CATALYTIC
ASYMMETRIC
SYNTHESIS

CATALYTIC ASYMMETRIC SYNTHESIS

Second Edition

Edited by
IWAO OJIMA

A John Wiley & Sons, Inc., Publication

New York • Chichester • Weinheim • Brisbane • Singapore • Toronto

Library of Congress Cataloging-in-Publication Data:
Ojima, Iwao, 1945–
    Catalytic asymmetric synthesis / edited by Iwao Ojima.—2nd ed.
        p. cm.
    Includes index.
    ISBN 0-471-29805-0 (alk. paper)
    1. Asymmetric synthesis. 2. Catalysis. I. Title.

QD262.O35 2000
541.3'9—dc21   99-053574

Printed in the United States of America.

10 9 8 7 6 5 4 3 2 1
ac^v

CONTENTS

PREFACE

The first edition of *Catalytic Asymmetric Synthesis*, published in the fall of 1993, was very warmly received by research communities in academia and industries from graduate students, research associates, faculty, staff, senior researchers, and others. The book was published at the very moment that the Food & Drug Administration (FDA) in the United States clarified the situation in "Chiral Drugs," the word "chirotechnology" was created, and chirotechnology industries were spawning in the United States and Britain.

As accurately predicted in the preface of the first edition, extensive research on new and effective catalytic asymmetric reactions have been continuing, in an explosive pace, and it is now obvious that these catalytic asymmetric processes promoted by man-made chiral catalysts will be the mainstream chemical technology in the 21st century. About five years from the publication of the original book, there was a clear demand in the synthetic community for an updated version of this book because advances in the field were accelerated during this period. Accordingly, I have agreed with the publisher to edit a second edition of this book.

In the second edition, I intended to incorporate all important reaction types that I am aware of, while keeping the monumental discovery and initial development of certain processes from the first edition, and highlighting recent advances in this field. The original book had 13 chapters (9 general-reaction types), which covered most of the important developments at that time. However, the second edition has 21 chapters (11 general-reaction types) (a total of 21 chapters for the 21st century is intriguing, isn't it?), reflecting the tremendous expansion in the scope of catalytic asymmetric synthesis in the past several years. In addition to the nine general-reaction types covered in the original book, the second edition includes "Asymmetric Carbometallations" (Chapter 4), "Asymmetric Amplification and Autocatalysis" (Chapter 9), and "Asymmetric Polymerization" (Chapter 11). "Cyclopropanation" in the original book has been replaced with "Asymmetric Carbene Reactions" (Chapter 5), which now includes powerful asymmetric intramolecular carbene insertion to C–H bonds. As the Table of Contents shows, there has been significant expansion and development in the asymmetric carbon–carbon bond-forming reactions (Chapter 8). Thus, this section consists of eight chapters dealing with cycloaddition reactions, aldol reactions, ene reactions, Michael reactions, allylic substitution reactions, cross-coupling reactions, and intramolecular Heck reactions. These processes provide very useful methods for the highly efficient synthesis of enantio-enriched or enantiopure compounds of biological, medicinal, agrochemical, and material science related interests.

Once again, the authors of these chapters are all world-leaders in this field, who outline and discuss the essence of each catalytic asymmetric reaction. Because the separate list of the chiral ligands in the original book was very well received, a convenient list of the chiral ligands with citation of relevant references appears in this book as an Appendix.

This book will, once again, serve as an excellent reference book for graduate students as well as chemists at all levels in both academic and industrial laboratories.

Iwao Ojima

PREFACE TO THE FIRST EDITION

Biological systems, in most cases, recognize a pair of enantiomers as different substances, and the two enantiomers will elicit different responses. Thus, one enantiomer may act as a very effective therapeutic drug whereas the other enantiomer is highly toxic. The sad example of thalidomide is well-known. It is the responsibility of synthetic chemists to provide highly efficient and reliable methods for the synthesis of desired compounds in an enantiomerically pure state, that is, with 100% enantiomeric excess (% ee), so that we shall not repeat the thalidomide tragedy. It has been shown for many pharmaceuticals that only one enantiomer contains all of the desired activity, and the other is either totally inactive or toxic. Recent movements of the Food & Drug Administration (FDA) in the United States clearly reflect the current situation in "Chiral Drugs", that is pharmaceutical industries will have to provide rigorous justification to obtain the FDA's approval of racemates. Several methods are used to obtain enantiomerically pure materials, which include classical optical resolution via diastereomers, chromatographic separation of enantiomers, enzymic resolution, chemical kinetic resolution, and asymmetric synthesis.

The importance and practicality of asymmetric synthesis as a tool to obtain enantiomerically pure or enriched compounds has been fully acknowledged to date by chemists in synthetic organic chemistry, medicinal chemistry, agricultural chemistry, natural products chemistry, pharmaceutical industries, and agricultural industries. This prominence is due to the explosive development of newer and more efficient methods during the last decade.

This book describes recent advances in catalytic asymmetric synthesis with brief summaries of the previous achievements as well as general discussions of the reactions. A previous book reviewing this topic, *Asymmetric Synthesis, Vol. 5—Chiral Catalysis*, edited by J. D. Morrison, Academic Press, Inc. (1985), compiles important contributions through 1982. Another book, *Asymmetric Catalysis*, edited by B. Bosnich, Martinus Nijhott (1986) also concisely covers contributions up to early 1984. In 1971, an excellent book, *Asymmetric Organic Reactions*, by J. D. Morrison and H. S. Mosher reviewed all earlier important work on the subject and compiled nearly 850 relevant publications through 1968, including some papers published in 1969. In the early 1980s, a survey of publications dealing with asymmetric synthesis (in a broad sense) indicated that the total number of papers in this area of research published in the 10 years after the Morrison/Mosher book, that is, 1971–1980, was almost the same as that of all the papers published before 1971. This doubling of output clearly indicates the attention paid to this important topic in 1970s. Since the 1980s, research on asymmetric synthesis has become even more important and popular when enantiomerically pure compounds are required for the total synthesis of natural products, pharmaceuticals, and agricultural agents. It would not be an exaggeration to say that the number of publications on asymmetric synthesis has been increasing exponentially every year.

Among the types of asymmetric reactions, the most desirable and the most challenging is *catalytic* asymmetric synthesis because one chiral catalyst molecule can create millions of chiral product molecules, just as enzymes do in biological systems. Among the significant achievements in basic research: (i) asymmetric hydrogenation of dehydroamino acids, a ground-breaking work by W.S. Knowles et al.; (ii) the Sharpless epoxidation by K. B. Sharpless et al.; and (iii) the second-generation asymmetric hydrogenation processes developed by R. Noyori et al. deserve particular attention because of the tremendous impact that these processes have made in synthetic organic chemistry. Catalytic asymmetric synthesis often has significant economic advantages over stoichiometric asymmetric synthesis for industrial-scale production of enantiomerically pure compounds. In fact, a number of catalytic asymmetric reactions, including the "Takasago Process" (asymmetric isomerization), the "Sumitomo Process" (asymmetric cyclopropanation), and the "Arco Process" (asymmetric Sharpless epoxidation) have been commercialized in the 1980s. These processes supplement the epoch-making "Monsanto Process" (asymmetric hydrogenation), established in the early 1970s. This book uncovers other catalytic asymmetric reactions that have high potential as commercial processes. Extensive research on new and effective catalytic asymmetric reactions will surely continue beyond the year 2000, and catalytic asymmetric processes promoted by man-made chiral catalysts will become mainstream chemical technology in the 21st century.

This book covers the following catalytic asymmetric reactions: asymmetric hydrogenation (Chapter 1); isomerization (Chapter 2); cyclopropanation (Chapter 3); oxidations (epoxidation of allylic alcohols as well as unfunctionalized olefins, oxidation of sulfides, and dihydroxylation of olefins) (Chapter 4); hydrocarbonylations (Chapter 5); hydrosilylation (Chapter 6); carbon-carbon bond-forming reactions (allylic alkylation, Grignard cross-coupling, and aldol reaction) (Chapter 7); phase-transfer reactions (Chapter 8); and Lewis acid-catalyzed reactions (Chapter 9). The authors of the chapters are all world-leaders in this field, who outline and discuss the essence of each catalytic asymmetric reaction. (In addition, a convenient list of the chiral ligands appearing in this book, with citation of relevant references, is provided as an Appendix.)

This book serves as an excellent reference for graduate students as well as chemists at all levels in both academic and industrial laboratories.

Iwao Ojima
March 1993

CONTRIBUTORS

Carsten Bolm, Institut für Organische Chemie, der RWTH Aachen, Professor-Pirlet-Str. 1, D-52056 Aachen, Germany

Erick M. Carreira, Laboratorium für Organische Chemie, Universitätstrasse 16, ETH-Z, CH-8092 Zürich, Switzerland

Yariv Donde, Department of Chemistry, University of California, Irvine, Irvine, CA 92717-2025

Michael P. Doyle, Vice President, Research Corporation, 101 North Wilmot Road, Suite 250, Tuscon, Arizona 85711

Tamio Hayashi, Department of Chemistry, Graduate School of Science, Kyoto University, Kyoto 606-01, JAPAN

Kenji Itoh, Department of Applied Chemistry, Graduate School of Engineering, Nagoya University, Chikusa, Nagoya 464-01, JAPAN

Henri B. Kagan, Laboratoire de Synthèse Asymétrique, Institut de Chimie Moleculaire d'Orsay, Batiment 420, Universite Paris-Sud, Centre d'Orsay, 91405 Orsay Cedex, FRANCE

Motomu Kanai, Faculty of Pharmaceutical Sciences, University of Tokyo, 7-3-1 Hongo, Bunkyo-ku, Tokyo 113, JAPAN

Tsutomu Katsuki, Department of Chemistry, Faculty of Science, Kyushu University, 6-10-1 Hakozaki, Higashi-ku, Fukouka 812-81, JAPAN

Masato Kitamura, Department of Chemistry, Faculty of Science, Nagoya University, Chikusa, Nagoya 464-01, JAPAN

Chulbom Lee, Department of Chemistry, Stanford University, Stanford, CA 94305

Keiji Maruoka, Department of Chemistry, Graduate School of Science, Hokkaido University, Sapporo 060, JAPAN

Koichi Mikami, Department of Chemical Technology, Tokyo Institute of Technology, 2-12-1 Ookayama, Meguro-ku, Tokyo 152, JAPAN

Tamaki Nakano, Department of Applied Chemistry, Graduate School of Engineering, Nagoya University, Chikusa, Nagoya 464-01, JAPAN

Ei-ichi Negishi, Department of Chemistry, Purdue University, 1393 Brown Laboratories, West Lafayette, IN 47907-1393

Hisao Nishiyama, School of Material Science, Toyohashi University of Technology, Tempaku-cho, Toyohashi 441, JAPAN

Ryoji Noyori, Department of Chemistry, Faculty of Science, Nagoya University, Chikusa, Nagoya 464-01, JAPAN

Kyoko Nozaki, Department of Material Chemistry, School of Industrial Chemistry, Kyoto University, Yoshida Hon-machi, Sakyo-ku, Kyoto 606-01, JAPAN

Martin J. O'Donnell, Department of Chemistry, Indiana University-Purdue University, 1125 East 38th Street, P.O. Box 647, Indianapolis, IN 46223

Masamichi Ogasawara, Department of Chemistry, Graduate School of Science, Kyoto University, Kyoto 606-01, JAPAN

Takesi Ohkuma, Department of Chemistry, Faculty of Science, Nagoya University, Chikusa, Nagoya 464-01, JAPAN

Iwao Ojima, Department of Chemistry, State University of New York at Stony Brook, Stony Brook, NY 11794-3400

Yoshio Okamoto, Department of Applied Chemistry, Graduate School of Engineering, Nagoya University, Chikusa, Nagoya 464-01, JAPAN

Larry E. Overman, Department of Chemistry, University of California, Irvine, Irvine, CA 92717-2025

Masakatsu Shibasaki, Faculty of Pharmaceutical Sciences, 7-3-1 Hongo, Bunkyo-ku, Tokyo 113, JAPAN

Takanori Shibata, Department of Applied Chemistry, Faculty of Science, Science University of Tokyo, Kagurazaka, Shinjuku-ku, Tokyo 162, JAPAN

Kenso Soai, Department of Applied Chemistry, Faculty of Science, Science University of Tokyo, Kagurazaka, Shinjuku-ku, Tokyo 162, JAPAN

Barry M. Trost, Department of Chemistry, Stanford University, Stanford, CA 94305

1

ASYMMETRIC HYDROGENATION

TAKESHI OHKUMA, MASATO KITAMURA, AND RYOJI NOYORI
Department of Chemistry and Research Center for Materials Science, Nagoya University, Chikusa, Nagoya, Japan 464-8602

1.1. INTRODUCTION

Chiral building blocks are indispensable for the syntheses of biologically active compounds such as pharmaceuticals, agrochemicals, flavors, and fragrances as well as in the creation of advanced materials. Asymmetric saturation of alkenes, ketones, and imines by hydrogen or organic hydrogen donors provides an ideal access to chiral alkanes, alcohols, and amines, respectively [1,2]. A small amount of a chiral catalyst repeatedly delivers hydrogen atoms to one of the enantiofaces of substrates, producing large amounts of optically active saturated compounds. The chirality multiplication is achieved largely in a homogeneous phase with chiral molecular catalysts consisting of a metallic element and an optically active organic compound(s). The molecular structures are highly diverse and, by choosing the handedness of catalyst, both enantiomers are available with equal ease. Furthermore, hydrogenation is economical and environment-conscious, producing no hazardous byproducts. Thus asymmetric hydrogenation has been increasingly important in practical organic syntheses ranging from research to production.

A high reaction rate and enantioselectivity are attainable by appropriate molecular architecture of the catalysts and suitable selection of reaction conditions. Historically, highly enantioselective hydrogenation was achieved mostly with substrates that have a functionality close to the unsaturated moiety and by use of Rh(I) or Ru(II) complexes that have a chiral diphosphine ligand [3]. Recently, various phosphine- and/or nitrogen-based ligands [3,4] have been developed for asymmetric hydrogenation or transfer hydrogenation of simple, unfunctionalized ketones and imines. Some simple olefins can be enantioselectively hydrogenated by early transition-metal complexes with a chiral cyclopentadienyl ligand or an Ir catalyst with a chiral amino phosphine ligand. This chapter presents recent advances in this important field. Some asymmetric heterogeneous catalyses are also discussed.

Catalytic Asymmetric Synthesis, Second Edition, Edited by Iwao Ojima
ISBN 0-471-29805-0   Copyright © 2000 Wiley-VCH, Inc.

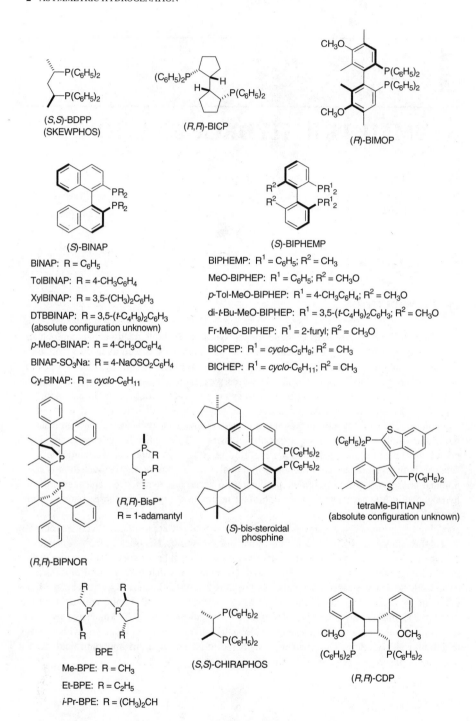

Figure 1.1. C_2-chiral diphosphine ligands (in alphabetical order for named compounds).

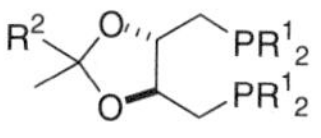

(S,S)-DIOP

DIOP: $R^1 = C_6H_5$; $R^2 = CH_3$

DIOP-OH: $R^1 = C_6H_5$; $R^2 = HOCH_2$

MOD-DIOP: $R^1 = 3,5\text{-}(CH_3)_2\text{-}4\text{-}(CH_3O)C_6H_2$; $R^2 = CH_3$

CyDIOP: $R^1 = cyclo\text{-}C_6H_{11}$; $R^2 = CH_3$

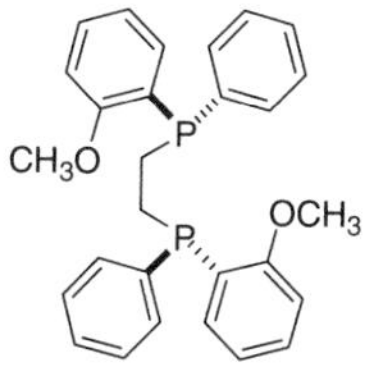

(S,S)-DIPAMP

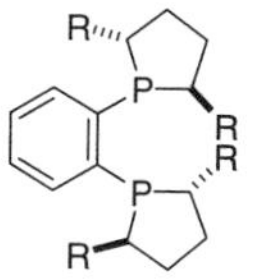

DuPHOS

Me-DuPHOS: R = CH_3

Et-DuPHOS: R = C_2H_5

i-Pr-DuPHOS: R = $(CH_3)_2CH$

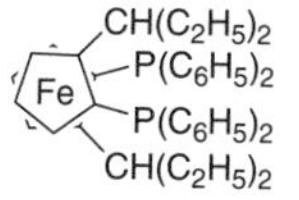

(S,S)-FerroPHOS

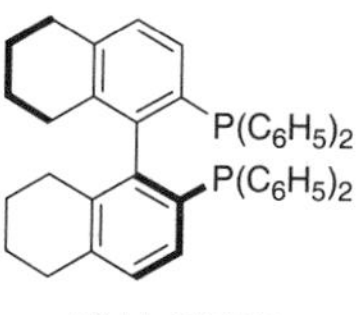

(S)-H$_8$-BINAP

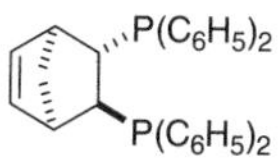

(S,S)-NORPHOS

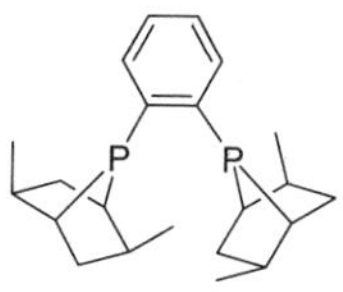

(R,S,R,S)-Me-PennPhos

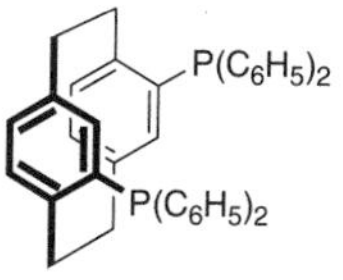

(S)-[2.2]PHANEPHOS

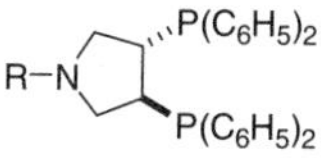

**(S,S)-PYRPHOS
(DEGUPHOS)**

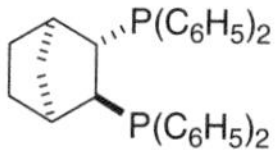

(S,S)-RENORPHOS

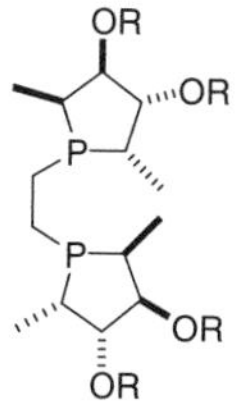

(S,S,S,S)-RoPHOS

R = $CH_2C_6H_5$ or $t\text{-}C_4H_9$

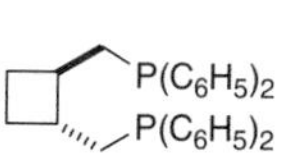

(R,R)-TBPC

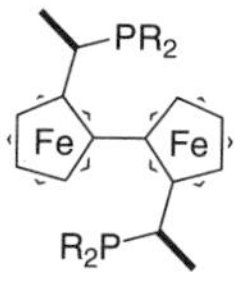

(R,R)-(S,S)-TRAP

EtTRAP: R = C_2H_5

i-BuTRAP: R = $(CH_3)_2CHCH_2$

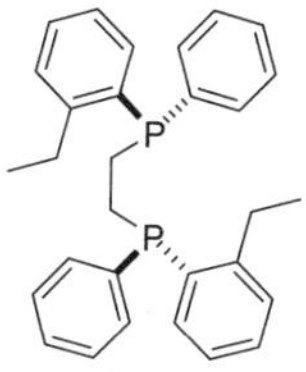

(S,S)-1

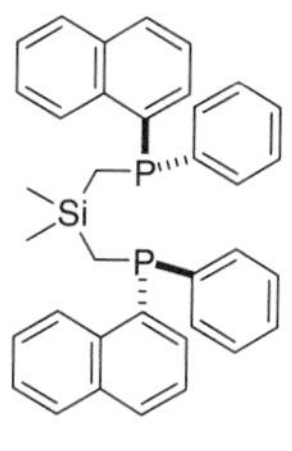

(S,S)-2

Figure 1.1. (*Continued*)

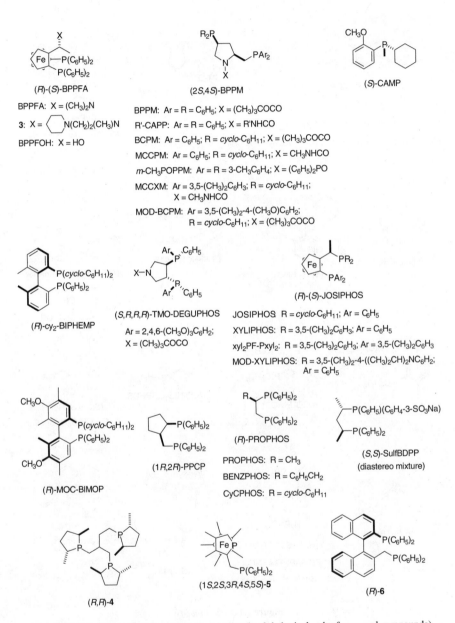

Figure 1.2. Phosphine ligands without C_2 chirality (in alphabetical order for named compounds).

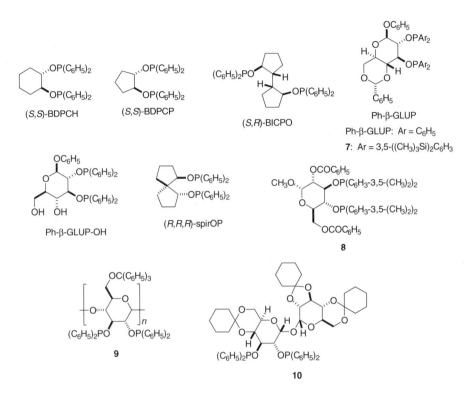

Figure 1.3. Bisphosphinite ligands.

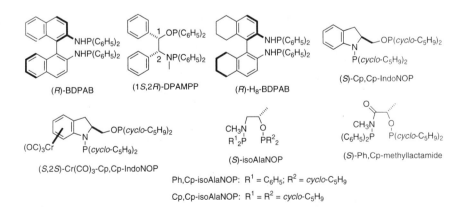

Figure 1.4. Amido- or aminophosphine ligands (in alphabetical order). (Continued on next page.)

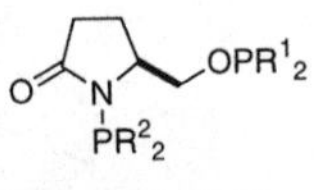

(S)-oxoProNOP

Ph,Ph-oxoProNOP: $R^1 = R^2 = C_6H_5$

Cp,Cp-oxoProNOP: $R^1 = R^2 = cyclo\text{-}C_5H_9$

Cy,Cy-oxoProNOP: $R^1 = R^2 = cyclo\text{-}C_6H_{11}$

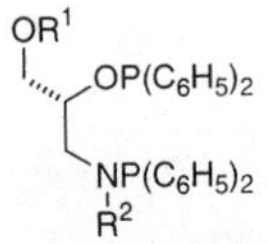

(R)-PINDOPHOS

PINDOPHOS: $R^1 = 7\text{-indolyl}$; $R^2 = (CH_3)_2CH$

11: $R^1 = 1\text{-naphthyl}$; $R^2 = cyclo\text{-}C_5H_9$

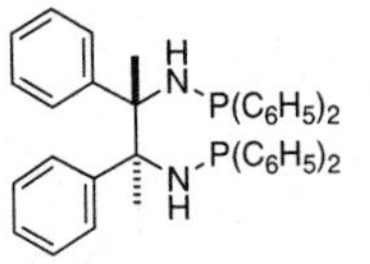

(S,S)-PNNP

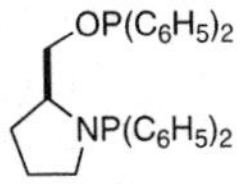

(S)-PROLOPHOS

Figure 1.4. (*Continued*)

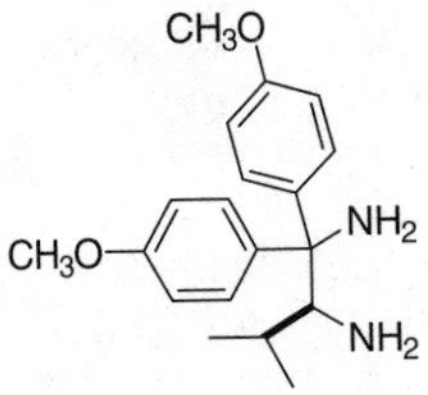

(S)-DAIPEN

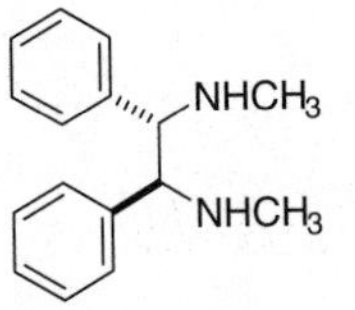

(S,S)-DMDPEN

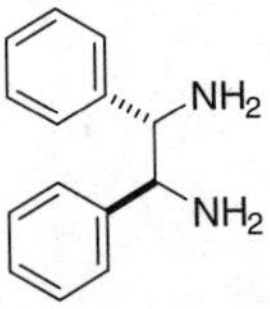

(S,S)-DPEN

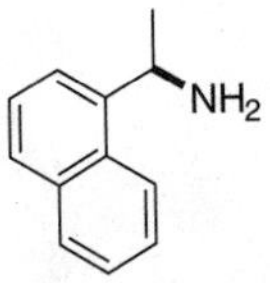

(R)-1-NEA

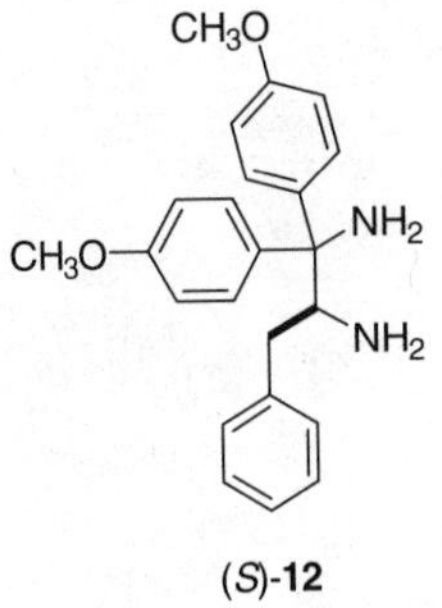

(S)-**12**

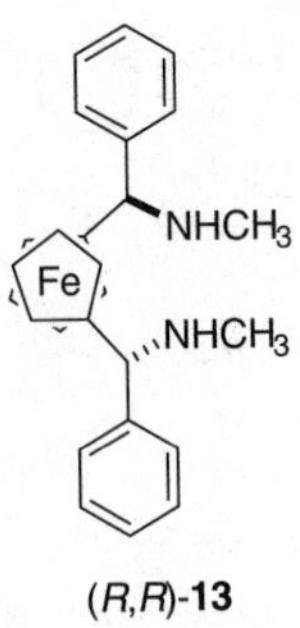

(R,R)-**13**

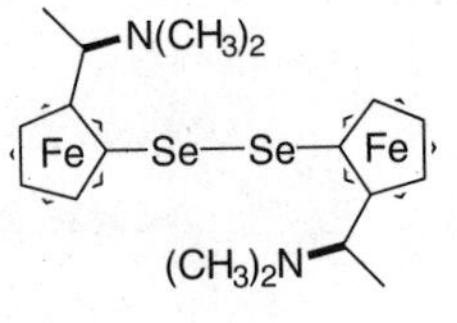

(R)-(S)-**14**

Figure 1.5. Amine ligands.

(*R*)-AMBOX

(*R*)-DHPPEI

(*R*,*R*)-PDPBI

(*S*)-PPEI

(*S*)-**15**

(*S*,*S*)-**16**

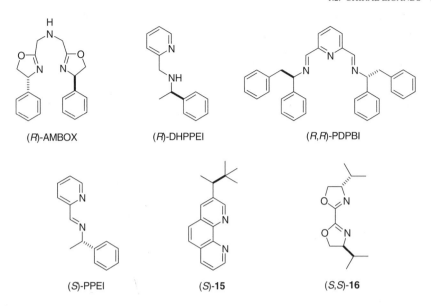

Figure 1.6. Pyridines, phenanthrolines, and oxazole ligands.

1.2. CHIRAL LIGANDS

A suitable combination of a metal species and chiral organic ligand [3,4] is the key factor to prepare high-performance catalysts for asymmetric hydrogenation. The electron-donating ligands often increase the hydride-donating ability of catalytic species. The stereo-regulation is normally achieved by utilizing repulsive interactions between substituents of the ligands and substrates, although bulky groups tend to decrease the reactivity. The stereochemical control with an attractive interaction is also important to achieve a high catalytic activity.

Development of optically active phosphine ligands, especially C_2-chiral diphosphines, used late transition-metal complexes as (pre)catalysts and provided a great advancement in asymmetric hydrogenation. Commonly used phosphorus-based ligands are listed in Figures 1.1–1.4. Chiral nitrogen-based ligands are also important, particularly for asymmetric transfer hydrogenation. The structures are classified in Figures 1.5–1.9. The chiral complexes with cyclopentadienyl ligands are given in the main text.

(*S*)-AMSO

(*R*,*R*)-TsDPEN

(*R*,*R*)-**17**

(*S*,*S*)-**18**: X = O
(*S*,*S*)-**19**: X = S

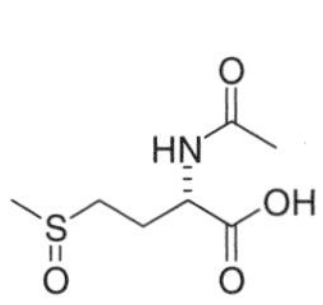

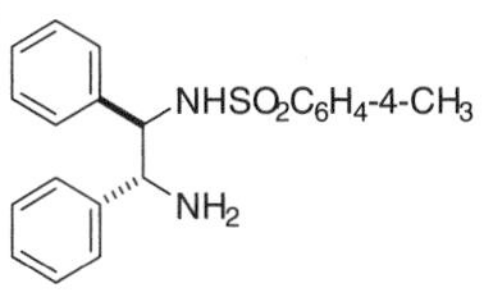

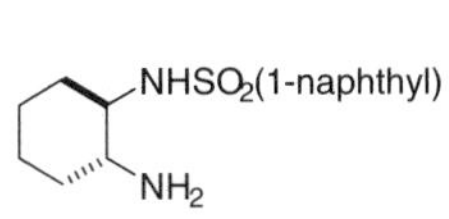

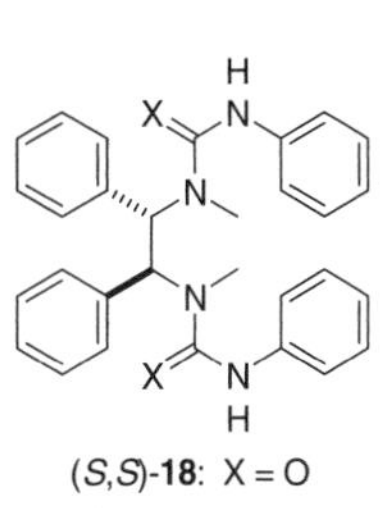

Figure 1.7. N-Substituted (di)amines.

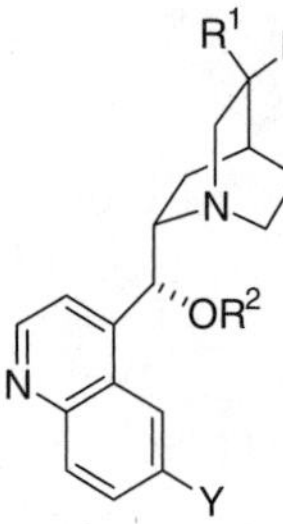

cinchona alkaloids

cinchonidine: $R^1 = CH_2=CH$; $R^2 = H$; $Y = H$

quinine: $R^1 = CH_2=CH$; $R^2 = H$; $Y = CH_3O$

MeOHCd: $R^1 = C_2H_5$; $R^2 = CH_3$; $Y = H$

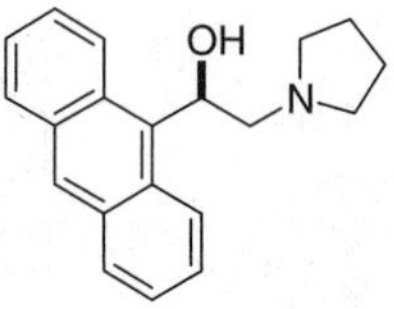

(R)-**20**

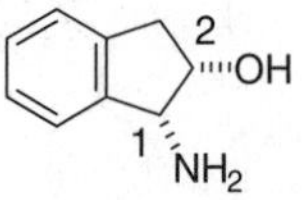

(1R,2S)-**21**

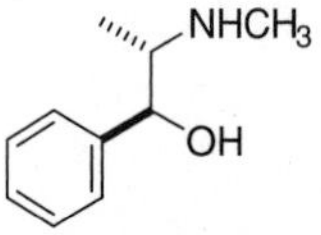

(S,S)-**22**

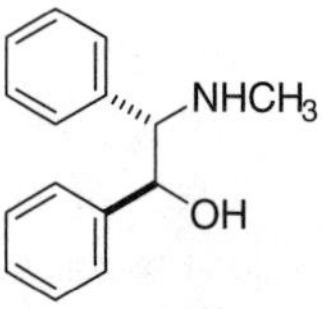

(S,S)-**23**

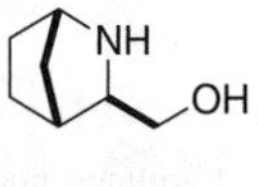

(S,R,R)-**24**

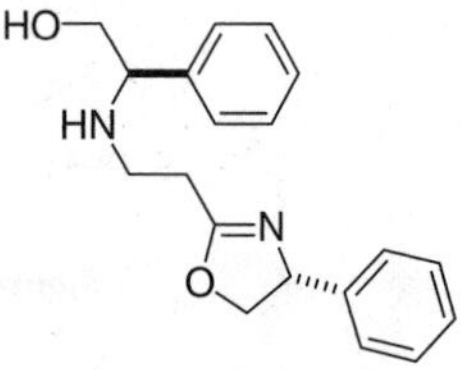

(R,R)-**25**

Figure 1.8. Amino alcohols.

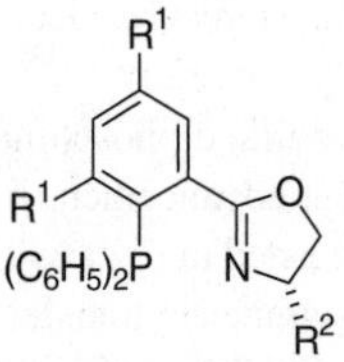

(S)-**26**: $R^1 = H$; $R^2 = (CH_3)_3C$
(S)-**27**: $R^1 = CH_3$; $R^2 = (CH_3)_2CH$

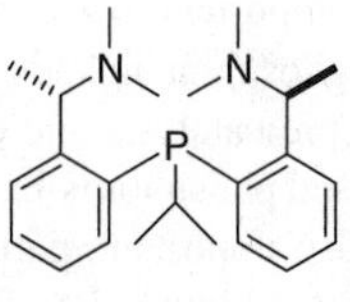

(S,S)-**28**

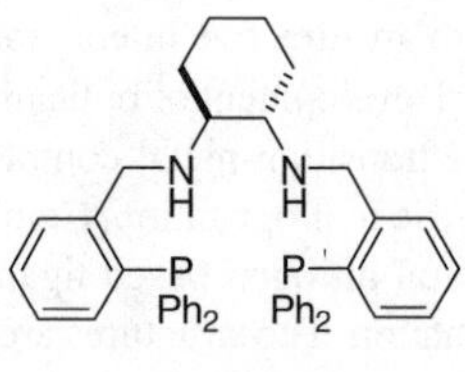

(S,S)-**29**

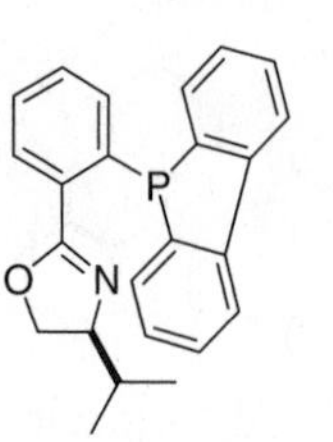

(S)-**30**

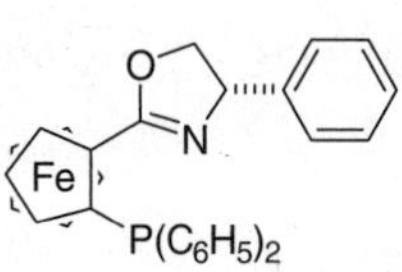

(S)-**31**

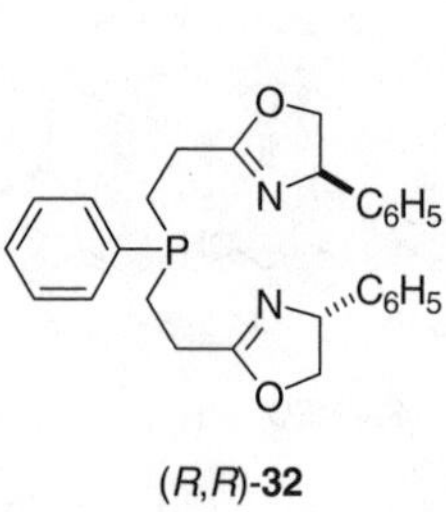

(R,R)-**32**

Figure 1.9. Amino and imino phosphine ligands.

1.3. HYDROGENATION OF OLEFINS

In the late 1960s, the discovery of the Wilkinson's hydrogenation catalyst, RhCl[P(C$_6$H$_5$)$_3$]$_3$, stimulated attempts to create tertiary stereogenic carbon atoms by enantioselective hydrogenation of olefins by using optically active transition-metal complexes [5]. In the beginning, only a very low enantioselectivity was obtained in hydrogenation of 2-phenylacrylic acid and 2-phenyl-1-propene with certain chiral tertiary phosphine-modified Rh complexes as catalysts [6]. However, the situation was soon dramatically changed by the invention of well-designed Rh complexes containing—among others—DIOP [7], CAMP [8], and DIPAMP [9] as ligands used for asymmetric hydrogenation of α-(acylamino)acrylic acids. This chemistry later became the standard method to synthesize optically active amino acids. In the mid-1980s the discovery of BINAP–Ru complexes significantly expanded the scope of olefinic substrates for asymmetric hydrogenation [1c]. Now a variety of natural and unnatural chiral compounds are accessible in a practical manner. Recently researchers have tried to extend the asymmetric catalysts from the late transition-metal complexes to the early transition-metal complexes.

1.3.1. Functionalized Olefins

Considerable success has been realized for asymmetric hydrogenation of functionalized alkenes. However, there is no universal catalyst that ensures generally high enantioselectivity for a wide range of substrates. The limitation has been overcome largely by combinatorial investigation of transition metals and chiral ligands together with mechanistic understanding of the multistep catalytic reaction.

1.3.1.1. Enamides Enantioselective hydrogenation of α-hydroxycarbonyl- or α-alkoxycarbonyl-substituted enamides is effectively catalyzed by a number of Rh complexes with a chiral ligand to produce the corresponding amino acid derivatives with >90% ee [3b,10]. One or two phosphorus atoms are incorporated into the phosphine, phosphinite, and aminophosphine ligands, where the chirality is installed at the phosphorus or carbon atoms or in the molecular framework. Chiral ligands may be immobilized onto polymers. In most cases, the efficiency of catalysts has been examined with (Z)-2-(acetamido)cinnamic acid, 2-(acetamido)acrylic acid, and their methyl or ethyl esters. The reaction is generally carried out in alcoholic solvents at 1–10 atm of hydrogen. The cationic complexes are most commonly used. Selected examples are compiled in Scheme 1.1.

High catalytic activities have been achieved by the PYRPHOS– [18], PPCP– [20], BICHEP– [21], Et-DuPHOS–Rh [19] complexes among others, allowing the reaction with a substrate-to-catalyst molar ratios (S/C) as high as 50,000. With a [2.2]PHANEPHOS–Rh complex, the reaction proceeds even at –45°C [27]. Supercritical carbon dioxide, a unique reaction medium, can be used in the DuPHOS and BPE–Rh-catalyzed hydrogenation [43]. A highly lipophilic counteranion such as tetrakis[3,5-bis(trifluoromethyl)phenyl]borate (BARF) or trifluoromethanesulfonate is used to enhance the solubility of the cationic Rh complexes. Under the most suitable reaction conditions of 102 atm of carbon dioxide, 1 atm of hydrogen, and 22°C, α-amino acid derivatives are produced with up to 99.7% ee.

Two-phase catalysis with water and organic solvent is industrially important because of the easy catalyst/product separation. For achieving sufficient water solubility of catalysts, a highly polar group such as amino, hydroxy, hydroxycarbonyl, or hydroxysulfonyl function is introduced into a chiral organic ligand. α-(Acetamido)cinnamic acid is hydrogenated in up to 94% optical yield by using a Rh complex of CHIRAPHOS bearing sulfonate groups or quarternary amino moieties in the *P*-phenyl rings [44]. Reaction of sodium α-(acylamino)cinnamate in

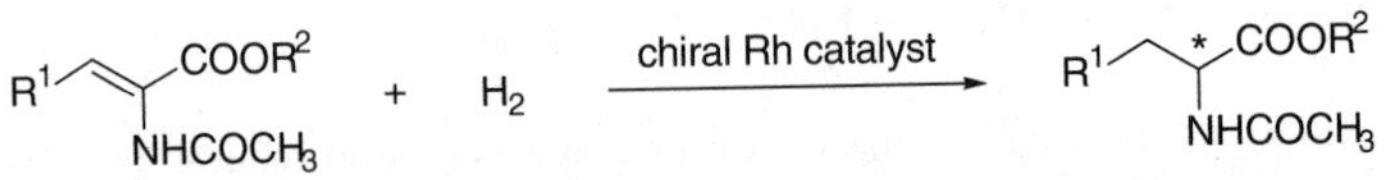

$$R^1\!\!-\!\!\overset{\displaystyle COOR^2}{\underset{\displaystyle NHCOCH_3}{}} + \ H_2 \ \xrightarrow{\text{chiral Rh catalyst}} \ R^1\!\!-\!\!\overset{*\,COOR^2}{\underset{NHCOCH_3}{}}$$

| | % ee of Product (config'n) | | | | |
Ligand	$R^1 = C_6H_5$ $R^2 = H$	$R^1 = C_6H_5$ $R^2 = CH_3$	$R^1 = H$ $R^2 = H$	$R^1 = H$ $R^2 = CH_3$	Ref
diphosphine					
(R,R)-DIOP	85 (R)		73 (R)		7
(R,R)-DIPAMP	96 (S)		94 (S)		9
(S)-(R)-BPPFA	93 (S)				11
(2S,4S)-BPPM	91 (R)[a]				12
(S,S)-CHIRAPHOS	99 (R)		91 (R)		13
(S,S)-NORPHOS[b]	95 (S)		90 (R)		14
(S)-BINAP	100 (R)		98 (R)[a]		15
(R)-CYCPHOS	88 (S)				16
(S,S)-BDPP	92 (R)				17
(R,R)-PYRPHOS	99 (S)				18
(S,S)-Et-DuPHOS	99 (S)		99.4 (S)		19
(1R,2R)-PPCP	96 (S)		87 (S)		20
(R)-BICHEP		98 (S)[c]			21
(R,R)-BICP	99 (S)[a]				22
(R,R)-Me-BPE		85 (R)			23
(R,R)-(S,S)-i-BuTRAP		92 (S)			24
(R,R)-(S,S)-EtTRAP				96 (R)	24
(S,S)-**1**	90 (R)	93 (R)	93 (R)	91 (R)	25
(S,S)-BIPNOR	>98 (S)				26
(R)-[2.2]PHANEPHOS		98 (R)		99.6 (R)	27
(1S,2S,3R,4S,5S)-**5**		87 (R)		87 (R)	28
(S,S)-**2**	97.1 (R)	97.7 (R)		64.1 (R)	29
(S,S)-BisP*	98.6 (R)	99.9 (R)		>99.9 (R)	30
(S,S)-FerroPHOS	98.9 (R)	97.6 (R)	98.2 (R)	97.5 (R)	31
(S,S,S,S)-RoPHOS	93.5 (S)	97.5 (S)			32
diphosphinite					
Ph-β-GLUP		91 (S)			33
7	99.0 (S)	97.6 (S)			34
8	97.0 (R)	96.3 (R)			34
(R,R,R)-spirOP	97.9 (R)	95.7 (R)	>99.9 (R)	99.0 (R)	35
10	84 (S)			66 (S)	36
(R,S)-BICPO	96.1 (S)				37

(*Continued on next page.*)

Scheme 1.1.

	% ee of Product (config n)				
	$R^1 = C_6H_5$	$R^1 = C_6H_5$	$R^1 = H$	$R^1 = H$	
Ligand	$R^2 = H$	$R^2 = CH_3$	$R^2 = H$	$R^2 = CH_3$	Ref
aminophosphine phosphinite					
(S)-PROLOPHOS			80 (S)		38
(R)-PINDOPHOS	91 (S)	90 (S)			39
(1S,2R)-DPAMPP		98.3 (R)			40
(S)-**11**	91 (R)^a	94 (R)^a			41
diaminophosphine					
(S,S)-PNNP	93.8 (S)			89.5 (S)	42

^{a} Hydrogenation of the N-benzoyl derivative. ^{b} The hydrogenation conditions form RENORPHOS.
^{c} Hydrogenation of the ethyl ester derivative.

Scheme 1.1. *(Continued)*

aqueous solution gives the saturated product in 90% optical yield by using a water-soluble Rh complex of a PYRPHOS-type ammonium ligand [45]. A Ph-β-GLUP-OH–Rh complex hydrogenates various α-(acetamido)cinnamates in water containing sodium dodecylsulfate (SDS) as a surfactant to yield phenylalanine derivatives with 96–98% ee [46]. Without SDS the enantioselectivities remain to be ca. 80% ee. In hydrogenation of methyl α-(acetamido)cinnamate with a DIOP-OH–Rh complex, the optical yield is increased from 1.5% to 76.6% [47]. This increase may be ascribed to the formation of micro-heterogeneous systems of colloidal dimensions on micelles. Use of polymeric surfactants generates swelled micelles that enclose catalysts, which can be separated by filtration [48]. These polymeric micelles represent a borderline case to polymer-attached catalysts.

Many trials on polymer fixation of homogeneous chiral Rh catalysts have been made since the pioneering work in 1976 [49]. Chiral diphosphines such as DIOP and BPPM are attached to a cross-linked polystyrene matrix containing hydrophilic hydroxyethyloxycarbonyl function. The immobilized Rh catalysts swell in an ethanol–benzene-mixed solvent and hydrogenate α-(acetamido)cinnamic acid with up to 90% ee [50]. Rh complexes of PYRPHOS bound to a polyethylene oxide-grafted styrene matrix and to a macroporous silica gel show higher reactivity and selectivity than the homogeneous complexes, giving N-acetylphenylalanine with 97% ee [51] and its methyl ester with 100% ee, respectively [52]. A nitrogen-based chiral ligand derived from natural proline is also anchored on silica and a modified USY zeolite. With these Rh complexes, N-benzoyl phenylalanine ethyl ester is obtained with 92–99% ee [53]. Cationic Rh complexes can be ionically bound to cation exchangers such as sulfonated styrene-divinylbenzene copolymer and arylsulfonic acid-functionalized silica gel [54]. Such immobilized Ph-β-GLUP–Rh complexes yield phenylalanine derivatives with ca. 95% ee. Water-soluble amine-functionalized BDPP and CHIRAPHOS derivatives are fixed on strongly acidic cation-exchange resin Nafion-H. The Rh complexes give similar enantioselectivities to the corresponding homogeneous catalysts [55]. In such ionic binding of the catalysts, however, a catalyst-leaching problem is always present.

In contrast to the high enantioselectivities obtained for the Z substrates, hydrogenation of the E isomers usually proceeds very slowly and in a poor optical yield partly due to the E/Z double-bond isomerization. The enantioselectivity in the BINAP–Rh-catalyzed hydrogenation of E enamides is enhanced in an aprotic solvent THF to minimize the isomerization [15,56].

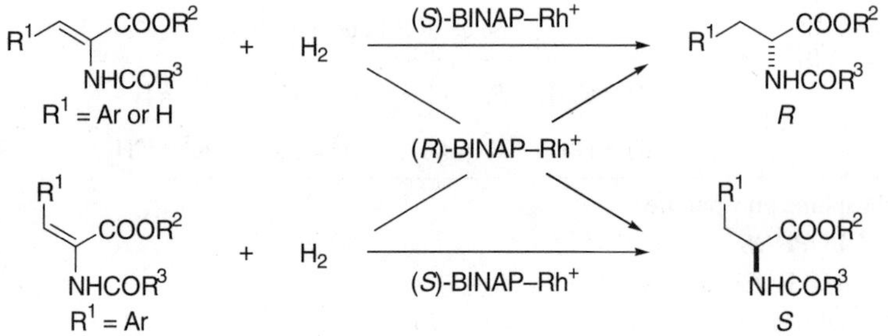

Scheme 1.2.

Scheme 1.2 illustrates the relationship between the double-bond geometry, the configuration of the BINAP ligand, and the stereochemistry of the products.

A DuPHOS–Rh catalyst reduces both E and Z enamides with a high enantioselectivity even in an alcoholic solvent without E/Z isomerization [57]. Notably, β,β-disubstituted α-enamides are also smoothly hydrogenated to β-branched α-amino acids [58]. Sterically less-hindered Me-DuPHOS– and Me-BPE–Rh catalysts provide a high enantioselectivity. The cationic Rh complexes of TRAP [24], BisP* [30], and [2.2]PHANEPHOS [27] are also active for hydro-

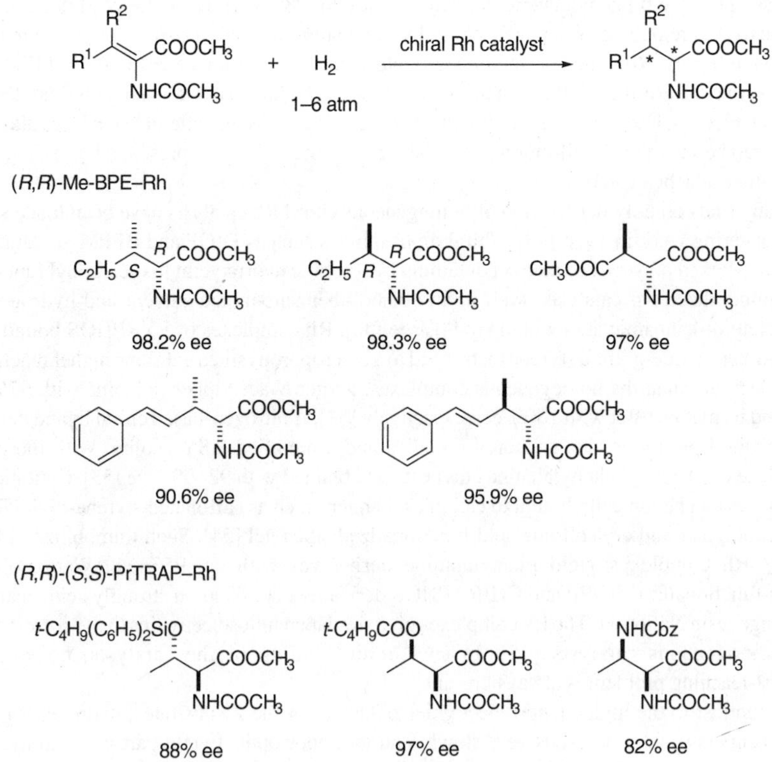

Scheme 1.3.

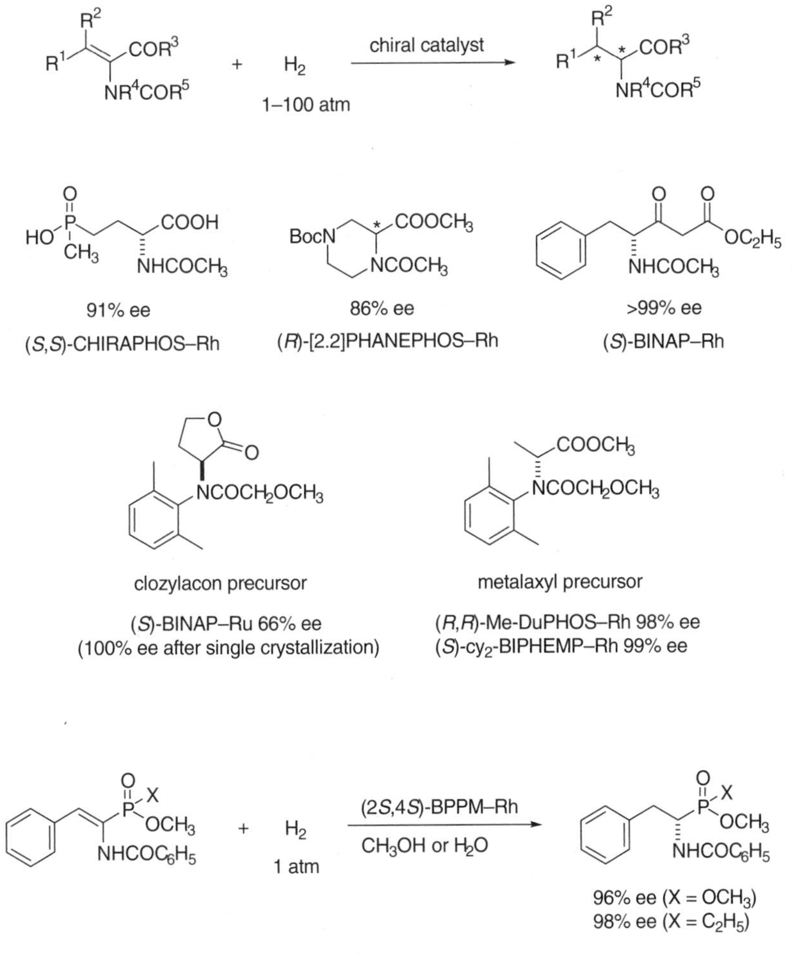

Scheme 1.4.

genation of β,β-disubstituted α-enamides. Hydrogenation of α-enamides possessing dissimilar β-substituents allows the simultaneous generation of an additional β-stereogenic center in the product. Both erythro and threo β-substituted amino acids are produced through hydrogenation of the *E* and *Z* enamides using Me-BPE–Rh complex (Scheme 1.3) [58]. Thus β-methyltryptophan with 97% ee can be prepared by use of Me-DuPHOS–Rh complex [59]. The cationic Et-DuPHOS– and Me-BPE–Rh complexes do not reduce isolated carbon–carbon double or triple bonds in the time frame for hydrogenation of the α-enamide moiety, allowing the synthesis of allylglycine derivatives [60]. Both *erythro*- and *threo*-β-hydroxy-α-amino acids with high enantiopurities are prepared by enantioselective hydrogenation of (*Z*)-β-siloxy-α-(acetamido)acrylates and (*E*)-β-pivaloyloxy-α-(acetamido)acrylates by using a TRAP–Rh complex [61]. The PrTRAP–Rh catalyst gives higher enantioselectivity than catalysts with Et- and BuTRAP. This method has been applied to the synthesis of 2,3-diamino carboxylic acids with 79–82% ee [62]. Use of the more easily removable N-protecting groups such as Cbz and Boc enhances the synthetic utility of asymmetric hydrogenation of enamides. [2.2]PHANEPHOS–, Et-DuPHOS–, and Pr-DuPHOS–Rh catalysts promote highly enantioselective hydrogenation of the N-Cbz or -Boc-protected substrates [27,39,63].

The chiral Rh catalytic systems are applicable to the synthesis of a variety of optically active α-amino acids, including (S)- and (R)-aromatic, -heteroaromatic, and -aliphatic alanine derivatives [10,34,64]. Functionalized substituents such as methylhydroxyphosphinylmethyl [65] and alkoxycarbonylamino group [27] may be introduced to the β-position of α-(acylamino)acrylates. The hydroxy- or alkoxycarbonyl group can be replaced by other electron-withdrawing groups [10] such as carbamoyl, keto, cyano, dialkoxyphosphoryl [32,66], and alkylhydroxyphosphinyl groups [67]. N-Aryl-substituted enamides are also usable as substrates, opening an enantioselective route to fungicidal agrochemicals such as clozylacon and metalaxyl [68]. Several examples recently reported are shown in Scheme 1.4.

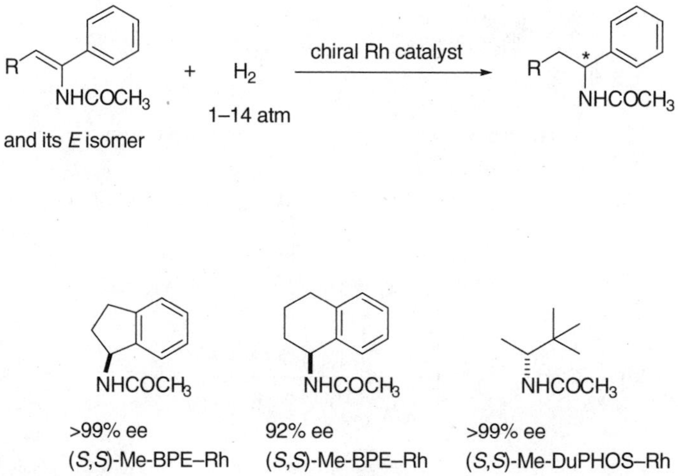

Substrate			Product	
R	E:Z	Ligand	% ee	Confign
H	—	(R,R)-Me-BPE	95.2	R
H	—	(R)-BDPAB	92.9	R
H	—	(R)-H$_8$-BDPAB	96.8	R
CH$_3$	0:100	(S,S)-DIOP	92	R
CH$_3$	50:50–20:80	(R,R)-Me-BPE	95.4	R
CH$_3$	65:35	(R,R)-CDP	92	R
CH$_3$OCH$_2$O	a	(R,R)-Me-DuPHOS	97	S
CH$_3$OCH$_2$O	a	(R,R)-BICP	94	S

a Unknown.

Scheme 1.5.

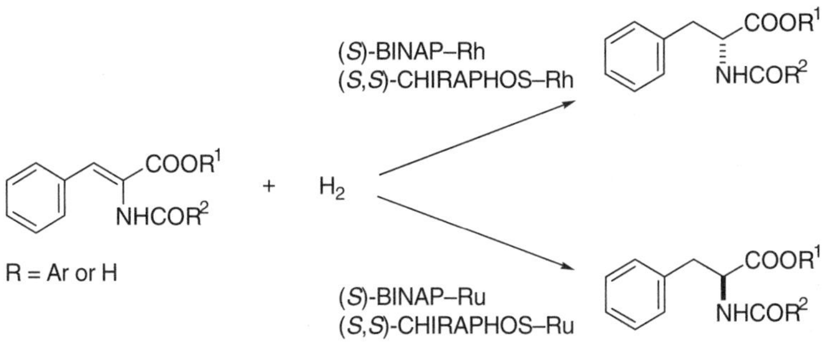

Scheme 1.6.

Existence of the α-carboxyl or alkoxycarbonyl function in α-(acylamino)acrylates is not a requisite condition for attaining a high reactivity and enantioselectivity. The chiral diphosphine– and bisaminophosphine–Rh complexes hydrogenate α-aryl-substituted α-enamides, providing an entry to α-arylamines as well as β-amino alcohol derivatives with >92% ee (Scheme 1.5) [69–73]. Both E and Z enamides are hydrogenated in a high optical yield with the same sense of enantioselectivity, thus allowing use of the E/Z mixture of β-substituted α-arylenamides. 1-Acetamido-2,3-benzocyclopentane and -2,3-benzocyclohexane are synthesized with up to >99% ee by Me-BPE–Rh-catalyzed hydrogenation [73]. Me-DuPHOS–Rh complex can be used for the hydrogenation of α-t-alkyl-substituted α-enamide [73].

Unlike the Rh-based hydrogenation of α-(acylamino)acrylates, the corresponding Ru chemistry has not been studied extensively. Ru complexes of (S)-BINAP and (S,S)-CHIRAPHOS catalyze the hydrogenation of (Z)-α-(acylamino)cinnamates to give the protected (S)-pheny-lalanine with 92% ee [74] and 97% ee [75], respectively. It is interesting that the Rh and Ru complexes with the same chiral diphosphines exhibit an opposite sense of asymmetric induction (Scheme 1.6) [13,15,56,74,75]. This condition is due primarily to the difference in the mechanisms; the Rh-catalyzed hydrogenation proceeds via Rh dihydride species [76], whereas the Ru-catalyzed reaction takes place via Ru monohydride intermediate [77]. The Rh-catalyzed reaction has been studied in more detail by kinetic measurement [78], isotope tracer experiments [79], NMR studies [80], and MO calculations [81]. The stereochemical outcome is understandable by considering the thermodynamic stability and reactivity of the catalyst–enamide complexes.

1-(Formamido)alkenylphosphonates and N-acyl-1-alkylidenetetrahydroisoquinolines are hydrogenated at 1–4 atm of hydrogen with almost perfect enantioselection by use of BINAP–Ru complexes [77a,82] (Schemes 1.7 and 1.8). The bulkier size of the sp^3-hybridized, tetrahedrally arranged phosphonic ester group in comparison with the sp^2-hybridized, planar carboxylic ester or the constrained cyclic system results in the high asymmetric induction. The BINAP–Ru method has provided a straightforward way to highly enantiomerically pure α-amino phos-

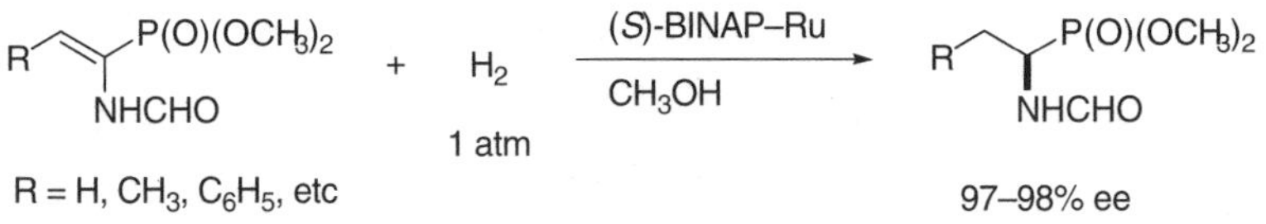

Scheme 1.7.

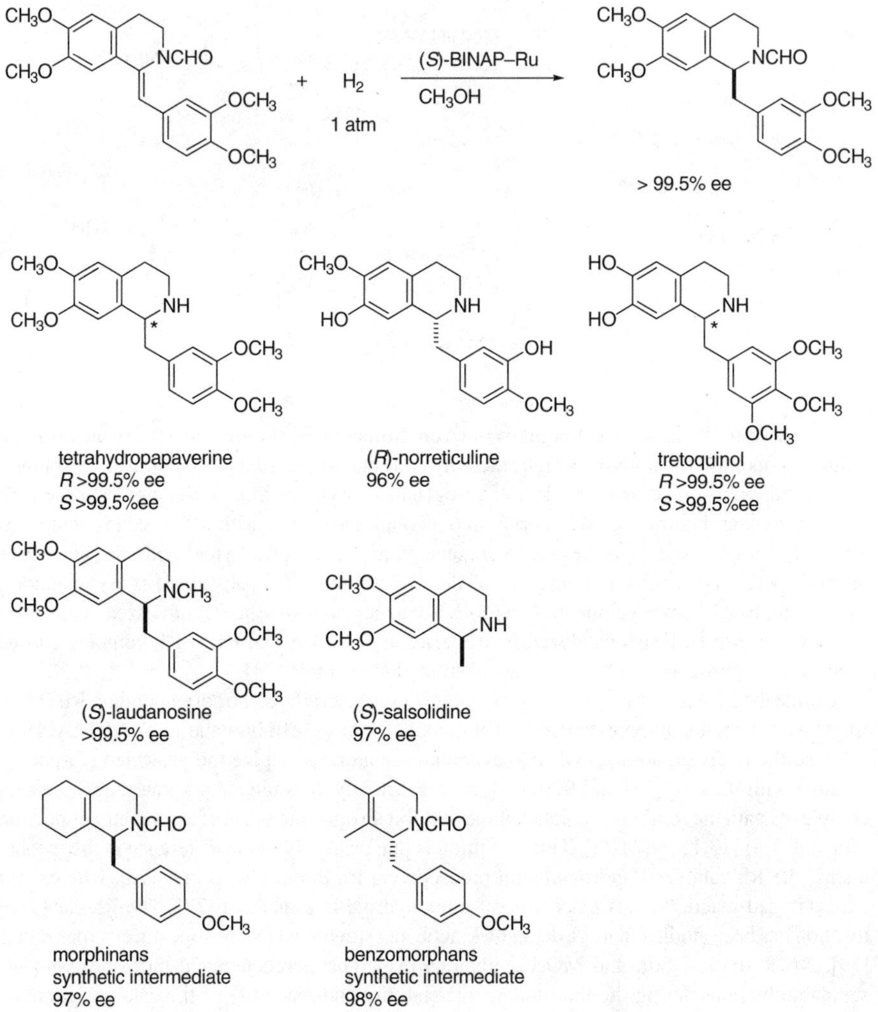

Scheme 1.8.

phonic acids and tetrahydroisoquinoline alkaloids. BIPHEMP–Ru-catalyzed hydrogenation is also effective for the asymmetric synthesis of 1-alkylated tetrahydroisoquinolines [83]. Use of the BINAP–Ru trifluoroacetate complexes at 100 atm of hydrogen allows the synthesis of optically active intermediates for benzomorphans and morphinans [84].

Hydrogenation of β-substituted (E)-β-(acylamino)acrylates catalyzed by the BINAP–Ru gives β-amino acid derivatives with high enantiopurities (Scheme 1.9) [85]. The Z double-bond isomers that have an intramolecular hydrogen bond between amide and ester groups are more reactive but are hydrogenated with poor enantioselectivity. The sodium or triethylammonium salts of the unsaturated acid are hydrogenated in water containing a sulfonated p-Tol-MeO-BIPHEP–Ru ditrifluoroacetate complex with an S/C of 1000–10,000 and an up to 99% optical

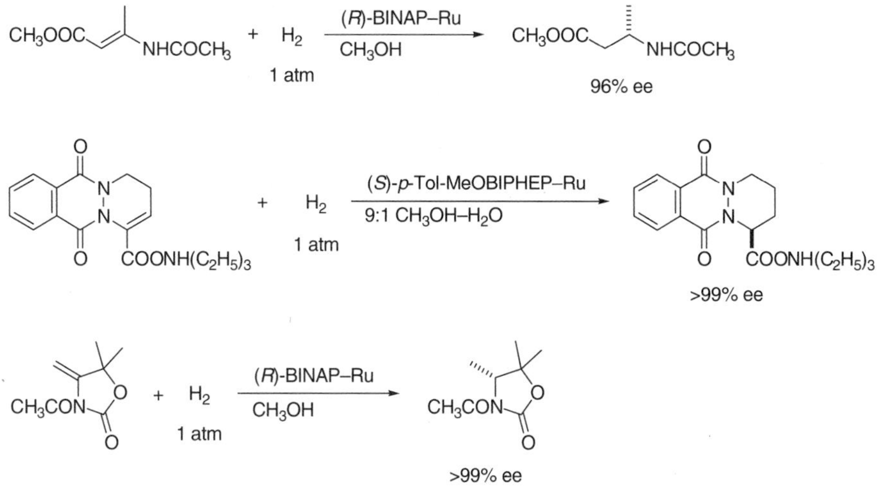

Scheme 1.9.

yield [86]. α-Methyl-*N*-acyloxazolidinones with high enantiomeric purities are prepared by the BINAP–Ru-catalyzed hydrogenation of the methylene substrates [87].

1.3.1.2. Enamines Removal of the *N*-acyl groups from the enamide substrates results in no hydrogenation with the phosphine–Rh or –Ru complexes. However, the titanocene catalyst **34**, generated by the addition of 2 equivalents of n-C_4H_9Li, followed by 2.5 equivalents of $C_6H_5SiH_3$ to a solution of **33** in THF under a hydrogen atmosphere, shows high reactivity toward 1,1-disubstituted enamines, giving the corresponding optically active tertiary amines [88] (Scheme 1.10). The enantioselectivity is independent of hydrogen pressure. For example, 1-(1-pyrrolidinyl)-1-(4-methoxyphenyl)ethene is reduced under the following conditions: 137 atm, 65°C; 6.4 atm, 65°C; 6.4 atm, rt; 2 atm, rt. In all cases, the ee of product is ca. 94%. The sterically crowded pyrrolidine enamine of pinacolone is unreactive. The catalyst system does not tolerate aromatic bromides. Titanium(III) hydride is believed to be the active catalyst.

1.3.1.3. Alkenyl Esters and Ethers Alkenyl carboxylates and enamides bear close resemblance in structure. Both possess a carbonyl oxygen atom, which is located three atoms from the olefin to be reduced and is properly aligned to facilitate chelation to a metal center. The topological analogy as well as the successful asymmetric hydrogenation of enamides imply the viability of enantioselective reduction of alkenyl carboxylates. In fact, some Rh and Ru complexes with chiral phosphines, including DIPAMP [89,90], PROPHOS [91], DuPHOS [19,92], and BINAP [90], are effective for this purpose (Scheme 1.11). DIPAMP–Rh and BINAP–Ru complexes hydrogenate (*Z*)-ethyl α-(acetyloxy)cinnamate and a 70:30 *E/Z* mixture of ethyl α-(acetyloxy)-β-(isopropyl)acrylate in 88–95% and 98% optical yields, respectively [90]. The sense of asymmetric induction in the DIPAMP–Rh-catalyzed hydrogenation of α-(acyloxy)acrylates and α-(acylamino)acrylates is the same. A wide range of α-(acyloxy)acrylates can be hydrogenated in the presence of a cationic Et-DuPHOS–Rh complex under 4 atm of hydrogen in alcoholic solvent to produce the protected α-hydroxy esters with enantiopurity as high as 93–99% ee [92]. Substrates bearing β-substituents can be used as *E/Z* isomeric

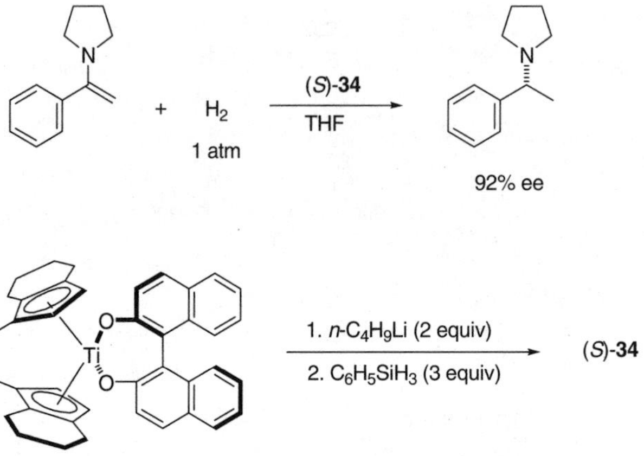

Scheme 1.10.

mixtures with no detrimental effect on the selectivity. Deuteration studies indicate that no substrate isomerization occurs during the reaction. Benzene as solvent inhibits the reaction through formation of a stable adduct between benzene and the cationic Et-DuPHOS–Rh fragment. Me-BPE–Rh complex is effective for asymmetric hydrogenation of β,β-disubstituted α-(acyloxy)acrylates (Scheme 1.12).

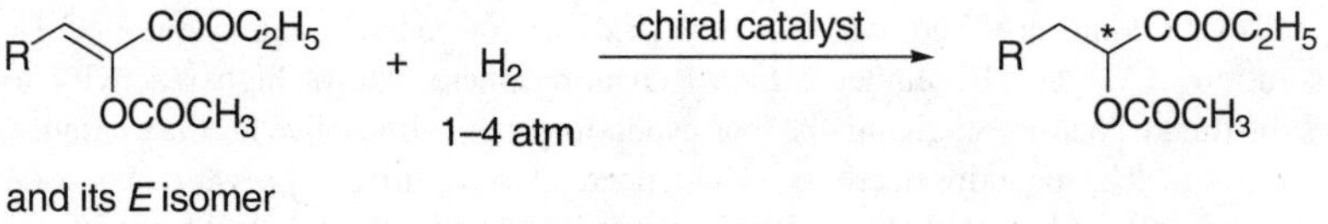

and its *E* isomer

	Substrate			Product	
R	E:Z	Catalyst	% ee	Config'n	
H	—	(R,R)-DIPAMP–Rh	89	S	
H	—	(R)-PROPHOS–Rh	81	S	
H	—	(S,S)-Et-DuPHOS–Rh	>99	S	
CH$_3$^{a}	75:25	(S,S)-Et-DuPHOS–Rh	99.3	S	
n-C$_5$H$_{11}$	78:22	(R,R)-Et-DuPHOS–Rh	>99	S	
n-C$_4$H$_9$	75:25	(R,R)-DIPAMP–Rh	92	S	
(CH$_3$)$_2$CH^a	86:14	(S,S)-Et-DuPHOS–Rh	96.9	S	
(CH$_3$)$_2$CH	70:30	(R,R)-DIPAMP–Rh	92	S	
(CH$_3$)$_2$CH	70:30	(R)-BINAP–Ru	98	S	
C$_6$H$_5$	0:100	(R,R)-DIPAMP–Rh	95	S	

^{a} Benzoyl enolate.

Scheme 1.11.

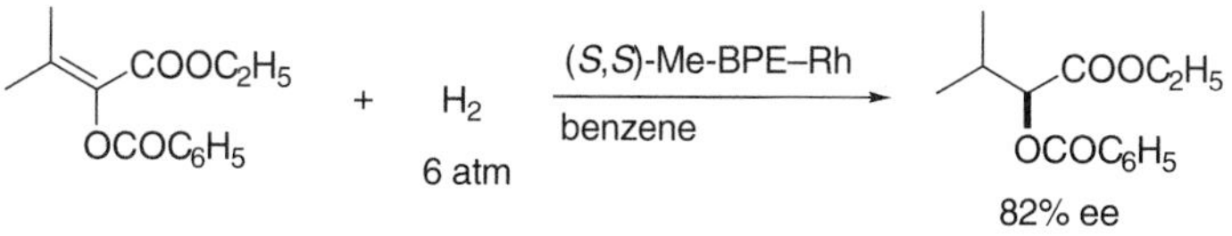

Scheme 1.12.

β-Acyloxy-substituted α,β-unsaturated carbonyl compounds are also the substrate of choice for asymmetric hydrogenation. Ethyl 3-(acetyloxy)-2-butenoate can be hydrogenated by DI-PAMP– [89] and PROPHOS–Rh complexes [91] in 89 and 81% optical yields, respectively.

Conjugation of the olefinic double bond to the alkoxycarbonyl or acyl function is not necessary for a high selectivity and reactivity. Thus, a DuPHOS–Rh [19], BPE–Rh [19], or BINAP–Ru complex [93] acts as an excellent catalyst for hydrogenations of such acyloxy-substituted olefins as shown in Schemes 1.13 and 1.14. With the DuPHOS–Rh complex, 1-phenyl-1-ethenyl acetate is efficiently converted, under 2 atm of hydrogen in methanol, to the phenethyl acetate with 89% ee. Higher enantioselectivities are achieved in the hydrogenation of 1-alkenyl- or 1-alkynyl-1-ethynyl acetate [94]. Under the reaction conditions, the C1 alkenyl group remains intact and the carbon–carbon triple bond is reduced to a double bond. With 1-alkyl-1-ethenyl acetate, the optical yield drops to ca. 30%. 1,1,1-Trifluoro-2-propenyl acetate and 1-naphthyl-1-ethenyl acetate are hydrogenated by a BPE–Rh complex to give the corresponding esters with 95% ee and 94% ee, respectively. A DIOP–Rh complex catalyzes the hydrogenation of 1-phenyl-1-ethenyl diphenylphosphinate in 80% optical yield [95]. Although little success has been reported with acyclic α-alkyl-substituted acyl enolates, four- and five-membered cyclic

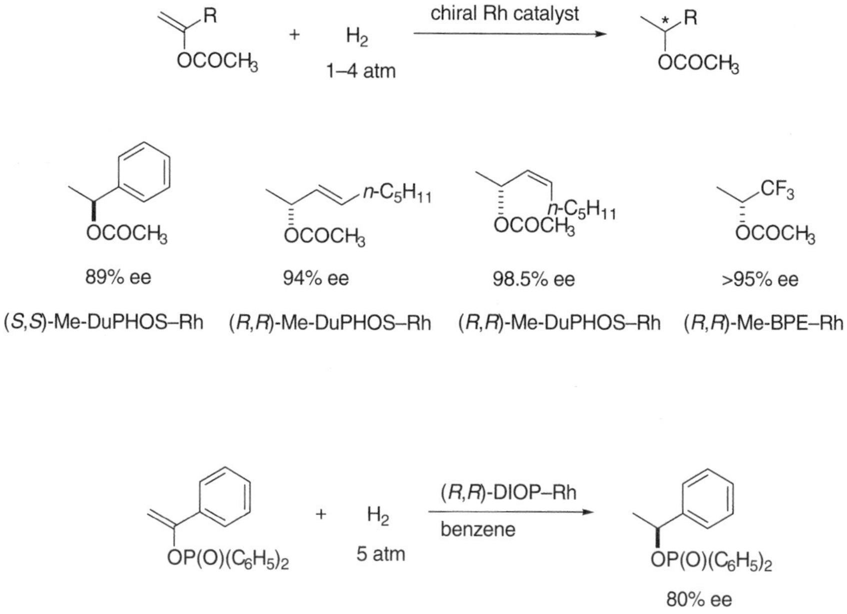

Scheme 1.13.

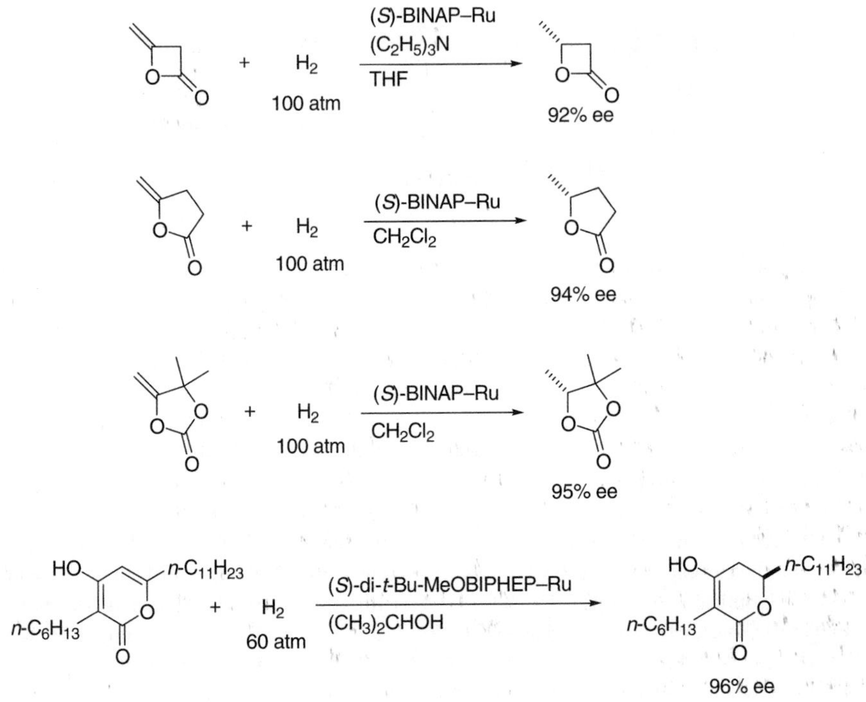

Scheme 1.14.

lactones or carbonates having an exocyclic methylene bond are hydrogenated in a highly enantioselective manner (Scheme 1.14). β-Methyl-β-propiolactone with 92% ee, γ-methyl-γ-butyrolactone with 95% ee [93], and the carbonate of 3-methyl-2,3-butanediol [96] with 95% ee are obtained by BINAP–Ru-catalyzed high-pressure hydrogenation in THF or CH_2Cl_2. Optically active β-methyl-β-propiolactone is a promising starting monomer of biodegradable polymers. With a six-membered substrate or an endo isomer of 4-methylene γ-lactone, the enantioselectivity is considerably decreased. The double chelation of olefin and oxygen atom to the Ru center may be important for high-enantioface differentiation [93]. A dicationic Ru complex of (S)-di-t-Bu-MeOBIPHEP allows a highly regio- and enantioselective hydrogenation of an α-pyrone. Selective saturation of the olefinic double bond in the acyl enolate part gives the corresponding R dihydropyrone with 96% ee [86].

BINAP–Ru complexes can catalyze the enantioselective hydrogenation of alkenyl ethers as shown in Scheme 1.15 [93]. 2-Methyltetrahydrofuran with 91% ee and 87% ee can be synthesized by BINAP–Ru-catalyzed hydrogenation of 2-methylenetetrahydrofuran and the endo-type substrate, 2-methyl-3,4-dihydrofuran, in CH_2Cl_2 under 100 atm of hydrogen, respectively. With the same Ru complex, phenyl 1-phenylethyl ether, an acyclic alkenyl ether, is reduced in a moderate optical yield.

Highly enantioselective hydrogenation of olefins with aprotic oxygen functionalities like esters and ethers has rarely been attained. Recent investigation with chiral transition-metal complexes, especially BINAP–Ru and DuPHOS–Rh complexes, has expanded the substrates to various alkenyl esters and ethers.

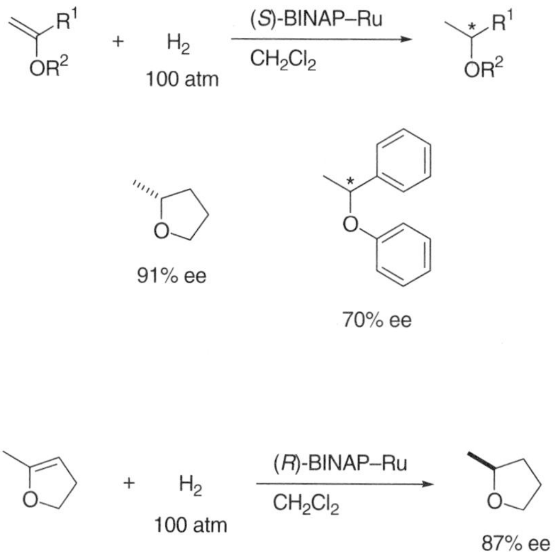

Scheme 1.15.

1.3.1.4. α,β- and β,γ-Unsaturated Carboxylic Acids

In the presence of a neomenthyl-diphenylphosphine–Rh complex, (E)-3-phenyl-2-butenoic acid was hydrogenated to the corresponding saturated acid with 61% ee [97]. This Morrison's result in 1971 had provided much hope for the further development of the asymmetric hydrogenation of simple acrylic acid analogues. However, the detailed investigation on the relationship between the substrate structures and selectivity has revealed that additional coordinating functionalities are required to attain a high enantioselectivity in Rh-catalyzed hydrogenation.

A breakthrough was provided by the discovery of the BINAP–Ru dicarboxylate complexes [82,98]. A range of substituted acrylic acids can be efficiently hydrogenated in alcoholic media to give saturated products with high enantiopurities (Scheme 1.16) [99]. The efficiency depends on the substitution pattern and reaction conditions, particularly the hydrogen pressure. With geranic acid, only the double bond closest to the carboxyl group is saturated. In the hydrogenation of tiglic acid by using the BINAP–Ru dicarboxylate complex, the operation of a monohydride mechanism is supposed on the basis of deuterium-labeling experiments and kinetics [100,101]. Other useful BINAP–Ru complexes and their derivatives include: [RuX(binap)(arene)]Y (X = halogen, Y = halogen or BF_4) [102], Ru(2-methallyl)$_2$(binap) [103], Ru(allyl)(acac-F_6)(binap) [104], [$NH_2(C_2H_5)_2$][{RuCl(binap)}$_2$(μ-Cl)$_3$] [74a,105,106], RuCl$_2$(ArCN)(binap)$_2$ [85], Ru(acac)(mnaa)(binap)(CH$_3$OH) (MNAA = 2-(6'-methoxy-naphth-2'-yl)acrylate anion) [107], RuH(binap)$_2$PF$_6$ [108], RuHCl(binap)$_2$ [108], and Ru(OCOCH$_3$)$_2$(bitianp) [109]. The hydrogenation of tiglic acid smoothly proceeds in supercritical carbon dioxide containing CF$_3$(CF$_2$)CH$_2$OH and Ru(OCOCH$_3$)$_2$[(S)-H$_8$-binap] under 25–35 atm of hydrogen and 175 atm of carbon dioxide at 50°C to give (S)-2-methylbutanoic acid in more than 99% yield and up to 89% ee [110].

Enantioselective hydrogenation of α-aryl-substituted acrylic acids has extensively been studied because of the pharmaceutical importance of the saturated products. Anti-inflammatory (S)-naproxen with 97% ee is obtained by the high-pressure hydrogenation of 2-(6'-methoxy-naphth-2'-yl)acrylic acid by using Ru(OCOCH$_3$)$_2$[(S)-binap] [99]. The hydrogenation rate is

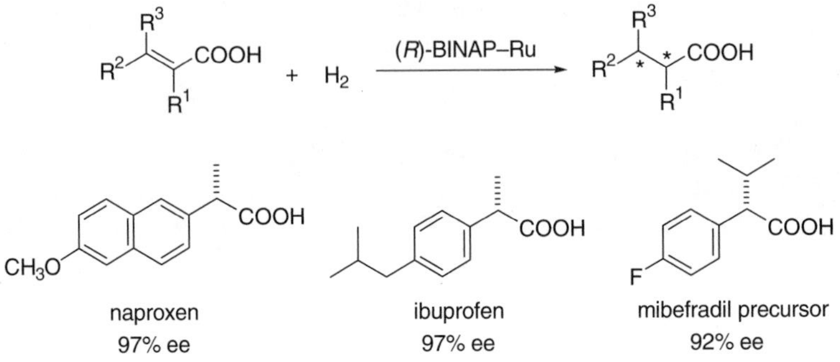

naproxen
97% ee

ibuprofen
97% ee

mibefradil precursor
92% ee

Substrate			Product	
R^1	R^2	R^3	% ee	Config'n
CH_3	CH_3	H	91 (97)[a]	2R
CH_3	C_2H_5	H	78 (97)[a]	2R
CH_3	H	H	35 (82)[a]	2R
C_6H_5	H	H	92	2R
$HOCH_2$	CH_3	H	95	_[b]
H	CH_3	C_6H_5	85	3S
H	$(CH_3)_2C{=}CHCH_2$	CH_3	87	3S
CH_3	CH_3OCOCH_2	H	83	2R
H	CH_3OCOCH_2	CH_3	95	3R
H	$HOCH_2$	CH_3	93	3S
CH_3	CH_3	H	89[c]	2S

[a] (R)-H_8-BINAP–Ru complex is used. [b] Not determined. [c] (S)-H_8-BINAP–Ru complex is used.

Scheme 1.16.

ca. 10-fold enhanced by use of Ru(acac)(mnaa)[(S)-binap](CH$_3$OH) [107]. For example, when an S/C of 1000 is used under 42 atm of hydrogen at rt, the reaction completes in 10 min to give (S)-naproxen with 88% ee. H_8-BINAP–Ru complexes also show higher reactivity and selectivity [111]. This finding presents a useful synthetic route to (S)-ibuprofen. The high efficiency may be due to the larger dihedral angle between the two aromatic rings of the tetralin moieties of H_8-BINAP than BINAP. A number of technical refinements have been tried for the significant reaction. 2-(4-Fluorophenyl)-3-methylbut-2-enoic acid, a trisubstituted acrylic acid, is hydrogenated in 92% optical yield under 270 atm of hydrogen by use of a continuously stirred tank reactor system, giving a calcium antagonist mibefradil synthetic intermediate [112]. A catalyst precursor [NH$_2$(C$_2$H$_5$)$_2$][{RuCl[(S)-binap]}$_2$(μ-Cl)$_3$] [74a,105,106] dissolved in 1-n-butyl-3-methylimidazolium tetrafluoroborate molten salt can hydrogenate 2-arylacrylic acids with enantioselectivities similar or higher than those obtained in homogeneous media [113]. The hydrogenation products can be quantitatively separated from the reaction mixture, and the recovered ionic liquid catalyst solution can be reused several times without any significant

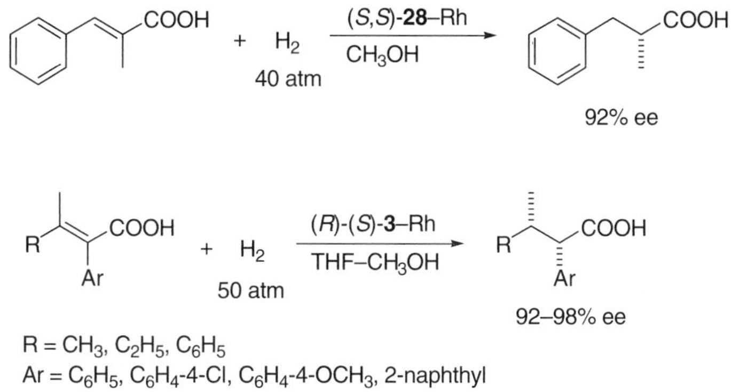

$R = CH_3, C_2H_5, C_6H_5$
$Ar = C_6H_5, C_6H_4\text{-}4\text{-}Cl, C_6H_4\text{-}4\text{-}OCH_3, 2\text{-naphthyl}$

Scheme 1.17.

changes in the catalytic reactivity or selectivity. $[RuCl(binap\text{-}4\text{-}SO_3Na)C_6H_6]Cl$ held in a film of ethylene glycol on a controlled porous hydrophilic support is used as a heterogeneous catalyst for the high-pressure hydrogenation of 2-(6′-methoxy-2′-naphthyl)acrylic acid in a 1:1 mixture of chloroform and cyclohexane. Naproxen with up to 96% ee is obtained at 100% conversion [114]. Asymmetric hydrogenation of 1-arylethenylphosphonic acid is also examined for the synthesis of phospho analog of naproxen-type drugs, though the ee values are moderate with BINAP– or MeOBIPHEP–Ru complexes [115].

Only limited successful examples of asymmetric hydrogenation of acrylic acids derivatives have included the use of chiral Rh complexes (Scheme 1.17). The diamino phosphine (**28**) utilizes selective ligation of the amino unit to a Rh center and also exerts electrostatic interaction with a substrate. Its Rh complex catalyzes enantioselective hydrogenation of 2-methylcinnamic acid in 92% optical yield [116]. Certain cationic Rh complexes can attain highly enantioselective hydrogenation of trisubstituted acrylic acids [117]. 2-(6′-Methoxynaphth-2′-yl)acrylic acid is hydrogenated by an (S,S)-BIPNOR–Rh complex in methanol at 4 atm to give (S)-naproxen with 98% ee but only in 30% yield [26].

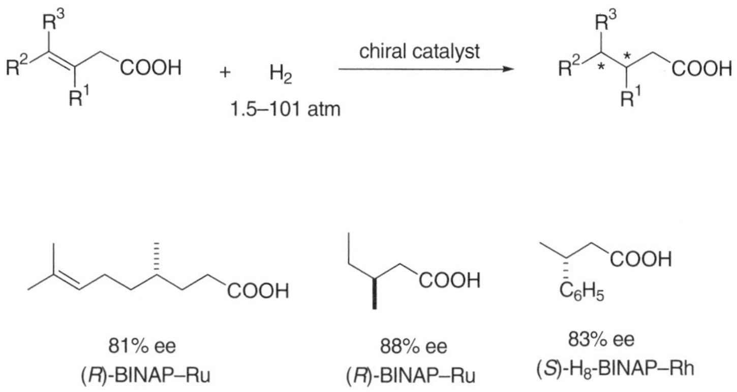

Scheme 1.18.

β,γ-Unsaturated carboxylic acids, shown in Scheme 18, are also enantioselectively hydrogenated with the aid of BINAP– or H_8-BINAP–Ru complexes to give the saturated acids with 80–90% ee [99,111].

Itaconic acid is classified as both α,β-unsaturated carboxylic acid and β,γ-unsaturated carboxylic acid. Usually it is conceived to behave as a β,γ-unsaturated carboxylic acid because the structure is viewed as the carbon analog of α-(acylamino)acrylates. In fact, hydrogenation of itaconic acid and its derivatives has been extensively studied [10]. The representative results are listed in Scheme 1.19. The first highly enantioselective hydrogenation of itaconic acid was achieved by using a cationic BPPM–Rh complex and an equimolar amount of triethylamine in methanol under 20 atm of hydrogen to give the saturated compound with 92% ee [118]. Endowing the original BPPM or DIOP diphosphine ligands with more electron-donating

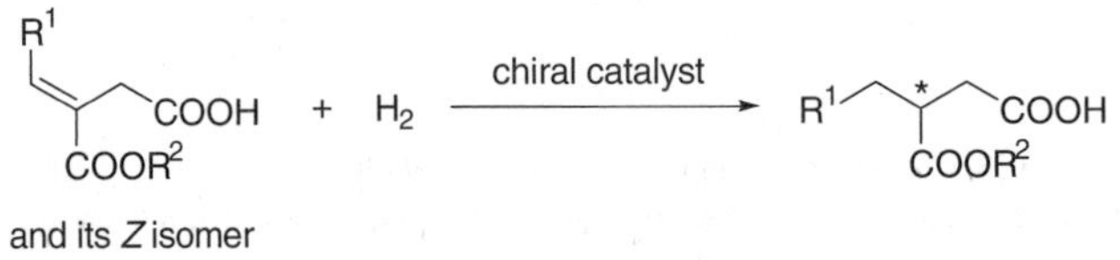

and its *Z* isomer

Substrate				Product	
R^1	R^2	*E:Z*	Catalyst	% ee	Confign
H	H	—	(2S,4S)-BPPM–Rh	92	S
H	H	—	(2S,4S)-Ph-CAPP–Rh	95	S
H	H	—	(2S,4S)-BCPM–Rh	92	S
H	H	—	(2S,4S)-MOD-BCPM–Rh	80.5	S
H	H	—	(R,R)-MOD-DIOP–Rh	91	S
H	H	—	(R,R)-BIPNOR–Rh	93	S
H	H	—	(R)-BICHEP–Rh	93	R
H	H	—	(S)-BICPEP–Rh	95	_[a]
H	H	—	(R)-BINAP–Ru	88	S
H	CH_3	—	(S,S)-Et-DuPHOS–Rh	97	S
H	CH_3	—	(R,R)-DIPAMP–Rh	88	R
C_6H_5	H	100:0	(R,R)-MOD-DIOP–Rh	96	S
C_6H_5	H	100:0	(R)-BINAP–Ru	90	S
[b]	H	100:0	(R)-BINAP–Ru	94	S
[b]	CH_3	100:0	(R,R)-MOD-DIOP–Rh	97	S
C_6H_5	CH_3	[c]	(S,S)-Et-DuPHOS–Rh	97	S
$(CH_3)_3C$	CH_3	[c]	(S,S)-Et-DuPHOS–Rh	99	S
$(CH_3)_2CH$	CH_3	[c]	(S,S)-Et-DuPHOS–Rh	99	S
C_2H_5	CH_3	[c]	(S,S)-Et-DuPHOS–Rh	99	S

[a] Not determined. [b] 3,4-Methylenedioxyphenyl. [c] A 67:33–90:10 mixture.

Scheme 1.19.

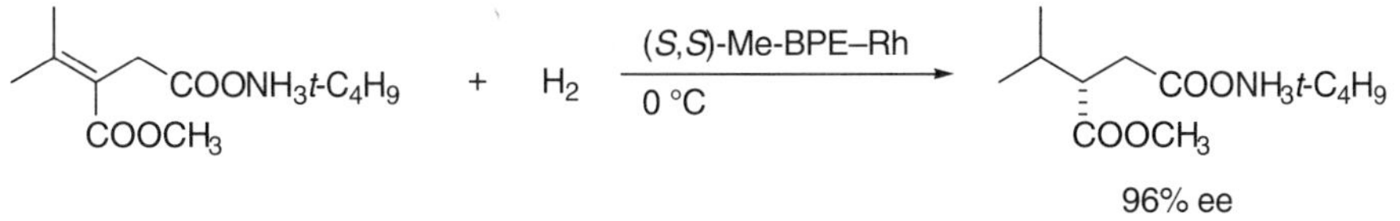

$$\text{Scheme 1.20.}$$

character results in rate enhancement. Rh complexes with CAPP [119], BCPM [120], MOD-BCPM [121], and MOD-DIOP [122] allow atmospheric pressure hydrogenation. Reaction by using a cationic BICHEP–Rh complex smoothly proceeds with an S/C of 10,000 under 5 atm of hydrogen, giving 2-methylsuccinic acid with 93% ee [123]. BICPEP– [124] and BIPNOR– Rh [26] complexes as well as a BINAP–Ru complex [74,125] are also effective, giving 2-methylsuccinic acid with 88–95% ee. A Ru complex with a BINAP derivative covalently bonded to an aminomethylated polystyrene resin is used for hydrogenation of itaconic acid, though both the rate and enantioselectivity are decreased [126]. Introduction of an aryl substituent at the β position of itaconic acid has little effect on the reactivity and selectivity in the MOD-DIOP–Rh- or BINAP–Ru-catalyzed hydrogenation [122,125]. Methyl itaconate, a β,γ-unsaturated carboxylic acid, is hydrogenated in up to 97% optical yield by use of DIPAMP– [127] and Et-DuPHOS–Rh complexes [128]. With a cationic Et-DuPHOS–Rh complex, methyl itaconate derivatives possessing a β-aryl or -alkyl substituent are usable. The substrates may be used as crude E/Z isomeric mixtures. Addition of a catalytic amount of a base such as sodium methoxide or primary or tertiary amines significantly enhances the reactivity. Given the sterically encumbered nature of the β,β-disubstituted itaconates, Me-BPE–Rh complex is effective in furnishing the products with up to 96% ee (Scheme 1.20) [128]. 2,2-Dimethylidene-succinic acid is hydrogenated by an (R)-BINAP–Ru complex at 100 atm to give a 99.8:1.2 mixture of (2S,3S)-dimethylsuccinic acid with 96% ee and the meso isomer (Scheme 21) [108].

1.3.1.5. α,β-Unsaturated Esters, Amides, Aldehydes, and Ketones The successful results with the title compounds are limited to a small range of substrates at the present stage. Some examples are collected in Schemes 1.22 and 1.23. Dimethyl itaconate is hydrogenated under low pressure by using Rh complexes coordinated by DIPAMP [127], a BPPM derivative [129], a triphospholane **4** [23], BICHEP [21], and so on, to give dimethyl 2-methylsuccinate with up to 99% ee. A Rh complex bearing a helical phosphine ligand, PHelix, is also used for asymmetric hydrogenation of dimethyl itaconate, although both the reactivity and enantioselectivity are low [130]. Dimethyl esters of racemic 3-substituted itaconic acids can be kinetically resolved with the k_f/k_s value of up to 16 by the DIPAMP–Rh-catalyzed hydrogenation [131]. Itaconic anhydride, though not an ester substrate, is hydrogenated by an (S)-BINAP–Ru complex under

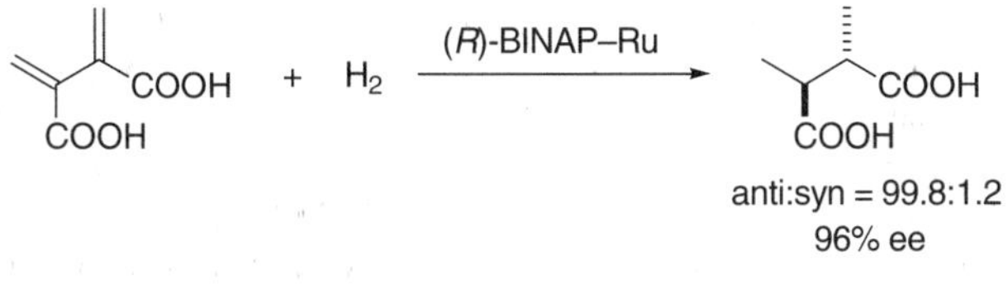

anti:syn = 99.8:1.2
96% ee

$$\text{Scheme 1.21.}$$

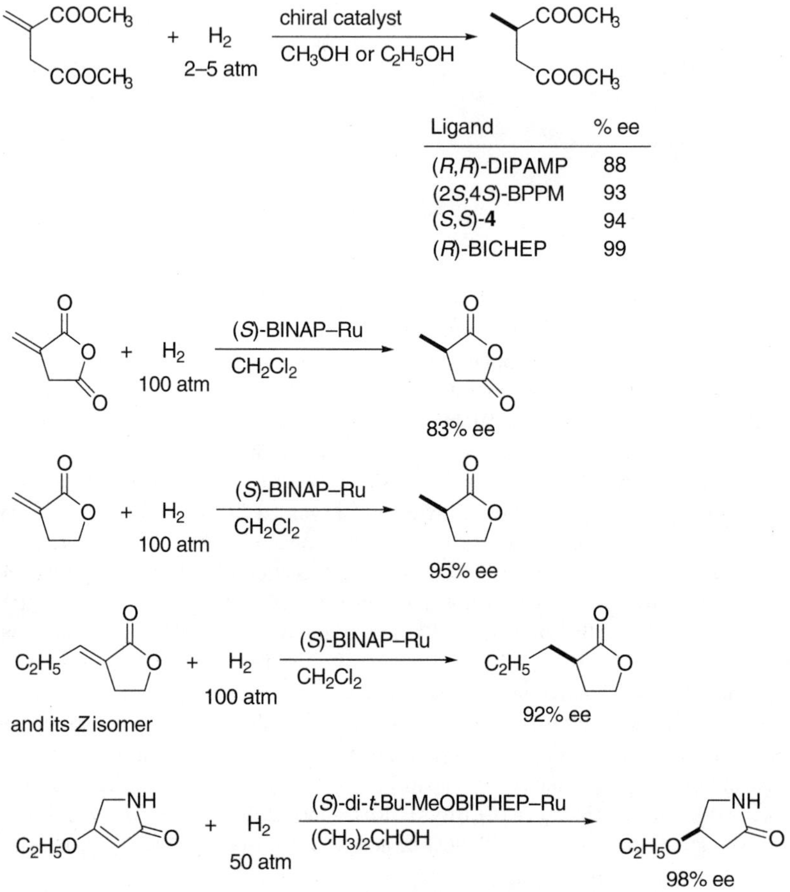

Ligand	% ee
(*R*,*R*)-DIPAMP	88
(2*S*,4*S*)-BPPM	93
(*S*,*S*)-**4**	94
(*R*)-BICHEP	99

Scheme 1.22.

100 atm of hydrogen to give the saturated anhydride with 83% ee [93]. Five-membered lactones with exocyclic C=C bonds are hydrogenated in a very high enantioselectivity by BINAP–Ru complexes [93]. Hydrogenation of 2-methylene- and -propylidene-γ-butyrolactones gives the corresponding γ-butyrolactones with 92% ee. Both the sense and degree of enantioselectivity are not affected by the olefin geometry. A dicationic (*S*)-di-*t*-Bu-MeOBIPHEP–Ru complex hydrogenates 3-ethoxypyrrolinone to give the β-alkoxy γ-lactam with 98% ee [86].

Moderate success has been achieved with the hydrogenation of acyclic α,β-unsaturated aldehydes—neral and geranial—with the aid of catalysts derived from TBPC and $Rh_6(CO)_{16}$ or $RhH(CO)[P(C_6H_5)_3]$ [132]. The reactions gave citronellal with 79% ee and 60% ee, respectively. In the presence of RuHCl(TBPC), endocyclic α,β-unsaturated ketones such as isophorone and 2-methyl-2-cyclohexenone are hydrogenated to give the chiral ketones with up to 62% ee [133], though the conversions are not satisfactory. The BINAP–Ru complexes, including [RuCl(binap)(benzene)]Cl, $[NH_2(C_2H_5)_2][\{RuCl(binap)\}_2(\mu\text{-}Cl)_3]$ [74a,105,106], or $Ru(OCOCH_3)_2$(binap), can hydrogenate 2-alkylidenecyclopentanones in up to 98% optical yield [93].

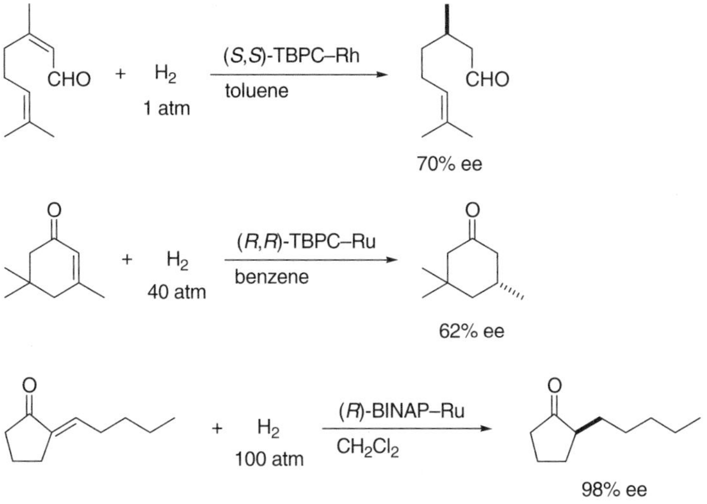

CHO + H₂ (1 atm) → (*S,S*)-TBPC–Rh / toluene → CHO (70% ee)

+ H₂ (40 atm) → (*R,R*)-TBPC–Ru / benzene → (62% ee)

+ H₂ (100 atm) → (*R*)-BINAP–Ru / CH₂Cl₂ → (98% ee)

Scheme 1.23.

1.3.1.6. Allylic and Homoallylic Alcohols In spite of many examples of diastereoselective hydrogenation of chiral allylic alcohols by using achiral Rh or Ir complexes [134], the enantioselective versions were unsuccessful. With a neutral Cy-BINAP–Rh complex, geraniol or nerol is hydrogenated in benzene to produce citronellol with only 50–60% ee [135]. The use of cationic BINAP–Rh complexes markedly reduces the enantioselectivity.

The invention of BINAP–Ru dicarboxylate complexes, however, changed the situation (Scheme 1.24). Prochiral allylic and homoallylic alcohols can be hydrogenated in a highly enantioselective manner [136]. Geraniol or nerol is quantitatively converted to citronellol with 96–99% ee in methanol at an initial hydrogen pressure higher than 30 atm. The S/C approaches 50,000 in the reaction by using the Ru bis(trifluoroacetate) catalyst. Only allylic alcohol double bond is hydrogenated, leaving the isolated C6–C7 double bond intact. In this catalytic system, the BINAP–Ru complex isomerizes geraniol to γ-geraniol, which is hydrogenated to citronellol of opposite absolute stereochemistry [137]. Therefore, the low-pressure hydrogenation that decreases the hydrogenation rate relative to the isomerization rate results in a low enantioselectivity. Nerol is insensitive to changes in pressure. The hydrogenation process has been successfully applied to the practical synthesis of optically active terpenes and related compounds, including (3*R*,7*R*)-3,7,11-trimethyldodecanol—a key synthetic intermediate of vitamins E and K1 [136] and dolicols [138]. Hydrogenation of homogeraniol occurs regioselectively at the C3–C4 double bond in a high optical yield with the same asymmetric orientation as observed with geraniol. Bishomogeraniol is not reduced. Similar dicarboxylate complexes with BIPHEMP and tetraMe-BITIANP ligands are also effective for asymmetric hydrogenation of allylic alcohols [83,109].

As shown in Figure 1.10, kinetic resolution of racemic acyclic and cyclic secondary alcohols can be achieved by the BINAP–Ru method with up to 74:1 differentiation between the enantiomers [139]. An application includes a practical resolution of a racemic 4-hydroxy-2-cyclopentenone, an important prostaglandin building block that is achievable on a multi-kilogram scale. Racemic methyl α-(hydroxyethyl)acrylate is reduced by hydrogen

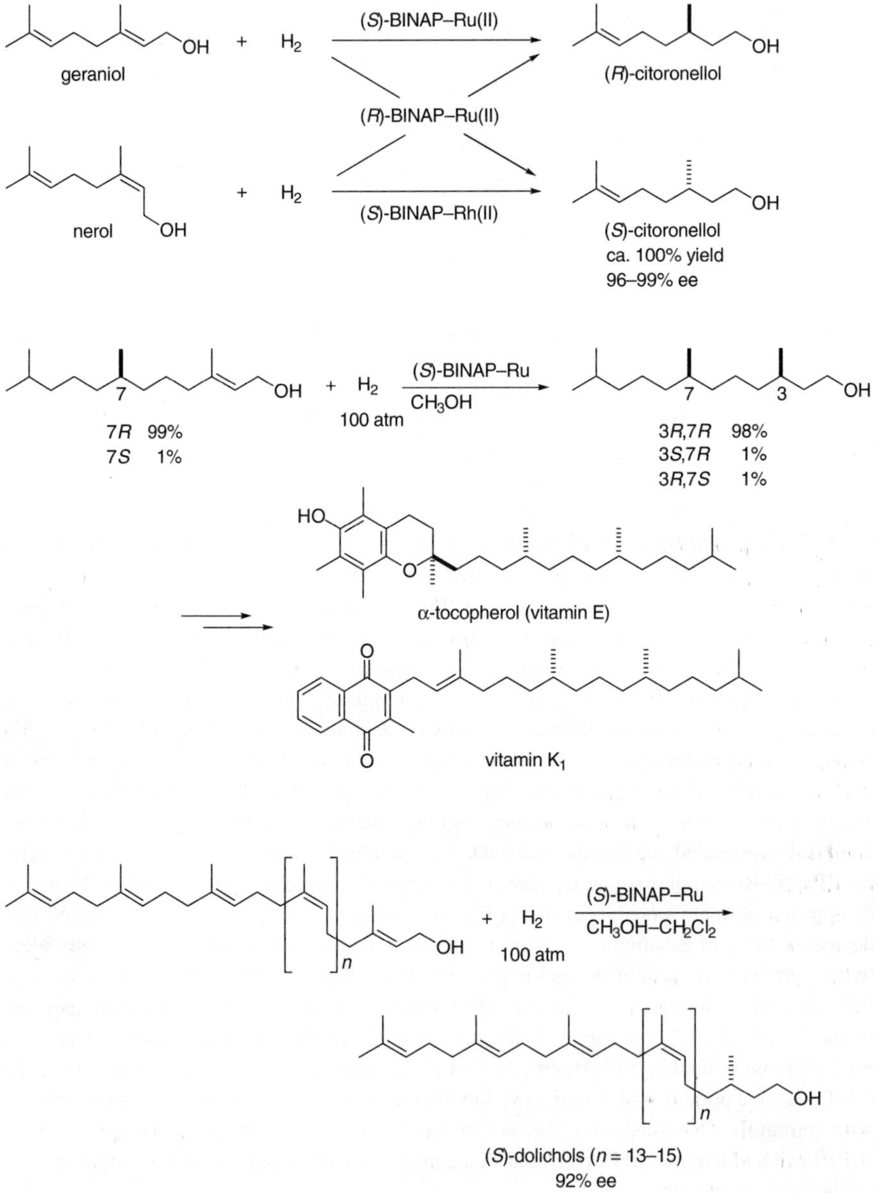
OH
geraniol
+ H2
(S)-BINAP–Ru(II)
(R)-BINAP–Ru(II)
(S)-BINAP–Rh(II)
OH
(R)-citoronellol
nerol
OH
+ H2
OH
(S)-citoronellol
ca. 100% yield
96–99% ee
7
OH
7R 99%
7S 1%
+ H2
100 atm
(S)-BINAP–Ru
CH3OH
7
3
OH
3R,7R 98%
3S,7R 1%
3R,7S 1%
HO
O
α-tocopherol (vitamin E)
O
O
vitamin K1
n
OH
+ H2
100 atm
(S)-BINAP–Ru
CH3OH–CH2Cl2
n
OH
(S)-dolichols (n = 13–15)
92% ee

Scheme 1.24.

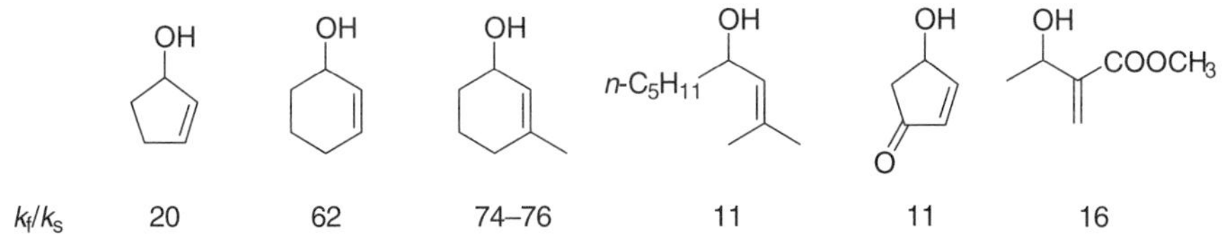

| k_f/k_s | 20 | 62 | 74–76 | 11 | 11 | 16 |

Figure 1.10. Kinetic resolution of racemic secondary alcohols by BINAP–Ru catalyzed hydrogenation.

in the presence of a BINAP–Ru or DIPAMP–Rh complex [134] with a k_f/k_s value of 16 and 7, respectively.

In the BINAP–Ru-catalyzed hydrogenation of 2-substituted 2-propen-1-ols, only a low level of enantioselectivity is observed. The selectivity is, however, pronouncedly enhanced by use of 2,4,6-trichlorobenzoate of the original allylic alcohol (Scheme 1.25) [140].

A cationic Ir complex possessing phosphanodihydrooxazole **26** is usable for asymmetric hydrogenation of allylic alcohols. (E)-2-Methyl-3-phenyl-2-propen-1-ol can be converted in CH$_2$Cl$_2$ containing 1 mol % of the Ir complex to the saturated product in 95% yield and 96% ee (Scheme 1.26) [141]. The process is used in the enantioselective synthesis of the artificial fragrance lilial.

1.3.2 Unfunctionalized Olefins

Historically, reaction of simple olefins in the presence of chiral phosphine–Rh complexes in 1968 marked the first examples of homogeneous asymmetric hydrogenation [6]. However, only a few successful results have been reported for asymmetric hydrogenation of unfunctionalized olefins. Some examples with late and early transition-metal complexes are illustrated in Schemes 1.27–28 and Schemes 1.29–30, respectively.

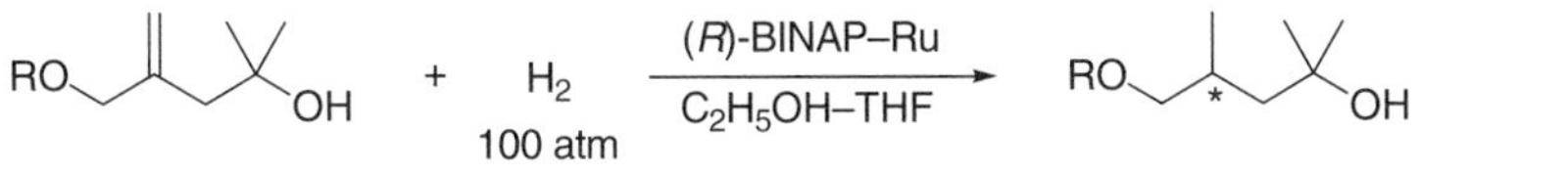

R	% ee	confign
H	9	R
2,4,6-Cl$_3$C$_6$H$_2$CO	85	S

Scheme 1.25.

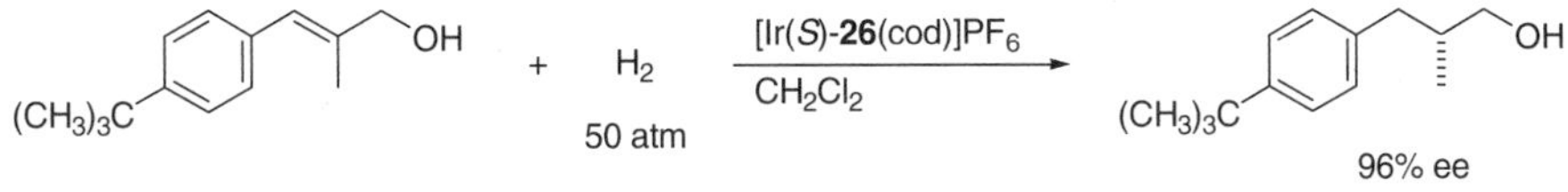

Scheme 1.26.

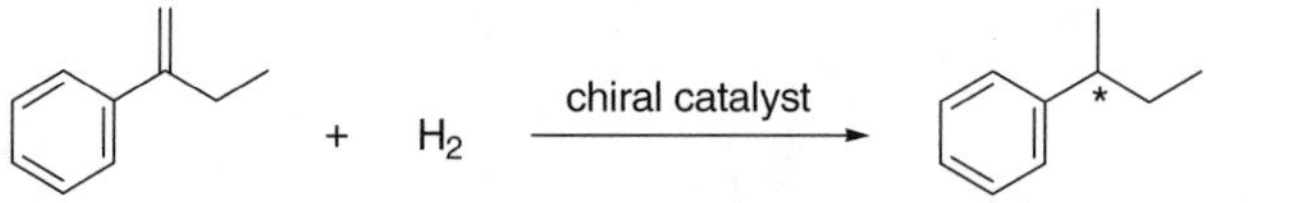

Ligand	% ee	Config'n
(S,S)-BDPCP	60	R
9	77	R
(R)-**6**	65	R
(S,S)-BDPP	54	S

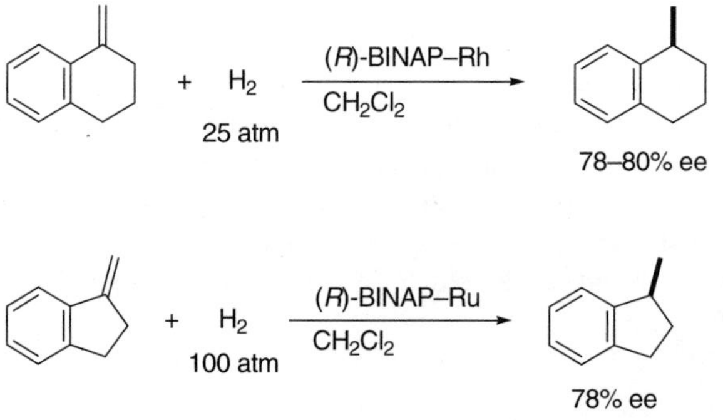

Scheme 1.27.

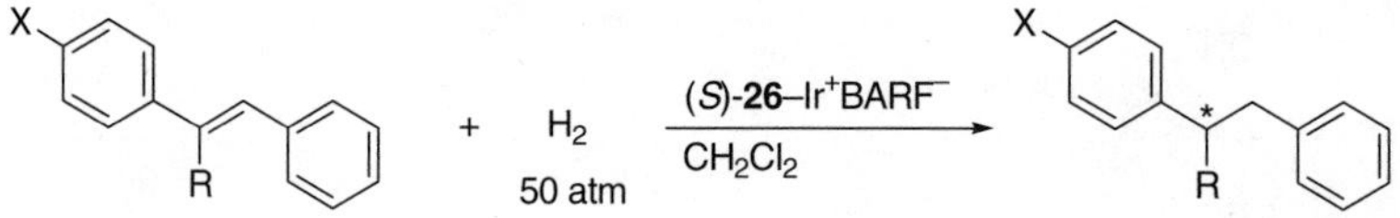

X	R	% ee (config'n)
H	CH$_3$	97 (−)
Cl	CH$_3$	95 (−)
CH$_3$O	CH$_3$	98 (R)
CH$_3$O	C$_2$H$_5$	95 (−)

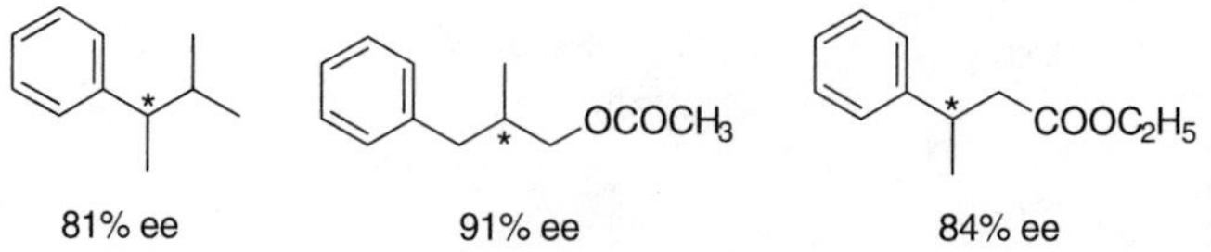

81% ee   91% ee   84% ee

Scheme 1.28.

Catalyst	R	% ee	Confign
35 + n-C$_4$H$_9$Li	C$_2$H$_5$	95	S
36	C$_2$H$_5$	96	S
37 + [Al(CH$_3$)O]$_n$	H^a	65	R

^{a} D$_2$ instead of H$_2$ is used.

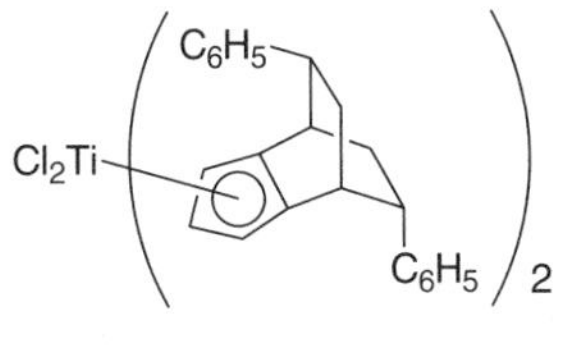

Scheme 1.29.

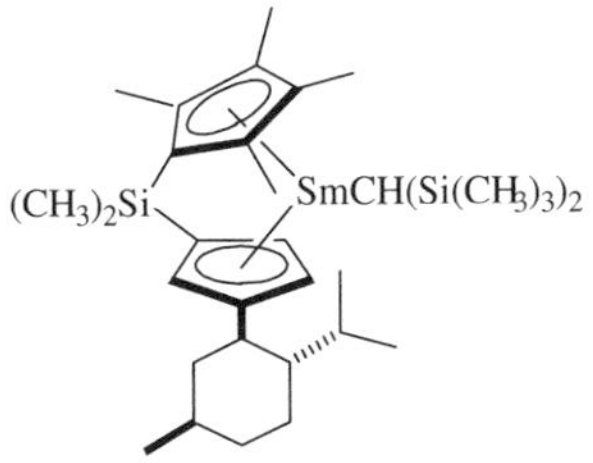

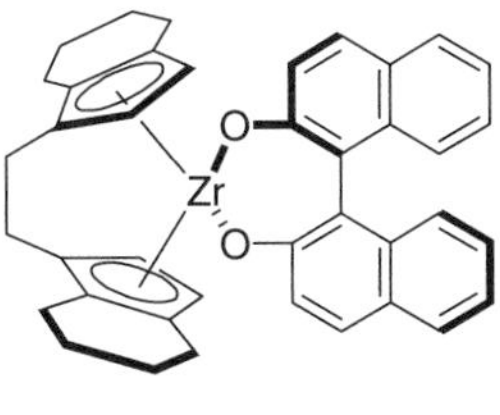

A neutral BDPCP–Rh complex promotes hydrogenation of 2-phenyl-1-butene at 50°C and 50 atm to give 2-phenylbutane with 60% ee [142]. Use of a Rh complex possessing BDPCH, a homologue of BDPCP phosphinite ligand, decreases the enantioselectivity to 33% ee [143]. By use of a Rh complex immobilized on 2,3-O-bis(diphenylphosphino)-6-O-triphenylmethylcellulose (**9**), the enantioselectivity is increased to 77% ee, although the conversion is only 1.2% even after 8 days at 50°C and 50 atm of hydrogen [144]. The hydrogen pressure can be reduced to 1 atm by use of the Rh complex of a bidentate phosphine, BDPP. α-Ethylstyrene is reduced to 3-phenylbutane with 54% ee [145]. With a Rh complex of 2-diphenylphosphino-2'-diphenyl-phosphinomethyl-1,1'-binaphthyl (**6**), α-ethylstyrene is saturated in 65% optical yield under 25 atm of hydrogen [146]. Using α-propylstyrene, the enantioselectivity increases to 77% ee, which is the highest value reported thus far in the hydrogenation of acyclic unfunctionalized olefins. Although the BINAP–Rh complexes are not effective for the hydrogenation of α-alkylstyrene, the same catalyst system affords ca. 80% ee for 1-methylenetetralin, a six-membered methylenebenzocycloalkane [147]. BINAP–Ru complexes hydrogenate 1-methyleneindane at a high pressure to give 1-methylindane with 78% ee [147]. With the same Ru complex, α-alkylstyrenes are hydrogenated in only 10–30% optical yield.

The cationic Ir complex with the phosphonodihydrooxazole ligands **26** and BARF as a counter anion smoothly hydrogenates a number of aryl-substituted unfunctionalized olefins in CH$_2$Cl$_2$ to give the saturated compound with >99% ee (Scheme 1.28) [141]. The loading amount of the catalysts is typically <1 mol %. Replacement of BARF with PF$_6^-$, SbF$_6^-$, BF$_4^-$, B(C$_6$H$_5$)$_4^-$, or TfO$^-$ anion significantly reduces the reactivity. (Z)-1,2-Diarylolefins are unreactive towards this catalyst, whereas both (E)- and (Z)-2-(4-methoxyphenyl)-2-butene give high conversions but only moderate enantioselectivities.

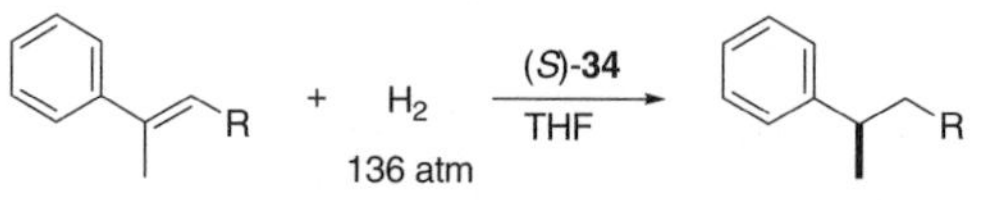

R	% ee
CH$_3$	95
C$_6$H$_5$	>99

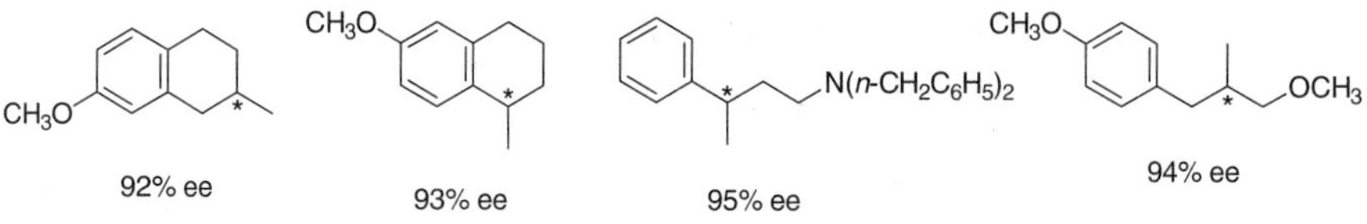

92% ee   93% ee   95% ee   94% ee

Scheme 1.30.

Considerable success has been accomplished by use of the early transition-metal complexes (Scheme 1.29). Optically active η^5-cyclopentadienes are generally used as ligands for Ti, Zr, or Sm on the basis of Kagan's earlier finding [148]. Ziegler-Natta-type Ti catalysts carrying a menthyl- or neomenthylcyclopentadienyl group reduces α-ethylstyrene under 1 atm of hydrogen in the presence of Li[H$_2$Al(OCH$_2$CH$_2$OCH$_3$)$_2$] as cocatalyst to give the saturated compound, although the optical yields do not exceed 15%. The metallocenes are also made chiral by fusion to naturally occurring terpenes such as pinene [149], camphor [150], and verbenone [151] or by use of bridged bis(cyclopentadiene) [152] and C_2-symmetric annulated cyclopentadienes [153]. The chiral Ti complex **35** catalyzes, after treatment with n-C$_4$H$_9$Li, the enantioselective hydrogenation of α-ethylstyrene. The reaction proceeds even at −75°C under hydrogen at atmospheric pressure to give 2-phenylbutane with 95% ee [153]. The chiral Zr complex **37** with an ethylenebis(tetrahydro-1-indenyl) ligand catalyzes the asymmetric deuteration of styrene at room temperature to give ethylbenzene-1,2-d_2 with 65% ee [154]. The chiral Sm complex **36** with neomenthyl-substituted cyclopentadienyl also brings about high enantioselectivity and high catalytic activity in hydrogenation of α-ethylstyrene [155]. The reaction proceeds in heptane at −78°C under 1 atm of hydrogen. The highest optical yield of 96% is obtained by use of a 70:30 mixture of S and R Sm complex.

As shown in Scheme 1.30, the chiral titanocene catalyst **34** hydrogenates unfunctionalized, disubstituted styrenes under 136 atm of hydrogen at 65°C to give the saturation products with ~83 to >99% ee [156]. A high enantioselectivity is now realized only with aryl-substituted olefins. The enantioselectivity of 41% ee attained 2-ethyl-1-hexene and **34** as catalyst is the highest for hydrogenation of non-aromatic olefins.

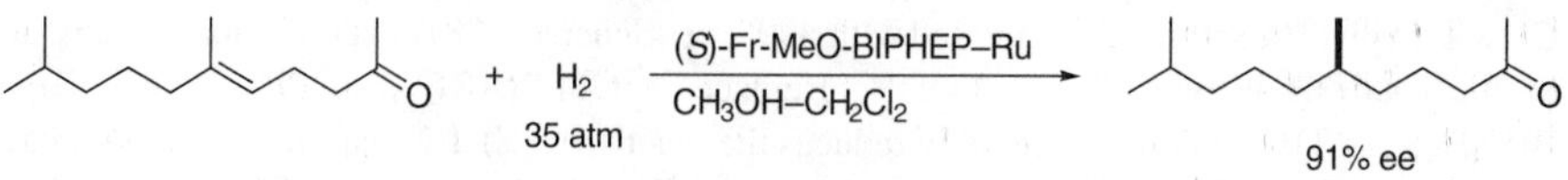

Scheme 1.31.

Dihydrogeranylacetone, though not a completely simple olefin, is chemoselectively hydrogenated at the C=C bond in the presence of a Ru complex with MeO-BIPHEP analogue containing four P-2-furyl groups to afford the saturated ketone with 91% ee (Scheme 1.31) [86]. Examples of hydrogenation of a trisubstituted olefin with an oxo or oxy substituent in the p-position are unknown.

1.3.3. Transfer Hydrogenation

Olefinic double bonds can be saturated by using manifold hydrogen donors other than hydrogen molecules. In the enantioselective reduction of olefins, formic acid, a 5:2 formic acid–triethylamine azeotrope, and 2-propanol are most frequently used in the presence of chiral Rh or Ru complexes. Other hydrogen donors such as ascorbic acid, benzyl alcohols, hydroaromatics, carbon monoxide/water combination have rarely been utilized. As shown in Scheme 1.32, transfer hydrogenation of itaconic acid with triethylammonium formate in DMSO containing a neutral BPPM–Rh complex gives 2-methylsuccinic acid with 82–92% ee [157]. Use of optically active phenethylamine instead of triethylamine increases the enantioselectivity to >97% ee [158]. Chiral Ru complexes of the general formula [Ru(acac-F$_6$)(η^3-allyl)(diphosphine)] effectively catalyze hydrogen transfer from formic acid-triethylamine azeotrope to itaconic acid in THF to afford the saturated carboxylic acids with up to 93% ee [159]. The most active and selective catalyst for this transformation is formed with BINAP. [RuH{(S)-binap}]$_2$PF$_6$, a cationic five-coordinate complex, catalyzes saturation of the same unsaturated carboxylic acids with 2-propanol in 97% optical yield [160]. In all cases, the sense of enantioselection is identical with that of reaction with molecular hydrogen. Use of DIOP, BPPM, or BINAP as ligand that forms a seven-membered metal chelate ring is crucial for obtaining high efficiency in the Ru-catalyzed reaction by using the HCOOH/R$_3$N system. On the contrary,

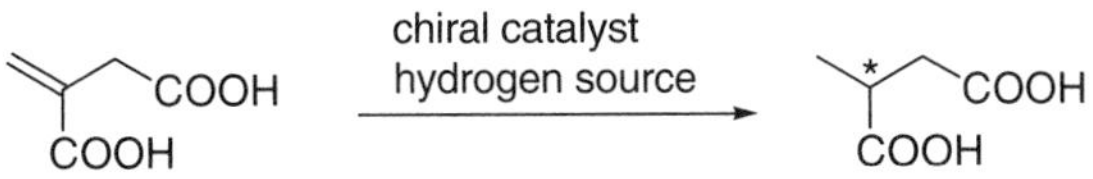

Catalyst	Hydrogen source	% ee	Config'n
(2S,4S)-BPPM–Rh	HCOOH/(C$_2$H$_5$)$_3$N	92	S
(2S,4S)-BPPM–Rh	HCOOH/(S)-C$_6$H$_5$CH(CH$_3$)NH$_2$	>97	S
(S)-BINAP–Ru	HCOOH/(C$_6$H$_5$)$_3$N	93	R
(S)-BINAP–Ru	(CH$_3$)$_2$CHOH	97	R

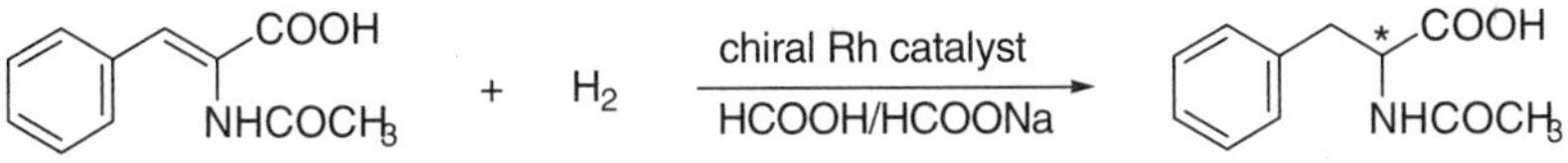

Ligand	% ee	Config'n
(R,R)-BDPP	92	R
(R,R)-PYRPHOS	91	S

Scheme 1.32.

PYRPHOS and BDPP forming a chiral five- and six-membered chelate ring are useful for Rh-catalyzed enantioselective reduction of (Z)-α-(acetamido)cinnamic acid with HCOOH/HCOONa [161].

1.4. HYDROGENATION OF KETONES

1.4.1 Functionalized Ketones

Highly enantioselective hydrogenation of functionalized ketones has been achieved with chiral phosphine–Rh(I) and –Ru(II) complexes [1,162]. The presence of a functional group close to the carbonyl moiety efficiently accelerates the reaction and also controls the stereochemical outcome. The heteroatom–metal interaction is supposed to effectively stabilize one of the diastereomeric-transition states and/or key intermediates in the hydrogenation.

1.4.1.1. Keto Esters Hydrogenation of α-keto esters and amides with chiral metal complexes can be achieved with a high enantioselectivity and a high reaction rate. Methyl pyruvate, the simplest substrate, is hydrogenated with an MCCPM–Rh complex to give methyl lactate quantitatively with 87% ee (Scheme 1.33) [163]. The electron-donating dicyclohexylphosphino group at the C4 position increases the activity of the catalyst, and the chirally arranged diphenylphosphino group on the C2 methylene acts effectively in differentiating the enantio-faces of the substrate. A Rh complex with Cy,Cy-oxoProNOP exhibits an excellent degree of

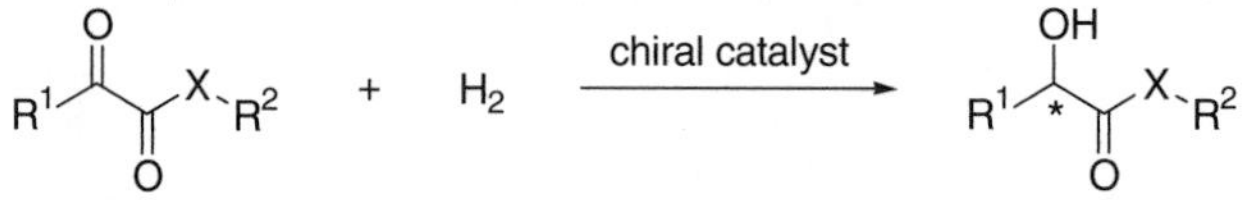

Substrate					Product	
R^1	XR^2	Catalyst	Solvent	H_2, atm	% ee	Config'n
CH_3	CH_3O	(2S,4S)-MCCPM–Rh	THF	20	87	R
CH_3	CH_3O	(−)-tetraMe-BITIANP–Ru	CH_3OH	97	88	S
CH_3	C_2H_5O	(S)-Cy,Cy-oxoProNOP–Rh	toluene	50	95	R
C_6H_5	CH_3O	(S)-MeO-BIPHEP–Ru	CH_3OH	20	86	S
C_6H_5	CH_3O	(R)-BICHEP–Ru	C_2H_5OH	5	>99	S
$4\text{-}CH_3C_6H_4$	CH_3O	(S)-BINAP–Ru	CH_3OH	100	93	S
$C_6H_5(CH_2)_2$	C_2H_5O	(S,S)-NORPHOS–Rh	CH_3OH	99	96	S
C_6H_5	$C_6H_5CH_2NH$	(S)-Ph,Cp-isoAlaNOP–Rh	toluene	1	88	S
C_6H_5	$C_6H_5CH_2NH$	(S)-Cy,Cy-oxoProNOP–Rh	toluene	50	95	S
C_6H_5	$C_6H_5CH_2NH$	(S)-Cp,Cp-IndoNOP–Rh	toluene	1	91	S
C_6H_5	$C_6H_5CH_2NH$	(S,2S)-Cr(CO)$_3$-Cp,Cp-IndoNOP–Rh	toluene	1	97	S
C_6H_5	$C_6H_5CH_2NH$	(S)-BICHEP–Ru	CH_3OH	40	96	R

Scheme 1.33.

enantioselection in hydrogenation of ethyl pyruvate and benzoylformamide derivatives [164]. The use of the $Cr(CO)_3$-complexed Cp,Cp-IndoNOP as ligand gives an even better enantiose-lectivity than the use of the original ligand (97% ee vs 91% ee) [165]. A Ph,Cp-isoAlaNOP–Rh complex is also usable [166]. Ethyl 2-oxo-4-phenylbutanoate is hydrogenated with a NOR-PHOS–Rh complex to give the alcohol with 96% ee while the reactivity is low [167]. Unlike Rh catalyses mentioned above, the Ru-catalyzed asymmetric hydrogenation is successfully achieved with C_2-chiral diphosphines. Hydrogenation of methyl 4'-methylbenzoylformate with a cationic BINAP–Ru complex in the presence of aqueous HBF_4 gives the hydroxy ester with 93% ee [168], whereas a neutral BINAP–Ru complex gives only 83% optical yield in hydro-genation of methyl pyruvate [169]. A BICHEP–Ru complex shows an excellent enantioselec-tivity in hydrogenation of methyl benzoylformate and its benzylamide derivative [170]. The optical yield reaches >99%. An electron-rich alkylphosphino group in place of the aryl-phosphino function increases the enantioselection. A MeO-BIPHEP–Ru complex also shows a high selectivity [171]. A tetraMe-BITIANP ligand with heteroaromatic rings is also effective [109].

Ketopantolactone is a standard substrate to test the enantioface-differentiating ability of chiral catalysts. The resulting pantoyl lactone serves as the key intermediate for the synthesis of pantothenic acid, a constituent of coenzyme A. A pioneering work has been done by using a neutral Rh complex with BPPM, a pyrrolidine-based diphosphine ligand, reaching 87% optical yield (Scheme 34) [172]. Similarly, a BCPM–Rh complex gives a 92% optical yield [173]. A m-CH$_3$POPPM–Rh complex promotes the hydrogenation to give the hydroxy lactone with 95% ee [174,175]. The turnover frequency (TOF), defined as moles of product per mole of catalyst per hour, reaches 50,000 with an S/C of 150,000. The reaction has been conducted on a 200-kg batch scale (Hoffmann-La Roche, Ltd). Rh complexes with an IndoNOP, ProNOP, and isoAlaNOP derivative also act as excellent asymmetric catalysts [165,176]. The use of $[Rh(OCOCF_3)(cp,cp-oxopronop)]_2$ achieves a 99% optical yield and a TOF as high as $3300\,h^{-1}$ under 1 atm of hydrogen [176b].

Highly enantioselective hydrogenation of β-keto esters is achieved by using Ru(II) catalysts [177] with a chiral diphosphine ligand. BINAP with an atropisomeric C_2 symmetric structure is the most effective ligand for this purpose [168,169,178–180]. A wide variety of β-keto esters are hydrogenated with the BINAP–Ru complexes, $RuX_2(binap)$ (X = Cl, Br, or I; empirical formula with a polymeric form) or $RuCl_2(binap)(dmf)_n$ (oligomeric form) [181], to give chiral

Ligand	Solvent	H$_2$, atm	Temp, °C	% ee	Config^n
(2S,4S)-BPPM	benzene	50	30	87	R
(2S,4S)-BCPM	THF	50	50	92	R
(2S,4S)-m-CH$_3$POPPM	toluene	12	40	95	R
(S)-Cp,Cp-IndoNOP	toluene	1	20	>99	R
(S)-Cp,Cp-oxoProNOP	toluene	1	rt	99	R
(S)-Cp,Cp-isoAlaNOP	toluene	1	rt	97	S
(S)-Ph,Cp-methyllactamide	toluene	1	rt	87	S

Scheme 1.34.

β-hydroxy esters in a nearly perfect optical yield (Scheme 1.35). For example, methyl 3-oxobutanoate is transformed by using the R complexes to (R)-methyl 3-hydroxybutanoate quantitatively with >99% ee at >20 atm of hydrogen with an S/C of up to 10,000 in an alcoholic solution [178]. β-Keto amides and thioesters are also hydrogenated with high enantioselection [169,182]. Because of the remarkable enantioselectivity and generality of the BINAP–Ru catalysis, many methods for preparation of the complex have been reported [74a,83,105,106,168,171,183]. The

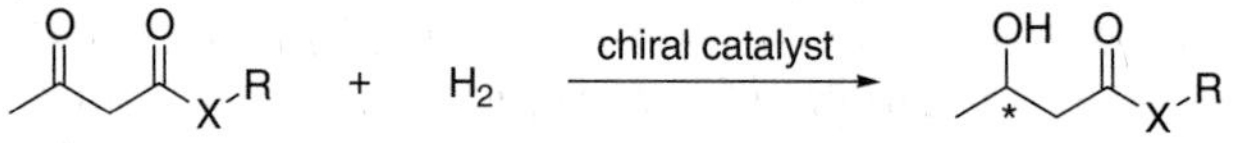

RX in substrate	Catalyst	Solvent	H_2, atm	Temp, °C	Product % ee	Config'n
CH₃O	RuCl₂[(R)-binap]	CH₃OH	100	23	>99	R
CH₃O	RuCl₂[(R)-binap](dmf)$_n$	CH₃OH	100	25	99	R
CH₃O	RuCl₂[(R)-binap](dmf)$_n$	CH₃OH	4	100	97	R
CH₃O	[NH₂(C₂H₅)₂][{RuCl[(R)-binap]}₂(μ-Cl)₃]	CH₃OH	100	25	>99	R
CH₃O	[RuI{(S)-binap}C₆H₆]I	CH₃OH	100	20	99	S
CH₃O	Ru[η³-CH₂C(CH₃)CH₂]₂-[(S)-binap] + HBr	CH₃OH	1	rt	97	S
CH₃O	(S)-bis-steroidal phosphine-Ru	CH₃OH	100	100	99	S
CH₃O	(S)-BIPHEMP–Ru	CH₃OH	5	50	>99	S
CH₃O	(R)-BIMOP–Ru	1:1 CH₂Cl₂-CH₃OH	10	30–40	100	R
CH₃O	(R,R)-i-Pr-BPE–Ru	9:1 CH₃OH-H₂O	4	35	99	S
CH₃O	Ru(OCOCF₃)₂[(S)-[2.2]-phanephos]^a	10:1 CH₃OH–H₂O	3	−5	96	R
C₂H₅O	(+)-tetraMe-BITIANP–Ru	CH₃OH	97	70	99	R
C₂H₅O	(R)-(S)-JOSIPHOS–Rh	CH₃OH	20	rt	97	S
(CH₃)₃CO	[NH₂(C₂H₅)₂][{RuCl[(R)-binap]}₂(μ-Cl)₃] + HCl	CH₃OH	3	40	>97	R
CH₃NH^b	RuCl₂[(R)-binap](dmf)$_n$	CH₃OH	14	100	100	S
C₆H₅NH	[NH₂(C₂H₅)₂][{RuCl[(R)-binap]}₂(μ-Cl)₃]	CH₃OH	30	60	>95	R
(CH₃)₂N	RuBr₂[(S)-binap]	C₂H₅OH	63	27	96	S
C₂H₅S	RuCl₂[(R)-binap]	C₂H₅OH	95	27	93	R

^{a} Reaction with (n-C₄H₉)₄NI. Ru:N = 1:6–13. ^{b} 3-Oxo-3-phenyl-propanoic acid N-methyl amide.

Scheme 1.35.

hydrogenation proceeds at 3 atm and at 40°C under strongly acidic conditions [183a,c]. Use of other C_2 symmetric biaryl diphosphines such as BIMOP [184], BIPHEMP [171], tetraMe-BI-TIANP [109], and bis-steroidal phosphine [185] also exhibits excellent enantioselection in hydrogenation of β-keto esters. A Ru complex with electron-rich i-Pr-BPE effectively promotes the hydrogenation under a low pressure [63,186]. Ru(OCOCF$_3$)$_2$([2.2]-phanephos) shows high activity in the presence of $(n$-C$_4$H$_9$)$_4$NI at a low temperature and a low hydrogen pressure without strong acids [187]. A Rh complex of C_1 chiral JOSIPHOS is also effective for asymmetric hydrogenation of ethyl 3-oxobutanoate [188]. BINAP–Ru complexes immobilized in a polydimethylsiloxane membrane matrix [189] or on polystyrene resin [126] are used for hydrogenation of methyl 3-oxopentanoate.

A BINAP–Ru catalyzed asymmetric hydrogenation of γ-keto esters and o-acylbenzoic esters gives γ-lactones and o-phthalides, respectively, with an excellent enantioselectivity (Scheme 1.36) [190,191]. The sense of enantioselection is the same as that in the hydrogenation of α- and β-keto esters.

The asymmetric hydrogenation of α-, β-, or γ-keto esters with BINAP–Ru complexes has been used for synthesis of a wide variety of natural chiral compounds as shown in Figure 1.11 [192]. The asymmetric reduction determines the stereocenter labeled by R or S.

The enantioselection in asymmetric hydrogenation of functionalized ketones stems from interaction of the heteroatom to a catalyst metal center. Therefore, in the reaction of substrates having two heteroatoms on both sides of the carbonyl moiety, the competitive interaction occurring at the enantioface-differentiating stage often causes a decrease in optical yield. The degree and sense of enantioselection are sensitively affected by the size and electronic characters of the functionalities. Hydrogenation of methyl 5-benzyloxy-3-oxobutanoate by using an (S)-BINAP–Ru complex affords the S alcohol with 99% ee, and the degree and sense of enantioselection are the same as those obtained with simple β-keto esters (Scheme 1.37) [169]. However, 4-benzyloxy- and 4-chloro-3-oxobutanoate are hydrogenated with the (S)-BINAP–Ru complex at room temperature to give the R alcohols with 78% ee and 56% ee, respectively. When the reaction is conducted at 100°C, the enantioselectivities are dramatically increased to 98% ee and 97% ee, respectively [1c,193]. Introduction of a bulky triisopropylsilyloxy group at the C4 position increases the selectivity to 95% ee at room temperature [169]. The trimethyl-ammonium chloride functionality does not interfere with enantioselection [171]. Hydrogenation of methyl 4-methoxy-3-oxobutanoate with an i-Pr-BPE–Ru complex gives the alcohol with

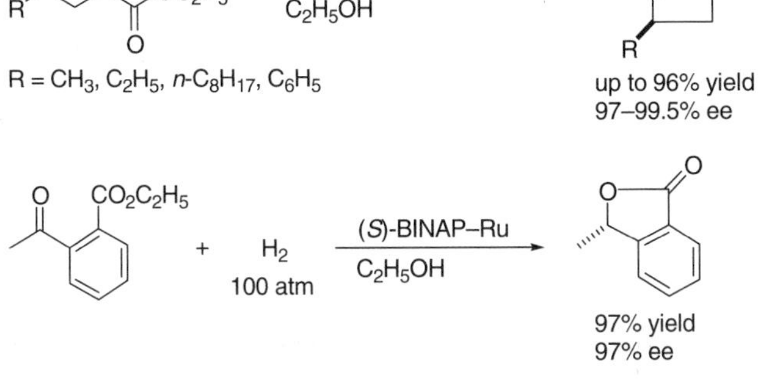

R = CH$_3$, C$_2$H$_5$, n-C$_8$H$_{17}$, C$_6$H$_5$

100 atm H$_2$
(R)-BINAP–Ru
C$_2$H$_5$OH

up to 96% yield
97–99.5% ee

(S)-BINAP–Ru
C$_2$H$_5$OH

97% yield
97% ee

Scheme 1.36.

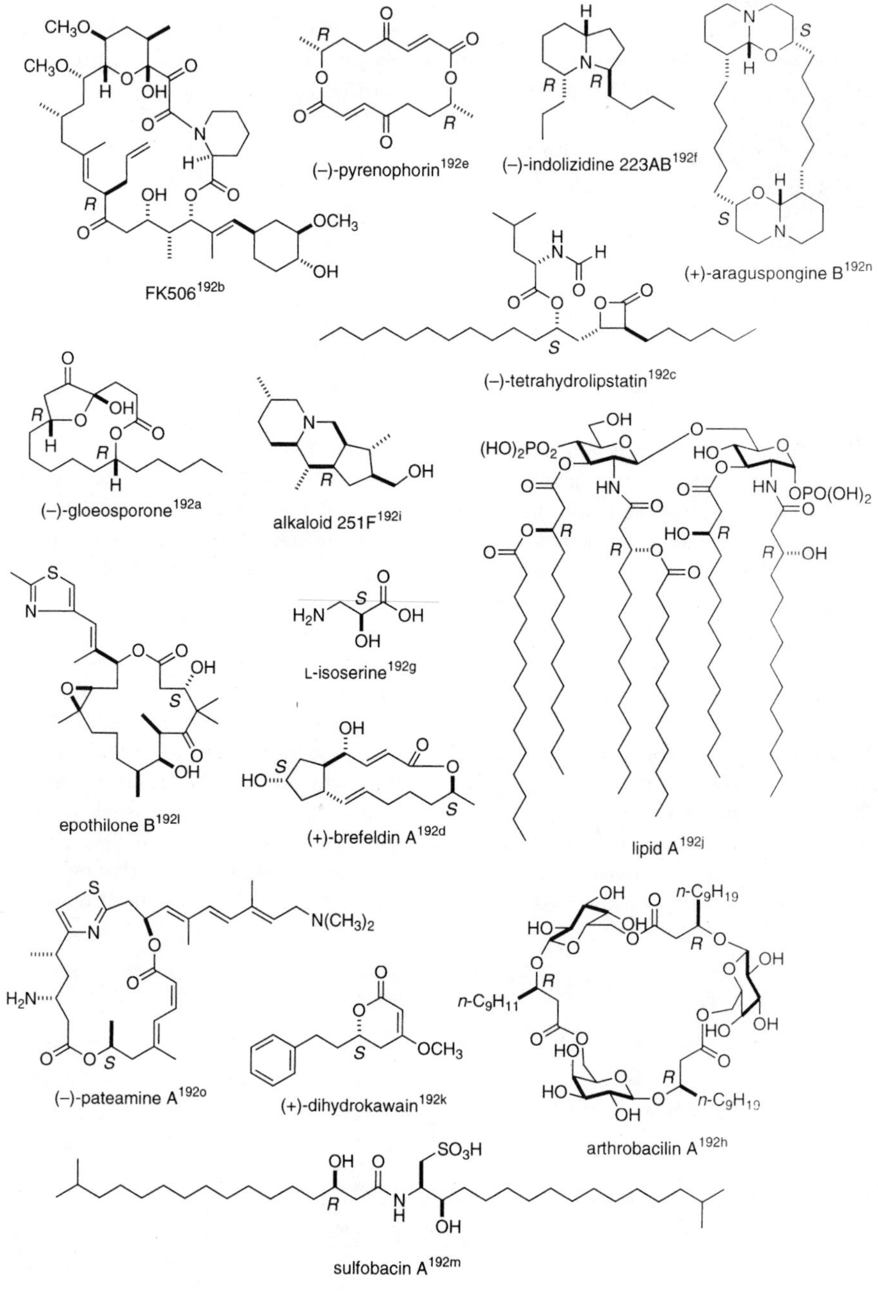

Figure 1.11. Examples of biologically active compounds obtainable via BINAP–Ru catalyzed hydrogenation of α-, β-, or γ-keto esters.

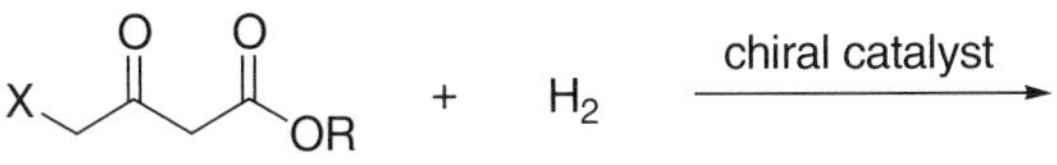

$$X \overset{O \quad\quad O}{\underset{OR}{\parallel}} \quad + \quad H_2 \quad \xrightarrow{\text{chiral catalyst}} \quad X \overset{OH \quad O}{\underset{*\quad\quad OR}{}}$$

Substrate			H₂,		Product	
X	R	Catalyst	atm	Temp, °C	% ee	Config'n
$C_6H_5CH_2OCH_2$	CH_3	(S)-BINAP–Ru	50	26	99	S
CH_3O	CH_3	(R,R)-i-Pr-BPE–Ru	4	35	96	R
$C_6H_5CH_2O$	C_2H_5	(S)-BINAP–Ru	100	28	78	R
$C_6H_5CH_2O$	C_2H_5	(S)-BINAP–Ru	100	100	98	R
$((CH_3)_2CH)_3SiO$	C_2H_5	(S)-BINAP–Ru	100	27	95	R
Cl	CH_3	(R,R)-i-Pr-BPE–Ru	4	35	76	R
Cl	C_2H_5	(S)-BINAP–Ru	77	24	56	R
Cl	C_2H_5	(S)-BINAP–Ru	100	100	97	R
Cl	C_2H_5	(S)-Ph,Ph-oxoProNOP–Ru	138	20	75	S
$Cl(CH_3)_2NH$	C_2H_5	(2S,4S)-MCCXM–Rh	20	50	85	S
$Cl(CH_3)_3N$	H	(R)-BINAP–Ru	100	25	96	S

Scheme 1.37.

96% ee, whereas the enantioselectivity decreases to 76% ee in reaction of its 4-chloro analogue [186]. An (S)-Ph,Ph-oxoProNOP–Ru complex shows a moderate selectivity for hydrogenation of ethyl 4-chloro-3-oxobutanoate [194]. A Rh complex with MCCXM gives an 85% optical yield for reduction of the 4-dimethylamino hydrochloride analogue [195].

The BINAP–Ru catalyzed asymmetric hydrogenation of difunctionalized ketones is applicable to synthesis of several biologically active compounds [1c,193,195,196]. In Figure 1.12, the stereocenter determined by the BINAP chemistry is labeled by R or S.

The diastereoselectivity in reaction of a chiral ketone with an optically active catalyst is determined by the intermolecular chirality-transfer ability of the catalyst and the extent of the intramolecular asymmetric induction based on the substrate structure. Scheme 1.38 exemplifies asymmetric hydrogenation via double stereodifferentiation [197]. The α-keto amide **A** derived from an (S)-amino ester is hydrogenated with an (R,R)-CyDIOP–Rh complex to give an 86:14 mixture of the (S,S)-hydroxy amide **B** and its R,S isomer [198]. (R)-BINAP–Ru catalyzed hydrogenation of a series of N-Boc-protected (S)-γ-amino β-keto esters **C** gives exclusively the syn amino alcohol **D** [199]. Reduction with the S catalyst selectively affords its anti-isomer. A tandem hydrogenation of N-acetyl- or N-boc-protected γ-amino γ,δ-unsaturated β-keto esters **E** with a mixture of an (S)-BINAP–Rh and –Ru catalyst gives predominantly (3R,4R)-**F** [200]. The BINAP–Rh catalyst selectively saturates the C=C bond of **E** under a low pressure of hydrogen, and then the BINAP–Ru catalyst hydrogenates the C=O linkage under high-pressure conditions. Hydrogenation of the N-boc-protected (S)-δ-amino β-keto ester **G** with an (R)-BINAP–Ru complex, followed by cyclization affords the trans-substituted lactone **H** and its cis isomer in a 96:4 ratio [201]. The products **D** and **F** are converted to a statin series, essential components of aspartic proteinase inhibitors [199,200]. The product **H** is also useful for the synthesis of theonellamide F, an antifungal agent [201].

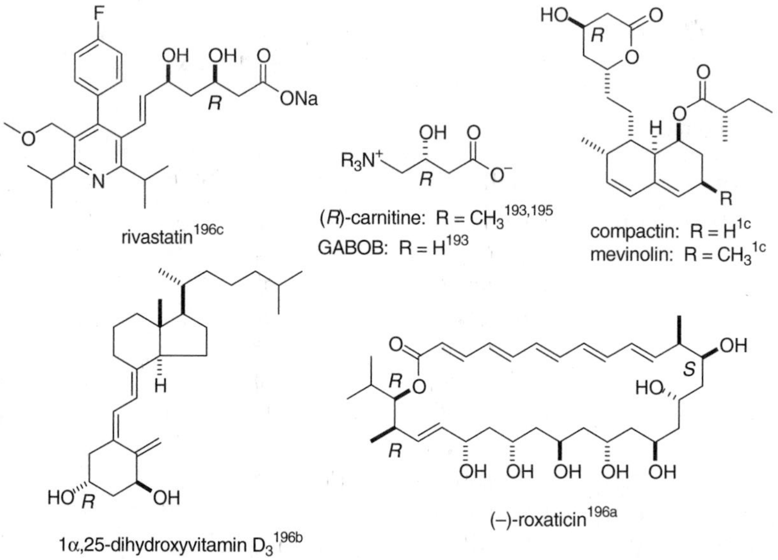

rivastatin[196c]

(R)-carnitine: R = CH$_3$ [193,195]
GABOB: R = H [193]

compactin: R = H[1c]
mevinolin: R = CH$_3$[1c]

1α,25-dihydroxyvitamin D$_3$[196b]

(−)-roxaticin[196a]

Figure 1.12. Examples of biologically active compounds obtainable via BINAP–Ru catalyzed hydrogena-tion of difunctionalized ketones.

Heterogeneous asymmetric hydrogenation of α-keto esters was first achieved by using a Pt/Al$_2$O$_3$ catalyst modified with an alkaloid [202–204]. Hydrogenation of methyl pyruvate catalyzed by Pt/Al$_2$O$_3$ in the presence of quinine in benzene affords (R)-methyl lactate with 87% ee (Scheme 1.39) [202b]. Ethyl benzoylformate is hydrogenated to the corresponding hydroxy ester with up to 90% ee [202c]. The enantioselectivity of hydrogenation of ethyl pyruvate is increased up to 97% ee by using a cinchonidine-modified catalyst with ultrasonic pretreatment [205]. Smaller metal-particle size (3.9 nm) of the catalyst may affect the increment of selectivity. When 10,11-dihydro-O-methylcinchonidine (MeOHCd) is used in hydrogenation in acetic acid, ethyl pyruvate is converted to (R)-ethyl lactate with up to 95% ee [206]. Hydrogenation of ethyl 2-oxo-4-phenylbutanoate also gives the alcohol in a high selectivity [167,204a]. This catalyst is also effective for asymmetric hydrogenation of α-keto acids [207]. Catalysts modified with simple chiral amines, (R)-1-(1-naphthyl)ethylamine [(R)-1-NEA] and (R)-1-(9-anthracenyl)-2-(1-pyrrolidinyl)ethanol [(R)-**20**], also show a good selectivity [208,209]. An extended aromatic π-system, with which the modifier adsorbs at the Pt surface, is necessary for a high enantioselectivity. The crucial structural elements are 1) the basic nitrogen center; 2) the flat aromatic ring, such as quinolyl or naphthyl; and 3) the stereogenic center(s) close to the nitrogen [204,209a,210]. For detailed discussion on this subject, see reviews and recent publications [137,203a,204,211]. A Pt-colloid catalyst stabilized by dihydrocinchonidine promotes the hydrogenation of ethyl pyruvate (91% ee) [212].

Most of the studies of Pt catalysts with cinchona alkaloids have focused on the hydrogenation of α-keto esters, especially ethyl pyruvate, as shown above. However, enantioselective hydro-genation of ketopantolactone and 1-ethyl-4,4-dimethylpyrrolidine-2,3,5-trione is attainable with a Pt catalyst modified by cinchonidine, giving the corresponding R alcohols with 92% ee and 91% ee, respectively (Scheme 1.40) [213]. These reactions can be performed with an S/C of up to 237,000 [213a].

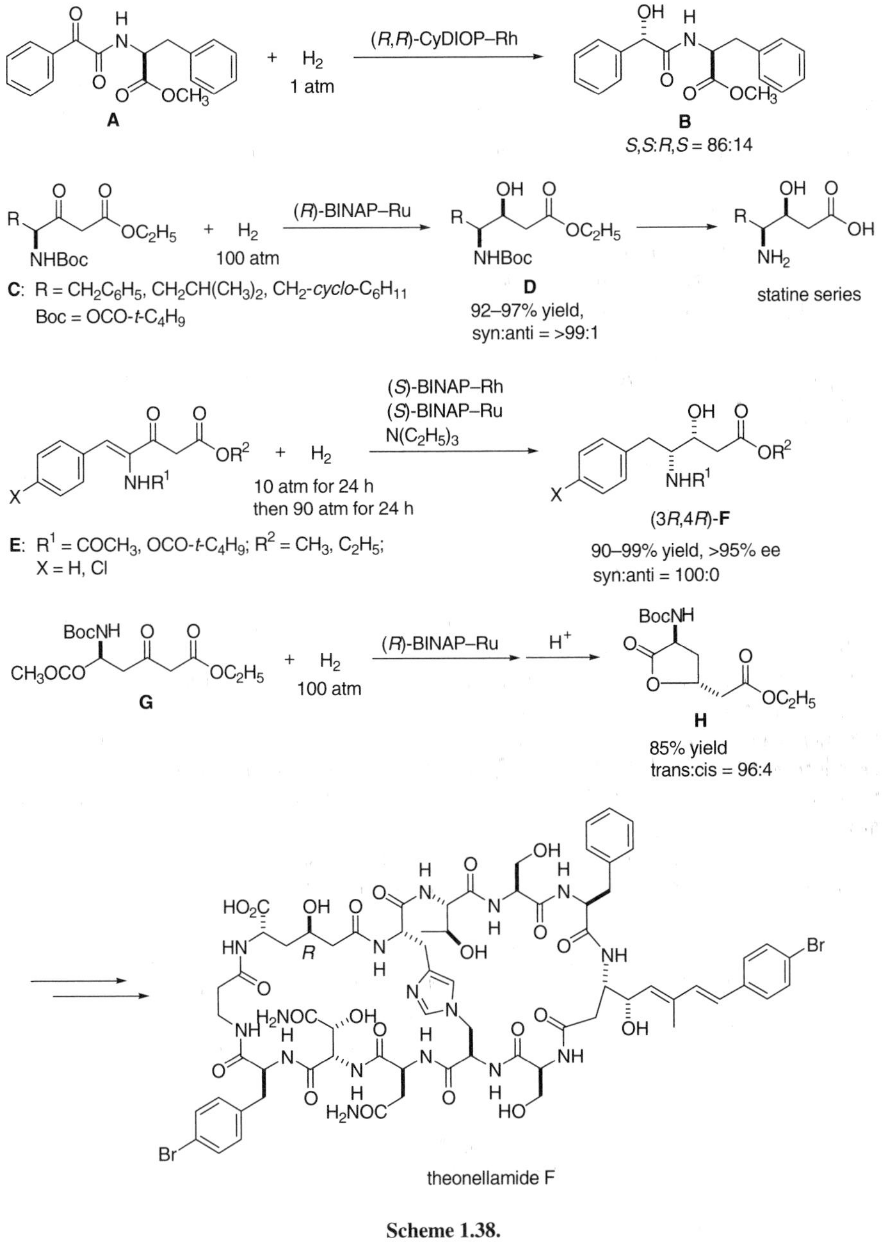
A
(R,R)-CyDIOP–Rh
H2
1 atm
B
S,S:R,S = 86:14
(R)-BINAP–Ru
H2
100 atm
C: R = CH2C6H5, CH2CH(CH3)2, CH2-cyclo-C6H11
Boc = OCO-t-C4H9
D
92–97% yield,
syn:anti = >99:1
statine series
(S)-BINAP–Rh
(S)-BINAP–Ru
N(C2H5)3
H2
10 atm for 24 h
then 90 atm for 24 h
E: R1 = COCH3, OCO-t-C4H9; R2 = CH3, C2H5;
X = H, Cl
(3R,4R)-F
90–99% yield, >95% ee
syn:anti = 100:0
G
(R)-BINAP–Ru
H+
H2
100 atm
H
85% yield
trans:cis = 96:4
theonellamide F

Scheme 1.38.

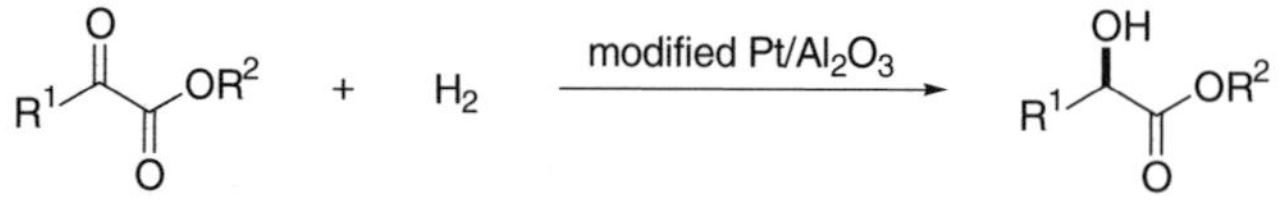

R^1	R^2	Modifier	Solvent	H$_2$, atm	Temp, °C	% ee of Product
CH$_3$	CH$_3$	quinine	benzene with quinine	68	rt	87
CH$_3$	C$_2$H$_5$	cinchonidine[a]	CH$_3$CO$_2$H	10	25	97
CH$_3$	C$_2$H$_5$	MeOHCd	CH$_3$CO$_2$H	99	20–25	95
CH$_3$	C$_2$H$_5$	(R)-1-NEA	CH$_3$CO$_2$H	8	9	82
CH$_3$	C$_2$H$_5$	(R)-**20**	CH$_3$CO$_2$H	69	10	87
C$_6$H$_5$(CH$_2$)$_2$	C$_2$H$_5$	MeOHCd	CH$_3$CO$_2$H	69	rt	92
CH$_3$	H	MeOHCd	9:1 C$_2$H$_5$OH–H$_2$O	99	20–30	79
C$_6$H$_5$(CH$_2$)$_2$	H	MeOHCd	9:1 C$_2$H$_5$OH–H$_2$O	99	20–30	85

[a] Ultrasonicated Pt/Al$_2$O$_3$ was used.

Scheme 1.39.

A Raney Ni catalyst modified by tartaric acid and NaBr is fairly effective for enantioselective hydrogenation of a series of β-keto esters (Scheme 1.41) [203a,214,215]. The enantio-discrimination ability of the catalyst is highly dependent on the preparation conditions such as pH (3–4), temperature (100°C), and concentration of the modifier (1%). Addition of NaBr as a second modifier is also crucial. Ultrasonic irradiation of the catalyst leads to even better activity and enantioselectivity up to 98% ee [214d–f]. The Ni catalyst is considered to consist of a stable, selective and weak, nonselective surface area, while the latter is selectively removed by ultrasonication.

Heterogeneous enantioselective hydrogenation of α- and β-keto esters is used for the synthesis of many biologically active compounds (Figure 1.13) [203b,204a,216]. Benazepril,

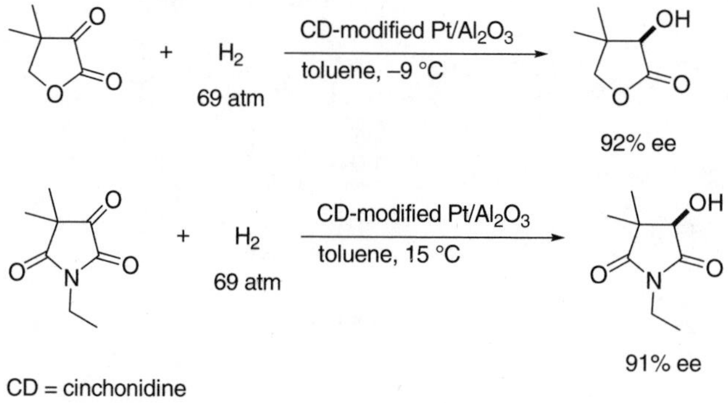

Scheme 1.40.

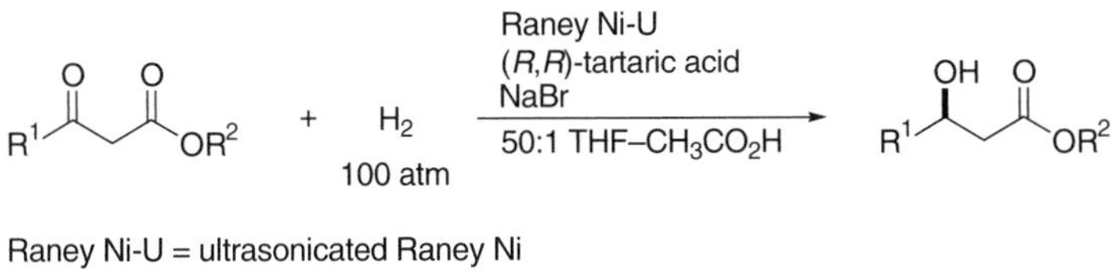

Raney Ni-U = ultrasonicated Raney Ni

| Substrate | | | | |
R[1]	R[2]	Temp, °C	Time, h	% ee of Product
CH_3	CH_3	100	4	85
C_2H_5	CH_3	60	34	94
$n\text{-}C_6H_{13}$	CH_3	60	52	90
$(CH_3)_2CH$	CH_3	60	71	96
$cyclo\text{-}C_3H_5$	CH_3	60	48	98
CH_3	$(CH_3)_2CH$	60	45	87
CH_3	$(CH_3)_3C$	60	40	88

Scheme 1.41.

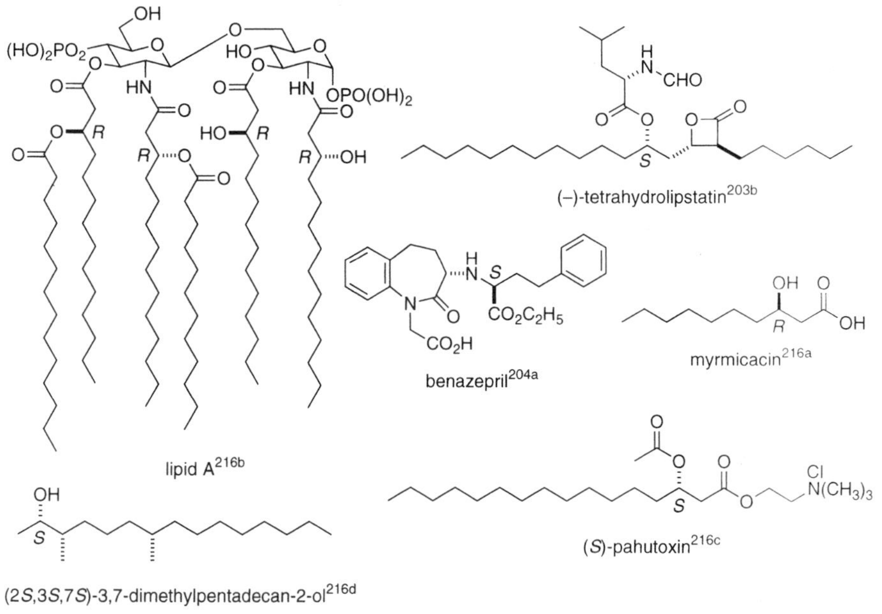

Figure 1.13. Examples of biologically active compounds obtainable via hydrogenation of α- or β-keto esters catalyzed by chirally modified Raney Ni or Pt/Al$_2$O$_3$.

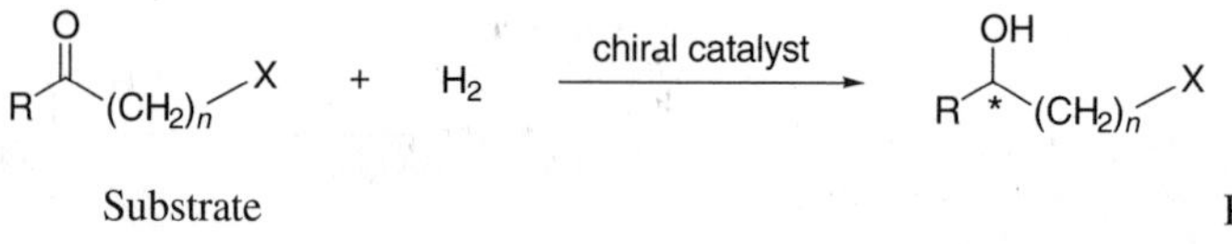

Substrate					Product	
R	n	X	Catalyst	H$_2$, atm	% ee	Config
C$_6$H$_5$	1	ClNH$_3$	(R)-MOC-BIMOP–Rh	90	93	R
C$_6$H$_5$	1	ClNH$_3$	(S)-Cy,Cy-oxoProNOP–Rh	50	93	S
C$_6$H$_5$	1	ClC$_6$H$_5$CH$_2$NH$_2$	(2S,4S)-MCCPM–Rh	20	93	S
3,4-(OH)$_2$-C$_6$H$_3$	1	ClCH$_3$NH$_2$	(R)-(S)-BPPFOH–Rh	50	95	R
CH$_3$	1	(CH$_3$)$_2$N	(S)-BINAP–Ru	102	99	S
C$_6$H$_5$	1	(CH$_3$)$_2$N	(S)-BINAP–Ru	100	95	S
2-naphthyl	1	(C$_2$H$_5$)$_2$N	(S,S)-DIOP–Rh	69	95	+
CH$_3$	1	Cl(CH$_3$)$_2$NH	(S)-Cy,Cy-oxoProNOP–Rh	50	97	S
C$_6$H$_5$	1	Cl(C$_2$H$_5$)$_2$NH	(2S,4S)-MCCPM–Rh	20	97	S
C$_6$H$_5$	1	Cl(CH$_3$)$_2$NH	(S)-Cp,Cp-IndoNOP–Rh	50	>99	S
C$_6$H$_5$	2	ClCH$_3$NH$_2$	(2S,4S)-MCCPM–Rh	30	80	R
C$_6$H$_5$	2	Cl(CH$_3$)$_2$NH	(S)-Cy,Cy-oxoProNOP–Rh	50	93[a]	R
C$_6$H$_5$	2	ClC$_6$H$_5$CH$_2$(CH$_3$)NH	(2S,4S)-MCCPM–Rh	30	91	R
C$_6$H$_5$	3	Cl(CH$_3$)$_2$NH	(S)-Cy,Cy-oxoProNOP–Rh	50	92	R
C$_6$H$_5$	3	ClC$_6$H$_5$CH$_2$(CH$_3$)NH	(2S,4S)-MCCPM–Rh	50	88	R
CH$_3$	1	HO	(R)-BINAP–Ru	93	92	R
n-C$_3$H$_7$	1	HO	(R)-BINAP–Ru	–	95	R
CH$_3$	2	HO	(R)-BINAP–Ru	70	98	R
CH$_3$	2	C$_6$H$_5$S	(S)-MeO-BIPHEP–Ru	30	98	S
C$_2$H$_5$	2	C$_6$H$_5$S	(S,S)-BDPP–Ru	30	95	S
CH$_3$	3	C$_6$H$_5$S	(S)-BINAP–Ru	113	70	S

[a] Contaminated with 5% of propiophenone.

Scheme 1.42.

an angiotensin-converting enzyme inhibitor (Novartis Services AG) [204a], and (−)-tetrahydrolipstatin, a pancreatic lipase inhibitor (Hoffmann-LaRoshe, Ltd) [203b] are produced on an industrial scale.

1.4.1.2. Amino, Hydroxy, and Phenylthio Ketones Asymmetric hydrogenation of amino ketones, in either a neutral or hydrochloride form, has extensively been studied. Both Rh(I) and Ru(II) complexes with an appropriate chiral diphosphine give a high enantioselectivity. As described in Scheme 1.42, α-aminoacetophenone hydrochloride is hydrogenated using a cationic Rh complex with (R)-MOC-BIMOP, an unsymmetrical biaryl diphosphine, to give the

R amino alcohol hydrochloride with 93% ee [217]. The use of a cationic Rh complex with Cy,Cy-oxoProNOP gives the same optical yield [218]. The mono-*N*-benzyl analogue is hydrogenated selectively in the presence of an MCCPM–Rh complex [219]. A cationic complex of BPPFOH, a chiral ferrocenyl diphosphine, effectively affords epinephrine hydrochloride with 95% ee [11,220]. In hydrogenation of α-dialkylamino ketones, a neutral BINAP–Ru [168,169] or DIOP–Rh complex [221] shows an excellent enantioselectivity. A series of α-, β-, and γ-amino ketone hydrochlorides are converted to the corresponding chiral alcohol hydrochlorides with up to 97% ee [218,219,222]. α-Dimethylaminoacetophenone hydrochloride is hydrogenated with an excellent enantioselectivity in the presence of a Cp,Cp-IndoNOP–Rh complex [165]. An (*R*)-BINAP–Ru complex is suitable for the hydrogenation of α- and β-hydroxy ketones to afford the *R* alcohols with up to 98% ee [169,223]. The sense of enantioface discrimination is the same as that observed in the hydrogenation of keto ester substrates. A Ru complex of BINAP, MeO-BIPHEP, or BDPP is also effective for hydrogenation of β-phenylthio ketones [224]. The γ-phenylthio analogue is somewhat less reactive.

Enantioselective hydrogenation of α,α′-bifunctionalized ketones is also known. Several 1-aryloxy-2-oxo-3-propylamines are hydrogenated with a (2*S*,4*S*)-MCCPM–Rh complex to give the (*S*)-amino alcohols with up to 97% ee (Scheme 1.43) [225]. The sense of enantioface differentiation is opposite to that obtained in the hydrogenation of α-monoamino ketones (see Table of Scheme 1.42). The BINAP–Ru catalyst possesses a high level of discrimination ability between a hydroxy group and an alkoxy or aryloxy group and, notably, between *n*-octadecyl and triphenylmethoxy groups [226].

Homogeneous enantioselective hydrogenation of amino or hydroxy ketones is useful for the synthesis of some biologically active compounds (Figure 1.14) [86,163b,179g,219,225,226c, 227,228]. The stereocenter determined by enantioselective hydrogenation is labeled by *R* or *S*. (*R*)-1,2-Propanediol formed by asymmetric hydrogenation of 1-hydroxy-2-propanone (50 tons/year at Takasago Int. Corp.) is used for the commercial synthesis of levofloxacin, an antibacterial agent (Dai-ichi Pharmaceutical Co.) [179g,228].

(*R*)-Amino ketone **A** in Scheme 1.44 is hydrogenated diastereoselectively with a neutral (*S*)-(*R*)-BPPFOH–[RhCl(cod)]$_2$ in ethyl acetate to give the *R*,*R* amino alcohol **B** [229]. The

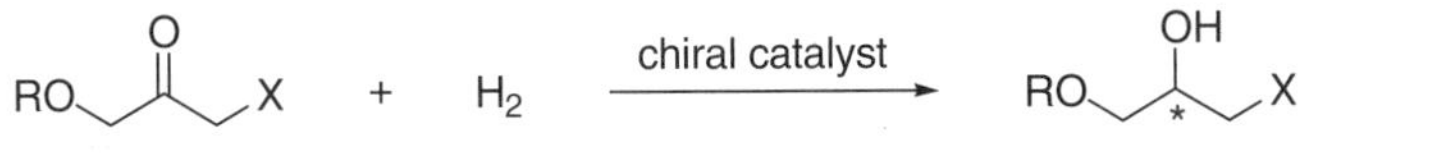

R	X	Catalyst	H$_2$, atm	% Yield	% ee	Config^n
C$_6$H$_5$	ClC$_6$H$_5$CH$_2$NH$_2$	(2*S*,4*S*)-MCCPM–Rh	20	100	97	*S*
Ar^a	HO	(*S*)-BINAP–Ru	94	86	>95	*R*
C$_6$H$_5$CH$_2$	HO	(*S*)-BINAP–Ru	97	>98	93	*R*
n-C$_{18}$H$_{37}$	(C$_6$H$_5$)$_3$CO	(*S*)-BINAP–Ru	97	>70	>96	*R*

^{a} ArO =

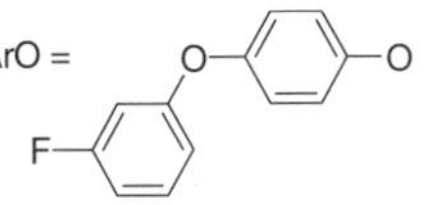

Scheme 1.43.

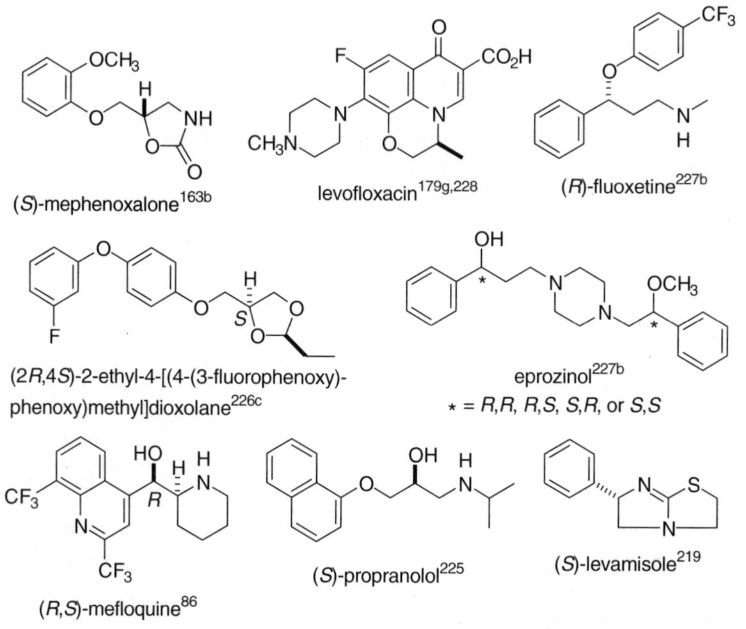

Figure 1.14. Examples of biologically active compounds obtainable via homogeneous asymmetric hydrogenation of amino or hydroxy ketones.

product is an isoproterenol analogue, a potent β-adrenoceptor agonist. When [Rh{(S)-(R)-bppfoh}]ClO$_4$, a cationic complex, is used in methanol, the S,R amino alcohol is obtained.

The BINAP–Ru complex effectively differentiates enantiomers of 1-hydroxy-1-phenyl-2-propanone. Hydrogenation of the racemic compounds catalyzed by an (R)-BINAP–Ru complex gives the corresponding 1S,2R diol with a 92% optical purity and unreacted R substrate with 92% ee at 50.5% conversion (Scheme 1.45) [1c]. The relative hydrogenation rate of the enantiomers, k_S/k_R, is calculated to be 64:1.

1.4.1.3. Diketones Asymmetric hydrogenation of α-diketones is rare. However, benzil is hydrogenated with a quinine–benzylamine–Co(dmg)$_2$ catalyst system to give (S)-benzoin with

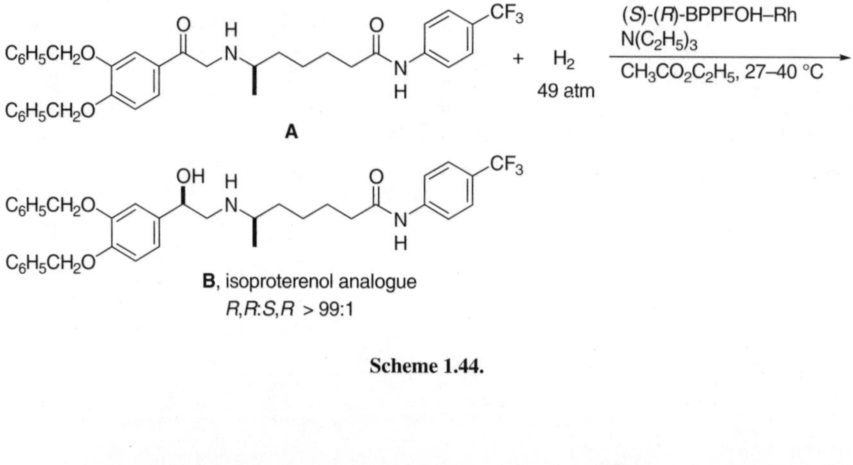

Scheme 1.44.

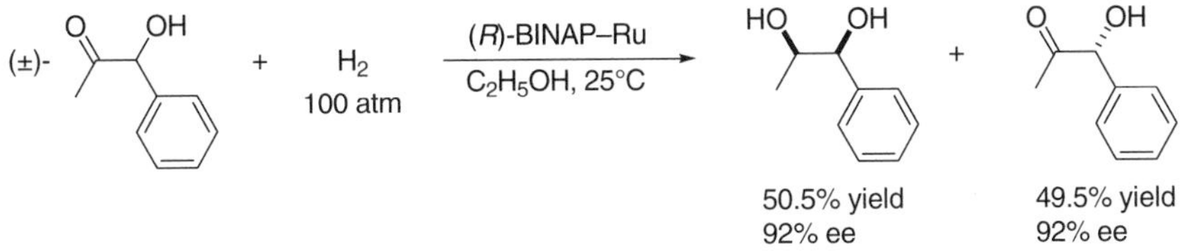

Scheme 1.45.

78% ee (Scheme 1.46) [230]. The addition of benzylamine as an achiral cocatalyst significantly enhances the reaction rate. A catalyst system with the BDM 1,3-pn ligand is also usable [231]. Hydrogenation of 2,3-butanedione catalyzed by an (R)-BINAP–Ru complex gives optically pure (R,R)-2,3-butanediol and the meso diol in a 26:74 ratio (Scheme 1.47) [169].

As shown in Scheme 1.48, asymmetric hydrogenation of β-diketones by using a C_2-chiral diphosphine–Ru complex gives the corresponding anti-diols with an excellent diastereo- and enantioselectivity. For example, 2,4-pentanedione is hydrogenated with an (R)-BINAP–Ru complex to give optically pure (R,R)-2,4-pentanediol in 99% yield [169]. With the same catalyst, hydrogenation of 5-methyl-2,4-hexanedione and 1-phenyl-1,3-butanedione affords the anti-diol in a high optical yield [169,232]. A BIPHEMP– [233] or BDPP–Ru complex [234] also shows a high enantioselectivity for hydrogenation of 2,4-pentanedione. Hydrogenation of methyl 3,5-dioxohexanoate with an (R)-BINAP–Ru complex affords an 81:19 mixture of an anti-(3S,5R, 78% ee) and syn-dihydroxy ester [235]. The absolute configuration of the product reveals that the C3 carbonyl function is more easily reduced than the C5 carbonyl group.

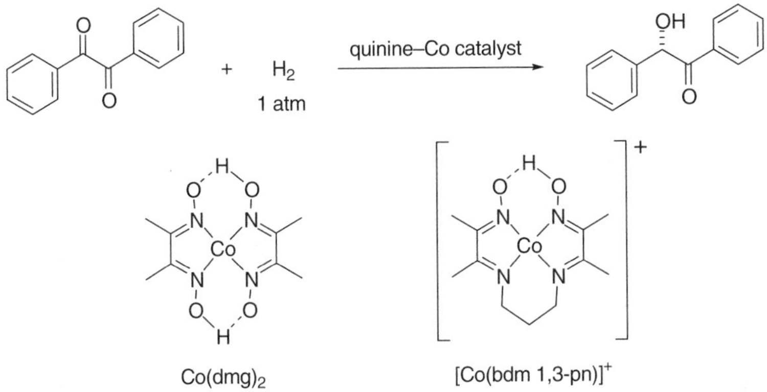

Catalyst system (molar ratio)	Solvent	Temp, °C	% ee of product
quinine–quinine•HCl–NH₂CH₂C₆H₅–Co(dmg)₂ (1:1:1)	benzene	−10	78
quinine–NH₂CH₂C₆H₅–H(BDM 1,3-pn)– CoCl₂•6H₂O (4:1:1.2:1)	1:9 (CH₃)₂CHOH– toluene	−8	79

Scheme 1.46.

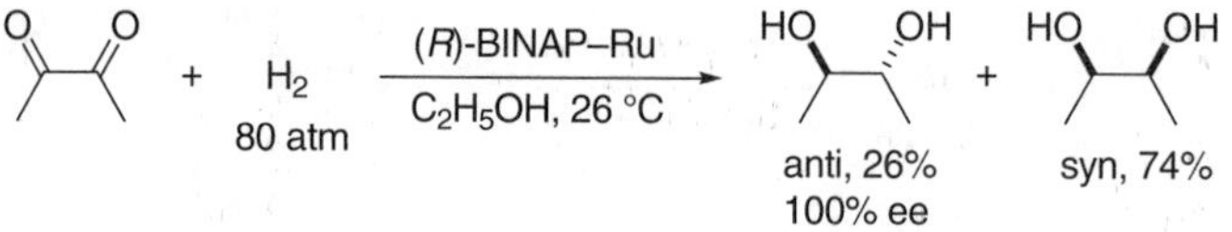

Scheme 1.47.

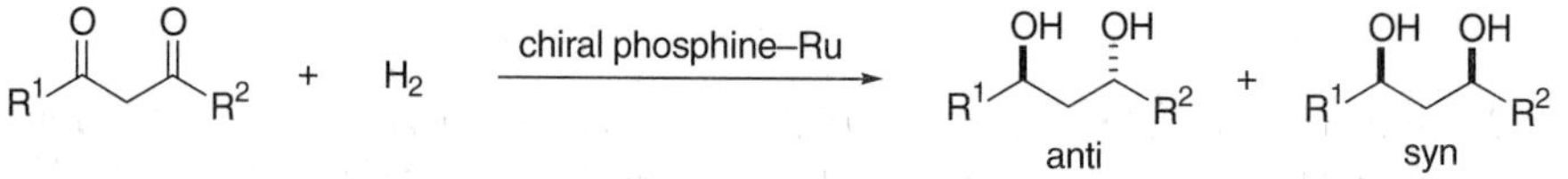

Substrate					Product		
R^1	R^2	Phosphine	H$_2$, atm	Temp, °C	% Yield	dr^a	% ee^b
CH$_3$	CH$_3$	(R)-BINAP	72	30	100	99:1	100
CH$_3$	CH$_3$	(R)-BIPHEMP	99	50	100	99:1	>99.9
CH$_3$	CH$_3$	(R,R)-BDPP	79	80	100	75:25	97
CH$_3$	(CH$_3$)$_2$CH	(R)-BINAP	48	50	92	97:3	98
CH$_3$	C$_6$H$_5$	(R)-BINAP	83	26	98	94:6	94
CH$_3$	CH$_3$OCOCH$_2$	(R)-BINAP	100	50	100^c	81:19	78
CH$_3$	C$_2$H$_5$OCO	(S)-MeO-BIPHEP	100	80	100^d	84:16	98^d
C$_6$H$_5$	C$_6$H$_5$	(R)-BIPHEMP	100	40	70	94:6	87
ClCH$_2$	ClCH$_2$	(R)-BINAP	85	102	—	—	92–94

^{a} Anti:syn diastereomer ratio. ^{b} % Ee of the anti diol. ^{c} A mixture of diol and δ-lactone. ^{d} (3R,5S)-3-Hy-droxy-5-methyltetrahydrofuran-2-one.

Scheme 1.48.

Hydrogenation of ethyl 2,4-dioxopentanoate with an (S)-MeO-BIPHEP–Ru complex followed by in situ cyclization gives (3R,5S)-3-hydroxy-5-methyltetrahydrofuran-2-one with 98% ee and the 3R,5R isomer with 87% ee in an 84:16 ratio [236]. Ethyl 2-hydroxy-4-oxopentanoate is the only detectable intermediate. 1,3-Diphenyl-1,3-propanediol with a BIPHEMP–Ru complex is transformed to an anti-diol selectively [237]. 1,5-Dichloro-2,4-pentanediol, a versatile chiral building block, is selectively obtained through the BINAP–Ru catalyzed hydrogenation of a

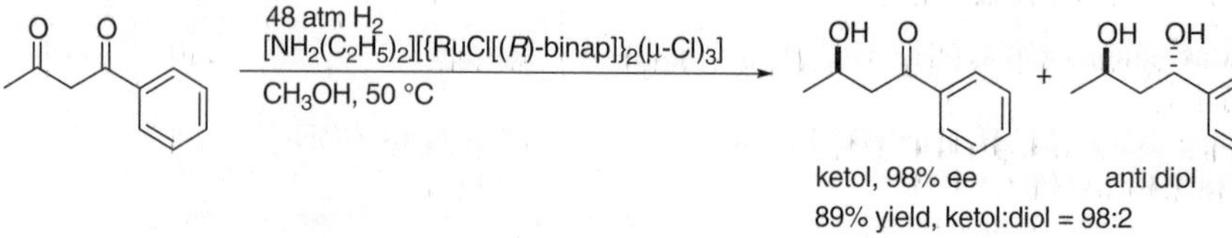

Scheme 1.49.

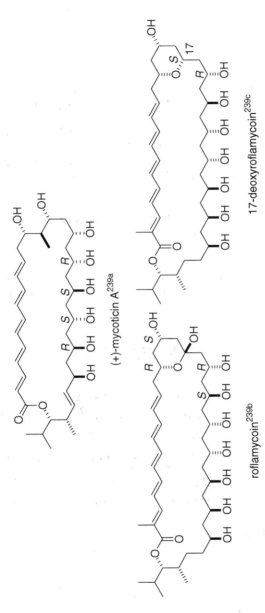

(+)-mycoticin A[239a]

roflamycoin[239b]

17-deoxyroflamycoin[239c]

Figure 1.15. Examples of biologically active compounds obtainable via BINAP–Ru catalyzed hydrogenation of β-diketones.

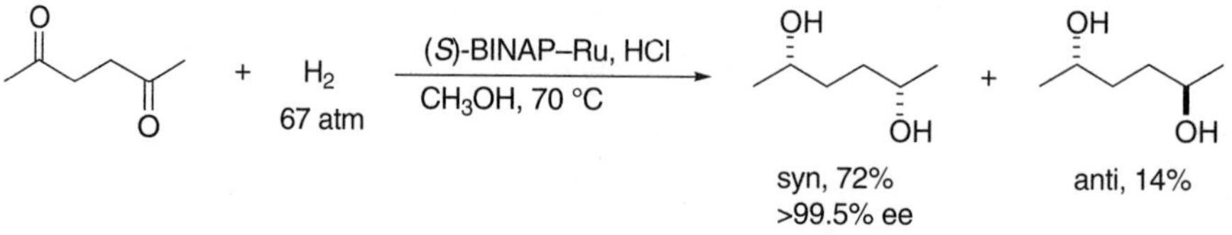

Scheme 1.50.

dichloro diketone [238]. 1-Phenyl-1,3-butanedione is hydrogenated selectively with [NH$_2$(C$_2$H$_5$)$_2$][{RuCl[(R)-binap]}$_2$(μ-Cl)$_3$] [74a,105,106] under appropriate conditions to give (R)-1-phenyl-3-hydroxybutan-1-one (Scheme 1.49) [232]. The BINAP–Ru catalyzed hydrogenation of β-diketones is useful for synthesis of naturally occurring compounds with many contiguous hydroxy groups (Figure 1.15) [239]. 2,5-Hexanedione, a γ-diketone, is also hydrogenated with a BINAP–Ru complex under acidic conditions to give optically pure *syn*-2,5-hexanediol in 72% yield (Scheme 1.50) [240]. Addition of HCl (Ru:HCl = 1:4) remarkably enhances the catalytic activity.

Heterogeneous asymmetric hydrogenation of 1,3-diketones is achieved by using a chirally modified Raney Ni catalyst (Scheme 1.51) [241]. Desired chiral diols are obtainable with about 90% ee. This method has been used for the synthesis of some naturally occurring compounds [242].

1.4.1.4. Keto Phosphonates, Keto Sulfonates, and Methylsulfones Asymmetric hydrogenation of β-keto phosphonates by using a BINAP–Ru complex affords β-hydroxy phosphonates

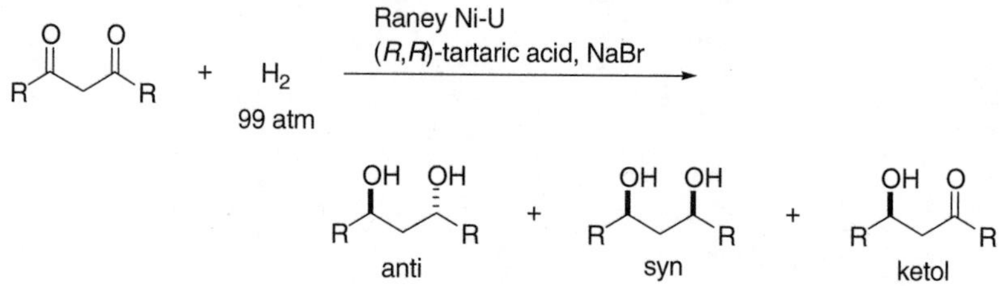

R	anti:syn:ketol	% ee of anti diol
CH$_3$	86:7:7	91
(CH$_3$)$_2$CH	72:22:6	90

examples of the application:

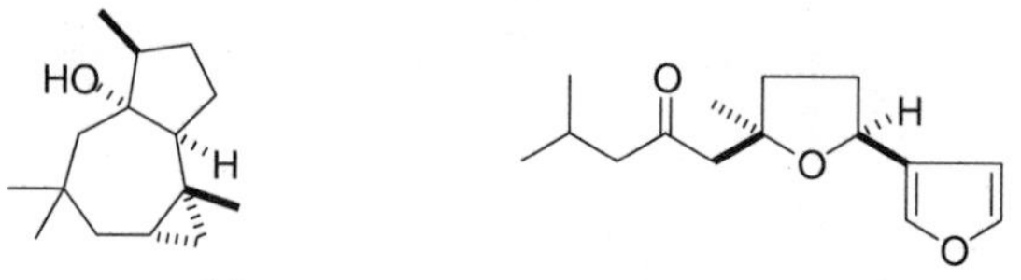

(+)-africanol[242a]

(–)-ngaione[242b]

Scheme 1.51.

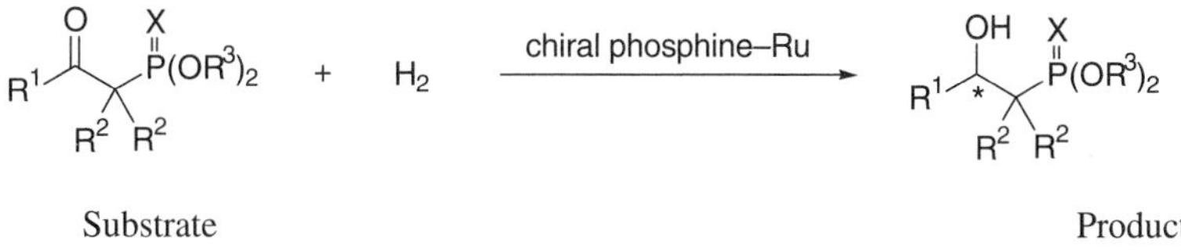

R^1	R^2	R^3	X	Phosphine	H$_2$, atm	Temp, °C	% ee	Config^n
CH$_3$	H	CH$_3$	O	(R)-BINAP	4	25	98	R
CH$_3$	H	C$_2$H$_5$	O	(S)-BINAP	1	50	99	S
CH$_3$	H	C$_2$H$_5$	O	(R,R)-BDPP	30	rt	95	R
CH$_3$	CH$_3$	CH$_3$	O	(R)-BINAP	4	50	98	R
n-C$_5$H$_{11}$	H	CH$_3$	O	(S)-BINAP	100	rt	98	S
(CH$_3$)$_2$CH	H	CH$_3$	O	(S)-BINAP	4	80	96	S
C$_6$H$_5$	H	CH$_3$	O	(R)-BINAP	4	60	95	R
n-C$_5$H$_{11}$	H	CH$_3$	S	(S)-MeO-BIPHEP	100	rt	94	S
(CH$_3$)$_2$CH	H	CH$_3$	S	(S)-MeO-BIPHEP	10	rt	93	S

Scheme 1.52.

with up to 99% ee (Scheme 1.52) [243]. The sense of enantioface discrimination is the same as that in hydrogenation of the corresponding β-keto carboxylic esters (see Table of Scheme 1.35). Notably, the reaction is much faster than that of keto carboxylic esters. β-Hydroxy phosphonates are obtainable quantitatively even at 1–4 atm of hydrogen and at room temperature. The enantioselectivity is not affected by hydrogen pressure ranging from 4 to 100 atm. A BDPP–Ru complex is also usable [224b]. Similarly, in the presence of a MeO-BIPHEP–Ru complex β-keto thiophosphates are transformed to the β-hydroxy thiophosphates in high optical yield [243b].

Enantioselective hydrogenation of sodium β-keto sulfonates can be conducted under atmospheric pressure and at 50°C in the presence of a BINAP–Ru catalyst and HCl (Ru:HCl = 1:50) to afford the β-hydroxy sulfonates with up to 97% ee (Scheme 1.53) [244]. Asymmetric

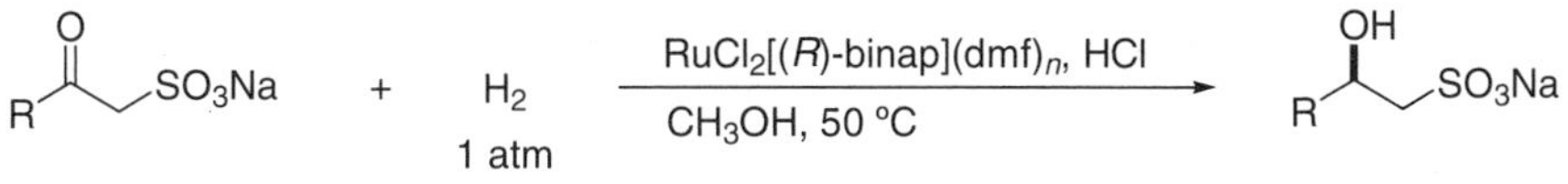

R	% Yield	% ee	Config^n
CH$_3$	100	97	R
n-C$_{15}$H$_{31}$	100	96	R
(CH$_3$)$_2$CH	100	97	R
C$_6$H$_5$	100	96	R

Scheme 1.53.

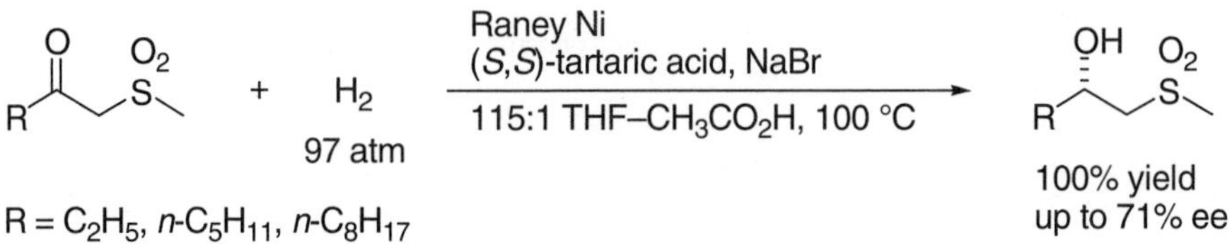

Scheme 1.54.

hydrogenation of β-keto methylsulfones catalyzed by an (S,S)-tartaric acid-modified Raney Ni gives the corresponding S alcohols with up to 71% ee (Scheme 1.54) [245].

1.4.1.5. Dynamic Kinetic Resolution Hydrogenation of racemic α-monosubstituted β-keto esters normally produces four possible stereoisomers of hydroxy esters. However, owing to the configurational ability at the α position, in principle, a single stereoisomer with two contiguous stereogenic centers is obtainable in 100% yield under suitable conditions. The rapid equilibration between the R and S enantiomers provides an opportunity for a chiral catalyst to reduce preferentially one of them. Then the combined effects of the catalyst-derived intermolecular chirality transfer and the substrate-controlled intramolecular asymmetric induction [197] determine kinetically the absolute configuration of the two stereocenters of the product. In fact, with [RuCl{(R)-binap}C$_6$H$_6$]Cl and in CH$_2$Cl$_2$, racemic 2-alkoxycarbonylcycloalkanones are hy-

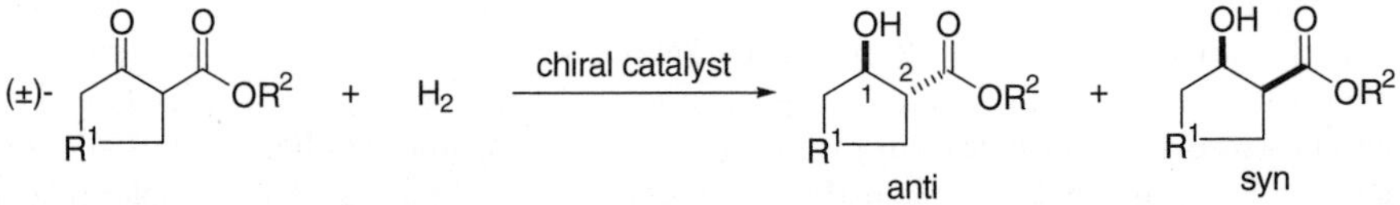

Substrate							Product	
R^1	R^2	Catalyst	Solvent	H$_2$, atm	dr^a	% ee^b	Config^b	
CH$_2$	CH$_3$	[RuCl{(R)-binap}C$_6$H$_6$]Cl	CH$_2$Cl$_2$	100	99:1	92	1R,2R	
CH$_2$	CH$_3$	[RuI{(S)-binap}-(p-cymene)]I	CH$_2$Cl$_2$^{c}	97	99:1	95	1S,2S	
CH$_2$	C$_2$H$_5$	Ru[η^3-CH$_2$C(CH$_3$)CH$_2$]$_2$-[(R)-binap] + HBr	CH$_3$OH	20	97:3	94	1R,2R	
CH$_2$	CH$_3$	(R,R)-i-Pr-BPE–Ru	9:1 CH$_3$OH-H$_2$O	4	96:4	98	1S,2S	
CH$_2$	CH$_3$	(+)-tetraMe-BITIANP–Ru	CH$_3$OH	97	93:7	99	1R,2R	
(CH$_2$)$_2$	C$_2$H$_5$	[RuCl{(R)-binap}C$_6$H$_6$]Cl	CH$_2$Cl$_2$	100	95:5	90	1R,2R	
(CH$_2$)$_2$	C$_2$H$_5$	Ru[η^3-CH$_2$C(CH$_3$)CH$_2$]$_2$-[(S)-binap] + HBr	CH$_3$OH	20	74:26	91	1S,2S	
(CH$_2$)$_3$	CH$_3$	[RuCl{(R)-binap}C$_6$H$_6$]Cl	CH$_2$Cl$_2$	100	93:7	93	1R,2R	

^{a} Anti:syn diastereomeric ratio. ^{b} Value of the anti alcohol. ^{c} Containing <1% of water.

Scheme 1.55.

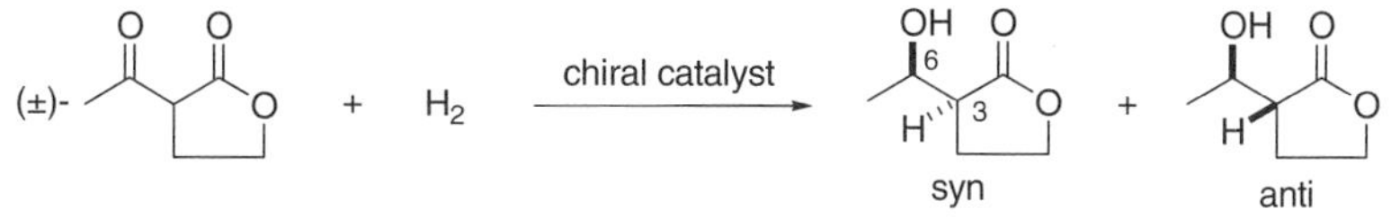

Catalyst	Solvent	H_2, atm	dr^a	% ee^b	$Confign^b$
[RuCl{(R)-binap}C$_6$H$_6$]Cl	CH$_3$OH	100	98:2	94	3S,6R
[RuI{(S)-binap}(p-cymene)]I	3:1 CH$_3$OH–CH$_2$Cl$_2$	97	99:1	97	3R,6S
(+)-tetraMe-BITIANP–Ru	CH$_3$OH	97	96:4	91	3S,6R

^{a} Syn:anti diastereomeric ratio. ^{b} Value of the syn alcohol.

Scheme 1.56.

drogenated to give 1R,2R hydroxy esters with a high anti-selectivity and optical yield through the dynamic kinetic resolution [246]. For example, 2-methoxycarbonylcyclopentanone is transformed to the anti-hydroxy ester in 99% yield and 92% ee (Scheme 1.55). The degree of enantio- and diastereoselectivity is highly dependent on the substrates, the preparation procedure of the BINAP–Ru catalysts, and the reaction conditions [168,247]. The selection of solvent is especially crucial. An i-Pr-BPE– [186] or tetraMe-BITIANP–Ru complex [109] also shows an excellent stereoselectivity. The kinetic behavior of the stereoselective hydrogenation of racemic 2-methoxycarbonylcycloheptanone with Ru(OCOCH$_3$)$_2$[(R)-binap] and 2 equiv of HCl is fully understood by a computer-aided quantitative analysis [248]. Thus, on the one hand, hydrogenation of the R keto ester in CH$_2$Cl$_2$ occurs 9.8 times faster than that of the S isomer, and the equilibration between the enantiomeric substrates is 4.4 times faster than hydrogenation of the slow-reacting S substrate. On the other hand, hydrogenation of racemic 3-acetyltetrahy-drofuran-2-one catalyzed by an (S)-BINAP–Ru complex gives the 3R,6S (syn) alcohol in 99% yield and up to 97% ee (Scheme 1.56) [168,246b]. A tetraMe-BITIANP–Ru complex is also usable [109].

A series of acyclic α-substituted β-keto esters is hydrogenated through dynamic kinetic resolution with an excellent enantio- and diastereoselectivity. α-Acylamino and α-amidomethyl substrates are converted to 2S,3R (syn) alcohols with up to 98% ee with an (R)-BINAP–Ru complex (Scheme 1.57) [168,246a,249]. The use of a sterically hindered DTBBINAP ligand, the α-amidomethyl keto ester is hydrogenated with almost perfect stereoselectivity, albeit at a lower rate [168]. It is interesting that hydrogenation of an α-chloro substrate in the presence of a BINAP–Ru(η^3-CH$_2$C(CH$_3$)CH$_2$)$_2$(cod) system gives exclusively the *anti*-chlorohydrin with 99% ee [250]. High diastereoselectivity is not accessible in hydrogenation of simple α-methyl β-keto esters, although the reaction proceeds with a high level of enantioselection [186,246].

Hydrogenation of α-acylamino or α-halogeno β-keto phosphonates with a BINAP–Ru complex gives the corresponding syn alcohols selectively with >98% ee (Scheme 1.58) [243a,251]. The sense of enantio- and diastereoface discrimination is the same as that in the case of α-substituted β-keto carboxylic esters (see Table of Scheme 1.57).

The stereoselective hydrogenation of α-monosubstituted β-keto carboxylates and phosphonates through dynamic kinetic resolution has been applied to the synthesis of a wide variety of useful bioactive compounds as well as some chiral diphosphines (Figure 1.16) [1c,20,162b,c,179,243,246,250,252]. The stereogenic center determined by the BINAP–Ru

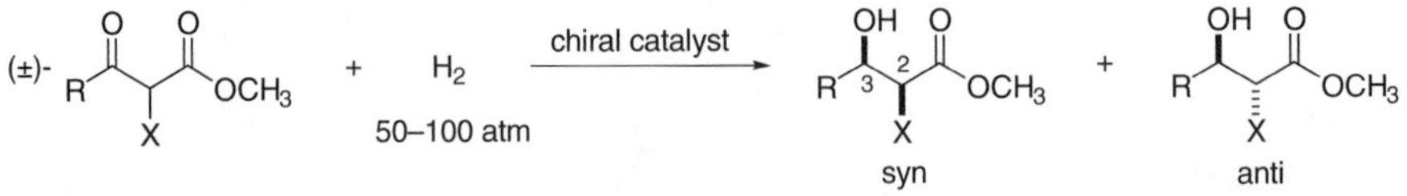

$$(\pm)\text{-} \quad \underset{X}{R-\overset{O}{C}-\overset{O}{C}-OCH_3} \quad + \quad H_2 \quad \xrightarrow[\text{50–100 atm}]{\text{chiral catalyst}} \quad \underset{X}{R\overset{OH}{\underset{3}{C}}\overset{}{\underset{2}{C}}\overset{O}{C}OCH_3} \quad + \quad \underset{X}{R\overset{OH}{C}\overset{O}{C}OCH_3}$$

syn   anti

	Substrate				Product	
R	X	Catalyst	Solvent	dr^a	% ee^b	Config^b
CH_3	CH_3CONH	$RuBr_2[(R)\text{-binap}]$	CH_2Cl_2	99:1	98	$2S,3R$
CH_3	CH_3CONH	$Ru[\eta^3\text{-}CH_2C(CH_3)CH_2]_2[(R)\text{-binap}] + HCl$	CH_3OH	76:24	95	$2S,3R$
CH_3	$(CH_3)_2CHCONH$	$Ru[\eta^3\text{-}CH_2C(CH_3)CH_2]_2[(R)\text{-binap}] + HBr$	CH_3OH	77:23	92	$2S,3R$
Ar^c	CH_3CONH	$RuBr_2[(R)\text{-binap}]$	CH_2Cl_2	99:1	94	$2S,3R$
CH_3	$C_6H_5CONHCH_2$	$[NH_2(C_2H_5)_2][\{RuCl\text{-}[(R)\text{-binap}]\}_2(\mu\text{-}Cl)_3]$	CH_2Cl_2	94:6	98	$2S,3R$
CH_3	$C_6H_5CONHCH_2$	$[RuI\{(S)\text{-binap}\}(p\text{-cymene})]I$	$CH_2Cl_2{}^d$	94:6	97	$2R,3S$
CH_3	$C_6H_5CONHCH_2$	$(+)\text{-DTBBINAP–Ru}$	7:1 CH_2Cl_2–CH_3OH	99:1	99	$2S,3R$
CH_3	Cl^e	$Ru[\eta^3\text{-}CH_2C(CH_3)CH_2]_2(cod)\text{–}(R)\text{-BINAP}$	CH_2Cl_2	1:99	99^f	$2R,3R^f$
CH_3	CH_3	$(R,R)\text{-}i\text{-Pr-BPE–Ru}$	9:1 CH_3OH–H_2O^g	58:42	96^f	$2R,3R^f$
CH_3	$CH_3{}^e$	$[RuCl\{(R)\text{-binap}\}C_6H_6]Cl$	CH_2Cl_2	32:68	94^f	$2R,3R^f$

^{a} Syn:anti diastereomeric ratio. ^{b} Value of the syn alcohol. ^{c} 3,4-Methylenedioxyphenyl. ^{d} Containing 0.5% of water. ^{e} Ethyl ester. ^{f} Value of the anti alcohol. ^{g} 4 atm of H_2.

Scheme 1.57.

chemistry is labeled by R or S. A very important application of this method is the industrial synthesis of the chiral 2-acetoxyazetidinone, a key intermediate for the synthesis of carbapenem antibiotics (Takasago Int. Corp) (Scheme 1.59) [1c,179,228,253].

1.4.2. Unfunctionalized Ketones

Despite fruitful results of asymmetric hydrogenation of functionalized ketones, only limited examples have been reported for reaction of ketonic substrates with no functionality near the carbonyl group [1,162,254]. Transition-metal catalysts with a bidentate chiral phosphine, successfully used for functionalized ketones, are often ineffective for reduction of simple ketones in terms of reactivity and enantioselectivity [162b,c]. However, a breakthrough in this subject has been provided by the invention of a new chiral Ru catalyst system.

1.4.2.1. Aromatic Ketones Although phosphine–Ru(II) dichlorides are poor catalysts for hydrogenation of unfunctionalized ketones [162b,255], a remarkably high reactivity emerges when the Ru compounds are further complexed with a 1,2-diamine ligand. Thus *trans-*

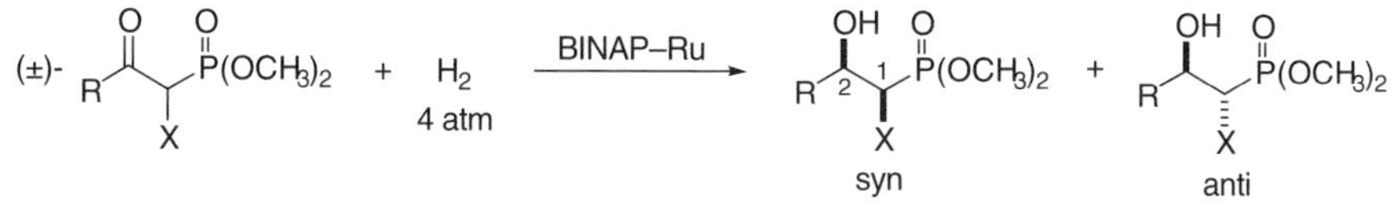

$$(\pm)\text{-} \quad R \overset{O}{\underset{X}{\|}} \overset{O}{\underset{}{\|}} P(OCH_3)_2 \;+\; H_2 \;\xrightarrow[\text{4 atm}]{\text{BINAP–Ru}}\; R \overset{OH}{\underset{2}{\,}} \overset{O}{\underset{1}{\,}} P(OCH_3)_2 \;+\; R \overset{OH}{\underset{}{\,}} \overset{O}{\underset{}{\,}} P(OCH_3)_2$$

syn   anti

	Substrate			Product		
R	X	BINAP	Temp, °C	dr[a]	% ee[b]	Confign[b]
CH₃	CH₃CONH	R	25	97:3	>98	1R,2R
C₆H₅	CH₃CONH	R	45	98:2	95	1R,2R
CH₃	Br	S	25	90:10[c]	98	1R,2S

[a] Syn:anti diastereomeric ratio. [b] Value of the syn alcohol. [c] Contaminated with 15% of a debrominated compound.

Scheme 1.58.

$RuCl_2(phosphine)_2(1,2\text{-diamine})$ combined with an inorganic base in 2-propanol is now recognized as the most reactive (pre)catalyst for homogeneous hydrogenation of ketones [162c,256,257]. Rapid, highly productive asymmetric hydrogenation of aromatic ketones has been realized by using $trans\text{-}RuCl_2[(S)\text{-tolbinap}][(S,S)\text{-dpen}]$ or the $R/R,R$ enantiomer (Scheme 1.60) [256]. For example, 601 g of acetophenone can be hydrogenated quantitatively by using only 2.2 mg of the Ru complex under 45 atm and at 30°C. The turnover number (TON, moles product per mole catalyst) reaches 2,400,000 and the TOF as high as 228,000 h^{-1} or 63 s^{-1}.

A wide variety of aromatic ketones can be hydrogenated quantitatively with an excellent enantioselectivity by using appropriate diphosphine/diamine Ru complexes (Scheme 1.61) [256–258]. $trans\text{-}RuCl_2[(S)\text{-xylbinap}][(S)\text{-daipen}]$ or its R,R isomer among many complexes exhibits the highest selectivity—up to 100%—and generality [258]. The reaction with an S/C up to 100,000 is performed under 1–10 atm of hydrogen. Many ring substituents, including F, Cl, Br, I, CF_3, CH_3O, $(CH_3)_2CHOCO$, NO_2, and NH_2, are not injured under the conditions. Aromatic ketones with an alkyl group such as methyl, ethyl, isopropyl, and cyclopropyl are hydrogenated with an excellent optical yield. A catalyst system prepared in situ from a BIPNOR–Ru complex, DPEN, and KOH is also effective for hydrogenation of 2′-acetonaphthone [26].

Examples of asymmetric hydrogenation of simple aromatic ketones with chiral Rh and Ir complexes are shown in Scheme 1.62. The reaction is conducted with an S/C of 100–200 under a relatively high pressure of hydrogen. Hydrogenation of aryl methyl ketones with a catalyst prepared from $[RhCl(nbd)]_2$, (S,S)-DIOP, and $N(C_2H_5)_3$ gives the corresponding chiral alcohols with up to 84% ee [259]. Addition of the organic base is supposed to promote the heterolytic cleavage of hydrogen on the metal center generating a reactive Rh hydride species. Acetophenone is hydrogenated with a BDPP–Rh catalyst to give 1-phenylethanol with 82% ee [145,260]. Several aromatic ketones are hydrogenated with an (R,S,R,S)-Me-PennPhos–Rh complex in the presence of 2,6-lutidine and KBr to give the corresponding S alcohols with up to 96% ee [261]. The presence of 2,6-lutidine and KBr increases reactivity and enantioselectivity. A catalyst system consisting of a cationic BINAP–Ir complex and an aminophosphine is used for hydrogenation of alkyl phenyl ketones with up to 84% optical yield [262].

A BINAP–Ir–aminophosphine system effectively promotes hydrogenation of cyclic aromatic ketones in a 5:1 dioxane–CH_3OH mixture under 50–57 atm of hydrogen at 90°C to give

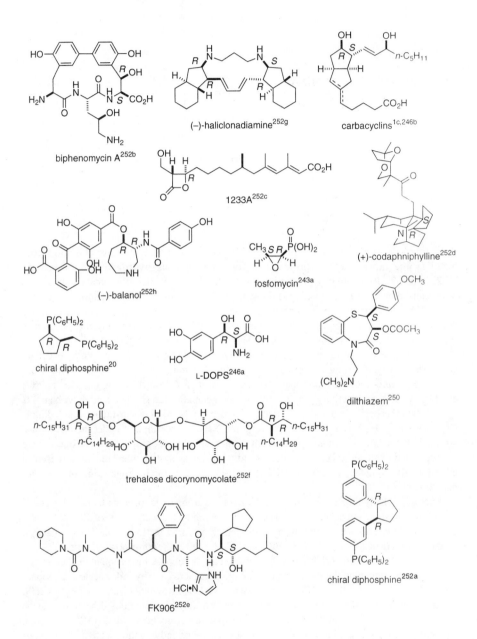

Figure 1.16. Examples of biologically active compounds and chiral diphosphines obtainable via BINAP–Ru catalyzed hydrogenation through dynamic kinetic resolution.

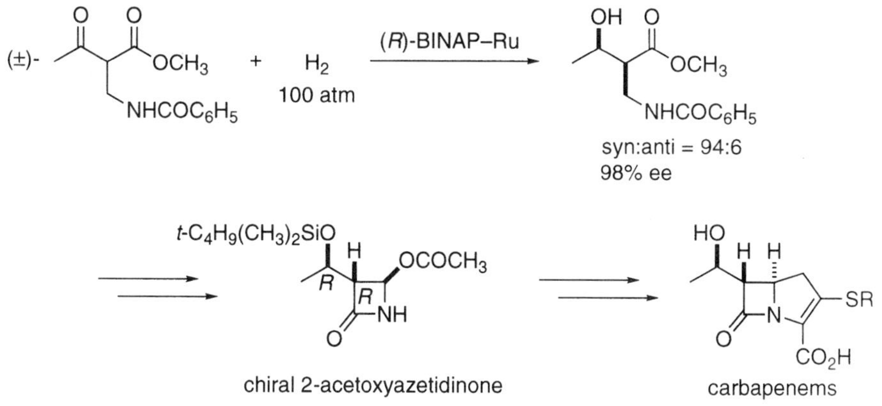
(±)-
O
O
OCH3
NHCOC6H5
+
H2
100 atm
(R)-BINAP–Ru
OH
O
OCH3
NHCOC6H5
syn:anti = 94:6
98% ee
t-C4H9(CH3)2SiO
H
OCOCH3
R
R
NH
O
chiral 2-acetoxyazetidinone
HO
H H
SR
N
O
CO2H
carbapenems

Scheme 1.59.

the corresponding alcohols with up to 95% ee (Scheme 1.63) [263]. β-Thiacycloalkanones are also hydrogenated stereoselectively with an added aminophosphine.

Hydrogenation of 2,2,2-trifluoroacetophenone and its derivatives with a mixture of *trans*-$RuCl_2$[(*S*)-xylbinap][(*S*)-daipen] and $(CH_3)_3COK$ in 2-propanol gives the *S* alcohols quantitatively with a high optical purity (Scheme 1.64) [258]. Unlike with many chiral borane reagents [264], the sense of enantioface discrimination is the same as in hydrogenation of acetophenone. The electronic effects of 4′-substituents on the enantioselectivity are small. These chiral fluorinated alcohols are useful as components of new functionalized materials [265].

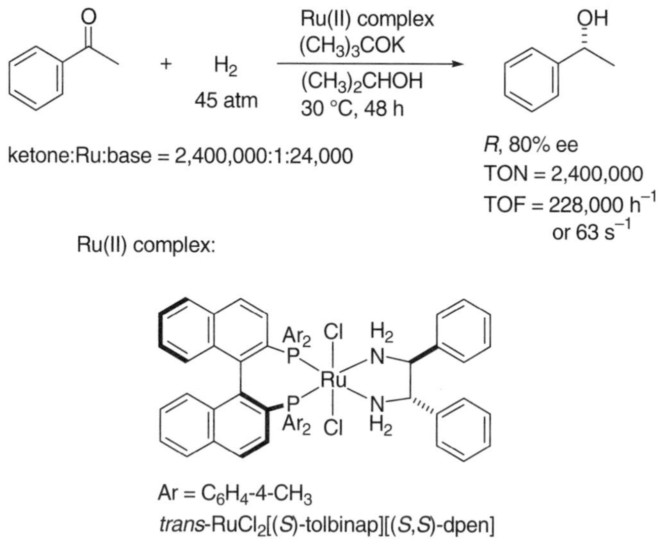
O
+
H2
45 atm
Ru(II) complex
(CH3)3COK
(CH3)2CHOH
30 °C, 48 h
OH
ketone:Ru:base = 2,400,000:1:24,000
R, 80% ee
TON = 2,400,000
TOF = 228,000 h−1
or 63 s−1
Ru(II) complex:
Ar2
Cl
H2
P
N
Ru
P
N
Ar2
Cl
H2
Ar = C6H4-4-CH3
trans-RuCl2[(S)-tolbinap][(S,S)-dpen]

Scheme 1.60.

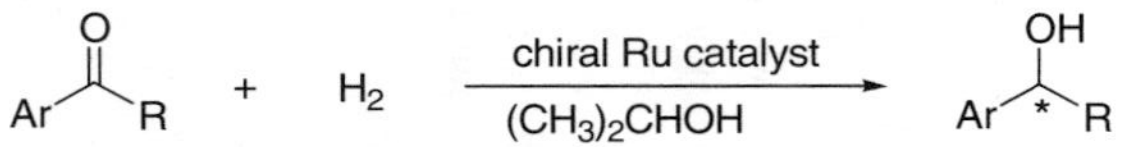

Substrate		Catalyst	Conditions			Product		
R	Ar		S/C[a]	H_2, atm	Time, h	% Yield	% ee	Config'n
CH_3	C_6H_5	trans-RuCl$_2$[(S)-xylbinap][(S)-daipen] + (CH$_3$)$_3$COK	100,000	8	60	97	99	R
CH_3	2-CH$_3$C$_6$H$_4$	trans-RuCl$_2$[(S)-tolbinap][(S)-daipen] + (CH$_3$)$_3$COK	100,000	10	48	94	99	R
CH_3	3-CH$_3$C$_6$H$_4$	trans-RuCl$_2$[(S)-xylbinap][(S)-daipen] + (CH$_3$)$_3$COK	10,000	10	48	98	100	R
CH_3	4-CH$_3$C$_6$H$_4$	trans-RuCl$_2$[(R)-xylbinap][(R,R)-dpen] + (CH$_3$)$_3$COK	2,000	4	7	100	98	S
CH_3	4-n-C$_4$H$_9$C$_6$H$_4$	trans-RuCl$_2$[(R)-xylbinap][(R,R)-dpen] + (CH$_3$)$_3$COK	2,000	4	5	100	98	S
CH_3	2,4-(CH$_3$)$_2$C$_6$H$_3$	trans-RuCl$_2$[(R)-xylbinap][(R)-daipen] + (CH$_3$)$_3$COK	2,000	4	8	99	99	S
CH_3	2-FC$_6$H$_4$	trans-RuCl$_2$[(S)-xylbinap][(S)-daipen] + (CH$_3$)$_3$COK	2,000	8	13	100	97	R
CH_3	3-FC$_6$H$_4$	trans-RuCl$_2$[(R)-xylbinap][(R,R)-dpen] + (CH$_3$)$_3$COK	2,000	4	6	99	98	S
CH_3	4-FC$_6$H$_4$	trans-RuCl$_2$[(R)-xylbinap][(R)-daipen] + (CH$_3$)$_3$COK	2,000	4	6	100	97	S
CH_3	2-ClC$_6$H$_4$	trans-RuCl$_2$[(R)-xylbinap][(R,R)-dpen] + (CH$_3$)$_3$COK	2,000	4	18	99	98	S
CH_3	4-ClC$_6$H$_4$	trans-RuCl$_2$[(R)-xylbinap][(R,R)-dpen] + (CH$_3$)$_3$COK	2,000	4	4	99.5	98	S
CH_3	2-BrC$_6$H$_4$	trans-RuCl$_2$[(R)-tolbinap][(R)-daipen] + (CH$_3$)$_3$COK	10,000	10	6	100	98	S
CH_3	2-BrC$_6$H$_4$	trans-RuCl$_2$[(R)-xylbinap][(R)-daipen] + (CH$_3$)$_3$COK	2,000	4	3	99	96	S
CH_3	3-BrC$_6$H$_4$	trans-RuCl$_2$[(R)-xylbinap][(R)-daipen] + (CH$_3$)$_3$COK	2,000	4	3	100	99.5	S
CH_3	4-BrC$_6$H$_4$	trans-RuCl$_2$[(S)-xylbinap][(S)-daipen] + (CH$_3$)$_3$COK	500	1	3	99.7	99.6	R
CH_3	4-IC$_6$H$_4$	trans-RuCl$_2$[(S)-xylbinap][(S)-daipen] + (CH$_3$)$_3$COK	2,000	8	4	99.7	99	R
CH_3	2-CF$_3$C$_6$H$_4$	trans-RuCl$_2$[(R)-xylbinap][(R)-daipen] + (CH$_3$)$_3$COK	2,000	4	3	99	99	S
CH_3	3-CF$_3$C$_6$H$_4$	trans-RuCl$_2$[(R)-xylbinap][(R)-daipen] + (CH$_3$)$_3$COK	2,000	4	4	100	99	S

CH_3	$4\text{-}CF_3C_6H_4$	*trans*-RuCl$_2$[(*S*)-xylbinap][(*S*)-daipen] + (CH$_3$)$_3$COK	10,000	10	20	100	99.6	*R*
CH_3	$2\text{-}CH_3OC_6H_4$	*trans*-RuCl$_2$[(*R*)-xylbinap][(*R*)-daipen] + (CH$_3$)$_3$COK	2,000	4	10	100	92	*S*
CH_3	$3\text{-}CH_3OC_6H_4$	*trans*-RuCl$_2$[(*R*)-xylbinap][(*R*)-daipen] + (CH$_3$)$_3$COK	2,000	4	4	99	99	*S*
CH_3	$4\text{-}CH_3OC_6H_4$	*trans*-RuCl$_2$[(*S*)-xylbinap][(*S*)-daipen] + (CH$_3$)$_3$COK	2,000	10	1	100	100	*R*
CH_3	$4\text{-}[(CH_3)_2CHOCO]C_6H_4$	*trans*-RuCl$_2$[(*S*)-xylbinap][(*S*)-daipen] + (CH$_3$)$_3$COK	2,000	8	3	100	99	*R*
CH_3	$4\text{-}NO_2C_6H_4$	*trans*-RuCl$_2$[(*S*)-xylbinap][(*S*)-daipen] + (CH$_3$)$_3$COK	2,000	8	15	100	99.8	*R*
CH_3	$4\text{-}NH_2C_6H_4$	*trans*-RuCl$_2$[(*S*)-xylbinap][(*S*)-daipen] + (CH$_3$)$_3$COK	2,000	8	4	100	99	*R*
CH_3	1-naphthyl	*trans*-RuCl$_2$[(*S*)-tolbinap][(*S,S*)-dpen] + (CH$_3$)$_3$COK	100,000	10	40	99.5	98	*R*
CH_3	1-naphthyl	*trans*-RuCl$_2$[(*R*)-xylbinap][(*R,R*)-dpen] + (CH$_3$)$_3$COK	2,000	4	4	99	99	*S*
CH_3	2-naphthyl	*trans*-RuCl$_2$[(*R*)-xylbinap][(*R,R*)-dpen] + (CH$_3$)$_3$COK	2,000	4	4	99	98	*S*
CH_3	2-naphthyl	RuBr$_2$[(*R,R*)-bipnor]–(*S,S*)-DPEN + KOH	500	5	10	65	81	*R*
C_2H_5	C_6H_5	*trans*-RuCl$_2$[(*R*)-xylbinap][(*R,R*)-dpen] + (CH$_3$)$_3$COK	2,000	4	15	100	99	*S*
C_2H_5	$4\text{-}FC_6H_4$	*trans*-RuCl$_2$[(*R*)-xylbinap][(*R,R*)-dpen] + (CH$_3$)$_3$COK	2,000	4	8	99	99	*S*
C_2H_5	$4\text{-}ClC_6H_4$	*trans*-RuCl$_2$[(*S*)-xylbinap][(*S*)-daipen] + (CH$_3$)$_3$COK	20,000	8	16	99.9	99	*R*
$(CH_3)_2CH$	C_6H_5	*trans*-RuCl$_2$[(*R*)-xylbinap][(*R*)-daipen] + (CH$_3$)$_3$COK	10,000	8	14	99.7	99	*S*
$cyclo\text{-}C_3H_5$	C_6H_5	*trans*-RuCl$_2$[(*S*)-xylbinap][(*S*)-daipen] + (CH$_3$)$_3$COK	2,000	8	14	99.7	96	*R*

[a] Substrate/catalyst molar ratio.

Scheme 1.61.

$$\underset{Ar}{\overset{O}{\|}}\!\!-\!\!R \;+\; H_2 \;\xrightarrow{\text{chiral catalyst}}\; \underset{Ar}{\overset{OH}{|}}\!\!\overset{*}{-}\!\!R$$

Substrate		Catalyst	Conditions				Product		
R	Ar		S/C	H$_2$, atm	Temp, °C	Time, h	% Yield	% ee	Config'n
CH$_3$	C$_6$H$_5$	[RhCl(nbd)]$_2$–(S,S)-DIOP + N(C$_2$H$_5$)$_3$	200	69	50	6	64	80	–
CH$_3$	C$_6$H$_5$	[RhCl(nbd)]$_2$–(S,S)-BDPP + N(C$_2$H$_5$)$_3$	100	69	50	24	72	82	S
CH$_3$	C$_6$H$_5$	[RhCl(cod)]$_2$–(R,S,R,S)-Me-PennPhos + 2,6-lutidine	100	30	rt	24	97	95	S
CH$_3$	C$_6$H$_5$	[Ir{(S)-binap}(cod)]BF$_4$–P[2-N(CH$_3$)$_2$C$_6$H$_4$]$_2$C$_6$H$_5$	100	54–61	60	126	63	54	S
CH$_3$	2-CH$_3$C$_6$H$_4$	[RhCl(nbd)]$_2$–(S,S)-DIOP + N(C$_2$H$_5$)$_3$	200	69	50	20	100	77	–
CH$_3$	4-CH$_3$OC$_6$H$_4$	[RhCl(cod)]$_2$–(R,S,R,S)-Me-PennPhos + 2,6-lutidine + KBr	100	30	rt	48	83	94	S
CH$_3$	1-naphthyl	[RhCl(nbd)]$_2$–(S,S)-DIOP + N(C$_2$H$_5$)$_3$	200	69	50	4	100	84	–
CH$_3$	1-furyl	[RhCl(cod)]$_2$–(R,S,R,S)-Me-PennPhos + 2,6-lutidine + KBr	100	30	rt	100	99	96	S
C$_2$H$_5$	C$_6$H$_5$	[RhCl(cod)]$_2$–(R,S,R,S)-Me-PennPhos + 2,6-lutidine + KBr	100	30	rt	88	95	93	S
(CH$_3$)$_2$CH	C$_6$H$_5$	[RhCl(cod)]$_2$–(R,S,R,S)-Me-PennPhos + 2,6-lutidine + KBr	100	30	rt	94	20	72	S
(CH$_3$)$_2$CH	C$_6$H$_5$	[Ir{(S)-binap}(cod)]BF$_4$–P[2-N(CH$_3$)$_2$C$_6$H$_4$]$_2$C$_6$H$_5$	200	54–61	90	160	78	84	R

Scheme 1.62.

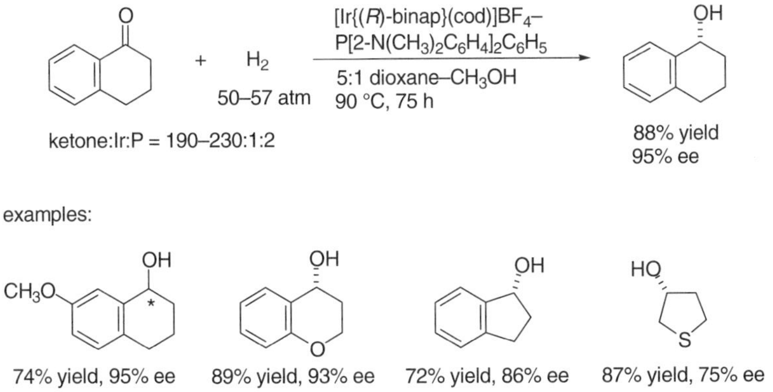

Scheme 1.63.

examples:

74% yield, 95% ee   89% yield, 93% ee   72% yield, 86% ee   87% yield, 75% ee

Cinchonidine-modified Pt/Al$_2$O$_3$ catalyzes hydrogenation of 2,2,2-trifluoroacetophenone to give the R alcohol with 56% ee [266].

1.4.2.2. Aliphatic Ketones Enantioselective hydrogenation of simple aliphatic ketones remains difficult, because at present there is no reliable way to differentiate two alkyl groups. Obviously methyl is different from other alkyl groups on the steric ground. Hydrogenation of pinacolone in CH$_3$OH containing an (R,S,R,S)-Me-PennPhos–Rh complex, 2,6-lutidine, and KBr gives the S alcohol with 94% ee (Scheme 1.65) [261]. Cyclohexyl methyl ketone and 2-hexanone are also reduced to the alcohols with 92% ee and 75% ee, respectively. A cationic (S,R,R,R)-TMO-DEGUPHOS–Rh is also effective for hydrogenation of pinacolone [267]. Cyclopropyl methyl ketone is hydrogenated in the presence of *trans*-RuCl$_2$[(S)-xylbinap][(S)-daipen] and (CH$_3$)$_3$COK to give the R alcohol with 95% ee without cleavage of the cyclopropyl ring [258]. The high level of enantio-distinction is partly due to the strong electron-donation ability of cyclopropyl group [268]. In comparison, cyclohexyl methyl ketone is transformed to the R alcohol with 85% ee with the same complex [258].

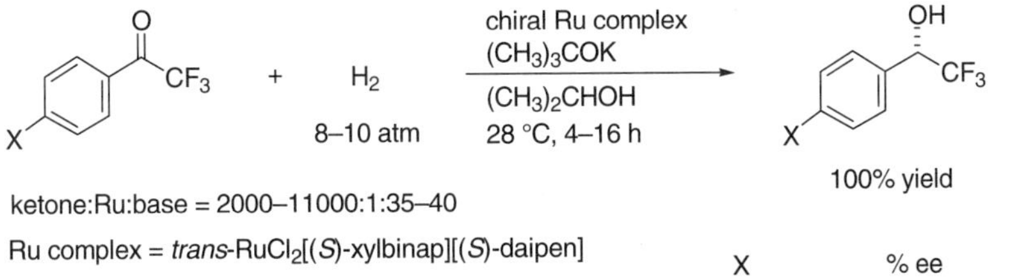

ketone:Ru:base = 2000–11000:1:35–40

Ru complex = *trans*-RuCl$_2$[(S)-xylbinap][(S)-daipen]

X	% ee
H	96
Cl	94
Br	94
CH$_3$O	96

Scheme 1.64.

$$R\text{-}\underset{O}{C}\text{-}CH_3 + H_2 \xrightarrow{\text{chiral catalyst}} R\overset{*}{-}\underset{OH}{C}H\text{-}CH_3$$

R in Substrate	Catalyst	S/C	Solvent	Conditions H₂, atm	Temp, °C	Time, h	Product % Yield	% ee	Config
n-C₄H₉	[RhCl(cod)]₂–(R,S,R,S)-Me-PennPhos + 2,6-lutidine + KBr	100	CH₃OH	30	rt	48	96	75	S
(CH₃)₂CHCH₂	[RhCl(cod)]₂–(R,S,R,S)-Me-PennPhos + 2,6-lutidine + KBr	100	CH₃OH	30	rt	75	66	85	S
(CH₃)₂CH	[RhCl(cod)]₂–(R,S,R,S)-Me-PennPhos + 2,6-lutidine + KBr	100	CH₃OH	30	rt	94	99	84	S
cyclo-C₃H₅	trans-RuCl₂[(S)-xylbinap][(S)-daipen] + (CH₃)₃COK	11,000	(CH₃)₂CHOH	10	28	12	96	95	R
cyclo-C₆H₁₁	[RhCl(cod)]₂–(R,S,R,S)-Me-PennPhos + 2,6-lutidine + KBr	100	CH₃OH	30	rt	106	90	92	S
cyclo-C₆H₁₁	trans-RuCl₂[(S)-xylbinap][(S)-daipen] + (CH₃)₃COK	10,000	(CH₃)₂CHOH	8	28	20	99	85	R
(CH₃)₃C	[RhCl(cod)]₂–(R,S,R,S)-Me-PennPhos + 2,6-lutidine + KBr	100	CH₃OH	30	rt	96	51	94	S
(CH₃)₃C	Rh{(S,R,R,R)-tmo-deguphos}(cod)]BF₄	1000	(CH3)₂CHOH	73	25	10	30	84	S

Scheme 1.65.

62

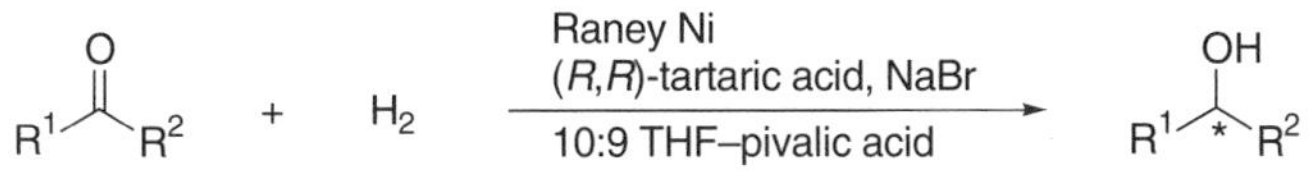

Reaction scheme:

R^1-C(=O)-R^2 + H_2 →[Raney Ni (*R,R*)-tartaric acid, NaBr; 10:9 THF–pivalic acid] R^1-CH(OH)-R^2 (with * at chiral center)

Substrate				Product		
R^1	R^2	H_2, atm	Temp, °C	% Yield	% ee	Config'n
C_2H_5	CH_3	90	50	>98	72	S
n-C_4H_9	CH_3	90	60	>98	80	S
n-C_6H_{13}	CH_3	90	60	>98	80	S
n-$C_{11}H_{23}$	CH_3	90	60	>98	75	S
$(CH_3)_2CHCH_2$	CH_3	90	60	>98	75	S
$(CH_3)_2CH$	CH_3	90	60	>98	85	S
n-C_6H_{13}	C_2H_5	90^a	100	>98	44	S

^{a} Fine Ni powder and 1-methyl-1-cyclohexanecarboxylic acid were used instead of Raney Ni and pivalic acid.

Scheme 1.66.

Heterogeneous asymmetric hydrogenation of simple aliphatic ketones with a Raney Ni catalyst modified by tartaric acid and NaBr shows fair-to-good enantioface differentiation [269]. In the presence of an excess amount of pivalic acid, 2-alkanones give the corresponding *S* configurated alcohols quantitatively with up to 85% ee (Scheme 1.66) [269a,b,f]. Distinction of methyl from even ethyl groups is possible when the NaBr/tartaric acid ratio is 22 [269d]. When fine Ni powder is used instead of Raney Ni, 3-octanone gives 3-octanol with up to 44% ee [269c,e].

4.2.3. Unsaturated Ketones

Enantioselective hydrogenation of unsaturated ketones giving chiral unsaturated alcohols is achievable by only limited catalysts. With most existing heterogeneous and homogeneous catalysts, saturation occurs at C=C bonds preferentially over C=O [270]. Thus development of carbonyl-selective and enantioselective hydrogenation is highly desirable.

This long-standing problem has been solved by applying *trans*-$RuCl_2$(binap)(1,2-diamine) as precatalyst [258]. The combination of the Ru catalyst, a diamine, and an inorganic base in 2-propanol promotes carbonyl-selective hydrogenation of enones, leaving the olefinic bond intact [162b,271]. For example, 1-phenyl-4-penten-1-one, an unconjugated enone, is hydrogen-

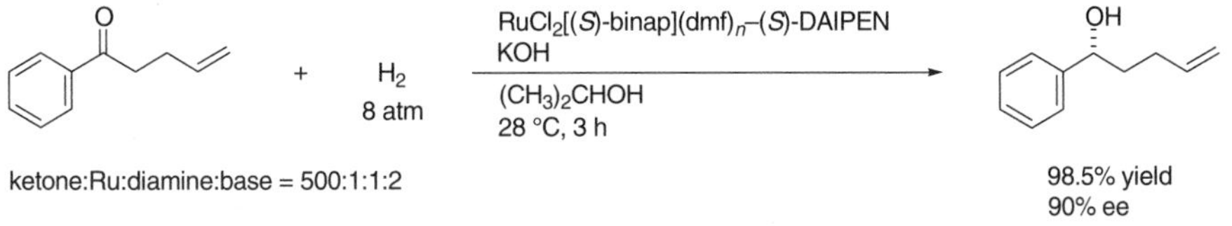

Reaction scheme:

ketone + H_2 (8 atm) →[$RuCl_2[(S)$-binap](dmf)$_n$–(S)-DAIPEN; KOH; $(CH_3)_2CHOH$; 28 °C, 3 h] product alcohol

ketone:Ru:diamine:base = 500:1:1:2

98.5% yield
90% ee

Scheme 1.67.

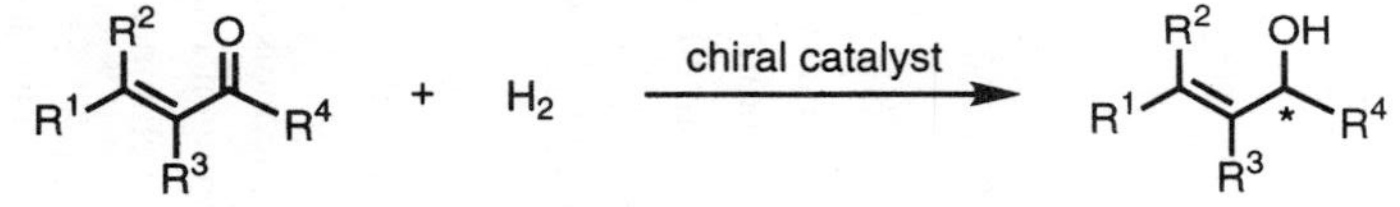

$$\text{ketone} + H_2 \xrightarrow{\text{chiral catalyst}} \text{chiral alcohol}$$

a: $R^1 = C_6H_5$; $R^2 = R^3 = H$; $R^4 = CH_3$

b: $R^1 = C_6H_5$; $R^2 = R^3 = H$; $R^4 = (CH_3)_2CH$

c: $R^1 = n\text{-}C_5H_{11}$; $R^2 = R^3 = H$; $R^4 = CH_3$

d: $R^1 = CH_3$; $R^2 = R^3 = H$; $R^4 = (CH_3)_2CHCH_2$

e: $R^1 = R^2 = R^4 = CH_3$; $R^3 = H$

f: $R^1 - R^3 = (CH_2)_4$; $R^2 = H$; $R^4 = CH_3$

g: $R^1 - R^3 = (CH_2)_5$; $R^2 = H$; $R^4 = CH_3$

h: $R^1 - R^3 = (CH_2)_3$; $R^2 = R^4 = CH_3$

i: $R^1 = 2,6,6\text{-trimethylcyclohexenyl}$; $R^2 = R^3 = H$; $R^4 = CH_3$ (β-ionone)

examples of the application:

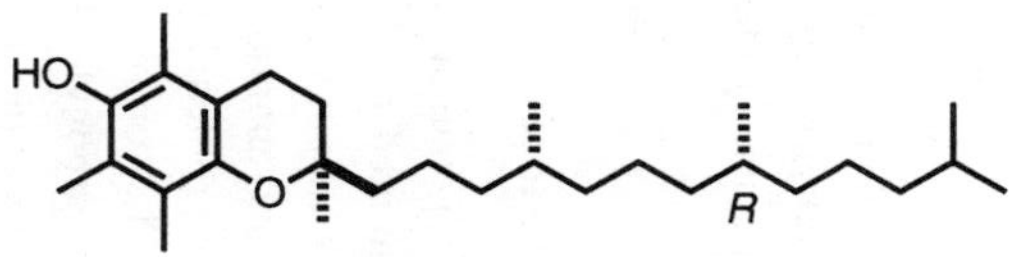

α-tocopherol[258,272a]

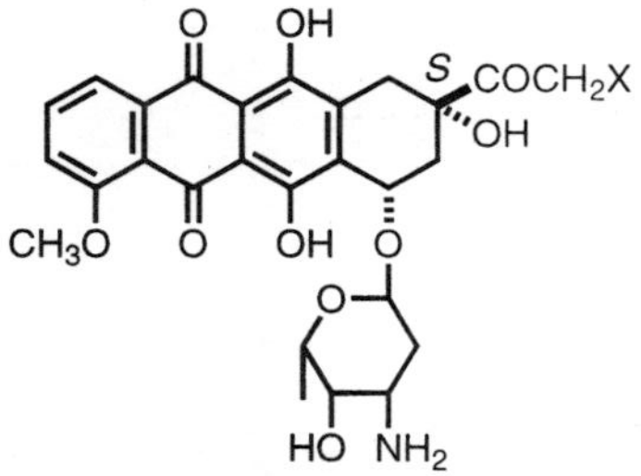

adriamycin: $X = OH$[271,272b]
daunorubicin: $X = H$[271,272b]

				Conditions			Product		
Substrate	Catalyst	S/C	Solvent	H$_2$, atm	Temp, °C	Time, h	% Yield	% ee	Config
a	trans-RuCl$_2$[(S)-xylbinap][(S)-daipen] + K$_2$CO$_3$	100,000	(CH$_3$)$_2$CHOH	80	30	43	100	97	R
a	trans-RuCl$_2$[(S)-xylbinap][(S)-daipen] + K$_2$CO$_3$	10,000	(CH$_3$)$_2$CHOH	10	28	15	100	96	R
a	[Ir{(R)-binap}(cod)]BF$_4$–P[2-N(CH$_3$)$_2$C$_6$H$_4$](C$_6$H$_5$)$_2$	120	THF	48	60	47	70	65	R
b	trans-RuCl$_2$[(S)-xylbinap][(S)-daipen] + K$_2$CO$_3$	2,000	(CH$_3$)$_2$CHOH	8	28	20	100	86	R
c	trans-RuCl$_2$[(S)-xylbinap][(S)-daipen] + K$_2$CO$_3$	2,000	(CH$_3$)$_2$CHOH	8	28	15	98	97	R
d	trans-RuCl$_2$[(R)-xylbinap][(R)-daipen] + K$_2$CO$_3$	2,000	(CH$_3$)$_2$CHOH	10	28	37	100	90	S
e	trans-RuCl$_2$[(S)-xylbinap][(S,S)-dpen] + K$_2$CO$_3$	10,000	(CH$_3$)$_2$CHOH	8	28	16	100	93	R
f	trans-RuCl$_2$[(S)-xylbinap][(S)-daipen] + (CH$_3$)$_3$COK	10,000	(CH$_3$)$_2$CHOH	10	28	16	99	100	R
g	trans-RuCl$_2$[(S)-xylbinap][(S)-daipen] + (CH$_3$)$_3$COK	2,000	(CH$_3$)$_2$CHOH	8	28	7	99.9	99	R
h	trans-RuCl$_2$[(S)-xylbinap][(S)-daipen] + (CH$_3$)$_3$COK	13,000	(CH$_3$)$_2$CHOH	10	28	15	100	99	R
i	trans-RuCl$_2$[(S)-xylbinap][(S)-daipen] + K$_2$CO$_3$	10,000	(CH$_3$)$_2$CHOH	8	28	22	99	94	R
i	[Ir{(S)-binap}(cod)]BF$_4$–P[2-N(CH$_3$)$_2$C$_6$H$_4$](C$_6$H$_5$)$_2$	110	3:2 THF–CH$_3$OH	48	60	106	97	19	S

Scheme 1.68.

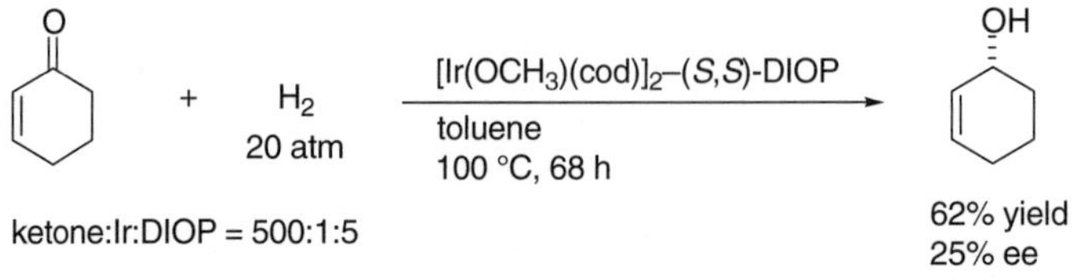

Scheme 1.69.

ated with a $RuCl_2[(S)\text{-binap}](dmf)_n-(S)\text{-DAIPEN-KOH}$ system to give the R olefinic alcohol in 98.5% yield and 90% ee (Scheme 1.67) [271].

Asymmetric hydrogenation of α,β-unsaturated ketones with different structural characteristics is also achievable with the diphosphine/diamine Ru catalyst. The use of K_2CO_3, a weak base cocatalyst, in place of KOH or $(CH_3)_3COK$, is suitable for reaction of base-sensitive substrates. As illustrated in Scheme 1.68, enones having various substitution patterns are quantitatively transformed to the corresponding allylic alcohols with high enantiopurities [256,258,271]. The reaction can be conducted even with an S/C of 100,000. Combination of (S)-XylBINAP and (S)-DAIPEN is suitable to obtain the R alcohols with high enantiopurities, in agreement with the sense of hydrogenation of aromatic ketones (see section 1.4.2.1) [258]. β-Ionone, a dienone, is convertible to β-ionol with a high enantiopurity. This method is applicable to the synthesis of some biologically active compounds such as α-tocopherol, adriamycin, and daunorubicin [258,271,272]. The stereocenter determined by this hydrogenation is labeled R or S. A BINAP–Ir–aminophosphine system shows a moderate enantioselectivity for hydrogenation of benzalacetone [273].

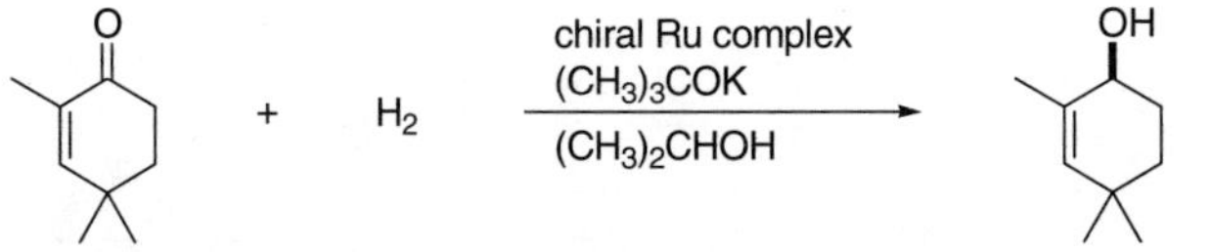

Ru complex = *trans*-$RuCl_2[(R)\text{-tolbinap}][(S,S)\text{-dpen}]$ 100% yield

ketone:Ru:base	H_2, atm	Temp, °C	Time, h	% ee
500:1:2	8	0	6	96
10,000:1:45	10	28	48	94

examples of the application:

α-damascone[277a]   dihydroactinidiolide[277a]

Scheme 1.70.

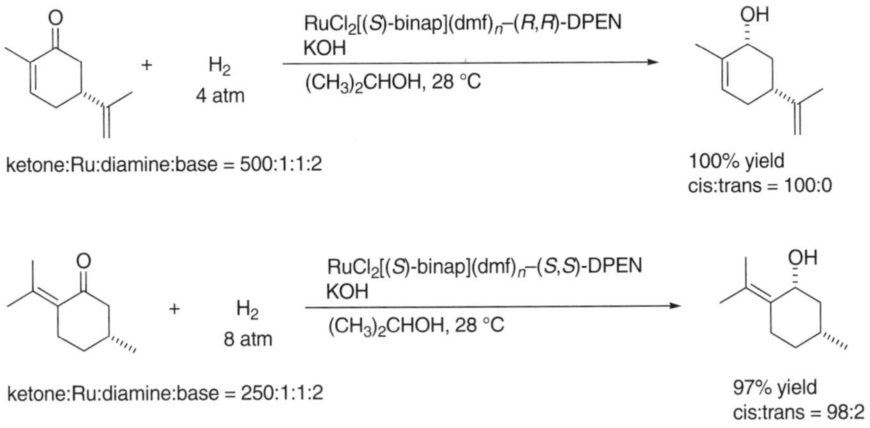

ketone:Ru:diamine:base = 500:1:1:2

100% yield
cis:trans = 100:0

ketone:Ru:diamine:base = 250:1:1:2

97% yield
cis:trans = 98:2

Scheme 1.71.

Asymmetric hydrogenation of simple 2-cyclohexenone is still difficult. A catalyst system prepared in situ from [Ir(OCH$_3$)(cod)]$_2$ and DIOP shows carbonyl-selectivity as high as 95% at 65% conversion, but with a poor enantioselectivity (Scheme 1.69) [274].

Hydrogenation of 2,4,4-trimethyl-2-cyclohexenone with *trans*-RuCl$_2$(tolbinap)(dpen) and (CH$_3$)$_3$COK under 8 atm of hydrogen gives 2,4,4-trimethyl-2-cyclohexenol quantitatively with 96% ee (Scheme 1.70) [256,275,276]. In this case, unlike in the reaction of aromatic ketones, the combination of the *R* diphosphine and *S,S* diamine most effectively discriminates the enantiofaces. The chiral allylic alcohol is a versatile intermediate in the synthesis of carotenoid-derived odorants and other bioactive terpens such as α-damascone and dihydroactinidiolide [277].

Hydrogenation of (*R*)-carvone, a chiral dienone, raises many selectivity problems, including 1,2 versus 1,4 selectivity in the conjugated enone moiety, chemoselectivity between the conjugated versus isolated C=C linkage, and cis versus trans diastereoselectivity with respect to the C5 substituent when the carbonyl function is reduced. With a catalyst in situ prepared from (*S*)-BINAP, (*R,R*)-DPEN, and KOH in 2-propanol, hydrogenation occurs selectively only at the C=O bond and with perfect diastereoselectivity resulting in the 1,5-cis relationship (Scheme 1.71) [275]. On the one hand, when enantiomeric (*R*)-BINAP and (*S,S*)-DPEN are used as chiral ligands, the reaction proceeds slowly to afford a 34:66 mixture of the cis and trans product in 98% yield. On the other hand, the (*S*)-BINAP/(*S,S*)-DPEN (not *S/R,R*) combination gives the highest reaction rate and diastereoselectivity in hydrogenation of (*R*)-pulegone, an *s*-cis chiral enone [275]. Racemic carvone can be resolved kinetically with a k_R/k_S ratio of 33:1 by hydrogenation with an (*S*)-BINAP–(*R,R*)-DPEN coupled system (Scheme 1.72) [275]. At

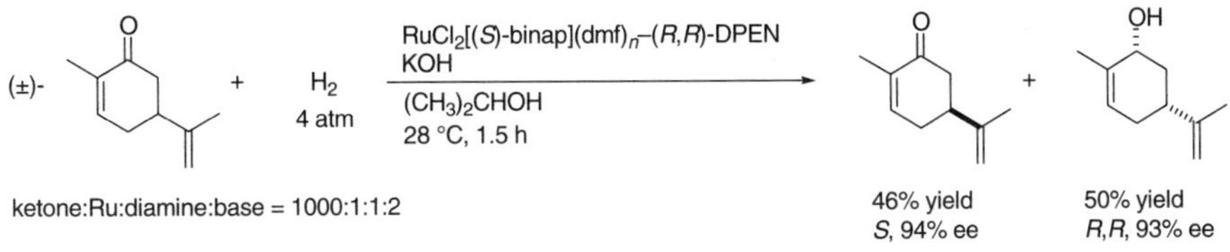

ketone:Ru:diamine:base = 1000:1:1:2

46% yield
S, 94% ee

50% yield
R,R, 93% ee

Scheme 1.72.

54% conversion, unreacted (*S*)-carvone and (1*R*,5*R*)-carveol are obtainable with 94% ee and 93% ee, respectively.

1.4.2.4. Dynamic Kinetic Resolution As described in Section 1.4.1.5, a single alcoholic compound among four possible stereoisomers is accessible from certain configurationally labile chiral ketones through dynamic kinetic resolution. This methodology is applicable to hydrogenation of α-substituted cycloalkanones. When racemic 2-isopropylcyclohexanone is hydrogenated with RuCl$_2$[(*S*)-binap](dmf)$_n$ and (*R*,*R*)-DPEN in 2-propanol in the presence of an excess amount of KOH promoting smooth substrate racemization, (1*R*,2*R*)-2-isopropylcyclohexanol with 93% ee is obtained with a cis:trans ratio of 99.8:0.2 (Scheme 1.73) [278]. The origin of the high enantio- and diastereoselectivity has been elucidated by a computer-aided analysis: Hydrogenation of the *R* ketone is 36 times faster than that of the *S* enantiomer, and stereochemical inversion at the α position occurs 47 times faster than hydrogenation of the less-reactive *S* substrate. Similarly, (−)-menthone possessing the configurationally stable C1 and the unstable C4 chiral center is hydrogenated with an (*R*)-BINAP–(*S*,*S*)-DPEN combined system to give (+)-neomenthol as a sole product [278]. The presence of the C1 stereogenic center favorably affects the diastereoface selection.

1.4.2.5. Asymmetric Activation Enantiomerically pure catalysts are often expensive. Thus, when a racemic metal complex can be activated as a chiral catalyst by addition of a cheap nonracemic auxiliary, this methodology is viable for practical asymmetric catalysis [279]. A racemic RuCl$_2$(tolbinap)(dmf)$_n$ is a feeble catalyst for hydrogenation of 2,4,4-trimethyl-2-cyclohexenone. However, the enone is hydrogenated quantitatively to give the *S* allylic alcohol with 95% ee when an equimolar amount of (*S*,*S*)-DPEN is added to the racemic complex in a 7:1 2-propanol–toluene mixture containing KOH (Scheme 1.74) [276]. The enantioselectivity is very close to the 96% ee attainable when optically pure (*R*)-TolBINAP and (*S*,*S*)-DPEN are used (see Scheme 1.70). Under such conditions, the highly enantioselective catalytic cycle involving an (*R*)-TolBINAP/(*S*,*S*)-DPEN complex occurs 121 times faster than the diastereomeric cycle with an *S*/*S*,*S* catalyst that gives the *R* alcohol with only 26% ee. The degree and sense of the resulting enantioselectivity are highly dependent on the structures of the diphosphine, diamine, and ketone substrate. 2′-Methylacetophenone, a simple aromatic ketone, is transformed to the *R* alcohol with 90% ee under the same catalytic conditions [276]. In this

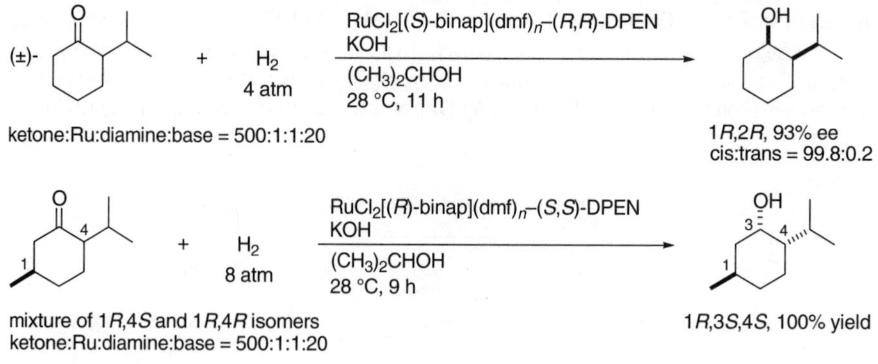

ketone:Ru:diamine:base = 500:1:1:20

1*R*,2*R*, 93% ee
cis:trans = 99.8:0.2

mixture of 1*R*,4*S* and 1*R*,4*R* isomers
ketone:Ru:diamine:base = 500:1:1:20

1*R*,3*S*,4*S*, 100% yield

Scheme 1.73.

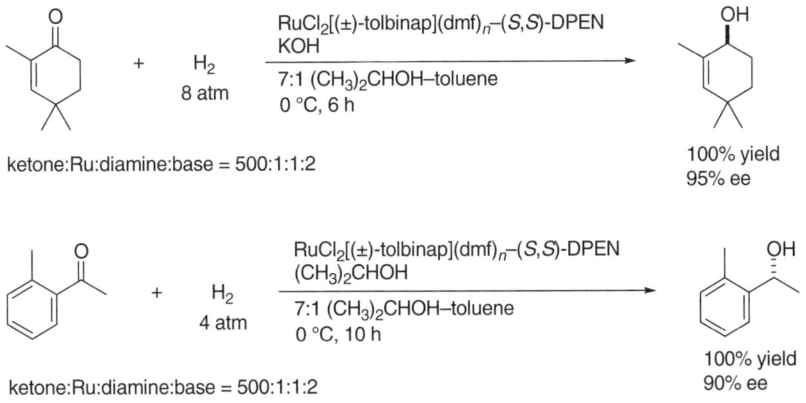

Scheme 1.74.

case, the $S/S,S$ cycle giving the R alcohol with 97.5% ee turns over 13 times faster than the $R/S,S$ cycle, affording the S alcohol with only 8% ee.

1.4.2.6. 1-Deuterio Aldehydes Asymmetric hydrogenation of 1-deuterio benzaldehydes with $Ru(OCOCH_3)_2[(R)$-binap] in the presence of 5 equiv. of HCl gives the corresponding chiral 1-deuterio alcohols with up to 89% ee (Scheme 1.75) [280]. The heteroatom substitution at the C2 position in the benzene ring tends to increase the enantioselectivity due to the heteroatom/metal interaction. This catalyst is less reactive in deuteration of benzaldehyde.

1.4.3. Transfer Hydrogenation

Transfer hydrogenation of ketones catalyzed by a transition-metal complex or a main group-metal alkoxide is a useful method to produce secondary alcohols. Pure organic compounds such as 2-propanol [2,281,282] and formic acid [283] are preferably used as hydrogen donors in place of hydrogen gas. This method is convenient for a small- or medium-scale reduction

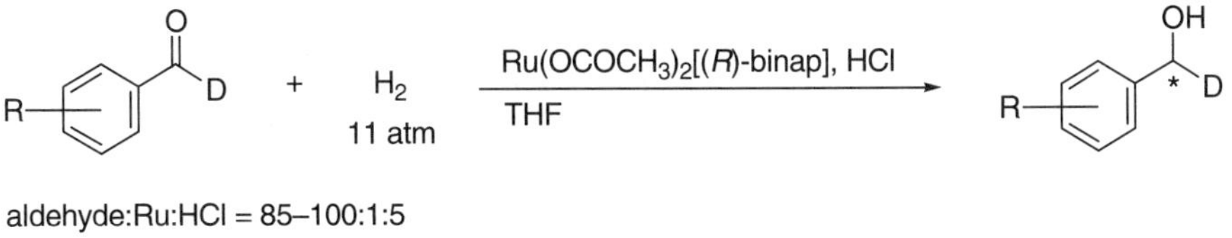

aldehyde:Ru:HCl = 85–100:1:5

R	% Yield	% ee	Config$n$
H	100	65	S
2-Br	100	89	S
3-Cl	67	73	–
4-Cl	100	70	–

Scheme 1.75.

because of the operational simplicity and no requirement of special apparatus. These advantages have prompted chemists to develop efficient asymmetric transfer hydrogenation [2,162c].

1.4.3.1. Meerwein–Ponndorf–Verley-Type Reduction

1.4.3.1. Meerwein–Ponndorf–Verley-Type Reduction Reduction of ketones by 2-propanol or related alcohols, known as Meerwein–Ponndorf–Verley (MPV) reduction, is promoted by various metal alkoxides, typically aluminum 2-propoxide [2a,d,281]. The C2 hydrogen of 2-propanol is transferred directly to the carbonyl carbon through a six-membered pericyclic transition state [284]. Earlier, a stoichiometric quantity of a metal alkoxide was required for this purpose, but recently, lanthanide [285] and aluminum [286] complexes acting as excellent catalysts have been reported.

The chiral Sm(III) complex **38** allows enantioselective MPV-type reduction of aryl methyl ketones in a 2:1 2-propanol–THF mixture giving an optical yield up to 97% (Scheme 1.76) [287]. The size of lanthanide metals affects the enantioselectivity, and the Sm(III) complex is the best catalyst. An electron-withdrawing nitro group at the C4 position in the benzene ring accelerates the reaction, whereas an electron-donating methoxy substituent decelerates the reduction. The presence of a hetero atom at the C2 position accelerates the reaction, perhaps owing to the chelation of the substrate to Sm. The degree of enantioselectivity is sensitive to the size of alkyl groups, 97% ee with 2′-chloroacetophenone versus 68% ee with 2′-chloropropiophenone. The optical purity of products is independent of the extent of conversion, in contrast to that in ordinary MPV reduction.

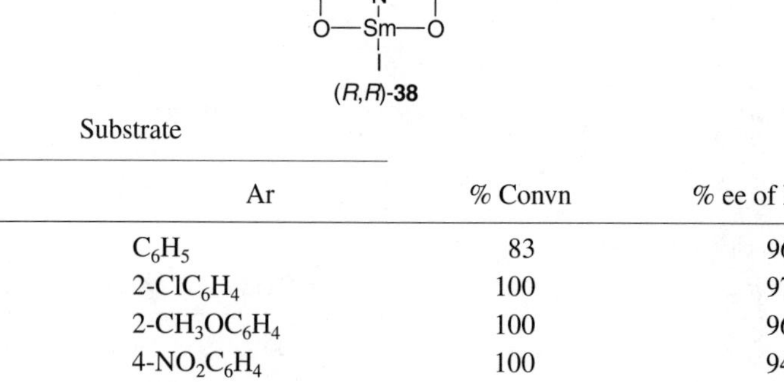

Substrate

R	Ar	% Convn	% ee of Product
CH_3	C_6H_5	83	96
CH_3	2-ClC_6H_4	100	97
CH_3	2-$CH_3OC_6H_4$	100	96
CH_3	4-$NO_2C_6H_4$	100	94
CH_3	4-$CH_3OC_6H_4$	43	92
CH_3	1-naphthyl	98	97
C_2H_5	2-ClC_6H_4	95	68

Scheme 1.76.

1.4.3.2. Transition-Metal–Catalyzed Reduction

1.4.3.2.1. Simple Ketones   Asymmetric transfer hydrogenation catalyzed by chiral Rh, Ir, and Ru complexes has been vigorously investigated. The efforts have mainly focused on the reaction of simple ketone substrates [2b,c]. In asymmetric hydrogenation, as described in Section 1.4.1 and 1.4.2, high reactivity and enantioselectivity in asymmetric hydrogenation are obtained by using chiral phosphine ligands. However enantioselective transfer hydrogenation largely uses chiral nitrogen-based ligands (Figure 1.5–9) [2c], although both reaction systems are believed to occur via transition-metal hydride species [288,289]. Most transfer hydrogenations require a strong base such as KOH, $(CH_3)_2CHOK$, or $(CH_3)_3COK$ as cocatalyst.

Ru(II) complexes with a chiral phosphine ligand promote rapid transfer-hydrogenation of aromatic ketones in 2-propanol with a strong base [289,290], however, the extent of enantioselectivity is unsatisfactory [290]. The situation changes dramatically when a nitrogen-based ligand, TsDPEN, is used for a $[RuCl_2(arene)]_2$ precursor [291–293]. For example, the Ru complex (*S,S*)-**39** effectively promotes reduction of acetophenone using a 0.1-M solution in 2-propanol containing KOH at room temperature to give (*S*)-1-phenylethanol in 95% yield and 97% ee (Scheme 1.77). The reaction rate and enantioselectivity are highly dependent on the electronic properties of substituents in the phenyl rings as well as the bulkiness around the carbonyl moiety. Similarly, an N-arenesulfonylated derivative of chiral cyclohexanediamine **17** can be used as a chiral auxiliary [294]. A Ru complex with a chiral ferrocenyl diamine **13** is reactive at –30°C [294,295]. When a catalyst prepared from $RuCl_2[P(C_6H_5)_3]_3$ and AMBOX (a tridentate nitrogen-based ligand) is used with $(CH_3)_2CHONa$, a series of alkyl aryl ketones is reduced at 82°C to alcohols with up to 98% ee [296]. The enantioselectivity is decreased by increasing the bulkiness of alkyl groups. Chiral β-amino alcohols are also effective for asymmetric transfer hydrogenation of aromatic ketones, where an appropriate pairing of the auxiliary and $[RuCl_2(arene)]_2$ [297] is crucial for high efficiency. For example, combination of $[RuCl_2\{C_6(CH_3)_6\}]_2$ and **23** affords 1-(1-naphthyl)ethanol with 93% ee [298]. In comparison with the reaction of using TsDPEN as a ligand, the optical yield is somewhat lower but the rate is higher. The amino alcohol ligands, **21** and **24**, also show a high enantioselectivity for reduction of some alkyl aryl ketones [299]. In these catalyst systems, an NH_2 or NH end in the chiral auxiliaries plays a crucial role to achieve a high reaction rate and enantioselectivity. A Ru complex with the chiral auxiliary **25** is effective for asymmetric reduction of sterically hindered pivalophenone [300]. The chiral bisthiourea **19** can be also used for reduction of alkyl aryl ketones [301]. This auxiliary is fairly useful for enantioselective reduction of isobutyrophenone.

Ru complexes with a suitable chiral amino or imino phosphine ligand effect highly reactive and enantioselective transfer hydrogenation of simple aromatic ketones. For example, preformed $RuCl_2(\textbf{30})[P(C_6H_5)_3]$ acts as an extremely active catalyst for reduction of acetophenone in 2-propanol containing NaOH to give a TOF of 42,600 h^{-1} at 82°C [302,303]. The degree of optical yield tends to increase by increasing of the size of alkyl groups in the ketonic substrates. Isobutyrophenone is reduced in up to 92% optical yield at 87% conversion. A (phosphinoferrocenyl)oxazoline **31** is also usable for asymmetric reduction of alkyl aryl ketones, providing the alcohols with up to 96% ee [304]. The presence of $P(C_6H_5)_3$ is crucial to achieve a high optical yield. Treatment of *trans*-$RuCl_2(dmso)_4$ with the C_2-symmetrical diphosphine/diamine tetradentate ligand **29** produces a hexacoordinate Ru complex **40** [305]. This unique chiral complex effectively promotes the conversion of aryl methyl ketones in 2-propanol containing $(CH_3)_2CHOK$ (Ru:base = 1:0.5) to 1-phenylethanol and its derivatives in up to 97% optical yield. The enantioselectivity is decreased only slightly even at a high conversion. Noticeably, the corresponding diphosphine/diimine Ru complex shows much lower reactivity, clearly indicating the significance of the NH function for the catalytic activity.

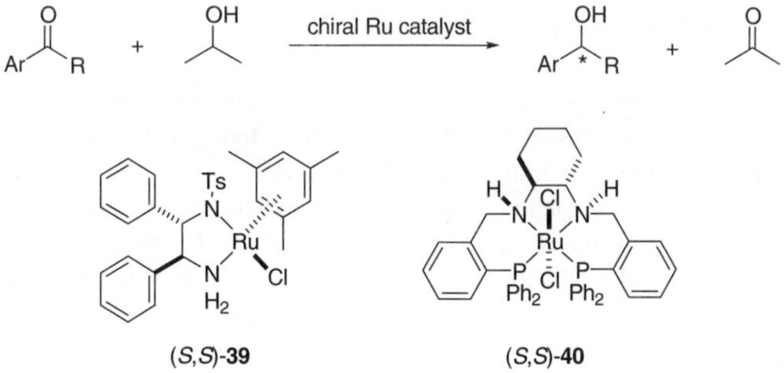

(S,S)-**39**   (S,S)-**40**

Substrate			Product			
R	Ar	Catalyst	Temp, °C	% Yield	% ee	Confign
CH_3	C_6H_5	(S,S)-**39** + KOH	rt	95	97	S
CH_3	C_6H_5	$RuCl_2[P(C_6H_5)_3]_3$–(R)-AMBOX + $(CH_3)_2CHONa$	82	80	98	S
CH_3	C_6H_5	$[RuCl_2(p\text{-cymene})]_2$– (R,R)-**17** + KOH	22	96	92	R
CH_3	C_6H_5	$[RuCl_2(p\text{-cymene})_2$–(1R,2S)- **21** + KOH	rt	70	91	S
CH_3	C_6H_5	$RuCl_2\{C_6(CH_3)_6\}]_2$–(S,R,R)- **24** + $(CH_3)_2CHOK$	rt	92	95	S
CH_3	C_6H_5	(S,S)-**40** + $(CH_3)_2CHOK$	45	93	97	R
CH_3	3-ClC_6H_4	(S,S)-**39** + KOH	rt	98	98	S
CH_3	2-$CH_3C_6H_4$	$RuCl_2[P(C_6H_5)_3]_3$–(R)-AMBOX + $(CH_3)_2CHONa$	82	96	98	S
CH_3	1-naphthyl	$[RuCl_2(p\text{-cymene})]_2$–(R,R)- **13** + KOH	−30	91	90	R
CH_3	1-naphthyl	$RuCl_2[P(C_6H_5)_3]_3$–(R)-AMBOX + $(CH_3)_2CHONa$	82	72	96	S
CH_3	1-naphthyl	$[RuCl_2(p\text{-cymene})]_2$–(1R,2S)- **21** + KOH	rt	79	94	S
CH_3	1-naphthyl	$[RuCl_2\{C_6(CH_3)_6\}]_2$–(S,S)- **23** + KOH	28	99	93	S
CH_3	1-naphthyl	$[RuCl_2\{C_6(CH_3)_6\}]_2$–(S,R,R)- **24** + $(CH_3)_2CHOK$	rt	92	97	S
CH_3	2-naphthyl	(S,S)-**39** + KOH	rt	93	98	S
C_2H_5	C_6H_5	$RuCl_2[P(C_6H_5)_3]_3$–(S)-**31** + $(CH_3)_2CHOK$	50	85	96	R
n-C_4H_9	C_6H_5	$[RuCl_2\{C_6(CH_3)_6\}]_2$–(S,R,R)- **24** + $(CH_3)_2CHOK$	rt	78	95	S
$(CH_3)_2CH$	C_6H_5	$[RuCl_2(C_6H_6)]_2$–(R,R)-**19** + $(CH_3)_3COK$	82	92	94	S
$(CH_3)_2CH$	C_6H_5	$RuCl_2[(S)\text{-}\mathbf{30}][P(C_6H_5)_3]$ + NaOH	82	87	92	R
$(CH_3)_3C$	C_6H_5	$[RuCl_2(C_6H_6)]_2$–(R,R)-**25** + $(CH_3)_2CHONa$	0	96	80	S

Scheme 1.77.

Asymmetric transfer hydrogenation of simple ketones using 2-propanol as a hydrogen donor [2c] was spurred by the invention of a catalyst prepared in situ from $[IrCl(cod)]_2$, the C_2 chiral tetrahydrobis(oxazole) ligand **16**, and KOH [306]. This catalyst system achieved for the first time an optical yield greater than 90% in reduction of isobutyrophenone at 70% conversion (Scheme 1.78), whereas the enantiopurity of the product is decreased at a higher conversion. The substrate structure strongly affects the degree of enantioselection. An Ir complex prepared in situ from $[IrCl(cod)]_2$ and TsDPEN exhibits a high enantioselectivity for reaction of acetophenone, although the scope is not clear [307]. A neutral Ir complex with a chiral diamine **12** with no N-alkyl substituents also promotes reduction of simple aromatic ketones [308]. The steric and electronic properties of ring substituents influence its reactivity and enantioselectivity. Pivalophenone, a sterically hindered aromatic ketone, is reduced enantioselectively with an IrI(cod)(ppei)–NaI–H_2O system in 2-propanol containing KOH [309]. The NaI effect is due probably to the increase in the concentration of a more selective neutral complex. On the contrary, a cationic DHPPEI–Ir(I) complex promotes asymmetric reduction of 1,4-diphenyl-1-butanone [2c]. Immobilization of chiral transition-metal complexes usually lowers catalytic activity and enantioselectivity in comparison with the original catalysts, however, the immobilized catalyst system **41** is much more reactive and enantioselective than the homogeneous catalyst in reaction of butyrophenone [2c]. Asymmetric reduction of acetophenone with a Rh complex of the chiral phenanthroline **15** and KOH in 2-propanol gives 1-phenylethanol with 62% ee with a high TOF of up to 12,000 h^{-1} [2c,310]. A Rh complex (S,S)-**42**, an analogue of Ru complex **39**, shows a high enantioselectivity in reduction of acetophenone in basic 2-propanol [311]. Reduction of propiophenone with a Rh complex with a chiral diurea **18** gives the alcohol with 80% ee [312]. A DMDPEN–Rh complex is usable for reduction of 4′-cyanoacetophenone [313]. An AMSO–Rh complex gives 75% optical yield in reduction of p-methylacetophenone [314]. A diphenylsilane–methanol system acts as a hydrogen donor in asymmetric reduction of alkyl aryl ketones [315]. In the presence of a complex prepared from $[RhCl(cod)]_2$ and the chiral bis(aminoethylferrocenyl) diselenide **14**, pivalophenone is transformed to the chiral alcohol with 95% ee but only in 11% yield.

Asymmetric transfer hydrogenation of simple aliphatic ketones remains difficult. A successful example has been reported only with pinacolone. A Ru complex prepared from $[RuCl_2(C_6H_6)]_2$ and a tridentate ligand **32** in 2-propanol containing a base gives the reduction product with 92% ee (Scheme 1.79) [316]. Under the same conditions, 5-methyl-3-hexanone is reduced quantitatively but in 63% optical yield. Cyclohexyl methyl ketone is reduced with a $[RuCl_2\{C_6(CH_3)_6\}]_2$–$(S,S)$-**22** [298] or $RuCl_2[(S)$-**30**][P(C_6H_5)_3]$ [302] system to give the S alcohol with 75% ee and 60% ee, respectively.

Studies on asymmetric reduction of ketones catalyzed by the chiral Ru(II) complex **39** and base in 2-propanol have fully elucidated the structure of the real catalytic intermediates and the reaction mechanisms (Scheme 1.80) [317]. Both catalyst **43** and reducing species **44** have been isolated in pure forms and the molecular structures have been determined by single-crystal X-ray analysis. The purple 16-electron complex **43** shows an eminent dehydrogenative activity for methanol, ethanol, 2-propanol, and other secondary alcohols at room temperature in the absence of any base, producing the orange Ru hydride species **44**. Reaction of the 18-electron complex **44** and a 10-fold excess of acetone regenerates instantaneously the original complex **43**. The forward and reverse processes are equally facile. The isolated complexes (S,S)-**43** and **44** show a reasonable activity for asymmetric reduction of acetophenone in 2-propanol without any base to give (S)-1-phenylethanol with 95% ee. These observations clearly remove the possibility that the reaction proceeds through metal alkoxide intermediate presumed for the MPV-type reduction. A strong base is necessary only for the generation of catalytic species **43** from the Ru chloride **39** by removal of HCl.

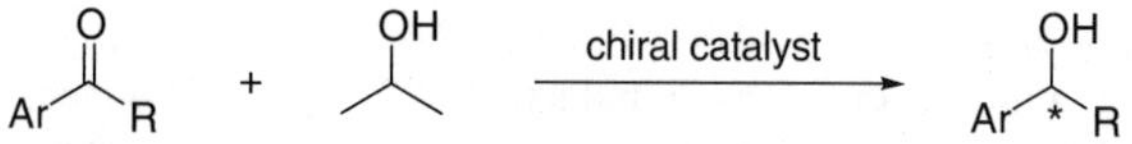

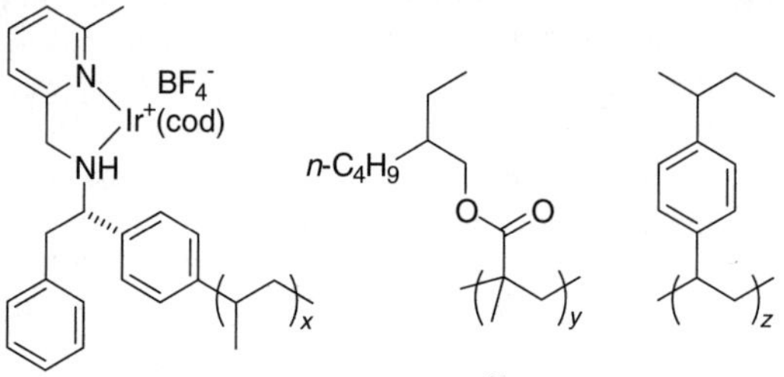

$$\frac{x}{x+y+z} = 0.02$$

(S)-**41**

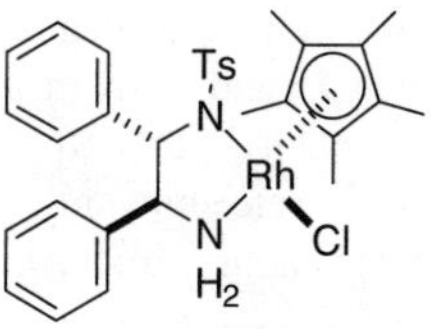

(S,S)-**42**

Substrate				Product		
R	Ar	Catalyst	Temp, °C	% Convn	% ee	Config'n
CH$_3$	C$_6$H$_5$	[IrCl(cod)]$_2$–(R,R)-TsDPEN + (CH$_3$)$_3$COK	rt	87	92	S
CH$_3$	C$_6$H$_5$	[RhCl(C$_6$H$_{10}$)]$_2$–(S)-**15** + KOH	60	94	62	S
CH$_3$	C$_6$H$_5$	(S,S)-**42** + KOH	rt	80	90	S
CH$_3$	4-CH$_3$C$_6$H$_4$	[RhCl(C$_6$H$_{10}$)]$_2$–(S)-AMSO + KOH	82	31	75	R
CH$_3$	4-CNC$_6$H$_4$	[RhCl(C$_6$H$_{10}$)]$_2$–(S,S)-DMDPEN + KOH	25	100	73	R
C$_2$H$_5$	C$_6$H$_5$	[IrCl(cod)]$_2$–(S)-**12** + KOH	rt	96	93	R
C$_2$H$_5$	C$_6$H$_5$	[RhCl(cod)]$_2$–(S,S)-**18** + KOC(CH$_3$)$_3$	60	87	80	R
n-C$_3$H$_7$	C$_6$H$_5$	(S)-**41** + KOH	60	92	84	S
C$_6$H$_5$(CH$_2$)$_3$	C$_6$H$_5$	[Ir(cod){(R)-dhppei}]BF$_4$ + KOH	60	–	>90	–
(CH$_3$)$_2$CH	C$_6$H$_5$	[IrCl(cod)]$_2$–(S,S)-**16** + KOH	80	70	91	R
(CH$_3$)$_3$C	C$_6$H$_5$	IrI(cod)[(S)-ppei]–NaI–H$_2$O + KOH	83	91	84	S

Scheme 1.78.

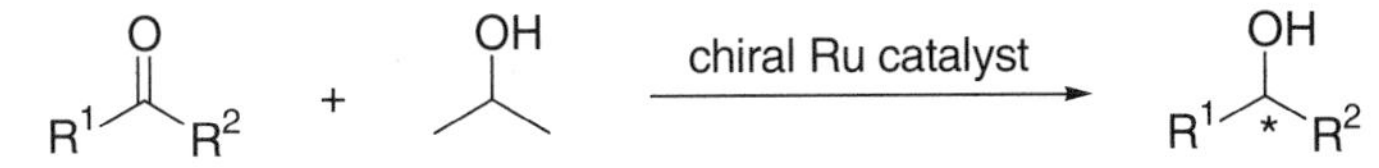

Substrate				Product		
R^1	R^2	Catalyst	Temp, °C	% Yield	% ee	Config'n
CH_3	cyclo-C_6H_{11}	$[RuCl_2\{C_6(CH_3)_6\}]_2-$ (S,S)-**22** + KOH	28	93	75	S
CH_3	cyclo-C_6H_{11}	$RuCl_2[(S)$-**30**][P(C$_6$H$_5$)$_3$] + NaOH	82	70	60	S
CH_3	$(CH_3)_3C$	$[RuCl_2(C_6H_6)]_2-(R,R)$-**32** + NaH	rt	85	92	S
CH_3CH_2	$(CH_3)_2CHCH_2$	$[RuCl_2(C_6H_6)]_2-(R,R)$-**32** + NaH	rt	100	63	S

Scheme 1.79.

2-Propanol is a convenient, most frequently used hydrogen donor for catalytic transfer hydrogenation of ketones. However, the inherent ketone/alcohol equilibrium prevents a high conversion, particularly from highly stable ketones to thermodynamically unfavorable alcohols [318]. Reduction of acetophenone with 2-propanol, for instance, requires the use of a dilute solution, as low as 0.1 M, to obtain a high yield. In asymmetric reduction, the occurrence of its reverse process frequently deteriorates the optical purity of the chiral alcohol, even if the catalyst has an excellent enantioface-differentiating ability. The use of formic acid, another inexpensive hydrogen donor, solves these thermodynamic problems [283]. Because formic acid is viewed as an adduct of H_2 and CO_2 [319], alcohols are irreversibly produced, in principle, in 100% yield. The enantioselection is made under fully kinetic control. In fact, with the Ru complex (S,S)-**39** and by using a 5:2 formic acid–N(C$_2$H$_5$)$_3$ azeotropic mixture [320] at 28°C, acetophenone is transformed quantitatively to (S)-1-phenylethanol with 98% ee (Scheme 1.81) [293,321]. Various alkyl aryl ketones are reduced quantitatively with a high enantioselectivity in a 2–10 M solution. The presence of N(C$_2$H$_5$)$_3$ is crucial to achieve a high reactivity, although alkaline bases are unnecessary. 2-Acetylfuran is reduced in a high optical yield without saturating the furan ring. Reduction of a benzophenone derivative with a methoxy and cyanide group at the 4- and 4'-positions gives the chiral alcohol with 66% ee [321]. Similarly, a Ru complex prepared from [RuCl$_2$(p-cymene)]$_2$ and **17** shows a high enantioselectivity for reduc-

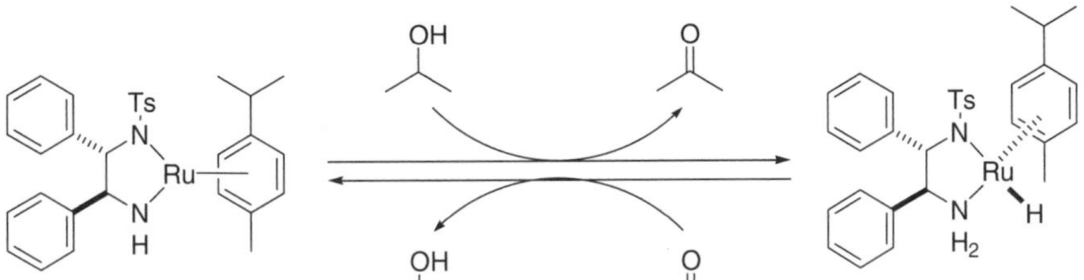

Scheme 1.80.

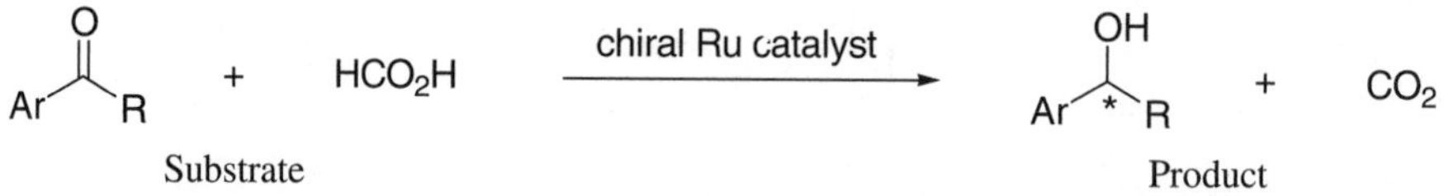

R	Ar	Catalyst	% Yield	% ee	Config^n
CH_3	C_6H_5	(S,S)-**39**	>99	98	S
CH_3	C_6H_5	[RuCl$_2$(p-cymene)]$_2$–(R,R)-**17**	>99	96	R
CH_3	$4-CH_3OC_6H_4$	(S,S)-**39**	>99	97	S
CH_3	2-naphthyl	(S,S)-**39**	>99	96	S
CH_3	2-furyl	(S,S)-**39**	>99	98	S
C_2H_5	C_6H_5	(S,S)-**39**	96	97	S
$4-CNC_6H_4$	$4-CH_3OC_6H_4$	(S,S)-**39**	54	66	S

Scheme 1.81.

tion of aryl methy ketones with a formic acid–$N(C_2H_5)_3$ mixture [294]. A Ru catalyst supported on a TsDPEN-containing polystyrene resin also gives a high optical yield in reduction of acetophenone [322]. The catalyst can be reused.

The multi-functionalized ketone **A** in Scheme 82 is reduced with an excellent ketone-selectivity with (R,R)-**39** and a formic acid–$(C_2H_5)_3N$ mixture, affording (R)-**B** with 92% ee [321]. The olefin, halogen atom, quinoline ring, and ester group are left intact. This product is a key intermediate in the synthesis of L-699,392 (LTD$_4$ antagonist) [323].

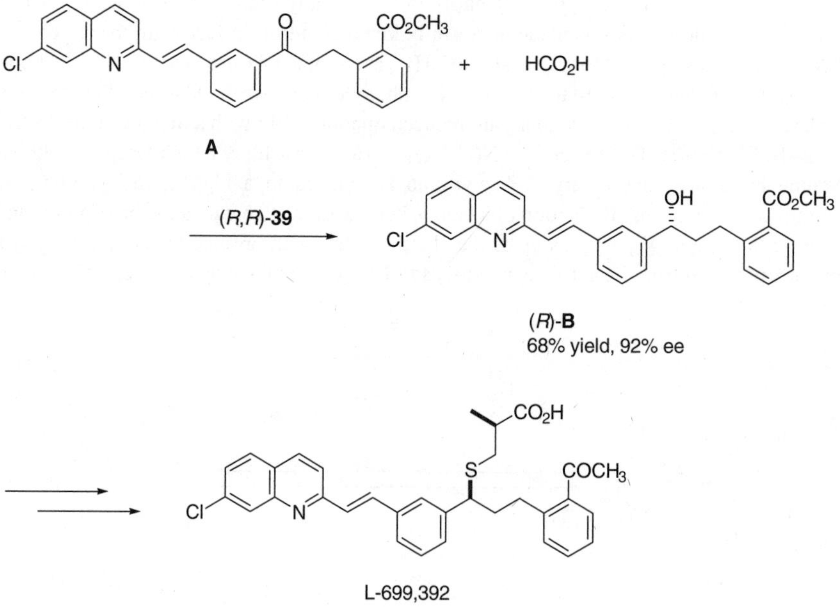

A

(R,R)-**39**

(R)-**B**
68% yield, 92% ee

L-699,392

Scheme 1.82.

High-yield transfer hydrogenation of α-tetralone or -indanone, which have low oxidation potential, is difficult under equilibrium conditions with 2-propanol as a hydrogen donor [292,318]. In fact, asymmetric reduction of α-tetralone with 2-propanol in the presence of a Ru catalyst with AMBOX, **21**, **22**, or prolinate as well as the Rh catalyst **42** gives α-tetralol only in a low-to-moderate yield, although an excellent enantioselectivity is accessible at a low conversion [296,298,299a,311,324]. In contrast, with the Ru complex **39** and a formic acid–$N(C_2H_5)_3$ mixture, α-tetralone and -indanone are reduced quantitatively with the Ru complex **39** and a formic acid–$N(C_2H_5)_3$ mixture to give the cyclic alcohols in 99% yield and up to 99% ee (Scheme 1.83) [321]. The sense of enantioface selection is the same as that in the reaction of acyclic analogues.

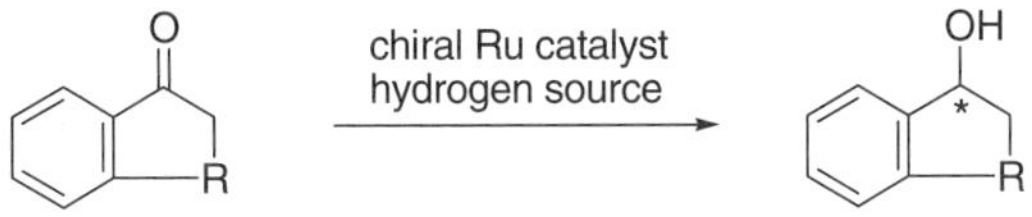

			Product		
R in Substrate	Hydrogen Source	Catalyst	% Yield	% ee	Config^n
CH_2	$(CH_3)_2CHOH$	(S,S)-**39** + KOH	45	91	S
CH_2	$(CH_3)_2CHOH$	(S,S)-**42** + KOH	47	99	S
CH_2	HCO_2H–$(C_2H_5)_3N$	(S,S)-**39**	>99	99	S
$(CH_2)_2$	$(CH_3)_2CHOH$	(S,S)-**39** + KOH	65	97	S
$(CH_2)_2$	$(CH_3)_2CHOH$	$RuCl_2[P(C_6H_5)_3]_3$–(R)-AMBOX + $(CH_3)_2CHONa$	42	95	S
$(CH_2)_2$	$(CH_3)_2CHOH$	$[RuCl_2(p\text{-cymene})]_2$–$(1R,2S)$-**21** + KOH	40	98	S
$(CH_2)_2$	$(CH_3)_2CHOH$	$[RuCl_2\{C_6(CH_3)_6\}]_2$–$(S,S)$-**22** + KOH	62	94	S
$(CH_2)_2$	$(CH_3)_2CHOH$	$RuCl[(S)\text{-prolinate}](p\text{-cymene})$ + KOH	8	92	R
$(CH_2)_2$	$(CH_3)_2CHOH$	(S,S)-**42** + KOH	79	97	S
$(CH_2)_2$	HCO_2H–$(C_2H_5)_3N$	(S,S)-**39**	>99	99	S

Scheme 1.83.

Reduction of the sulfur-containing cyclic ketones **A** and **B** with (R,R)-**39** gives the *R* alcohols **C** and **D** with 99% ee and 98% ee, respectively (Scheme 1.84) [321]. The chiral products are important intermediates in the synthesis of MK-0417, which is a carbonic anhydrase inhibitor [325].

1.4.3.2.2. α,β-Olefinic and α,β-Acetylenic Ketones Only very limited catalytic systems are available for the chemoselective and enantioselective transfer hydrogenation of unsaturated ketones. As shown in Scheme 1.85, an (R,R)-PDPBI–Ir(I) complex catalyzes reduction of benzalacetone with 2-propanol containing KOH and H_2O to give the *S* allylic alcohol with 82% ee at 43% conversion. The carbonyl-selectivity is 94% [326]. At a higher conversion, the optical yield is decreased, whereas the chemoselectivity is unaffected.

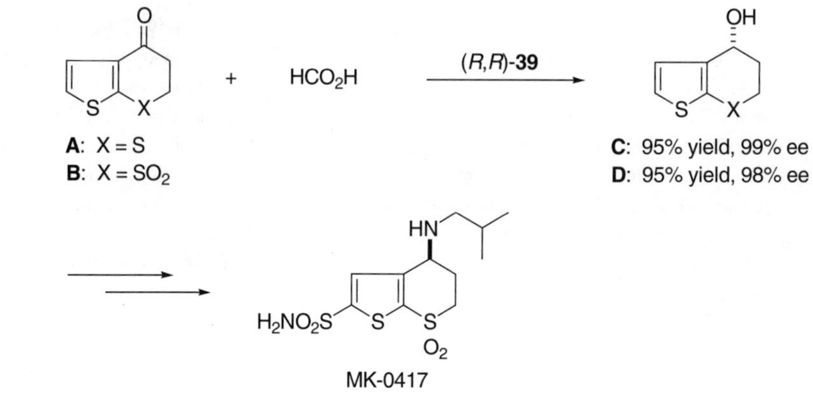

HCO2H
(R,R)-39
A: X = S
B: X = SO2
C: 95% yield, 99% ee
D: 95% yield, 98% ee
H2NO2S
O2
MK-0417

Scheme 1.84.

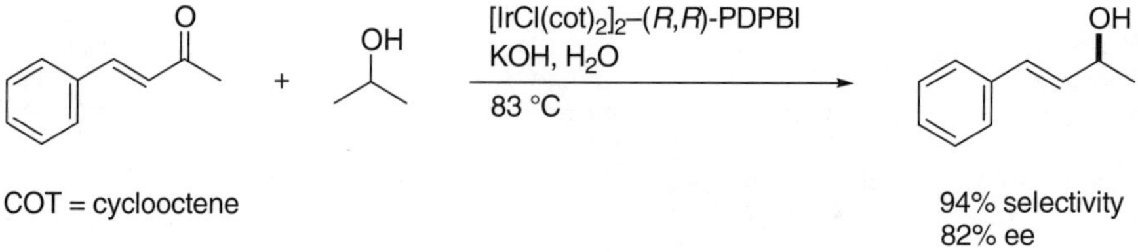

[IrCl(cot)2]2–(R,R)-PDPBI
KOH, H2O
83 °C
COT = cyclooctene
94% selectivity
82% ee

Scheme 1.85.

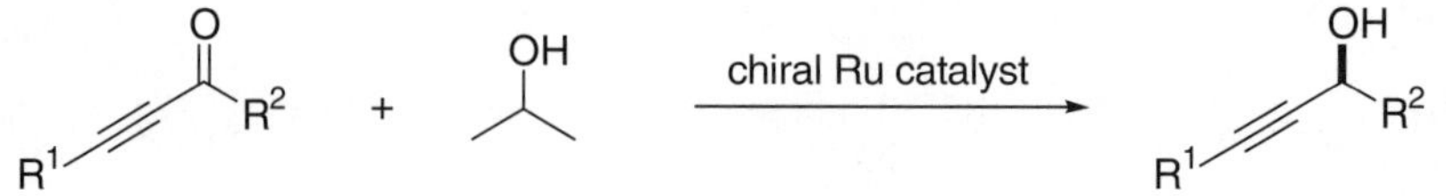

chiral Ru catalyst
R1
R2

Substrate			Product	
R^1	R^2	Catalyst	% Yield	% ee
C_6H_5	CH_3	(*S*,*S*)-**39** + KOH	>99	97
C_6H_5	CH_3	(*S*,*S*)-**43**	>99	97
C_6H_5	C_2H_5	(*S*,*S*)-**39** + KOH	97	97
C_6H_5	$(CH_3)_2CH$	(*S*,*S*)-**39** + KOH	98	99
C_6H_5	$(CH_3)_3C$	(*S*,*S*)-**39** + KOH	84	98
n-C_4H_9	$(CH_3)_2CH$	(*S*,*S*)-**39** + KOH	90	>99
$(CH_3)_3Si$	CH_3	(*S*,*S*)-**43**	>99	98
$(CH_3)_3Si$	$(CH_3)_2CH$	(*S*,*S*)-**43**	>99	99

Scheme 1.86.

The use of the chiral Ru(II) complex **39** and KOH, or the isolated catalyst **43**, has resulted in the highly enantioselective transfer hydrogenation of α,β-acetylenic ketones in 2-propanol [327]. Regardless of the size of alkyl groups in the substrates, various propargylic alcohols are formed in >99% yield and up to 98% ee (Scheme 1.86). Unlike in the reduction of alkyl aryl ketones, 2-propanol is the best hydrogen donor. The favorable ynone/ynol thermodynamic balance leads to high conversion with a 0.1–1 M ynone solution. The Ru catalyst **43** is less reactive in a 1:1 mixture of formic acid and $N(C_2H_5)_3$.

Diastereoselective transfer hydrogenation of the chiral ketone (S)-**A** with (R,R)-**43** in 2-propanol gives (3S,4S)-**B** in >97% yield (Scheme 1.87) [327]. Reaction with (S,S)-**43** affords the 3R,4S alcohol predominantly. The degree and sense of diastereoface differentiation are mostly controlled by the chirality of the Ru catalyst.

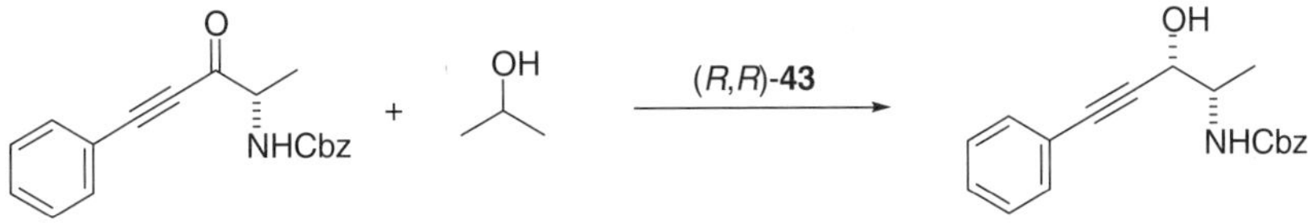

(S)-**A** (98% ee) (3S,4S)-**B**
  >97% yield, >99% ee

Scheme 1.87.

Diastereoselective reduction of the chiral silyloxy ynone **A** with (R,R)-**43** and 2-propanol predominantly gives (7R,9R)-**B** (Scheme 1.88) [328]. The chirality of the C9 position is controlled by the BINAP–Ru catalyzed asymmetric hydrogenation. The chiral product is a key intermediate in the synthesis of taurospongin A, a potent inhibitor of DNA polymerase β and HIV reverse transcriptase.

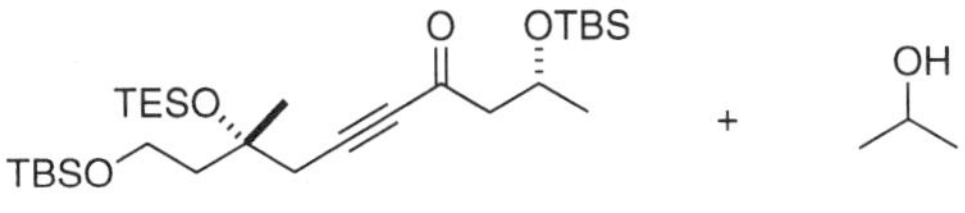

A
TBS = t-$C_4H_9(CH_3)_2Si$; TES = $(C_2H_5)_3Si$

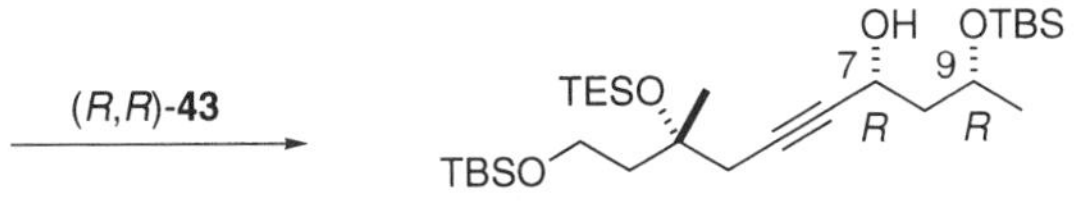

(7R,9R)-**B**
94% yield, 7R,9R:7S,9R > 95:5

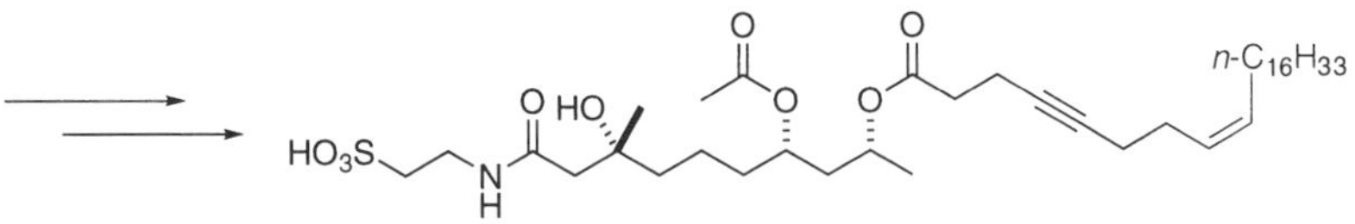

taurospongin A

Scheme 1.88.

1.4.3.2.3. Keto Esters Asymmetric transfer hydrogenation of functionalized ketones is rare. However, an excellent optical yield is obtainable in reduction of methyl benzoylformate by using 2-propanol and with a catalyst system consisting of [RhCl(C_6H_{10})]$_2$, (S,S)-DMDPEN, and KOH (Scheme 1.89) [313].

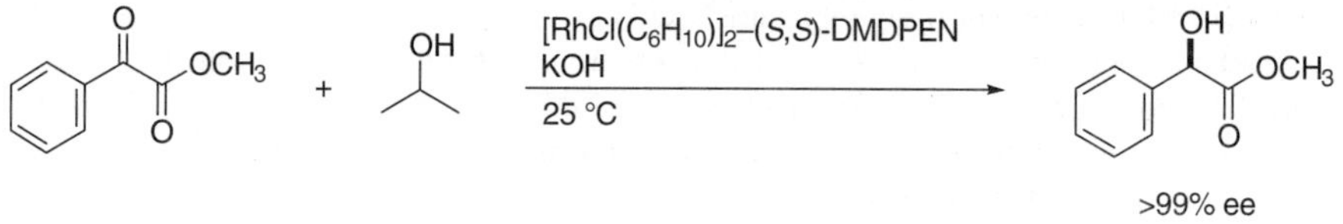

>99% ee

Scheme 1.89.

Reduction of some aromatic keto esters by using the Ru complex (S,S)-**39** and a formic acid–N(C_2H_5)$_3$ mixture gives the corresponding chiral alcohols with up to 95% ee (Scheme 1.90) [293,321]. The optical yield is increased in the order of α-, β-, and δ-keto esters. A Ru complex prepared from [RuCl$_2$(p-cymene)]$_2$ and (S,R)-**22** promotes reduction of ethyl 3-phenyl-3-oxopropanoate in 2-propanol containing a base to afford the S alcohol with 94% ee [329].

1.4.3.3. Kinetic Resolution of Racemic Alcohols Kinetic resolution of racemic secondary alcohols is another useful approach to obtain optically active alcohols. Dehydrogenative oxidation of secondary alcohols with acetone is the reverse process of transfer hydrogenation of ketones by using 2-propanol [281b,289b]. When the reaction is promoted by a chiral metal complex, a racemic alcohol can be converted to a mixture of an unreacted chiral alcohol and a ketonic product. In fact, various racemic aromatic or unsaturated alcohols are efficiently resolved in acetone containing the diamine TsDPEN–Ru(II) complex, **43** or **45** [330]. Thanks to the excellent enantiomer-discriminating ability of the catalyst, the chiral alcohols can be recovered with up to 99% ee at nearly 50% conversion (Scheme 1.91). The efficiency of resolution, k_f/k_s, is >100:1. This process is well-suited for the resolution of racemic aromatic alcohols, which have a high reduction potential [318]. These alcohols are hardly obtainable with a high optical purity by asymmetric transfer hydrogenation of ketones in 2-propanol. These two redox methods are complementary. Notably, racemic 2-cyclohexenol, a simple cyclic allylic alcohol, is also resolved by this process in 93% optical yield at 43% conversion. The resolution of 4-phenyl-3-butene-2-ol, a flexible allylic alcohol, gives only moderate enantioselectivity, however. Desymmetrization of the meso unsaturated diol **A** with the aid of Ru complex (S,S)-**45** and acetone affords the hydroxy enone **B** in 70% yield and 96% ee (Scheme 1.92) [330]. This product is a versatile intermediate for the syntheses of bioactive compounds such as (−)-conduritol C and eutypoxide B, in which the R stereocenters are determined by the asymmetric reduction [331].

Enzymatic resolution of racemic secondary alcohols by enantiomer-selective acylation gives optically pure compounds with up to 50% yield [332]. When this method is coupled with the principle of dynamic kinetic resolution (see Section 1.4.1.5), the theoretical yield increases to 100%. Thus a reaction system consisting of an achiral transition-metal catalyst for racemization, a suitable enzyme, acetophenone, and an acetyl donor allows the transformation of racemic 1-phenylethanol to the R acetates with an excellent ee (Scheme 1.93) [333]. The presence of one equiv. of acetophenone is necessary to promote the alcohol racemization catalyzed by the

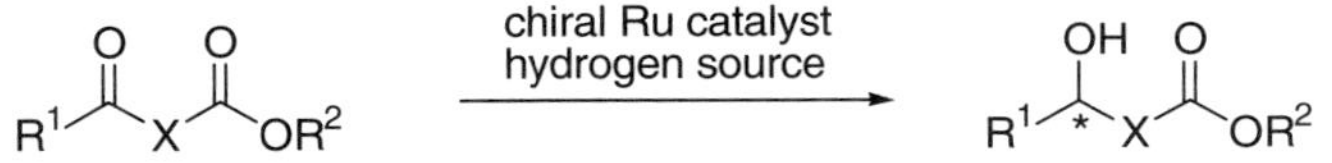

$$R^1 \underset{O}{\overset{O}{\|}} X \underset{O}{\overset{O}{\|}} OR^2 \xrightarrow[\text{hydrogen source}]{\text{chiral Ru catalyst}} R^1 \overset{OH}{\underset{*}{\|}} X \underset{O}{\overset{O}{\|}} OR^2$$

Substrate			Hydrogen Source	Catalyst	Temp, °C	Product		
R^1	X	R^2				% Yield	% ee	Config
C_6H_5	—	$(CH_3)_2CH$	$HCO_2H–(C_2H_5)_3N$	(S,S)-39	28	94	75	R
CH_3	CH_2	C_2H_5	$(CH_3)_2CHOH$	$[RuCl_2(p\text{-cymene})]_2$-$(S,R)$-22 + $KOCH(CH_3)_2$	50	100	15	R
C_6H_5	CH_2	C_2H_5	$(CH_3)_2CHOH$	$[RuCl_2(p\text{-cymene})]_2$-$(S,R)$-22 + $KOCH(CH_3)_2$	50	85	94	S
C_6H_5	CH_2	C_2H_5	$HCO_2H–(C_2H_5)_3N$	(S,S)-39	28	98	93	S
C_6H_5	$(CH_2)_3$	C_2H_5	$HCO_2H–(C_2H_5)_3N$	(S,S)-39	28	99	95	S

Scheme 1.90.

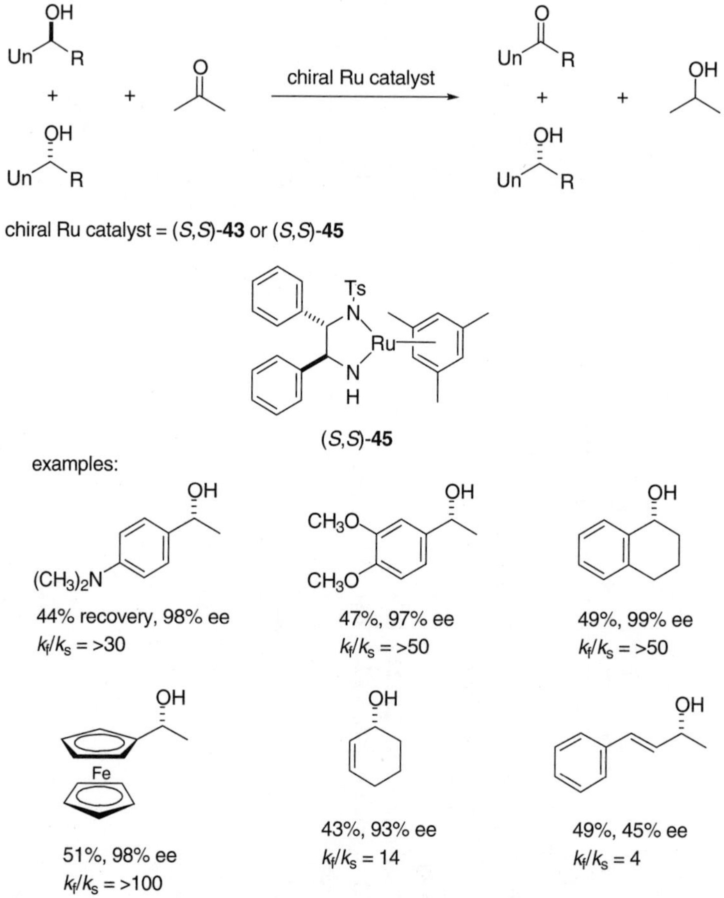

chiral Ru catalyst = (*S,S*)-**43** or (*S,S*)-**45**

(*S,S*)-**45**

examples:

44% recovery, 98% ee
$k_f/k_s = >30$

47%, 97% ee
$k_f/k_s = >50$

49%, 99% ee
$k_f/k_s = >50$

51%, 98% ee
$k_f/k_s = >100$

43%, 93% ee
$k_f/k_s = 14$

49%, 45% ee
$k_f/k_s = 4$

Scheme 1.91.

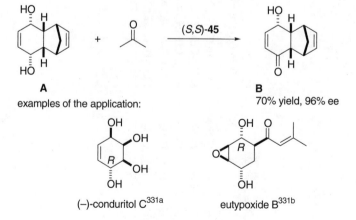

A

B
70% yield, 96% ee

examples of the application:

(−)-conduritol C[331a]

eutypoxide B[331b]

Scheme 1.92.

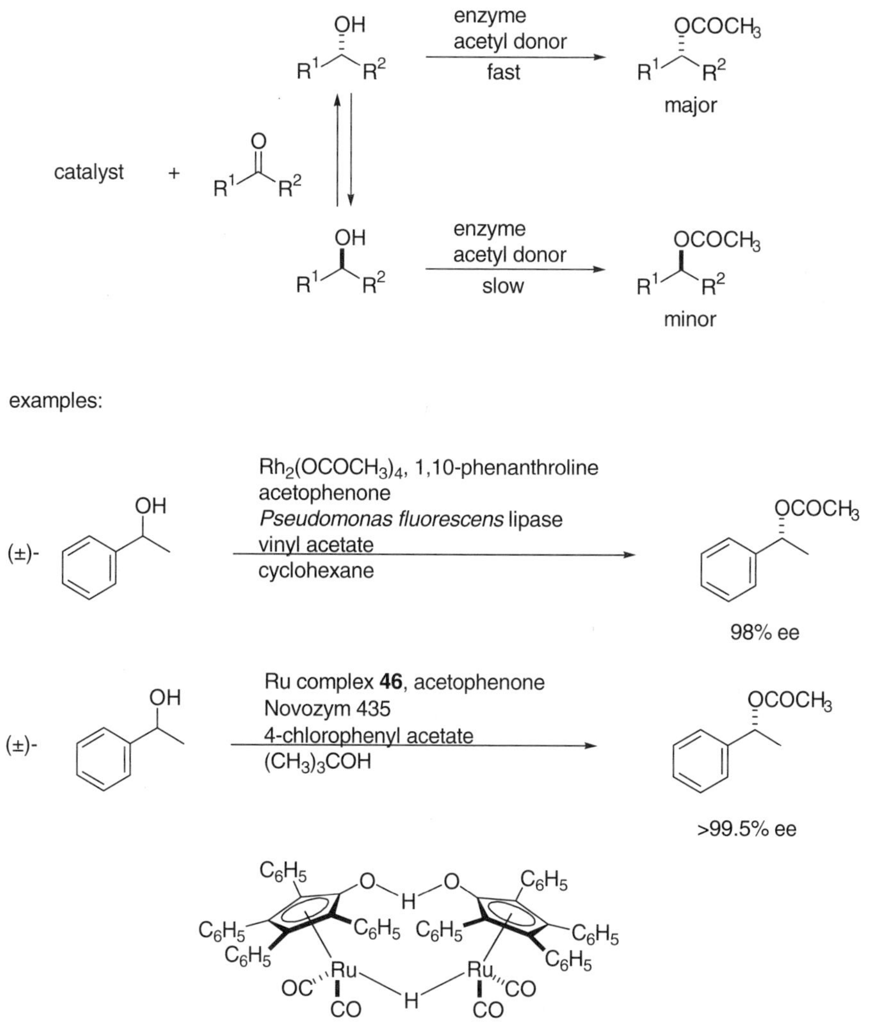

examples:

98% ee

>99.5% ee

46

Scheme 1.93.

Ru complex **46** [333b,334]. 4-Chlorophenyl acetate is a suitable acetyl donor because the produced 4-chlorophenol does not interfere with the catalytic racemization.

1.5. HYDROGENATION OF IMINES

As in asymmetric hydrogenation of olefins and ketones, chiral diphosphine–Rh or –Ir complexes have frequently been used as catalysts [1,162,335]. Recently, a chiral titanocene catalyst has been developed as well. Furthermore, highly enantioselective transfer hydrogenation has been achieved by using chiral amine–Ru complexes.

1.5.1. Acyclic Imines

In solution, acyclic imines exist as a mixture of the easily interconvertible E and Z isomers [336]. Therefore, to achieve a high enantioselectivity, chiral catalysts are required to reduce one stereoisomer selectively or to hydrogenate both isomers with the same sense of enantioselection. Hydrogenation has been studied largely with chiral Rh or Ir complexes as catalysts and N-benzylimines of acetophenone and its analogues as substrates. A neutral Rh complex formed from [RhCl(nbd)]$_2$ and BENZPHOS or CyCPHOS promotes the hydrogenation of a ketimine (Ar = 4-CH$_3$OC$_6$H$_4$) in up to 91% optical yield (Scheme 1.94) [337,338]. The presence of a

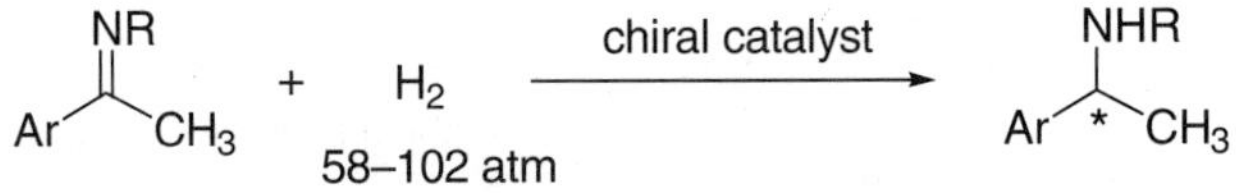

Substrate				Product	
Ar	R	Catalyst	Solvent	% ee	Config[n]
C$_6$H$_5$	C$_6$H$_5$CH$_2$	(R)-BENZPHOS–Rh	1:1 CH$_3$OH–benzene	72	S
C$_6$H$_5$	C$_6$H$_5$CH$_2$	(R)-CyCPHOS–Rh + KI	1:1 CH$_3$OH–benzene	79	S
C$_6$H$_5$	C$_6$H$_5$CH$_2$	(S,S)-BDPP–Rh + (C$_2$H$_5$)$_3$N	CH$_3$OH	73	R
C$_6$H$_5$	C$_6$H$_5$CH$_2$	(S,S)-BDPP–Rh	benzene–ASS^a + 15-crown-5	89	R
C$_6$H$_5$	C$_6$H$_5$CH$_2$	(S,S)-BDPP–Rh + (n-C$_4$H$_9$)$_4$NI	benzene	80	S
C$_6$H$_5$	C$_6$H$_5$CH$_2$	(S,S)-SulfBDPP^b–Rh	1:1 H$_2$O–CH$_3$CO$_2$C$_2$H$_5$	96	R
C$_6$H$_5$	C$_6$H$_5$CH$_2$	(S,S)-SulfBDPP–Rh	1:1 H$_2$O-CH$_3$CO$_2$C$_2$H$_5$	94	R
C$_6$H$_5$	C$_6$H$_5$CH$_2$	(S)-TolBINAP–Ir + C$_6$H$_5$CH$_2$NH$_2$	CH$_3$OH	70	R
C$_6$H$_5$	C$_6$H$_5$CH$_2$	(S)-27–Ir	CH$_2$Cl$_2$	76	R
C$_6$H$_5$	C$_6$H$_5$	(S)-27–Ir	CH$_2$Cl$_2$	89	R
4-CH$_3$C$_6$H$_4$	C$_6$H$_5$CH$_2$	(S)-27–Ir	CH$_2$Cl$_2$	79	R
4-ClC$_6$H$_4$	C$_6$H$_5$CH$_2$	(S,S)-SulfBDPP–Rh	1:1 H$_2$O–CH$_3$CO$_2$C$_2$H$_5$	92	R
2-CH$_3$OC$_6$H$_4$	C$_6$H$_5$CH$_2$	(S,S)-SulfBDPP^b–Rh	1:1 H$_2$O–CH$_3$CO$_2$C$_2$H$_5$	91	R
4-CH$_3$OC$_6$H$_4$	C$_6$H$_5$CH$_2$	(R)-CyCPHOS–Rh + KI	2:1 CH$_3$OH–toluene	91	S
4-CH$_3$OC$_6$H$_4$	C$_6$H$_5$CH$_2$	(S,S)-BDPP–Rh	benzene–ASS^a	92	R
4-CH$_3$OC$_6$H$_4$	C$_6$H$_5$CH$_2$	(S,S)-SulfBDPP^b–Rh	1:1 H$_2$O–CH$_3$CO$_2$C$_2$H$_5$	95	R
4-CH$_3$OC$_6$H$_4$	C$_6$H$_5$CH$_2$	(S,S)-SulfBDPP–Rh	1:1 H$_2$O–CH$_3$CO$_2$C$_2$H$_5$	92	R

^{a} ASS = aggregate of bis(2-ethylhexyl) 2-sulfosuccinate sodium salt. ^{b} Contaminated with 1,3-bis(diphenyl(3-sulfophenyl)phosphino)pentane disodium salt.

Scheme 1.94.

halide anion, especially I^-, is crucial to obtain a high enantioselectivity. The use of methanol as solvent is also important. Rh complexes of BDPP and its analogues are also useful. Hydrogenation of a ketimine (Ar = C_6H_5) in the presence of a BDPP–Rh complex and $(C_2H_5)_3N$ in methanol gives the amine with 73% ee [145]. The use of [Rh{(S,S)-bdpp}(nbd)]ClO$_4$ with an excess amount of aggregate of bis(2-ethylhexyl) 2-sulfosuccinate sodium salt in benzene gives the R amine with 92% ee [339]. Other sulfonate salts show a similar effect. It is interesting that addition of $(n\text{-}C_4H_9)_4NI$ to the system instead of a sulfonate salt gives rise to the opposite enantioface selection with the same chiral source. Interaction of the sulfonate function with a Rh center may play an important role for the enantioselection. A Rh complex with a monosulfonated BDPP (SulfBDPP) promotes the asymmetric hydrogenation of some acyclic ketimines in a water–ethyl acetate two-phase system [340]. The optical yield reaches 96% [341], the highest value for hydrogenation of acyclic ketimines. The degree of enantioface differentiation is little affected by the electronic properties of C4 substituents in the benzene ring [342]. Contamination of a less-selective 1,3-disulfonated BDPP–Rh complex is not serious because of its lower catalytic activity. However, because SulfBDPP is a mixture of two diastereomers due to the P-chiral center, the origin of high enantioselection is puzzled [341,342].

The use of a neutral (S)-TolBINAP–Ir complex and excess benzylamine gives the R amine with 70% ee [343]. A cationic Ir complex combined with the chiral iminophosphine **27** is highly active in weakly coordinating solvents such as CH_2Cl_2 [344]. Hydrogenation of an N-phenyl substrate at 99 atm exhibits a TON of 5,000 and an initial TOF of 3 s^{-1}. Under high dilution conditions, an 89% optical yield is achievable. Kinetic resolution of a racemic α-methylbenzyl imine of 2′-methoxyacetophenone is attainable by hydrogenation with an (S,S)-BDPP–Rh complex with $k_R/k_S = 5.7$, leaving the less-reactive S substrate with 98% ee at 72% conversion [345].

As shown in Scheme 1.95, the chiral titanocene catalyst **34** (see Scheme 1.10) prepared from **33**, $n\text{-}C_4H_9Li$, and $C_6H_5SiH_3$ shows a moderate-to-good enantioselectivity in the hydrogenation of N-benzyl imines of aryl methyl ketones, whereas the catalytic activity is rather low even at 137 atm [346]. The ketimine with $R^1 = 4\text{-}CH_3OC_6H_4$ is hydrogenated with (R)-**34** to give the R amine with 86% ee. The E:Z of the imine substrate affects the enantioselection. The optical

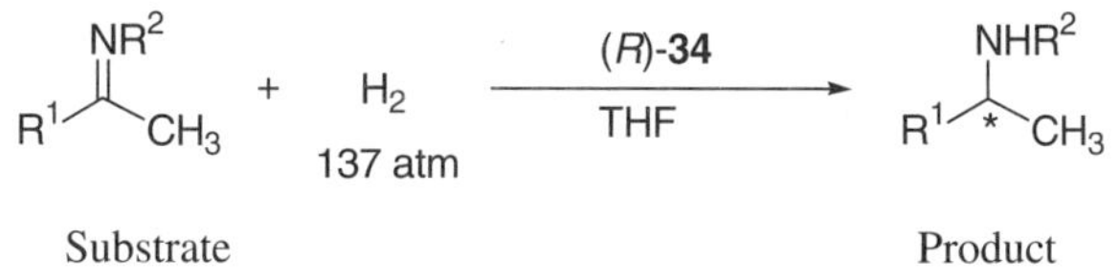

$$\underset{\text{Substrate}}{\underset{R^1}{\overset{NR^2}{\|}}\underset{CH_3}{\diagup}} \quad + \quad \underset{\substack{137\ \text{atm}}}{H_2} \quad \xrightarrow[\text{THF}]{(R)\text{-}\mathbf{34}} \quad \underset{\text{Product}}{\underset{R^1}{\overset{NHR^2}{\diagup}}\underset{*}{\underset{CH_3}{\diagdown}}}$$

R^1	R^2	E:Z Ratio	% Yield	% ee	Config'n
C_6H_5	$C_6H_5CH_2$	17:1	93	85	R
4-CH$_3$OC$_6$H$_4$	$C_6H_5CH_2$	17:1	86	86	R
2-naphthyl	$C_6H_5CH_2$	44:1	82	70	+^{a}
n-C$_4$H$_9$	$C_6H_5CH_2$	3.3:1	68	58	−^{a}
cyclo-C$_6$H$_{11}$	CH$_3$	11:1	85	92^b	R
cyclo-C$_6$H$_{11}$	$C_6H_5CH_2$	11:1	93	76	R

^{a} Sign of rotation. ^{b} At 35 atm of H$_2$.

Scheme 1.95.

yield is significantly decreased under a lower hydrogen pressure. The catalyst effectively discriminates between an alkyl and methyl group. For example, an N-methyl substrate derived from cyclohexyl methyl ketone is hydrogenated in 92% optical yield [346b]. The optical yield decreases to 76% when its N-benzyl analogue is used. With the same catalyst system, the imine of 2-hexanone is converted in 58% optical yield.

High catalytic activity is necessary for industrial hydrogenation [175]. (S)-Metolachlor, an herbicide, is provided in a >10,000-ton quantity per year through asymmetric hydrogenation of the phenyl imine of 1-methoxy-2-propanone by the Novartis Services AG (Scheme 1.96) [68,335f,347,348]. An Ir complex prepared from [IrCl(cod)]$_2$ and (R)-(S)-XYLIPHOS with added (n-C$_4$H$_9$)$_4$NI in acetic acid shows a remarkably high catalytic activity. The reaction is conducted with an S/C of 1,000,000 at 79 atm of hydrogen and at 50°C [68,347]. The TOF number reaches 500 s^{-1}. This homogeneous catalyst is one of the most active for hydrogenation. The enantioselectivity of the amine product, 79% ee, is not excellent, however, the tremendous reactivity offsets the disadvantage. The same catalyst is applicable to the thiophenyl-imine derivative with an S/C of 100 [335f]. An MOD-XYLIPHOS–Ir catalyst achieves higher optical yield in reaction of the phenyl imine but with a lower activity than that of the XYLIPHOS–Ir catalyst [335f]. Other chiral Ir complexes have been used for laboratory-scale research. A neutral BDPP–Ir complex with (n-C$_4$H$_9$)$_4$NI promotes highly enantioselective hydrogenation of an analogous imine, albeit with a low activity [349]. [Ir(diop)HI$_2$]$_2$, a dimeric Ir(III) complex, shows a relatively high activity [350]. A 90% optical yield is obtained by using Ir(OCOCF$_3$)$_3$(bdpp) [351].

1.5.2. Cyclic Imines

Enantiocontrol of hydrogenation of geometry-fixed cyclic imines seems easier than that with flexible, acyclic imines. The chiral titanocene catalyst **34** exhibits excellent enantioselection in hydrogenation of various cyclic imines [346, 352]. For example, 2-phenyl-1-pyrroline (R^1 = C$_6$H$_5$, R^2 = CH$_2$), a five-membered imine, is hydrogenated with an R catalyst to give (R)-2-phenylpyrrolidine in 83% yield and 99% ee (Scheme 1.97) [346b,352]. The optical yield is not affected by hydrogen pressure. 2-Alkylated substrates are also reduced selectively. Under these conditions, many functional groups such as trisubstituted olefins, hydroxy, t-C$_4$H$_9$(CH$_3$)$_2$SiO, and N-benzylpyrrolyl are tolerated, although mono- or 1,2-disubstituted olefins are reducible. Six- and seven-membered cyclic analogues are also hydrogenated under the standard conditions with a similar enantioselectivity [346a,b,352]. A TolBINAP–Ir complex with added benzylimine also shows a high enantioselectivity in hydrogenation of a six-membered substrate [R^1 = C$_6$H$_5$, R^2 = (CH$_2$)$_2$] [343].

The chiral titanocene catalyst **34** is very effective for the kinetic resolution of racemic 2,5-disubstituted 1-pyrrolines. When hydrogenation of racemic 5-methyl-2-phenyl-1-pyrroline with (S)-**34** is interrupted at ca. 50% conversion, unreacted R substrate with 99% ee is obtainable with a (2S,5S)-cis-pyrrolidine derivative with 99% ee (Scheme 1.98) [353]. As summarized in the table, some other racemic substrates can be resolved in >95% optical yield.

The activity and enantioselectivity of chiral Ir catalysts have been tested by using 2,3,3-trimethylindolenine as a model substrate. Hydrogenation of the cyclic imine with [Ir(bdpp)HI$_2$]$_2$ gives the corresponding chiral amine with 80% ee (Scheme 1.99) [350]. The stereoselectivity is somewhat better than that with acyclic substrates (see Scheme 1.94). A neutral BCPM–Ir complex with BiI$_3$ effects asymmetric hydrogenation in 91% optical yield [354]. A complex of MCCPM shows similar enantioselection [354]. These complexes are not applicable to the reaction of other acyclic and six-membered cyclic imines. An MOD-DIOP–Ir complex is also usable with the aid of (n-C$_4$H$_9$)$_4$NI [355]. An Ir complex of BICP with phthalimide effectively

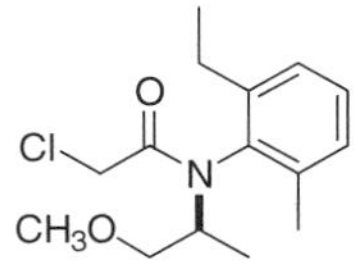

(S)-metolachlor

Ar in Imine	Catalyst	S/C[a]	Solvent	H$_2$, atm	Temp, °C	Product % ee	Product Config'n
2,6-(CH$_3$)$_2$C$_6$H$_3$	(S,S)-BDPP–[IrCl(cod)]$_2$ + (n-C$_4$H$_9$)$_4$NI	100	1:1 CH$_3$OH–benzene	20	0	84	S
2,6-(CH$_3$)$_2$C$_6$H$_3$	[Ir{(S,S)-diop}HI$_2$]$_2$	2000	3:1 THF–CH$_2$Cl$_2$	99	20	63	S
2,6-(CH$_3$)$_2$C$_6$H$_3$	Ir(OCOCF$_3$)$_3$[(S,S)-bdpp]	500	3:1 THF–CH$_2$Cl$_2$	39	0	90	R
2-CH$_3$-6-C$_2$H$_5$C$_6$H$_3$	(R)-(S)-XYLIPHOS–[IrCl(cod)]$_2$ + (n-C$_4$H$_9$)$_4$NI	1,000,000	CH$_3$CO$_2$H	79	50	79	S
2-CH$_3$-6-C$_2$H$_5$C$_6$H$_3$	(R)-(S)-MOD-XYLIPHOS–[IrCl(cod)]$_2$ + I$^-$	5,000	CH$_3$CO$_2$H	79	–10	87	S
3-(2,4-dimethyl)thiophenyl	(R)-(S)-XYLIPHOS–[IrCl(cod)]$_2$ + I$^-$	100	CH$_3$CO$_2$H	59	5	80	S

[a] Substrate/catalyst molar ratio.

Scheme 1.96.

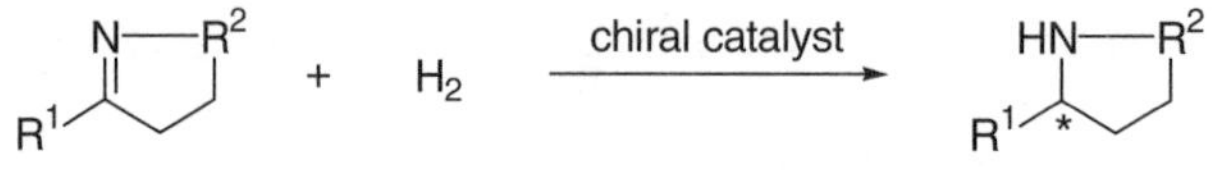

Substrate | Product

R^1	R^2	Catalyst	H_2, atm	% Yield	% ee	Confign
n-C_6H_{13}	CH_2	(*R*)-**34**	137	81	98	+[a]
$(CH_3)_2C{=}CH(CH_2)_2$	CH_2	(*R*)-**34**	6	79	99	+[a]
$HO(CH_2)_7$	CH_2	(*R*)-**34**	6[b]	84	99	+[a]
$TBDMSO(CH_2)_4$[c]	CH_2	(*R*)-**34**	6	82	99	+[a]
C_6H_5	CH_2	(*R*)-**34**	6	83	99	*R*
2-(*N*-benzyl)pyrrolyl	CH_2	(*R*)-**34**	6	72	99	*R*
C_6H_5	$(CH_2)_2$	(*R*)-**34**	35	78	98	*R*
C_6H_5	$(CH_2)_2$	(*S*)-TolBINAP–Ir + $C_6H_5CH_2NH_2$	58	100	90	*R*
C_6H_5	$(CH_2)_3$	(*R*)-**34**	35	71	98	+[a]

[a] Sign of rotation. [b] With 1.1 equiv of phenylsilane to imine. [c] TBDMS = t-$C_4H_9(CH_3)_2Si$.

Scheme 1.97.

promotes hydrogenation to produce the chiral amine quantitatively with 95% ee [356]. The catalyst system is not effective for enantioselective hydrogenation of any other imines. A high enantioselectivity is also achievable in acetic acid containing a xyl_2PF-$Pxyl_2$–Ir catalyst [335f]. An (2*S*,4*S*)-BPPM–Ir catalyzed hydrogenation of a 3-methyl-2*H*-1,4-benzoxadine gives the *S* product with 90% ee (Scheme 1.100) [357].

Asymmetric hydrogenation of Scheme 1.101 provides a general route to isoquinoline alkaloids (see Section 1.3.1.1). An imine substrate is hydrogenated with the chiral titanocene (*R*)-**34** to give the *S* product with 98% ee [346a,b,352]. A neutral BCPM–Ir complex with phthalimide in toluene also shows high enantioselection [358]. The choice of a weakly polar

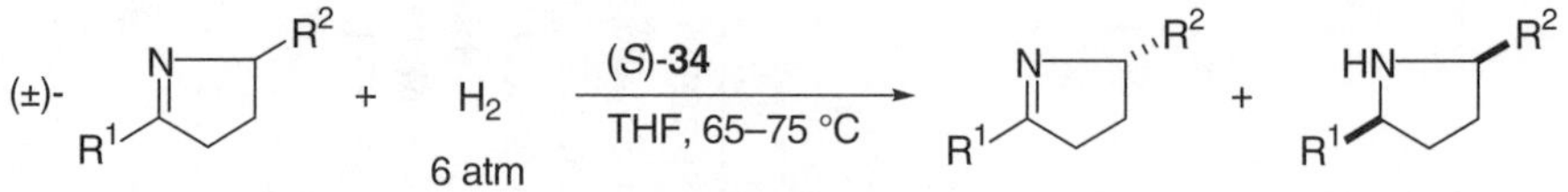

Substrate | Recovered Imine | Amine Product

R^1	R^2	Recovered Imine % Yield[a]	Recovered Imine % ee	Amine Product % Yield[a]	Amine Product % ee
n-$C_{11}H_{23}$	CH_3	41	>95	41	>95
C_6H_5	CH_3	37	99	34	99
C_6H_5	$TIPSOCH_2$[b]	43	98	41	98
2-(*N*-benzyl)pyrrolyl	CH_3	42	96	44	98

[a] Isolated yield. [b] TIP = $((CH_3)_2CH)_3Si$

Scheme 1.98.

Catalyst	Solvent	H$_2$, atm	Temp, °C	Product	
				% ee	Sign[a]
[Ir{(S,S)-bdpp}HI$_2$]$_2$	3:1 THF–CH$_2$Cl$_2$	39	30	80	+
(2S,4S)-BCPM–[IrCl(cod)]$_2$ + BiI$_3$	1:1 CH$_3$OH–benzene	100	−30	91	+
(2S,4S)-MCCPM–[IrCl(cod)]$_2$ + BiI$_3$	1:1 CH$_3$OH–benzene	100	−10	90	+
(R,R)-MOD-DIOP–[IrCl(cod)]$_2$ + (n-C$_4$H$_9$)$_4$NI	1:1 CH$_3$OH–benzene	100	20	81	+
(R,R)-BICP–[IrCl(cod)]$_2$ + phthalimide	CH$_2$Cl$_2$	68	0	95	+
(R)-(S)-xyl$_2$PF-Pxyl$_2$–[IrCl(cod)]$_2$ + I$^-$	CH$_3$CO$_2$H	79	5	94	b

[a] Sign of rotation. [b] Not indicated.

Scheme 1.99.

solvent is important. With the same complex, 1-arylmethyl, 1-(2-arylethyl), and 1-(2-arylvinyl) analogues are also hydrogenated in a high optical yield [359]. In some cases, addition of 3,4,5,6-tetrafluorophthalimide gives a slightly better optical yield than that with an original phthalimide. A BINAP–Ir complex gives a similar selectivity [359]. The benzyloxymethyl derivative is hydrogenated enantioselectively with a BINAP–Ir complex in the presence of 3,4,5,6-tetrafluorophthalimide [360].

Double hydrogenation of 2-methylquinoxaline with (R)-**47** in methanol gives (S)-2-methyl-1,2,3,4-tetrahydroquinoxaline in 54% yield and 90% ee (Scheme 1.102) [361]. The two C=N bonds are reduced at comparable rates. The degree of enantioselection tends to increase by decreasing hydrogen pressure. Diastereoselective hydrogenation of folic acid with a (2S,4S)-BPPM–Rh complex supported on silica gel in phosphate buffer gives 5,6,7,8-tetrahydrofolic acid with a syn:anti ratio of 96:4 (Scheme 1.103) [362]. Medium particle size (40 μm) of silica gel is crucial to achieve a high stereoselectivity.

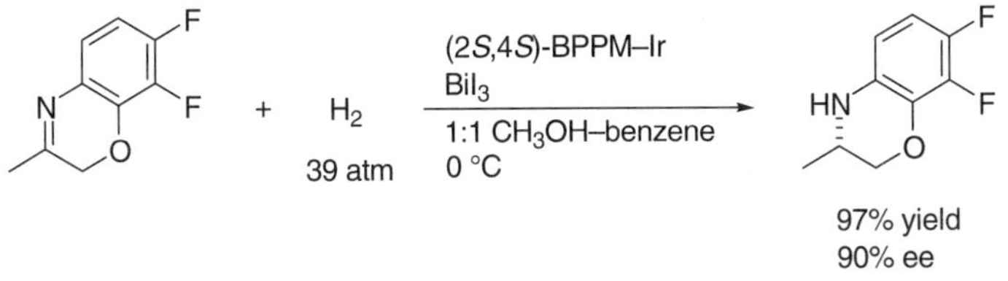

97% yield
90% ee

Scheme 1.100.

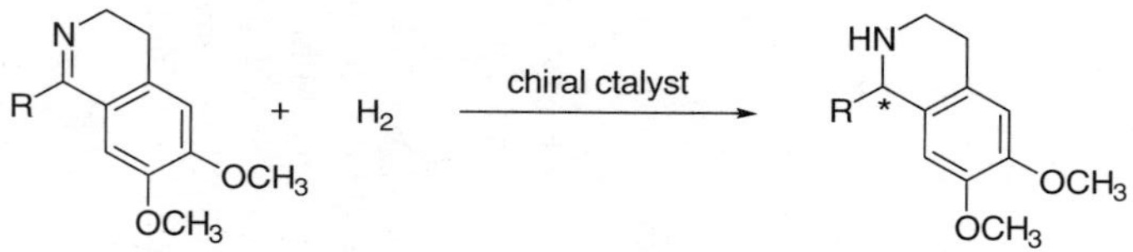

imine + H$_2$ →(chiral ctalyst) product

salsolidine: R = CH$_3$

norlaudanosine: R = 3,4-(CH$_3$O)$_2$C$_6$H$_3$CH$_2$

tetrahydrohomopapaverine: R = 3,4-(CH$_3$O)$_2$C$_6$H$_3$(CH$_2$)$_2$

calycotomine: R = HOCH$_2$

R in Imine	Catalyst	Solvent	H$_2$, atm	Temp, °C	% Yield	% ee	Config
						Product	
CH$_3$	(R)-34	THF	137	65	82	98	S
CH$_3$	(2S,4S)-BCPM–Ir + phthalimide	toluene	100	2–5	95	93	S
3,4-(CH$_3$O)$_2$C$_6$H$_3$CH$_2$	(2S,4S)-BCPM–Ir + phthalimide-F$_4$[a]	toluene	100	5	84	88	S
3,4-(CH$_3$O)$_2$C$_6$H$_3$(CH$_2$)$_2$	(2S,4S)-BCPM–Ir + phthalimide	toluene	100	2	75	87	S
3,4-(CH$_3$O)$_2$C$_6$H$_3$(CH$_2$)$_2$	(S)-BINAP–Ir + phthalimide-F$_4$[a]	CH$_3$OH–toluene	100	2	89	86	S
(E)-3,4-(CH$_3$O)$_2$C$_6$H$_3$CH$_2$CH=CH	(2S,4S)-BCPM–Ir + phthalimide	toluene	100	2	79	86	S
C$_6$H$_5$CH$_2$OCH$_2$	(R)-BINAP–Ir + phthalimide-F$_4$[a]	CH$_3$OH–toluene	100	2–5	85	86	S

[a] 3,4,5,6-Tetrefluorophthalimide.

Scheme 1.101.

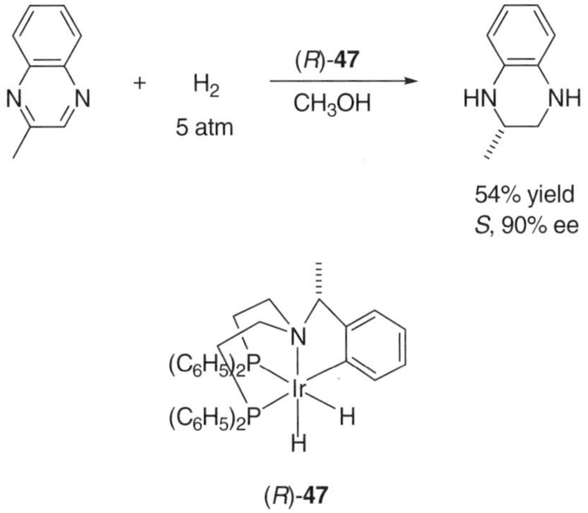

Scheme 1.102.

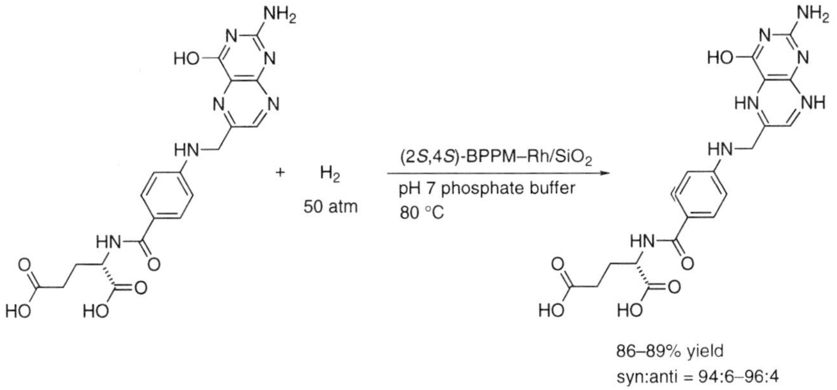

Scheme 1.103.

1.5.3. *N*-Acylhydrazones and Sulfonimides

Studies on asymmetric hydrogenation of C=N bonds have focused on the *N*-alkyl or -aryl compounds. When a functionality is introduced to the nitrogen atom instead of the alkyl or aryl group, the C=N bond is often activated for hydrogenation. In fact, a variety of *N*-benzoylhydrazones are hydrogenated with [Rh(et-duphos)(cod)]CF$_3$SO$_3$ to afford chiral *N*-benzoylhydrazines with a fair-to-excellent optical yield [363]. For example, hydrogenation of a substrate containing phenyl and methyl group (R^1 = C$_6$H$_5$, R^2 = CH$_3$) with an (*R,R*)-Et-DuPHOS–Rh complex in 2-propanol at −10°C under atmospheric pressure gives the *S* product with 95% ee (Scheme 1.104) [363]. The reaction proceeds smoothly with <4 atm of hydrogen, and the highest enantioselectivity is obtained in 2-propanol. The use of a neutral Rh complex decreases the enantioselectivity. Noticeably, replacement of NH function with NCH$_3$ group results in a loss of the optical yield (8%). Hydrogenation of a substrate (R^1 = C$_6$H$_5$, R^2 = C$_2$H$_5$) gives a somewhat

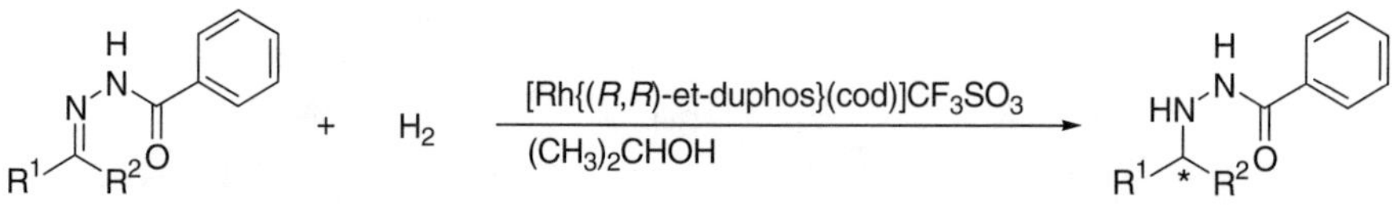

Substrate				Product	
R^1	R^2	H_2, atm	Temp, °C	% ee	Config'n
C_6H_5	CH_3	1	−10	95	S
C_6H_5	C_2H_5	4	−10	85	S
$4\text{-}CH_3OC_6H_4$	CH_3	4	0	88	S
$4\text{-}NO_2C_6H_4$	CH_3	4	0	97	S
2-naphthyl	CH_3	4	0	95	S
$(CH_3)_2CH$	CH_3	4	−10	73	S
C_2H_5	CH_3	4	−10	43	S
CH_3OCO	C_6H_5	4	0	91	$-^a$
$(C_2H_5O)_2PO$	C_6H_5	4	−10	90	$-^a$

^{a} Absolute configuration is unknown.

Scheme 1.104.

lower selectivity. An electron-withdrawing substituent, such as the nitro group, in the benzene ring tends to increase the degree of enantioface differentiation. The catalyst can distinguish methyl from other alkyl groups, even from ethyl (43% ee). A substrate derived from an α-keto ester ($R^1 = CH_3OCO$, $R^2 = C_6H_5$) or an α-keto phosphonate ($R^1 = (C_2H_5O)_2PO$, $R^2 = C_6H_5$) is also hydrogenated in a high optical yield [363]. This catalyst system shows a high chemose-lectivity. Carbonyl compounds, nitriles, imines, organic halides, and nitro compounds are tolerated under the standard conditions. Unfunctionalized alkenes and alkynes are less reactive. The obtained chiral *N*-benzoylhydrazines are easily converted to the primary amines by SmI$_2$-induced reductive cleavage of the N–N bond without loss of optical purity.

Enantioselective hydrogenation of sulfonimides is achievable by a BINAP–Ru complex. A *p*-toluenesulfonimide derived from propiophenone is hydrogenated with Ru(OCOCH$_3$)$_2$[(*R*)-binap] in THF to give the *R* product with 84% ee, albeit with a very low activity (Scheme 1.105) [364]. The degree of enantioface differentiation is highly dependent on the structure of the substrate. A cyclic sulfonimide is also hydrogenated with [NH$_2$(C$_2$H$_5$)$_2$][{RuCl[(*R*)-binap]}$_2$(μ-Cl)$_3$] under 4 atm of hydrogen to give the nearly enantiomerically pure *R* sultam [365].

1.5.4. Transfer Hydrogenation

Asymmetric transfer hydrogenation of imines by using a stable organic hydrogen donor is a useful synthetic method. Despite its importance, no efficient chiral catalysts have been reported until recently [366]. However, highly enantioselective transfer hydrogenation is now accessible by using a unique chiral Ru catalyst and formic acid as a hydrogen source. As illustrated in

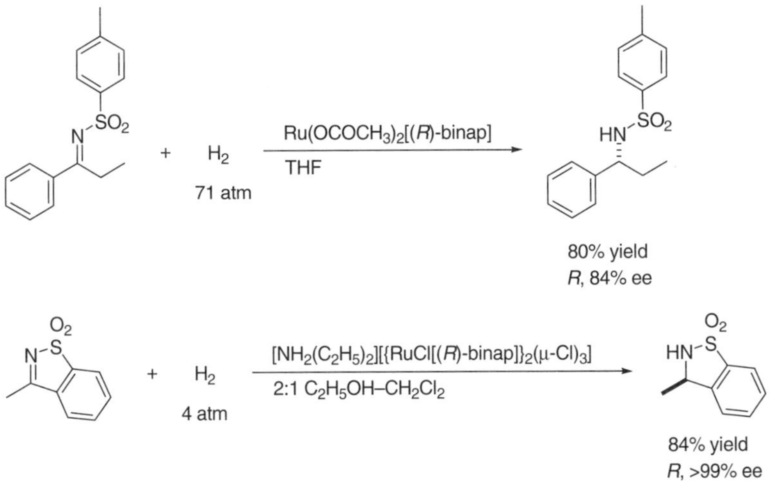

Scheme 1.105.

Scheme 1.106, a six-membered cyclic imine with R = CH_3 is reduced in the presence of (S,S)-**48a** and a 5:2 formic acid–$(C_2H_5)_3N$ azeotropic mixture [320] in CH_3CN at 28°C to give quantitatively (R)-salsolidine with 95% ee [293,367]. 2-Propanol, commonly used for reduction of ketones, cannot be used as hydrogen source for this transformation (see Section 1.4.3). The reaction proceeds in aprotic polar solvents such as CH_3CN, DMF, DMSO, and CH_2Cl_2, whereas the reaction is very slow in a neat formic acid–$(C_2H_5)_3N$ mixture. The reactivity and enantioselectivity are highly sensitive to the structures of the η^6-arene and 1,2-diamine ligands in **48**. The presence of NH_2 and $ArSO_2$ groups is crucial to achieve a high reactivity. The structure of Ar group and substitution pattern of the $ArSO_2$ group can be flexibly changed toward the imine substrates. Cyclic imines substituted by alkyl, benzyl, and aryl groups are transformed to the amines in a high optical yield (Scheme 1.106) [293,367]. This method has opened a general route to natural and unnatural isoquinoline alkaloids (see Section 1.3.1.1 and 1.5.2). An indol with 97% ee is obtainable by reduction of the corresponding imine with the complex **48a** [293,367]. The benzyl imine of acetophenone (E:Z = 94:6) is reduced in 77% optical yield. The imine derived from α-tetralone also gives the chiral product with 89% ee. A remarkable feature of this reduction is the excellent chemoselectivity for C=N bond. Reaction of a cyclic imine is >1000 times faster than that of a structurally related ketone [293,367]. The C=N/C=O selectivity is even higher than that observed in $NaBH_3CN$ reduction (98:1) [368]. Structurally similar aromatic olefins such as α-methylstyrene are inert under the standard conditions.

This methodology is applicable to the synthesis of biologically important compounds. For example, imines **A** and **B** in Scheme 1.107 having a heterocyclic ring are reduced with (S,S)-**48d** and a formic acid–$(C_2H_5)_3N$ mixture to give the amines **C** and **D** with 85% ee and 88% ee, respectively, useful intermediates for the synthesis of MK-0417 (see Section 1.4.3.2.1) [325].

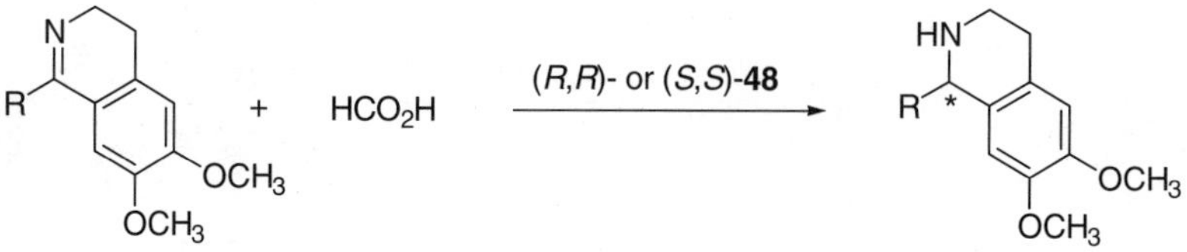

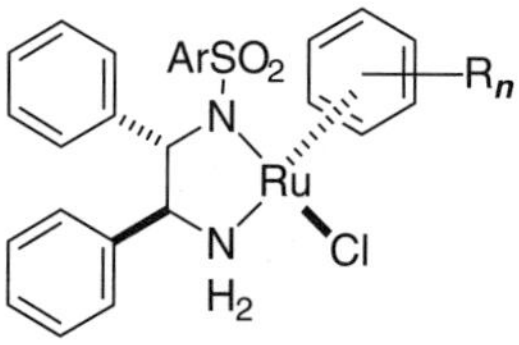

a: η^6-arene = p-cymene; Ar = 4-CH$_3$C$_6$H$_4$
b: η^6-arene = p-cymene; Ar = 2,4,6-(CH$_3$)$_3$C$_6$H$_2$
c: η^6-arene = benzene; Ar = 2,4,6-(CH$_3$)$_3$C$_6$H$_2$
d: η^6-arene = benzene; Ar = 1-naphthyl

other examples:

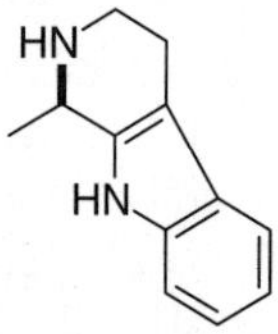

97% ee
with (S,S)-**48a** in DMF

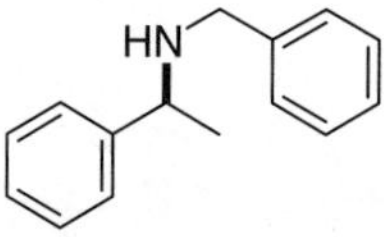

77% ee
with (S,S)-**48c** in CH$_2$Cl$_2$

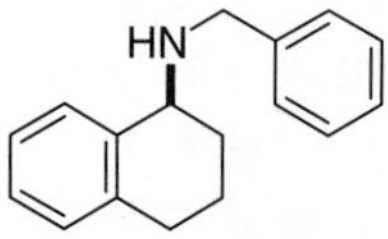

89% ee
with (S,S)-**48d** in CH$_2$Cl$_2$

			Product		
R in Imine	Catalyst	Solvent	% Yield	% ee	Config^n
CH$_3$	(S,S)-**48a**	CH$_3$CN	>99	95	R^a
3,4-(CH$_3$O)$_2$C$_6$H$_3$CH$_2$	(R,R)-**48b**	(CH$_3$)$_2$NCHO	90	95	S^b
3,4-(CH$_3$O)$_2$C$_6$H$_3$(CH$_2$)$_2$	(R,R)-**48b**	CH$_2$Cl$_2$	99	92	S^c
C$_6$H$_5$	(S,S)-**48d**	CH$_2$Cl$_2$	99	84	R
3,4-(CH$_3$O)$_2$C$_6$H$_3$	(R,R)-**48d**	CH$_2$Cl$_2$	>99	84	S^d

^{a} Salsolidine. ^{b} Tetrahydropapaverine. ^{c} Norhomolaudanosine. ^{d} Norcryptostyline.

Scheme 1.106.

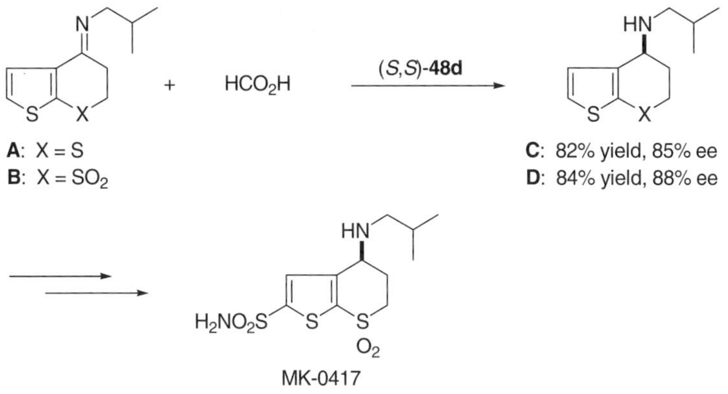

A: X = S
B: X = SO₂

C: 82% yield, 85% ee
D: 84% yield, 88% ee

MK-0417

Scheme 1.107.

1.6. CONCLUSION

Asymmetric hydrogenation provides a straightforward method to obtain nonracemic alkanes, alcohols, and amines from achiral olefins, ketones, and imines, respectively. This chapter has focused on the reaction with homogeneous catalysts that are constructed with an early or late transition metal and a chiral ligand(s). Molecular catalysts are, in principle, obtainable as a single species endowed with any chemical function and chiral environment. The chirality of catalyst can be repeatedly transferred to a large number of achiral unsaturated molecules, resulting in chiral multiplication. Due to the high potential of such molecular catalysts, asymmetric hydrogenation is, and will remain, a major subject in synthetic chemistry. The efficiency of asymmetric hydrogenation can be evaluated by the following items among others: (1) the enantioselectivity leading to nonracemic products in a high ee; (2) the reactivity of catalysts represented by TON and TOF; (3) the generality or scope of reaction; (4) the operational simplicity, including safety and environmental factors; (5) the cost of the catalyst and substrate; (6) the ease of recovery of the precious metal and ligand; and (7) the value of a specific product added by the asymmetric reaction. Some of these factors are purely scientific in nature but are strongly connected with the utility in research and even commercial production of chiral substances. Chemistry is characterized by generality. Although a variety of chiral molecular catalysts has been devised for highly enantioselective hydrogenation, the scope is still limited. Because of a diverse array of unsaturated organic compounds, further development of well-shaped and well-functionalized chiral ligands is desirable to expand the scope of this process. Flexible molecular design based on accumulated chemical knowledge will lead to the discovery of original high-performance chiral catalysts. The rapidly growing combinatorial approach that allows high-throughput screening is of great help, particularly for optimizing catalysts. In addition, computational methodology presents a powerful aid for this purpose. Cost performance is clearly a major issue in industry. High-performance catalysts must not only be reactive, but also robust. Some hydrogenation catalysts already exhibit a TON of >1,000,000 and a TOF of >100 per second, making the reaction truly practical. To devise many other efficient hydrogenations, introduction of unique chemical and physical concepts is essential. Properties of reaction media are also important from both scientific and technical points of view. It is obvious that environmentally unconscious solvents are to be avoided. Certain unorthodox media lead to enormous benefits in hydrogenation. For example, hydrogenation in supercritical fluids has been shown to display unique selectivity and unusually high reactivity in some cases [369]. The latter is largely due to the high concentration of H_2 in a supercritical fluid, which overcomes the mass-transfer problem encountered in conventional liquid-phase hydrogenation. Because chiral catalysts are often expensive, their reuse is highly desirable. Organic/aqueous biphasic reaction by using a water-soluble catalyst allows for easy separation of the catalyst from products [370]. Immobilization of catalysts on polymer or gel matrices [371] or the use in membrane [372] provides another way of the repeated use. Thus, there remains a variety of problems to be solved or improved. Nevertheless, it is certain that this field has a bright future.

REFERENCES

1. (a) Noyori, R.; Kitamura, M. In *Modern Synthetic Methods*; Scheffold, R. (Ed.); Springer: Berlin, 1989; Vol. 5, pp. 115. (b) Takaya, H.; Ohta, T.; Noyori, R. In *Catalytic Asymmetric Synthesis*; Ojima, I. (Ed.); VCH: New York, 1993; Chapter 1. (c) Noyori, R. In *Asymmetric Catalysis in Organic Synthesis*; Wiley: New York, 1994; Chapter 2. (d) Genêt, J.-P. In *Reductions in Organic Synthesis: Recent Advances and Practical Applications (ACS symposium series 641)*; Abdel-Magid, A. F. (Ed.); American Chemical Society: Washington, DC, 1996; Chapter 2. (e) Albrecht, J.; Nagel, U. *Angew. Chem. Int. Ed.* **1996**, *35*, 407.

2. (a) Morrison, J. D.; Mosher, H. S. In *Asymmetric Organic Reactions*; Prentice-Hall: New Jersey, 1971; Chapter 5. (b) Matteoli, U.; Frediani, P.; Bianchi, M.; Botteghi, C.; Gladiali, S. *J. Mol. Catal.* **1981**, *12*, 265. (c) Zassinovich, G.; Mestroni, G.; Gladiali, S. *Chem. Rev.* **1992**, *92*, 1051. (d) de Graauw, C. F.; Peters, J. A.; van Bekkum, H.; Huskens, J. *Synthesis* **1994**, 1007. (e) Gladiali, S.; Mestroni, G. In *Transition Metals for Organic Synthesis*; Beller, M.; Bolm, C. (Eds.); Wiley-VCH: Weinheim, 1998; Vol. 2, Chapter 1.3.

3. (a) Kagan, H. B. In *Asymmetric Synthesis*; Morrison, J. D. (Ed.); Academic Press: Orlando, 1985; Vol. 5, Chapter 1. (b) Brunner, H. *Top. Stereochem.* **1988**, *18*, 129. (c) Blaser, H.-U. *Chem. Rev.* **1992**, *92*, 935. (d) Brunner. H.; Zettlmeier, W. *Handbook of Enantioselective Catalysis*; VCH: Weinheim, 1993. (e) Schwink, L.; Knochel, P. *Chem. Eur. J.* **1998**, *4*, 950. (f) Richards, C. J.; Locke, A. J. *Tetrahedron: Asymmetry* **1998**, *9*, 2377.

4. Lucet, D.; Le Gall, T.; Mioskowski, C. *Angew. Chem. Int. Ed.* **1998**, *37*, 2580.

5. Evans, D.; Osborn, J. A.; Jardine, F. H.; Wilkinson, G. *Nature* **1965**, *208*, 1203.

6. (a) Horner, L.; Siegel, H.; Büthe, H. *Angew. Chem. Int. Ed.* **1968**, *7*, 942. (b) Knowles, W. S.; Sabacky, M. J. *J. Chem. Soc., Chem. Commun.* **1968**, 1445.

7. Dang, T. P.; Kagan, H. B. *J. Chem. Soc., Chem. Commun.* **1971**, 481.

8. Knowles, W. S.; Sabacky, M. J.; Vineyard, B. D. *J. Chem. Soc., Chem. Commun.* **1972**, 10.

9. Vineyard, B. D.; Knowles, W. S.; Sabacky, M. J.; Bachman, G. L.; Weinkauff, D. J. *J. Am. Chem. Soc.* **1977**, *99*, 5946.

10. (a) Koenig, K. E. In *Catalysis of Organic Reactions*; Kosak, J. R. (Ed.); Marcel Dekker: New York, 1984; Chapter 3. (b) Koenig, K. E. In *Asymmetric Synthesis*; Morrison, J. D. (Ed.); Academic Press: Orlando, 1985; Vol. 5, Chapter 3. (c) Pfaltz, A.; Brown, J. M. In *Methods of Organic Chemistry (Houben-Weyl)*, 4th ed.: Helmchen, G.; Hoffmann, R. W., Mulzer, J., Schaumann, E. (Ed.); Thieme: Stuttgart, 1995; Vol. E21d, pp. 4334. (d) Nagel, U.; Albrecht, J. *Top. Catalysis* **1998**, *5*, 3. (e) Burk, M. J.; Bienewald, F. In *Transition Metals for Organic Synthesis*; Beller, M., Bolm, C. (Ed.); Wiley-VCH: Weinheim, 1998; Vol. 2, Chapter 1.1.2.

11. Hayashi, T.; Kumada, M. *Acc. Chem. Res.* **1982**, *15*, 395.

12. Achiwa, K. *J. Am. Chem. Soc.* **1976**, *98*, 8265.

13. Fryzuk, M. D.; Bosnich, B. *J. Am. Chem. Soc.* **1977**, *99*, 6262.

14. Brunner, H.; Pieronczyk, W. *Angew. Chem. Int. Ed.* **1979**, *18*, 655.

15. Miyashita, A.; Yasuda, A.; Takaya, H.; Toriumi, K.; Ito, T.; Souchi, T.; Noyori, R. *J. Am. Chem. Soc.* **1980**, *102*, 7932.

16. Oliver, J. D.; Riley, D. P. *Organometallics* **1983**, *2*, 1032.

17. MacNeil, P. A.; Roberts, N. K.; Bosnich, B. *J. Am. Chem. Soc.* **1981**, *103*, 2273.

18. (a) Nagel, U. *Angew. Chem. Int. Ed.* **1984**, *23*, 435. (b) Nagel, U.; Kinzel, E.; Andrade, J.; Prescher, G. *Chem. Ber.* **1986**, *119*, 3326.

19. Burk, M. J. *J. Am. Chem. Soc.* **1991**, *113*, 8518.

20. Inoguchi, K.; Achiwa, K. *Synlett* **1991**, 49.

21. Miyashita, A.; Karino, H.; Shimamura, J.; Chiba, T.; Nagano, K.; Nohira, H.; Takaya, H. *Chem. Lett.* **1989**, 1849.

22. Zhu, G.; Cao, P.; Jiang, Q.; Zhang, X. *J. Am. Chem. Soc.* **1997**, *119*, 1799.

23. Burk, M. J.; Feaster, J. E.; Harlow, R. L. *Tetrahedron: Asymmetry* **1991**, *2*, 569.

24. Sawamura, M.; Kuwano, R.; Ito, Y. *J. Am. Chem. Soc.* **1995**, *117*, 9602.

25. Imamoto, T.; Tsuruta, H.; Wada, Y.; Masuda, H.; Yamaguchi, K. *Tetrahedron Lett.* **1995**, *36*, 8271.

26. Robin, F.; Mercier, F.; Ricard, L.; Mathey, F.; Spagnol, M. *Chem. Eur. J.* **1997**, *3*, 1365.

27. Pye, P. J.; Rossen, K.; Reamer, R. A.; Tsou, N. N.; Volante, R. P.; Reider, P. J. *J. Am. Chem. Soc.* **1997**, *119*, 6207.

28. Qiao, S.; Fu, G. C. *J. Org. Chem.* **1998**, *63*, 4168.

29. Stoop, R. M.; Mezzetti, A. *Organometallics* **1998**, *17*, 668.

30. Imamoto, T.; Watanabe, J.; Wada, Y.; Masuda, H.; Yamada, H.; Tsuruta, H.; Matsukawa, S.; Yamaguchi, K. *J. Am. Chem. Soc.* **1998**, *120*, 1635.

31. Kang, J.; Lee, J. H.; Ahn, S. H.; Choi, J. S. *Tetrahedron Lett.* **1998**, *39*, 5523.

32. Holz, J.; Quirmbach, M.; Schmidt, U.; Heller, D.; Stürmer, R.; Börner, A. *J. Org. Chem.* **1998**, *63*, 8031.

33. (a) Selke, R.; Pracejus, H. *J. Mol. Catal.* **1986**, *37*, 213. (b) Selke, R.; Ohff, M.; Riepe, A. *Tetrahedron* **1996**, *52*, 15079.

34. RajanBabu, T. V.; Ayers, T. A.; Halliday, G. A.; You, K. K.; Calabrese, J. C. *J. Org. Chem.* **1997**, *62*, 6012.

35. Chan, A.S.C.; Hu, W.; Pai, C.-C.; Lau, C.-P. *J. Am. Chem. Soc.* **1997**, *119*, 9570.

36. Yonehara, K.; Hashizume, T.; Ohe, K.; Uemura, S. *Bull. Chem. Soc. Jpn.* **1998**, *71*, 1967.

37. Zhu, G.; Zhang, X. *J. Org. Chem.* **1998**, *63*, 3133.

38. Cesarotti, E.; Chiesa, A.; Ciani, G.; Sironi, A. *J. Organomet. Chem.* **1983**, *251*, 79.

39. Kreuzfeld, H.-J.; Schmidt, U.; Döbler, C.; Krause, H. W. *Tetrahedron: Asymmetry* **1996**, *7*, 1011.

40. Mi, A.; Lou, R.; Jiang, Y.; Deng, J.; Qin, Y.; Fu, F.; Li, Z.; Hu, W.; Chan, A.S.C. *Synlett* **1998**, 847.

41. Krause, H. W.; Schmidt, U.; Taudien, S.; Costisella, B.; Michalik, M. *J. Mol. Catal.* **1995**, *104*, 147.

42. (a) Fiorini, M.; Giongo, G. M. *J. Mol. Catal.* **1979**, *5*, 303. (b) Fiorini, M.; Giongo, G. M. *J. Mol. Catal.* **1980**, *7*, 411.

43. Burk, M. J.; Feng, S.; Gross, M. F.; Tumas, W. *J. Am. Chem. Soc.* **1995**, *117*, 8277.

44. (a) Alario, F.; Amrani, Y.; Colleuille, Y.; Dang, T.-P.; Jenck, J.; Morel, D.; Sinou, D. *J. Chem. Soc., Chem. Commun.* **1986**, 202. (b) Tóth, I.; Hanson, B. E.; Davis, M. E. *Tetrahedron: Asymmetry* **1990**, *1*, 913.

45. Nagel, U.; Kinzel, E. *Chem. Ber.* **1986**, *119*, 1731.

46. Kumar, A.; Oehme, G.; Roque, J. P.; Schwarze, M.; Selke, R. *Angew. Chem. Int. Ed.* **1994**, *33*, 2197.

47. Selke, R.; Holz, J.; Riepe, A.; Börner, A. *Chem. Eur. J.* **1998**, *4*, 769.

48. Flach, H. N.; Grassert, I.; Oehme, G. *Macromol. Chem. Phys.* **1994**, *195*, 3289.

49. Takaishi, N.; Imai, H.; Bertelo, C. A.; Stille, J. K. *J. Am. Chem. Soc.* **1976**, *98*, 5400.

50. Baker, G. L.; Fritschel, S. J.; Stille, J. R.; Stille, J. K. *J. Org. Chem.* **1981**, *46*, 2954.

51. Nagel, U.; Leipold, J. *Chem. Ber.* **1996**, *129*, 815.

52. Nagel, U.; Kinzel, E. *J. Chem. Soc., Chem. Commun.* **1986**, 1098.

53. (a) Corma, A.; Iglesias, M.; del Pino, C.; Sanchez, F. *J. Chem. Soc., Chem. Commun.* **1991**, 1253. (b) Corma, A.; Iglesias, M.; del Pino, C.; Sanchez, F. *Stud. Surf. Sci. Catal.* **1993**, *75*, 2293.

54. (a) Selke, R. *J. Mol. Catal.* **1986**, *37*, 227. (b) Selke, R.; Capka, M. *J. Mol. Catal.* **1990**, *63*, 319.

55. (a) Tóth, I.; Hanson, B. E.; Davis, M. E. *J. Organomet. Chem.* **1990**, *397*, 109. (b) Tóth, I.; Hanson, B. E. *J. Mol. Catal.* **1992**, *71*, 365.

56. Miyashita, A.; Takaya, H.; Souchi, T.; Noyori, R. *Tetrahedron* **1984**, *40*, 1245.

57. Burk, M. J.; Feaster, J. E.; Nugent, W. A.; Harlow, R. L. *J. Am. Chem. Soc.* **1993**, *115*, 10125.

58. Burk, M. J.; Gross, M. F.; Martinez, J. P. *J. Am. Chem. Soc.* **1995**, *117*, 9375.

59. Hoerrner, R. S.; Askin, D.; Volante, R. P.; Reider, P. J. *Tetrahedron Lett.* **1998**, *39*, 3455.

60. Burk, M. J.; Allen, J. G.; Kiesman, W. F. *J. Am. Chem. Soc.* **1998**, *120*, 657.

61. Kuwano, R.; Okuda, S.; Ito, Y. *J. Org. Chem.* **1998**, *63*, 3499.

62. Kuwano, R.; Okuda, S.; Ito, Y. *Tetrahedron: Asymmetry* **1998**, *9*, 2773.

63. (a) Burk, M. J.; Gross, M. F.; Harper, T.G.P.; Kalberg, C. S.; Lee, J. R.; Martinez, J. P. *Pure Appl. Chem.* **1996**, *68*, 37. (b) Burk, M. J. *CHEMTRACTS–ORGANIC CHEMISTRY* **1998**, *11*, 787.

64. Bozell, J. J.; Vogt, C. E.; Gozum, J. *J. Org. Chem.* **1991**, *56*, 2584.

65. Zeiss, H.-J. *J. Org. Chem.* **1991**, *56*, 1783.

66. (a) Schöllkopf, U.; Hoppe, I.; Thiele, A. *Liebigs Ann. Chem.* **1985**, *1*, 555. (b) Schmidt, U.; Oehme, G.; Krause, H. *Synth. Commun.* **1996**, *26*, 777. (c) Schmidt, U.; Krause, H. W.; Oehme, G.; Michalik, M.; Fischer, C. *Chirality* **1998**, *10*, 564.

67. Dwars, T.; Schmidt, U.; Fischer, C.; Grassert, I.; Kempe, R.; Fröhlich, R.; Drauz, K.; Oehme, G. *Angew. Chem. Int. Ed.* **1998**, *37*, 2851.

68. Blaser, H.-U.; Spindler, F. *Top. Catalysis* **1997**, *4*, 275.

69. Burk, M. J.; Wang, Y. M.; Lee, J. R. *J. Am. Chem. Soc.* **1996**, *118*, 5142.

70. Zhang, F.-Y.; Pai, C.-C.; Chan, A.S.C. *J. Am. Chem. Soc.* **1998**, *120*, 5808.

71. (a) Kagan, H. B.; Langlois, N.; Dang, T. P. *J. Organomet. Chem.* **1975**, *90*, 353. (b) Sinou, D.; Kagan, H. B. *J. Organomet. Chem.* **1976**, *114*, 325.

72. Hayashi, M.; Hashimoto, Y.; Takezaki, H.; Watanabe, Y.; Saigo, K. *Tetrahedron: Asymmetry* **1998**, *9*, 1863.

73. Burk, M. J.; Casy, G.; Johnson, N. B. *J. Org. Chem.* **1998**, *63*, 6084.

74. (a) Ikariya, T.; Ishii, Y.; Kawano, H.; Arai, T.; Saburi, M.; Yoshikawa, S.; Akutagawa, S. *J. Chem. Soc., Chem. Commun.* **1985**, 922. (b) Kawano, H.; Ikariya, T.; Ishii, Y.; Saburi, M.; Yoshikawa, S.; Uchida, Y.; Kumobayashi, H. *J. Chem. Soc., Perkin Trans. 1* **1989**, 1571.

75. James, B. R.; Pacheco, A.; Rettig, S. J.; Thorburn, I. S.; Ball, R. G.; Ibers, J. A. *J. Mol. Catal.* **1987**, *41*, 147.

76. (a) Brown, J. M. *Chem. Soc. Rev.* **1993**, 25. (b) Brown, J. M.; Guiry, P. J.; Wienand, A. In *Principle of Molecular Recognition*; Buckingham, A. D., Legon, A. C., Roberts, S. M. (Ed.); Blackie: Glasgow, 1993; pp. 79. (c) Halpern, J. *Precious Met.* **1995**, *19*, 411.

77. (a) Kitamura, M.; Yoshimura, M.; Tsukamoto, M.; Noyori, R. *Enantiomer* **1996**, *1*, 281. (b) Wiles, J. A.; Bergens, S. H.; Young, V. G. *J. Am. Chem. Soc.* **1997**, *119*, 2940. (c) Wiles, J. A.; Bergens, S. H. *Organometallics* **1998**, *17*, 2228. (d) Eisenberg, R.; Eisenschmid, T. C.; Chinn, M. S.; Kriss, R. U. In *Homogeneous Transition Metal Catalyzed Reactions*; Moser, W. R., Slocum, D. W. (Ed.); American Chemical Society: Washington, DC, 1992; pp. 47.

78. Landis, C. R.; Halpern, J. *J. Am. Chem. Soc.* **1987**, *109*, 1746.

79. (a) Brown, J. M.; Parker, D. *Organometallics* **1982**, *1*, 950. (b) Landis, C. R.; Brauch, T. W. *Inorg. Chim. Acta* **1998**, *270*, 285. (c) Bakos, J.; Karaivanov, R.; Laghmari, M.; Sinou, D. *Organometallics* **1994**, *13*, 2951.

80. (a) Brown, J. M.; Chaloner, P. A. *J. Am. Chem. Soc.* **1980**, *102*, 3040. (b) Brown, J. M.; Chaloner, P. A.; Morris, G. A. *J. Chem. Soc., Chem. Commun.* **1983**, 664. (c) Bircher, H.; Bender, B. R.; von Philipsborn, W. *Magn. Reson. Chem.* **1993**, *31*, 293. (d) Bender, B. R.; Koller, M.; Nanz, D.; von Philipsborn, W. *J. Am. Chem. Soc.* **1993**, *115*, 5889. (e) Ramsden, J. A.; Claridge, T.D.W.; Brown, J. M. *J. Chem. Soc., Chem. Commun.* **1995**, 2469. (f) Harthun, A.; Kadyrov, R.; Selke, R.; Bargon, J. *Angew. Chem. Int. Ed.* **1997**, *36*, 1103. (g) Giernoth, R.; Huebler, P.; Bargon, J. *Angew. Chem. Int. Ed.* **1998**, *37*, 2473.

81. (a) Brown, J. M.; Evans, P. L. *Tetrahedron* **1988**, *44*, 4905. (b) Bogdan, P. L.; Irwin, J. J.; Bosnich, B. *Organometallics* **1989**, *8*, 1450. (c) Giovannetti, J. S.; Kelly, C. M.; Landis, C. R. *J. Am. Chem. Soc.* **1993**, *115*, 4040.

82. (a) Noyori, R.; Ohta, M.; Hsiao, Yi; Kitamura, M.; Ohta, T.; Takaya, H. *J. Am. Chem. Soc.* **1986**, *108*, 7117. (b) Kitamura, M.; Hsiao, Yi; Ohta, M.; Tsukamoto, M.; Ohta, T.; Takaya, H.; Noyori, R. *J. Org. Chem.* **1994**, *59*, 297.

83. Heiser, B.; Broger, E. A.; Crameri, Y. *Tetrahedron: Asymmetry* **1991**, *2*, 51.

84. Kitamura, M.; Hsiao, Yi; Noyori, R.; Takaya, H. *Tetrahedron Lett.* **1987**, *28*, 4829.

85. Lubell, W. D.; Kitamura, M.; Noyori, R. *Tetrahedron: Asymmetry* **1991**, *2*, 543.

86. Schmid, R.; Broger, E. A.; Cereghetti, M.; Crameri, Y.; Foricher, J.; Lalonde, M.; Müller, R. K.; Scalone, M.; Schoettel, G.; Zutter, U. *Pure Appl. Chem.* **1996**, *68*, 131.

87. Gendre, P. L.; Thominot, P.; Bruneau, C.; Dixneuf, P. H. *J. Org. Chem.* **1998**, *63*, 1806.

88. Lee, N. E.; Buchwald, S. L. *J. Am. Chem. Soc.* **1994**, *116*, 5985.

89. Koenig, K. E.; Bachman, G. L.; Vineyard, B. D. *J. Org. Chem.* **1980**, *45*, 2362.

90. Schmidt, U.; Langner, J.; Kirschbaum, B.; Braun, C. *Synthesis* **1994**, 1138.

91. Merrill, R. E. *CHEMTECH* **1981**, 118.

92. Burk, M. J.; Kalberg, C. S.; Pizzano, A. *J. Am. Chem. Soc.* **1998**, *120*, 4345.

93. Ohta, T.; Miyake, T.; Seido, N.; Kumobayashi, H.; Takaya, H. *J. Org. Chem.* **1995**, *60*, 357.

94. Boaz, N. W. *Tetrahedron Lett.* **1998**, *39*, 5505.

95. Hayashi, T.; Kanehira, K.; Kumada, M. *Tetrahedron Lett.* **1981**, *22*, 4417.

96. Gendre, P. L.; Braun, T.; Bruneau, C.; Dixneuf, P. H. *J. Org. Chem.* **1996**, *61*, 8453.

97. Morrison, J. D.; Burnett, R. E.; Aguiar, A. M.; Morrow, C. J.; Phillips, C. *J. Am. Chem. Soc.* **1971**, *93*, 1301.

98. (a) Kitamura, M.; Tokunaga, M.; Noyori, R. *J. Org. Chem.* **1992**, *57*, 4053. (b) Takaya, H.; Inoue, S.; Tokunaga, M.; Kitamura, M.; Noyori, R. *Org. Synth.* **1993**, *72*, 74. (c) Ohta, T.; Takaya, H.; Noyori, R. *Inorg. Chem.* **1988**, *27*, 566.

99. Ohta, T.; Takaya, H.; Kitamura, M.; Nagai, K.; Noyori, R. *J. Org. Chem.* **1987**, *52*, 3174.

100. Ohta, T.; Takaya, H.; Noyori, R. *Tetrahedron Lett.* **1990**, *31*, 7189.

101. Ashby, M. T.; Halpern, J. *J. Am. Chem. Soc.* **1991**, *113*, 589.

102. Mashima, K.; Kusano, K.; Ohta, T.; Noyori, R.; Takaya, H. *J. Chem. Soc., Chem. Commun.* **1989**, 1208.

103. Genêt, J. P.; Mallart, S.; Pinel, C.; Juge, S.; Laffitte, J. A. *Tetrahedron: Asymmetry* **1991**, *2*, 43.

104. Alcock, N. W.; Brown, J. M.; Rose, M.; Wienand, A. *Tetrahedron: Asymmetry* **1991**, *2*, 47.

105. Shao, L.; Takeuchi, K.; Ikemoto, M.; Kawai, T.; Ogasawara, M.; Takeuchi, H.; Kawano, H.; Saburi, M. *J. Organomet. Chem.* **1992**, *435*, 133.

106. (a) King, S. A.; DiMichele, L. In *Catalysis of Organic Reactions*; Scaros, M. G., Prunier, M. L. (Ed.); Dekker: New York, 1995; pp. 157. (b) Ohta, T.; Tonomura, Y.; Nozaki, K.; Takaya, H.; Mashima, K. *Organometallics* **1996**, *15*, 1521.

107. Chen, C.-C.; Huang, T.-T.; Lin, C.-W.; Cao, R.; Chan, A.S.C.; Wong, W. T. *Inorg. Chim. Acta* **1998**, *270*, 247.

108. Saburi, M.; Takeuchi, H.; Ogasawara, M.; Tsukahara, T.; Ishii, Y.; Ikariya, T.; Takahashi, T.; Uchida, Y. *J. Organomet. Chem.* **1992**, *428*, 155.

109. Benincori, T.; Brenna, E.; Sannicolò, F.; Trimarco, L.; Antognazza, P.; Cesarotti, E.; Demartin, F.; Pilati, T. *J. Org. Chem.* **1996**, *61*, 6244.

110. Xiao, J.; Nefkens, S.C.A.; Jessop, P. G.; Ikariya, T.; Noyori, R. *Tetrahedron Lett.* **1996**, *37*, 2813.

111. (a) Zhang, X.; Uemura, T.; Matsumura, K.; Sayo, N.; Kumobayashi, H.; Takaya, H. *Synlett* **1994**, 501. (b) Uemura, T.; Zhang, X.; Matsumura, K.; Sayo, N.; Kumobayashi, H.; Ohta, T.; Nozaki, K.; Takaya, H. *J. Org. Chem.* **1996**, *61*, 5510.

112. Wang, S.; Kienzle, F. *Organic Process Research & Development* **1998**, *2*, 226.

113. Monteiro, A. L.; Zinn, F. K.; De Souza, R. F.; Dupont, J. *Tetrahedron: Asymmetry* **1997**, *8*, 177.

114. Wan, K. T.; Davis, M. E. *Nature* **1994**, *370*, 449.

115. Henry, J.-C.; Lavergne, D.; Ratovelomanana-Vidal, V.; Genêt, J.-P.; Beletskaya, I. P.; Dolgina, T. M. *Tetrahedron Lett.* **1998**, *39*, 3473.

116. Yamada, I.; Yamaguchi, M.; Yamagishi, T. *Tetrahedron: Asymmetry* **1996**, *7*, 3339.

117. (a) Hayashi, T.; Kawamura, N.; Ito, Y. *J. Am. Chem. Soc.* **1987**, *109*, 7876. (b) Hayashi, T.; Kawamura, N.; Ito, Y. *Tetrahedron Lett.* **1988**, *29*, 5969.

118. Ojima, I.; Kogure, T. *Chem. Lett.* **1978**, 567.

119. Ojima, I.; Yoda, N. *Tetrahedron Lett.* **1980**, 1051.

120. Takahashi, H.; Achiwa, K. *Chem. Lett.* **1987**, 1921.

121. Takahashi, H.; Yamamoto, N.; Takeda, H.; Achiwa, K. *Chem. Lett.* **1989**, 559.

122. Morimoto, T.; Chiba, M.; Achiwa, K. *Tetrahedron Lett.* **1989**, *30*, 735.

123. Chiba, T.; Miyashita, A.; Nohira, H.; Takaya, H. *Tetrahedron Lett.* **1991**, *32*, 4745.

124. Ohshima, S.; Matsumoto, T.; Aoki, Y.; Hirose, T.; Miyashita, A.; Nohira, H. *Enantiomer* **1998**, *3*, 191.

125. (a) Kawano, H.; Ishii, Y.; Ikariya, T.; Saburi, M.; Yoshikawa, S.; Uchida, Y.; Kumobayashi, H. *Tetrahedron Lett.* **1987**, *28*, 1905. (b) Shao, L.; Miyata, S.; Muramatsu, H.; Kawano, H.; Ishii, Y.; Saburi, M.; Uchida, Y. *J. Chem. Soc., Perkin Trans. I* **1990**, 1441.

126. Bayston, D. J.; Fraser, J. L.; Ashton, M. R.; Baxter, A. D.; Polywka, M.E.C.; Moses, E. *J. Org. Chem.* **1998**, *63*, 3137.

127. Christopfel, W. C.; Vineyard, B. D. *J. Am. Chem. Soc.* **1979**, *101*, 4406.

128. Burk, M. J.; Bienewald, F.; Harris, M.; Zanotti-Gerosa, A. *Angew. Chem. Int. Ed.* **1998**, *37*, 1931.

129. Inoguchi, K.; Morimoto, T.; Achiwa, K. *J. Organomet. Chem.* **1989**, *370*, C9.

130. Reetz, M. T.; Beuttenmüller, E. W.; Goddard, R. *Tetrahedron Lett.* **1997**, *38*, 3211.

131. Brown, J. M.; James, A. P. *J. Chem. Soc., Chem. Commun.* **1987**, 181.

132. Dang, T.-P.; Aviron-Violet, P.; Colleuille, Y.; Varagnat, J. *J. Mol. Catal.* **1982**, *16*, 51.

133. Massonneau, V.; Le Maux, P.; Simonneaux, G. *J. Organomet. Chem.* **1987**, *327*, 269.

134. Brown, J. M. *Angew. Chem. Int. Ed.* **1987**, *26*, 190.

135. Inoue, S.; Osada, M.; Koyano, K.; Takaya, H.; Noyori, R. *Chem. Lett.* **1985**, 1007.

136. Takaya, H.; Ohta, T.; Sayo, N.; Kumobayashi, H.; Akutagawa, S.; Inoue, S.; Kasahara, I.; Noyori, R. *J. Am. Chem. Soc.* **1987**, *109*, 1596, 4129.

137. (a) Sun, Y.; LeBlond, C.; Wang, J.; Blackmond, D. G. *J. Am. Chem. Soc.* **1995**, *117*, 12647. (b) Sun, Y.; Landau, R. N.; Wang, J.; LeBlond, C.; Blackmond, D. G. *J. Am. Chem. Soc.* **1996**, *118*, 1348. (c) Sun, Y.; Wang, J.; LeBlond, C.; Landau, R. N.; Blackmond, D. G. *J. Catal.* **1996**, *161*, 759.

138. (a) Imperiali, B.; Zimmerman, J. W. *Tetrahedron Lett.* **1988**, *29*, 5343. (b) Kitamura, M.; Noyori, R. unpublished result.

139. Kitamura, M.; Kasahara, I.; Manabe, K.; Noyori, R.; Takaya, H. *J. Org. Chem.* **1988**, *53*, 708.

140. Shimizu, H.; Shimada, Y.; Tomita, A.; Mitsunobu, O. *Tetrahedron Lett.* **1997**, *38*, 849.

141. Lightfoot, A.; Schnider, P.; Pfaltz, A. *Angew. Chem. Int. Ed.* **1998**, *37*, 2897.

142. Hayashi, T.; Tanaka, M.; Ogata, I. *Tetrahedron Lett.* **1977**, 295.

143. Tanaka, M.; Ogata, I. *J. Chem. Soc., Chem. Commun.* **1975**, 735.

144. Kawabata, Y.; Tanaka, M.; Ogata, I. *Chem. Lett.* **1976**, 1213.

145. Bakos, J.; Tóth, I.; Heil, B.; Markó, L. *J. Organomet. Chem.* **1985**, *279*, 23.

146. Inagaki, K.; Ohta, T.; Nozaki, K.; Takaya, H. *J. Organomet. Chem.* **1997**, *531*, 159.

147. Ohta, T.; Ikegami, H.; Miyake, T.; Takaya, H. *J. Organomet. Chem.* **1995**, *502*, 169.

148. Cesarotti, E.; Ugo, R.; Kagan, H. B. *Angew. Chem. Int. Ed.* **1979**, *18*, 779.

149. Paquette, L. A.; McKinney, J. A.; McLaughlin, M. L.; Rheingold, A. L. *Tetrahedron Lett.* **1986**, *27*, 5599.

150. Halterman, R. L.; Vollhardt, K.P.C. *Organometallics* **1988**, *7*, 883.

151. Paquette, L. A.; Sivik, M. R.; Bzowej, E. I.; Stanton, K. J. *Organometallics* **1995**, *14*, 4865.

152. Gibis, K.-L.; Helmchen, G.; Huttner, G.; Zsolnai, L. *J. Organomet. Chem.* **1993**, *445*, 181.

153. Halterman, R. L.; Vollhardt, P. C.; Welker, M. E. *J. Am. Chem. Soc.* **1987**, *109*, 8105.

154. Waymouth, R.; Pino, P. *J. Am. Chem. Soc.* **1990**, *112*, 4911.

155. Conticello, V. P.; Brard, L.; Giardello, M. A.; Tsuji, Y.; Sabat, M.; Stern, C. L.; Marks, T. J. *J. Am. Chem. Soc.* **1992**, *114*, 2761.

156. Broene, R. D.; Buchwald, S. L. *J. Am. Chem. Soc.* **1993**, *115*, 12569.

157. Brunner, H.; Leitner, W. *Angew. Chem. Int. Ed.* **1988**, *27*, 1180.

158. Brunner, H.; Graf, E.; Leitner, W.; Wutz, K. *Synthesis* **1989**, 743.

159. Brown, J. M.; Brunner, H.; Leitner, W.; Rose, M. *Tetrahedron: Asymmetry* **1991**, *2*, 331.

160. Saburi, M.; Ohnuki, M.; Ogasawara, M.; Takahashi, T.; Uchida, Y. *Tetrahedron Lett.* **1992**, *33*, 5783.

161. Gonsalves, A. M. d'A. R.; Bayón, J. C.; Pereira, M. M.; Serra, M.E.S.; Pereira, J.P.R. *J. Organomet. Chem.* **1998**, *553*, 199.

162. (a) Brunner, H. In *Methods of Organic Chemistry (Houben-Weyl)*, 4th ed.: Helmchen, G.; Hoffmann, R. W.; Mulzer, J.; Schaumann, E. (Ed.); Thieme: Stuttgart, 1995; Vol. E21d, pp. 3945. (b) Ohkuma, T.; Noyori, R. In *Transition Metals for Organic Synthesis*; Beller, M., Bolm, C. (Ed.); Wiley-VCH: Weinheim, 1998; Vol. 2, Chapter 1.1.3. (c) Ohkuma, T.; Noyori, R. In *Comprehensive Asymmetric Catalysis*; Jacobsen, E. N., Pfaltz, A., Yamamoto, H. (Ed.); Springer; Berlin, 1999; Vol. 1, Chapter 6.1.

163. (a) Takahashi, H.; Morimoto, T.; Achiwa, K. *Chem. Lett.* **1987**, 855. (b) Inoguchi, K.; Sakuraba, S.; Achiwa, K. *Synlett* **1992**, 169.

164. Carpentier, J.-F.; Mortreux, A. *Tetrahedron: Asymmetry* **1997**, *8*, 1083.

165. Pasquier, C.; Naili, S.; Pelinski, L.; Brocard, J.; Mortreux, A.; Agbossou, F. *Tetrahedron: Asymmetry* **1998**, *9*, 193.

166. Agbossou, F.; Carpentier, J.-F.; Hatat, C.; Kokel, N.; Mortreux, A.; Betz, P.; Goddard, R.; Krüger, C. *Organometallics* **1995**, *14*, 2480.

167. Blaser, H.-U.; Jalett, H.-P.; Spindler, F. *J. Mol. Catal. A: Chemical* **1996**, *107*, 85.

168. Mashima, K.; Kusano, K.; Sato, N.; Matsumura, Y.; Nozaki, K.; Kumobayashi, H.; Sayo, N.; Hori, Y.; Ishizaki, T.; Akutagawa, S.; Takaya, H. *J. Org. Chem.* **1994**, *59*, 3064.

169. Kitamura, M.; Ohkuma, T.; Inoue, S.; Sayo, N.; Kumobayashi, H.; Akutagawa, S.; Ohta, T.; Takaya, H.; Noyori, R. *J. Am. Chem. Soc.* **1988**, *110*, 629.

170. Chiba, T.; Miyashita, A.; Nohira, H.; Takaya, H. *Tetrahedron Lett.* **1993**, *34*, 2351.

171. Genêt, J. P.; Pinel, C.; Ratovelomanana-Vidal, V.; Mallart, S.; Pfister, X.; Bischoff, L.; Caño De Andrade, M. C.; Darses, S.; Galopin, C.; Laffitte, J. A. *Tetrahedron: Asymmetry* **1994**, *5*, 675.

172. (a) Ojima, I.; Kogure, T.; Terasaki, T.; Achiwa, K. *J. Org. Chem.* **1978**, *43*, 3444. (b) Ojima, I.; Kogure, T.; Yoda, Y. *Org. Synth.* **1985**, *63*, 18.

173. Takahashi, H.; Hattori, M.; Chiba, M.; Morimoto, T.; Achiwa, K. *Tetrahedron Lett.* **1986**, *27*, 4477.

174. (a) Broger, E. A.; Crameri, Y. EP 0218970, **1987**. (b) Broger, E. A.; Crameri, Y. US 5,142,063, **1992**. (c) Schmid, R. *CHIMIA*, **1996**, *50*, 110.

175. Blaser, H.-U.; Pugin, B.; Spindler, F. In *Applied Homogeneous Catalysis with Organometallic Compounds*; Cornils, B.; Herrmann, W. A. (Ed.); VCH: Weinheim, 1996; Vol. 2, Chapter 3.3.1.

176. (a) Roucoux, A.; Devocelle, M.; Carpentier, J.-F.; Agbossou, F.; Mortreux, A. *Synlett* **1995**, 358. (b) Roucoux, A.; Thieffry, L.; Carpentier, J.-F.; Devocelle, M.; Méliet, C.; Agbossou, F.; Mortreux, A.; Welch, A. J. *Organometallics* **1996**, *15*, 2440.

177. Naota, T.; Takaya, H.; Murahashi, S. *Chem. Rev.* **1998**, *98*, 2599.

178. Noyori, R.; Ohkuma, T.; Kitamura, M.; Takaya, H.; Sayo, N.; Kumobayashi, H.; Akutagawa, S. *J. Am. Chem. Soc.* **1987**, *109*, 5856.

179. (a) Noyori, R. *Chem. Soc. Rev.* **1989**, *18*, 187. (b) Noyori, R. *Science* **1990**, *248*, 1194. (c) Noyori, R.; Takaya, H. *Acc. Chem. Res.* **1990**, *23*, 345. (d) Noyori, R. *CHEMTECH* **1992**, *22*, 360. (e) Noyori, R. *Tetrahedron* **1994**, *50*, 4259. (f) Noyori, R. In *Stereocontrolled Organic Synthesis*; Trost, B. M. (Ed.); Blackwell Scientific Publications: Oxford, 1994; pp. 1. (g) Noroyi, R. *Acta Chem. Scand.* **1996**, *50*, 380.

180. Ager, D. J.; Laneman, S. A. *Tetrahedron: Asymmetry* **1997**, *8*, 3327.

181. (a) Kitamura, M.; Tokunaga, M.; Ohkuma, T.; Noyori, R. *Tetrahedron Lett.* **1991**, *32*, 4163. (b) Kitamura, M.; Tokunaga, M.; Ohkuma, T.; Noyori, R. *Org. Synth.* **1993**, *71*, 1.

182. (a) Huang, H.-L.; Liu, L. T.; Chen, S.-F.; Ku, H. *Tetrahedron: Asymmetry* **1998**, *9*, 1637. (b) Gendre, P. L.; Offenbecher, M.; Bruneau, C.; Dixneuf, P. H. *Tetrahedron: Asymmetry* **1998**, *9*, 2279.

183. (a) Taber, D. F.; Silverberg, L. J. *Tetrahedron Lett.* **1991**, *32*, 4227. (b) Mashima, K.; Hino, T.; Takaya, H. *J. Chem. Soc., Dalton Trans.* **1992**, 2099. (c) King, S. A.; Thompson, A. S.; King, A. O.; Verhoeven, T. R. *J. Org. Chem.* **1992**, *57*, 6689. (d) Hoke, J. B.; Hollis, L. S.; Stern, E. W. *J. Organomet. Chem.* **1993**, *455*, 193. (e) Pathak, D. D.; Adams, H.; Bailey, N. A.; King, P. J.; White, C. *J. Organomet. Chem.* **1994**, *479*, 237. (f) Genêt, J. P.; Ratovelomanana-Vidal, V.; Caño de Andrade, M. C.; Pfister, X.; Guerreiro, P.; Lenoir, J. Y. *Tetrahedron Lett.* **1995**, *36*, 4801. (g) Doucet, H.; Gendre, P. L.; Bruneau, C.; Dixneuf, P. H.; Souvie, J.-C. *Tetrahedron: Asymmetry* **1996**, *7*, 525.

184. Murata, M.; Morimoto, T.; Achiwa, K. *Synlett* **1991**, 827.

185. Enev, V.; Ewers, C.L J.; Harre, M.; Nickisch, K.; Mohr, J. T. *J. Org. Chem.* **1997**, *62*, 7092.

186. Burk, M. J.; Harper, T. G. P.; Kalberg, C. S. *J. Am. Chem. Soc.* **1995**, *117*, 4423.

187. Pye, P. J.; Rossen, K.; Reamer, R. A.; Volante, R. P.; Reider, P. J. *Tetrahedron Lett.* **1998**, *39*, 4441.

188. Togni, A.; Breutel, C.; Schnyder, A.; Spindler, F.; Landert, H.; Tijani, A. *J. Am. Chem. Soc.* **1994**, *116*, 4062.

189. Tas, D.; Thoelen, C.; Vankelecom, I.F. J.; Jacobs, P. A. *Chem. Commun.* **1997**, 2323.

190. Ohkuma, T.; Kitamura, M.; Noyori, R. *Tetrahedron Lett.* **1990**, *31*, 5509.

191. Nishi, T.; Kataoka, M.; Morisawa, Y. *Chem. Lett.* **1989**, 1993.

192. (a) Schreiber, S. L.; Kelly, S. E.; Porco, Jr., J. A.; Sammakia, T.; Suh, E. M. *J. Am. Chem. Soc.* **1988**, *110*, 6210. (b) Nakatsuka, M.; Ragan, J. A.; Sammakia, T.; Smith, D. B.; Uehling, D. E.; Schreiber, S. L. *J. Am. Chem. Soc.* **1990**, *112*, 5583. (c) Case-Green, S. C.; Davies, S. G.; Hedgecock, C. J. R. *Synlett* **1991**, 781. (d) Taber, D. F.; Silverberg, L. J.; Robinson, E. D. *J. Am. Chem. Soc.* **1991**, *113*, 6639. (e) Baldwin, J. E.; Adlington, R. M.; Ramcharitar, S. H. *Synlett* **1992**, 875. (f) Taber, D. F.; Deker, P. B.; Silverberg, L. J. *J. Org. Chem.* **1992**, *57*, 5990. (g) Nozaki, K.; Sato, N.; Takaya, H. *Tetrahedron: Asymmetry* **1993**, *4*, 2179. (h) Garcia, D. M.; Yamada, H.; Hatakeyama, S.; Nishizawa, M. *Tetrahedron Lett.* **1994**, *35*, 3325. (i) Taber, D. F.; You, K. K. *J. Am. Chem. Soc.* **1995**, *117*, 5757. (j) Keegan, D. S.; Hagen, S. R.; Johnson, D. A. *Tetrahedron: Asymmetry* **1996**, *7*, 3559. (k) Spino, C.; Mayes, N.; Desfossés, H. *Tetrahedron Lett.* **1996**, *37*, 6503. (l) Balog, A.; Harris, C.; Savin, K.; Zhang, X.-G.; Chou, T. C.; Danishefsky, S. J. *Angew. Chem. Int. Ed.* **1998**, *37*, 2675. (m) Irako, N.;

Shioiri, T. *Tetrahedron Lett.* **1998**, *39*, 5793. (n) Baldwin, J. E.; Melman, A.; Lee, V.; Firkin, C. R.; Whitehead, R. C. *J. Am. Chem. Soc.* **1998**, *120*, 8559. (o) Romo, D.; Rzasa, R. M.; Shea, H. A.; Park, K.; Langenhan, J. M.; Sun, L.; Akhiezer, A.; Liu, J. O. *J. Am. Chem. Soc.* **1998**, *120*, 12237.

193. Kitamura, M.; Ohkuma, T.; Takaya, H.; Noyori, R. *Tetrahedron Lett.* **1988**, *29*, 1555.

194. Hapiot, F.; Agbossou, F.; Mortreux, A. *Tetrahedron: Asymmetry* **1997**, *8*, 2881.

195. Takeda, H.; Hosokawa, S.; Aburatani, M.; Achiwa, K. *Synlett* **1991**, 193.

196. (a) Rychnovsky, S. D.; Hoye, R. C. *J. Am. Chem. Soc.* **1994**, *116*, 1753. (b) Trost, B. M.; Hanson, P. R. *Tetrahedron Lett.* **1994**, *35*, 8119. (c) Beck, G.; Jendralla, H.; Kesseler, K. *Synthesis* **1995**, 1014.

197. Masamune, S.; Choy, W.; Petersen, J. S.; Sita, L. R. *Angew. Chem. Int. Ed.* **1985**, *24*, 1.

198. (a) Tani, K.; Tanigawa, E.; Tatsuno, Y.; Otsuka, S. *Chem. Lett.* **1986**, 737. (b) Tani, K.; Suwa, K.; Tanigawa, E.; Ise, T.; Yamagata, T.; Tatsuno, Y.; Otsuka, S. *J. Organomet. Chem.* **1989**, *370*, 203.

199. Nishi, T.; Kitamura, M.; Ohkuma, T.; Noyori, R. *Tetrahedron Lett.* **1988**, *29*, 6327.

200. Doi, T.; Kokubo, M.; Yamamoto, K.; Takahashi, T. *J. Org. Chem.* **1988**, *63*, 428.

201. Tohdo, K.; Hamada, Y.; Shioiri, T. *Synlett* **1994**, 105.

202. (a) Orito, Y.; Imai, S.; Niwa, S. *Nippon Kagaku Kaishi* **1979**, 1118. (b) Orito, Y.; Imai, S.; Niwa, S. *Nippon Kagaku Kaishi* **1980**, 670. (c) Niwa, S.; Imai, S.; Orito, Y. *Nippon Kagaku Kaishi* **1982**, 137.

203. (a) Webb, G.; Wells, P. B. *Catal. Today* **1992**, *12*, 319. (b) Blaser, H.-U.; Pugin, B. In *Chiral Reactions in Heterogeneous Catalysis*; Jannes, G., Dubois, V. (Ed.); Plenum: New York, 1995; pp. 33.

204. (a) Blaser, H.-U.; Jalett, H.-P.; Müller, M.; Studer, M. *Catal. Today* **1997**, *37*, 441. (b) Baiker, A. *J. Mol. Catal. A: Chemical* **1997**, *115*, 473. (c) Baiker, A.; Blaser, H. U. In *Handbook of Heterogeneous Catalysis*; Ertl, G., Knözinger, H., Weitkamp, J. (Ed.); VCH: Weinheim, 1997; Vol. 5, Chapter 4.14. (d) Wells, P. B.; Wilkinson, A. G. *Topics Catal.* **1998**, *5*, 39.

205. Török, B.; Felföldi, K.; Szakonyi, G.; Balázsik, K.; Bartók, M. *Catal. Lett.* **1998**, *52*, 81.

206. Blaser, H. U.; Jalett, H. P.; Wiehl, J. *J. Mol. Catal.* **1991**, *68*, 215.

207. Blaser, H. U.; Jalett, H. P. In *Heterogeneous Catalysis and Fine Chemicals III*; Guisnet, M. et al. (Ed.); Elsevier: Amsterdam, 1993; pp 139.

208. Minder, B.; Schürch, M.; Mallat, T.; Baiker, A.; Heinz, T.; Pfaltz, A. *J. Catal.* **1996**, *160*, 261.

209. (a) Pfaltz, A.; Heinz, T. *Topics Catal.* **1997**, *4*, 229. (b) Schürch, M.; Heinz, T.; Aeschimann, R.; Mallat, T.; Pfaltz, A.; Baiker, A. *J. Catal.* **1998**, *173*, 187.

210. Blaser, H. U.; Jalett, H. P.; Monti, D. M.; Baiker, A.; Wehrli, J. T. In *Structure-Activity and Selectivity Relationships in Heterogeneous Catalysis*; Grasselli, R. K., Sleight, A. W. (Ed.); Elsevier: Amsterdam, 1991; pp. 147.

211. (a) Garland, M.; Blaser, H.-U. *J. Am. Chem. Soc.* **1990**, *112*, 7048. (b) Simons, K. E.; Meheux, P. A.; Griffiths, S. P.; Sutherland, I. M.; Johnston, P.; Wells, P. B.; Carley, A. F.; Rajumon, M. K.; Roberts, M. W.; Ibbotson, A. *Recl. Trav. Chim. Pays-Bas* **1994**, *113*, 465. (c) Augustine, R. L.; Tanielyan, S. K. *J. Mol. Catal. A: Chemical* **1996**, *112*, 93. (d) Margitfalvi, J. L.; Hegedüs, M.; Tfirst, E. *Stud. Surf. Sci. Catal.* **1996**, *101*, 241. (e) Margitfalvi, J. L.; Hegedüs, M.; Tfirst, E. *Tetrahedron: Asymmetry* **1996**, *7*, 571. (f) Blaser, H.-U.; Jalett, H.-P.; Garland, M.; Studer, M.; Thies, H.; Wirth-Tijani, A. *J. Catal.* **1998**, *173*, 282.

212. (a) Bönnemann, H.; Braun, G. A. *Angew. Chem. Int. Ed.* **1996**, *35*, 1992. (b) Bönnemann, H.; Braun, G. A. *Chem. Eur. J.* **1997**, *3*, 1200. (c) Köhler, J. U.; Bradley, J. S. *Catal. Lett.* **1997**, *45*, 203.

213. (a) Schürch, M.; Künzle, N.; Mallat, T.; Baiker, A. *J. Catal.* **1998**, *176*, 569. (b) Künzle, N.; Szabo, A.; Schürch, M.; Wang, G.; Mallat, T.; Baiker, A. *Chem. Commun.* **1998**, 1377.

214. (a) Izumi, Y. *Adv. Catal.* **1983**, *32*, 215. (b) Tai, A.; Harada, T. In *Tailored Metal Catalysts*; Iwasawa, Y. (Ed.); Reidel: Dordrecht, 1986; pp. 265. (c) Osawa, T.; Harada, T.; Tai, A. *J. Catal.* **1990**, *121*, 7. (d) Tai, A.; Kikukawa, T.; Sugimura, T.; Inoue, Y.; Abe, S.; Osawa, T.; Harada, T. *Bull. Chem. Soc. Jpn.* **1994**, *67*, 2473. (e) Sugimura, T.; Osawa, T.; Nakagawa, S.; Harada, T.; Tai, A. *Stud. Surf. Sci. Catal.* **1996**, *101*, 231. (f) Nakagawa, S.; Sugimura, T.; Tai, A. *Chem. Lett.* **1997**, 859. (g) Osawa, T.; Mita, S.; Iwai, A.; Miyazaki, T.; Takayasu, O.; Harada, T.; Matsuura, I. *Chem. Lett.* **1997**, 1131.

215. (a) Petrov, Y. I.; Klabnovskii, E. I.; Balandin, A. A. *Kinet. Katal.* **1967**, *8*, 814. (b) Nitta, Y.; Utsumi, T.; Imanaka, T.; Teranishi, S. *J. Catal.* **1986**, *101*, 376. (c) Fu, L.; Kung, H. H.; Sachtler, W.M.H. *J. Mol. Catal.* **1987**, *42*, 29. (d) Wittmann, G.; Bartók, G. B.; Bartók, M.; Smith, G. V. *J. Mol. Catal.* **1990**, *60*, 1. (e) Brunner, H.; Muschiol, M.; Wischert, T. *Tetrahedron: Asymmetry* **1990**, *1*, 159. (f) Webb, G. In *Chiral Reactions in Heterogeneous Catalysis*; Jannes, G., Dubois, V. (Ed.); Plenum: New York, 1995; pp. 61.

216. (a) Schildknecht, H.; Koob, K. *Angew. Chem.* **1971**, *83*, 110. (b) Shiba, T.; Kusumoto, S. *J. Synth. Org. Chem. Jpn.* **1988**, *46*, 501. (c) Yoshikawa, M.; Sugimura, T.; Tai, A. *Agric. Biol. Chem.* **1989**, *53*, 37. (d) Tai, A.; Morimoto, N.; Yoshikawa, M.; Uehara, K.; Sugimura, T.; Kikukawa, T. *Agric. Biol. Chem.* **1990**, *54*, 1753. (e) Kikukawa, T.; Tai, A. *Shokubai*, **1992**, *34*, 182.

217. Yoshikawa, K.; Yamamoto, N.; Murata, M.; Awano, K.; Morimoto, T.; Achiwa, K. *Tetrahedron: Asymmetry* **1992**, *3*, 13.

218. Devocelle, M.; Agbossou, F.; Mortreux, A. *Synlett* **1997**, 1306.

219. Takeda, H.; Tachinami, T.; Aburatani, M.; Takahashi, H.; Morimoto, T.; Achiwa, K. *Tetrahedron Lett.* **1989**, *30*, 363.

220. Hayashi, T.; Katsumura, A.; Konishi, M.; Kumada, M. *Tetrahedron Lett.* **1979**, 425.

221. Törös, S.; Kollár, L.; Heil, B.; Markó, L. *J. Organomet. Chem.* **1982**, *232*, C17.

222. (a) Sakuraba, S.; Achiwa, K. *Synlett* **1991**, 689. (b) Sakuraba, S.; Nakajima, N.; Achiwa, K. *Synlett* **1992**, 829.

223. Tombo, G. M. R.; Belluš, D. *Angew. Chem. Int. Ed.* **1991**, *30*, 1193.

224. (a) Tranchier, J.-P.; Ratovelomanana-Vidal, V.; Genêt, J.-P.; Tong, S.; Cohen, T. *Tetrahedron Lett.* **1997**, *38*, 2951. (b) Blanc, D.; Henry, J.-C.; Ratovelomanana-Vidal, V.; Genêt, J.-P. *Tetrahedron Lett.* **1997**, *38*, 6603.

225. Takahashi, H.; Sakuraba, S.; Takeda, H.; Achiwa, K. *J. Am. Chem. Soc.* **1990**, *112*, 5876.

226. (a) Cesarotti, E.; Mauri, A.; Pallavicini, M.; Villa, L. *Tetrahedron Lett.* **1991**, *32*, 4381. (b) Cesarotti, E.; Antognazza, P.; Pallavicini, M.; Villa, L. *Helv. Chim. Acta* **1993**, *76*, 2344. (c) Buser, H.-P.; Spindler, F. *Tetrahedron: Asymmetry* **1993**, *4*, 2451.

227. (a) Sakuraba, S.; Takahashi, H.; Takeda, H.; Achiwa, K. *Chem. Pharm. Bull.* **1995**, *43*, 738. (b) Sakuraba, S.; Achiwa, K. *Chem. Pharm. Bull.* **1995**, *43*, 748.

228. Noyori, R.; Hashiguchi, S. In *Applied Homogeneous Catalysis with Organometallic Compounds*; Cornils, B., Herrmann, W. A. (Ed.); VCH: Weinheim, 1996; Vol. 1, Chapter 2.9.

229. Märki, H. P.; Crameri, Y.; Eigenmann, R.; Krasso, A.; Ramuz, H.; Bernauer, K.; Goodman, M.; Melmon, K. L. *Helv. Chim. Acta* **1988**, *71*, 320.

230. (a) Ohgo, Y.; Natori, Y.; Takeuchi, S.; Yoshimura, J. *Chem. Lett.* **1974**, 1327. (b) Ohgo, Y.; Takeuchi, S.; Natori, Y.; Yoshimura, J. *Bull. Chem. Soc. Jpn.* **1981**, *54*, 2124.

231. Waldron, R. W.; Weber, J. H. *Inorg. Chem.* **1977**, *16*, 1220.

232. Kawano, H.; Ishii, Y.; Saburi, M.; Uchida, Y. *J. Chem. Soc., Chem. Commun.* **1988**, 87.

233. Mezzetti, A.; Tschumper, A.; Consiglio, G. *J. Chem. Soc., Dalton Trans.* **1995**, 49.

234. Brunner, H.; Terfort, A. *Tetrahedron: Asymmetry* **1995**, *6*, 919.

235. Shao, L.; Kawano, H.; Saburi, M.; Uchida, Y. *Tetrahedron* **1993**, *49*, 1997.

236. Blandin, V.; Carpentier, J.-F.; Mortreux, A. *Tetrahedron: Asymmetry* **1998**, *9*, 2765.

237. Pini, D.; Mandoli, A.; Iuliano, A.; Salvadori, P. *Tetrahedron: Asymmetry* **1995**, *6*, 1031.

238. Rychnovsky, S. D.; Griesgraber, G.; Zeller, S.; Skalitzky, D. J. *J. Org. Chem.* **1991**, *56*, 5161.

239. (a) Poss, C. S.; Rychnovsky, S. D.; Schreiber, S. L. *J. Am. Chem. Soc.* **1993**, *115*, 3360. (b) Rychnovsky, S. D.; Khire, U. R.; Yang, G. *J. Am. Chem. Soc.* **1997**, *119*, 2058. (c) Rychnovsky, S. D.; Yang, G.; Hu, Y.; Khire, U. R. *J. Org. Chem.* **1997**, *62*, 3022.

240. Fan, Q.; Yeung, C.; Chan, A.S.C. *Tetrahedron: Asymmetry* **1997**, *8*, 4041.

241. (a) Tai, A.; Kikukawa, T.; Sugimura, T.; Inoue, Y.; Osawa, T.; Fujii, S. *J. Chem. Soc., Chem. Commun.* **1991**, 795. (b) Brunner, H.; Amberger, K.; Wiehl, J. *Bull. Soc. Chim. Belg.* **1991**, *100*, 571.

242. (a) Sugimura, T.; Futagawa, T.; Tai, A. *Chem. Lett.* **1990**, 2295. (b) Sugimura, T.; Tai, A.; Koguro, K. *Tetrahedron* **1994**, *50*, 11647.

243. (a) Kitamura, M.; Tokunaga, M.; Noyori, R. *J. Am. Chem. Soc.* **1995**, *117*, 2931. (b) Gautier, I.; Ratovelomanana-Vidal, V.; Savignac, P.; Genêt, J.-P. *Tetrahedron Lett.* **1996**, *37*, 7721.

244. Kitamura, M.; Yoshimura, M.; Kanda, N.; Noyori, R. *Tetrahedron* **1999**, *55*, 8769.

245. Hiraki, Y.; Ito, K.; Harada, T.; Tai, A. *Chem. Lett.* **1981**, 131.

246. (a) Noyori, R.; Ikeda, T.; Ohkuma, T.; Widhalm, M.; Kitamura, M.; Takaya, H.; Akutagawa, S.; Sayo, N.; Saito, T.; Taketomi, T.; Kumobayashi, H. *J. Am. Chem. Soc.* **1989**, *111*, 9134. (b) Kitamura, M.; Ohkuma, T.; Tokunaga, M.; Noyori, R. *Tetrahedron: Asymmetry* **1990**, *1*, 1.

247. Genêt, J. P.; Pfister, X.; Ratovelomanana-Vidal, V.; Pinel, C.; Laffitte, J. A. *Tetrahedron Lett.* **1994**, *35*, 4559.

248. (a) Kitamura, M.; Tokunaga, M.; Noyori, R. *J. Am. Chem. Soc.* **1993**, *115*, 144. (b) Kitamura, M.; Tokunaga, M.; Noyori, R. *Tetrahedron* **1993**, *49*, 1853. (c) Noyori, R.; Tokunaga, M.; Kitamura, M. *Bull. Chem. Soc. Jpn.* **1995**, *68*, 36.

249. Genêt, J. P.; Pinel, C.; Mallart, S.; Juge, S.; Thorimbert, S.; Laffitte, J. A. *Tetrahedron: Asymmetry* **1991**, *2*, 555.

250. Genêt, J.-P.; Caño de Andrade, M. C.; Ratovelomanana-Vidal, V. *Tetrahedron Lett.* **1995**, *36*, 2063.

251. Kitamura, M.; Tokunaga, M.; Pham, T.; Lubell, W. D.; Noyori, R. *Tetrahedron Lett.* **1995**, *36*, 5769.

252. (a) Fukuda, N.; Mashima, K.; Matsumura, Y.; Takaya, H. *Tetrahedron Lett.* **1990**, *31*, 7185. (b) Schmidt, U.; Leitenberger, V.; Griesser, H.; Schmidt, J.; Meyer, R. *Synthesis* **1992**, 1248. (c) Wovkulich, P. M.; Shankaran, K.; Kiegiel, J.; Uskoković, M. R. *J. Org. Chem.* **1993**, *58*, 832. (d) Heathcock, C. H.; Kath, J. C.; Ruggeri, R. B. *J. Org. Chem.* **1995**, *60*, 1120. (e) Ohtake, H.; Yonishi, S.; Tsutsumi, H.; Murata, M. *Abstracts of Papers, 69th National Meeting of the Chemical Society of Japan*, Kyoto, Chemical Society of Japan, Tokyo; 1995, p. 1030, 1H107. (f) Nishizawa, M.; García, D. M.; Minagawa, R.; Noguchi, Y.; Imagawa, H.; Yamada, H.; Watanabe, R.; Yoo, Y. C.; Azuma, I. *Synlett* **1996**, 452. (g) Taber, D. F.; Wang, Y. *J. Am. Chem. Soc.* **1997**, *119*, 22. (h) Coulon, E.; Cristina, M.; Caño de Andrade, M. C.; Ratovelomanana-Vidal, V.; Genêt, J.-P. *Tetrahedron Lett.* **1998**, *39*, 6467.

253. Akutagawa, S. In *Chirality in Industry*; Collins, A. N., Sheldrake, G. N., Crosby, J. (Ed.); Wiley: Chichester, 1992; Chapter 17.

254. Fehring, V.; Selke, R. *Angew. Chem. Int. Ed.* **1998**, *37*, 1827.

255. Bennett, M. A.; Matheson, T. W. In *Comprehensive Organometallic Chemistry*; Wilkinson, G., Stone, F.G.A., Abel, E. W. (Ed.); Pergamon Press: Oxford, 1982; Vol. 4, Chapter 32.9.

256. Doucet, H.; Ohkuma, T.; Murata, K.; Yokozawa, T.; Kozawa, M.; Katayama, E.; England, A. F.; Ikariya, T.; Noyori, R. *Angew. Chem. Int. Ed.* **1998**, *37*, 1703.

257. Ohkuma, T.; Ooka, H.; Hashiguchi, S.; Ikariya, T.; Noyori, R. *J. Am. Chem. Soc.* **1995**, *117*, 2675.

258. Ohkuma, T.; Koizumi, M.; Doucet, H.; Pham, T.; Kozawa, M.; Murata, K.; Katayama, E.; Yokozawa, T.; Ikariya, T.; Noyori, R. *J. Am. Chem. Soc.* **1998**, *120*, 13529.

259. Törös, S.; Heil, B.; Kollár, L.; Markó, L. *J. Organomet. Chem.* **1980**, *197*, 85.

260. Bakos, J.; Tóth, I.; Heil, B.; Szalontai, G.; Párkányi, L.; Fülöp, V. *J. Organomet. Chem.* **1989**, *370*, 263.

261. Jiang, Q.; Jiang, Y.; Xiao, D.; Cao, P.; Zhang, X. *Angew. Chem. Int. Ed.* **1998**, *37*, 1100.

262. Zhang, X.; Kumobayashi, H.; Takaya, H. *Tetrahedron: Asymmetry* **1994**, *5*, 1179.

263. Zhang, X.; Taketomi, T.; Yoshizumi, T.; Kumobayashi, H.; Akutagawa, S.; Mashima, K.; Takaya, H. *J. Am. Chem. Soc.* **1993**, *115*, 3318.

264. Ramachandran, P. V.; Teodorović, A. V.; Brown, H. C. *Tetrahedron* **1993**, *49*, 1725.

265. Iseki, K. *Tetrahedron* **1998**, *54*, 13887.

266. Mallat, T.; Bodmer, M.; Baiker, A. *Catal. Lett.* **1997**, *44*, 95.

267. Nagel, U.; Roller, C. Z. *Naturforsch., Ser. B* **1998**, *53*, 267.

268. Corey, E. J.; Helal, C. J. *Tetrahedron Lett.* **1995**, *36*, 9153.

269. (a) Osawa, T. *Chem. Lett.* **1985**, 1609. (b) Osawa, T.; Harada, T.; Tai, A. *J. Mol. Catal.* **1994**, *87*, 333. (c) Osawa, T.; Tai, A.; Imachi, Y.; Takasaki, S. In *Chiral Reactions in Heterogeneous Catalysis*; Jannes, G., Dubois, V. (Ed.); Plenum: New York, 1995; pp. 75. (d) Harada, T.; Osawa, T. In *Chiral Reactions in Heterogeneous Catalysis*; Jannes, G., Dubois, V. (Ed.); Plenum: New York, 1995; pp. 83. (e) Osawa, T.; Harada, T.; Tai, A.; Takayasu, O.; Matsuura, I. In *Heterogeneous Catalysis and Fine Chemicals IV*; Blaser, H. U., Baiker, A., Prins, R. (Ed.); Elsevier: Amsterdam, 1997; pp. 199. (f) Osawa, T.; Harada, T.; Tai, A. *Catal. Today* **1997**, *37*, 465.

270. (a) Augustine, R. L. *Adv. Catal.* **1976**, *25*, 56, (b) Birch, A. J.; Williamson, D. H. *Org. React. (NY)* **1976**, *24*, 1. (c) Rylander, P. *Catalytic Hydrogenation in Organic Syntheses*; Academic Press: New York, 1979. (d) Siegel, S. In *Comprehensive Organic Synthesis*; Trost, B. M., Fleming, I. (Ed.); Pergamon: Oxford, 1991; Vol. 8, Chapter 3.1. (e) Takaya, H.; Noyori, R. In *Comprehensive Organic Synthesis*; Trost, B. M., Fleming, I. (Ed.); Pergamon: Oxford, 1991; Vol. 8, Chapter 3.2. (f) Keinan, E.; Greenspoon, N. In *Comprehensive Organic Synthesis*; Trost, B. M., Fleming, I. (Ed.); Pergamon: Oxford, 1991; Vol. 8, Chapter 3.5. (g) Hudlický, M. *Reductions in Organic Chemistry*, 2nd ed.; American Chemical Society: Washington, DC, 1996; Chapter 13.

271. Ohkuma, T.; Ooka, H.; Ikariya, T.; Noyori, R. *J. Am. Chem. Soc.* **1995**, *117*, 10417.

272. (a) Chan, K.-K.; Cohen, N.; De Noble, J. P.; Specian, Jr., A. C.; Saucy, G. *J. Org. Chem.* **1976**, *41*, 3497. (b) Terashima, S.; Tanno, N.; Koga, K. *Tetrahedron Lett.* **1980**, *21*, 2753.

273. Mashima, K.; Akutagawa, T.; Zhang, X.; Takaya, H.; Taketomi, T.; Kumobayashi, H.; Akutagawa, S. *J. Organomet. Chem.* **1992**, *428*, 213.

274. Spogliarich, R.; Vidotto, S.; Farnetti, E.; Graziani, M.; Gulati, N. V. *Tetrahedron: Asymmetry* **1992**, *3*, 1001.

275. Ohkuma, T.; Ikehira, H.; Ikariya, T.; Noyori, R. *Synlett* **1997**, 467.

276. Ohkuma, T.; Doucet, H.; Pham, T.; Mikami, K.; Korenaga, T.; Terada, M.; Noyori, R. *J. Am. Chem. Soc.* **1998**, *120*, 1086.

277. (a) Mori, K.; Puapoomchareon, P. *Liebigs Ann. Chem.* **1991**, 1053. (b) Croteau, R.; Karp, F. In *Perfumes: Art, Science and Technology*; Müller, P. M., Lamparsky, D. (Ed.); Blackie Academic & Professional: London, 1991; Chapter 4.

278. Ohkuma, T.; Ooka, H.; Yamakawa, M.; Ikariya, T.; Noyori, R. *J. Org. Chem.* **1996**, *61*, 4872.

279. (a) Matsukawa, S.; Mikami, K. *Enantiomer* **1996**, *1*, 69. (b) Mikami, K.; Matsukawa, S. *Nature* **1997**, *385*, 613. (c) Matsukawa, S.; Mikami, K. *Tetrahedron: Asymmetry* **1997**, *8*, 815. (d) Volk, T.; Korenaga, T.; Matsukawa, S.; Terada, M.; Mikami, K. *CHIRALITY* **1998**, *10*, 717.

280. Ohta, T.; Tsutsumi, T.; Takaya, H. *J. Organomet. Chem.* **1994**, *484*, 191.

281. (a) Wilds, A. L. *Org. React. (N.Y.)* **1944**, *2*, 178. (b) Djerassi, C. *Org. React. (N.Y.)* **1951**, *6*, 207. (c) Krohn, K. In *Methods of Organic Chemistry (Houben-Weyl)*, 4th ed.; Helmchen, G., Hoffmann, R. W., Mulzer, J., Schaumann, E. (Ed.); Thieme: Stuttgart, 1995; Vol. E21d, Chapter 2.3.5.2.

282. Chaloner, P. A.; Esteruelas, M. A.; Joó, F.; Oro, L. A. *Homogeneous Hydrogenation*; Kluwer Academic Publishers: Dordrecht, 1994; Chapter 3.

283. (a) Watanabe, Y.; Ohta, T.; Tsuji, Y. *Bull. Chem. Soc. Jpn.* **1982**, *55*, 2441. (b) Nakano, T.; Ando, J.; Ishii, Y.; Ogawa, M. *Tech. Rep. Kansai Univ.* **1987**, *29*, 69.

284. (a) Moulton, W. N.; van Atta, R. E.; Ruch, R. R. *J. Org. Chem.* **1961**, *26*, 290. (b) Shiner, Jr., V. J.; Whittaker, D. *J. Am. Chem. Soc.* **1969**, *91*, 394. (c) Nasipuri, D.; Gupta, M. D.; Banerjee, S. *Tetrahedron Lett.* **1984**, *25*, 5551.

285. (a) Namy, J. L.; Souppe, J.; Collin, J.; Kagan, H. B. *J. Org. Chem.* **1984**, *49*, 2045. (b) Lebrun, A.; Namy, J.-L.; Kagan, H. B. *Tetrahedron Lett.* **1991**, *32*, 2355. (c) Huskens, J.; de Graauw, C. F.; Peters, J. A.; van Bekkum, H. *Recl. Trav. Chim. Pays-Bas* **1994**, *113*, 488.

286. Ooi, T.; Miura, T.; Maruoka, K. *Angew. Chem. Int. Ed.* **1998**, *37*, 2347.

287. Evans, D. A.; Nelson, S. G.; Gagné, M. R.; Muci, A. R. *J. Am. Chem. Soc.* **1993**, *115*, 9800.

288. Morton, D.; Cole-Hamilton, D. J.; Utuk, I. D.; Paneque-Sosa, M.; Lopez-Poveda, M. *J. Chem. Soc., Dalton Trans.* **1989**, 489.

289. (a) Chowdhury, R. L.; Bäckvall, J.-E. *J. Chem. Soc., Chem. Commun.* **1991**, 1063. (b) Bäckvall, J.-E.; Chowdhury, R. L.; Karlsson, U.; Wang, G. In *Perspectives in Coordination Chemistry*; Williams, A. F., Floriani, C., Merbach, A. E. (Ed.); VHCA: Basel, 1992; pp. 463.

290. (a) Bianchi, M.; Matteoli, U.; Frediani, P.; Menchi, G.; Piacenti, F. *J. Organomet. Chem.* **1982**, *236*, 375. (b) Chauvin, R. *J. Mol. Catal.* **1990**, *62*, 147. (c) Genêt, J.-P.; Ratovelomanana-Vidal, V.; Pinel, C. *Synlett* **1993**, 478. (d) Bhaduri, S.; Sharma, K.; Mukesh, D. *J. Chem. Soc., Dalton Trans.* **1993**, 1191. (e) Jiang, Q.; van Plew, D.; Murtuza, S.; Zhang, X. *Tetrahedron Lett.* **1996**, *37*, 797. (f) Barbaro, P.; Bianchini, C.; Togni, A. *Organometallics* **1997**, *16*, 3004.

291. Krasik, P.; Alper, H. *Tetrahedron* **1994**, *50*, 4347.

292. Hashiguchi, S.; Fujii, A.; Takehara, J.; Ikariya, T.; Noyori, R. *J. Am. Chem. Soc.* **1995**, *117*, 7562.

293. Noyori, R.; Hashiguchi, S. *Acc. Chem. Res.* **1997**, *30*, 97.

294. Püntener, K.; Schwink, L.; Knochel, P. *Tetrahedron Lett.* **1996**, *37*, 8165.

295. Schwink, L.; Ireland, T.; Püntener, K.; Knochel, P. *Tetrahedron: Asymmetry* **1998**, *9*, 1143.

296. Jiang, Y.; Jiang, Q.; Zhang, X. *J. Am. Chem. Soc.* **1998**, *120*, 3817.

297. (a) Silverthorn, W. E. *Adv. Organomet. Chem.* **1975**, *13*, 47. (b) Muetterties, E. L.; Bleeke, J. R.; Wucherer, E. J.; Albright, T. A. *Chem. Rev.* **1982**, *82*, 499.

298. Takehara, J.; Hashiguchi, S.; Fujii, A.; Inoue, S.; Ikariya, T.; Noyori, R. *Chem. Commun.* **1996**, 233.

299. (a) Palmer, M.; Walsgrove, T.; Wills, M. *J. Org. Chem.* **1997**, *62*, 5226. (b) Alonso, D. A.; Guijarro, D.; Pinho, P.; Temme, O.; Andersson, P. G. *J. Org. Chem.* **1998**, *63*, 2749.

300. Jiang, Y.; Jiang, Q.; Zhu, G.; Zhang, X. *Tetrahedron Lett.* **1997**, *38*, 6565.

301. Touchard, F.; Gamez, P.; Fache, F.; Lemaire, M. *Tetrahedron Lett.* **1997**, *38*, 2275.

302. Langer, T.; Helmchen, G. *Tetrahedron Lett.* **1996**, *37*, 1381.

303. Yang, H.; Alvarez, M.; Lugan, N.; Mathieu, R. *J. Chem. Soc., Chem. Commun.* **1995**, 1721.

304. Sammakia, T.; Stangeland, E. L. *J. Org. Chem.* **1997**, *62*, 6104.

305. Gao, J.-X.; Ikariya, T.; Noyori, R. *Organometallics* **1996**, *15*, 1087.

306. Müller, D.; Umbricht, G.; Weber, B.; Pfaltz, A. *Helv. Chim. Acta* **1991**, *74*, 232.

307. ter Halle, R.; Bréhéret, A.; Schulz, E.; Pinel, C.; Lemaire, M. *Tetrahedron: Asymmetry* **1997**, *8*, 2101.

308. Inoue, S.; Nomura, K.; Hashiguchi, S.; Noyori, R.; Izawa, Y. *Chem. Lett.* **1997**, 957.

309. Zassinovich, G.; Bettella, R.; Mestroni, G.; Bresciani-Pahor, N.; Geremia, S.; Randaccio, L. *J. Organomet. Chem.* **1989**, *370*, 187.

310. Gladiali, S.; Pinna, L.; Delogu, G.; De Martin, S.; Zassinovich, G.; Mestroni, G. *Tetrahedron: Asymmetry* **1990**, *1*, 635.

311. (a) Mashima, K.; Abe, T.; Tani, K. *Chem. Lett.* **1998**, 1199. (b) Mashima, K.; Abe, T.; Tani, K. *Chem. Lett.* **1998**, 1201.

312. Gamez, P.; Dunjic, B.; Lemaire, M. *J. Org. Chem.* **1996**, *61*, 5196.

313. (a) Gamez, P.; Fache, F.; Mangeney, P.; Lemaire, M. *Tetrahedron Lett.* **1993**, *34*, 6897. (b) Gamez, P.; Fache, F.; Lemaire, M. *Tetrahedron: Asymmetry* **1995**, *6*, 705.

314. Kvintovics, P.; James, B. R.; Heil, B. *J. Chem. Soc., Chem. Commun.* **1986**, 1810.

315. Nishibayashi, Y.; Singh, J. D.; Arikawa, Y.; Uemura, S.; Hidai, M. *J. Organomet. Chem.* **1997**, *531*, 13.

316. Jiang, Y.; Jiang, Q.; Zhu, G.; Zhang, X. *Tetrahedron Lett.* **1997**, *38*, 215.

317. Haack, K.-J.; Hashiguchi, S.; Fujii, A.; Ikariya, T.; Noyori, R. *Angew. Chem. Int. Ed.* **1997**, *36*, 285.

318. (a) Adkins, H.; Elofson, R. M.; Rossow, A. G.; Robinson, C. C. *J. Am. Chem. Soc.* **1949**, *71*, 3622. (b) Hach, V. *J. Org. Chem.* **1973**, *38*, 293.

319. (a) Jessop, P. G.; Ikariya, T.; Noyori, R. *Chem. Rev.* **1995**, *95*, 259. (b) Leitner, W. *Angew. Chem. Int. Ed.* **1995**, *34*, 2207.

320. (a) Wagner, V. K. *Angew. Chem. Int. Ed.* **1970**, *9*, 50. (b) Narita, K.; Sekiya, M. *Chem. Pharm. Bull.* **1977**, *25*, 135.

321. Fujii, A.; Hashiguchi, S.; Uematsu, N.; Ikariya, T.; Noyori, R. *J. Am. Chem. Soc.* **1996**, *118*, 2521.

322. Bayston, D. J.; Travers, C. B.; Polywka, M.E.C. *Tetrahedron: Asymmetry* **1998**, *9*, 2015.

323. King, A. O.; Corley, E. G.; Anderson, R. K.; Larsen, R. D.; Verhoeven, T. R.; Reider, P. J.; Xiang, Y. B.; Belley, M.; Leblanc, Y.; Labelle, M.; Prasit, P.; Zamboni, R. J. *J. Org. Chem.* **1993**, *58*, 3731.

324. Ohta, T.; Nakahara, S.; Shigemura, Y.; Hattori, K.; Furukawa, I. *Chem. Lett.* **1998**, 491.

325. Jones, T. K.; Mohan, J. J.; Xavier, L. C.; Blacklock, T. J.; Mathre, D. J.; Sohar, P.; Jones, E. T. T.; Reamer, R. A.; Roberts, F. E.; Grabowski, E. J. J. *J. Org. Chem.* **1991**, *56*, 763.

326. De Martin, S.; Zassinovich, G.; Mestroni, G. *Inorg. Chim. Acta* **1990**, *174*, 9.

327. Matsumura, K.; Hashiguchi, S.; Ikariya, T.; Noyori, R. *J. Am. Chem. Soc.* **1997**, *119*, 8738.

328. Lebel, H.; Jacobsen, E. N. *J. Org. Chem.* **1998**, *63*, 9624.

329. Everaere, K.; Carpentier, J.-F.; Mortreux, A.; Bulliard, M. *Tetrahedron: Asymmetry* **1998**, *9*, 2971.

330. Hashiguchi, S.; Fujii, A.; Haack, K.-J.; Matsumura, K.; Ikariya, T.; Noyori, R.*Angew. Chem. Int. Ed.* **1997**, *36*, 288.

331. (a) Takano, S.; Moriya, M.; Higashi, Y.; Ogasawara, K. *J. Chem. Soc., Chem. Commun.* **1993**, 177. (b) Takano, S.; Moriya, M.; Ogasawara, K. *J. Chem. Soc., Chem. Commun.* **1993**, 614.

332. (a) Jones, J. B. *Tetrahedron* **1986**, *42*, 3351. (b) Crout, D.H.G.; Christen, M. In *Modern Synthetic Methods*; Scheffold, R. (Ed.); Springer: Berlin, 1989; Vol. 5, pp. 1. (c) Santaniello, E.; Ferraboschi, P.; Grisenti, P.; Manzocchi, A. *Chem. Rev.* **1992**, *92*, 1071. (d) Schoffers, E.; Golebiowski. A.; Johnson, C. R. *Tetrahedron* **1996**, *52*, 3769.

333. (a) Dinh, P. M.; Howarth, J. A.; Hudnott, A. R.; Williams, J.M.J.; Harris, W. *Tetrahedron Lett.* **1996**, *37*, 7623. (b) Larsson, A.L.E.; Persson, B. A.; Bäckvall, J.-E. *Angew. Chem. Int. Ed.* **1997**, *36*, 1211.

334. Koh, J. H.; Jeong, H. M.; Park, J. *Tetrahedron Lett.* **1998**, *39*, 5545.

335. (a) Bolm, C. *Angew. Chem. Int. Ed.* **1993**, *32*, 232. (b) Martens, J. In *Methods of Organic Chemistry (Houben-Weyl)*, 4th ed.: Helmchen, G.; Hoffmann, R. W., Mulzer, J., Schaumann, E. (Ed.); Thieme: Stuttgart, 1995; Vol. E21d, Chapter 2.4.1. (c) James, B. R. *Chem. Ind.* **1995**, *62*, 167. (d) Johansson, A. *Contemp. Org. Synth.* **1995**, 393. (e) Yurovskaya, M. A.; Karchava, A. V. *Tetrahedron: Asymmetry* **1998**, *9*, 3331. (f) Spindler, F.; Blaser, H.-U. In *Transition Metals for Organic Synthesis*; Beller, M., Bolm, C. (Ed.); Wiley-VCH: Weinheim, 1998; Vol. 2, Chapter 1.1.4.

336. (a) McCarty, C. G. In *The Chemistry of the Carbon–Nitrogen Double Bond*; Patai, S. (Ed.); Wiley: London, 1970; Chapter 9. (b) Bjørgo, J.; Boyd, D. R.; Watson, C. G.; Jennings, W. B. *J. Chem. Soc., Perkin Trans. 2* **1974**, 757. (c) Johnson, G. P.; Marples, B. A. *Tetrahedron Lett.* **1984**, *25*, 3359.

337. Vastag, S.; Bakos, J.; Törös, S.; Takach, N. E.; King, R. B.; Heil, B.; Markó, L. *J. Mol. Catal.* **1984**, *22*, 283.

338. (a) Kang, G.-J.; Cullen, W. R.; Fryzuk, M. D.; James, B. R.; Kutney, J. P. *J. Chem. Soc., Chem. Commun.* **1988**, 1466. (b) Becalski, A. G.; Cullen, W. R.; Fryzuk, M. D.; James, B. R.; Kang, G.-J.; Rettig, S. J. *Inorg. Chem.* **1991**, *30*, 5002.

339. Buriak, J. M.; Osborn, J. A. *Organometallics* **1996**, *15*, 3161.

340. Longley, C. J.; Goodwin, T. J.; Wilkinson, G. *Polyhedron* **1986**, *5*, 1625.

341. (a) Amrani, Y.; Lecomte, L.; Sinou, D.; Bakos, J.; Toth, I.; Heil, B. *Organometallics* **1989**, *8*, 542. (b) Bakos, J.; Orosz, Á.; Heil, B.; Laghmari, M.; Lhoste, P.; Sinou, D. *J. Chem. Soc., Chem. Commun.* **1991**, 1684.

342. (a) Lensink, C.; de Vries, J. G. *Tetrahedron: Asymmetry* **1992**, *3*, 235. (b) Lensink, C.; Rijnberg, E.; de Vries, J. G. *J. Mol. Catal. A: Chemical* **1997**, *116*, 199.

343. Tani, K.; Onouchi, J.; Yamagata, T.; Kataoka, Y. *Chem. Lett.* **1995**, 955.

344. Schnider, P.; Koch, G.; Prétôt, R.; Wang, G.; Bohnen, F. M.; Krüger, C.; Pfaltz, A. *Chem. Eur. J.* **1997**, *3*, 887.

345. Lensink, C.; de Vries, J. G. *Tetrahedron: Asymmetry* **1993**, *4*, 215.

346. (a) Willoughby, C. A.; Buchwald, S. L. *J. Am. Chem. Soc.* **1992**, *114*, 7562. (b) Willoughby, C. A.; Buchwald, S. L. *J. Am. Chem. Soc.* **1994**, *116*, 8952. (c) Willoughby, C. A.; Buchwald, S. L. *J. Am. Chem. Soc.* **1994**, *116*, 11703.

347. Bader, R. R.; Blaser, H.-U. In *Heterogeneous Catalysis and Fine Chemicals IV*; Blaser, H. U., Baiker, A., Prins, R. (Ed.); Elsevier: Amsterdam, 1997; pp. 17.

348. Togni, A. *Angew. Chem. Int. Ed.* **1996**, *35*, 1475.

349. Spindler, F.; Pugin, B.; Blaser, H.-U. *Angew. Chem. Int. Ed.* **1990**, *29*, 558.

350. Cheong, Y. N.; Osborn, J. A. *J. Am. Chem. Soc.* **1990**, *112*, 9400.

351. Sablong, R.; Osborn, J. A. *Tetrahedron: Asymmetry* **1996**, *7*, 3059.

352. Willoughby, C. A.; Buchwald, S. L. *J. Org. Chem.* **1993**, *58*, 7627.

353. Viso, A.; Lee, N. E.; Buchwald, S. L. *J. Am. Chem. Soc.* **1994**, *116*, 9373.

354. Morimoto, T.; Nalajima, N.; Achiwa, K. *Synlett* **1995**, 748.

355. Morimoto, T.; Nalajima, N.; Achiwa, K. *Chem. Pharm. Bull.* **1994**, *42*, 1951.

356. Zhu, G.; Zhang, X. *Tetrahedron: Asymmetry* **1998**, *9*, 2415.

357. Yukimoto, J.; Kanai, K.; Inanaga, M.; Achiwa, K. *Jpn. Kokai Tokyo Koho* **1993**, JP 05, *246*, 995. *Chem. Abstr.* **1994**, *120*, 134275y.

358. Morimoto, T.; Achiwa, K. *Tetrahedron: Asymmetry* **1995**, *6*, 2661.

359. Morimoto, T.; Suzuki, N.; Achiwa, K. *Heterocycles* **1996**, *43*, 2557.

360. Morimoto, T.; Suzuki, N.; Achiwa, K. *Tetrahedron: Asymmetry* **1998**, *9*, 183.

361. Bianchini, C.; Barbaro, P.; Scapacci, G.; Farnetti, E.; Graziani, M. *Organometallics* **1998**, *17*, 3308.

362. (a) Brunner, H.; Huber, C. *Chem. Ber.* **1992**, *125*, 2085. (b) Brunner, H.; Bublak, P.; Helget, M. *Chem. Ber.* **1997**, *130*, 55.

363. (a) Burk, M. J.; Feaster, J. E. *J. Am. Chem. Soc.* **1992**, *114*, 6266. (b) Burk, M. J.; Martinez, J. P.; Feaster, J. E.; Cosford, N. *Tetrahedron* **1994**, *50*, 4399.

364. Charette, A. B.; Giroux, A. *Tetrahedron Lett.* **1996**, *37*, 6669.

365. Oppolzer, W.; Wills, M.; Starkemann, C.; Bernardinelli, G. *Tetrahedron Lett.* **1990**, *31*, 4117.

366. (a) Basu, A.; Bhaduri, S.; Sharma, K.; Jones, P. G. *J. Chem. Soc., Chem. Commun.* **1987**, 1126. (b) Wang, G.-Z.; Bäckvall, J.-E. *J. Chem. Soc., Chem. Commun.* **1992**, 980.

367. Uematsu, N.; Fujii, A.; Hashiguchi, S.; Ikariya, T.; Noyori, R. *J. Am. Chem. Soc.* **1996**, *118*, 4916.

368. Borch, R. F.; Bernstein, M. D.; Durst, H. D. *J. Am. Chem. Soc.* **1971**, *93*, 2897.

369. (a) Jessop, P. G.; Ikariya, T.; Noyori, R. *Science* **1995**, *269*, 1065. (b) Jessop, P. G.; Ikariya, T.; Noyori, R. *Chem. Rev.* **1999**, *99*, 475. (c) Jessop, P. G.; Leitner, W. In *Chemical Synthesis Using Supercritical Fluids*, Jessop, P. G.; Leitner, W. (Ed.); Wiley-VCH: Weinheim, 1999, Chapter 4.7.

370. (a) Cornils, B.; Herrmann, W. A. In *Applied Homogeneous Catalysis with Organometallic Compounds*; Cornils, B., Herrmann, W. A. (Ed.); VCH: Weinheim, 1996; Vol. 2, Chapter 3.1.1.1. (b) Sinou, D. In *Transition Metals for Organic Synthesis*; Beller, M., Bolm, C. (Ed.); Wiley-VCH: Weinheim, 1998; Vol. 2, Chapter 3.2. (c) *Aqueous-Phase Organometallic Catalysis*, Cornils, B., Herrmann, W. A. (Ed.); Wiley-VCH: Weinheim, 1998. See also: Horváth, I. T. In *Applied Homogeneous Catalysis with Organometallic Compounds*: Cornils, B., Herrmann, W. A. (Ed.); VCH: Weinheim, 1996; Vol. 2, Chapter 3.1.1.2.

371. (a) Panster, P.; Wieland, S. In *Applied Homogeneous Catalysis with Organometallic Compounds*; Cornils, B., Herrmann, W. A. (Ed.); VCH: Weinheim, 1996; Vol. 2, Chapter 3.1.1.3. (b) Basset, J.-M.: Niccolai, G. P. In *Applied Homogeneous Catalysis with Organometallic Compounds*; Cornils, B., Herrmann, W. A. (Ed.); VCH: Weinheim, 1996; Vol. 2, Chapter 3.1.1.4. (c) Shuttleworth, S. J.; Allin,

S. M.; Sharma, P. K. *Synthesis* **1997**, 1217. (d) Basset, J. M.; Candy, J. P.; Santini, C. C. In *Transition Metals for Organic Synthesis*; Beller, M., Bolm, C. (Ed.); Wiley-VCH: Weinheim, 1998; Vol. 2, Chapter 3.1.

372. Kragl, U.; Dreisbach, C.; Wandrey, C. In *Applied Homogeneous Catalysis with Organometallic Compounds*; Cornils, B., Herrmann, W. A. (Ed.); VCH: Weinheim, 1996; Vol. 2, Chapter 3.2.3.

2

ASYMMETRIC HYDROSILYLATION AND RELATED REACTIONS

HISAO NISHIYAMA

School of Materials Science, Toyohashi University of Technology, Toyohashi 441-8580, Japan

KENJI ITOH

Department of Molecular Design and Engineering, Graduate School of Engineering, Nagoya University, Nagoya 464-8603, Japan

2.1. INTRODUCTION

Asymmetric hydrometallation of ketones and imines with H–M (M = Si, B, Al) catalyzed by chiral transition-metal complexes followed by hydrolysis provides an effective route to optically active alcohols and amines, respectively. Asymmetric addition of metal hydrides to olefins provides an alternative and attractive route to optically active alcohols or halides via subsequent oxidation of the resulting metal–carbon bonds (Scheme 2.1).

In this chapter, recent advances in asymmetric hydrosilylations promoted by chiral transition-metal catalysts will be reviewed, which attained spectacular increase in enantioselectivity in the 1990s [1]. After our previous review in the original "Catalytic Asymmetric Synthesis," which covered literature through the end of 1992 [2], various chiral P_n, N_n, and P–N type ligands have been developed extensively with great successes. In addition to common rhodium and palladium catalysts, other new chiral transition-metal catalysts, including Ti and Ru complexes, have emerged. This chapter also discusses catalytic hydrometallation reactions other than hydrosilylation such as hydroboration and hydroalumination.

2.2. ASYMMETRIC HYDROSILYLATION OF KETONES AND IMINES WITH RH AND RU CATALYSTS

Since the catalytic activity of the Wilkinson complex, $RhCl(PPh_3)_3$, for the hydrosilylation of ketones was discovered, the application of this reaction to asymmetric synthesis has been studied

Catalytic Asymmetric Synthesis, Second Edition, Edited by Iwao Ojima
ISBN 0-471-29805-0 Copyright © 2000 Wiley-VCH, Inc.

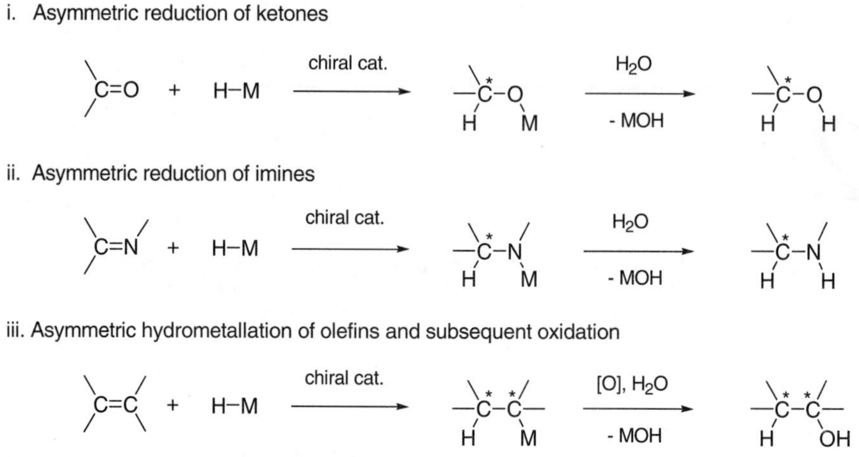

i. Asymmetric reduction of ketones

$$\text{C=O} + \text{H—M} \xrightarrow{\text{chiral cat.}} \overset{*}{\underset{\underset{M}{|}}{\underset{\underset{H}{|}}{\text{C}}}}\text{-O} \xrightarrow[\text{- MOH}]{\text{H}_2\text{O}} \overset{*}{\underset{\underset{H}{|}}{\underset{\underset{H}{|}}{\text{C}}}}\text{-O}$$

ii. Asymmetric reduction of imines

$$\text{C=N} + \text{H—M} \xrightarrow{\text{chiral cat.}} \overset{*}{\text{C}}\text{-N} \xrightarrow[\text{- MOH}]{\text{H}_2\text{O}} \overset{*}{\text{C}}\text{-N}$$

iii. Asymmetric hydrometallation of olefins and subsequent oxidation

$$\text{C=C} + \text{H—M} \xrightarrow{\text{chiral cat.}} \overset{*}{\text{C}}\text{-}\overset{*}{\text{C}} \xrightarrow[\text{- MOH}]{\text{[O], H}_2\text{O}} \overset{*}{\text{C}}\text{-}\overset{*}{\text{C}}$$

Scheme 2.1.   Asymmetric Hydrometallation (M = Si, B, Al)

by using Rh catalysts with chiral phosphine ligands [3,4]. Typically, catalytic asymmetric hydrosilylation of prochiral ketone is carried out with a dihydrosilane, for example diphenylsilane (Ph$_2$SiH$_2$, **2**) or 1-naphthylphenylsilane (1-NpPhSiH$_2$, **3**), in the presence of a chiral catalyst by using acetophenone (**1**) as the standard substrate in aromatic or ether solvent or sometimes without solvent at 0°C ~ room temperature (Scheme 2.2). After hydrolysis of the reaction mixture, optically active 1-phenylethanol (**4**) is obtained.

In the 1970s, various chiral phosphine-Rh catalysts were used for this reaction, which gave only low-to-moderate enantioselectivity. Two representative chiral catalysts that achieved relatively good enantioselectivity at that time were (+)-DIOP(**5**)/[Rh(COD)Cl]$_2$ [5] and [Rh(Glucophinite, **6**) (COD)]BF$_4$ [6], which afforded 1-phenylethanol (**4**) with 58% ee (S) and 65% ee (S), respectively, by using 1-NpPhSiH$_2$ (**3**) as the hydrosilane.

In the early 1980s, nitrogen-containing chiral ligands such as Iminopyridine **7** [7], Amphos **8** [8], or Aminphos **9** [9] were developed for Rh catalysts, which gave **4** with 50–79% ee. However, it was necessary to use a large excess of ligand, up to 10-fold excess with respect to Rh atom, to achieve high enantioselectivity. For these chiral nitrogen-containing ligands,

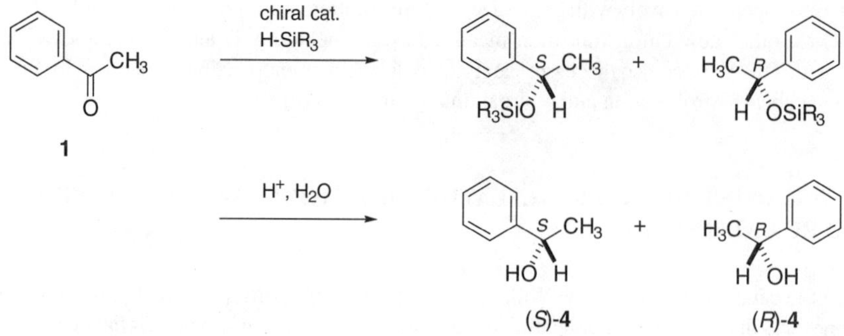

Scheme 2.2.

diphenylsilane (**2**) was the hydrosilane of choice to achieve the best results in contrast to the cases of DIOP and Glucophinite ligands wherein 1-NpPhSiH$_2$ (**3**) was the best hydrosilane.

A chiral nitrogen ligand, pyridine-thiazolidine (Pythia, **10**), synthesized from pyridinecarboaldehyde and L-cysteine ethyl ester in 1983, brought about a breakthrough in this field [10]. Acetophenone (**1**) was reduced to **4** with 87.2~97.6% ee by using **2** in the presence of [Rh(COD)Cl]$_2$ (**C**) and Pythia **10** (8–13 fold excess to Rh) (S/C = 140–450, S = substrate, C = catalyst) at –20°C without solvent. The *R*-absolute configuration of the secondary alcohol product was derived from the *R*-configuration at the 4′-position of the thiazolidine ring of **10**.

Pyridine-(2-mono-oxazoline), Pymox **11**, appeared for the Cu-catalyzed asymmetric monophenylation of diols with bismuth reagent in 1986 [11]. The chirality of the oxazoline ring is derived from enantiopure aminoalcohols, which are readily obtained by the reduction of enantiopure amino acids. Pymox can be synthesized via the imidate route from 2-cyanopyridine or through amide-formation/cyclization from pyridine-2-carboxylic acid [12,13]. Pymox-*t*-Bu **11c** in the presence of [Rh(COD)Cl]$_2$ and diphenylsilane achieved 83–91% ee for acetophenone. The use of trityl ligand **11d** and **3** as the hydrosilane gave 80% ee [14]. Modification of the oxazolin ring by introduction of two phenyl rings, as shown in **11e** and **11f**, increased enantioselectivity from 60% ee with **11b** to 83–89% ee [15]. An appropriate amount of Pymox may be 4–5 fold excess to Rh atom. In the case of Pymox, the use of tetrachloromethane leads to a remarkable increase in the enantioselectivity as compared with aromatic solvents or dichloromethane [12]. This increase is probably due to a change in the catalytically active species arising from the oxidative addition of tetrachloromethane. It is interesting that the absolute configuration of the product alcohol was reverse between the cases of Pymox and Pythia, that is the chirality transfer from the catalysts to the substrate runs opposite in these systems. It is curious that the introduction of a substituent at the 6-position of pyridine skeleton of Pymox led to the formation of **4** with absolute configuration opposite that obtained by using the parent Pymox.

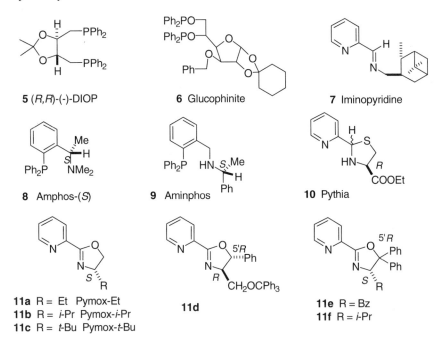

5 (*R,R*)-(-)-DIOP

6 Glucophinite

7 Iminopyridine

8 Amphos-(*S*)

9 Aminphos

10 Pythia

11a R = Et Pymox-Et
11b R = *i*-Pr Pymox-*i*-Pr
11c R = *t*-Bu Pymox-*t*-Bu

11d

11e R = Bz
11f R = *i*-Pr

The C_2 symmetric design was first introduced in 1989 to a chiral nitrogen ligand, 2,6-bis(oxazolinyl)pyridine, Pybox **12**, for the asymmetric hydrosilylation of ketones [13,16]. Pybox can be readily synthesized by the condensation of pyridine-2,6-dicarboxylic acid with enantiopure aminoalcohols [16] or by BF_3-catalyzed cyclization of intermediate amidoalcohols [17]. A combination of Pybox-*i*-Pr **12a** (3 mol%) and [Rh(COD)Cl]$_2$ (**C**) (1 mol%) with diphenylsilane reduced **1** to **4** with 76% ee (*S*) at 0°C. Remarkable improvement of enantioselectivity up to 95% ee (*S*) was realized by using RhCl$_3$(Pybox-*i*-Pr) complex (**D**) (1 mol%) and AgBF$_4$ (2 mol%) in the presence of a 4-fold excess of Pybox-*i*-Pr. Although an excess Pybox was necessary to achieve the best result, the ligand-exchange experiments indicated that the catalytically active Rh species should possess only one Pybox. The reaction is usually performed in a small amount of THF and **2** is the best choice as hydrosilane. It should be noted that the absolute configuration of the product alcohol **4** was again reverse from that obtained using Pymox **11** even though both Pybox **12** and Pybox **11** were derived from the same aminoalcohols with *S* configuration.

The electronic effects of the remote substituents at the 4-position of the pyridine ring of Pybox were shown to affect both the reaction rate and enantioselectivity in the Pybox-Rh–catalyzed reactions [18]. For example, the Rh complex with a Pybox bearing an electron-withdrawing substituent, Pybox-*i*-Pr(4-CO$_2$Me) **12c**, catalyzed the reduction of **1** with **2** affording **4** with 96% ee, at −5°C for 4 h [19], whereas Pybox having an electron-donating group such as NMe$_2$ resulted in retarding the reaction (20°C for 16 h to complete) to give **4** with 90% ee.

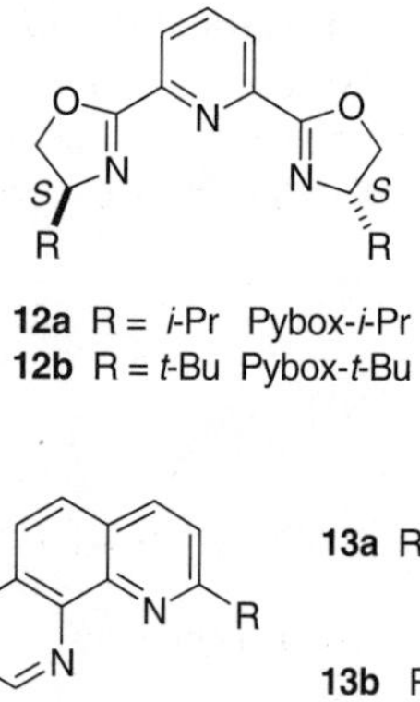

12a R = *i*-Pr   Pybox-*i*-Pr
12b R = *t*-Bu   Pybox-*t*-Bu

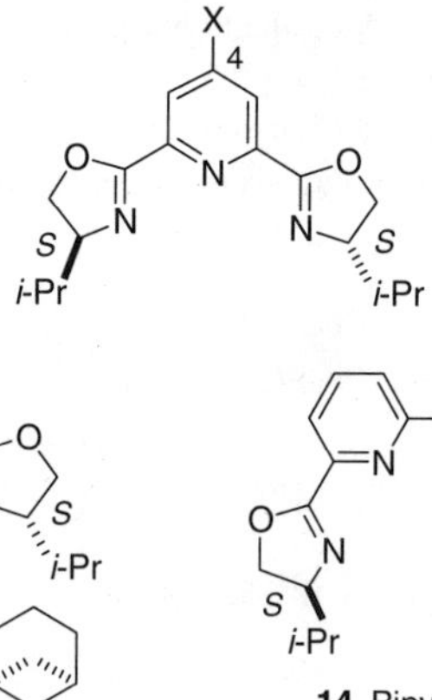

12c X = CO$_2$Me
12d X = NMe$_2$

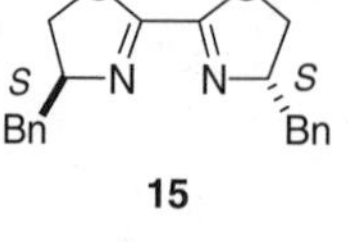

13a R =

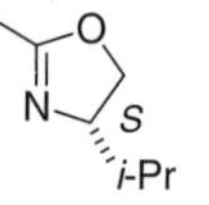

13b R =

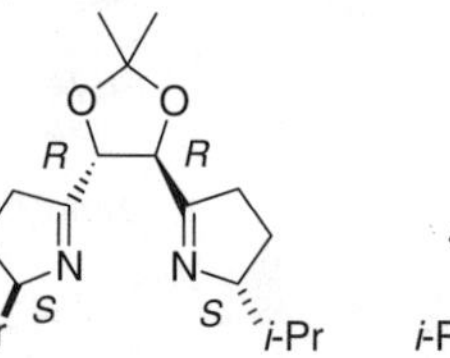

14 Bipymox-*i*-Pr

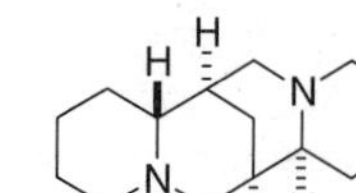

15

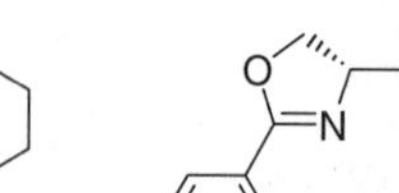

16

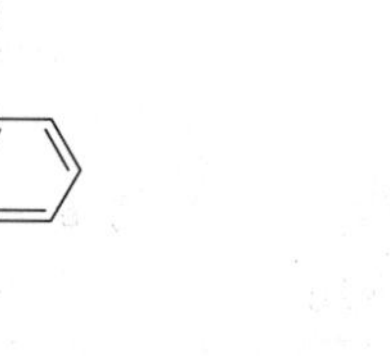

17a

17b

18 (−)-Sparteine

19

Among several dipyridine derivatives developed as chiral ligands, a Rh catalyst with phenanthroline **13b** having a pinane skeleton gave **4** with 76% ee in the reduction of **1**, which was higher than that obtained with Rh-phenanthroline-oxazoline **13a** [20], whereas the tetradentate dipyridine-bis(oxazoline) Bipymox **14** in combination with RhCl$_3$ gave **4** with 90% ee (S) [21].

Since the late 1980s to 1990s, several bidentate chiral bis-oxazoline ligands derived from oxalic acid or malonic acid were developed for asymmetric cyclopropanation and allylic substitution [22]. Among these bis-oxazoline ligands, only the benzylic ligand **15** gave a better result in the reduction of **1**, giving **4** with 84% ee (R) [23]. The difference in the chiral backbone structure between **17a** and **17b** influenced the enantioselectivity, that is 64% ee for the former and 49% ee for the latter, with the same absolute configuration [24]. Use of 4–10 fold excess of a naturally occurring chiral amine, (+)-spartein (**18**), as a bidentate ligand in the same reaction attained 65–75% yields with 34–37% ee [25]. A tetradentate pyridyl bis-oxazoline **19** having dual chirality in the backbone was applied to the reduction of **1** to give **4** with 50% ee (R) [26]. Thus, the attempts with chiral bis-oxazoline ligands resulted in realizing only fairly good-to-good enantioselectivity.

In contrast to the remarkable developments of the chiral nitrogen-containing ligands as described above, chiral phosphines did not show noteworthy progress until 1994. In 1994, however, a bis-phosphinoferrocene derivative, TRAP **20**, possessing 164° for the bite angle of P-Rh-P, achieved a high enantioselectivity (>90% ee) for the first time as chiral diphosphine ligand [27]. The combination of TRAP-n-Bu **20a** with [Rh(COD)Cl]$_2$ (**C**) reduced **1** at –40°C with **2** to give **4** with 92% ee (S). Monodentate phosphonite ligands TADDOL-Phosphonite **21** derived from tartrates exhibited good-to-excellent enantioselectivity (62–87% ee) in the Rh-catalyzed reduction of **1** [28]. Similar cyclic monophosphonites were also investigated in detail for the effects of their backbone structures and temperature on enantioselectivity [28b].

20a R = n-Bu (R,R)-(S,S)-TRAP-n-Bu
20b R = Et (R,R)-(S,S)-TRAP-Et

(R,R)-TADDOL-Phosphonites:

21a R = Ph, R' = Ph-4-Me
21b R = 2-Np, R' = Ph
21c R = 2-Np, R' = 2-Np

In the mid 1990s, new N-P type ligands similar to Amphos and Aminphos were developed as good ligands for the hydrosilylation. For example, a rather simple Cystophos **22** having methylthio group gave **4** with 64% ee (S) in the Rh-catalyzed reduction of **1** [29]. Also, ferrocenylphosphineimine **23** [30] and the ferrocenylphosphine-oxazoline (DIPOF) **24** [31] attained 87% ee and 91% ee, respectively, for the same reaction. Modification of phosphine-oxazoline **24** [32], which was developed in 1993 as an efficient ligand for palladium-catalyzed allylic substitution, was further performed [33]. The isopropyl derivative **25a** gave **4** with 82% ee. The *tert*-butyl derivative **25b** decreased enantioselectivity to 40% ee, whereas the bis(3,5-bistrifluoromethylphenyl)phosphino derivative **25c** increased the selectivity to 86% ee, and, finally, the indane derivative **25d** achieved enantioselectivity of 94% ee for the same reaction [34]. It should be noted that the absolute configuration of **4** obtained by using phosphine-oxazoline ligands, Phos-Oxazole **25**, is the same as that by pyridine-oxazoline ligands, Pymox **11**, bearing the oxazoline moieties with the same absolute configuration. In 1997, bis(phosphinophenyl-oxazoline), Phos-Biox **26**, exhibited extremely high enantioselectivity, giving **4** with 97% ee (*S*) in the reduction of **1** at 0°C in THF with an S/C of 400 [35]. In addition, a unique bis(aminoferrocenyl)diselenide **27** was developed for the hydrosilylation of ketones, reported as 85% ee in the reduction of **1** [36]. Heterocyclic carbene-Rh complexes **28** and **29** were also examined for their efficiency in the asymmetric reduction of **1**, but resulted in only moderate enantioselectivity (38–40% ee) [37,38].

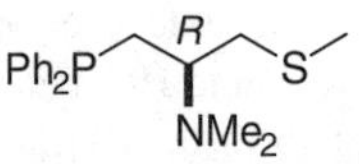

22 (*R*)-Cystphos

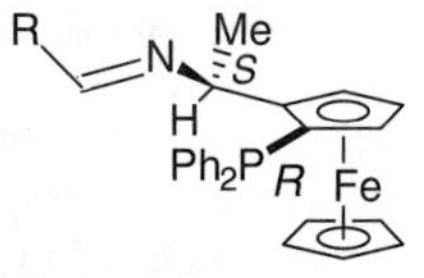

Ferrocenylphosphine-Imines
23a-(*R,S*) R = Ph
23b-(*R,S*) R = 3-CF$_3$- C$_6$H$_4$
23c-(*R,S*) R = 4-CF$_3$- C$_6$H$_4$

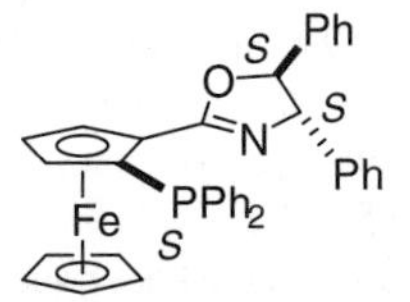

Ferrocenylphosphine-oxazoline
24 R = Ph (*S,S,S*)-DIPOF

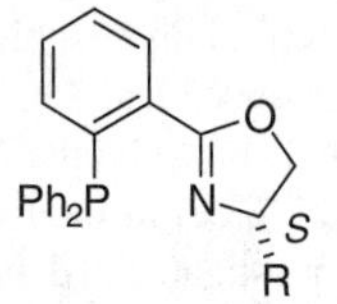

25a R = *i*-Pr
25b R = *t*-Bu

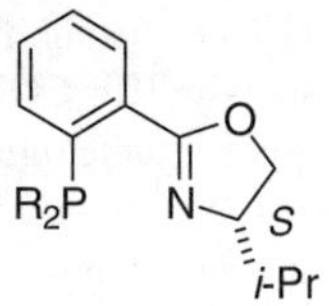

25c R = 3,5-(CF$_3$)$_2$C$_6$H$_3$

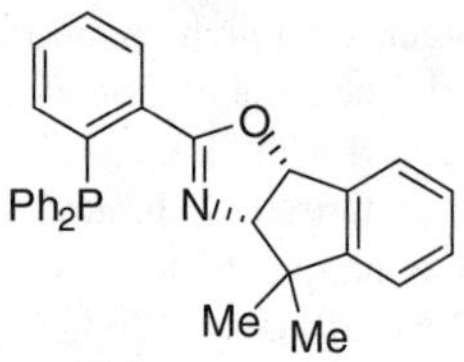

25d

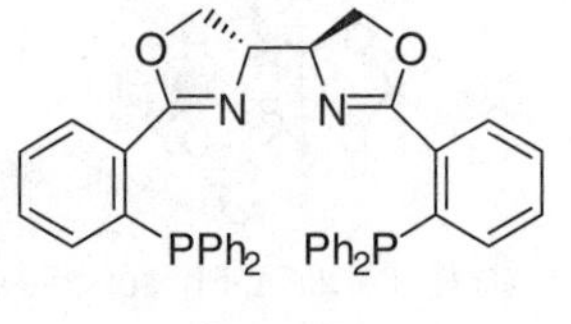

26 Phos-Biox

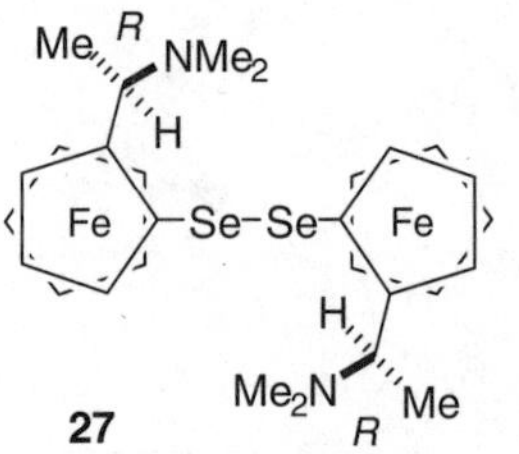

27

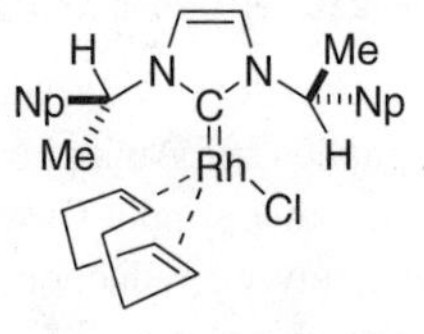

28

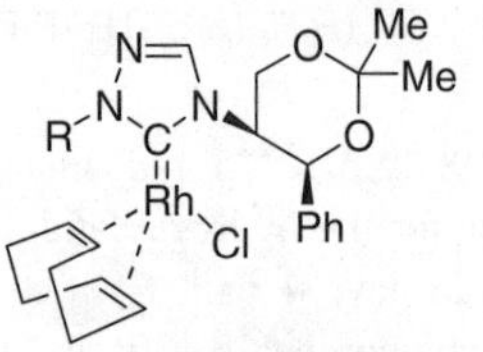

29a R = Me
29b R = t-Bu

The results on the asymmetric reduction of acetophenone discussed above are summarized in Table 2.1. The chiral catalyst systems described above have also been applied to a variety of ketones other than acetophenone. The results are summarized in Table 2.2.

The chiral Rh catalyst systems described above for the reduction of **1** have been shown to achieve high enantioselectivity (>90% ee) for aryl alkyl ketones **30–37**. Propiophenone **30** and

TABLE 2.1. Asymmetric Hydrosilylation of Acetophenone with Chiral Rhodium Catalysts

Ligand	Rh-Cat.[a]	L/Rh	S/C[b]	Silane[c]	Silane/Ketone	Solvent	Temp/Time (°C/h)	Yield (%)	Product: 1-phenylethanol (4) %ee	abs config	Ref.
(+)-DIOP 5	A	1	50	3	2	benzene	rt/20	100	58	S	5
Glucophinite 6	B	1	50	3	1	benzene	rt/12	65	65	—	6
	B	1	50	2	1	benzene	rt/12	65	55	—	6
Iminopyridine 7	C	9	100	2	1.1	no solv.	0–25/48	99	78.8	S	7
Amphos-(S) 8	A	1	300	2	1.1	benzene	rt/72	98	72	S	8
Aminphos 9	C	2.2	500	2	1.1	THF	–10/96	93	52.7	S	9
Pythia 10	C	8	450	2	1.1	no solv.	–20/100	86	87.2	R	10
	C	13	140	2	1.1	no solv.	–20/120	99	97.6	R	10
Pymox-Et-(R) 11a	C	5	200	2	1	no solv.	0–20/18	60	39.8	S	12
	C	5	200	2	1	benzene	0–20/18	60	31.0	S	12
Pymox-i-Pr-(S) 11b	C	5	200	2	1	CCl$_4$	0–20/18	93	56.6	S	12
Pymox-t-Bu-(S) 11c	C	5	200	2	1	CCl$_4$	0–20/18	85	62.2	R	12
	C	5	200	2	1	CCl$_4$	0–20/18	90	83.4	R	12
Pymox-(CH$_2$OTr)(Ph) 11d	C	5	100	2	1.6	no solv.	–5/24	90	91	R	13
	A	4	171	2	1.1	no solv.	0/72	96	63.2	R	14
	A	4	171	3	1.1	no solv.	0/72	100	80	R	14
Pymox-(i-Pr)(Ph$_2$) 11e	C	5	100	2	1	CCl$_4$	0–rt/18	80	89.1	R	15
Pybox-i-Pr-(S,S) 12a	D	5	100	2	1.6	THF	0/3	94	95	S	13,16
Pybox-t-Bu-(S,S) 12b	D	5	100	2	1.6	THF	0/18	92	83	S	13,16
Pybox-i-Pr(4-CO$_2$Me) 12c	D	5	100	2	1.6	THF	–5/4	90	96	S	19
Pybox-i-Pr(4-NME$_2$) 12d	D	5	100	2	1.6	THF	20/16	83	90	S	18
Phenanthroline 13a	C	5	200	2	1	toluene	0–rt/18	100	10.8	S	20
phenanthroline 13b	C	5	200	2	1	toluene	0–rt/18	100	75.6	R	20
Bipymox-i-Pr-(S,S) 14	E	4	100	2	1.5	THF	5/2	98	90	S	21
Bisoxazole-Bn-(S) 15	C	10	100	2	1.2	CCl$_4$	0	59	84	R	23

(continued)

TABLE 2.1. Continued

Ligand	Rh-Cat.[a]	L/Rh	S/C[b]	Silane[c]	Silane/Ketone	Solvent	Temp/Time (°C/h)	Product: 1-phenylethanol (4)			
								Yield (%)	% ee	abs config	Ref.
Bisoxazole-i-Pr-(S) 16	C	10	100	2	1.2	CCl4	rt	—	12	R	23
Bisoxazole-(S,S:R,R) 17a	C	8	100	2	1.6	CCl4	-5/72	90	64	R	24
Bisoxazole-(S,S:S,S) 17b	C	8	100	2	1.6	CCl4	-5/72	72	49	R	24
(−)-Sparteine 18	C	4	1000	2	1.1	no solv.	25	65	34	R	25
Pyridyl-bisoxazole 19	C	4.3	75	2	1.1	no solv.	8/30	75.5	50	R	26
TRAP-n-Bu-(R,R)(S,S) 20a	F	1.1	100	3	1.5	THF	-40/11	88	92	S	27
TADDOL-Phos 21a	C	10	100	3	1.2	benzene	0–20/10–15	91	82	R	28
TADDOL-Phos 21b	C	10	100	2	1.2	benzene	0–20/10–15	99	84	R	28
Cystphos-(R) 22	C	2.2	100	2	1.07	THF	0/20	99	64	S	29
Fc-phosphine-Imine 23a	G	1.5	100	2	1.25	THF	20/1	90	87	S	30
Fc-phosphine-Imine 23b	G	1.5	100	2	1.25	THF	20/0.2	90	90	S	30
DIPOF-(S,S,S) 24	C	1	200	2	1.5	Et2O	25/15–25	100	91	R	31
Phos-Oxazole-i-Pr-(S) 25a	C	10	50	2	4	THF	-78/–	86	82	R	32
	C	10	50	3	4	THF	-78/–	85	86	R	32
Phos-Oxazole-t-Bu-(S) 25b	C	1.3	333	2	1.1	THF	rt/70	65	40	R	32
Phos-Oxazole 25c	C	1.3	333	2	1.1	THF	10/70	90	85	R	32
Phos-Oxazole 25d	C	2	50	2	2	toluene	rt/24	84	94	R	34
Phos-Biox 26	C	2	400	2	1.6	THF	0/7	98	97	R	35
Amino-Fc-Se/2 27	C	1	100	2	1.5	THF	0/48	31	85	R	36
Carbene-Rh 28	—	1	100	2	1	THF	-34/48	90	32	—	37
Carbene-Rh 29a	—	1	100	2	1	THF	11/144	60	40	S	38

[a]Rh cat.: A, [Rh(C2H4)2Cl]2; B, [Rh(COD)L]BF4; C, [Rh(COD)Cl]2; D, RhCl3(L); E, RhCl3(L)/AgBF4; F, [Rh(COD)2]BF4; G, [Rh(NBD)Cl]2.
[b]S/C (substrate/catalyst) = ketone/Rh.
[c]Silane: 2, Ph2SiH2; 3, 1-NpPhSiH2.

TABLE 2.2. Asymmetric Hydrosilylation of Other Ketones with Rhodium Catalysts[a]

Ketone	Ligand	Rh-cat.[c]	L/Rh	S/C[d]	Solvent	Temp/Time (°C/h)	Yield (%)	% ee	abs config	Ref.
PhCOEt **30**	Glucophinite **6**[b]	B	1	50	benzene	rt/12	65	61	—	6
	Pythia **10**	C	8	150	no solv.	0–20/18	90	76.7	R	10
	Pybox-i-Pr **12a**	D	5	100	THF	5/4	73	91	S	16
	Fc-Phos-Imine **23a**	G	1.5	100	THF	20/<1	97	86	S	30
	Phos-Oxazole **25d**	C	2	50	toluene	rt/24	91	91	R	34
1-NpCOMe **31**	Pythia **10**	C	8	150	no solv.	0–20/18	95	83.3	R	10
	Pybox-i-Pr **12a**	D	5	100	THF	–5/5	87	94	S	16
	TADDOL-Phos **21c**	C	10	100	benzene	0–20/10–15	92	87	R	28
	Phos-Oxazole **25d**	C	2	50	toluene	rt/24	90	92	R	34
2-NpCOMe **32**	Pythia **10**	C	8	150	benzene	0–20/10–15	90	77.9	R	10
	Pybox-i-Pr **12a**	D	5	100	THF	–5/5	93	97	S	16
	DIPOF **24**	C	1	200	Et$_2$O	25/15–25	99	88	R	31
2-PyCOMe **33**	Pythia **10**	C	8	150	no solv.	0–20/18	90	89.6	R	10
4-MeOPhCOMe **34**	Pythia **10**	C	8	150	no solv.	–15/192	95	93.3	R	10
2-(MeO$_2$C)PhCOMe **35**	Pybox-i-Pr **12a**	D	7	100	THF	0/14	95	96	S	16
PhCOCH$_2$Cl **36**	Amino-Fc-Se/2 **27**	C	1.5	100	THF	0/120	85	88	R	36
1-Tetralone **37**	Pythia **10**	C	8	150	no solv.	0–20/18	85	82.6	R	10
	Pybox-i-Pr **12a**	D	5	100	THF	0/2	92	99	S	16
	Phos-Oxazole **25d**	C	2	50	toluene	rt/24	89(97)[e]	89(92)	R(S)	34
n-C$_6$H$_{13}$COMe **38**	Pybox-i-Pr **12a**	D	5	100	THF	0/2	85	63	S	16

(continued)

TABLE 2.2. Continued

Ketone	Ligand	Rh-cat.[c]	L/Rh	S/C[d]	Solvent	Temp/Time (°C/h)	Product: alcohol		abs config	Ref.
							Yield (%)	% ee		
$n\text{-}C_7H_{15}COMe$ **39**	Phos-Oxazole **25d**	C	2	50	toluene	rt/24	90	52	R	34
$t\text{-}C_4H_9COMe$ **40**	DIPOF **24**	C	1	200	Et_2O	25/15−25	92	60	R	31
	DIPOF **24**	C	1	200	Et_2O	25/15−25	100	87	R	31
Adamantyl-COMe **41**	TRAP-n-Bu **20a**	F	1.1	100	THF	−40/48	92	91	S	27
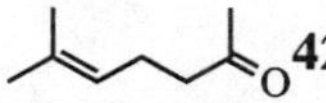 (structure) **42**	Pybox-i-Pr **12a**	D	6	100	THF	20/20	94	70	S	16
2-Ph-cyclohexanone **43**	Pybox-i-Pr **12a**	D	5	100	THF	0/4	92 (51:49)	99	$1S,2R$	39
								96	$1S,2S$	
$MeCOCO_2\text{-}n\text{-}Pr$ **44**	(−)-DIOP **5**[b]	C	1.5	300	benzene	0−rt/6	90	85.4	R	40
$MeCOCO_2\text{-}Et$ **45**	TRAP-Et **20b**	F	1.5	100	THF	0/4	60	80	S	27
$MeCOC(CH_3)_2CO_2Et$ **47**	TRAP-Et **20b**	F	1.5	100	THF	−30/11	69	93	S	27
$MeCO(CH_2)_2CO_2\text{-}i\text{-}Bu$ **48**	(−)-DIOP **5**[b]	C	1.5	150	benzene	20/12	96	84.4	S	40
$MeCO(CH_2)_2CO_2Et$ **49**	Pybox-i-Pr **12a**	D	5	100	THF	0/7	91	95	S	16
$MeCO(CH_2)_2CO_2Me$ **50**	TRAP-Et **20b**	F	1.5	100	THF	−30/31	74	88	S	27
$MeCOC(CH_3)_2COMe$ **51**	TRAP-Et **20b**	F	1.5	100	THF	−30/58	58(96:4)	99	$2S,4S$	27

[a]Silane = Ph_2SiH_2 (1−2 equiv), unless otherwise commented.

[b]Silane = 1-$NpPhSiH_2$.

[c]Rh cat.: **A**, $[Rh(C_2H_4)_2Cl]_2$; **B**, $[Rh(COD)Cl]_2$; **C**, $[Rh(COD)L]BF_4$; **D**, $RhCl_3(L)$; **E**, $RhCl_3(L)/AgBF_4$; **F**, $[Rh(COD)_2]BF_4$; **G**, $[Rh(NBD)Cl]_2$.

[d]S/C (substrate/catalyst) = ketone/Rh.

[e]use of (+)-**25d**.

acetonaphthones **31** and **32** were reduced to the corresponding alcohols with 91–97% ee using Pybox-*i*-Pr **12a** [16] and Phos-Oxazole **25d** [34]. Fc-Phos-Imine **23a** [30], TADDOL-Phos **21c** [28], and DIPOF **24** [31] gave a little lower enantioselectivity than that obtained using Pybox-*i*-Pr. Pythia **10** exhibited high enantioselectivity for the reduction of 2-pyridyl methyl ketone **33** and 4-methoxyphenyl methyl ketone **34** [10]. 1-Tetralone was reduced to the corresponding alcohol with 99% ee (*S*) by using Pybox-Rh catalyst [16].

In the hydrosilylation-reduction of dialkyl ketones by using chiral phosphine ligands, enantioselectivity did not exceed 50% ee [2]. In contrast, Rh catalysts with Pybox **12a**, Phos-Oxazole **25d**, and DIPOF **24** realized 52–63% ee in the reaction of linear 2-alkanones **38** and **39** [16,34,31]. The reduction of 4-phenyl-2-butanone by using 4-Cl-Pybox-*i*-Pr **12d** gave 4-phenyl-2-butanol with 80% ee (*S*) at −5°C for 3 h in THF [18]. For the reaction of *tert*-butyl methyl ketone, a Rh catalyst with DIPOF (**24**) achieved 87% ee, whereas the same reactions with Amphos **8** [8] and the methyl ester of Pythia **10** [10] gave 3,3-dimethyl-2-butanol with 68.5% ee and 72% ee, respectively. Bulky adamantyl methyl ketone (**41**) was reduced to 1-adamantylethanol with 91% ee (*S*) by using a TRAP-Rh catalyst at −40°C for 48 h [27]. The reaction of 6-methyl-5-hepten-2-one (**42**) by using a Pybox-Rh catalyst gave (*S*)-sulcatol with 70% ee [16].

Racemic 2-phenylcyclohexanone (**43**) was reduced with **2** by using a Pybox-Rh catalyst to the corresponding diastereomeric alcohols [(1*S*,2*R*):(1*S*,2*S*) = cis:trans = 51:49] with 99% ee for (1*S*,2*R*)-isomer and 96% ee for (1*S*,2*S*)-isomer [39]. It is noteworthy that the ketone moiety at C-1 was reduced to the alcohols with *S* configuration regardless of the absolute configuration at C-2 bearing a bulky phenyl group. The reaction of 4-*tert*-butylcyclohexanone with **2** by using a Pybox-Rh catalyst to the corresponding alcohols (cis:trans = 33/67) in 92% isolated yield. It should be noted that bulky reducing agents usually give *cis*-alcohol as the major product via an equatorial attack in the reduction of this substrate. On the contrary, the Pybox-Rh catalyst afforded *trans*-alcohol as the major product through thermodynamic control via an axial attack of hydride species.

The first enantioselective hydrosilylation of α-keto esters was carried out in 1977 by using DIOP-Rh catalysts, giving α-hydroxy esters after hydrolysis [40]. *i*-Propyl pyruvate **44** was reduced to *i*-propyl (*R*)-lactate with 85.4% ee by using 1-NpPhSiH$_2$ (**3**) and a (–)-DIOP-Rh catalyst.

Although β-keto esters such as acetoacetate **46** or benzoylacetate did not give enantioselectivity better than 70% ee with DIOP-Rh catalysts [40], 2,2-dimethylacetoacetate **47** was reduced with **2** using a TRAP-Rh catalyst to give the corresponding β-hydroxyacetate with 93% ee (*S*) at –30°C [27].

The reaction of a γ-keto ester, *i*-butyl levulinate (**49**), with **3** using a DIOP-Rh catalyst, gave methyl-γ-lactone with 84.4% ee via the corresponding γ-hydroxy ester [40]. The reaction of ethyl levulinate (**49**) by using a Pybox-Rh catalyst and **2** as the hydrosilane gave ethyl 4-hydroxypentanoate with 95% ee [16]. Reactions of methyl levurinate (**50**) and 3,3-dimethylpentane-2,4-dione (**51**) with **2** using a TRAP-Rh catalyst gave the corresponding alcohols with 88% ee and 99% ee, respectively [27].

43

44 R = *n*-C$_3$H$_7$
45 R = C$_2$H$_5$

46 R = H
47 R = CH$_3$

48 R = *i*-Bu
49 R = C$_2$H$_5$
50 R = CH$_3$

51

Hydrosilylation of α,β-unsaturated ketones with Rh-phosphine catalysts is known to bring about 1,2-addition with dihydrosilanes such as **2** and **3** [41]. Reduction of 2-methyl-2-cyclohexenone with a DIOP-Rh catalyst gave 2-methylcyclohexan-1-ol with 52% ee [31]. The reactions of benzylideneacetone (**52**) with Rh catalysts with Pythia **10**, Pybox **12**, and Phos-Oxazole **25d** gave low enantioselectivity, that is, 13.8–22% ee, whereas the reaction of chalcone (**53**) by using a Pybox-Rh catalyst gave the corresponding alcohol with 71% ee [16]. The reduction of cyclohexenyl methyl ketone (**54**) by using Rh-TRAP afforded 1-cyclohexenylethanol with 95% ee (*S*) at –40°C [27].

52 R = CH$_3$
53 R = Ph

54

Intramolecular hydrosilylation of siloxyacetone **55** catalyzed by a cationic Rh complex with DuPHOS-*i*-Pr (**56**), [Rh(COD)(DuPHOS-*i*-Pr)]OTf, to give the corresponding cyclic silyl ether with 93% ee (*S*) [42]. The product was converted to 1,2-diol **57**, which can also be prepared by asymmetric dihydroxylation of propene. In the same reaction, the use of BINAP **58** gave only 45% ee.

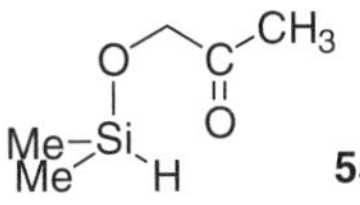

55

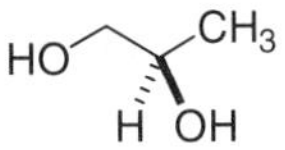

57

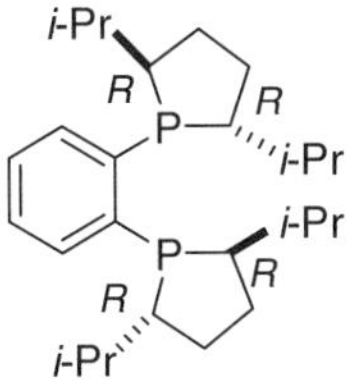

56 R = *i*-Pr, DuPHOS-*i*-Pr

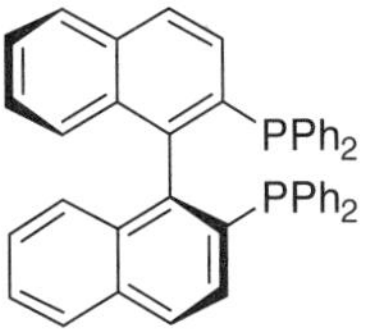

58 (*S*)-(-)-BINAP

In the 1990s, no improvement has been reported for the asymmetric hydrosilylation of imines (Scheme 2.3) by using chiral Rh-catalysts. Imines **59** and **60** were reduced with **2** by using DIOP-Rh catalysts to the corresponding secondary amines **61** and **62** with 65% ee and 66% ee, respectively [43]. Chiral Ru-catalyst, however, achieved higher enantioselectivity in the asymmetric reduction of a particular imine (see below).

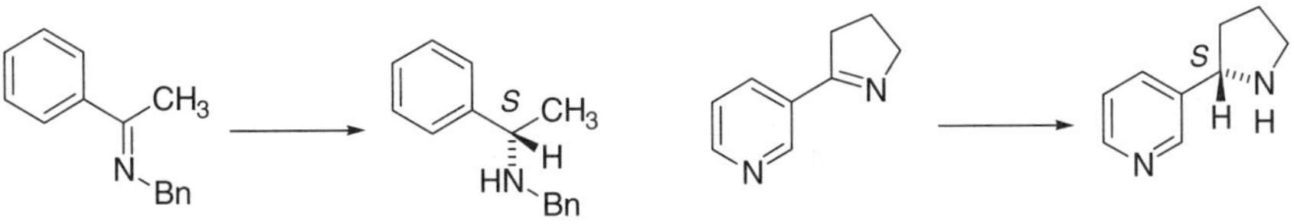

59

61

60

62

The use of Ru complexes in place of rhodium complexes as catalysts has been investigated. The first Ru complex–catalyzed asymmetric hydrosilylation of ketones appeared in 1997 [44]. The reaction of **1** with **2** (1.6 equiv) by using a chiral Ru-catalyst generated from [RuCl$_2$(C$_6$H$_6$)]$_2$ (0.5 mol%) and (*R*)-BINAP [**58**-(*R*)] (2.2 mol%) in the presence of AgOTf (2 mol%) gave **4** in 95% yield at room temperature, but the enantioselectivity was only 5% ee. Under similar conditions the same reaction catalyzed by a Ru complex with Pybox-*i*-Pr (**12a**) proceeded sluggishly to give **4** in only 16% yield. Accordingly, a new chiral tridentate ligand **63** containing two phosphine groups and one pyridine moiety was developed in the hope of solving the problem. Indeed the Ru-catalyst with ligand **63** in the presence of AgOTf improved the enantioselectivity of this reaction, giving **4** with 54% ee (*R*) [44]. The reduction of 2-acetonaphthone (**32**) by using the same chiral Ru-catalyst gave 1-naphthylethanol with 66% ee (*R*) in 98% yield. Next, a Ru-complex with ferrocenylphosphine-oxazoline **64** was developed, which exhibited extremely high catalytic activity and enantioselectivity in asymmetric hydrosilylation of ketones and imines [45]. The reaction of **1** with **2** (2 equiv) by using the chiral Ru-complex with **64** (1 mol%) in the presence of AgOTf (1 mol%) at 0°C gave **4** with 93% ee (*R*) in 53% yield. The reaction of propiophenone (**30**) catalyzed by the same chiral Ru-complex in the presence of Cu(OTf)$_2$ afforded 1-phenylpropanol with 97% ee. It is worthy of note that

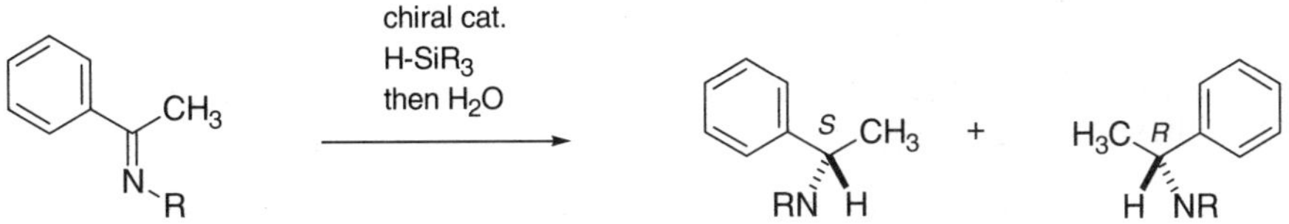

Scheme 2.3.

this chiral Ru-complex with **64** is also highly effective for the reduction of imine **65** to chiral amine **66** with 88% ee (*S*) in 60% yield.

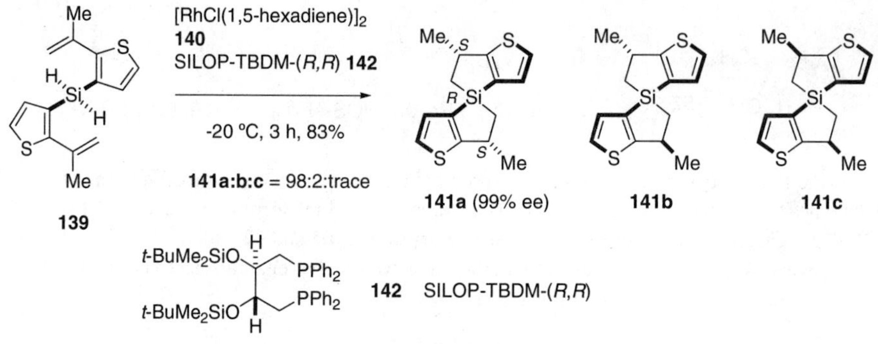

The hydrosilylation of several nitrones **67–69** with diphenylsilane catalyzed by Ru_2Cl_4(Tol-BINAP, **70**)$_2$(Et$_3$N) gives the corresponding hydroxyamines **71–73** with 63–91% ee [46]. This reaction provides a new route to chiral amines from racemic amines by means of nitrones. Although an iron complex is rarely used as catalyst for hydrosilylation, [Cp$_2$Fe$_2$(PHMen$_2$, **74**)(CO)$_2$] acts as a catalyst in the asymmetric hydrosilylation of **1** under irradiation to give **4** with 33% ee [47].

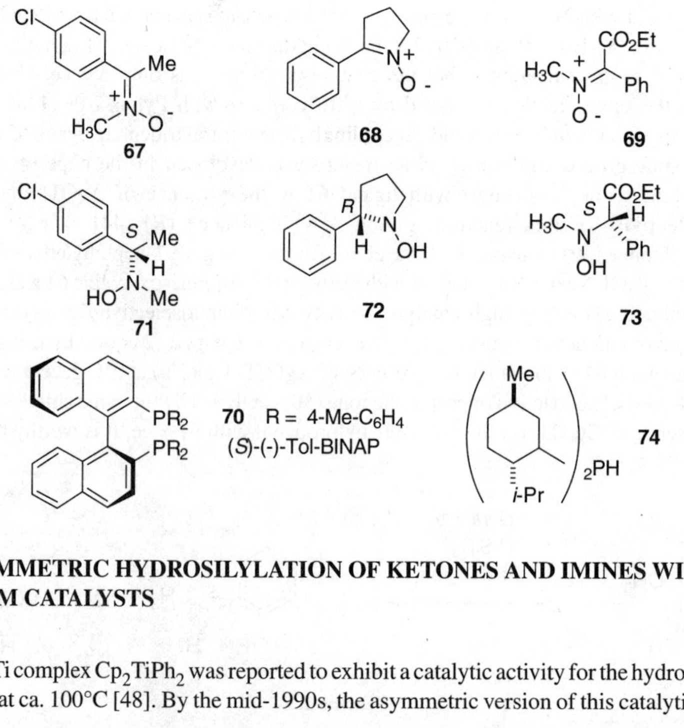

2.3. ASYMMETRIC HYDROSILYLATION OF KETONES AND IMINES WITH TITANIUM CATALYSTS

In 1988, a Ti complex Cp$_2$TiPh$_2$ was reported to exhibit a catalytic activity for the hydrosilylation of ketones at ca. 100°C [48]. By the mid-1990s, the asymmetric version of this catalytic process was developed. Binaphthyl(indenyl)titanium chloride **75** (4.5 mol%) catalyzed the asymmetric hydrosilylation-reduction of **1** and 2-acetonaphthone (**31**) with triethoxysilane in THF at −78°C ~ room temperature for 24 h to give the corresponding secondary alcohols with 14–40% ee in 90–100% yield [49]. Asymmetric reduction of ketones by using bis(indenyl)titanium binaphth-diolate **76** (4.5 mol%) as the catalyst and polymethylhydrosiloxane (PMHS) (5 equiv) in benzene at room temperature for 22–84 h achieved exceedingly high enantioselectivity, that is, 97% ee (*S*) for **1**, 95% ee (*S*) for **30**, and 95% ee (*S*) for **31** [50,51,52] (Scheme 2.4). Dialkyl ketones, however, were reduced to the corresponding alcohols with low enantiopurity. In this catalytic system, it is necessary to add alkyllithium reagents such as *n*-BuLi to the Ti pre-cata-lysts to form active low-valent Ti hydride species.

Bis(indenyl)titanium chloride **77** also exhibited good enantioselectivity in the reactions of ethyl isopropyl ketone (65% ee, *R*) and ethyl cyclopentyl ketone (70% ee, *R*) [53]. However, the reduction of **1** with the Ti catalyst **76** activated by MeLi resulted in giving **4** with 12% ee in 90% yield [53]. The reaction of **1** with triethylsilane catalyzed by (*R*)-BI-NOL-Ti(O-*i*-Pr)$_2$ (**79**) (10 mol%) in ether at 50°C gave **4** with 54% ee (*R*) [54]. Asymmetric hydrosilylation of imines was successfully carried out by using bis(indenyl)titanium fluoride **78** (1–2 mol%) and PhSiH$_3$ (1.5 equiv) in THF, that is, the reactions of imines **65** and **80–83** were carried out at room temperature ~ 35°C to give the corresponding chiral amines **66** and **84–87** in 80–96% yield with extremely high enantioselectivity, that is, up to 99% ee (*S*) for **66**, 97% ee for **84**, 96% ee for **85**, 93% ee for **86**, and 98% ee for **87** [55].

Addition of isobutylamine in combination with the use of PMHS as the hydrosilane

75

76-(*R*,*R*)
X$_2$ = 1,1'-binaphth-2,2'-diolate

77-(*S*,*S*) X = Cl
78-(*S*,*S*) X = F

79-(*R*)

accelerated the reduction of *N*-benzyl-1-indanimine **88** to give the corresponding amine **89** with 92% ee [56]. The catalyst **78** was initially activated by treatment with PhSiH$_3$, piperidine, and methanol in THF at 60°C. Then, to this catalyst solution was added *N*-benzyl-1-indanimine **88** in the presence of *i*-BuNH$_2$ and PMHS to initiate the reaction. It is interesting that acyclic imines, even as a mixture of *Z* and *E* isomers (1.8:1~3.5:1), were also reduced by this catalytic system to the corresponding amines with >90% ee [56].

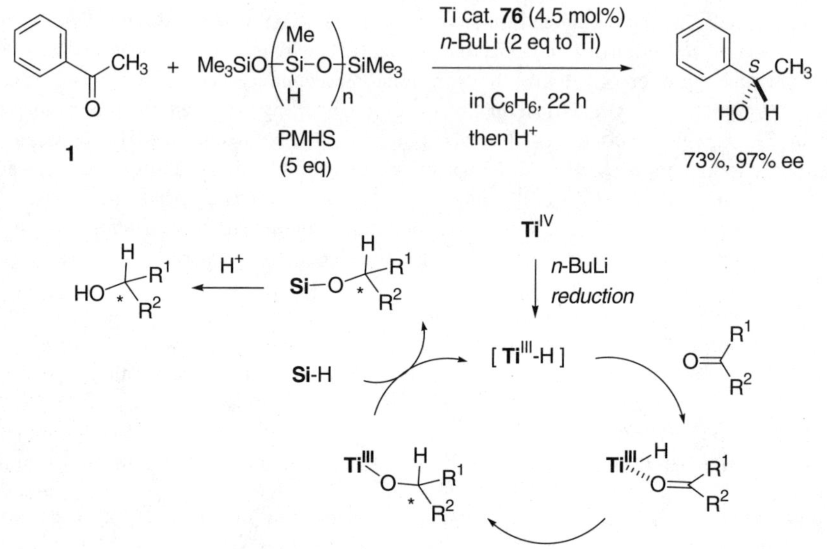

Ti cat. **76** (4.5 mol%)
n-BuLi (2 eq to Ti)

PMHS
(5 eq)

in C$_6$H$_6$, 22 h
then H$^+$

73%, 97% ee

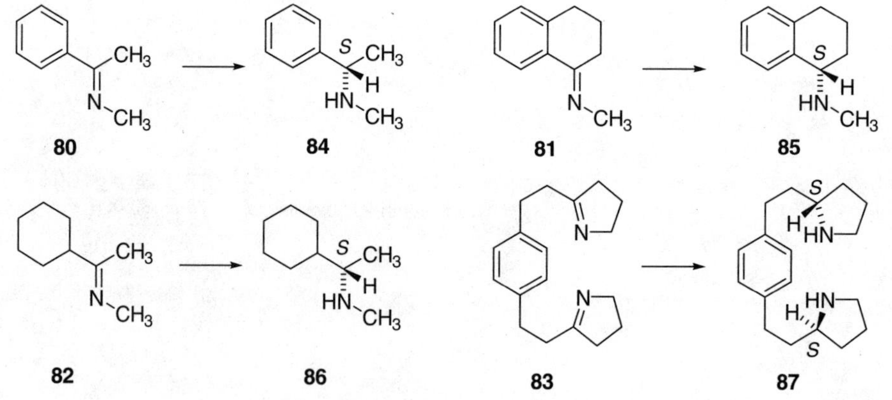

1

Scheme 2.4.

80

84

81

85

82

86

83

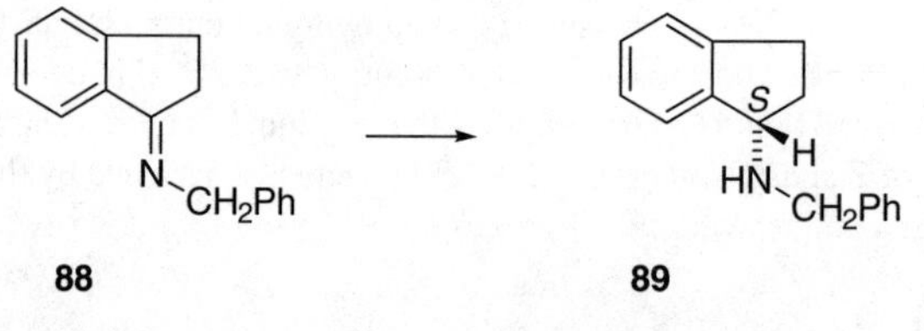

87

88

89

2.4. ASYMMETRIC HYDROSILYLATION OF OLEFINS WITH TRANSITION-METAL CATALYSTS

Hydrosilylation of olefins provides efficient and versatile routes not only to organosilanes, but also to alcohols and halides by using the oxidation procedure developed by Tamao [57]. The asymmetric hydrosilylation of styrene catalyzed by Pt complex with chiral P(Bn)MePh ligand was first reported in 1971 to afford the corresponding chiral silane **90** with 5.2% ee [58]. A moderate enantioselectivity (52–53% ee) was realized by using PPFA(**91**)-PdCl$_2$ as catalyst (0.01 mol%) for the reactions of styrene and norbornene (**92**) with trichlorosilane followed by oxidation and bromination to give **93** (52% ee) and **94** (53% ee), respectively [59] (Scheme 2.5). The reaction of 1-phenyl-1,3-butadiene (**95**) with trichlorosilane catalyzed by a modified PPFA(**96**)-PdCl$_2$ was carried out at 80°C, followed by oxidation to give *cis*-allylic alcohol **97** with 66% ee (*R*) in 52% yield [60]. The use of fluorosilanes such as HSiPh$_2$F for the reaction catalyzed by a combination of modified binaphthyl ligand **98** and (η^3-C$_3$H$_5$PdCl)$_2$ (0.3 mol%) increased the yield of the hydrosilylation product **99** (66% ee) up to 96% [61]. The reaction of cyclic diene **100** with F$_2$PhSiH catalyzed by Pd complex with PPF-OH (**101**) or PPF-OAc (**102**), followed by difurylation of the silyl moiety, gave allylsilane **103** with 62–72% ee [62].

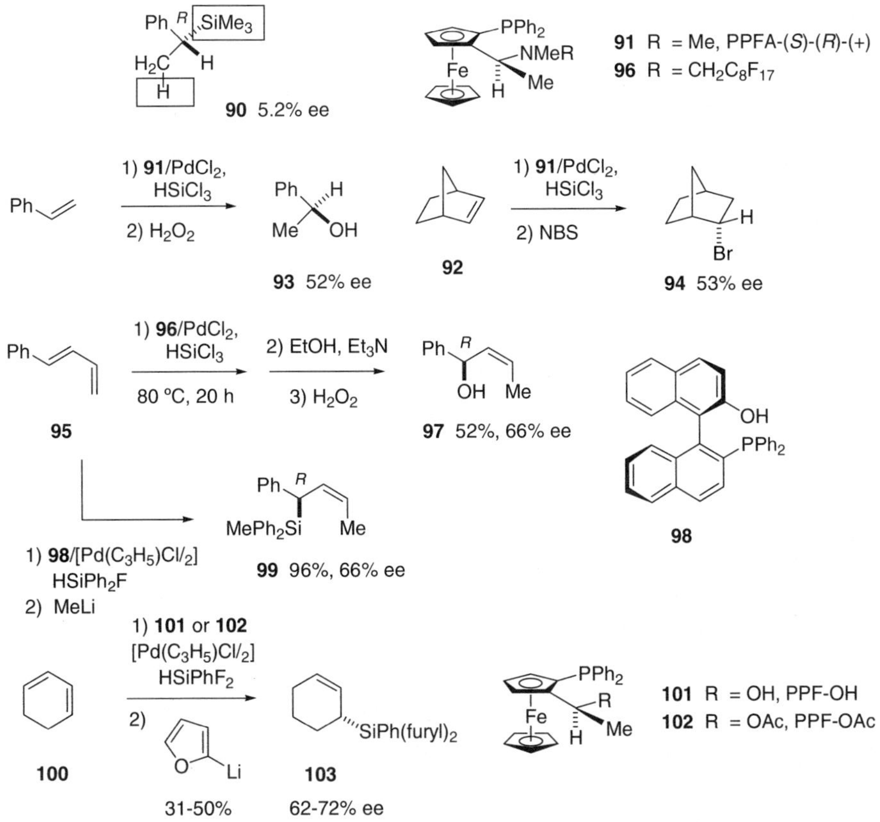

Scheme 2.5.

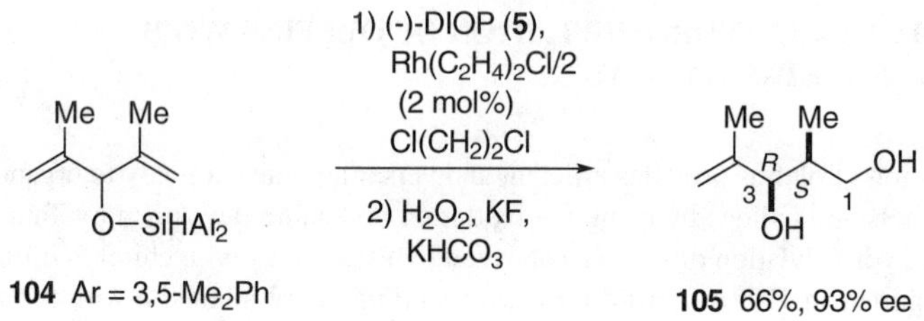

Scheme 2.6.

In 1990, the enantioselective intramolecular hydrosilylation of allylic alcohols was success-fully applied to the synthesis of chiral 1,3-diol [63] (Scheme 2.6). The reaction of 3-diarysiloxy-1,4-pentadiene (**104**) catalyzed by (−)-DIOP-[Rh(C$_2$H$_4$)Cl]$_2$ (2 mol%), followed by Tamao oxidation, gave (2S,3R)-1,3-diol **105** (syn:anti = 98:2) with 93% ee [63].

Rh-diphosphine complex, [Rh(Chiraphos, **106**)](ClO$_4$)$_2$, was used as a catalyst for the intramolecular hydrosilylations of homoallylic silane **107** and silyl allyl ether **108** in acetone at 25°C to give the corresponding 1,4-diol **109** (60% ee, R) in 84% yield and 1,3-diol **110** (with 56% ee, R) in 96% yield, respectively [64] (Scheme 2.7). The Rh-Chiraphos catalyzed reaction of 1-(3-phenylpropen-2-yloxy)silacyclohexane (**111a**) gave diol **112** with 74% ee (R) in 61%

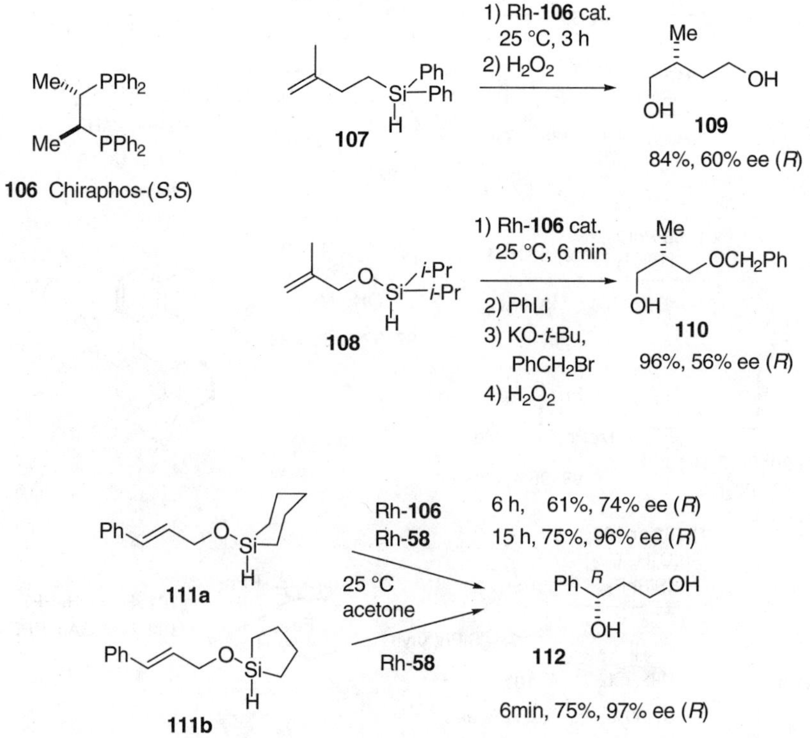

Scheme 2.7.

TABLE 2.3. Asymmetric Hydrosilylation of Ketones with Titanium Catalysts[a]

Titanium cat. (RLi)[b]	Ketone	Silane	S/C[c]	Silane/Ketone	Solvent	Temp/Time (°C/h)	Yield (%)	Product: alcohol		
								% ee	abs config	Ref.
75 (n-BuLi)	PhCOMe **1**	**S1**	200	2.5	THF	−78–rt/24	100	14	S	51
	2-NpCOMe **32**	**S1**	200	2.5	THF	−78–rt/24	90	40	S	51
76 (n-BuLi)	PhCOMe **1**	**S2**	23	5	benzene	rt/22	73	97	S	52
	1-Tetralone **37**	**S2**	23	5	benzene	rt/84	92	91	S	52
	2-NpCOMe **32**	**S2**	23	5	benzene	rt/72	84	95	S	52
	1-C$_6$H$_9$-COMe **54**	**S2**	23	5	benzene	rt/24	72	90	S	52
	PhCOEt **30**	**S2**	23	5	benzene	rt/24	96	95	S	52
77 (MeLi)	PhCOMe **1**	**S3**	100	1	no solv.	rt/300	90	12	R	53
	i-PrCOMe	**S4**	100	1	no solv.	rt/230	91	65	R	53
79 (no)	PhCOMe **1**	**S1**	10	6	Et$_2$O	50/5	98	55	R	55
	1-NpCOMe **31**	**S1**	10	6	Et$_2$O	50/5	98	54	R	55

[a]**S1** = (EtO)$_3$Si, **S2** = Me$_3$Si(SiHMeO)$_n$SiMe$_3$ (PMHS), **S3** = Ph$_2$SiH$_2$, **S4** = PhMeSiH$_2$.
[b]RLi, 2 equiv. to Ti.
[c]S/C, substrate ketone/catalyst Ti.

yield, whereas the reactions of **111a** and 1-(3-phenylpropen-2-yloxy)silacyclopentane (**111b**) catalyzed by [Rh(BINAP, **58**)](ClO$_4$)$_2$ achieved exceedingly high enantioselectivity in yielding **112**, that is, 97% ee (R) for **111a** and 97% ee (R) for **111b**.

The mechanism of the intramolecular hydrosilylation catalyzed by Rh and Pt complexes was investigated by using deuterated silanes, which indicated the operation of both the traditional Chalk-Harrod "hydrometallation" and "silylmetallation" pathways accompanied by rapid β-hydride elimination [65]. This intramolecular reaction was applied to the syntheses of natural products [66].

Breakthrough in the asymmetric hydrosilylation of olefins was brought about by using a Pd catalyst (0.1 mol%) with a chiral monodentate phosphine ligand MOP (**116**) [67], which was readily synthesized from enantiopure binaphthol in several steps [68]. It is worth mentioning that Pd catalysts with chelating chiral diphosphines did not exhibit catalytic activity in the hydrosilylation of olefins at temperatures <80°C [67]. This observation gave rise to the design of the novel chiral monodentate phosphine ligand, MOP. With this ligand, the regioselectivity for the branched product is >90%. The asymmetric hydrosilylation of simple 1-alkenes such as 1-octene (**113**) and 4-phenyl-1-butene (**114**) catalyzed by Pd complex **115** with **116** at 40°C followed by Tamao oxidation gives optically active 2-alkanols with >90% ee, for example, 95% ee for **113** and 97% ee for **114** (Scheme 2.8). Cyclic olefins, norbornene (**92**), its diester derivative **119**, and 2,5-dihydrofuran (**122**), were converted to the corresponding alcohols **120**, **121**, and **123**, respectively, in 94–96% ee [69].

The regioselectivity (branched:linear) for the addition of trichlorosilane to 1-alkenes is in a range of 80:20 ~ 93:7 [67]. In the case of 1-aryl-1-alkenes such as (2-chlorophenyl)ethene (**124**), 1-phenylprop-1-ene (**125**), and indene, the regioselectivity reaches as high as 99:1, although the enantioselectivity for the formation of the correspond-

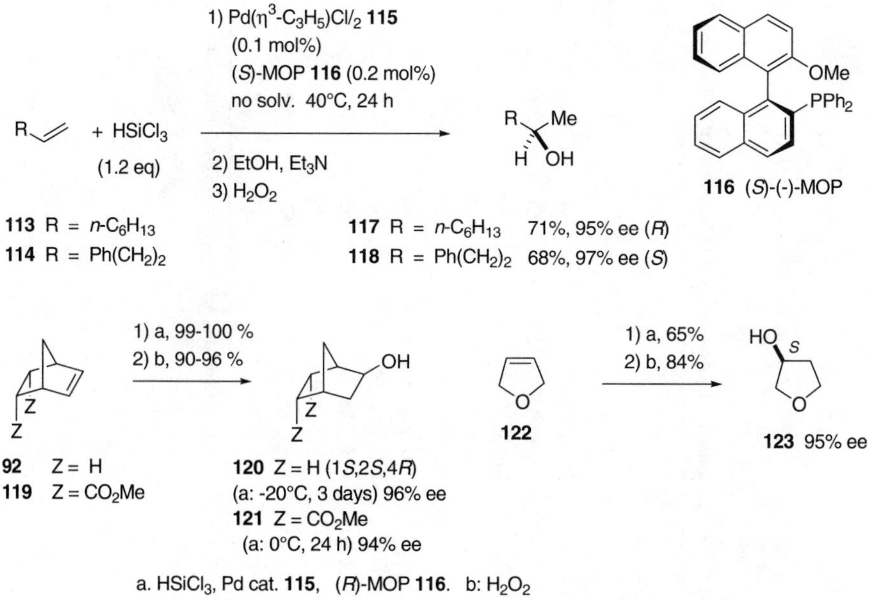

Scheme 2.8.

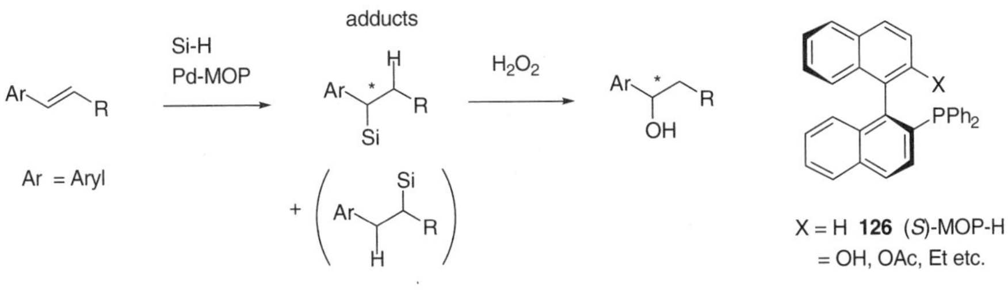

Scheme 2.9.

ing (R)-benzylic aldohols is in a range of 71–85% ee of using R-**116** as the chiral ligand (Scheme 2.9) (Table 2.4) [70]. A variety of MOP ligands having H, OH, OAc, Et, and COOMe groups in place of MeO group of MOP (**116**) has also been developed (Scheme 2.9) [71]. It is rather surprising and thus worth mentioning that among these MOP ligands the simplest, (S)-MOP-H (X = H, S-**126**), attained the highest enantioselectivity (94–96% ee) in the hydrosilylation of styrene and substituted styrenes **127**–**129** [72] (Table 2.4). In contrast to this finding, other MOP ligands that bear CN, Et, COOMe, and OH as the substituent X resulted in low enantioselectivity for the same reactions (18–34% ee). (R)-MOP-phen (**131**) is so far the best ligand for the asymmetric hydrosilylation-oxidation of diene **130** to give alcohol **133** with 80% ee via **132** [9] (Scheme 2.10). For this reaction, **116** and **126** induce only 28–39% ee.

TABLE 2.4. Asymmetric Hydrosilylation of Styrene Derivatives with MOP-Pd Catalysts[a]

			Product: alcohol			
		Temp/				
		Time	Yield		abs	
Ligand	Styrene	(°C/h)	(%)[b]	% ee	config	Ref.
(R)-(−)-MOP	styrene	5/44	97	71	R	70
116	2-ClC$_6$H$_4$CH=CH$_2$ **124**	5/13	94	81	R	70
	PhCH=CHCH$_3$ **125**	40/48	78	82	R	70
	indene	5/72	83	85	R	70
(S)-MOP-H	styrene	−10/32	100[c]	94	R	72
125	4-CF$_3$C$_6$H$_4$CH=CH$_2$ **127**	0/120	98[c]	96	R	72
	3-ClC$_6$H$_4$CH=CH$_2$ **128**	0/36	68[c]	95	R	72
	4-ClC$_6$H$_4$CH=CH$_2$ **129**	0/120	80[c]	94	R	72

[a]HSiCl$_3$ (1.2 equiv. to styrenes), Pd(η^3-C$_3$H$_5$)Cl/$_2$ (olefin/Pd = 1000), Ligand (2 equiv to Pd), no solvent.
[b]yield from olefin via a mixture of the adduct isomers.
[c]yield of adduct; subsequent oxidation, >90%.

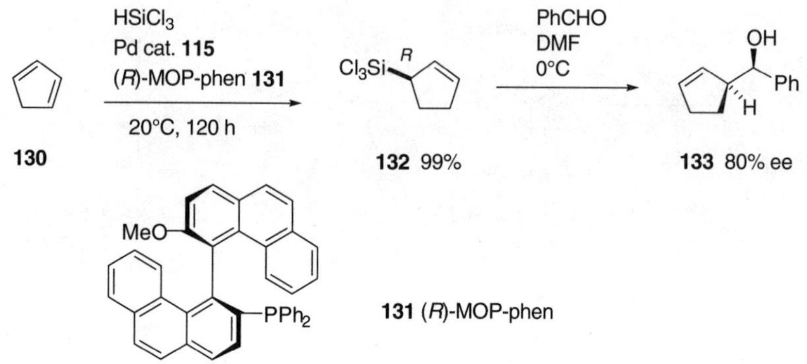

Scheme 2.10.

2.5. ASYMMETRIC SYNTHESIS OF CHIRAL SILICON COMPOUNDS

Hydrosilylation of ketones with prochiral dihydrosilanes provides a useful method to yield optically active silanes. In 1974, the asymmetric synthesis of 1-NpPhMeSiH (46% ee, R) (**134**) was reported through hydrosilylation of diethyl ketone with 1-NpPhSiH$_2$ (**3**) catalyzed by (+)-DIOP-Rh complex followed by methylation (Scheme 2.11, eq 1) [74]. Use of (−)-menthone (**135**), **3**, and an achiral catalyst, RhCl(PPh$_3$)$_3$, gave 1-NpPhEtSiH (67% ee, R) (**136**) (Scheme 2.11, eq 2) [75]. The combination of (+)-DIOP-Rh catalyst and (−)-menthone improved the enantiopurity of **136** to 82% ee (R) through double-asymmetric induction (matching pair), whereas the reaction by using (−)-DIOP-Rh catalyst and (−)-menthone (mismatching pair) gave **136** with 46% ee (R) [75].

In 1994, extremely high enantioselectivity was achieved for this type of reactions Cybinap (**137**)/Rh catalyst [76]. The hydrosilylation of **3** with acetone catalyzed by

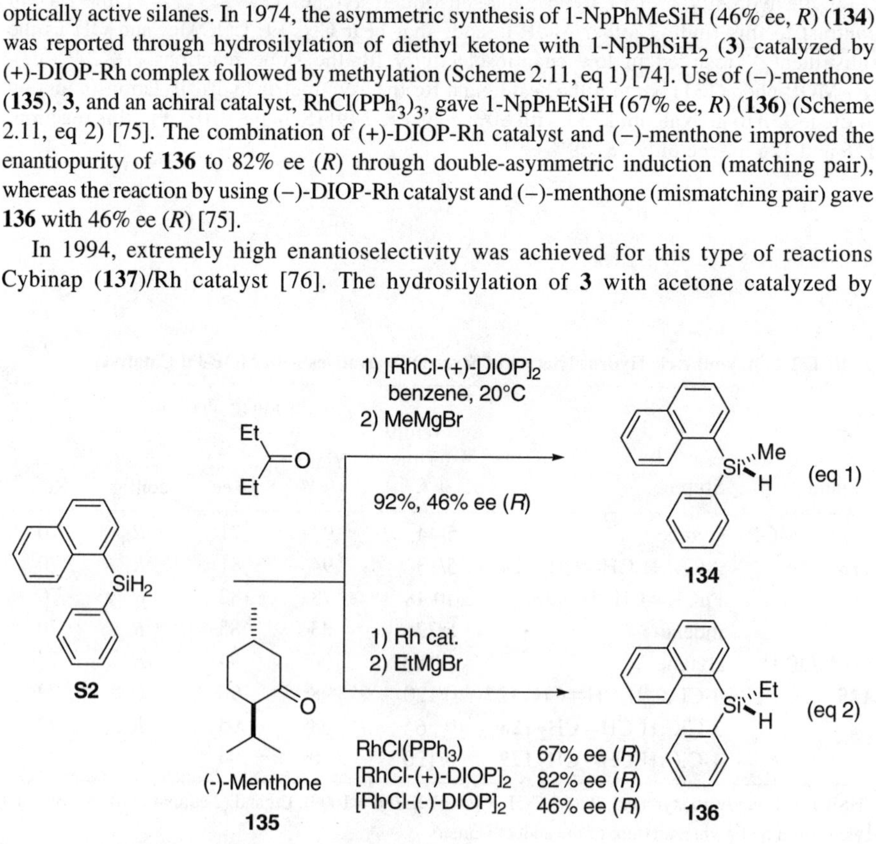

Scheme 2.11.

(R)-Cybinap/[Rh(COD)Cl]$_2$ (0.5 mol%) in THF at –20°C to give 1-NpPhSi(OCHMe$_2$)H (**138**) in 81% yield, whose enantiopurity was determined >99% ees (R) on the basis of chiral HPLC analysis. The use of diethyl ketone and diisopropyl ketone for this reaction also achieved excellent enantioselectivity (98–99 % ee). Under the same reaction conditions, (R)-BINAP (**58**)/Rh catalyst exhibited somewhat lower enantioselectivity (83–95% ee).

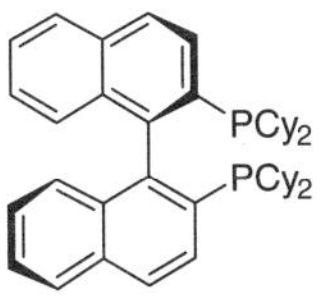

137 (R)-Cybinap 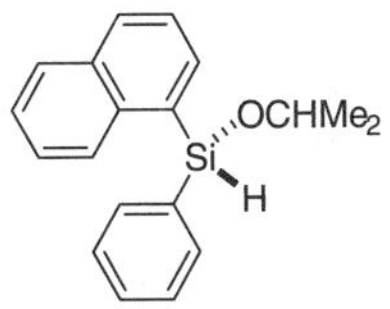   **138** 99% ee (R)

Hydrosilylation of olefins was successfully applied to the synthesis of unique axially chiral spirosilanes (Scheme 2.12) [77]. Intramolecular hydrosilylation of bis(alkenyl)dihydrosilane **139** promoted by a catalyst in situ generated from [Rh(1,5-hexadiene)Cl]$_2$ (0.3–0.5 mmol) **140** and a chiral phosphine (1.1–1.3 equiv to Rh) to give a mixture of spirosilanes **141a–c** in 61–83% yields, wherein for the C$_2$-symmetric product **141a** is predominant. The reactions catalyzed by BINAP-Rh and DIOP-Rh gave **141a** with 58% ee and 83% ee, whereas the reaction by using disiloxy-diphosphine SILOP (**142**) as the chiral ligand afforded **141a** with 99% ee and excellent product, that is, **141a:141b:141c** = 98:2:trace, in 83% yield.

2.6. ASYMMETRIC HYDROBORATION WITH TRANSITION-METAL CATALYSTS

Catalytic hydroboration of olefins with catecholborane (**143**) in the presence of Wilkinson catalyst was first reported in 1985 [78,79,80]. Although the reaction takes place without the catalyst, it requires high temperature. The Rh-catalyzed reaction proceeds smoothly at room

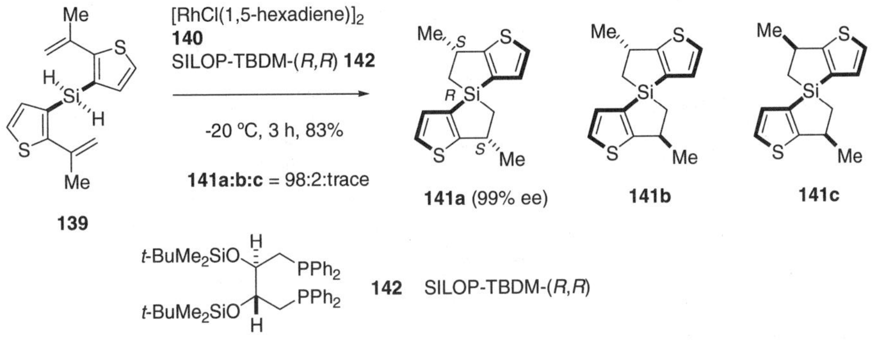

Scheme 2.12.

temperature to give the catecholborane adducts, which are subsequently treated with an alkaline hydrogen peroxide solution to produce the corresponding alcohols. In general, 1-alkenes are regioselectively converted to primary alcohols. Mechanistic aspects for Rh-catalyzed hydroboration of olefins have been studied in detail based on deuterium-labeling experiments [79a, 80].

In 1988, the first catalytic asymmetric hydroboration of olefins was reported by using norbornene and 2-*tert*-butylpropene as substrates and DIOP (**5**)/[Rh(COD)Cl]$_2$ (2 mol%) as the catalyst to give 2-hydroxynorbornane (**120**) (57% ee) and 2,3,3-trimethylbutanol (**144**) (69% ee) (Scheme 2.13) (Table 2.5) [81a]. Under the same reaction conditions, the use of BDPP-*S,S* (**145**) and 2-MeODIOP (**146**) as chiral ligand attained higher enantioselectivity (80–82% ee) for the reaction of norbornene [81b].

In 1989, the hydroboration of styrene catalyzed by Rh-BINAP complex (2 mol%) was found to proceed regioselectively in DME at –30°C, followed by oxidation to give 1-phenylethanol (**4**) with 70% ee in 90% yield [82] (Scheme 2.14). The enantioselectivity was increased to 96% ee when the reaction was carried out at –78°C. P,N-type ligand **148**, an aza analog of MOP, was found to serve as an effective chiral ligand for the Rh-catalyzed reaction of *p*-MeO-styrene at room temperature to give **149** with 94% ee [83]. A similar ligand **151** was also developed, which realized better enantioselectivity for the reactions of dihydronaphthalene (84% ee, *R*) and 1-phenylpropene (91% ee, *R*) [83c,84,85].

Double asymmetric induction was studied in the reaction of *p*-MeO-styrene catalyzed by Rh-BINAP complex by using chiral borane **152** derived from ephedrine. The Rh-(*R*)-BINAP/**152** combination (matching pair) gave **149** with 86% ee, whereas the enantioselectivity observed for the Rh-(*S*)-BINAP/**152** combination (mismatching pair) was only 8% ee [86]. A ferrocenyldiphosphine, Josiphos (**153**), is also an effective ligand, which gave **4** with 91.5% ee in the reaction of styrene at –70°C [87]. Pyrazole-containing ferrocenylphosphines, **154a** and **154b**, achieved even higher enantioselectivity than **153** in the reaction of styrene although regioselectivity was not impressive [88]. The reaction of styrene catalyzed by **154a**/[Rh(COD)$_2$]BF$_4$ (1 mol%) gave **4** with 95.1% ee (*R*) and 66% regioselectivity (total yield, 91%) [88]. The use of 4-CF$_3$-phenyl analog **154b** as ligand increased the enantioselectivity to 98% ee for the same reaction.

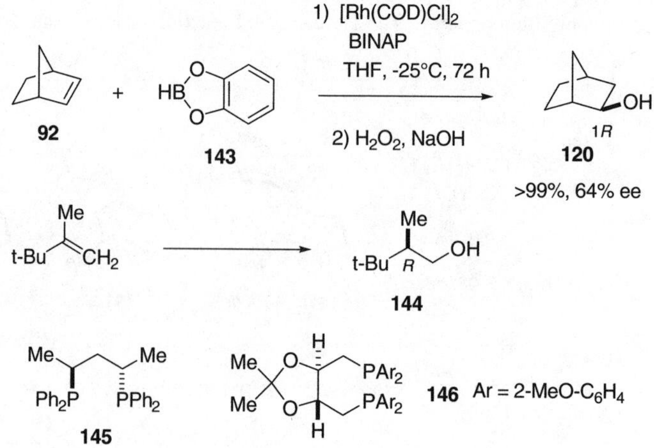

Scheme 2.13.

TABLE 2.5. Asymmetric Hydroboration of Olefins with Rhodium Catalysts and Catecholborane[a]

Ligand	Olefin	Rh-cat.[b]	L/Rh	S/C[c]	Solvent	Temp/ Time (°C/h)		Product: Alcohol Yield (%)	% ee	abs config	Ref.
(R,R)-(−)-DIOP **5**	norbornene	**C**	1	50	THF	−25/72	**120**	>99	57	*R*	81a
(R)-(+)-BINAP **58**	norbornene	**C**	1	50	THF	−25/72	**120**	>99	64	*R*	81a
(R,R)-(−)-DIOP **5**	t-BuMeC=CH₂	**C**	1	50	THF	−25/72	**144**	>99	69	*R*	81a
BDPP-*S,S* **145**	norbornene	**C**	1	50	THF	−25/6	**120**	>95	80	*R*	81b
2-MeODIOP-*R,R* **146**	norbornene	**C**	1	50	toluene	−25/6	**120**	>95	82	*R*	81b
(R)-(+)-BINAP **58**	styrene	**F**	1.1	50	THF	−30/0.5	**4**	90	70	*R*	82a
	styrene	**F**	1.1	50	DME	−78/2	**4**	91	96	*R*	82a
	p-Me-styrene	**F**	1.1	50	DME	−78/6	**147**	77	94	*R*	82a
P-N ligand **148**	*p*-MeO-styrene	**H**	1	100	THF	20/0.25	**149**	57	94	*S*	83a
	styrene	**H**	1	100	THF	20/0.25	**4**	69	88	*S*	83a
	indene	**H**	1	100	THF	20/0.25	**150**	58	91	*S*	83a
Josiphos **153**	styrene	**I**	1	50	DME	−70/10	**4**	65	91.5	*R*	87
	indene	**I**	1.1	100	DME	−70/10	**150**	—	42	*R*	87
P-N ligand **154a**	styrene	**F**	1.1	100	THF	20/3–5	**4**	60	95.1	*R*	88
P-N ligand **154b**	styrene	**F**	1.1	100	THF	20/3–5	**4**	37	98	*R*	88

[a]Catecholborane (1.0–1.2 equiv.).
[b]Rh cat.: **C**, [Rh(COD)Cl]₂; **F**, [Rh(COD)₂]BF₄; **H**, [Rh(COD)(**148**)]OTf; **I**, [Rh(NBD)₂]BF₄.
[c]S/C (substrate/catalyst) = olefin/Rh.

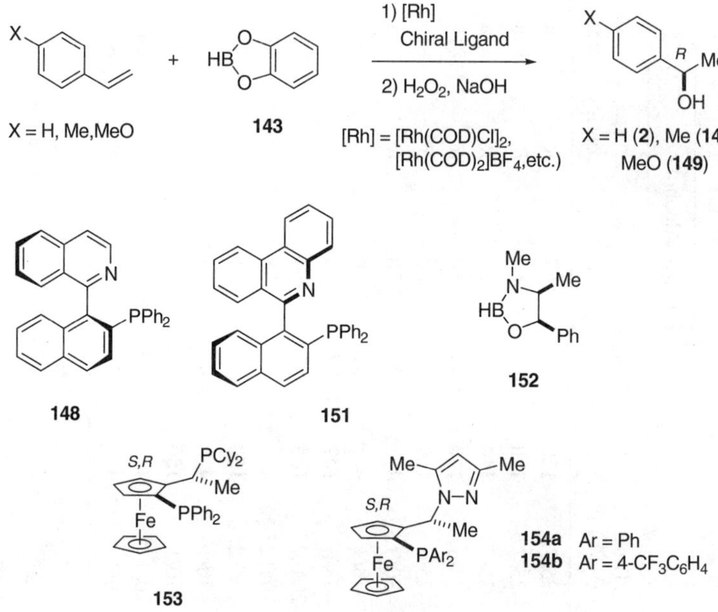

Scheme 2.14.

Asymmetric hydroboration of 1-phenyl-1,3-butadiene (**95**) catalyzed by Rh-BINAP gave the corresponding optically active 1,3-diol **155** with 72% ee [89,90] (Scheme 2.15). Palladium-MOP complex also exhibited catalytic activity for the asymmetric hydroboration of but-1-en-3-yne (**156**), giving an optically active allenyl borane **157** [91].

Transformation of alkylcatecholborane intermediates, thus formed after asymmetric hydroboration prior to oxidation, to the corresponding optically active amines with retention of configuration provides a very useful synthetic method. In fact, this process was realized through the reaction of an alkylcatecholborane with Grignard reagent and HN_2OSO_3H, for example, benzylic amine **159**, was obtained from **158** in 56% yield with retention of configuration [92] (Scheme 2.16).

Sodium borohydride (**160**) was found to serve as a hydrogen donor in the asymmetric reduction of the presence of an α,β-unsaturated ester or amide **162** catalyzed by a cobalt-Semicorrin **161** complex, which gave the corresponding saturated carbonyl compound **163** with 94–97% ee [93]. The β-hydrogen in the products was confirmed to come from sodium borohydride, indicating the formation of a metal enolate intermediate via conjugate addition of cobalt-hydride species (Scheme 2.17).

A combination of chiral cobalt-catalyst and sodium borohydride was successfully applied to the asymmetric reduction of aromatic ketones. A chiral cobalt complex **164** (5 mol%), prepared from the corresponding salen-type chiral bisketoaldimine and cobalt(II) chloride, catalyzed the reduction of dimethylchromanone **165** in the presence of sodium borohydride (1.5 equiv to ketone) in chloroform, including a small amount of ethanol at −20°C for 120 h to give alcohol **166** 92% ee (*S*) in 94% yield (Scheme 2.18) [94]. Addition of tetrahydrofurfuryl alcohol (THFFA) to the reaction system or the use of pre-modified borohydride, $NaBH_2(THFFA)_2$, improved the catalyst activity, that is, using this protocol, the reactions of ketone **165** and

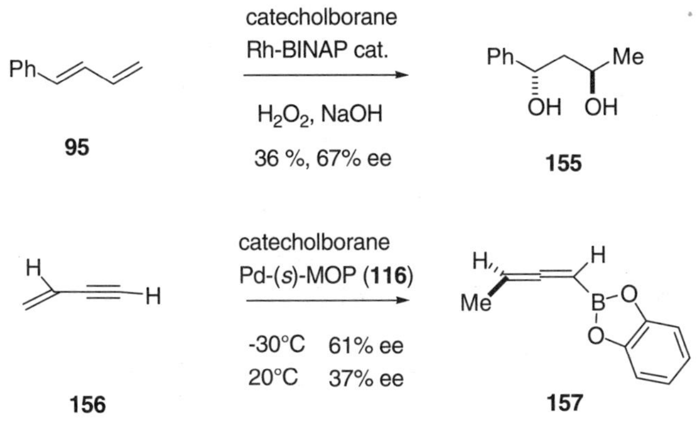

Scheme 2.15.

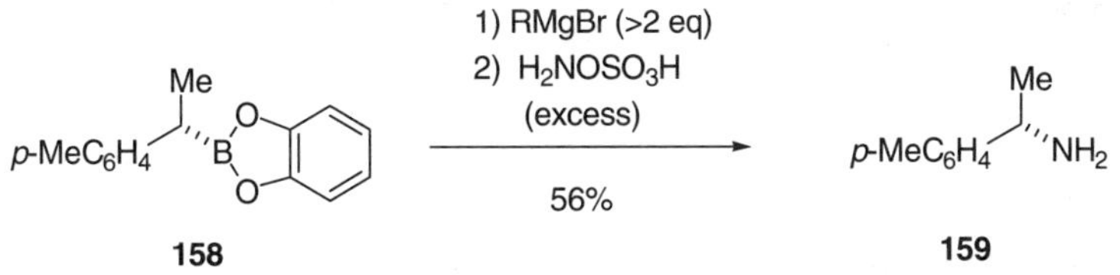

Scheme 2.16.

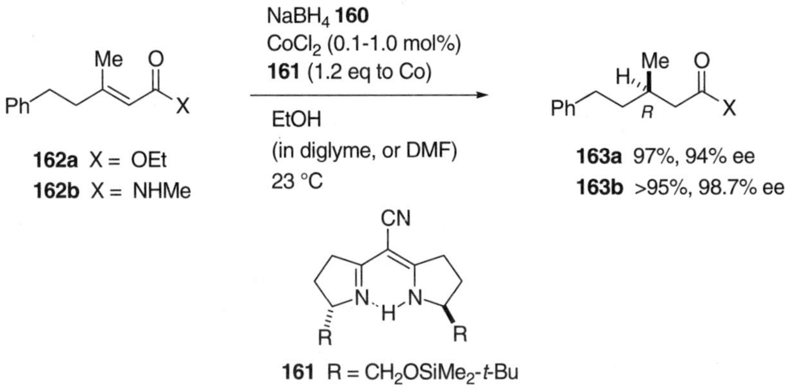

Scheme 2.17.

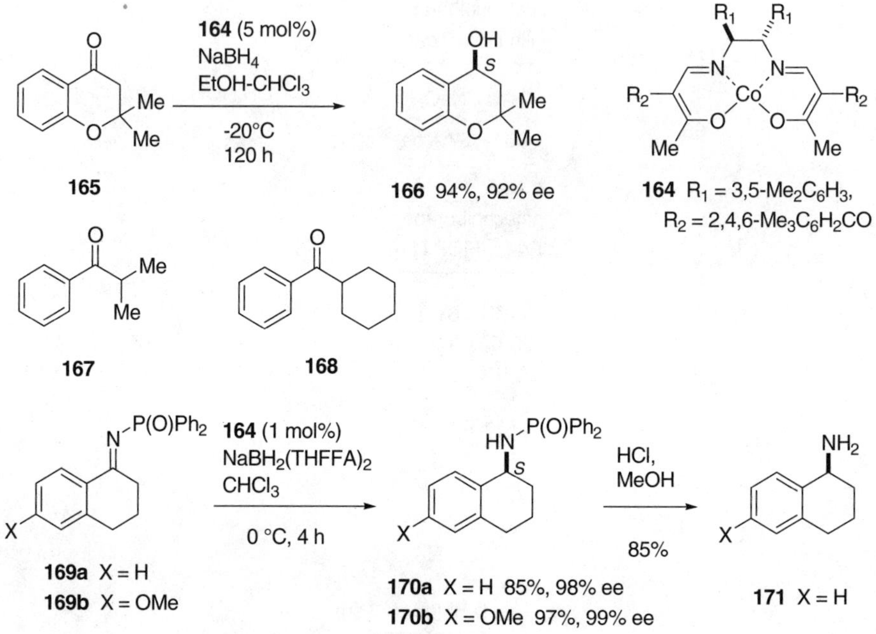

Scheme 2.18.

propiophenone (**30**) by using only 0.1~1 mol% of **164** completed within 1 h to give **166** with 92% ee in 98% yield and 1-phenylpropanol with 97% ee in 98% yield [95a,b]. In a similar manner, the addition of methanol to the reaction system was effective, achieving high enantioselectivity for the reactions of isopropyl phenyl ketone (**167**) (95% ee) and cyclohexyl ketone (**168**) (95% ee) [95c]. This catalytic reduction system was also applied to the asymmetric reduction of *N*-phosphinylketimines **169** to *N*-phosphinylamines **170** with >98% ee, which can be readily hydrolyzed to the corresponding primary amines **171** [96].

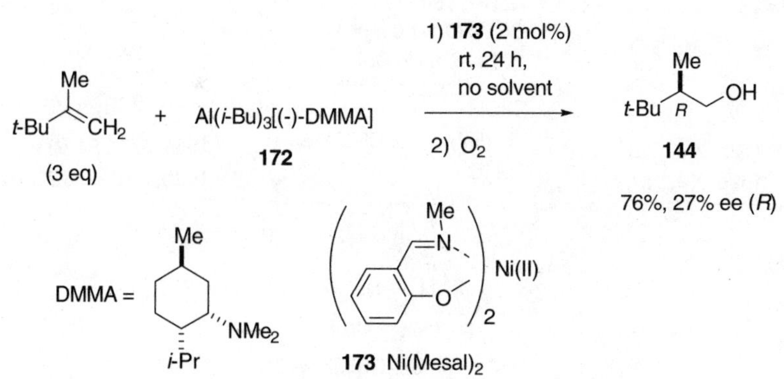

Scheme 2.19.

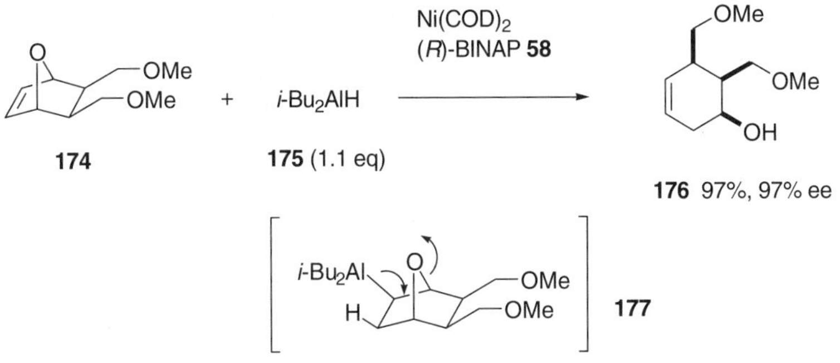

Scheme 2.20.

2.7. ASYMMETRIC HYDROALUMINATION WITH TRANSITION-METAL CATALYSTS

Nickel catalysts are well-known to promote hydroalumination of olefins, and the resulting organoaluminum compounds can be converted to the corresponding alcohols through oxidation with molecular oxygen [97].

In 1981 the first attempt for the asymmetric hydroalumination with a chiral alane reagent took place. The reaction of a chiral amine-trialkylaluminum complex, $Al(i\text{-}Bu)_3$-[(−)-DMMA] (**172**), with 1,1-dialkylethylenes (excess) catalyzed by Ni(II)[Mesal]$_2$ (**173**) (2 mol%) without solvent, followed by exposure of the resulting adduct to the oxygen atmosphere gave the corresponding alcohols with low enantiopurity, for example, the reaction of 2-*tert*-butylpropene with **172** in the presence of **173** gave **144** with 27% ee (*R*) (Scheme 2.19) [98].

Highly enantioselective hydroalumination was realized in 1995 in the reaction of **174** with $i\text{-}Bu_2AlH$ (DIBAL, **175**) promoted by the catalyst generated in situ from (*R*)-BINAP (**58**) (21 mol%) and Ni(COD)$_2$ (14 mol%) at −78°C, which gave **176** with 97% ee in 97% yield through the ring-opening of the initially formed adduct **177** (Scheme 2.20) [99].

2.8. CONCLUSION

In the past decade, transition-metal catalyzed asymmetric hydrosilylation, hydroboration, and hydroalumination have reached a sufficiently high level of catalytic efficiency in terms of turnover number and enantioselectivity, which are comparable with other catalytic asymmetric processes, such as asymmetric hydrogenation, oxidation, and cyclopropanation. These developments have been brought about largely based on the design and synthesis of various new and unique chiral ligands such as multidentate nitrogen ligands, phosphine-nitrogen and phosphine-oxygen ligands, monodentate phosphines, and cyclopentadiene derivatives, which have been combined exactly with appropriate metals such as Rh, Pd, and Ti. In spite of ambiguity in mechanistic details, the evolutionary construction of enantiotopic environments around these catalysts has been realized. In the next decade, it will be necessary to shed light on the mechanistic details and to identify true active catalyst species in these catalytic asymmetric reactions. These future efforts will open an avenue for the discovery and development of new methodology and new processes based on asymmetric hydrosilylation and related reactions for the efficient production of fine and specialty chemicals.

REFERENCES

1. (a) Ojima, I.; Yamamoto, K.; Kumada, M. In *Aspects of Homogeneous Catalysis*; Ugp, R. (Ed.); Reider Dordrecht: 1977; Vol. 3, pp. 185. (b) Ojima, I.; Hirai, K. In *Asymmetric Synthesis*; Morrison, J. D. (Ed.); Academic Press: Orlando, FL, 1985; Vol. 5, pp. 103. (c) Brunner, H. *Synthesis*, **1988**, 645. (d) Brunner, H. *Top. Stereochem.* **1988**, *18*, 129. (e) Ojima, I.; Clos, N.; Bastos, C. *Tetrahedron*, **1989**, *45*, 6901. (f) Ojima, I. In *The Chemistry of Organic Silicon Compounds*, Part 2; Patai, S.; Rappoport, Z. (Eds.); Wiley: New York, 1989; pp. 1479. (g) Aitken, R. A.; Kilényi, S. N. (Eds.); *Asymmetric Synthesis*; Chapmann & Hall: London, 1992. (h) Noyori, R. *Asymmetric Catalysis in Organic Synthesis*; John Wiley & Sons Inc.: New York, 1994. (i) Stephenson, G. R. (Ed.); *Advanced Asymmetric Synthesis*; Chapmann & Hall: London, 1996.

2. Brunner, H.; Nishiyama, H.; Itoh, K. In *Catalytic Asymmetric Synthesis*; Ojima, I. (Ed.); VCH Publisher, Inc.: New York, 1993: Chap. 6, pp. 303.

3. (a) Ojima, I.; Nihonyanagi, M.; Nagai, Y. *J. Chem. Soc., Chem. Commun.* **1972**, 938. (b) Ojima, I.; Kogure, T.; Nihonyanagi, M.; Nagai, T. *Bull. Chem. Soc. Jpn.* **1972**, *45*, 3506.

4. Yamamoto, K.; Uramoto, Y.; Kumada, M. *J. Organomet. Chem.* **1971**, *31*, C9.

5. Dumont, W.; Poulin, J. C.; Dang, T.-P.; Kagan, H. B. *J. Am. Chem. Soc.* **1973**, *95*, 8295.

6. Johnson, T.; Klein, K.; Thomen, S. *J. Mol. Catal.* **1981**, *12*, 37.

7. (a) Brunner, H.; Riepl, G. *Angew. Chem., Int. Ed.* **1982**, *21*, 377; *Angew. Chem. Suppl.* **1982**, 769. (b) Brunner, H.; Reiter, B.; Riepl, G. *Chem. Ber.* **1984**, *117*, 1330. (c) Botteghi, C.; Schionato, A.; Chelucci, G.; Brunner, H.; Kürzinger, A.; Obermann, U. *J. Organomet. Chem.* **1989**, *370*, 17.

8. Payne, N. C.; Stephan, D. W. *Inorg. Chem.* **1982**, *21*, 182.

9. Brunner, H.; Rahman, A. M. F. *Chem. Ber.* **1984**, *117*, 710.

10. (a) Brunner, H.; Riepl, G.; Weitzer, H. *Angew. Chem., Int. Ed.* **1983**, *22*, 331; *Angew. Chem. Suppl.* **1983**, 445. (b) Brunner, H.; Becker, R.; Riepl, G. *Organometallics* **1984**, *3*, 1354. (c) Brunner, H.; Kürzinger, A. *J. Organomet. Chem.* **1988**, *346*, 413.

11. Brunner, H.; Obermann, U.; Wimmer, P. *J. Organomet. Chem.* **1986**, *316*, C1.

12. (a) Brunner, H.; Obermann, U. *Chem. Ber.* **1989**, *122*, 499. (b) Brunner, H.; Brandl, P. *J. Organomet. Chem.* **1990**, *390*, C81. (c) Brunner, H.; Brandl, P. *Tetrahedron: Asymmetry* **1991**, *2*, 919.

13. Nishiyama, H.; Sakaguchi, H.; Nakamura, T.; Horihata, M.; Kondo, M.; Itoh, K. *Organometallics* **1989**, *8*, 846.

14. Balavoine, G.; Client, J. C.; Lellouche, I. *Tetrahedron Lett.* **1989**, *30*, 5141.

15. Brunner, H.; Henrichs, C. *Tetrahedron: Asymmetry* **1995**, *6*, 653.

16. Nishiyama, H.; Kondo, M.; Nakamura, T.; Itoh, K. *Organometallics* **1991**, *10*, 500.

17. Daives, I. W.; Gerena, L.; Lu, N.; Larsen, R. D.; Reider, P. J. *J. Org. Chem.* **1996**, *61*, 9629.

18. Nishiyama, H.; Yamaguchi, S.; Kondo, M.; Itoh, K. *J. Org. Chem.* **1992**, *57*, 4306.

19. Nishiyama, H. unpublished result. Cf., Park, S. B.; Murata, K.; Matsumoto, H.; Nishiyama, H. *Tetrahedron: Asymmetry* **1995**, *6*, 2487.

20. Gladiali, S.; Pinna, L.; Delogu, G.; Graf, E.; Brunner, H. *Tetrahedron: Asymmetry* **1990**, *1*, 937.

21. Nishiyama, H.; Yamaguchi, S.; Park, S. B.; Itoh, K. *Tetrahedron: Asymmetry* **1993**, *4*, 143.

22. (a) Lowental, R. E.; Abiko, A.; Masamune, S. *Tetrahedron Lett.* **1990**, *31*, 6005. (b) Evans, D. A.; Woerpel, K. A.; Hinman, M. M.; Faul, M. M. *J. Am. Chem. Soc.* **1991**, *113*, 726. (c) Pfaltz, A. *Acc. Chem. Res.* **1993**, *26*, 339.

23. Helmchen, G.; Krotz, A.; Ganz, K. T.; Hansen, D. *Synlett* **1991**, 257.

24. Imai, Y.; Zhang, W.; Kida, T.; Nakatsuji, Y.; Ikeda, I. *Tetrahedron: Asymmetry* **1996**, *7*, 2453.

25. Alper, H.; Goldberg, Y. *Tetrahedron: Asymmetry* **1992**, *3*, 1055.

26. Lee, S.-G.; Lim, C. W.; Song, C. E.; Kim, I. O.; Jun, C.-H. *Tetrahedron: Asymmetry* **1997**, *8*, 2927.

27. (a) Sawamura, M.; Kuwano, R.; Ito, Y. *Angew. Chem., Int. Ed.* **1994**, *33*, 111. (b) Sawamura, M.; Kuwano, R.; Shirai, J.; Ito, Y. *Synlett* **1995**, 347. (c) Kuwano, R.; Sawamura, M.; Shirai, J.; Takahashi, M.; Ito, Y. *Tetrahedron Lett.* **1995**, *36*, 5239.

28. (a) Sasaki, J.; Schweizer, W. B.; Seebach, D. *Helv. Chim. Acta* **1993**, *76*, 2654. (b) Haag, D.; Runsink, J.; Scharf, H.-D. *Organometallics* **1998**, *17*, 398.

29. Faller, J. W.; Chase, K. J. *Organometallics* **1994**, *13*, 989.

30. Hayashi, T.; Hayashi, C.; Uozumi, Y. *Tetrahedron: Asymmetry* **1995**, *6*, 2503.

31. Nishibayashi, Y.; Segawa, K.; Ohe, K.; Uemura, S. *Organometallics* **1995**, *14*, 5486.

32. (a) Newman, L. M.; Williams, J. M. J.; McCague, R.; Potter, G. A. *Tetrahedron: Asymmetry* **1996**, *7*, 1597. (b) Langer, T.; Janssen, J.; Helmchen, G. *Tetrahedron: Asymmetry* **1996**, *7*, 1599.

33. (a) Sprinz, J.; Helmchen, G. *Tetrahedron Lett.* **1993**, *34*, 1769. (b) von Matt, P.; Pfaltz, A. *Angew. Chem., Int. Ed.* **1993**, *32*, 566. (c) Dawson, G. J.; Frost, C. G.; Williams, J. M. J.; Coote, S. J. *Tetrahedron Lett.* **1993**, *34*, 3149.

34. Sudo, A.; Yoshida, H.; Saigo, K. *Tetrahedron: Asymmetry* **1997**, *8*, 3205.

35. Lee, S.-G.; Lim, C. W.; Song, C. E.; Kim, I. O. *Tetrahedron: Asymmetry* **1997**, *8*, 4027.

36. (a) Nishibayashi, Y.; Singh, J. D.; Segawa, K.; Fukuzawa, S.; Uemura, S. *J. Chem. Soc., Chem. Commun.* **1994**, 1375. (b) Nishibayashi, Y.; Segawa, K.; Singh, J. D.; Fukuzawa, S.; Ohe, K.; Uemura, S. *Organometallics* **1996**, *15*, 370.

37. Herrmann, W. A.; Goossen, L. J.; Köcher, C.; Artus, G. R. *Angew. Chem., Int. Ed.* **1996**, *35*, 2805.

38. Enders, D.; Gielen, H.; Breuer, K. *Tetrahedron: Asymmetry* **1997**, *8*, 3571.

39. Nishiyama, H.; Park, S.-B.; Itoh, K. *Tetrahedron: Asymmetry* **1992**, *3*, 1029.

40. (a) Ojima, I.; Kogure, T.; Kumagai, M. *J. Org. Chem.* **1977**, *42*, 1671. (b) Ojima, I.; Tanaka, T.; Kogure, T. *Chem. Lett.* **1981**, 823.

41. Ojima, I.; Kogure, T. *Organometallics* **1982**, *1*, 1390.

42. Burk, M. J.; Feaster, J. E. *Tetrahedron Lett.* **1992**, *33*, 2099.

43. (a) Ojima, I.; Kogure, T.; Nagai, Y. *Tetrahedron Lett.* **1973**, 2475. (b) Langlois, N.; Dang, T.-P.; Kagan, H. B. *Tetrahedron Lett.* **1973**, 4865. (c) Kagan, H. B.; Langlois, N.; Dang, T.-P. *J. Organomet. Chem.* **1975**, *90*, 353. (d) Becker, R.; Brunner, H.; Mahboobi, S.; Wiegrebe, W. *Angew. Chem., Int. Ed.* **1985**, *24*, 996. (e) Brunner, H.; Kürzinger, A.; Mahboobi, S.; Wiegrebe, W. *Arch. Pharm.* **1988**, *321*, 73. (f) Kokel, N.; Mortreux, A.; Petit, F. *J. Mol. Catal.* **1989**, *57*, L5. (g) Brunner, H.; Becker, R. *Angew. Chem., Int. Ed.* **1984**, *23*, 222. (h) Brunner, H.; Becker, R.; Grauder, S. *Organometallics* **1986**, *5*, 739.

44. Zhu, G.; Terry, M.; Zhang, X. *J. Organomet. Chem.* **1997**, *547*, 97.

45. Nishibayashi, Y.; Takei, I.; Uemura, S.; Hidai, M. *Organometallics* **1998**, *17*, 3420.

46. Murahashi, S.; Watanabe, S.; Shiota, T. *J. Chem. Soc., Chem. Commun.* **1994**, 725.

47. Brunner, H.; Rötzer, M. *J. Organomet. Chem.* **1992**, *425*, 119.

48. Nakano, T.; Nagai, Y. *Chem. Lett.* **1988**, 481.

49. Halterman, R. L.; Ramsey, T. M.; Chen, Z. *J. Org. Chem.* **1994**, *59*, 2642.

50. Carter, M. B.; Schiøtt, B.; Gutiérrez, A.; Buchwald, S. L. *J. Am. Chem. Soc.* **1994**, *116*, 11667.

51. For reduction of esters; see, (b) Berk, S. C.; Kreutzer, K. A.; Buchwald, S. L. *J. Am. Chem. Soc.* **1991**, *113*, 5093. (c) Barr, K. J.; Berk, S. C.; Buchwald, S. L. *J. Org. Chem.* **1994**, *59*, 4323.

52. Rahimian, K.; Harrod, J. F. *Inorg. Chim. Acta* **1998**, *270*, 330.

53. Xin, S.; Harrod, J. F. *Can. J. Chem.* **1995**, *73*, 999.

54. Verdaguer, X.; Lange, U. E. W.; Reding, M. T.; Buchwald, S. L. *J. Am. Chem. Soc.* **1996**, *118*, 6784.

55. Imma, H.; Mori, M.; Nakai, T. *Synlett* **1996**, 1229.

56. Verdaguer, X.; Lange, U. E. W.; Buchwald, S. L. *Angew. Chem., Int. Ed.* **1998**, *37*, 1103.

57. (a) Tamao, K.; Kakui, T.; Kumada, M. *J. Am. Chem. Soc.* **1978**, *100*, 2268. (b) Tamao, K.; Ishida, N.; Tanaka, T.; Kumada, M. *Organometallics* **1983**, *2*, 1694.

58. Yamamoto, K.; Hayashi, T.; Kumada, M. *J. Am. Chem. Soc.* **1971**, *93*, 5301.

59. Hayashi, T.; Tamao, K.; Katsuro, Y.; Nakae, I.; Kumada, M. *Tetrahedron Lett.* **1980**, *21*, 1871.

60. (a) Hayashi, T.; Kabeta, K.; Yamamoto, T.; Tamao, K.; Kumada, M. *Tetrahedron Lett.* **1983**, *24*, 5661. (b) Hayashi, T.; Kabeta, K. *Tetrahedron Lett.* **1985**, *26*, 3023. (c) Hayashi, T.; Matsumoto, Y.; Morikawa, I.; Ito, Y. *Tetrahedron Lett.* **1990**, *2*, 601.

61. Hatanaka, Y.; Goda, K.; Yamashita, F.; Hiyama, T. *Tetrahedron Lett.* **1994**, *35*, 7981.

62. Ohmura, H.; Matsuhashi, H.; Tanaka, M.; Kuroboshi, M.; Hiyama, T.; Hatanaka, Y.; Goda, K. *J. Organomet. Chem.* **1995**, *499*, 167.

63. Tamao, K.; Tohma, T.; Inui, N.; Nakayama, O.; Ito, Y. *Tetrahedron Lett.* **1990**, *31*, 7333.

64. Bergens, S. H.; Noheda, P.; Whelan, J.; Bosnich, B. *J. Am. Chem. Soc.* **1992**, *114*, 2121.

65. (a) Bergens, S. H.; Noheda, P.; Whelan, J.; Bosnich, B. *J. Am. Chem. Soc.* **1992**, *114*, 2128. (b) Tamao, K.; Nakagawa, Y.; Ito, Y. *Organometallics* **1993**, *12*, 2297.

66. (a) Curtis, N. R.; Holmes, A. B.; Looney, M. G. *Tetrahedron Lett.* **1992**, *33*, 671. (b) Curtis, N. R.; Holmes, A. B. *Tetrahedron Lett.* **1992**, *33*, 675.

67. (a) Hayashi, T.; Uozumi, Y. *J. Am. Chem. Soc.* **1991**, *113*, 9887. (b) Uozumi, Y.; Kitayama, K.; Hayashi, T.; Yanagi, K.; Fukuyo, E. *Bull. Chem. Soc. Jpn.* **1995**, *68*, 713.

68. Uozumi, Y.; Tanahashi, A.; Lee, S.-Y.; Hayashi, T. *J. Org. Chem.* **1993**, *58*, 1945.

69. (a) Uozumi, Y.; Lee, S.-Y.; Hayashi, T. *Tetrahedron Lett.* **1992**, *33*, 7185. (b) Uozumi, T.; Hayashi, T. *Tetrahedron Lett.* **1993**, *34*, 2335.

70. Uozumi, T.; Kitayama, K.; Hayashi, T. *Tetrahedron: Asymmetry* **1993**, *4*, 2419.

71. Uozumi, Y.; Suzuki, N.; Ogiwara, A.; Hayashi, T. *Tetrahedron* **1994**, *50*, 4293.

72. Kitayama, K.; Uozumi, Y.; Hayashi, T. *J. Chem. Soc., Chem. Commun.* **1995**, 1533.

73. Kitayama, K.; Tsuji, H.; Uozumi, Y.; Hayashi, T. *Tetrahedron Lett.* **1996**, 4169.

74. (a) Corriu, R. J. P.; Moreau, J. J. E. *J. Organomet. Chem.* **1974**, *64*, C51. (b) Corriu, R. J. P.; Moreau, J. J. E. *J. Organomet. Chem.* **1975**, *85*, 19.

75. (a) Corriu, R.J.P.; Moreau, J. J. E. *J. Organomet. Chem.* **1975**, *91*, C27. (b) Corriu, R. J. P.; Moreau, J. J. E. *Nouv. J. Chim.* **1977**, *1*, 71.

76. Ohta, T.; Ito, M.; Tsuneto, A.; Takaya, H. *J. Chem. Soc., Chem. Commun.* **1994**, 2525.

77. Tamao, K.; Nakamura, K.; Ishii, H.; Yamaguchi, S.; Shiro, M. *J. Am. Chem. Soc.* **1996**, *118*, 12469.

78. Männig, D.; Nöth, H. *Angew. Chem., Int. Ed.* **1985**, *24*, 878.

79. (a) Fu, G. C.; Evans, D. A.; Muci, A. R. In *Advances in Catalytic Processes*; Doyle, M. P. (Ed.); JAI Press Inc.: London, 1995; Vol. 1, pp. 95. (b) Burgess, K.; Van der Donk, W. A. In *Advances Asymmetric Synthesis*; Stephenson, G. R. (Ed.); Chapman & Hall: London, 1996; pp. 181. (c) Burgess, K.; Ohlmeyer, J. M. *Chem. Rev.* **1991**, *91*, 1179.

80. (a) Evans, D. A.; Fu, G. C. *J. Org. Chem.* **1990**, *55*, 5178. (b) Evans, D. A.; Fu, G. C.; Anderson, B. A. *J. Am. Chem. Soc.* **1992**, *114*, 6679.

81. (a) Burgess, K.; Ohlmeyer, M. J. *J. Org. Chem.* **1988**, *53*, 5178. (b) Burgess, K.; Donk, W. A.; Ohlmeyer, M. J. *Tetrahedron: Asymmetry* **1991**, *2*, 613. (c) Burgess, K.; Ohlmeyer, M. J.; Whitmire, K. H. *Organometallics* **1992**, *11*, 3588.

82. (a) Hayashi, T.; Matsumoto, Y.; Ito, Y. *J. Am. Chem. Soc.* **1989**, *111*, 3426. (b) Hayashi, T.; Matsumoto, Y.; Ito, Y. *Tetrahedron: Asymmetry* **1991**, *2*, 601.

83. (a) Brown, J. M.; Hulmes, D. I.; Layzell, T. P. *J. Chem. Soc., Chem. Commun.* **1993**, 1673. (b) Alcock, N. W.; Brown, J. M.; Hulmes, D. I. *Tetrahedron Lett.* **1993**, *4*, 743. (c) Valk, J. M.; Whitlock, G. A.; Layzell, T. P.; Brwon, J. M. *Tetrahedron: Asymmetry* **1995**, *6*, 2593.

84. Sato, M.; Miyaura, N.; Suzuki, A. *Tetrahedron Lett.* **1990**, *31*, 231.

85. Zhang, J.; Lou, B.; Guo, G.; Dai, L. *J. Org. Chem.* **1991**, *56*, 1670.

86. Brown, J. M.; Lloyd-Jones, G. C. *Tetrahedron: Asymmetry* **1990**, *1*, 869.

87. Togni, A.; Breutel, C.; Schnyder, A.; Spindler, F.; Landert, H.; Tijani, A. *J. Am. Chem. Soc.* **1994**, *116*, 4062.

88. (a) Schnyder, A.; Hintermann, L.; Togni, A. *Angew. Chem., Int. Ed.* **1995**, *34*, 931. (b) Schnyder, A.; Togni, A.; Wiesli, U. *Organometallics* **1997**, *16*, 255. (c) Abbenhuis, H. C. L.; Burckhardt, U.; Gramlich, V.; Martelleci, A.; Spencer, J.; Steiner, I.; Togni, A. *Organometallics* **1996**, *15*, 1614.

89. Matsumoto, Y.; Hayashi, T. *Tetrahedron Lett.* **1991**, *32*, 3387.

90. Wiesauer, C.; Weissensteiner, A. *Tetrahedron: Asymmetry* **1996**, *7*, 5.

91. Matsumoto, Y.; Naito, M.; Hozumi, Y.; Hayashi, T. *J. Chem. Soc., Chem. Commun.* **1993**, 1468.

92. (a) Fernandez, E.; Hooper, M. W.; Knight, F. I.; Brown, J. M. *Chem. Commun.* **1997**, 173. (b) Knight, F. I.; Brown, J. M.; Lazzari, D.; Ricci, A.; Blacker, A. J. *Tetrahedron* **1997**, *53*, 11411.

93. (a) Leutenegger, A.; Madin, A.; Pfaltz, A. *Angew. Chem., Int. Ed.* **1989**, *28*, 60. (b) von Matt, P.; Pfaltz, A. *Tetrahedron: Asymmetry* **1991**, *2*, 691. (c) Pfaltz, A. *Acta Chemica Scandinavica* **1996**, *50*, 189.

94. Nagata, T.; Yorozu, K.; Yamada, T.; Mukaiyama, T. *Angew. Chem., Int. Ed.* **1995**, *34*, 2145.

95. (a) Sugi, K. D.; Nagata, T.; Yamada, T.; Mukaiyama, T. *Chem. Lett.* **1996**, 737 and 1081. (b) Nagata, T.; Sugi, K. D.; Yamada, T.; Mukaiyama, T. *Synlett* **1996**, 1076.

96. Sugi, K. D.; Nagata, T.; Yamada, T.; Mukaiyama, T. *Chem. Lett.* **1997**, 493.

97. Eisch, J. J. In *Comprehensive Organic Synthesis*; Trost, B. M. (Ed.); Pergamon Press: Oxford, 1991; Vol. 8, pp. 733.

98. Giacomelli, G.; Bertero, L.; Lardicci, L. *Tetrahedron Lett.* **1981**, *22*, 883.

99. Lautens, M.; Chiu, P.; Ma, S.; Rovis, T. *J. Am. Chem. Soc.* **1995**, *117*, 532.

3

ASYMMETRIC ISOMERIZATION OF ALLYLAMINES

SUSUMU AKUTAGAWA
Takasago International Corporation, Nissay Aroma Square 18F, 37-1, Kamata 5-chome Ohta-ku, Tokyo, Japan 144-8721

KAZUHIDE TANI
Department of Chemistry, Faculty of Engineering Science, Osaka University, Toyonaka, Osaka, Japan 560

3.1. INTRODUCTION

Olefinic double-bond isomerization is probably one of the most commonly observed and well-studied reactions that uses transition metals as catalysts [1]. However, prior to our first achievement of asymmetric isomerization of allylamine by optically active Co(I) complex catalysts [2], there were only a few examples of catalytic asymmetric isomerization, and these were characterized by very low asymmetric induction (<4% ee) [3]. In 1978 we reported that an enantioselective hydrogen migration of a prochiral allylamine such as *N,N*-diethylgeranylamine, (**1**) or *N,N*-diethylnerylamine (**2**) gave optically active citronellal (*E*)-enamine **3** with about 32% ee utilizing Co(I)-DIOP [DIOP = 2,3-*O*-isopropylidene-2,3-dihydroxy-1,4-bis(diphenylphosphino)butane] complexes as the catalyst (eq 3.1).

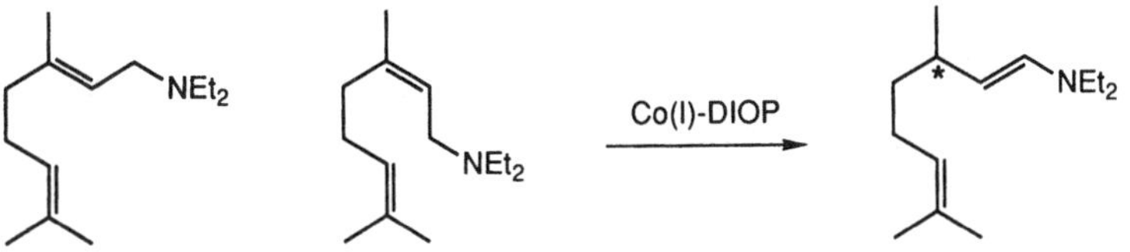

$$(3.1)$$

Catalytic Asymmetric Synthesis, Second Edition, Edited by Iwao Ojima
ISBN 0-471-29805-0   Copyright © 2000 Wiley-VCH, Inc.

Optically active citronellal, which can be quantitatively obtained from the optically active citronellal enamine, is a useful intermediate for the synthesis of many optically active natural products. Thus, the discovery of the asymmetric isomerization, although both chemo- and enantioselectivities were insufficient, opened the door to the synthesis of optically active terpenoids. A great improvement in the chemo- and enantioselectivities for the reaction in Equation 1 (up to 99% yield for **3** and 98% ee) has been achieved by using cationic Rh(I)-BINAP (BINAP = 2,2'-bis(diphenylphosphino)-1,1'-binaphthyl) complexes, $[Rh(BINAP)(L_2)]^+$ (L = diene or solvent), as the catalyst precursor. Further studies of the Rh(I)-BINAP species led to the discovery of the thermally stable $[Rh(BINAP)_2]^+$ complex, which is suitable for industrial uses, achieving a very high turnover number (TON: enamine moles produced by one mole of catalyst during 18 h) ($\sim 3 \times 10^5$) without impairing chemo- and enantioselectivities. Since 1983 Takasago International Corporation has been successfully manufacturing (−)-menthol on a commercial basis ($\sim$1500 t/year) by using this asymmetric isomerization process, the "Takasago process."

This chapter describes the discovery of a new asymmetric isomerization process, improvements in the catalyst, some practical aspects, and the scope and limitation of the process, with emphasis on developments during commercial manufacturing [4].

3.2. ASYMMETRIC ISOMERIZATION OF ALLYLAMINES

Although various transition-metal complexes have reportedly been active catalysts for the migration of inner double bonds to terminal ones in functionalized allylic systems (Eq. 3.2) [5], prochiral allylic compounds with a multisubstituted olefin (R^1, $R^2 \neq$ H in eq 2) are not always susceptible to catalysis or they show only a low reactivity [1d]. Choosing allylamines **1** and **2** as the substrates for enantioselective isomerization has its merits: (1) optically pure citronellal, which is an important starting material for optically active terpenoids such as (−)-menthol, cannot be obtained directly from natural sources [6], and (2) both (E)-allylamine **1** and (Z)-allylamine **2** can be prepared in reasonable yields from myrcene or isoprene, respectively. The (E)-allylamine **1** is obtained from the reaction of myrcene and diethylamine in the presence of lithium diethylamide under Ar in an almost quantitative yield (Eq. 3.3) [7]. The (Z)-allylamine **2** can also be prepared with high selectivity ($\sim$90%) by Li-catalyzed telomerization of isoprene using diethylamine as a telomer (Eq. 3.4) [8]. Thus, natural or petroleum resources can be selected.

$$R^2 \underset{\quad}{\overset{R^1}{\diagup}} X \longrightarrow R^2 \underset{\quad}{\overset{R^1}{\diagup}} X \qquad (3.2)$$

X=OH, OR, NRR', etc.

Migration of the multisubstituted inner double bonds of **1** or **2** has shown to be effected by various metal catalysts (Table 3.1). As Table 3.1 shows, the product selectivity varies dramatically with the catalyst used. With the exception of cationic Rh(I)-tertiary phosphine complexes, the selectivity for the desired citronellal enamine **3** is not sufficient. In the case of Ti and Co complexes, the 6-double bond also migrates to give a considerable amount of the undesired conjugated dienamine **4**. A secondary allylamine, N-cyclohexylgeranylamine, (**5**), is an effective substrate that can be isomerized cleanly to give the corresponding imine **6** even with Co(I) catalysts. Cationic Rh(I) complexes with phosphine or bisphosphine are very effective catalysts for the isomerization of **1** and **2** to give the same (E)-enamine **3**; the undesired dienamine **4** is virtually absent.

TABLE 3.1. Metal Catalyzed Isomerization of Allylamine[*]

Catalyst	Substrate	Product (Selectivity %)
$Cp_2TiCl_2/^iPrMgBr$	1	**4** (100)
$CoH(N_2)(PPh_3)_3$	1	**3** (85), **4** (15)
$Co(acac)_2/PPh_3/^iBu_2AlH$	2	**3** (81), **4** (19)
$Co(acac)_2/PPh_3/^iBu_2AlH$	5	**6** (~100)
$[Rh(PPh_3)_2(COD)]^+$	1	**3** (~100)
$[Rh(DIPHOS)(Solvent)_n]^+$	1	**3** (>96)
$[Rh(BINAP)(COD)]^+$	1	**3** (>96)

[*]Reaction was carried out in THF at 60°C under N_2, [Substrate]/[Catalyst] = 100, for 15 h.

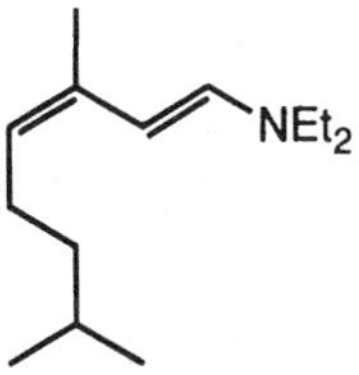

4 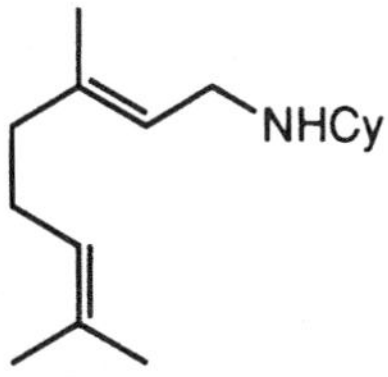   **5** 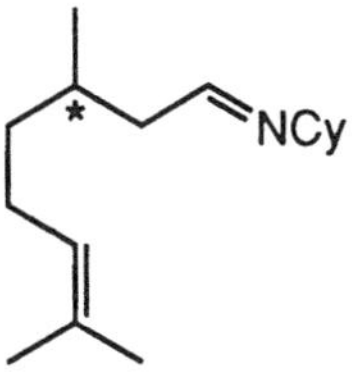   **6** 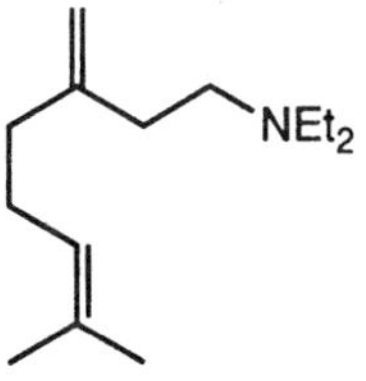   **7**

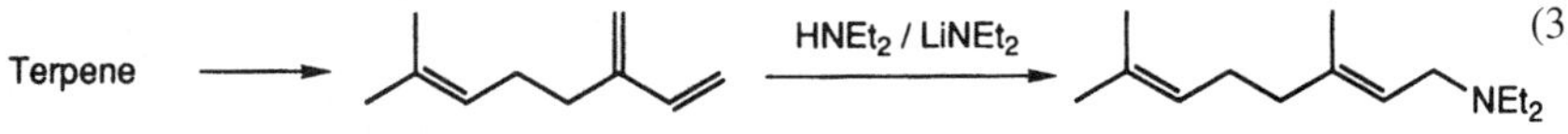

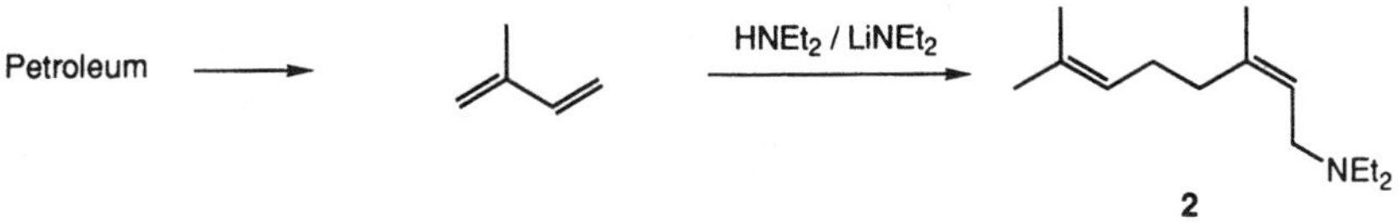

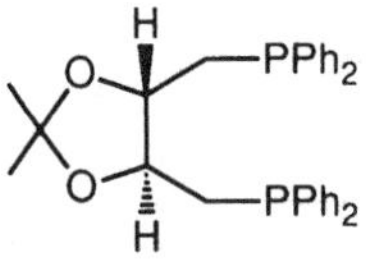

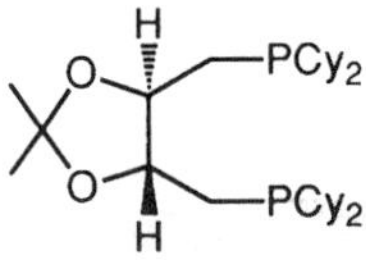

The first asymmetric isomerization of an allylamine was achieved with the optically active cobalt catalysts as mentioned above. The isomerization of **1** with $Co(acac)_2/(+)$-DIOP/$^iBu_2AlH$ catalyst gave $(3R)$-**3** with 32% ee (39% chemical yield) accompanied by a considerable amount of undesired dienamine **4**. Although with the same catalyst system, a higher enantioselectivity (57% ee) and chemical yield (60%) were obtained by using the secondary allylamine **5** as the substrate; these values, i.e., 57% ee and 60% yield, were insufficient for a commercial process. As mentioned earlier, we have found that the cationic Rh(I)-phosphine complexes, which are effective as hydrogenation catalysts, showed very high selectivities and catalytic activities in the isomerization of **1** or **2** to the (*E*)-enamine **3**. Because many chiral diphosphines had already been developed, several typical optically active diphosphines were tested as the chiral ligand for asymmetric isomerization. Results are listed in Table 3.2. Among the DIOP-based ligands, only CyDIOP gave an exceptionally high enantioselectivity of 77% ee, but the catalytic activity was not sufficient with such a diphosphine where all the phosphorus substituents were aliphatic groups. The DIOP complex showed a fairly high catalytic activity but only a low asymmetric induction of about 22% ee; DIOP was also known to give low enantioselectivities in the asymmetric hydrogenation of some olefins [9]. In contrast to these conventional optically active diphosphines, BINAP, in which all the substituents on both phosphorus atoms are aromatic groups, showed excellent enantioselectivity as well as chemoselectivity and catalytic activity for the isomerization. One of the favorable features of this asymmetric isomerization is the desirable stereochemical correlation between the substrate geometries, product (*E*)-enamine configurations, and the BINAP chirality, as shown in Scheme 3.1. This correlation is established with an almost perfect enantioselectivity and a practically quantitative chemical yield in all routes.

(*R*)(+)-BINAP   (*S*)(-)-BINAP

Because BINAP enantiomers and (*E*)- and (*Z*)-allylamine substrates are easily obtained, this stereochemical relation provides the following economical advantages: (1) the option of taking the starting material from either a natural resource (renewable terpene) or petroleum, and (2) easy access to both enantiomers of citronellal from a single intermediate.

TABLE 3.2. Asymmetric Isomerization of Allylamine[*]

Catalyst	Substrate	Conversion (%)	Product (Selectivity %)	% ee (Configuration)
Co(acac)$_2$/(+)-DIOP/^{i}Bu$_2$AlH	1	45	**3** (87), **4** (13)	35(*R*)
Co(acac)$_2$/(+)-DIOP/^{i}Bu$_2$AlH	5	62	**6** (97)	57(*R*)
[Rh{(+)-DIOP}(COD)]$^+$	1	71	**3** (100)	22(*R*)
[Rh{(−)-CyDIOP}(Solvent)$_2$]$^+$	1	18	**3** (80)	77(*S*)
[Rh{(*R*)-BINAP}(COD)]$^+$	1	100	**3** (100)	97(*S*)
[Rh{(*R*)-BINAP}(COD)]$^+$	2	97	**3** (100)	96(*R*)
[Rh{(*S*)-BINAP}(COD)]$^+$	1	100	**3** (100)	97(*R*)
[Rh{(*S*)-BINAP}(COD)]$^+$	5	100	**6** (100)	98(*R*)

[*]Reaction was carried out in THF at 60°C with the s/c of 100 for 15 h.

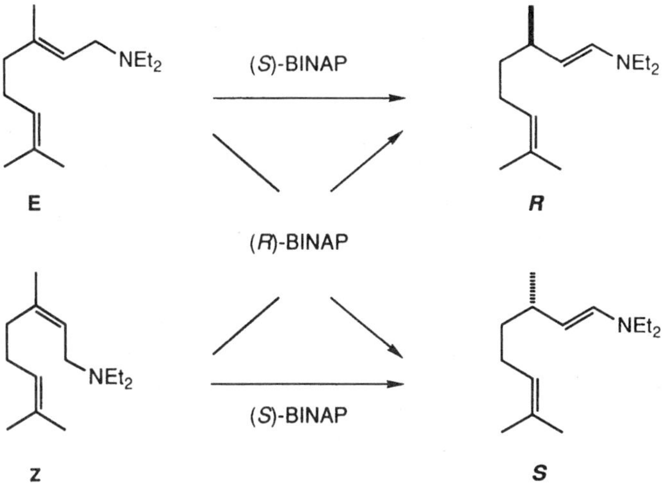

Scheme 3.1.

3.3. ISOMERIZATION PROCESS DEVELOPMENT

In spite of the remarkably high selectivities, the drawback of the original asymmetric isomerization process was the high price of the Rh–BINAP catalyst (cf. RhCl$_3$, $42.50/500 mg, (+)- or (−)-BINAP, $25.40/100 mg: *Aldrich*, 1991). For the industrial production of (+)-citronellal in the 1000 t/year scale, we had to raise the TON to an optimum level >50,000 while maintaining high regio- and stereo selectivities. The following process development has enabled us to fulfill these requirements.

3.3.1. Great Improvement in the Procedure for Synthesizing Optically Active BINAP

The chiral ligand BINAP was originally prepared from 2,2′-binaphthol and resolved by complexation with an optically active Pd complex [10]. A new method starting from 2-naphthol was developed (Scheme 3.2) [11]. In this method optical resolution was achieved at the stage of BINAP dioxide (BINAPO) by using inexpensive optically active acids such as camphorsulfonic acid and dibenzoyltartaric acid.

3.3.2. Removal of Catalyst Inhibitors

The Rh–BINAP catalysts are very sensitive to impurities such as oxygen, moisture, and carbon dioxide. If an excess of water ([H$_2$O]/[Rh] = 15) is present in the reaction mixture, the isomerization is stopped after a few turnovers with precipitation of air-stable red-brown crystals, which were found to be [{Rh(BINAP)}$_3$(μ_3-OH)$_2$]ClO$_4$. X-Ray analysis of this complex (H$_2$O was replaced by D$_2$O) revealed a unique structure of a triangular Rh$_3$ core capped with two triply bridging OH groups (Fig. 3.1). This Rh(I)-trinuclear complex was totally inactive as an isomerization catalyst, which suggests a catalyst deactivation mechanism [12]. The effect of several additives on the isomerization of **2** with [Rh(BINAP)(COD)]$^+$ has been examined, and the results are summarized in Table 3.3.

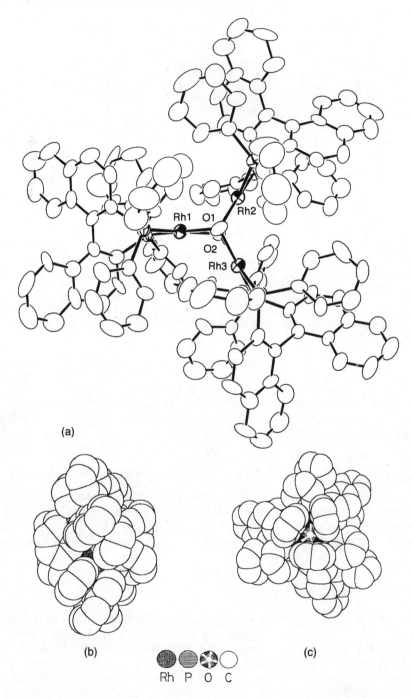

Figure 3.1. Molecular structure of the cationic part in [{Rh((*R*)(+)-BINAP)}₃(OD)₂]ClO₄:
ORTEP drawing (a) and space-filling representations of the side view (b) and top view (c).
Rh–Rh = 3.086–3.102 Å, Rh–P = 2.194–2.222 Å, Rh–O = 2.122–2.163 Å; Rh–Rh–Rh = 59.75–60.25°.

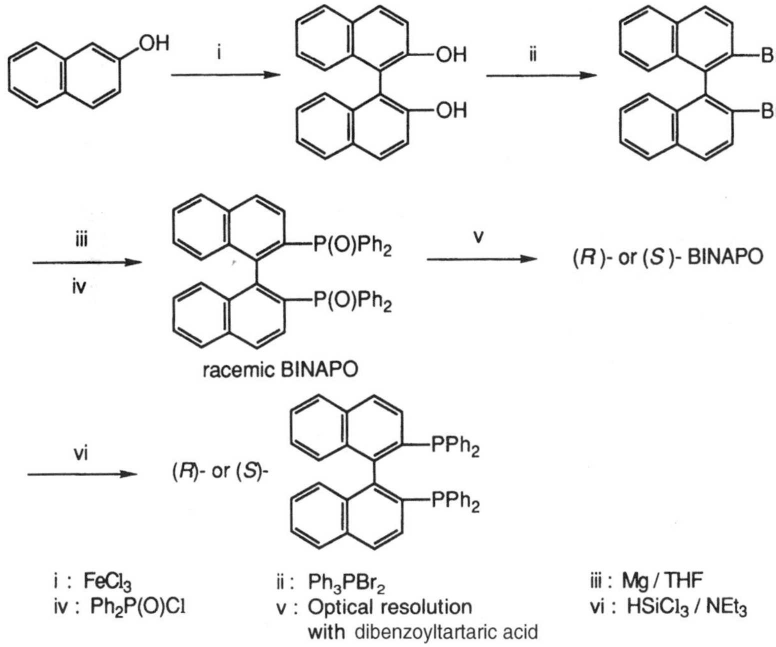

OH
i
OH
OH
ii
Br
Br
iii
iv
P(O)Ph2
P(O)Ph2
racemic BINAPO
v
(R)- or (S)- BINAPO
vi
(R)- or (S)-
PPh2
PPh2
i : FeCl3
iv : Ph2P(O)Cl
ii : Ph3PBr2
v : Optical resolution
with dibenzoyltartaric acid
iii : Mg / THF
vi : HSiCl3 / NEt3

Scheme 3.2.

Although donor substances retard the reaction rate, significant deactivation of the catalyst by the conjugated dienamine **4** was observed. In connection with this observation, we have found that an isomeric dienamine of the substrate 2-[2-(*N,N*-diethylamino)ethyl]-6-methyl-1,5-heptadiene (**7**), which is always present in small amounts (0.5–0.7%) in the crude commercial products **1**, also acts as a deactivator (see Table 3.4). Thus, a careful and exhaustive pretreatment of the substrate and the reaction system is necessary to attain a high TON.

TABLE 3.3. Effect of Additive on the Isomerization of 1 with Rh-BINAP Catalyst

Substrate	Additive to $[Rh\{(+)\text{-BINAP}\}(COD)]^+$	Additive Catalyst	k_{obs}[a] mol/*l*/min
1	none	—	37.
1	NEt_3	4	20.
1	COD	2	8.5
1	dienamine(**4**)	2	1.8

[a]THF, $[Rh\{(+)\text{-BINAP}\}(COD)]^+$, [S] = 0.24 mol/*l*, [S] / [C] = 100, 60°C

TABLE 3.4. Advancement of TON for the Isomerization of 1

Improvement	TON
[Rh(BINAP)(COD)]+,	
1 without special pretreatment	100
[Rh(BINAP)(COD)]+,	
1 treated with Vitride®	1,000
[Rh(BINAP)(COD)]+,	
1 after removal of amine isomer 7	8,000
Reuse of 8	
(10% loss)	80,000
Reuse of 8	
(2% loss*)	400,000

* Owing to total quality control.

3.3.3. Development of New Catalyst Systems

As the catalyst precursor for the present isomerization, $[Rh(BINAP)(COD)]^+$ was conveniently used in the original process. However, further improvement in TON was achieved by the discovery of a new catalyst, $[Rh(BINAP)_2]^+$(8) [13]. It was a surprise for us to find that the bis(BINAP) complex showed considerable catalytic activity at a high temperature (>80°C), because $[Rh(DIPHOS)_2]ClO_4$ was completely inactive for the isomerization of 1 and 2, even at 120°C. The enantioselectivity was maintained as high (>96% ee) as that obtained with [Rh(BI-NAP)(COD)]+ at 40°C. Through the use of the bis(BINAP) complex, it became feasible to reuse the catalyst and raise the TON >400,000. Progress in the improvement of TON is summarized in Table 4. An X-ray structural analysis of complex 8 [13] indicated the existence of significant distortion toward a tetrahedron and a considerable elongation of Rh–P distances (2.368(6)-2.388(6)Å) due to steric congestion. The observed Rh–P distance is perhaps the longest among the known bisphosphine–Rh(I) complexes. These features may explain the unusual catalytic activity of 8.

The stereochemical correlation observed with the $[Rh(BINAP)_2]^+$ catalyst was the same as that obtained with $[Rh(BINAP)(COD)]^+$ or $[Rh(BINAP)(solvent)_n]^+$ (Scheme 3.1), implying that an identical catalyst species determines the enantioselection. This requires dissociation of one of the BINAP ligands of $[Rh(BINAP)_2]^+$ during the catalysis. Supporting evidence for the dissociation of BINAP from $[Rh(BINAP)_2]^+$ has been obtained from a ^{31}P NMR study of the reaction mixture of 8 with a pseudosubstrate, triethylamine, at 90°C [14]. The greatest advantage of using $[Rh(BINAP)_2]^+$ as the catalyst is the resulting thermal and chemical stability, which enabled the reuse of the recovered catalyst. A further improvement of the catalyst was realized by employing the p-TolBINAP ligand [11b]. The new catalyst, $[Rh(p\text{-TolBINAP})_2]^+$, also showed excellent stability, allowing multiple reuses of the recovered catalyst. Both chemo- and enantioselectivities for the isomerization of 1 also remained high enough (> 97% ee) with the new catalyst. Another merit found in the new catalyst was a better solubility in organic solvent, a valuable feature in industrial processes.

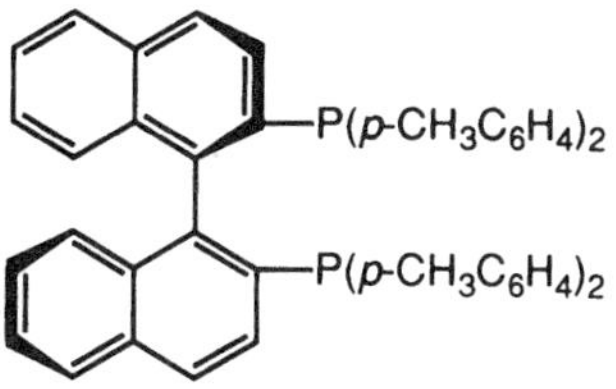

(*R*)(+)-*p*-Tol BINAP

3.4. COMMERCIAL MANUFACTURE

As described above, hydrolysis of the optically active enamine **3** proceeds without racemization and produces an optically active aldehyde, citronellal, with a very high optical purity (>98% ee). The optical purity of citronellal) available from natural sources is known to be no more than 80% ee [5]. The present asymmetric isomerization of the allylamine **1** is utilized as the key step for the industrial production of (–)-menthol (Scheme 3.3).

The ZnBr$_2$-catalyzed cyclization of (+)-citronellal, an intramolecular ene reaction, proceeds stereospecifically to give almost quantitatively (–)-isopulegol, where all the substituents are in the equatorial position. This forms a contrast to ordinary Lewis acid catalyzed cyclization of citronellal, which gives only 65% of (–)-isopulegol. The enantiomeric purity of (–)-isopulegol was raised to 100% ee by recrystallization at –50°C, and each step proceeds with high chemical yield (>92%), therefore the synthesis of pure (–)-menthol by the present procedure became a commercial process. With an estimated worldwide consumption of 4500 tons per year, (–)-menthol is widely used in many consumer products including cigarettes, chewing gum, toothpaste, and pharmaceutical products. At present, natural (–)-menthol is obtained mostly from *Mentha arvensis* cultivated in China, while synthetic materials are produced in Germany, the United States, and Japan. Takasago's menthol synthesis is now yielding the isomer with (1*R*, 3*R*, 4*S*)-configuration, the only useful isomer, among the eight possible isomers, in its enantiomerically pure form without any optical resolution process. Since 1984 Takasago has also

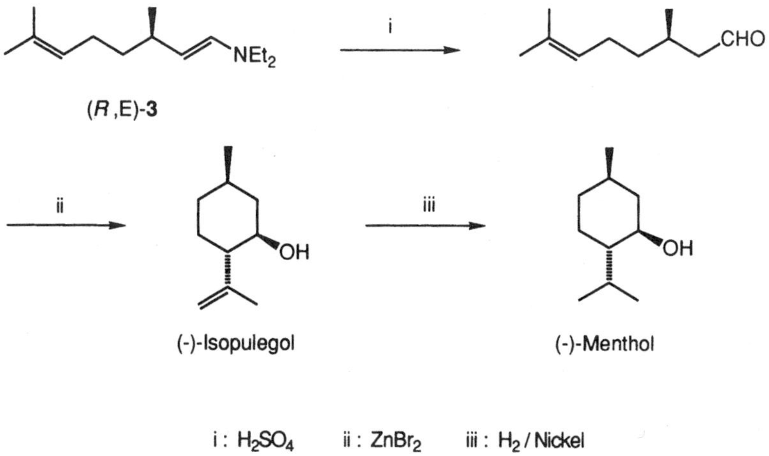

(*R*,E)-**3**

(-)-Isopulegol (-)-Menthol

i : H$_2$SO$_4$ ii : ZnBr$_2$ iii : H$_2$ / Nickel

Scheme 3.3.

TABLE 3.5. Optically Active Terpenoids Produced Based on the Asymmetric Isomerization of Allylamines by Takasago

Name	Formula	Chem. Purity (%)	% ee	Production (tons/year)	Use
(+)-Citronellal		98.0	97±1	1,500	Intermediate
(−)-Isopulegol		100	100	1,100	Intermediate
(−)-Menthol		100	100	1,000	Pharmaceuticals Tobacco Household Products
(−)-Citronellol		99	98	20	Fragrances
(+)-Citronellol		99	98	20	Fragrances
(−)-7-Hydroxy-citronellal		99	98	40	Fragrances
S-7-Methoxy-citronellal		99	98	10	Insect Growth Regulator
S-3,7-Dimethyl-1-octanal		99	98	7	Insect Growth Regulator

been producing a number of optically active terpenoids besides (–)-menthol based on the asymmetric isomerization of allylamines (Table 3.5).

3.5. SCOPE AND LIMITATIONS

The present enantioselective isomerization process requires prochiral allylamines free of the geometrical isomer. If such an allylamine is at hand, high asymmetric induction can be realized for a wide range of tertiary or secondary amines with alkyl substituents on the nitrogen atom. Thus, with [Rh{(+)-BINAP}(COD)]$^+$ the secondary allylamine **5** gives quantitatively the corresponding (S)-imine **6** with 96% ee. Allylamines with a styrene-type conjugation (**9**) (though slowly reacting substrates) and those having a hydroxy group (**10**) and a methoxy group (**11**) also act as effective substrates to give the corresponding enamines with high enantiomeric purities (>95% ee) (see Table 3.5). For all these substrates the stereochemical relationship shown in Scheme 1 remain unaffected. Asymmetric isomerization has successfully been applied to the synthesis of the optically active α-tocopherol side chain [15]. Thus the C-15 allylamine (E)(7R)-1-dimethylamino-3,7,11-trimethyl-2-dodecene (**12**) was isomerized to the corresponding C-15 (R,R)-enamine (**13**) with a very high enantiomeric purity (97% ee) in excellent chemical yield (98%) (eq 3.5).

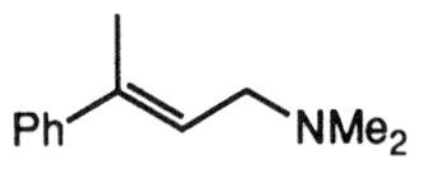

9	**10**	**11**

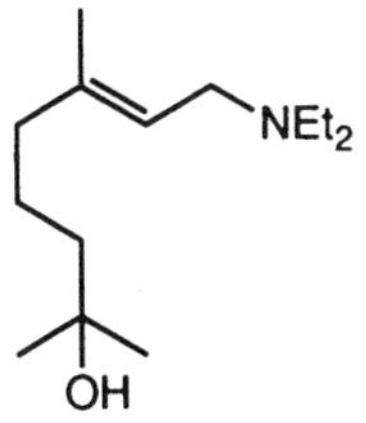

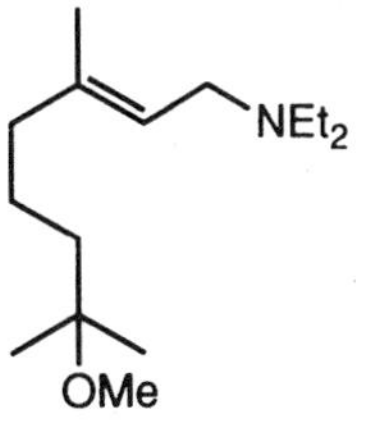

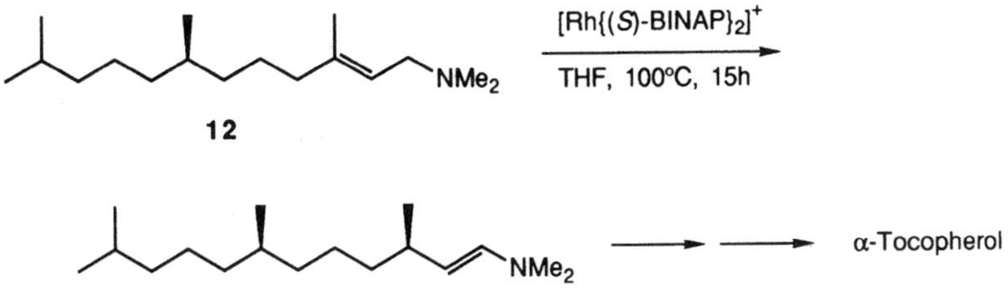

The basicity of the amine nitrogen appears to be an important factor for an effective asymmetric induction. Phenyl substituents on the nitrogen atom greatly retard the reaction rate. Thus, N-phenyl- and N,N-diphenylgeranylamine are inert at 40°C and 24 h reaction time. Few characteristic features are worth noting. If an allylamine is secondary, the product is the corresponding imine, a more stable valence tautomer of the enamine, which cannot be detected in the reaction mixture. The exclusive formation of an (E)-enamine regardless of the double-bond geometry of the starting substrate is another noticeable feature of isomerization. Homoallylamines such as **7** or *N,N*-dimethyl-3-butenylamine (**14**) are inactive substrates. Besides acyclic allylamines, cyclic allylamine **15**, which should yield the (*Z*)-enamine **16** on isomerization, is also an active substrate. We have examined the isomerization of several tetrahydropyridine derivatives (**17a–e**) with the Rh-BINAP catalyst [15]. In these reactions, however, the corresponding (*Z*)-enamines of type **16** are not isolated as monomeric forms, only the dimer **18** (from the secondary allylamine) or trimer **20** (from the tertiary allylamine) has been isolated in optically active forms when prochiral substrates are used (eqs 3.6 and 3.8). The configuration and the enantiomeric purity of these enamines (**18** and **20**) have not been determined. The achiral system **17a,b** yielded only racemic dimers **19a,b** (eq 3.7). Thus, it is likely that the oligomerization of the enamine intermediate is a noncatalyzed thermal process.

Allylamides are slow-reacting substrates [16]. The isomerization of *N*-acetylgeranylamide (**21**) and *N*-acetylnerylamide (**22**) with [Rh{(+)-BINAP}(COD)]$^+$ needs a temperature of 150°C to give the corresponding allylamide **23** with high enantiomeric purity (>95% ee), and the yield

14 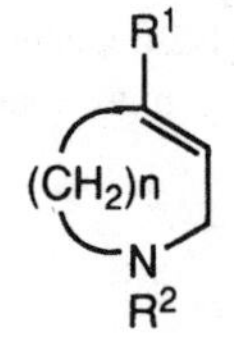   **15** 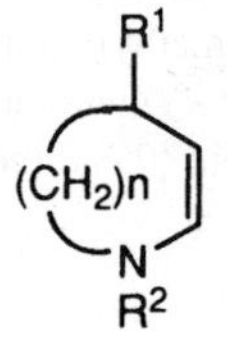   **16**

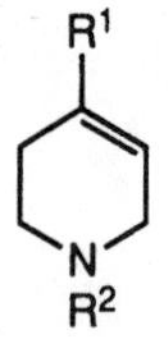

17

a	R^1=H,	R^2=Me
b	R^1=H,	R^2=CH$_2$Ph
c	R^1=Et,	R^2=Me
d	R^1=Me,	R^2=CH$_2$Ph
e	R^1=Me,	R^2=H

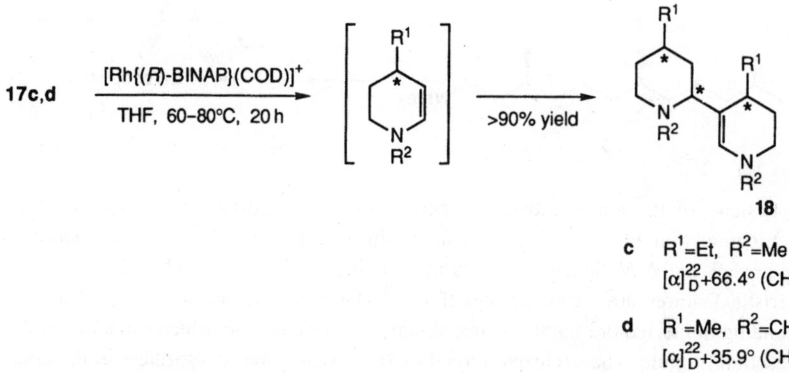

17c,d $\xrightarrow[\text{THF, 60–80°C, 20 h}]{\text{[Rh{(}R\text{)-BINAP}(COD)]}^+}$

>90% yield

18

$$\qquad(3.6)$$

c R^1=Et, R^2=Me
$[\alpha]_D^{22}$+66.4° (CHCl$_3$)

d R^1=Me, R^2=CH$_2$Ph
$[\alpha]_D^{22}$+35.9° (CHCl$_3$)

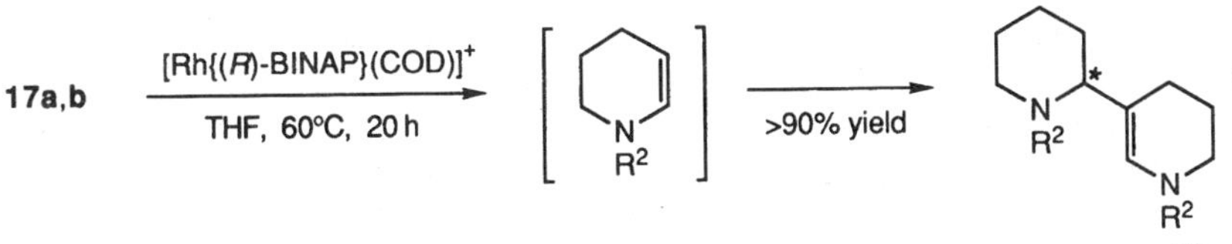

$$17a,b \xrightarrow[\text{THF, 60°C, 20 h}]{[\text{Rh}\{(R)\text{-BINAP}\}(\text{COD})]^+} \left[\text{structure} \right] \xrightarrow{>90\% \text{ yield}} \text{structure } \mathbf{19} \qquad (3.7)$$

a R^2=Me, $[\alpha]_D$ 0°

b R^2=CH$_2$Ph, $[\alpha]_D$ 0°

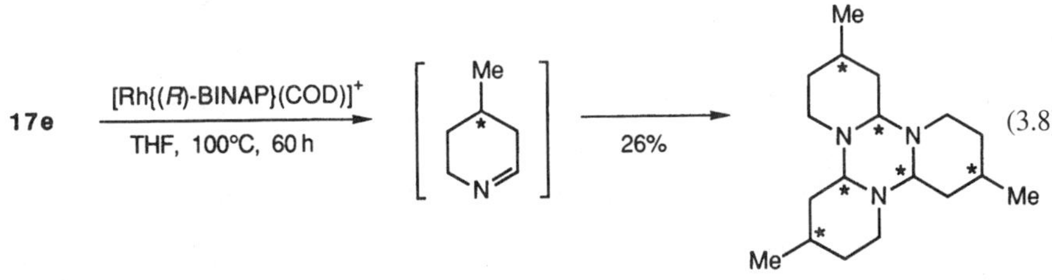

$$17e \xrightarrow[\text{THF, 100°C, 60 h}]{[\text{Rh}\{(R)\text{-BINAP}\}(\text{COD})]^+} \left[\text{structure} \right] \xrightarrow{26\%} \text{structure} \qquad (3.8)$$

20 $[\alpha]_D^{22}$ +14.7° (CHCl$_3$)

is low (<30%) due to the formation of a considerable amount of dienamide **24**. The stereochemical relationship is the same as that found with allylamine (see Scheme 3.1). A cyclic allylamide (**25**) is also isomerized selectively at 150°C to the corresponding enamide (**26**) with high enantiomeric purity (98% ee)(eq 3.9).

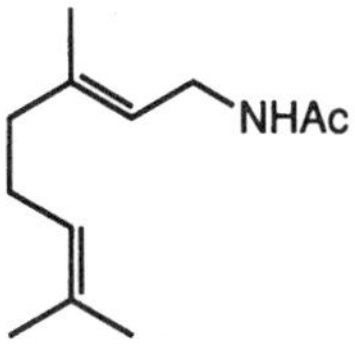

21 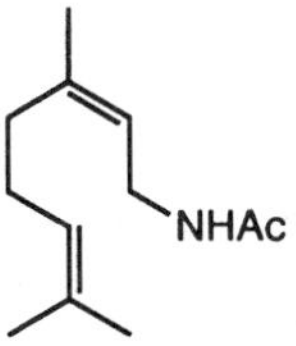   **22** 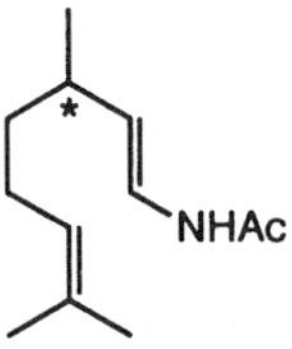   **23** 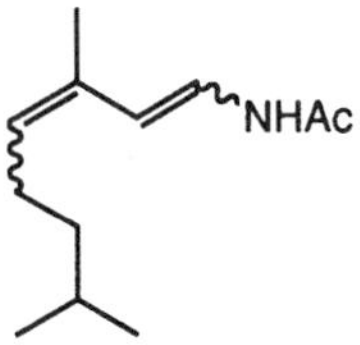   **24**

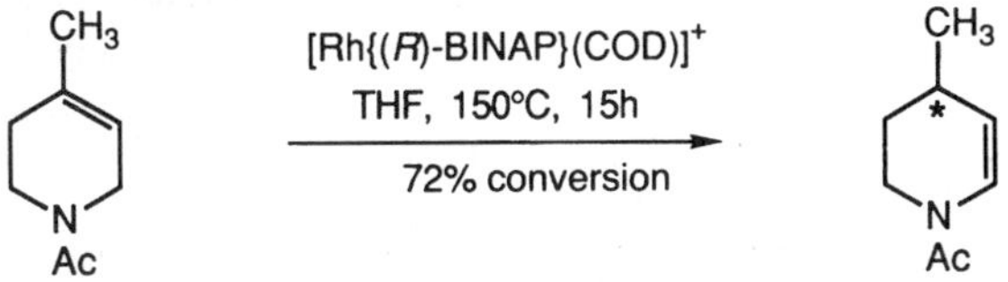

$$\text{structure }\mathbf{25} \xrightarrow[\substack{\text{THF, 150°C, 15h} \\ \text{72\% conversion}}]{[\text{Rh}\{(R)\text{-BINAP}\}(\text{COD})]^+} \text{structure }\mathbf{26} \qquad (3.9)$$

98% ee

The enantioselective isomerization of the prochiral alcohols **27** and **28** has also been achieved with 1 mol % of Rh–BINAP catalyst (eqs 3.10 and 3.11) [17]. However, neither chemo- nor enantioselectivity was sufficiently high, though the enantiomeric excesses of the products were much higher than the values hitherto reported [3a]. The formation of the *S*-configuration at C(3) of the aldehyde produced from the *E*-double bond of the starting alcohol is again the same as the situation observed for the isomerization of allylamine (Scheme 3.1).

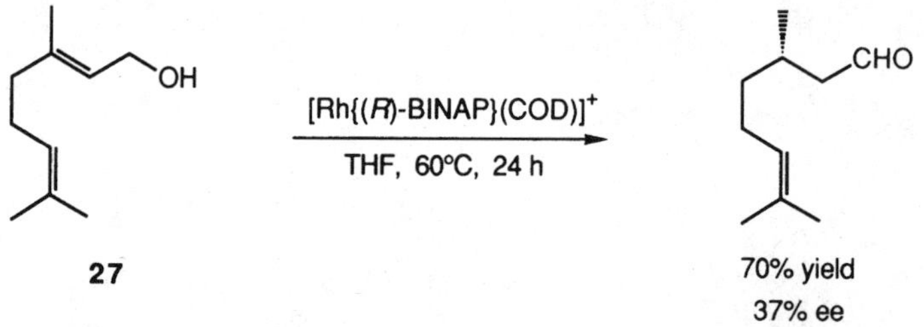

$$[Rh\{(R)\text{-BINAP}\}(COD)]^+$$
THF, 60°C, 24 h

27

70% yield
37% ee

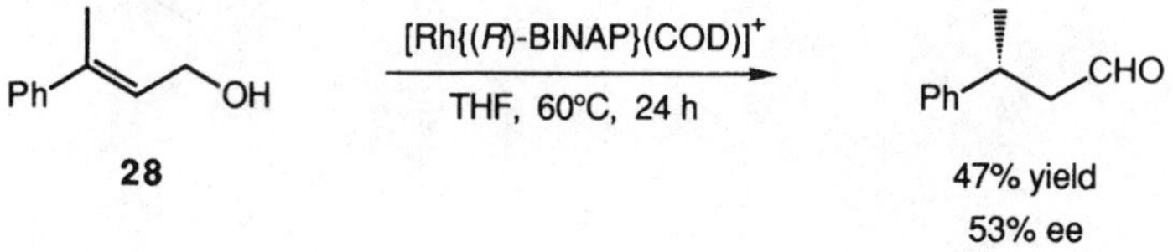

(3.10)

$$[Rh\{(R)\text{-BINAP}\}(COD)]^+$$
THF, 60°C, 24 h

28

47% yield
53% ee

(3.11)

The isomerization of cyclic allyl alcohols to produce ketones proceeds more cleanly [17]. Effective kinetic resolution of racemic cyclic allylic alcohols has been reported [18]. The isomerization of racemic 4-hydroxy-2-cyclopentanone (**29**) in the presence of 0.5 mol % of $\{Rh[(R)\text{-BINAP}](MeOH)_2\}^+$ in THF proceeded with 5:1 enantiomeric discrimination at 0°C to give 1,3-cyclopentadione (**31**) via enol ketone **30**, leaving the *R*-starting allylic alcohol (91% ee and 27% recovery yield) at 72% conversion after 14 days (eq 3.12). (*R*)-4-Hydroxy-2-cyclopentenone is a key building block for prostaglandin synthesis [19].

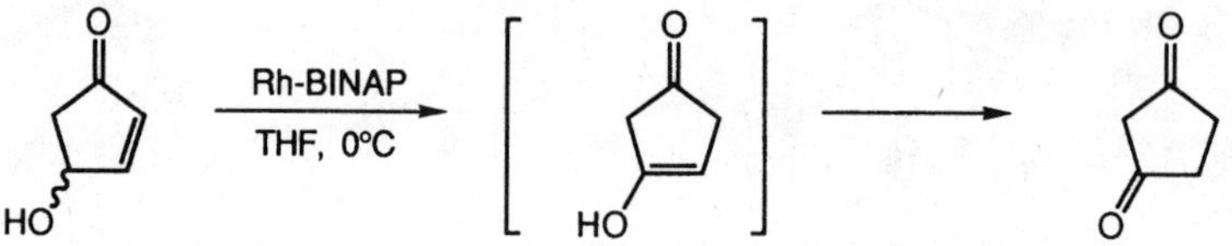

Rh-BINAP
THF, 0°C

29   **30**   **31**

(3.12)

Recently it has been reported that the catalytic isomerization of allylic alcohols is promoted by $[Rh(diphosphine)(solvent)_2]^+$ at 25°C yields synthetically useful quantities of the corresponding simple enols and that the transformation of allylic alcohols to enols and thereby to ketonic products proceeds catalytically via hydrido-π-allylic and hydrido-π-oxy-allylic intermediates, respectively [20]. Consistently observed, enantioselection has been in the process of conversion of a prochiral enol to a chiral aldehyde. Thus, the prochiral substrate **32** is transformed to the optically active aldehyde **34** with 18% ee by using $[Rh(BINAP)]^+$ catalyst (eq 3.13). Accordingly, this isomerization proceeds via a different mechanism from that of the isomerization of allylamine. For the reaction mechanism of the asymmetric isomerization of allylamine, we propose a unique, nitrogen-triggered mechanism, where the nitrogen coordination of the allylamine to the Rh$^+$ center plays an important role in the reaction process as well as in the chiral recognition [4b,14]. A proposed mechanism for the isomerization of allylamine is illustrated in Scheme 3.4.

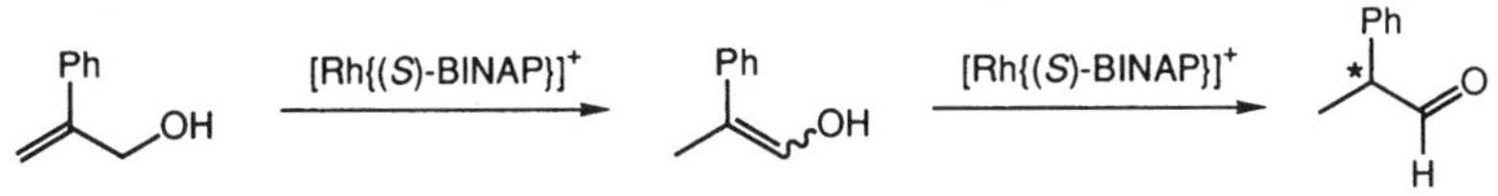

$$(3.13)$$

32   **33**   **34**

S = solvent, P⌒P = BINAP

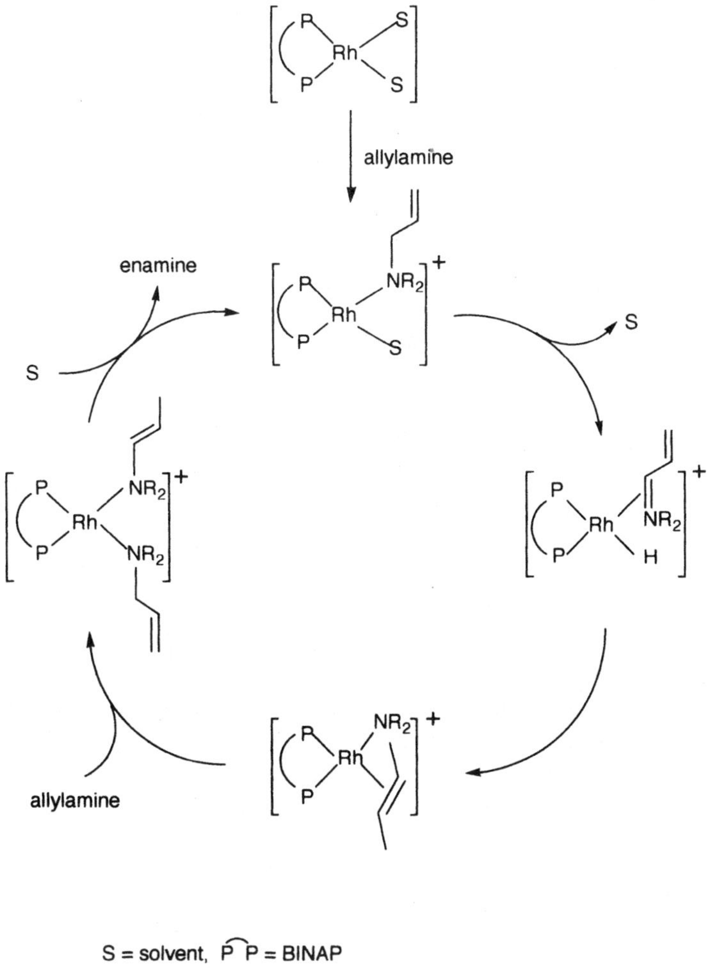

Scheme 3.4. A nitrogen-triggered mechanism for the isomerization of allylamine with cationic Rh(I)-BI-NAP complexes.

Regarding the catalyst precursor among cationic Rh(I)-bisphosphane complexes, the solvent complex [Rh(BINAP)(solvent)n]$^+$ is the most active. For example, [Rh(BINAP)(MeOH)$_2$] ClO$_4$ is effective for the isomerization of 1 even at temperatures below $-20°C$, but it is very sensitive to impurities. The diene complex [Rh(BINAP)(diene)]$^+$ is most conveniently used in laboratory-scale experiments and is sufficiently active around ambient temperature. The bis(BINAP) complex [Rh(BINAP)$_2$]$^+$ is a less-active catalyst and needs temperatures >90°C for practical use, but this complex is the most robust and the most adequate for industrial purposes, as mentioned above. All three catalyst types show equally high chemo- and stereoselectivity with the same stereochemical relationship for the isomerization. In contrast to the very active cationic Rh(I) complexes, the neutral Rh(I) complex, "Rh(BINAP)Cl" prepared in situ from [Rh(diene)Cl]$_2$ with two molecules of the BINAP ligand, was not effective for the isomerization of allylamine. This finding may imply that the Lewis acidity of the Rh metal is an important factor for catalytic activity. Peraryldiphosphine ligands, where all substituents on the phosphorus atoms are aromatic groups, form the most active catalysts.

Besides BINAP or p-TolBINAP, optically active peraryldiphosphines with axial chirality based on the biphenyl groups (6,6′-dimethylbiphenyl-2,2′-diyl)bis(diphenylphosphine) and its analog are also effective ligands for the asymmetric isomerization as expected [21].

3.6. CONCLUSION

During the past decade, metal-catalyzed asymmetric reactions have become one of the indispensable synthetic methodologies in academic and industrial fields. The asymmetric isomerization of allylamine to an optically active enamine is a typical example of the successful application of basic research to an industrial process. We believe that Takasago's successful development of large-scale asymmetric catalysis will have a great impact on both synthetic chemistry and the fine chemical industries. The Rh–BINAP catalysts, though very expensive, have become one of the cheapest catalysts in the chemical industry through extensive process development.

ACKNOWLEDGMENT

The authors express their deepest gratitude to the following people for their collaboration: Professors J. Tanaka (Shizuoka University) and S. Watanabe (Chiba University)—terpenoid amine synthesis; Professors R. Noyori (Nagoya University) and the late H. Takaya (Kyoto University)—BINAP synthesis; Professor S. Otsuka and Dr. T. Yamagata (Osaka University) and Dr. H. Kumobayashi (Takasago)—asymmetric isomerizations; and T. Sakaguchi, M. Yagi, H. Nagashima, and N. Murakami (Takasago)—process development.

REFERENCES

1. (a) Jolly, P. W.; Wilke, G. *The Organic Chemistry of Nickel*; Academic Press: Orlando, FL, **1975**; *Vol. 2*, Chapter 1. (b) Houghton, R. P. *Metal Complexes in Organic Chemistry*; Cambridge University Press: London, New York, **1979**; pp. 258. (c) Parshall, G. W. *Homogeneous Catalysis*; Wiley: New York, **1980**; Chapter 3.3. (d) Davies, S. G. *Organotransition Metal Chemistry:*

Applications to Organic Synthesis; Pergamon: Oxford, **1982**; Chapter 7. (e) Colquhoun, H. M.; Holton, J.; Thompson, D. J.; Twigg, M. V. In *New Pathways for Organic Synthesis*; Plenum: New York, **1984**; Chapter 5.

2. Kumobayashi, H.; Akutagawa, S.; Otsuka, S. *J. Am. Chem. Soc.* **1978**, *100*, 3949.

3. (a) Botteghi, C.; Giacomelli, G. *Gazz. Chim. Ital.* **1976**, *106*, 1131. (b) Carlini, C.; Politi, D.; Ciardelli, F. *J. Chem. Soc., Chem. Commun.* **1970**, 1260. (c) Giacomelli, G.; Bertero, L.; Lardicci, L.; Menicagli, R. *J. Org. Chem.* **1981**, *46*, 3707.

4. (a) Otsuka, S.; Tani, K. In *Asymmetric Synthesis*; Morrison, J. D. (Ed.); Academic Press: Orlando, FL, **1985**; *Vol. 5,* Chapter 6. (b) Otsuka, S.; Tani, K. *Synthesis* **1991**, 665.

5. For examples, see (a) Baudry, D.; Ephritikhine, M.; Felkin, H. *Nouveau J. Chim.* **1978**, *2*, 355. (b) Golborn, P.; Scheinmann, F. *J. Chem. Soc. Perkin Trans. 1* **1973**, 2870. (c) Baudry, D.; Ephritikhine, M.; Felkin, H. *J. Chem. Soc. Chem. Commun.* **1978**, 694. (d) Stille, J. K.; Becker, Y. *J. Org. Chem.* **1980**, *45*, 2139. (e) Suzuki, H.; Koyama, Y.; Moro-Oka, Y.; Ikawa, T. *Tetrahedron Lett.* **1979**, 1415.

6. Sully, B. D.; Williams, P. L. *Perfum. Essent. Oil Rec.* **1968**, *59*, 365.

7. (a) Takabe, K.; Katagiri, T.; Tanaka, J. *Bull. Chem. Soc. Jpn.* **1973**, *46*, 222. (b) Fujita, T.; Suga, K.; Watanabe, S. *Chem. Ind. (London)* **1973**, 231. (c) Takabe, K.; Katagiri, T.; Tanaka, J.; Fugita, T.; Watanabe, S.; Suga, K. *Org. Synth.* **1989**, *67*, 44.

8. (a) Takabe, K.; Katagiri, T.; Tanaka, J. *Chem. Lett.* **1977**, 1025. (b) Takabe, K.; Katagiri, T.; Tanaka, J. *Tetrahedron Lett.* **1972**, 4009. (c) Takabe, K.; Yamada, T.; Katagiri, T.; Tanaka, J. *Org. Synth.* **1989**, *67*, 48.

9. Koenig, K. E. In *Asymmetric Synthesis*; Morrison, J. D. (Ed.); Academic Press: Orlando, FL, **1985**; *Vol. 5*, Chapter 3.

10. (a) Miyashita, A.; Takaya, H.; Souchi, T.; Noyori, R. *Tetrahedron* **1984**, *40*, 1245. (b) Miyashita, A.;Yasuda, A.; Takaya, H.; Toriumi, K.; Ito, T.; Souchi, T.; Noyori, R. *J. Am. Chem. Soc.* **1980**, *102*, 7932.

11. (a) Takaya, H.; Akutagawa, S.; Noyori, R. *Org. Synth.* **1989**, *67*, 20. (b) Takaya, H.; Mashima, K.; Koyano, K.; Yagi, M.; Kumobayashi, H.; Taketomi, T.; Akutagawa, S.; Noyori, R. *J. Org. Chem.* **1986**, *51*, 629.

12. Yamagata, T.; Tani, K.; Tatsuno, Y.; Saito, T. *J. Chem. Soc., Chem. Commun.* **1988**, 466.

13. Tani, K.; Yamagata, T.; Tatsuno, Y.; Yamagata, Y.; Tomita, K.; Akutagawa, S.; Kumobayashi, H.; Otsuka, S. *Angew. Chem.* **1989**, *85*, 232; *Angew. Chem., Int. Ed.* **1985**, *24*, 217.

14. Inoue, S.-I.; Takaya, H.; Tani, K.; Otsuka, S.; Saito, T.; Noyori, R. *J. Am. Chem. Soc.* **1990**, *112*, 4897.

15. Takabe, K.; Uchiyama, Y.; Okisaka, K.; Yamada, T.; Katagiri, T.; Okazaki, T.; Oketa, Y.; Kumobayashi, H.; Akutagawa, S. *Tetrahedron Lett.* **1985**, *26*, 5153.

16. Tani, K. Unpublished results.

17. Tani, K. *Pure Appl. Chem.* **1985**, *57*, 1845.

18. Kitamura, M.; Manabe, K.; Noyori, R.; Takaya, H. *Tetrahedron Lett.* **1987**, *28*, 4719.

19. (a) Noyori, R.; Suzuki, M. *Angew. Chem., Int. Ed.* **1984**, *23*, 847. (b) Suzuki, M.; Yanagisawa, A.; Noyori, R. *J. Am. Chem. Soc.* **1985**, *107*, 3348.

20. Bergens, S. T.; Bosnich, B. *J. Am. Chem. Soc.* **1991**, *113*, 958.

21. Schmid, R.; Cereghetti, M.; Heiser, B.; Schönholzer, P.; Hansen, H.-J. *Helv. Chim. Acta* **1988**, *71*, 897.

CHAPTER 3 ADDENDUM—1999

XIN WEN AND IWAO OJIMA
Department of Chemistry, State University of New York at Stony Brook

SECTION 3.5

Asymmetric isomerization of *meso*-1,4-enediol and its derivatives was studied by using [Rh{(*S*)-BINAP}(cod)]ClO$_4$ (Scheme 3A.1) [1]. Reaction of *meso*-diol **1** (0.1 M in THF) catalyzed by (*S*)-BINAP/Rh(I) (5 mol %) produced hydroxyketone (+)-**2a** with 43% ee in quantitative yield. The bis-ether derivatives of diol **3** underwent asymmetric isomerization with the same chiral catalyst in 1,2-dichloroethane to afford ketones (–)-**2** with the opposite configuration and much higher enantiopurities in good-to-excellent yields. For example, the reactions of silyl ethers achieved excellent enantioselectivities (93.5–97.5% ee). The reason for the observed inversion of enantioselectivity on going from the free diol **1** to the bis-ethers **3** in the presence of the same chiral catalyst is not clear. Nevertheless, the results of bis-ethers **3** can be rationalized by a mechanism involving a suprafacial 1,3-hydrogen shift.[1]

Asymmetric desymmetrization of *meso*-cycloheptene-1,4-diol bis-silyl ether **4** by using the Rh(I)-(*S*)-BINAP catalyst gave (*S*)-4-hydroxycycloheptanone (**5**) with 71% ee (Scheme 3A.2).[2]

Asymmetric isomerization of 4-*tert*-butyl-1-vinylcyclohexane (**6**) catalyzed by bis(indenyl)-titanium complex (**8**) bearing a chiral bridging moiety afforded (*S*)-4-*tert*-butyl-1-ethylidenecyclohexane (**7**) with up to 80% ee (Scheme 3A.3).[3,4]

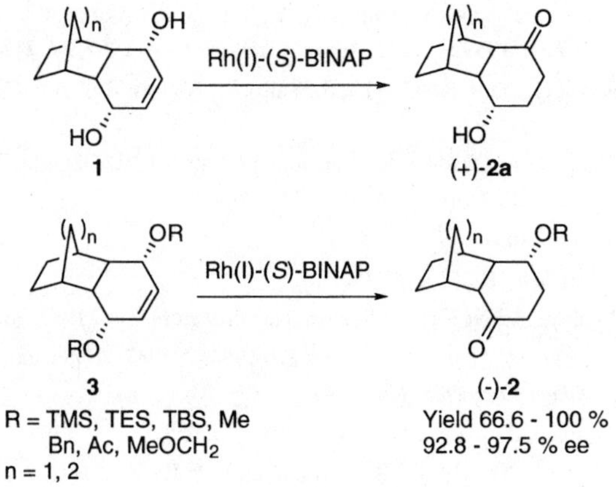

R = TMS, TES, TBS, Me
    Bn, Ac, MeOCH$_2$
n = 1, 2

Yield 66.6 - 100 %
92.8 - 97.5 % ee

Scheme 3A.1.

Catalytic Asymmetric Synthesis, Second Edition, Edited by Iwao Ojima
ISBN 0-471-29805-0 Copyright © 2000 Wiley-VCH, Inc.

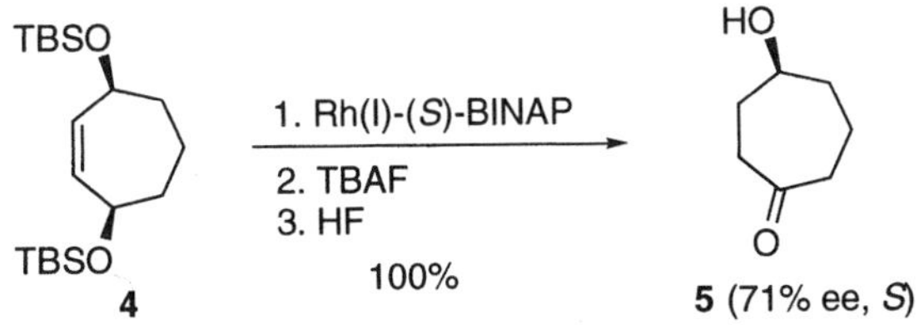

Scheme 3A.2.

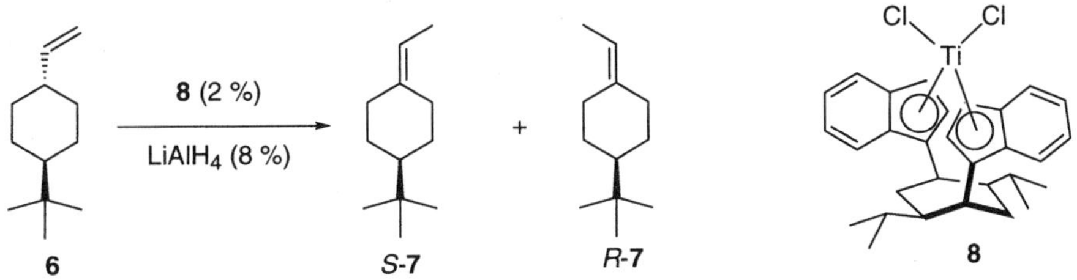

Scheme 3A. 3.

Ab Initio MO calculations of a model complex $Rh(PH_3)_2(NH_3)(CH_2=CHCH_2NH_2)$ were carried out to shed light to the detailed mechanism of Rh(I)-catalyzed isomerization of allylic amines to enamines.[5] This study suggests that the square-planar $[Rh(PH_3)_2(NH_3)(CH_2=CHCH_2NH_2)]^+$ complex is transformed to $[Rh(PH_3)_2(NH_3)((E)-CH_3CH=CHNH_2)]^+$ via intramolecular oxidative addition of the C(1)-H bond to the Rh(I) center, giving a distorted-octahedral Rh(III) hydride intermediate, followed by reductive elimination accompanied by allylic transposition.

REFERENCES

1. Hiroya, K.; Kurihara, Y.; Ogasawara, K. *Agnew. Chem., Int. Ed.* **1995**, *34*, 2287.

2. Hiroya, K.; Ogasawara, K. *J. Chem. Soc., Chem. Commun.* **1995**, 2205.

3. Chen, Z.; Halterman, R. L. *J. Am. Chem. Soc.* **1992**, *114*, 2276.

4. Halterman, R. L.; Chen, Z.; Khan, M. A. *Organometallics* **1996**, *15*, 3957.

5. Yamakawa, M.; Noyori, R. *Organometallics* **1992**, *11*, 3167.

4

ASYMMETRIC CARBOMETALLATIONS

EI-ICHI NEGISHI

Department of Chemistry, Purdue University, West Lafayette, Indiana

4.1. INTRODUCTION

"Carbometallation" is a term coined for describing chemical processes involving net addition of carbon–metal bonds to carbon–carbon π-bonds [1] (Scheme 4.1). It represents a class of insertion reactions. Whereas the term "insertion" per se does not imply anything chemical, the term carbometallation itself not only explicitly and clearly indicates carbon–metal bond addition but also is readily modifiable to generate many additional, more specific terms such as carboalumination, arylpalladation, and so on. In principle, carbometallation may involve addition of carbon–metal double and triple bonds, that is, carbene- and carbyne-metal bonds, as well as those of metallacycles. Inasmuch as alkene- and alkyne-metal π-complexes can also be represented as three-membered metallacycles, their ring expansion reactions via addition to alkenes and alkynes may also be viewed as carbometallation processes (Scheme 4.1).

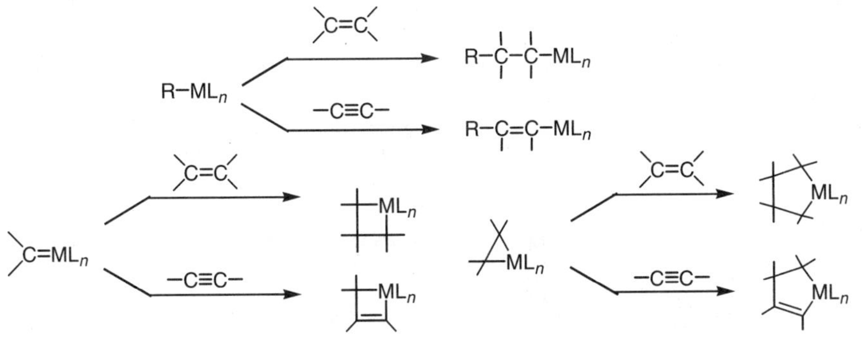

Scheme 4.1.

Catalytic Asymmetric Synthesis, Second Edition, Edited by Iwao Ojima
ISBN 0-471-29805-0   Copyright © 2000 Wiley-VCH, Inc.

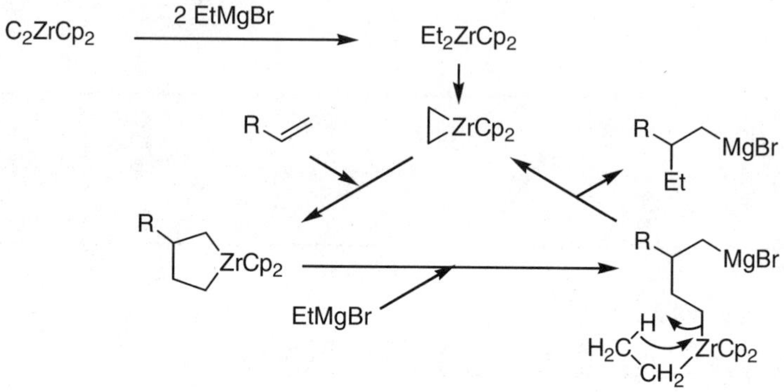

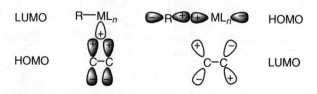

Scheme 4.2.

It is important to note that the term carbometallation primarily represents starting material–product relationships with little or no actual mechanistic implications. In fact, many mechanistically intricate relationships among seemingly discrete processes shown in Scheme 4.1 have been observed. Thus, for example, a seemingly straightforward addition reaction of ethylmagnesiums to 1-alkenes catalyzed by Cp_2ZrCl_2 commonly known as the Dzhemilev reaction [2] was shown to proceed via a cyclic mechanism shown in Scheme 4.2 [3]. Even a more intricate cyclic mechanism has recently been shown to be operative in some Zr-catalyzed ethylalumination reactions of alkynes [4]. It is also well established that alkyne polymerization reactions catalyzed by carbene–metal complexes proceed via a series of formation and cleavage (metathesis) of four-membered metallacycles [5].

Despite mechanistic complications, however, it appears very likely that most, if not all, of the facile and synthetically attractive carbometallation reactions involve, at a critical moment, concerted addition of carbon–metal bonds where the synergistic HOMO-LUMO interactions shown in Scheme 4.3, akin to those for the concerted hydrometallation reactions, provide a plausible common mechanism. This mechanism requires the ready availability of a metal empty orbital. It also requires that addition of carbon–metal bonds be strictly syn, as has generally been observed. Perhaps more important in the present discussion is that concerted syn carbometallation must proceed via a transition state in which a carbon–metal bond and a carbon–carbon bond become coplanar. Under such constraints, one can readily see how chirally discriminated carbon–metal bonds can select either *re* or *si* face of alkenes. In principle, the mechanistic and stereochemical considerations presented above are essentially the same as for related concerted syn hydrometallation. In reality, however, carbometallation is generally less facile than the corresponding hydrometallation, which may be largely attributable to more demanding steric and

Scheme 4.3.

stereoelectronic requirements of carbon groups relative to those of hydrogen. This condition may, in turn, be responsible for the fact that the development of simple asymmetric carbometallations has lagged behind that of similar hydrometallation reactions.

4.2. HISTORICAL BACKGROUND AND SCOPE

Despite its inherent difficulties, carbometallation has, in fact, played important roles in catalytic asymmetric carbon–carbonal bond formation. Isotactic and syndiotactic alkene polymerization involving both heterogeneous and homogeneous Ti and Zr catalysts must involve a series of face-selective carbometallation processes, although the main stereochemical concern in poly(alkene) formation is diastereoselectivity rather than enantioselectivity. This fascinating topic, however, is outside the scope of this chapter, and the readers are referred to Chapter 11 and other previous reviews [6].

One of the earliest enantioselective carbon–carbon bond-forming processes catalyzed by chiral transition-metal complexes is asymmetric cyclopropanation discussed in Chapter 5, which can proceed via face-selective carbometallation of carbene–metal complexes. Some other more recently developed enantioselective carbon–carbon bond forming reactions, such as Pd-catalyzed enantioselective alkene-CO copolymerization (Chapter 7) and Pd-catalyzed enantioselective alkene cyclization (Chapter 8.7), are thought to involve face-selective carbometallation of acyl-Pd and carbon-Pd bonds, respectively (Scheme 4.4). Similarly, the asymmetric Pauson-Khand reaction catalyzed by chiral Co complexes most likely involves face-selective cyclic carbometallation of chiral alkyne-Co complexes (Chapter 8.7).

In this chapter, attention is primarily focused on simple single-stage catalytic enantioselective carbometallation reactions leading to the formation of acyclic products in most cases regardless of their actual mechanisms. However, some closely related cyclic processes are also discussed. At present, the scope of such processes appears to be largely limited to those involving a few early transition metals, especially Zr.

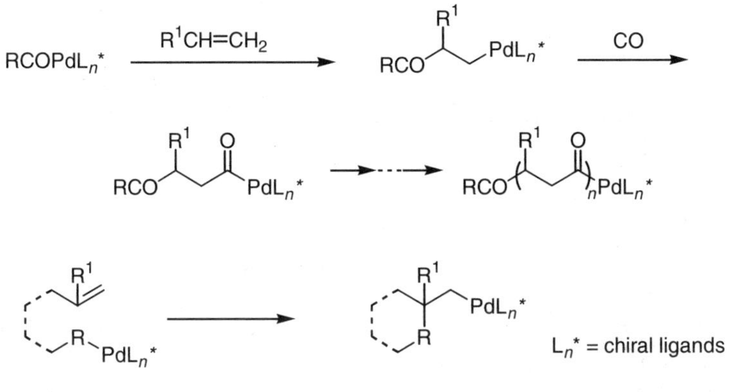

Scheme 4.4.

4.3. ZIRCONIUM-CATALYZED ENANTIOSELECTIVE MONOCARBOALUMINATION OF UNACTIVATED ALKENES

4.3.1. Background

Since the discovery of the Zr-catalyzed controlled monocarboalumination of alkynes in 1978 [1] (Eq. 4.1), its application to the development of a related Zr-catalyzed enantioselective carboalumination of alkenes as shown in Equation 4.2 has been an attractive but rather elusive synthetic goal. Although enantioselective hydroboration of 1,1-disubstituted alkenes produces related organoboron products, hydroboration of 1,1-disubstituted alkenes has exhibited very low levels (usually <10–20% ee) of enantioselectivity [7] (Eq. 4.3) (Scheme 4.5).

As detailed later, the Zr-catalyzed enantioselective monocarbometallation of simple, unactivated alkenes as shown in Equation 4.2 was finally developed during the past few years [8,9]. Before the discussion, however, it is useful to become familiar with various potentially competing reactions, patterns, and mechanisms involving mutually related reagents and catalysts to fully appreciate the highly intricate relationships among them and apply the knowledge.

(a) Polymerization and Oligomerization. In parallel with the development of the Zr-catalyzed monocarboalumination of alkynes, alkene polymerization of high tacticity has been developed using methylalumoxane-zirconocene reagent systems [6b]. In the single-state carboalumination of alkynes shown in Equation 4.1 (Scheme 4.5), an alkylaluminum, for example, Me_3Al, is converted to an alkenyldialkylaluminum. Both starting material and product exist as doubly bridged dimers in the absence of a strongly coordinating solvent. However, the alkenyl double bridge is energetically more favorable, making the product less reactive, and the observed single-stage carboalumination results. In the desired alkene carboalumination, however, other factors must be exploited to control the degree of polymerization.

In the presence of hydrogen, for example, optically active saturated trimers of propene were obtained, albeit in low yields, by using $(EBTHI)ZrCl_2$ (**1**) [10] and methylalumoxane [11]. At low relative concentrations of alkenes, their oligomers, mainly dimers through pentamers, were

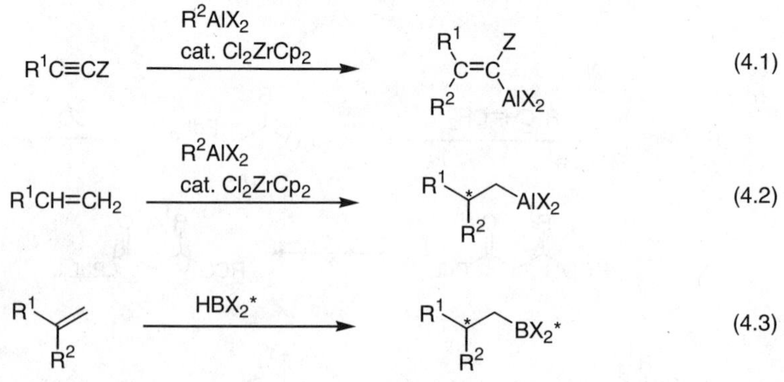

$$R^1C\equiv CZ \xrightarrow[\text{cat. } Cl_2ZrCp_2]{R^2AlX_2} \begin{array}{c} R^1 \quad Z \\ C=C \\ R^2 \quad AlX_2 \end{array} \qquad (4.1)$$

$$R^1CH=CH_2 \xrightarrow[\text{cat. } Cl_2ZrCp_2]{R^2AlX_2} \quad R^1 \overset{*}{\underset{R^2}{\diagdown}} AlX_2 \qquad (4.2)$$

$$R^1 \overset{}{\underset{R^2}{\diagdown}} \xrightarrow{HBX_2^*} \quad R^1 \overset{*}{\underset{R^2}{\diagdown}} BX_2^* \qquad (4.3)$$

$Cp = \eta^5\text{-}C_5H_5$. Cp^* = chiral Cp derivative. R^1, R^2 = C group. X = C or halogen ligand. X^* = chiral ligand. Z = H, C, Si group.

Scheme 4.5.

obtained as 1,1-disubstituted alkenes by using (*S*)-(EBTHI)Zr[OCOCH(OAc)Ph-(*R*)]$_2$ and methylalumoxane (Scheme 4.6). It is interesting that the trimer of 1-butene containing one asymmetric carbon atom was only 27% ee, but generation of the second chiral center in the formation of the tetramer proceeded in 80% ee [12]. The results suggest that, in the asymmetric alkene polymerization, the enantioselectivity in the initial phase may be low but that the chiral growing chain may lead to much higher enantioselectivity levels. From the viewpoint of the development of a Zr-catalyzed enantioselective single-stage carboalumination of alkenes, these results presented a mixed blessing in that they showed both feasibility and inherent difficulty in achieving such a goal.

(b) β-Dehydrometallation Leading to Zr-Catalyzed Hydroalumination. It was observed in 1978 that, under the conditions of the Zr-catalyzed monomethylalumination of alkynes with Me$_3$Al and Cl$_2$ZrCp$_2$ (Eq. 4.1, Scheme 4.5), terminal alkenes did react but did not produce either the desired methylalumination product or alkene polymers in significant yields. Recent reinvestigation by using 1-octene indicated that the two major products were 2-(*n*-octyl)-1-octene and 2-methyl-1-octene which were obtained in ~60 and 20% yields, respectively. Evidently, the desired 2-methyloctylmetal must have been formed, but it must have undergone hydrogen-transfer hydrometallation to produce 2-methyl-1-octene and an *n*-octylmetal, which would then undergo a similar carbometallation and a hydrogen-transfer hydrometallation, as shown in Scheme 4.7 [8]. Clearly, successful development of an enantioselective single-stage carboalumination of alkenes must somehow avoid dehydrometallation leading to hydrogen-transfer hydrometallation as well.

(c) Zr-Catalyzed Cyclic Carbometallation of Alkenes. As discussed earlier and shown in Scheme 4.2, net addition of carbon–metal bonds to alkenes can proceed via cyclic carbometallation, which must be clearly recognized as such and distinguished from acyclic processes, especially in connection with enantioselective carbometallation. It may be readily envisioned

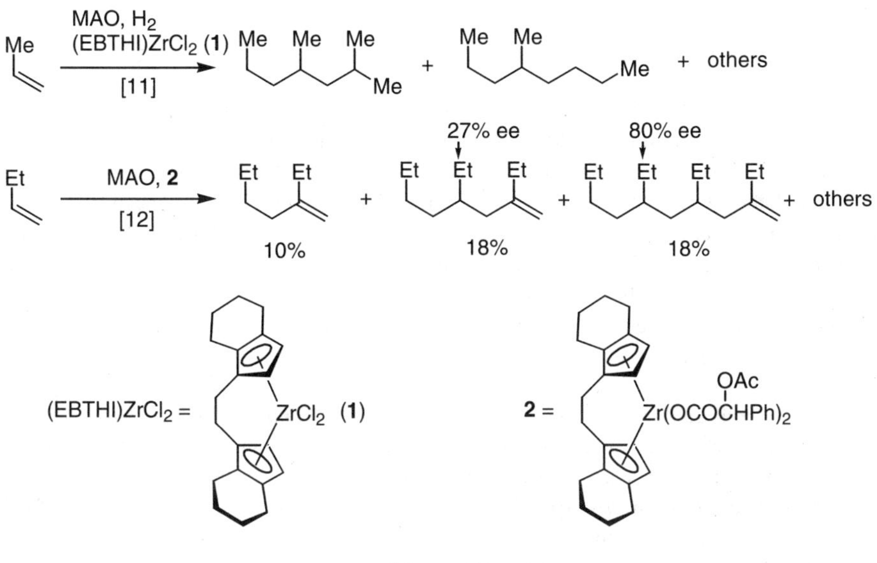

Scheme 4.6.

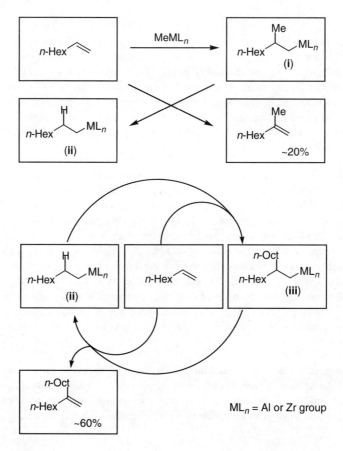

Scheme 4.7.

that various factors affecting the enantioselective aspects of acyclic and cyclic carbometallation processes could be significantly different. Until recently, the Dzhemilev ethylmagnesation of alkenes [2] had been essentially the only single-stage and high-yielding Zr-catalyzed alkene carbometallation. Consequently, efforts have been made to develop its enantioselective versions. Although highly enantioselective processes have been recently developed with proximally heteroatom-substituted alkenes, as discussed in Section 4.4, few results obtained with simple, unactivated alkenes have been reported. Furthermore, the highest enantioselectivity thus far reported is 33% ee for the reaction shown in Scheme 4.8 [9]. Further developments along this line are highly desirable.

(d)   Zr-Centered Carbometallation vs. Al- or Other Metal-Centered Carbometallation.
Chiral ligands are generally expensive. It is therefore highly desirable to use chiral ligands as catalyst components rather than those of stoichiometric regents. In the desired Zr-catalyzed enantioselective carboalumination of alkenes, for example, chiral ligands should be part of the Zr catalysts. Furthermore, it appears desirable to devise Zr-centered carbometallation processes rather than Al-centered ones. This factor can be potentially serious as both Zr- and Al-centered

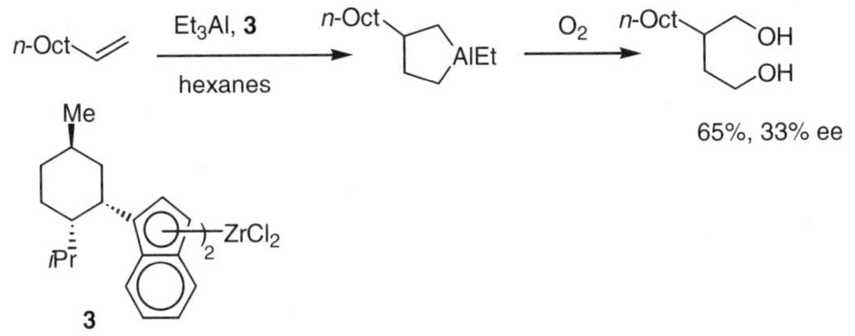

Scheme 4.8.

carbometallation processes have been observed in the Zr-catalyzed carboalumination of alkynes [13]. Specifically, the currently available data indicate that the Zr-catalyzed carboalumination of alkynes with trialkylaluminums must be a Zr-centered process, whereas the corresponding reaction with dialkylchloroaluminums is most likely an Al-centered process [13].

4.3.2. Zirconium-Catalyzed Enantioselective Methylalumination of Unactivated Alkenes

(a) Discovery and Exploration of Synthetic Scope. Despite various difficulties discussed above, the reaction of unactivated terminal alkenes with Me_3Al in the presence of chiral zirconocene derivatives was investigated. With 8 mol% of dichlorobis(1-neomenthylin-denyl)zirconium (**3**), $Cl_2Zr(NMI)_2$ [14] hereafter, as a catalyst, various monosubstituted alkenes reacted with one molar equivalent of Me_3Al at 22°C to give the corresponding 2-methyl-1-al-kanols in 77–92% yields, after oxidation with O_2, except in the case of styrene where the yield was 30% [8]. Only minor amounts, if any, of polymers and hydrogen-transfer hydrometallation products were present in the product mixtures. Furthermore, ^{1}H and ^{13}C NMR analysis of the esters derived from (+)- and (−)-MTPA, that is, α-methoxy-α-(trifluoromethyl)phenylacetic acid, indicated that the reaction proceeded with good-to-high enantioselectivity (65–75% ee) [8] (Scheme 4.9 and Table 4.1).

Several other chiral catalysts have also been tested, but they have been less satisfactory. Specifically, the reaction of 1-octene with Me_3Al in the presence of 8 mol % of dichlorobis(1-neoisomenthylindenyl)zirconium, dichlorobis(1-neoisomenthyl-4,5,6,7-tetrahydroindenyl)zir-conium, (*R,R*)-dichloroethylenebis(4,5,6,7-tetrahydro-1-indenyl)zirconium, and (*R,R*)-

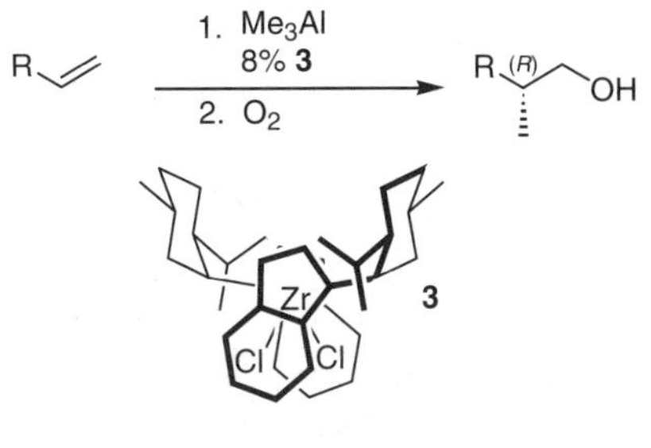

Scheme 4.9.

TABLE 4.1. Zirconium-Catalyzed Methylalumination of Monosubstituted Alkenes[a]

Substrate	Time, h	Product	Yield,[b] %	% ee
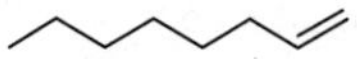 (structure)	12	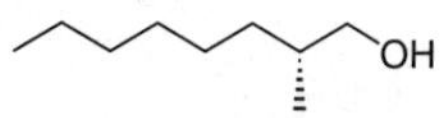 (structure) OH	88	72
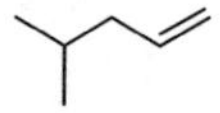 (structure)	12	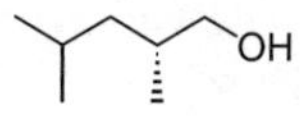 (structure) OH	92	74
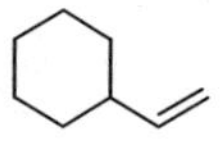 (structure)	12	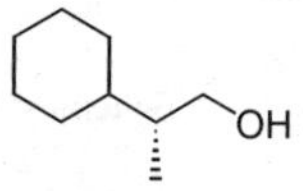 (structure) OH	80	65
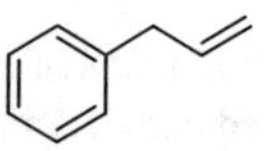 (structure)	24	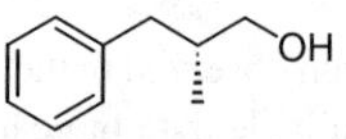 (structure) OH	77	70
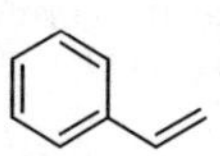 (structure)	528	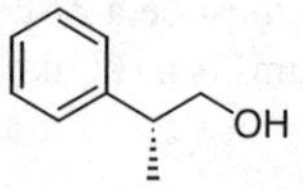 (structure) OH	30	85
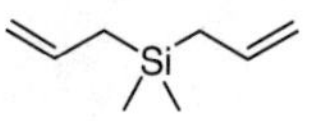 (structure)	12	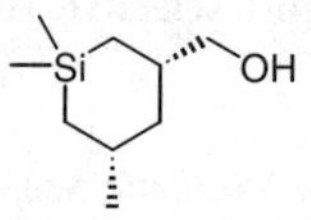 (structure) OH	81	74
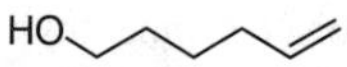 HO—(structure)	12[c]	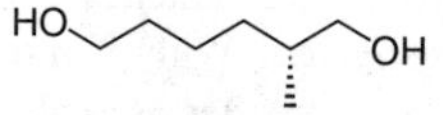 HO—(structure) OH	79	75
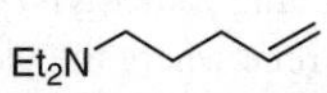 Et$_2$N—(structure)	96[d]	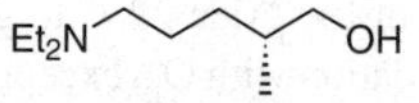 Et$_2$N—(structure) OH	68	71

[a]The reactions were run by using 8 mol % of **3** and 1 equiv. of Me$_3$Al in 1,2-dichloroethane at 22°C. [b]Isolated yields. [c]Three-fold excess of Me$_3$Al was used. [d]Two-fold excess of Me$_3$Al was used.

ethylenebis(4,5,6,7-tetrahydro-1-indenyl)zirconium (R)-1′,1″-bi-2-naphtholate gave 2-methyl-1-octanol, after oxidation, in 67% (50% ee, S), 60% (6% ee, S), 45% (8% ee, R), ad 53% (6% ee, R) yields, respectively, the absolute configuration and enantiopurity of each product being indicated in parentheses [8]. In a brief study for optimizing enantioselectivity, it was found that the methylalumination of 1-octene with Me$_3$Al in the presence of **3** could be readily achieved to give 2-methyl-1-octanol with 81% ee in 1,1-dichloroenthane at 0°C. Clearly, further optimization, especially with respect to chiral zirconocene catalysts, is desirable. Nonetheless, the results presented above have clearly established the following.

1. Controlled single-stage carboalumination of 1-alkenes by using a zirconocene catalyst can be achieved selectively and in high yields without the production of unwanted oligomers, polymers, and hydrometallation products in significant yields.

2. Under unoptimized conditions, the reaction uniformly gives (R)-2-methyl-1-alkanols with ~65–75% ee. In the case of 1-octene, running the reaction at 0°C in 1,1-dichloroethane has improved the enantioselectivity from 72 to 81% ee.

3. Only monosubstituted terminal alkenes, including some terminal dienes, have so far been successfully methylaluminated. Within this restriction, however, the reaction appears to be reasonably general, as shown in Table 4.1. Specifically, n-alkyl, isoalkyl, and secondary alkyl substituents can be accommodated, but tertially alkyl-substituted ethylenes, such as t-BuCH=CH$_2$, fail to react under the same conditions. In contrast with styrene, allylbenzene reacts normally.

To apply this reaction to the synthesis of natural products, it is essential to accommodate various functional groups. To probe this point, the effects of hydroxy, amino, and alkenyl groups have been briefly examined. As shown in Table 4.1, both 5-hexenol and 4-pentenyldiethylamine react normally provided that 3 and 2 equivalents, respectively, of Me$_3$Al are used, and the enantioselectivity is also in the normal range. Preliminary results indicate that 8-nonenol, 4-pentenol and 3-butenol also react normally to give the desired diols within 10–12 h at 22°C, after oxidation, in 60, 55, and 72% yields, respectively [15]. However, the reaction of the t-BuPh$_2$Si-protected 3-butenol was very sluggish, requiring 7 days for producing the methylaluminated product in 44% yield. The reaction of 4-(phenyldimethylsilyl)-1-butene, however, gave the methylalumination product in 69% yield [15].

The effects of the second double bond have also been examined. As might be expected from the case of styrene, the reaction of (E)-1,3-decadiene and 2-(n-hexyl)-1,3-butadiene was very sluggish and did not proceed detectably over 12 h. Although 1,4-pentadiene itself has not yet been tested, (E)-5-methyl-1,4-undecadiene, (E)-4-(n-butyl)-1,4-nonadiene, and (Z)-4-(trimethylsilyl)-1,4-nonadiene have provided the desired 2-methylated alcohols in 79, 84, and 69% yields, respectively [15]. Similarly, (E)-6-methyl-1,5-dodecadiene and (E)-5-(n-butyl)-1,5-decadiene acted as if they were simple monosubstituted terminal alkenes, producing the desired 2-methylated alcohols in 55 and 71% yields, respectively [15]. However, 1,5- and 1,6-dienes containing two terminal vinyl groups preferentially undergo cyclic carboalumination following the desired methylmetallation. Thus, 1,5-hexadiene and 1,6-heptadiene give 3-methylcyclopentylmethanol and 3-methylcyclohexylemethanol in 65 and 72% yields, respectively. Because these alcohols are only 84 and 74% cis, respectively, their enantiopurities have not been measured. However, the reaction of dimethyldiallylsilane gives (3,3,5-trimethyl-3-silacyclohexyl)methanol, which was >95% cis and 74% ee, in 81% yield [8].

To further probe the potential synthetic utility of the Zr-catalyzed enantioselective carboalumination of alkenes, the C$_{14}$ sidechain units of vitamin E, that is, (2R,6R)-2,6,10-trimethyl-1-undecanol (4) and the corresponding iodide 5, have been synthesized as summarized in Scheme 4.10 [15]. Whereas the enantioselectivity in each methylalumination step must be significantly improved, the four-step synthesis of 4 and 5 summarized in Scheme 4.10 appears to provide one of the most efficient enantioselective routes to the C$_{14}$ sidechain units of vitamin E [16].

(b) Mechanism. Because Me$_3$Al lacks β-CH bonds, the Zr-catalyzed methylalumination of 1-alkenes described above cannot proceed by cyclic mechanisms, such as those shown in Scheme 2. The moderate-to-high enantioselectivity and the uniform R configuration of the products in cases where 3 is used as a chiral catalyst are difficult to explain in terms of an Al-centered carbometallation process. However, they are consistent with Zr-centered carbometallation. Initially, it was thought that Me$_3$Al would not only monomethylate 3 but also polarize the second Zr-Cl bond to generate a cationic Zr intermediate 6 so as to maintain C$_2$ symmetry about the Zr-C bond in line with the widely believed mechanism for the homogeneous Ziegler-Natta-type polymerization [17] (Scheme 4.11). However, by virtue of the C$_2$ symmetry of 3, substitution of either of the two Cl atoms with Me leads to a single species 7 of C$_1$ symmetry, which may then undergo methylalumination most probably assisted by an aluminum species

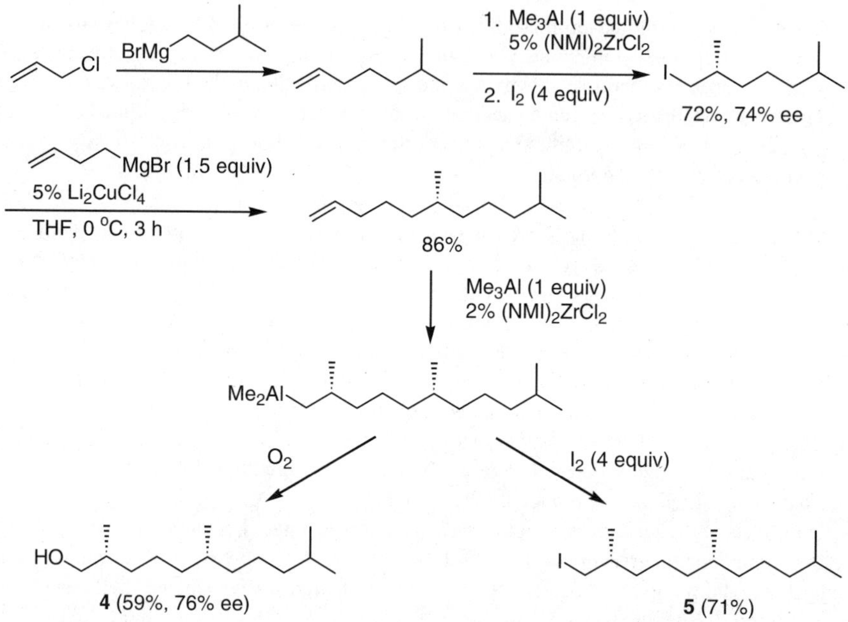

Scheme 4.10.

(Scheme 4.11). It is important to note that only one side or one locale of an empty orbital is available for olefin coordination. The mechanistic dichotomy presented above is not unlike the S_N1 versus S_N2 dichotomy, and their distinction through kinetic studies is very desirable.

Another mechanistically intriguing aspect of the reaction is the virtual absence of β-dehydrometallation of the methylmetallated species. This absence must be attributable to one

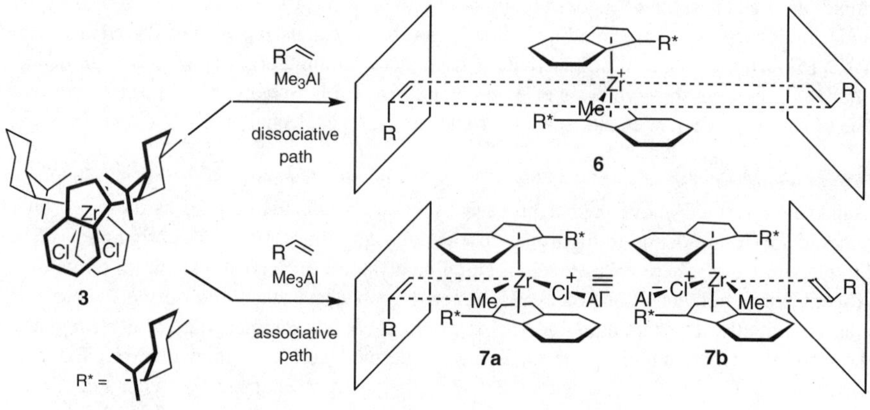

Scheme 4.11.

or more of the following three differences between Cp_2ZrCl_2 and $(NMI)_2ZrCl_2$, that is, (1) steric requirements, (2) chirality, and (3) electronic factors due to the indene ring. To probe this point, Cp_2ZrCl_2, $(1\text{-Ind})_2ZrCl_2$, and $(Me_5C_5)_2ZrCl_2$ were compared in the reaction of 1-decene with Me_3Al. As indicated by the results summarized in Scheme 4.12, indene is a satisfactory ligand leading to the formation of the desired product in high yield without inducing β-dehydrometallation. However, no reaction was observed with $(Me_5C_5)_2ZrCl_2$. Clearly, ligand chirality is not essential in blocking β-dehydrometallation. However, it is not clear at present if the benzene ring of indene exerts any beneficial electronic influence.

4.3.3. Zirconium-Catalyzed Enantioselective Ethyl- and Higher Alkylalumination of Unactivated Alkenes

The reaction of monosubstituted alkenes with Et_3Al in the presence of a catalytic amount of Cp_2ZrCl_2 was reported to give the corresponding 3-alkyl-substituted aluminacycles [16], but the corresponding reaction of 1-decene in the presence of $(NMI)_2ZrCl_2$ in hexanes proceeded with only 33% ee as shown in Scheme 4.8. It is interesting that the reaction with 1 molar equiv of Et_3Al in the presence of 8 mol % of Cp_2ZrCl_2 in $(CH_2Cl)_2$ in place of hexanes produced, after deuterolysis, a 37% yield of 3-(deuteriomethyl)undecane, which contained D in the C-1 position only to the extent of 9%. The extent of D incorporation in the deuteriomethyl group was >90%. 2-Ethyl-1-decene and 1-deuteriodecane were also obtained in 20% yield each (Scheme 4.13). Clearly, acyclic ethylalumination was favored over its cyclic counterpart in $(CH_2Cl)_2$, even though hydrogen-transfer hydrometallation is a serious side reaction. Encouraged by this promising lead, the reaction was carried out in CH_3CHCl_2 in the presence of 8 mol % of $(NMI)_2ZrCl_2$ (**3**). Under these conditions, 1-decene gave, after oxidation, (R)-2-ethyl-1-decanol in 63% yield and 92% ee (Scheme 4.13) [9]. The results indicate that the mechanism must have completely switched from cyclic to acyclic as the solvent was changed from hexanes to CH_3CHCl_2. Of several chlorinated hydrocarbon solvents tested, CH_3CHCl_2 led to the highest enantioselectivity, although CH_2Cl_2 was almost equally satisfactory. However, 1,2-dichloroethane,

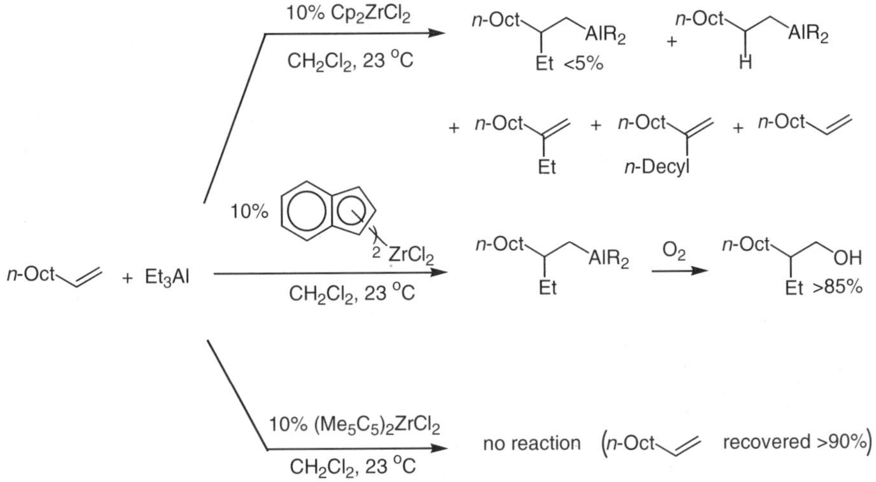

Scheme 4.12.

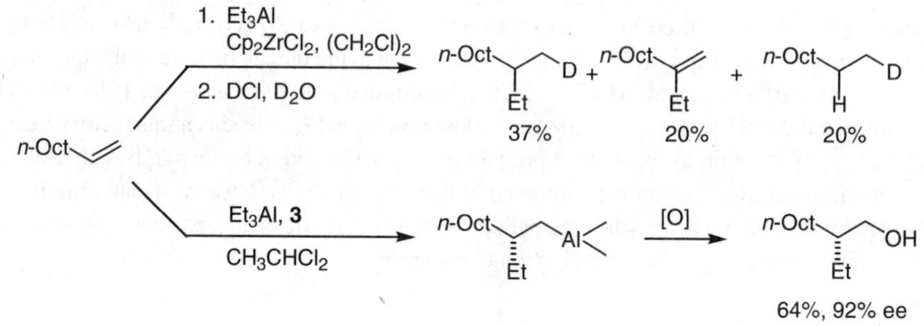

Scheme 4.13.

chlorobenzene, and 1,2-dichlorobenzene appeared to be somewhat inferior. The generally higher carbometallation rates observed with ethyl- and higher alkylaluminums relative to methylalumination permit the reaction to go to completion at 0°C usually within 24 h. Under these conditions, the ethylalumination reaction has been shown to proceed in reasonable yields and ≥90% ee in most cases. The experimental results are summarized in Table 4.2. It should be pointed out that not only t-BuCH=CH$_2$ but c-HexCH=CH$_2$ also failed to undergo ethylalumination. The chemical yields of ethylalumination shown in Table 4.2 are generally 10–15% lower

TABLE 4.2. Zirconium-Catalyzed Ethyl- and n-Propylalumination of Monosubstituted Alkenes[a]

Substrate	R of R$_3$Al	Time, h	Product	Yield,[b] %	% ee
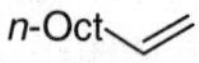 n-Oct	Et	12	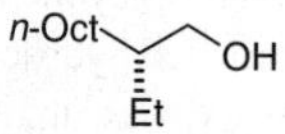 n-Oct (OH, Et)	64	92
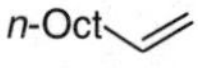 n-Oct	n-Pr	12	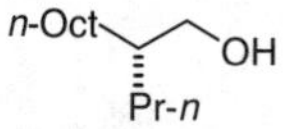 n-Oct (OH, Pr-n)	62	91
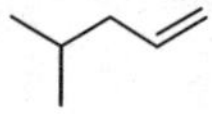	Et	24	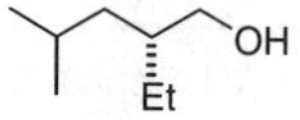 (OH, Et)	77	90
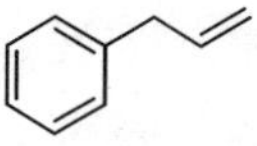	Et	24	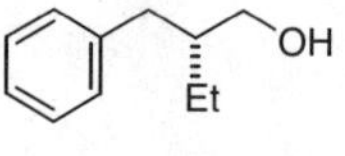 (OH, Et)	69	93
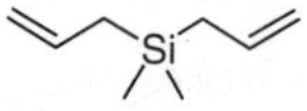 Si	Et	24	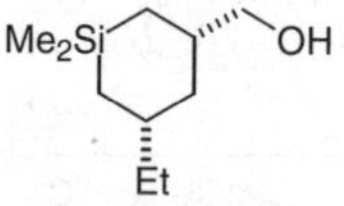 Me$_2$Si (OH, Et)	76	96
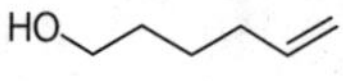 HO	n-Pr	24	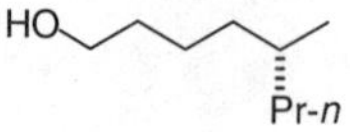 HO (Pr-n)	90	91
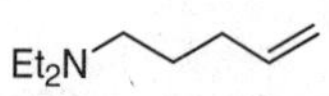 Et$_2$N	Et	72[d]	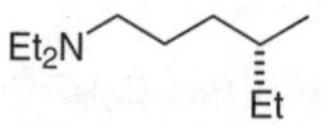 Et$_2$N (Et)	88	90

[a]The reactions were run by using 8 mol % of **3** and 1 equiv. of R$_3$Al at 0°C in CH$_3$CHCl$_2$, unless otherwise mentioned. [b]Isolated yields. [c]Run at 10°C by using a three-fold excess of n-Pr$_3$Al. [d]Run at 25°C by using a two-fold excess of Et$_3$Al.

than those of methylalumination due to competitive hydrometallation. Although a full-fledged investigation of the synthetic scope with respect to the alkyl groups of alkylaluminums is yet to be performed, n-propylalumination appears to proceed equally well (Table 4.2) [9].

4.4. ZIRCONIUM-CATALYZED ENANTIOSELECTIVE CYCLIC CARBOMETALLATION OF ALLYLIC ETHERS, ALCOHOLS, AMINES, AND SULFIDES

4.4.1. Background

In 1985 the development of the Zr-promoted bicyclization reaction of enynes shown in Eq. 4.4 [19,20] was a milestone in the Cp$_2$Zr(II) chemistry. Coupled with a simple and convenient method for in situ generation of Cp$_2$Zr(II) in the form of Cp$_2$Zr(II)-alkene complexes shown in Eq. 4.5 [21,22], it has led to the development of a wide variety of cyclization reactions applicable to the synthesis of simple and complex organic compounds. In addition to enynes, various other π-compounds including dienes (Eq. 4.6) [23,24] and monoenes (Eq. 4.7) [25,26] have also been successfully used (Scheme 4.14). It should be noted that in the reactions shown in Scheme 4.14, one or more asymmetric carbon centers are generated and that many of these reactions have displayed very high diastereoselectivities, which often approach 100%. Another significant milestone is the finding that some of these reactions can be catalytic in Zr [2–4,18,27,28]. Also important is the recognition that some seemingly straightforward carbometallation reactions actually proceed by cyclic mechanisms, as shown earlier in Scheme 4.2.

Despite many such promising results, development of Zr-catalyzed enantioselective reactions based on these Zr-promoted or Zr-catalyzed cyclic carbometallation processes has not been straightforward. Thus, there has been no report of Zr-catalyzed cyclic carbometallation of simple, unactivated alkenes exhibiting enantioselectivity >33% [9,29].

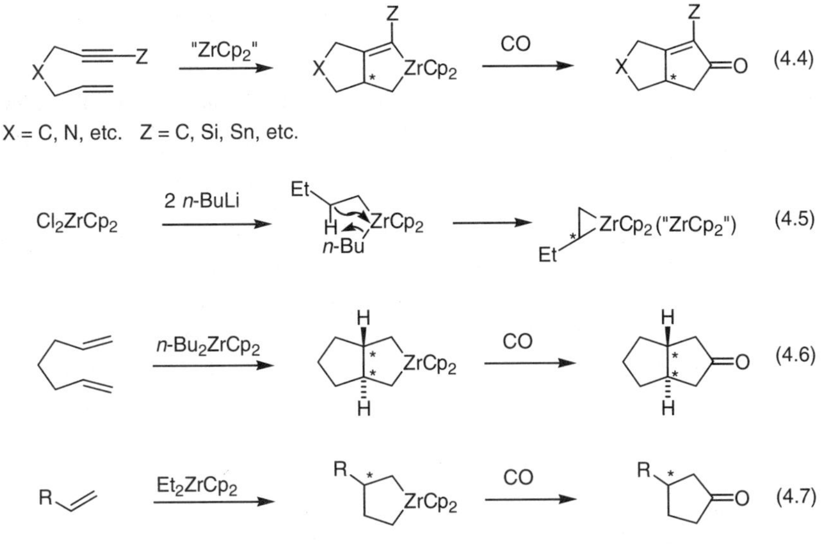

$$X = C, N, \text{ etc.}\quad Z = C, Si, Sn, \text{ etc.}$$

Scheme 4.14.

A significant breakthrough in this area was a finding made in 1993 that allylic ethers, alcohols, and amines can undergo Zr-catalyzed highly enantioselective cyclic carbometallation [30]. Since then, several groups of workers have developed a number of related reactions of high enantioselectivity and potential synthetic utility, as detailed below. It is striking that all of the examples displaying high ee figures are heteroatom-substituted allylic derivatives.

4.4.2. Zirconium-Catalyzed Enantioselective Carbomagnesation-Elimination Tandem Reaction of Allylic Derivatives

(a) *Asymmetric Synthesis of ω-Alkenols and ω-Alkenylamines.* The reaction of five-through seven-membered allylic ethers and amines with a several-fold excess of alkylmagnesium halides in the presence of 10 mol % of (EBTHI)ZrCl$_2$ (**1**) gives allylically alkylated ω-alkenols and ω-alkenylamines, respectively, in good yields and >90% ee [30,31] (Table 4.3). A plausible mechanism involves (a) formation of a zirconacyclopropane intermediate, (b) its face-selective ring expansion, (c) regioselective transmetallation, and (d) β-H abstraction to give the product and the zirconacyclopropane in its original form for completion of a catalytic cycle (Scheme 4.15). The allylic substrate approaches Zr in a manner to minimize steric interactions with the cyclohexene ring. Although this mechanism is plausible and consistent with the observed stereochemistry, it does not appear to explain the seemingly crucial role of the allylic O or N atom. The strategically positioned heteroatom must intimately interact with Zr and/or Mg.

Despite the severe limitations in scope, this reaction is of considerable synthetic potential, as suggested by the synthesis of part of a structure of Sch 38516 [32] (Scheme 4.16). It should also be noted that the required cyclic allyl derivatives may be readily synthesized via metathesis cyclization [33].

(b) *Kinetic Resolution of Racemic Allyl Ethers.* The Zr-catalyzed enantioselective carbomagnesation-elimination reaction has also been applied to the kinetic resolution of racemic cyclic allyl ethers as well as acyclic ethers containing cyclic allyl groups, as indicated by the results shown in Table 4.4 [33,34]. At about 60% conversion, the unreacted allylic ethers with >90% ee may be recovered in most cases.

The generally higher reactivity of dihydrofurans makes them difficult to resolve in the manner discussed above. However, the two enantiomeric dihydrofurans react with EtMgCl in the presence of (*R*)-(EBTHI)ZrCl$_2$ via two different paths, as exemplified by the results shown in Scheme 4.17 [35].

4.4.3. Zirconium-Catalyzed Enantioselective Carboalumination-Elimination Tandem Reaction of Allylic Derivatives and Its Application to Kinetic Resolution

In view of the Zr-catalyzed enantioslective carbomagnesation-elimination tandem reaction of allylic derivatives discussed earlier, a similar process with Et$_3$Al might be expected and has indeed been developed recently [29]. As a representative example, the reaction of 2,5-dihydrofuran with 3 equiv. of Et$_3$Al in the presence of (*R*)-(EBTHI)Zr[BINOL-(*S*)] (**8**) and (NMTHI)ZrCpCl$_2$ (**9**) produced, after hydrolysis, (*S*)-2-ethyl-3-buten-1-ol in 90 and 67% yields, respectively. The enantioselectivity observed with **8** was >99% ee, whereas that observed with **9** was 85–90% ee. Upon deuterolysis of the organoaluminum products, a mixture of monodeuterated and nondeuterated products was obtained and the extent of D incorporation increased to 94% with neat Et$_3$Al without any solvent. The results indicate that the reaction must produce two organoaluminum products, **10** and **11** (Scheme 4.18). On oxidation with O$_2$ only

TABLE 4.3. Zr-Catalyzed Enantioselective Carbomagnesation-Elimination Reaction of Allylic Derivatives[a]

Substrate	R of RMgCl	Product	Yield, %	% ee
	Et	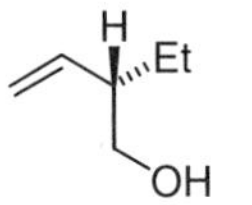	65	>97
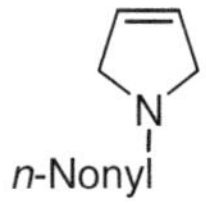	Et	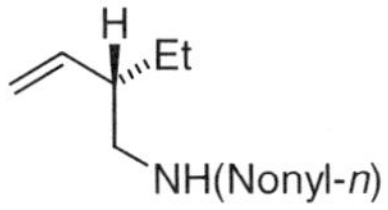	75	>95
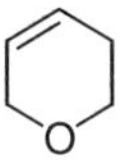	Et	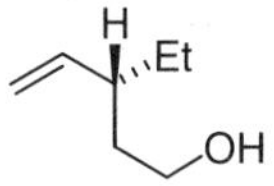	73	95
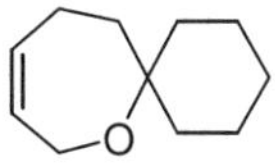	Et	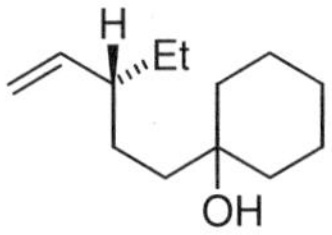	75	92
	n-Pr[b]	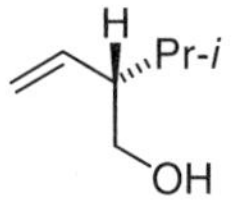	35–40	94
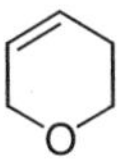	n-Pr	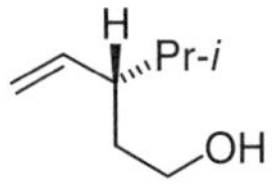	35–40	98
	n-Bu[b]	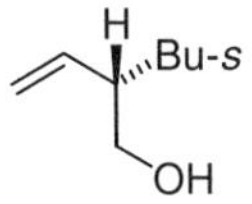	35–40	90
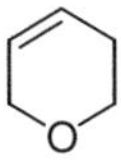	n-Bu	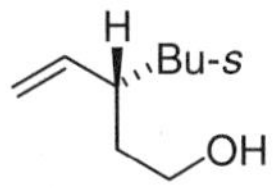	35–40	>95

[a]The reactions were run by using 5 equiv. of RMgCl and 10 mol % of (R)-**1** at 22°C for 16 h. [b]Run at 70°C.

10 can give (S)-2-vinyl-1,4-butanediol. The two catalysts, that is, **8** and **9**, led to the formation of the diol in 45 and 68% yields, respectively. However, the enantioselectivities in these cases were 99 and 90% ee, respectively. So, the overall merits and demerits of the two classes of chiral ligands used are unclear.

As in the case of ethylmagnesation shown in Scheme 4.17, kinetic resolution of 2-aryl-2,5-dihydrofurans as indicated in Scheme 4.19 can be achieved by using Et$_3$Al and **8** or **9**. Whereas the enantioselectivities observed with **9** were 64–84% ee, those observed with **8** were 96% ee for either product, indicating that **8** is superior to **9** [29] (Scheme 4.19).

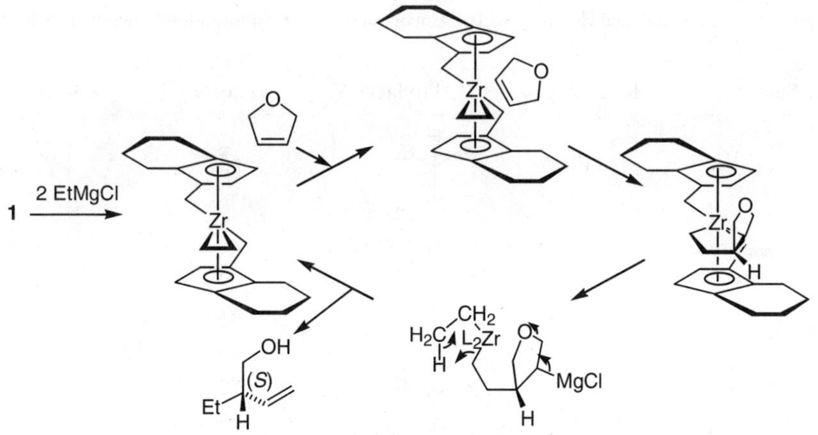

Scheme 4.15.

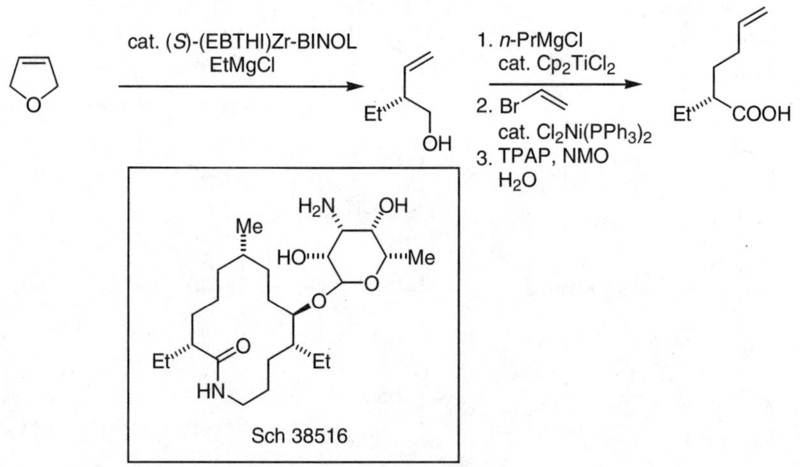

Scheme 4.16.

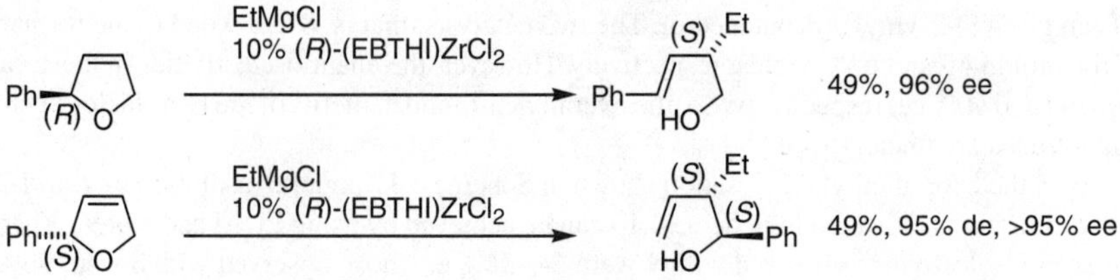

Scheme 4.17.

TABLE 4.4. Zr-Catalyzed Kinetic Resolution of Allylic Ethers[a]

Substrate	Conversion, %	% ee	Configuration
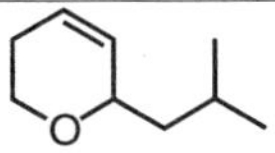	60	96	R
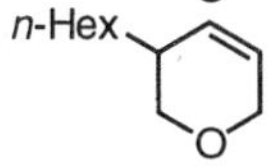	58	99	R
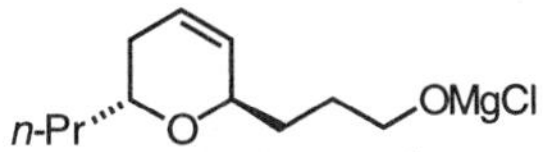 (racemic *trans*)	56	>99	R
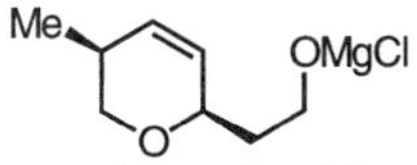 (racemic *cis*)	63	>99	R
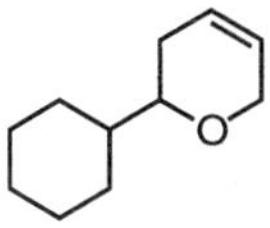	60	41	S
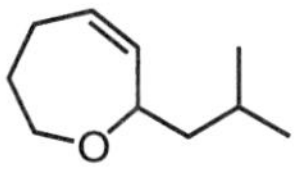	58	>99	R
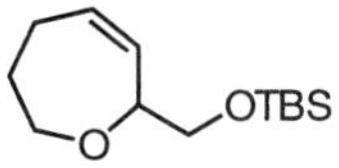	63	96	R
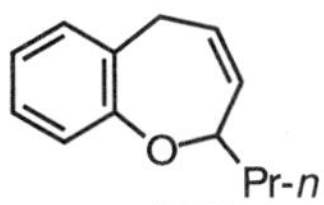	60	60	R
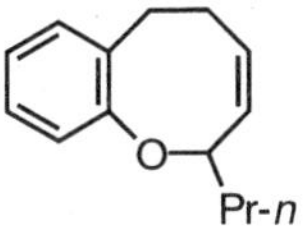	63	79	R
 (racemic *cis*)	55	98	1S,4R
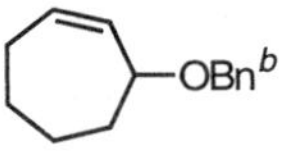	60	>99	S
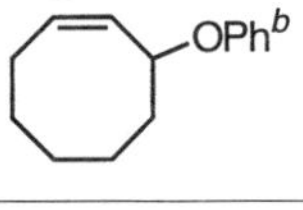	59	81	S

[a]The reaction was carried out with 5 equiv. of EtMgCl in the presence of 10 mol % of (R)-(EBTHI)ZrCl$_2$ at 70°C in THF. [b](R)-(EBTHI)Zr-Binol was used as a catalyst.

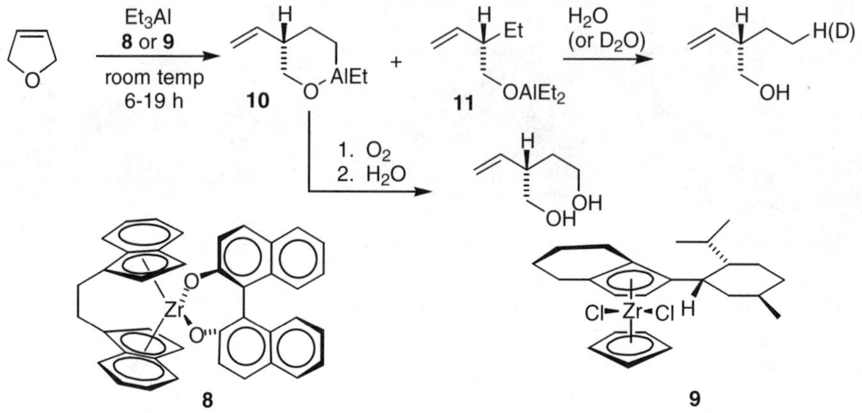

Scheme 4.18.

4.4.4. Zirconium-Catalyzed Enantioselective Ethylmagnesation and Ethylalumination of Allyl Derivatives

Recent results indicate that Zr-catalyzed enantioselective ethylmetallation of allyl derivatives not accompanied by β-elimination is feasible with various allylamines, allyl sulfides, and even with allyl alcohol [29,36] (Table 4.5). Although both chemical yields and enantioselectivity are still generally modest, the following favorable examples appear to indicate that further improvements may be expected. In some cases (NMTHI)ZrCpCl$_2$ (9) of C$_1$ symmetry and smaller overall steric requirements provide distinct advantages over sterically more demanding C$_2$-symmetric complexes, such as 1, 3, and 8.

4.4.5. Zirconium-Catalyzed Enantioselective Cyclic Carbomagnesation of Diallylamines

Whereas diallyl ethers do not readily undergo Zr-promoted cyclization due to competitive oxidative addition [23a], diallylamines have been shown to readily undergo Zr-promoted cyclization, even in cases where the corresponding all-C dienes fail to do so [23]. Since the development of the Zr-catalyzed diene cyclization with n-butylmagnesiums, such as n-BuMgBr

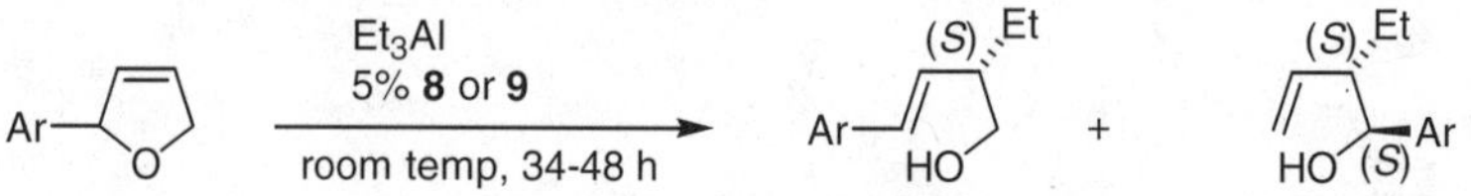

Ar	8 or 9	yield, % (% ee)	yield, % (% ee)
Ph	8	39 (>99)	32 (97)
Ph	9	50 (65)	45 (84)
p-FC$_6$H$_4$	8	34 (99)	31 (96)
p-FC$_6$H$_4$	9	40 (65)	32 (81)

Scheme 4.19.

TABLE 4.5. Zirconium-Catalyzed Enantioselective Ethylmagnesation and Ethylalumination of Allyl Derivatives

$$Z \diagdown \diagup \quad \xrightarrow{\text{8 or 9}} \quad Z \diagdown \overset{\overset{\text{Et}}{|}}{\diagup} \diagdown M \quad \xrightarrow{E^+} \quad Z \diagdown \overset{\overset{\text{Et}}{|}}{\diagup} \diagdown E$$

Z of $Z \diagdown \diagup$	EtM	Catalyst	Electrophile	Solvent	Yield, %	% ee
NHPh	EtMgCl	9	MeSSMe	Et$_2$O	90	81
NHPh	Et$_2$Mg	9	MeSSMe	THF	95	75
NHPh	Et$_2$Mg	8	MeSSMe	THF	39	26
NHCy	EtMgCl	9	MeSSMe	Et$_2$O	87	83
SPh	Et$_3$Al^a	9	H$_2$O	hexane	76	64
OH	Et$_2$Mg	9	H$_2$O	THF	75	56
OH	Et$_2$Mg	8	H$_2$O	THF	35	27

^{a}The product is 1-ethyl-3-phenylthiomethylaluminacyclopentane.

and n-Bu$_2$Mg [27,28], Zr-catalyzed enantioselective diene cyclization has become an attractive synthetic target. To date, only diallylamines have been successfully used for this purpose. Little or no favorable information appears to be available on the corresponding reaction of all-C dienes at this time. Judging from the results discussed in the preceding sections, it appears likely that allylic N atoms exert favorable influences, rendering highly enantioselective cyclization feasible.

The reaction of N-benzyldiallylamine with Bu$_2$Mg (8 equiv.) in the presence of 10 mol % of **8** in refluxing THF was sluggish. After 64 h, protonolysis with 10% HCl afforded *cis*- and *trans*-3,4-dimethyl-N-benzylpiperidines in 13 and 6% yields, respectively. The enantiopurity of the chiral *trans* isomer was only 13% ee [37]. Another discouraging case was the stoichiometric reaction of a diallylamine **12** with the dibutyl derivative of **8** (1.15 equiv.), which, after protonolysis with 10% HCl, gave **13** in 59% yield but with only 9% ee [37]. In sharp contrast, however, the reaction of **12** with n-Bu$_2$Mg (8 equiv.) in the presence of 10 mol % of **8** in refluxing THF was complete in several hours, and the protonolysis product **13** was obtained in 84% yield and 61% ee [37]. NMR analysis of the oxidation products indicated that **13** was derived from a mixture of mono- and dimagnesated derivatives and that the zirconabicyclic intermediate **14** must be *cis*-fused (Scheme 4.20) [37].

The results presented above suggest that, under the catalytic conditions, **14** and *ent*-**14** interconvert and that kinetic resolution must occur in their subsequent reaction with n-butyl-magnesium reagents. Comparison of n-BuMgCl with n-Bu$_2$Mg has indicated that n-BuMgCl is a more satisfactory reagent than n-Bu$_2$Mg. Using 4 equiv. of n-BuMgCl and 10 mol % of **8**, several dienes containing one or two allylamine moieties have been enantioselectively cyclized to give products with 85–95% ee, as shown in Table 4.6 [37].

4.4.6. Zirconium-Catalyzed Enantioselective Cocyclization of Alkenes with 2-Picoline

Chiral Zr(η^2-6-Me-pyrid-2-yl) complexes **15** react readily with terminal alkenes to give the corresponding azazirconacyclopentenes **16** in high yields. Even internal alkenes react similarly with **15b** [38]. Propene and 1-hexene react with **15a** so as to place the alkyl substituent in the β position, whereas styrene and trimethylsilylethylene place the Ph and Me$_3$Si groups in the α

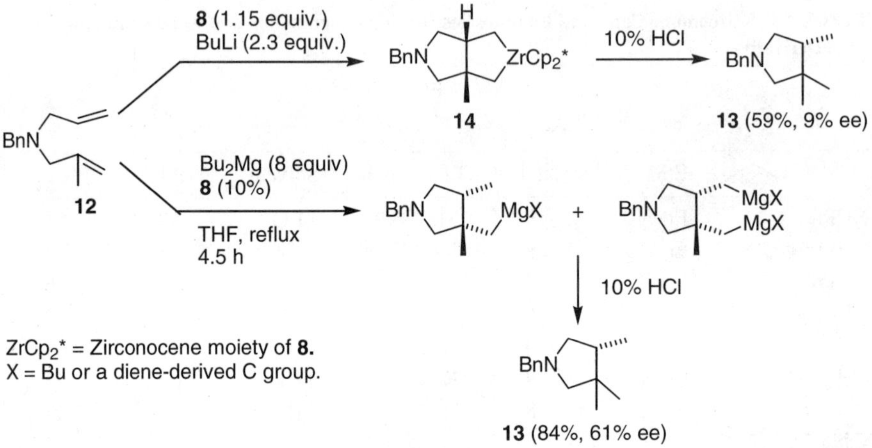

Scheme 4.20.

position. Stereochemically, Ph and Me₃Si point away from the benzene ring of indene, as might be expected, whereas the alkyl groups of propene and 1-hexene point toward the benzene ring, providing an interesting puzzle to be solved. The α-Me group derived from either E or Z 2-butene points away from the benzene ring. In cases where there is an α-substitutent, the diastereoselectivity of cyclization is >98% de whereas, the placement of a β-substituent in the absence of an α-substitutent is 83% de for either Me or Bu group (Scheme 4.21) [38].

The feasibility of devising a catalytic version of the reaction discussed above was demonstrated by the results shown in Scheme 4.22. Thus, the reaction of 1-hexene with an excess of

TABLE 4.6.   Zr-Catalyzed Enantioselective Cyclization of Dienes Containing Allylamine Moieties Induced by n-BuMgCl

Substrate	Time, (h)	Electrophile	Product	Yield, %	% ee
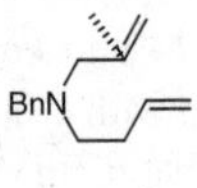	45	O_2	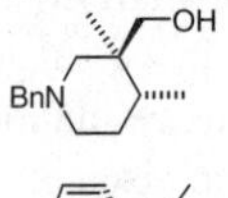	41	93
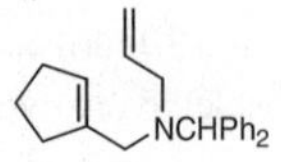	6	H_2O	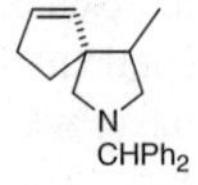	70	87 and 84 (*cis/trans* = 2)
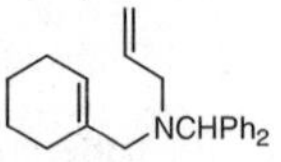	1.5	H_2O	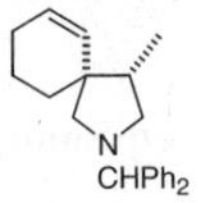	47	94
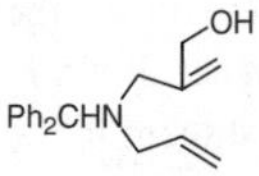	9.5	H_2O	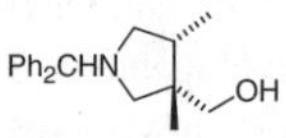 + isomer	40	95

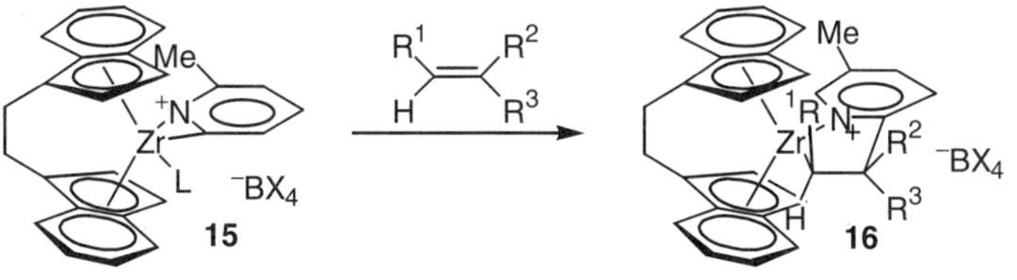

a: L = picoline. $^-BX_4 = {}^-BPh_4$.

b: L = none. $^-BX_4 = Me^-B(C_6F_5)_3$.

alkene	**15a** or **15b**	R^1	R^2	R^3	de (%)
propene	**15a**	H	H	Me	83
1-hexene	**15a**	H	H	n-Bu	83
styrene	**15a**	Ph	H	H	>98
Me₃SiCH=CH₂	**15a**	Me₃Si	H	H	>98
cis-2-butene	**15b**	Me	Me	H	>98
trans-2-butene	**15b**	Me	H	Me	>98

Scheme 4.21.

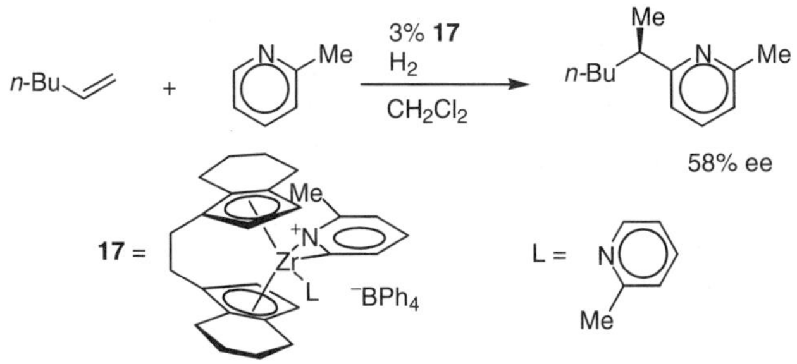

Scheme 4.22.

2-picoline catalyzed by 3 mol % of **17** under H_2 provided (R)-2-methyl-6-(2-hexyl)pyridine with 58% ee. The TON was 6.2 [38].

4.5. TITANIUM-CATALYZED ENANTIOSELECTIVE ENYNE BICYCLIZATION-CARBONYLATION

Since the advent of the stoichiometric Zr-promoted enyne bicyclization-carbonylation [19,20], development of its catalytic and enantioselective version has been an attractive but elusive target. Although its catalytic but racemic version has recently been developed in the form of Zr-catalyzed cyclic carboalumination [39], there does not appear to be any successful report on Zr-catalyzed enantioselective enyne bicyclization-carbonylation, the closest process being that discussed in

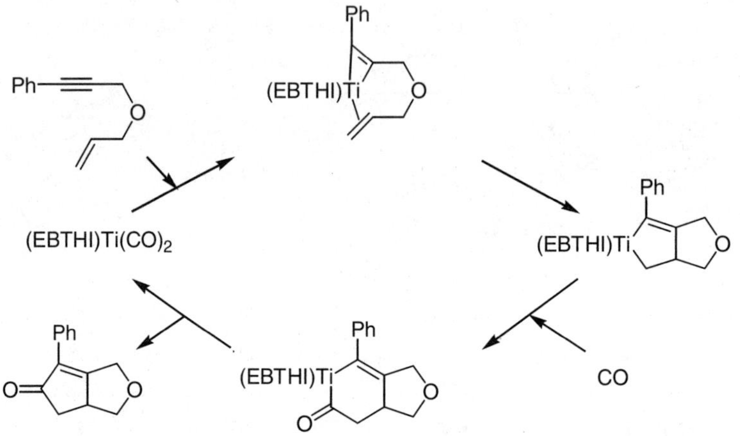

Scheme 4.23.

TABLE 4.7. Ti-Catalyzed Enantioselective Enyne Bicyclization-Carbonylation

Enyne	Amount of **17**	Product	Yield, %	% ee
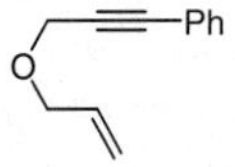	20	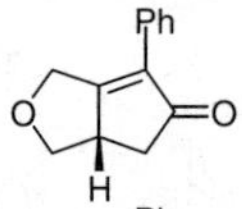	85	96
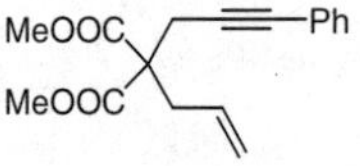	7.5	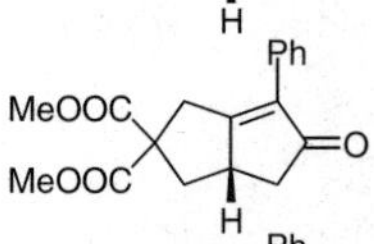	92	94
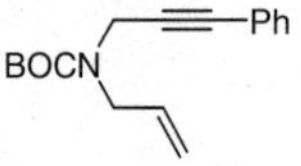	10	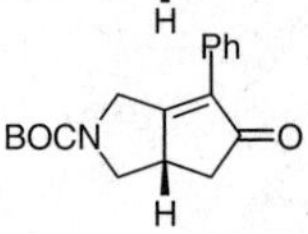	84	74
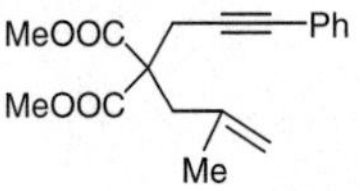	20	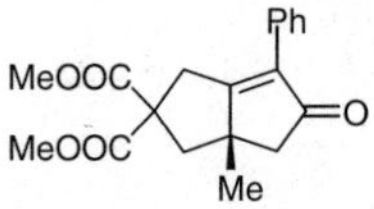	90	72

Sect. 4.4.5. However, a major breakthrough was made recently with chiral Ti complexes, as shown in Scheme 4.23 [40]. Both chemical yield and enantioselectivity are generally satisfactory, as indicated by the results summarized in Table 4.7 [40].

4.6. CONCLUSION

Catalytic enantioselective single-stage carbometallation of alkenes had until recently been an elusive synthetic goal, even though its principle and potential feasibility were recognized. It is a highly desirable chemical transformation, as it would not only provide an attractive, selective, and efficient tool for the synthesis of natural products and related complex organic compounds but also serve as single-stage models of stereoselective alkene polymerization reactions.

Over the past several years, several Zr- and Ti-catalyzed reactions that fall into the above-defined category have been discovered and developed. These reactions can be divided into two general classes, that is, those involving acyclic carbometallation and others involving cyclic carbometallation, regardless of the overall outcome.

In general, Zr- or Ti-catalyzed processes proceeding via acyclic carbometallation appear to require bimetallic activation [41] involving Al- or B-containing Lewis acids, as in the Ziegler-Natta polymerization [6]. At present, the Zr-catalyzed enantioselective alkene carboalumination discussed in Sect. 4.3. appears to represent the only example of this class of reaction of reasonable scope and synthetic potential. Although its applications to the syntheses of natural products have just been initiated, its potential appears to be considerable. It also appears that this reaction has provided a clear and positive answer to the controversial question of whether the initial stage of the stereoregular polymerization of propene and other unactivated alkenes catalyzed by Zr or Ti can be highly enantioselective.

To the best of this author's knowledge, all of the other known Zr- or Ti-catalyzed enantiose-lective single-stage carbometallation reactions involve cyclic mechanisms clarified, for the first time, in 1991. Moreover, all of the currently known Zr-catalyzed cyclic carbometallation reactions exhibiting high enantioselectivity (>60–70% ee) are strictly limited to the cases of allylically hetero-substituted alkenes. Although some of these reactions appear to be also bimetallic, metallacyclopropanes and metallacyclopropenes containing Zr and Ti are known to be capable of undergoing cyclic carbometallation with alkenes without the assistance of other Lewis acids. So, even their catalytic version may be monometallic.

Regardless of their mechanistic details, however, (i) Zr-catalyzed enantioselective carbomagnesation-elimination (Sect. 4.4.2.) and its Al analogue (Sect. 4.4.3.), (ii) Zr-catalyzed enantioselctive carbomagnesation and carboalumination followed by trapping with electrophiles (Sect. 4.4.4), and (iii) Zr-catalyzed enantioselective cyclic carbometallation of diallylamines (Sect. 4.4.5.) promise to provide collectively many attractive asymmetric transformations. An important pending question is whether other classes of alkenes can be satisfactorily carbometallated by these methods through their appropriate modifications. In this connection, the Ti-catalyzed enantioselective enyne bicyclization-carbonylation reaction (Sect. 4.5) is noteworthy, as it represents the only known highly enantioselective cyclic carbometallation reaction catalyzed by early transition metals.

REFERENCES

1. Van Horn, D. E.; Negishi, E. *J. Am. Chem. Soc.* **1978**, *100*, 2252.

2. Dzhemilev, U. M.; Vostrikova, O. S.; Sultanov, R. M. *Izv. Akad. Nauk SSSR, Ser. Khim.* **1983**, 218.

3. Takahashi, T.; Seki, T.; Nitto, Y.; Saburi, M.; Rousset, C. J.; Negishi, E. *J. Am. Chem. Soc.* **1991**, *113*, 6266.

4. Negishi, E.; Kondakov, D. Y.; Choueiry, D.; Kasai, K.; Takahashi, T. *J. Am. Chem. Soc.* **1996**, *118*, 9577.

5. (a) Ivin, K. J.; Mol, J. C. *Olefin Metathesis and Metathesis Polymerization*; Academic Press: New York, 1997. (b) Grubbs, R. H.; Miller, S. J.; Fu, G. C. *Acc. Chem. Res.* **1995**, *28*, 446.

6. (a) Boor, J. *Ziegler-Natta Catalysts and Polymerizations*; Academic Press: New York, 1979. (b) Janiak, C. In *Metallocenes*; Togni, A..; Halterman, R. L. (Ed.); 2 Vols., Wiley-VCH: Weinheim, 1998, Chap. 9, pp. 547–623.

7. Pelter, A.; Smith, K.; Brown, H. C., *Borane Reagents*; Academic Press: London; 1985, pp. 503.

8. Kondakov, D. Y.; Negishi, E. *J. Am. Chem. Soc.* **1995**, *117*, 10771.

9. Kondakov, D. Y.; Negishi, E. *J. Am. Chem. Soc.* **1996**, *118*, 1577.

10. (a) Wild, F.R.W.P.; Zsolnai, L.; Huttner, G.; Brintzinger, H. H. *J. Organomet. Chem.* **1982**, *232*, 233. (b) Wild, F.R.W.P.; Wasiucionek, M.; Huttner, G.; Brintzinger, H. H. *J. Organomet. Chem.* **1985**, *288*, 63. (c) Grossman, R. B.; Davis, W. M.; Buchwald, S. L. *J. Am. Chem. Soc.* **1991**, *113*, 2321. (d) Chin, B.; Buchwald, S. L. *J. Org. Chem.* **1997**, *62*, 2267.

11. Pino, P.; Cioni, P.; Wei, J. *J. Am. Chem. Soc.* **1987**, *109*, 6189.

12. Kaminsky, W.; Ahlers, A.; Moller-Lindenhof, N. *Angew. Chem., Int. Ed.* **1989**, *28*, 1216.

13. (a) Negishi, E.; Yoshida, T. *J. Am. Chem. Soc.* **1981**, *103*, 4985. (b) Negishi, E.; Van Horn, D. E.; Yoshida, T. *J. Am. Chem. Soc.* **1985**, *107*, 6639.

14. Erker, G.; Aulbach, M.; Knickmeier, M.; Wingbermuhle, D.; Kruger, C.; Nolte, M.; Werner, S. *J. Am. Chem. Soc.* **1993**, *115*, 4590.

15. Negishi, E.; Huo, S.; Alimardanov, A. Unpublished results.

16. (a) Cohen, N.; Eichel, W. F.; Lopresti, R. J.; Neukom, C.; Saucy, G. *J. Org. Chem.* **1976**, *41*, 3505. (b) Trost, B. M.; Klun, T. P. *J. Am. Chem. Soc.* **1981**, *103*, 1864. (c) Fujisawa, J.; Fukutani, Y.; Hasegawaa, M.; Maruoka, K.; Yamamoto, H. *J. Am. Chem. Soc.* **1984**, *106*, 5004.

17. (a) Sinn, H.; Kaminsky, W. *Adv. Organomet. Chem.* **1980**, *18*, 99. (b) Brintzinger, H. H.; Fisher, D.; Mulhaupt, R.; Rieger, B.; Waymouth, R. M. *Angew. Chem., Int. Ed.* **1995**, *34*, 1143.

18. Dzhemilev, U. M.; Ibragimov, A. G.; Zoltarev, A. P.; Muslukhov, R. R.; Tolstikov, G. A. *Izv. Akad. Nauk SSSR, Ser. Khim.* **1989**, 115; **1991**, 2570.

19. (a) Negishi, E.; Holmes, S. J.; Tour, J. M.; Miller, J. A. *J. Am. Chem. Soc.* **1985**, *107*, 2568. (b) Negishi, E.; Holmes, S. J.; Tour, J. M.; Miller, J. A.; Cederbaum, F. E.; Swanson, D. R.; Takahashi, T. *J. Am. Chem. Soc.* **1989**, *111*, 3336. (c) Negishi, E. In *Comprehensive Organic Synthesis; Vol. 5*, Paquette, L. A. (Ed.); Pergamon Press: 1991, 1163. (d) Takahashi, T.; Xi, Z.; Nishihara, Y.; Huo, S.; Kasai, K.; Aoyagi, K.; Denisov, V.; Negishi, E. *Tetrahedron* **1997**, *53*, 9123.

20. (a) RajanBabu, T. V.; Nugent, W. A.; Taber, D. F.; Fagan, P. J. *J. Am. Chem. Soc.* **1988**, *110*, 7128. (b) Jensen, M.; Livinghouse, T. *J. Am. Chem. Soc.* **1989**, *111*, 4495. (c) Lund, E.; Livinghouse, T. *J. Org. Chem.* **1989**, *54*, 4487.

21. (a) Negishi, E.; Cederbaum, F. E.; Takahashi, T. *Tetrahedron Lett.* **1986**, *27*, 2829. (b) Negishi, E.; Nguyen, T.; Maye, J. P.; Choueiri, D.; Suzuki, N.; Takahashi, T. *Chem. Lett.* **1992**, 2367. (c) Negishi, E.; Takahashi, T. *Bull. Chem. Soc. Jpn.* **1998**, *71*, 755.

22. Takahashi, T.; Murakami, M.; Kaunishage, M.; Saburi, M.; Uchida, Y.; Kozawa, K.; Uchida, T.; Swanson, D. R.; Negishi, E. *Chem. Lett.* **1989**, 761.

23. (a) Rousset, C. J.; Swanson, D. R.; Lamaty, F.; Negishi, E. *Tetrahedron Lett.* **1989**, *30*, 5105. (b) Negishi, E.; Maye, J. P.; Choueiry, D. *Tetrahedron* **1995**, *51*, 4447.

24. Nugent, W. A.; Taber, D. F. *J. Am. Chem. Soc.* **1989**, *111*, 6435.

25. Swanson, D. R.; Rousset, C. J.; Negishi, E.; Takahashi, T.; Seki, T.; Saburi, M.; Uchida, Y. *J. Org. Chem.* **1989**, *54*, 3521.

26. Negishi, E.; Takahashi, T. *Acc. Chem. Res.* **1994**, *27*, 124.

27. (a) Negishi, E.; Rousset, C. J.; Choueiry, D.; Maye, J. P.; Suzuki, N.; Takahashi, T. *Inorg. Chim. Acta* **1998**, *280/1–2*, 8.

28. (a) Knight, K. S.; Waymouth, R. M. *J. Am. Chem. Soc.* **1991**, *113*, 6268. (b) Knight, K. S.; Wang, D.; Waymouth, R. M.; Ziller, J. *J. Am. Chem. Soc.* **1994**, *116*, 1845.

29. Dawson, G.; Durrant, C. A.; Kirk, G. G.; Whitby, R. J.; Jones, R.V.H.; Standen, C. H. *Tetrahedron Lett.* **1997**, *38*, 2335.

30. Morken, J. P.; Didiuk, M. T.; Hoveyda, A. H. *J. Am. Chem. Soc.* **1993**, *115*, 6997.

31. Didiuk, M. T.; Johannes, C. W.; Morken, J. P.; Hoveyda, A. H. *J. Am. Chem. Soc.* **1995**, *117*, 7097.

32. Houri, A. F.; Xu, Z. M.; Cogan, D. A.; Hoveyda, A. H. *J. Am. Chem. Soc.* **1995**, *117*, 2943.

33. Visser, M. S.; Heron, N. M.; Didiuk, M. T.; Sagal, J. F.; Hoveyda, A. H. *J. Am. Chem. Soc.* **1996**, *118*, 4291.

34. Morken, J. P.; Didiuk, M. T.; Visser, M. S.; Hoveyda, A. H. *J. Am. Chem. Soc.* **1994**, *116*, 3123.

35. Visser, M. S.; Hoveyda, A. H. *Tetrahedron* **1995**, *51*, 4383.

36. (a) Bell, L.; Whitby, R. J.; Jones, R.V.H.; Standen, M.C.H. *Tetrahedron Lett.* **1996**, *37*, 7139. (b) Bell, L.; Brookings, D. C.; Dawson, G. J.; Whitby, R. J.; Jones, R.V.H.; Standen, M.C.H. *Tetrahedron* **1998**, *54*, 14617.

37. Yamaura, Y.; Hyakutake, M.; Mori, M. *J. Am. Chem. Soc.* **1997**, *119*, 7615.

38. Rodewald, S.; Jordan, R. F. *J. Am. Chem. Soc.* **1994**, *116*, 4491.

39. (a) Negishi, E.; Montchamp, J.-L.; Anastasia, L.; Elizarov, A.; Choueiry, D. *Tetrahedron Lett.* **1998**, *39*, 2503. (b) Montchamp, J.-L.; Negishi, E. *J. Am. Chem. Soc.* **1998**, *120*, 5345.

40. Hicks, F. A.; Buchwald, S. L. *J. Am. Chem. Soc.* **1996**, *118*, 11688.

41. Negishi, E. *Chem. Eur. J.* **1999**, *5*, 411.

5

ASYMMETRIC ADDITION AND INSERTION REACTIONS OF CATALYTICALLY-GENERATED METAL CARBENES

MICHAEL P. DOYLE
Department of Chemistry, University of Arizona, Tucson, Arizona 85721, U.S.A.

5.1. INTRODUCTION

The first chiral transition-metal catalyst designed for an enantioselective transformation was applied to the reaction between a diazo ester and an alkene to form cyclopropanes [1]. In that application Nozaki and coworkers used a Schiff base-Cu(II) complex (**1**), whose chiral ligand was derived from α-phenethylamine, to catalyze the cyclopropanation of styrene with ethyl diazoacetate (Eq. 5.1) [2].

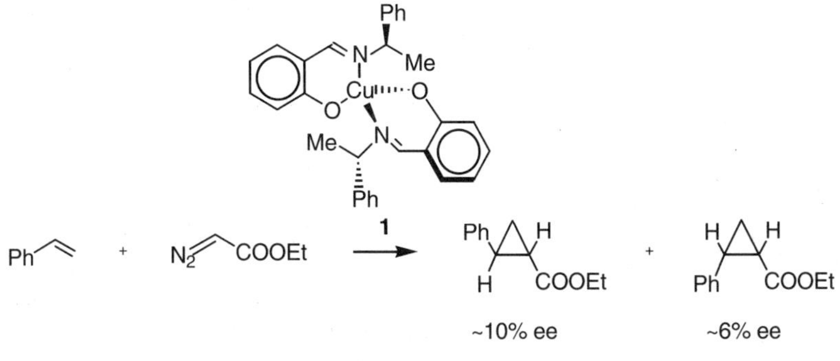

$$Ph\diagup\!\!\!\diagdown \quad + \quad N_2\!\!\diagdown\text{COOEt} \quad \longrightarrow \quad \text{(Ph cyclopropane COOEt)} \quad + \quad \text{(Ph cyclopropane COOEt)} \tag{5.1}$$

~10% ee   ~6% ee

Although enantioselectivity was low, the principle of chiral catalyst development for generating and distinguishing between diastereomeric transition states involving transition metals was established. From that first report in 1966, elaboration on ligand designs for chiral

Catalytic Asymmetric Synthesis, Second Edition, Edited by Iwao Ojima
ISBN 0-471-29805-0   Copyright © 2000 Wiley-VCH, Inc.

Schiff base-Cu(II) complexes, **2**, by Aratani and co-workers [3–6] led to high enantioselectivities in selected intermolecular cyclopropanation reactions, one of which was the critical asymmetrization step in the Sumotomo–Merck synthesis of cilastatin (Eq. 5.2).

A = CH$_3$, CH$_2$Ph   B = C$_4$H$_9$, C$_8$H$_{17}$
(R)-**2**

$$(5.2)$$

92% ee

cilastatin

Cyclopropanation reactions are one set in an array of C–C bond-forming transformations attributable to metal carbenes (Scheme 5.1) and are often mistakenly referred to by the nonspecific term "carbenoid." Both cyclopropanation and cyclopropenation reactions, as well as the related aromatic cycloaddition process, occur by addition. Ylide formation is an association transformation, and "insertion" requires no further definition. All of these reactions occur with diazo compounds, preferably those with at least one attached carbonyl group. Several general reviews of diazo compounds and their reactions have been published recently and serve as valuable references to this rapidly expanding field [7–10]. The book by Doyle, McKervey, and Ye [7] provides an intensive and thorough overview of the field through 1996 and part of 1997.

Like electrophilic addition to diazo compounds [7] from which diazonium ions and, subsequently, carbocations are generated, transition-metal compounds that can act as Lewis acids are potentially effective catalysts for metal carbene transformations. These compounds possess an open coordination site that allows the formation of a diazo carbon–metal bond with a diazo compound and, after loss of dinitrogen, affords a metal carbene (Scheme 5.2).

Lewis bases (B:) that can occupy the open coordination site inhibit catalytic activity. The electrophilic nature of the metal carbene is seen in its subsequent reactions with nucleophiles (S:), which occur with the transfer of the carbene entity from the metal to a nucleophile without ever having generated or transformed an actual "free" carbene.

Among transition-metal compounds that are effective for metal carbene transformations, those of Cu and Rh have received the most attention [7–10]. Cu catalysis for reactions of diazo compounds with olefins has been known for more than 90 years [11], but the first report of Rh catalysis, in the form of dirhodium(II) tetraacetate, has been recent [12]. Although metal carbene intermediates with catalytically active Cu or Rh compounds have not yet been observed, those

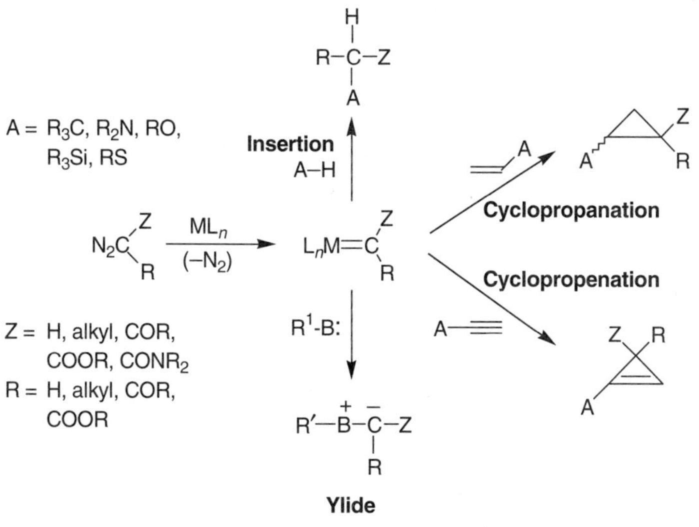

$$
\begin{array}{c}
\text{H} \\
| \\
\text{R}-\text{C}-\text{Z} \\
| \\
\text{A}
\end{array}
$$

A = R$_3$C, R$_2$N, RO,
R$_3$Si, RS

Insertion
A–H ↑

$$
N_2C\overset{Z}{\underset{R}{\diagdown}} \xrightarrow[(-N_2)]{ML_n} L_nM=C\overset{Z}{\underset{R}{\diagdown}}
$$

Cyclopropanation

Cyclopropenation

Z = H, alkyl, COR,
COOR, CONR$_2$
R = H, alkyl, COR,
COOR

R^1-B: |
↓

$$
\overset{+}{R'}-\overset{}{B}-\overset{-}{C}-Z \\
| \\
R
$$

Ylide

Scheme 5.1.

derived from a Ru-pybox catalyst have been characterized [13,14]. In addition, correlations in relative reactivities and diastereomeric selectivities for cyclopropanation between pentacarbonyltungsten carbenes [(CO)$_5$W=CHR, R = Ph, COOEt] and those issued from dirhodium(II) tetraacetate, Rh$_2$(OAc)$_4$, [15,16] strongly suggest their involvement in catalytic reactions. Apparent trapping of a Rh carbene or its diazonium ion precursor by iodide [17] from the reaction of Rh(TTP)I with ethyl diazoacetate (Eq. 5.3, TTP = tetra-p-tolylporphyrin) also supports the catalytic cycle that is outlined in Scheme 5.2.

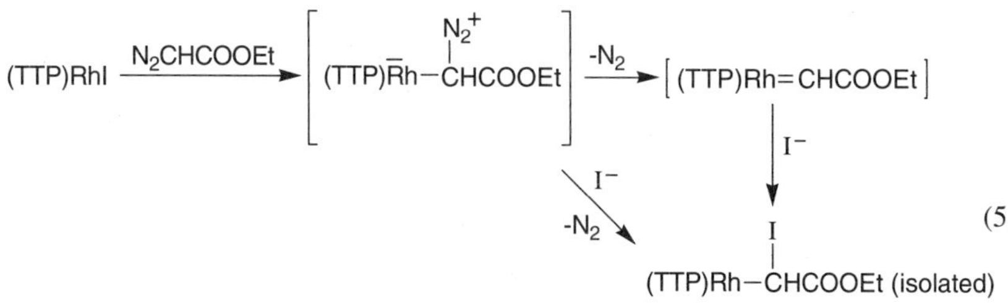

$$
(TTP)RhI \xrightarrow{N_2CHCOOEt} \left[(TTP)\overset{N_2^+}{\overset{|}{\bar{R}h-CHCOOEt}} \right] \xrightarrow{-N_2} \left[(TTP)Rh=CHCOOEt \right]
$$

$$
\searrow \; \overset{I^-}{-N_2} \qquad\qquad \downarrow I^-
$$

$$
\overset{I}{\underset{}{|}} \\
(TTP)Rh-CHCOOEt \text{ (isolated)}
$$

(5.3)

Two resonance-contributing structures (**3a** and **3b**), in the formalism of ylide structures, can be used to describe metal carbene intermediates. The highly electrophilic character of those derived from Cu and Rh catalysts suggests that the contribution from the metal-stabilized carbocation **3b** is important in the overall evaluation of the reactivities and selectivities of these metal carbene intermediates. Emphasis on the metal carbene structure **3a** has led to the subsequently discounted proposal that cyclopropane formation from reactions with alkenes occurs through the intervention of a metallocyclobutane intermediate [18]. The metal-stabilized carbocation structure **3b** is consistent with the cyclopropanation mechanism in which L$_n$M dissociates from the carbene as bond-formation occurs between the carbene and the reacting alkene (Eq. 5.4) [7,15].

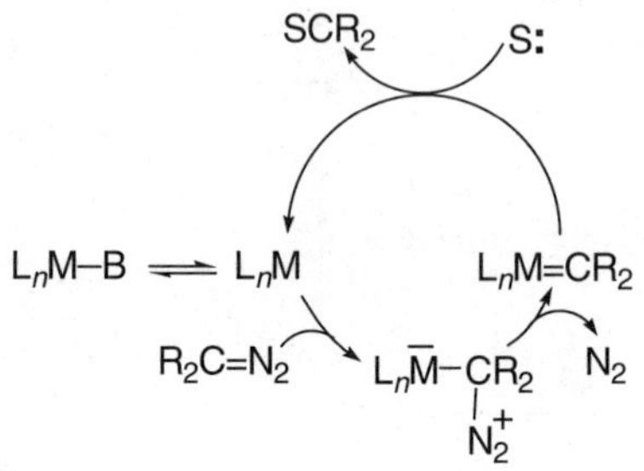

$$L_nM-B \;\rightleftharpoons\; L_nM \qquad L_nM=CR_2$$

Scheme 5.2.

$$L_nM=CR_2 \;\longleftrightarrow\; L_n\overline{M}-\overset{+}{C}R_2$$

3a **3b**

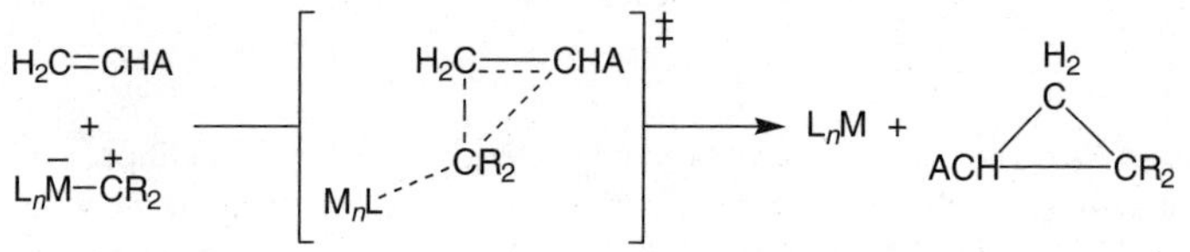

$$H_2C=CHA \;+\; L_n\overset{-}{M}-\overset{+}{C}R_2 \;\longrightarrow\; \left[\begin{array}{c} H_2C\cdots CHA \\ | \quad \\ \cdot CR_2 \\ M_nL \end{array} \right]^{\ddagger} \;\longrightarrow\; L_nM \;+\; \underset{ACH\quad CR_2}{\overset{H_2}{\triangle}} \qquad (5.4)$$

Diazocarbonyl compounds, especially diazo ketones and diazo esters [19], are the most suitable substrates for metal carbene transformations catalyzed by Cu or Rh compounds. Diazoalkanes are less useful owing to more pronounced "carbene dimer" formation that competes with, for example, cyclopropanation [7]. This competing reaction occurs by electrophilic addition of the metal-stabilized carbocation to the diazo compound followed by dinitrogen loss and formation of the alkene product that occurs with regeneration of the catalytically active metal complex (Eq. 5.5) [20].

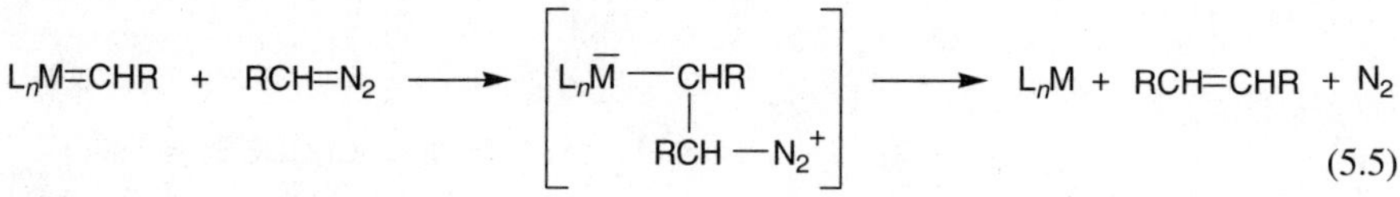

$$L_nM=CHR \;+\; RCH=N_2 \;\longrightarrow\; \left[\begin{array}{c} L_n\overline{M}-CHR \\ | \\ RCH-N_2^{+} \end{array} \right] \;\longrightarrow\; L_nM \;+\; RCH=CHR \;+\; N_2 \qquad (5.5)$$

Among the transition-metal catalysts that have been used, only those of Pd(II) are productive with diazomethane, which may be the result in cyclopropanation reactions [7,9,21] of a mechanism whereby the Pd-coordinated alkene undergoes electrophilic addition to diazomethane rather than by a metal carbene transformation; in any case, asymmetric induction does not occur by using Pd(II) complexes of chiral bis-oxazolines [22].

The development of chiral transition-metal catalysts for these transformations has taken place during the evolution of mechanistic understanding for metal carbene generation. Concepts applied to the design of chiral catalysts have been influenced by these mechanistic developments, and further refinements can be expected.

5.2. ASYMMETRIC INTERMOLECULAR CYCLOPROPANATION

Cyclopropane formation occurs from reactions between diazo compounds and alkenes, catalyzed by a wide variety of transition-metal compounds [7–9], that involve the addition of a carbene entity to a C–C double bond. This transformation is stereospecific and generally occurs with "electron-rich" alkenes, including substituted olefins, dienes, and vinyl ethers, but not α,β-unsaturated carbonyl compounds or nitriles [23,24]. Relative reactivities portray a highly electrophilic intermediate and an "early" transition state for cyclopropanation reactions [15,25], accounting in part for the relative difficulty in controlling selectivity. For intermolecular reactions, the formation of geometrical isomers, regioisomers from reactions with dienes, and enantiomers must all be taken into account.

5.2.1. Chiral Salicylaldimine Cu Catalysts

Although the degree of enantioselection for Nozaki's Cu(II)-Schiff base catalyst (**1**) was low [2], extensive systematic screening of salicylaldimine ligands by Aratani and co-workers nearly 10 years later brought about design changes in the chiral salicylaldimine Cu catalysts that gave dramatic improvements in optical yields for select intermolecular cyclopropanation reactions [3,4]. The ligands offering the greatest advantage were those prepared from salicylaldehyde and amino alcohols derived from alanine [3]. Those derived from phenylalanine, valine, and leucine were also evaluated [6], but lower enantioselectivities were obtained with them for cyclopropanation reactions of 2,5-dimethyl-2,4-hexadiene (Eq. 5.6).

$$N_2CHCOOR \quad + \quad \text{(Me, Me, Me diene)} \quad \xrightarrow{(R)\text{-}2} \quad \text{(cyclopropane product)} \quad (5.6)$$

When the carbinol substituents (R) were the bulky 5-*tert*-butyl-2-(n-octyloxy)phenyl group, optimum enantioselectivities were achieved with the catalytic use of the corresponding Cu(II) complex (**2**) in both enantiomeric forms. Specific applications of the Aratani catalysts have included the synthesis of chrysanthemic acid esters (Eq. 5.6) and a precursor to permethrinic acid, both potent units of pyrethroid insecticides, and for the commercial preparation of ethyl (*S*)-2,2-dimethylcyclopropanecarboxylate (Eq. 5.2), which is used for constructing cilastatin. Several other uses of these catalysts and their derivatives for cyclopropanation reactions have been reported albeit, in most cases, with only moderate enantioselectivities [26–29].

Enantioselection with the Aratani catalysts is influenced by the steric bulk of the ligand R group (see **2**) and by the size of the alkyl substituent derived from the amino alcohol (A = CH$_3$ > CH$_2$Ph > i-Pr > i-Bu). In addition, the diazo ester has a significant influence on both enantioselectivity and on the *trans:cis* ratio of the resulting cyclopropane product [4]. As reported in Table 5.1 for the production of chrysanthemic acid esters, increasing the size of the diazo ester substituent increases the percentage of cyclopropane product having the trans geometry and also increases the enantiomeric excess of this product. Enantioselectivity for the cis isomer decreases with increasing size of the diazoacetate, at least for the examples reported in Table 5.1. A higher diastereomeric ratio for the trans isomer and a lower diastereomeric ratio for the cis isomer are obtained with *l*-menthyl diazoacetate and (**R**)-2 (A = Me), which also produces a higher *trans:cis* product ratio.

TABLE 5.1. Cyclopropanation of 2,5-Dimethyl-2,4-hexadiene with Diazo Esters (Eq. 5.6) Catalyzed by the Aratani Catalysts [4][a]

		N$_2$CHCOOR			% ee[b]	
R', B =	A	R =	Yield (%)	trans:cis	trans	cis
n-Octyl	CH$_3$	CH$_2$CH$_3$	54	51:49	68	62
n-Octyl	CH$_3$	Cyclohexyl	71	58:42	70	58
n-Octyl	CH$_3$	C(CH$_3$)$_3$	74	75:25	75	46
n-Octyl	CH$_3$	1-adamantyl	82	84:16	85	46
n-Octyl	CH$_3$	C(i-Pr)$_2$CH$_3$	64	92:8	88	n.r.[c]
n-Octyl	CH$_3$	d-menthyl	64	72:28	90	59
n-Octyl	CH$_3$	l-menthyl[d]	c	93:7	94	46
n-Butyl	CH$_3$	l-menthyl[d]	67	89:11	87	25
n-Heptyl	PhCH$_2$	l-menthyl[d]	42	91:9	86	22

[a]Results for (R)-**2** are reported. [b]% de for menthyl esters. [c]Not reported. [d]l-Menthyl = (1R,2S,5R)-2-isopropyl-5-methylcyclohexyl.

With a series of substituted ethylenes, diastereoselectivities for cyclopropanation with chiral menthyl diazoacetates generally increase with increasing substitution about the C–C double bond (Table 5.2).

TABLE 5.2. Diastereoselective Cyclopropanation of Alkenes with l-menthyl Diazoacetate (l-MDA) Catalyzed by the Aratani Catalyst 2 (A = CH$_3$) [5][a]

				Diastereomer Ratio[b]	
Entry	Alkene	Catalyst Configuration	trans:cis	trans	cis
1	styrene	R	86:14	84:16(1S,2S)	77:33(1S,2R)
2		S	82:18	91:9(1R,2R)	89:11(1R,2S)
3	1-octene	R	83:17	88:12(1S,2S)	73:27(1S,2S)
4	trans-4-octene	R		91:9(2S,3S)	
5	α-methylstyrene	R	60:40	84:16	93:7
6	1,1-diphenylethylene	R		87:13(1S)	
7	Cl$_3$CCH$_2$CH=CMe$_2$ [c]	S	15:85	60:40(1R,2R)	96:4(1R,2S)
8	BrCH$_2$CH$_2$CH=CMe$_2$	S	17:83	62:38(1R,2R)	98:2(1R,2S)
9	Me$_2$C=CH–CH=CMe$_2$	R	93:7	97:3(1S,2S)	73:27(1S,2R)

[a]l-Menthyl = (1R,2S,5R)-2-isopropyl-5-methylcyclohexyl. [b]Configuration of carbon atoms 1 and 2 for major isomer in parentheses. [c]With ethyl diazoacetate the same trans:cis product ratio was obtained, and % ee's were 91(cis) and 11(trans).

However, structural effects from the olefin on stereoselection are relatively minor, and only with trisubstituted olefins do diastereoisomer ratios exceed 95:5. 5-Halo-2-pentenes (entries 7 and 8) are exceptional in this series with both an inverted preference for formation of the cis-cyclopropane derivative and the high selectivity for this stereoisomer. Only one other example has been reported to have this unexpectedly high stereochemical preference (Eq. 5.7) [27], which may be the result of a dipolar interaction between the polar substituent of the reacting alkene and the metal carbene, but none show the relative independence of enantio-selectivity on the diazoacetate as do the halogenated alkenes.

$$(5.7)$$

	trans:cis	% ee cis:	% ee trans:
R = l-menthyl	44:56	> 95%	> 95%
Et	37:63	80%	29%

With modified Aratani catalysts (**2**, R = Ph and A = CH$_2$Ph), Reissig observed moderate enantioselectivities (30–40% ee for the trans cyclopropane isomer) for reactions between trimethylsilyl vinyl ethers and methyl diazoacetate [26], but vinyl ethers are the most reactive olefins towards cyclopropanation and also the least selective [30,31]. Other chiral Schiff bases have been examined for enantio-selection by using the in situ method for catalyst preparation that was pioneered by Brünner, but enantioselectivities were generally low [32].

Methyl diazoacetates were used extensively in earlier work because menthyl esters were amenable to separation by chromatographic methods. However, advances in chiral separation technologies have significantly reduced the need for chiral auxiliaries today, and enantiomeric separations of even simple esters are now routine. Selectivities obtained with menthyl diazoacetates are now of mainly historical interest.

According to Aratani, the results obtained are consistent with a reduced mononuclear Cu(I) complex, existing in a tetrahedral configuration, that forms a metal carbene intermediate [6]. Alkene addition occurs by displacement of an oxygen ligand, which allows the production of a metallacyclobutane intermediate that collapses to cyclopropane product with regeneration of the catalytically active Cu compound. Approach of the alkene to the carbene occurs from the less hindered side. However, although this explanation predicts the observed predominant cyclopropane configuration, it does not adequately account for the preferential trans stereochemistry that is observed in most cases (Table 5.2). An alternative explanation, based on metal carbenes as metal-stabilized carbocations, has also been proposed to explain these results [33], and its predictions better represent observed selectivities.

5.2.2. C$_2$-Symmetric Ligands: Chiral Semicorrin/Bis-oxazoline Cu Catalysts

The most significant advances in the development of chiral Cu catalysts for enantioselective cyclopropanation reactions were initiated by Pfaltz with his report in 1986 [34] of semicorrin Cu(II) complexes (**4**) and extended to their aza analogues **5** [35]. Of the catalysts with chiral semicorrin ligands that were initially evaluated [36], the Cu complex **4a** afforded the highest selectivities [37] [L = a second semicorrin ligand for the Cu(II) complex or a carbene ligand in a Cu(I) complex]. For the cyclopropanation of two monosubstituted alkenes, styrene and 1-heptene, enantioselectivities were significantly higher than those obtained with the use of the Aratani catalysts (Table 5.3).

a R = CMe₂OH
b R = CH₂OSiMe₂ᵗBu
c R = COOMe

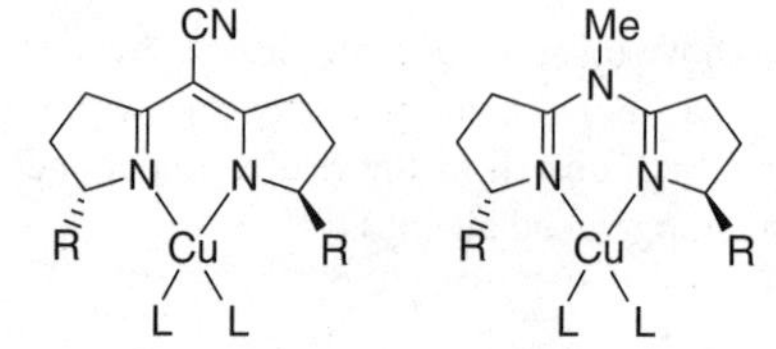

a R = CH₂OSiMe₂ᵗBu
b R = CMe₂OSiMe₃
c R = CMe₂OSiMe₂ᵗBu

4   **5**

As with the Aratani catalysts, enantioselectivities for cyclopropane formation with **4** and **5** are responsive to the steric bulk of the diazo ester, are higher for the trans isomer than for the cis form, and are influenced by the absolute configuration of a chiral diazo ester (*d*- and *l*-menthyl diazoacetate), although not to the same degree as reported for **2** in Tables 5.1 and 5.2. 1,3-Butadiene and 4-methyl-1,3-pentadiene, whose higher reactivities for metal carbene addition result in higher product yields than do terminal alkenes, form cyclopropane products with 97% ee in reactions with *d*-menthyl diazoacetate (Eq. 5.8). Regiocontrol is complete, but diastereocontrol (*trans:cis* selectivity) is only moderate.

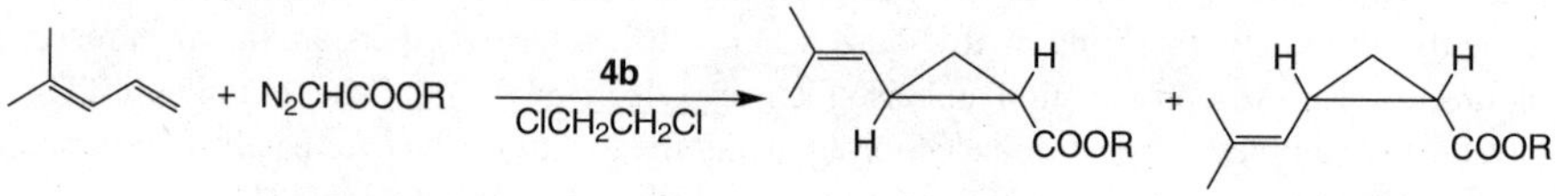

$$+ \ N_2CHCOOR \xrightarrow[\text{ClCH}_2\text{CH}_2\text{Cl}]{\textbf{4b}}$$

R = *d*-menthyl

67%
(97% ee)

33%
(97% ee)

(5.8)

Whereas the (semicorrinato)Cu catalysts are exceptionally effective with mono-substituted alkenes, their enantiocontrol with 1,2-di- and trisubstituted alkenes are less than those achieved

TABLE 5.3. Enantioselective Cyclopropanation of Alkenes with Diazo Compounds Catalyzed by the Pfaltz Semicorrin Catalysts 4 and 5

Catalyst	Alkene	N₂CHCOOR[a] R =	trans:cis	% ee/de trans[b]	% ee/de cis[c]
4a	styrene	*l*-menthyl	85:15	91	90
4a	styrene	*d*-menthyl	82:18	97	95
5b	styrene	*d*-menthyl	84:16	98	99
5c	styrene	*d*-menthyl	84:16	98	99
4a	styrene	*tert*-butyl	81:19	93	93
4a	styrene	ethyl	73:27	92	79
5b	styrene	ethyl	75:25	94	68
5c	styrene	ethyl	77:23	95	90
4a	1-heptene	*d*-menthyl	82:18	92	92
4a	1,3-butadiene	*d*-menthyl	63:37	97	97
4a	4-methyl-1,3-butadiene	*d*-menthyl	63:37	97	97

[a]*d*-Menthyl = (1*S*,2*R*,5*S*)-2-isopropyl-5-methylcyclohexyl. [b]Absolute configuration for cyclopropane product is (1*S*,2*S*). [c]Absolute configuration of cyclopropane product is (1*S*,2*R*).

with the use of the Aratani catalysts. Chrysanthemic acid esters, for example, are prepared in much lower yields and with significantly reduced enantiocontrol with the Pfaltz catalysts than with the Aratani catalysts [38]. In addition, alkenes less reactive than conjugated olefins like styrene or 1,3-butadiene are sluggish in their reactions with the intermediate metal carbene; for example, cyclopropanation of 1-heptene gave only a 30% product yield (92% ee) whereas with styrene or conjugated dienes, cyclopropanation yields were more than double this value. Pfaltz has described the active catalyst as a Cu(I) derivative with only one semicorrin ligand, and he has provided a mechanistic description that portrays rational transition-state geometries [37].

Cu complexes with bis-oxazoline ligands **6** that were first reported by Masamune and co-workers [39] have incited considerable interest because of the exceptional enantiocontrol that can be achieved with their use as catalysts for cyclopropanation reactions. Concurrent investigations by Evans [40], who added **7**; Masamune, who provided **8** and **9** [41]; and Pfaltz [42], who investigated a similar series, established that the C_2-symmetric bis-oxazoline ligands are suitable alternatives to semicorrin ligands for Cu in creating a highly enantioselective environment for intermolecular cyclopropanation. For the first time, diazoacetates with ester substituents as small as ethyl could be used to achieve enantioselectivity >90% ee in reactions with styrene (Table 5.4).

6		**7**		**8**		**9**	
a	R = Ph	**a**	R = ^{i}Pr	**a**	R = Me, R′ = Ph	**a**	R = Me, R′ = PH
b	R = Me$_2$CH	**b**	R = ^{t}Bu	**b**	R = CH$_2$OH, R′ = Ph	**b**	R = Et, R′ = ^{t}Bu
c	R = (S)-EtCH(Me)					**c**	R = R′ = Ph
d	R = Et						
e	R = PhCH$_2$						
f	R = ^{t}Bu						
g	R = CH$_2$OH						
h	R = Me$_2$C(OH)						
i	R = CH$_2$OSiMe$_2$^{t}Bu						

As expected, increasing the size of the R group increases enantioselection, and the buttressing effect on the bis-oxazoline ring caused by the geminal disubstitution in **10** provides further enhancement of enantiocontrol. From the results in Table 5.4, however, the ligand's R substituent has only a minor influence on the *trans:cis* of cyclopropane products. To increase product diastereoselectivity, Evans [40] increased the size of the ester substituent from ethyl to *tert*-butyl and then to the bulky BHT ester, previously reported by Doyle [43] to provide exceptional diastereocontrol in catalytic cyclopropanation reactions. Applications of these catalysts to alkenes other than styrene have demonstrated the potential generality of their uses for asymmetric intermolecular cyclopropanation (Table 5.5).

The *trans:cis* (or *E/Z*) ratios of cyclopropane products are higher than those obtained with styrene, and the diastereomer ratio or enantiomer ratio for the trans(*E*) isomer is generally >90:10. Although these reactions are usually performed with 1.0 mol % of catalyst, Evans has optimized the cyclopropanation of isobutylene to a 0.25 mol scale [40], by using only 0.1 mol % of catalyst, and obtained a 91% yield of the (*S*)-cyclopropane enantiomer whose enantiopurity

TABLE 5.4. Enantioselective Cyclopropanation of Styrene with Diazo Esters using Bis-oxazoline Copper Catalysts

Ligand	R=	$N_2CHCOOR'$ R'=	trans:cis	% ee/de trans[a]	cis[b]	Ref.
6a	Ph	Et	70:30	60	52	[39]
6e	PhCH$_2$	Et	71:29	36	15	[39]
6b	i-Pr	Et	71:29	46	31	[39]
6b	t-Pr	Et	64:36	64	48	[40]
6f	t-Bu	Et	75:25	90	77	[39]
6f	t-Bu	Et	77:23	98	93	[40]
6f	t-Bu	d-menthyl	84:16	98	80	[39]
6f	t-Bu	l-menthyl	84:16	98	96	[39]
6f	t-Bu	l-menthyl	87:13	96	97	[42]
6h	Me$_2$C(OH)[c]	d-menthyl	83:17	90[d]	90[e]	[42]
8a	Me	Et	71:29	28[d]	30[e]	[39]
7b	t-Bu	Et	73:27	99	97	[40]
7b	t-Bu	t-Bu	81:19	96	93	[40]
7b	t-Bu	BHT[f]	94:6	99		[40]
13d	TMS	t-Bu	86:14	92	98	[48]
14	—	Et	76:24	73	44	[49]
14	—	t-Bu	86:14	87	82	[49]
14	—	l-menthyl	85:15	89	84	[49]
14	—	BHT[f]	96:4	94		[49]
15	—	Et	74:26	86	58	[50]
15	—	l-menthyl	91:9	94		[50]

[a]Product from styrene has the $(1R,2R)$-configuration. [b]Product from styrene has the $(1R,2S)$-configuration. [c]Catalyst has opposite configuration to **6** (R = t-Bu). [d]$(1S,2S)$-configuration. [e]$(1S,2R)$-configuration. [f]2,6-Di-tert-butyl-4-methylphenyl.

was greater than 99% ee, which is significantly greater than that for the same reaction performed with the Aratani catalyst (Eq. 5.2) [6]. Results obtained by Masamune and Lowenthal [41] for the cyclopropanation of 2,5-dimethyl-2,4-hexadiene (Eq. 5.6) demonstrate that high enantio-control and diastereocontrol can be achieved with this system (Table 5.5), but that such results require the use of **9**; no single bis-oxazoline fits all applications.

Various chiral bis-oxazolines have been prepared in attempts to supercede results achieved with **6–9** and, especially, **7b**. These ligands have utilized linkers such as pyridine, as in pybox structures **10** [45], 1,2-ethanediol, as in **11** [46], and biphenyl, as in **12** [47]. However, from results obtained mainly with styrene, these less conveniently accessed ligands are inferior to **7b** in enantiocontrol and diastereocontrol.

Cu catalysts for metal carbene transformations are active as Cu(I) complexes and not as Cu(II). Although in the distant past there was some disagreement with this proposition, bis-oxazoline, semicorrin, and even the Aratani catalysts are active only when Cu is in its +1 oxidation state [6,34,39,40]. The chiral Cu(I) catalysts have been produced from the correspond-

TABLE 5.5. Enantioselective Cyclopropanation of Alkenes with Diazo Esters using Bis-oxazoline Copper Catalysts

Alkene	Ligand	$N_2CHCOOR'$ R' =	Yield (%)	trans:cis	% ee[a] trans	% ee[a] cis	Ref.
1-octene	**6f**	*l*-menthyl	76	94:6	99	30	[39]
	13d	*t*-Bu	65	85:15	91	n.d.	[48]
	6f	*d*-menthyl	72	90:10	75	45	[39]
	14	BHT	64	93:7	90	80	[49]
Ph(Me)C=CH$_2$	**6f**	*l*-menthyl	78	89:11	92	79	[39]
	6f	*d*-menthyl	72	85:15	83	77	[39]
trans-4-octene	**6f**	*l*-menthyl	52	—	88		[39]
TMSOCH=CH$_2$	**7b**	Me	55	66:34	73	76	[44]
	9c	Me	50	>97:3	11	n.d.[b]	[44]
Et$_3$SiCH=CH$_2$	**14**	BHT	64	99:1	95	64	[49]
PhCH$_2$CH=CH$_2$	**14**	BHT	78	94:6	91	78	[49]
p-MeOC$_6$H$_4$CH=CH$_2$	**13d**	*t*-Bu	73	90:10	83	>99	[48]
	14	BHT	90	94:6	87	90	[49]
Me$_3$CC(Me)=CH$_2$	**6f**	*l*-menthyl	60	95:5	80	91	[39]
	6f	*d*-menthyl	55	98:2	77	n.d.[b]	[39]
Me$_2$C=CH$_2$	**7b**	Et	91	—	>99		[40]
Ph$_2$C=CH$_2$	**7b**	Et	75	—	99		[40]

[a]% de values for reactions with menthyl diazoacetate. [b]Not determined.

ing chiral Cu(II) complex by reduction with phenylhydrazine [6,34,39] prior to use (Eq. 5.9) or generated in situ by ligand replacement from Cu(I) triflate [40] or, less commonly, Cu(I) *tert*-butoxide [34] (Eq. 5.10, X = OTf or O-*t*-Bu). The method of catalyst preparation does influence enantioselection in product formation. With **6f**, for example, the catalyst generated according to Eq. 5.9 [39] provided 90% ee and 77% ee for the *trans*- and *cis*-2-phenylcyclo-propanecarboxylate esters, respectively (Table 5.4), whereas the presumed same catalyst generated in situ (Eq. 5.10, X = OTf) gave 98% ee and 93% ee, respectively, for the same products [40]. Similar results were observed by Pfaltz [34] with semicorrin ligated copper, but the catalyst generated in situ gave lower enantioselectivity than that produced by phenylhy-

10

a R = Et
b R = Ph

11

a R = PhCH$_2$
b R = Me$_2$CH
c R = ^{t}Bu
d R = Ph

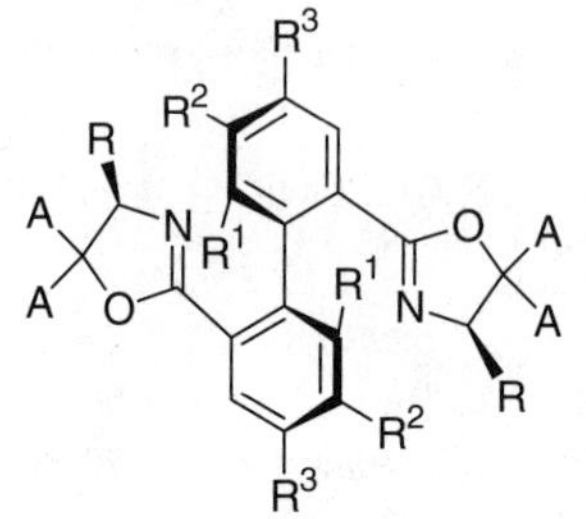

12

a $R^1 = R^2 = R^3 = OMe$, $A = H$, $R = {}^iPr$
b $R^1 = R^2 = R^3 = OMe$, $A = H$, $R = {}^tBu$

drazine reduction. This finding does seem to be related to the uncertainty in redox reactions, and today catalyst preparation via Eq. 5.10 is virtually exclusive.

$$CuL_2^* \xrightarrow{PhNHNH_2} CuL^* + L^*H \tag{5.9}$$

$$CuX + L^*H \longrightarrow CuL^* + HX \tag{5.10}$$

5.2.3. Other Chiral Ligated Cu Catalysts

The number of chiral ligands designed for bidentate attachment to copper has grown rapidly during the 1990s, but only three have shown substantial promise: Katsuki's chiral bipyridines **13** [48], Fu's C_2-symmetric bisazaferrocene **14** [49], and chiral 1,2-diamine **15** [50].

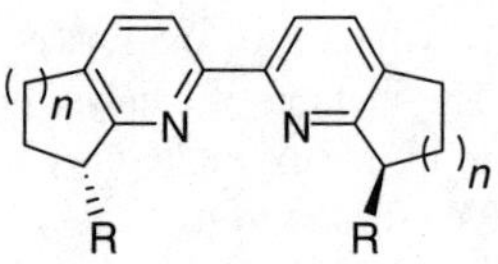

13

a $n = 1$, $R = {}^tBu$
b $n = 2$, $R = {}^tBu$
c $n = 2$, $R = CMe_2(OMe)$
d $n = 2$, $R = TMS$

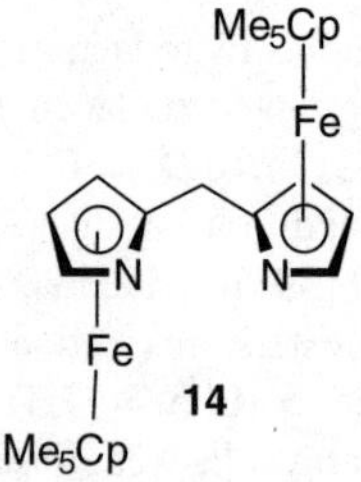

14

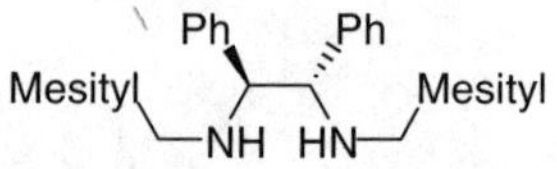

15

Selected results from using these catalysts are given in Tables 5.4 and 5.5. It is interesting that **13** in combination with CuOTf provides exceptional enantiocontrol in the cyclopropanation of E-olefins (Eq. 5.11). Additional information that might describe other advantages is lacking, however, so these ligands await a "champion" that can further promote their unique characteristics.

$$\text{(5.11)}$$

trans:cis (to Me) = 40:60
>99% ee (1S,2S,3S)
24% ee (1S,2R,3R)

5.2.4. Chiral Dirhodium(II) Carboxylate Catalysts

Rh(II) carboxylates, especially $Rh_2(OAc)_4$, have emerged as the most generally effective catalysts for metal carbene transformations [7–10] and thus interest continues in the design and development of dirhodium(II) complexes that possess chiral5ligands. They are structurally well-defined, with D_{2h} symmetry [51] and axial coordination sites at which carbene formation occurs in reactions with diazo compounds. With chiral dirhodium(II) carboxylates the asymmetric center is located relatively far from the carbene center in the metal carbene intermediate. The first of these to be reported with applications to cyclopropanation reactions was developed by Brunner [52], who prepared 13 chiral dirhodium(II) tetrakis(carboxylate) derivatives (**16**) from enantiomerically pure carboxylic acids $R^1R^2R^3CCOOH$ with substituents that were varied from H, Me, and Ph to OH, NHAc, and CF_3. However, reactions performed between ethyl diazoacetate and styrene yielded cyclopropane products whose enantiopurities were less than 12% ee, a situation analogous to that encountered by Nozaki [2] in the first applications of chiral Schiff base-Cu(II) catalysts.

About the same time McKervey reported N-sulfonamidoprolinate catalysts [53] (**17**), and later Ikegami, Hashimoto, and co-workers described uses of dirhodium(II) catalysts with ligands that were phthalimide derivatives of phenylalanine [54] (**18a**), tert-leucine [55] (**18b**), and alanine (**18c**), but they were similarly unselective in intermolecular cyclopropanation reactions of ethyl diazoacetate. Only when Davies applied chiral prolinate **17a** and those that he reported for the first time (**17c**, X = ^{t}Bu, $C_{12}H_{25}$) to cyclopropanation reactions of vinyldiazocarboxylates (Eq. 5.12) did the value of these catalysts for cyclopropanation reactions become fully expressed (Table 5.6) [56,57]. The advantage of **17c** (X = $C_{12}H_{25}$) is its solubility in pentane, even at −78°C.

16

In addition, reactions of methyl phenyldiazoacetate with alkenes exhibit similar selectivities [58,59]. The use of pentane, rather than dichloromethane, as the solvent has a significant influence on enantioselectivities, increasing enantiopurities of the products to ≥90% ee. The

TABLE 5.6. Enantioselective Cyclopropanation of Alkenes with Methyl Phenyldiazoacetate (A) and Methyl Cinnamyldiazoacetate (B) in Pentane

Catalyst	R^1	R^2	% ee A	Ref.	% ee B	Ref.
17c (X = ^{t}Bu)	Ph	H	87	[59]	90	[57]
17c (X = ^{t}Bu)	p-ClC$_6$H$_4$	H	85	[59]	89	[57]
17c (X = ^{t}Bu)	p-MeOC$_6$H$_4$	H	83	[59]	88	[57]
17c (X = ^{t}Bu)	Ph	Ph	97	[58]	—	
17c (X = C$_{12}$H$_{25}$)	EtO	H	66	[59]	59	[57]
17c (X = C$_{12}$H$_{25}$)	n-Bu	H	77	[59]	<90	[57]

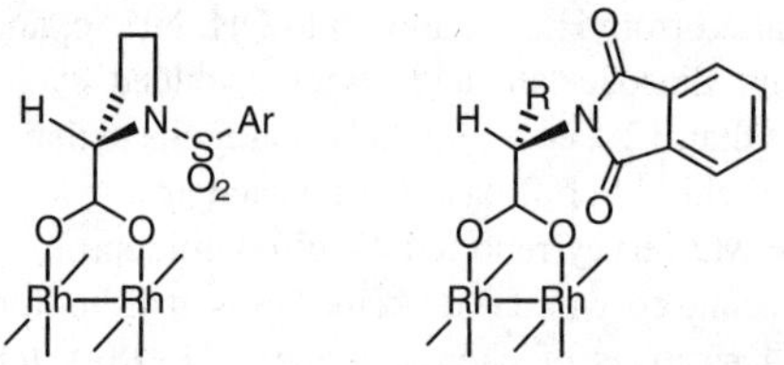

17a Ar = Ph
17b Ar = Nap
17c Ar = C$_6$H$_4$X

18a R = PhCH$_2$
18b R = ^{t}Bu
18c R = Me

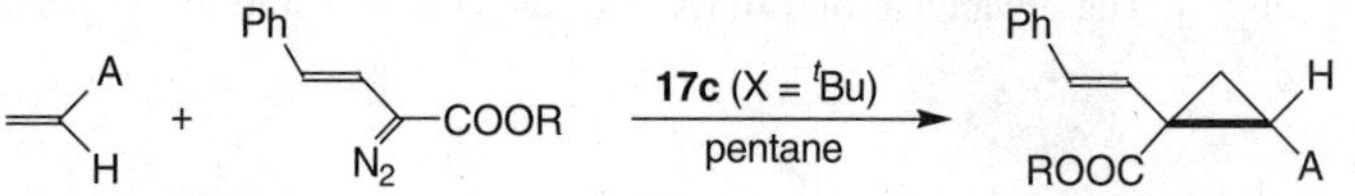

$$(5.12)$$

cause for this solvent influence is the change in the relative conformations of prolinate ligands [58], depicted below.

Conformational orientations of prolinate ligands:
A = ArSO$_2$

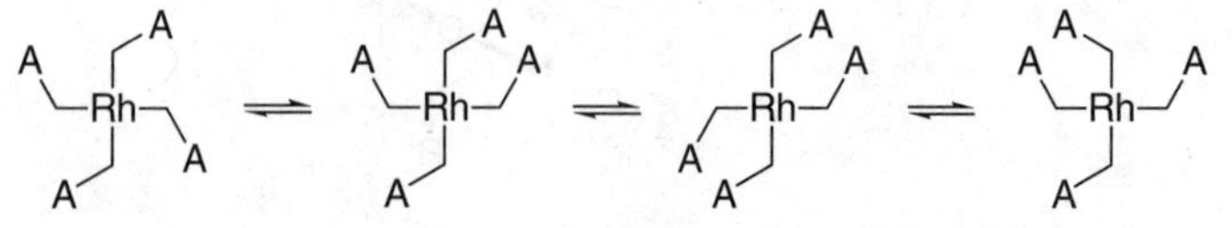

(−)-Sertraline [60], chiral 1,4-cycloheptadienes [61], and select cyclopentenes [62] have been prepared by using these catalysts and vinyldiazocarboxylates, and this approach has also been applied to the enantioselective synthesis of functionalized tropanes [63] and of the four stereoisomers of 2-phenylcyclopropane-1-amino acid [64].

(-)-sertraline

98% ee

97% ee

86% ee

5.2.5. Chiral Dirhodium(II) Carboxamidates

In contrast to Rh(II) carboxylates, Rh(II) carboxamidates provide placement of a chiral center adjacent to nitrogen in close proximity to the carbenoid center. However, unlike the Rh(II) carboxylates, for which only one isomer derived from the ligand is possible, Rh(II) carboxamidates can conceivably be constructed from chiral carboxamides in any one or all of four possible configurations. In spite of this, the isomer possessing four bridging amide ligands and configured so that two oxygen and two nitrogen donor atoms are bonded to each Rh, with the nitrogen donor atoms in a cis geometry (**19**), is preferred among the other arrangements. Rh(II) carboxamidates are less reactive towards diazo compounds and generally provide higher selectivities in carbenoid reactions than do Rh(II) carboxylates [7].

19

The preparation and uses of chiral dirhodium(II) tetrakis(methyl 2-oxapyrrolidine-5-carboxylate) (**20**) and those with oxazolidinone, imidiazolidinone, or azetidinone ligands (**21–23**) for intermolecular cyclopropanation reactions have been reported [7]. These catalysts are prepared in high yield and, except for imidazolidinone-ligated **22** [65], only in the cis geometry by ligand exchange in refluxing chlorobenzene with Rh(II) acetate, and the mechanism for this selective conversion has been established [66]. Their design places the chiral center adjacent to the Rh-bound nitrogen and, like the salicylaldimine and semicorrin/bis-oxazoline copper catalysts, interactions from the protruding alkyl or carboxylate substituents control enantioselectivity. However, these dirhodium(II) catalysts block approach by the alkene in a manner that is significantly different from previous catalyst designs so that a higher level of stereocontrol occurs in the formation of the *cis*-cyclopropane derivative from reactions with styrene than in the formation of the *trans*-cyclopropane derivative [33,67].

Enantiocontrol with **21–23** is lower than that achieved with chiral copper catalysts for reactions of diazoacetates with styrene and a few other alkenes examined thus far [68], but the carboxamidates display far greater stereocontrol than do the dirhodium(II) carboxylates for the same reactions [69]. However, Hashimoto has reported the use of chiral piperidinonate **24** and found exceptional enantiocontrol in the cyclopropanation of styrene and both mono- and

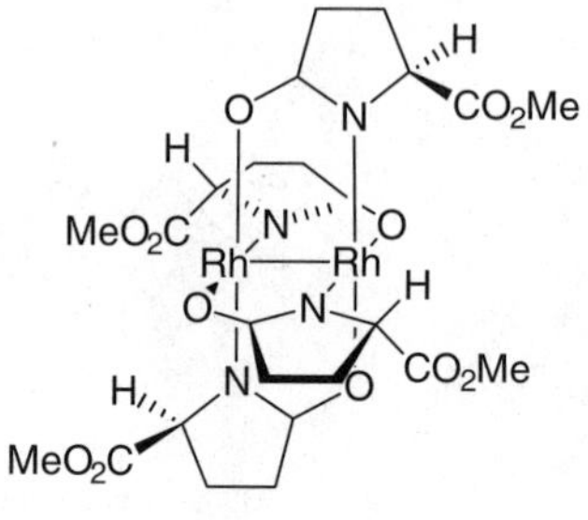

20: Rh₂(5S-MEPY)₄

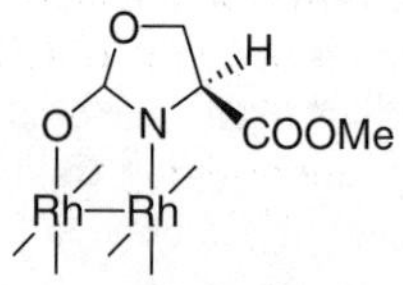

21

Rh₂(4S-MEOX)₄

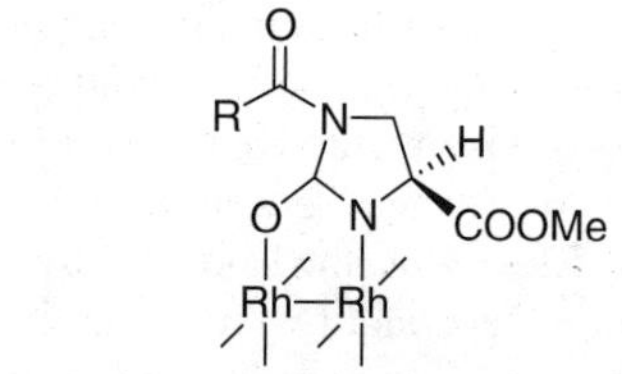

22

a. R = Me : Rh₂(4S-MACIM)₄
b. R = Ph : Rh₂(4S-MBOIM)₄
c. R = PhCH₂CH₂ : Rh₂(4S-MPPIM)₄

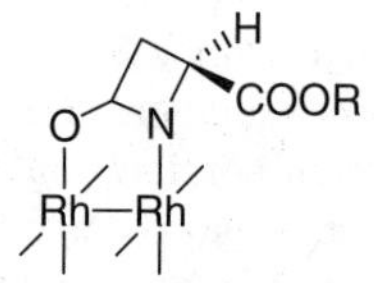

23

a. R = ⁱBu : Rh₂(4S-IBAZ)₄
b. R = Bn : Rh₂(4S-BNAZ)₄

1,1-disubstituted alkenes with 2,3,4-trimethyl-3-pentyl diazoacetate although diastereocontrol was low [70]. With **24** solvent effects are important, and the use of ether increases enantiocontrol to optimum values. Reactions of methyl phenyldiazoacetate with alkenes catalyzed by Rh₂(4S-IBAZ)₄ result in levels of enantiocontrol comparable with those obtained by the catalysis of the prolinate complex **17c** [58].

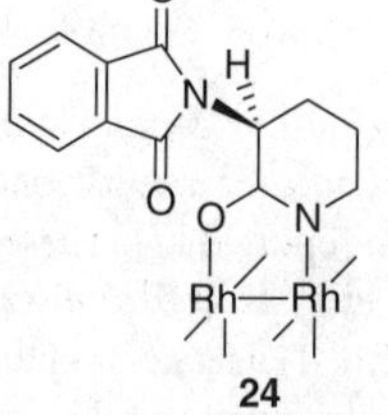

24

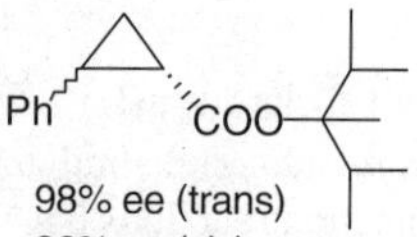

98% ee (trans)
96% ee (cis)
trans:cis = 74:26

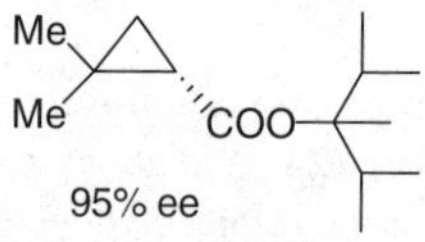

95% ee

The carboxylate attachment to the carboxamidate ligand is essential for high levels of enantiocontrol. Replacement of the COOMe moiety by benzyl (**25a**), isopropyl (**25b**), or phenyl (**25c**) group, which is of similar size, almost always results in significantly lower selectivities [68,71].

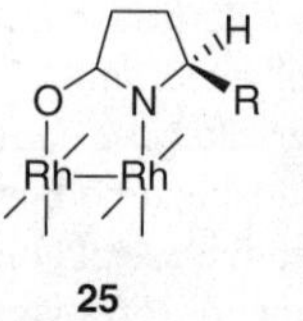

a R = Bn : Rh₂(5R-BNOX)₄
b R = ⁱPr : Rh₂(5R-IPOX)₄
c R = Ph : Rh₂(5R-PHOX)₄

25

The selectivity enhancement brought about by the carboxylate group has been attributed [11] to the ability of this group to orient and stabilize the Rh-bound carbene and thereby direct incoming nucleophiles to backside attack on the side of the carbene opposite to the stabilizing substituents (Scheme 5.3), wherein electronic interaction of the p-orbital of the metal carbene with the polar carboxylate substituent of the ligand (δ^-) provides the required stabilization and orientation. Because the carboxamide's two nitrogen ligands place two polar substituents in a cis geometry on each face of the dirhodium(II) compound, two limiting carbene orientations are possible when the carbene substituent R is assumed to take the optimum electronic or less sterically congested position (**26A, 26B**).

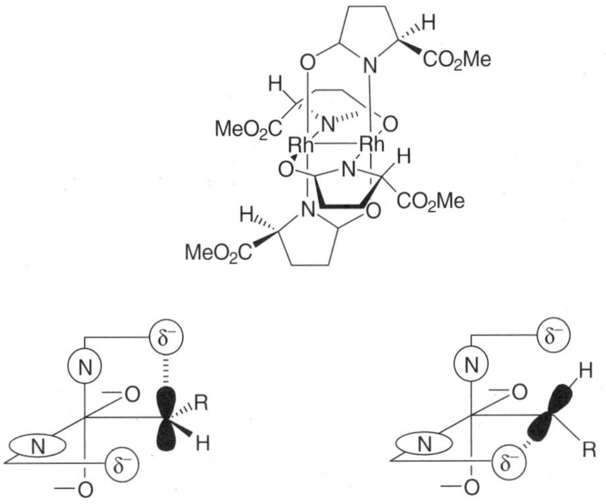

Scheme 5.3.

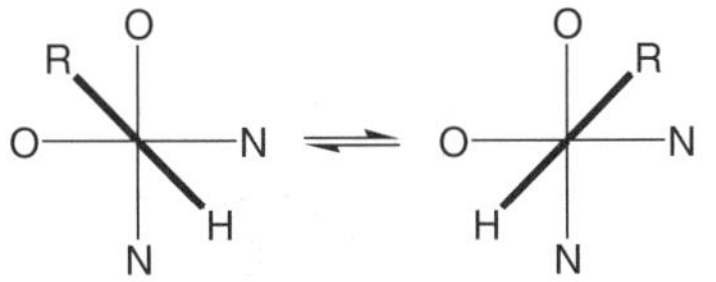

26A **26B**

5.2.6. Chiral Rh(III) Porphyrin Catalysts

Callot and co-workers established in 1982 that iodorhodium(III) porphyrin complexes could be used as cyclopropanation catalysts with diazo esters and alkenes; with *cis*-disubstituted alkenes these catalysts provide preferential production of cis(syn) disubstituted cyclopropanes (syn/anti up to 3.3 with 1,4-cyclohexadiene) [72]. More recently, chiral porphyrins have been designed and prepared by Kodadek and co-workers [73], and their iodorhodium(III) complexes have been examined for asymmetric induction in catalytic cyclopropanation reactions [74,75]. The intent here has been to affix chiral attachments onto the four porphyrin positions that are occupied in tetraphenylporphyrin by a phenyl group. Iodorhodium(III) catalysts with chiral binaphthyl (**27**, called "chiral wall" porphyrin [74]) and the structurally analogous chiral pyrenyl-naphthyl (**28**, called "chiral fortress" porphyrin [75]) have been prepared, but their ability to control enantioselectivity is low to moderate (10–60% ee) with ethyl diazoacetate. However, these catalysts have an exceptional propensity for the production of the cis(syn) stereoisomer to an even greater extent than that reported by Callot and co-workers for the corresponding tetramesitylporphyrin catalyst [72]. The oxidation state of the catalytically active Rh compound is unknown and could be Rh(II).

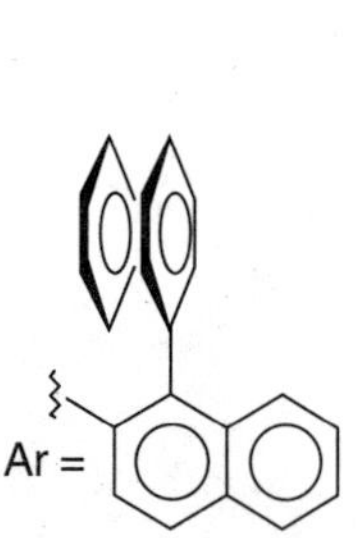

27 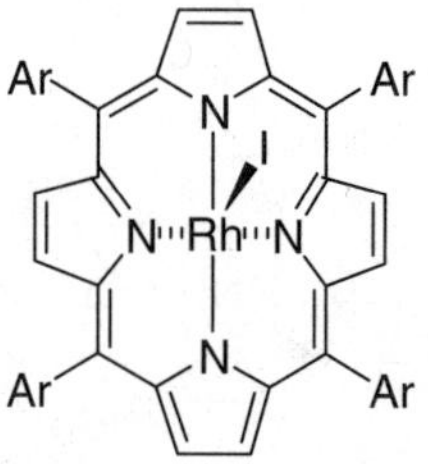 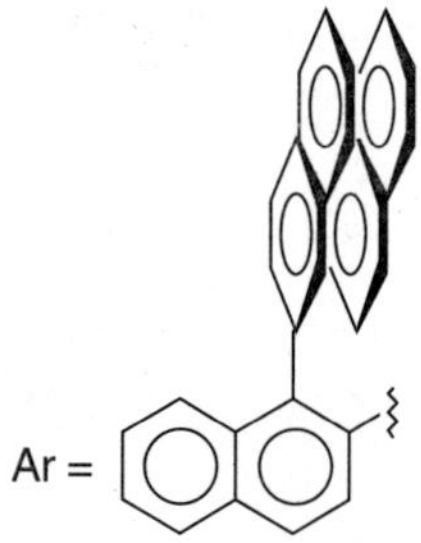   **28**

5.2.7. Chiral Cobalt Catalysts

Even before Aratani introduced his chiral salicylaldimine copper catalysts, Nakamura and Otsuka reported in 1974 that chiral bis(α-camphorquinonedioximato)cobalt(II) (**29**) and related complexes were effective enantioselective cyclopropanation catalysts [76], and they more fully described the preparation, characteristics, and uses of these catalysts in 1978 [77]. Optical yields as high as 88% were achieved for the cyclopropanation of styrene with neopentyl diazoacetate, and chemical yields greater than 90% were obtained in several cases.

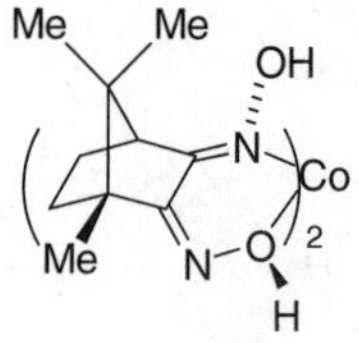

29

Dioximato-cobalt(II) catalysts are unusual in their ability to catalyze cyclopropanation reactions that occur with conjugated olefins (e.g., styrene, 1,3-butadiene, and 1-phenyl-1,3-butadiene) and, also, certain α,β-unsaturated esters (e.g., methyl α-phenylacrylate, Eq. 5.13), but not with simple olefins and vinyl ethers. In this regard they do not behave like metal carbenes formed with Cu or Rh catalysts that are characteristically electrophilic in their reactions towards alkenes (vinyl ethers > dienes > simple olefins >> α,β-unsaturated esters) [7], and this divergence has not been adequately explained. However, despite their ability to attain high enantioselectivities in cyclopropanation reactions with ethyl diazoacetate and other diazo esters, no additional details concerning these Co(II) catalysts have been published since the initial reports by Nakamura and Otsuka.

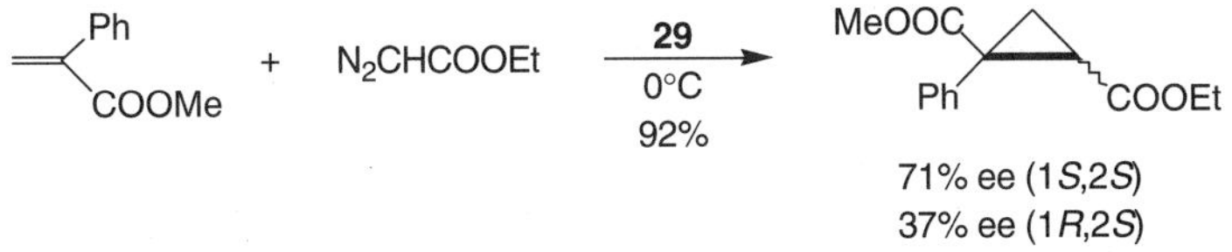

$$\begin{array}{c}\text{71\% ee }(1S,2S)\\\text{37\% ee }(1R,2S)\end{array}\qquad(5.13)$$

Although enantioselectivities attained by the Co catalyst **29** are relatively high (75% ee for ethyl (1S,2S)-*trans*-2-phenylcyclopropanecarboxylate and 67% ee for the (1S,2R)-*cis*-isomer resulting from the reaction of styrene with ethyl diazoacetate), diastereoselectivities are low. However, at least for reactions with styrene, the *trans:cis* ratio changes markedly with small changes in the ester group of diazoacetates [77], and enantioselectivities are similarly responsive. In reactions of ethyl diazoacetate with *cis*-1,2-dideuteriostyrene, scrambling of deuterium in the cyclopropane product is observed, which is extensive, but not complete. Free rotation about the C_1-C_2 bond of the original 1,2-dideuteriostyrene in a reaction intermediate is inferred.

The extensive data accumulated by Nakamura and Otsuka, although interpreted by them as being due to the intervention of metal carbene and metallocyclobutane intermediates, can also be rationalized by an alternative mechanism in which coordination of the chiral Co(II) catalyst with the alkene activates the alkene for electrophilic addition to the diazo compound (Scheme 5.4). Subsequent ring closure can be envisioned to occur via a diazonium ion intermediate, without involving at any stage a metal carbene intermediate.

Reactions of transition-metal coordinated olefins with diazo compounds as a route to cyclopropane products have not yet been rigorously established. Catalysts that should be effective in this pathway are those that are more susceptible to olefin coordination than to association with a diazo compound and also those whose coordinated alkene is sufficiently electrophilic to react with diazo compounds, especially diazomethane. Pd(II), Pt(II), and Co(II) compounds appear to be capable of olefin coordination-induced cyclopropanation reactions, but further investigations will be required to unravel this mechanistic possibility.

Katsuki has recently constructed a series of chiral Co(III)-salen complexes **30** and used them as catalysts for cyclopropanation of styrene using *tert*-butyl diazoacetate [78]. Not only is enantiocontrol for addition greater than 90% ee, but diastereoselectivity favors the *trans*-

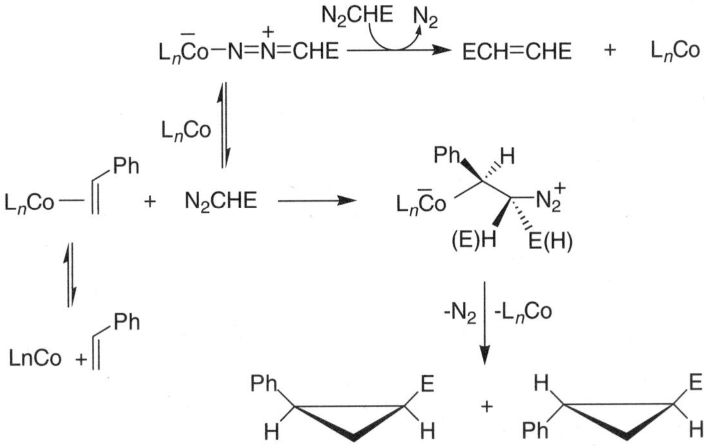

Scheme 5.4.

cyclopropane isomer almost exclusively (96:4). Further details concerning the activities of these catalysts and their selectivities are needed.

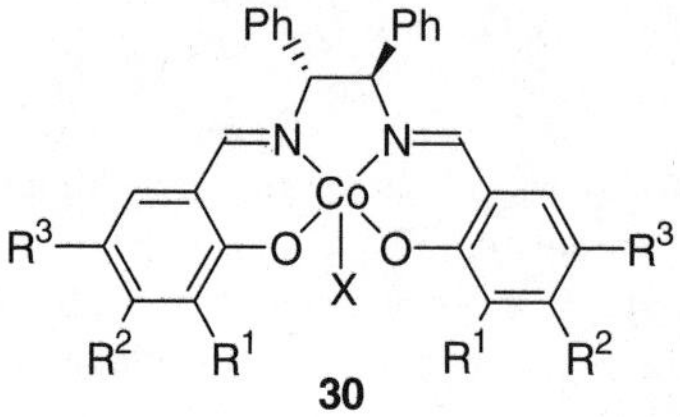

a. X = Br, $R^1 = R^3$ = H, R^2 = ^{t}Bu
b. X = Br, $R^1 = R^2$ = H, R^3 = OMe

30

5.2.8. Chiral Ru Catalysts

The C_2-symmetric 2,6-bis(2-oxazolin-2-yl)pyridine (pybox) ligand was originally applied with Rh for enantioselective hydrosilylation of ketones [79], but Nishiyama, Itoh, and co-workers have used the chiral pybox ligands with Ru(II) as an effective cyclopropanation catalyst **31** [80]. The advantages in the use of this catalyst are the high enantiocontrol in product formation (≥95% ee) and the exceptional diastereocontrol for production of the trans-cyclopropane isomer (>92:8) in reactions of diazoacetates with monosubstituted olefins. Electronic influences from 4-substituents of pyridine in **31** affect relative reactivity (ρ = +1.53) and enantioselectivity, but not diastereoselectivity [81]. The disadvantage in the use of these catalysts, at least for synthetic purposes, is their sluggish reactivity. In fact, the stability of the intermediate metal carbene has allowed their isolation in two cases [82].

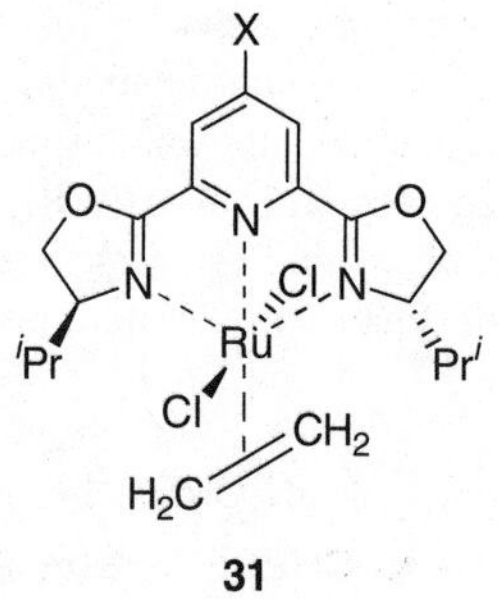

31

A chiral Ru porphyrin complex based, in part, on the Kodadek design for Rh has recently been reported to give good-to-high (80–91% ee) enantiocontrol for the cyclopropanation of styrenes and 1,1-disubstituted alkenes [83]. The advantage here is the high *trans:cis* ratio, which seems to be characteristic of Ru, and, like other metalloporphyrin catalysts, turnover numbers are large.

5.3. ASYMMETRIC INTRAMOLECULAR CYCLOPROPANATION

Intramolecular cyclopropanation reactions of alkenyl diazo carbonyl compounds are among the most useful catalytic metal carbene transformations, and the diversity of their applications for organic syntheses is substantial [7,10,24,84]. Their catalytic asymmetric reactions, however, have only recently been reported. An early application of the Aratani catalyst **2** (A = PhCH$_2$) to the asymmetric synthesis of dihydrochrysanthemolactone (Eq. 5.14, in refluxing cyclohexane) showed low enantiocontrol (23% ee) and relatively low yield (47%) [85]. Dauben and co-workers subsequently examined enantioselectivity in the intramolecular cyclopropanation of 1-diazo-5-hexen-2-one and its homolog using a similar Aratani catalyst [86], but found limited improvement in enantiocontrol.

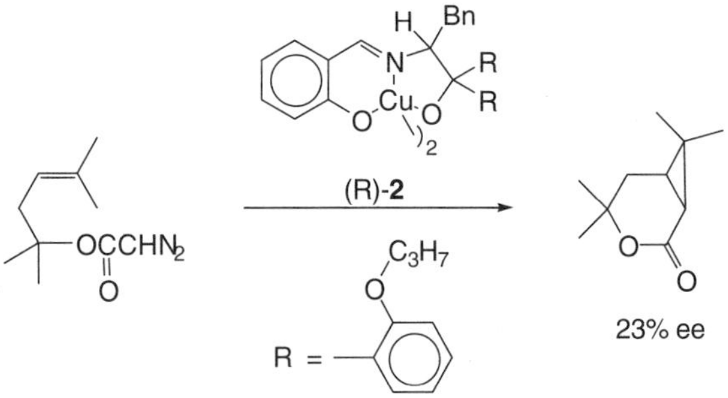

$$(5.14)$$

As may now be expected from Cu(II) compounds, whose reduction to Cu(I) is a requirement for catalyst activation, product yield and enantioselectivity depend on the reagents and conditions used. Other diazo carbonyl compounds, including a β-keto-α-diazo esters and α-diazo esters, underwent intramolecular cyclopropanation in moderate-to-good yields, but with low enantiocontrol.

Pfaltz has also examined enantiocontrol in the intramolecular cyclopropanation of diazo ketones (Eq. 5.15), and found relatively high enantioselectivity with the use of his semicorrin Cu(I) catalyst [38]. This catalyst is obviously superior to the salicylaldimine catalysts for intramolecular cyclopropanation reactions and, furthermore, enantiocontrol increases with an increase in the ring size from five to six.

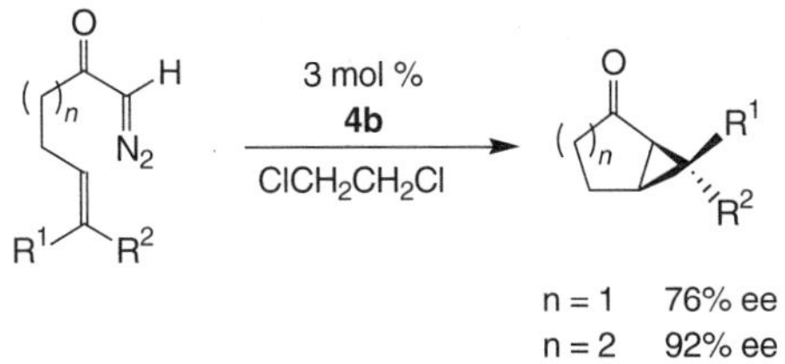

$$(5.15)$$

5.3.1. Allylic and Homoallylic Cyclopropanation

Doyle, Martin, Müller, and co-workers communicated exceptional enantiocontrol for intramolecular cyclopropanation of a series of allyl diazoacetates (Eq. 5.16) by using dirhodium(II) tetrakis(methyl 2-oxopyrrolidine-5-carboxylates), $Rh_2(MEPY)_4$, in either their R- or S-configurations [87], and they have fully elaborated these results in a subsequent report [88].

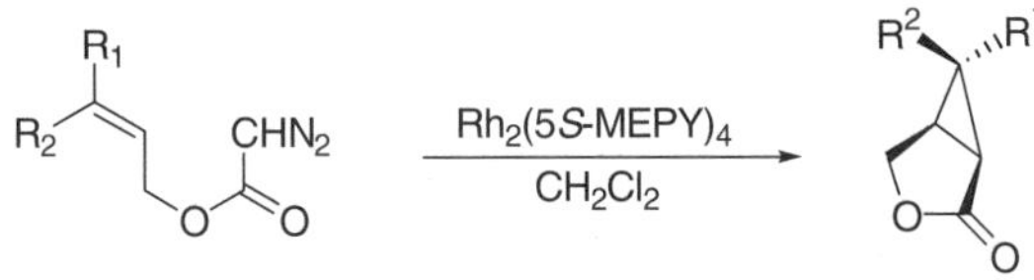

$$(5.16)$$

TABLE 5.7. Enantioselective Intramolecular Cyclopropanation of Allyl Diazoacetates [87,88,91]

31	R'	R^2	Catalyst	Yield (%)	% ee	Configuration
(a)	H	H	$Rh_2(5S\text{-MEPY})_4$	75	95	$(1R,5S)$
	H	H	$Rh_2(5R\text{-MEPY})_4$	72	95	$(1S,5R)$
	H	H	$CuPF_6/\mathbf{7b}$	61	20	$(1R,5S)$
(b)	Ph	H	$Rh_2(5S\text{-MEPY})_4$	78	68	$(1R,5S)$
	Ph	H	$Rh_2(4S\text{-MPPIM})_4$	61	96	$(1R,5S)$
(c)	H	Ph	$Rh_2(5S\text{-MEPY})_4$	70	≥94	$(1R,5S)$
(d)	^{n}Pr	H	$Rh_2(5S\text{-MEPY})_4$	93	85	$(1R,5S)$
	^{n}Pr	H	$Rh_2(4S\text{-MPPIM})_4$	83	95	$(1R,5S)$
	^{n}Pr	H	$CuPF_6/\mathbf{7b}$	74	29	$(1S,5R)$
	^{n}Pr	H	$\mathbf{30}$ (X = H)	68	78	$(1R,5S)$
(e)	H	^{n}Pr	$Rh_2(5S\text{-MEPY})_4$	88	≥94	$(1R,5S)$
	H	^{n}Pr	$CuPF_6/\mathbf{7b}$	82	37	$(1S,5R)$
	H	^{n}Pr	$\mathbf{30}$ (X = H)	54	21	$(1R,5S)$
(f)	H	Bn	$Rh_2(5S\text{-MEPY})_4$	80	≥94	$(1R,5S)$
(g)	H	^{i}Bu	$Rh_2(5S\text{-MEPY})_4$	73	≥94	$(1R,5S)$
(h)	H	^{i}Pr	$Rh_2(5S\text{-MEPY})_4$	85	≥94	$(1R,5S)$
(i)	H	$(^n\text{Bu})_3\text{Sn}$	$Rh_2(5S\text{-MEPY})_4$	79	≥94	$(1R,5S)$
(j)	H	I	$Rh_2(5S\text{-MEPY})_4$	78	≥94	$(1R,5S)$
(k)	Me	Me	$Rh_2(5S\text{-MEPY})_4$	89	98	$(1S,5R)$
	Me	Me	$CuPF_6/\mathbf{7b}$	61	20	$(1R,5S)$
	Me	Me	$\mathbf{30}$ (X = H)	91	76	$(1R,5S)$
(l)	$Me_2C=CH(CH_2)_2$	Me	$Rh_2(5S\text{-MEPY})_4$	88	95	$(1S,5R)$
(m)	Me	$Me_2C=CH(CH_2)_2$	$Rh_2(5S\text{-MEPY})_4$	79	93	$(1S,5R)$

Cis-disubstituted olefins lead to higher enantioselectivities than do the corresponding trans-disubstituted olefins (Table 5.7), even with substituents as bulky as tri-n-butyltin.

However, the use of $Rh_2(MPPIM)_4$ provides enhanced enantiocontrol for cyclopropanations of trans-disubstituted double bonds, up to 96% ee in the cases examined [89]. Trisubstituted allylic double bonds, even that in farnesyl diazoacetate (**33**, Eq. 5.17) [90], undergo effective, efficient, and highly enantioselective cyclopropanation (Eq. 5.17). Product yields are high except for those cases in which steric factors appear to limit olefin approach to the metal carbene center.

$$\text{(5.17)}$$

33   **34**, 94% ee

Chiral Rh(II) oxazolidinones $Rh_2(BNOX)_4$ and $Rh_2(IPOX)_4$ (**25a,b**) were not as effective as $Rh_2(MEPY)_4$ for enantioselective intramolecular cyclopropanation, even though the steric bulk of their chiral ligand attachments (COOMe versus *i*-Pr or CH_2Ph) are similar. Significantly lower yields and lower enantioselectivities resulted from dinitrogen extrusion from prenyl diazoacetate catalyzed by either $Rh_2(4S\text{-}IPOX)_4$ or $Rh_2(4S\text{-}BNOX)_4$. This difference, and those associated with butenolide formation [91], can be attributed to the ability of the carboxylate substituents to stabilize the carbocation form of the intermediate metal carbene (**3b**), thus limiting the $Rh_2(MEPY)_4$-catalyzed reaction to concerted carbene addition onto both carbon atoms of the C–C double bond.

35

The directional orientation of chiral ligand substituents on $Rh_2(5S\text{-}MEPY)_4$ establishes relatively unimpeded pathways for intramolecular cyclization from two limiting configurations of which **35** is preferred. According to this model Z-olefins should afford higher enantioselectivities than do E-olefins, whose substituent is buttressed against the ligand's carboxylate group, and substituents on the internal carbon position of the C–C double bond should give rise to very low levels of enantioselectivity. The absolute configuration of the bicyclic cyclopropane product is consistent with this interpretation.

The predicted low enantiocontrol from reactions performed with methallyl diazoacetate (Eq. 5.18) was borne out in reactions catalyzed by $Rh_2(MEPY)_4$ and $Rh_2(MEOX)_4$, but when chiral imidazolidinone-ligated dirhodium(II) was used, enantioselectivity rose to 89% ee (Table 5.8) [89]. The use of $CuPF_6/\textbf{7b}$ also caused relatively high enantiocontrol (87% ee) [92] which, however, decreases to 82% ee when the methyl group of **36** was replaced by *n*-butyl, whereas with $Rh_2(4S\text{-}MPPIM)_4$ the enantiopurity of the product corresponding to **37** was 93% ee. The *N*-3-phenylpropanoyl substituents of $Rh_2(4S\text{-}MPPIM)_4$ help to create a more conformationally restrictive environment that leads to enhanced enantiocontrol.

TABLE 5.8. Enantioselective Intramolecular Cyclopropanation of Methallyl Diazoacetate [87,88,91]

Catalyst	Yield (%)	% ee	Configuration
$Rh_2(5S\text{-}MEPY)_4$	72	7	(1*R*,5*S*)
$Rh_2(4S\text{-}MPPIM)_4$	75	89	(1*S*,5*R*)
$Rh_2(4S\text{-}MEOX)_4$	84	1	(1*S*,5*R*)
$Rh_2(4S\text{-}MACIM)_4$	90	78	(1*S*,5*R*)
$CuPF_6/\textbf{7b}$	58	87	(1*S*,5*R*)

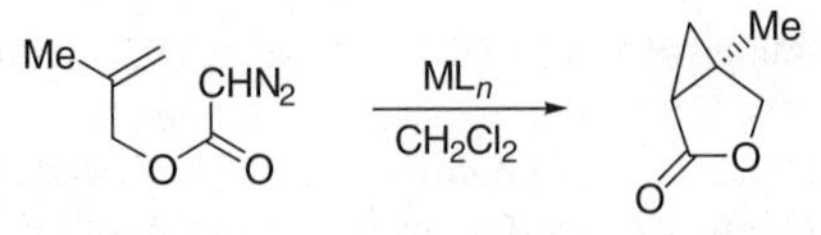

$$(5.18)$$

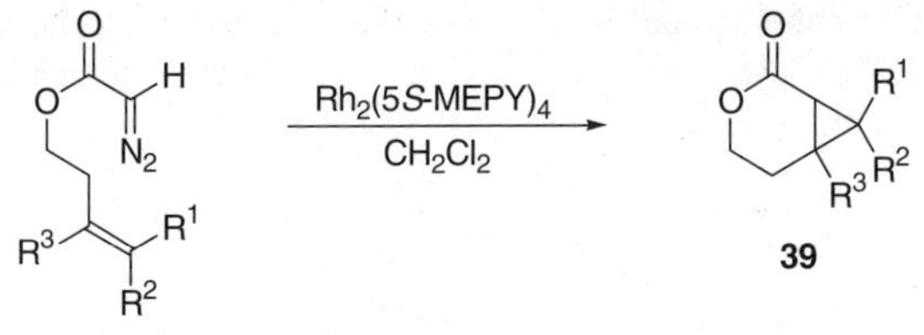

$$(5.19)$$

Intramolecular cyclopropanation of the next higher homologs of the allyl diazoacetates (Eq. 5.19) catalyzed by $Rh_2(MEPY)_4$ give moderate-to-high percent of ee's for the addition product, and isolated yields are also high (Table 5.9) [88].

TABLE 5.9. Enantioselective Intramolecular Cyclopropanation of Homoallyl Diazoacetates (38) Catalyzed by Rh$_2$(5S-MEPY)$_4$ [87]

38	R^1	R^2	R^3	Yield (%)	% ee	Configuration
(a)	H	H	H	80	71	(1R,6S)
(b)	CH$_3$	CH$_3$	H	74	77	(1S,6R)
(c)	H	H	CH$_3$	76	83	(1R,6S)
(d)	CH$_3$CH$_2$	H	H	80	90	(1S,6R)
(e)	H	CH$_3$CH$_2$	H	65	82	(1S,6R)
(f)	Ph	H	H	73	88	(1S,6R)
(g)	H	Ph	H	55	73	(1S,6R)
(h)	c-C$_6$H$_{11}$CH$_2$	H	H	77	80	(1S,6R)
(i)	PhCH$_2$	H	H	68	80	(1S,6R)
(j)	TMS	H	H	65	86	(1S,6R)

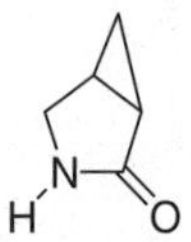

98% ee
40% yield
Rh$_2$(4S-MEOX)$_4$
40

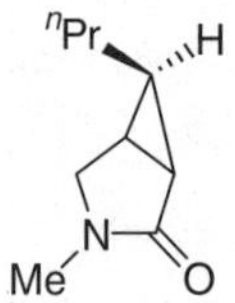

95% ee
88% yield
Rh$_2$(4S-MPPIM)$_4$
41

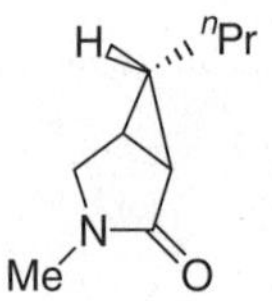

92% ee
93% yield
Rh$_2$(4S-MPPIM)$_4$
42

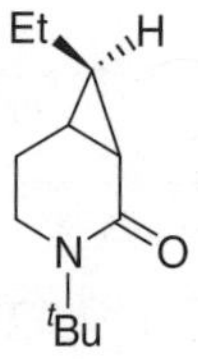

90% ee
94% yield
Rh$_2$(5S-MEPY)$_4$
43

Once again, cis-disubstituted olefins lead to higher enantioselectivities than do trans-disubstituted olefins, but here the differences are not as great as they were with allyl diazoacetates. Both allylic and homoallylic diazoacetamides also undergo highly enantioselective intramolecular cyclopropanation (**40–43**) [93,94]. However, with allylic α-diazopropionates enantiocontrol is lower by 10–30% ee [95]. The composite data suggest that chiral dirhodium(II) carboxamide catalysts are superior to chiral Cu or Ru catalysts for intramolecular cyclopropanation reactions of allylic and homoallylic diazoacetates.

5.3.2. Asymmetric Macrocyclic Cyclopropanation

Entropy effects in the formation of macrocyclic compounds by cyclopropanation lie between allylic/homoallylic cyclopropanation and intermolecular cyclopropanation. Yet despite extensive investigations of intramolecular reactions, the feasibility of the macrocyclization process for cyclopropanation has only recently been established [96,97]. The asymmetric versions of this transformation have been reported for the first time only in 1996 [98], but certain control features of these reactions are already evident. First, one of the most surprising is that these reactions can be performed in high yield without the requirement of high dilution. Second, Cu(MeCN)$_4$PF$_6$/**7b** gives consistently high enantioselectivities, independent of ring size, with methallyl systems (**44–46**). Third, in a competition that would produce either **45**, or a product from addition to the cis-disubstituted double bond of the diazo compound from which **45** originates, the more reactive methallyl double bond is the only site for addition.

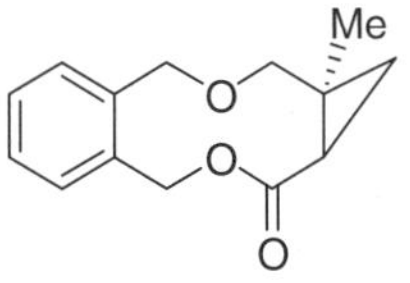

90% ee
44

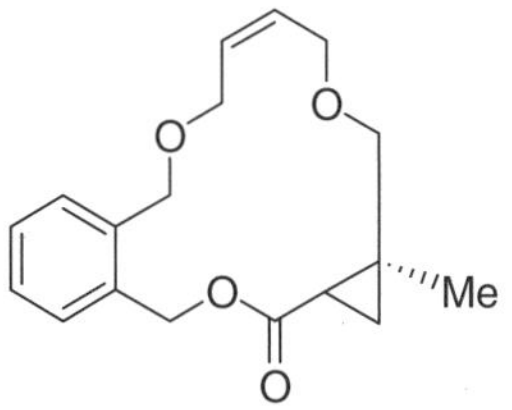

90% ee
45

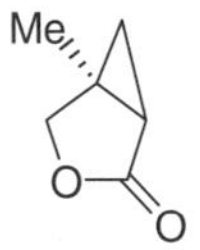

87% ee
37

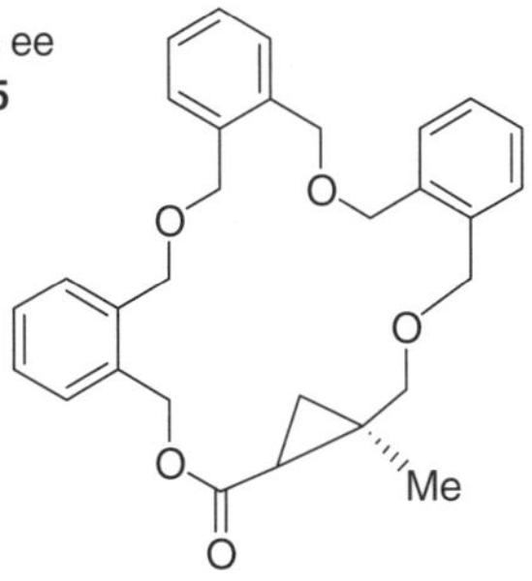

86% ee
46

In the competition between allylic intramolecular cyclopropanation and macrocyclization (Eq. 5.20), the more electrophilic catalyst favors macrocyclization. Doyle has explained this differential selectivity as due to the formation of an intermediate π-complex between the C–C double bond and the carbene center. The more electrophilic the carbene carbon or the more electron-rich the double bond, the more that this π-complex is favored and the more favorable the pathway to macrocyclization [97]. However, thus far few systems have been examined for enantioselective macrocyclic cyclopropanation, so one can expect a few surprises for the future.

$$(5.20)$$

5.4. ASYMMETRIC CYCLOPROPENATION OF ALKYNES

5.4.1. Intermolecular Cyclopropenation

Functionalized cyclopropenes are viable synthetic intermediates whose applications [99,100] extend to a wide variety of carbocyclic and heterocyclic systems. However, advances in the synthesis of cyclopropenes, particularly through Rh(II) carboxylate—catalyzed decomposition of diazo esters in the presence of alkynes [100–102], has made available an array of stable 3-cyclopropenecarboxylate esters. Previously, copper catalysts provided low to moderate yields of cyclopropenes in reactions of diazo esters with disubstituted acetylenes [103], but the higher temperatures required for these carbenoid reactions often led to thermal or catalytic ring opening and products derived from vinylcarbene intermediates (104–107).

In the presence of catalytic amounts of $Rh_2(5R\text{-MEPY})_4$, reaction between ethyl diazoacetate and representative alkynes results in the formation of ethyl cyclopropene-3-carboxylates (Eq. 5.21) with enantiopurities ranging from 54% ee to 98% ee in good yields (70–85%) [108].

$$(5.21)$$

Virtually identical results, except in the opposite stereocontrol sense, are obtained with the use of $Rh_2(5S\text{-MEPY})_4$. With d-menthyl diazoacetate and the same series of alkynes, selectivities as high as ≥97:3 diastereomer ratio have been achieved (Table 5.10), but diastereoselectivities with diazoacetates having other chiral auxiliaries did not show any improvement in diastereocontrol [109].

Enantioselectivities increase with the steric size of the diazo ester, and the polarity of the alkyne substituent also appears to influence enantiocontrol. For instance, reactions using N,N-dimethyldiazoacetamide showed a 20% or more improvement in selectivity. The fact that the enantiopurities of cyclopropenes arising from reactions with propargyl methyl ether are higher than those from reactions with 1-hexyne and 3,3-dimethyl-2-butyne suggest that polar interactions of the methoxy substituent of this alkyne with ligands of the catalyst may be operative. Indeed, cyclopropenation of propargyl acetate by l-menthyl diazoacetate in the presence of $Rh_2(5S\text{-MEPY})_4$ also resulted in ≥97:3 diastereomer ratio for the cyclopropenecarboxylate product. The use of these chiral cyclopropenes as precursors to cis-disubstituted cyclopropanes is an effective methodology [108,110].

TABLE 5.10. Enantioselective Cyclopropenation of Alkynes with Diazoacetates (Eq. 5.21) [108]

Diazoacetate R =	1-Alkyne R′ =	Catalyst	Yield (%)	% ee (or % de)	Config.[a]
Me	CH(OEt)$_2$	Rh$_2$(5S-MEPY)$_4$	42	≥98	S
^{t}Bu	CH$_2$OMe	Rh$_2$(5S-MEPY)$_4$	52	78	S
^{t}Bu	CH$_2$OMe	Rh$_2$(5R-MEPY)$_4$	56	78	R
Et	CH$_2$OMe	Rh$_2$(5R-MEPY)$_4$	73	69	R
d-menthyl	CH$_2$OMe	Rh$_2$(5R-MEPY)$_4$	43	≥94	R
l-menthyl	CH$_2$OMe	Rh$_2$(5R-MEPY)$_4$	45	44	R
Et	^{n}Bu	Rh$_2$(5R-MEPY)$_4$	70	54	R
d-menthyl	^{n}Bu	Rh$_2$(5R-MEPY)$_4$	46	86	R
Et	^{t}Bu	Rh$_2$(5R-MEPY)$_4$	85	57	R
d-menthyl	^{t}Bu	Rh$_2$(5R-MEPY)$_4$	51	78	R

[a]Absolute configuration of the cyclopropene.

5.4.2. Intramolecular Cyclopropenation

What was evident for macrocyclization involving metal carbene addition to alkenes is even more so for addition to alkynes [110]. However, here the chiral dirhodium(II) catalyst Rh$_2$(4S-IBAZ)$_4$ exhibits the highest degree of enantiocontrol, superior even to Cu(MeCN)$_4$PF$_6$ (Eq. 5.22).

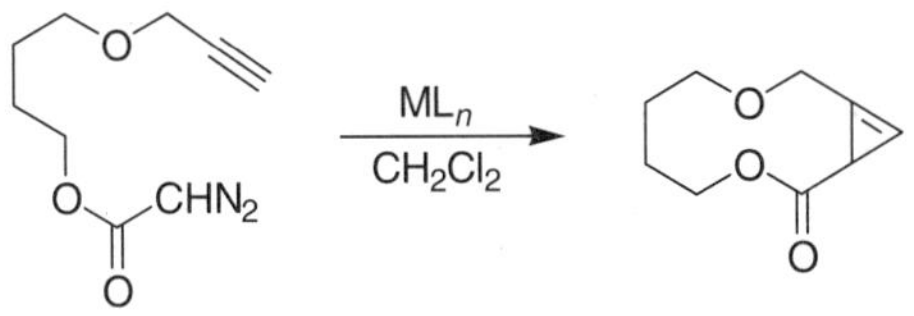

ML$_n$	
Rh$_2$(4S-IBAZ)$_4$	99% ee
CuPF$_6$/**7b**	82% ee
Rh$_2$(5S-MEPY)$_4$	78% ee

(5.22)

In addition, a high degree of chemoselectivity can be achieved with the proper selection of catalyst, as shown in Eq. 5.23, which describes exceptional control of product selectivity based

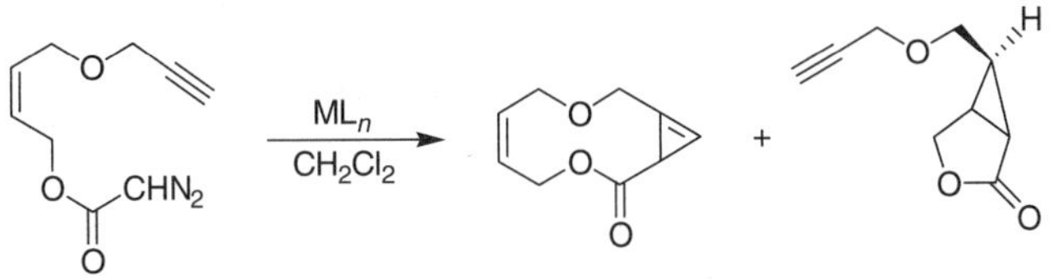

ML$_n$	yield,%		
Rh$_2$(4S-IBAZ)$_4$	80	84 (97% ee)	16 (88% ee)
Rh$_2$(5S-MEPY)$_4$	76	4	96 (96% ee)
CuPF$_6$/**7b**	54	31 (75% ee)	69 (46% ee)
17c (Ar = C$_6$H$_4$^{t}Bu)	87	94 (12% ee)	4 (20% ee)

(5.23)

solely on the ligand of the dirhodium(II) catalyst [111]. Ring sizes up to 15 have been reported, and in all cases addition to the C–C triple bond is preferred over addition to a C–C double bond with $Rh_2(4S\text{-}IBAZ)_4$.

5.5. ASYMMETRIC INTRAMOLECULAR C–H INSERTION

Although lesser-known than the addition transformations of metal carbenes, C–H insertion processes are among the most important asymmetric C–C bond-forming reactions available to synthetic chemists [7,10,112,113]. They result from cleavage of C–H and metal–C bonds with concurrent formation of C–C and C–H bonds (Scheme 5.5). The construction of five-membered rings is most common, but six-, four- and even eleven-membered rings [96] have been formed by C–H insertion.

Dirhodium(II) compounds are reported to be the most suitable catalysts for insertion. Selectivity is higher and yields are greater with dirhodium(II) carboxylates or carboxamidates than with copper catalysts, whereas Ru catalysts are not known to facilitate C–H insertion. As expected by a process that is basically electrophilic, electron-donating substituents that are adjacent to the site of insertion activate that center for C–H insertion [114]. In addition to electronic influences, however, conformational effects that are basically steric in origin can also control reaction selectivity [115].

Asymmetric induction in intramolecular C–H insertion reactions was first reported by McKervey and co-workers [53], who used chiral Rh(II) prolinate **17a** (Eq. 5.24). Although enantiocontrol was low, this report established the feasibility of the methodology and left open advances that were subsequently made by Ikegami and Hashimoto, who were able to convert α-diazo-β-ketoester **47** into cyclopentanone **48** with **18a** (Eq. 5.25) with 32–76% ee, dependent on the substituent Z and the size of the ester alkyl group [54,116].

$$(5.24)$$

~ 12% ee

$$(5.25)$$

32-76% ee

5.5.1. Chiral Dirhodium(II) Carboxylate Catalysts

Since the early reports by the research groups led by McKervey and Hashimoto and Ikegami, major advances have been made in the applications of these catalysts. McKervey has reported [117] a highly diastereoselective synthesis of chromanones with enantiocontrol up to 82% ee (Eq. 5.26), and Hashimoto [118] recently reported a clever asymmetric synthesis of β-lactams (Eq. 5.27), previously prepared by Ponsford and Southgate with $Rh_2(OAc)_4$ as the first published

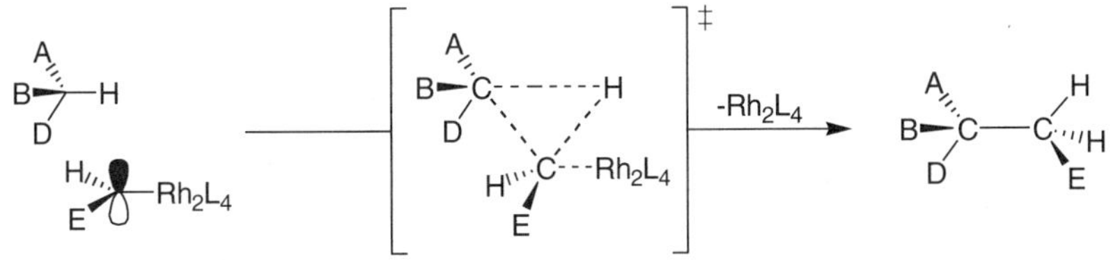

Scheme 5.5.

example of a C–H insertion reaction catalyzed by the dirhodium(II) catalyst [119]. Using chiral carboxylate **18a**, Hashimoto had previously shown that similar diazo compounds formed *N-tert*-butyl-β-lactams in high yield and 56–74% ee [120]. More recently, efforts to prepare γ-lactams were reported by using **18b**, but enantiocontrol was only moderate [121].

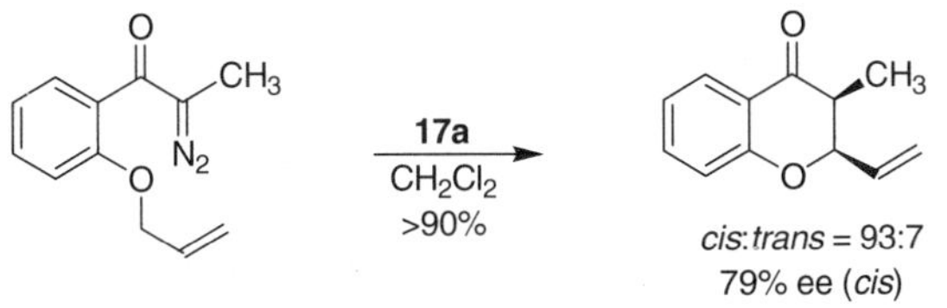

$$cis:trans = 93:7$$
$$79\% \text{ ee } (cis)$$

(5.26)

96% ee

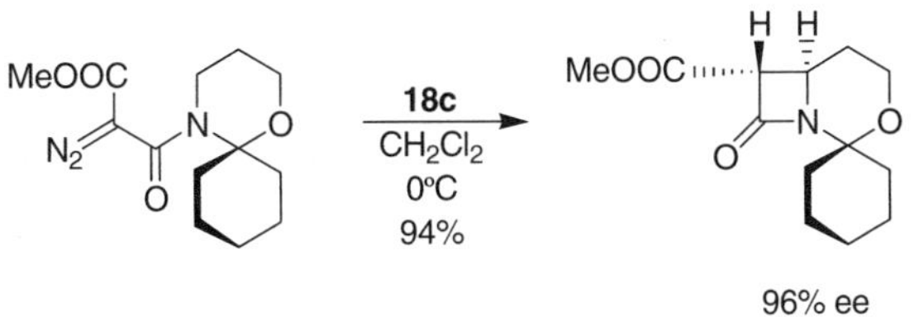

(5.27)

The use of chiral dirhodium carboxylate, **17** or **18**, is preferred over chiral dirhodium carboxamidates for chemical transformations of α-diazo-β-ketocarbonyl compounds primarily because of reactivity considerations, that is, these diazo compounds do not undergo dinitrogen loss with the carboxamidate catalysts even at elevated temperatures. In addition, the orientation of the chiral ligands in **17** and **18** provides closer access to bulky diazo compounds. When the two attachments to the diazomethane unit are vastly unequal in size, high levels of enantiocontrol can result.

5.5.2. Chiral Dirhodium(II) Carboxamidate Catalysts

By far the greatest advances in enantiocontrolled C–H insertion reactions have been provided by Doyle and co-workers with chiral dirhodium(II) carboxamidate catalysts [7,10]. The key development here is the creation of chiral imidazolidinone-ligated dirhodium catalysts **22** to control diastereoselectivity and enhance enantiocontrol [122]. A significant example of the power of this methodology is the insertion reactions of cycloalkyl diazoacetates. With cyclohexyl diazoacetate, for example, four products are possible via C–H insertion constituted in two pairs of diastereoisomers (Eq. 5.28).

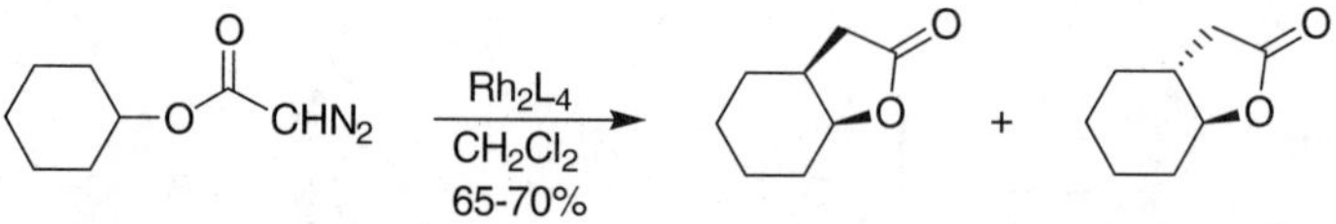

Rh$_2$(OAc)$_4$	40	60
Rh$_2$(4S-MEOX)$_4$	55 (96% ee)	45 (95% ee)
Rh$_2$(5S-MEPY)$_4$	75 (97% ee)	25 (91% ee)
Rh$_2$(4S-MACIM)$_4$	99 (97% ee)	1 (65% ee)

$$(5.28)$$

Enantiocontrol is high throughout, but diastereocontrol is markedly dependent on the structure of the ligand, that is, the ligand providing the most restrictive cavity is the one that influences the conformational equilibrium of the metal carbene intermediate (Scheme 5.6).

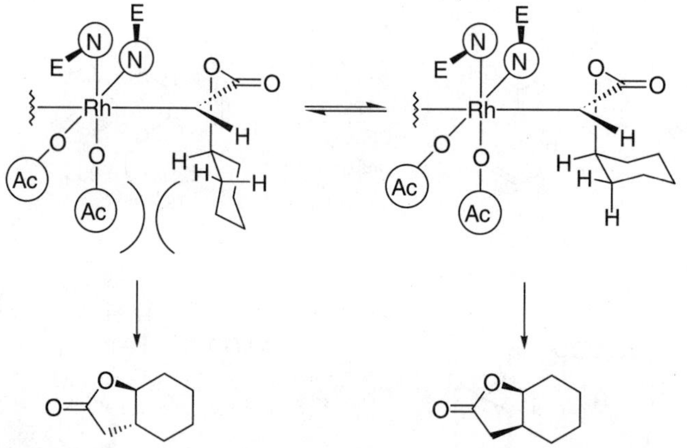

Scheme 5.6.

The protruding N-acyl group is buttressed against the cyclohexane ring in the more stable conformation causing insertion to occur with a lower energy from the less stable conformation. This level of enantiocontrol and diastereoselectivity is general for ring sizes of five through eight, but results with larger rings have not been reported. Similar success in achieving high stereocontrol is evident with acyclic substrates (Eq. 5.29) by using imidazolidinone-ligated **22c** [89].

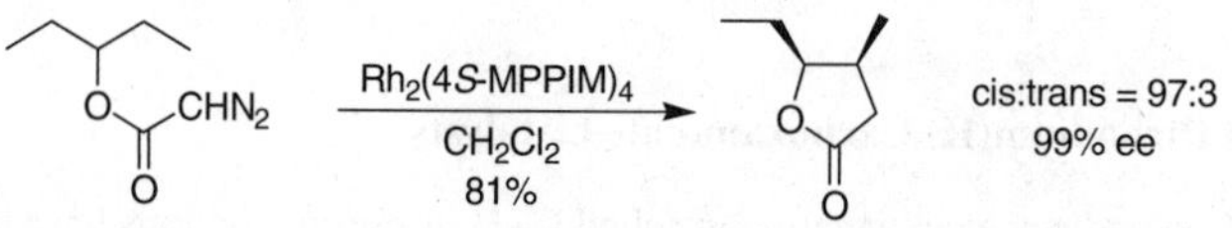

$$(5.29)$$

The synthesis of 2-deoxyxylolactone in high yield and 94% ee from 1,3-dichloro-2-propanol (Scheme 5.7) requires the exclusive use of Rh$_2$(5R-MEPY)$_4$ catalysts [123]. Insertion results in the predominant formation of one stereoisomer that is the thermodynamically less stable one. Here again, the success of the synthesis is determined as much by diastereocontrol as by enantiocontrol. As little as 0.1 mol % of catalyst is required to bring about complete reaction.

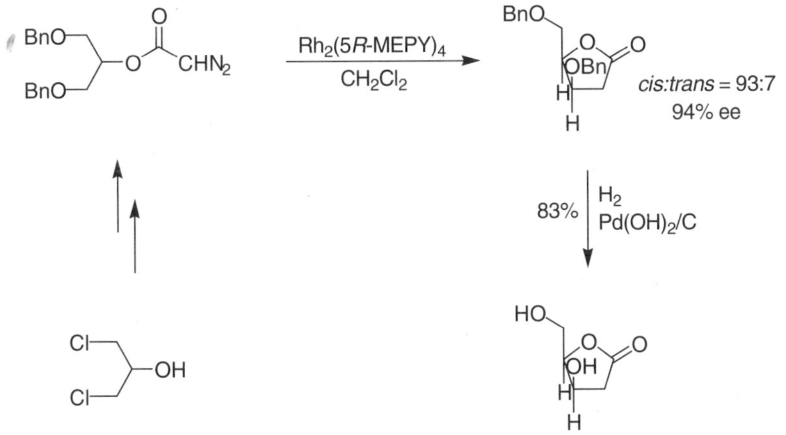

Scheme 5.7.

Match and mismatch of chiral catalyst with chiral substrate are distinctive, that is, matched combination generally brings high yield and high stereocontrol/regiocontrol. For example, treatment of (1S,2R)-cis-2-methylcyclohexyl diazoacetate with Rh_2(4R-MPPIM)$_4$ resulted virtually completely in the production of the all-cis bicyclic insertion product of Scheme 5.8 [124]. Similarly, treatment of (1R,2S)-cis-2-methylcyclohexyl diazoacetate with Rh_2(4S-MPPIM)$_4$ gave the same result, except for the opposite stereochemical sense. However, the mismatch of substrate and catalyst configurations results in a low yield overall and in a mixture of products (Scheme 5.8).

On the other hand, matched and mismatched combinations can produce distinctively different C–H insertion products (Eq. 5.30) with high stereochemical purity. When the reactant is a racemic mixture, the chiral catalyst directs one enantiomer to a single enantiomer of a given C–H insertion product while the enantiomer forms a distinctively different product (Eq. 5.31) [125].

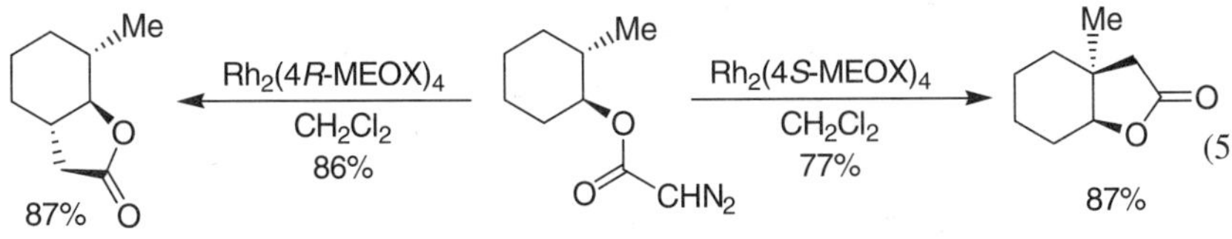

(5.30)

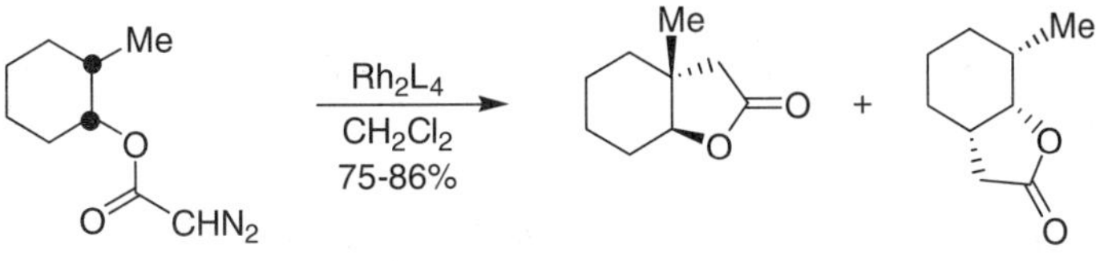

(5.31)

Rh_2(5S-MEPY)$_4$	45 (91% ee)	49 (98% ee)
Rh_2(4S-MEOX)$_4$	40 (99% ee)	47 (99% ee)
Rh_2(4S-MACIM)$_4$	11 (87% ee)	66 (77% ee)

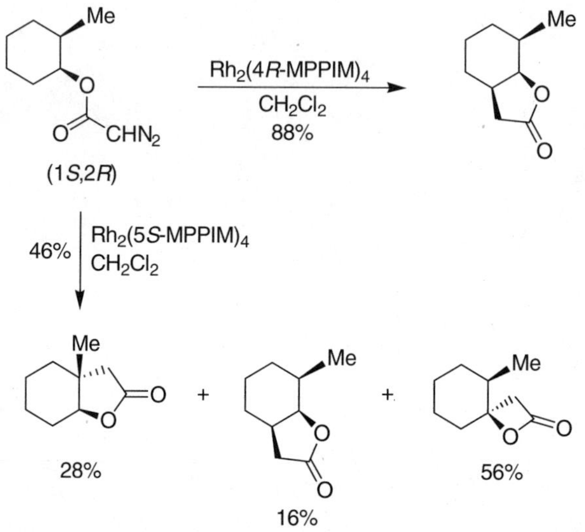

Scheme 5.8.

Other examples suggest the generality of this selectivity for cyclic diazoacetates and a limited number of acyclic analogs [124], and the composite provides, in an absolute sense, the reality of diastereotopic reaction considerations.

A critical development in efforts to achieve high enantiocontrol in C–H insertion reactions was the synthesis and applications of Rh$_2$(MPPIM)$_4$ [89]. This catalyst provided enhanced enantiocontrol in virtually all cases examined, but especially with 3-substituted-1-propyl diazoacetates (Eq. 5.32). Results obtained with various substrates are given in Table 5.11 [126,127], which show the unique ability of Rh$_2$(MPPIM)$_4$ to increase enantioselectivity. Notice also that the S-configured catalyst produces the S-configured product and that the R-catalyst produces the R-product.

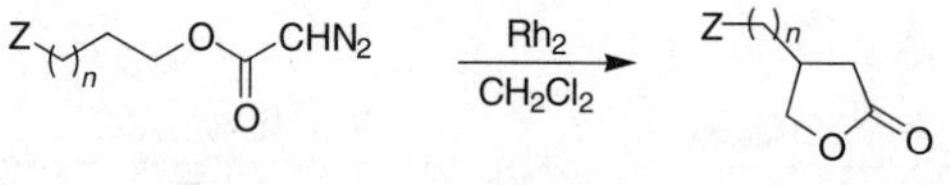

$$(5.32)$$

As a result of these developments, Doyle and co-workers have synthesized several lignans, among which are (−)-enterolactone, (+)-isodeoxypodophyllotoxin, and (−)-arctigenin [126].

Selected examples have been reported of impressive results from C–H insertion reactions of diazoacetamides that result in β-lactams. For example, β-lactam formation was the sole C–H insertion process to occur with **51** (Eq. 5.33) [128] or other seven-membered ring diazoacetamides.

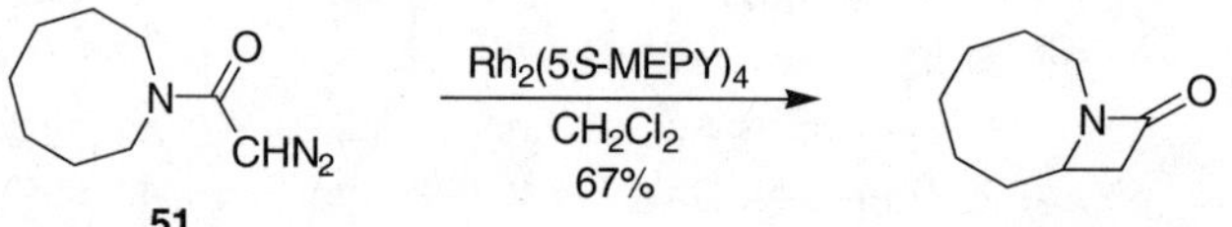

$$(5.33)$$

TABLE 5.11. Enantiocontrol in C–H Insertion Reactions of 3-Substituted-1-propyl Diazoacetates (49) [126,127]

49, Z =	n =	Catalyst	50 Yield (%)	50 % ee	Config.
Me	1	Rh$_2$(4S-MPPIM)$_4$	52	96	S
^{i}Pr	1	Rh$_2$(4S-MPPIM)$_4$	60	95	S
Ph	1	Rh$_2$(4S-MEOX)$_4$	76	51	S
Ph	1	Rh$_2$(5R-MEPY)$_4$	49	72	R
Ph	1	Rh$_2$(4R-MPPIM)$_4$	76	91	R
m-MeOC$_6$H$_4$	1	Rh$_2$(4R-MPPIM)$_4$	63	93	R
3,4-CH$_2$(O)OC$_6$H$_4$	1	Rh$_2$(4R-MPPIM)$_4$	57	95	R
3,4-(MeO)$_2$C$_6$H$_4$	1	Rh$_2$(4S-MPPIM)$_4$	62	94	S
MeO	0	Rh$_2$(4S-MPPIM)$_4$	>95	93	S
MeO	0	Rh$_2$(4R-MEPY)$_4$	56	91	R
BnO	0	Rh$_2$(4R-MEPY)$_4$	69	87	R

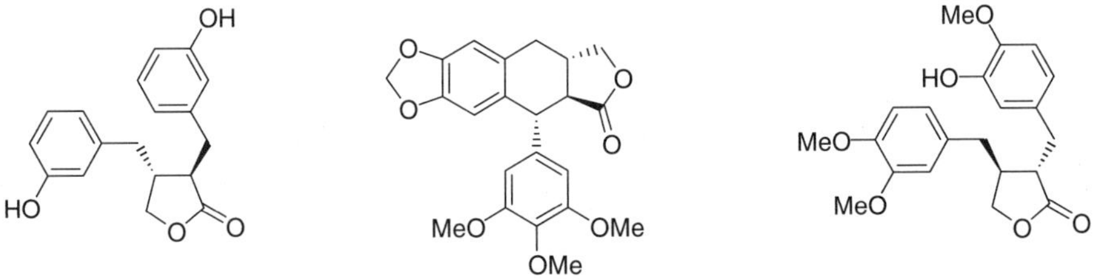

(-)-enterolactone   (+)-isodeoxypodophyllotoxin   (-)-arctigenin

However, the higher homolog proved to be too flexible, and mixtures of β- and γ-lactams resulted from catalytic diazo decomposition. In these cases the γ-lactams had higher enantiopurity. More modest selectivities were reported from reactions of their acyclic counterparts [129].

A novel methodology for the synthesis of optically pure pyrrolizidine bases also utilized the C–H insertion process [130]. Diazoacetamide **53** generated from readily available chiral reactants underwent Rh$_2$(4S-MPPIM)$_4$-catalyzed conversion to bicyclic product **54** with high diastereocontrol that was not possible with achiral dirhodium(II) catalysts or with chiral Cu(I) bis-oxazolines (Eq. 5.34) [130]. The Rh$_2$(4S-MPPIM)$_4$ catalyst, with its restrictions to conformational arrangements suitable for C–H insertion, provides easy access to this direct precursor of (−)-heliotridane, the thermodynamically less-stable isomer. With Rh$_2$(OAc)$_4$ or Cu catalysts the isomer formed in greatest yield was the thermodynamically more-stable one.

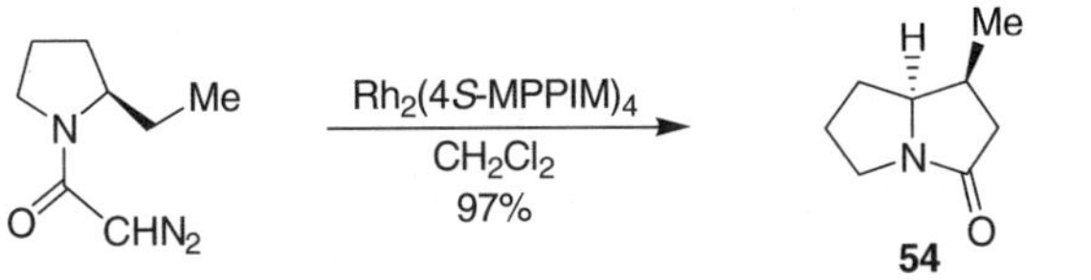

(5.34)

53

54
(endo:exo = 96:4)

Although exceptional diastereocontrol and enantiocontrol can now be achieved in C–H insertion reactions of catalytically generated metal carbenes, further improvements are needed. Insertion into tertiary C–H bonds occurs with diminished enantiocontrol and regiocontrol [131,132]. In addition, chiral dirhodium carboxamidates do not react with α-diazo-β-ketocarbonyl compounds. Thus, the potential for their impact on a broad range of C–H insertion processes is yet to be tested.

5.6. SUMMARY

Significant progress has been made recently in the design and development of chiral transition-metal catalysts for metal carbene transformations. Applications now extend beyond cyclopropanation to cyclopropenation and C–H insertion. The major focus of these efforts has been intermolecular reactions, but more recent results have demonstrated highly enantioselective intramolecular counterparts. Cu and dirhodium(II) catalysts that possess chiral ligands are still the most versatile, but Ru and Co catalysts show promise. From a practical point of view, the catalyst effectiveness is related to high turnover numbers and, especially, to high turnover rates, and these considerations are just now being refined.

ACKNOWLEDGMENT

Support from the National Science Foundation and the National Institutes of Health (GM-46503) for investigations of the design and development of chiral dirhodium(II) carboxamidate catalysts for asymmetric metal carbene transformations is gratefully acknowledged.

REFERENCES

1. Noyori, R. *Science* **1990**, *248*, 1194.

2. Nozaki, H.; Moriuti, S.; Takaya, H.; Noyori, R. *Tetrahedron Lett.* **1966**, 5239.

3. Aratani, T.; Yoneyoshi, Y.; Nagase, T. *Tetrahedron Lett.* **1975**, 1707.

4. Aratani, T.; Yoneyoshi, Y.; Nagase, T. *Tetrahedron Lett.* **1977**, 2599.

5. Aratani, T.; Yoneyoshi, Y.; Nagase, T. *Tetrahedron Lett.* **1982**, *23*, 685.

6. Aratani, T. *Pure Appl. Chem.* **1985**, *57*, 1839.

7. Doyle, M. P.; McKervey, M. A.; Ye, T. *Modern Catalytic Methods for Organic Synthesis with Diazo Compounds*; John Wiley & Sons, Inc.: New York, 1998.

8. Ye, T.; McKervey, M. A. *Chem. Rev.* **1994**, *94*, 1091.

9. Nefedov, O. M.; Shapiro, E. A.; Dyatkin, A. B. In *Supplement B: The Chemistry of Acid Derivatives*; Patai, S. (Ed.); Wiley: New York 1992; Chapter 25.

10. Doyle, M. P.; Forbes, D. C. *Chem. Rev.* **1998**, *98*, 911.

11. Silberrad, O.; Roy, C. S. *J. Chem. Soc.* **1906**, *89*, 179.

12. (a) Paulissen, R.; Reimlinger, H.; Hayez, E.; Hubert, A. J.; Teyssie, Ph. *Tetrahedron Lett.* **1973**, 2233. (b) Hubert, A. J.; Noels, A. F.; Anciaux, A. J.; Teyssie, Ph. *Synthesis* **1976**, 600.

13. Park, S.-B.; Nishiyama, H.; Itoh, Y.; Itoh, K. *J. Chem. Soc., Chem. Commun.* **1994**, 1315.

14. Park, S.-B.; Sakata, N.; Nishiyama, H. *Chem. Eur. J.* **1996**, *2*, 303.

15. Doyle, M. P.; Griffin, J. H.; Bagheri, V.; Dorow, R. L. *Organometallics* **1984**, *3*, 53.

16. Doyle, M. P.; Griffin, J. H.; Conceicao, J. da *J. Chem. Soc., Chem. Commun.* **1985**, 328.

17. Maxwell, J.; Kodadek, T. *Organometallics* **1991**, *10*, 4.

18. Brookhart, M.; Studabaker, W. B. *Chem. Rev.* **1987**, *87*, 411.

19. Regitz, M.; Maas, G. *Aliphatic Diazo Compounds—Properties and Synthesis*; Academic Press: New York, 1987.

20. (a) Shankar, B. K. R.; Shechter, H. *Tetrahedron Lett.* **1982**, *23*, 2277. (b) Müller, P.; Pautex, N.; Doyle, M. P.; Bagheri, V. *Helv. Chim. Acta* **1990**, *73*, 1233.

21. (a) Suda, M. *Synthesis* **1981**, 714. (b) Vallgarda, J.; Hacksell, U. *Tetrahedron Lett.* **1991**, *32*, 5625.

22. Denmark, S. E.; Stavenger, R. A.; Faucher, A.-M.; Edwards, J. P. *J. Org. Chem.* **1997**, *62*, 3375.

23. (a) Doyle, M. P. *Chem. Rev.* **1986**, *19*, 348. (b) Doyle, M. P. *Acc. Chem. Res.* **1986**, *19*, 348.

24. Doyle, M. P.; Protopopova, M. N. *Tetrahedron* **1998**, *54*, 7919.

25. Anciaux, A. J.; Hubert, A. J.; Noels, A. F.; Petiniot, N.; Teyssié, Ph. *J. Org. Chem.* **1980**, *45*, 695.

26. (a) Kunz, T.; Reissig, H.-U. *Tetrahedron Lett.* **1989**, *30*, 2079. (b) Reissig, H.-U.; Dammast, F. *Chem. Ber.* **1993**, *126*, 2727.

27. Becalski, A.; Cullen, W. R.; Fryzuk, M. D.; Herb, G.; James, B. R.; Kutney, J. P.; Piotrowska, K.; Tapiolas, D. *Can. J. Chem.* **1988**, *66*, 3108.

28. Laidler, D. A.; Milner, D. J. *J. Organometal. Chem.* **1984**, *270*, 121.

29. Baldwin, J. E.; Barden, T. C. *J. Am. Chem. Soc.* **1984**, *106*, 6364.

30. Doyle, M. P.; Dorow, R. L.; Buhro, W. E.; Griffin, J. H.; Tamblyn, W. H.; Trudell, M. L. *Organometallics* **1984**, *3*, 44.

31. Doyle, M. P.; van Leusen, D. *J. Org. Chem.* **1982**, *47*, 5326.

32. Brunner, H.; Miehling, W. *Monatsh. Chem.* **1984**, *115*, 1237.

33. (a) Doyle, M. P. In *Catalytic Asymmetric Synthesis*; Ojima, I. (Ed.); VCH Publishers: New York, 1992; Ch. 3. (b) Doyle, M. P. *Rec. Trav. Chim. Pays-Bas* **1991**, *110*, 305.

34. Fritschi, H.; Leuteneggar, U.; Pfaltz, A. *Angew. Chem., Int. Ed.* **1986**, *25*, 1005.

35. Leutenegger, U.; Umbricht, G.; Fahrni, C.; von Matt, P.; Pfaltz, A. *Tetrahedron* **1992**, *48*, 2143.

36. Fritschi, H.; Leutenegger, U.; Siegmann, K.; Pfaltz, A.; Keller, W.; Kratky, Ch. *Helv. Chim. Acta* **1988**, *71*, 1541.

37. Fritschi, H.; Leutenegger, U.; Pfaltz, A. *Helv. Chim. Acta* **1988**, *71*, 1553.

38. (a) Pfaltz, A. *Modern Synthetic Methods* 1989: Schelfold, R. (Ed.); Springer: Berlin-Heidelberg, 1989; pp. 199–248. (b) Pfaltz, A. *Acc. Chem. Res.* **1993**, *26*, 339.

39. Lowenthal, R. E.; Abiko, A.; Masamune, S. *Tetrahedron Lett.* **1990**, *31*, 6005.

40. Evans, D. A.; Woerpel, K. A.; Hinman, M. M. *J. Am. Chem. Soc.* **1991**, *113*, 726.

41. Lowenthal, R. E.; Masamune, S. *Tetrahedron Lett.* **1991**, *32*, 7373.

42. Müller, D.; Umbricht, G.; Weber, B.; Pfaltz, A. *Helv. Chim. Acta* **1991**, *74*, 232.

43. Doyle, M. P.; Bagheri, V.; Wandless, T. J.; Harn, N. K.; Brinker, D. A.; Eagle, C. T.; Loh, K.-L. *J. Am. Chem. Soc.* **1990**, *112*, 1906.

44. Schumacher, R.; Reissig, H.-U. *Chem. Eur. J.* **1997**, *3*, 614.

45. Gupta, A. D.; Bhuniya, D.; Singh, V. K. *Tetrahedron* **1994**, *50*, 13725.

46. (a) Bedeker, A. V.; Andersson, P. G. *Tetrahedron Lett.* **1996**, *37*, 4073. (b) Harm, A. M.; Knight, J. G.; Stemp, G. *Synlett* **1996**, 677.

47. Rippert, A. J. *Helv. Chim. Acta* **1998**, *81*, 676.

48. (a) Ito, K.; Katsuki, T. *Tetrahedron Lett.* **1993**, *34*, 2661. (b) Ito, K.; Katsuki, T. *Synlett* **1993**, 638.

49. Lo, M. M.-C.; Fu, G. C. *J. Am. Chem. Soc.* **1998**, *120*, 10270.

50. Kanemasa, S.; Hamura, S.; Harada, E.; Yamamoto, H. *Tetrahedron Lett.* **1994**, *35*, 7985.

51. (a) Cotton, F. A.; Walton, R. A. *Multiple Bonds Between Metal Atoms*; Wiley: New York, 1982; Ch. 7. (b) Felthouse, T. R. *Prog. Inorg. Chem.* **1982**, *29*, 73. (c) Boyar, E. B.; Robinson, S. D. *Coord. Chem. Rev.* **1983**, *50*, 109.

52. Brunner, H.; Kluschanzoff, H.; Wutz, K. *Bull. Chem. Soc. Belg.* **1989**, *98*, 63.

53. Kennedy, M.; McKervey, M. A.; Maguire, A. R.; Roos, G. H. P. *J. Chem. Soc., Chem. Commun.* **1990**, 361.

54. Hashimoto, S.; Watanabe, N.; Ikegami, S. *Tetrahedron Lett.* **1990**, *31*, 5173.

55. Watanabe, N.; Ogawa, T.; Ohtake, Y.; Ikegami, S.; Hashimoto, S. *Synlett* **1996**, 85.

56. Davies, H. M. L.; Hutcheson, D. K. *Tetrahedron Lett.* **1993**, *34*, 7243.

57. Davies, H. M. L.; Bruzinski, P. R.; Lake, D. H.; Kong, N.; Fall, M. J. *J. Am. Chem. Soc.* **1996**, *118*, 6897.

58. Doyle, M. P.; Zhou, Q.-L.; Charnsangavej, C.; Longoria, M. A.; McKervey, M. A.; Garcia, C. F. *Tetrahedron Lett.* **1996**, *37*, 4129.

59. Davies, H. M. L.; Bruzinski, P. R.; Fall, M. J. *Tetrahedron Lett.* **1996**, *37*, 4133.

60. Corey, E. J.; Grant, T. G. *Tetrahedron Lett.* **1994**, *35*, 5373.

61. Davies, H. M. L.; Stafford, D. G.; Doan, B. D.; Houser, J. H. *J. Am. Chem. Soc.* **1998**, *120*, 3326.

62. Davies, H. M. L.; Kong, N.; Churchill, M. R. *J. Org. Chem.* **1998**, *63*, 6586.

63. Davies, H. M. L.; Matasi, J. J.; Hodges, L. M.; Huby, N. J. S.; Thornley, C.; Kong, N.; Houser, J. H. *J. Org. Chem.* **1997**, *62*, 1095.

64. Davies, H. M. L.; Bruzinski, P. R.; Lake, D. H.; Konig, N.; Fall, M. J. *J. Am. Chem. Soc.* **1996**, *118*, 6897.

65. Doyle, M. P.; Zhou, Q.-L.; Raab, C. E.; Roos, G. H. P.; Simonsen, S. H.; Lynch, V. *Inorg. Chem.* **1996**, *35*, 6064.

66. Doyle, M. P.; Raab, C. E.; Roos, G. H. P.; Lynch, V.; Simonsen, S. H. *Inorg. Chim. Acta* **1997**, *266*, 13.

67. Doyle, M. P.; Brandes, B. D.; Kazala, A. P.; Pieters, R. J.; Jarstfer, M. B.; Watkins, L. M.; Eagle, C. T. *Tetrahedron Lett.* **1990**, *31*, 6613.

68. Müller, P.; Baud, C.; Ené, D.; Motallebi, S.; Doyle, M. P.; Brandes, B. D.; Dyatkin, A. B.; See, M. M. *Helv. Chim. Acta* **1995**, *78*, 459.

69. Davies, H. M. L.; Hutcheson, D. K. *Tetrahedron Lett.* **1993**, *34*, 7243.

70. Kitagaki, S.; Matsuda, H.; Watanabe, N.; Hashimoto, S. *Synlett* **1997**, 1171.

71. Doyle, M. P.; Winchester, W. R.; Hoorn, J. A. A.; Lynch, V.; Simonsen, S. H.; Ghosh, R. *J. Am. Chem. Soc.* **1993**, *115*, 9968.

72. Callot, H. J.; Metz, F.; Piechoki, C. *Tetrahedron* **1982**, *38*, 2365.

73. O'Malley, S.; Kodadek, T. *J. Am. Chem. Soc.* **1989**, *111*, 9116.

74. O'Malley, S.; Kodadek, T. *Tetrahedron Lett.* **1991**, *32*, 2445.

75. O'Malley, S.; Kodadek, T. *Organometallics* **1992**, *11*, 2299.

76. Tatsuno, Y.; Konishi, A.; Nakamura, A.; Otsuka, S. *J. Chem. Soc., Chem. Commun.* **1974**, 588.

77. (a) Nakamura, A.; Konishi, A.; Tatsuno, Y.; Otsuka, S. *J. Am. Chem. Soc.* **1978**, *100*, 3443, 6544. (b) Nakamura, A.; Konishi, A.; Tsujitani, R.; Kudo, M.; Otsuka, S. *J. Am. Chem. Soc.* **1978**, *100*, 3449.

78. (a) Fukuda, T.; Katsuki, T. *Synlett* **1995**, 825. (b) Fukuda, T.; Katsuki, T. *Tetrahedron* **1997**, *53*, 7201.

79. Brunner, H.; Nishiyama, H.; Itoh, K. In *Catalytic Asymmetric Synthesis*; Ojima, I. (Ed.); VCH Publishers: New York, 1993; Ch. 6.

80. Nishiyama, H.; Itoh, Y.; Sugawara, Y.; Matsumoto, H.; Aoki, K.; Itoh, K. *Bull. Chem. Soc. Jpn.* **1995**, *68*, 1247.

81. Park, S.-B.; Murata, K.; Matsumoto, H.; Nishiyama, H. *Tetrahedron: Asymmetry* **1995**, *6*, 2487.

82. (a) Park, S.-B.; Sakata, N.; Nishiyama, H. *Chem. Eur. J.* **1996**, *2*, 303. (b) Park, S.-B.; Nishiyama, H.; Itoh, Y.; Itoh, K. *J. Chem. Soc., Chem. Commun.* **1994**, 1315.

83. Lo, W.-C.; Che, C.-M.; Cheng, K.-F.; Mak, T. C. W. *J. Chem. Soc., Chem. Commun.* **1997**, 1205.

84. Burke, S. T.; Grieco, P. A. *Org. Reactions (N.Y.)* **1979**, *26*, 361.

85. Hirai, H.; Matsui, M. *Agr. Biol. Chem.* **1974**, *40*, 169.

86. Dauben, W. G.; Hendricks, R. T.; Luzzio, M. J.; Ng, H. P. *Tetrahedron Lett.* **1990**, *31*, 6969.

87. Doyle, M. P.; Pieters, R. J.; Martin, S. F.; Austin, R. E.; Oalmann, C. J.; Müller, P. *J. Am. Chem. Soc.* **1991**, *113*, 1423.

88. Doyle, M. P.; Austin, R. E.; Bailey, A. S.; Dwyer, M. P.; Dyatkin, A. B.; Kalinin, A. V.; Kwan, M. M. Y.; Liras, S.; Oalmann, C. J.; Pieters, R. J.; Protopopova, M. N.; Raab, C. E.; Roos, G. H. P.; Zhou, Q.-L.; Martin, S. F. *J. Am. Chem. Soc.* **1995**, *117*, 5763.

89. Doyle, M. P.; Zhou, Q.-L.; Dyatkin, A. B.; Ruppar, D. A. *Tetrahedron Lett.* **1995**, *36*, 7579.

90. Rogers, D. H.; Yi, E. C.; Poulter, C. D. *J. Org. Chem.* **1995**, *60*, 941.

91. Doyle, M. P.; Winchester, W. R.; Hoorn, J. A. A.; Lynch, V.; Simonsen, S. H.; Ghosh, R. *J. Am. Chem. Soc.* **1993**, *115*, 9968.

92. Doyle, M. P.; Peterson, C. S.; Zhou, Q.-L.; Nishiyama, H. *J. Chem. Soc., Chem. Commun.* **1997**, 211.

93. Doyle, M. P.; Eismont, M. Y.; Protopopova, M. N.; Kwan, M. M. Y. *Tetrahedron* **1994**, *50*, 1665.

94. Doyle, M. P.; Kalinin, A. V. *J. Org. Chem.* **1996**, *61*, 2179.

95. Doyle, M. P.; Zhou, Q.-L. *Tetrahedron: Asymmetry* **1995**, *6*, 2157.

96. Doyle, M. P.; Protopopova, M. N.; Poulter, C. D.; Rogers, D. H. *J. Am. Chem. Soc.* **1995**, *117*, 7281.

97. Doyle, M. P.; Peterson, C. S.; Protopopova, M. N.; Marnett, A. B.; Parker, D. L., Jr.; Ene, D. G.; Lynch, V. *J. Am. Chem. Soc.* **1997**, *119*, 8826.

98. Doyle, M. P.; Peterson, C. S.; Parker, D. L., Jr. *Angew. Chem., Int. Ed.* **1996**, *35*, 1334.

99. (a) Baird, M. S. *Top. Curr. Chem.* **1988**, *144*, 137. (b) Salaün, J. *Chem. Rev.* **1989**, *89*, 1247. (c) Binger, P.; Buch, H. M. *Top. Curr. Chem.* **1986**, *135*, 77.

100. Protopopova, M. N.; Shapiro, E. A. *Russ. Chem. Rev.* **1989**, *58*, 667.

101. Petiniot, N.; Anciaux, A. J.; Noels, A. F.; Hubert, A. J.; Teyssie, P. *Tetrahedron Lett.* **1978**, 1239.

102. Shapiro, E. A.; Kalinin, A. V.; Ugrak, B. I.; Nefedov, O. M. *J. Chem. Soc., Perkin Trans. 2* **1994**, 709.

103. (a) Maier, G.; Hoppe, M.; Reisenauer, H. P.; Kruger, C. *Angew. Chem., Int. Ed.* **1982**, *21*, 437. (b) Wenkert, E.; Alonso, M. E.; Buckwalter, B. L.; Sanchez, E. L. *J. Am. Chem. Soc.* **1983**, *105*, 2021.

104. Müller, P.; Gränicher, C. *Helv. Chim. Acta* **1995**, *78*, 129.

105. Müller, P.; Pautex, N.; Doyle, M. P.; Bagheri, V. *Helv. Chim. Acta* **1990**, *73*, 1233.

106. Padwa, A.; Weingarten, M. D. *Chem. Rev.* **1996**, *96*, 223.

107. Hoye, T. R.; Vyvyan, J. R. *J. Org. Chem.* **1995**, *60*, 4184.

108. Doyle, M. P.; Protopopova, M. N.; Müller, P.; Ene, D.; Shapiro, E. A. *J. Am. Chem. Soc.* **1994**, *116*, 8492.

109. Doyle, M. P.; Protopopova, M. N.; Brandes, B. D.; Davies, H. M. L.; Huby, N. J. S.; Whitesell, J. K. *Synlett* **1993**, 151.

110. Imogai, H.; Bernardinelli, G.; Granicher, C.; Moran, M.; Rossier, J.-C.; Müller, P. *Helv. Chim. Acta* **1998**, *81*, 1754.

111. Doyle, M. P.; Ene, D. G.; Peterson, C. S.; Lynch, V. *Angew. Chem., Int. Ed.*, in press.

112. Doyle, M. P. In *Comprehensive Organometallic Chemistry II*; Hegedus, L. S. (Ed.); Pergamon Press: New York, 1995; Vol. 12, Ch. 5.2.

113. (a) Taber, D. F. In *Comprehensive Organic Synthesis: Selectivity, Strategy, and Efficiency in Modern Organic Chemistry*; Trost, B. M.; Fleming, I. (Ed.); Pergamon Press: New York, 1991; Vol. 3, Ch. 4.2. (b) Taber, D. F. In *Stereoselective Synthesis*; Helmchen, G.; Hoffmann, R. W.; Mulzer, J.; Schaumann, E. (Ed.); Verlag: New York, Vol. E21a of *Methods of Organic Chemistry*; 1995; Ch. 1.2.

114. Adams, J.; Spero, D. M. *Tetrahedron* **1991**, *47*, 1765.

115. Doyle, M. P.; Westrum, L. J.; Wolthuis, W. N. E.; See, M. M.; Boone, W. P.; Bagheri, V.; Pearson, M. M. *J. Am. Chem. Soc.* **1993**, *115*, 958.

116. Hashimoto, S.; Watanabe, N.; Sato, T.; Shiro, M.; Ikegami, S. *Tetrahedron Lett.* **1993**, *34*, 5109.

117. Ye, T.; Garcia, C. F.; McKervey, M. A. *J. Chem. Soc., Perkin Trans. 1* **1995**, 1373.

118. Anada, M.; Watanabe, N.; Hashimoto, S. *J. Chem. Soc., Chem. Commun.* **1998**, 1517.

119. Ponsford, R. J.; Southgate, R. *J. Chem. Soc., Chem. Commun.* **1979**, 846.

120. Watanabe, N.; Anada, M.; Hashimoto, S.; Ikegami, S. *Synlett* **1994**, 1031.

121. Anada, M.; Hashimoto, S. *Tetrahedron Lett.* **1998**, *39*, 79.

122. Doyle, M. P.; Dyatkin, A. B.; Roos, G. H. P.; Cañas, F.; Pierson, D. A.; van Basten, A.; Müller, P.; Polleux, P. *J. Am. Chem. Soc.* **1994**, *116*, 4507.

123. Doyle, M. P.; Dyatkin, A. B.; Tedrow, J. S. *Tetrahedron Lett.* **1994**, *35*, 3853.

124. Doyle, M. P.; Kalinin, A. V.; Ene, D. G. *J. Am. Chem. Soc.* **1996**, *118*, 8837.

125. Doyle, M. P.; Kalinin, A. V. *Russ. Chem. Bull.* **1995**, *44*, 1729.

126. Bode, J. W.; Doyle, M. P.; Protopopova, M. N.; Zhou, Q.-L. *J. Org. Chem.* **1996**, *61*, 9146.

127. Doyle, M. P.; van Oeveren, A.; Westrum, L. J.; Protopopova, M. N.; Clayton, T. W., Jr. *J. Am. Chem. Soc.* **1991**, *113*, 8982.

128. Doyle, M. P.; Kalinin, A. V. *Synlett* **1995**, 1075.

129. Doyle, M. P.; Protopopova, M. N.; Winchester, W. R.; Daniel, K. L. *Tetrahedron Lett.* **1992**, *33*, 7819.

130. Doyle, M. P.; Kalinin, A. V. *Tetrahedron Lett.* **1996**, *37*, 1371.

131. Doyle, M. P.; Zhou, Q.-L.; Raab, C. E.; Roos, G. H. P. *Tetrahedron Lett.* **1995**, *36*, 4745.

132. Doyle, M. P.; Dyatkin, A. B. *J. Org. Chem.* **1995**, *60*, 3035.

6
ASYMMETRIC OXIDATIONS AND RELATED REACTIONS

6A

CATALYTIC ASYMMETRIC EPOXIDATION OF ALLYLIC ALCOHOLS

ROY A. JOHNSON
The Upjohn Company, Kalamazoo, Michigan

K. BARRY SHARPLESS
The Scripps Research Institute, La Jolla, California

6A.1. INTRODUCTION

Efforts to achieve asymmetric induction in the epoxidation of olefins commenced in 1965 with a report by Henbest that a low level of enantioselectivity (8%) was achieved in epoxidations by using percamphoric acid [1]. A useful level of asymmetric induction remained an elusive goal for 15 years until Katsuki and Sharpless reported that the combination of a Ti (IV) alkoxide, an optically active tartrate ester, and *t*-butyl hydroperoxide was capable of epoxidizing a wide variety of allylic alcohols in good yield and with an ee usually greater than 90% [2]. The Ti-tartrate complex was shown to perform as a true catalyst; the most reactive allylic alcohols were completely oxidized when only 5–10 mol % of the complex was used for the reaction and the level of enantioselectivity achieved under these catalytic conditions was within 1 or 2% of that obtained when a stoichiometric amount of the complex was used. However, with many slower-reacting allylic alcohols, a complete reaction was difficult to achieve under catalytic conditions and, as a consequence, a stoichiometric quantity of the complex was used for most asymmetric epoxidations before 1986. In 1986, the addition of activated molecular sieves to the asymmetric epoxidation process was found to have an extremely beneficial effect on the reaction, such that nearly all epoxidations are completed efficiently with only 5–10 mol % of the catalyst [3,4]. In addition to an economy of reagents, the advantages of the catalytic reaction include mild conditions, easier isolations, increased yields especially of very reactive epoxy alcohol products, and application of in situ derivatization of the product. These aspects of the reaction are all discussed further in this chapter.

Catalytic Asymmetric Synthesis, Second Edition, Edited by Iwao Ojima
ISBN 0-471-29805-0   Copyright © 2000 Wiley-VCH, Inc.

Success in the use of Ti tartrate catalyzed asymmetric epoxidation depends on the presence of the hydroxyl group of the allylic alcohol. The hydroxyl group enhances the rate of the reaction, thereby providing selective epoxidation of the allylic olefin in the presence of other olefins; it also is essential for the achievement of asymmetric induction. The role played by the hydroxyl group in this reaction is described in a later section of this chapter. The need for a hydroxyl group necessarily limits the scope of this asymmetric epoxidation to a fraction of all olefins. Fortunately, allylic alcohols are easily introduced into synthetic intermediates and are very versatile in organic synthesis. The Ti tartrate catalyzed asymmetric epoxidation of allylic alcohols has been applied extensively as documented in the literature and in this review. The development of methods aimed at catalytic asymmetric epoxidation of unfunctionalized olefins is described in Chapter 6B, whereas the catalytic asymmetric dihydroxylation of olefins, which provides an alternate method for olefin functionalization, is described in Chapter 6D.

The literature was reviewed through early 1990 for the purposes of preparing this chapter. The following sections of this review are reprinted with only very minor changes from an earlier version [5]. Other reviews have covered various aspects of asymmetric epoxidation, including an extensive discussion of the mechanism of the reaction [6], synthetic applications through 1984 [7], a thorough compilation of uses through 1987 [8], and an excellent review of synthetic applications utilizing nonracemic glycidol and related 2,3-epoxy alcohols [9]. Use of enantiomerically pure epoxy alcohols in the synthesis of sugars and other polyhydroxylated compounds [10] and for the preparation of various synthetic intermediates has been reviewed [11]. A personal account of the discovery of Ti-catalyzed asymmetric epoxidation has been recorded [12]. A comprehensive review of Ti-catalyzed asymmetric epoxidation has appeared [13].

6A.2. FUNDAMENTAL ELEMENTS OF Ti-TARTRATE CATALYZED ASYMMETRIC EPOXIDATION

The essence of Ti-catalyzed asymmetric epoxidation is illustrated in Figure 6A.1. As Figure 6A.1 shows, the four essential components of the reaction are the allylic alcohol substrate, a Ti(IV) alkoxide, a chiral tartrate ester, and an alkyl hydroperoxide. The asymmetric complex formed from these reagents delivers the peroxy oxygen to one face or the other of the allylic alcohol depending on the absolute configuration of the tartrate used. If D-(−)-tartrate is used, oxygen delivery will be from the top face of the allylic alcohol when drawn in the orientation shown in Figure 6A.1, and if L-(+)-tartrate is used, oxygen delivery will be from the bottom face. The enantioselectivity of this reaction approaches 100% as measured by the enantiomeric purities of the epoxy alcohol products. An enantiomeric excess (ee) of 94%, a degree of enantiomeric purity attained in many of the epoxide products, reflects an enantioselectivity of 97:3 for epoxidation of one face of the allylic alcohol over the other.

The enantioselectivity principles portrayed in Figure 6A.1 have been followed without exception in all epoxidations of prochiral allylic alcohols reported to date, and one may use these principles to assign absolute configurations to the epoxy alcohols prepared by the method. On the other hand, epoxidation of allylic alcohols with chiral substituents at C-1, C-2, and/or C-3 does not always follow these principles, and assignment of absolute configuration to the products must be made with care. Even in the latter cases, reliable assignments usually can be made if the outcome (diastereomeric ratio) of epoxidation with both the (+) and (−)-tartrate ester ligands is compared.

A structural variant of the allylic alcohol not shown in Figure 6A.1 is encountered when a substituent is placed on C-1 as illustrated in Figure 6A.2. Such an allylic alcohol is a racemate

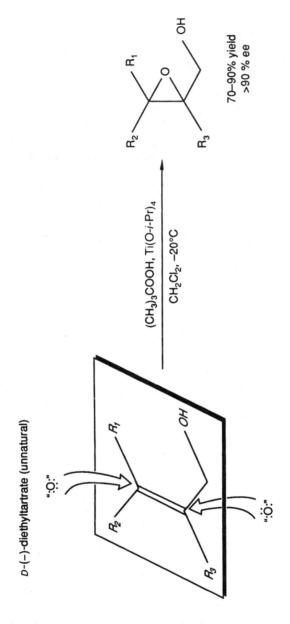

Figure 6A.1. Enantiofacial selectivity in the epoxidation of prochiral allylic alcohols with titanium/tartrate/TBHP.

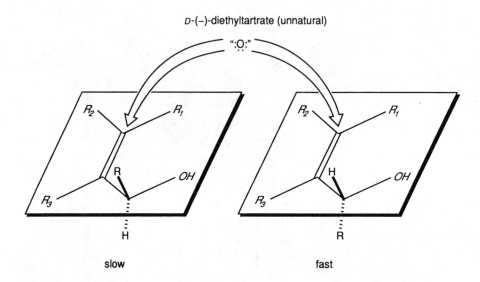

Figure 6A.2. Diastereofacial selectivity in the epoxidation of 1-substituted allylic alcohols with titanium/tartrate/TBHP.

(unless it has been resolved) in which one enantiomer will have the R group oriented in the direction of oxygen delivery whereas the other enantiomer will have the R group oriented away from the direction of oxygen delivery. The enantioselective principles of asymmetric epoxidation remain in force for epoxidation of this type of substrate, but now oxygen is delivered at different rates to the two enantiomers depending on the orientation of the R group.

Experimental results have shown that the difference in these rates is of sufficient magnitude that one enantiomer of the allylic alcohol will remain largely unoxidized whereas the other undergoes complete epoxidation, the net result is a kinetic resolution of the enantiomers [14]. Experience has further shown that the slow-reacting enantiomer will always be the one with the R group oriented in the direction of "oxygen" delivery. For the example illustrated in Figure 6A.2, the titanium/D-(–)-diethyl tartrate complex will deliver oxygen to the top face in preference to the bottom face of the substrate in accordance with the rules implied in Figure 6A.1, and this delivery will be more rapid when the R group is oriented toward the bottom face of the molecule. Opposite results will be obtained with the titanium/L-(+)-diethyl tartrate complex. Additional details for using this reaction in the kinetic resolution mode may be found in a later section.

An important aspect of asymmetric epoxidation not shown in Figures 6A.1 and 6A.2 is the fact that the allylic alcohol is coordinated to Ti as the alkoxide during the epoxidation process (see Section 6C.1.). Not only does this coordination play a key role in orientation of allylic alcohol during the epoxidation process, but it also accounts for the selectivity of the process for allylic and homoallylic alcohols in preference to nearly all other olefins. This effect is most clearly seen in comparison of allylic alcohols with the analogous allylic ethers. The latter are essentially unchanged by the $Ti(OR)_4$/tartrate/TBHP system during the same time required for epoxidation of the allylic alcohol. The $Ti(OR)_4$/tartrate/TBHP reagent thereby exhibits selectivity for allylic and homoallylic alcohols while being compatible with other olefinic groups. Use of the reagent is compatible with many other functional groups as well (see Table 6A.1).

TABLE 6A.1. Compatibility of Functional Groups with the Asymmetric Epoxidation Reaction

Compatible Functional Groups		Incompatible Groups
Acetals, ketals	Nitriles	Amines (most)
Acetylenes	Nitro	Carboxylic Acids
Alcohols (remote)	Olefins	Mercaptans
Aldehydes	Pyridines	Phenols (most)
Amides	Silyl ethers	Phosphines
Azides	Sulfones	
Carboxylic Esters	Sulfoxides	
Epoxides	Tetrazoles	
Ethers	Ureas	
Hydrazides	Urethans	
Ketones		

An important improvement in the asymmetric epoxidation process is the finding, reported in 1986, that by adding molecular sieves to the reaction medium virtually all reactions can be performed with a catalytic amount (5–10 mol %) of the Ti-tartrate complex [3]. Previously, only a few structural classes of allylic alcohols were efficiently epoxidized by less than stoichiometric amounts of the complex, and most reactions were routinely performed with stoichiometric quantities of the reagent.

In situ derivatization of the crude epoxy alcohol product becomes a viable alternative to isolation when 5–10 mol % of catalyst is used for the epoxidation. This procedure is especially useful when the product is reactive or is difficult to isolate because of solubility in an aqueous extraction phase [15,16]. Low-molecular-weight epoxy alcohols such as glycidol are readily extracted from the reaction mixture after conversion to ester derivatives such as the *p*-nitrobenzoate or 3-nitrobenzenesulfonate [4,17]. This derivatization not only facilitates isolation of the product but also preserves the epoxide in a synthetically useful form.

6A.3. REACTION VARIABLES FOR Ti-TARTRATE CATALYZED ASYMMETRIC EPOXIDATION

This section presents a summary of the currently preferred conditions for performing Ti-catalyzed asymmetric epoxidations and is derived primarily from the detailed account of Gao et al. [4]. We wish to draw the reader's attention to several aspects of the terminology used here and throughout this chapter. The terms Ti-tartrate *complex* and Ti-tartrate *catalyst* are used interchangeably. The term *stoichiometric reaction* refers to the use of the Ti-tartrate complex in a stoichiometric ratio (100 mol %) relative to the substrate (allylic alcohol). The term *catalytic reaction* (or *quantity*) refers to the use of the Ti-tartrate complex in a catalytic ratio (usually 5–10 mol %) relative to the substrate.

6A.3.1. Stoichiometry

Two aspects of stoichiometry are important in asymmetric epoxidation: one is the ratio of Ti to tartrate used for the catalyst and the other is the ratio of catalyst to substrate. With regard to the catalyst, it is crucial in obtaining the highest possible enantioselectivity that at least a 10% excess of tartrate ester to Ti(IV) alkoxide be used in all asymmetric epoxidations. This condition is important whether the reaction is being done with a stoichiometric or a catalytic quantity of the complex. There appears to be no need to increase the excess of tartrate ester beyond 10–20% and, in fact, a larger excess has been shown to slow the epoxidation reaction unnecessarily [4].

The second stoichiometry consideration is the ratio of catalyst to substrate. As noted in the preceding section, virtually all asymmetric epoxidations can be performed with a catalytic amount of Ti-tartrate complex if molecular sieves are added to the reaction milieu. A study of catalyst/substrate ratios in the epoxidation of cinnamyl alcohol revealed a significant loss in enantioselectivity (Table 6A.2) below the level of 5 mol % catalyst. At this catalyst level, the reaction rate also decreases, with the consequence that incomplete epoxidation of the substrate may occur. Presently, the recommended catalyst stoichiometry is from 5% Ti and 6% tartrate ester to 10% Ti and 12% tartrate ester [4].

6A.3.2. Concentration

The concentration of substrate used in the asymmetric epoxidation must be given consideration because competing side reactions may increase with increased reagent concentration. The use of catalytic quantities of the Ti-tartrate complex has greatly reduced this problem. The epoxidation of most substrates under catalytic conditions may be performed at a substrate concentration up to 1 M. By contrast, epoxidations using stoichiometric amounts of complex are best run at substrate concentrations of 0.1 M or lower. Even with catalytic amounts of the complex, a concentration of 0.1 M may be maximal for substrates such as cinnamyl alcohol, which produce sensitive epoxy alcohol products [4].

6A.3.3. Preparation and Aging of the Catalyst

Proper preparation of the catalyst is essential for optimal reaction rates and enantioselectivity. The preparation and storage of stock solutions of the Ti-tartrate catalyst should not be attempted because the complex is not sufficiently stable for long-term storage. Best results are obtained when the catalyst is prepared by mixing the Ti(IV) alkoxide and the tartrate in solvent at –20°C, adding either *t*-butylhydroperoxide (TBHP) *or* the allylic alcohol, and aging the system at this temperature for 20–30 min. This aging period is critical to the success of the reaction and must not be eliminated. On the rare occasion that a bulky Ti(IV) alkoxide such as the *t*-butoxide is used, the aging period should be increased to 1 hr [18]. After the aging period, the temperature

TABLE 6A.2. Dependence of Enantioselectivity on Catalyst Stoichiometry

Entry	Ti(O-*i*-Pr)$_4$, mole %	(+)-DIPT mole %	% ee
1	5.0	6.0	92
2	4.0	5.2	87
3	2.0	2.5	69

is adjusted to the desired level and the last reagent, either the allyl alcohol or the hydroperoxide, is added.

6A.3.4. Oxidant and Epoxidation Solvent

TBHP is used as the oxidant for nearly all Ti-catalyzed asymmetric epoxidations. Exceptions are for allyl alcohol and methallyl alcohol, where cumylhydroperoxide is used to advantage for the epoxidation [4]. Cumylhydroperoxide can be used for other epoxidations and is reported to result in slightly faster reaction rates than are observed with TBHP [4]. Trityl hydroperoxide can also serve as an effective replacement for TBHP [6]. TBHP is generally preferred, however, because product isolation is significantly easier when this oxidant is used. The most economical source of TBHP is the commercially available 70% solution in water, in which case steps must be taken to obtain anhydrous material. Detailed instructions for obtaining dry solutions of TBHP have been published elsewhere [3,4]. For smaller laboratory-scale reactions, anhydrous solutions of TBHP in 2,2,4-trimethylpentane (isooctane) are available commercially. Storage of TBHP solutions over molecular sieves is not recommended, but brief drying over sieves (ca. 30 min) of the required amount of the solution just before use is a good practice.

Because the preparation and storage of stock quantities of TBHP is a convenient way to deal with this reagent, compatibility with the solvent is essential. Much care has gone into finding the optimum solvent for storage of TBHP, and recommendations have changed as additional experience has been gained. The current solvent of choice is isooctane, with the favored alternates being dichloromethane or toluene [4]. Dichloroethane should not be used [19]. Dichloromethane solutions of TBHP require storage at 0°C, and toluene solutions occasionally develop a contaminant that inhibits the catalytic reaction. Safety considerations (chance of slight pressurization) make high-density polyethylene bottles preferable to glass bottles for storage of TBHP solutions. However, both dichloromethane and toluene, but not isooctane, permeate through such bottles, with the result that the concentration of the contents slowly changes with time. If the published instructions [3,4] for preparation of anhydrous TBHP in isooctane are followed, a relatively concentrated solution (5–6 M) is obtained. Aliquots of this solution are briefly dried over sieves and added directly to the epoxidation reaction without concern for removal of the isooctane. The use of dilute solutions of TBHP in isooctane should be avoided, because the additional isooctane involved in transfer will have an inhibitory effect on the rate of epoxidation and can lead to solubility problems with some substrates. Solutions of 5.5 M TBHP in isooctane are available commercially and should *always* be used instead of the 3.0 M solution.

For the asymmetric epoxidation reaction, dry alcohol-free dichloromethane (the use of dichloromethane stabilized with methanol must be avoided) is usually the solvent of choice: It is inert to the reagents, has good solvent power for the components of the reaction, and supports good epoxidation rates. A fortunate consequence of the asymmetric epoxidation process is that ligation of the allylic alcohol to the Ti center aids in solubilization of the substrate. Substrates that normally may be only modestly soluble in the above-mentioned solvents will be brought into solution as they complex with the Ti-tartrate catalyst.

6A.3.5. Tartrate Esters

Optically active tartrate esters are the source of chirality for the asymmetric epoxidation process. With a few subtle exceptions, the esters used conventionally—dimethyl (DMT), diethyl (DET), and diisopropyl tartrate (DIPT)—are equally effective at inducing asymmetry during the crucial epoxidation event. The minor exceptions that have been noted include (a) a slight improvement in enantioselectivity (from 93 to 95% ee) when changing from DIPT to DET in the epoxidation of E-substituted allylic alcohols such as (E)-2-hexen-l-ol (having only a primary alkyl chain at C-3) and (b) a higher product yield (but no change in enantioselectivity) when changing from DET to DIPT in the epoxidation of allyl alcohol [4]. Other subtle variations such as these may exist, but their discovery awaits execution of the appropriate comparative experiments. If optimal conditions are desired for a specific asymmetric epoxidation, variation of the tartrate ester is likely to be a useful exercise.

In the kinetic resolution of chiral 1-substituted allylic alcohols, there clearly is benefit to be gained in the choice of tartrate ester used for the reaction. In these reactions, the efficiency of kinetic resolution increases as the size of the tartrate alkyl ester group increases. Data for DMT, DET, and DIPT are summarized below (see Table 6A.8 [6]), and the trend shown there continues with the use of the crystalline dicyclohexyl and dicyclododecyl tartrates [4].

The nonconventional tartrate esters **1–3** have been used to probe the mechanism of the asymmetric epoxidation process [20a]. These chain-linked bistartrates when complexed with 2 equiv. of Ti(O-t-Bu)$_4$ catalyze asymmetric epoxidation with good enantiofacial selectivity.

1, n = 3
2, n = 4
3, n = 5

A number of tartrate-like ligands have been studied as potential chiral auxiliaries in the asymmetric epoxidation and kinetic resolution processes [6,20b]. Although on occasion a ligand has been found that has the capability to induce high enantioselectivity into selected substrates (see Section, 6A.7.3.) none has exhibited the broad scope of effectiveness seen with the tartrate esters.

Polymer-linked tartrate esters have been prepared and used for asymmetric epoxidation in efforts to simplify reaction workup procedures and to allow recycling of the chiral tartrate [21]. The tartrates were linked through an ester bond to either a hydroxymethyl or a hydroxyethyl group on the polymer backbone to form **4** and **5**, respectively.

4, n = 1
5, n = 2

Epoxidation catalysts were prepared from these polymer-linked tartrates by combination with 0.5 equiv. of Ti(O-i-Pr)$_4$, based on the weight of tartrate ester, which had been added to the polymer. Epoxidation of geraniol with **4** or **5** gave epoxy alcohol with 49 and 65% ee, respectively. Recycling of the polymer-linked tartrate was possible, but the subsequent epoxidation suffered from significant loss in enantioselectivity [21].

6A.3.6. Ti Alkoxides

Ti(IV) isopropoxide [*Chemical Abstracts* nomenclature: 2-propanol, Ti(4+) salt], is the Ti species of choice for preparation of the Ti-tartrate complex in the asymmetric epoxidation process. The use of Ti(IV) *t*-butoxide has been recommended for reactions in which the epoxy alcohol product is particularly sensitive to ring opening by the alkoxide [18]. The 2-substituted epoxy alcohols are one such class of compounds. Ring opening by *t*-butoxide is much slower than by *i*-propoxide. With the reduced amount of catalyst that now is sufficient for all asymmetric epoxidations, the use of Ti(O-*t*-Bu)$_4$ appears to be unnecessary in most cases, but the concept is worth noting.

6A.3.7. Molecular Sieves

The addition of activated molecular sieves (zeolites) to the asymmetric epoxidation milieu has the beneficial effect of permitting virtually all reactions to be carried out with only 5–10 mol % of the Ti-tartrate catalyst [3,4]. Without molecular sieves, only a few of the more reactive allylic alcohols are epoxidized efficiently with less than an equivalent of the catalyst. The role of the molecular sieves is thought to be protection of the catalyst from (a) adventitious water and (b) water that may be generated in small amounts by side reactions during the epoxidation process.

There are several important guidelines to be followed in using activated molecular sieves for the asymmetric epoxidation process [4]. Stock solutions of TBHP should not be stored over molecular sieves (the sieves catalyze the slow decomposition of TBHP), but the amount of TBHP solution required for a reaction should be placed over sieves briefly (10–60 min) before use. Likewise, neither the tartrate ester nor the Ti(IV) isopropoxide should be stored over sieves. Addition of the sieves at the time of mixing the tartrate ester with the Ti(O-*i*-Pr)$_4$ followed by the normal "aging" of the catalyst is sufficient to dry these reagents, provided they initially are of good quality (see Ref. 4, p. 276). Powdered, activated 4A molecular sieves, commercially available in preactivated form, are preferred; 3A, 4A, and 5A molecular sieves in pellet form are also effective. Only 3A sieves are effective in the case of allyl alcohol, because this substrate is small enough to be sequestered by 4A or 5A sieves. Unactivated sieves can be activated by heating at 200°C under high vacuum for at least 3 hr.

6A.4. SOURCES OF ALLYLIC ALCOHOLS

One of the amenities of present-day organic synthesis is the availability of intermediates from the many chemical supply companies. More than 100 allylic alcohols (excluding extensive listings of phorbol esters and prostaglandin structures) are offered for sale from these sources. Two concerns about such supplies should be noted. The first is the *E/Z* composition of acyclic allylic alcohols, which should be checked when it is not specified, and second is the optical purity of allylic alcohols offered in optically active form, which likewise should be checked.

When the allylic alcohol needed for asymmetric epoxidation is unavailable from a commercial source, reasonably general synthetic routes have been developed to allylic alcohols of several different substitution patterns. Good methods are available for the preparation of 3-substituted allylic alcohols, whereas synthesis of 2-substituted allylic alcohols is more problematic. The substrates for kinetic resolution, 1-substituted allylic alcohols, frequently can be derived by addition of alkenyl or alkynyl organometallic reagents to aldehydes followed by modification of the resulting product as required.

The Homer-Emmons addition of dialkyl carboalkoxymethylenephosphonates to aldehydes [22] has been widely used to generate α,β-unsaturated esters which, in turn, can be reduced to allylic alcohols. Under the original conditions of the Homer-Emmons reaction, the stereochemistry of the α,β-unsaturated ester is predominantly trans and therefore the *trans* allylic alcohol is obtained upon reduction. Still and Gennari have introduced an important modification of the Homer-Emmons reaction, which shifts the stereochemistry of the α,β-unsaturated ester to predominantly cis [23]. Diisobutylaluminum hydride (DIBAL) has frequently been used for reduction of the alkoxycarbonyl to the primary alcohol functionality. The aldehyde needed for reaction with the Homer-Emmons reagent may be derived via Swern oxidation [24] of a primary alcohol. The net result is that one frequently sees the reaction sequence shown in Eq. 6A.1 used for the net preparation of $3E$ and $3Z$ allylic alcohols.

$$(6A.1)$$

The propargylic alcohol group may be exploited as an allylic alcohol precursor (Eq. 6A.2) and may be generated by nucleophilic addition to an electrophile [25] or by addition of a formaldehyde equivalent to a preexisting terminal acetylene group [26]. Once in place, reduction of the propargylic alcohol with lithium aluminum hydride or, preferably, with sodium bis(2-methoxyethoxy)aluminum hydride (Red-Al) [27] will produce the *trans* allylic alcohol. Alternately, catalytic reduction over Lindlar catalyst can be used to obtain the *cis* allylic alcohol [28]. The addition of other lithium acetylides to ketones produces chiral secondary alcohols, which also can be reduced by the preceding methods to the *cis* or *trans* allylic alcohols. Additional synthetic approaches to allylic alcohols may be found in the various references cited in this chapter.

$$(6A.2)$$

6A.5. ASYMMETRIC EPOXIDATION BY SUBSTRATE STRUCTURE

The scope of allylic alcohol structures that are subject to asymmetric epoxidation was foreshadowed in the first report of this reaction. Examples of nearly all the possible substitution patterns were shown to be epoxidized in good yield and with high enantiofacial selectivity [2]. The numerous results that have appeared since the initial report have confirmed and extended the scope of the structures that have been epoxidized. This section of the chapter illustrates the structural scope without being exhaustive in coverage of the literature. Examples were chosen as much as possible from the reports in the literature that provide experimentally determined yield and % ee data. When there are limitations to the structural scope as reflected by lower enantiofacial selectivity, these cases are noted. The results presented in this section are divided according to the substitution patterns of the allylic alcohol substrates. This organization is intended to provide easy access to precedent when the synthetic chemist is contemplating asymmetric epoxidation of a new substrate.

Before commencing, the attention of the reader is drawn to the terms "enantiofacial selectivity" and "diastereoselectivity." The usage in this chapter does not conform to the strictest possible definitions of these terms. In particular, *enantiofacial selectivity* is used with reference to the selection and delivery of oxygen by the epoxidation catalyst to one face of the olefin in preference to the other. This usage extends to chiral allylic alcohols (primarily the 1-substituted allylic alcohols) when the focus of the discussion is on face selection in the epoxidation process. *Diastereoselectivity* is used in the discussion of kinetic resolution when the generation of diastereomeric compounds is emphasized.

6A.5.1. Allyl Alcohol

Glyceraldehyde derivatives [29], asymmetrically substituted glycerol [30], and glycidol [31] are three-carbon molecules which, especially in their optically active forms, find widespread use in organic synthesis. In the past, these compounds in optically active form came almost exclusively from the degradation of natural products such as mannitol. Efficient, multistep routes from the natural products provide access to either enantiomer of these three-carbon compounds. Since the discovery of asymmetric epoxidation in 1980, the potential has existed for a convenient one-step synthesis of optically active glycidol (**7**) from allyl alcohol (**6**) [2]. However, because glycidol is one of the epoxy alcohols more sensitive to ring-opening reactions and also is a water-soluble molecule, isolation from the stoichiometric asymmetric epoxidation is difficult, and very little glycidol has been prepared in this way. Now with the use of catalytic epoxidation in the presence of molecular sieves, it is possible to isolate optically active glycidol with 88–92% ee in yields of 50–60% [4]. As a result of these improvements, both enantiomers of glycidol are available commercially.

An attractive alternative to isolation of glycidol is in situ derivatization of the crude product during workup [15]. Two distinct applications of this method have been described. In the first, ring opening of glycidol (*R*)-**7** with a nucleophile such as sodium 1-naphthoxide produces an intermediate (**8**) that can be carried on to useful products—for example, for the synthesis of β-adrenergic blocking agents [15a] and antidepressants [32]. In the second, esterification of the hydroxyl group of glycidol improves the extraction of the glycidol moiety from the reaction mixture and at the same time generates a synthon in which all three carbon centers are differentiated for further reaction. Another benefit is that with certain derivatives, such as the 3-nitrobenzenesulfonate ester (9), recrystallization can be used to upgrade the enantiomeric purity (to >99% ee) [4,16a].

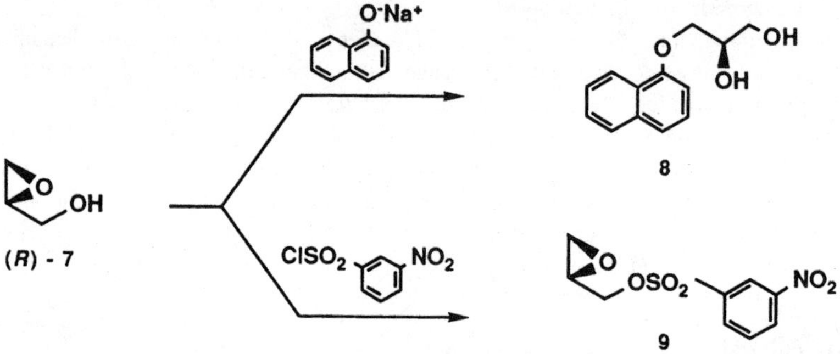

As an industrial process, production of optically active glycidol is at an early stage of development, with additional improvements and economies certain to occur. As a chemical intermediate, optically active glycidol is the most versatile epoxy alcohol prepared by asymmetric epoxidation and is poised for exploitation in organic synthesis [9,16].

6A.5.2. 2-Substituted Allyl Alcohols

The epoxides (11) derived from 2-substituted allylic alcohols (10) are particularly susceptible to nucleophilic attack at C-3, a reaction that is promoted by Ti(IV) species [18]. When stoichiometric amounts of Ti-tartrate complex are used in these epoxidations, considerable product is lost via opening of the epoxide before it can be isolated from the reaction.

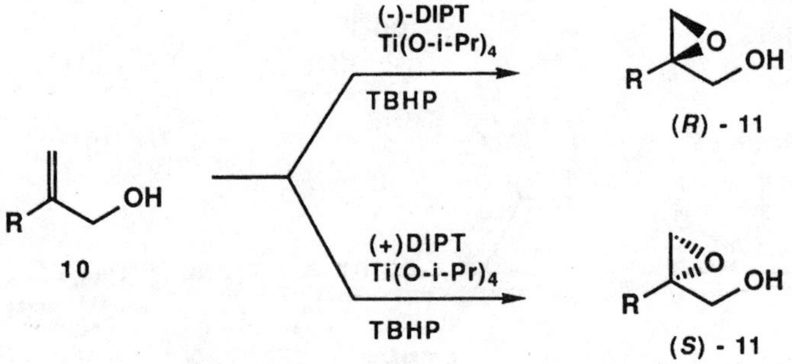

The primary nucleophilic culprit is the isopropoxide ligand of the Ti(O-i-Pr)$_4$. The use of Ti(O-t-Bu)$_4$ in place of Ti(O-i-Pr)$_4$ has been prescribed as a means to reduce this problem (the t-butoxide being a poorer nucleophile) [18]. Fortunately a better solution now exists in the form of the catalytic version of the reaction that uses only 5–10 mol % of Ti-tartrate complex and greatly reduces the amount of epoxide ring opening. Some comparisons of results from reactions run under the two sets of conditions are possible from the epoxidations summarized in Table 6A.3 [2,4,18,33–38].

TABLE 6A.3. Epoxides from 2-Substituted Allylic Alcohols

	Epoxide		Catalyst				
Entry	R_1	R_2	%Ti/%Tart.	Tartrate	Yield(%)	% ee	Ref.
1	H	Me	100/100[a]	(+)-DET	—	85	33
2	H	Me	27/27	(−)-DET	32	94	34
3	H	Me	7.6/10[a]	(−)-DET	47	>95	35
4	PNB	Me	5/6	(+)-DIPT	78	92(98)[b]	4
5	Tos	Me	5/6	(+)-DIPT	69	95	4
6	Nps[c]	Me	5/6	(+)-DIPT	60	(92)[b]	4
7	H	n-Pr	4.7/5.9	(+)-DET	88	95	4
8	H	N-Nonyl	100/110	(+)-DET	53	>96	36
9	H	n-Tetradecyl	100/110[a]	(+)-DET	51	95	18
10	H	n-Tetradecyl	10/13	(+)-DET	91	96	4
11	H	i-Pr	65/120	(+)-DET	56	86	37
12	H	t-Bu	120/150[a]	(+)-DET	42	86	38
13	H	Cyclohexyl	100/100	(+)-DET	81	>95	2
14	H	−CH$_2$OBn	7.6/10[a]	(−)-DET	74	>95	35

[a]Ti(O-t-Bu)$_4$ used in this reaction. [b]Value in parenthesis is after recrystallization. [c]Nps = 2-Naphthalene-sulfonyl.

The prototype for this structural class is 2-methyl-2-propen-1-ol (methallyl alcohol), from which asymmetric epoxidation generates optically active 2-methyloxiranemethanol. Like glycidol, 2-methyloxiranemethanol has been difficult to obtain by stoichiometric asymmetric epoxidation, but with the use of the catalytic version, reasonable quantities, are now produced [4] and the compound has become commercially available. In situ derivatization also can be used to recover this epoxy alcohol from the epoxidation reaction. Progress in the isolation of 2-methyloxiranemethanol is reflected in entries 1–3 of Table 6A.3, and the results of in situ derivatization are revealed by entries 4–6. The enantiomeric purity of 2-methyloxiranemethanol produced in this way is very good (92–95% ee), and improvement to 98% ee is observed after recrystallization of the 4-nitrobenzoate derivative.

Several other allylic alcohols with primary C-2 substituents have been epoxidized with very good results (entries 7–10, 14). Epoxy alcohols have been obtained with 95–96% ee and, when the catalytic version of the reaction is used, as in entry 10, the yield is excellent. When the C-2 substituent is more highly branched, as in entries 11–13, there may be some interference with high enantiofacial selectivity by the bulky group, because the enantioselectivity in two cases (entries 11 and 12) is 86%. Another example that supports this possibility of steric interference to selective epoxidation is summarized in Eq. 6A.3a [39]. In this case the optically active allylic alcohol **12**, (3R)-3,7-dimethyl-2-methylene-6-penten-1-ol, was subjected to epoxidation with both antipodes of the Ti-tartrate catalyst. With (+)-DIPT, enantiofacial selectivity was 96:4

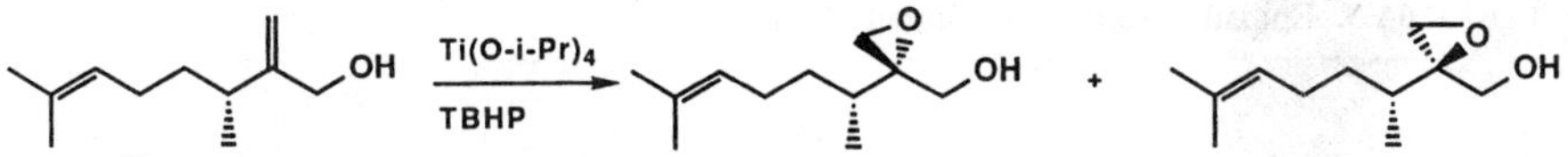

(6A.3a)

(+)-DIPT   96 : 4

(−)-DIPT   1 : 3

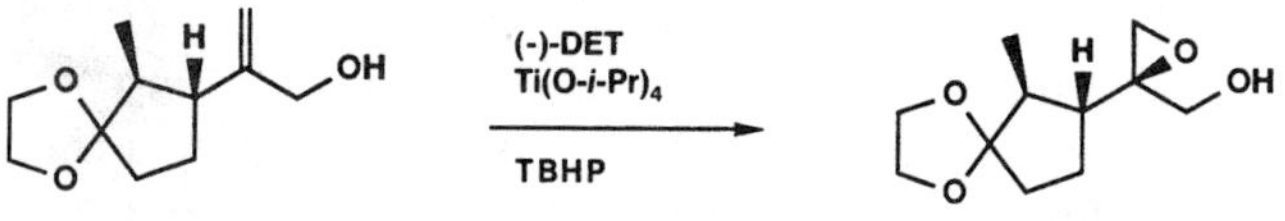

(6A.3b)

("matched pair" [40a]), but with (−)-DIPT selectivity fell to only 1:3 ("mismatched pair"), a further indication that a secondary C_2 substituent can perturb the fit of the substrate to the active catalyst species.

In the epoxidation of the allylic alcohol shown in Eq. 6A.3b, the epoxy alcohol is obtained in 96% yield and with a 14:1 ratio of enantiofacial selectivity [40b]. An interesting alternate route to the epoxide of entry 12 (Table 6A.3) has been described in which 2-t-butylpropene is first converted to an allylic hydroperoxide via photooxygenation and then, in the presence of Ti-tartrate catalyst, undergoes asymmetric epoxidation (79% yield, 72% ee) [38b]. The intermediate hydroperoxide serves as the source of oxygen for the epoxidation step.

6A.5.3. (3*E*)-Substituted Allyl Alcohols

Several factors contribute to the frequent use of (3*E*)-substituted allylic alcohols (**13**) for asymmetric epoxidation: (a) The allylic alcohols are easily prepared; (b) conversion to epoxy alcohol normally proceeds with good chemical yield and with better than 95% ee; (c) a large variety of functionality in the (3*E*) position is tolerated by the epoxidation catalyst. Representative epoxy alcohols (**14**) are summarized in Table 6A.4 [2,4,18,41–53] and Figure 6A.3 (4,54–61], with results divided arbitrarily according to whether the (3*E*) substituent is a hydrocarbon (Table 6A.4) or otherwise (Fig. 6A.3). The versatility of these and other 3-substituted epoxy alcohols for organic synthesis is illustrated with several examples in the following discussion.

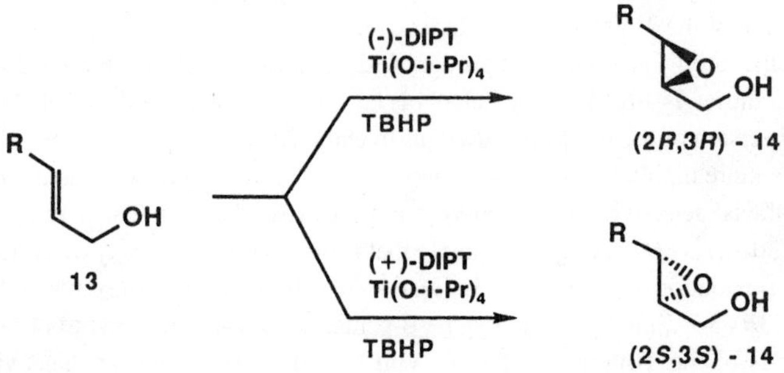

Compatibility of asymmetric epoxidation with acetals, ketals, ethers, and esters has led to extensive use of allylic alcohols containing these groups in the synthesis of polyoxygenated natural products. One such synthetic approach is illustrated by the asymmetric epoxidation of **15**, an allylic alcohol derived from (S)-glyceraldehyde acetonide [59,62]. In the epoxy alcohol (**16**) obtained from **15**, each carbon of the five-carbon chain is oxygenated, and all stereochemistry has been controlled. The structural relationship of **16** to the pentoses is evident, and methods leading to these carbohydrates have been described [59,62a].

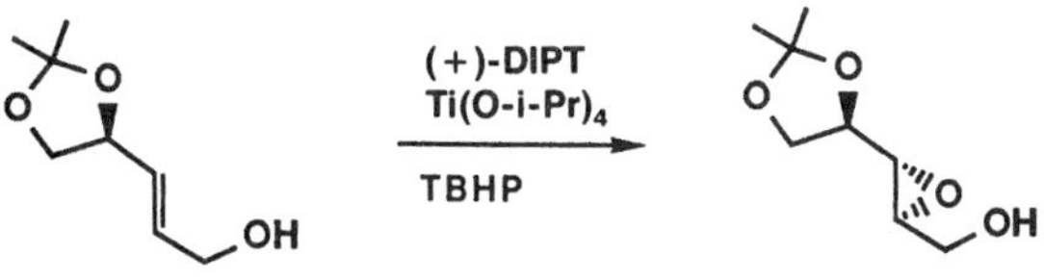

$$(+)\text{-DIPT}$$
$$Ti(O\text{-i-}Pr)_4$$
$$\xrightarrow{\hspace{2cm}}$$
$$TBHP$$

15   **16**

This synthetic methodology has been extended by the development of an efficient series of reactions that can transform one allylic alcohol into a second, which is two carbons longer than the first. Repetition of the reaction sequence can, in principle, be continued to any desired chain length.

The key steps in this "reiterative two-carbon extension cycle" are illustrated in Figure 6A.4, which shows a synthetic route leading from an achiral alkoxyacetaldehyde (**17**) to L-allose, one of the eight possible (L)-hexoses [63]. In practice, all eight L-hexoses were synthesized by taking advantage of branch points in the scheme and by using both antipodes of the Ti-tartrate catalyst to generate epimeric epoxides. The sequence of reactions begins with the two-carbon benzyloxyacetaldeyde **17**, which can be converted to the four-carbon intermediate **18** by means of a Wittig reaction (step a). In the actual synthesis, intermediate **18** was the starting point and was obtained by an alternate method. The carboxylic acid ester of **18** is reduced with DIBAL (step b) to the (3E) allylic alcohol **19** which, by asymmetric epoxidation (step c) is converted to epoxy alcohol **20**. Base-catalyzed (Payne) rearrangement of **20** establishes the equilibrium shown between **20** and the 1,2-epoxy alcohol **21**. Phenylthiolate reacts regioselectively to open the 1,2-epoxide (step d), leading to the dihydroxysulfide **22**. The diol is protected by conversion (step e) to the acetonide **23**, which, upon oxidation of the sulfide to a sulfoxide followed by Pummerer rearrangement (step f), is converted to the acetoxythioacetal **24**. Reduction of the latter (**24**) with one equivalent of DIBAL (step g) produces aldehyde **25**. At this point the synthetic sequence can be branched by converting a portion of the aldehyde **25** to the epimeric aldehyde (not shown) by epimerization with potassium carbonate in methanol. Both these new aldehydes can now be chain-extended by repeating steps a–g, which in the case of **25** leads to the hexose derivative **26**. To obtain all eight hexoses, a further branching during the second cycle is initiated at step c with part of the material (an allylic alcohol) being subjected to asymmetric epoxidation with (–)-DET. Both these branches are carried on through step f or g, thereby producing all eight L-hexose derivatives. Deprotection of the derivatives completes the syntheses as shown for L-allose (**27**) in Figure 6A.4.

1,2-Epoxy-3-alcohols can be derived from 2,3-epoxy-1-alcohols by the base-catalyzed Payne rearrangement as illustrated in step d of Figure 6A.4 [59,64]. The rearrangement is completely stereospecific but, because it is reversible, it usually results in an equilibrium mixture of the two epoxy alcohols for which the relative proportions are structure-dependent. Practical synthetic applications of this rearrangement therefore depend on methods that will shift the equilibrium completely in the direction desired. Nucleophiles such as thiolates and amines are

TABLE 6A.4. Epoxides from (3E)-Substituted Allylic Alcohols (Hydrocarbon Substituents)

Entry	Epoxide R	Catalyst Mole %Ti/%Tart.	Tartrate	Epoxide Config.	Yield (%)	% ee	Ref.
1	CH$_3$	100/100	(−)-DIPT	R,R	40–58	95	41–43
2	CH$_3$	5/6	(+)-DIPT	S,S	70	92	4
3	C$_2$H$_5$	stoich.	(−)-DIPT	R,R	80	>95	44
4	n-C$_3$H$_7$	100/104	(+)-DET	S,S	64	93	45
5	i-C$_3$H$_7$	5/6	(+)-DET	S,S	85	94	4
6	i-C$_3$H$_7$	100/104	(+)-DET	S,S	66	98	45
7	s-C$_4$H$_9$	—	(−)-DET	R,R	a	a	46
8	t-C$_4$H$_9$	120/150	(+)-DET	S,S	52	>95	38
9	CH$_2$=CH	5/6	(+)-DIPT	S,S	56	>91	47
10	CH$_3$CH=CHCH$_2$	8/10	(+)-DET	S,S	81	a	48
11	n-C$_5$H$_{11}$	a	(+)-DET	S,S	78	95	49
12	CH$_2$=CH(CH$_2$)$_3$	100/100	(+)-DET	S,S	80	>95	41
13	n-C$_7$H$_{15}$	5/7.3	(+)-DET	S,S	99	96	4
14	n-C$_8$H$_{17}$	5/6	(+)-DET	S,S	78	94	4
15	C$_2$H$_5$CH=CHCH$_2$CH=CHCH$_2$	a	(+)-DET	S,S	82	>95	46
16	C$_2$H$_5$C≡CCH$_2$C≡CCH$_2$	a	(−)-DET	R,R	76	a	50
17	n-C$_{10}$H$_{21}$	100/100	(+)-DET	S,S	79	>95	2
18	n-C$_{12}$H$_{25}$	a	(+)-DET	S,S	a	a	51
19	C$_{14}$H$_{29}$	a	(+)-DIPT	S,S	77	a	52
20	C$_{15}$H$_{31}$	120/160	(−)-DET	R,R	88	>95	53

a: Not reported.

246

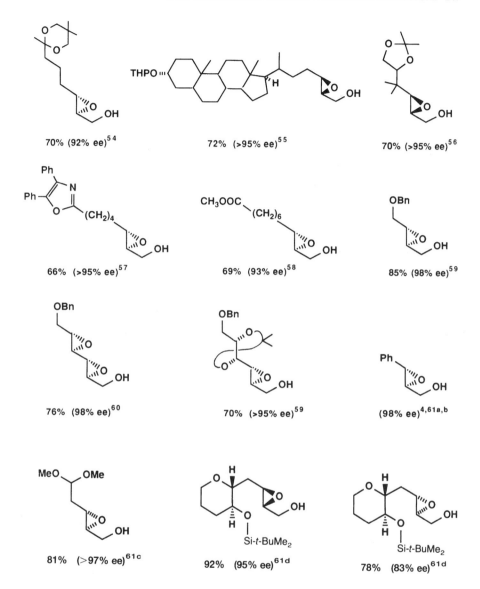

Figure 6A.3.   Epoxy alcohols from asymmetric epoxidation of (3*E*)-monosubstituted allylic alcohols.

sufficiently selective to react preferentially at C-1 of the 1,2-epoxy-3-alcohol and thereby shift the equilibrium completely in that direction. However, many other nucleophiles are incompatible with the reaction conditions required for the Payne rearrangement, and the approach of trapping the 1,2-epoxide cannot be used. To circumvent this problem and increase the scope of the Payne rearrangement/opening process, methods have been developed that lead to isolation of the terminal 1,2-epoxy-3-alcohols [11,65].

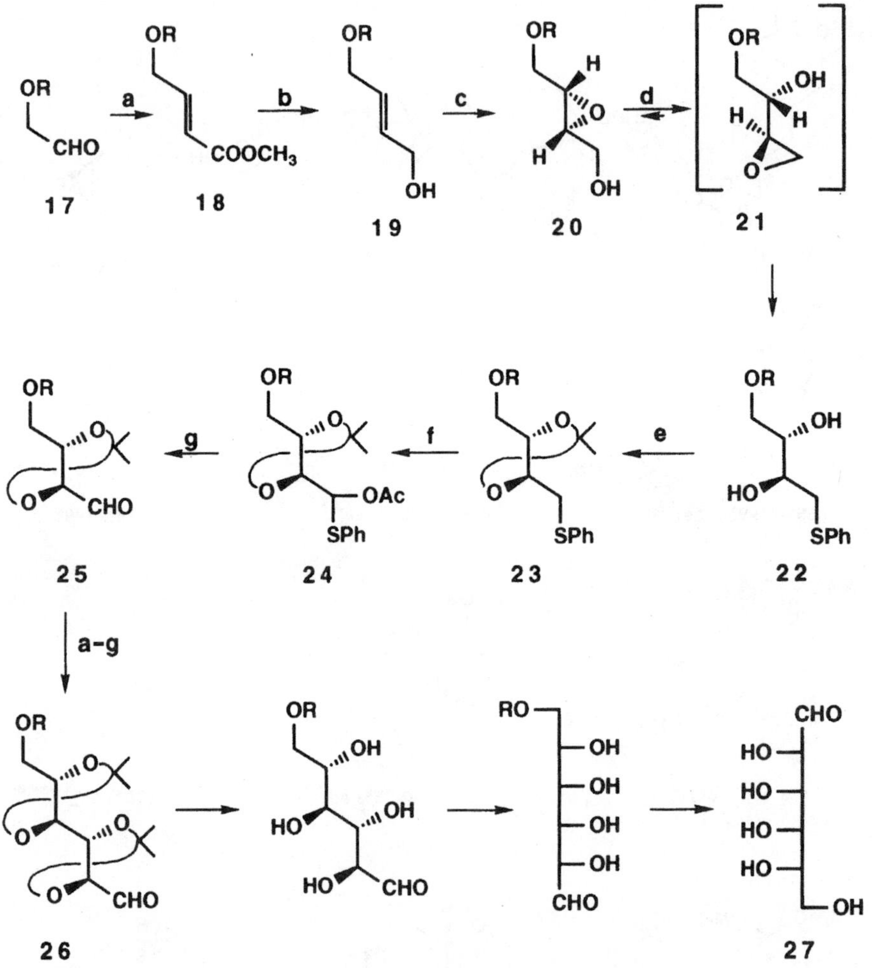

Figure 6A.4. Synthetic route to L-allose illustrating a reiterative two-carbon extension cycle: (a) $(CH_3O)_3P(O)CHCOOCH_3$, (b) DIBAL, (c) (+)-DET/Ti(O-i-Pr)$_4$, (d) PhSH/OH$^-$, (e) $CH_3C(=CH_2)OCH_3/H^+$, (f) m-CPBA, and (g) DIBAL (1 equiv.); H_2O.

One method uses the 2,3-diol-1-sulfide **30**, produced by thiolate trapping of the 1,2-epoxide from the Payne rearrangement equilibrium between **28** and **29** [11,65]. The sulfide is alkylated with Me$_3$OBF$_4$ to produce a good leaving group in **31**. Then base-promoted ring closure gives the 1,2-epoxide **32** in complete preference to formation of any 1,3-oxetane. The *erythro*-epoxy alcohol precursor **31** requires sodium hydride as the base to ensure that the Payne rearrangement is not reversed back to the starting 2,3-epoxy alcohol **28**. The analogous *threo*-epoxy alcohol precursor can be closed with sodium hydroxide.

In the second method, the 2,3-epoxy-1-alcohol **28** is first converted to a mesylate (or a tosylate) and then the epoxide is opened hydrolytically with inversion at C$_3$ to give the diol-mesylate **33**. The slight loss of optical purity observed in this process is due to lack of complete regioselectivity for C-3 opening. Mild base is sufficient to effect ring closure of the diol-mesylate **33** to give the 1,2-epoxide **34** [11,65].

The two methods are complementary in terms of stereochemistry, such that if a 2,3-epoxy alcohol of the same absolute configuration is used to start each sequence, the *erythro*-1,2-epoxy-3-ols produced will have opposite configurations at C-2 and C-3. This result is because inversion occurs at C-2 during the Payne rearrangement, whereas in the epoxy-mesylate opening, inversion occurs at C-3. Detailed discussions of these Payne rearrangement processes as well as of further synthetic transformations of the 1,2-epoxy alcohols have been presented elsewhere [11,65].

When two allylic alcohols are contained in a symmetrical molecule, asymmetric epoxidation proceeds with interesting consequences for stereochemical purity. The results were first described for the asymmetric epoxidation of (2Z, 6E, 10Z)-dodeca-2,6,10-trien-1,12-diol (**35**) [66]. The first epoxidation of **35** produces the major and minor enantiomers **36** and **37**. Because the stereogenic centers in these compounds are remote from the second allylic alcohol, each enantiomer undergoes a second epoxidation with essentially the same enantiofacial selectivity as in the first epoxidation. Three bisepoxides result, **38**, **39** (a meso compound), and **38′** (the mirror image of **38**). The overall consequence is that most of the epoxidation resulting from the undesired enantiofacial attack leads to the meso compound **39**, which is in principle separable from the major product. Very little of the mirror-image compound **38′** is formed and therefore the enantiomeric purity of the major product will be very high. In the example cited, enantiomeric purity could not be determined directly but was calculated according to the expression $(A_1 + B_1)(A_2 + B_2)$, where the first term in parentheses gives the enantiofacial selectivity of the first epoxidation and the second, the enantiofacial selectivity of the second epoxidation. In the example being discussed, an enantiofacial selectivity of 19:1 (90% ee) was assumed for both steps. The ratio of the three products therefore should be $(19 + 1)(19 + 1)$ or 361:38:1, and the enantiomeric purity of **38** should be 99.45% ee [66].

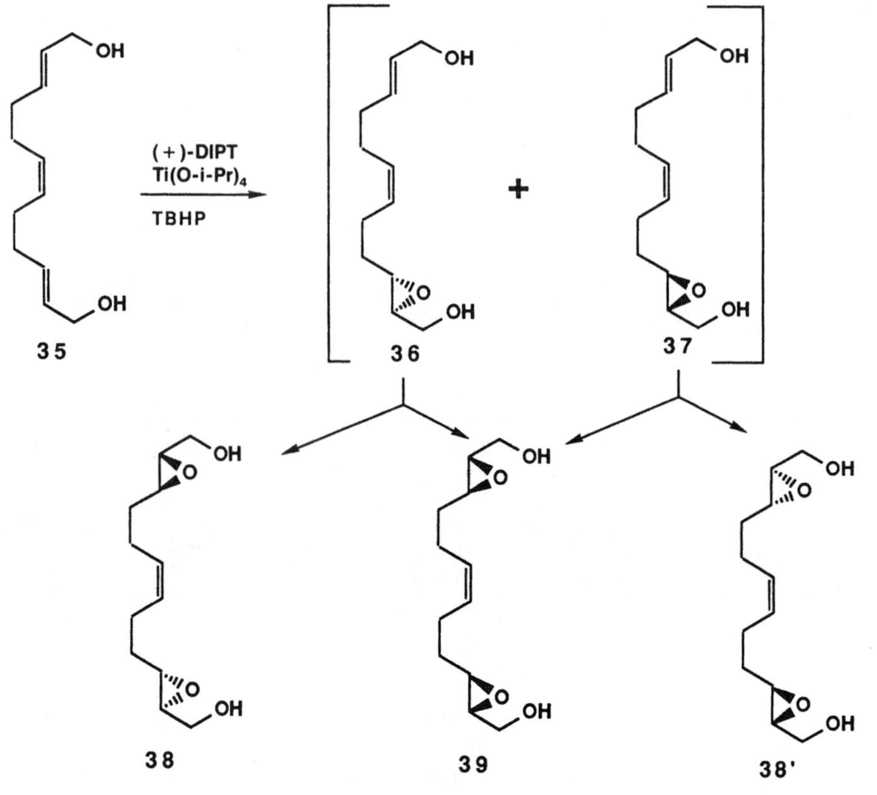
OH
OH
35
(+)-DIPT
Ti(O-i-Pr)4
TBHP
OH
O
OH
36
+
OH
O
OH
37
OH
O
O
OH
38
OH
O
O
OH
39
OH
O
O
OH
38'

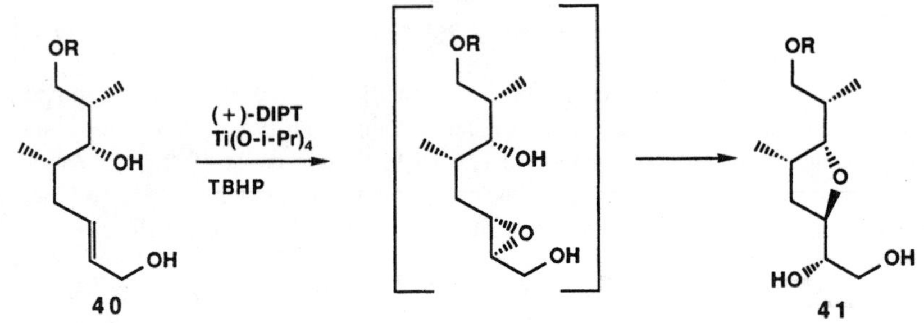
OR
OH
OH
40
(+)-DIPT
Ti(O-i-Pr)4
TBHP
OR
OH
O
OH
OR
O
HO
OH
41

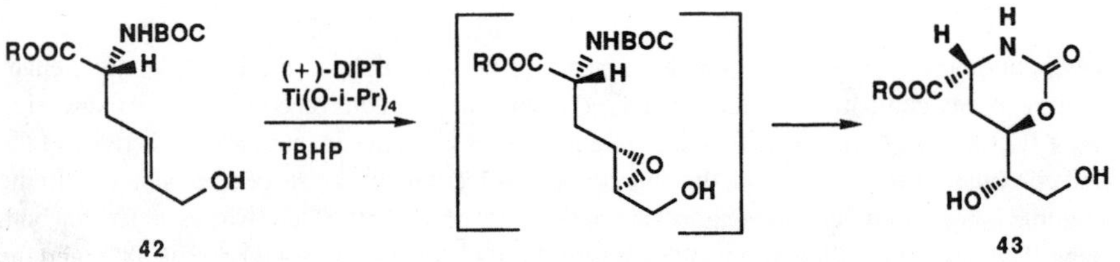
ROOC
NHBOC
H
OH
42
(+)-DIPT
Ti(O-i-Pr)4
TBHP
ROOC
NHBOC
H
O
OH
H
H
N
O
ROOC
O
HO
OH
43

Fortunately, a wide variety of functionality is compatible with the Ti-tartrate catalyst (see Table 6A.1), but the judicious placement of functional groups relative to the allylic alcohol can lead to further desirable reactions following epoxidation. For example, in **40**, asymmetric epoxidation of the allylic alcohol is followed by intramolecular cyclization under the reaction conditions to give the tetrahydrofuran **41** [67]. Likewise, in the epoxidation of **42**, cyclization of the intermediate epoxy alcohol occurs under the reaction conditions and leads to the cyclic urethane **43** [68].

Ti(IV) isopropoxide is an effective reagent for promoting regioselective attack by nucleophiles at the 3-position of 2,3-epoxy alcohols [69], 2,3-epoxy acids [70], and 2,3-epoxy amides [70]. It has been proposed that this process involves coordination to the metal center in the bidentate manner shown for a 2,3-epoxy alcohol in structure **44**. Such Ti-assisted nucleophilic opening of epoxides is thought to play a role in the in situ reactions leading to **41** and **43**.

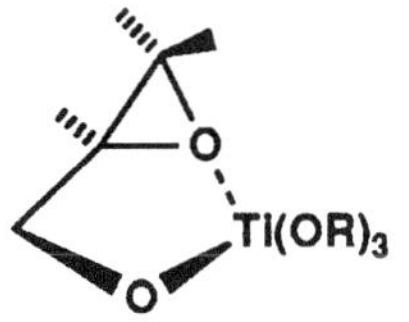

44

6A.5.4. (3Z)-Monosubstituted Allyl Alcohols

Allylic alcohols with a cis-3-substituent (**45**) are the slowest to be epoxidized, and they give the most variable enantiofacial selectivity. Both these characteristics suggest that allylic alcohols of this structure have the poorest fit to the requirements of the active epoxidation catalyst. Nevertheless, asymmetric epoxidation of these substrates is still effective and in most cases gives an enantiomeric purity of at least 80% ee and often as high as 95% ee. Patience with the slower reaction rate usually is rewarded with chemical yields of epoxy alcohols comparable with those obtained with other allylic alcohols. A number of representative examples are collected in Table 6A.5 [2,4,38,59,62a,71–78].

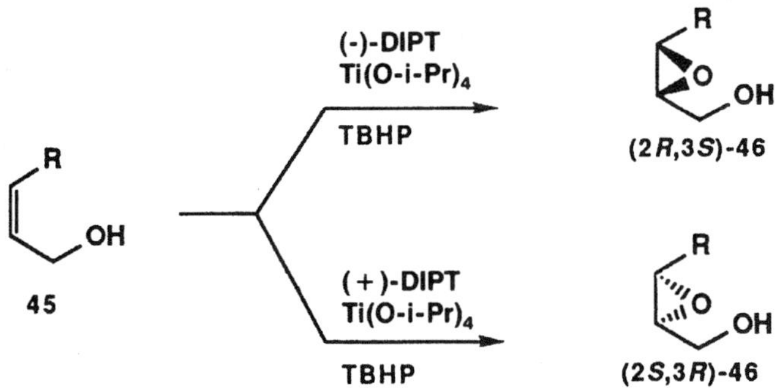

There is a rough correlation between the enantiomeric purity observed for these epoxy alcohols and the steric complexity at the α-carbon of the C-3 substituent. When the C-3 substituent is a primary group (Table 6A.5, entries 1, 2, 4, 6–12, 19–21), enantiofacial selectivity is highest and 80–95% ee are observed for these compounds. When the substituent is secondary

TABLE 6A.5. Epoxides from (3Z)-Substituted Allylic Alcohols

Entry	Epoxide R	Catalyst mole %Ti/%Tart.	Tartrate	Epoxide config.	Yield (%)	% ee	Ref.
1	CH_3	5/6	(+)-DIPT	S,R	68	92	4
2	C_2H_5	a	(+)-DET	S,R	60	80	71
3	$CH(CH_3)_2$	a	(+)-DET	S,R	54	66	72
4	$CH_2CH(CH_3)_2$	a	(+)-DET	S,R	80	95	73
5	$C(CH_3)_3$	120/150	(+)-DET	S,R	77	25	38
6	$C_9H_{19}(n)$	100/100	(−)-DET	R,S	80	91	2
7	$C_7H_{15}(n)$	10/14	(+)-DET	S,R	74	86	4
8	$C_8H_{17}(n)$	5/7.4	(+)-DIPT	S,R	63	>80	4
9	$CH_2CH=CHC_5H_{11}$	110/110	(+)-DMT	S,R	70	94	74
10	$(CH_2)_3COOCH_3$	a	(+)-DET	S,R	57	95	75
11	CH_2OBn	100/100	(−)-DET	R,S	84	92	59
12	CH_2OBn	14/14	—	—	—	95	4
13	Ph	100/120	(+)-DET	S,R	61	78	76
14	$CH(CH_3)Ph$	a	(+)-DIPT	S,R	a	a	77
15	(structure)	100/100	(+)-DET	S,R	55,57	93,84	59,62a
16	as above	100/100	(−)-DET	R,S	a	20	62a
17	$CH(CH_3)CH_2OBn$	100/100	(+)-DET	S,R	a	66	78a
18	as above	100/100	(−)-DET	R,S	a	0	78a
19	CH_2CH_2OBn	a	(+)-DIPT	S,R	75	92	78b
20	$C_{11}H_{23}(n)$	120/130	(+)-DIPT	S,R	83	92	78c
21	$(CH_2CH=CH)_2CH=CH_2$	100/148	(+)-DET	S,R	59	89	78d

a: Not reported.

(entries 3, 15–18) or tertiary (entry 5), enantiofacial selectivity is much more variable. When the substituent is asymmetric, enantiofacial selectivity depends on the absolute configuration, as is evident in comparison of entries 15 with 16 and of 17 with 18 in Table 6A.5. Epoxidation of these chiral allylic alcohols with one antipode of catalyst yields moderate-to-good diastereoselectivity, whereas the other antipode, diastereoselectivity is virtually lacking.

6A.5.5. (2,3E)-Disubstituted Allyl Alcohols

Extensive use in synthesis has been made of the asymmetric epoxidation of (2,3E)-disubstituted allylic alcohols. With few exceptions enantiofacial selectivity is excellent (90–95% ee). The results for a number of epoxidations of allylic alcohols with smaller substituents are collected in Table 6A.6 [2,4,41,61b,79–84], while a variety of other compounds with larger groups are illustrated by structures **47–60**.

TABLE 6A.6. Epoxides from (2,3E)-Disubstituted Allylic Alcohols

Entry	R_1	R_2	Tartrate	Epoxide Config.	Yield (%)	% ee	Ref.
				Epoxide			
1	CH_3	CH_3	(+)-DET	2S,3S	77	94	61b
2	CH_3	C_2H_5	(+)-DMT	2S,3S	79	95	41,78c
3	$-(CH_2)_3-$		(+)-DET	2S,3S	38	>95	79
4	$-(CH_2)_4-$		(+)-DET	2S,3S	77	93	4
5	CH_3	CH_2OBn	(−)-DIPT	2R,3R	87	90	80
6	CH_3	$CH(CH_3)CH_2OBn$	—	—	93	>95	81
7	CH_3	C_6H_5	(+)-DIPT	2S,3S	79	>98	4
8	C_6H_5	C_6H_5	(+)-DET	2S,3S	70	>95[a]	2,4
9	CH_3	$CH_2CH_2CH=CH2$	(+)-DET	2S,3S	71	96	82
10	CH_3	-CH2 (OMe, MeO)	(−)-DET	2R,3R	87	>95	83
11	CH_3	$CH_2CH=C(CH_3)_2$	(+)-DET	2S,3S	64	>90	84a
12	CH_3	$CH_2CH=C(CH_3)_2$	(−)-DET	2R,3R	59	>91	84a
13	CH_3	$(CH_2)_3OSi(CH_3)_2Bu$	(−)-DET	2R,3R	89	93	84b

[a]Ee after crystallization.

The epoxy alcohol **47** is a squalene oxide analog that has been used to examine substrate specificity in enzymatic cyclizations by baker's yeast [85]. The epoxy alcohol **48** provided an optically active intermediate used in the synthesis of 3,6-epoxyauraptene and marmine [86], and epoxy alcohol **49** served as an intermediate in the synthesis of the antibiotic virantmycin [87]. In the synthesis of the three stilbene oxides **50**, **51**, and **52**, the presence of an *o*-chloro group in the 2-phenyl ring resulted in a lower enantiomeric purity (70% ee) when compared with the analogs without this chlorine substituent [88a]. The very efficient (80% yield, 96% ee) formation of **52a** by asymmetric epoxidation of the allylic alcohol precursor offers a synthetic entry to optically active 11-deoxyanthracyclinones [88b], whereas epoxy alcohol **52b** is one of several examples of asymmetric epoxidation used in the synthesis of brevitoxin precursors [88c]. Diastereomeric epoxy alcohols **54** and **55** are obtained in combined 90% yield (>95% ee each) from epoxidation of the racemic alcohol **53** [89]. Diastereomeric epoxy alcohols, **57** and **58**, also are obtained with high enantiomeric purity in the epoxidation of **56** [44]. The epoxy alcohol obtained from substrate **59** undergoes further intramolecular cyclization with stereospecific formation of the cyclic ether **60** [90].

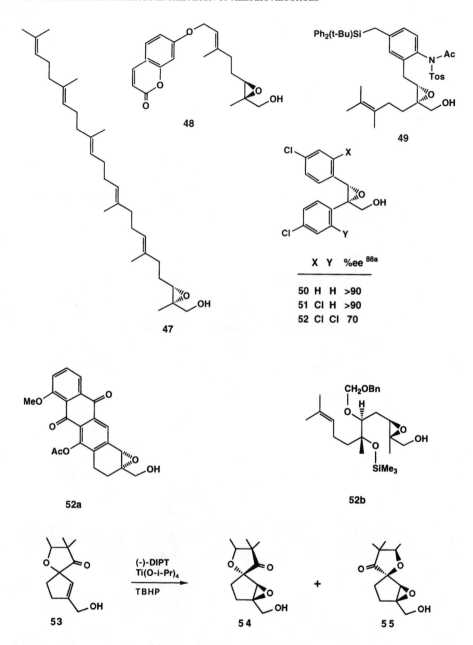

48

49

X	Y	%ee [88a]
50	H	H	>90
51	Cl	H	>90
52	Cl	Cl	70

52a

52b

53   (-)-DIPT Ti(O-i-Pr)₄   TBHP   54   +   55

47

6A.5.6. (2,3Z)-Disubstituted Allyl Alcohols

A limited number of allylic alcohols of the (2,3Z)-disubstituted type have been subjected to asymmetric epoxidation. With one exception, the C-2 substituent in these substrates has been a methyl group, the exception being a *t*-butyl group [38]. The (3Z)-substituents have been more varied, as illustrated by structures **61–64**, which show the epoxy alcohols derived from the corresponding allylic alcohol substrates.

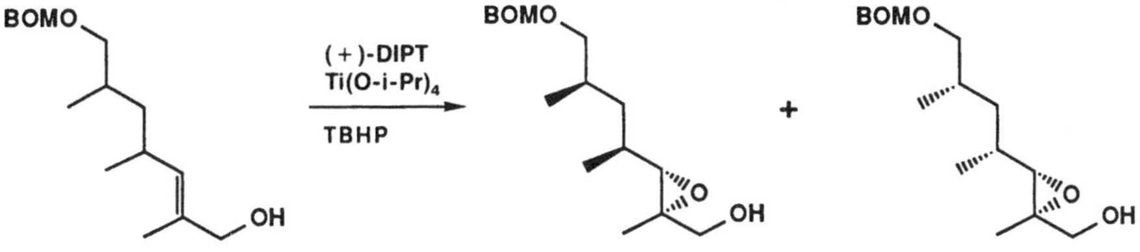

Epoxidation of (Z)-2-methyl-2-hepten-1-ol gave epoxy alcohol **61** (80% yield, 89% ee) [2], of (Z)-2-methyl-4-phenyl-2-buten-1-ol gave **62** (90%, 91% ee) [77], and of (Z)-1-hy-droxysqualene gave **63** (93%, 78% ee) [85]. The epoxy alcohol **64** had >95% ee after recrystallization [91]. In the epoxidation of (Z)-2-*t*-butyl-2-buten-1-ol, the allylic alcohol with a C-2 *t*-butyl group, the epoxy alcohol was obtained in 43% yield and with 60% ee [38]. These results lead one to expect that other 2,3Z-disubstituted allylic alcohols will be epoxidized in good yield and with enantioselectivity similar to that observed for the 3Z-monosubstituted allylic alcohols (*i.e.*, 80–95% ee).

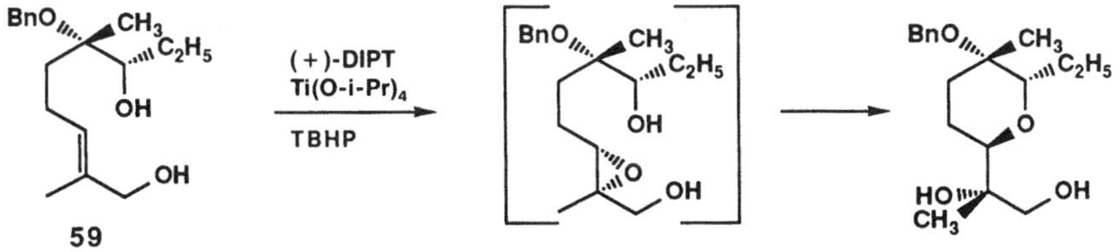

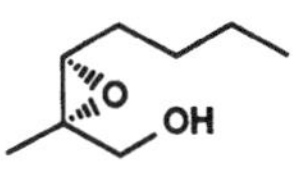

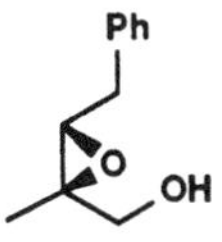

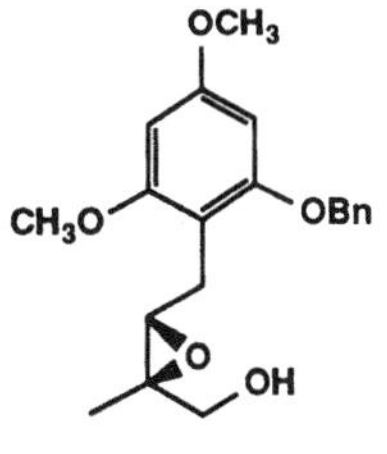

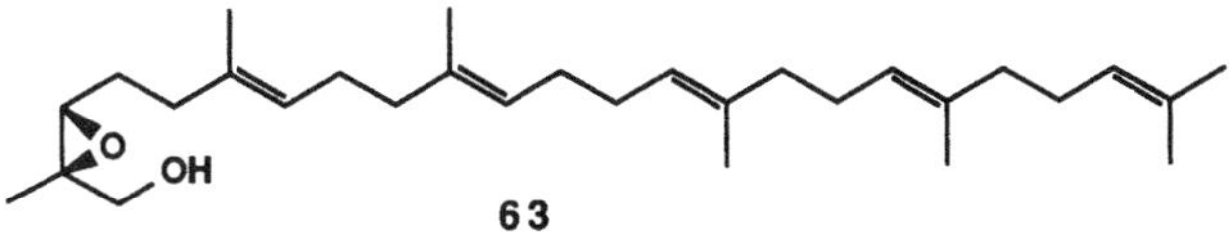

6A.5.7. 3,3-Disubstituted Allyl Alcohols

The 3,3-disubstituted allyl alcohols are substrates that combine a 3*E*-substituent with a 3*Z*-sub-stituent in the same molecule. Allylic alcohols with only a 3*E*-substituent generally are epoxidized with excellent enantioselectivity, whereas those with only a 3*Z*-substituent are epoxidized with enantioselectivity in the range of 80–95% ee. In combination, many of the reported examples have a methyl substituent at the 3*Z*-position, and all are epoxidized with 90–95% ee (see Table 6A.7, entries 1–4, 6) [2,4,92–97]. Only a limited number of examples

TABLE 6A.7. Epoxides from 3,3-Disubstituted Allylic Alcohols

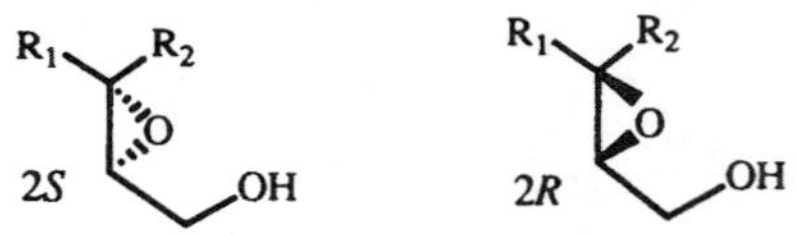

Entry	Epoxide R₁	R₂	Catalyst mol %Ti/%Tart.	Tartrate	Epoxide Config.	Yield (%)	% ee	Ref.
1	CH_3	CH_3	100/100	(–)-DBT	2R	25	>90	92
2	$CH_2CH=C(CH_3)_2$	$CH3$	200/200	(+)-DET	2S,3S	67	95	93
3	$(CH_2)_2CH=C(CH_3)_2$	$CH3$	100/100	(+)-DET	2S,3S	77	95	2
4	$(CH_2)_2CH=C(CH_3)_2$	CH_3	5/7.4	(+)-DET	2S,3S	95	91	4
5	CH_3	$(CH_2)_2CH=C(CH_3)_2$	100/100	(+)-DET	2S,3R	79	94	2
6	$(CH_2)_2OSi(CH_3)_2C(CH_3)_3$	CH_3	105/157	(–)-DET	2R,3R	81	>95	94
7	CH_3	$(CH_2)_2OSi(CH_3)_2C(CH_3)_3$	a	(+)-DET	2S,3R	98	90	95
8	CH_3	$(CH_2)_2OSi(CH_3)_2C(CH_3)_3$	a	(–)-DET	2R,3S	98	86	95
9	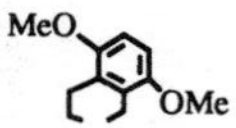 MeO—OMe (aromatic)		10/15	(+)-DET	2S,3R	97	93	96
10	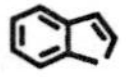 (indane/indene structure)		100/110	(+)-DET	2S,3S	a	94	97a
11	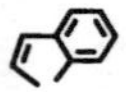 (naphthalene structure)		100/110	(+)-DET	2S,3R	a	88	97a
12	$CH_2CH(CH_3)CH_2OMEM$	$(CH_2)_4OBn$	10/12	(–)-DIPT	2R,3S	83	91	97b

a: Not reported.

256

with larger groups at the 3Z-position have been reported (entries 5, 7–12) and in these the enantioselectivity ranges from 84 to 94% ee.

3,3-Dimethylallyl alcohol was epoxidized with better than 90% ee (entry 1) but in low yield when a stoichiometric amount of the Ti-tartrate complex was used. However, when a catalytic amount of the complex was used and an in situ derivatization employed, the p-nitrobenzoate (>98% ee after recrystallization) and p-toluenesulfonate (93% ee) were isolated in yields of 70 and 55%, respectively.

Likewise, the epoxidation of geraniol with a stoichiometric amount of the complex gave epoxide (entry 3) in 77% yield (95% ee), which was improved to 95% yield (91% ee) when a catalytic amount of complex was used (entry 4).

6A.5.8. 2,3,3-Trisubstituted Allyl Alcohols

Interesting structural diversity is present in the limited examples of trisubstituted allyl alcohols (equivalent to tetrasubstituted olefins) to which asymmetric epoxidation has been applied. The epoxides **65–70** [98–103] have been obtained from the corresponding allylic alcohols with yield and enantiomeric purity as indicated when such data have been reported. The lower enantiomeric purity observed for epoxy alcohol **69** may result from disruption of the catalyst structure by the phenolic groups or from alternate modes of binding of substrate to catalyst, again because of the phenolic groups. Phenols bind strongly to Ti(IV), which may account for the large excess (6 equiv.) of the Ti-tartrate complex that was required to achieve the yield and enantiomeric purity reported in the case of **69**.

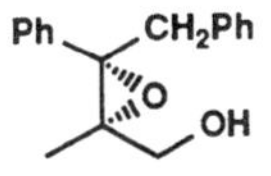

65
90% (94%ee)[98]

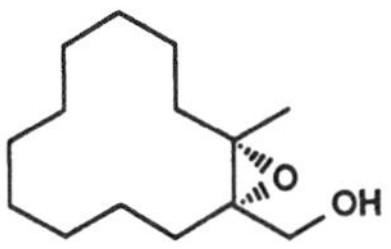

66
72% (94%ee)[99]

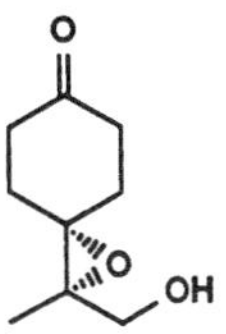

67
(>90%ee)[100]

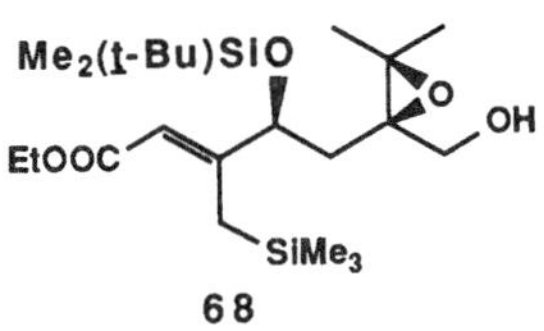

68
(>90%ee)[101]

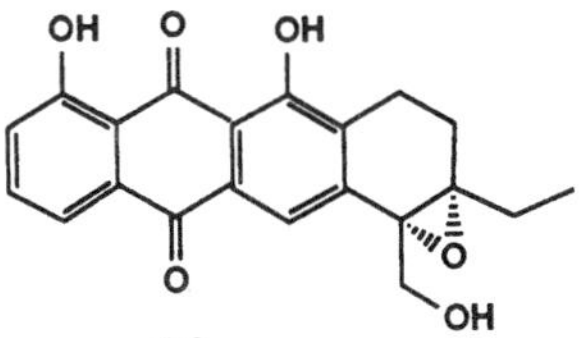

69
85% (53%ee)[102]

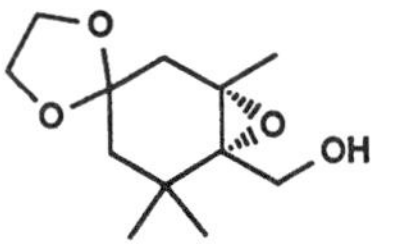

70
95% (95%ee)[103]

6A.5.9. 1-Substituted Allyl Alcohols: Kinetic Resolution

The presence of a stereogenic center at C_1 of an allylic alcohol introduces an additional factor into the asymmetric epoxidation process in that now both enantiofacial selectivity and diastereoselectivity must be considered. It is helpful in these cases to examine epoxidation of each enantiomer of the allylic alcohol separately. Epoxidation of one enantiomer proceeds normally and produces an erythro epoxy alcohol in accord with the rules shown in Figure 6A.1.

Epoxidation of the other enantiomer proceeds at a reduced rate because contact between the C-1 substituent and the catalyst seriously impedes the necessary approach of olefin to oxidant (see Fig. 6A.2). The difference in epoxidation rates for the two enantiomers is usually of sufficient magnitude that either the epoxy alcohol or the recovered allylic alcohol can be produced with high enantiomeric purity. The net result is a kinetic resolution [14]. In the case of a homochiral C-1 substituted allylic alcohol, asymmetric epoxidation will be fast and highly diastereoselective with one antipode of the Ti-tartrate catalyst but not with the other, according to the guidelines of Figure 6A.2. Although kinetic resolution is most frequently encountered and applied to chiral C-1 substituted allylic alcohols, the rationale also is applicable to allylic alcohols with chiral substituents at other positions, examples of which have been given in several preceding subsections.

The ratio of the rates of epoxidation of the two enantiomers, k_{fast}/k_{slow}, has been defined as the relative rate (k_{rel}) and is related to the percent conversion of allylic alcohol to epoxy alcohol and the enantiomeric purity of the remaining allylic alcohol. A mathematical relationship between these variables exists and can be represented graphically as shown in Figure 6A.5 [14].

If values are known for two of the three variables, the third can be predicted by use of this graph. Inspection of the graph reveals that relative rates of 25 or more are very effective for achieving kinetic resolution of 1-substituted allylic alcohols. With a relative rate of 25, the epoxidation need be carried to <60% conversion to achieve essentially 100% ee for the unreacted alcohol. A convenient method for limiting the extent of epoxidation to 60% is simply by controlling the amount of oxidant used in the reaction. However, for some substrates (see Table 6A.8, entries 1, 9, or 10) even k_{fast} is extremely slow and the epoxidation takes several days [2,13,104–106]. To shorten the time needed for such reactions, an alternate practice is to use an

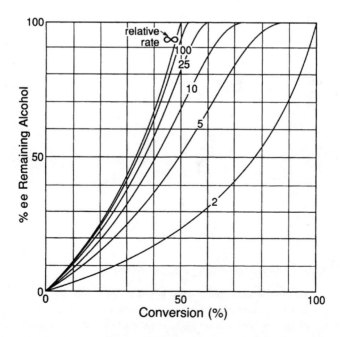

Figure 6A.5. Dependence of enantiomeric excess on relative rate in the epoxidation of 1-substituted allylic alcohols.

TABLE 6A.8. Relative Rate (k_{rel}) Data for Kinetic Resolution of 1-Substituted Allylic Alcohols

Entry	Allylic Alcohol	Reaction Time	Rel. Rate at −20°C	% ee	Rel. rates at 0°C			Ref.
					DIPT	DET	DMT	
1	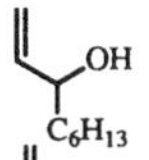 OH, C_6H_{13}	12 da	83	>96	60			2,13
2	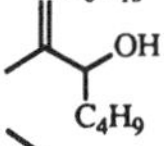 OH, C_4H_9	15 hr	138	>96	96	52		2,13
3	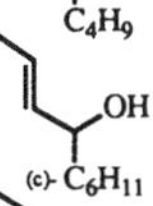 OH, (c)-C_6H_{11}	15 hr	104	>96	74	28	15	2,13
4	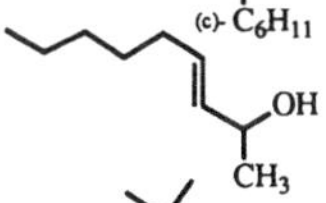 OH, CH_3		160					104
5	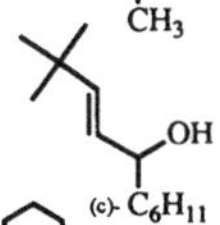 OH, (c)-C_6H_{11}		300					104
6	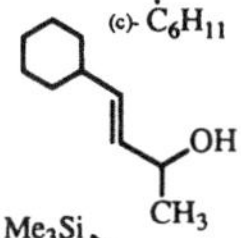 OH, CH_3		330					104
7	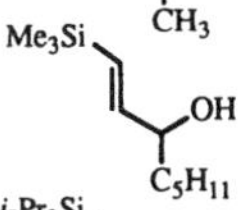 Me_3Si, OH, C_5H_{11}		700					104–106
8	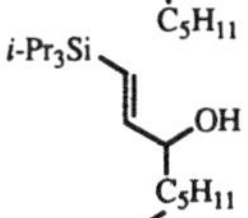 i-Pr_3Si, OH, C_5H_{11}		300					104
9	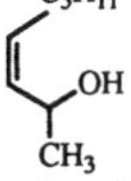 OH, CH_3	6 da	20	91				2,13
10	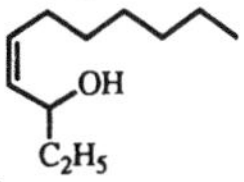 OH, C_2H_5	2da	16	82	13			2,13
11	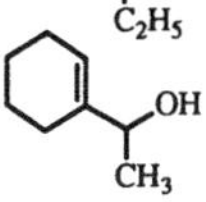 OH, CH_3	15 hr	83	>96	60	38		2,13

excess of oxidant and to monitor the extent of epoxidation by an appropriate analytical method. If the optically active epoxy alcohol is the desired reaction product, high enantioselectivity can be ensured by running the reaction to approximately 45% completion.

Relative rate data for the kinetic resolution/epoxidation of 1-substituted allylic alcohols of varying structure are summarized in Table 6A.8. The k_{rel} values at −20°C for all entries in Table 6A.8 were determined using DIPT as the chiral ligand. Additionally, for several entries (1–3, 10, 11) the dependence of k_{rel} on temperature, 0 versus −20°C, and on steric bulk of the tartrate ester (DIPT vs. DET vs. DMT) has been measured. Lower reaction temperature and larger tartrate ester groups are factors that clearly increase the magnitude of k_{rel} and, therefore, improve the efficiency of the kinetic resolution process. Although the results summarized in Table 6A.8

are from experiments in which stoichiometric quantities of Ti-tartrate complex were used, the catalytic version of the reaction also may be used for kinetic resolution [4]. When comparing results with the same tartrate ester, a slight loss in enantioselectivity is seen in the catalytic mode relative to the stoichiometric reaction. The trend toward higher enantioselectivity with bulkier tartrate esters can be used to advantage in the catalytic reaction by using dicyclohexyltartrate (DCT), which gives higher enantioselectivity than DIPT, or dicyclododecyltartrate, which gives yet higher enantioselectivity than DCT [4].

The efficiency of kinetic resolution is even greater when there is a silicon or iodo substituent in the 3E position of the C-1 chiral allylic alcohols. The compatibility of silyl substituents with asymmetric epoxidation conditions was first shown by the conversion of (3E)-3-trimethylsilyl-allyl alcohol into (2R,3R)-3-trimethylsilyloxiranemethanol in 60% yield with more than 95% ee [107a] and further exploited by the conversion of (E)-3-(triphenylsilyl)-2-[2,3-^{2}H$_2$]propenol into (2R,3R)-3-triphenylsilyl [2,3-^{2}H$_2$]oxiranemethanol in 96% yield with 94% ee [107b,c]. With an n-pentyl group at C-1, the k_{rel} for asymmetric epoxidation of the enantiomeric allylic alcohols is 700 (Table 6A.8, entry 7), and both epoxy alcohol and optically active recovered allylic alcohol are obtained in 42% yield with more than 99% ee (see Table 6A.9, entry 1). Equally good yields and enantiomeric purities are observed with other substituents in the C-1 position as is shown by entries 2–9 in Table 6A.9 [105,108,109]. Good yields with high enantioselectivities also are reported in the kinetic resolution of (3E)-iodo analogs (entries 10–14) and of a (3E)-chloro analog (entry 15). (3E)-Stannyl substituents (entries 16–18) appear to be similar to carbon substituents in their effect on kinetic resolution.

The influence of both the steric and electronic properties of the silyl group on the rate of epoxidation have been examined experimentally [104]. Two rate effects were considered. First, the overall rate of epoxidation of the silyl allylic alcohols was found to be one-fifth to one-sixth that of the similar carbon analogs. This rate difference was attributed to electronic differences between the silicon and carbon substituents. Second, the increase in k_{rel} to 700 for silyl allylic alcohols compared with carbon analogs (e.g., 104 for entry 3, Table 6A.8) was attributed to the steric effect of the large trimethylsilyl group. As expected, when a bulky t-butyl group was placed at C-3, k_{rel} increased to 300 [104].

At the end of 1989, the number of 1-substituted allylic alcohols that had been used in kinetic resolution/asymmetric epoxidation experiments exceeded 75. In slightly more than half of these experiments, the desired product was the kinetically resolved allylic alcohol, whereas in the remainder the epoxy alcohol was desired. In addition to the compounds in Table 6A.8, experimental results for other kinetically resolved alcohols are summarized in Table 6A.10 [38,77,110–115a–d]. From these results, it appears that kinetic resolution is successful regardless of the nature of the (3E) substituent and is successful with any except the most bulky substituents at C-2.

When the allylic alcohol is the desired product of the kinetic resolution process, the accompanying epoxy alcohol also may be converted to the desired allylic alcohol by the two-step sequence shown in Scheme 6A.1. The epoxy alcohol, after separation from the allyl alcohol, is mesylated and then subjected to reaction with sodium telluride, which effects the transformation of epoxy mesylate to the allylic alcohol with inversion at the asymmetric carbinol center [115e]. Preliminary results suggest that the rearrangement follows this pathway only when the epoxy alcohol is unsubstituted at the 3-position.

A small, structurally distinct class of 1-substituted allylic alcohols consists of those that are conformationally restricted by incorporation into a ring system. These allylic alcohols may be further subdivided into two types, depending on whether the double bond is endocyclic or exocyclic. For allylic alcohols with endocyclic double bonds, kinetic resolution gives 2-cyclohexen-1-ol (**71**) with 30% ee [14], (4aS,2R)-4a-methyl-2,3,4,4a,5,6,7,8-octahydro-

TABLE 6A.9. Kinetic Resolution of 3-Silyl-, Halo-, and Stannyl-Substituted Allylic Alcohols

$$R_2 \diagdown \!\!\!\!\! \diagup^{OH} \diagdown R_1$$

Entry	Allylic Alcohol		Allylic Alcohol		Epoxy Alcohol		Ref.
	R_1	R_2	Yield(%)[a]	% ee	Yield(%)[a]	% ee	
1	C_5H_{11} (n)	SiMe$_3$	42	>99	42	>99	105
2	C_3H_7 (i)	SiMe$_3$	40	>99	41	99	105b
3	C_6H_5	SiMe$_3$	44	>99	42	97	105b
4	CH_2OPh	SiMe$_3$	47	>99	46	>99	105b
5	CH_2OCH_2Ph	SiMe$_3$	43	>99	48	>99	105b
6	$CH_2CH=CHC_5H_{11}$	SiMe$_3$	44	>99	43	>99	105b
7	$CH_2CH_2OCH_2Ph$	SiMe$_3$	43	>99	45	>99	105b
8	CH_2COOBu (n)	SiMe$_3$	44	>99			105c
9	$(CH_2)_3COOCH_3$	SiMe$_3$	43	>99	45	>99	105b
10	C_5H_{11} (n)	1	49	>99	49	>99	108
11	C_2H_5	I	40	>98			108
12	$CH_2C_6H_{11}$ (c)	I	42	>99			108
13	C_5H_9 (c)	I	44	>99			108
14	C_6H_5	I	43	>98			108
15	C_5H_{11} (n)	Cl	43	>99			108
16	C_5H_{11} (n)	SnBu$_3$	40	>99	[b]	84	109
17	C_6H_{11} (c)	SnBu$_3$	41	>99			109
18	CH_2OPh	SnBu$_3$	40	>99			109

[a]Maximum yield is 50%. [b]Not reported.

naphthalen-2-ol (**72**) with 55% ee [116], and (R)-2-cyclohepten-1-ol (**73**) with 80% ee [14]. The epoxy alcohols (1S,2S,3R)-2,3-epoxycyclopenten-1-ol (**74**) [117], (1S,2S,4aR)-4a-decahydronaphthalen-2-ol (**75**) [116], and (1R,2S,3R)-2,3-epoxy-6-cyclononen-1-ol (**76**) [118] are obtained with 60, 61, and 90% ee, respectively.

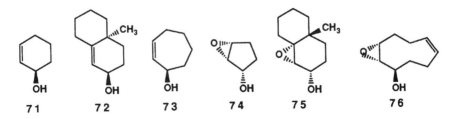

(R)-trans-Verbenol (**77**) is epoxidized five times as fast as (S)-trans-verbenol when (+)-DIPT is used in the catalyst [77]. For allylic alcohols with an exocyclic double bond, kinetic resolution gives 2-methylenecyclohexanol (**78**) with 80% ee in 46% yield when (−)-DIPT is used [119].

TABLE 6A.10. Representative Kinetic Resolutions of 1-Substituted Allylic Alcohols

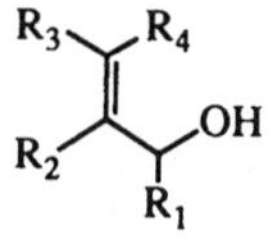

	Allylic Alcohol				Yield		
Entry	R_1	R_2	R_3	R_4	$(\%)^a$	% ee	Ref.
1	CH_3	H	$CH_2CH=CH_2$	H	39	90	110
2	C_6H_{11} (c)	H	H	H	32	>98	111
3	C_2H_5	CH_3	H	H	b	>98	112
4	C_4H_9 (n)	H	H	H	43	>90	113
5	C_6H_3 (2,4-diCl)	C_6H_3 (2,4-diCl)	H	H	42	90	114
6	C_2H_5	H	Ph	H	b	99	77
7	CH_2CH_2Ph	H	H	H	b	99	77
8	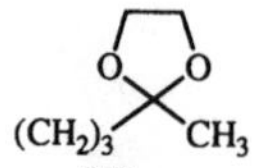 $(CH_2)_3$ CH_3	H	H	H	b	99	77
9	$CH_2CCH=CH_2$	H	CH_3	CH_3	10	>99	77
10	$C_{12}H_{25}$ (n)	H	CH_3	H	44	97	115a
11	CH_3	$C(CH_3)_3$	H	H	b	30	38
12	$C(CH_3)_3$	H	CH_3	H	b	5	38
13	C_4H_9 (n)	H	$CH=CH_2$	H	40	90	115b
14	CH_2COOEt	CH_3	H	H	11	>95	115c
15	$C\equiv CC_6H_{13}$ (n)	H	H	H	35	95	115d

^{a}Maximum yield is 50%. ^{b}Not reported.

Epoxidation of the enantiomerically pure 4-methylene-5α-cholestan-3β-ol (**79**) is reported to be much faster with catalyst derived from (+)-DET than from (−)-DET [120].

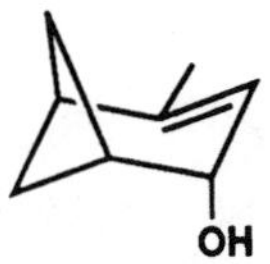

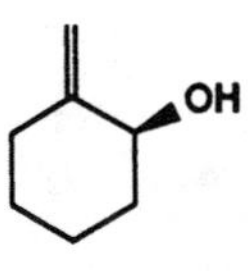

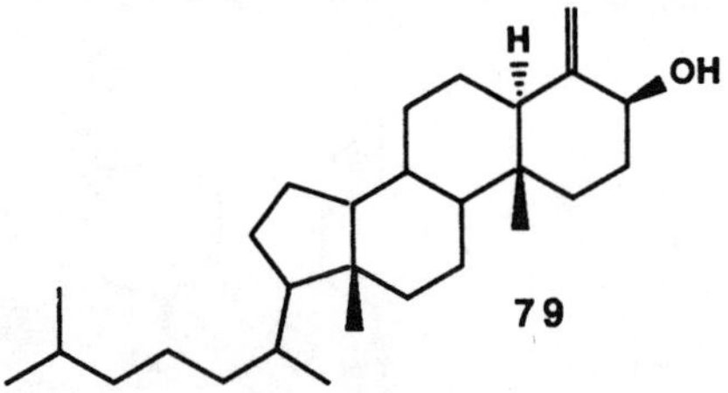

The variable enantioselectivities seen in these results likely stem from conformational restraints imposed by the cyclic structures, which prevent the allylic alcohols from attaining an ideal conformation for the epoxidation process (see Fig. 6A.9, below, for the proposed ideal conformation).

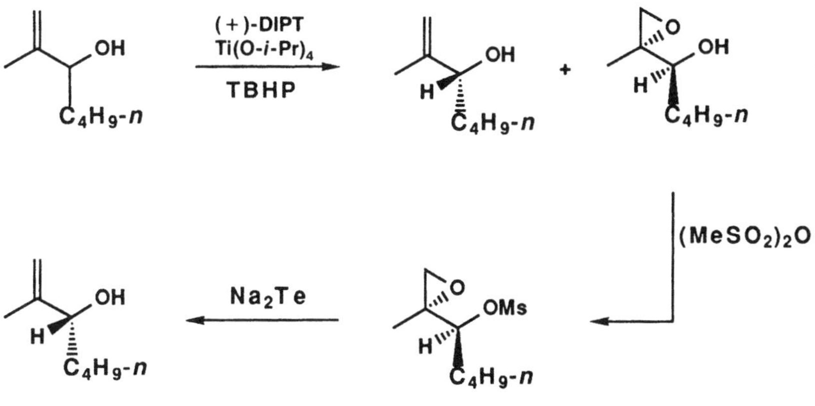

Scheme 6A.1.

One especially interesting kinetic resolution/ asymmetric epoxidation substrate is (R,S)-2,4-hexadien-3-ol (**80**) [77]. The racemic diene has eight different olefinic faces at which epoxidation can occur and thereby presents an interesting challenge to the selectivity of the epoxidation catalyst. The selectivity can be tested by using slightly less than 0.5 equiv. of oxidant (because the substrate is a racemate, the maximum yield of any one product is 50%). When the reaction was run under these conditions, the only product that was formed was the $(1R,2R,3R)$-epoxy alcohol **81**.

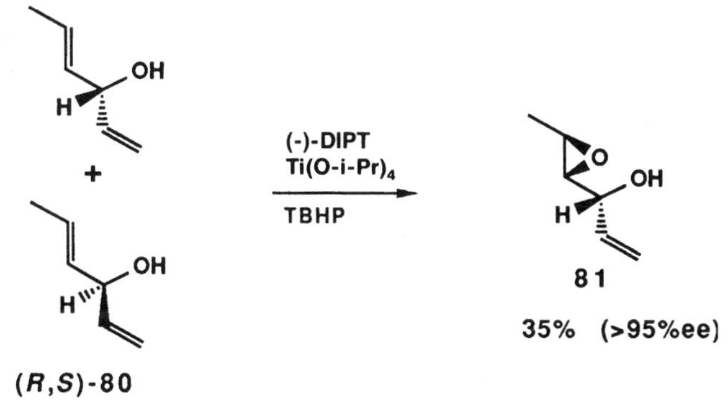

Three different principles of selectivity are required to achieve this result. First, the difference in rate of epoxidation by the catalyst of a disubstituted versus a monosubstituted olefin must be such that the propenyl group is epoxidized in complete preference to the vinyl group. The effect of this selectivity is to reduce the choice of olefinic faces to four of the two propenyl groups. Second, the inherent enantiofacial selectivity of the catalyst as represented in Figure 6A.1 will narrow the choice of propenyl faces from four to two. Finally, the steric factor responsible for kinetic resolution of 1-substituted allylic alcohols (Fig. 6A.2) will determine the final choice between the propenyl groups in the enantiomers of **80**. The net result is the formation of epoxy alcohol **81** and enrichment of the unreacted allylic alcohol in the $(3S)$-enantiomer.

trans-1,2-Dialkylcycloalkenes (**82**) have helical chirality and can be resolved if flipping of the ring from one face of the olefin to the other is restricted. When appropriately substituted, these compounds also serve as synthetic precursors to the betweenanenes. The asymmetric epoxidation approach to kinetic resolution is ideally suited for the resolution of the cycloalkenes when a hydroxymethyl group is one of the substituents on the double bond, as shown for **82**. The epoxidation of **82** with Ti(O-*i*-Pr)$_4$/(+)-DET and 0.6 equiv. of TBHP was complete within 10 minutes and gave resolved allylic alcohol **83** in 41% isolated yield with no detectable enantiomeric impurity and epoxy alcohol **84** in 50% yield (the maximum yields possible for **83** and **84** are 50%) [121a]. A variety of analogs of **82**, including different ring sizes, have been resolved by this method and have been used for the synthesis of optically active betweenanenes [121b].

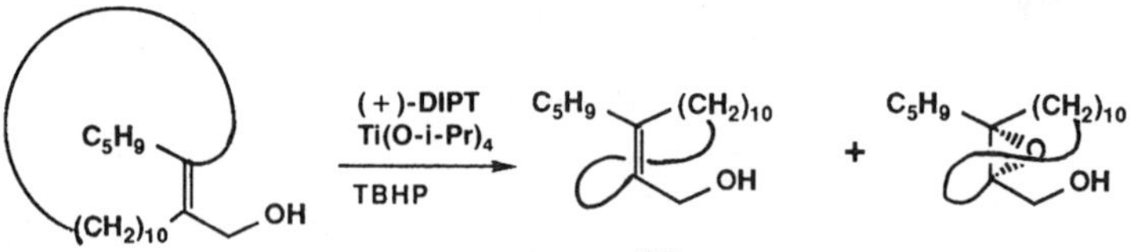

A final subclass of 1-substituted allylic alcohols is made up of carbinol derivatives with two identical olefinic substituents, the simplest example is 1,4-pentadien-3-ol or divinylcarbinol, **85**. Although these compounds per se are achiral, once they have bound to the chiral Ti complex, the two vinyl groups become stereochemically nonequivalent (diastereotopic). Asymmetric epoxidation will now occur selectively at one of the two vinyl groups, and the choice is controlled by factors identical to those in effect during the kinetic resolution process. The similarity can be seen by comparison of the Ti-allylic alcohol complex portrayed in Figure 6A.6 with the kinetic resolution process depicted in Figure 6A.2. The pro-*S* and pro-*R* conformations shown will be sterically favored and disfavored, respectively, for the same reasons that the enantiomers of chiral C$_1$ allylic alcohols are distinguished during kinetic resolution. Therefore, epoxidation of **85** produces (2*R*,3*S*) epoxy alcohol **86** [122].

Further analysis of the asymmetric epoxidation of divinylcarbinol **85**, including the minor products, has led to recognition of the second factor that influences the optical purity of the major product **86** [26,123]. One of the minor epoxy alcohols is enantiomeric to **86** and therefore is responsible for lowering the enantiomeric purity of **86**. However, in this minor isomer the configuration of the remaining allylic alcohol group favors a rapid second epoxidation, and this isomer is quickly converted to a diepoxide. As a consequence, the enantiomeric purity of the major epoxy alcohol **86** increases as the reaction progresses. A mathematical equation relating enantiomeric purity to the various rates of epoxidation for these divinylcarbinols has been derived. This analysis can also be applied to asymmetric epoxidation of prochiral compounds such as **87** [124].

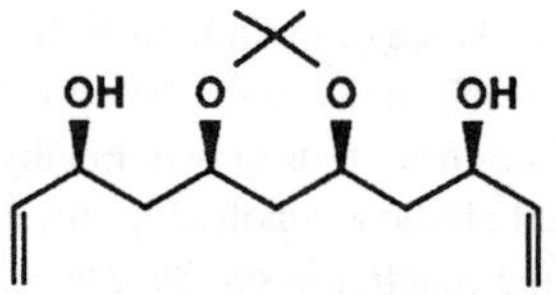

8 7

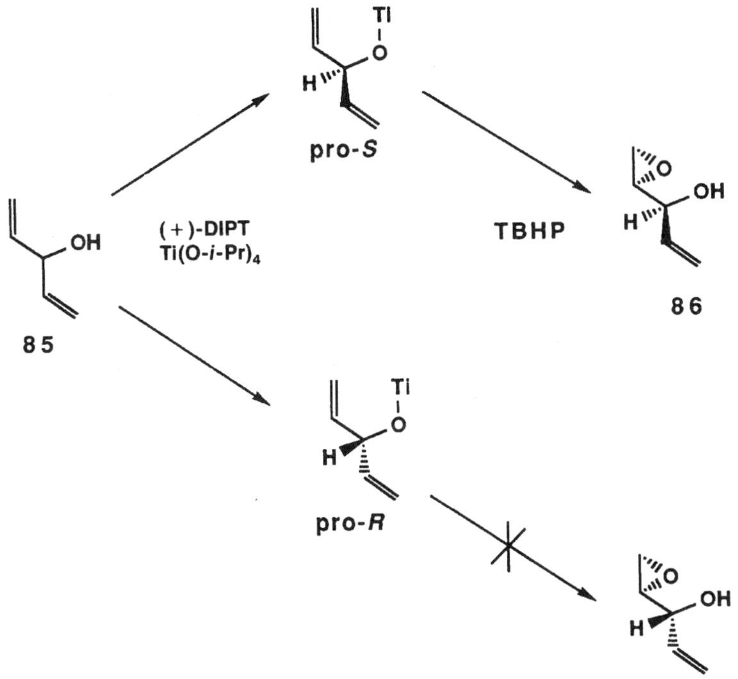

Figure 6A.6. Asymmetric epoxidation of 1,4-pentadien-3-ol.

As noted earlier in this section on C-1 substituted compounds, preparation of the epoxy alcohol has been the synthetic objective nearly as often as has been the optically active allylic alcohol. The principles outlined in Figure 6A.2 can again be used to guide the choice of tartrate ester needed to obtain the erythro epoxy alcohol of desired absolute configuration. By limiting the amount of oxidant (TBHP) used for the epoxidation to 0.4 equiv. (relative to substrate), optimum enantiomeric purity of the epoxy alcohol can be ensured and, in most cases, will be excellent. A few representative examples of epoxides prepared in this way are summarized in Table 6A.11 [90,105b,115c,125–130b].

In the special case in which the substrate is already enantiomerically pure (as in entry 5), it should be clear from Figure 6A.2 that asymmetric epoxidation will be successful (with regard to diastereomeric purity) only when the choice of catalyst directs delivery of oxygen to the face of the olefin opposite that of the C-1 substituent. Such choice of catalyst is further illustrated in Scheme 6A.2, wherein the two sequential epoxidations each proceed with better than 97% diastereoselectivity. The bisepoxide is obtained in an overall yield of 80% [130c].

6A.5.10. 1,1-Disubstituted Allyl Alcohols

The rationale that explains the kinetic resolution of the 1-monosubstituted allylic alcohols predicts that a 1,1-disubstituted allylic alcohol will be difficult to epoxidize with the Ti-tartrate catalyst. In practice, the epoxidation of 1,1-dimethylallyl alcohol (**88**) with a stoichiometric quantity of the Ti-tartrate complex is very slow, and no epoxy alcohol is isolated

TABLE 6A.11. Epoxides from 1-Substituted Allylic Alcohols

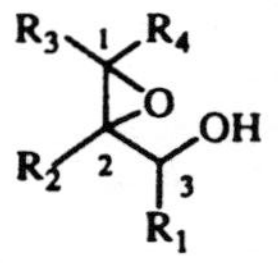

Entry	Epoxide R$_1$	R$_2$	R$_3$	R$_4$	Tartrate	Config.[a]	Yield (%)[b]	% ee	Ref.
1	C$_2$H$_5$	H	H	H	(+)-DIPT	2R,3S	c	d	125
2	C$_5$H$_{11}$ (n)	H	H	H	(−)-DIPT	2S,3R	47	91	126
3	(CH$_2$)$_6$COOMe	H	H	H	(+)-DIPT	2R,3S	36	>95	127
4	CH$_2$C≡CSi(iPr)$_3$	H	H	H	(−)-DIPT	2S,3R	40	>90	128
5	C$_2$H$_5$	CH$_3$	H	H	(+)-DIPT	2R,3S	82[e]	92	90
6	CH$_2$CH=CH$_2$	H	CH$_3$	H	(−)-DIPT	1R,2R,3R	27	>95	129a–129c
7	CH(OBn)CH=CHCH$_3$	H	CH$_3$	H	(+)-DIPT	1S,2S,3S	35	>95	130a
8	CH$_3$	H	CH$_2$CH=CH$_2$	H	(−)-DIPT	1R,2R,3R	40	90	129b
9	C$_5$H$_{11}$	H	SiMe$_3$	H	(+)-DIPT	1S,2S,3S	40	99	105b
10	CH$_2$COOEt	CH$_3$	H	H	(−)-DET	1S,2S,3R	d	>95	115c
11	CH$_3$	–CH$_2$CH=CCH$_2$– Me		H	(+)-DIPT	1S,2S,3S	37	95	130b

[a]Note that the arbitrary numbering used here may not coincide in all cases (e.g., entries 7, 10, 11) with correct *Chem. Abstr.* numbering. [b]Maximum yield is 50%, except for entry 5. [c]The epoxy alcohol was converted without isolation to the ethoxyethyl derivative. [d]Not reported. [e](3S)-2-Methylpent-1-en-3-ol was used as the substrate for this epoxidation.

266

[131]. Clearly, the rate of epoxidation of this substrate is slower than the subsequent reaction(s) of the epoxide.

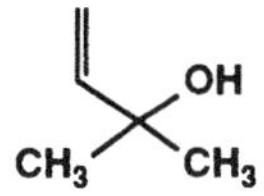

88

6A.5.11. Homoallylic, Bishomoallylic, and Trishomoallylic Alcohols

In contrast to allylic alcohols, the asymmetric epoxidation of homoallylic alcohols shows the following three general characteristics [132]: (a) the rates of epoxidation are slower, (b) enantiofacial selectivity is reversed (i.e., oxygen is delivered to the opposite face of the olefin when the same tartrate ester is used), and (c) the degree of enantiofacial selectivity is lower, with enantiomeric purities of the epoxy alcohols ranging from 20 to 55% ee. A series of seven model homoallylic alcohols, including all but one of the possible substitution patterns, has been subjected to epoxidation by using the stoichiometric version of the reaction with the results providing the basis for the preceding generalizations. An analogous complex composed of Zr(IV) isopropoxide [Zr(O-i-Pr)$_4$] and (+)-dicyclohexyltartramide has been found to catalyze asymmetric epoxidation of homoallylic alcohols with the same sense of enantiofacial selectivity as the Ti-tartrate ester complex. An improvement in enantiomeric purity was noted for epoxy alcohols derived from Z-homoallylic alcohols (to 77% ee), whereas other epoxy alcohols were obtained with enantiomeric purities comparable with those achieved with Ti [133].

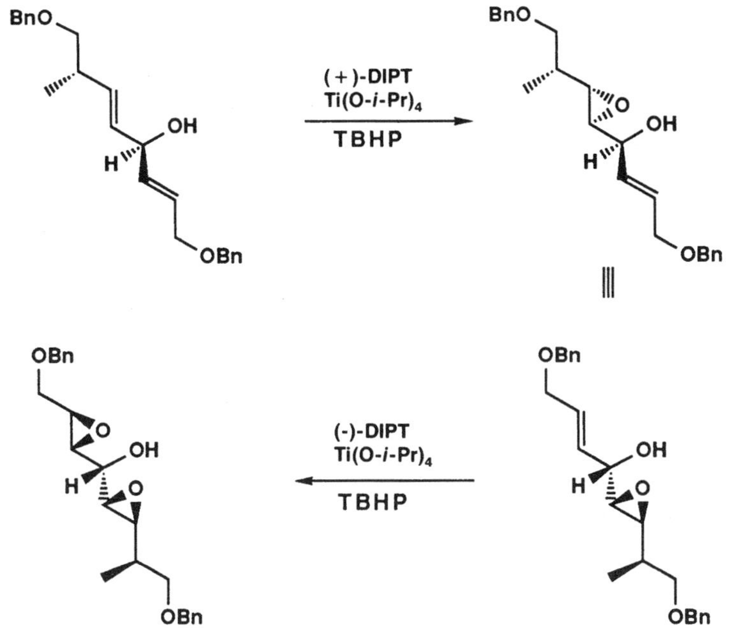

Scheme 6A.2.

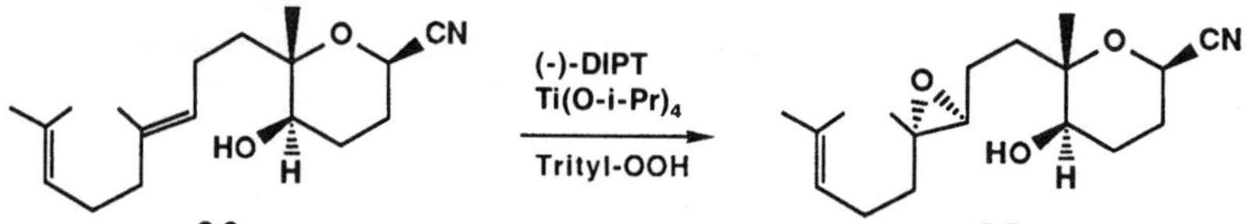

The trishomoallylic alcohol **89** undergoes asymmetric epoxidation in a yield of 74% and with "high" diastereofacial selectivity to give **90**. Trityl hydroperoxide, which had been shown to be effective in the asymmetric epoxidation of allylic alcohols [6], was required to attain enantiofacial selectivity in the epoxidation of **89** [134a]. The Ti/tartrate/TBHP-catalyzed conversions of the bishomoallylic phenol (**90a**, $n = 1$, R = H) and the trishomoallylic analog (**90a**, $n = 2$, R = Me) into dihydrobenzofuran **90b** (22% yield, 29% ee) and dihydrobenzopyran **90c** (49 and 56% ee), respectively, is assumed to occur via the intermediate epoxides [134b]. The dihydrofuran **90b** is assigned the (2*S*,1′*R*)-configuration, whereas the configurations of the dihydropyran **90c** are unspecified.

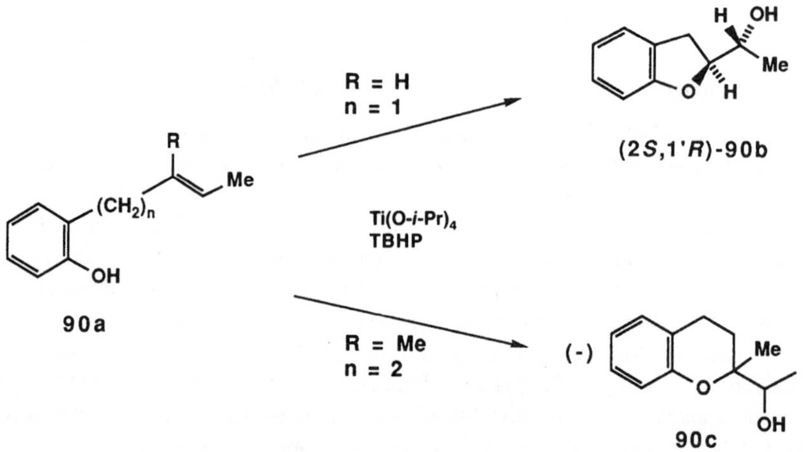

6A.6. MECHANISM OF Ti-TARTRATE CATALYZED ASYMMETRIC EPOXIDATION

The hallmark of Ti–tartrate catalyzed asymmetric epoxidation is the high degree of enantiofacial selectivity seen for a wide range of allylic alcohols. It is natural to inquire into what the mechanism of this reaction might be and what structural features of the catalyst produce these desirable results. These questions have been studied extensively, and the results have been the subject of considerable discussion [6,135,136]. For the purpose of this chapter, we review the aspects of the mechanistic-structural studies that may be helpful in devising synthetic applications of this reaction.

Of fundamental importance to an understanding of the reaction and its mechanism is the fact that in solution there is rapid exchange of Ti ligands [6,135,136]. Thus, when equimolar solutions of a Ti alkoxide and a dialkyl tartrate are mixed, the equilibrium represented by Eq. 6A.4 will be quickly reached with all but the most sterically demanding alkoxides.

$$Ti(OR)_4 + tartrate \rightleftarrows Ti(tartrate)(OR)_2 + 2\ ROH \qquad (6A.4)$$

This equilibrium is shifted far to the right because a chelating diol (i.e., the tartrate) has a much higher binding constant for Ti than do monodentate alcohols. The binding of tartrate is also enhanced by the increased acidity of its hydroxyl groups (due to the inductive effect of the esters). Spectroscopic evidence clearly reveals that two moles of free monodentate alcohol are present at equilibrium. Rapid ligand exchange continues as the hydroperoxide oxidant and the allylic alcohol substrate are added to the reaction medium. Pseudo-first-order kinetic experiments have shown a first-order rate dependence on the Ti-tartrate complex, the hydroperoxide, and the allylic alcohol and an inverse second-order dependence on the nonolefinic alcohol ligands (i.e., the isopropyl alcohol). The rate law derived from these results is expressed in Eq. 6A.5. The mechanistic pathway outlined below in Figure 6A.7 is consistent with Eq. 6A.5 and clearly illustrates the ligand-exchange processes essential for catalytic epoxidation.

$$Rate = k\ \frac{[Ti(tartrate)(OR)_2][TBHP][allylic\ alcohol]}{[ligand\ alcohol]^2} \qquad (6A.5)$$

After formation of the Ti(tartrate)(OR)$_2$ complex, the two remaining alkoxide ligands are replaced in reversible exchange reactions by the hydroperoxide (TBHP) and the allylic alcohol to give the "loaded" complex Ti(tartrate)(TBHP)(allylic alcohol). Now, in the rate-controlling step of the process, oxygen-transfer from the coordinated hydroperoxide to the allylic alcohol gives the complex Ti(tartrate)(t-OBu)(epoxy alcohol). The product alkoxides are replaced by more allylic alcohol and TBHP to regenerate the "loaded" complex and complete the catalytic cycle.

An alternate mechanism invoking an ion-pair transition-state assembly has been proposed to account for the enantioselectivity of the asymmetric epoxidation process [137]. In this proposal, two additional alcohol species are required in the transition-state complex. This

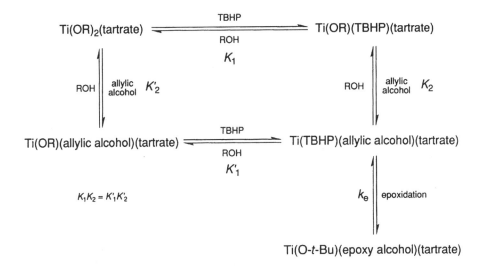

Figure 6A.7. Ligand exchange on titanium during the asymmetric epoxidation catalytic cycle.

requirement is inconsistent with the kinetic studies of this reaction that have led to the rate law expressed in Eq. 6A.5 and, therefore, this proposal must be considered to be incorrect.

Much of the experimental success of asymmetric epoxidation lies in exercising proper control of Eq. 6A.4 [6]. Both $TI(OR)_4$ and $Ti(tartrate)(OR)_2$ are active epoxidation catalysts, and because the former is achiral, any contribution by that species to the epoxidation will result in loss of enantioselectivity. The addition to the reaction of more than one equivalent of tartrate, relative to Ti, will have the effect of minimizing the leftward component of the equilibrium and will suppress the amount of $Ti(OR)_4$ present in the reaction. The excess tartrate, however, forms $Ti(tartrate)_2$, which has been shown to be a catalytically inactive species and will cause a decrease in reaction rate that is proportional to the excess tartrate added. The need to minimize $Ti(OR)_4$ concentration and, at the same time, to avoid a drastic reduction in rate of epoxidation is the basis for the recommendation of a 10–20 mol % excess of tartrate over Ti for formation of the catalytic complex. After the addition of hydroperoxide and allylic alcohol to the reaction, the concentration of ROH will increase accordingly, and this will increase the leftward pressure on the equilibrium shown in Eq. 6A.4. Fortunately, in most situations this shift apparently is extremely slight and is effectively suppressed by the use of excess tartrate. A shift in the equilibrium does begin to occur, however, when the reaction is run in the catalytic mode and the amount of catalyst used is less than ~5 mol % relative to allylic alcohol substrate. Loss in enantioselectivity then may be observed. This factor is the basis of the recommendation for use of 5–10 mol % of Ti-tartrate complex when the catalytic version of asymmetric epoxidation is used.

Comparison of the epoxidation rates of several parasubstituted cinnamyl alcohols reveals that the olefin acts as a nucleophile toward the activated peroxide oxygen in the epoxidation reaction [136]. Relative to unsubstituted cinnamyl alcohol (relative rate = 1), an electron-withdrawing *p*-nitro group decreases the rate of epoxidation (0.42), whereas an electron-releasing group such as *p*-methoxy increases the rate (4.39). These results are consistent with the notion of the olefin acting as a nucleophile. Additional support for this conclusion arises from comparison of the rates for epoxidation of less substituted allylic alcohols with those for more highly substituted analogs. A clear example of this substituent effect is seen in the epoxidation of (*R,S*)-2,4-hexadien-3-ol (**80**), described in the preceding section, where the propenyl group is epoxidized in nearly complete preference to the vinyl group [77]. Another example is seen with the allylic-homoallylic alcohol **91**, where epoxidation occurs preferentially at the tetrasubstituted homoallylic olefin to give **92** [99]. The preferential epoxidation of the more highly substituted olefin in these compounds is consistent with a nucleophilic role for the olefin.

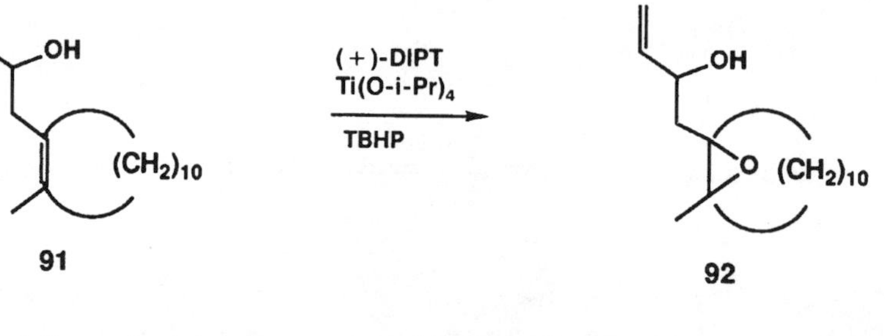

Although the mechanistic scheme portrayed in Figure 6A.7 provides important insight into the experimental aspects of asymmetric epoxidation, it sheds little light on the structure of the catalyst and on the features of the catalyst responsible for the concurrent high stereoselectivity and broad generality. The rapid ligand exchange, so crucial to the success of the reaction, makes characterization of the catalyst structure extremely difficult. Some reliable structural information has been obtained from spectroscopic measurements on the complex in solution [6,136b]. These data clearly support the conclusion that the major molecular species formed in solution is the dimeric composite, $Ti_2(tartrate)_2(OR)_4$. Efforts to isolate this complex, ideally as a crystalline solid, have so far been fruitless. Therefore, assignment of a structure to the dimeric complex has depended on information provided by the X-ray crystallographic structure obtained for the closely related complex, $Ti_2(dibenzyltartramide)_2(OR)_4$ [138]. The assumption of a similarity of structure for these two complexes receives some support from the fact that both catalyze the epoxidation of α-phenylcinnamyl alcohol with the same enantiofacial selectivity. From this analogy, the structure shown in Figure 6A.8 has been proposed for the $Ti_2(tar-trate)_2(OR)_4$ complex. This structure has a C_2 axis of symmetry with the two Ti atoms in identical stereochemical environments. To account for the sameness of all the tartrate ester groups in the room-temperature NMR spectrum, a fluxional equilibrium between the two structurally degenerate complexes shown in Figure 6A.8 has been proposed. Catalysis of the epoxidation process is thought to involve only one of the two Ti atoms, but the possibility that both are required has not yet been ruled out.

"Loading" of the catalyst with hydroperoxide and substrate can now be considered in terms of the proposed structure [6]. Orientation of these two ligands on the catalyst becomes a crucial issue. Three coordination sites, two axial and one equatorial, become available by exchange of two isopropoxides and dissociation of the coordinated ester carbonyl group. These processes can occur with minimal perturbation of the remaining catalyst structure. The three coordination sites are in a semicircular (i.e., meridional) array around one edge of the catalyst surface. In the reactive mode, coordination of the hydroperoxide is assumed to be bidentate by analogy to the precedent of bidentate TBHP coordination to vanadium [6,139]. The hydroperoxide must occupy the equatorial and one of the two available axial coordination sites, with the allylic alcohol in the remaining axial site. To achieve the necessary proximity for transfer of oxygen (the distal peroxide oxygen is assumed to be transferred) to the olefin, the distal oxygen is placed in the equatorial site (Fig. 6A.9) and the proximal oxygen is placed in the axial site. The axial site on the lower face of the complex (as drawn in Fig. 6A.9) is chosen for the peroxide because of the larger steric demands of the *t*-butyl, or especially of the trityl group when trityl hydroperoxide is used, compared with the allylic alcohol.

The allylic alcohol binds to the remaining axial coordination site, where stereochemical and stereoelectronic effects dictate the conformation shown in Figure 6A.9 [6]. The structural model of catalyst, oxidant, and substrate shown in Figure 6A.9 illustrates a detailed version of the formalized rule presented in Figure 6A.1. Ideally, all observed stereochemistry of epoxy alcohol and kinetic resolution products can be rationalized according to the compatibility of their binding with the stereochemistry and stereoelectronic requirements imposed by this site [6]. A

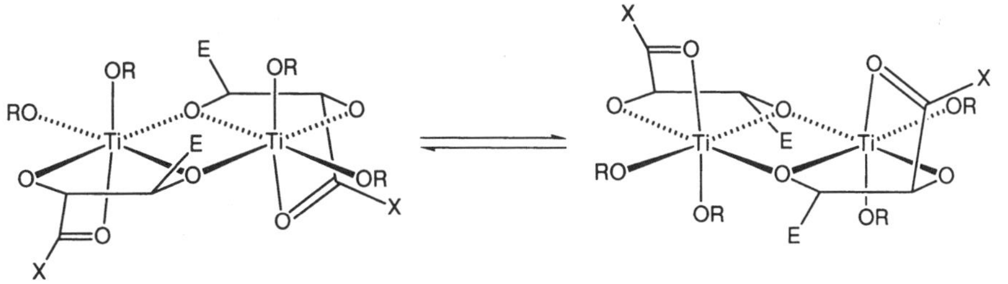

Figure 6A.8. Fluxional equilibrium proposed for the titanium-tartrate complex in solution.

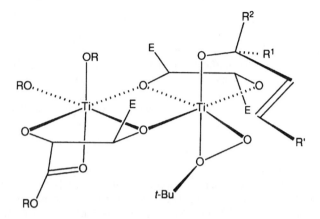

Figure 6A.9.   Proposed structure of "loaded" catalyst at the time of oxygen transfer.

transition-state model for the asymmetric epoxidation complex has been calculated by a frontier orbital approach and is consistent with the formulation portrayed in Figure 6A.9 [140].

6A.7. OTHER ASYMMETRIC EPOXIDATIONS AND OXIDATIONS CATALYZED BY TITANIUM COMPLEXES

6A.7.1. Dititanium-Ditartrate Complex

The discussion to this point has focused entirely on the epoxidation of allylic (and homoallylic) alcohols catalyzed by the $Ti(OR)_2$(tartrate) complex. The role of the olefin as a nucleophile toward the activated peroxide oxygen in this reaction has been established (see discussion of mechanism). If the olefin of the allylic alcohol is replaced by another nucleophilic group then, in principle, oxidation of that group may occur (Eq. 6A.6) [141].

$$G\text{—}(C)_n\text{—}\overset{|}{\underset{|}{C}}\text{—OH} \longrightarrow \overset{O}{\underset{|}{\overset{\|}{G}}}\text{—}(C)_n\text{—}\overset{|}{\underset{|}{C}}\text{—OH} \tag{6A.6}$$

In practice, oxidations of this type have been observed and generally have been carried out with a substrate bearing a racemic secondary alcohol so that kinetic resolution is achieved. Although these oxidations are not strictly within the scope of this chapter, they are summarized briefly in Eqs. 6A.7–6A.9 to acquaint the reader with other potential uses for the Ti-tartrate catalytic complex. In the kinetic resolutions shown in Eqs. 6A.7 and 6A.8, the oxidations are controlled by limiting the amount of oxidant used to 0.6 equiv. Only modest resolution was attained for the acetylenic alcohol (Eq. 6A.7, 21% ee) [77] and the allenic alcohol (Eq. 6A.8, 40% ee) [77]. Resolutions of the furanols [142] or the thiophene alcohols [143] of Eq. 6A.9 generally are excellent (90–98% ee except when R_1 is a t-butyl group). Only in the kinetic resolution of the furanols has the oxidation product been identified and, in that case, is a dihydropyranone.

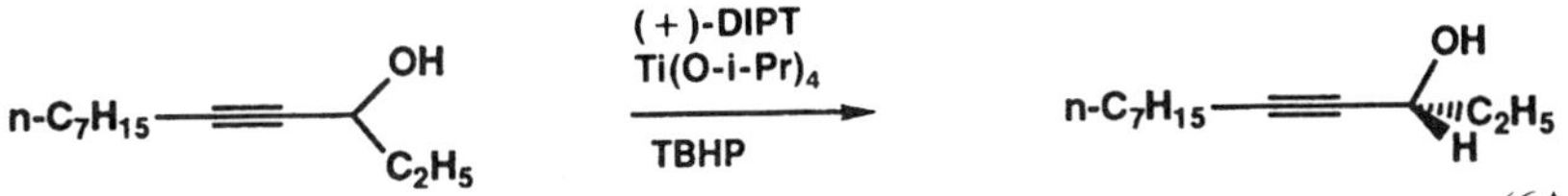

(6A.7)

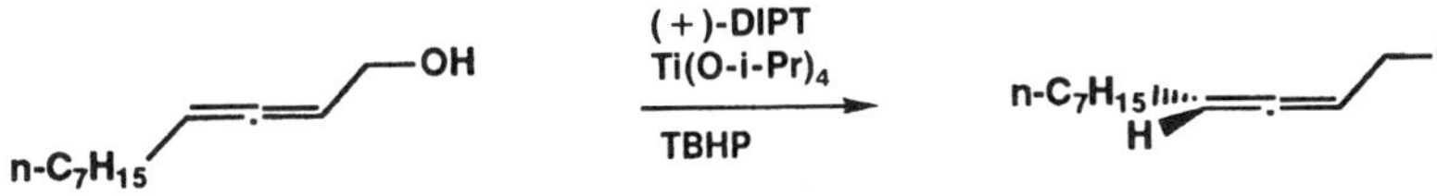

(6A.8)

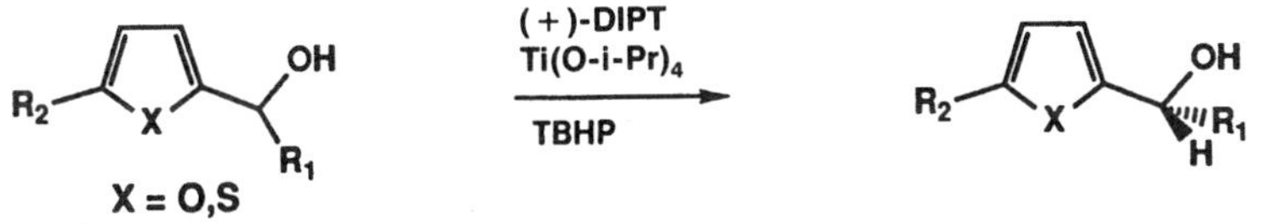

(6A.9)

The asymmetric epoxidation of an allylic alcohol in which the carbinol has been replaced by a silanol has been described [144]. As shown in Eq. 6A.10 [144], (3*E*)-phenylethenyldimethylsilanol is converted to an epoxysilanol in 50% yield with 85–95% ee. Note that here the longer Si–C bonds appear to overcome the restriction to epoxidation associated with a fully substituted C-1 atom in the allylic alcohol series. Fluoride cleavage of the silanol group gives (*S*)-styrene oxide.

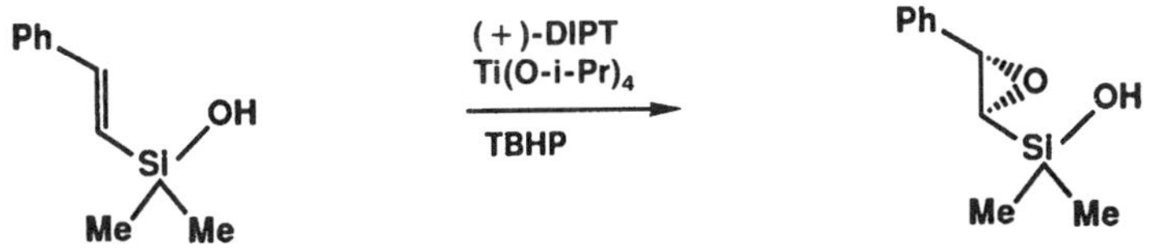

(6A.10)

6A.7.2. Dititanium-Tartrate Complex

The β-hydroxy amines are a class of compounds falling within the generic definition of Eq. 6A.6. When the alcohol is secondary, the possibility for kinetic resolution exists if the Ti-tartrate complex is capable of catalyzing the enantioselective oxidation of the amine to an amine oxide (or other oxidation product). The use of the "standard" asymmetric epoxidation complex (i.e., $T_2(\text{tartrate})_2$) to achieve such an enantioselective oxidation was unsuccessful. However, modification of the complex so that the stoichiometry lies between $Ti_2(\text{tartrate})_1$ and $Ti_2(\text{tartrate})_{1.5}$ leads to very successful kinetic resolutions of β-hydroxyamines. A representative example is shown in Eq. 6A.11 [141b,c]. The oxidation and kinetic resolution of more than 20 secondary β-hydroxyamines [141,145a] provides an indication of the scope of the reaction and of some

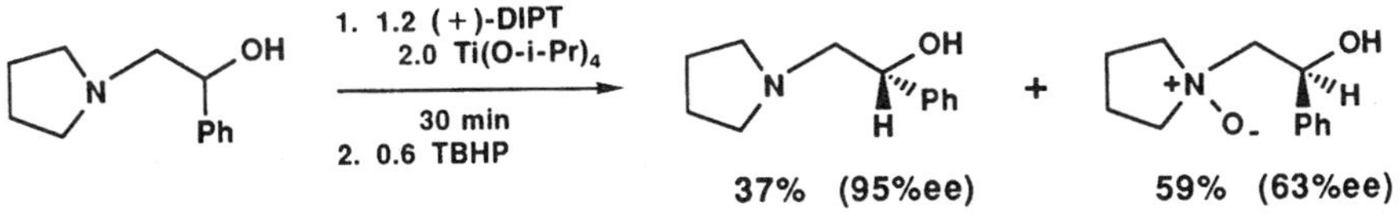

37% (95%ee)   59% (63%ee)

(6A.11)

structural limitations to good kinetic resolution. These results also show a consistent correlation of absolute configuration of the resolved hydroxyamine with the configuration of tartrate used in the catalyst. This correlation is as shown in Eq. 6A.11, where use of (+)-DIPT results in oxidation of the (S)-β-hydroxyamine and leaves unoxidized the (R)-enantiomer.

6A.7.3. Titanium-Tartramide Complexes

A number of derivatives of the tartaric acid structure have been examined as substitutes for the tartrate ester in the asymmetric epoxidation catalyst. These derivatives have included a variety of tartramides, some of which are effective in catalyzing asymmetric epoxidation (although none display the broad consistency of results typical of the esters). One notable example is the dibenzyltartramide, which in a 1:1 ratio (in reality, a 2:2 complex as shown by an X-ray crystallographic structure determination [138]) with Ti(O-i-Pr)$_4$ catalyzes the epoxidation of allylic alcohols with the same enantiofacial selectivity as does the Ti-tartrate ester complex [18]. It is remarkable that, when the ratio of dibenzyltartramide to Ti is changed to 1:2, epoxidation is catalyzed with *reversed* enantiofacial selectivity. These results are illustrated for the epoxidation of α-phenylcinnamyl alcohol (Eq. 6A.12a).

(6A.12a)

(α-Phenylcinnamyl alcohol is a particularly felicitous substrate for asymmetric epoxidation; epoxidation of other allylic alcohols with the 1:2 dibenzyltartramide-Ti complex does not give as high enantioselectivities, but the reversed selectivity is consistent throughout [18].) An extensive listing of tartramides used in the epoxidation of α-phenylcinnamyl alcohol with both 1:1 and 1:2 catalysts has been tabulated elsewhere [6].

6A.7.4. Ti(O-i-Pr)$_2$Cl$_2$-tartrate Complexes

As described in earlier sections of this chapter, certain epoxy alcohols (e.g., the 2-monosubstituted epoxy alcohols) are particularly susceptible to ring-opening processes. The epoxidation catalyst was modified by the use of Ti(O-i-Pr)$_2$Cl$_2$ in place of Ti(O-i-Pr)$_4$, with the intent of controlling the ring-opening reaction. The idea was to open the ring with chloride to produce a chlorodiol [18]. This modification was successful, with 3-chloro-1,2-diols formed in yields of 60–80% with good regioselectivity. Epoxy alcohols were assumed to be intermediates in these reactions, which can be regenerated from the chlorodiols by base-promoted ring closure. Unfortunately, the enantioselectivity of the process is variable, with enantiomeric purities generally in the range of 20–70% ee. A point of interest concerning the chlorohydroxylation process is that the reversal of enantiofacial selectivity from that of the normal asymmetric epoxidation process is not altered by changing the ratio of Ti(O-*i*-Pr)$_2$Cl$_2$ to tartrate from 1:1 to 2:1. Chlorohydroxylation of 2-(6-chloropyridin-2-yl)-2-propen-1-ol (shown in Eq. 6A.12b), followed by closure of the epoxide ring has provided a useful route to the optically active epoxy alcohol in 50% yield and with 90% ee [145b].

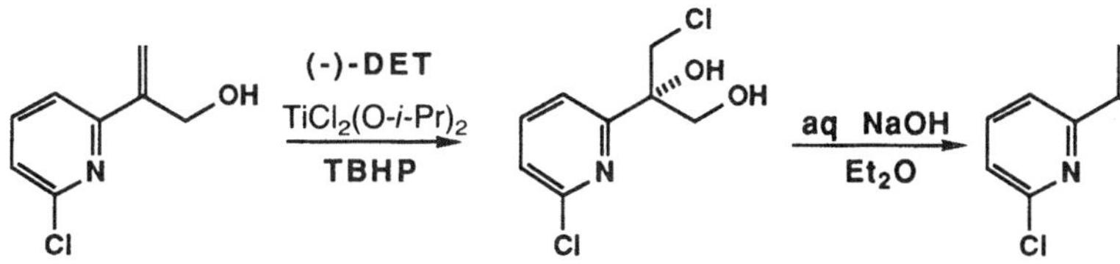

(6A.12b)

6A.8. CONCLUSION

Asymmetric epoxidation of allylic alcohols is a very reliable chemical reaction. More than a decade of experience has confirmed that the Ti-tartrate catalyst is extremely tolerant of structural diversity in the allylic alcohol substrate for epoxidation yet is highly selective in its ability to discriminate between the enantiofaces of the prochiral olefin. Today the practitioner of organic chemistry need provide only the allylic alcohol to perform the reaction. All other reagents and materials required for the reaction are available from supply houses and usually are sufficiently pure as received to be used directly in the asymmetric epoxidation process. [When purchasing *t*-butyl hydroperoxide in prepared solutions, however, the more concentrated 5.5-M solution in isooctane (2,2,4-trimethylpentane) should always be chosen over the 3.0-M solution.] If the considerations presented in this chapter are observed, with attention to the moderately stringent technique outlined, no difficulty should be encountered in performing this reaction.

Before 1986, asymmetric epoxidations frequently were performed by using a stoichiometric amount of the Ti-tartrate catalyst relative to the amount of allylic alcohol. This technique was necessary to obtain a reasonable reaction rate for the epoxidation as well as to drive the reaction to completion. The report in 1986 recommending the addition of activated molecular sieves to the reaction milieu makes it possible to use only catalytic amounts of the Ti-tartrate complex for nearly all asymmetric epoxidations.

The essence of the asymmetric epoxidation process, including correlation of enantiofacial selectivity with tartrate ester stereochemistry, is outlined in Figure 6A.1. No exceptions to the face-selectivity rules shown in Figure 6A.1 have been reported to date. Consequently, one can use this scheme with considerable confidence to predict and assign absolute configuration to the epoxides obtained from *prochiral allylic alcohols*. When allylic alcohols with chiral substituents at C-1, C-2, and/or C-3 are used in the reaction, the assignment of stereochemistry to the newly introduced epoxide group must be done with considerably more care.

A remaining goal related to asymmetric epoxidation is to obtain additional structural information about the Ti-tartrate catalyst as well as about the catalyst loaded with substrate and oxidant.

REFERENCES

1. (a) Henbest, H. B. *Chem. Soc., Spec. Publ.* **1965**, *19*, 83. (b) Ewins, R. C.; Henbest, H. B.; McKervey, M. A. *J. Chem. Soc., Commun.* **1967**, 1085.

2. Katsuki, T.; Sharpless, K. B. *J. Am. Chem. Soc.* **1980**, *102*, 5974.

3. Hanson, R. M.; Sharpless, K. B. *J. Org. Chem.* **1986**, *51*, 1922.

4. Gao, Y.; Hanson, R. M.; Klunder, J. M.; Ko, S. Y.; Masamune, H.; Sharpless, K. B. *J. Am. Chem. Soc.* **1987**, *109*, 5765.

5. Reprinted with permission from *Comprehensive Organic Synthesis*, Vol. 7; Trost, B. M.; Fleming, I. (Ed.); Pergamon:Oxford, 1991; pp. 389–436.

6. Finn, M. G.; Sharpless, K. B. In *Asymmetric Synthesis*, Morrison, J. D. (Ed.); Academic Press: Orlando, FL, 1985; Vol. 5, p. 247.

7. (a) Rossiter, B. In *Asymmetric Synthesis*; Morrison, J. D. (Ed.); Academic Press: Orlando, FL, 1985; Vol. 5, pp. 193. (b) Pfenniger, A. *Synthesis*, **1986**, 89.

8. Zeller, K. P. In *Houben-Weyl, Methoden der organische Chemie*, Vol. E13, Part 2; *Organische Peroxo-verbindungen*; Kropf, H. (Ed.); Thieme: Stuttgart, 1988; pp. 1210–1250.

9. Hanson, R. M. *Chem. Rev.* **1991**, *91*, 437.

10. McGarvey, G. J.; Kimura, M.; Oh, T.; Williams, J. M. *Carbohydr. Chem.* **1984**, *3*, 125.

11. Behrens, C. H.; Sharpless, K. B. *Aldrichim. Acta*, **1983**, *16*, 67.

12. (a) Sharpless, K. B. *Proc. Robert A. Welch Found. Conf. Chem. Res.* **1984**, *27*, 59. (b) Sharpless, K. B. *ChemTech*, **1985**, *15*, 692. (c) Sharpless, K. B. *Chem. Br.* **1986**, *22*, 38.

13. Katsuki, T.; Martin, V. S. Organic Reactions **1996**, *48*, 1.

14. Martin, V. S.; Woodard, S. S.; Katsuki, T.; Yamada, Y.; Ikeda, M.; Sharpless, K. B. *J. Am. Chem. Soc.* **1981**, *103*, 6237.

15. (a) Klunder, J. M.; Ko, S. Y.; Sharpless, K. B. *J. Org. Chem.* **1986**, *51*, 3710. (b) Ko, S. Y.; Sharpless, K. B. *J. Org. Chem.* **1986**, *51*, 5413.

16. (a) Klunder, J. M.; Onami, T.; Sharpless, K. B. *J. Org. Chem.* **1989**, *54*, 1295. (b) Burgos, C. E.; Ayer, D. E.; Johnson, R. A. *J. Org. Chem.* **1988**, *53*, 4973. (c) Guivisdalsky, P. N.; Bittman, R. *Tetrahedron Lett.* **1988**, *29*, 4393. (d) Johnson, R. A.; Burgos, C. E.; Nidy, E. G. *Chem. Phys. Lipids*, **1989**, *50*, 119. (e) Burgos, C. E.; Nidy, E. G.; Johnson, R. A. *Tetrahedron Lett.* **1989**, *30*, 5081. (f) Guivisdalsky, P. N.; Bittman, R. *J. Am. Chem. Soc.* **1989**, *111*, 3077. (g) Byun, H.-S.; Bittman, R. *Tetrahedron Lett.* **1989**, *30*, 2751. (h) Guivisdalsky, P. N.; Bittman, R. *J. Org. Chem.* **1989**, *54*, 4637. (h) Guivisdalsky, P. N.; Bittman, R. *J. Org. Chem.* **1989**, *54*, 4643. (i) (*R*)- and (*S*)-glycidol are available in commercial quantity from ARCO Chemical Company, Newton Square, PA.

17. Ko, S. Y.; Masamune, H.; Sharpless, K. B. *J. Org. Chem.* **1987**, *52*, 667.

18. Lu, L.D.L.; Johnson, R. A.; Finn, M. G.; Sharpless, K. B. *J. Org. Chem.* **1984**, 49, 728.

19. Hill, J. G.; Rossiter, B. E.; Sharpless, K. B. *J. Org. Chem.* **1983**, *48*, 3607.

20. (a) Carlier, P. R.; Sharpless, K. B. *J. Org. Chem.* **1989**, *54*, 4016. (b) Burns, C. J.; Martin, C. A.; Sharpless, K. B. *J. Org. Chem.* **1989**, *54*, 2826.

21. Farrall, M. J.; Alexis, M.; Trecarten, M. *Nouv. J. Chim.* **1983**, *7*, 449.

22. (a) Wadsworth, W. S. *Org. React.*, **1977**, *25*, 73. (b) Walker, B. J. In *Organophosphorus Reagents in Organic Synthesis*; Cadogan, J. I. G. (Ed.); Academic Press:Orlando, FL, 1979; pp. 155–205.

23. Still, W. C.; Gennari, C. *Tetrahedron Lett.* **1983**, *24*, 4405.

24. Mancuso, A. J.; Huang, S.-L.; Swern, D. *J. Org. Chem.* **1978**, *43*, 2480.

25. Nicolaou, K. C.; Daines, R. A.; Uenishi, J.; Li, W. S.; Papahatjia, D. P.; Chakraborty, T. K. *J. Am. Chem. Soc.* **1987**, *109*, 2205; **1988**, *110*, 4672.

26. Schreiber, S. L.; Schreiber, T. S.; Smith, D. B. *J. Am. Chem. Soc.* **1987**, *109*, 1525.

27. Denmark, S. E.; Jones, T. K. *J. Org. Chem.* **1982**, *47*, 4595.

28. Marvell, E. N.; Li, T. *Synthesis*, **1973**, 457.

29. Jurczak, J.; Pikul, S.; Bauer, T. *Tetrahedron*, **1986**, *42*, 447.

30. Breitgoff, D.; Laumen, K.; Schneider, M. P. *J. Chem. Soc., Chem. Commun.* **1986**, 1523.

31. Kleemann, A.; Wagner, R. *Glycidol*; Huthig: *New York, 1981.*

32. Gao, Y.; Sharpless, K. B. *J. Org. Chem.* **1988**, *53*, 4081.

33. Dung, J. S.; Armstrong, R. W.; Anderson, O. P.; Williams, R. M. *J. Org. Chem.* **1983**, *48*, 3592.

34. Meister. C.; Scharf, H. D. *Justus Liebigs Ann. Chem.* **1983**, 913.

35. Tanner, D.; Somfai, P. *Tetrahedron*, **1986**, *42*, 5985.

36. Giese, B.. Rupaner, R. *Justus Liebigs Ann. Chem.* **1987**, 231.

37. Mori, K.; Ebata, T; Takechi, S. *Tetrahedron*, **1984**, *40*, 1761.

38. (a) Schweiter, M. J.; Sharpless, K. B. *Tetrahedron Lett.* **1985**, *26*, 2543. (b) Adam, W.; Griesbeck, A.; Staab, E. *Tetrahedron Lett.* **1986**, *27*, 2839. (c) Adam, W.; Braun, M.; Griesbeck, A.; Lucchini, V.: Staab, E.; Will, B. *J. Am. Chem. Soc.* **1989**, *111*, 203.

39. (a) White, J. D.; Jayasinghe, L. R. *Tetrahedron Lett.* **1988**, *29*, 2138. (b) White, J. D.; Amedio, J. C., Jr.; Gut, S.: Jayasinghe, L. *J. Org. Chem.* **1989**, *54*, 4268.

40. (a) Masamune, S.: Choy, W.; Petersen, J. S.; Sita, L. R. *Angew. Chem., Int. Ed. Engl.* **1985**, *24*, 1. (b) Nishikimi, Y.; Iimori, 1; Sodeoka, M.; Shibasaki, M. *J. Org. Chem.* **1989**, *54*, 3354.

41. Rossiter, B. E.; Katsuki, T.; Sharpless K. B. *J. Am. Chem. Soc.* **1981**, *103*, 464.

42. Baker, R.; Cummings, W. J.; Hayes, J. F.; Kumar, A. *J. Chem. Soc., Chem. Commun.* **1986**, 1237.

43. Kuroda, C.; Theramongkol, P.; Engebrecht, J. R.; White, J. D. *J. Org. Chem.* **1986**, *51*, 956.

44. Honda, M.; Katsuki, T.; Yamaguchi, M. *Tetrahedron Lett.* **1984**, *25*, 3857.

45. Gorthy, L. A.; Vairamani, M.; Djerassi, C. *J. Org. Chem.* **1984**, *49*, 1511.

46. Hanessian, S.; Ugolini, A.; Dube, D.; Hodges, P. J.; Andre, C. *J. Am. Chem. Soc.* **1986**, *108*, 2776.

47. Wershofen. S.; Scharf, H. D. *Synthesis*, **1988**, 854.

48. Tung, R. D.; Rich. D. H. *Tetrahedron Lett.* **1987**, *28*, 1139.

49. Molander, G. A.; Hahn, G. *J. Org. Chem.* **1986**, *51*, 2596.

50. Corey, E. J., Pyne, S. G.; Su, W. G. *Tetrahedron Lett.* **1983**, *24*, 4883.

51. Furukawa, J.; Iwasaki, S.; Okuda, S. *Tetrahedron Lett.* **1983**, *24*, 5257.

52. Kitano, Y.; Kobayashi, Y.; Sato, F. *J. Chem. Soc., Chem. Commun.* **1985**, 498.

53. Roush, W. R.; Adam, M. A. *J. Org. Chem.* **1985**, *50*, 3752.

54. Oehlschlager, A. C.; Johnston, B. D. *J. Org. Chem.* **1987**, *52*, 940.

55. Lai, C. K.; Gut, M. *J. Org. Chem.* **1987**, *52*, 685.

56. Dolle, R. E.; Nicolaou, K. C. *J. Am. Chem. Soc.* **1985**, *107*, 1691.

57. Pridgen, L. N.; Shilcrat, S. C.; Lantos, I. *Tetrahedron Lett.* **1984**, *25*, 2835.

58. Baker, S. R.; Boot, J. R.; Morgan, S. E.; Osborne, D. J.; Ross, W. J.; Shrubsall, P. R. *Tetrahedron Lett.* **1983**, *24*, 4469.

59. Katsuki, T.; Lee, A.W.M.; Ma, P.; Martin, V. S.; Masamune, S.; Sharpless, K. B.; Tuddenham,. D.; Walker, F. J. *J. Org. Chem.* **1982**, *47*, 1373.

60. Ma, P.; Martin, V. S.; Masamune, S.; Sharpless, K. B.; Viti, S. M. *J. Org. Chem.* **1982**, *47*, 1378.

61. (a) Brunner, H.; Sicheneder, A. *Angew Chem., Int. Ed.* **1988**, *27*, 718. (b) Evans, D. A.; Williams, J. M. *Tetrahedron Lett.* **1988**, *29*, 5065. (c) Hughes, P.; Clardy, J. *J. Org. Chem.* **1989**, *54*, 3260. (d) Nicolaou, K. C.; Prasad, C.V.C.; Somers, P. K.; Hwang, C.-K. *J. Am. Chem. Soc.* **1989**, *111*, 5330.

62. (a) Minami, N.; Ko, S. S.; Kishi, Y. *J. Am. Chem. Soc.* **1982**, *104*, 1109. (b) Roush, W. R.; Adam, M. A.; Walts, A. E.; Harris, D. J. *J. Am. Chem. Soc.* **1986**, *108*, 3422.

63. (a) Ko, S. Y.; Lee, A.W.M.; Masamune, S.; Reed, L. A., III; Sharpless, K. B.; Walker, F. J. *Science* **1983**, *220*, 949. (b) Ko, S. Y.; Lee, A.W.M.; Masamune, S.; Reed, L. A., III; Sharpless, K. B.; Walker, F. J. *Tetrahedron*, **1990**, *46*, 245. (c) Reed, L. A., III; Ito, Y.; Masamune, S.; Sharpless, K. B. *J. Am. Chem. Soc.* **1982**, *104*, 6468.

64. Wrobel, J. E.; Ganem, B. *J. Org. Chem.* **1983**, *48*, 3761.

65. Behrens, C. H.; Ko, S. Y.; Sharpless, K. B.; Walker, F. J. *J. Org. Chem.* **1985**, *50*, 5687.

66. (a) Hoye, T. R.; Suhadolnik, J. C. *J. Am. Chem. Soc.* **1985**, *107*, 5312. (b) Hoye, T. R.; Suhadolnik, J. C. *Tetrahedron* **1986**, *42*, 2855.

67. Doherty, A. M.; Ley, S. V. *Tetrahedron Lett.* **1986**, *27*, 105.

68. Baldwin, J. E.; Flinn, A. *Tetrahedron Lett.* **1987**, *28*, 3605.

69. Caron, M.; Sharpless, K. B. *J. Org. Chem.* **1985**, *50*, 1557.

70. Chong, J. M.; Sharpless, K. B. *J. Org. Chem.* **1985**, *50*, 1560.

71. Baker, R.; Swain, C. J.; Head, J. C. *J. Chem. Soc., Chem. Commun.* **1986**, 874.

72. Wood, R. D.; Ganem, B. *Tetrahedron Lett.* **1982**, *23*, 707.

73. Takahashi, T.; Miyazawa, M.; Veno, H.; Tsuji, J. *Tetrahedron Lett.* **1986**, *27*, 3881.

74. Mills, L. S.; North, P. C. *Tetrahedron Lett.* **1983**, *24*, 409.

75. (a) Suzuki, M.; Morita, Y.; Yanagisawa, A.; Noyori, R.; Baker, B. J.; Scheur, B. J. *J. Am. Chem. Soc.* **1986**, *108*, 5021. (b) Suzuki, M.; Morita, Y.; Yanagisawa, A.; Baker, B. J.; Scheuer, P. J.; Noyori, R. *J. Org. Chem.* **1988**, *53*, 286.

76. Denis, J.-N.; Greene, A. E.; Serra, A. A.; Luche, M.-J. *J. Org. Chem.* **1986**, *51*, 46.

77. Sharpless, K. B.; Behrens, C. H.; Katsuki, T.; Lee, A.W.M.; Martin, V. S.; Takatani, M.; Viti, S. M.; Walker, F. J.; Woodard, S. S. *Pure Appl. Chem.* **1983**, *55*, 589.

78. (a) Nagaoka, H.; Kishi, Y. *Tetrahedron* **1981**, *37*, 3873. (b) Hirai, Y.; Chintani, M.; Yamazaki, T.; Momose, T. *Chem. Lett.* **1989**, 1449. (c) Ebata, T.; Mori, K. *Agric. Biol. Chem.* **1989**, *53*, 801. (d) Mori, K.; Takeuchi, T. *Justus Liebigs Ann. Chem.* **1989**, 453. (e) Boeckman, R. K., Jr.; Pruitt, J. R. *J. Am. Chem. Soc.* **1989**, *111*, 8286.

79. Still, W. C.; Ohmizu, H. *J. Org. Chem.* **1981**, *46*, 5242.

80. Garner, P.; Park, J. M.; Rotello, V. *Tetrahedron Lett.* **1985**, *26*, 3299.

81. Meyers, A. I.; Hudspeth, J. P. *Tetrahedron Lett.* **1981**, *22*, 3925.

82. Niwa, N.; Miyachi, Y.; Uosaki, Y.; Yamada, K. *Tetrahedron Lett.* **1986**, *27*, 4601.

83. Takabe, K.; Okisaki, K.; Uchiyama, Y.; Katagiri, T.; Yoda, H. *Chem. Lett.* **1985**, 561.

84. (a) Mori, K.; Ueda, H. *Tetrahedron* **1981**, *37*, 2581. (b) Nicolaou, K. C.; Prasad, C.V.C.; Hwang, C.-K.; Duggan, M. E.; Veale, C. A. *J. Am. Chem. Soc.* **1989**, *111*, 5321.

85. Medina, J. C.; Kyler, K. S. *J. Am. Chem. Soc.* **1988**, *110*, 4818.

86. Aziz, M.; Rouessac, F. *Tetrahedron* **1988**, *44*, 101.

87. Morimoto, Y.; Oda, K.; Shirahama, H.; Matsumoto, T.; Omura, S. *Chem. Lett.* **1988**, 909.

88. (a) Takahashi, K.; Ogata, M. *J. Org. Chem.* **1987**, *52*, 1877. (b) Naruta, Y.; Nishigaichi, Y.; Maruyama, K. *Tetrahedron Lett.* **1989**, *30*, 3319. (c) Nicolaou, K. C.; Duggan, M. E.; Hwang, C.-K. *J. Am. Chem. Soc.* **1989**, *111*, 6676.

89. Rastetter, W. H.; Adams, J. *Tetrahedron Lett.* **1982**, *23*, 1319.

90. Evans, D. A.; Bender, S. L.; Morris, J. *J. Am. Chem. Soc.* **1988**, *110*, 2506.

91. Reddy, K. S.; Ko, O. H.; Ho, D.; Persons, P. E.; Cassidy, J. M. *Tetrahedron Lett.* **1987**, *28*, 3075.

92. Yamada, S.; Shiraishi, M.; Ohmori, M.; Takayama, H. *Tetrahedron Lett.* **1984**, *25*, 3347.

93. Mori, K.; Okada, K. *Tetrahedron* **1985**, *41*, 557.

94. Roush, W. R.; Blizzard, T. A. *J. Org. Chem.* **1984**, *49*, 4332.

95. Bonadies, F.; Rossi, G.; Bonini, C. *Tetrahedron Lett.* **1984**, *25*, 5431.

96. Sodeoka, M.; Iimori, T.; Shibasaki, M. *Tetrahedron Lett.* **1985**, *26*, 6497.

97. (a) Meyers, A. G.; Porteau, P. J.; Handel, T. M. *J. Am. Chem. Soc.* **1988**, *110*, 7212. (b) Williams, D. R.; Brown, D. L.; Benbow, J. W. *J. Am. Chem. Soc.* **1989**, *111*, 1923.

98. Erickson, T. J. *J. Org. Chem.* **1986**, *51*, 934.

99. Marshall, J. A.; Jenson, T. M. *J. Org. Chem.* **1984**, *49*, 1707.

100. Hamon, D.P.G.; Shirley, N. J. *J. Chem. Soc., Chem. Commun.* **1988**, 425.

101. Pettersson, L.; Frejd, T.; Magnusson, G. *Tetrahedron Lett.* **1987**, *28*, 2753.

102. Rizzi, J. P.; Kende, A. S. *Tetrahedron* **1984**, *40*, 4693.

103. Acemoglu, M.; Uebelhart, P.; Rey, M.; Eugster, C. H. *Helv. Chim. Acta* **1988**, *71*, 931.

104. Carlier, P. R.; Mungall, W. S.; Schroder, G.; Sharpless, K. B. *J. Am. Chem. Soc.* **1988**, *110*, 2978.

105. (a) Kitano, Y.; Matsumoto, T.; Takeda, Y.; Sato, F. *J. Chem. Soc., Chem. Commun.* **1986**, 1323. (b) Kitano, Y.; Matsumoto, T.; Sato, F. *Tetrahedron* **1988**, *44*, 4073. (c) Kitano, Y.; Okamoto, S.; Sato, F. *Chem. Lett.* **1989**, 2163.

106. (a) Russell, A. T.; Procter, G. *Tetrahedron Lett.* **1987**, *28*, 2041. (b) Procter, G.; Russell, A. T.; Murphy, P. J.; Tan, T. S.; Mather, A. N. *Tetrahedron* **1988**, *44*, 3953.

107. (a) Katsuki, T. *Tetrahedron Lett.* **1984**, *25*, 2821. (b) Schwab, J. M.; Ho, C.-K. *J. Chem. Soc., Chem. Commun.* **1986**, 872. (c) Schwab, J. M.; Ray, T.; Ho, C.-K. *J. Am. Chem. Soc.* **1989**, *111*, 1057.

108. Kitano, Y.; Matsumoto, T.; Wakasa, T.; Okamoto, S.; Shimazaki, T.; Kobayashi, Y.; Sato, F.; Miyaji, K.; Arai, K. *Tetrahedron Lett.* **1987**, *28*, 6351.

109. Kitano, Y.; Matsumoto, T.; Okamoto, S.; Shimazaki, T.; Kobayashi, Y.; Sato, F. *Chem. Lett.* **1987**, 1523.

110. Roush, W. R.; Spada, A. P. *Tetrahedron Lett.* **1983**, *24*, 3693.

111. Aristoff, P. A.; Johnson, P. D.; Harrison, A. W. *J. Am. Chem. Soc.* **1985**, *107*, 7967.

112. Overman, L. E.; Lin, N.-H. *J. Org. Chem.* **1985**, *50*, 3670.

113. Aggarwal, S. K.; Bradshaw, J. S.; Eguchi, M.; Parry, S.; Rossiter, B. E.; Markides, K. E.; Lee, M. L. *Tetrahedron* **1987**, *43*, 451.

114. Ogata, M.; Matsumoto, H.; Takahashi, K.; Shimizu, S.; Kida, S.; Murahayashi, A.; Shiro, M.; Tawara, K. *J. Med. Chem.* **1987**, *30*, 1054.

115. (a) Sugiyama, S.; Honda, M.; Komori, T. *Justus Liebigs Ann. Chem.* **1988**, 619. (b) Tanaka, A.; Suzuki, H.; Yamashita, K. *Agric. Biol. Chem.* **1989**, *53*, 2253. (c) Chamberlin, A. R.; Dezube, M.; Reich, S. H.; Sall, D. J. *J. Am. Chem. Soc.* **1989**, *111*, 6247. (d) Rama Rao, A. V.; Khrimian, A. P.; Radha Krishna, P.; Yagadiri, P.; Yadav, J. S. *Synth. Commun.* **1989**, *18*, 2325. (e) Discordia, R. P.; Dittmer, D. C. *J. Org. Chem.* **1990**, *55*, 1414.

116. Marshall, J. A.; Flynn, K. E. *J. Am. Chem. Soc.* **1982**, *104*, 7430.

117. Mihelich, E. D. Unpublished results.

118. Alvarez, E.; Manta, E.; Martin, J. D.; Rodriquez, M. L.; Ruiz-Perez, C. *Tetrahedron Lett.* **1988**, *29*, 2093.

119. Ronald, R. C.; Ruder, S. M.; Lillie, T. S. *Tetrahedron Lett.* **1987**, *28*, 131.

120. Ekhato, I. V.; Silverton, J. V.; Robinson, C. H. *J. Org. Chem.* **1988**, *53*, 2180.

121. (a) Marshall, J. A.; Flynn, K. E. *J. Am. Chem. Soc.* **1984**, *106*, 723. (b) Marshall, J. A.; Audia, V. H. *J. Org. Chem.* **1987**, *52*, 1106.

122. (a) Hatakeyama, S.; Sakurai, K.; Takano, S. *J. Chem. Soc., Chem. Commun.* **1985**, 1759. (b) Hafele, B.; Schroter, D.; Jager, V. *Angew. Chem., Int. Ed. Engl.* **1986**, *25*, 87. (c) Babine, R. E. *Tetrahedron Lett.* **1986**, *27*, 5791. (d) Schreiber, S. L.; Schreiber, T. S.; Smith, D. B. *J. Am. Chem. Soc.* **1987**, *109*, 1525. (e) Askin, D.; Volante, R. P.; Reamer, R. A.; Ryan, K. M.; Shinkai, I. *Tetrahedron Lett.* **1988**, *29*, 277.

123. Bergens, S.; Bosnich, B. *Comments Inorg. Chem.* **1987**, *6*, 85.

124. Schreiber, S. L.; Goulet, M. T.; Schule, G. *J. Am. Chem. Soc.* **1987**, *109*, 4718.

125. Mori, K.; Seu, Y. B. *Tetrahedron* **1985**, *41*, 3429.

126. Mori, K.; Otsuka, T. *Tetrahedron* **1985**, *41*, 553.

127. Lewis, M. D.; Duffy, J. P.; Blough, B. E.; Crute, T. D. *Tetrahedron Lett.* **1988**, *29*, 2279.

128. Corey, E. J.; Tramontano, A. *J. Am. Chem. Soc.* **1984**, *106*, 462.

129. (a) Roush, W. R.; Brown, R. J. *J. Org. Chem.* **1982**, *47*, 1373. (b) Roush, W. R.; Brown, R. J. *J. Org. Chem.* **1983**, *48*, 5093. (c) Bulman-Page, P. C.; Carefull, J. F.; Powell, L. H.; Sutherland, I. O. *J. Chem. Soc., Chem. Commun.* **1985**, 822.

130. (a) Kufner, U.; Schmidt, R. R. *Angew Chem., Int. Ed. Engl.* **1986**, *25*, 89. (b) Frater, G.; Muller, J. *Helv. Chim. Acta.* **1989**, *72*, 653. (c) Ibuka, T.; Tanaka, M.; Yamamoto, Y. *J. Chem. Soc., Chem. Commun.* **1989**, 967.

131. Katsuki, T.; Sharpless, K. B. Unpublished results.

132. Rossiter, B. E.; Sharpless, K. B. *J. Org. Chem.* **1984**, *49*, 3707.

133. Ikeqami, S.; Katsuki, T.; Yamaguchi, M. *Chem. Lett.* **1987**, 83.

134. (a) Corey, E. J.; Ha, D.-C. *Tetrahedron Lett.* **1988**, *29*, 3171. (b) Hosokawa, T.; Kono, T.; Shinohara, T.; Murahashi, S.-I. *J. Organometal. Chem.* **1989**, 370, C13.

135. Sharpless, K. B.; Woodard, S. S.; Finn, M. G. *Pure Appl. Chem.* **1983**, *55*, 1823.

136. (a) Woodard, S. S.; Finn, M. G.; Sharpless, K. B. *J. Am. Chem. Soc.* **1991**, *113*, 106. (b) Finn, M. G.; Sharpless, K. B. *J. Am. Chem. Soc.* **1991**, *113*, 113.

137. Corey, E. J. *J. Org. Chem.* **1990**, *55*, 1693.

138. Williams, I. D.; Pedersen, S. F.; Sharpless, K. B.; Lippard, S. J. *J. Am. Chem. Soc.* **1984**, *106*, 6430.

139. Mimoun, H.; Chaumette, P.; Mignard, M.; Saussine, L.; Fischer, J.; Weiss, R. *Nouv. J. Chim.* **1983**, *7*, 467.

140. (a) Jorgensen, K. A.; Wheeler, R. A.; Hoffmann, R. *J. Am. Chem. Soc.* **1987**, *109*, 3240. (b) Jorgensen, K. A. *Chem. Rev.* **1989**, *89*, 431.

141. (a) Katsuki, T.; Sharpless, K. B. U.S. Patent 4,471,130, Sept. 11, 1984; *Chem. Abstr.* **1985**, *102*, 24872m. (b) Miyano, S.; Lu, L. D.-L.; Viti, S. M.; Sharpless, K. B. *J. Org. Chem.* **1983**, *48*, 3611. (c) Miyano, S.; Lu, L. D.-L.; Viti, S. M.; Sharpless, K. B. *J. Org. Chem.* **1985**, *50*, 4350.

142. (a) Kobayashi, Y.; Kusakabe, M.; Kitano, Y.; Sato, F. *J. Org. Chem.* **1988**, *53*, 1586. (b) Kamatani, T.; Tsubuki, M.; Tatsuzaki, Y.; Honda, T. *Heterocycles* **1988**, *27*, 2107. (c) Kusakabe, M.; Kitano, Y.; Kobayashi, Y.; Sato, F. *J. Org. Chem.* **1989**, *54*, 2085. (d) Kusakabe, M.; Sato, F. *J. Org. Chem.* **1989**, *54*, 3486. (e) Kamatani, T.; Tatsuzaki, Y.; Tsubuki, M.; Honda, T. *Heterocycles* **1989**, *29*, 1247.

143. Kitano, Y.; Kusakabe, M.; Kobayashi, Y.; Sato, F. *J. Org. Chem.* **1989**, *54*, 994.

144. Chan, T. H.; Chen, L. M.; Wang, D. *J. Chem. Soc., Chem. Commun* **1988**, 1280.

145. (a) Kihara, M.; Ohnishi, K.; Kobayashi, S. *J. Heterocyclic Chem.* **1988**, *25*, 161. (b) Jikihara, T.; Katsurada, M.; Ikeda, O.; Yoneyama, K.; Takematsu, T. Presentation ORGN 220; 198th National Meeting of the American Chemical Society; Sept. 15–20, 1989.

CHAPTER 6A ADDENDUM—1999

XIN WEN AND IWAO OJIMA

Department of Chemistry, State University of New York at Stony Brook

ADDENDUM TO SECTION 6A.5.3.

Sharpless epoxidation was used to synthesize a key intermediate for the synthesis of 2-keto-3-deoxy-D-gluconic acid (KDG).[1] Epoxide **1** was obtained in 80% yield and >90% ee (Eq. 6AA.1).

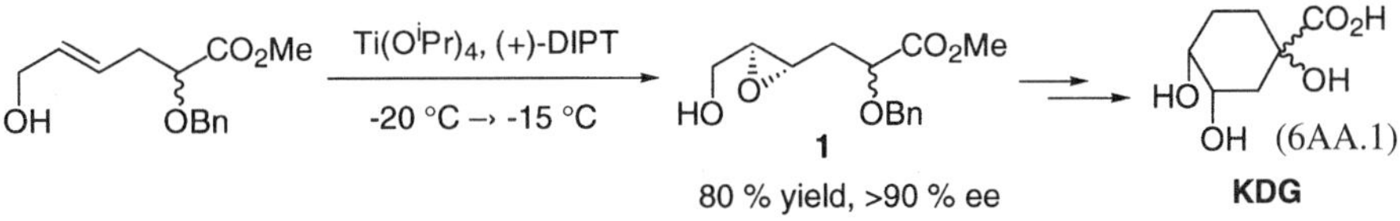

80 % yield, >90 % ee   **KDG**

ADDENDUM TO SECTIONS 6A.5.3. AND 6A.5.9.

An anorectic agent, (S)-fenfluramine [(S)-**5**], as well as alcohols, (R)-**4** and (S)-**4**, were synthesized by using Sharpless asymmetric epoxidation.[2] The asymmetric epoxidation of

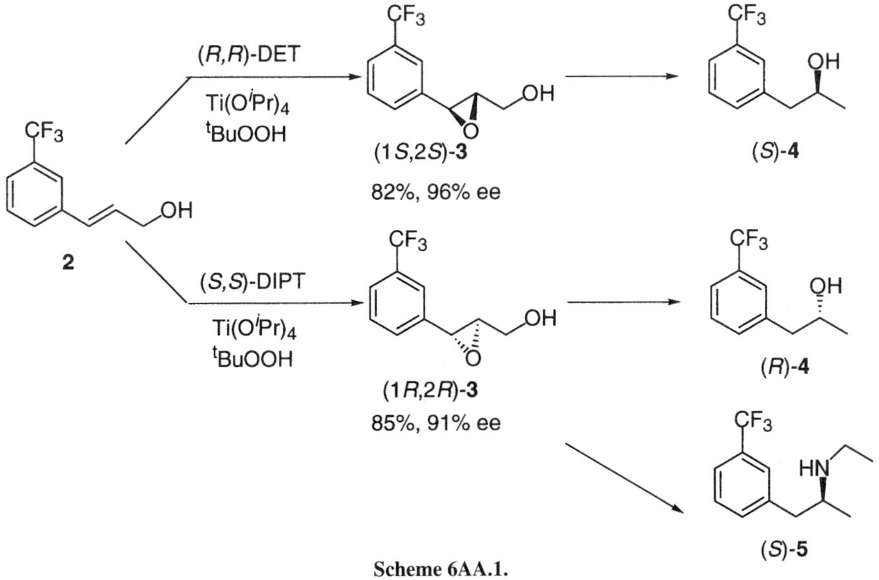

Scheme 6AA.1.

Catalytic Asymmetric Synthesis, Second Edition, Edited by Iwao Ojima
ISBN 0-471-29805-0   Copyright © 2000 Wiley-VCH, Inc.

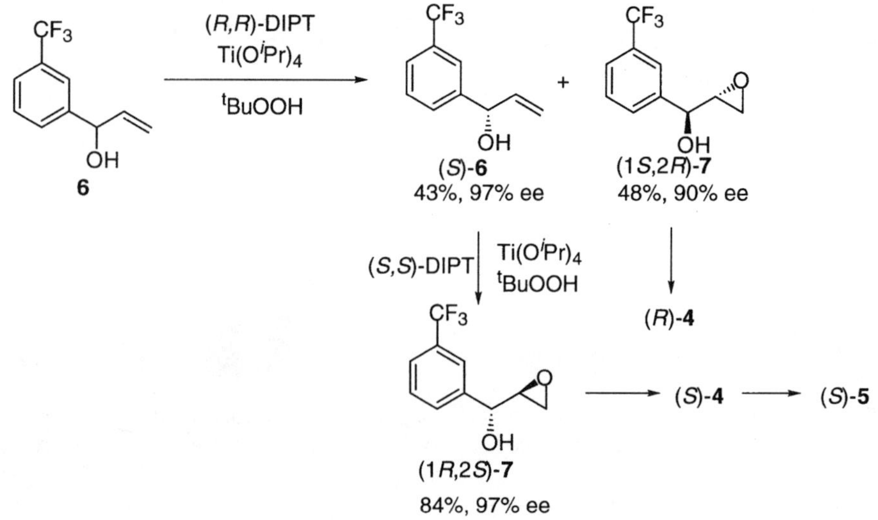

Scheme 6AA.2.

primary allylic alcohol **2** by using (*R,R*)-DET gave epoxide (1*S*,2*S*)-**3** in 82% yield and 96% ee. In the same manner, (1*R*,2*R*)-**3** was obtained in 85% yield and 91% ee by using (*S,S*)-DIPT (Scheme 6AA.1).[2] Epoxide (1*R*,2*R*)-**3** was converted to (*S*)-**5** (90% ee) via an aziridine in 5 steps. Kinetic resolution of secondary allylic alcohol **6** afforded 43% of alcohol (*S*)-**6** (97% ee) and 48% of epoxyalcohol (1*S*,2*R*)-**7** (99% erythro, 90% ee). Second Sharpless epoxidation of (*S*)-**6** afforded (1*R*,2*S*)-**7** in 84% yield and 97% ee, which was converted to (*S*)-**4** (95% ee) (Scheme 6AA.2).[2] Synthesis of (*S*)-**5** from (*S*)-**4** was performed previously.

ADDENDUM TO SECTION 6A.5.5.

Double Sharpless epoxidation was applied to the syntheses of both enantiomers of 2,3:22,23-diepoxysqualene-1,24-diol, **8a** and **8b**, in 58% and 37% yields, respectively.[3] Tetradeuterated

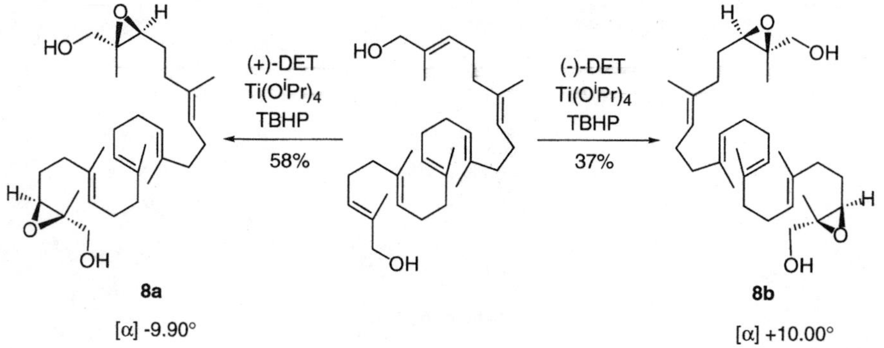

Scheme 6AA.3.

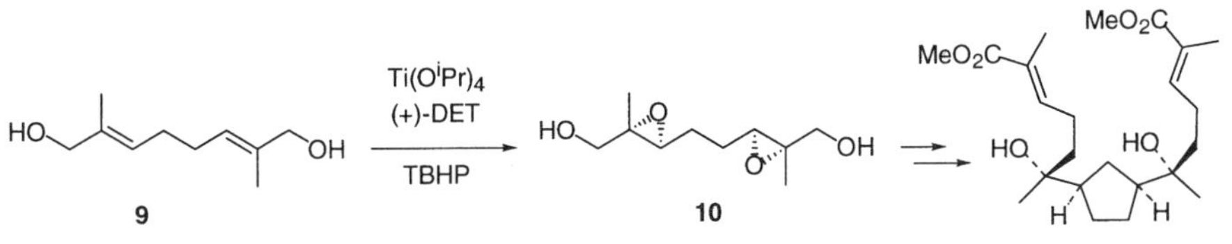

Scheme 6AA.4.

derivative of the diepoxydiol **8b** (the positions of deuterium labeling are indicated by asterisks) was also obtained in 60% yield (Scheme 6AA.3).[3] These compounds are useful to the study of postsqualene steroid biosynthesis.

Double Sharpless epoxidation was also applied to the synthesis of a key intermediate **10** to the meso-compound **11** that is related to teurilene, a bioactive polycyclic triterpene isolated from red alga *Laurencia obtusa*.[4] The reaction of (E,E)-Bisallylic alcohol **9** gave the bisglycidic alcohol **10** with 80% de and 89% ee for each epoxidation (Scheme 6AA.4).

Sharpless epoxidation of electrophilic (E)-2-cyanoallylic alcohols was successfully carried out using reaction temperature at $-20°C \rightarrow 5°C$ instead of the standard $-20°C$ to give the corresponding epoxides in 49–64% yields and 96–98% ee's (Eq. 6AA.2).[8] The Z isomer was completely unreactive under these conditions.

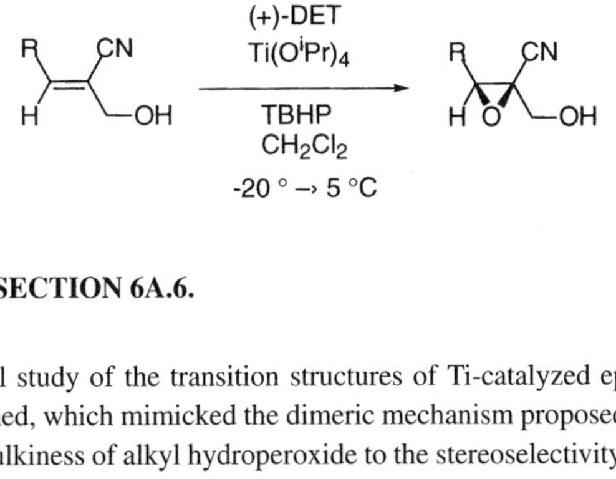

(6AA.2)

ADDENDUM TO SECTION 6A.6.

A density functional study of the transition structures of Ti-catalyzed epoxidation of allylic alcohol was performed, which mimicked the dimeric mechanism proposed by Sharpless et al.[5] Importance of the bulkiness of alkyl hydroperoxide to the stereoselectivity, the conformational features of tartrate esters in the epoxidation transition structure, and the loading of allylic alcohol in the dimeric transition structure model were pointed out.

ADDENDUM TO SECTION 6A.7.1.

Sharpless epoxidation of a variety of E-alkenylsilanols was investigated. The reactions gave the corresponding silylepoxides with 85–95% ee except for the case of phenylethenyldiphenylsilanol due to the steric effects of this bulky group on asymmetric induction (7–20% ee) (Scheme 6AA.5).[6]

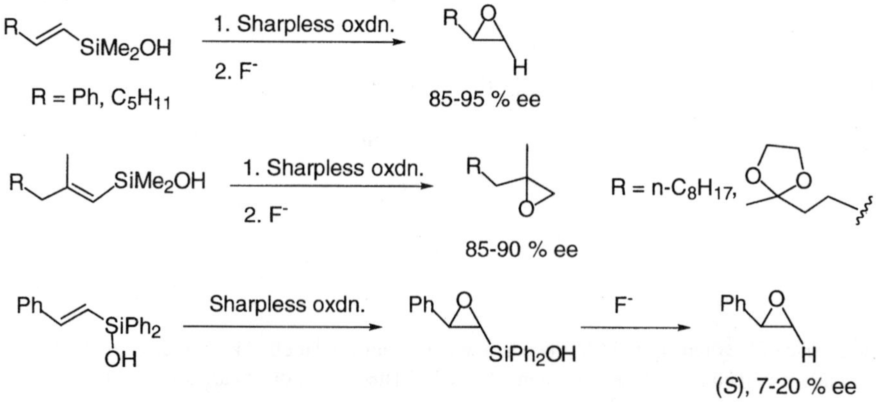

Scheme 6AA.5.

Sharpless epoxidation of (E)-(1,2-dialkyl)vinylsilanols **13**, prepared from hydrolysis of (E)-(1,2-dialkyl)vinyldimethylbutoxysilanes **12**, gave silylepoxides **14**, which were treated with Et$_4$NF in MeCN to afford epoxides **15** in 62–70% overall yield and 44–70% ee (Scheme 6AA.6).[7] The overall transformation can be considered as asymmetric epoxidation of simple internal alkenes. This approach was applied to the synthesis of a naturally occurring insect sex pheromone (+)-disparlure.[7]

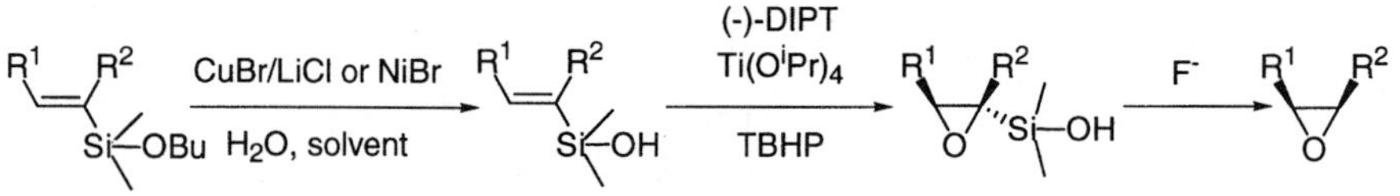

Scheme 6AA.6.

ADDENDUM TO SECTION 6A.7.

The kinetic resolution of racemic tertiary hydroperoxides via catalytic Sharpless epoxidation with various allylic alcohols was investigated (Eq. 6AA.3).[9] The reaction of 1-cyclohexyl-1-phenylethyl hydroperoxide gave partially resolved hydroperoxide with up to 29% at about 50% conversion.

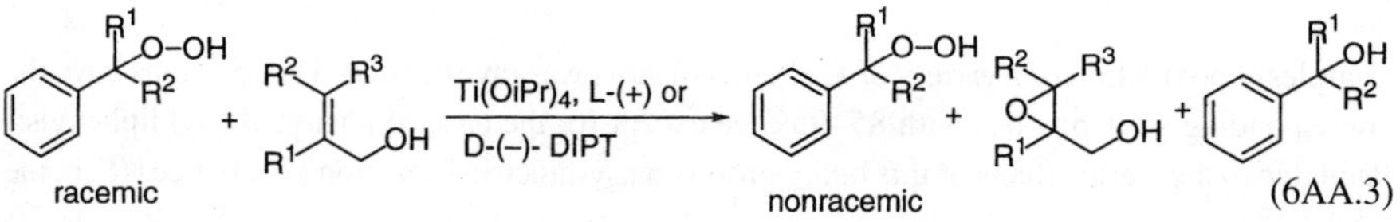

(6AA.3)

REFERENCES

1. Shimizu, M.; Yoshida, A.; Mikami, K. *Syn Lett* **1996**, 1112.

2. Goument, B.; Duhamel, L.; Mauge, R. *Tetrahedron* **1994**, *50*, 171.

3. Hauptfleisch, R.; Franck, B. *Tetrahedron Lett.* **1997**, *38*, 383.

4. Lindel, T.; Franck, B. *Tetrahedron Lett.* **1995**, *36*, 9465.

5. Wu, Y.; Lai, D.K.W. *J. Am. Chem. Soc.* **1995**, *117*, 11327.

6. Chan, T. H.; Chen, L. M.; Wang, D.; Li, L. H. *Can. J. Chem.* **1993**, *71*, 60.

7. Li, L. H.; Wang, D.; Chan, T. H. *Tetrahedron Lett.* **1997**, *38*, 101.

8. Aiai, M; Robert, A.; Baudy-Floeh, M; Le Grel, P. *Tetrahedron (A)* **1995**, *6*, 2249.

9. (a) Hoft, E.; Hamann, H.-J.; Kunath, A. *J. Prakt. Chem.* **1994**, *336*, 534. (b) Hoft, E.; Hamann H.-J.; Kunath, A.; Ruffer, L. *Tetrahedron (A)* **1992**, *3*, 507.

6B

ASYMMETRIC EPOXIDATION OF UNFUNCTIONALIZED OLEFINS AND RELATED REACTIONS

TSUTOMU KATSUKI

Department of Chemistry, Faculty of Science, Kyushu University, 6-10-1 Hakozaki, Higashi-ku, Fukuoka 812-8581, Japan

6B.1. INTRODUCTION

Three-membered heterocyclic ring compounds such as oxiranes (epoxides) and aziridines are highly strained and react with various nucleophiles to provide the ring-opening products stereospecifically [1]. The absolute and relative configurations of the resulting two functionalized carbons in the ring-opening products are determined by the configurations of the starting heterocyclic compounds (Scheme 6B.1). Therefore, three-membered heterocyclic compounds, especially optically active ones, serve as useful building blocks for the synthesis of a wide range of organic compounds and the asymmetric synthesis of these heterocyclic compounds is one of the major objectives in organic synthesis.

Although many methodologies are available for the asymmetric synthesis of this class of compounds, oxidation of olefins with electrophilic chiral oxene, nitrene, or their equivalents is

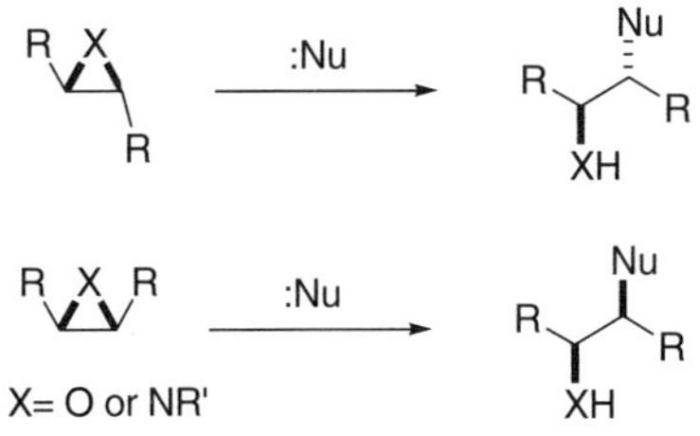

Scheme 6B.1.

Catalytic Asymmetric Synthesis, Second Edition, Edited by Iwao Ojima
ISBN 0-471-29805-0 Copyright © 2000 Wiley-VCH, Inc.

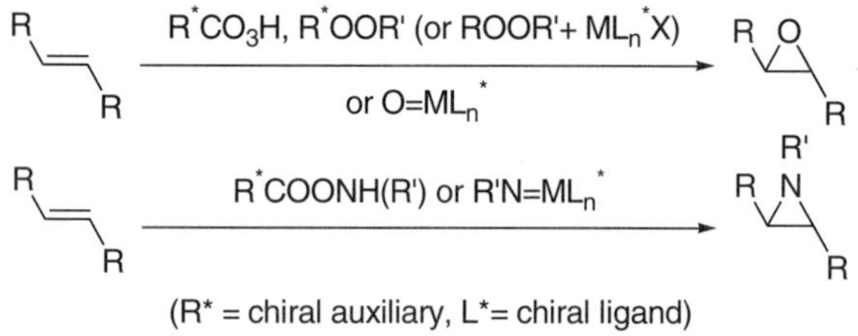

Scheme 6B.2.

the most important one from the practical point of view: i) a single-step synthesis, ii) ready availability of various olefins and oxidants, and iii) mild reaction conditions (high chemoselectivity). Representative chiral oxene equivalents are optically active peracids, peroxides (in some cases, a combination of achiral peroxide and chiral-metal catalyst), and oxo-metal species (metal oxenoids) bearing a chiral ligand. Nitrene equivalents are chiral acyloxy amine derivatives and chiral imino-metal species (metal nitrenoids) (Scheme 6B.2) [2].

In this chapter, the recent development of catalytic asymmetric epoxidation and aziridination of simple olefins bearing no pre-coordinating substituent is discussed.

6B.2. ASYMMETRIC EPOXIDATION WITH CHIRAL PEROXIDES

In 1967, Henbest et al. reported the first asymmetric epoxidation with an optically active peracid as the chiral oxidant [3]. Since then, many optically active peracids have been used for this purpose but enantioselectivity remains low (<20% ee). This is probably because the substrate and the asymmetric center in the peracid are distant from each other in the transition state of the epoxidation as shown in Figure 6B.1 [4].

Differing from optically active peracids, dioxiranes and oxaziridines carry asymmetric center(s) next to the electrophilic oxygen atom that is nucleophilically attacked by olefins. Therefore, the incoming olefin interacts more strongly with the asymmetric center in the oxidants (Figure 6B.2) and higher enantioselectivity can be expected in the epoxidation. Epoxidation with chiral oxaziridine was first reported by Davis et al. and, as expected, higher enantioselectivity (65%) was realized in the epoxidation of trans-β-methylstyrene by using an N-sulfonyloxaziridine as the chiral oxidant (Scheme 6B.3) [5].

Recently, oxaziridinium salts derived in situ from chiral iminium salts (1 and 2) and Oxone were found to catalyze epoxidation with moderate-to-good enantioselectivity (up to 73% ee) (Scheme 6B.4) [6]. Although the substrates are limited to conjugated olefins, this reaction has an advantage in being catalytic with respect to chiral iminium salts.

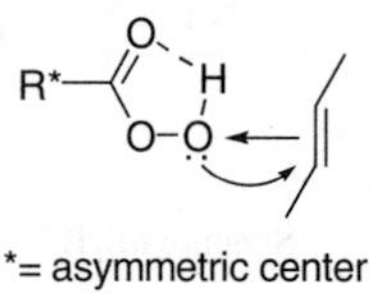

*= asymmetric center

Figure 6B.1. Transition-state model for epoxidation with optically active peracids.

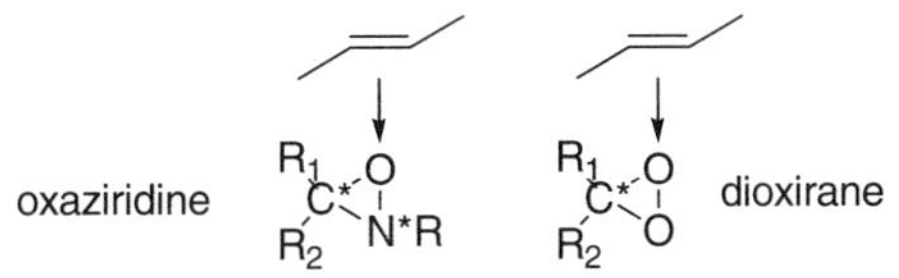

Figure 6B.2. Transition-state models for epoxidation with optically active oxaziridines and dioxiranes.

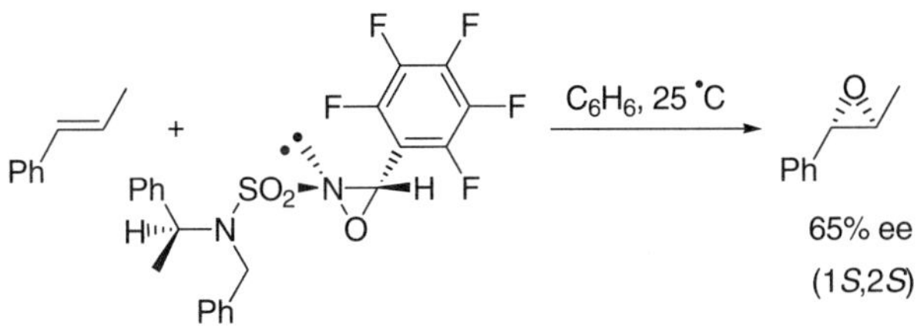

Scheme 6B.3.

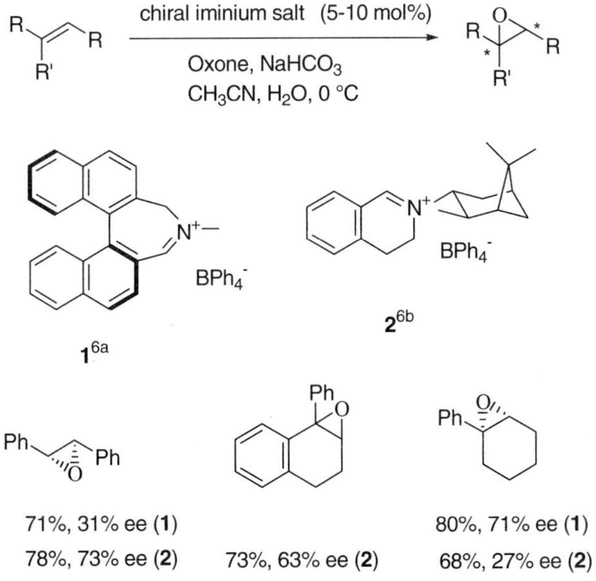

Scheme 6B.4.

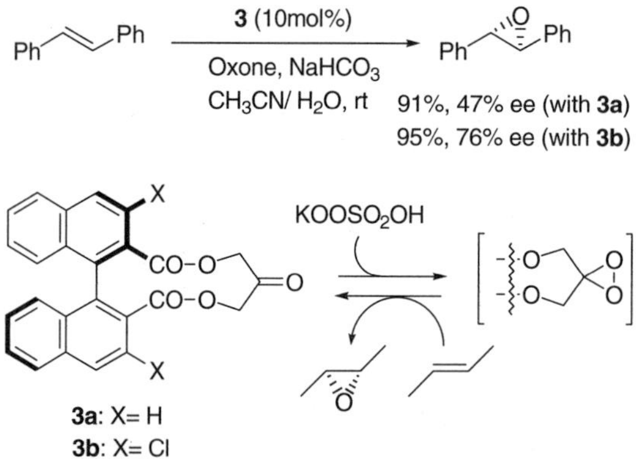

Scheme 6B.5.

Chiral dioxirane that was also generated in situ from the corresponding ketone and Oxone was first used for catalytic asymmetric epoxidation by Curci et al., although enantioselectivity was low [7]. Later, Yang et al. disclosed that this approach had a bright prospect if used with a combination of Oxone and chiral ketone **3** [8]. Ketone **3** is converted into the corresponding dioxirane in situ, which epoxidizes olefins (Scheme 6B.5).

Shi et al. have recently reported that the fructose-derived ketone **4** is an excellent catalyst for asymmetric epoxidation (Scheme 6B.6) [9]. High enantioselectivity is realized in the epoxidation of *trans*-di- and tri-substituted olefins regardless of the presence or absence of functional groups, whereas the reaction of *cis*-disubstituted and terminal olefins exhibits moderate enantioselectivity (Fig. 6B.3) [10]. Chemoselectivity of the reaction is very high so that the functional groups such as acetal, silyl ether, hydroxy group, and halo-substituent are tolerated under the reaction conditions [9,10]. Epoxidations of hydroxylated olefins such as allylic, homoallylic and bishomoallylic alcohols are accelerated probably through H-bond formation with dioxirane [10].

When the reaction is carried out at pH 7–8, however, ketone **4** is consumed rapidly because of the undesired Bayer-Villiger (B.V.) reaction of the intermediary hydroxy peroxysulfonate, which competes with the desired dioxirane formation (Scheme 6B.7). The suppression of the undesired B. V. reaction is indispensable for performing the desired epoxidation in a catalytic manner. Shi et al. have found that alkaline reaction conditions accelerate the conversion of the hydroxy peroxysulfonate to dioxirane and the reaction can be carried out with a catalytic amount (0.3 equiv.) of chiral ketone (Fig. 6B.3) [11]. Alkaline conditions also bring about enhance-

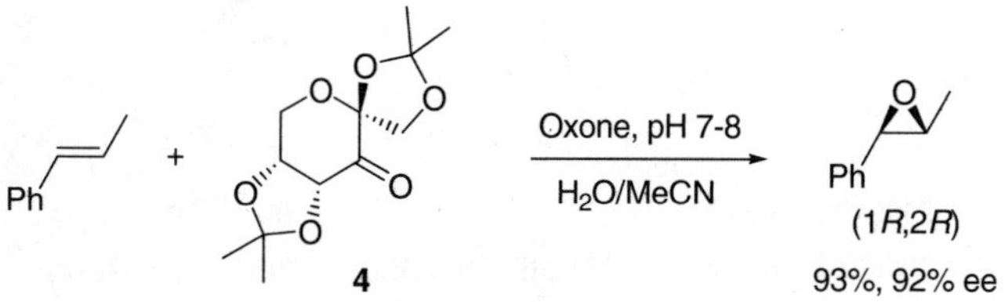

Scheme 6B.6.

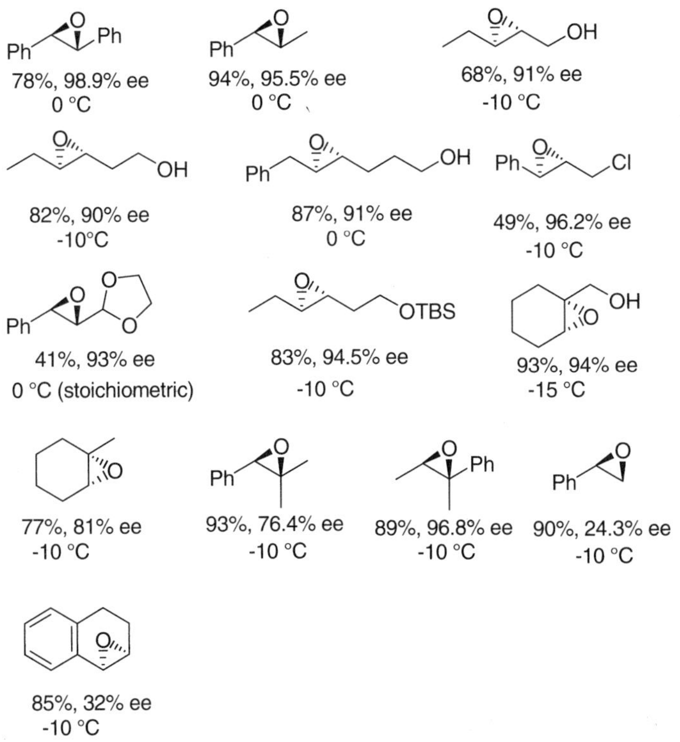
Ph
Ph
78%, 98.9% ee
0 °C
Ph
94%, 95.5% ee
0 °C
OH
68%, 91% ee
-10 °C
OH
82%, 90% ee
-10°C
Ph
OH
87%, 91% ee
0 °C
Ph
Cl
49%, 96.2% ee
-10 °C
Ph
41%, 93% ee
0 °C (stoichiometric)
OTBS
83%, 94.5% ee
-10 °C
OH
93%, 94% ee
-15 °C
77%, 81% ee
-10 °C
Ph
93%, 76.4% ee
-10 °C
Ph
89%, 96.8% ee
-10 °C
Ph
90%, 24.3% ee
-10 °C
85%, 32% ee
-10 °C

chiral ketone=0.3 equiv; Oxone=1.4 eq

Figure 6B.3. Epoxidation with chiral ketone **4** as a catalyst.

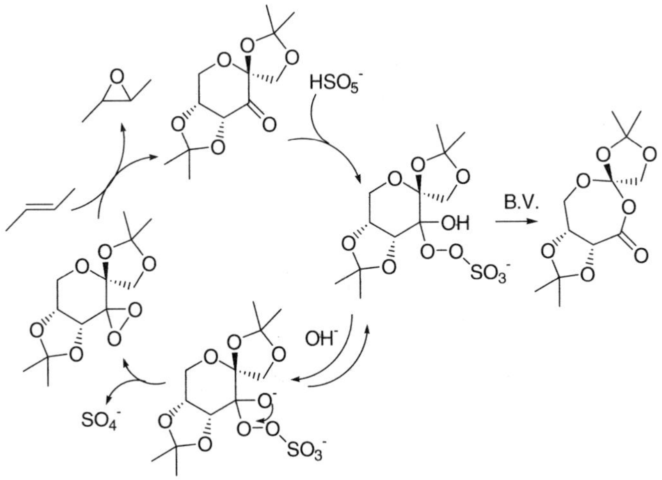
HSO5-
OH
B.V.
SO3-
OH-
SO4-

Scheme 6B.7.

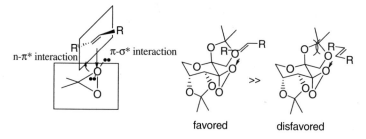

Figure 6B.4. Mechanism for asymmetric induction by chiral dioxiranes.

ment of the enantioselectivity in the epoxidation because the epoxidation of olefins with free potassium peroxysulfonate, which gives racemic epoxides, is suppressed under the conditions [10].

The stereochemistry of epoxidation by using chiral ketones (**3** and **4**) as catalysts can be explained by *spiro* transition state, in which π-electrons of the olefin attack the σ*-orbital of O–O bond and the lone-paired electrons concurrently attack the π*-orbital of double bond to give the epoxide (Figure 6B.4) [12,13]. The observed effect of the size and location of the olefinic substituents on enantioselectivity (Figure 6B.3) is compatible with the proposed transition-state model [10a].

6B.3. METAL-CATALYZED ASYMMETRIC EPOXIDATION

Active species such as alkylperoxo-, peroxo- and oxo-metal species, which epoxidize olefins, can be produced by treating appropriate metal complexes with oxidants such as alkyl hydroperoxide, hydrogen peroxide, and iodosylbenzene (Figure 6B.5) [2]. These epoxidations occur in the coordination sphere of the metal complexes. Therefore, if the metal complex carries chiral bystander ligand(s), epoxidation proceeds in an asymmetric environment to give optically active epoxides. As has been described in the first edition of this volume [14], Sharpless and Katsuki discovered that the Ti complex bearing dialkyl tartrate as a chiral auxiliary was an excellent catalyst for the asymmetric epoxidation of allylic alcohols [15]. With the use of this catalyst, most of allylic alcohols can be converted to the corresponding epoxy alcohols with high enantioselectivity. However, asymmetric epoxidation of simple olefins is still a major challenge in organic synthesis. Thus, continued efforts have been directed to the design of suitable catalysts for the epoxidation of this class of olefins.

$$L^*_n{\cdot}M\underset{O}{\overset{O{-}R}{\langle}} \qquad L^*_n{\cdot}M\underset{O}{\overset{O}{\langle}} \qquad L^*_n{\cdot}M{=}O$$

alkylperoxo-metal species   peroxo-metal species   oxo-metal species

L*= chiral bystander ligand

Figure 6B.5. Active species for metal-catalyzed oxygen-transfer reactions.

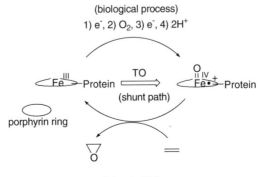

Scheme 6B.8.

6B.3.1. Molybdenum-Catalyzed Epoxidation

Diperoxo(oxo)molybdenum(IV) complex bearing (*S*)-lactic acid piperidineamide as a chiral ligand has been used for the epoxidation of *E*-2-butene (Scheme 6B.8) and moderate enantiose-lectivity (49%) is achieved wherein the reaction is stoichiometric [16]. Two possible mecha-nisms have been proposed for this reaction. One mechanism includes coordination of an olefin prior to epoxidation, which makes the olefin electrophilic and facilitates the nucleophilic attack of the proximal oxygen atom of the peroxide on the olefin. The other one is that an olefin nucleophilically attacks the peroxo group of the molybdenum complex.

6B.3.2. Metalloporphyrin-Catalyzed Epoxidation

Oxo-metal species participate in a wide range of biological and chemical oxidation reactions. Representative oxidizing enzyme, cytochrome P-450, which carries iron(III)-porphyrin com-plex as its active site, catalyzes various O-atom transfer reactions such as epoxidation, hydroxy-lation of C–H bond, and oxidation of sulfides. These reactions have been proven to proceed through cationic oxoiron(IV)-porphyrin species, which are generated by the oxidation of Fe(III) complex with molecular oxygen. This conversion from Fe(III) to O=Fe(IV) species is a

(biological process)
1) e⁻, 2) O₂, 3) e⁻, 4) 2H⁺

TO

(shunt path)

porphyrin ring

Scheme 6B.9.

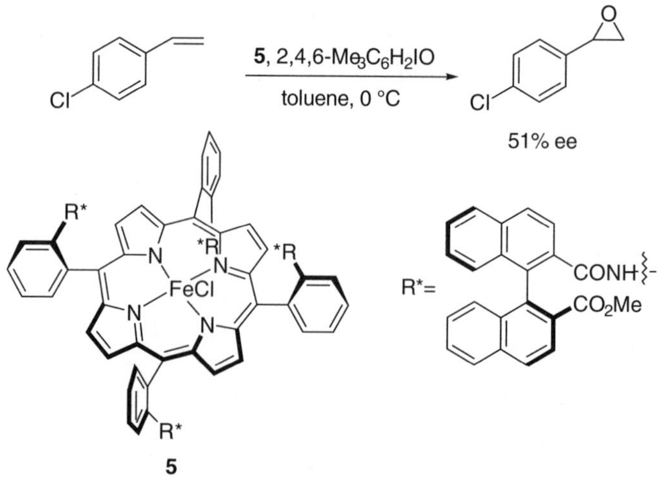

Scheme 6B.10.

complex process, which includes electron and proton transfers (Scheme 6B.9). In 1979, however, Groves et al. found that treatment of Fe(III)-porphyrin complex with terminal oxidant (TO) such as iodosylbenzene directly provided cationic oxoiron(IV)-porphyrin complexes that epoxidized simple olefins (shunt path) [17]. It should be noted that *cis*-olefins reacted much faster than *trans*-olefins in this porphyrin-catalyzed epoxidation. This substrate-specificity has been explained by assuming that an olefin approaches the oxo-metal bond from its side parallel to the porphyrin ring [18]. The proposed side-on approach is now widely accepted for the metalloporphyrin- and metallosalen-catalyzed epoxidation, which will be discussed below.

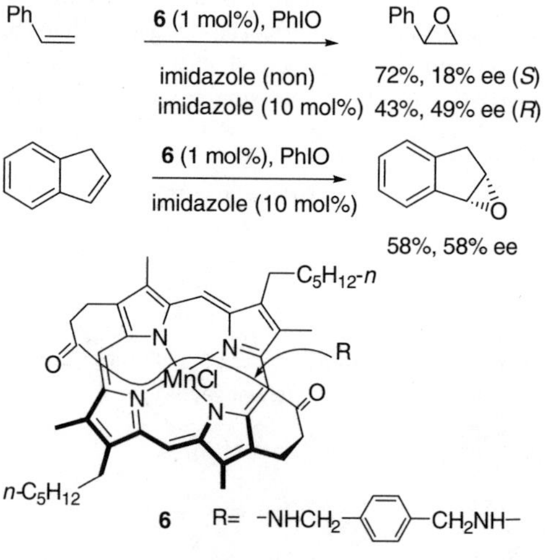

Scheme 6B.11.

These findings prompted the development of chiral metalloporphyrin-catalyzed epoxidation of simple olefins.

In 1983, Groves synthesized the first chiral Fe(III)-porphyrin complex (**5**), which had chiral auxiliaries on the *meso*-carbons and achieved moderate enantioselectivity (51%) in the epoxidation of *p*-chlorostyrene (Scheme 6B.10) [19].

In the wake of this report, many chiral iron(III)- and Mn(III)-porphyrin complexes have been synthesized and applied to the epoxidation of styrene derivatives [20]. Because these asymmetric epoxidations are discussed in the first edition of this book [21], the discussion on metalloporphyrin-catalyzed epoxidation here is limited to some recent examples. Most chiral metalloporphyrins bear chiral auxiliaries such as the one derived from α-amino acid or binapthol. Differing from these complexes is complex **6**, which has no chiral auxiliary but is endowed with facial chirality by introducing a strap and has been reported by Inoue et al. [20f]. Epoxidation of styrene by using only **6** as the catalyst shows low enantioselectivity, but the selectivity is remarkably enhanced when the reaction is performed in the presence of imidazole (Scheme 6B.11). This result can be explained by assuming that imidazole coordinates to the unhindered face of the complex and the reaction occur on the strapped face [20f].

Collman et al. have recently reported that threitol-strapped manganese(III)-porphyrin complex **7** shows high enantioselectivity in the epoxidation of a wide range of olefins when the reaction is carried out in the presence of 1,5-dicyclohexylimidazole (Scheme 6B.12) [22]. The substituted imidazole in this reaction appears to play the same role as the imidazole in Inoue's reaction. Detailed understanding of the mechanism of asymmetric induction by **7** needs further investigation, the major pathway should accommodate the steric interaction between the olefinic substituent and the inner oxygen atom of the threitol strap (Figure 6B.6). Thus, the pathway **a** is likely to be the major pathway.

The scope of the porphyrin-catalyzed epoxidations is still limited, and there is room for improvement in their enantioselectivity. However, metalloporphyrins have an advantage in their robustness against various terminal oxidants, especially when the ligand carries electron-withdrawing substituents. High turnover number (TON) exceeding thousands has been realized in porphyrin-catalyzed epoxidation [21]. Another advantage of porphyrin-catalyzed epoxidation is the fact that molecular oxygen can be used as oxidant. Different from the biological pathway, most of metalloporphyrin-catalyzed epoxidations so far reported use the terminal oxidants other than molecular oxygen (shunt path in Scheme 6B.9). However, molecular oxygen is an economical and environmentally acceptable oxidant. Groves et al. has found that dioxoruthenium(VI)-porphyrin complex catalyzes aerobic epoxidation even in the absence of a reducing agent [23]. Recently, enantioselective version of the aerobic epoxidation by using the dioxoruthenium(VI)-complex **8** as catalyst was reported by Che et al. (Scheme 6B.13) [24]. The epoxidation of *cis*-β-methylstyrene under 8 atm of oxygen proceeds with a good enantioselectivity (73% ee) though TON is not high (up to 21).

6B.3.3. Metallosalen-Catalyzed Epoxidation

Metallosalen complex [salen = N,N-ethylenebis(salicyldeneaminato)] has a structure similar to metalloporphyrin, and these two complexes catalyze the epoxidation of olefins. For example, Kochi et al. have found that metallosalen complexes such as (salen)manganese(III) [25] and (salen)chromium(III) complexes [26] (hereafter referred to as Mn- and Cr-salen complexes, respectively) serve as catalysts for the epoxidation of unfunctionalized olefins by using iodosylbenzene [25] or sodium hypochlorite [27]. In particular, cationic Mn-salen complex is a good catalyst for epoxidation of unfunctionalized olefins, which proceeds through an oxo(salen)manganese(V) species (Scheme 6B.14) [25,28]. The presence of oxo-Mn(V)-salen

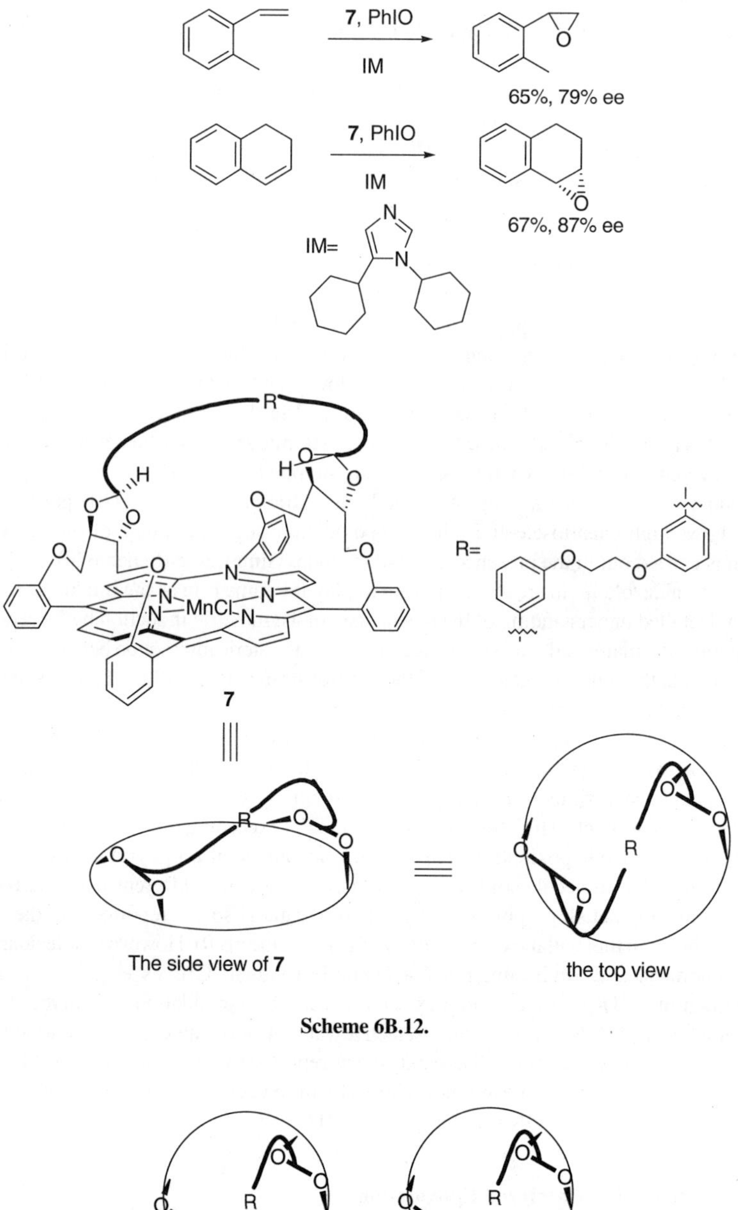

7, PhIO
→
IM

65%, 79% ee

7, PhIO
→
IM

67%, 87% ee

IM=

R=

7

The side view of **7**   ≡   the top view

Scheme 6B.12.

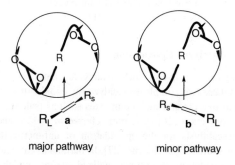

R_L **a**

major pathway

b R_L

minor pathway

Figure 6B.6. Reasonable explanation of asymmetric induction by porphyrin complex **7**.

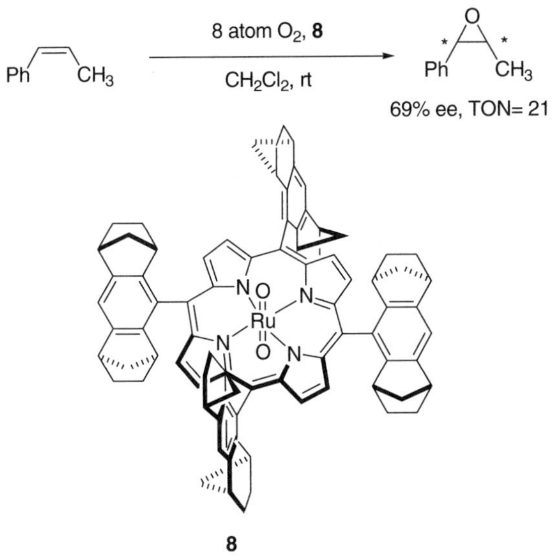

8 atom O2, 8
CH2Cl2, rt
Ph
CH3
O
69% ee, TON= 21
N
O
Ru
8

Scheme 6B.13.

PhIO
PhI
N
Mn+
O
PF6-

Scheme 6B.14.

sp3 carbons
N
M
O
1
3
1'
3'
open space
metalloporphyrin
metallosalen

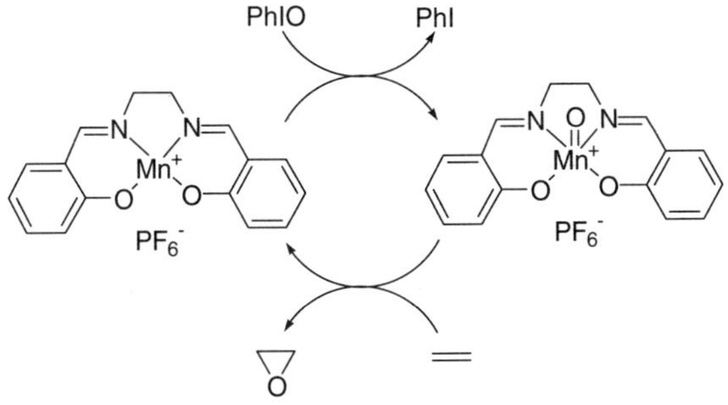

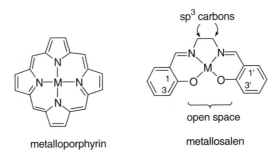

Figure 6B.7. Basic structures of metallo-porphyrin and -salen complexes.

species was recently proved by MS/MS study of μ-oxo Mn(IV)-salen complexes [28]. However, porphyrin and salen complexes have some structural differences. On the one hand, porphyrin ligand encircles a central metal ion and its peripheral carbons are all sp^2. On the other hand, salen ligand surrounds a central metal partly and it includes two sp^3 carbons in its ethylenediamine moiety (Figure 6B.7). When these structural features were used to advantage, various chiral metallosalen complexes were synthesized to construct efficient reaction sites for asymmetric epoxidation of unfunctionalized olefins.

In 1990, Jacobsen et al. and Katsuki et al. independently reported asymmetric epoxidation of conjugated olefins by using complexes **9** and **10**, respectively, as catalysts [29]. These Mn-salen complexes were further improved to complexes **11** [30], **12** [31], and **13** [32]. The common features of these first-generation Mn-salen complexes are i) they possess C$_2$-symmetry, ii) two sp^3 carbons at the ethylenediamine moiety are replaced with chiral ones, and iii) they have *tert*-butyl groups or enantiopure 1-phenylpropyl groups at the C3 and C3′ positions.

These catalysts, **11–13**, show good enantioselectivity ranging from 80 to 95% ee in the epoxidation of conjugated *cis*-di- and tri-substituted olefins. Epoxidation of "good" substrates such as 2,2-dimethylchromene derivatives proceeds with excellent enantioselectivity (>95% ee). Since the results obtained with these first-generation Mn-salen catalysts have been reviewed [21,33], only typical examples are shown in Table 6B.1. These reactions are usually carried out in the presence of donor ligand [34] such as 4-phenylpyridine *N*-oxide with terminal oxidants such as iodosylbenzene and sodium hypochlorite as described above. However, the use of some other terminal oxidants under well-optimized conditions expands the scope of the Mn-salen-

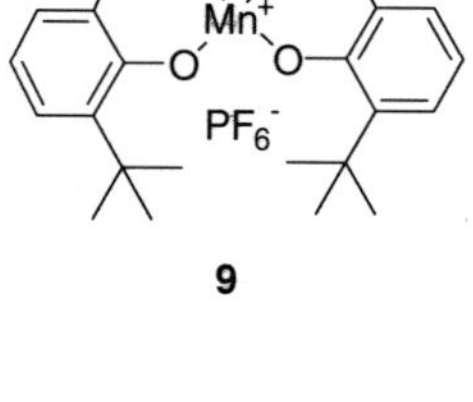

9

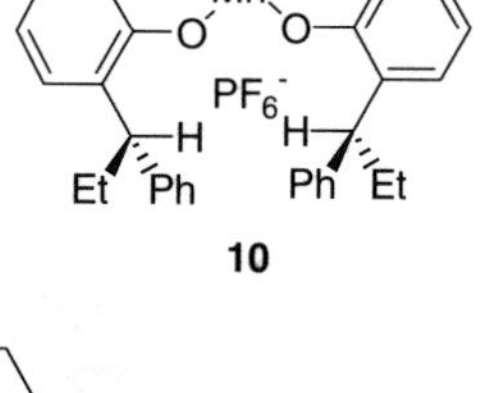

10

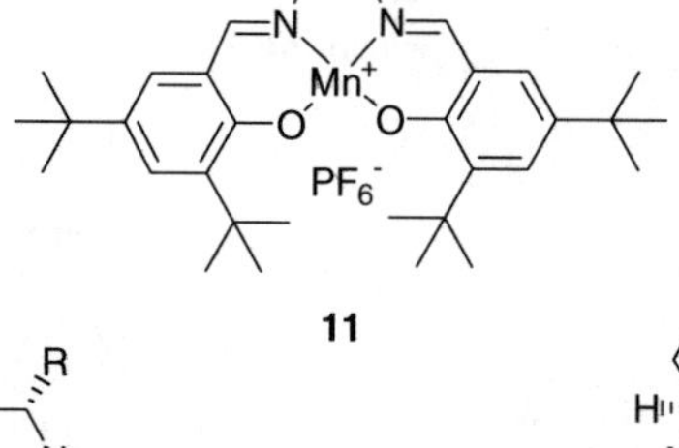

11

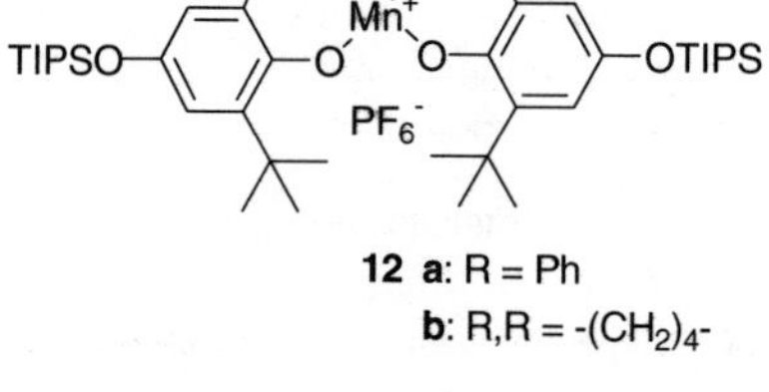

12 a: R = Ph
b: R,R = -(CH$_2$)$_4$-

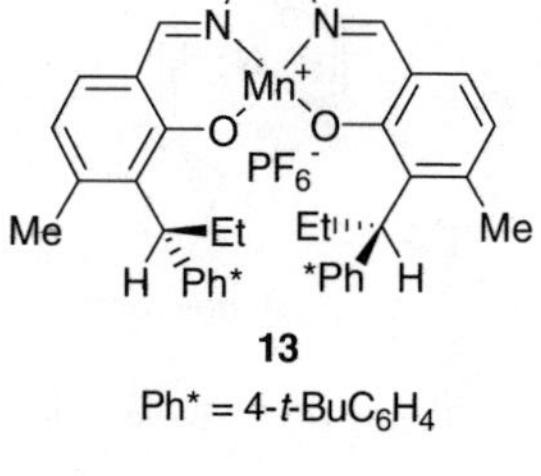

13

Ph* = 4-*t*-BuC$_6$H$_4$

catalyzed epoxidation. Epoxidation of conjugated olefins is considered to proceed through a radical intermediate in which the carbon–carbon bond between the radical and its vicinal carbons can rotate, and this bond rotation leads to the formation of stereoisomeric epoxides. For example, epoxidation of *cis*-β-methylstyrene (R = Me) gives a mixture of *cis*- and *trans*-epoxides (Scheme 6B.15) [25,29b]. When the substrate is styrene (R = H), the bond rotation leads to racemization and thus diminishes the enantioselectivity of the reaction. It has been shown that the epoxidation at low temperature by using *m*-chloroperbenzoic acid in the presence of *N*-methylmorpholine *N*-oxide suppresses the bond rotation, achieving high enantioselectivity in the epoxidation of styrene derivatives (Table 6B.1, entries 1 and 2) [35].

On the other hand, the addition of a quaternary ammonium salt to the reaction medium accelerates the isomerization of the radical intermediate [36]. Thus, the epoxidation of *cis*-stilbene in the presence of *N*-benzylquinine salt gives *trans*-stilbene oxide with 90% ee as major product (Table 6B.1, entry 24). This protocol provides an effective method for the synthesis of *trans*-epoxides. In contrast to the epoxidation of *cis*-di- and tri-substituted olefins for which complexes **11–13** are the catalysts of choice, the best catalyst for the epoxidation of tetra-substituted conjugated olefins varies with substrates (Table 6B.1, entries 27 and 28) [37]. The asymmetric epoxidation of 6-bromo-2,2,3,4-tetramethylchromene is well-promoted by complex **14** and that of 2-methyl-3-phenylindene, by complex **12a**.

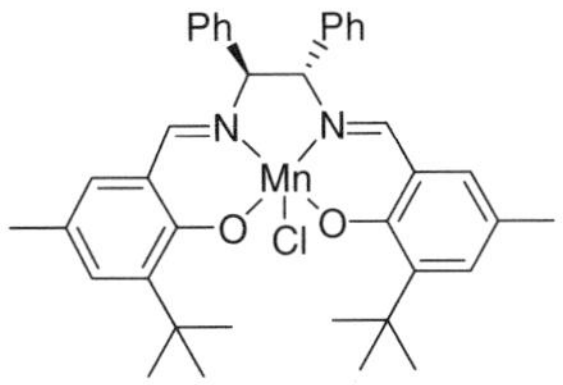

14

Enantioface selection of olefins by chiral Mn-salen complexes is induced through the interaction between the chiral salen ligand and the incoming olefin molecule. To achieve high enantioselectivity, the trajectory of the olefin and its orientation must be regulated strictly (Figure 6B.8). Complexes **9** and **10** have been designed based on the following premises: i) the salen ligand has a planar structure, ii) an olefin approaches oxo-metal bond from its side parallel

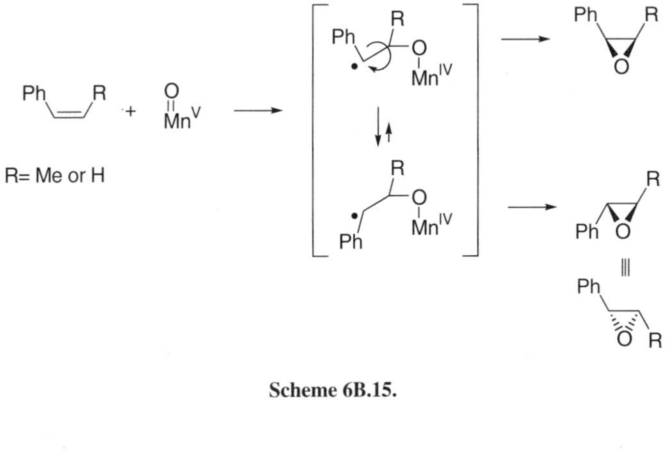

R= Me or H

Scheme 6B.15.

TABLE 6B.1. Epoxidation of Conjugated Olefins With the First-Generation Mn-salen Catalyst

Entry	Substrate	Catalyst	Solvent	Oxidant	Temp.	Yield (%)	% ee	Config.	Ref.
1	Ph	12a	CH$_2$Cl$_2$	m-CPBA^a	−78°C	88	86	—	[34a]
2	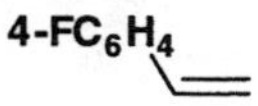 4-FC$_6$H$_4$	12a	CH$_2$Cl$_2$	m-CPBA^a	−78°C	83	85	—	[34b]
3		ent-12b	Et$_2$O	NaOCl^b	4°C	30	65	—	[31]
4	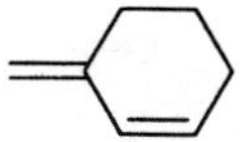	13	CH$_3$CN	PhIO	rt	10	87	—	[38]
5	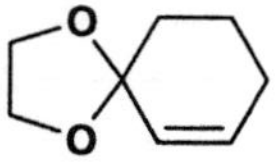	ent-11	CH$_2$Cl$_2$	NaOCl	4°C	63	94	—	[39]
6	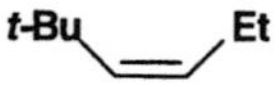 t-Bu / Et	ent-12b	ClC$_6$H$_5$	NaOCl^c	4°C	−(69:31)^d	84^e	—	[36]
7	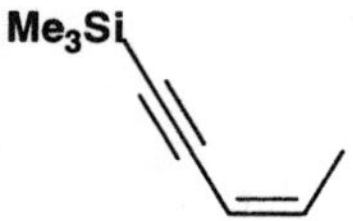 Me$_3$Si	ent-11	CH$_2$Cl$_2$	NaOCl	rt	84 (2.5:1)^d	90^e (78)^f	3R,4R	[40]
8	 Ph / Me	13	CH$_2$Cl$_2$	NaOCl	4°C	36	86	1S,2R	[32]
9	"	ent-11	CH$_2$Cl$_2$	NaOCl	4°C	81	92	1S,2R	[41]
10	"	12a	CH$_2$Cl$_2$	m-CPBA^a	−78°C	89	96	—	[34b]

(continued)

300

TABLE 6B.1. Continued

Entry	Substrate	Catalyst	Solvent	Oxidant	Temp.	Yield (%)	% ee	Confign.	Ref.
11	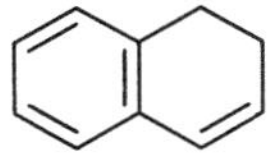	**13**	CH_2Cl_2	$NaOCl^g$	rt	38	91	1S,2R	[32]
12	"	**ent-11**	CH_2Cl_2	NaOCl	0°C	67	86	1S,2R	[33d]
13	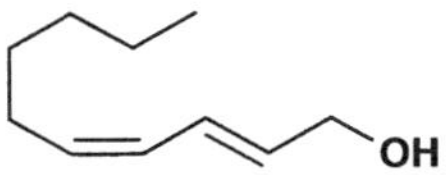	**11**	CH_2Cl_2	$NaOCl^b$	0°C	16 (2.3:1)^{d,h}	83^e	—	[41]
14	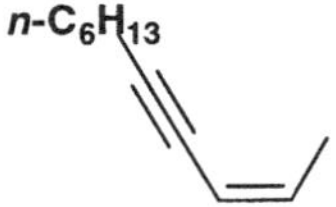	**ent-11**	CH_2Cl_2	NaOCl	rt	65 (1.6:1)^d	90^e (72)^f	3R,4R	[40]
15	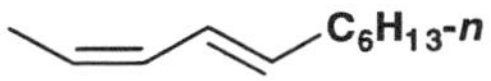	**11**	CH_2Cl_2	$NaOCl^b$	0°C	50 (1.1:1)^d	92^e	—	[41]
16	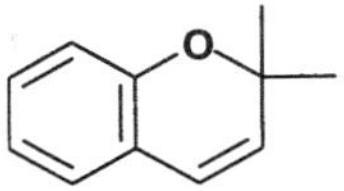	**11**	CH_2Cl_2	NaOCl	4°C	72	98	3R,4R	[39]
17	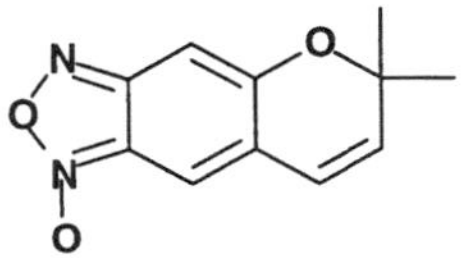	**13**	CH_3CN	PhIO		63	94	3S,4S	[32]
18	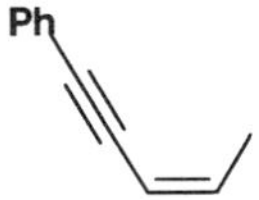	**ent-11**	CH_2Cl_2	NaOCl	rt	85 (2:1)^d	93^e (81)^f	3R,4R	[40]

(continued)

TABLE 6B.1. Continued

Entry	Substrate	Catalyst	Solvent	Oxidant	Temp.	Yield (%)	% ee	Config.	Ref.
19	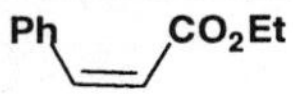 Ph⌣CO₂Et	11	CH₂Cl₂	NaOCl[b]	4°C	56 (13:87)[d]	95–97[i]	1R,2R	[42]
20	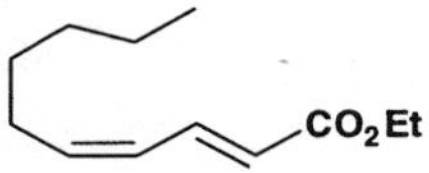 CO₂Et (diene)	11	CH₂Cl₂	NaOCl[b]	0°C	81 (9:1)[d]	87[e]	—	[41]
21	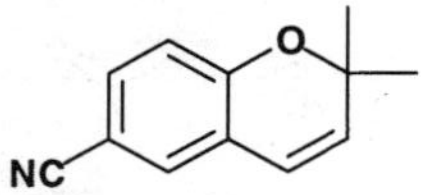 NC-chromene	11	CH₂Cl₂	NaOCl	4°C	96	97	3R,4R	[39]
22	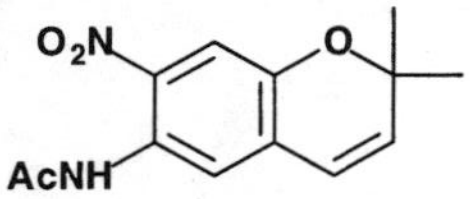 O₂N / AcNH-chromene	13	CH₃CN	PhIO	rt	78	96	—	[32]
23	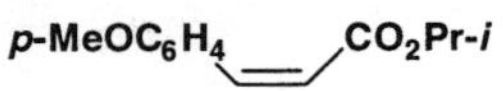 p-MeOC₆H₄⌣CO₂Pr-i	ent-12b	ClC₆H₅	NaOCl[c]	4°C	–(89:11)[d]	86[e]	2S,3S	[36]
24	 Ph⌣Ph	ent-12b	ClC₆H₅	NaOCl[c]	4°C	–(>96:4)[d]	90[e]	1S,2S	[36]
25	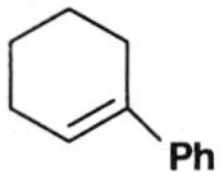 Ph-cyclohexene	11	CH₂Cl₂	NaOCl[b]	0°C	69	93	1S,2S	[43a]
26	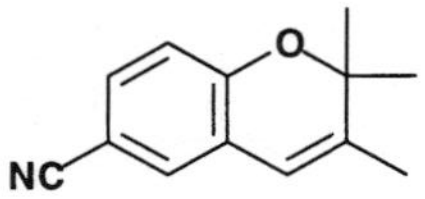 NC-chromene	11	CH₂Cl₂	NaOCl	0°C	82	>98	3R,4R	[43b]
27	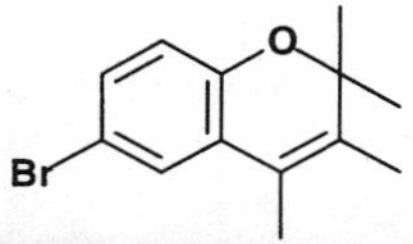 Br-chromene	14	CH₂Cl₂	NaOCl[a]	0°C	84	96	3S,4R	[37]

(continued)

TABLE 6B.1. Continued

Entry	Substrate	Catalyst	Solvent	Oxidant	Temp.	Yield (%)	% ee	Config.	Ref.
28	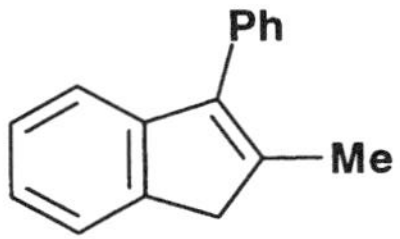	**12a**	CH₂Cl₂	NaOCl[b]	0°C	90	90	—	[37]

[a]Reaction was carried out in the presence of excess *N*-methylmorpholine *N*-oxide.

[b]Reaction was carried out in the presence of 4-phenylpyridine *N*-oxide.

[c]Reaction was carried out in the presence of *N*-benzylquinine salt.

[d]Product is a mixture of *trans*- and *cis*-epoxides. Numbers in parentheses are the ratio of *trans*- and *cis*-epoxides.

[e]The number stands for the % ee of *trans*-epoxide.

[f]The number in parentheses stands for the face selectivity. Face selectivity = % ee $_{trans}$ × %*trans* + %ee$_{cis}$ × %cis (reference 44).

[g]Reaction was carried out in the presence of 4-(*N*,*N*-dimethylamino)pyridine *N*-oxide.

[h]The allylic alcohol was oxidized to the corresponding aldehyde.

[i]The number stands for the % ee of *cis*-epoxide.

303

to the salen ligand, and iii) an olefin approaches along the sterically less-congested pathway. According to these premises, an olefin is expected to approach the oxo-metal bond along the pathway **a**, directing their bulkier substituents away from the bulky substituents at C3 and C3′ in the epoxidation with complexes **9** and **10** (R′= H). However, the pathway **b** is proposed for the epoxidation by using **11** [R′ = *t*-Bu, R″ = R″ = –(CH$_2$)$_6$-] as catalyst [39]. *tert*-Butyl groups introduced at C5 and C5 should block the pathway **a**. In the proposed pathways, **a** and **b**, the orientation of the incoming olefin should be dictated by the steric repulsion between the substituent of the olefin and the salen ligand. However, these pathways do not give a reasonable answer to the question of why "good" substrates for this epoxidation are limited to conjugated olefins and the reaction of non-conjugated olefins is slow and less enantioselective [38]. Furthermore, in the epoxidation of enynes, alkynyl substituents behave as larger groups than the methyl group (Table 6B.1, entries 14 and 18) though the steric requirements (A values) of alkynyl groups are much smaller than that of the methyl group. These results suggest that there should be another factor contributing to the determination of the orientation of the incoming conjugated olefin in addition to steric repulsion. The answer to this question is obtained by the comparison of epoxidations of 3-methylene-1-cyclohexene (87% ee, Table 6B.1, entry 4) and of 2-methyl-1,3-cyclohexadiene (63% ee) [38]. The results suggest that the π–π repulsive interaction and the steric repulsion between an unsaturated substituent of an olefin (L = unsaturated group) and the salen ligand synergistically work to direct the substituent of an olefin to the ethylenediamine side, when the substrate approach the oxo species along the pathway **c**.

This hypothesis that the π–π repulsion plays an important role in asymmetric induction by the Mn-salen catalyst has led to the development of a new type of Mn-salen complexes **15**, in which the salen ligand contains a binaphthyl unit as the chiral auxiliary instead of a chiral substituent at C3 (Figure 6B.9) [45]. The phenyl group on the naphthyl moiety at C3′ is expected to protrude toward the incoming olefins along the pathway **a** or **c**, and to increase both the steric and π–π repulsion, which would strictly regulate the orientation of the substrate. In accord with the expectation, the epoxidation of conjugated *cis*-olefins catalyzed by complex **15** [Ar = 3,5-(CH$_3$)$_2$C$_6$H$_4$] shows remarkably improved enantioselectivity (Table 6B.2) [45,46].

Discussions to this point rely on the hypothesis that the ligands of oxo(salen)manganese(V) complexes have planar structures by analogy to metalloporphyrin complexes and Mn(III)-salen complexes **11** and **12**, the structures of which were determined by the X-ray crystallographic analysis [29a,47]. However, the assumption that the ligand of the oxo-Mn-salen species is planar failed to give a satisfactory explanation for the following stereochemistry observed in the recent study. *Trans-cis* selectivity in the epoxidation of 1-alkylindenes usually improves as the steric bulk of the catalyst increases. However, the epoxidation of 1-methylindene with the smallest

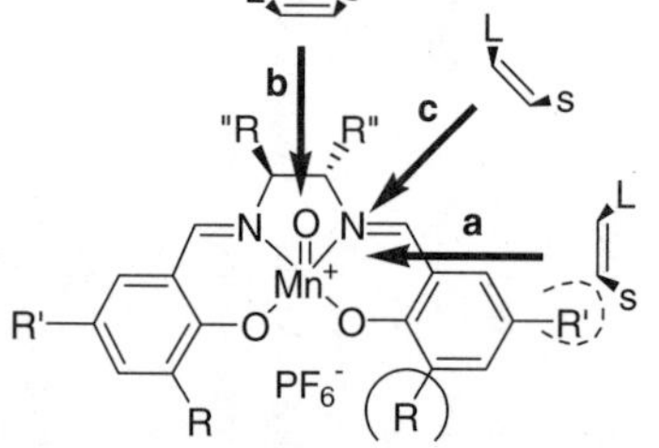

Figure 6B.8. Plausible olefin's approaches [only the incoming olefin of the favored orientation is des cribed in each approach (**a**, **b**, or **c**)].

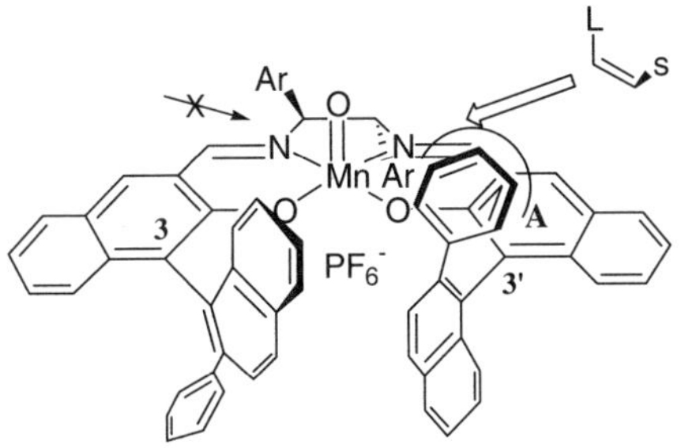

L
S
Ar
O
N
Mn
Ar
O
A
3
3'
PF6-

15

Figure 6B.9. Role of the chiral binaphthyl unit of **15** in its asymmetric induction.

catalyst **16a** shows the lowest selectivity, but the reactions using catalysts **16b** and **16c** exhibit a similar level of moderate selectivity, suggesting that the *tert*-butyl groups at C5 and C5′ have little effect on the selectivity [48]. These results do not match with the assumption that the ligand of the oxo-metal species is planar. However, the results can be rationalized by assuming that the ligands of oxo(salen)manganese(V) complexes take a non-planar stepped conformation (Scheme 6B.16), that is, an olefin approaches the oxo-metal species over the downward benzene ring in the ligand so that the substituent of the olefin does not interact with the *tert*-butyl group at C5 of the benzene ring.

The proposed non-planar ligand conformation agrees with the structure of oxo(salen)chromium(V) complex characterized by Srinivasan and Kochi, who isolated an oxo(salen)chromium(V) complex as well as its pyridine *N*-oxide adduct, and unambiguously demonstrated that the salen ligands of these two complexes took a non-planar stepped conformation on the basis of X-ray crystallographic analyses [49]. The hypothesis that the ligand of oxo-Mn-salen

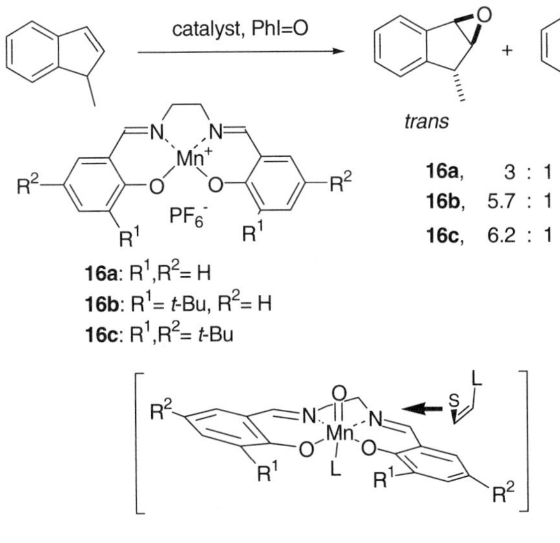

catalyst, PhI=O
trans
cis
16a, 3 : 1
16b, 5.7 : 1
16c, 6.2 : 1
R2
R1
Mn+
PF6-
16a: R1,R2= H
16b: R1= t-Bu, R2= H
16c: R1,R2= t-Bu
L
S

Scheme 6B.16.

TABLE 6B.2. Epoxidation of Conjugated *cis*-di- and Tri-substituted Olefins with Mn-Salen Complex 15 as Catalyst

Entry	Olefin	Oxidant	Solvent	Temp	Yield	% ee	(% ee)[a]
1	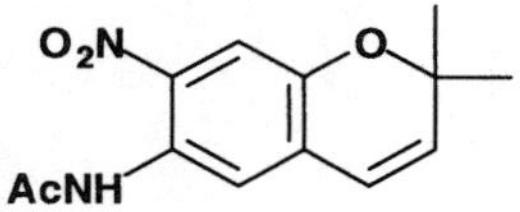 (structure: O_2N, AcNH-substituted 2,2-dimethylchromene)	PhIO	CH_2Cl_2	0°C	80	>99	
2	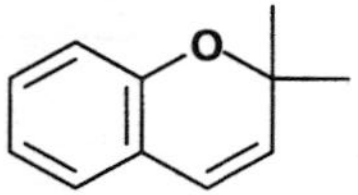 (structure: 2,2-dimethylchromene)	PhIO	CH_3CN	−20°C	60	>99	98[b]
3	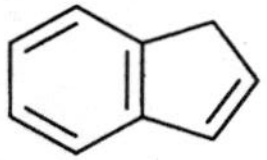 (structure: indene)	NaOCl	CH_2Cl_2	0°C	55	98	88[b]
4	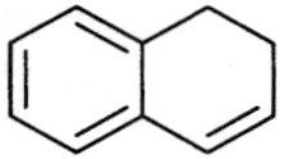 (structure: 1,2-dihydronaphthalene)	NaOCl	CH_2Cl_2	0°C	78	98	
5	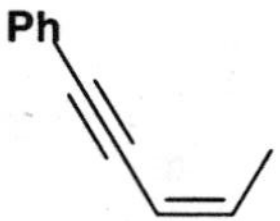 (structure: Ph-substituted enyne)	NaOCl	CH_2Cl_2	0°C	80[c]	96 (trans) 92 (cis)[e]	93[d] (81)[b,e]
6	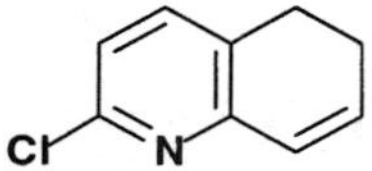 (structure: Cl-substituted dihydroquinoline)	NaOCl	CH_2Cl_2	0°C	77	96	67[b]
7	 (structure: cyclopentadiene)	NaOCl	CH_2Cl_2	−18°C	40	93	64[f]

(continued)

306

TABLE 6B.2. Continued

Entry	Olefin	Oxidant	Solvent	Temp	Yield	% ee	(% ee)[a]	
8	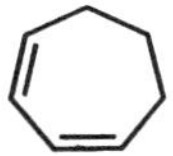	NaOCl	CH_2Cl_2	−18°C	54	94	70[f]	—
9		NaOCl	CH_2Cl_2	−18°C	37 (68)	88	65[f]	(3S,4R)
10	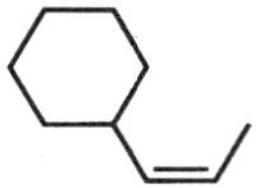	NaOCl	CH_2Cl_2	−18°C	23	82	50[g]	—
11	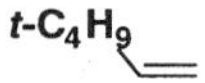	NaOCl	CH_2Cl_2	−18°C	5	70	—	
12	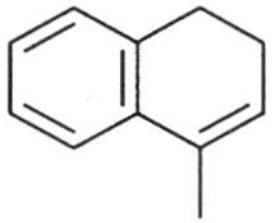	PhIO	CH_3CN	−20°C	41	96	—	(1S,2R)
13	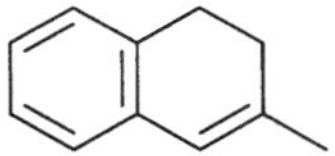	PhIO	"	−20°C	48	92	—	
14	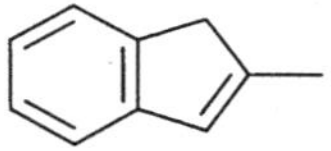	NaClO	CH_2Cl_2	0°C	91	88	—	
15	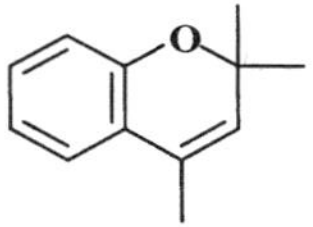	NaClO	CH_2Cl_2	0°C	88	96	—	

(continued)

TABLE 6B.2. Continued

Entry	Olefin	Oxidant	Solvent	Temp	Yield	% ee	(% ee)[a]
16	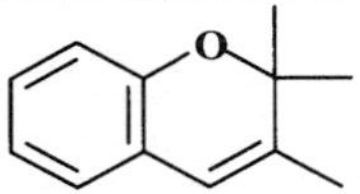	PhIO	CH_3CN	$-20°C$	81	>99	—
17	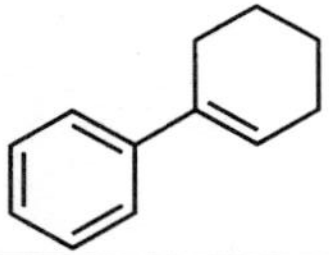	PhIO	CH_3CN	$-20°C$	26	83	93[b] (1R,2S)

[a]The highest % ee reported with the first generation Mn-salen catalysts.

[b]With Jacobsen's catalyst **11**.

[c]A mixture of *cis*- and *trans*-epoxides in a ratio of 2:1.

[d]With Jacobsen's catalyst **11**. The enantiomeric excess of *cis*-epoxide was 58% ee.

[e]The number in parentheses stands for the face selectivity. Face selectivity = ee $_{trans}$ × %*trans* + ee$_{cis}$ × %cis (reference 44).

[f]With modified Jacobsen's catalyst **12b** (4°C): Chang, S.; Heid, R.M.; Jacobsen, E.N. *Tetrahedron Lett.*, **1994**, *35*, 669.

[g]With complex **13**.

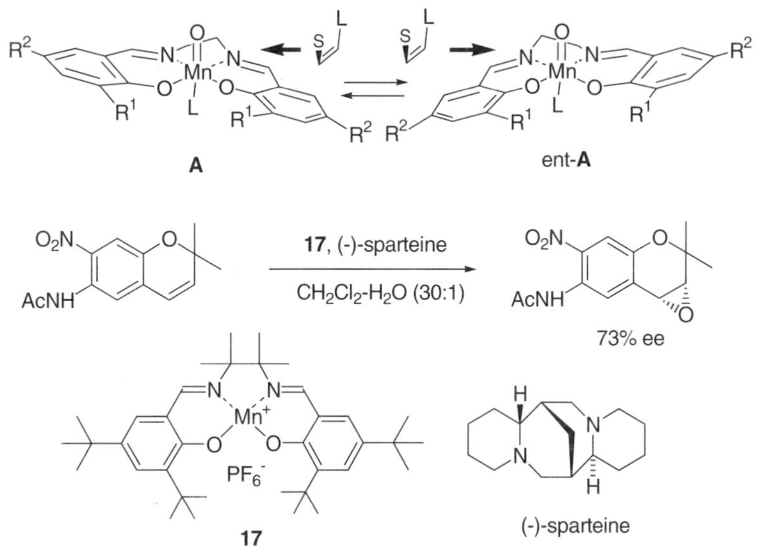

Scheme 6B.17.

complex also takes a stepped conformation has led to the new asymmetric epoxidation system using an achiral Mn-salen complex as the catalyst (Scheme 6B.17) [50]. If salen ligand takes a stepped conformation, achiral oxo-Mn-salen complex should exist as an equilibrium mixture of enantiomeric conformers **A** and **ent-A**, and an olefin should approach these enantiomeric oxo-metal species from their downward benzene ring side to give racemic epoxides. If this equilibrium is shifted to one side by some means, epoxides should be formed in optically active form. For example, the equilibrium is expected to be shifted to one side by the coordination of a chiral ligand at the apical position (L = optically active ligand), which makes **A** and **ent-A** diastereomeric. In fact, the epoxidation of 2,2-dimethylchromene derivative with achiral complex **17** in the presence of (–)-sparteine and a small amount of water gives the corresponding epoxide with good enantioselectivity (73% ee).

It is reasonable to assume that the oxo-Mn-salen complexes derived from the usual C_2-symmetric Mn-salen complexes preferentially exist as the conformer eq-**B** in which two substituents (R) at the ethylenediamine moiety take pseudo-equatorial positions because the conformer ax-**B** bearing two pseudo-axial substituents suffers from steric repulsion with the oxene atom and the apical ligand (L) (Scheme 6B.18). Thus, olefins approach the conformer eq-**B** from its downward benzene ring side to give the major enantiomer of the epoxides. However, when the substituent R is a carboxylate, the pseudo-axial conformer ax-**C** should be preferred due to the stabilization by the coordination of the carboxylate group to the manganese ion, and the equilibrium should be shifted to ax-**C** with the reversal of the ligand conformation. To prove this hypothesis, Mn-salen complex **18**, in which the configuration of the ethylenediamine part is opposite to that of **15**, was synthesized (Scheme 6B.19). It should be noted that complex **18** carries only one carboxylate group at the ethylenediamine part and is not C_2-symmetric. If the complex is C_2-symmetric, another substituent (enclosed with a dotted line) should also take pseudo-axial position with reversal of the ligand conformation, which would cause undesired interaction with the incoming olefin of the desired orientation (Scheme 18, ax-**C**) [51].

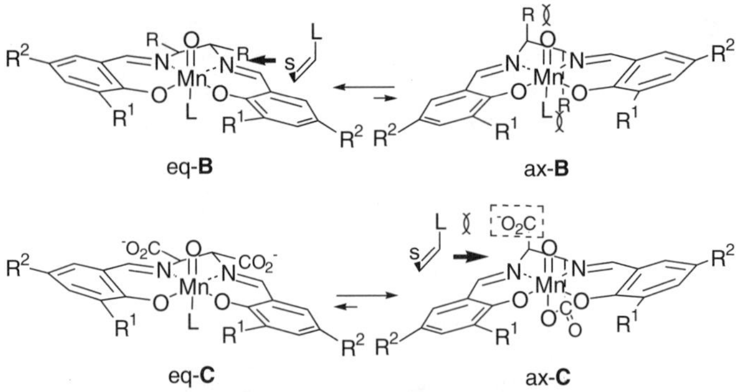
eq-B
ax-B
eq-C
ax-C

Scheme 6B.18.

As a matter of fact, complexes **15** and **18** show the same sense of asymmetric induction with high enantioselectivity (≥99% ee) in the epoxidation of 2,2-dimethylchromenes, though their configurations at the ethylenediamine moieties are opposite to each other (Scheme 6B.19) [51]. Another advantage of epoxidation with complex **18** is that the reaction can be carried out in the absence of donor ligands, because the carboxylate group serves as the apical ligand. Therefore, one of the two apical coordination sites of Mn(III) complex **18** is always free. On the other hand, epoxidation with the C2-symmetric Mn(III)-salen complexes is usually performed in the presence of a donor ligand in which the two apical sites of the regenerated Mn(III) complex are occupied by two molecules of the donor ligands [34]. This condition suggests that reoxidation of the regenerated **18** to the oxo-metal species is much faster than the reoxidation of usual Mn(III)-salen complexes such as **15**. In fact, the TON as high as 9,200 has been realized in the epoxidation by using **18** as the catalyst [51].

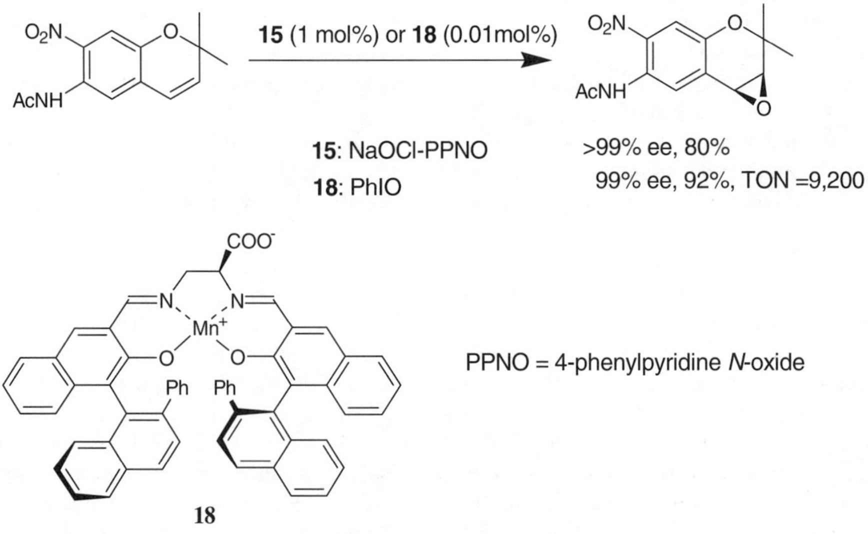
15 (1 mol%) or 18 (0.01mol%)
15: NaOCl-PPNO
18: PhIO
>99% ee, 80%
99% ee, 92%, TON =9,200
18
PPNO = 4-phenylpyridine N-oxide

Scheme 6B.19.

Recently it has been demonstrated that the ligand of **15** is structurally pliable and takes a stepped conformation based on the single crystal X-ray analysis of **15a** and **15b** (Figure 6B.10) [52]. Because it is expected that the bond between manganese ion and equatorial oxygen atom becomes shorter as the oxidation state of manganese ion increases and this bond-shortening

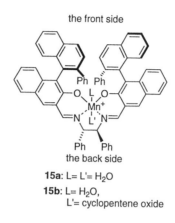

15a: L= L'= H_2O

15b: L= H_2O,
        L'= cyclopentene oxide

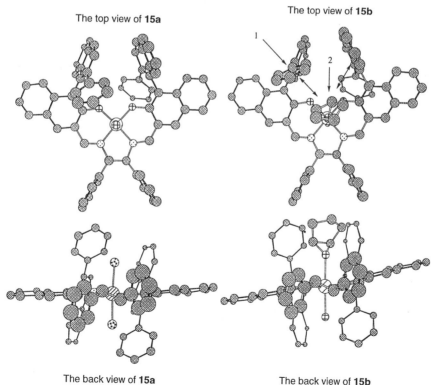

The top view of **15a**

The top view of **15b**

The back view of **15a**

The back view of **15b**

Figure 6B.10. X-ray structures of complexes **15a** and **15b**.

amplifies the non-planarity of the ligand, it is reasonable to assume that the ligand of the oxo-Mn-salen complexes takes a more deeply folded stepped conformation.

Mechanism of the transfer of oxygen atom from the oxo species to an olefin is still in controversy. There have been two mechanisms proposed for this process: i) concerted mechanism and ii) stepwise mechanism (Scheme 6B.20). From the studies of electronic effect on enantioselectivity and *cis-trans* isomerization, kinetic isotope effect, and temperature effect, Jacobsen et al. proposed the concerted mechanism (route **a**) [53]. However, Norrby et al. and Katsuki et al. have independently proposed a stepwise mechanism in which oxygen atom transfer proceeds through a metallaoxetane intermediate that decomposes to the epoxide directly (route **b**) or by way of a radical intermediate (route **c**) [54,55]. Calculation by using Macromodel/MM3 supports the intermediacy of the metallaoxetane intermediate [54]. However, unusual non-linear phenomena in electronic and temperature effects [56] on enantioselectivity are observed in the epoxidation by using **15** or its derivative as catalyst [55]. These results strongly suggest the presence of an intermediate in the oxygen atom transfer process [56]. A recent paper on the stereochemistry in the oxidation of enol ethers also supports the intermediary of the metallaoxetane complex [57].

The X-ray crystallographic analysis of complex **15b** suggests that the second-generation salen ligand may recognize the substrate through an attractive interaction, probably CH–π interaction, between the ligand and the substrate. As Figure 6B.10 illustrates, the cyclopentane ring of the apical epoxide ligand is located close to the sterically most-congested area of the salen ligand and the distance between the 2-phenylnaphthyl groups (arrows 1 and 2) and the cyclopentane ring was only ca. 3.8Å at the nearest point. This finding suggests that the second-generation Mn-salen complexes can also be used as catalysts for asymmetric oxidation of non-olefinic substrate such as C-H oxidation [52]. Actually, high enantioselectivity has been realized in the oxidation of compounds bearing activated methylene moieties using the second-generation Mn-salen complexes [58]. The asymmetric desymmetrization of *meso*-tetrahydrofurans [58c] and pyrrolidines [58e] has also been successfully carried out by enantiotopos-selective C–H oxidation by using **19** and **20** as catalysts, respectively, with high selectivity (Scheme 6B.21).

Thus far, a variety of oxidants such as iodosylbenzene, sodium hypochlorite, Oxone [59], dioxirane [60], sodium periodate [61], peracid, and hydrogen peroxide [62] has been used as terminal oxidants. An electrochemical method has been shown to be useful for manganese recycling [63]. Molecular oxygen can be used as a terminal oxidant in the presence of an aldehyde [64]. In this process, aldehyde serves as a reducing agent to generate acylperoxo-Mn(IV) species **21**, which epoxidizes olefins (Scheme 6B.22). It should be noted that the sense of enantioselection in this reaction depends on the presence or absence of a donor ligand. It is

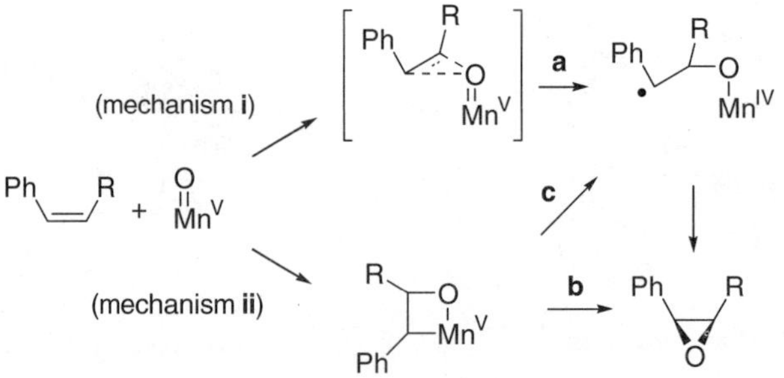

Scheme 6B.20.

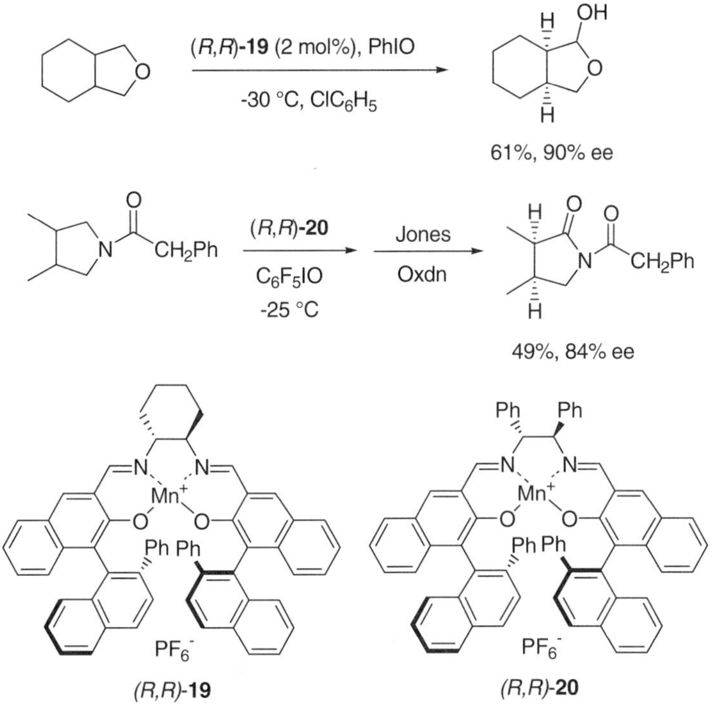

Scheme 6B.21.

proposed that the acylperoxo-Mn(IV) species **21** is converted to the oxo-Mn(V) species in the presence of a donor ligand (Scheme 6B.22).

Gilheany et al. have reported that Cr-salen complex **23** shows better enantioselectivity in the epoxidation of *trans*-β-methylstyrene than *cis*-β-methylstyrene (Scheme 6B.23) [65]. In contrast to Mn-salen catalysts in which the introduction of an electron-withdrawing group to their salen ligands diminishes asymmetric induction [66], Cr-salen complex **23** bearing an electron-withdrawing chlorine substituent shows better enantioselectivity than the parent non-substituted Cr-salen complex when the epoxidation is carried out in the presence of triphenylphosphine oxide. It has also been reported that the sense of enantioface selection in the epoxidation of 2,2-dimethylchromenes with Cr-salen complex **24** is affected by the polarity of the solvent used. For example, the epoxidation of 6-acetoamido-2,2-dimethyl-7-nitrochromene in acetonitrile gives the corresponding (*R,R*)-epoxide with 78% ee, whereas the epoxidation in toluene affords the (*S,S*)-epoxide with 65% ee [67]. The scope of the current Cr-salen-catalyzed epoxidation is very limited, but this process warrants further investigation because of its unique features.

The syntheses of most chiral Mn-salen complexes are simple, and thus their recovery and reuse have not been studied extensively. Only a few studies of epoxidation with a polymer-bound Mn-salen complex or a Mn-salen complex embedded in nano-porous materials as catalyst has been performed. It has been disclosed, however, that the microenvironment provided by the macromolecule adversely affects the asymmetric induction by the Mn-salen catalyst to a considerable extent although reuse of the catalysts for several cycles are realized [68].

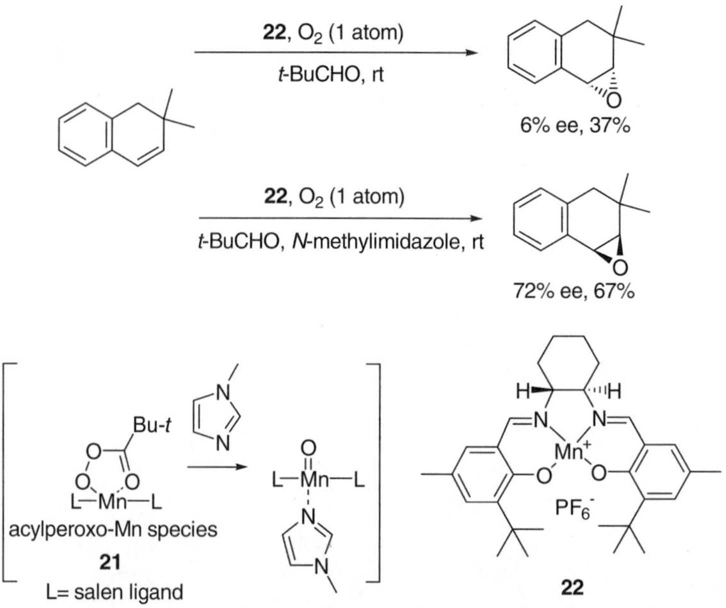

Scheme 6B.22.

6B.3.4. Other Metal-Catalyzed Asymmetric Epoxidation

In connection with the study on aerobic enantioselective epoxidation, N,N-'bis(3-oxobutyl-idene)diaminatomanganese(III) complex (**25**) is reported to be an effective catalyst for the reaction (Scheme 6B.24) [69]. It is noteworthy that epoxidations using **25** as the catalyst show higher enantioselectivity in the *absence* of a donor ligand rather than that in the presence of the ligand, which is different from the reaction using Mn-salen catalyst **22** (vide supra). This reaction

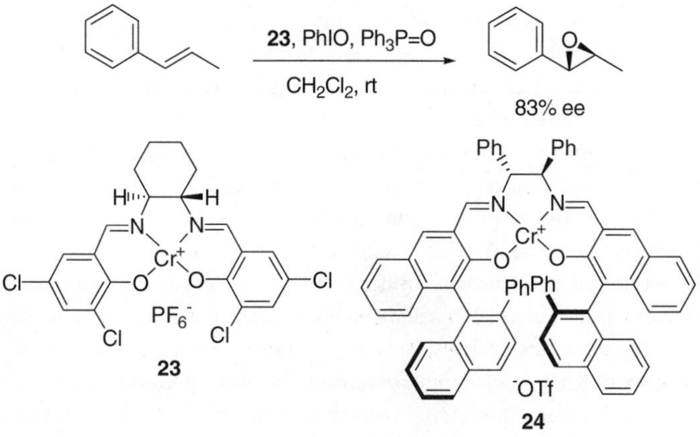

Scheme 6B.23.

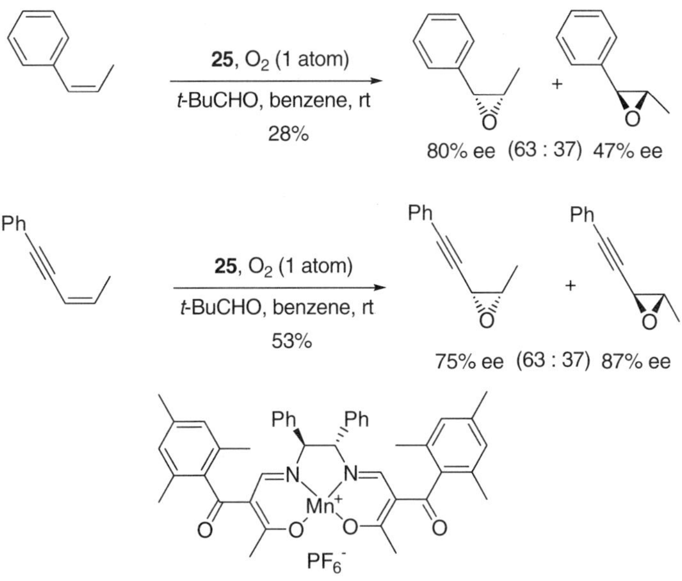

Scheme 6B.24.

is considered to proceed through the corresponding acylperoxo-Mn(V) species as proposed in the case of **22**.

End and Pfalz have reported that chiral ruthenium-bisamide (**26**) complex is a useful catalyst for epoxidation of *E*-olefins with sodium periodate as a terminal oxidant, although enantiose-lectivity is moderate (Scheme 6B.25) [70]. It is noteworthy that *trans*-β-methylstyrene is a better substrate than *cis*-β-methylstyrene for this reaction in terms of enantioselectivity. Competitive

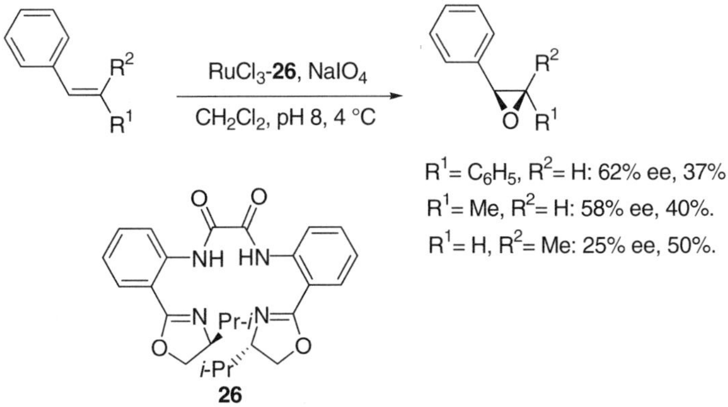

$R^1 = C_6H_5$, $R^2 = H$: 62% ee, 37%
$R^1 = Me$, $R^2 = H$: 58% ee, 40%.
$R^1 = H$, $R^2 = Me$: 25% ee, 50%.

Scheme 6B.25.

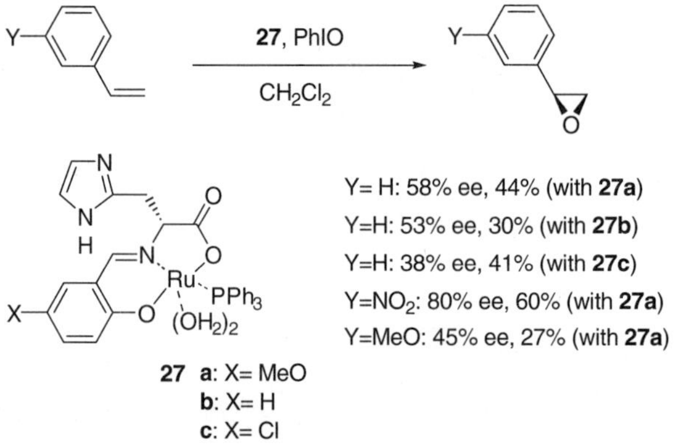

Y= H: 58% ee, 44% (with **27a**)
Y=H: 53% ee, 30% (with **27b**)
Y=H: 38% ee, 41% (with **27c**)
Y=NO$_2$: 80% ee, 60% (with **27a**)
Y=MeO: 45% ee, 27% (with **27a**)

27 a: X= MeO
b: X= H
c: X= Cl

Scheme 6B.26.

oxidative bond cleavage to give benzaldehyde is found to be a side reaction, but it can be considerably suppressed under the optimized conditions.

Kureshy et al. have reported that ruthenium-Schiff base complexes **27** serve as a catalyst for enantioselective epoxidation of styrene derivatives (Scheme 6B.26) [71]. An electronic effect similar to that described in the Mn-salen-catalyzed epoxidation (vide supra) is observed in this epoxidation, that is, an electron-donating group on the catalyst and electron-withdrawing group on the substrate lead to higher enantioselectivity. For example, the epoxidation of styrene with **27c** shows modest enantioselectivity (38% ee), whereas that of *m*-nitrostyrene with **27a** exhibits much higher enantioselectivity (80% ee).

It was recently found that (ON$^+$)(salen)ruthenium(II) complex **28** was an efficient catalyst for the epoxidation of conjugated *trans*-, *cis*- and terminal olefins. These reactions were remarkably accelerated by irradiation. It is noteworthy that *trans*- and *cis*-olefins gave the corresponding *trans*- and *cis*-epoxides, respectively (Scheme 6B.27) [72].

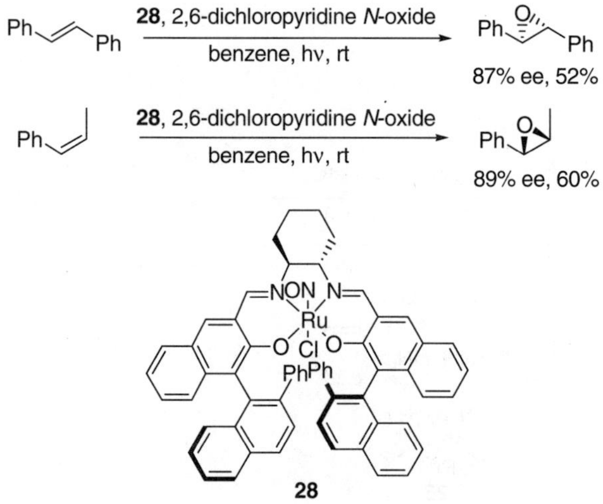

87% ee, 52%

89% ee, 60%

28

Scheme 6B.27.

6B.4. METAL-CATALYZED ASYMMETRIC AZIRIDINATION

As discussed above, iodosylbenzene (PhI=O) oxidizes various transition-metal ions (M^{n+}) such as manganese, iron, ruthenium, and chromium ions to the corresponding oxo-metal species ($O=M^{(n+2)+}$) that are potent oxene-transfer agents. Likewise, [N-(p-toluenesulfonyl)imino]-phenyliodinane (PhI=NTs) also oxidizes these metal ions to give the corresponding tosylimino-metal species ($TsN=M^{(n+2)+}$) that undergo nitrene-transfer reaction such as aziridination (Scheme 6B.28) [73].

In 1991, Evans et al. reported that cationic Cu(I) ions catalyzed the nitrene-transfer reaction smoothly (Scheme 6B.29) [74]. Since then, many studies on asymmetric aziridination have been carried out with chiral copper(I) complexes as catalysts.

Evans and co-workers were the first to report that 4,4′-disubstituted bisoxazolines **29** are excellent chiral ligands for enantioselective aziridination (Scheme 6B.30) [74,75]. Aryl-substituted olefins, especially cinnamate esters, are "good" substrates for this aziridination. The best reaction conditions, however, vary with the substrates used. For the reactions of cinnamate esters, bisoxazoline **29a** and benzene are the ligand and solvent of choice. Under these conditions, enantioselectivity up to 97% ee is observed. For the aziridination of styrene, bisoxazoline **29b** and acetonitrile are the appropriate ligand and solvent.

Lowenthal and Masamune also reported that the copper complex bearing a bisoxazoline ligand **30** was an effective catalyst for aziridination of styrene (88% ee) (Scheme 6B.31) [76]. However, Evans et al. later claimed that this result was not reproducible [75].

Enantioselectivity of copper-catalyzed aziridination is dependent on the nitrene precursor used (Scheme 6B.32) [77]. Although the precursor of choice varies with the substrates, p-MeOC$_6$H$_4$SO$_2$N=IPh or p-O$_2$NC$_6$H$_4$SO$_2$N=IPh is superior to TsN=IPh in many cases. For example, the aziridination of styrene in the presence of copper-bisoxazoline complex **29b** gives the product with 78% ee using p-MeOC$_6$H$_4$SO$_2$N=IPh as the nitrene precursor, whereas the enantioselectivity is 52% ee when TsN=IPh is used as the precursor.

Another bisoxazoline ligand **31** was prepared from tartrate and applied to asymmetric aziridination. However, enantioselectivity observed was modest (Scheme 6B.33) [78]. Bisaziridine ligand **32** was prepared, but its copper complex showed only modest enantioselectivity in the aziridination of styrene (Scheme 6B.34) [79].

Jacobsen and co-workers have reported that chiral diimine **33a** serves as an effective chiral auxiliary for the copper-catalyzed aziridination of aryl-substituted Z-olefins (Scheme 6B.35) [80]. For example, the aziridination of 6-cyanochromene proceeds with high enantioselectivity (>98% ee). Comparison of ligands **33a**–**33c** has revealed that the o-substituents in the ligands sterically and electronically influence the enantioselectivity of the reaction, that is, the introduction of chlorines at o-positions not only prolongs catalyst lifetime but also enhances enantioselectivity. The reactions of other Z-substrates and cinnamate esters catalyzed by **33a** show moderate-to-high enantioselectivity, whereas that of E-stilbene gave low enantioselectivity (Table 6B.3).

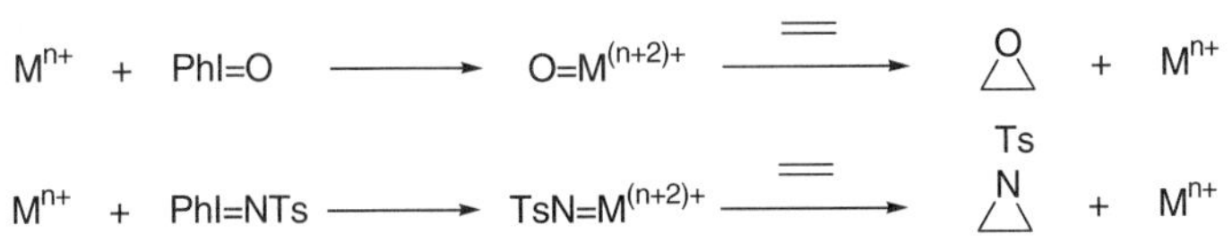

Scheme 6B.28.

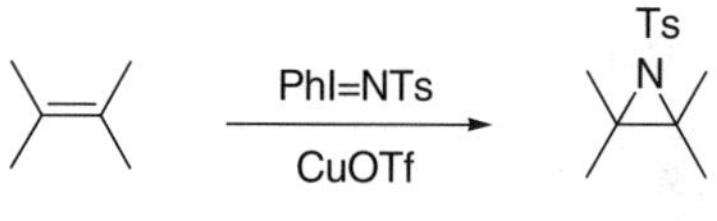

Scheme 6B.29.

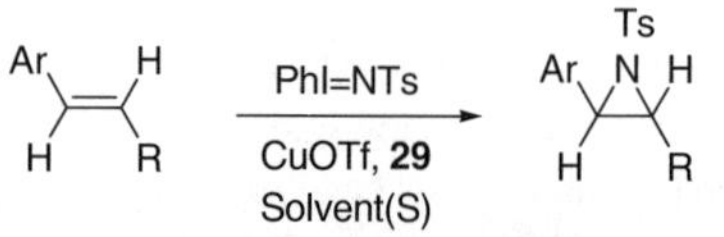

Ar = Ph, R = CO₂Me, **29a** (S = C₆H₆): 63%, 94% ee

Ar = Ph, R = CO₂Ph, **29a** (S = C₆H₆): 64%, 97% ee

Ar = α-C₁₀H₇, R = CO₂Me, **29a** (S = C₆H₆): 76%, 95% ee

Ar = Ph, R = Me, **29b** (S = MeCN): 62%, 70% ee

Ar = Ph, R = H,   **29b** (S = styrene): 89%, 63% ee

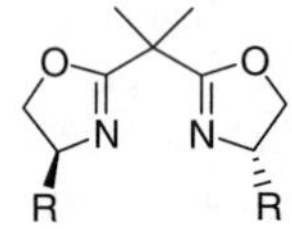

29a, R = C₆H₅

29b, R = CMe₃

Scheme 6B.30.

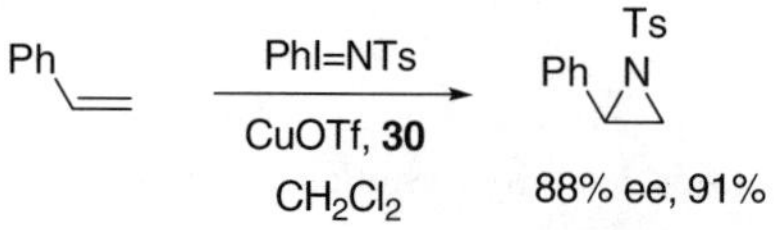

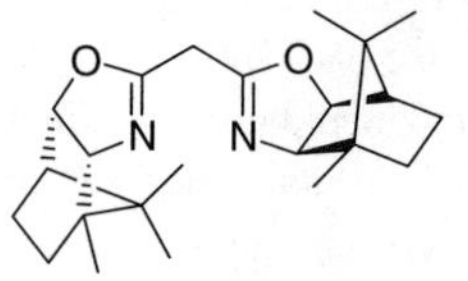

88% ee, 91%

30

Scheme 6B.31.

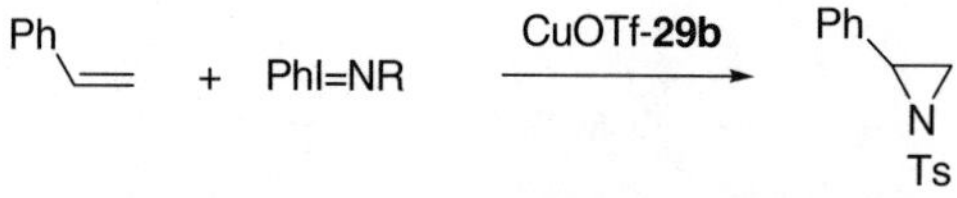

R= Ts: 52% ee, 77%

R= p-MeOC₆H₄SO₂: 78% ee, 86%

Scheme 6B.32.

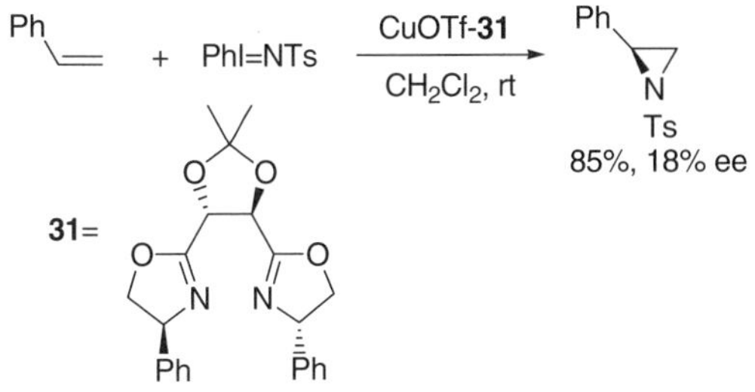

Scheme 6B.33.

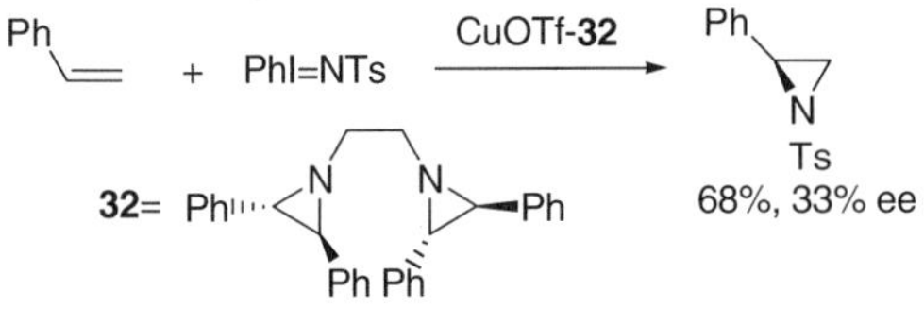

Scheme 6B.34.

Evans et al. proposed that an imino-copper species in the 3+ oxidation state (Cu^{3+}=NTs) should be the key intermediate in copper-catalyzed aziridinations [75b]. This proposal was supported by Jacobsen's study on the dependence of enantioselectivity on the nitrene precursors and/or the substrate structures with two iminoiodoarenes, PhI=NTs and 2,3,4-Me_3-6-(*t*-Bu)C_6HI=NTs), in the presence of $CuPF_6$-**33a** complex and four olefins [80b]. This study disclosed that enantioselectivity did not depend on the iminoiodoarene, but on the olefins used, that is, the finding excludes the possibility that a Cu-ArI=NTs adduct is a key intermediate. It has also been observed that the photochemical aziridination with tosyl azide (TsN$_3$) catalyzed

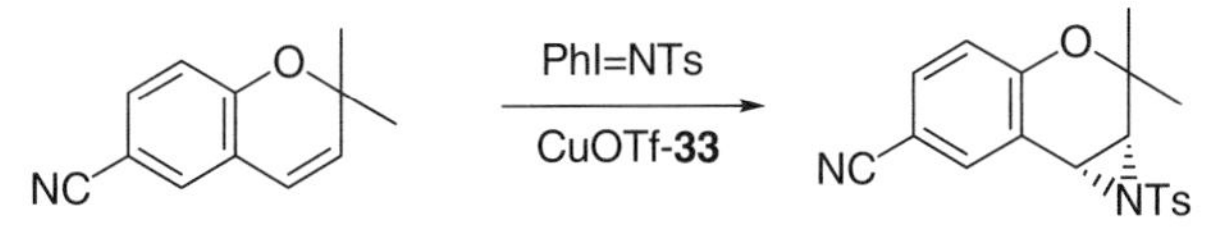

	33a	33b	33c
% ee:	>98	50	92
catalyst turnovers:	16	10	6.1

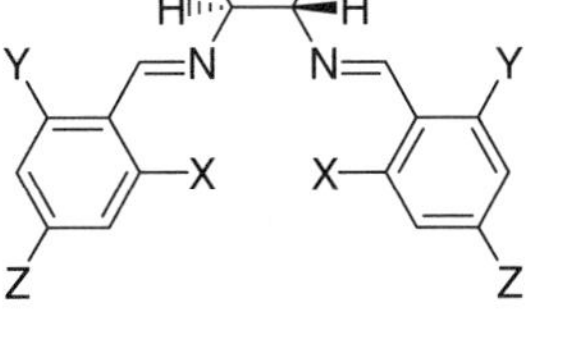

33a: X=Y= Cl, Z= H
33b: X=Y=Z= H
33c: X=Y=Z= CH_3

Scheme 6B.35.

TABLE 6B.3. Asymmetric Aziridination with Copper-diimine Complex 33a as a Catalyst

$$\text{alkene} \xrightarrow[\text{CuOTf (or CuPF}_6)\text{-}\textbf{33a}]{\text{PhI=NTs}} \text{aziridine}$$

Substrate	Cu(I)	Product(s)	Yield (%)	% ee
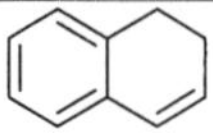	CuOTf	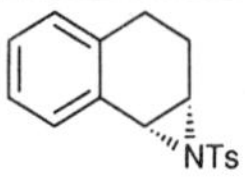	70	87
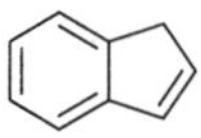	CuOTf	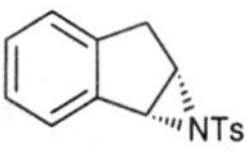	50	58
	CuOTf	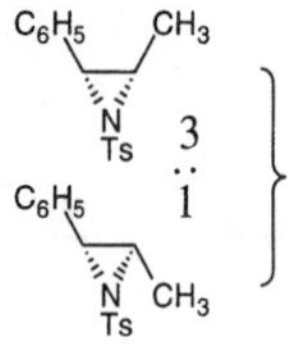	79	67 / 81
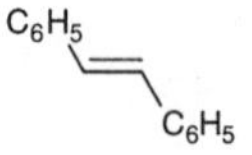	CuOTf	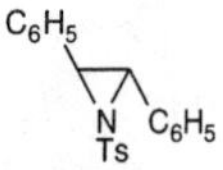	—	30
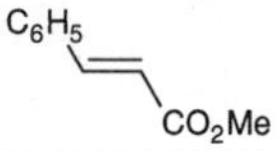	CuOTf	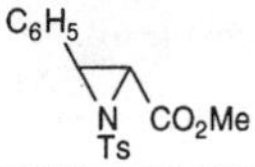	—	71

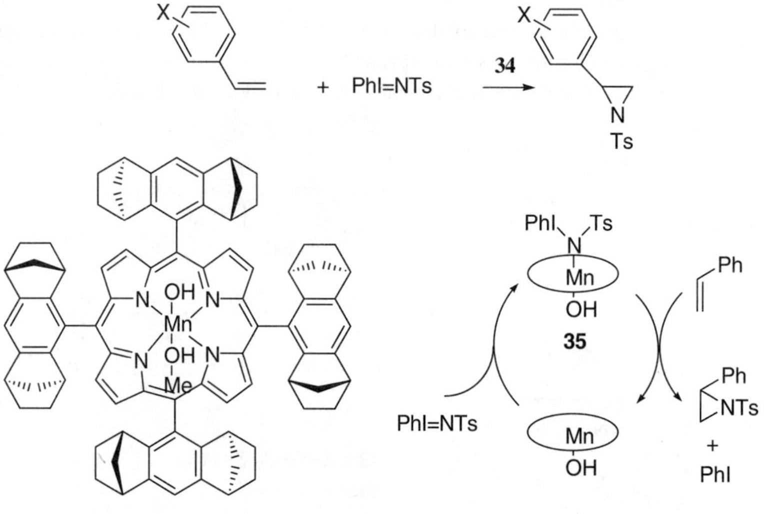

34

Scheme 6B.36.

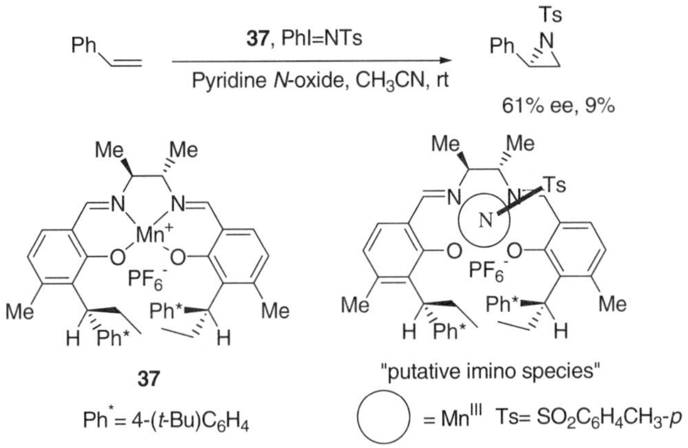

Scheme 6B.37.

by $CuPF_6$-**33a** complex shows the same enantioselectivity as the reaction with PhI=NTs. Because TsN_3 is known to generate a free nitrene intermediate under photochemical conditions, this result also supports the intermediacy of Cu^{3+}=NTs species.

Metalloporphyrin complexes serve as catalysts for aziridination in the presence of PhI=NTs [73]. Che et al. have reported the chiral version of metalloporphyrin-catalyzed aziridination (Scheme 6B.36) [81]. The reaction of styrene derivatives with a D_4-manganese(III) porphyrin complex **34** proceeds with fairly good enantioselectivity, up to 68% ee. This reaction is proposed to proceed through a Mn(IV)-PhINTs adduct **35** on the basis of EPR analysis.

In 1992, Burrows et al. reported asymmetric epoxidation and aziridination by using Mn-salen complex **36** as a catalyst [82]. Some enantioselectivity was observed in the epoxidation, but no asymmetric induction took place in the aziridination.

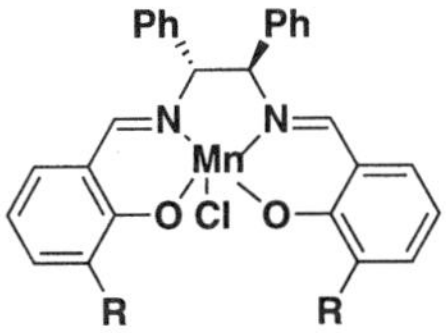

36 a: R= Me_3Si
 b: R= t-BuMe$_2$Si

Most of the first-generation Mn-salen complexes such as **16** show poor enantioselectivity (<25% ee) [83]. The Mn-salen complex **37**, which has a small substituent at the ethylenediamine moiety shows moderate enantioselectivity, but chemical yield is poor (Scheme 6B.37). Differing from the oxo-Mn species, the imino-Mn species that is the putative active catalyst species in aziridination has a substituent on its nitrogen atom. It is reasonable to assume that the size of the substituents at the ethylenediamine moiety directs the orientation of the substituent at the nitrogen atom, which influences enantioselectivity.

The second-generation Mn-salen complexes such as **18** show higher catalytic activity than the first-generation Mn-salen complexes, probably due to the attractive interaction between the

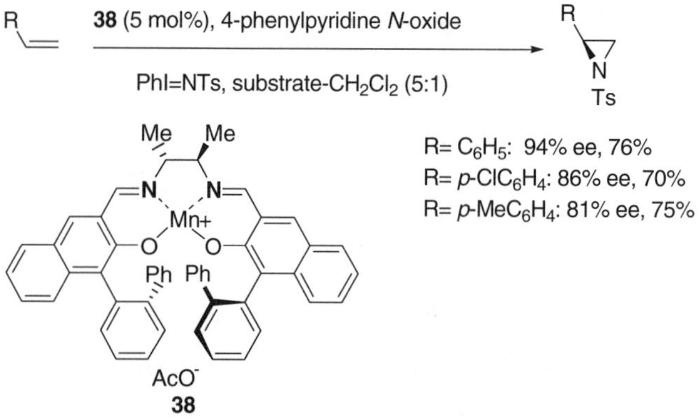

Scheme 6B.38.

ligand and substrates [84]. Nishikori and Katsuki took advantage of this fact and achieved high enantioselectivity (>90% ee) and chemical yield in the aziridination of styrene derivatives by using a slightly modified Mn-salen complex **38** as the catalyst (Scheme 6B.38) [85]. However, "good" substrates for this reaction are limited to styrene derivatives at present.

6B.5. CONCLUSION

As described above, introduction of chiral oxene or nitrene equivalents such as well-designed dioxiranes and oxo- and imino-metal species has realized highly enantioselective epoxidation and aziridination reactions of unfunctionalized olefins. Although many approaches are available for these reactions, this chapter deals mainly with the reaction systems that have achieved high enantioselectivity due to page limitation. It should be noted that various approaches have not realized sufficient enantioselectivity, but they include interesting concepts. Further development of such approaches may bring about a breakthrough in the epoxidation and aziridination. It is also worth mentioning that besides asymmetric oxidation are other excellent methods for the preparation of optically active epoxides and aziridines, such as catalytic asymmetric epoxidation by using chiral sulfide [86] and stereospecific conversion of chiral diols to epoxides [87], which are out of the scope of this chapter.

REFERENCES

1. Rossiter, B. "Synthetic Aspects and Applications of Asymmetric Epoxidation." In *Asymmetric Synthesis*; Morrison, J.D. (Ed.); Academic Press, Inc.: Orland, 1985; Vol. 5, pp. 193–246.

2. *Comprehensive Organic Synthesis*. Trost, B.M. (Ed.); Pergamon Press: Oxford 1991; Vol. 7.

3. Ewins, R. C.; Henbest, H. B.; Mckarvey, M. A. *J. Chem. Soc., Chem. Commun.* **1967**, 1085–1086.

4. Sharpless, K. B.; Verhoeven, T. R. *Aldrichimica Acta* **1979**, *12*, 63.

5. a) Davis, F. A.; Harabal, M. E.; Awad, S. B. *J. Amer. Chem. Soc.* **1983**, *105*, 3123–3126. b) Davis, F. A.; Sheppard, A. C. *Tetrahedron* **1989**, *45*, 5703–5742.

6. a) Aggarwal, V. A.; Wang, M. F. *Chem. Commun.* **1996**, 191–192. b) Page, P. C. B.; Rassias, G. A. R.; Bethel, D.; Schilling, M. B. *J. Org. Chem.* **1998**, *63*, 2774–2777.

7. Curci, R.; Fiolentino, M.; Seri, M. R. *J. Chem. Soc., Chem. Commun.* **1984**, 155–156.

8. Yang, D.; Yip, Y. C.; Tang, M. W.; Wong, M. K.; Zheng, J. H.; Cheung, K. K. *J. Am. Chem. Soc.* **1996**, *118*, 491–492.

9. Tu, Y.; Wang, Z.-X.; Shi, Y. *J. Am. Chem. Soc.* **1996**, *118*, 9806–9807.

10. a) Wang, Z-X.; Tu, Y.; Frohn, M.; Zhang, J.-R.; Shi, Y. *J. Am. Chem. Soc.* **1997**, *119*, 11224–11235. b) Wang, Z-X.; Shi, Y. *J. Org. Chem.* **1998**, *63*, 3099–3104.

11. Wang, Z-X.; Tu, Y.; Frohn, M.; Shi, Y. *J. Org. Chem.* **1997**, *62*, 2328–2329.

12. Yang, D.; Yip, Y.-C.; Tang, M.-W.; Wong, M.-K.; Zheng, J.-H.; Cheung, K.-K. *J. Am. Chem. Soc.* **1998**, *120*, 5943–5952.

13. Cao, G-A.; Wang, Z-X.; Shi, Y. *Tetrahedron Lett.* **1998**, *39*, 4425–4428.

14. Johnson, R. A.; Sharpless, K. B. "Catalytic asymmetric epoxidation of allylic alcohols." In *Catalytic Asymmetric Synthesis*; Ojima, I. (Ed.); VCH publishers: New York, 1993; pp. 103–158.

15. a) Katsuki, T.; Sharpless, K. B. *J. Am. Chem. Soc.* **1980**, *102*, 5974–5976. b) Katsuki, T.; Martin, V. S. *Organic Reactions* **1996**, *48*, 1–299.

16. Schurig, V.; Hintzer, K.; Leyree, U.; Mark, C.; Pitchen, P.; Kagan, H. *J. Organometal. Chem.* **1989**, *370*, 81–96.

17. Groves, J. T.; Nemo, T. E.; Myers, R. S. *J. Am. Chem. Soc.* **1979**, *101*, 1032–1033.

18. Groves, J. T.; Nemo, T. E. *J. Am. Chem. Soc.* **1983**, *105*, 5786.

19. Groves, J. T.; Myers, R. S. *J. Am. Chem. Soc.* **1983**, *105*, 5791–5796.

20. a) Mansuy, D.; Battioni, P.; Renaud, J. P.; Guerin, P. *J. Chem. Soc. Chem. Commun.* **1985**, 155–156. b) Groves, J. T.; Viski, P. *J. Org. Chem.* **1990**, *55*, 3628–3634. c) Halterman, R. L.; Jan, S.-T. *J. Org. Chem.* **1991**, 56, 5253–5254. d) Naruta, Y.; Tani, F.; Maruyama, K. *Chem. Lett.* **1989**, 1269–1272. e) Naruta, Y.; Tani, F.; Ishihara, N.; Maruyama, K. *J. Am. Chem. Soc.* **1991**, *113*, 6865–6872. f) Konishi, K.; Oda, K.; Nishida, K.; Aida, T.; Inoue, S. *J. Am. Chem. Soc.* **1992**, *114*, 1313–1317. g) Berkessel, A.; Frauenkron, M. *J. Chem. Soc., Perkin Trans.* **1997**, *1*, 2265–2266.

21. Jacobsen, E. N. "Asymmetric catalytic epoxidation of unfuctionalized olefins." In *Catalytic Asymmetric Synthesis*; Ojima, I. (Ed.); VCH publishers: New York, 1993; pp. 159–202.

22. Collman, J. P.; Lee, V. J.; Kellen-Yuen, C.J.; Zhang, X.; Ibers, J. A.; Brauman, J. I. *J. Am. Chem. Soc.* **1995**, *117*, 692–703.

23. Groves, J. T.; Quinn, R. *J. Am. Chem. Soc.* **1985**, *107*, 5790.

24. Lai, T.-S.; Zhang, R.; Cheung, K.-K.; Kwong, H.-L.; Che, C.-M. *Chem. Commun.* **1998**, 1583–1584.

25. Srinivasan, K.; Michaud, P.; Kochi, J. K. *J. Am. Chem. Soc.* **1986**, *108*, 2309–2320.

26. a) Samsel, E. G.; Srinivasan, K.; Kochi, J. K. *J. Am. Chem. Soc.* **1985**, *107*, 7606–7617. b) Siddall, T. L.; Miyaura, N.; Huffman, J. C.; Kochi, J. K. *J. Chem. Soc., Chem. Commun.* **1983**, 1185–1186.

27. a) Meunier, B.; Guilmet, E.; Decalvalho, M.-E.; Poilblanc, R. *J. Am. Chem. Soc.* **1984**, *106*, 6668–6676. b) Jacobsen, E. N.; Zhang, W. *J. Org. Chem.* **1991**, *56*, 2296–2298.

28. Feichtinger, D.; Plattner, D. A. *Angew. Chem., Int. Ed.* **1997**, *36*, 1718–1719.

29. a) Zhang, W.; Loebach, J. L.; Wilson, S. R.; Jacobsen, E. N. *J. Am. Chem. Soc.* **1990**, *112*, 2801–2803. b) Irie, R.; Noda, K.; Ito, Y.; Matsumoto, N.; Katsuki, T. *Tetrahedron Lett.* **1990**, *31*, 7345–7348.

30. Jacobsen, E. N.; Zhang, W.; Muci, L. C.; Ecker, J. R.; Deng, L. *J. Am. Chem. Soc.* **1991**, *113*, 7063–7064.

31. Chang, S.; Heid, R. M.; Jacobsen, E. N. *Tetrahedron Lett.* **1994**, *35*, 669–672.

32. Hosoya, N.; Irie, R.; Katsuki, T. *Synlett* **1993**, 261–263.

33. a) Katsuki, T. *Coord. Chem. Rev.* **1995**, *140*, 189–214. b) Katsuki, T. *J. Mol. Cat. A, Chem.* **1996**, *113*, 87–107. c) Dalton, C. T.; Ryan, K. M.; Wall, V. M.; Bousquet, C.; Gilheany, D. G. *Topics in Catalysis* **1998**, *5*, 75–91.

34. a) Irie, R.; Ito, Y.; Katsuki, T. *Synlett* **1991**, 265–266. b) Irie, R.; Noda, K.; Ito, Y.; Matsumoto, N.; Katsuki, T. *Tetrahedron:Asymmetry* **1991**, *2*, 481–494.

35. a) Palucki, M.; Pospisil, P. J.; Zhang, W.; Jacobsen, E. N. *J. Am. Chem. Soc.* **1994**, *116*, 9333–9334. b) Palucki, M.; McCormick, G. J.; Jacobsen, E. N. *Tetrahedron Lett.* **1995**, *35*, 5457–5460.

36. Chang, S.; Galvin, J. M.; Jacobsen, E. N. *J. Am. Chem. Soc.* **1994**, *116*, 6937–6938.

37. Brandes, B. D.; Jacobsen, E. N. *Tetrahedron Lett.* **1995**, *36*, 5123–5126.

38. Hamada, T.; Irie, R.; Katsuki, T. *Synlett* **1994**, 479–481.

39. Jacobsen, E. N.; Zhang, W.; Muci, L. C.; Ecker, J. R.; Deng, L. *J. Am. Chem. Soc.* **1991**, *113*, 7063–7064.

40. Lee, N. H.; Jacobsen, E. N. *Tetrahedron Lett.* **1991**, *32*, 6533–6536.

41. Chang, S.; Lee, N. H.; Jacobsen, E. N. *J. Org. Chem.* **1993**, *58*, 6939–6941.

42. Deng, L.; Jacobsen, E. N. *J. Org. Chem.* **1992**, *57*, 4320–4323.

43. a) Brandes, B. D.; Jacobsen, E. N. *J. Org. Chem.* **1994**, *59*, 4378–4380. b) Lee, N. H.; Muci, A. R.; Jacobsen, E. N. *Tetrahedron Lett.* **1991**, *32*, 5055–5058.

44. Zhang, W.; Lee, N. H.; Jacobsen, E. N. *J. Am. Chem. Soc.* **1994**, *116*, 425–426.

45. a) Hosoya, N.; Hatayama, A.; Irie, R.; Sasaki, H.; Katsuki, T. *Tetrahedron* **1994**, *50*, 4311–4322. b) Sasaki, H.; Irie, R.; Katsuki, T. *Synlett* **1994**, 356–358. c) Sasaki, H.; Irie, R.; Hamada, T.; Suzuki, K.; Katsuki, T. *Tetrahedron* **1994**, *50*, 11827–11838.

46. Fukuda, T.; Irie, R.; Katsuki, T. *Synlett* **1995**, 197–198.

47. a) Rispens, M. T.; Meetsma, A.; Feringa, B. L. *Recl. Trav. Chim. Pays-Bas.* **1994**, *113*, 413. b) Pospisil, P. J.; Carsten, D. H.; Jacobsen, E. N. *Chem. Eur. J.* **1996**, *2*, 974–980. c) Finney, N. S.; Pospisil, P. J.; Chan, S.; Palucki, M.; Konsler, R. G.; Hansen, K. B.; Jacobsen, E. N. *Angew. Chem., Int. Ed.* **1997**, *36*, 1720–1723.

48. Noguchi, Y.; Irie, R.; Fukuda, T.; Katsuki, T. *Tetrahedron Lett.* **1966**, *37*, 4533–4536.

49. Srinivasan K.; Kochi, J. K. *Inorg. Chem.* **1985**, *24*, 4671–4679.

50. a) Hashihayata, T.; Ito, Y.; Katsuki, T. *Synlett* **1996**, 1079–1081. b) Hashihayata, T.; Ito, Y.; Katsuki, T. *Tetrahedron* **1997**, *53*, 9541–9552. c) Miura, K.; Katsuki, T. *Synlett* **1999**, 783–785.

51. Ito, Y. N.; Katsuki, T. *Tetrahedron Lett.* **1998**, *39*, 4325–4328.

52. a) Irie, R.; Hashihayata, T.; Katsuki, T.; Akita, M.; Moro-oka, Y. *Chem. Lett.* **1998**, 1041–1042. b) Punniyamurthy, T.; Irie, R.; Katsuki, T.; Akita, M.; Moro-oka, Y. *Synlett* **1999**, 1049–1052..

53. Palucki, M.; Finny, N.: Pospisill, P. J.; Guler, M. J.; Ishida, T.; Jacobsen, E. N. *J. Am. Chem. Soc.* **1998**, *120*, 948–954.

54. Norrby, P.-O.; Linde, C.; Åkermark, B. *J. Am. Chem. Soc.* **1995**, *34*, 11035–11036.

55. Hamada, T.; Fukuda, T.; Imanishi, H.; Katsuki, T. *Tetrahedron* **1996**, *52*, 515–530.

56. Buschmann, H.; Scharf, H.-D.; Hoffmann, N.; Esser, P. *Angew. Chem., Int. Ed.* **1991**, *30*, 477–515.

57. Adam, W.; Fell, R. T.; Stegmann, V. R.; Saha-Moller, C. R. *J. Am. Chem. Soc.* **1998**, *120*, 708–714.

58. a) Hamachi, K.; Irie, R.; Katsuki, T. *Tetrahedron Lett.* **1996**, *37*, 4979–4982. b) Miyafuji, A.; Katsuki, T. *Synlett* **1997**, 836–838. c) Miyafuji, A.; Katsuki, T. *Tetrahedron* **1998**, *54*, 10339–10348. d) Hamada, T.; Irie, R.; Mihara, J.; Hamachi, K.; Katsuki, T. *Tetrahedron* **1998**, *54*, 10017–10028. e) Punniyamurthy, T.; Miyafuji, A.; Katsuki, T. *Tetrahedron Lett.* **1998**, *39*, 8295–8298.

59. Gurjar, M. K.; Sarma, E.; Rama Rao, A. V. *Indian J. Chem., Sect. B: Org. Chem. Incl. Med. Chem.* **1997**, *36B*, 213–215.

60. Adam, W.; Jeko, J.; Levai, A.; Nemes, C.; Patonay, T.; Sebok, P. *Tetrahedron Lett.* **1995**, *36*, 3669–3672.

61. Pietikäinen, P. *Tetrahedron Lett.* **1995**, *36*, 319–332.

62. a) Pietikäinen, P. *Tetrahedron Lett.* **1994**, *35*, 941–944. b) Irie, R.; Hosoya, N.; Katsuki, T. *Synlett* **1994**, 255–256.

63. Torii, S. *Electrochem. Soc. Interface, Winter* **1997**, 46–48.

64. a) Yamada, T.; Nagata, T.; Imagawa, K.; Mukaiyama, T. *Chem. Lett.* **1992**, 2231–2234. b) Yamada, T.; Imagawa, K.; Nagata, T.; Mukaiyama, *Bull. Chem. Soc. Jpn.* **1994**, *67*, 2248–2526.

65. Bousquet, C.; Gilheany, D. G. *Tetrahedron Lett.* **1995**, *36*, 7739–7742.

66. Jacobsen, E. N.; Zhang, W.; Guller, M. L. *J. Am. Chem. Soc.* **1991**, *113*, 6703–6704.

67. Imanishi, H.; Katsuki, T. *Tetrahedron Lett.* **1997**, *38*, 251–254.

68. a) De, B. B.; Lohray, B. B.; Sivaram, S.; Dhal, P. K. *Tetrahedron: Asymmetry* **1995**, *6*, 2105–2108. b) Minutoro, F.; Pini, D.; Petri, A.; Salvadori, P. *Tetrahedron, Asymmetry* **1996**, *7*, 2293–2302. c) Minutoro, F.; Pini, D.; Salvadori, P. *Tetrahedron Lett.* **1996**, *37*, 3375–3378. d) Sabater, M. J.; Corma, A.; Domench, A.; Fornes, V.; Garcia, H. *Chem. Commun.* **1997**, 1285–1286.

69. a) Mukaiyama, T.; Yamada, T.; Nagata, T.; Imagawa, K. *Chem. Lett.* **1993**, 327–330. b) Nagata, T.; Imagawa, K.; Yamada, T.; Mukaiyama, T. *Bull. Chem. Soc. Jpn.* **1995**, *68*, 1455–1465.

70. a) End, N.; Pfaltz, A. *Chem. Commun.* **1998**, 589–590. b) End, N.; Marko, L. Zehnder, M.; Pfaltz, A. *Chem. Eur. J.* **1998**, *4*, 818–824.

71. Kureshy, R. I.; Khan, N. H.; Abdi, S.H.R. *J. Mol. Cat. A: Chem.* **1995**, *96*, 117–122.

72. Takeda, T.; Irie, R.; Shinoda, Y.; Katsuki, T. *Synlett*, **1999**, 1157–1159.

73. a) Mansuy, D.; Battioni, P.; Mahy, J.-P. *J. Am. Chem. Soc.* **1982**, *104*, 4487–4489. b) Groves, J. T.; Takahashi, T. *J. Am. Chem. Soc.* **1983**, *105*, 2073–2074.

74. Evans, D. A.; Faul, M. M.; Bilodeau, M. T. *J. Org. Chem.* **1991**, *56*, 6744–6746.

75. a) Evans, D. A.; Woerpel, K. A.; Hinman, M. M.; Faul, M. M. *J. Am. Chem. Soc.* **1991**, *113*, 726–728. b) Evans, D. A.; Faul, M. M.; Bilodeau, M. T.; Anderson, B. A. *J. Am. Chem. Soc.* **1993**, *115*, 5328–5329.

76. Lowenthal, R. E.; Masamune, S. *Tetrahedron Lett.* **1991**, *32*, 7373–7376.

77. Södergren, M. J.; Alonso, D. A.; Andersson, P. G. *Tetrahedron: Asymmetry* **1997**, *8*, 3563–3564.

78. a) Harm, A. M.; Knight, J. G.; Stemp, G. *Synlett* **1996** 677–678. b) *Idem*, *Tetrahedron Lett.* **1996**, *37*, 6189–6192.

79. a) Tanner, D.; Anderson, P. G.; Harden, A.; Somfai, B. *Tetrahedron Lett.* **1994**, *35*, 4631–4634. b) Tanner, D.; Harden, A.; Johansson, F.; Wyatt, P.; Anderson, P. G. *Acta Chemica Scandinavica* **1996**, *50*, 361–368.

80. a) Li, Z.; Conser, K. R.; Jacobsen, E. N. *J. Am. Chem. Soc.* **1993**, *115*, 5326–5327. b) Li, Z.; Quan, R. W.; Jacobsen, E. N. *J. Am. Chem. Soc.* **1995**, *117*, 5889–5890.

81. Lai, T-S.; Kwong, H.-L.; Che, C.-M.; Peng, S.-M. *J. Chem. Soc. Chem. Commun.* **1997**, 2373–2374.

82. O'Connor, K. J.; Wey, S. J.; Burrows, C. J. *Tetrahedron Lett.* **1992**, *33*, 1001–1004.

83. Noda, K.; Hosoya, N.; Irie, R.; Ito, Y.; Katsuki, T. *Synlett* **1993**, 469–471.

84. a) Noguchi, Y.; Irie, R.; Fukuda, T.; Katsuki, T. *Tetrahedron Lett.* **1996**, *37*, 4533–4536. b) Hamada, T.; Irie, R.; Mihara, J.; Hamachi, K.; Katsuki, T. *Tetrahedron* **1998**, *54*, 10017–10028.

85. Nishikori, H.; Katsuki, T. *Tetrahedron Lett.* **1996**, *37*, 9245–9248.

86. a) Aggarwal, V. K.; Ford, J. G.; Thompson, A.; Jones, R.V.H.; Standen, M.C.H. *J. Am. Hem. Soc.* **1996**, *118*, 7004–7005. b) Li, A.-H.; Dai, L.-X.; Aggarwal, V. K. *Chem. Rev.* **1997**, *97*, 2341–2372.

87. a) Keinan, E.; Sinha, S. C.; Sinha-Bagchi, A.; Wang, Z.-M.; Zhang, X.-L.; Sharpless, K. B. *Tetrahedron Lett.* **1992**, *33*, 6411–6414. b) Kolb, H. C.; Sharpless, K. B. *Tetrahedron* **1992**, *48*, 10515–10530.

6C

ASYMMETRIC OXIDATION OF SULFIDES

HENRI B. KAGAN

Laboratoire de Synthèse Asymétrique, Institut de Chimie Moléculaire d'Orsay, Université Paris-Sud, 91405 Orsay, France

6C.1. INTRODUCTION

Chiral sulfoxides are an important class of compounds that find increasing use as chiral auxiliaries in asymmetric synthesis [1–6]. Current interest in chiral sulfoxides also reflects the existence of products with biological properties that need a sulfinyl group with a defined configuration. Some materials can also be based on a chiral sulfoxide structure, for example, liquid crystals. For all these reasons, it is very important to develop efficient methods to prepare chiral sulfoxides of high enantiomertic purity [3,5–7]. Among various approaches to enantiomerically pure sulfoxides, the most practical one is the Andersen method, which uses a diastereomerically and enantiomerically pure sulfinate [7]. The main difficulty associated with this method, however, lies in achieving high diastereoselectivity, except for suitable cases such as menthyl p-tolylsulfinate, where crystallization can be combined with in situ epimerization [1,7]. Asymmetric oxidation of achiral R^1-S-R^2 sulfides is, in principle, a very straightforward route to chiral sulfoxides with much flexibility in the choice of R^1 and R^2 groups. Unfortunately, for a long time the enantioselectivity of such reactions remained very low, and this route was devoid of synthetic interest [8]. Renewed interest came in the early 1980s with progress obtained in various approaches. Among these methods, oxidation by hydroperoxides in the presence of chiral complexes, the use of chiral oxaziridines [9], electrochemichal oxidation with chiral electrodes [10], and enzymatic or microbial reactions seem to be the most attractive for the synthesis. This chapter covers the progress in these processes by selecting reactions that are catalytic with respect to the source of chirality. Surveys of the asymmetric oxidation of sulfides (catalytic or stoichiometric) can be found in articles devoted to various aspects of sulfoxide chemistry [5–8,11–13].

In this chapter, the oxidations with hydroperoxides mediated or catalyzed by chiral titanium alcoholates will be discussed first. Then, results obtained with various chiral metal-Schiff bases

Catalytic Asymmetric Synthesis, Second Edition, Edited by Iwao Ojima
ISBN 0-471-29805-0   Copyright © 2000 Wiley-VCH, Inc.

or metal-porphyrins as catalysts will be presented. Heterogeneous chiral catalysts will be briefly discussed. Oxidations from using chiral catalysts or chiral complexing agents such as flavins, imines, cyclodextrin, or bovin serum albumin (BSA) will also be included. Finally, the most efficient enzymatic oxidation of thio ethers will be summarized.

6C.2. OXIDATIONS IN THE PRESENCE OF CHIRAL TITANIUM ALCOHOLATES

Sharpless asymmetric epoxidation of allylic alcohols with hydroperoxides in the presence of a chiral titanium-complex consisting of $Ti(Oi\text{-}Pr)_4$ and diethyl tartrate (DET) or diisopropyl tartrate (1:1), discovered in 1980, soon evolved into a major methodology for synthesis of enantiomerically pure compounds [14,15]. The success of this reaction attracted much attention to the titanium/chiral alcoholate combinations as potential mediators or catalysts in various reactions. In 1984 Kagan et al. in Orsay attempted to oxidize simple thio ethers with t-butyl hydroperoxide in the presence of the Sharpless reagent. At first, racemic mixtures were isolated, but it was soon discovered that the addition of 1 mole equiv. of water provided sulfoxides with very high enantiomeric purity [16,17]. Under these conditions, the epoxidation of allylic alcohols is completely blocked. Independently that same year, Modena et al. in Padua reported the beneficial effect of using a large excess of diethyl tartrate (4 mol equiv.) with respect to $Ti(Oi\text{-}Pr)_4$ [18]. This stoichiometry also blocks the epoxidation of allylic alcohols [16]. The Orsay and Padua procedures seem to involve closely related complexes because the results are very similar for comparable examples. These systems are described in Sections 6C.2.1. and 6C.2.2. Originally the titanium complexes were used in stoichiometric amounts [16–18]. Later a procedure that allows the use of the titanium complex in catalytic amounts was devised. For that reason, the results obtained in stoichiometric reactions are presented because they are closely related to the catalytic reaction.

6C.2.1. Complexes Based on Titanium Tartrate–Water Combination

The Orsay group found serendipitously that methyl p-tolyl sulfide was oxidized to methyl p-tolyl sulfoxide with high enantiomeric purity (80–90% ee) when the Sharpless reagent was modified by addition of 1 mole equiv. of water [16,17]. The story of this discovery was described in a review [19]. Sharpless conditions gave racemic sulfoxide and sulfone. Careful optimization of the stoichiometry of the titanium complex in the oxidation of p-tolyl sulfide led to the selection of $Ti(OiPr)_4/(R,R)\text{-}DET/H_2O$ (1:2:1) combination as the standard system [17]. In the beginning of their investigations, the standard conditions implied a stoichiometric amount of the chiral titanium complex with respect to the prochiral sulfide [16,17,20–23]. Later, proper conditions were found, which decreased the amount of the titanium complex without too much alteration of the enantioselectivity [24,25].

Table 6C.1 lists representative results for the asymmetric oxidation of thio ethers with t-butyl hydroperoxide under the standard conditions (in dichloromethane at −20°C). Enantioselectivities are especially good (80–95% ee) for the oxidation of aryl methyl sufoxides (Table 6C.1). A substantial decrease in enantioselectivity is observed for oxidation of aryl-S-alkyl–type sulfides in which an alkyl group is larger than methyl such as n-propyl an n-butyl.

In the case of the oxidation of methyl alkyl sulfides, enantioselectivity remains in the range of 50–60% ee (Table 6C.2). Disufides R-S-S-R′, sulfenamides R-S-NR′$_2$, and sulfenates R-S-O-R′ were oxidized to the corresponding chiral thiosulfinates, sulfinamides, and sulfinates, respectively (<52% ee, Table 6C.3) [21].

TABLE 6C.1. Asymmetric Oxidation of Sulfide Ar-S-R by t-BuOOH in the Presence of Ti(O-i-Pr)$_4$/(R,R)-DET/H$_2$O in a 1:2:1 Ratio[a]

Entry	Ar	R	Isolated Yield	(%) %ee[b]	Ref.
1	p-Tolyl	Methyl	90	89	[17]
2	p-Tolyl	Ethyl	71	74	[17]
			60	83	[26]
3	p-Tolyl	n-Butyl	75	75	[17]
4	1-Naphthyl	Methyl	98	89	[20]
5	2-Naphthyl	Methyl	88	90	[17]
6	2-Naphthyl	n-Propyl	78	24	[17]
7	9-Anthracenyl	Methyl	33	86	[20]
8	o-Tolyl	Methyl	77	89	[20]
9	p-OMeC$_6$H$_4$	Methyl	72	86	[17]
10	o-OMeC$_6$H$_4$	Methyl	70	84	[17]
11	Phenyl	CH$_2$Cl	60	47	[20]
12	Phenyl	CH$_2$CN	85	34	[20]
13	2-Pyridyl	Methyl	63	77	[17]

[a]Reaction performed at 5 mmol scale. [Sulfide] = [reagent] = $2*10^{-1}$ M in CH$_2$Cl$_2$ at $-20°$C.
[b]Measured by ^{1}H NMR with Eu(hfc)$_3$ or (R)-(3,5-dinitrobenzoyl)- 1-phenylethylamine [27,28]. All sulfoxides have (R) configuration.

Enantioselectivity is highly dependent on the solvent employed. A screening of appropriate solvents for the oxidation of methyl p-tolyl sulfide showed a dramatic solvent effect (Table 6C.4) [22]. The best solvents were dichloromethane and 1,2-dichloroethane, which have similar dielectric constants, that is, 1.6 and 1.44, respectively.

The nature of the hydroperoxides is important in the enantioselective oxidation of sulfides [23–25]. The best hydroperoxide is cumene hydroperoxide, a readily available compound.

TABLE 6C.2. Asymmetric Oxidation of Dialkyl Sulfides (R^1-S-R^2) by t-BuOOH and the Reagent Ti(O-i-Pr)$_4$/(R,R)-DET/H$_2$O in a Ratio of 1:2:1[a]

Entry	R^1	R^2	Isolated Yield (%)	% ee[b]	Ref.
1	t-Butyl	Methyl	72	53	[17]
2	n-Octyl	Methyl	77	53	[24]
3	Cyclohexyl	Methyl	67	54	[17]
4	PhCH$_2$	Methyl	88	35	[24]
5	PhCH$_2$CH$_2$CH$_2$	Methyl	84	50	[17]
6	Methyl	CH$_2$CO$_2$Et	84	63	[24]
7	Methyl	CH$_2$CH$_2$CO$_2$Me	85	64	[24]

[a]Reaction performed at 5 mmol scale. [Sulfide] = [reagent] = $2 * 10^{-1}$ M in CH$_2$Cl$_2$ at $-20°$C.
[b]Measured by proton NMR spectroscopy with Eu(hfc)$_3$ or (R)-(3,5-dinitrobenzoyl)1-phenylethylamine [27,28]. All sulfoxides have (R) configuration.

TABLE 6C.3. Asymmetric Oxidation of R-S-X by *t*-BuOOH and the Reagent Ti(O-*i*-Pr)$_4$/(*R,R*)-DET/H$_2$O in a Ratio of 1:2:1^a

Entry	R	X	Isolated Yield (%)	%ee^b	Config.
1	Methyl	S-Methyl	60	41	(S)
2	*i*-Propyl	S-*i*-Propyl	43	52	(S)
3	*t*-Butyl	S-*t*-Butyl	34	41	(S)
4	Phenyl	O-Methyl	86	29	(R)
5	*p*-Tolyl	O-Methyl	88	36	(R)
6	*p*-Tolyl	NH-*i*-Propyl	28	24	(S)
7	*p*-Tolyl	N-Diethyl	60	35	(S)

^{a}Reaction performed at 5 mmol scale. [Sulfide] = [reagent] = $2 * 10^{-1}$ M in CH$_2$Cl$_2$ at -20°C.
^{b}Measured by proton NMR spectroscopy with Eu(hfc)$_3$ or (*R*)-(3,5-dinitrobenzoyl)-1-phenylethylamine [27,28].
Source: Reference 21.

Table 6C.5 shows comparisons of *t*-BuOOH, PhC(Me)$_2$OOH, and Ph$_3$COOH in the oxidation of typical sulfides.

As is often found in asymmetric synthesis, *temperature* is an experimental parameter that can increase enantioselectivity. Here, however, a decrease in the reaction temperature does not always increase the enantioselectivity. An optimum temperature was found to be -20°C to -25°C for the oxidation of methyl *p*-tolyl sulfide [17], and this temperature range was retained for the standard oxidations. In the case of the monooxidation of dithianes, the maximum enantioselectivity was obtained at ca. -40°C [29] (Scheme 6C.1).

Mechanistic studies have provided some information on the titanium-tartrate combination, although many aspects of the reaction remain obscure [17]. The infrared spectroscopic analysis of the titanium-tartrate complex in dichloromethane showed the carbonyl stretching bands of tartrate at 1745 cm^{-1} (free) and 1675 cm^{-1} (chelated). The latter is more intense than that in the Sharpless reagent (which has absorptions at 1735 and 1635 cm^{-1}) [30]. The Hammett plot on *p*-R-C$_6$H$_4$-S(O)-Me gave a very good linear correlation with the ρ-value of -1.02, indicative of an electrophilic attack on sulfur [17]. No evidence (IR or polarimetric) has been found for the precoordination of sulfides to the titanium complex. It has been hypothesized that water addition is beneficial for building Ti-O-Ti units as it is known in controlled hydrolysis of titanium alcoholates [31], although other explanations are possible. The selective hydrolysis of one ester

TABLE 6C.4. Enantioselectivity in Asymmetric Oxidation of Methyl *p*-Tolyl Sulfide by Various Hydroperoxides in the Presence of Ti(O-*i*-Pr)$_4$/(*R,R*)-DET/H$_2$O in a Ratio of 1:2:1 in Various Solvents

Solvent	CCl$_4$	CHCl$_3$	CH$_2$Cl$_2$	ClCH$_2$CH$_2$Cl	Toluene	Acetone
% ee	4.5	70	85	86	26	62
Config.	(S)	(R)	(R)	(R)	(R)	(R)

Source: Reference 22.

TABLE 6C.5. Enantiomeric Excess in Asymmetric Oxidation by Various Hydroperoxides in the Presence of Ti(O-i-Pr)$_4$/(R,R)-DET/H$_2$O in a Ratio 1:2:1^a

Sulfide	Cumene Hydroperox.	t-BuOOH	Ph$_3$COOH
Me-S-(p-tolyl)	96	89	16
Me-S-(o-anisyl)	93	74	
Me-S-phenyl	93	88	
Me-S-(n-octyl)	80	53	32
Me-S-benzyl	61	35	

^{a}Reactions performed at $-20°C$ in CH$_2$Cl$_2$.
Source: Reference 24.

function of diethyl tartrate or dimethyl tartrate could be excluded. An excellent correlation exists between the absolute configuration of tartrate and that of the resulting sulfoxides. Scheme 6C.2 provides excellent predictions, by taking L and S groups (on steric grounds), as large and small, respectively.

The aromatic rings have a directing effect, which is a combination of steric and polar effects. Polar effects were seen in the oxidation of (p-R)-C$_6$H$_4$-S-Me, that is, enantioselectivity decreases with σ_R values (for example, R = p-NO$_2$, 17% ee; R = p-OMe, 90% ee). A triple bond behaves like a phenyl group in Scheme 6C.2. For example, n-Bu-C≡C-S-Me gave the corresponding (R)-sulfoxide with 75% ee [24]. If one assumes a tridentate tartrate around the titanium peroxide moiety in a Ti-O-Ti binuclear complex, one can propose Figure 6C.1 to be the preferred transition state. This mechanism is based on the hypothesis that the nucleophilic attack of sulfide takes place along the O–O bond of the coordinated peroxide.

Diethyl tartrate is the best tartaric acid derivative for enantioselective oxidation of thioethers. This finding was established for the asymmetric oxidation of methyl p-tolyl sulfide with cumene hydroperoxide, that is, 96% ee (DET); 87% ee (diisopropyl tartrate); 62% ee (dimethyl tartrate) [24]; and 1.5% ee (bis N,N-dimethyltartramide, t-BuOOH as the oxidant) [17].

A well-defined procedure for the preparation of the Orsay reagent (obtained by the sequential addition in dichloromethane of DET, Ti(O-i-Pr)$_4$, and water) improved the enantioselectivity up to 99% ee in the oxidation of various sulfides (p-R)-C$_6$H$_4$-S-Me (for example, R = p-Me, 99.5% ee; R = Ph, 99.2% ee; R = p-OMe, 99.5% ee; R = p-NO$_2$, 99.3% ee) [32].

Catalytic reactions in Sharpless epoxidation were achieved in 1986 by addition of molecular sieves, which suppress the formation of nonenantioselective complexes by moisture already present in the medium or produced during the reaction [33]. Similar problems needed to be solved in the asymmetric oxidation of sulfides because a decrease in the concentration of a

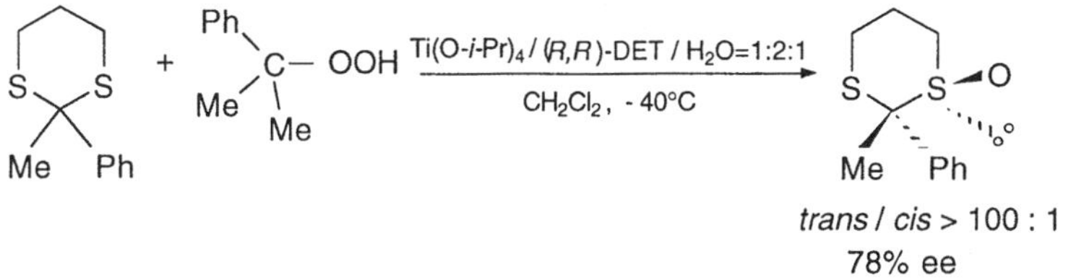

trans / cis > 100 : 1
78% ee

Scheme 6C.1. (From Ref. 29).

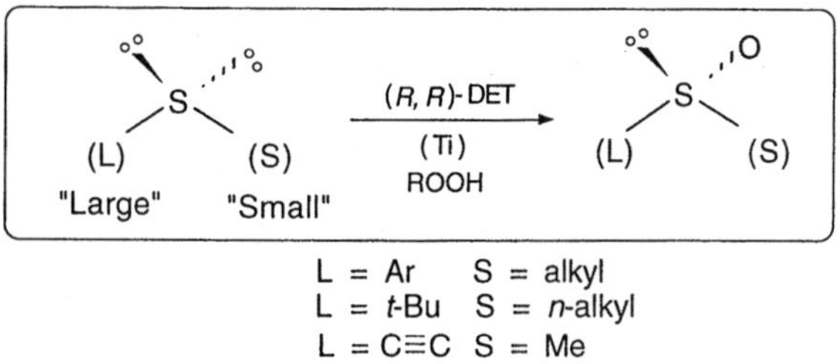

L = Ar   S = alkyl
L = t-Bu   S = n-alkyl
L = C≡C   S = Me

Scheme 6C.2. (From Refs. 17, 24).

titanium complex parallels a decrease in the enantiomeric purity of the resulting sulfoxide. The first difficulty might be due to an increase of the uncatalyzed pathway. It was estimated that the latter reaction was almost 200 times slower than the titanium-mediated oxidation with t-BuOOH [17]. The sluggishness of direct oxidation under the standard conditions left room to set up a catalytic process. Because of the various equilibria involved, many titanium species are potential catalysts. After many variations in experimental conditions, it was found that cumene hydroperoxide allowed for catalytic conditions [24]. Table 6C.5 shows typical results for the enantioselective oxidation of methyl p-tolyl sulfide. The enantioselectivity remains high (85% ee) until the amount of titanium complex is 0.2 mol equiv. A substantial decrease in enantioselectivity starts from 0.1 mol equiv. or below, although the chemical yield remains good. Obviously undesired catalytic species are quite active at low catalyst concentrations. It is curious that the addition of molecular sieves (pellets) helps to maintain good enantioselectivity, perhaps by efficient regulation of the amount of water. For preparative work at the 5–20 mmol scale, it is very convenient to use 0.5 mol equiv. of the titanium complex. A detailed procedure has been published [25]. It has been applied, for example, to the oxidation of 2,2-disubstituted 1,3-dithianes [29].

6C.2.2. Complexes Based on the Titanium-Excess Tartrate Combination

In 1984 the Padua group described the asymmetric oxidation of some sulfides with t-butyl hydroperoxide in the presence of 1 mole equiv. of Ti(O-i-Pr)$_4$ / (R,R)-DET (1:4) combination [18]. The reactions were mostly performed at −20°C in toluene or 1,2-dichloroethane. Results,

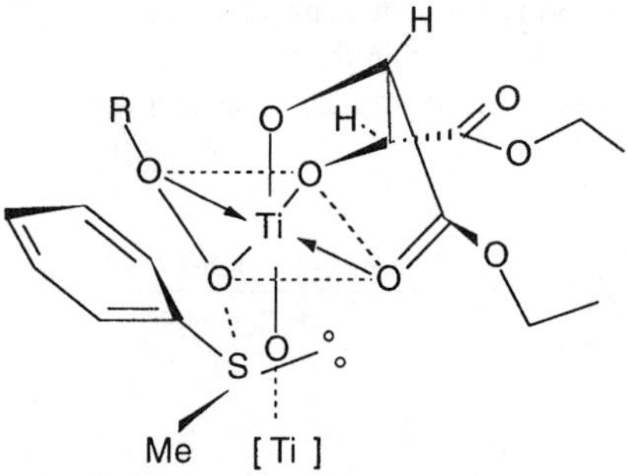

Figure 6C.1. Proposed preferred transition state for the asymmetric oxidation of phenyl methyl sulfide with binuclear titanium(peroxide)-tartrate complex.

listed in Table 6C.6, are similar to those given by using the water-modified reagent (cf. data in Tables 6C.1, 6C.2, and 6C.6). For example, (R)-methyl p-tolyl sulfoxide with virtually the same enantiomeric purity was obtained in both cases, that is, the Orsay complex, 89% ee and the Padua complex, 88% ee. Also, there is a substantial decrease in enantioselectivity upon replacing dichloromethane or 1,2-dichloroethane with toluene in both systems. It was hypothesized that identical active species could be involved in both systems [24, 136]. The procedure with excess diethyl tartrate was applied to the oxidation of 1,3-dithiolanes (Table 6C.7) [34–36].

The oxidation of the corresponding 1,3-dithianes or 1,3-oxathiolanes gave much lower enantioselectivity. An interesting application is the resolution of racemic ketones through their transformation into 1,3-dithiolanes by using asymmetric S-monooxidation followed by the separation of diastereomers and regeneration of the carbonyl group. This procedure has been applied to the resolution of *dl*-menthone wherein (−)-menthone was recovered with 93% ee. β-Hydroxysulfides of structure $PhCH(OH)CH_2$-S-R (racemic mixture) gave some kinetic resolution by partial conversion to the corresponding sulfoxide [37]. Stereoselectivity was much improved by protection of the hydroxyl group via silylation or acylation [38].

The Orsay group discovered that addition of 4 mol equiv. of isopropanol to the Padua complex (Ti(O-*i*-Pr)$_4$/(R,R)-DET = 1:4) gave a complex that can be used in catalytic quantities (10 mol % with respect to sulfide) in the sulfoxidations with cumene hydroperoxide [39,40]. With this catalyst system, which needs the presence of 4A molecular sieves, enantioselectivities in the range of 90% ee have been achieved for the preparation of various aryl methyl sulfoxides. Even benzyl methyl sulfoxide was obtained with 90% ee.

Thanks to the sign and size of the corresponding nonlinear effects, the use of diethyl tartrate of various enantiomeric excesses allowed them to trace the differences between the various combinations of Ti(O-*i*-Pr)$_4$/(R,R)-DET and additives such as water, isopropanol, and molecular sieves [41].

An interesting behavior of the Padua reagent (Ti(O-*i*-Pr)$_4$/(R,R)-DET = 1:4) was described by Scretti et al. [42,43], who used racemic furylhydroperoxides **1** instead of cumyl hydroperoxide as oxidant. The enantioselectivities in the oxidation of methyl aryl sulfides are very good. For example, methyl p-tolyl sulfoxide was obtained in 75% yield and >95% ee together with about 15% of sulfone by using hydroperoxide **1** (R^1 = OEt, R^2 = *i*-Pr and R^3 = Me)· Simultaneously there is a kinetic resolution of the racemic hydroperoxide takes place; is used in excess (2 mol equiv. with respect to sulfide). Thus in the example mentioned above, the enantiopurity of the residual hydroperoxide was 81% ee. It has also been established that some kinetic resolution of

TABLE 6C.6. Asymmetric Oxidation of R^1-S-R^2 by *t*-BuOOH in the Presence of Ti(O-*i*-Pr)$_4$/(R,R)-DET in a Ratio of 1:4^a

Entry	R^1	R^2	Isolated Yield	(%) %ee^b	Config.
1	p-Tolyl	Methyl	46	64	R
2	p-Tolyl	Methyl	60^b	88	R
3	p-Tolyl	t-Butyl	99	34	(+)
4	p-ClC$_6$H$_4$	CH$_2$CH$_2$OH	41	14	(−)
5	Benzyl	Methyl	70^c	46	(+)

^{a}1 mol equiv. of Ti complex, at −20°C. in toluene unless stated.

^{b}In 1,2-dichloroethane.

^{c}In dichloromethane at −77°C.

Source: Reference 18

TABLE 6C.7. Asymmetric Monooxidation of 1,3-Dithiolanes by *t*-BuOOH in the Presence of Ti(O-*i*-Pr)₄/(R,R)-DET in a Ratio of 1:4[a]

Substrate	Yield (%)[a]	Diastereomer Ratio	%ee
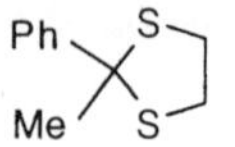 Ph, Me (1,3-dithiolane)	66	97:3	64
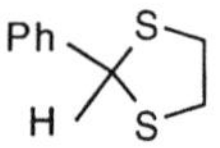 Ph, H (1,3-dithiolane)	76	94:6	88
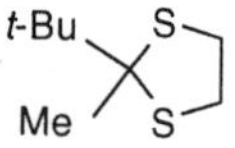 *t*-Bu, Me (1,3-dithiolane)	61	99:1	34
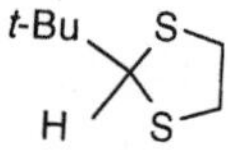 *t*-Bu, H (1,3-dithiolane)	82	99:1	14
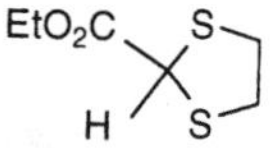 EtO₂C, H (1,3-dithiolane)	62	85:15	46

[a]Reaction at –20°C, in 1,2-dichloroethane. The relative stereochemistry of the S-monooxide is preferentially *trans* (between oxygen and the bulky group); the absolute configuration has not been established.
Source: References 34–36.

racemic sulfoxides is possible, enhancing the enantiomeric excess of sulfoxides arising from the enantioselective sulfide oxidations.

6C.2.3. Complexes Based on the Titanium-1,2-Diol Combination

Yamamoto et al. found, in 1989, that methyl *p*-tolyl sulfide gave the corresponding sulfoxide with 84% ee when used with 1,2-bisaryl 1,2-ethanediol [44] in the presence of stoichiometric amounts of Ti(O-*i*-Pr)₄/diol/H₂O (1:2:1) combination. It is curious that an inversion of configuration changing from bis(*o*-anisyl) ligand **2** to bis(*p*-anisyl) ligand **3** of the same absolute configuration (Scheme 6C.3). The same phenomenon was observed in the oxidation of methyl benzyl sulfide with ligands **2** and **3**, which gave the corresponding sulfoxides with 66% ee (*S*) and 43% ee (*R*), respectively.

Rosini et al. recently investigated the use of Ti(O-*i*-Pr)₄/(R,R)-diphenylethane-1,2-diol **4**/H₂O combination [45]. In a preliminary screening [45a] they defined the standard procedure for the oxidation with *t*-BuOOH (70% in water) in CCl₄ using the Ti(O-*i*-Pr)₄/**4**/H₂O (1:2:20) combination as the catalyst (5 mol %). This procedure provided (S)-methyl p-tolyl sulfoxide (80% ee) in 62% yield, minimizing the amount (8%) of sulfone produced. In the subsequent work, Rosini et al. extended the procedure to oxidation of various sulfides [45b] (Scheme 6C.3). The enantioselectivity of 80% ee was observed in the oxidation of p-R-C$_6$H$_4$-S-Me, wherein R is an electron-donor group. Enantioselectivity decreases with an electron-withdrawing group for R. An interesting feature of this system is the high enantioselectivity (~98% ee) in the oxidation of aryl benzyl sulfides.

Enantiopure 2,2,5,5-tetramethyl-3,4-hexanediol was prepared by Yamanoi and Imamoto [46]. A combination of Ti(O-i-Pr)$_4$ with this diol (1:2) gives a chiral catalyst for sulfide oxidation with cumyl hydroperoxide in the presence of 4A molecular sieves in toluene. At –20°C p-tolyl methyl sulfoxide (95% ee) was obtained in 42% yield together with 40% sulfone. A kinetic resolution increased, to some extent, the enantiomeric excess of the product, that is, at lower conversion (20% yield) the enantiopurity of the resulting sulfoxide was only 40% ee. This catalytic system is ineffective for the enantioselective oxidation of dialkyl sulfides.

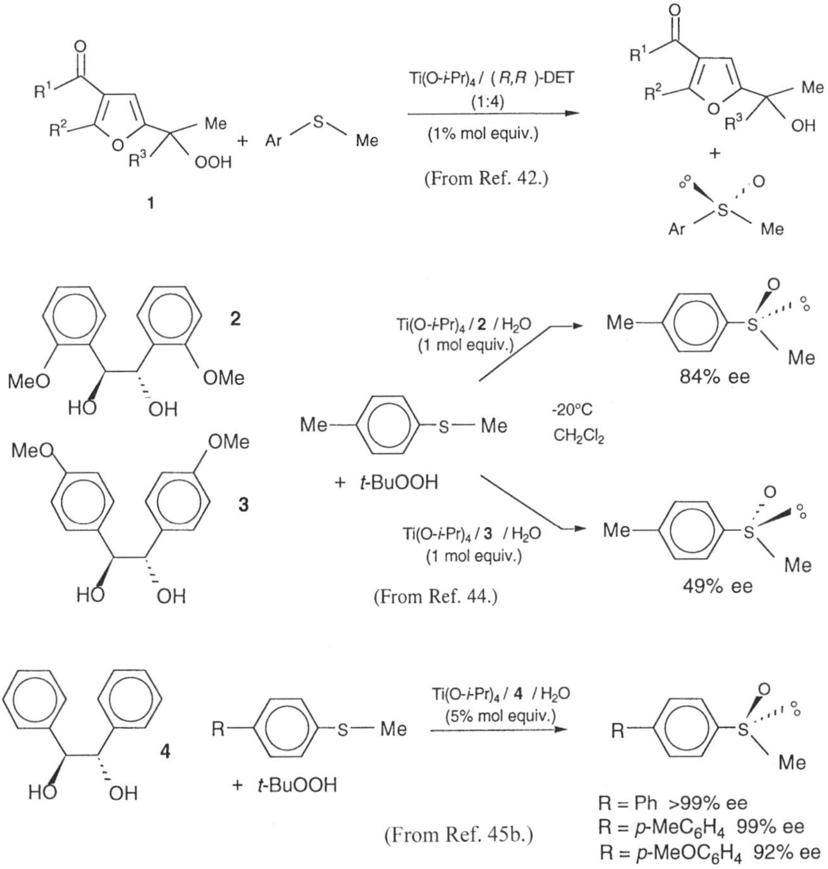

Scheme 6C.3.

6C.2.4. Complexes Based on the Titanium-Trialkanolamines Combination

Licini et al. studied titanium complexes arising from trialkanolamines **5** (Scheme 6C.4) and $Ti(O\text{-}i\text{-}Pr)_4$ [47]. In this study, complex **6** was identified by proton NMR spectroscopy in $CDCl_3$. Subsequent addition of *t*-BuOOH to **6** generated peroxo complexes **7** that were also clearly identified (equilibrium constant = 3.5 at 22°C). It should be noted that the X-ray structure of an achiral analogue of **7** has been determined [48].

Catalytic oxidations of sulfides were carried out in 1,2-dichloroethane with cumyl hydroperoxide by using 10 mol % of the catalyst. The best enantioselectivity was achieved with complex **6c**. However, sulfone was always produced as byproduct of the reaction. Even with a limited amount of hydroperoxide, the sulfone formation could not be avoided. For example, the reaction of methyl *p*-tolyl sulfide using 0.5 mol equiv. of cumyl hydroperoxide with respect to sulfide gave a 62:38 mixture of the corresponding (*S*)-sulfoxide and sulfone. The reaction of benzyl phenyl sulfide led to the formation of (*S*)-sulfoxide (84% ee) and sulfone ([sulfoxide]/[sulfone] = 77:23). It was established that sulfone was produced from the early stages of the reaction. It was also demonstrated that some kinetic resolution of the sulfoxide cooperated with the enantioselective oxidation of the sulfide. A unique feature of this oxidation system, as compared to those using various Ti(IV)/(DET) complexes, is the insensitivity of the enantioselectivity (40–60% ee at 0°C) to the nature of the alkyl group of sulfides Ar-S-alkyl.

6C.2.5. Complexes Based on the Titanium-Binol Combination

Uemura et al. [49] found that (*R*)-1,1′-binaphthol could replace (*R,R*)-diethyl tartrate in the water-modified catalyst, giving good results (up to 73% ee) in the oxidation of methyl *p*-tolyl sulfoxide with *t*-BuOOH (at –20°C in toluene). The chemical yield was close to 90% with the use of a catalytic amount (10 mol %) of the titanium complex ($Ti(O\text{-}i\text{-}Pr)_4$/(*R*)-binaphthol/$H_2O$ = 1:2:20). They studied the effect of added water and found that high enantioselectivity was obtained when using 0.5–3.0 equivalents of water with respect to the sulfide. In the absence of water, enantioselectivity was very low. The beneficial effect of water is clearly established here, but the amount of water needed is much higher than that in the case of the catalyst with diethyl tartrate. They assumed that a mononuclear titanium complex with two binaphthol ligands was involved, in which water affects the structure of the titanium complex and its rate of formation.

More recently, Uemura et al. studied the scope and limitations of their catalyst system [50]. The various parameters affecting the enantioselectivity were studied. A protocol using 2.5 mol

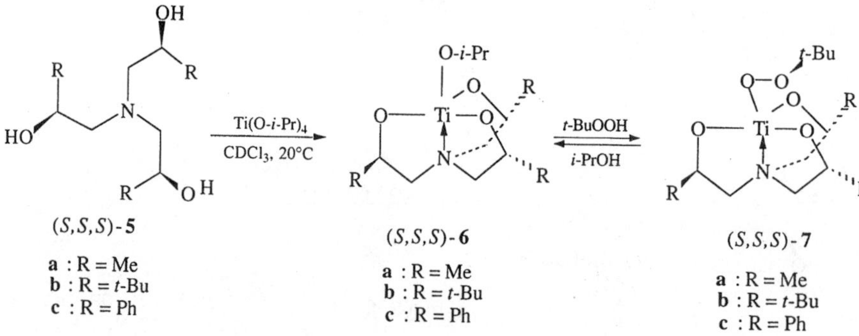
OH
R
R
HO
N
R
OH
Ti(O-i-Pr)4
CDCl3, 20°C
O-i-Pr
O
R
O
Ti
O
N
R
R
t-BuOOH
i-PrOH
t-Bu
O
O
O
R
O
Ti
O
N
R
R
(S,S,S)-5
a : R = Me
b : R = t-Bu
c : R = Ph
(S,S,S)-6
a : R = Me
b : R = t-Bu
c : R = Ph
(S,S,S)-7
a : R = Me
b : R = t-Bu
c : R = Ph

Scheme 6C.4.

% of titanium complex and a large proportion of water gave methyl p-tolyl sulfoxide with 96% ee in 24% yield. It was found that the high enantioselectivity originated from the simultaneous occurrence of the kinetic resolution of the resulting sulfoxide in the initial stage of the enantioselective oxidation of the sulfide. It was estimated that the enantiopurity of (R)-methyl p-tolyl sulfoxide initially formed was only 50% ee. The oxidation of racemic methyl p-tolyl sulfoxide to the corresponding sulfone, in a controlled experiment under the same conditions, clearly established the existence of kinetic resolution [51,52]. The applicability of the enantioselective oxidation catalyzed by the titanium/binol system is limited to aryl alkyl sulfides, and the enantioselectivity is quite low for the oxidation of dialkyl sulfides.

Reetz et al. [53] prepared analogues of (R)-binol as ligands for the titanium complex in the presence of water under the same conditions as Uemura's mentioned above. In this study, (R)-octahydrobinol and its dinitro derivative were synthesized. The reaction using (R)-dinitro-octahydrobinol ligand gave (S)-methyl p-tolyl sulfoxide (86% ee) [53], which makes a sharp contrast to the reaction using (R)-binol wherein (R)-methyl p-tolyl sulfoxide was formed [50]. It is probable that kinetic resolution is involved, giving some asymmetric amplification.

6C.2.6. Some Applications

Most of the applications described in the literature deal with the water-modified titanium system. Some examples are illustrated in Scheme 6C.5 [54–57]. Beckwith et al., during their investigation of homolytic substitution at sulfur, prepared sulfoxides **8** and **9** by oxidation with t-BuOOH in the presence of $Ti(O-i-Pr)_4/(R,R)$-DET/H_2O. Absolute configurations of **8** and **9** were assigned on the basis of the rule shown in Scheme 6C.2 [17,24] and were later confirmed by X-ray crystallographic analysis of sulfoxide **8**. Total synthesis of Itomanindole A (**10**), an indolic compound isolated from red algae, was affected by oxidation with cumene hydroperoxide in the presence of $Ti(O-i-Pr)_4/(R,R)$-DET/H_2O combination, and (R)-configuration was assigned to the product based on the rule shown in Scheme 6C.2. A Syntex team prepared both enantiomers of p-anisyl methyl sulfoxide (95% ee) by oxidation with cumene hydroperoxide in the presence of the titanium complex and used these sulfoxides as chiral auxiliary as well as a building block for the synthesis of a cardiovascular drug [56]. The Orsay complex was also useful to prepare some metallocene sulfoxides. Examples are shown in Scheme 6C.5. Asymmetric oxidation of 2-aryl 1,3-dithiolane or 2-aryl-1,3-oxathiolane with t-BuOOH in the presence of the same chiral titanium complex gave a superior level of diastereoselectivity and enantioselectivity as compared to the oxidation catalyzed by flavin-containing monooxygenase [57].

The $Ti(O-i-Pr)_4/(R,R)$-DET (1:4) combination has been used for resolution of 2,2'-dimethylthio-[1,1'-binaphthalene]. Through demethylation it is possible to recover both enantiomers of 2,2'-dithiol-[1,1'-binaphthalene], a useful chiral auxiliary [59,60].

Aggarwal et al. performed an extensive investigation into the usefulness of 1,3-dithiolane-1,3-dioxide and 1,3-dithiane-1,3-dioxide in asymmetric synthesis. For that purpose they prepared these compounds by asymmetric bis-oxidation of the corresponding 1,3-dithiolanes and 1,3-dithianes. The Padua protocol was very efficient to prepare 1,3-dithiolane-1,3-dioxide **12** (Scheme 6C.6), which was then transformed to the powerful chiral dienophile **14** [61]. They compared the Padua and Orsay protocols in the asymmetric oxidation of a range of 1,3-dithianes [62]. It was found that the Orsay reagent promoted efficient enantioselective monooxidation only when there is a substituent, for example, carbethoxy group (vide supra) [29], at the 2-position. Page et al. applied the same procedure for the monooxidation of 2-acyl-1,3-dithiane with excellent enantioselectivity [63].

Aggarwal et al. studied the bis-oxidation process in detail. In principle, the level of enantioselectivity at the first stage is the key factor for isolation of the resulting dioxide with

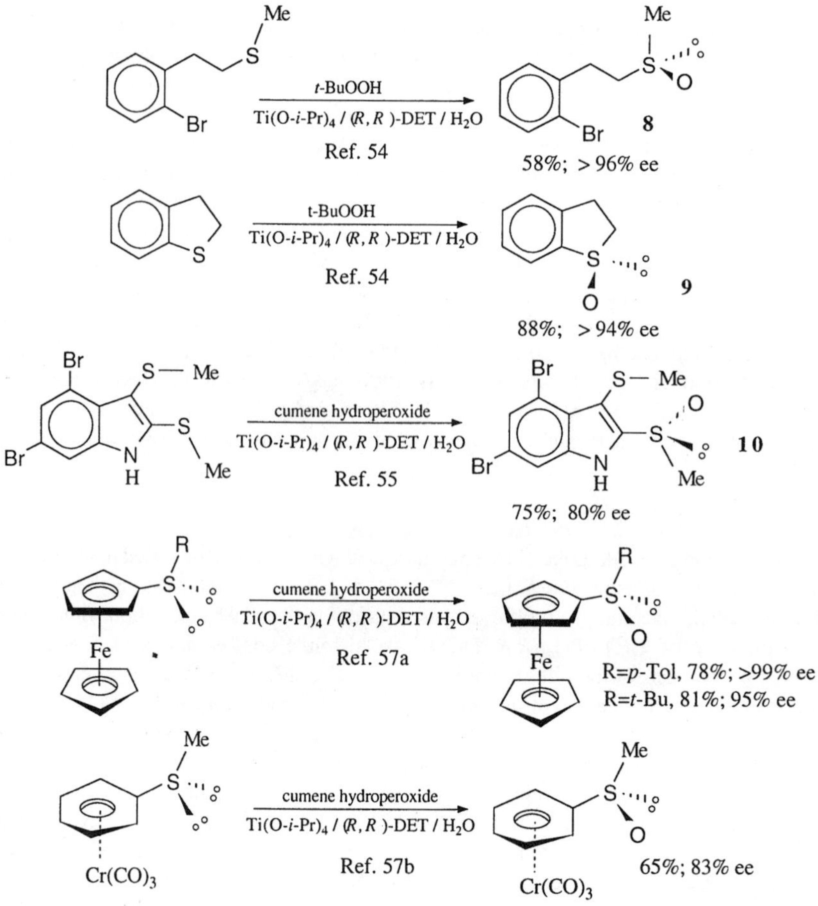

Scheme 6C. 5.

high enantiomeric excess. However, some asymmetric amplification may occur following the "Horeau principle," that is, the minor enantiomer of the first step is mostly eliminated through the minor *meso* dioxide formation even if the asymmetric induction at sulfur remains the same in the second oxidation step. For example, 1,3-dithiane ester **11** was oxidized with cumyl hydroperoxide at −35°C in the presence of the Orsay reagent, giving monooxide (85% yield,

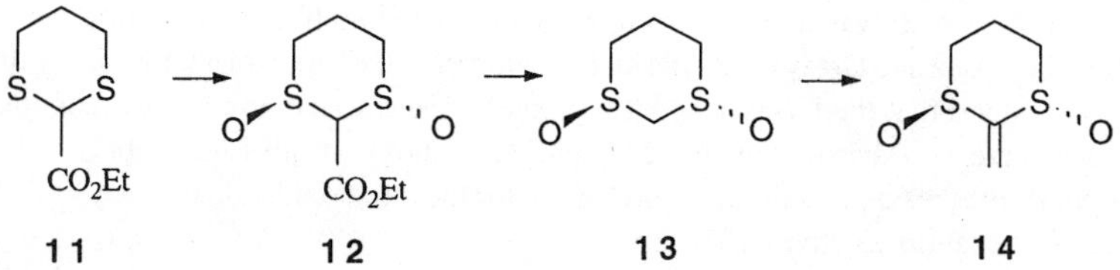

Scheme 6C.6.

85% ee) and *trans*-dioxide **12** (10% yield, >97% ee). After the systematic screening of experimental conditions for improving the yield of dioxide **12**, it was found that the presence of water in the reagent gave a mild oxidant that was quite selective for the monooxidation. The use of an excess of cumyl hydroperoxide afforded some dioxide **12** mixed with an overoxidation product, sulfoxide-sulfone. The Padua reagent was found to be an excellent oxidant for the selective synthesis of dioxide **12**, for example, the reaction at −22°C gave the desired **12** (>97% ee) in 60% yield together with monoxide (8%) and sulfoxide-sulfone (19%). It was proven that the high enantioselectivity was not due in part to kinetic resolution during overoxidation to sulfoxide-sulfone. The synthetic utility of this reaction was exemplified by the easy transformation of dioxide **12** to disulfoxide **14** that is a versatile chiral building block in asymmetric synthesis.

It was also found that the Padua reagent was useful for the oxidation of three isomeric bis(methylsulfinyl)benzenes with *t*-BuOOH, yielding the corresponding almost enantiopure bis-sulfoxides [64]. As mentioned above, the enantioselectivity of the first oxidation was enhanced (Horeau principle), because a minor amount of *meso*-disulfoxide was formed and separated from the chiral disulfoxide. It was reported that the Orsay reagent with cumyl hydroperoxide oxidized an *N*-protected 2-*S*-methylindole with moderate enantioselectivity, whereas the Padua protocol afforded the desired sulfoxide with 80% ee [65].

Finally, it is worth mentioning an enantioselective sulfoxidation in a kilogram scale performed by a team at Rhône-Poulenc Rorer [66]. This process involves the oxidation of an imidazolyl methyl sulfide with cumyl hydroperoxide in the presence of a half equivalent of Ti(O-*i*-Pr)$_4$/DET (1:2) combination. The corresponding sulfoxide was obtained with enantioselectivities in the range of 90–95% ee, depending on the protecting group of the imidazolyl nitrogen.

6C.3. CHIRAL TITANIUM-SCHIFF BASE CATALYSTS

The high enantioselectivity achieved by the chiral titanium alcoholates described in the preceding section is often counterbalanced by the low catalytic efficiency. Pasini et al. have developed chiral oxotitanium(IV)-Schiff bases **15** (Scheme 6C.7), which are good catalysts (catalyst:substrate ratio = 1:1000 to 1:1500) for the oxidation of methyl phenyl sulfide with 35% H$_2$O$_2$ in aqueous ethanol or dichloromethane [67]. Unfortunately, the enantioselectivity of this process is low (<20% ee), and some sulfone is formed. They proposed that sulfide would coordinate to titanium (that is, asymmetric induction takes place at sulfur) prior to the attack by hydrogen peroxide (instead of forming peroxotitanium species). A low-level asymmetric induction was also observed by Colonna et al. using chiral titanium complexes of *N*-salicylidene-*L*-amino acids **16** (Scheme 6C.7) [68]. These catalysts (0.1 mol equiv.) gave enantioselectivities below 25% ee in the oxidation of methyl *p*-tolyl sulfide and various sulfides with *t*-BuOOH in benzene at room temperature.

Fujita et al. reported a more promising approach in which active catalysts were prepared by reacting a Schiff base of (*R,R*)-1,2-cyclohexanediamine **17** with TiCl$_4$ in pyridine. The isolated complex served as a catalyst (4 mol %) for the asymmetric oxidation of methyl phenyl sulfide with trityl hydroperoxide in methanol at 0°C [69]. The reaction gave (*R*)-sulfoxide with 53% ee in good yield. Other hydroperoxides, for example, TBHP and cumene hydroperoxide, gave inferior results. The X-ray crystal structure of the isolated titanium-Schiff base complex revealed, surprisingly, the existence of an additional oxygen (see **18**, Scheme 6C.7) acting as a bridge between two titanium metals (Ti-O-Ti unit). Each titanium metal has an octahedral

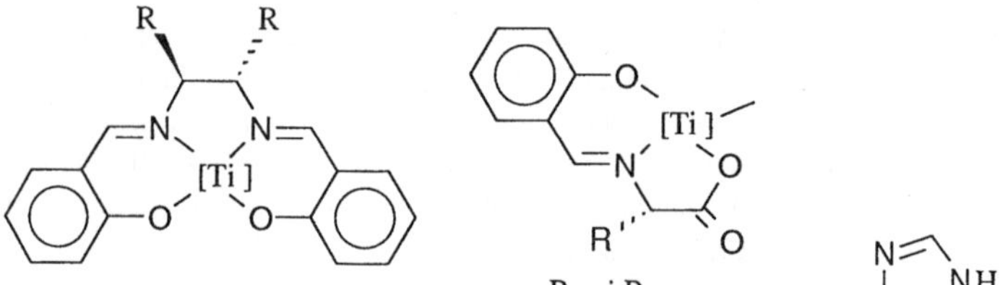

15 R = Me or Ph
Ref. [67]

16 a R = i-Pr
b R = t-Bu
c R = CH₂—

Ref. [68]

[Ti]: Ti = O (presumably as polymeric complex with Ti-O-Ti chains and pseudo octahedral structure).

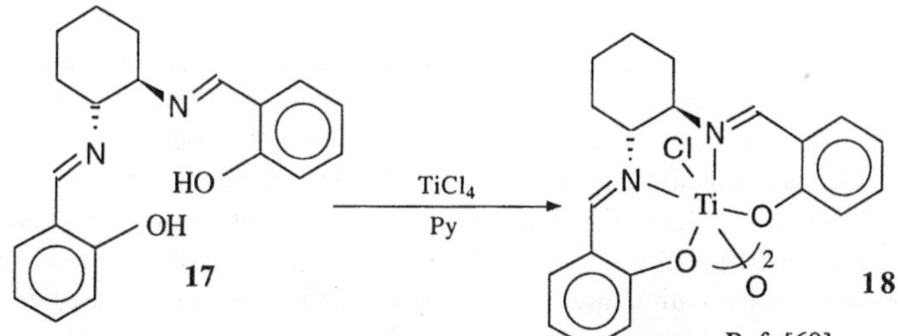

Scheme 6C.7.

configuration; the two planes of the Ti and the Schiff base are almost parallel to each other. The authors believe that the solution structure is different from the structure in the solid state.

6C.4. CHIRAL VANADIUM (IV)-SCHIFF BASE CATALYSTS

The Schiff base-oxovanadium(IV) complex formulated as **19** was found to catalyze the asymmetric oxidation of sulfides with cumene hydroperoxide (Scheme 6C.8) [70]. Various aryl methyl sulfides were used for this process (room temperature in dichloromethane and 0.1 mol equiv. of the catalyst). Chemical yields were excellent, but enantioselectivities were not higher than 40% for the resulting methyl phenyl sulfoxide. Complex **16a**, where [Ti] was replaced by VO, was also examined in the oxidation of sulfides, but the reactions gave only racemic sulfoxides [68].

Bolm and Bienewald discovered in 1995 that some chiral vanadium (IV)-Schiff base complexes were efficient catalysts (1 mol %) for sulfoxidation [71a]. The catalyst **20** was prepared in situ by reacting VO(acac)₂ with the Schiff base of a β-aminoalcohol (Scheme 6C.8). Reactions were conveniently performed in air at room temperature by slow addition of 1.1 mol equiv. of aqueous hydrogen peroxide (30%). Under these experimental conditions the reaction of methyl phenyl sulfide gave the corresponding sulfoxide in 94% yield and 70% ee. The best enantioselectivity was obtained in the formation of sulfoxide **21** (85% ee). Many structural analogues of catalyst **20** were screened for their efficacy, but none of

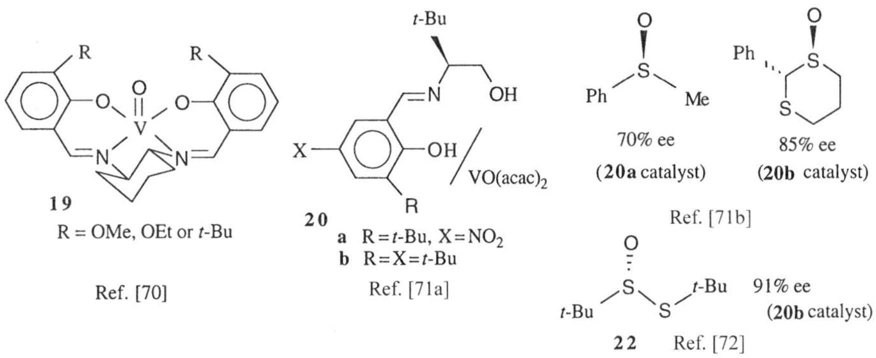

Scheme 6C.8.

these analogues gave good enantioselectivity. The scope of this process for the oxidation of various dithioketals that are mono- or disubstitued at the C-2 position has also been investigated [71b].

The Bolm protocol was recently used by Ellman et al. for the enantioselective oxidation of *t*-butyl disulfide **22** [72]. Excellent result was achieved in the formation of thiosulfinate **22** (91% ee, 93% yield) by using catalyst **20** (0.25 mol %) in a 0.5 mmol scale. In spite of extensive screening of chiral Schiff bases related to catalyst **20**, better enantioselectivity was not realized. Chiral thiosulfinate **22** is a convenient starting material for the preparation of *t*-butyl sulfinamides and *t*-butyl sulfoxides. Vetter and Berkessel modified the structure of the Schiff base moiety of catalyst **20** by replacing the aryl ring with a 1,1'-binaphthyl system [73]. The corresponding vanadium catalyst realized 78% ee in the oxidation of thioanisol, which was better than that attained by the Bolm catalyst (59% ee).

6C.5. MANGANESE-SCHIFF BASE CATALYSTS

Chiral (salen)Mn(III)Cl complexes are useful catalysts for the asymmetric epoxidation of isolated bonds. Jacobsen et al. used these catalysts for the asymmetric oxidation of aryl alkyl sulfides with unbuffered 30% hydrogen peroxide in acetonitrile [74]. The catalytic activity of these complexes was high (2–3 mol %), but the maximum enantioselectivity achieved was rather modest (68% ee for methyl *o*-bromophenyl sulfoxide). The chiral salen ligands used for the catalysts were based on **23** (Scheme 6C.9) bearing substituents at the ortho and meta positions of the phenol moiety. Because the structures of these ligands can easily be modified, substantial improvements may well be made by changing the steric and electronic properties of the substituents. Katsuki et al. reported that cationic chiral (salen)Mn(III) complexes **24** and **25** were excellent catalysts (1 mol %) for the oxidation of sulfides with iodosylbenzene, which achieved excellent enantioselectivity [75,76]. The best result in this catalyst system was given by complex **24** in the formation of orthonitrophenyl methyl sulfoxide that was isolated in 94% yield and 94% ee [76].

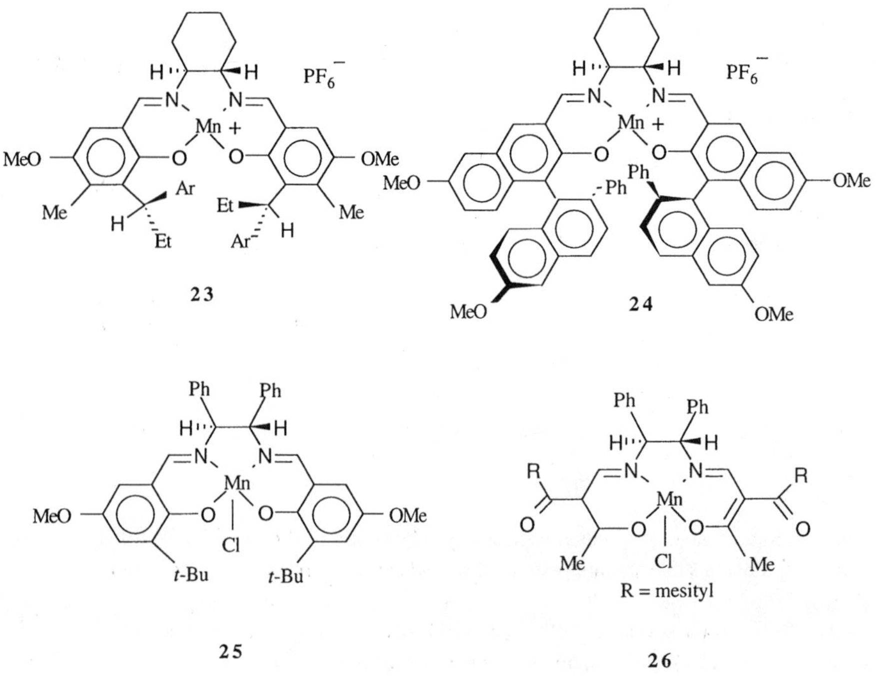

Scheme 6C.9.

6C.6. β-OXO-ALDIMINATOMANGANESE (III) CATALYSTS

A new family of chiral manganese catalysts, related to the salen manganese complexes, has been devised by Mukaiyama et al. [77,78]. An example of these β-oxo aldiminatomanganese (III) complexes **26** is shown in Scheme 6C.9. The oxidant in this system consists of molecular oxygen (3 atm) and pivalaldehyde. Presumably the peracid is formed in situ from the aldehyde, which reacts with the Mn(III) complex to generate an acylperoxomanganese complex as the actual oxidant. The oxidation of sulfides by using these catalysts (2 mol %) gives the corresponding sulfoxides with up to 70% ee in excellent yields with almost no overoxidation to sulfones.

6C.7. IRON- OR MANGANESE-PORPHYRIN CATALYSTS

Oxometalloporphyrins were taken as models of intermediates in the catalytic cycle of cytochrome P-450 and peroxidases. The oxygen transfer from iodosyl aromatics to sulfides with metalloporphyrins Fe(III) or Mn(III) as catalysts is very clean, giving sulfoxides. The first examples of asymmetric oxidation of sulfides to sulfoxides with significant enantioselectivity were published in 1990 by Naruta et al. who used chiral "twin coronet" iron porphyrin **27** as the catalyst (Figure 6C.2) [79]. This C_2 symmetric complex efficiently catalyzed the oxidation of sulfides with iodosylbenzene [turnover number (TON) up to 290] with enantioselectivity up to 73% ee (Table 6C.8). Addition of 1-methylimidazole, which acts as an axial ligand of iron, is necessary to achieve good enantioselectivitiy. For example, in the absence of 1-methylimidazole pentafluoro methyl sulfoxide is formed with only 31% ee (compare this result with entry 5, Table 6C.8). It was proposed that asymmetric induction should be mainly controlled by the steric hindrance around sulfur rather than by electronic effects. In 1991 the same authors disclosed full details on their catalytic system and proposed a mechanism for explaining asymmetric induction [80]. This mechanism is based on the steric-approach control of a sulfide to the oxo-iron center in the molecular cleft.

Groves and Viski reported similar results with binaphthyl iron (III)-tetraphenylporphyrin **28** as the catalyst (0.1 mol %) in the asymmetric oxidation of sulfides with iodosylbenzene, and enantioselectivity up to 48% was achieved [81]. Halterman et al. provided new examples of asymmetric oxidation of sulfides with iodosylbenzene, catalyzed by a D_4 symmetric manga-

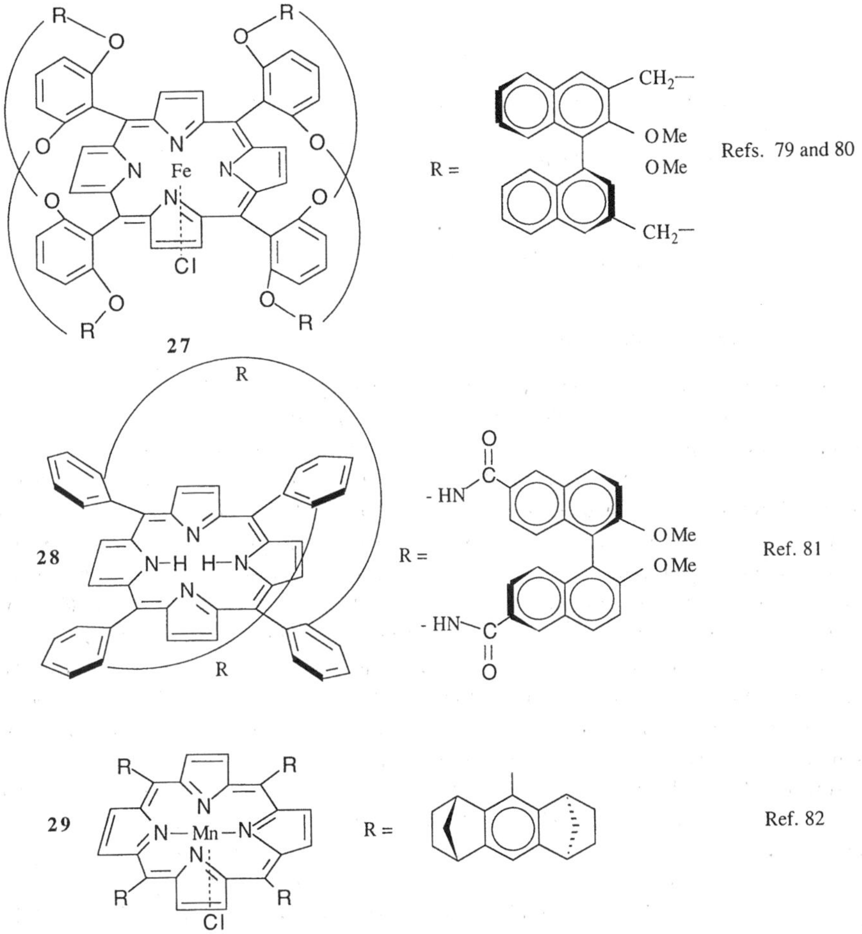

Figure 6C.2. Metal-porphyrin catalysts with bulky stereo-controlling chiral substituents

TABLE 6C.8. Asymmetric Oxidation of Ar-S-Me by PhIO Catalyzed by Iron (II)-Porphyrin 27

	Ar	Temp. (°C)	Turnover Number[b]	% ee	Config.
1	C_6H_5	−15	139	46	S
2	$2\text{-}NO_2C_6H_4$	−5	88	24	S
3	$3\text{-}NO_2C_6H_4$	−15	128	45	S
4	$4\text{-}NO_2C_6H_4$	0	120	53	S
5	C_6F_5	−15	55	73	S
6	$4\text{-}MeC_6H_4$	−15	144	54	S
7	2-Naphthyl	−15	168	34	R

[a]Reaction performed in CH_2Cl_2, PhIO = 260 μmol, **27** = 1 μmol, sulfide = 500 μmol, 1-methylimidazole = 100 μmol.

[b]Based on the amount of isolated sulfoxides.

Source: Reference 79

nese-tetraphenylporphyrin complex **29** (Figure 6C.2) [82]. Catalytic activity of **29** was excellent in dichloromethane at 20°C (catalyst/PhIO/sulfides = 1:200:400), and methyl phenyl sulfoxide was produced with 55% ee and methyl *o*-bromophenyl sulfoxide with 68% ee.

6C.8. HETEROGENEOUS CATALYSTS

6C.8.1. Montmorillonite Support

It has been shown that an ion-exchanged adduct of a clay and a chiral metal complex can be useful in resolution of a racemic mixture or in asymmetric synthesis. Usually when one enantiomer of a chiral chelate is adsorbed on a clay, it leaves half of the surface unoccupied (whereas the racemic mixture occupies all the active sites). The empty sites can be occupied by prochiral sulfides, which enables an asymmetric photooxidation to occur [83]. In this process photoexcited Λ or Δ-(2,2′-bipy)$_3$RuCl$_2$ first reacts with O_2 to provide O_2^- and [Ru(bipy)$_3$]$^{3+}$. Attack of an oxygen molecule on the cation radical of a sulfide yields the corresponding sulfoxide. Reaction is performed in methanol/water (1:4) by stirring the clay-chelate adduct (0.5 mol equiv.) under bubbling oxygen gas and irradiation with a 500-W tungsten lamp. There is a complete conversion of sulfides to sulfoxides after 2 h, and no sulfones are detected. Unfortunately, the sulfoxides thus formed have low enantiomeric purities (i.e., 15–20% ee for a variety of alkyl groups R). The ruthenium complex itself under homogeneous conditions led to racemic sulfoxides, in agreement with the hypothesis of an asymmetric control during the oxidation of adsorbed sulfides. Clay-chiral chelate adducts were also used as templates in the presence of an oxidant such as sodium metaperiodate, MCPBA or $K_2O_8S_2$ in water [84]. Δ-Ni (phen)$_3^{2+}$ -montmorillonite clay showed an appreciable enantioselectivity. Cyclohexyl-S-Ph at room temperature gave the corresponding sulfoxide in 78% ee (90% yield) by using NaIO$_4$ and 62% ee (90% yield) by using MCPBA. Similar results were obtained in oxidation of *n*-Bu-S-Ph or Bn-S-Ph. Unfortunately, the sulfide has to be preadsorbed on montmorillonite-clay, with the latter in large excess, making the transition into a practical catalytic process difficult.

6C.8.2. Chiral Electrodes

Electrooxidation of sulfides to sulfoxides on electrodes chemically modified with optically active compounds is an attractive approach. However, the first attempts were disappointing (ee <2%) [85]. High enantioselectivity was reported by Komori and Nonaka in 1984 [86]. These authors prepared various types of poly amino acid-coated electrodes. The electrodes were platinum or graphite plates, and in some cases polypyrrole-films were coated or covalently bound to the base electrode surface. Oxidations were carried out by means of a controlled-potential method in acetonitrile containing $(n\text{-Bu}_4)\text{NBF}_4$ and water. Reactions of alkyl aryl sulfides were investigated in detail with several modified electrodes. The reaction of Ph-S-Me gave very low enantioselectivity. The best results were observed for sulfides bearing bulky alkyl groups. The most appropriate electrode is the one prepared by dip-coating a platinum electrode modified chemically with polypyrrole and then poly(L-valine). With the use of this electrode, the following results were obtained for the oxidation of Ph-S-R :

R = i-Bu	44% ee
R = cyclohexyl	54% ee
R = i-Pr	73% ee
R = t-Bu	93% ee

Sulfoxides yielded have the (S)-configuration. A detailed mechanism of the origin of asymmetric induction is unknown. These chiral electrodes are reusable without loss of enantioselectivity. This approach is synthetically promising, but preparation of coated electrodes is a delicate undertaking, making the method difficult in practice unless robust chiral electrodes become commercially available.

6C.9. CHIRAL FLAVINS AS THE CATALYSTS

Biological oxidation of sulfides involves cytochromes P-450 or flavin-dependent oxygenases. A chiral flavin model was prepared by Shinkai et al. and used as the catalyst in the oxidation of aryl methyl sulfides [87]. Flavinophane **30** (Scheme 6C.10) is a compound with planar chirality. It catalyzes the oxidation of sulfides with 35% H_2O_2 in aqueous methanol at −20°C in the dark.

Flavinophane **30** acts as a true catalyst with TON number up to 800. The results in the oxidation of $p\text{-R-C}_4\text{H}_4\text{-S-Me}$ were

R = H	47% ee
R = Me	65% ee
R = t-Bu	42% ee
R = CN	25% ee

The reaction is much slower for R being CN as compared to R being H, indicative of an electrophilic character in the oxygen transfer at sulfur, presumably through the intermediate depicted in Scheme 6C.10.

6C.10. CHIRAL IMINES AND IMINIUM SALTS AS THE CATALYST

Imines and iminiums salts are oxidized by peracids into oxazidines and oxaziridinium ions, which are good reagents for oxidation of sulfides. The imine or iminium salt can, in principle,

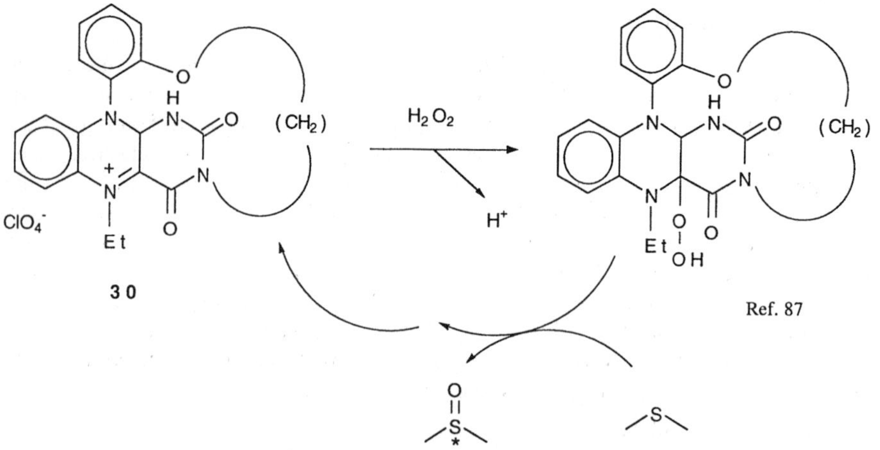

Scheme 6C.10.

be used in a catalytic amount if the terminal oxidant reacts faster on them than sulfide. The observation of some asymmetric induction (32% ee) was reported by Lusinchi et al. in the oxidation of methyl p-tolyl sulfide with oxone in alkaline medium in the presence of a stoichiometric amount of a chiral oxaziridinium salt derived from an iminium salt [88]. Page et al. developed a system by using aqueous hydrogen peroxide in basic conditions as the stoichiometric oxidant in the presence of a chiral imine derived from camphor [89,90]. The chiral imine is present in a stoichiometric amount (in principle, it could be used in a catalytic amount), but reaction slows down. It was established that the reaction does not involve a transient oxaziridine, but occurs with an α-hydroxyperoxyamine in a mechanism similar to that of the Payne reaction in which the oxidant system is the combination of a nitrile and hydrogen peroxide. This approach nicely complements other systems in the literature, as it has been proven particularly effective for the asymmetric oxidation of non-aryl sulfides. Enantioselectivity up to 98% ee was observed in the monooxidation of 2-phenyl-1,3-dithiane.

6C.11. TEMPLATE EFFECTS

6C.11.1. Sulfoxidations in the Presence of Cyclodextrins

It is known that cyclodextrins have a hydrophobic cavity (a binding site for aromatics) and a hydrophilic external surface. A "template-directed" asymmetric sulfoxidation has been attempted with various aryl alkyl sulfides [91]. Oxidations were performed by using metachloroperbenzoic acid in water in the presence of an excess of β-cyclodextrin. The best ee (33%) was attained for *meta*-(t-butyl)phenyl ethyl sulfoxide. The decrease in the amount of β-cyclodextrin below 1 mol equiv. causes a sharp decrease in enantioselectivity because of competition with oxidation of free substrate by the oxidant. Similarly, Drabowicz and Mikolajczyk observed modest asymmetric induction (27% ee) in the oxidation of Ph-S-n-Bu with H_2O_2 in the presence of β-cyclodextrin [92].

6C.11.2. Sulfoxidations in the Presence of Bovin Serum Albumin (BSA)

Sugimoto et al. found that BSA, a carrier protein in biological systems, is a host for aromatic sulfides. Based on this observation, the oxidation of sulfides with NaIO$_4$ was attempted in aqueous solution (pH 9.2) in the presence of BSA (0.3–2.0 mm, for example, 0.06 to 0.5 mol equiv. with respect to sulfoxide) [93,94]. The best results were obtained by using 0.3 mol equiv. of BSA. Results are shown in Table 6C.9.

There is a strong dependence of enantioselectivity on the structure of the aromatic moiety (reminiscent of enzymatic reactions). For example, Ph-S(O)-i-Pr and p-Tol-S(O)-i-Pr were isolated in about 80% yields with 81% ee (R) and 34% ee (S), respectively. The enantioselectivity is highly pH-dependent, for example, enantioselectivity vanishes at pH < 5. This property could be ascribed to conformational changes in the gross protein structure. With MPCBA or H$_2$O$_2$ as oxidant, lower enantiomeric excess or overoxidation to sulfones were unavoidable. The authors found that 30% hydrogen peroxide at pH 9.2 in the presence of isobutyl phenyl sulfide and BSA (0.33 mol equiv.) gave the corresponding sulfoxide with enantiomeric purity that increased with conversion. This finding is indicative of a kinetic resolution process [95]. Indeed, kinetic resolution of racemic sulfoxides was realized by using BSA. For example, enantiopurities of i-Pr-S(O)-Ph, t-Bu-S(O)-Ph, PhCH$_2$-S(O)-Ph, and i-Pr-S(O)-p-Tol at 50% conversion were 18, 33, 21, and 6% ee, respectively. By the simultaneous combination of asymmetric oxidation and kinetic resolution, it should be possible to increase substantially the enantiomeric excesses of phenyl alkyl sulfoxides. In fact, the oxidation of i-Pr-S-Ph gave (R)-sulfoxide with 62% ee in 78% yield, whereas through overoxidation the (R)-sulfoxide with 93% ee was recovered in 47% yield. This result was interpreted as a two-stage process in the binding domain of BSA, with preferential formation of the (R)-sulfoxide and the subsequent preferential destruction of the (S)-sulfoxide. However no such effect was observed with p-tolylsulfides, where overoxidation rather decreased the enantiomeric purity of the sulfoxides.

Asymmetric oxidation of formaldehyde dithioacetals with aqueous NaIO$_4$ was realized by Ogura et al. in the presence of a catalytic amount of BSA (0.005–0.02 mol equiv.) [96]. Under conditions the authors used, the starting sulfide was virtually insoluble in water (pH 9.2), and the best results were obtained at low concentrations of BSA. This result clearly indicates that the BSA/sulfide ratio is not the controlling factor of enantioselectivity. With this protocol, p-Tol-S-CH$_2$-S-p-Tol could be transformed into monosulfoxide with 60% ee. The same protocol gave isopropyl phenyl sulfoxide with 60% ee.

TABLE 6C.9. Asymmetric Oxidation of Sulfides Ar-S-R by NaIO$_4$ Containing BSA (0.3 mol equiv.)

Entry	Ar	R	Isolated Yield (%)	% ee	Config.
1	Phenyl	Methyl	47	7	R
2	Phenyl	Ethyl	58	29	R
2	Phenyl	i-Propyl	78	81	R
3	Phenyl	n-Butyl	87	36	R
4	Phenyl	i-Butyl	86	22	S
5	Phenyl	t-Butyl	86	75	R
6	Phenyl	Benzyl	52	49	R
7	p-Tolyl	i-Propyl	82	34	S

TABLE 6C.10. Asymmetric Oxidation of Sulfides Ar-S-R by Dioxiranes Catalyzed by BSA at 4°C

Entry	Ar	Me	Dioxirane Precursor	Yield (%)	% ee	Config.
1	Ph	Me	Acetone[a]	98	7	S
2	Ph	i-Pr	Acetone[a]	56	79	R
3	Ph	t-Bu	Acetone[a]	70	73	R
4	p-Tolyl	i-Pr	Acetone[a]	50	29	S
5	Ph	i-Pr	CH_3COCF_3 [b]	67	89	R
6	Ph	CH_2Ph	CH_3COCF_3 [c]	56	67	R

[a]Sulfide/KHSO$_5$/ketone/BSA = 1:2:13:0.05.
[b]Sulfide/KHSO$_5$/ketone/BSA = 1:2:0.44:0.05.
[c]Sulfide/KHSO$_5$/ketone/BSA = 1:0.5:0.11:0.0125.
Source: References 99 and 100.

Colonna et al. later investigated a wide range of sulfides to study periodate oxidation catalyzed by BSA [97,98]. The reactions were performed by stirring a heterogeneous mixture of sulfides, NaIO$_4$, and BSA (0.05 mol equiv.). In this study, lowering or increasing the amount of BSA had a detrimental effect on enantioselectivity. The enantiomeric purities of the sulfoxides thus obtained are as follows: n-butyl t-butyl sulfoxide 33% ee; mesityl phenyl sulfoxide, 40% ee; p-Tol-S-CH_2-S(O)-p-Tol, 40% ee. Oxidation to the sulfur of racemic sulfides with a stereocenter α or β gave two diastereomeric sulfoxides, each with some enantiomeric excess. Spectral data and circular dichroism of a mixture of BSA and some sulfides indicated that the latter are not tightly bound to BSA under the reaction conditions.

Dioxiranes were also generated in situ from ketones and caroate (KHSO$_5$) in the presence of BSA and sulfides at pH 7.5–8.0 [99,100]. Reactions were performed at 4°C for ~15 to 180 min, depending on the substrates and dioxiranes. The involvement of dioxiranes as actual oxidant is well-supported by the significant differences in enantioselectivity and absolute configurations, which were found by changing the structure of ketones that were the precursors of the dioxiranes. Some representative examples are listed in Table 6C.10. Yields are satisfactory and enantioselectivity is up to 89% ee. Moreover, the catalyst amount (1.25–5 mol %) is noteworthy, although the high molecular weight of BSA (170,000) necessitates a decrease in the quantity of BSA if the method is to be synthetically valuable.

6C.12. ENZYMATIC REACTIONS

Biooxidation of chiral sulfides was initially investigated in the 1960s, especially through the pioneering work of Henbest et al. [101]. Since then, many developments have been reported and are summarized in reviews [102,103]. It would be helpful to reveal some structural or mechanistic details of enzymes involved in the oxidation processes. Biotransformations are also of great current interest for the preparation of chiral sulfoxides, which are useful as synthetic intermediates and chiral auxiliaries. Because extensive review of these transformations is beyond the scope of this chapter, only highlights are discussed in comparison with the abiotic enantioselective oxidations described earlier. Biooxidations by microorganisms and by isolated enzymes are discussed in Sections 6C.12.1. and 6C.12.2.

6C.12.1. Microbiological Oxidations

Microbiological oxidation is the easiest procedure because it uses the intact cells. Scheme 6C.11 shows results obtained by using *Aspergillus Niger* [101]. Enantioselectivity can be very high but experiments are performed on a small scale, which results in a low yield of sulfoxides. Both enantiomers of methyl *p*-tolyl sulfoxide were prepared by Sih et al. with *Mortierella isabellina* NRRL 1757, giving (*R*)-sulfoxide with 100% ee in 60% yield or with *Helminthosporium sp* NRRL 4671, giving (*S*)-sulfoxide with 100% ee in 50% yield [104]. A similar result was obtained for ethyl *p*-tolyl sulfide. A predictive model for sulfoxidation by *Helminthosporium sp* NRRL 4671 was proposed by Holland et al. [105], which was based on the analysis of more than 90 biotransformations of sulfides.

Corynebacterium equi IF0 3730 gave high enantioselectivity in the oxidation of aryl alkyl sulfides [106]. The results listed in Scheme 6C.11 arise from experiments with no formation of sulfone, which occurs quite easily in several cases. Thio ketals and thio acetals were oxidized into mono S-oxides by various fungal species with enantioselectivity up to 70% ee [107]. *Corynebacterium equi* was very successfully used in the oxidation of formaldehyde dithioacetals to mono S-oxide or sulfone-sulfoxide, depending on the substrates. Thus *n*-Bu-S-CH-S-*n*-Bu was transformed to *n*-Bu-SO_2-CH_2-S(O)-*n*-Bu with more than 95% ee in 70% yield.

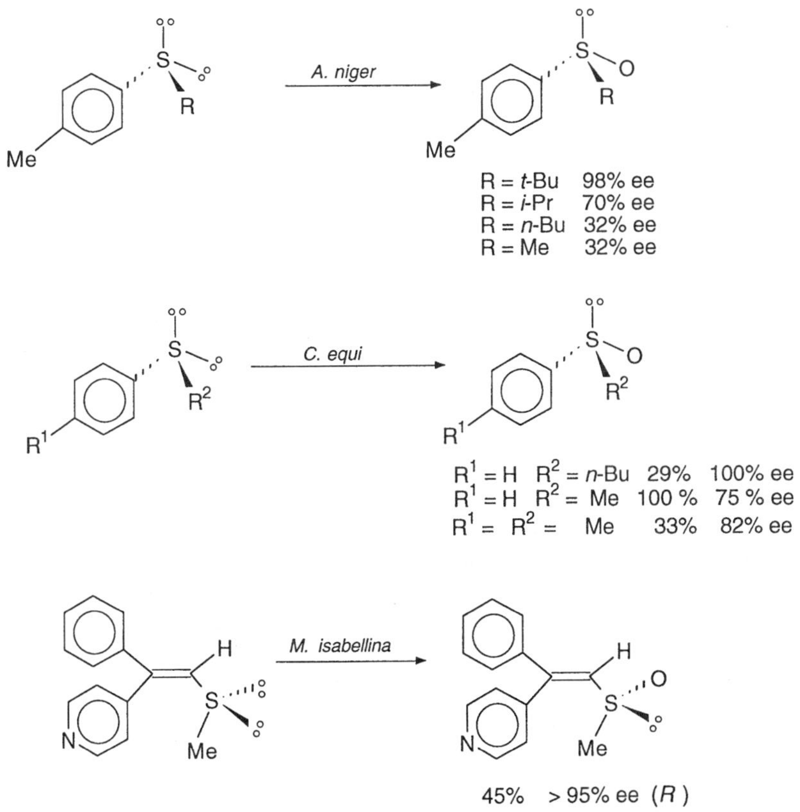
A. niger
Me
R
R = t-Bu 98% ee
R = i-Pr 70% ee
R = n-Bu 32% ee
R = Me 32% ee
C. equi
R1
R2
R1 = H R2 = n-Bu 29% 100% ee
R1 = H R2 = Me 100 % 75 % ee
R1 = R2 = Me 33% 82% ee
M. isabellina
H
N
Me
O
45% > 95% ee (R)

Scheme 6C.11.

Saccharomycies cerevisiae (bakers' yeast) has a desaturase that can oxidize some sulfides such as methyl 9-thiasterate [108]. To obtain a less symmetrical sulfoxide, the oxidation of PhCH$_2$-S-(CH$_2$)$_7$-CO$_2$Me was carried out with *S. cerevisiae* to give the corresponding (*S*)-sulfoxide with 70% ee. The stereochemical course of the oxidation of 9-thiostereate with cultures of bakers' yeast was later reinvestigated, and it was demonstrated that the biosulfoxidation on this quasi-symmetrical substrate was highly enantioselective (>95% ee) [109]. *Mortierella isabellina* oxidizes a vinylic sulfide to the corresponding sulfoxide with very high enantiomeric purity, as shown in Scheme 6C.11 [110]. Oxidation of various methyl β-arylvinyl sulfides with *Helminthosporium* sp. and by *Fusarium oxysporum* were also performed [111], wherein the corresponding sulfoxides were sometimes obtained with enantioselectivity greater than 98% ee. Comparisons were made between various chemical oxidations, including the catalysis of the water-modified titanium complex [24] and microbiological oxidations. In general, the latter method gave higher enantioselectivities but lower chemical yields.

Selected strains of the bacterium *Pseudomonas putida* were found to yield chiral sulfoxides Ph-S(O)-R (*R* or *S*-configuration) without sulfone formation. Similar results were obtained with an *Escherichia coli* clone [112].

Whole-cell enantioselective oxidation of sulfides is in fast development and of preparative interest. Additional examples with *Acinetobacter* sp NCIMB 9871, *Pseudomonas* sp. NCIMB 9872, and *Xanthobacter autotrophicus* DSM 431 have been reported [113,114].

6C.12.2. Oxidations with Isolated Enzymes

Cytochrome P-450 is a monooxygenase present in mammalian tissues. Takata et al. investigated the oxidation of sulfides with rabbit liver microsomes, and substantial asymmetric induction (up to 54% ee) was observed with some simple prochiral sulfides [115]. The horseradish peroxidase (HRP) has some similarities with cytochrome P450 and was investigated by Ortiz de Montellano and Ozaki in the oxidation of alkyl phenyl sulfides [116]. Methyl phenyl sulfoxide and *n*-butyl phenyl sulfoxide were isolated with 77% ee and 12% ee, respectively. The active site of HRP involves a catalytic histidine residue close to Phe-41. Replacement of Phe-41 with smaller amino acids, leucine or threonine, gave mutants F41L and F41T, respectively, which were screened for sulfoxidation. The mutant F41L exhibited an enhanced catalytic activity and higher enantioselectivity than the native enzyme, for example, 97% ee for methyl phenyl sulfoxide and 94% ee for *n*-butyl sulfoxide. It is noteworthy that interesting perspectives for catalyst-tuning are revealed by this work.

The structure of cytochrome P450cam (CYP101), a soluble bacterial enzyme, has been solved. This enzyme was studied for its catalytic properties in the oxidation of methyl phenyl sulfide and methyl *p*-tolyl sulfide [117]. Molecular dynamics calculations were used to predict the absolute configuration and enantioselectivity of the reaction.

Dopamine β-hydroxylase (DBH), a copper monooxygenase, catalyzes the benzylic oxidation of dopamine to norepinephrine. Replacement of benzylic carbon by a sulfur atom was investigated by May and Philipps [118]. The model sulfide PhSCH$_2$CH$_2$NH$_2$ was treated with oxygen in the presence of DBH at pH 5.0. The (*S*)-aminosulfoxide was produced with a very high enantiomeric excess, which—interestingly—has the same stereochemistry as the phenylethanolamines arising from the hydroxylation of phenyethylamines catalyzed by DBH. Walsh et al. studied the oxidation in air of *p*-tolyl ethyl sulfide catalyzed by two cytochrome P-450 isoenzymes, which gave (*S*)-sulfoxide (up to 80% ee) [119]. Oxidation of *p*-tolyl methyl sulfide with a monooxygenase containing flavin adenine dinucleotide (FAD) and purified from pig liver microsomes gave the corresponding (*R*)-sulfoxide (95% ee) [120]. Cashman et al. made a careful comparison of the oxidation of 2-(*p*-methoxyphenyl)-1,3-dithiolane and 2-(*p*-cyano-phenyl)-1,3-oxathiolane with chemical system ($NaIO_4$/BSA, chiral titanium complex/t-BuOOH) described in Section 6C.2 and with the purified microsomal flavin-containing monooxygenase from hog liver as well as from cytochrome P450 from rate or mouse liver [58]. In all the enzymatic oxidations, very high enantioselectivities (>90% ee) were observed, with the *trans*-sulfoxide being the major product as in the chemical oxidation. 2-Methyl-1,3-benzodithiole has been used by Cashman et al. as a stereochemical probe to study sulfur oxidation catalyzed by some pure enzymes as well as intact fungal and bacterial systems [121].

Pseudomonas oleovorans contains *P. oleovorans* monooxygenase (POM), which is a typical "ω-hydroxylase" for hydroxylation of the terminal methyl of alkanes as well as epoxidation of terminal olefins. The ω-hydroxylation system of *P. oleovorans* was reconstituted from purified components, POM, rubredoxin, and a flavoprotein reductase [122]. In the presence of NADH and oxygen, it oxidizes a wide range of aliphatic methyl alkyl sulfides. Enantioselectivities are very much dependent of the length of the alkyl chain of Me-S(O)-R, as exemplified by the following results:

R = *n*-Pr   80% ee
R = *n*-Bu   80% ee
R = *n*-pentyl   60% ee
R = *n*-hexyl   30% ee
R = *n*-heptyl   70% ee

Chloroperoxidase, a heme protein isolated from the mold *Caldariomyces fumago*, was used by Colonna et al. as the catalyst for the oxidation of thio ethers to sulfoxides [123–125]. *t*-Butyl hydroperoxide and hydrogen peroxide are the preferred oxidants. Reactions were performed in water at pH 5 at 25°C with 1.6×10^{-5} mol equiv. of the enzyme. The following (*R*)-sulfoxides, among others, were prepared in good-to-excellent yields: $PhCH_2S(O)Me$ (90% ee), PhS(O)Me (98% ee), *p*-TolS(O)Me (91% ee), *p*-ClC_6H_4S(O)Me (90% ee), and *p*-$MeOC_6H_4$S(O)Me (90% ee). Several dialkyl sulfoxides were also produced with quite high enantioselectivities [125]. Wong et al. performed similar experiments with chloroperoxidase, but they used racemic 1-arylethyl hydroperoxide as the oxidant [126]. In this sytem, aryl methyl sulfides were transformed to the corresponding (*R*)-sulfoxides, whereas kinetic resolution of the chiral hydroperoxide occurred simultaneously. For example, 1 equiv. of racemic 1-phenylethyl hydroperoxide oxidized phenyl methyl sulfide at 4°C (48% conversion of the sulfide) to give (*R*)-sulfoxide (91% ee) together with the unreacted chiral hydroperoxide (74% ee) and 1-phenylethanol (80% ee).

Allenmark and Andersson investigated the scope of the sulfoxidation with chloroperoxidase by using a series of rigid and aromatic sulfides of well-defined geometry in the presence of hydrogen peroxide at pH 5 and 20°C) [127]. The almost planar 1-thiaindane is an excellent substrate, giving the corresponding (*R*)-sulfoxide (>96% ee) in quantitative yield. 1-Thiatetrahydro-naphthalene is less reactive, but the sulfoxide was formed with very high enantiopurity (>96%).

Sulfoxidations catalyzed by cyclohexanone monooxygenases from *Acinetobacter calcoaceticus* and by other flavin monooxygenases have recently been actively investigated by several groups. Colonna et al. reviewed most of this area [128]. These oxidant systems use molecular oxygen and a cofactor, NADPH, which is enzymatically regenerated by glucose-6-phosphate. High enantioselectivities (*R*-configuration) were observed for the formation of cycloalkyl methyl sulfoxides and methyl sulfoxides with a branched alkyl chain of up to four carbon atoms [129]. A predictive active site model was proposed, which was based on cubic-space descriptors and on the results obtained with more than 30 different sulfides [130]. A refinement of the model has been proposed for taking into account the influence of the electronic properties of para-substituents in benzyl methyl sulfides [131]. Unsubstituted 1,3-dithioacetals were transformed to the corresponding monooxides in good yields and enantioselectivities over 98% [128]. The introduction of a substituent in the 2-position decreases the enantioselectivity in most cases [132].

Two diketocamphane monooxygenases from *Pseudomonas putida* were found to catalyze the oxidation of a wide range of sulfides with modest enantioselectivities, and predictive active-site models were proposed [133].

A rabbit flavin-containing monooxygenase (FMO2) was used to oxidize prochiral sulfides Ar-S-alkyl in the presence of oxygen and NADPH [134]. It is interesting that when phenyl *n*-alkyl sulfides were used, a complete reversal of the absolute configuration of the resulting sulfoxide occurred by increasing the length of the alkyl chain, that is, 99% ee (*R*) for Me; 97% ee (*S*) for *n*-heptyl. The results were discussed in terms of an active-site model, which contains two binding domains.

6C.13. CONCLUSION

Asymmetric oxidation of sulfides has been widely expanded in the past decade, both by chemical and biochemical methods. Chemical methods can now give rise to a wide range of chiral sulfoxides with enantioselectivities often over 90% with the use of, for example, hydroperoxides and chiral Ti complexes. However, the catalyst efficiency of these systems is moderate. On the other hand, highly active catalyst systems, for example, chiral porphyrins and Schiff base-metal complexes, have not reached the level to achieve very high enantioselectivity. Accordingly, one can expect, in both systems, significant improvements in the near future. Biooxidations of sulfides are now readily available with many microorganisms, giving very satisfactory results. Molecular engineering of some enzymes have been made, which could enhance the scope of biosulfoxidations. Some catalytic antibodies have been used in sulfoxidation [135], but no information on enantioselectivity is presently available.

REFERENCES

1. Solladié, G. *Synthesis* **1981**, 185–196.

2. Barbachyn, M. R.; Johnson, C. R. In *Asymmetric Synthesis*; Morrison, J. D. (Ed.); Academic Press: Orlando, FL, 1984; Vol. 4, pp. 227–261.

3. Posner, G. H. *Acc. Chem. Res.* **1987**, *20*, 72–78.

4. Posner, G. H. In *The Chemistry of Sulphones and Sulphoxides*; Patai, S.; Rappoport, Z.; Stirling, C. J. M. (Ed.); John Wiley and Sons: Chichester, UK; 1988, Chapter 16, pp. 823–848.

5. Drabowicz, J.; Kielbasinski, P.; Mikolajczyk, M. In *The Chemistry of Sulphones and Sulphoxides*; Patai, S.; Rappoport, Z.; Stirling, C. J. M. (Ed.); John Wiley and Sons: Chichester, UK; 1988, Chapter 8, pp. 233–253.

6. Carreno, M. C. *Chem. Rev.* **1995**, 95, 1717–1760.

7. Andersen, K. K. In *The Chemistry of Sulphones and Sulphoxides*; Patai, S.; Rappoport, Z.; Stirling, C. J. M. (Ed.); John Wiley and Sons: Chichester, UK; 1988, Chapter 3, pp. 55–92.

8. Early attempts are reported in Morrison, J. D.; Mosher, H. S. *Asymmetric Organic Reactions*; Prentice Hall: Englewood Cliffs, NJ; 1971, pp. 336–351.

9. Davis, F. A.; Mc Cauley, J. P. Jr.; Harakal, M. E. *J. Org. Chem.* **1984**, *49*, 1465–1467.

10. Komori T.; Nonaka, T. *J. Am. Chem. Soc.* **1983**, *105*, 5690–5691.

11. Madesclaire, M. *Tetrahedron* **1986**, *42,* 5459–5495.

12. Kagan, H. B.; Diter, P. In *Organosulfur Chemistry*; Page, P. C. B. (Ed.); Academic Press: New York; 1998, Vol. 2, Chapter 1, pp. 1–39.

13. Kagan, H. B.; Luukas, T. In *Transition Metals for Organic Synthesis*; Beller, M.; Bolm, C. (Ed.); John Wiley and Sons, Weinheim; 1998, Chapter 2.13, pp. 361–373.

14. Katsuki, T.; Sharpless, K. B. *J. Am. Chem. Soc.* **1980**, *102*, 5974–5976.

15. Rossiter, B. E.; Katsuki, T.; Sharpless, K. B. *J. Am. Chem. Soc.* **1981**, *103,* 464–465.

16. Pitchen, P.; Kagan H. B. *Tetrahedron Lett.* **1984**, *25*, 1049–1052.

17. Pitchen, P.; Deshmukh, M.; Dunach, E.; Kagan, H. B. *J. Am. Chem. Soc.* **1984**, *106,* 8188–8193.

18. Di Furia, F.; Modena, G.; Seraglia, R. *Synthesis* **1984**, 325–326.

19. Kagan, H. B.; Rebiere, F. *Synlett* **1990**, 643–650.

20. Dunach, E.; Kagan H. B. *New J. Chem.* **1985**, *9,* 1–3.

21. Nemecek, C.; Dunach, E.; Kagan, H. B. *New J. Chem.* **1986**, *10,* 761–764.

22. Kagan H. B.; Dunach, E.; Nemecek, C.; Pitchen, O.; Samuel, O.; Zhao, S. H. *Pure Appl. Chem.* **1985**, *57,* 1911–1916.

23. Zhao, S.; Samuel, O.; Kagan, H. B. *C. R. Acad. Sci. Paris, Ser. B.* **1987**, *304*, 273–275.

24. Zhao, S.; Samuel, O.; Kagan, H. B. *Tetrahedron* **1987**, *43*, 5135–5144.

25. Zhao, S.; Samuel, O.; Kagan, H. B. *Organic Syntheses* **1989**, *68,* 49–56.

26. Glahsl, G.; Herrmann, R. *J. Chem. Soc. Perkin Trans 1* **1988**, 1753–1757.

27. Deshmukh, M.; Dunach, E.; Jugé, S.; Kagan, H. B. *Tetrahedron Lett.* **1984**, *25*, 3467–3470.

28. Corrigendum, Deshmukh, M.; Dunach, E.; Jugé, S.; Kagan, H. B. *Tetrahedron Lett.* **1985**, *26*, 402.

29. Samuel, O.; Ronan, B.; Kagan, H. B. *J. Organomet. Chem.* **1989**, *370,* 43–50.

30. Finn, M. G.; Sharpless, K. B. *J. Am. Chem. Soc.* **1991**, *113*, 113–126.

31. Bradley, D. C.; Mehrota, R. C.; Gaur, D. P. In *Metal Alkoxide*; Academic Press: London, UK; 1978.

32. Brunel, J.-M.; Diter, P.; Duetsch, M.; Kagan, H. B. *J. Org. Chem.* **1995**, *60*, 8086–8088.

33. Hanson, R. M.; Sharpless, K. B. *J. Org. Chem.* **1986**, *51,* 1922–1925.

34. Bortolini, O.; Di Furia, F.; Licini, G.; Modena, G.; Rossi, M. *Tetrahedron Lett.* **1986**, *27,* 6257–6260.

35. Bortolini, O.; Di Furia, F.; Licini, G.; Modena, G. In *The Role of Oxygen in Chemistry and Biochemistry*; Ando, W.; Moro-oka, Y. (Ed.); *Studies in Organic Chemistry*; Elseviers: Amsterdam; 1988, Vol. 33, pp. 193–200 .

36. Bortolini, O.; Di Furia, F.; Licini, G.; Modena, G. *Rev. Heteroatom Chem.* **1988**, *1,* 66–79.

37. Bortolini, O.; Di Furia, F.; Licini, G.; Modena, G. *Phosphorus Sulfur* **1988**, *37,* 171–174.

38. Conte, V.; Di Furia, F.; Licini, G.; Modena, G. *Tetrahedron Lett.* **1989**, *30,* 4859–4862.

39. Brunel, J.-M.; Kagan, H. B. *Synlett* **1996**, 404–406.

40. Brunel, J.-M.; Kagan, H. B. *Bull. Soc. Chim. Fr.* **1996**, *133*, 1109–1115.

41. Brunel, J.-M.; Luukas, T.; Kagan, H. B. *Tetrahedron: Asymmetry* **1998**, *9*, 1941–1946.

42. Scretti, A.; Bonadies, F.; Lattanzi, A. *Tetrahedron: Asymmetry* **1996**, *7*, 629–632.

43. Scretti, A.; Bonadies, F.; Lattanzi, A. *Tetrahedron: Asymmetry* **1998**, *9*, 1817–1822.

44. Yamamoto, K.; Ando, H.; Shuetake, T.; Chikamatsu, H. *J. Chem. Soc. Chem. Commun.* **1989**, 754–755.

45. (a) Superchi, S.: Rosini, C. *Tetrahedron:Asymmetry* **1997**, *8,* 349–352. (b) Donnoli, M. I.; Superchi, S.; Rosini, C. *J. Org. Chem.* **1998**, *63*, 9392–9395.

46. Yamanoi, Y.; Imamoto, T. *J. Org. Chem.* **1997**, *62*, 8560–8564.

47. Di Furia, F.; Licini, G.; Modena, G.; Motterle, R.; Nugent, W. A. *J. Org. Chem.* **1996**, *61*, 5175–5177.

48. Boche, G.; Möbus, K.; Harms, K.; Marsh, M. *J. Am. Chem. Soc.* **1996**, *118*, 2770–2771.

49. Komatsu, K.; Nishibayashi, Y.; Suguta, T.; Uemura, S. *Tetrahedron Lett.* **1986**, *33*, 5391–5394.

50. Komatsu, N.; Hashizume, M.; Sugita, T.; Uemura, S. *J. Org. Chem.* **1993**, *56*, 6341–634.

51. Komatsu, N.; Hashizume, M.; Sugita, T.; Uemura, S. *J. Org. Chem.* **1993**, *56*, 4529–4533.

52. Komatsu, N.; Hashizume, M.; Sugita, T.; Uemura, S. *J. Org. Chem.* **1993**, *56*, 7624–7626.

53. Reetz, M. T.; Merk, C.; Naberfeld, G.; Rudolph, J.; Griebenow, N.; Goddard, R. *Tetrahedron Lett.* **1997**, *38*, 5273–52.

54. Beckwith, A. L. J.; Boate, D. R. *J. Chem. Soc. Chem. Commun.* **1986**, 189–190.

55. Tanaka, J.; Higa, T.; Bernardinelli, G.; Jefford, C. W. *Tetrahedron* **1989**, *45*, 7301–7310.

56. Davis, F.; Kern, J. R.; Kurtz, L. J.; Pfister, J. R. *J. Am. Chem. Soc.* **1988**, *110*, 7873–7874.

57. (a) Diter, P.; Samuel, O.; Taudien, S.; Kagan, H. B. *Tetrahedron: Asymmetry* **1994**, *5*, 549–552. (b) Griffiths, S. L.; Perrio, S.; Thomas, S. E. *Tetrahedron: Asymmetry* **1994**, *5*, 1847–1864.

58. Cashman, J. R.; Olsen, L. D.; Bornheim, L. M. *J. Am. Chem. Soc.* **1990**, *112*, 3191–3195.

59. Di Furia, F.; Licini, G.; Modena, G.; Valle, G. *Bull. Soc. Chim. Fr.* **1990**, 734–744.

60. Di Furia, F.; Licini, G.; Modena, G. *Tetrahedron Lett.* **1989**, *30*, 2575–2576.

61. Aggarwal, V. K.; Drabowicz, J.; Grainger, R. S.; Gültekin, Z.; Lightowler, M.; Spargo, P. L. *J. Org. Chem.* **1995**, *60*, 4962–4963.

62. Aggarwal, V. K.; Esquivel-Zamora, B. N.; Evans, G. R.; Jones, E. *J. Org. Chem.* **1998**, *63*, 7306–7310.

63. Page, P. C. B.; Gareh, M. T.; Porter, R. A. *Tetrahedron: Asymmetry* **1993**, *4*, 2139–2142.

64. Bendazzoli, P; Di Furia, F.; Licini, G.; Modena, G. *Tetrahedron Lett.* **1993**, *34*, 2975–2978.

65. Marino, J. P.; Bogdan, S.; Kimura, K. *J. Am. Chem. Soc.* **1992**, *114*, 5566–5672.

66. Pitchen, P.; France, C. J.; McFarlane, I. M.; Newton, C. G.; Thompson, D. M. *Tetrahedron Lett.* **1994**, *35*, 485–488.

67. Colombo, A.; Marturano, G.; Pasini, A. *Gazz. Chim. Ital.* **1986**, *116*, 35–40.

68. Colonna, S.; Manfredi, A.; Spadoni, M.; Casella, L.; Gullotti, M. *J. Chem. Soc., Perkin Trans 1* **1987**, 71–73.

69. Nakajima, K.; Sasaki, C.; Kojima, M.; Aoyama, T.; Ohba, S.; Saito, Y.; Fujita, J. *Chem. Lett.* **1987**, 2189–2192.

70. Nakajima, K.; Kojima, M.; Fujita, J. *Chem. Lett.* **1986**, 1483–1486.

71. (a) Bolm, C.; Bienewald, F. *Angew. Chem. Int. Ed.* **1995**, *34*, 2640–2642. (b) Bolm, C.; Bienewald, F. *Synlett* **1998**, 1327–1328.

72. Cogan, D. E.; Liu, G.; Kim, K.; Backes, B. J.; Ellman, J. A. *J. Am. Chem. Soc.* **1998**, *120*, 8011–8019.

73. Vetter, A. H.; Berkessel, A. *Tetrahedron Lett.* **1998**, *39*, 1741–1744.

74. Palucki, M.; Hanson, P.; Jacobsen, E. N. *Tetrahedron Lett.* **1992**, *33*, 7111–7114.

75. Noda, K.; Hosoya, N.; Irie, R.; Yamashita, Y.; Katsuki, T. *Tetrahedron* **1994**, *50*, 9609–9618.

76. Kokubo, C.; Katsuki T. *Tetrahedron* **1996**, *52*, 13895–13900.

77. Imagawa, K.; Nagata, T.; Yamada, T.; Mukaiyama, T. *Chem. Lett.* **1995**, 335–336.

78. Nagata, T.; Imagawa, K.; Yamada, T.; Mukaiyama, T. *Bull. Chem. Soc. Jpn.* **1995**, *68*, 3241–3246.

79. Naruta, Y.; Tani, F.; Maruyama, K. *J. Chem. Soc. Chem. Commun.*, **1990**, 1378–1380.

80. Naruta, Y.; Tani, F.; Maruyama, K. *Tetrahedron: Asymmetry* **1991**, *2*, 533–542.

81. Groves, J. T.; Viski, P. *J. Org. Chem.* **1990**, *55*, 3628–3634.

82. Halterman, R. L.; Jan, S. T.; Nimmens, H. L. *Synlett* **1991**, 791–792.

83. Hikita, T.; Tamaru, K.; Yamagishi, A.; Iwamoto, T. *Inorg. Chem.* **1989**, *28*, 2221–2223.

84. Yamagishi, A. *J. Chem. Soc., Chem. Commun.* **1986**, 290–291.

85. Firth, B. E.; Miller, L. L. *J. Am. Chem. Soc.* **1976**, *98* 8272–8273.

86. Komori, T.; Nonaka, T. *J. Am. Chem. Soc.* **1984**, *106*, 2656–2659.

87. Shinkai, S.; Yamaguchi, T.; Manabe, O.; Toda, F. *J. Chem. Soc., Chem. Commun.* **1988**, 1399–1401.

88. Bohé, L.; Hanquet, G.; Lusinchi, M.; Lusinchi, X. *Tetrahedron Lett.* **1993**, *34*, 9629–9632.

89. Page, P. C. B.; Heer, J. P.; Bethell, D.; Collington, E. W.; Andrews, D. M. *Tetrahedron Lett.*, **1994**, *35*, 9629–9632.

90. Page, P. C. B.; Heer, J. P.; Bethell, D.; Collington, E. W.; Andrews, D. M. *Synlett* **1995**, 773–775.

91. Czarnik, A. W. *J. Org. Chem.* **1984**, *49,* 924–927.

92. Drabowicz, J.; Mikolajczyk, M. *Phosphorus Sulfur* **1984**, *21,* 245–250.

93. (a) Sugimoto, T.; Kokubo, T.; Miyazaki, J.; Tanimoto, S.; Okano, M. *J. Chem. Soc., Chem. Commun.* **1989**, 402–404. (b) Sugimoto, T.; Kokubo, T.; Miyazaki, J.; Tanimoto, S.; Okano, M. *J. Chem. Soc., Chem. Commun.* **1989**, 1052–1053.

94. Sugimoto, T.; Kokubo, T.; Miyazaki, J.; Tanimoto, S.; Okano, M. *Bioorg. Chem.* **1981**, *10,* 311–323.

95. Kagan, H. B.; Fiaud, J. C. *Top. Stereochem.* **1988**, *18,* 249–339.

96. Ogura, K.; Fujita, M.; Iida, H. *Tetrahedron Lett.* **1980**, *21,* 2233–2236.

97. Colonna, S.; Banfi, S.; Sommaruga, M. *J. Org. Chem.* **1985**, *50*, 769–771.

98. Colonna, S.; Banfi, S.; Annunziata, R.; Casella, L. *J. Org. Chem.* **1986**, *51,* 891–895.

99. Colonna, S.; Gaggero, N. *Tetrahedron Lett.*, **1989**, *30*, 6233-6236.

100. Colonna, S.; Gaggero, N.; Leone, M.; Pasta, P. *Tetrahedron* **1991**, *47*, 8385–8398.

101. Auret, B. J.; Boyd, D. R.; Henbest, H. B.; Ross, S. *J. Chem. Soc. C* **1968**, 2371–2376.

102. Madesclaire, M. *Tetrahedron* **1986**, *42,* 5459–5495.

103. Holland, H. L. *Chem. Rev.* **1988**, *88,* 473–485.

104. Abushanab, E.; Reed, D.; Suzuki, F.; Sih, C. J. *Tetrahedron Lett.* **1978**, *37*, 3415–3418.

105. Holland, H. L.; Brown, F. M.; Lakshmaiah, G.; Larsen, B. G.; Patel, M. *Tetrahedron: Asymmetry* **1997**, *8*, 683–697.

106. Ohta, H.; Okamoto, Y.; Tsuchihashi, G. *Chem. Lett.* **1984**, 205–208.

107. (a) Auret, B. J.; Boyd, D. R.; Cassidy, E. S.; Hamilton, R.; Turley, F.; Drake, A. F. *J. Chem. Soc., Perkin Trans 1* **1985**, 1547–1552. (b) Auret, B. J.; Boyd, D. R.; Dunlop, R.; Drake, A. F. *J. Chem. Soc., Perkin Trans 1* **1988**, 2827–2829.

108. Buist, P. H.; Marecak, D. M.; Partington, E. T.; Skala, P. *J. Org. Chem.* **1990**, *55*, 5667–5669.

109. Buist, P. H.; Marecak, D. M. *J. Am. Chem. Soc.* **1991**, *113,* 5877–5878.

110. Madesclaire, M.; Fauve, A.; Metin, J.; Carpy, A. *Tetrahedron: Asymmetry* **1990**, *1*, 311–314.

111. Rossi, C.; Fauve, A.; Madesclaire, M.; Roche, D.; Davis, F. A.; Thimma Reddy, R. *Tetrahedron: Asymmetry* **1992**, *3,* 629–636.

112. Allen, C. C. R.; Boyd, D. R.; Dalton, H.; Sharma, N. D.; Haughey, S. A.; McMordie, R. A. S.; McMurray, B. T.; Sheldrake, G. N.; Sproule, K. *J. Chem. Soc., Chem. Commun.* **1995**, 119–120.

113. Kelly, D. R.; Knowles, C. J.; Mahdi, J. G.; Taylor, I. N.; Wright, M. A. *Tetrahedron: Asymmetry* **1996**, *7,* 365–368.

114. Alphand, V.; Gaggero, N.; Colonna, S.; Furstoss, R. *Tetrahedron: Lett.* **1996**, *37,* 6117–6120.

115. Takata, T.; Yamazaki, M.; Fujimori, K.; Kim, Y. H.; Oae, S.; Iyanagi, T. *Chem. Lett.* **1980**, 1441–1444.

116. Ozaki, S.; Ortiz de Montellano, P. R. *J. Am. Chem. Soc.* **1995**, 117, 7056–7064.

117. Fruetel, J.; Chang, Y.-T.; Collins, J.; Loew, G.; Ortiz de Montellano, P. R. *J. Am. Chem. Soc.* **1994**, *116*, 11643–11648.

118. May, S. W.; Phillips, R. S. *J. Am. Chem. Soc.* **1980**, *102,* 5981–5983.

119. Light, D. R.; Waxman, D. J.; Walsh, C. *Biochemistry* **1982**, *21,* 2490–2493.

120. Waxman, D. J.; Light, D. R.; Walsh, C. *Biochemistry* **1982**, *21,* 2499–2502.

121. Cashman, J. R.; Olsen, L. D.; Boyd, D. R.; McMordie, R. A. S.; Dunlop, R.; Dalton, H. *J. Am. Chem. Soc.* **1992**, *114,* 8772–8777.

122. Katopodis, A. G.; Smith, H. A. Jr. ; May, S. W. *J. Am. Chem. Soc.* **1988**, *110,* 897–899.

123. Colonna, S.; Gaggero, N.; Manfredi, A.; Casella, L.; Gullotti, M. *J. Chem. Soc., Chem. Commun.* **1988**, 1451–1452.

124. Colonna, S.; Gaggero, N.; Casella, L.; Carrea, G.; Pasta, P. *Tetrahedron: Asymmetry* **1992**, *3*, 95–106.

125. Colonna, S.; Gaggero, N.; Manfredi, A.; Carrea, G.; Pasta, P. *J. Chem. Soc., Chem. Commun.* **1997**, 439–440.

126. Fu, H.; Kondo, H.; Ichikawa, Y.; Loo, G. C.; Wong, C.-H. *J. Org. Chem.* **1992**, *57*, 7265–7270.

127. Allenmark, S. G.; Andersson, M. A. *Tetrahedron: Asymmetry* **1996**, *7*, 1089–1094.

128. Colonna, S.; Gaggero, N.; Pasta, P.; Ottolina, G. *J. Chem. Soc., Chem. Commun.* **1996**, 2303–2307.

129. Colonna, S.; Gaggero, N.; Carrea, G.; Pasta, P. *J. Chem. Soc., Chem. Commun.* **1997**, 2303–2307.

130. Ottolina, G.; Pasta, P.; Carrea, C.; Colonna, S.; Dallavalle, S.; Holland, H. L. *Tetrahedron: Asymmetry* **1995**, *6*, 1375–1386.

131. Ottolina, G.; Pasta, P.; Varley, D.; Holland, H. L. *Tetrahedron: Asymmetry* **1996**, *7*, 3427–3430.

132. Colonna, S.; Gaggero, N.; Carrea, G.; Pasta, P. *Tetrahedron: Asymmetry* **1996**, *7*, 565–570.

133. Beecher, J.; Willets, A. *Tetrahedron: Asymmetry*, **1998**, *9*, 1899–1916.

134. Fisher, M. B.; Rettie, A. E. *Tetrahedron: Asymmetry* **1997**, *8*, 613–618.

135. Hsieh, L. C.; Stephans, J. C.; Schultz, P. G. *J. Am. Chem. Soc.* **1994**, *116*, 2167–2168.

136. Potvin, P. G.; Fieldhouse, B. G. *Tetrahedron: Asymmetry* **1999**, *10*, 1661–1672.

6D

CATALYTIC ASYMMETRIC DIHYDROXYLATION—DISCOVERY AND DEVELOPMENT

ROY A. JOHNSON
The Upjohn Company, Kalamazoo, Michigan

K. BARRY SHARPLESS
The Scripps Research Institute, La Jolla, California

6D.1. INTRODUCTION

The cis dihydroxylation of olefins mediated by osmium tetroxide represents an important general method for olefin functionalization [1,2]. For the purpose of introducing the subject of this chapter, it is useful to divide osmium tetroxide mediated cis dihydroxylations into four categories: (1) the stoichiometric dihydroxylation of olefins, in which a stoichiometric equivalent of osmium tetroxide is used for an equivalent of olefin; (2) the catalytic dihydroxylation of olefins, in which only a catalytic amount of osmium tetroxide is used relative to the amount of olefin in the reaction; (3) the stoichiometric, asymmetric dihydroxylation of olefins, in which osmium tetroxide, an olefinic compound, and a chiral auxiliary are all used in equivalent or stoichiometric amounts; and (4) the catalytic, asymmetric dihydroxylation of olefins. The last category is the focus of this chapter. Many features of the reaction are common to all four categories, and are outlined briefly in this introductory section.

In the reaction with osmium tetroxide (**1**), the olefin (**2**) is first osmylated to form an osmium(VI) monoglycolate ester for which Criegee proposed structure **3** [1] (see Scheme 6D.1) in analogy to the cyclic ester postulated by Boeseken and van Giffen for permanganate glycolates [3]. The osmylation process has been proposed to proceed by either [3+2] cycloaddition leading directly to the monoglycolate ester or a reversible [2+2] cycloaddition leading to a metallaoxetane intermediate (**4**) [4], which undergoes irreversible rearrangement to the monoglycolate ester (**3**). Arguments favoring each of these pathways have been put forward and have been reviewed recently [5]. In the absence of hydrolytic conditions, the monoglycolate

Catalytic Asymmetric Synthesis, Second Edition, Edited by Iwao Ojima
ISBN 0-471-29805-0   Copyright © 2000 Wiley-VCH, Inc.

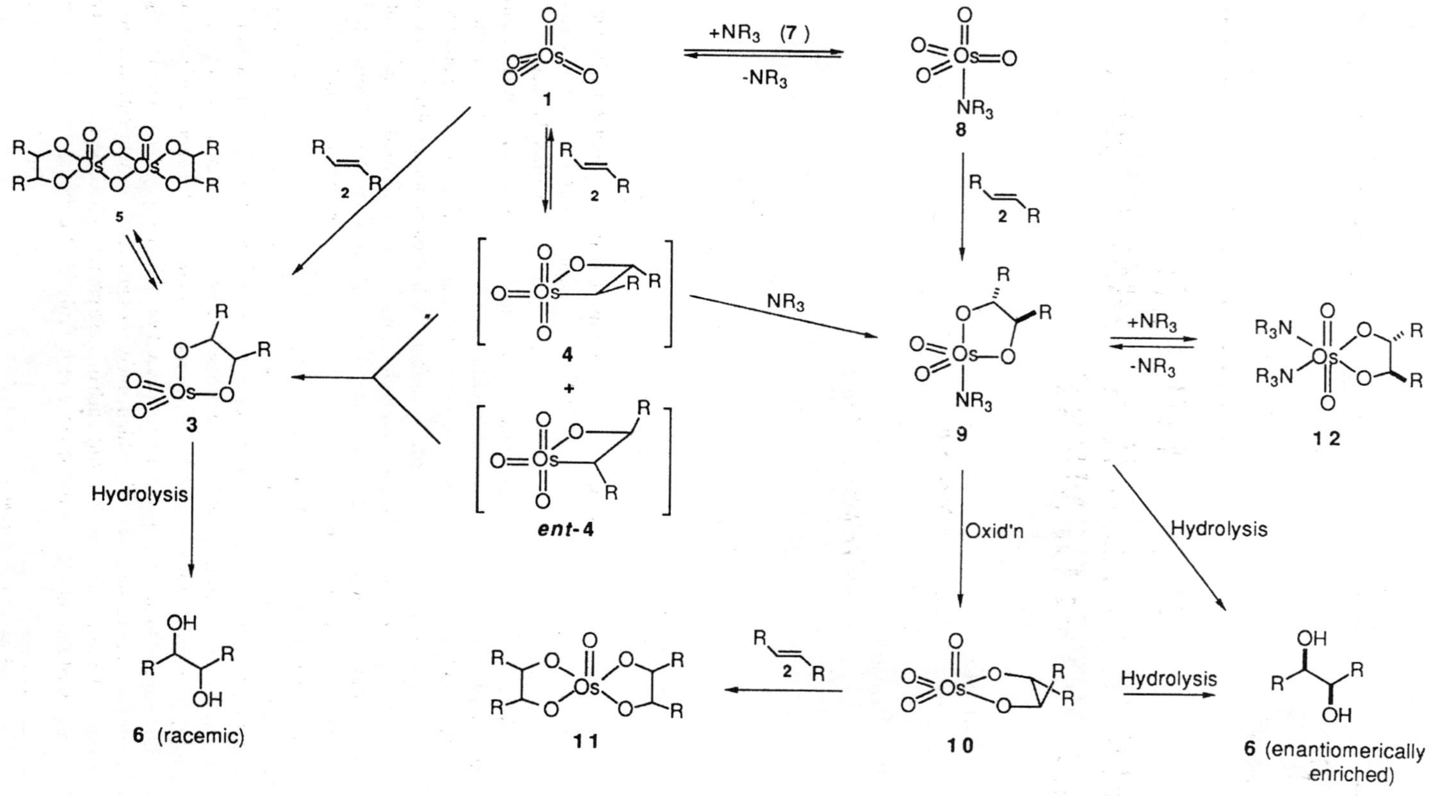

Scheme 6D.1. Complexation and reaction of osmium tetroxide with tetiary amines and olefins.

ester **3** may form a dimeric complex (**5**) [6]; under hydrolytic conditions; whereas hydrolysis of the glycolate ester(s) occurs with breaking of the osmium-oxygen bonds, releasing the diol **6** and an osmium(VI) oxo complex.

Inclusion in the reaction of a cooxidant serves to return the osmium to the osmium tetroxide level of oxidation and allows for the use of osmium in catalytic amounts. Various cooxidants have been used for this purpose; historically, the application of sodium or potassium chlorate in this regard was first reported by Hofmann [7]. Milas and co-workers [8,9] introduced the use of hydrogen peroxide in *t*-butyl alcohol as an alternative to the metal chlorates. Although catalytic cis dihydroxylation by using perchlorates or hydrogen peroxide usually gives good yields of diols, it is difficult to avoid overoxidation, which with some types of olefins becomes a serious limitation to the method. Superior cooxidants that minimize overoxidation are alkaline t-butylhydroperoxide, introduced by Sharpless and Akashi [10], and tertiary amine oxides such as *N*-methylmorpholine-*N*-oxide (NMO), introduced by VanRheenen, Kelly, and Cha (the Upjohn process) [11]. A new, important addition to this list of cooxidants is potassium ferricyanide, introduced by Minato, Yamamoto, and Tsuji in 1990 [12].

Acceleration of osmylation by the addition of pyridine to the reaction system was first observed by Criegee et al. [13] and is a property shared by other tertiary amines as well. The reactions and equilibria included in Scheme 6D.1 can be postulated on the basis of spectral and crystallographic characterization of various reaction components. Osmium tetroxide (**1**) and tertiary amines (**7**) have been observed to form monoamine complexes such as **8** [14]. Osmium monoglycolate-amine complexes such as **9** have also been isolated and characterized [15]. However, the mechanism by which complex **9** forms is not yet clear. Whether **9** arises from osmylation of the olefin (**2**) by complex **8** or via the metallaoxetane **4** remains to be determined. Oxidation of complex **9** gives the putative trioxoglycolate **10** [1], which can react with a second olefin and form the osmium(VI) bisglycolate ester **11**, an event with important consequences to the catalytic asymmetric dibydroxylation process (see Section 6D.2.2.) [16]. Alternatively, complex **9** can add a second amine ligand giving complex **12**, which is found to be unreactive to further oxidation or hydrolysis [1,17]. High binding constants for complex **12** are observed when the amine is pyridine or a chelating diamine, a somewhat lower binding constant is seen for quinuclidine, and weak binding constants are seen for derivatives of the cinchona alkaloids dihydroquinidine (**13**) and dihydroquinine (**14**) (see Chart 6D.1), which are used as chiral ligands in asymmetric dihydroxylation. Hydrolysis of the glycolate esters **9** and **10** releases the diols.

Amine complexation by osmium tetroxide opened the door to asymmetric dihy-droxylation of olefins when Hentges and Sharpless found that osmylation of olefins in the presence of dihydroquinidine acetate (**15**) or dihydroquinine acetate (**16**) under stoichiometric conditions gave, after hydrolysis, optically active diols with 25–94% ee [18]. The crystallographically determined structure shown in Figure 6D.1 for the complex of dioxo[(3S,4S)-2,2,5,5-tetra-methyl-3,4-hexanediolatolosmium(VI) with dihydroquinine *p*-chlorobenzoate clearly shows the osmium(VI) monoglycolate complexed to the quinuclidine nitrogen of the alkaloid [19]. Following the report of Hentges and Sharpless, several other groups have described stoichiomet-ric asymmetric dihydroxylation systems. Generally, these systems use chiral, chelating diamines as the chiral auxiliary with osmium tetroxide for induction of asymmetry in the diol products. Diamines used as chiral auxiliaries include 1,4-piperidinylbutan-2,3-diol ketals and acetals (**17**) [20], (−)-(R,R)-N,N,N′N-tetramethylcyclohexane-1,2-*trans*-diamine (**18**) [21], 1,2-(*trans*-3,4-diaryl)pyrro-lidinylethanes (**19**) [22], N,N′-dialkyl-2,2′-bipyrrolidines (**20**) [23], and 1,2-di-phenyl-N,N′-bis(2,4,6-trimethylbenzylidene)-1,2-diaminoethane (**21**) [24].

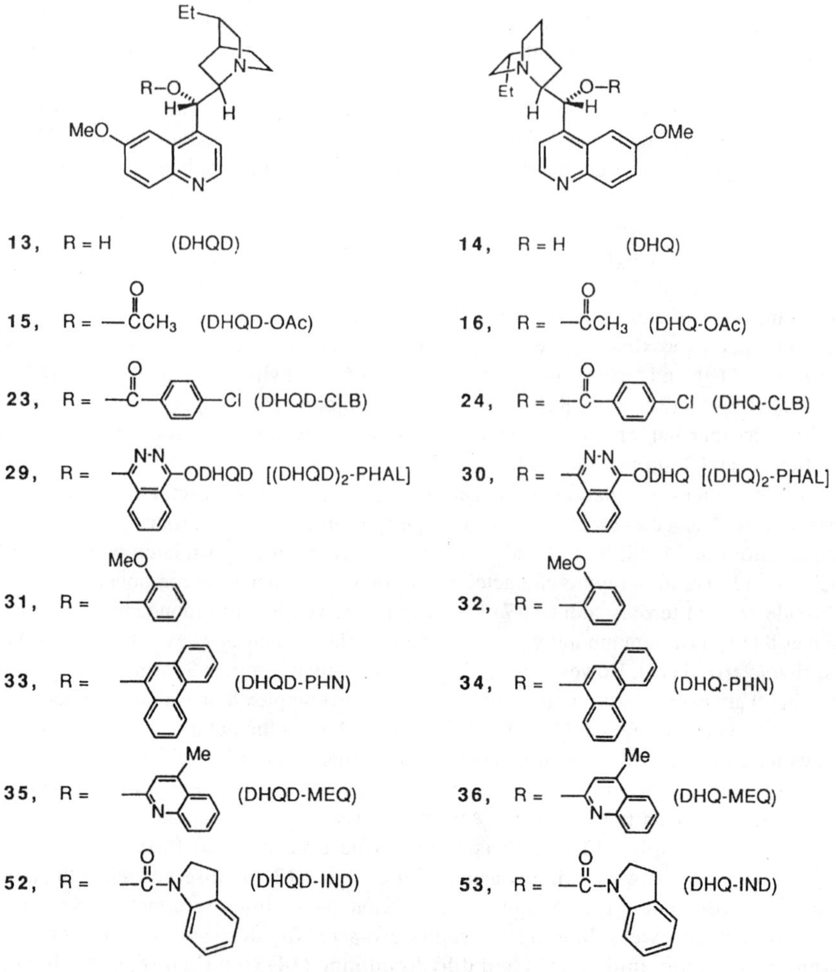

13, R = H   (DHQD)

14, R = H   (DHQ)

15, R = —$\overset{O}{\overset{\|}{C}}$CH₃ (DHQD-OAc)

16, R = —$\overset{O}{\overset{\|}{C}}$CH₃ (DHQ-OAc)

23, R = —$\overset{O}{\overset{\|}{C}}$—⟨⟩—Cl (DHQD-CLB)

24, R = —$\overset{O}{\overset{\|}{C}}$—⟨⟩—Cl (DHQ-CLB)

29, R = —ODHQD [(DHQD)₂-PHAL]

30, R = —ODHQ [(DHQ)₂-PHAL]

31, R =

32, R =

33, R = (DHQD-PHN)

34, R = (DHQ-PHN)

35, R = (DHQD-MEQ)

36, R = (DHQ-MEQ)

52, R = —$\overset{O}{\overset{\|}{C}}$—N⟨⟩ (DHQD-IND)

53, R = —$\overset{O}{\overset{\|}{C}}$—N⟨⟩ (DHQ-IND)

Chart 6D.1. Structures of chiral ligands.

Asymmetric induction also occurs during osmium tetroxide mediated dihydroxylation of olefinic molecules containing a stereogenic center, especially if this center is near the double bond. In these reactions, the chiral framework of the molecule serves to induce the diastereoselectivity of the oxidation. These diastereoselective reactions are achieved with either stoichiometric or catalytic quantities of osmium tetroxide. The possibility exists for pairing or "matching" this diastereoselectivity with the face selectivity of asymmetric dibydroxylation to achieve enhanced or "double diastereoselectivity" [25], as discussed further later in the chapter.

This brief outline of historical developments in osmium tetroxide-mediated olefin hydroxy-lation brings us to our main subject, catalytic asymmetric dihydroxylation. The transition from *stoichiometric* to *catalytic* asymmetric dihydroxylation was made in 1987 with the discovery by Sharpless and co-workers that the stoichiometric process became catalytic when *N*-methyl-

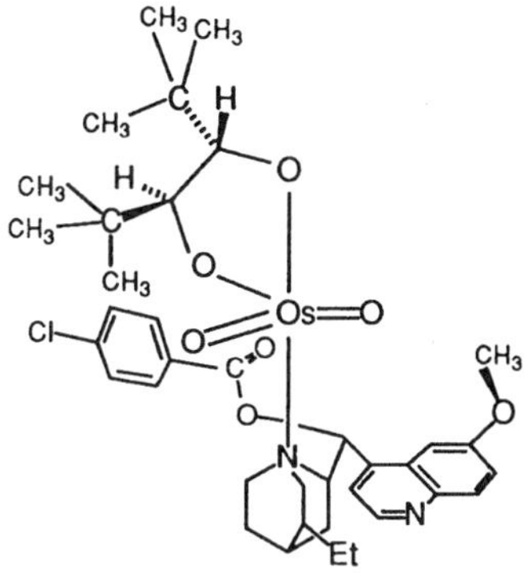

Figure 6D.1. Structure of the dioxo[(3S,4S)-2,2,5,5-tetramethyl-3,4-hexanediolato]osmium(VI) complex with dihydroquinine *p*-chlorobenzoate (DHQ-CLB).

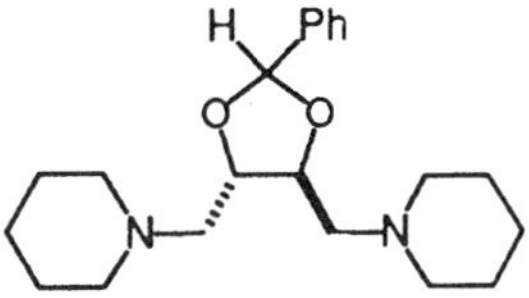

17

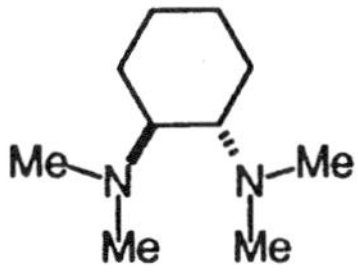

18

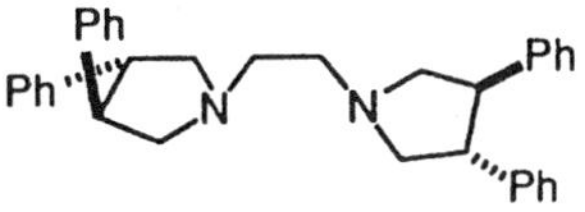

19

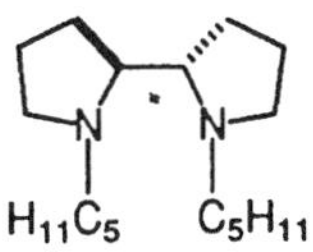

20

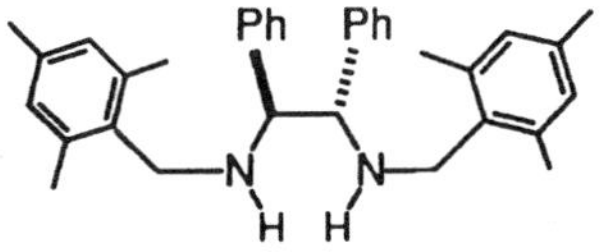

21

morpholine-*N*-oxide (**22**, NMO) was used as the cooxidant in the reaction [26]. DHQD *p*-chlorobenzoate (**23**) and DHQD *p*-chlorobenzoate (**24**) (see Chart 6D.1) also were introduced as new chiral ligands with improved enantioselective properties. By using 0.002 mol of osmium tetroxide and 0.134 mol of a chiral auxiliary (0.033 mol when the reaction was performed at 0°C), one mole of (*E*)-stilbene (**25**) was converted to (*R,R*)-(+)-dihydrobenzoin (**26**) with 88% ee in 80% yield using 1.2 mol of NMO as the oxidant for this catalytic process.

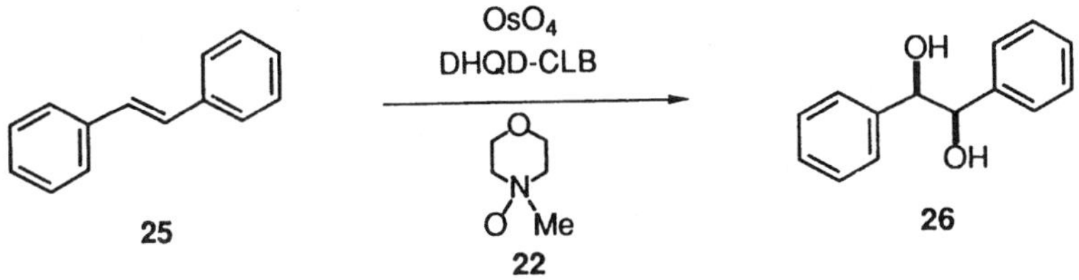

25

OsO$_4$
DHQD-CLB

22

26

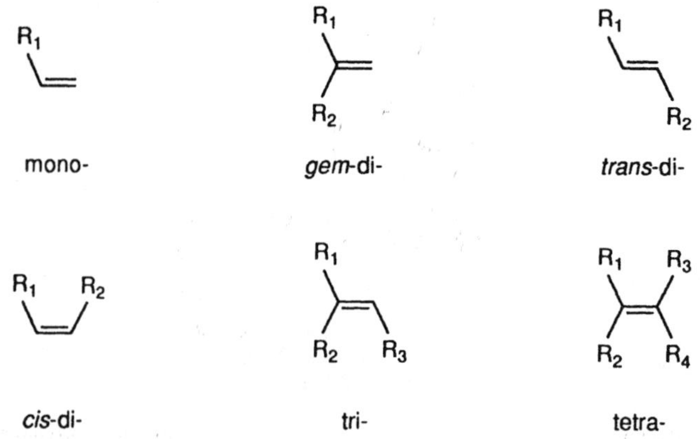

Chart 6D.2. Olefin substitution patterns.

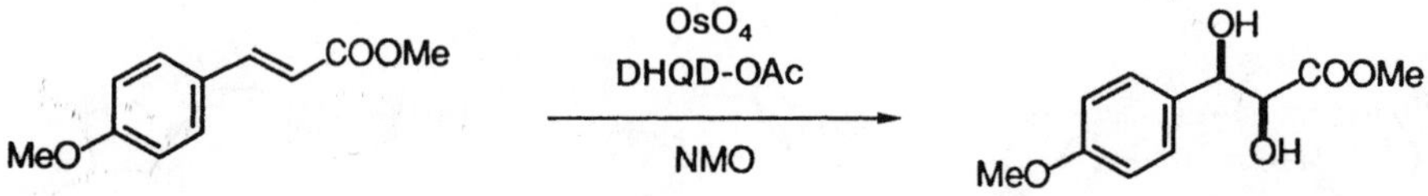

The original communication also described the asymmetric dihydroxylation of seven other olefins, among which were representatives from three of the six possible substitution patterns (see Chart 6D.2). The observation of a catalytic asymmetric dihydroxylation of methyl (*E*)-*p*-methoxycinnamate (**27**) with NMO as oxidant to give diol **28** (70% ee) was described independently by Gredley in a patent application published in 1989 [27] and reported in a communication in 1990 [28]. Since 1987, the catalytic asymmetric dihydroxylation process has undergone intensive development resulting in a number of improvements, including several very recent modifications of reaction conditions and of the chiral auxiliary, which significantly improve enantioselectivity over a broad range of olefin structures. Development activity continues and will lead to additional improvements.

The focus of this chapter is to acquaint the reader with details of catalytic asymmetric dihydroxylation with osmium tetroxide and the scope of results that one can expect to achieve with current optimum conditions. The literature through mid-1992 has been reviewed in compiling this chapter. Osmium tetroxide catalyzed hydroxylations of olefins and acetylenes are the subject of an extensive review by Schröder published in 1980 [2a]. A comprehensive review of research and industrial applications of asymmetric dihydroxylations is in preparation [2b].

6D.2. GENERAL FEATURES OF OSMIUM-CATALYZED ASYMMETRIC DIHYDROXYLATION

Catalytic asymmetric dihydroxylations (ADs) are easy reactions to perform. The reaction actually requires water and is insensitive to oxygen and so can be carried out without fear of exposure to the atmosphere. The reaction uses a multicomponent reagent system, which allows for considerable variation of each component. In the years since the discovery of the catalytic asymmetric process, an evolution of improvements in each of the variables has occurred. These improvements led to the publication of the following "recipe" [29] for the catalytic asymmetric dihydroxylation of one millimole of an olefin: potassium ferricyanide (0.94 g, 3.0 mmols), potassium carbonate (0.41 g, 3.0 mmols), either 1,4-bis(9-O-dihydroquinidine) phthalazine, (DHQD)$_2$-PHAL (**29**), or 1,4-bis(9-O-dihydroquinine)phthalazine, (DHQ)$_2$-PHAL (**30**) (see Chart 6D.1), (0.0078 g. 0.010 mmol), and potassium osmate(VI) dihydrate (0.00074 g, 0.0020 mmol) in a 1:1 mixture of t-butyl alcohol and water 95 mL of each). [*However, see the next paragraph for an improved version of this recipe.*] The recommended temperature for the reaction is 0°C [29]. The addition of a sulfonamide to the reaction should also be considered according to the guidelines outlined in Section 6D.2.5. One requirement for this set of reaction conditions is that the reaction mixture be stirred vigorously, because addition of the inorganic salts causes separation of the solvent into two phases. This separation of phases is itself thought to be important to the reaction (see Section 6D.2.2.).

The total weight of the four solid ingredients used for the one-millimole-scale reaction is 1.36 g, of which less than a milligram is potassium osmate(VI) dihydrate. Ready-made mixtures of the four solid components, for which the names AD-mix-β [containing (DHQD)$_2$-PHAL, see Section 6D.2.6 for derivation of names] and AD-mix-α [containing (DHQ)$_2$-PHAL] have been coined, are available from the Aldrich Chemical Company. AD-mix-α and AD-mix-β were used for many of the entries in the tables of results discussed below [29]. The AD-mixes offer a convenient alternative to the need to weigh out each component for one or more small-scale reactions. The stability of the mixture shows promise for a practical shelf-life [29]. The original formula [29] for one kilogram of AD-mix is K$_3$Fe(CN)$_6$, 699.96 g; K$_2$CO$_3$, 294.00 g; (DHQD)$_2$- or (DHQ)$_2$-PHAL, 5.52 g; and K$_2$OsO$_2$(OH)$_4$, 0.52 g; and the ratio of ligand to Os is 5:1. *This formula is improved for general use by increasing the amount of* $K_2OsO_2(OH)_4$ *fivefold, to 2.60 g*. The advantage of this formula is that olefins such as α,β-unsaturated amides, which are hydroxylated only sluggishly with the original formula, are now hydroxylated at a practical rate.

In the following sections, the reaction components and variables for catalytic AD are discussed in greater detail.

6D.2.1. Osmium

Descriptions in the literature of osmium tetroxide frequently include several qualifying statements that express the essential emotions accompanying the use of this reagent: (1) The *cis* dihydroxylation of olefins with osmium tetroxide is a reliable and powerful synthetic method. (2) Osmium tetroxide is an expensive reagent. (3) Osmium tetroxide is regarded as a toxic material. Judging from its frequent use, the utility of osmium tetroxide clearly outweighs the drawbacks of cost and danger. At the same time, the problems of cost and toxicity exert pressure for the development of efficient methods for the use of osmium tetroxide at the catalytic level. Fortunately, in this regard, the discovery early in the history of osmium tetroxide that the molecule is very easily reduced and reoxidized has led to extensive use of the reagent in a catalytic mode.

Osmium tetroxide is the traditional osmium species used in the dihydroxylation of olefins. For large-scale reactions, osmium tetroxide may be weighed and transferred as the solid. For many catalytic applications on a laboratory scale, the amount of osmium tetroxide required is too small to be weighed conveniently. In these cases, advantage can be taken of the solubility of osmium tetroxide in organic solvents by the preparation of a stock solution of known concentration and the use of an aliquot for the small-scale reaction.

Alternate sources of osmium can be used and should be considered, as they offer advantages in ease of handling. Both osmium(III) chloride hydrate [OsCl·xH$_2$O] [30] and potassium osmate(VI) dihydrate [K$_2$OsO$_2$(OH)$_4$] [31] have been used as replacements for osmium tetroxide, and the latter reagent is specified in the preceding standard set of conditions for catalytic AD.

Curiously, osmium tetroxide's reputation for chronic toxicity is not supported in the toxicology literature [32]. In fact, an aqueous solution of osmium tetroxide is used to treat refractory rheumatoid arthritis in humans by direct injection into the knee joint [33,34].

Osmium tetroxide, like the skunk, carries its own warning system-a strong odor described variously as resembling chlorine, bromine, or ozone. Clearly, a commonsense approach to the use of this reagent in an efficient hood is called for, and under these conditions osmium tetroxide should be regarded as no more dangerous than many other reagents found in daily use in the laboratory.

6D.2.2. Oxidants

Either amine oxides (usually NMO) [11,26] or potassium ferricyanide/potassium carbonate [12,35] are used as cooxidants for catalytic AD. The choice of oxidant carries with it the choice of solvent for the reaction, and the details of the catalytic cycle appear to be quite different depending on which oxidant–solvent combination is used. When potassium ferricyanide/potassium carbonate is used as the oxidant, the solvent used for the reaction is a 1:1 mixture of t-butyl alcohol and water [35,36]. This solvent mixture, normally miscible, separates into two liquid phases upon addition of the inorganic reagents. The sequence of reactions summarized below in Eqs. 6D.1–6D.5 has been postulated as occurring under these conditions. This reaction sequence is further illustrated in the reaction cycle shown in Scheme 6D.2, which also emphasizes the role of the two-phase solvent system.

First, osmylation of the olefin proceeds to form the osmium(VI) monoglycolate-amine complex (**9**) as shown in Eq. 6D.1. Formation of **9** is presumed to occur in the organic phase in which all the involved species are soluble. Next, at the organic-aqueous interface, hydrolysis of the glycolate ester releases the diol into the organic phase and the reduced osmium into the aqueous phase as the hydrated osmate(VI) dianion (shown as the potassium salt in Eq. 6D.2). Oxidation of osmate(VI) by potassium ferricyanide regenerates osmium tetroxide via an intermediate perosmate(VIII) dianion (Eq. 6D.3). Loss of two hydroxide groups from the perosmate(VIII) ion gives osmium tetroxide, which then migrates back into the organic phase to restart the cycle (Eq. 6D.4) [36]. Summation of Eqs. 6D.1–6D.4 gives Eq. 6D.5, showing that in this catalytic cycle both oxygens of the diol are provided by water.

A new development is that electrochemical oxidation of ferrocyanide to ferricyanide can be coupled with AD to give a very efficient electrocatalytic process [37]. Under these conditions, the amount of potassium ferricyanide needed for the reaction becomes catalytic and Eqs. 6D.6 and 7 can be added following Eq. 6D.4. Summation of Eq. 6D.1–6D.4, 6D.6, and 6D.7 gives 6D.8, showing that only water in addition to electricity is needed for the conversion of olefins to asymmetric diols and that hydrogen gas, released at the cathode, is the only byproduct of this process. In practice, sodium ferrocyanide is used in the reaction and the amount of this reagent used in comparison with the potassium ferricyanide method mentioned above has been reduced from 3.0 equiv. to 0.15 equiv. (relative to an equivalent of olefin).

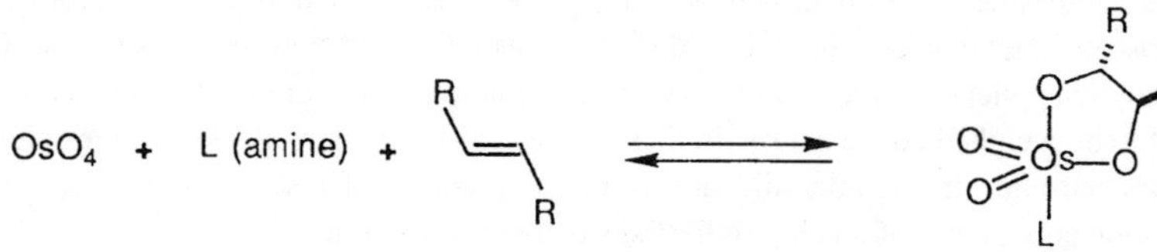

$$\text{OsO}_4 \; + \; \text{L (amine)} \; + \quad \text{olefin} \quad \rightleftharpoons \quad \text{osmium glycolate complex} \tag{6D.1}$$

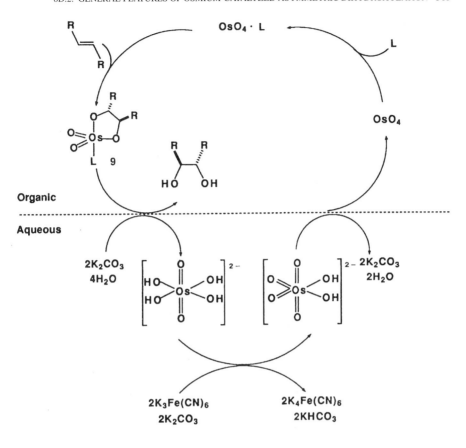

Scheme 6D.2. Catalytic cycle for asymmetric dihydroxylation using potassium ferricyanide as cooxidant.

$$\text{(from the structure)} \quad + \; 4H_2O \; + \; 2K_2CO_3 \;\longrightarrow\; \underset{HO \quad OH}{R \quad R} \; + \; K_2OsO_2(OH)_4 \qquad (6D.2)$$
$$+ \; 2KHCO_3 \; + \; L$$

$$K_2OsO_2(OH)_4 \; + \; K_3Fe(CN)_6 \; + \; 2K_2CO_3 \;\longrightarrow\; K_2OsO_4(OH)_2 \; + \; 2K_4Fe(CN)_6$$
$$+ \; 2KHCO_3$$
$$(6D.3)$$

$$OsO_4(OH)_2 \; + \; 2KHCO_3 \;\rightleftharpoons\; OsO_4 \; + \; 2K_2CO_3 \; + \; 2H_2O \qquad (6D.4)$$

$$\underset{R}{\overset{R}{\diagdown}} \; + \; 2K_3Fe(CN)_6 \; + \; 2K_2CO_3 \;\longrightarrow\; \underset{HO \quad OH}{R \quad R} \; + \; 2K_4Fe(CN)_6 \; + \; 2KHCO_3$$
$$+ \; 2H_2O$$
$$(6D.5)$$

$$2K_4Fe(CN)_6 \quad - \quad 2e^- \quad \longrightarrow \quad 2K_3Fe(CN)_6 \quad + \quad 2K^+ \qquad (6D.6)$$

$$2H_2O \quad + \quad 2e^- \quad \longrightarrow \quad 2OH^- \quad + \quad H_2 \qquad (6D.7)$$

$$(6D.8)$$

For laboratory-scale reactions, this electrocatalytic AD generally is performed in a glass H-type cell in which the anode and cathode compartments are separated by a semipermeable Nafion cation-exchange membrane and platinum electrodes are used. A 5% aqueous solution of phosphoric acid is used in the cathode compartment, and the reaction in the anode compartment is stirred vigorously. Under a controlled anode potential of 0.4 V (vs. Ag/AgCl) and with $(DHQD)_2$-PHAL as chiral ligand, α-methylstyrene was converted to R-2-phenyl-1,2-propanediol in 15 h with the electrical consumption of 2.1 F/mol. The product was isolated in 100% yield with 92% ee [37].

When NMO is used as the oxidant, typically with acetone-water (10:1) as a solvent, the reaction mixture is essentially homogeneous and a different catalytic cycle is observed [16,36]. The cycle can be broken down into Eqs. 6D.9–6D.13 and can be illustrated as shown in Scheme 6D.3. The first step (Eq. 6D.9) is the same as above, that is, formation of the osmium(VI) monoglycolate-amine complex **9**. Next the organic soluble oxidant NMO is postulated to bind reversibly to the osmium(VI) glycolate ester (**9**), as shown in Eq. 6D.10. Transfer of oxygen to osmium leads to formation of an osmium(VIII) trioxoglycolate (**10**) as shown in Eq. 6D.11, and hydrolysis of this trioxoglycolate (Eq. 6D.12) gives the diol and regenerated osmium tetroxide. Equation 6D.13 simply indicates the equilibrium between complexed and free osmium tetroxide to complete the cycle. Summation of Eqs. 6D.9–6D.13 gives 14 and shows that as this catalytic system recycles, NMO and water provide one oxygen each to the newly formed diol.

In addition to participating in the primary catalytic cycle just outlined, the osmium(VIII) trioxoglycolate (**10**) shown in Eq. 6D.11 has access to a secondary reaction cycle, which has a devastating effect on the overall enantioselectivity of the dihydroxylation process [16,36]. The trioxoglycolate (**10**) can add to (osmylate) a second olefin molecule, forming a bisglycolate (**11**) as shown for the secondary cycle in Scheme 6D.3. Completion of this secondary cycle involves hydrolysis of the bisglycolate (**11**) to the monoglycolate (**9**) followed by reoxidation leading back to the osmium (VIII) trioxoglycolate (**10**) intermediate [36]. By modification of experimental conditions, the enantioselectivity of the secondary cycle can be determined and is found to be very low [16]. In fact, in the specific case examined the diol produced in this way has a slight preponderance of the configuration opposite to that of the primary cycle. Clearly, avoidance of this secondary cycle is necessary to attain the highest degree of enantioselectivity possible for a given catalytic AD. The extent to which the secondary cycle will be followed depends on the rate of hydrolysis of the trioxoglycolate ester, the concentration of olefin available for reaction to form the bisglycolate, and the rate of the latter process. Consistent with this hypothesis is the finding that slow addition of olefin to the reaction is accompanied by an increase in the enantiomeric purity of the diol product [16]. These findings with NMO have added much insight into the processes by

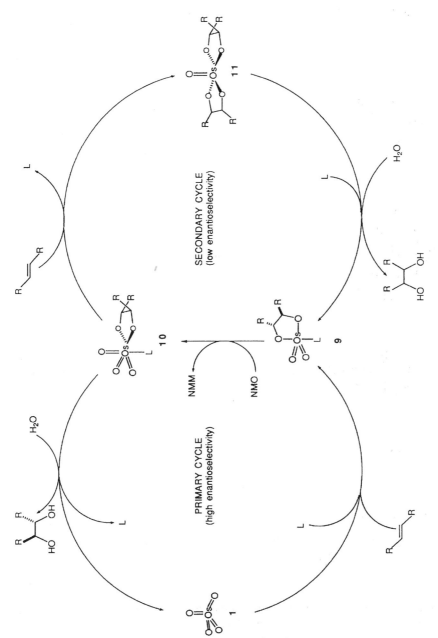

Scheme 6D.3. Catalytic cycle for asymmetric dihydroxylation using *N*-methylmorpholine-*N*-oxide as cooxidant.

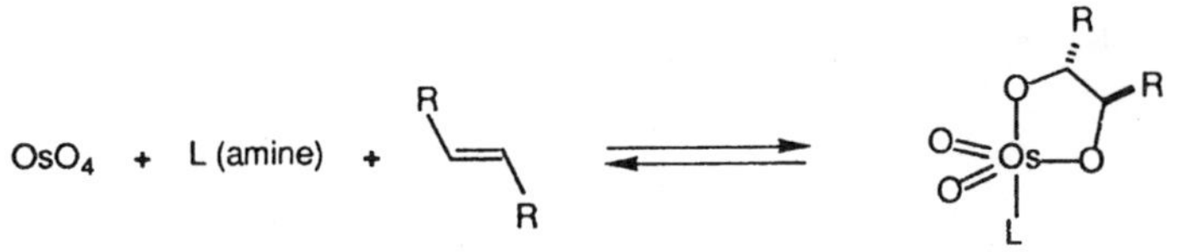

$$\text{OsO}_4 \; + \; \text{L (amine)} \; + \; \text{(olefin)} \;\; \rightleftharpoons \;\; \text{(osmate ester)} \tag{6D.9}$$

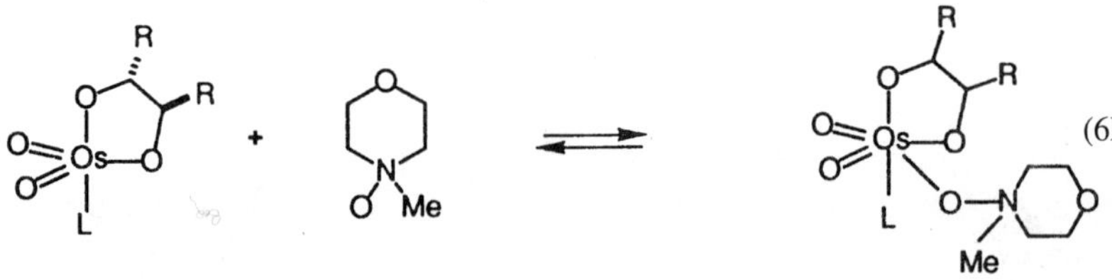

$$\text{(osmate ester)} \; + \; \text{(N-methylmorpholine-N-oxide)} \;\; \rightleftharpoons \;\; \text{(adduct)} \tag{6D.10}$$

$$\text{(adduct)} \;\; \longrightarrow \;\; \text{(osmium species)} \; + \; \text{(N-methylmorpholine)} \tag{6D.11}$$

$$\text{(osmium species)} \; + \; \text{H}_2\text{O} \;\; \longrightarrow \;\; \text{OsO}_4 \cdot \text{L} \; + \; \text{(diol)} \tag{6D.12}$$

$$\text{OsO}_4 \cdot \text{L} \;\; \rightleftharpoons \;\; \text{OsO}_4 \; + \; \text{L} \tag{6D.13}$$

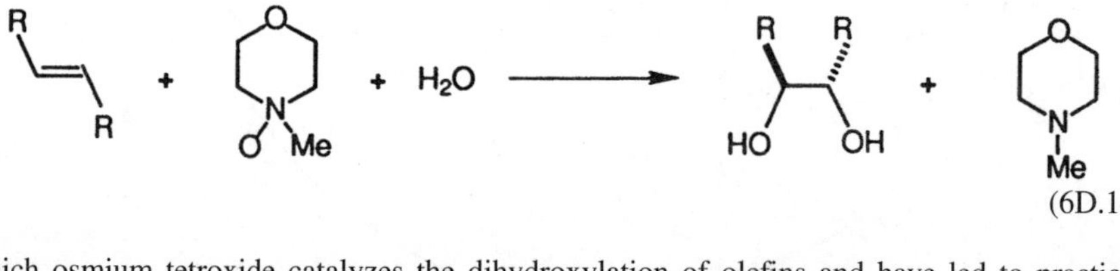

$$\text{(olefin)} \; + \; \text{(N-methylmorpholine-N-oxide)} \; + \; \text{H}_2\text{O} \;\; \longrightarrow \;\; \text{(diol)} \; + \; \text{(N-methylmorpholine)} \tag{6D.14}$$

which osmium tetroxide catalyzes the dihydroxylation of olefins and have led to practical improvements in the enantiomeric purity of products obtained under this set of conditions. However, with the advent of the potassium ferricyanide/potassium carbonate cooxidant modification, which appears to allow for hydrolysis of the osmium(VI) monoglycolate ester (**9**) before further oxidation can occur [36], a considerably easier method is in hand for avoiding the secondary reaction cycle and obtaining optimum enantioselectivities.

6D.2.3. Chiral Ligand (Auxiliary)

The chiral ligands for catalytic AD are derived from dthydroquinidine (DHQD) (**13**, Chart 1) and dihydroquinine (DHQ) (**14**). DHQD and DHQ are minor components of the naturally occurring cinchona alkaloids and are separated in sufficient quantity from the more abundant quinidine and quinine to satisfy current commercial demand. They may also be obtained by catalytic reduction of quinidine and quinine. The cinchona alkaloids have an illustrious history in the fields of chemistry and medicine, areas that have been frequently reviewed [38]. DHQD and DHQ become very effective ligands for AD when derivatized at the C_9 hydroxyl group (new results indicate that derivatives of the parent quinidine and quinine alkaloids are almost as effective when used as ligands [29]). The best such derivatives (to date) are discussed further in the next paragraph. In their role as chiral ligands, these pairs of derivatives function almost as if they are enantiomers, although they are actually diastereoisomers because of differences in the attachment of the ethyl group (cf. **13** and **14** in Chart 6D.1). Because they operate like enantiomers in AD, the term "pseudoenantiomers" has been applied to these pairs [39]. In the preparation of enantiomeric diols, the enantiomeric purity of the diol obtained, by using by the DHQD derivative usually exceeds by 2–10% that of the diol obtained by using the DHQD derivative. This difference is generally seen regardless of the derivative used.

More than 250 derivatives, mainly of the cinchona alkaloids, have been made and tested as chiral ligands in the catalytic AD process. (Recall that the acetate was used in the original stoichiometric AD work.) The initial catalytic AD used the *p*-chlorobenzoate derivative, and this pair of ligands (**15** and **16**) was used in much subsequent work through 1990 [16,17,26,35,40]. By that time several ethers of DHQD and DHQ were found to give improved enantioselectivities when compared with the *p*-chlorobenzoates [41]. Particularly notable in this regard was the *o*-methoxyphenyl ethers (**31** and **32**), the 9′-phenanthryl ethers (**33** and **34**), and the 4′-methyl-2′-quinolyl ethers (**35** and **36**). Most recently, a new pair of ligands, the bis-DHQD and bis-DHQ ethers of phthalazine-1,4-diol (**29** and **30**), show significant improvement over the earlier ethers in the AD of most olefins [29].

Attachment of the alkaloid derivatives to a polymer support has been examined as a way in which to recycle the AD catalyst [42]. Polymers **37–40** were prepared by copolymerization of the appropriate alkaloid olefin monomer with acrylonitrile. After complexation with osmium tetroxide, the polymers were used as heterogeneous catalysts in the AD of (*E*)-stilbene. The reaction time with polymer **37**, in which the quinuclidine ring of the alkaloid is attached directly to the polymer chain, was unacceptably long (7 days). The other three polymers (**38–40**) with spacer groups between the alkaloid and polymer chain were effective catalysts for AD, although reaction times are still longer than those with homogeneous catalysts. At room temperature with 25 mol % alkaloid and 1 mol % osmium tetroxide, reaction times of 18–48 h gave stilbene diol in yields of 75–96%. The enantiomeric purities of the diols ranged from 85% ee to 93% ee with polymer **38** and were 80–82% ee with polymers **39** and **40** when NMO was used as cooxidant. With polymers **39** and **40** an improvement in enantioselectivity to 86–87% ee was observed when $K_3Fe(CN)_6/K_2CO_3$ was used as cooxidant. Recovery and reuse of polymer **38** in a second hydroxylation gave diol with very slightly reduced enantiomeric purity, indicating a potential for continuous use [42].

37

38

40

39

6D.2.4. Stoichiometry

From the practical perspective of laboratory use, the catalytic AD process requires a minimum of concern over stoichiometry. The new osmium tetroxide-chiral ligand complexes are so efficient that for most olefins, 0.2 mol % of osmium will provide a satisfactory rate of reaction at 0°C [29]. In the occasional case where hydroxylation is slow under these conditions, the quantity of osmium in the catalyst should be increased to 1 mol % and the reaction temperature kept at 0 °C. In the rare case where hydroxylation is still slow under these conditions, the temperature may be raised to 25°C and, to ensure no loss in enantioselectivity, the ligand concentration may be increased from 1 mol % to 2 mol %.

There are two main hindrances to the achievement of 100% ee in the catalytic AD reaction, namely, the nonenantioselective catalysis of dihydroxylation by species other than the desired complex (such as osmium tetroxide itself or the trioxoglycolate **10**) and less than a 100% exclusive fit of one face of the prochiral olefin to the chiral catalyst at the transition state. The dependence of enantioselectivity on the chiral ligand:osmium tetroxide ratio has been examined carefully in reaching the prescribed 5:1 stoichiometry of these reagents. First, the effect on enantioselectivity by changing this ratio was examined by using *trans*-5-decene as the substrate for catalytic AD. The results are shown in Table 6D.1 (note that the DHQ-derived PHAL is used in this experiment) and show a near-maximum 93% ee for the diol when the reaction is run at 0°C with a 5:1 ratio of ligand to OsO_4. The catalytic efficiency of the PHAL ligand-osmium tetroxide system is dramatically portrayed in the AD of (*E*)-stilbene [29]. Under the original conditions (a 5:1 ligand/OsO_4 ratio) and at 0°C, (*E*)-stilbene is dihydroxylated with better than 99.5% ee. The effect on enantioselectivity of lowering this ratio was examined, and it was found that even when the ratio of ligand to OsO_4 was only 1:20, diol with 96% ee was obtained from AD of stilbene. In other words, under the conditions of this experiment, there were 10,000 molecules of stilbene and 20 molecules of OsO_4 for every molecule of chiral ligand, yet the ligand-OsO4 complex imparted optical activity to 9600 molecules of olefin, leaving only 400 molecules to be hydroxylated by OsO_4 alone (Table 6D.1).

TABLE 6D.1. Relationship of Ligand Concentration to Enantiometric Excess

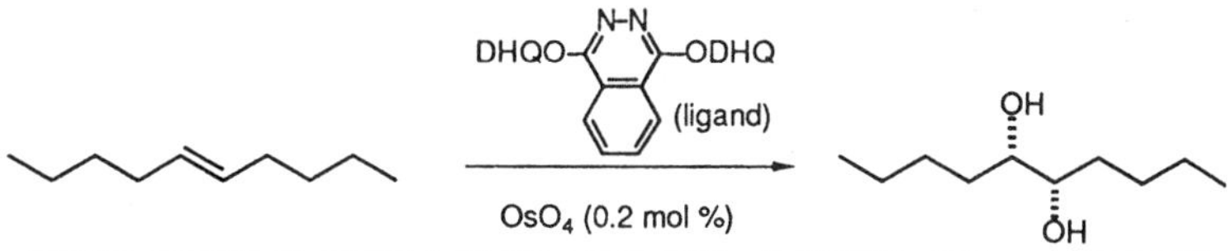

Ligand (mol %)	% ee	
	rt	0°C
10	93	
8	92	
6	92	
4	92	
2	90	94
1	89	93
0.5		93
0.25		91

The chiral ligand is the crucial component controlling fit of the olefin on the catalyst and, clearly, for E-stilbene the fit must be nearly perfect. Efforts at modeling the chiral ligand–OsO$_2$–olefin complex with the intent to understand olefin fit have suggested some possible features of importance [43], but improvements in the chiral ligand have thus far depended largely on a trial-and-error approach to design.

6D.2.5. Additives

When the secondary reaction cycle shown in Scheme 6D.3 was discovered, it became clear that an increase in the rate of hydrolysis of trioxoglycolate **10** should reduce the role played by this cycle. The addition of nucleophiles such as acetate (tetraethylammonium acetate is used) to osmylations is known to facilitate hydrolysis of osmate esters. Addition of acetate ion to catalytic ADs by using NMO as cooxidant was found to improve the enantiomeric purity for some diols, presumably as a result of accelerated osmate ester hydrolysis [16]. The subsequent change to potassium ferricyanide as cooxidant appears to result in nearly complete avoidance of the secondary cycle (see Section 4.4.2.2.), but the turnover rate of the new catalytic cycle may still depend on the rate of hydrolysis of the osmate ester **9**. The addition of a sulfonamide (usually methanesulfonamide) has been found to enhance the rate of hydrolysis for osmate esters derived from 1,2-disubstituted and trisubstituted olefins [29]. However, for reasons that are not yet understood, addition of a sulfon-amide to the catalytic AD of terminal olefins (i.e., monosubstituted and 1,1-disubstituted olefins) actually slows the overall rate of the reaction. Therefore, when called for, the sulfonamide is added to the reaction at the rate of one equivalent per equivalent of olefin. This enhancement in rate of osmate hydrolysis allows most sluggish dihydroxylation reactions to be run at 0°C rather than at room temperature [29].

6D.2.6. Enantioselectivity Mnemonic

The asymmetric complex formed by osmium tetroxide, chiral ligand, and the olefinic substrate delivers two oxygen atoms to one face or the other of the olefin depending on which antipode of the chiral auxiliary is used in the reaction (see Section 6D.2.3 for a discussion of the pseudoenantiomeric character of DHQ and DHQD). Figure 6D.2 presents a mnemonic showing olefin orientation and face selectivity [26]. In the mnemonic, the olefin is oriented to fit the size constraints, where R_L = largest substituent, R_M = medium-sized substituent, and R_S = smallest substituent other than hydrogen. The oxygens will then be delivered from the upper or β-face if a DHQD-derived chiral auxiliary is used (this is the source of the β designation used in the name AD-mix-β) and from the lower or α-face if a DHQ-derived auxiliary is used (AD-mix-α). Any olefin can be oriented four ways in the plane of this mnemonic (the number of orientations is reduced for olefins that are symmetrically substituted). Two orientations are possible for each face of an olefin, and they are related by a 180° rotation in the plane of the paper. Both orientations of one face yield the same diol upon *cis* dihydroxylation. Consider the two orientations (**41** and **41'**) shown for one face of *trans*-3-decene, which are related by a simple 180° rotation. Hydroxylation of either with osmium tetroxide and (DHQD)₂-PHAL (**29**) as chiral ligand will produce (*R,R*)-decane-3,4-diol (**42**).

The mnemonic shown in Figure 6D.2 is empirical and is based primarily on the enantiose-lectivities observed in the synthesis of the diols derived from the olefins listed in Tables 6D.2–6D.5 in Section 6D.3. Conversely, the absolute configurations of the diols obtained from these olefins can be assigned from the mnemonic, and from there it is a small step to the use of Figure 6D.2 in the prediction of enantioselectivities for new dihydroxylations. However, this step is one that must be taken with caution. Unlike the case of asymmetric epoxidation (Chapter 6.1), where a similarly styled mnemonic can be safely used to predict absolute configurations of epoxides, the mnemonic for AD is only suggestive of new diol configurations. The reasons for this lie in the much greater structural diversity (all olefins vs. only allylic alcohols) available for AD as well as in the nature of the interactions between substrate and catalyst (allylic alcohols coordinate as the alkoxides to the metal of the asymmetric epoxidation catalyst). For some classes of olefins, such as the trans-disubstituted olefins, Figure 6D.2 should correctly predict the enantioselectivity of dihydroxylation for all prochiral substrates for reasons such as those given in the preceding discussion of **41**. The prediction of enantioselectivities in the dihydroxy-lation of monosubstituted olefins should also be quite reliable for similar reasons. In other cases,

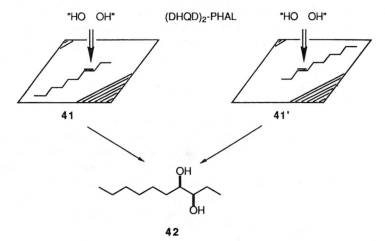

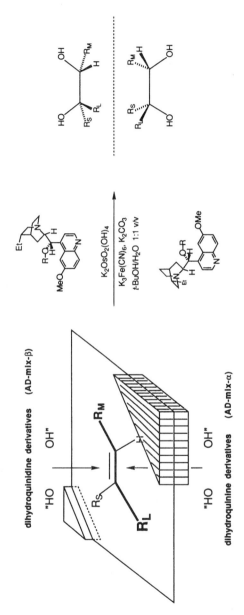

Figure 6D.2. Enantioselectivity mnemonic scheme.

particularly the 1,1-disubstituted olefins, the difficulty in applying the mnemonic when the two substituents are very similar in nature quickly becomes apparent. Figure 6D.2, therefore, should be regarded as a useful tool in the analysis of AD reactions.

6D.3. CATALYTIC ASYMMETRIC DIHYDROXYLATIONS BY OLEFIN SUBSTITUTION PATTERN

The challenges facing the development of any catalytic process for the asymmetric functionalization of olefins are large when one considers, first, the number and variety of olefins that are potential substrates for the reaction and, second, a useful asymmetric synthesis should yield an enantioselectivity of at least 90% ee, with 100% ee as the ultimate goal. When the catalytic version of AD was announced in 1988, the hydroxylation of only one olefin, *E*-stilbene, approached this standard for enantioselectivity. At the start of 1992, examples from four of the six substitution classes of olefins can be hydroxylated with 94% ee or higher, and the number of examples meeting this standard is limited primarily by the number of olefins that have been subjected to the catalytic process. Nevertheless, it is clear that this catalytic system will not be effective for every olefin, even within these four substitution classes. It is important therefore to examine carefully the structural nature of the olefin when considering application of AD. Examination of an olefin with the mnemonic (Fig. 6D.2) in mind should be of help in this regard.

A summary of results from experiments exploring the scope of catalytic AD is presented in Tables 6D.2–6D.6. The results are divided according to the five substitution patterns for which useful levels of enantioselectivity have thus far been achieved so that the synthetic chemist can easily compare a potential candidate for AD with the existing precedent. Our emphasis in this chapter is on the best conditions currently available for a general catalytic AD process, because this is very likely of the most practical interest to the reader. However, Tables 6D.2–6D.6 include most of the published data from the Sharpless laboratory for AD under a variety of conditions. These data are included to permit the reader to place in context results reported in several publications and to illustrate the variety of olefins that have been used at least once for an AD reaction. Additional applications of catalytic AD reported in the literature are integrated into the discussions of the different olefin categories. Examples illustrating "double diastereoselectivity" are collected and discussed in a separate section.

Please note the following features of Tables 6D.2–6D.5. The tables are arranged in nine columns each; the numerical data refer to the enantiomeric excess of the diol obtained from the olefin shown in Column 2. Yields are generally in the range of 75–95%. The absolute configurations of the diols are known and can be found in the references cited or deduced with the aid of the enantioselectivity mnemonic of Figure 6D.2. References are found in the footnotes to the tables. The absence of entries for an olefin in a column means simply that the experiment has not been performed. Several olefins have a nearly complete series of entries across the table (see Table 6D.2, entry 6; Table 6D.3, entries 16 and 17; and Table 6D.5, entry 3), which serve to illustrate the degree of improvement attained as beneficial modifications of the reaction were discovered. Note that numbers *without parentheses* refer to experiments with DHQD ligands and numbers *within parentheses* are for experiments with DHQ ligands. Columns 1 and 2 list entry number and show olefin structure, respectively. Column 3 lists the enantiomeric purity (% ee) obtained under *stoichiometric* AD conditions by using *p*-chlorobenzoate (CLB) derivatives of the alkaloid ligands. Column 4 lists results obtained under initial catalytic AD conditions with NMO as cooxidant. Column 5 lists results obtained following discovery of the secondary cycle shown in Scheme 6D.2 and introduction of the method of slow addition of olefin to

TABLE 6D.2. Asymmetric Dihydroxylation of Monosubstituted Olefins (% ee)

1	2	3[a]	4[b]	5[c]	6[d]	7[e]	8[f]	9[g]
Entry	Olefin	CLB stoich.; 0°C	CLB cat.; NMO 0°C/25°C	CLB cat.; NMO slo/OAc⁻	CLB cat.; Fe 0°C/25°C	MEQ cat.; Fe 0°C	PHN cat.; Fe 0°C	PHAL cat.; Fe 0°C
1					–/45	65	74	84(80)
2	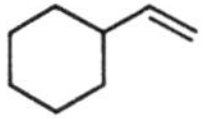		46/–				80	
3	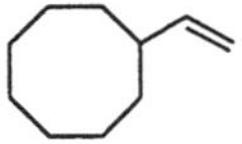				–/64	85	93	
4	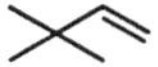				–/44	79	79	97(97)
5	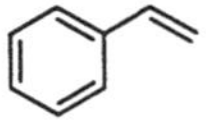		62(60)/–	60/–	73/74	87	78	
6	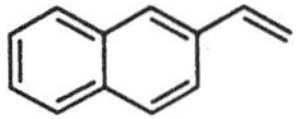				–/88	93	83	
7	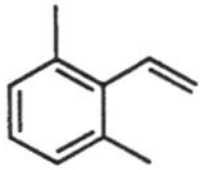		65/--					
8	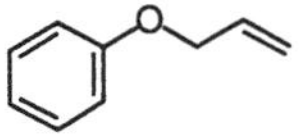					66		88(77)
9	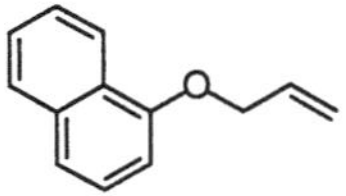			40/–	–/60	44		91(88)

(*continued*)

375

TABLE 6D.2. Continued

1	2	3[a]	4[b]	5[c]	6[d]	7[e]	8[f]	9[g]
		CLB stoich.; 0°C	CLB cat.; NMO 0°C/25°C	CLB cat.; NMO slo/OAc⁻	CLB cat.; Fe 0°C/25°C	MEQ cat.; Fe 0°C	PHN cat.; Fe 0°C	PHAL cat.; Fe 0°C
Entry	Olefin							
10	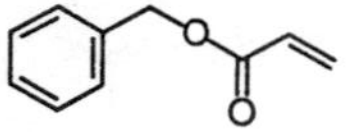						50(49)	77(70)
11	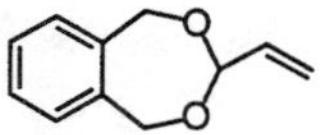						84	67
12	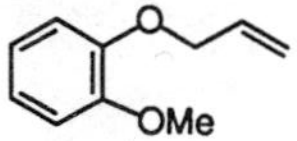							
13							38	63(56)
14	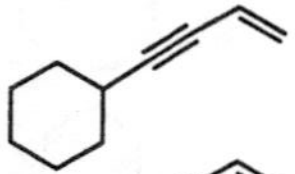						44	72
15	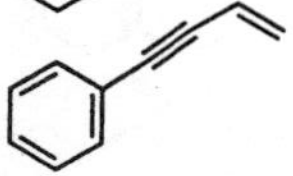						53	73

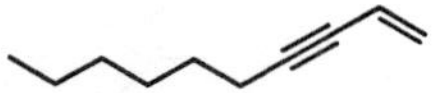

[a]Reference for column 3: entry 5 [40]. [b]Refs for column 4: entries 2,5,7 [26]. [c]Refs for column 5: entry 5 [16], entry 9 [49]. [d]Refs for column 6: entries 1,3,4,6 [31]; entry 5 [31,35]; entry 9 [49]. [e]Refs for column 7: entries 1, 3–6 [31]; entries 8,9 [44]. [f]Refs for column 8: entries 1,3,4,5,6 [31]; entries 2,10 [44]; entry 11 [45]; entries 13–15 [46]. [g]Refs for column 9: entries 1,5,10 [29]; entry 4,8,9,11,12 [45]; entries 13–15 [46].

376

TABLE 6D.3. Asymmetric Dihydroxylation of trans-Disubstituted Olefins (% ee)

Entry	Olefin	3[a] CLB stoich.; 0°C	4[b] CLB cat.; NMO 0°C/25°C	5[c] CLB cat.; NMO slo/OAc⁻	6[d] CLB cat.; Fe 0°C/25°C	7[e] MEQ cat.; Fe 0°C	8[f] PHN cat.; Fe 0°C	9[g] PHAL cat.; Fe 0°C
1	 (trans-alkene)	69–71		70/–				97(93)
2	 (trans-alkene)	71–73	20	–/69	74–79/–	90	95	
3	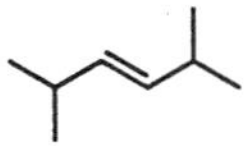 (branched diene)	79–80	12	46/76				
4	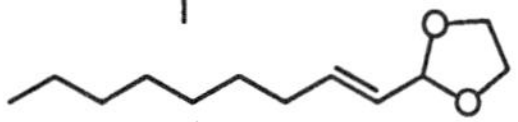 (dioxolane alkene)	83		56/66				
5	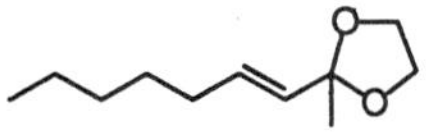 (dioxolane alkene)	50		46/50				
6	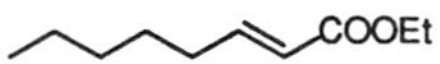 (alkene COOEt)	66–67				85	94	99(96)
7	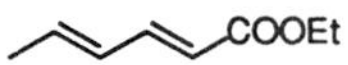 (alkene COOEt)							92
8	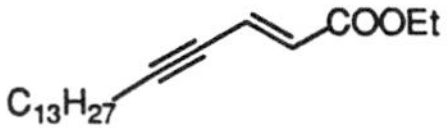 C₁₃H₂₇ (enyne COOEt)							93 / 97de
9	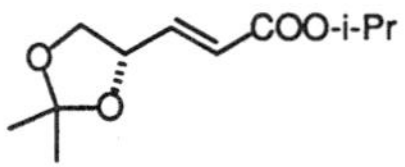 (acetonide CH=CH COO-i-Pr)	77		76de/–				
10	(acetonide CH=CH COOMe)		52de/–					
11	(acetonide CH=CH COOMe)	54						

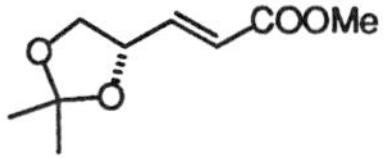

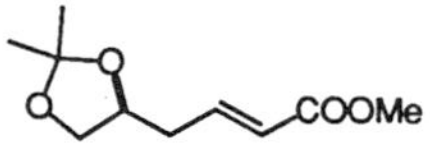

(continued)

TABLE 6D.3. Continued

1 Entry	2 Olefin	3[a] CLB stoich.; 0°C	4[b] CLB cat.; NMO 0°C/25°C	5[c] CLB cat.; NMO slo/OAc⁻	6[d] CLB cat.; Fe 0°C/25°C	7[e] MEQ cat.; Fe 0°C	8[f] PHN cat.; Fe 0°C	9[g] PHAL cat.; Fe 0°C
12	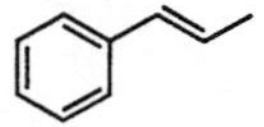	87	66(55)	86/–	–/91			(97)
13	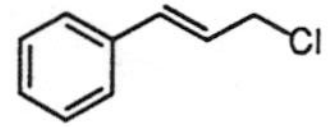	79		78/–				
14	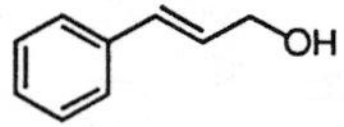	66	76	66/–				
15	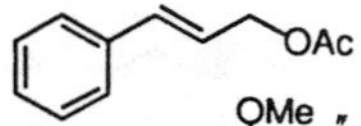	82		79/–	–/91			
16	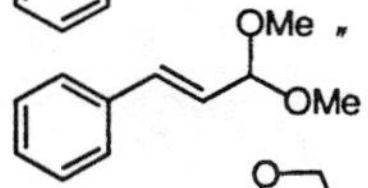	82		75/83				
17	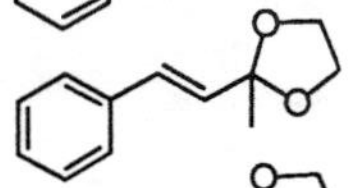	85		60/89				
18	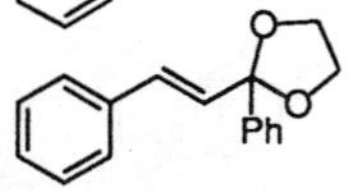	86		84/87				
19	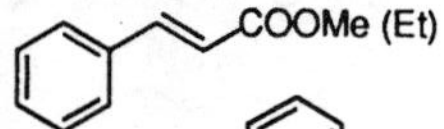	89	60	86-88/–	91/95	98	98	97(95)
20	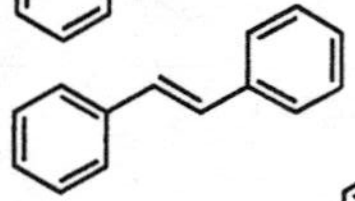	99	88(75)	95/–	–/99	98	99	>99.5(>99.5)
21	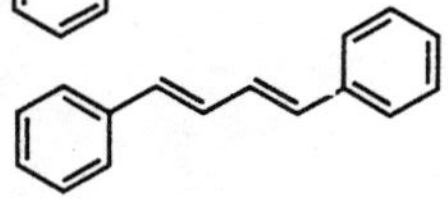							97

(continued)

378

TABLE 6D.3. Continued

1	2	3[a]	4[b]	5[c]	6[d]	7[e]	8[f]	9[g]
		CLB stoich.; 0°C	CLB cat.; NMO 0°C/25°C	CLB cat.; NMO slo/OAc⁻	CLB cat.; Fe 0°C/25°C	MEQ cat.; Fe 0°C	PHN cat.; Fe 0°C	PHAL cat.; Fe 0°C
Entry	Olefin							
22	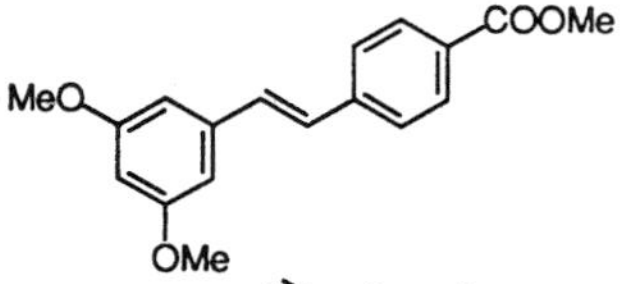							>99(>99)
23	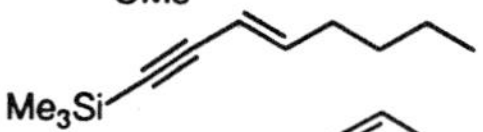						93	
24	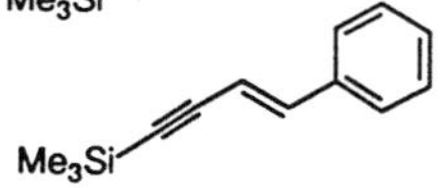						94	
25	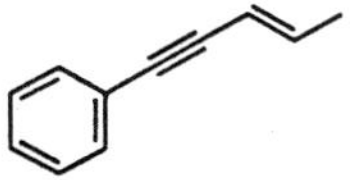						73	
26	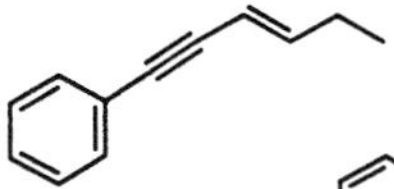						90	
27	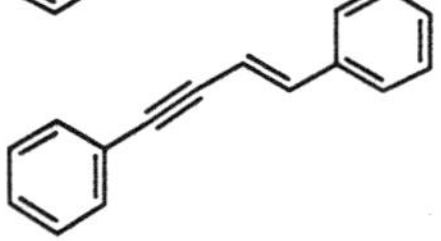						97	

[a]References for column 3: entries 1–3,6 [40,41]; entries 4,5,10–20 [40]. [b]Refs for column 4: entries 1,12,15, [26]; entries 2, 19 [35]; entry 3 [16]; entry 20 [47]; [c]Refs for column 5: entries 1–5,10–19 [40]; entry 20 [35,40]; entry 20 [35,40]. [d]Refs for column 6: entries 2,19 [31,35]; entries 12,15,20 [35]. [e]Refs for column 7: entries 2,6,19,20 [31]. [f]Refs for column 8: entries 2,6,19,20 [31]; entries 23–27 [46]. [g]Refs for column 9: entries 2,6–9,19–21 [29]; entry 22 [48].

379

TABLE 6D.4. Asymmetric Dihydroxylation of 1,1-Disubstituted Olefins (% ee)

1	2	3	4[a]	5	6[b]	7[c]	8[d]	9[e]
		CLB stoich.; 0°C	CLB cat.; NMO 0°C/25°C	CLB cat.; NMO slo/OAc⁻	CLB cat.; Fe 0°C/25°C	MEQ cat.; Fe 0°C	PHN cat.; Fe 0°C	PHAL cat.; Fe 0°C
Entry	Olefin							
1								78(76)
2	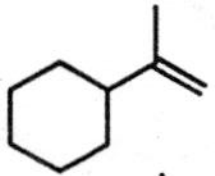				–/37	73	82	
3	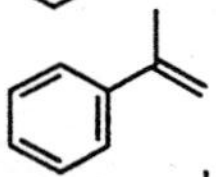		33					94(93)
4	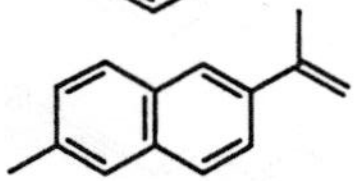				–/74	88	69	
5	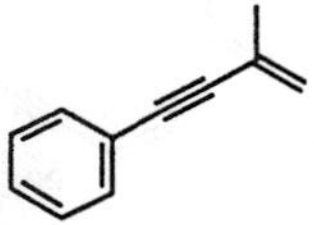						48	79

[a]Reference for column 4: entry 3 [26]. [b]Refs. for column 6: entries 2,4 [31]. [c]Ref. for column 7: entries 2,4 [31]. [d]Refs. for column 8: entries 2,4 [31]. [e]entry 5 [46]. Refs. for column 9: entries 1,3 [29]; entry 5 [46].

TABLE 6D.5. Asymmetric Dihydroxylation of Trisubstituted Olefins (% ee)

1 Entry	2 Olefin	3[a] CLB stoich.; 0°C	4 CLB cat.; NMO 0°C/25°C	5[b] CLB cat.; NMO slo/OAc⁻	6[c] CLB cat.; Fe 0°C/25°C	7[d] MEQ cat.; Fe 0°C	8[e] PHN cat.; Fe 0°C	9[f] PHAL cat.; Fe 0°C
1	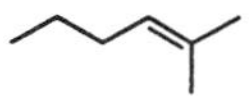	55		–/54				
2								98(95)
3	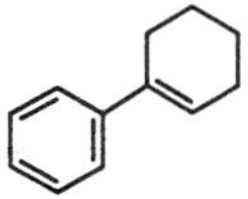	79		78/81	91/–	92	93	99(97)
4	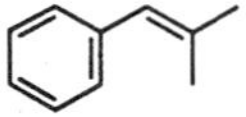	55		53/–	74/–	81	84	
5	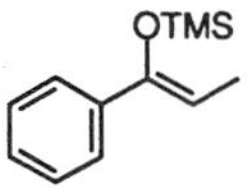	85		72/78				
6	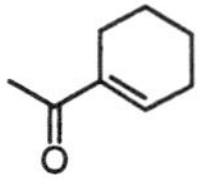	56		–/54				

[a]Reference for column 3: entries 1,3–6 [40]. [b]Ref. for column 5: entries 1,3–6 [40]. [c]Ref. for column 6: entries 3,4 [31]. [d]Ref. for column 7: entries 3,4 [31]. [e]Ref. for column 8: entries 3,4 [31]. [f]Ref. for column 9: entries 2,3 [29].

TABLE 6D.6. Asymmetric Dihydroxylation of *cis*-Disubstituted Olefins

Entry	Olefin	% ee, rt	% ee, 0°C
1	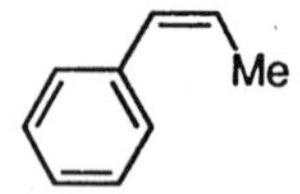 Me	65	72
2	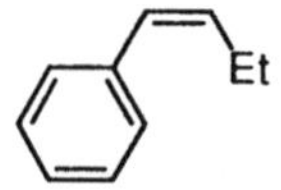 Et	67	
3	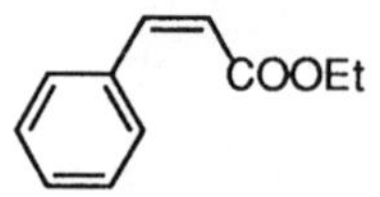 COOEt	75	78
4	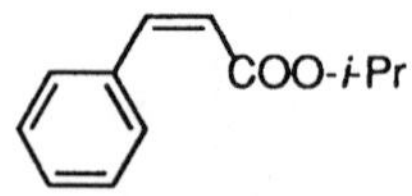 COO-*i*-Pr	74	80
5	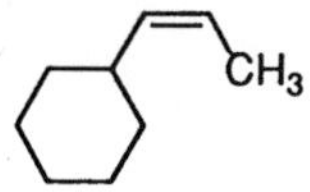 CH₃	49	
6	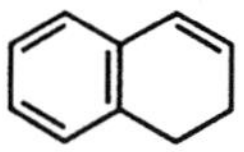	14	

suppress this cycle. Results in Column 5 are divided between slow addition without/with acetate added to the reaction. Column 6 lists results obtained by using potassium ferricyanide as cooxidant, and data are divided between reactions run at 0°C/25°C. Column 7 lists results obtained by using the 4′-methyl-2′-quinolyl ether (MEQ) derivative of DHQD as the chiral ligand. Column 8 lists results obtained by using the 9′-phenanthryl ether (PHN) derivative of DHQD as the chiral ligand. Column 9 lists results obtained with the bis-DHQD and bis-DHQ ethers of phthalazine-1,4-diol (PHAL) as the chiral ligands.

6D.3.1. Monosubstituted Olefins

The enantiomeric excesses obtained to this point for the catalytic AD of monosubstituted olefins (see Table 6D.2 [16,26,29,31,40,44–46,49]) are lower than for *trans*-disubstituted olefins (Table 6D.3). The entries in Column 9 show enantiomeric purities ranging from 54% ee to 97% ee for dihydroxylations with the (DHQD)₂-PHAL and (DHQ)₂-PHAL pair of chiral ligands. Several monosubstituted olefins with branching at the α-position (e.g., entries 2–4 and 11) are dihydroxylated with higher enantioselectivities when DHQD-PHN is used as the chiral ligand instead of (DHQD)₂-PHAL. Recently, a new ligand for terminal olefins has been discovered [48b].

43, R = H
44, R = Tos

45

The diol (**43**) obtained from dihydroxylation of acrolein benzene-1,2-dimethanol acetal (entry 11) is a masked glyceraldehyde and has the potential to be a very useful synthon. Although the enantiomeric purity of the crude diol formed in this reaction is 84% ee, one recrystallization from ethyl acetate improves it to 97% ee in 55% recovery yield. The masked glyceraldehyde **43** is converted via the tosylate **44** to the masked glycidaldehyde **45** in an overall yield of 85%. Both these masked aldehydes are superior to the free aldehydes in terms of handling ease, stability, and safety. The aldehydes can be released from the acetal under the mild conditions of catalytic hydrogenolysis [45].

Diols obtained from the catalytic AD of aryloxyallyl ethers, such as those of entries 8, 9, and 12, are useful as precursors for the synthesis of the pharmacologically important β-blockers [49].

6D.3.2. *trans*-Disubstituted Olefins

Olefins of the *trans*-disubstituted type have given diols with excellent enantiomeric purities when dihydroxylated with the (DHQD)$_2$-PHAL/(DHQ-PHAL pair of chiral ligands with osmium tetroxide (see Table 6D.3 [16,26,29,31,35,40,41,46–48]). All the entries but one in Column 9 for diols obtained with these ligands exceed 90% ee (or 90% de). From other entries in Table 6D.3, particularly those of Column 3 for earlier stoichiometric ADs with the DHQD-CLB ligand, good enantioselectivities are anticipated for the dihydroxylation of most *trans*-disubstituted olefins when the PHAL ligands are used.

Included in this class of olefins is (*E*)-stilbene (entry 20), which throughout studies of AD has usually been the olefin dihydroxylated with the highest degree of enantioselectivity. Availability of (*R,R*) or (*S,S*)-1,2-diphenyl-1,2-ethanediol (also referred to as stilbenediol or dihydrobenzoin) with high enantiomeric purities has led to reports of a number of applications, including incorporation into chiral dioxaphospholanes [50], chiral boronates [51], chiral ketene acetals [52], chiral crown ethers [53], and conversion into 1,2-diphenylethane-1,2-diamines [54]. Dihydroxylation of the substituted *trans*-stilbene **46** with OsO$_4$/NMO and DHQD-CLB gives the *R,R*-diol **47** with 82% ee in 88% yield [55].

46

47

Details for the large-scale synthesis of (*R,R*)-1,2-diphenyl-1,2-ethanediol by using the DHQD-CLB/NMO variation of catalytic AD have been published [47]. Under these conditions the crude diol is produced with 90% ee and upon crystallization, essentially enantiomerically pure diol is obtained in 75% yield. Subsequent improvements in the catalytic AD process now allow this dihydroxylation to be achieved with >99.8% ee (entry 20, Column 9); however, the *Organic Synthesis* procedure [47] is still an excellent choice for preparing large amounts of the enantiomerically pure stilbene diols as it is fast and needs little solvent (the reaction is run at 2 molar in stilbene).

The asymmetric dihydroxylation of dienes has been examined, originally with the use of NMO as the cooxidant for osmium [56a] and, more recently, with potassium ferricyanide as the cooxidant [56b]. Tetraols are the main product of the reaction when NMO is used, but with $K_3Fe(CN)_6$, ene-diols are produced with excellent regioselectivity. The example of dihydroxylation of *trans,trans*-1,4-diphenyl-1,3-butadiene is included in Table 6D.3 (entry 21). One double bond of this diene is hydroxylated in 84% yield with 99% ee when the amounts of $K_3Fe(CN)_6$ and K_2CO_3 are limited to 1.5 equiv. each. Unsymmetrical dienes are also dihydroxylated with excellent regioselectivity. In these dienes, preference is shown for (a) a bans over a cis olefin, (b) the terminal olefin in $\alpha,\beta,\gamma,\delta$-unsaturated esters, and (c) the more highly substituted olefin [56b].

A class of *trans*-disubstituted olefin that is encountered frequently in organic chemistry is the α,β-unsaturated ester. Catalytic AD of several examples may be found in Table 6D.3 (entries 6–11, 19) and in these cases, diols with high enantiomeric purities are produced. The olefin of entry 7 is an interesting variation of this class and is $\alpha,\beta,\gamma,\delta$-bis-unsaturated ester. Catalytic AD of this diene under controlled conditions (so that only enough oxidant is present to allow dihydroxylation of one double bond) gives the γ,δ-dihydroxy-α,β-unsaturated ester with 88% ee (by using $(DHQ)_2$-PHAL) in 76% yield. Dihydroxylation of methyl *p*-methoxycinnamate has been reported, and the $2R,3S$-diol was taken on in a synthesis of (+)-diltiazem, a vasodilating agent [28]. The catalytic AD of other α,β-unsaturated esters found in chiral molecules is described in Section 6D.3.7.

6D.3.3. 1,1-Disubstituted Olefins

A limited number of 1,1-disubstituted olefins have been subjected to catalytic AD, and these are listed in Table 6D.4 [26,29,31,46]. The results shown there indicate that selected members of this class of olefins can be dihydroxylated with high enantioselectivity but it should be recognized that these olefins all have relatively dissimilar substituents. As the two substituents becomes more alike, the discriminatory capability of the catalyst is expected to lessen and lower enantioselectivities to be observed. Such a trend may be found in the comparison of entries 1 and 5 with entry 3. The difference in size of substituents clearly is greater in the case of entry 3 and the diol derived from this olefin has the highest enantiomeric purity.

6D.3.4. Trisubstituted Olefins

A listing of trisubstituted olefins that have been subjected to catalytic AD is given in Table 6D.5 [29,31,40]. Only three compounds of this type have been dihydroxylated by using the newer, more efficient chiral ligands, and all three are converted to diols with high enantiomeric purities under these conditions.

Squalene (**48**) presents an interesting array of trisubstituted double bonds with which to test the selectivity of catalytic AD. Osmium tetroxide catalyzed dihydroxylation of squalene in the *absence* of a chiral ligand generates a mixture of the 2,3-diol (**49**), 6,7-diol (**50**), and 10,11-diol (**51**) in a ratio of 1:1:1. (Squalene is used in excess in these experiments to minimize multiple-dihydroxylations of the squalene molecule.) Dihydroxylation with $(DHQD)_2$-PHAL as the chiral ligand produces the diols **49**, **50**, and **51** in a ratio of 46:35:19. Stereochemical analysis of the 2,3-diol (**49**) indicates formation of the 2,3R-diol with 96% ee [57a]. Perhydroxylation of squalene has also been achieved with 98% ee or de for each of the six dihydroxylation events required in the process [57b].

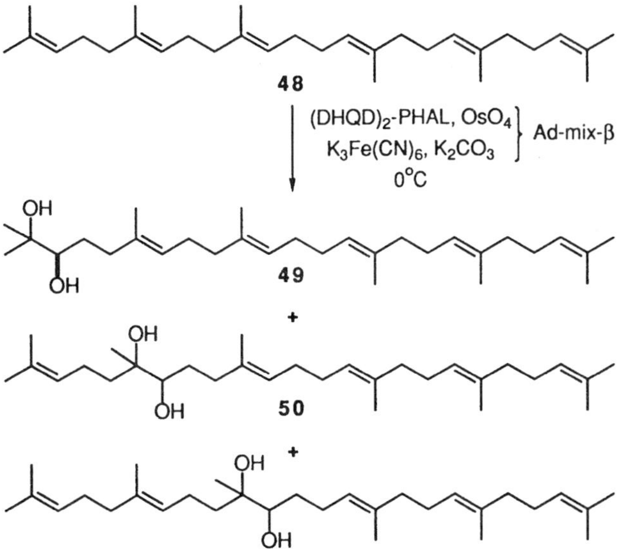

6D.3.5. *cis*-Disubstituted Olefins

cis-Disubstituted olefins have proven to be the most difficult class of substrates from which to obtain diols with high enantiomeric purities. With any of the previously discussed chiral ligands, enantioselectivity in the AD of these olefins has been low. Using (Z)-1-phenyl-propene (*cis*-β-methylstyrene) as an example, AD with DHQD-CLB gave 1-phenylpropan-1,2-diol with 35% ee, DHQD-MEQ gave 26% ee, DHQD-PHN gave 22% ee, and (DHQD)$_2$-PHAL gave 29% ee. Continued searching for better chiral ligands for this class of olefins has recently led to the discovery of a new pair of ligands that imparts significantly higher enantiomeric purity to the diols [57c]. The new ligands are 9-*O*-indolinylcar-bamoyldihydroquinidine (**52**, Chart 6D.1) and the pseudoenantiomeric dihydroquinine derivative (**53**) (DHQD-IND and DHQ-IND, respectively). Although work with these new ligands is still in progress, current results for AD by using DHQD-IND are listed in Table 6D.6. *cis*-β-Methylstyrene now is converted to (1*R*,2*S*)-1-phenylpropan-1,2-diol with 72% ee (entry 1) and *i*-propyl *cis*-cinnamate gives isopropyl-(2*R*,3*R*)-2,3-dihydroxy-3-phenyl-propionate with 80% ee (entry 4).

6D.3.6. Tetrasubstituted Olefins

Using the AD recipe given in Section 6D.2, tetrasubstituted silyl enol ethers give good yields of α-hydroxyketones after workup with enantioselectivities ranging from 60% to 97% ee [57d]. The AD of other tetrasubstituted olefins is also possible by running the reaction at room temperature with a more powerful AD recipe consisting of 1 mol % K$_2$OsO$_2$(OH)$_4$, 5 mol % ligand, and 3 equiv. of methanesulfonamide [57e]. The use of this more active recipe is recommended whenever a very unreactive (in the turnover sense) olefin is encountered.

6D.3.7. Double Diastereoselectivity

A diastereoselective osmylation must have occurred in the first reaction of osmium tetroxide with a chiral molecule, which may have been of pinene by Hofmann in 1912 [7]. However, at that time even the cis nature of the dihydroxylation was unknown and was not determined until the careful work of Criegee in the 1930s. By the late 1930s, diastereoselectivity was clearly recognized in the osmium tetroxide catalyzed dihydroxylations of natural products and, for example, played an important role in structural assignments and synthetic elaborations of the corticosteroid side chain. A discussion of these diastereoselective reactions and references to the original literature can be found in *Steroids,* the classical account by Fieser and Fieser [58] of the early decades of steroid chemistry. Increased efforts at exercising stereocontrol in synthesis, particularly in the past 10–15 years, have led to frequent observations of diastereoselectivity in olefin osmylation and to efforts at developing models for use in predictions of this selectivity.

With the discovery of AD and subsequent catalytic modifications came the possibility for "matching and mismatching" [25] diastereoselectivity with the enantiofacial selectivity of the asymmetric process. Knowledge of the diastereoselectivity of a given olefin hydroxylation is needed to match substrate with the correct choice of chiral catalyst. The best way to determine the diastereoselectivity for dihydroxylation of a chiral olefin is to perform the osmylation in the absence of a chiral ligand and then determine the ratio (de) of the resulting diols. A much less rigorous alternative is the inspection of molecular models, which may suggest that there is a diastereoselective preference for the dihydroxylation of a particular chiral olefin. Such predictions of diastereoselectivity may be reliable for some rigid cyclic molecules but will be difficult for olefins found in acyclic environments. The most frequently studied acyclic olefins are those in which the double bond is attached to a chiral carbon carrying a heteroatom substituent. Several analyses of the diastereoselectivity of osmylation of these allylic systems, including the closely related γ-substituted α,β-unsaturated esters, have been put forward [59]. From consideration of the rationale behind these results, it appears that as a first approximation, the osmylation of such olefins will occur from the sterically more accessible face rather than in response to the electronic nature of the allylic system.

quinuclidine	2.5 : 1
DHQD-OAc	40 : 1
DHQ-OAc	1 : 16

The matching and mismatching of chiral olefin **54** and catalyst was examined briefly by using *stoichiometric* quantities of osmium tetroxide with achiral and chiral ligands [60]. The monothio acetal derived from camphor (**54**) was dihydroxylated with osmium tetroxide in the presence of quinuclidine, DHQD-OAc, or DHQ-OAc. With the achiral quinuclidine as ligand, the ratio of (2*S*,3*R*) to (2*R*,3*S*) diastereomers **55** and **56** was 2.5:1. With DHQD-OAc as the chiral ligand, catalyst and substrate are matched and the ratio is enhanced to 40:1 while with DHQ-OAc catalyst and substrate are mismatched and a reversed selectivity of 1:16 is observed.

A series of α,β-unsaturated esters (**57a–e**) has been matched and mismatched with catalyst in a study of the catalytic AD of octuronic acid derivatives [61]. In this study, hydroxylations

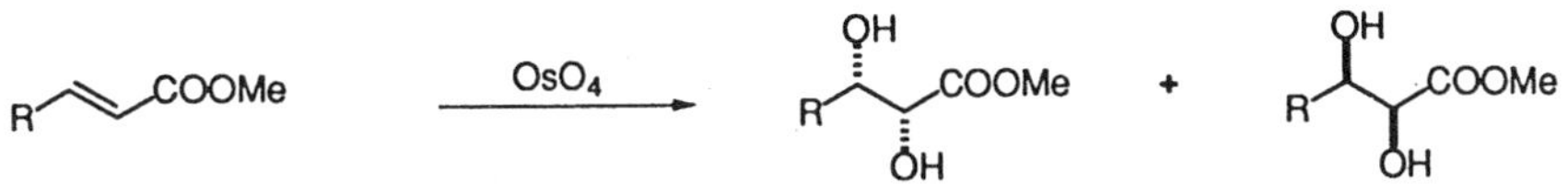

R	Reagent	Ratio 58 : 59
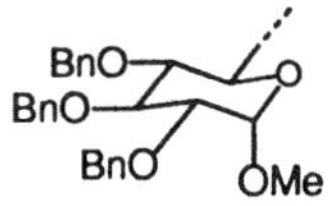 **57a**	OsO$_4$ only	10.3 : 1
	DHQD-CLB	1.3 : 1
	DHQ-CLB	20.5 : 1
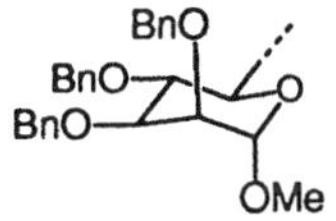 **57b**	OsO$_4$ only	7.4 : 1
	DHQD-CLB	3.4 : 1
	DHQ-CLB	15.9 : 1
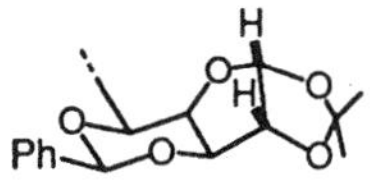 **57c**	OsO$_4$ only	1 : 2.2
	DHQD-CLB	1 : 5.3
	DHQ-CLB	1 : 1.6

with osmium tetroxide alone were compared with hydroxylations with DHQD-CLB and DHQ-CLB as the chiral ligands. Olefins **57a** and **57b** are matched with DHQ-CLB and olefin **57c** is matched with DHQD-CLB. In all three eases, matching of olefin diastereoseleetivity with catalyst enantioselectivity enhances the overall selectivity of the dihydroxylation. The authors of this work observed greater enhancement of selectivity with stoichiometric reagents in comparison with catalytic amounts of reagents, a discrepancy that may be reduced with application of the ferricyanide procedure for the catalytic process, if indeed it does not disappear.

With the newest chiral ligands, (DHQD)$_2$-PHAL and (DHQ)$_2$-PHAL, the effect of matching and mismatching on the dihydroxylation of olefin **60** has been tested [62].

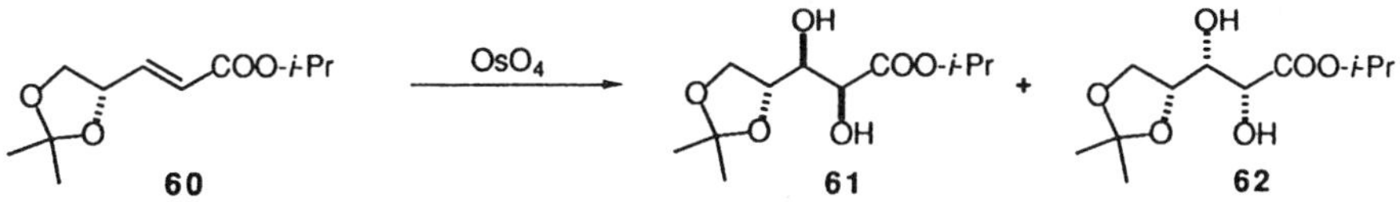

Reagent	Ratio 61 : 62
no ligand	2.8 : 1
(DHQD)$_2$-PHAL (Ad-mix-β)	39 : 1
(DHQ)$_2$-PHAL (Ad-mix-α)	1 : 1.3

The ratio of diols **61:62** from dihydroxylation with osmium tetroxide alone is 2.8:1, whereas with (DHQD)$_2$-PHAL the ratio is 39:1, and with (DHQ)$_2$-PHAL the ratio is 1:1.3, which shows the cumulative effect that can be achieved by matching diastereoselectivities and also reveals that AD may not be as reliable as is asymmetric epoxidation (Chapter 6A) for attaining good results when diastereoselectivities are mismatched.

Other comparisons of diastereoselective osmylations in the absence and in the presence of chiral ligands have been reported. A study of the dihydroxylation of unsaturated side chains attached to various steroid nuclei allows some comparisons of matching and mismatching of diastereoselectivities [63].

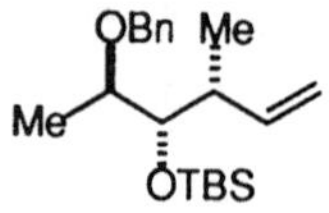

Dihydroxylation of the double bond in the synthetic intermediate **63** gives a 3.5:1 ratio of diastereoisomers in the absence of a chiral catalyst and a 5:1 ratio when matched with DHQ-CLB [64]. Synthetic intermediate **64** is dihydroxylated with a 12:1 ratio of diastereoselectivity without a chiral catalyst and, under otherwise identical conditions, in a 20:1 ratio with DHQ-CLB [65]. It should be noted that dihydroxylation of **64**, the subject of extensive process development, is very sensitive to solvent composition, and an improved diastereoselective ratio (39:1) was obtained by altering this variable. Dihydroxylation of **65** is an example of the stereochemistry required for a synthetic goal necessitating the use of a mismatched system. The ratio of diastereomeric products in the absence of chiral ligand was not given but was stated to be "reversed" from the 2.5:1 ratio observed with DHQD-CLB [66]. Finally, both olefins in the synthetic intermediate **66** were dihydroxylated at the same time with DHQ-CLB; the diastereo-facial selectivity at each olefin was determined to be 12.9:1 [67]. The use of the newer chiral ligands, which are capable of inducing higher levels of enantioselectivity, is expected to give improved results in future applications of AD for achieving double diastereoselectivity.

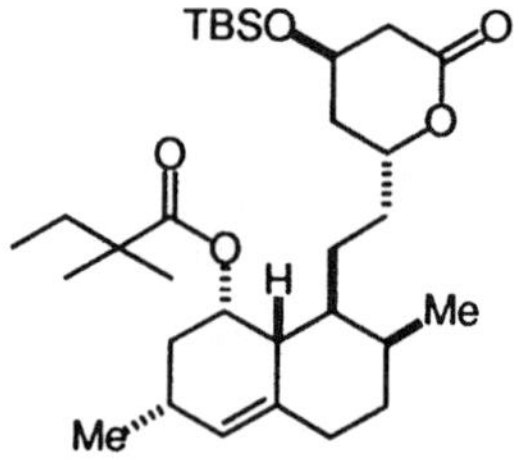

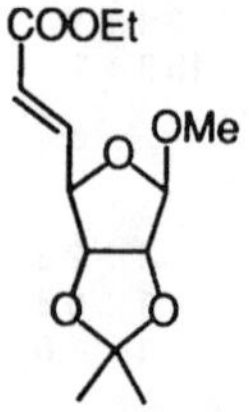

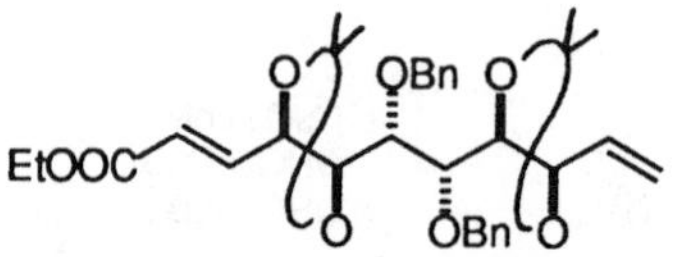

6D.4. DIOL ACTIVATION

This section outlines three chemical transformations designed to allow further synthetic elaboration of the diols obtained from AD. The first and most broadly applicable method is the conversion of the diols into cyclic sulfates, a functionality that has reactive properties like an epoxide but is even more electrophilic than an epoxide [68]. The second approach to diol activation is the regioselective conversion of one of the hydroxyl groups into a sulfonate ester [69]. This approach requires that the diol be substituted in a way that leads to regioselective derivatization of one of the two hydroxyl groups, and diol esters are a prime example of such an arrangement. The third and very convenient approach currently under development is the conversion of diols into acetoxy halides from which epoxides are easily obtained.

6D.4.1. Cyclic Sulfates, Sulfites, and Sulfamidates

The use of cyclic sulfates in synthetic applications has been limited in the past because, although cyclic sulfites are easily prepared from diols, a convenient method for oxidation of the cyclic sulfites to cyclic sulfates had not been developed. The experiments of Denmark [70] and of Lowe and co-workers [71] with stoichiometric ruthenium tetroxide oxidations and of Brandes and Katzenellenbogen [72a] and Gao and Sharpless [68] with catalytic ruthenium tetroxide and sodium periodate as cooxidant have led to an efficient method for this oxidation step. Examples of the conversion of several diols (**67**) to cyclic sulfites (**68**) followed by oxidation to cyclic sulfates (**69**) are listed in Table 6D.7. The cyclic sulfite/cyclic sulfate sequence has been applied to 1,2-, 1,3-, and 1,4-diols with equal success. Cyclic sulfates, like epoxides, are excellent electrophiles and, as a consequence of their stereoelectronic makeup, are less susceptible to the elimination reactions that usually accompany attack by nucleophiles at a secondary carbon. With the development of convenient methods for their syntheses, the reactions of cyclic sulfates have been explored. Most of the reactions have been nucleophilic displacements with opening of the cyclic sulfate ring. The variety of nucleophiles used in this way is already extensive and includes H$^-$ [68], N$_3^-$ [68,73–76], F$^-$ [68,72,74], PhCOO$^-$ [68,73,74], NO$_3^-$ [68], SCN$^-$ [68], PhCH$_2^-$ [68], (ROOC)$_2$CH$^-$ [68], RNH$_2$ [76], (RS)$_2$CH$^-$ and (RS)$_3$C$^-$ [77], RCC≡C$^-$ [78], PhC(NH$_2$)=NH [79], and RPH$_2$ [80], a list that is sure to grow. Upon opening of the cyclic sulfate (**69**), one obtains a sulfate ester (**70**), itself a versatile functional group. Hydrolysis of the sulfate ester (**70**) with aqueous acid regenerates the hydroxyl group, giving **71**; conditions suitable for both the generation of the cyclic sulfate and the hydrolysis of the sulfate ester in the presence of other acid-sensitive groups have been developed [73].

Alternatively, the remaining sulfate ester of **70** may serve as a leaving group for a second nucleophilic displacement reaction. When this displacement is by an intramolecular nucleophile, a new ring is formed, as was first shown in the synthesis of a cyclopropane with malonate as the nucleophile [68] and of aziridines with amines as the nucleophiles [76]. The concept is further illustrated in the double displacement on (*R,R*)-stilbenediol cyclic sulfate (**72**) by benzamidine (**73**) to produce the chiral imidazoline **74** [79]. Conversion of the imidazoline (**74**) to (*S,S*)-stilbenediamine **75** demonstrates an alternative route to optically active 1,2-diamines. Acylation of **75** with chloroacetyl chloride forms a bisamide, which, after reduction with diborane, is cyclized to the enantiomerically pure *trans*-2,3-diphenyl-1,4-diazabicyclo[2.2.2]octane (**76**) [81].

Cyclic sulfites (**68**) also are opened by nucleophiles, although they are less reactive than cyclic sulfates and require higher reaction temperatures for the opening reaction. Cyclic sulfite **77**, in which the hydroxamic ester is too labile to withstand ruthenium tetroxide oxidation of the sulfite, is opened to **78** in 76% yield by reaction with lithium azide in hot DMF [82]. Cyclic sulfite **79** is opened with nucleophiles such as azide ion [83] or bromide ion [84], by using elevated temperatures in polar aprotic solvents. Structures such as 80 generally are not isolated but as in the case of **80** are carried on (when X = N$_3$) to amino alcohols [83] or (when X = Br) to maleates [84] by reduction. Yields are good and for compounds unaffected by the harsher conditions needed to achieve the displacement reaction, use of the cyclic sulfite eliminates the added step of oxidation to the sulfate.

In the same vein as the cyclic sulfate activation of diols, cyclic sulfamidates have been prepared from 1,2- and 1,3-amino alcohols for the purpose of activating the carbinol toward nucleophilic attack [85–88]. (*S*)-Prolinol [85], *N*-benzyl serine *t*-butyl ester [86], and 2-(2-hy-

TABLE 6D.7. Cyclic Sulfates (69) from Diols (67)

Entry	R_1	R_2	Yield (%)
1	COO-i-Pr	COO-i-Pr	90
2	COOEt	COOEt	69
3	COOMe	COOMe	63
4	n-C$_8$H$_{17}$	H	92
5	c-C$_6$H$_{11}$	H	97
6	n-C$_4$H$_9$	n-C$_4$H$_9$	89
7	n-C$_{15}$H$_{31}$	COOMe	90
8	c-C$_6$H$_{11}$	COOEt	95
9	H	COO-c-C$_6$H$_{11}$	88
10	H	CONHCH$_2$Ph	64
11	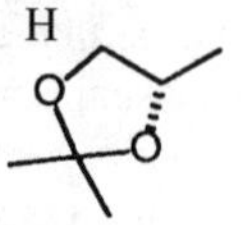 [structure]	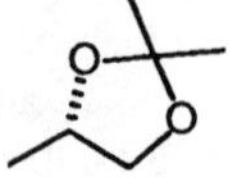 [structure]	94
12	H	CH$_2$OSiMe$_2$-t-Bu	87
13	COOMe	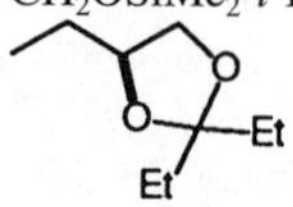 [structure] —Et	95

droxyethyl)piperidine [87] react with thionyl chloride to form the corresponding cyclic sulfamidite, and the latter are oxidized with RuO$_4$/NaIO$_4$ to the cyclic sulfamidates **81**, **82** and **83**, respectively. Nucleophilic displacements on the oxy carbon of the cyclic sulfamidate have been achieved with RNH$_2$ and ROH [85]; with H$_2$O, N$_3^-$, SCN$^-$, and CN$^-$ [86]; and with F$^-$ [88].

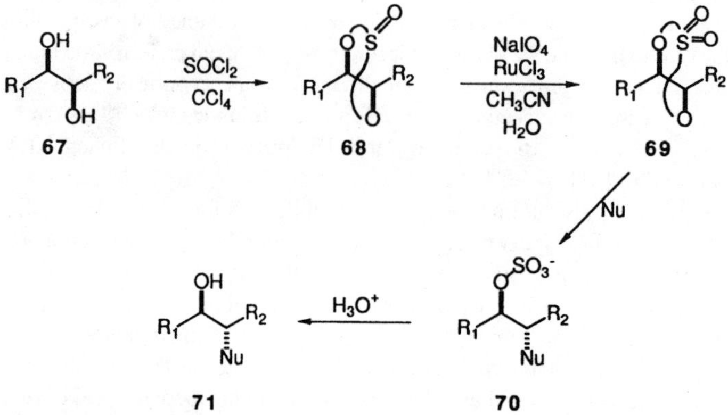

6D.4.2. Regioselective Sulfonylation

2,3-Dihydroxy esters, such as **84**, are monosulfonylated in good yield and with high regioselectivity for the 2-hydroxyl group with either p-toluenesulfonyl chloride or p-nitrosulfonyl chloride [69]. Minor side products formed in the reaction are the bissulfonate ester and the

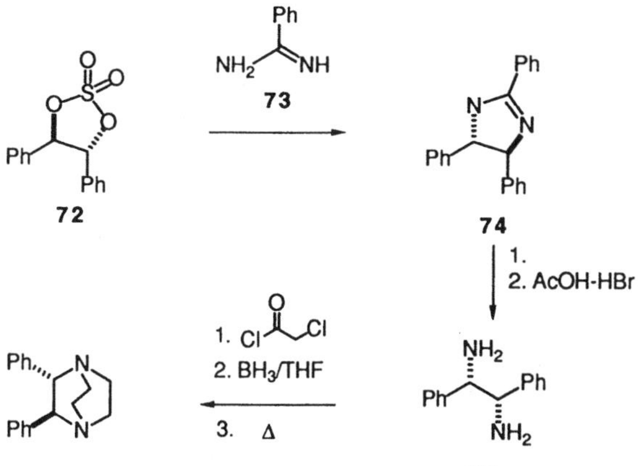

72 74 75 76

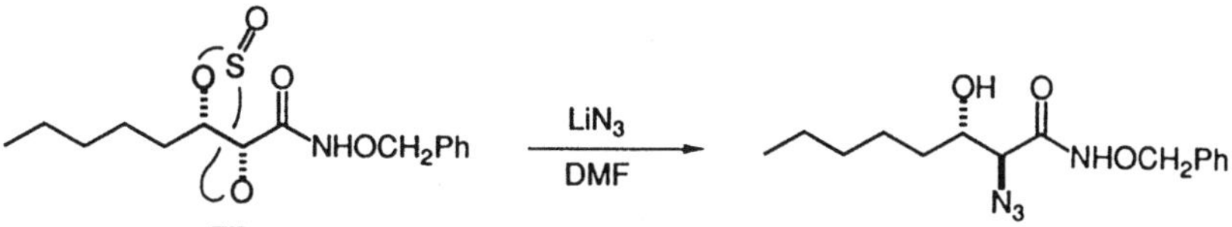

77 78

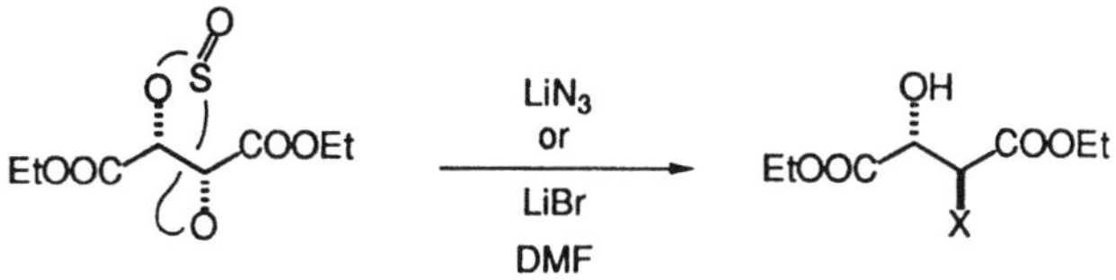

79 80 $X = N_3$; Br

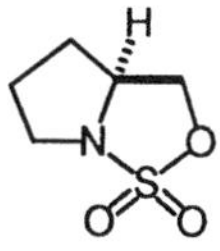

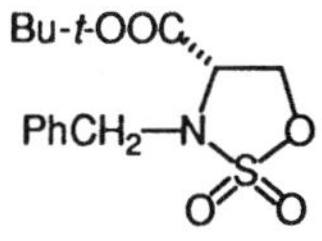

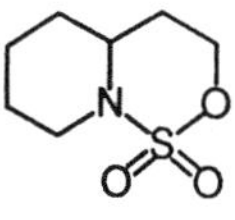

81 82 83

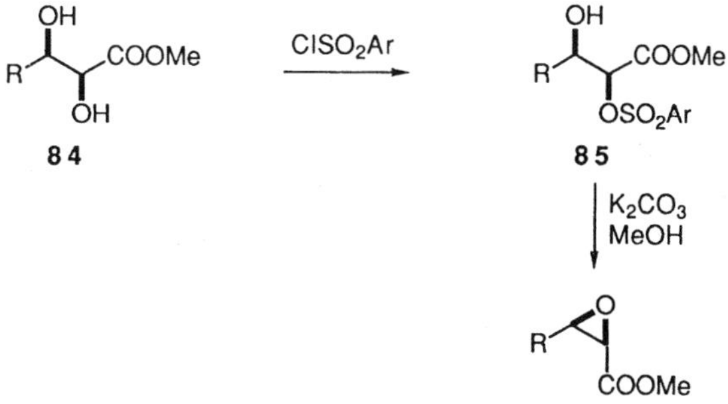

84 85

86

α-sulfonyloxy-α,β-unsaturated ester, and the formation in both cases are kept to a minimum by carrying out the reaction under relatively dilute conditions. Under mild alkaline conditions, the monosulfonates (**85**) are converted to glycidic esters (**86**) in high yield. The nosylates, but not the tosylates, undergo displacement by azide ion to give α-azido-β-hydroxy esters, which are envisioned as precursors to β-hydroxyamino acids [69].

6D.4.3. Regioselective Acetoxyhalogenation and Conversion into Epoxides

1,2-Diols such as **87** are converted in excellent yields [89] into acetoxy chlorides (**88**) by treatment with trimethyl orthoacetate and trimethylsilyl chloride [90] or into acetoxybromides (**89**) with trimethyl orthoacetate and acetyl bromide [91]. These reactions proceed through nucleophilic attack on an intermediate 1,3-dioxolan-2-ylium cation [91] with inversion of configuration. In the presence of an aryl substituent as in **87**, displacement occurs exclusively at the benzylic position. With aliphatic diols such as **90**, the halide is introduced mainly at the less hindered position and acetoxybromides **91** and **92** are formed in a ratio of 7:1. Treatment of the acetoxy halides **88** or **89** under mildly alkaline conditions affords epoxide **93** in 84–87% yield, while the mixture of **91** and **92** is converted to epoxide **94** in 94% yield. Because both steps can be performed in the same reaction vessel, this reaction sequence constitutes an extremely efficient method for the direct conversion of 1,2-diols into epoxides with overall retention of configuration [89].

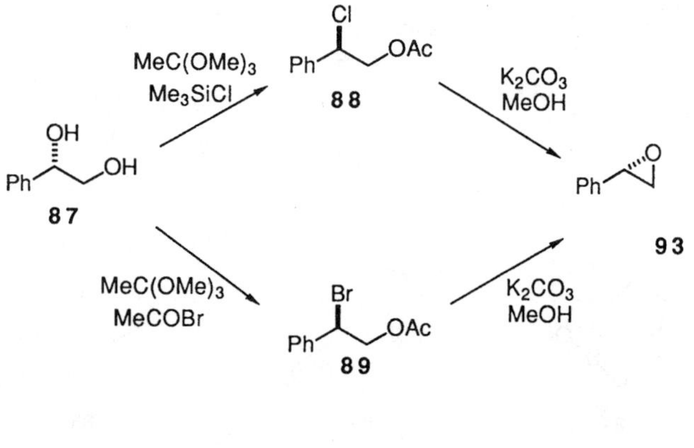

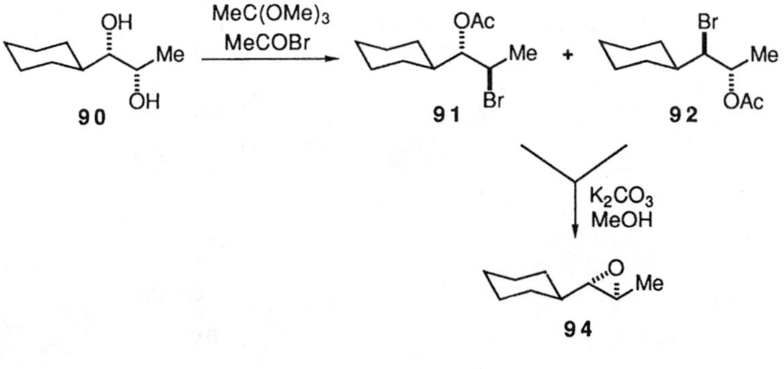

6D.5. CONCLUSION

With the very recent discoveries of the phthalazine class of ligands (**29** and **30**) and the acceleration of osmate ester hydrolysis in the presence of organic sulfonamides, the osmium-catalyzed AD process has reached a level of effectiveness and simplicity unique among asymmetric catalytic transformations. A striking characteristic of this system, which is shared with Mn(III) salen catalyzed asymmetric epoxidation (see Chapter 6.2), is the ability to deliver very high enantioselectivities without the need for an ancillary functional group to act as binding tether. This latter requirement is essential for titanium tartrate catalyzed asymmetric epoxidation (AE, see Chapter 6A) and catalytic asymmetric hydrogenation (see Chapter 1) and has serious consequences for the scope of these two otherwise excellent systems. This point stands out clearly in Chart 6D.3, which compares the outcome of the catalytic AD and AE processes for a closely related family of olefins.

Broad scope and high enantioselectivity are important for any catalytic asymmetric transformation, but they are not sufficient to ensure that a process will be widely used, especially on a commercial scale. To achieve this latter goal the process must also be economical and easy to perform. From this point of view it is hard to imagine a selective catalytic process that could have been more responsive to improvement than has catalytic AD. The adjustable parameters of the catalytic AD reaction were amenable to improvement at virtually every turn and have resulted in a procedure requiring only trace amounts of the expensive alkaloid ligand and the osmium catalyst. The key responsible variables are the large ligand acceleration effect and the binding constant ($L + OsO_4 \rightleftarrows L \cdot OsO_4$). Both these variables are dramatically temperature dependent, such that everything related to the enantioselectivity of the process favors operating

	AD	AE
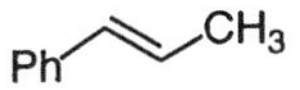 Ph⌒⌒CH$_3$	>95% ee	NR
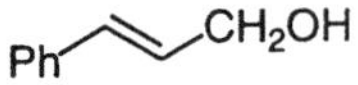 Ph⌒⌒CH$_2$OH	>95% ee	>95% ee
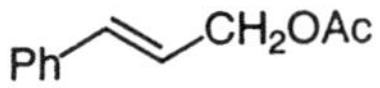 Ph⌒⌒CH$_2$OAc	>95% ee	NR
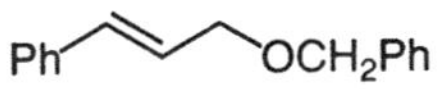 Ph⌒⌒OCH$_2$Ph	>95% ee	NR
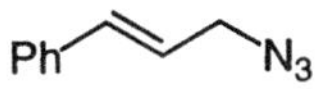 Ph⌒⌒N$_3$	>95% ee	NR
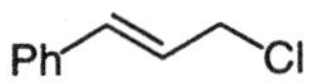 Ph⌒⌒Cl	>95% ee	NR
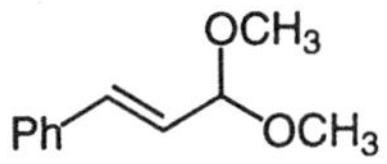 Ph⌒⌒CH(OCH$_3$)OCH$_3$	>95% ee	NR
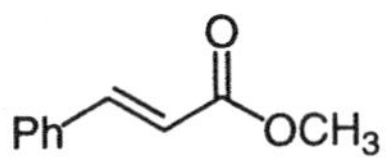 Ph⌒⌒C(O)OCH$_3$	>95% ee	NR

Chart 6D.3. Comparison of asymmetric dihydroxylation with asymmetric epoxidation of a series of olefins.

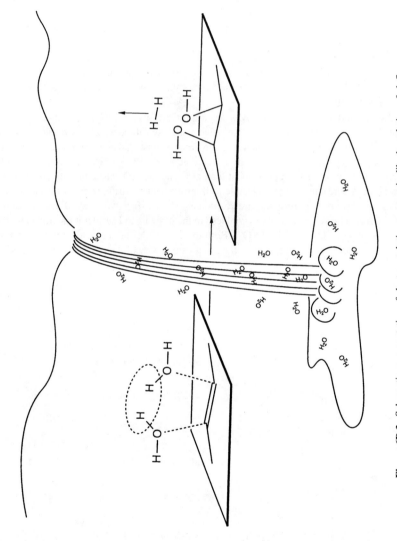

Figure 6D.3. Schematic representation of electrocatalytic asymmetric dihydroxylation of olefins.

at lower temperature. The slow rate of osmate ester hydrolysis prevented use of lower temperature with the more substituted olefins, but now the sulfonamide effect has overcome this limitation.

Last but not least, from the process improvement point of view, a research group at Sepracor led by Yun Gao has developed an extremely effective electrocatalytic version of the catalytic AD process (see Section 6D.2.2) [37]. As symbolized in Figure 6D.3, this electrocatalytic approach to AD appears to be ideal. Enantiomerically pure diols arise from electricity, water, and olefins compounds, with hydrogen gas and a little water over the dam as the only byproducts.

ACKNOWLEDGMENT

We thank David J. Berrisford for the use of his excellent bibliography of the asymmetric dihydroxylation literature.

REFERENCES

1. Criegee, R. *Justus Liebigs Ann. Chem.* **1936**, 75.

2. (a) Schröder, M. *Chem. Rev.* **1980**, *80,* 187. (b) Kolb, H. C.; Van Nieuwenhze, M. S.; Sharpless, K. B. *Chem. Rev.* **1994**, *94*, 2487.

3. Böeseken, J.; van Giffen, J. *Recl. Trav. Chim. Pays-Bas* **1921**, *39*, 183.

4. Sharpless, K. B.; Teranishi, A. Y.; Böackval, J.-E. *J. Am. Chem. Soc.* **1977**, *99*, 3120.

5. Jorgensen, K. A.; Schiott, B. *Chem. Rev.* **1990**, *90*, 1483.

6. For an X-ray crystallographic structure determination of such a dimeric complex, see: Coll in, R.; Griffith, W. P.; Phillips, F. L.; Skapski, A. C. *Biochim. Biophys. Acta* **1973**, *320*, 745.

7. Hofmann, K. A. *Chem. Ber.* **1912**, *45*, 3329.

8. Milas, N. A.; Sussmani. S. *J. Am. Chem. Soc.* **1936**, *58,* 1302.

9. Milas, N. A.; Trepagnier, J. H.; Nolan, J. T., Jr.; Iliopulos, M. 1. *J. Am. Chem. Soc.* **1959**, *81*, 4730.

10. Sharpless, K. B.; Akashi, K. *J. Am. Chem. Soc.* **1976**, *98*, 1986.

11. VanRheenen, V.; Kelly, R. C.; Cha, D. Y. *Tetrahedron Lett.* **1976**, 1973, wherein the invention of this method is attributed to: Schneider, W. P.; Mclntosh, A. V. U.S. Patent 2,769,824 Nov. 6, 1956.

12. Minato, M.; Yamamoto, K.; Tsuji, J. *J. Org. Chem.* **1990**, *55*, 766.

13. Criegee, R.; Marchand, B.; Wannowius, H. *Justus Liebigs Ann. Chem.* **1942**, *550*, 99.

14. For an X-ray crystallographic structure determination of one such monoamine complex, see: Svendsen, J. S.; Markó, I.; Jacobsen, E. N.; Pulla Rao, C.; Bott, S.; Sharpless. K. B. *J. Org. Chem.* **1989**, *54*, 2263.

15. For X-ray crystallographic structure determinations of complexes of this type, see Ref. 19 a nd: Cartwright, B. A.; Griffith, W. P.; Schroder, M.; Skapski, A. C. *J. Chem. Soc., Chem. Commun.* **1978**, 853.

16. Wai, J.S.M.; Markó, I.; Svendsen, J. S.; Finn, M. G.; Jacobsen, E. N.; Sharpless, K. B. *J. Am. Chem. Soc.* **1989**, *111*, 1123.

17. Jacobsen, E. N.; Markó, I.; France, M. B.; Svendsen, J. S.; Sharpless, K. B. *J. Am. Chem. Soc.* **1989**, *111*, 737.

18. Hentges, S. G.; Sharpless, K. B. *J. Am. Chem. Soc.* **1980**, 4263.

19. Pearlstein, R. M.; Blackburn, B. K.; Davis, W. M.; Sharpless, K. B. *Angew. Chem., Int. Ed.* **1990**, *29*, 639.

20. Yamada, T.; Narasaka, K. *Chem. Lett.* **1986**, 131.

21. Tokles, M.; Snyder, J. K. *Tetrahedron Lett.* **1986**, *34*, 3951.

22. Tomioka, K.; Nakajima, M.; Koga, K. *J. Am. Chem. Soc.* **1987**, *109*, 6213.

23. Hirama, M.; Oishi, T.; Ito, S. *J. Chem. Soc., Chem. Commun.* **1989**, 665.

24. Corey, E. J.; Jardine, P. D.; Virgil, S.; Yuen, P.-W.; Connell, R. D. *J. Am. Chem. Soc.* **1989**, *111*, 9243.

25. Masamune, S.; Choy, W.; Petersen, J. S.; Sita, L. R. *Angew. Chem., Int. Ed.* **1985**, *24*, 1.

26. Jacobsen, E. N.; Markó, I.; Mungall, W. S.; Schröder, G.; Sharpless. K. B. *J. Am. Chem. Soc.* **1988**, *110*, 1968.

27. Gredley, M. PCT Int. Appl. WO 89 02,428; *Chem. Abstr.* **1989**, *111,* 173782v.

28. Watson, K. G.; Fung, Y. M.; Gredley, M.; Bird, G. J.; Jackson, W. R.; Gountzos, H.; Matthews, B. R. *J. Chem. Soc., Chem. Commun.* **1990**, 1018.

29. Sharpless, K. B.; Åmberg, W.; Bennani, Y. L.; Crispino, G. A.; Hartung, J.; Jeong, K.-S.; Kwong, H.-L.; Morikawa, K.; Wang, Z.-M.; Xu. D.; Zhang, X.-L. *J. Org. Chem.* **1992**, *57*, 2768.

30. See footnote 6 of Ref. 26.

31. Sharpless, K. B.; Amberg, W.; Beller, M.; Chen, H.; Hartung, J.; Kawanami, Y.; Lübben, D.; Manoury, E.; Ogino, Y.; Shibata, T.; Ukita, T. *J. Org. Chem.* **1991**, *56*, 4585.

32. McLaughlin, A.I.G.; Milton, R.; Perry, K. M. A. *Br. J. Ind. Med.* **1946**, *3*, 183; *Chem. Abstr.* **1946**, *40*, 5841.

33. Nissila, M. *Scand. J. Rheumatol.* **1978**, *Suppl.* 29, 1–44.

34. Hinckley, C. C.; Bemiller, J. N.; Strack, L. E.; Russell, L. D. *"Platinum, Gold, and Other Metal Chemotherapeutic Agents,"* Lippard, S. J. (Ed.), ACS Symp. Series, No. 209, **1983**, 421–437.

35. Kwong, H.-L.; Sorato, C.; Ogino, Y.; Chen. H.; Sharpless, K. B. *Tetrahedron Lett.* 1990, *31*, 2999.

36. Ogino, Y.; Chen, H.; Kwong, H.-L.; Sharpless, K. B. *Tetrahedron Lett.* **1991**, *32*, 3965.

37. Gao, Y.; Zepp, C. M. Sepracor, Inc. Unpublished results.

38. Verpoorte, R.; Schripsema, J.; van der Leer, T. In *The Alkaloids;* Brossi, A. (Ed.); Academic Press: Orlando, FL; **1988**; Vol. 34, pp. 331–398.

39. See Ref. 30, footnote 20.

40. (a) Lohray, B. B.; Kalantar, T. H.; Kim, B. M.; Park, C. Y.; Shibata, T.; Wai, J. S.; Sharpless, K. B. *Tetrahedron Lett.* **1989**, *30*, 2041. (b) Hashiyama, T.; Morikawa, K.; Sharpless, K. B. *J. Org. Chem.* **1992**, *57*, 5069.

41. Shibata, T.; Gilheany, D. G.; Blackburn, B. K.; Sharpless, K. B. *Tetrahedron Lett.* **1990**, *31*, 3817.

42. Kim, B. M.; Sharpless, K. B. *Tetrahedron Lett.* **1990**, *31*, 3003.

43. Ogino, Y.; Chen, H.; Manoury, E.; Shibata, T.; Beller, M.; Lübben, D.; Sharpless, K. B. *Tetrahedron Lett.* **1991**, *32*, 5761.

44. Wang, Z.-M.; Zhang, X.-L.; Sharpless, K. B. *Tetrahedron Lett.* **1993**, *34*, 2267.

45. Oi, R.; Sharpless, K. B. *Tetrahedron Lett.* **1992**, *33*, 2095.

46. Jeong, K.-S.; Sjö, P.; Sharpless, K. B. *Tetrahedron Lett.* **1992**, *33*, 3833.

47. McKee, B. H.; Gilheany, D. G.; Sharpless, K. B. *Org. Synth.* **1991**, *70*, 47.

48. (a) Keinan, E.; Sinha, S. C.; Sharpless, K. B. Unpublished results. (b) Crispino, G. A.; Jeong, K. S.; Kolb, H.; Wang, Z.-M.; Xu, D.; Sharpless, K. B. *J. Org. Chem.* **1993**, *58*, 3785.

49. Rama Rao, A. V.; Gurjar, M. K.; Joshi, S. V. *Tetrahedron: Asymmetry* **1990**, *1*, 697.

50. Wink, D. J.; Kwok, T. J.; Yee, A. *Inorg. Chem.* **1990**, *29*, 5006.

51. (a) Hoffmann, R. W.; Ditrich, K.; Köster, G.; Stürmer, R. *Chem. Ber.* **1989**, *122*, 1783. (b) Stürmer, R. *Angew, Chem., Int. Ed.* **1990**, *29*, 59. (c) Stürmer, R.; Hoffman, R. W. *Synlett* **1990**, 759.

52. (a) Konopelski, J. P.; Boehler, M. A.; Tarasow, T. M. *J. Org. Chem.* **1989**, *54*, 4966. (b) Eid, C. N., Jr.; Konopelski, J. P. *Tetrahedron Lett.* **1991**, *32*, 461.

53. Crosby, J.; Fakley, M. E.; Gemmell, C.; Martin, K.; Quick, A.; Slawin, A. M. Z.; Shahriai-Zavareh, H.; Stoddart, J. F.; Williams, D. J. *Tetrahedron Lett.* **1989**, *30*, 3849.

54. Pini, D.; Iuliano, A.; Rosini, C.; Salvadori, P. *Synthesis* **1990**, 1023.

55. Hirsenkorn, R.; *Tetrahedron Lett.* **1990**, *31*, 7591; **1991**, *32*, 1775.

56. (a) Xu, D.; Crispino, G. A.; Sharpless, K. B. *J. Am. Chem. Soc.* **1992**, *114*, 7570. (b) Park, C. Y.; Kim, B. M.; Sharpless, K. B. *Tetrahedron Lett.* **1991**, *32*, 1003.

57. (a) Crispino, G. A.; Sharpless, K. B. *Tetrahedron Lett.* **1992**, *33*, 4273. (b) Crispino, G. A.; Ho, P.-T.; Sharpless, K. B. *Science* **1993**, *259*, 64. (c) Wang, L.; Sharpless, K. B. *J. Am. Chem. Soc.* **1992**, *114*, 7568. (d) Morikawa, K.; Hashiyama, T.; Sharpless, K. B. Unpublished results. (e) Morikawa, K.; Andersson, P.; Park, J.; Hashiyama, T.; Sharpless, K. B. *J. Org. Chem.* **1992**, *57*, 2768.

58. Fieser, L. F.; Fieser, M. *Steroids*; Reinhold: New York, 1959; pp. 612–632.

59. *Cf.* (a) Cha, J. K.; Christ, W. J.; Kishi, Y. *Tetrahedron* **1984**, *40*, 2247. (b) Dent, W. H., III; Vedejs, E. *J. Am. Chem. Soc.* **1989**, *111*, 6861. (c) Houk, K. N.; Duh, H.-Y.; Wu, Y.-D.; Moses, S. *J. Am. Chem. Soc.* **1986**, *108*, 2754.

60. Annuziata, R.; Cinquini, M.; Cozzi, F.; Raimondi, L.; Stefanelli, S. *Tetrahedron Lett.* **1987**, *28*, 3139.

61. Brimacombe, J. S.; McDonald, G.; Rahman, M. A. *Caroohydr. Res.* **1990**, *205*, 422.

62. Sharpless, K. B.; Morikawa, K.; Kim, B.-M. Unpublished results.

63. (a) Sun, L.-Q.; Zhou, W.-S.; Pan, X.-F. *Tetrahedron: Asymmetry* **1991**, *2*, 973. (b) Zhou, W.-S.; Huang, L.-F.; Sun, L.-Q.; Pan, X.-F. *Tetrahedron Lett.* **1991**, *32*, 6745.

64. Ireland, R. E.; Wipf, P.; Roper, T. D. *J. Org. Chem.* **1990**, *55*, 2284.

65. DeCamp, A. E.; Mills, S. G.; Kawaguchi, A. T.; Desmond. R.; Reamer, R. A.; DiMichele, L.; Volante, R. P. *J. Org. Chem.* **1991**, *56*, 3564.

66. Cooper, A. J.; Salomon, R. G. *Tetrahedron Lett.* **1990**, *31*, 3813.

67. Ikemoto, N.; Schreiber, S. L. *J. Am. Chem. Soc.* **1990**, *112*, 9657.

68. Gao, Y.; Sharpless, K. B. *J. Am. Chem. Soc.* **1988**, *110*, 7538.

69. Fleming, P. R.; Sharpless, K. B. *J. Org. Chem.* **1991**, *56*, 2869.

70. Denmark, S. E. *J. Org. Chem.* **1981**, *48*, 3144.

71. (a) Lowe, G.; Salamone, S. J. *J. Chem. Soc., Chem. Commun.* **1983**, 1392. (b) Lowe, G.; Parratt, M. J. *Bioorg. Chem.* **1988**, *16*, 283.

72. (a) Brandes, S. J.; Katzenellenbogen, J. A. *Mol. Pharmacol.* **1991**, *32*, 391. (b) Liu, A.; Katzenellenbogen, J. A.; VanBrocklin, H. F.; Mathias, C. J.; Welch, M. J. *J. Nucl. Med.* **1991**, *32*, 81.

73. Kim, B. M.; Sharpless, K. B. *Tetrahedron Lett.* **1989**, *30*, 655.

74. Vanhessche, K.; Van der Eycken, E.; Vandewalle, M. *Tetrahedron Lett.* **1990**, *31*, 2337.

75. (a) Machinaga, N.; Kibayashi, C. *Tetrahedron Lett.* **1990**, *31*, 3637. (b) Machinaga, N.; Kibayashi, C. *J. Chem. Soc., Chem. Commun.* **1991**, 405. (c) Machinaga, N.; Kibayashi, C. *J. Org. Chem.* **1991**, *56*, 1386.

76. Lohray, B. B.; Gao, Y.; Sharpless, K. B. *Tetrahedron Lett.* **1989**, *30*, 2623.

77. (a) van der Klein, P.; Boons, G.J.P.H.; Veeneman, G. H.; van der Marel, G. A., van Boom, J. H. *Synthesis* **1990**, 311. (b) van der Klein, P.: de Nooy, A.E.J.; van der Marel, G. A.; van Boom, J. H. *Synthesis* **1991**, 347.

78. Bates, R. W.; Fernández-Moro, R.; Ley, S. V. *Tetrahedron Lett.* **1991**, *32*, 2651.

79. Oi, R.; Sharpless, K. B. *Tetrahedron Lett.* **1991**, *32*, 999.

80. Burk, M. J. *J. Am. Chem. Soc.* **1991**, *113*, 8518.

81. Oi, R.; Sharpless, K. B. *Tetrahedron Lett.* **1991**, *32*, 4853.

82. Kim, B. M.; Sharpless, K. B. *Tetrahedron Lett.* **1991**, *31*, 4317.

83. Lohray, B. B.; Ahuja, J. R. *J. Chem. Soc., Chem. Commun.* **1991**, 95.

84. Gao, Y.; Zepp, C. M. *Tetrahedron Lett.* **1991**, *32*, 3155.

85. Alker, D.; Doyle, K. J.; Harwood, L. M.; McGregor, A. *Tetrahedron: Asymmetry* **1990**, *1*, 877.

86. Baldwin, J. E.; Spivey, A. C.; Schofield, C. J. *Tetrahedron: Asymmetry* **1990**, *1*, 881.

87. Lowe, G.; Reed, M. A. *Tetrahedron: Asymmetry* **1990**, *1*, 885.

88. White, G. J.; Garst, M. E. J. *Org. Chem.* **1991**, *56*, 3177.

89. Kolb, H.; Sharpless, K. B. *Tetrahedron* **1992**, *48*, 10515.

90. (a) Newman, M. S.; Olson, D. R. *J. Org. Chem.* **1973**, *38*, 4203. (b) Dansette, P.; Jerina, D. M. *J. Am. Chem. Soc.* **1974**, *94*, 1224.

91. Harmann, W.; Heine, H.-G.; Wendisch, D. *Tetrahedron Lett.* **1977**, 2263.

6E

RECENT ADVANCES IN ASYMMETRIC DIHYDROXYLATION AND AMINOHYDROXYLATION

CARSTEN BOLM, JENS P. HILDEBRAND, KILIAN MUÑIZ
Institut für Organische Chemie der RWTH Aachen, Prof.-Pirlet-Str. 1, D-52056 Aachen, Germany

6E.1. ASYMMETRIC DIHYDROXYLATION

6E.1.1. Introduction

The catalytic asymmetric dihydroxylation (AD) of olefins has emerged as one of the most general methods in asymmetric catalysis and has become a powerful tool for synthetic chemists today. After the first edition of this book appeared (see Section 6D.1), substantial improvements have been achieved in both enhancing enantioselectivity and in understanding the mechanistic basis [1]. The AD has also been successfully applied to other substrate classes, and the as-yet-unsolved debate about the precise reaction path in the initial steps of the AD has led to a more mechanism-based design of new ligands. Due to its high applicability the AD has been used to construct a number of synthetically valuable building blocks and key intermediates in natural product synthesis.

An important improvement with regard to both substrate tolerance and enantioselectivity of the AD was the introduction of the anthraquinone-bridged ligands $(DHQ)_2AQN$ and $(DHQD)_2AQN$ [2]. These ligands are easily accessible from phthalic anhydride and 1,4-difluoro benzene, and in many aspects their catalytic behavior reflects that of the standard bis-cinchona alkaloids described in Section 6D.1.

For a number of substrate classes these new AQN ligands are superior to the phthalazine-derived ones, that is, most vinylic and allylic compounds are dihydroxylated with significantly higher enantioselectivities. For example, using $(DHQD)_2AQN$ in the AD of allyl tosylate gives

Catalytic Asymmetric Synthesis, Second Edition, Edited by Iwao Ojima
ISBN 0-471-29805-0

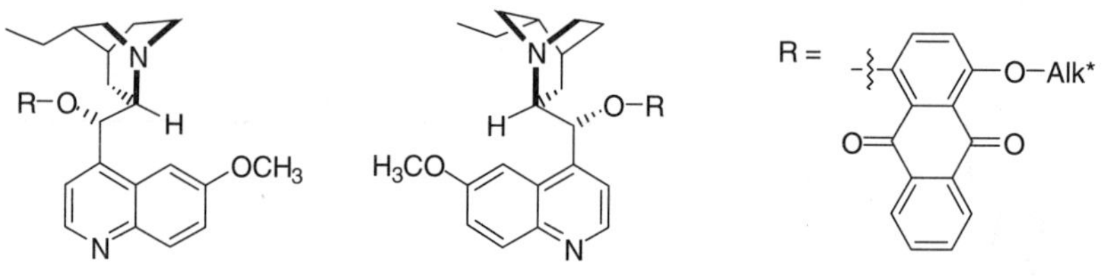

| DHQD Ligands | DHQ Ligands | AQN Ligands |
| Dihydroquinidine (R = H) | Dihydroquinine (R = H) | Alk* = DHQ or DHQD |

Chart 6E.1.

the corresponding diol with 83% ee (94% ee after recrystallization) whereas the same reaction with phthalazine-derived (DHQD)$_2$PHAL affords the product with only 40% ee [2].

The use of the AQN ligands in the AD of alkyl-substituted terminal olefins and di- or trisubstituted alkenes improves the enantioselectivity only slightly compared to reactions with (DHQD)$_2$PHAL. It is interesting, however, that the AD proceeds at a higher rate in certain cases. For example, the dihydroxylation of 1-decene is about 10% faster [2]. Thus, using the AQN ligands overcomes several limitations of the earlier ligand generations.

In general, (DHQ)$_2$AQN and (DHQD)$_2$AQN are most-suited for terminal and some branched aliphatic olefins whereas PYR ligands remain the best for alkenes bearing sterically demanding groups. For most *cis*-1,2-disubstituted olefins and for aromatic substrates, IND ligands and the DPP system, respectively, give the best results [1].

AD reactions of oligoprenyl compounds [3] can be problematic because of position selectivity (see also Section 6D.1.3.4). For example, by using pyridazine-bridged **7** as ligand, 2-*E*-geranyl acetate (**1**) was oxidized to give (*R*)-6,7-dihydroxy derivative **2** with >95% ee in 76% yield [4], whereas 2,6-*E*,*E*-farnesyl acetate (**3**) gave diol **5** in only a 34% yield (92% ee) along with remaining starting material, tetraols, and other regioisomeric diols [5]. Previously, Sharpless had described a slight position selectivity in the oxidation of squalene (**4**) with (DHQD)$_2$PHAL to give **6** as the major diol with 96% ee [3].

To improve the position selectivity in the AD of oligoprenyl compounds bis-cinchona alkaloid ligand **8** was introduced by Corey [5,6]. Its design was based on the [3+2]-cycloaddition model for the AD mechanism, which will be discussed in Section 6E.1.2. The two 4-heptyl ether substituents of the quinolines are supposed to assist fixation of the substrate in the binding cleft. Additionally, the *N*-methylquinuclidinium unit and the linking naphthopyridazine were introduced to rigidify the osmium tetroxide complex of **8** [6].

AD of 2,6-*E*,*E*-farnesyl acetate (**3**) with **8** as ligand occurred selectively at the terminal double bond to give **5** with 96% ee (position selectivity of about 120:1). In a special solvent system consisting of *tert*-butanol-water-methylcyclohexane, at ca. 50% conversion of squalene (**4**) diol **6** was obtained in 32% yield with 90% ee. Here, the position selectivity was 8:1 [5].

Mono-cinchona alkaloid **9** is another ligand that was designed on the basis of Corey's mechanistic model. It was expected to adopt a U-shaped geometry and to have similar properties as **7**. Steric requirements would lead to a *trans*-arrangement of the *tert*-butyl group and the pyridazine linker forcing the latter to adopt a perpendicular position with regard to the anthryl moiety [7]. Indeed, **9** performed similar to **7** affording the diols of stilbene and styrene with >98% ee and 91% ee, respectively.

The scope of the AD was further broadened by the discovery that allyl 4-methoxy-benzoates are excellent substrates for this reaction [8]. The 4-methoxybenzoyl substituent is supposed to interact with the aryl groups of the ligand-binding pocket and to suppress acyl migration from

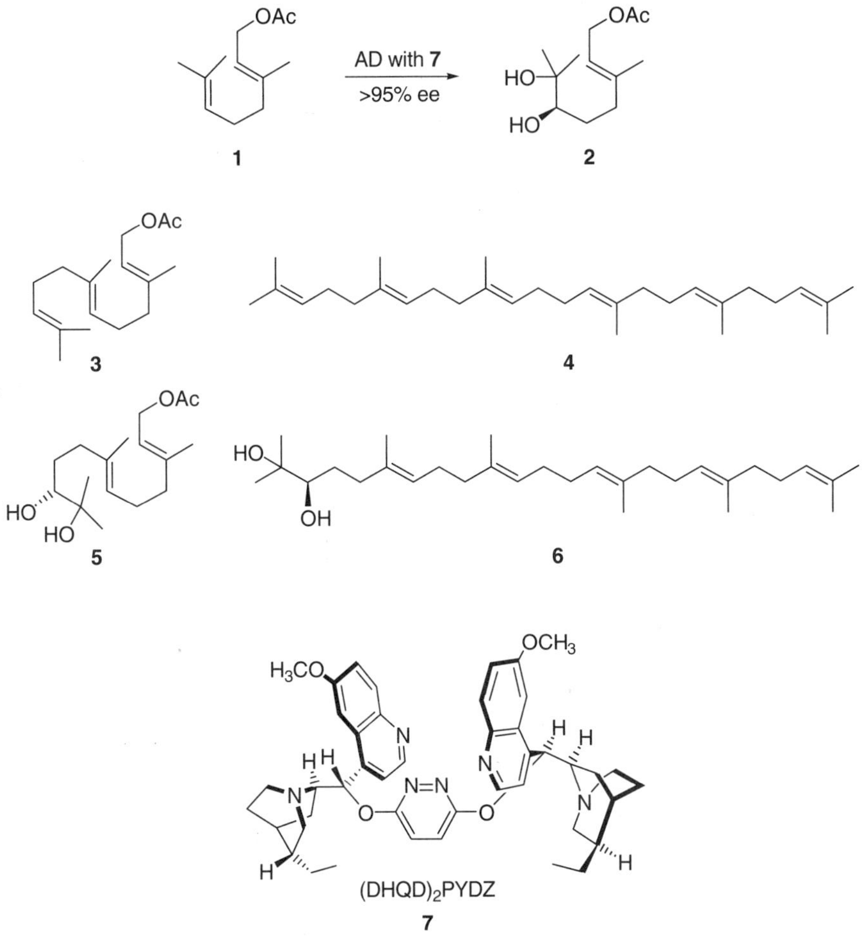

Scheme 6E.1.

one oxygen to another as observed with acetyl- or benzoyl-containing substrates. Excellent enantiomeric excesses were achieved in dihydroxylations of *trans*-1,2-disubstituted and trisubstituted alkenes, and, surprisingly, even substrates like terminal and 1,1-disubstituted allyl 4-methoxybenzoates gave diols with high enantioselectivity. This discovery was useful for the synthesis of (−)-ovalicin (see Section 6E.2.3).

Aryl homoallyl ketones and 4-methoxy phenyl ethers are also good substrates for the AD [8], whereas the structurally related allyl amides and thioesters give products with insufficient enantioselectivity. Homoallylic 4-methoxy benzoates perform relatively poor, which is consistent with Corey's proposed mechanistic model [9].

Various studies focused on applying the AD for kinetic resolutions of racemates. One of the most impressive examples was described by Hawkins, who discovered a kinetic resolution of a chiral fullerene [10]. The isolation of both enantiomers of this carbon allotrope became possible by a kinetic resolution using pseudo-enantiomeric ligands **10** and **11**. Thus, performing the AD in the presence of **10** yielded (−)-C_{76} and the OsO_4 adduct of (+)-C_{76}. Reductive removal of the

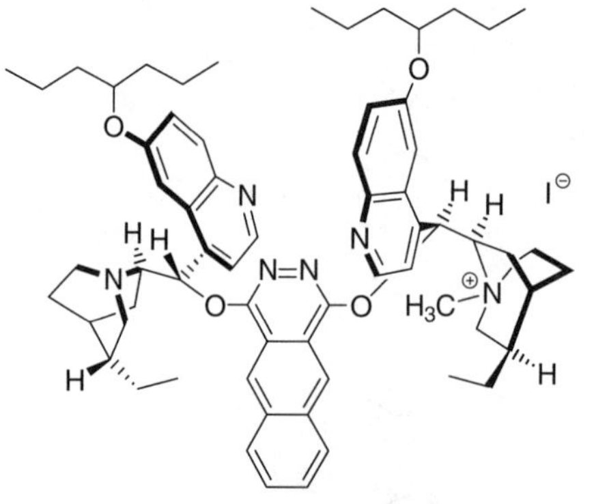

8 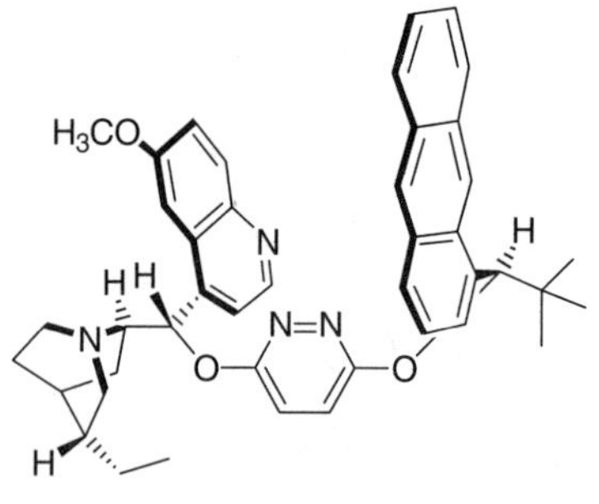   **9**

metal oxide from the latter with stannous dichloride released (+)-C_{76}. This dihydroxylation is particularly interesting considering the fact that the fullerene contains 30 types of double bonds that *a priori* could all react with the osmium reagent.

Moreover, kinetic resolutions of racemic allylic acetates [11], olefins [12], and allyl 4-methoxybenzoates [13] have been investigated [14]. In the resolution of allylic acetates a selectivity factor (S) of up to 25 was reached with a bridged C_2-symmetric bis-cinchona alkaloid ligand. Axially dissymmetric alkenes were oxidized to their corresponding diols with ratios of rate constants for the fast- versus slow-reacting enantiomers of up to 32 [12]. The use of anthryl-substituted mono-cinchona alkaloid ligand **9** in the kinetic resolution of allyl 4-methoxybenzoates gave relative rate factors of up to 79 [13].

With the intention to use AD reactions in combinatorial chemistry, polymer-attached substrates were dihydroxylated [15–18]. In general, these transformations worked well and products with high enantiomeric excesses were obtained. The selectivity trends remained the same as in solution.

Multiple sequential AD gave access to interesting dendritic structures as shown by Sharpless [19]. Thus, dihydroxylations of 3- and 4-vinylbenzyl chlorides followed by coupling of the products with other chiral diols in a double-exponential manner afforded novel dendrimers with chiral polyether subunits.

6E.1.2. Mechanistic Considerations

In September 1997, *Chemical and Engineering News* summarized the ongoing discussion about the precise mechanism of the initial steps of the osmium-catalyzed olefin dihydroxylation in an

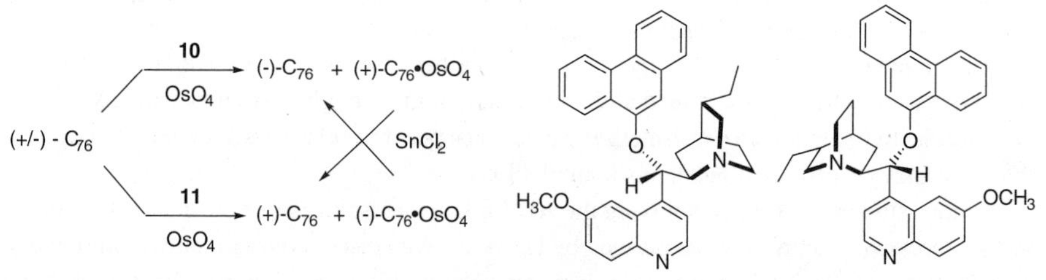

10   **11**

Scheme 6E.2.

Stepwise [2+2] Mechanism

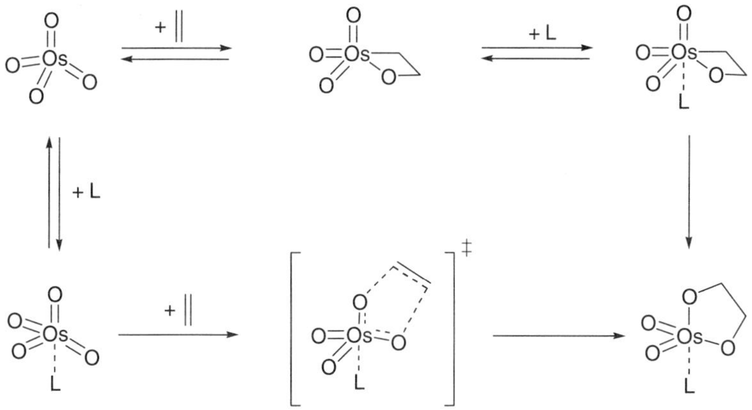

Concerted [3+2] Mechanism

Scheme 6E.3.

article entitled "A Reaction Under Scrutiny" [20]. In fact, the first mechanistic proposal dates back to the early 1920s when Böeseken stated that the osmylations might proceed in a similar fashion as permanganate oxidations of alkenes to diols [21]. Thus, he and later Criegee [22] proposed that the dihydroxylation was initiated by a direct and concerted cycloaddition of the olefin with two oxygens of osmium tetroxide. Since then a vast amount of information on mechanistic details has been collected, and the discussion has focused on the question whether the AD proceeds via a concerted [3+2] mechanism or a stepwise [2+2] reaction pathway (see Section 6D and [1]).

The following section highlights facts supporting each mechanism. To date no final decision about an exclusive involvement of one of these two models seems possible, and the reader is encouraged to evaluate the data independently.

The [2+2] Mechanism Already in 1977 Sharpless proposed a stepwise [2+2] mechanism for the osmylation of olefins in analogy to related oxidative processes with d^0-metals such as alkene oxidations with CrO_2Cl_2 [23, 24]. Metallaoxetanes [25] were suggested to be formed by suprafacial addition of the oxygens to the olefinic double bond. In the case of osmylation the intermediate osmaoxetane would be derived from an olefin–osmium(VIII) complex that subsequently would rearrange to the stable osmium(VI) ester.

To explain the high enantioselectivity of the AD with the PHAL ligand in such a stepwise process, Sharpless proposed the formation of arrangement **12** with an L-shaped binding pocket that is built up by the phthalazine and one of the methoxyquinolines [26].

Arrangement **12** is particularly well-suited for olefinic substrates to fit into, especially if additional interactions from the olefin substituents help stabilize the resulting conformation and the transition state for osmaoxetane formation [27]. Responsible factors for the excellent enantioselectivities in the AD are, first, favorable interactions between one of the osmaoxetane

12   **13**

groups and the methoxyquinoline unit in **13**, and, second, repulsions between the hydrogens of the metallaoxetane and those of the binding cleft, which destabilize the transition state. The observed ligand acceleration (see below) is proposed to arise from an attractive π-stacking between substrate and binding domain, which is particularly pronounced in dihydroxylations of aromatic substrates.

One of the most striking arguments in favor of a stepwise mechanism was obtained from variable-temperature studies of stoichiometric osmylation reactions [28]. Here, Sharpless found Eyring-type diagrams that showed two linear areas and one characteristic inversion point for every catalyst olefin combination. The finding of non-linear behavior in these modified Eyring plots convincingly points toward a path involving a reaction intermediate like an osmaoxetane. The appearance of inversion points is well-known for a number of asymmetric catalytic transformations [29], and it indicates that the reaction path includes at least two enantioselective steps that differ in their temperature-dependence.

Evidence for a [2+2]-type mechanism was provided by Gable who reported on related alkene oxidations with Cp^*ReO_3 [30]. In this study, the equilibrium between alkene oxidation by Cp^*ReO_3 and alkene extrusion was investigated, and activation parameters for both processes were obtained. It was found that the level of olefin strain, which has a large effect on the activation enthalpy of the diolate formation, did not substantially influence the activation enthalpy of olefin extrusion. This finding and further evidence from other investigations, including rate measurements, favor diolate formation via a rhenaoxetane intermediate. Given the similarities between osmium and rhenium in oxidation chemistry, it could be concluded that a closely related pathway should also operate in osmylation reactions.

The [3+2] Mechanism In 1993, Corey described his detailed investigations on the origin of the enantioselectivity in the AD [31] by using pyridazine-bridged bis-cinchona alkaloid **7** as ligand [32]. First, the involvement of a μ-oxo-bridged bis-osmium species was suggested [31–33], but later, key experiments using ligand **14** excluded this intermediate from consideration. The two nitrogens of the quinuclidines in **14**, which represents an adipate bridged analog of **7**, appeared to be too distant from each other to allow the reaction to proceed via the proposed μ-oxo-bridged bis-osmium intermediate [34].

Instead, evidence now favors the intermediacy of the highly organized mono-osmium complex **15**. Various factors could contribute to its excellent enantioselectivity, the most important one being the formation of a binding site consisting of the two methoxy quinolines and the pyridazine linker. As confirmed by X-ray analysis and NMR studies [35,36], this pocket adopts a U-shaped conformation and is capable of perfectly binding aromatic substrates such as styrene through attractive interactions with the methoxyquinolines. In contrast, bulky

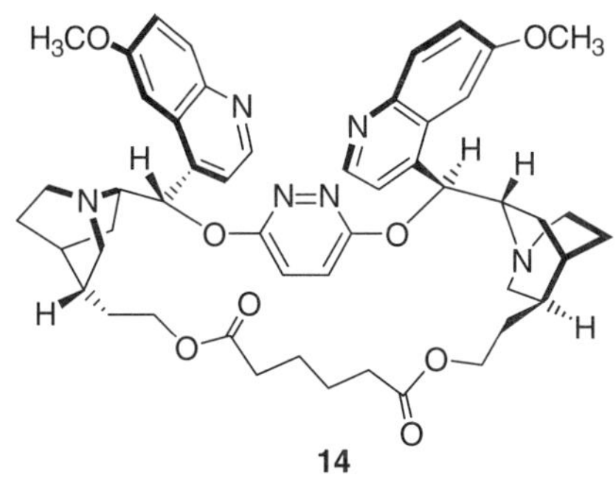

14

substrates like *tert*-butyl ethylene cause severe steric constraints in the binding domain, resulting in low enantioselectivity.

Coordination of the olefin to osmium results in additional substrate-catalyst binding and leads to arrangement **16** with one axial and one equatorial oxygen in proximity to the olefinic double bond [37]. Consequently, a minimal motion pathway from this arrangement via a [3+2]-cycloaddition would directly produce the pentacoordinate osmium(VI) ester in the energetically most-favorable geometry.

Corey also pointed out that **16** reflects the transition-state of an enzyme–substrate complex. Its formation was later supported by the observation of Michaelis–Menten-type kinetics in dihydroxylation reactions and in competitive inhibition studies [37]. This kinetic behavior was held responsible for the non-linearity in the Eyring diagrams, which would otherwise be inconsistent with a concerted mechanism. Contrary, Sharpless stated that the observed Michaelis–Menten behavior in the catalytic AD would result from a step other than osmylation. Kinetic studies on the stoichiometric AD of styrene under conditions that replicate the organic phase of the catalytic AD had revealed that the rate expression was clearly first-order in substrate over a wide range of concentrations [38].

Sharpless, Singleton, and Houk obtained additional data favoring a [3+2] mechanism from studies involving experimental kinetic isotope effects (KIE) and transition-structure/KIE calculations [39]. Their results predict a highly symmetrical transition-state and a [3+2] cycloaddition as the rate-determining step instead of a rate-limiting ring expansion of an osmaoxetane to the osmate ester [40].

Corey collected further evidence against a [2+2] mechanism by investigating the rate acceleration in stoichiometric dihydroxylations with chiral 1,2-diamines [36]. The X-ray structure of a highly reactive heptacoordinate—formally 20-electron bisamine-OsO$_4$ com-

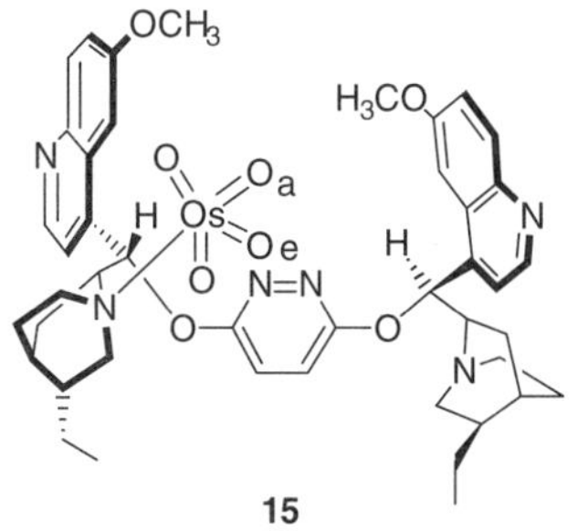

15 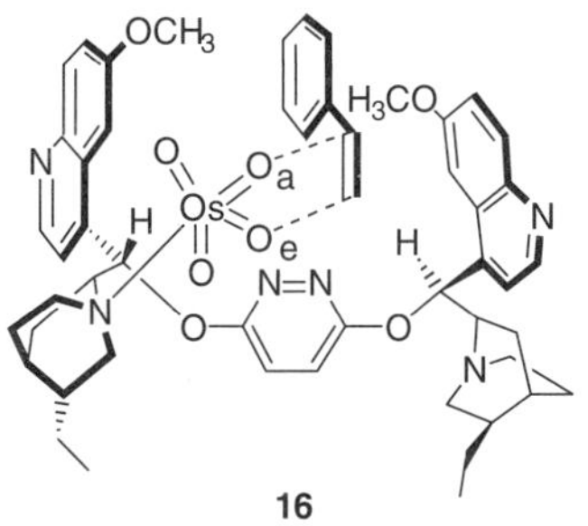   **16**

plex—was obtained, and a pathway involving a rate-limiting olefin π-complexation to osmium and subsequent fast rearrangement to a dioxo-osmate ester was proposed. On the basis of both, the observed absolute stereochemical course in transformations with various diamines [41] and quantum mechanical calculations [42] a reaction path via a [2+2] cycloaddition was excluded in this system.

Evidence for a possibility that both the [2+2] and the [3+2] mechanism might well be involved in the AD process was detailed by Sharpless [38]. Nonlinear Hammett plots for the AD in the presence of DHQD-CLB revealed that under the usual AD conditions a metallaoxetane pathway is favored, whereas for certain substrate ligand combinations a [3+2] mechanism is evidenced.

Ligand Acceleration and Non-Linear Effects One of the most intriguing characteristics of the catalytic asymmetric dihydroxylation of olefins is the significant increase in reaction rate caused by the ligand. Thus, the AD (besides the asymmetric epoxidation of allylic alcohols discovered by Sharpless; see Section 6A) became the most prominent example of a ligand-accelerated catalysis (LAC) [43]. In general, if the presence of a ligand modifies the reaction rate of a catalytic process two cases must be distinguished: (i) the ligand accelerates the rate of product formation ($k_1 > k_0$) or (ii) the ligand slows it down ($k_1 < k_0$). The LAC concept is illustrated below for the osmium tetroxide addition to an olefin.

The ligand acceleration is particularly useful in catalysis with *chiral* ligands. Here, this phenomenon helps a stereoselective reaction mode to dominate over competing unselective pathways, leading to a highly efficient asymmetric catalysis.

Ligand acceleration in the reaction of olefins with OsO_4 has been known since the 1930s from observations by Criegee [22], who reported that the addition of tertiary amines including pyridine led to a significant rate increase. This effect—termed "Criegee effect"—is also observed in the AD with bis-cinchona alkaloid ligands in both the catalytic and the stoichiometric versions [43].

Although the ligand-acceleration phenomenon positively influences the reaction rate of osmate ester formation, the turnover of the catalytic AD can be retarded, or even blocked, in cases where the hydrolysis of the intermediate osmate ester is the rate-limiting step. For obtaining a high overall reaction rate it is therefore important to establish a fine-tuned system that includes appropriate rates for all individual steps of the catalytic cycle. To measure the rate constants that depend only on the ligand-acceleration effects and not on the rate-determining osmate ester hydrolysis, Sharpless investigated the dihydroxylation of a variety of olefins under pseudo first-order conditions, that is, the reaction proceeded completely on the ligand-accelerated pathway under these conditions (Table 6E.1).

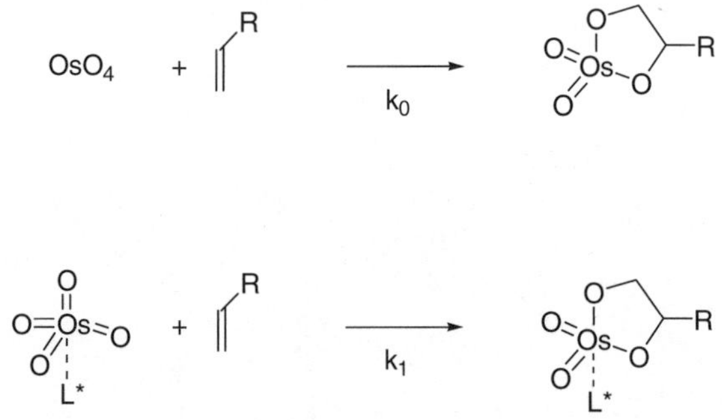

Scheme 6E.4.

TABLE 6E.1. Ratio of Rate Constants k_c/k_0 in *tert*-Butanol at 25°C for the Osmylation of Several Terminal Olefins by Using Various Alkaloid Ligands (k_c = rate as ligand concentration $\to \infty$, k_0 = rate with no ligand present) [43]

Ligand	![naphthyl alkene]	![styrene]	C_8H_{17} ![alkene]	![cyclohexyl alkene]
(DHQD)$_2$PHAL	14771	4182	770	341
DHQD-CLB	791	622	238	139
quinuclidine	61	75	55	—

The following facts can be deduced from Sharpless' study: a) the ligands have a substantial accelerating effect on the reaction rate; b) the bis-cinchona alkaloids are uniformly more effective than the monodentate quinuclidine; and c) the increase in rate is substrate-dependent.

The ligand-acceleration effect in the catalytic AD can also reach the maximum enantioselectivity with a much lower amount of ligand than necessary to reach the onset of rate saturation. Consequently, very low ligand-to-metal loadings can be used without loss of selectivity. A very illustrative description of this behavior was used by Sharpless [43]: "Expressed in a different way, in a mixture of 10,000 molecules of stilbene, 20 molecules of OsO$_4$ and 1 molecule of chiral ligand, only 400 stilbene molecules are dihydroxylated by unbound OsO$_4$. The remaining 9600 stilbene molecules are enantioselectively oxidized through the OsO$_4$/ligand-assisted process."

During the course of his studies of non-linear effects in asymmetric catalysis [44], Kagan investigated the influence of competing pseudo-enantiomeric cinchona alkaloid ligands in AD reactions [45]. As a test system the AD of 2,2'-dibromo stilbene in the presence of (DHQ)$_2$PHAL and (DHQD)$_2$PHAL was chosen. Both catalysts gave the corresponding diols with 98% ee and opposite absolute configurations [(S,S)- and (R,R)-diol, respectively]. Next, catalytic dihydroxylations in the presence of mixtures of these pseudo-enantiomeric ligands were studied. Assuming identical reaction rates and negligible mutual interactions between catalysts, a *linear* correlation between the enantiopurity of the product and the composition of the mixture of the two ligands should result. In fact, only a minor deviation from linearity was detected, indicating a slight superiority in rate for the (DHQD)$_2$PHAL-based catalyst (Figure 6E.1, left).

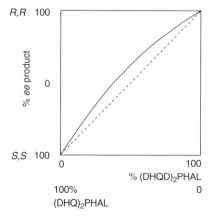 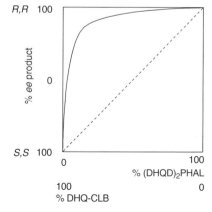

Figure 6E.1.   Plots indicating the relationship between the ee of the product and the catalyst composition.

However, when a mixture of (DHQD)$_2$PHAL and the first-generation ligand DHQ-CLB was used in the catalysis, a substantial departure from linearity was found (Figure 6E.1, right). Apparently, the catalyst derived from (DHQD)$_2$PHAL is acting at a much higher rate. Even a 1:9 mixture of (DHQD)$_2$PHAL and DHQ-CLB afforded the (R,R)-configuration product with 77% ee.

6E.1.3. Polymer-Supported Asymmetric Dihydroxylations

Owing to the high cost of osmium salt and ligand, efficient and simple recovery of both is of major interest for large-scale applications and industrial processes. Thus, to facilitate the reuse of the alkaloids, the standard AD ligands have been attached to organic polymers or inorganic support [46,47]. In heterogeneous systems, the ligands can easily be recovered by simple filtration. This concept generally suffers from a severe decrease in efficiency as indicated by lower enantioselectivity and catalyst activity. Careful optimization is then required to enable a successful adoption of the original AD systems.

As an alternative, the use of soluble supported ligands has emerged. This approach is advantageous in a sense that most characteristics of the original homogeneous AD are maintained. After the reaction, ligand recovery is easily accomplished by precipitation upon addition of an appropriate solvent followed by filtration.

The first example of the use of a polymer-bound cinchona alkaloid in the AD was described in 1990 by Sharpless [48,49]. The polymer was readily obtained by radical co-polymerization of 9-(p-chlorobenzoyl)quinidine acrylate with acrylonitrile. First applications in dihydroxylations of trans-stilbene using NMO as co-oxidant yielded products with enantioselectivities in the range of 85–93% ee. It is interesting that a repetitive use of the polymer was possible without great loss of reactivity, indicating that the metal was retained in the polymeric array.

Similar systems were reported by Salvadori [50] and Lohray [51], who prepared different polyacrylonitrile- and polystyrene-supported 9-O-acylquinine derivatives. However, application of these systems afforded products with significantly lower enantiomeric excesses. In the case of Lohray's ligands, reuse of the polymeric ligands led to a decrease in enantioselectivity, and addition of osmium salt was necessary to maintain the catalytic activity. Despite Lohray's original report [51], one of his polymeric ligands was found by Song to be excellent for the oxidation of trans-stilbene with K$_3$[Fe(CN)$_6$] as secondary oxidant [52]. Later, these results were critically evaluated by Sherrington [53].

A major development in this area was brought about by the invention of crosslinked polystyrene-supported 9-(p-chlorobenzoyl)quinine ligands **17** [54] and **18** [55]. The salient feature of this invention is the connection of the quinine unit to the polymer backbone through a sterically undemanding spacer. Thereby, the quinuclidine, which in catalysis coordinates to osmium, is free of steric interaction with the polymeric side chain. Dihydroxylation of trans-stilbene in the presence of **17** and NMO as co-oxidant gave stilbene diol with 87% ee. However, changing the terminal oxidant to K$_3$[Fe(CN)$_6$] led to full inhibition of the reaction. This result was explained by a possible collapse of the polymer in the required protic solvent, which prevented substrate penetration.

Introducing a polymeric backbone containing additional free hydroxyl groups like in **18** could suppress this inhibitory effect due to the ability of the polymer to swell in polar protic solvents. Thus, by using **18** the dihydroxylation of several olefins became highly efficient yielding the desired diols with up to 95% ee. In agreement with the results reported for the original homogeneous AD, use of NMO was generally less-efficient than that of K$_3$[Fe(CN)$_6$].

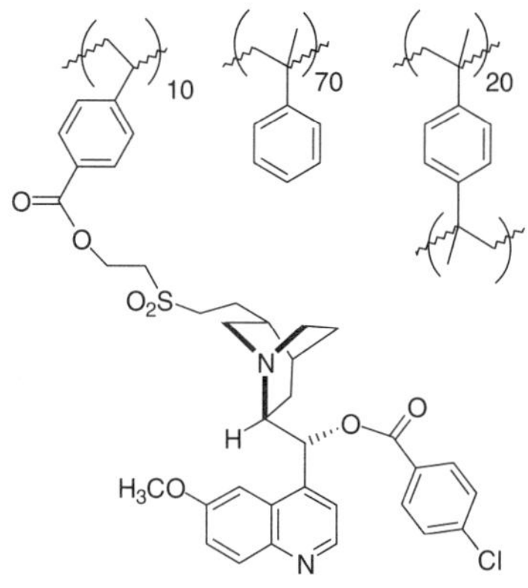

17 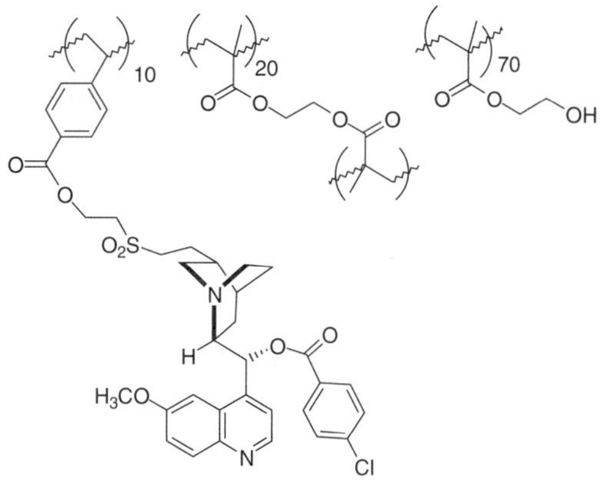   **18**

In 1994, Lohray introduced the first polymer-supported (DHQ)$_2$PYDZ ligand [56]. Being very active, it gave stilbene diol with 98% ee in the oxidation of *trans*-stilbene at room temperature within 20 h. This result is comparable with that obtained in the original homogeneous system (99% ee).

To date, the best ligand is polymer-bound (DHQD)$_2$PHAL **19**, reported by Salvadori [57]. Enantioselectivities in the range of 91~99% ee have been achieved with the use of this ligand and K$_3$[Fe(CN)$_6$] as secondary oxidant. After the reaction, **19** could be recovered quantitatively by filtration, although a loss of osmium during this operation required a further addition of 0.2

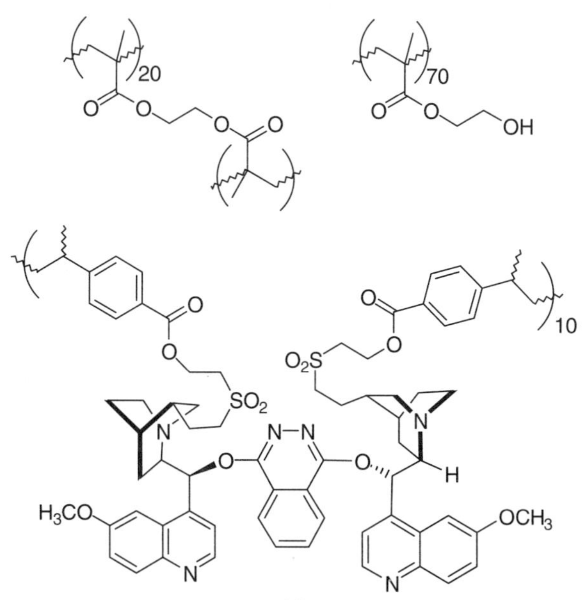

19

20

mol % of osmium salt to prevent a rate decrease [58]. Further contributions in this area include various other polymer-supported PHAL and PYR ligands [59,60].

In 1996, Lohray reported on the first AD with cinchona alkaloids immobilized on silica gel [61]. This heterogeneous system was expected to be superior to other systems because the cinchona alkaloids are located on the surface of the silica exposed to the reactants. Thus, an encapsulation in the polymer matrix should occur to a much smaller extent. The first catalytic attempts, however, resulted only in unsatisfactory enantioselectivities, which were even lower than those reported for the systems with purely organic polymers. The major breakthrough was achieved by Song, who applied silica gel-supported (DHQ)$_2$PHAL **20** to AD [62]. With this ligand, identical enantioselectivities as those with the original Sharpless system were obtained for a variety of aromatic di- and trisubstituted olefins. However, reuse of the recovered catalyst proved problematic, and a loss of catalytic activity was still observed. This loss could be overcome only by addition of a further amount of OsO$_4$, indicating a loss of metal during the workup.

A different approach was introduced by Bolm [63]. Whereas in the ligands described above the solid support was attached to the alkaloid through modification of the olefinic double bond of quinuclidine, in **21** the nitrogen-containing heterocycle was used for the binding to porous glass (**21a**) and silica gel (**21b**). Thereby, a minimization of negative steric interactions between the quinuclidine part of the ligand and the polymer backbone was expected. In accordance with this assumption, an excellent enantioselectivity (97% ee) was achieved in the oxidation of styrene with **21a** [63]. In addition, the system was found to be active under the given conditions for four consecutive runs without loss of enantioselectivity. A slight drop in enantioselectivity was observed in the subsequent cycles, which was attributed to a partial hydrolysis of the ester linker under the slightly basic reaction conditions. In fact, replacement of the ester linker by an ether functionality improved the stability. Thus, in ADs of styrene and *trans*-stilbene with ligand **21b** optically active diols with excellent enantiopurities (98 and 99% ee, respectively) were obtained, and the ligand proved stable during several consecutive runs. However, to maintain this catalytic activity the addition of a further 1 mol % of osmium salt was necessary prior to reuse of the recovered ligand.

Finally, the AD of terminal aliphatic olefins was accomplished by the use of silica-bound PYR ether **22** [63b]. For example, the oxidation of 1-decene provided the corresponding diol with 84% ee. This result is well-comparable with the reported 87% ee for the AD with the original homogeneous system.

Another recent development in the area of polymer-supported ligands has been the introduction of polyethylene glycol (PEG) tethers. Although soluble PEG-modified ligands have been known since the pioneering work by Bayer and Schurig [64], it took until 1996 that the first PEG-modified alkaloid was used in AD oxidation chemistry by Janda [65]. In this case, alkaloids were bound to polyethylene glycol monomethyl ethers (MeO-PEG) of an average molecular

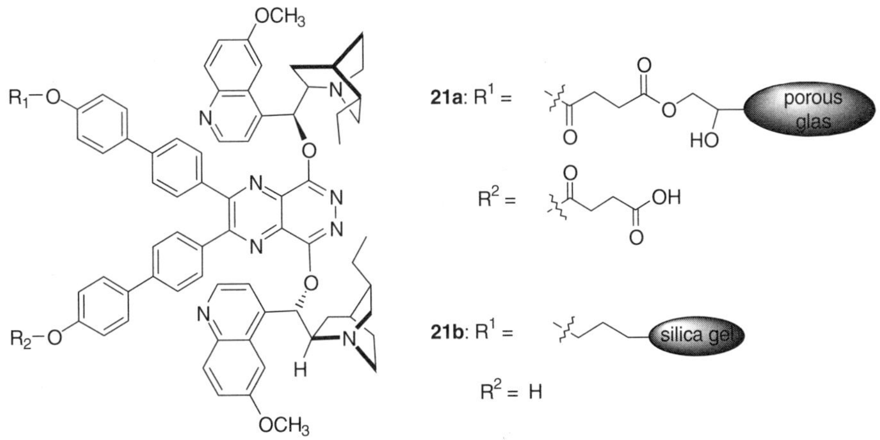

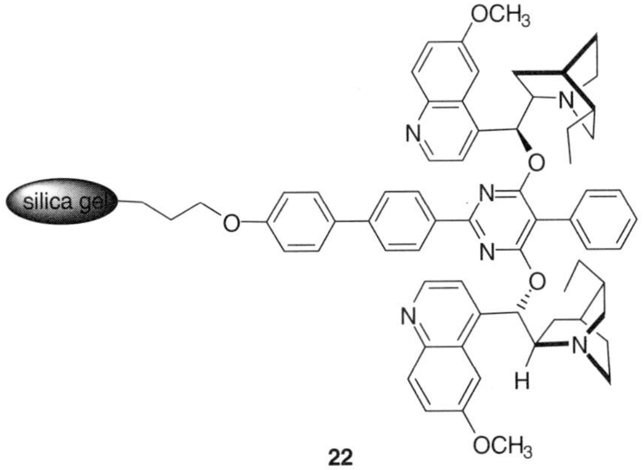

21a

21b

22

weight of 5000. Due to the fact that these MeO-PEG-modified ligands are completely soluble under the reaction conditions [NMO as co-oxidant, acetone/water (1:1, v/v) as solvent], the positive effects of the original Sharpless system remain largely unaffected. In this early study, Janda used a "first-generation" alkaloid and the enantioselectivities remained below 90% ee. After the reaction the MeO-PEG ligands could be recovered almost quantitatively, although, again, their reuse in catalysis was possible only upon addition of further osmium tetroxide. Later, the unsatisfying enantioselectivities were improved by using a MeO-PEG-bound phthalazine ligand and potassium ferricyanide as co-oxidant [66]. Janda also described the use of this ligand in the AD of polymer-bound substrates [16,67].

MeO–PEG-modified DPP and PYR ligands **23** and **24** were introduced by Bolm [68]. Whereas **23** showed the expected excellent enantioselectivities in the oxidation of various aromatic olefins (ee_{max} = 99%), the catalysis of **24** gave an efficient conversion of terminal aliphatic olefins into their respective diols (ee_{max} = 90% for 3,3-dimethyl-1-butene). As with the corresponding silica gel-bound ligand **21**, sequential use of **23a** led to a slight decrease in enantioselectivity in consecutive runs. This result was again attributed to partial hydrolysis of the ester moiety. After replacement of the ester by an ether function, the enantiomeric excess of the product remained constant over a period of several runs [69].

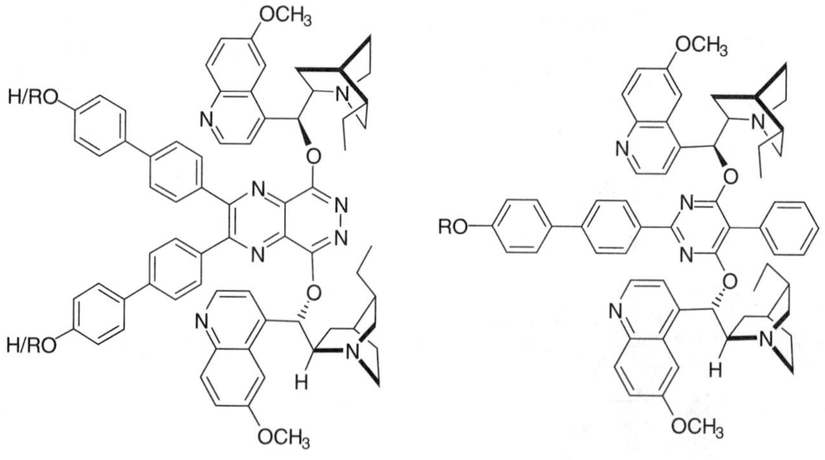

23a,b   **24a,b**

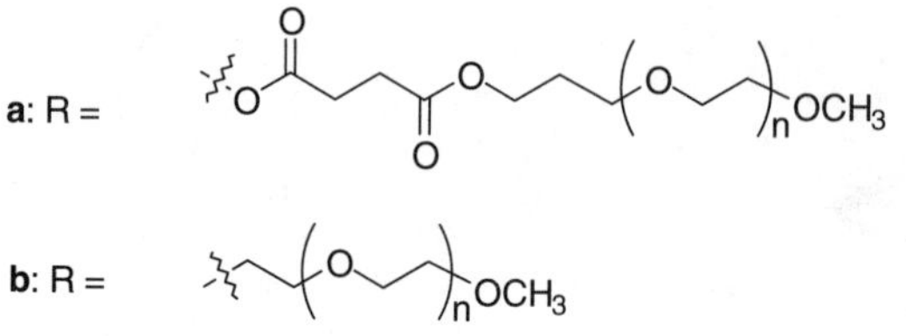

a: R =

b: R =

6E.2. ASYMMETRIC AMINOHYDROXYLATION

6E.2.1. Introduction and Mechanism

In the asymmetric aminohydroxylation (AA) an olefin is converted into a vicinal amino alcohol by means of an osmium(VIII)–mediated suprafacial addition of a nitrogen and an oxygen atom to the double bond. Like the AD, the AA has been developed by modifying an originally stoichiometric, achiral version. Although the first aminohydroxylations were reported in 1976 [70], the asymmetric catalytic protocol is still under development [71].

The AA is initiated by the in situ formation of imidotrioxoosmium(VIII) (**25**) from osmium tetroxide and an appropriate nitrene source. Coordination of a chiral ligand L* gives 18 electron species **26**, which adds to the olefin in a suprafacial fashion. In analogy to the AD mechanism, Sharpless suggests the formation of an intermediate osmaazetidine **27**, which is in agreement with the observations that in the product the nitrogen atom is generally located in the β-position to the most electron-withdrawing group R′. Rearrangement of complex **27** to the formal [3+2]-adduct **28** and dissociation of the chiral ligand L* followed by addition of another nitrogen results in the formation of imidodioxoosmium azaglycolate **29**. Hydrolysis then releases the 1,2-amino alcohol and regenerates **25**.

The second possible catalytic cycle, which is in competition with the first one, involves the reaction of **29** with another olefin. Because the resulting intermediate has no chiral ligand, the formation of the bisazaglycolate complex must occur with little or even no asymmetric induction. Fortunately, this competing catalytic cycle can be suppressed by simply carrying out

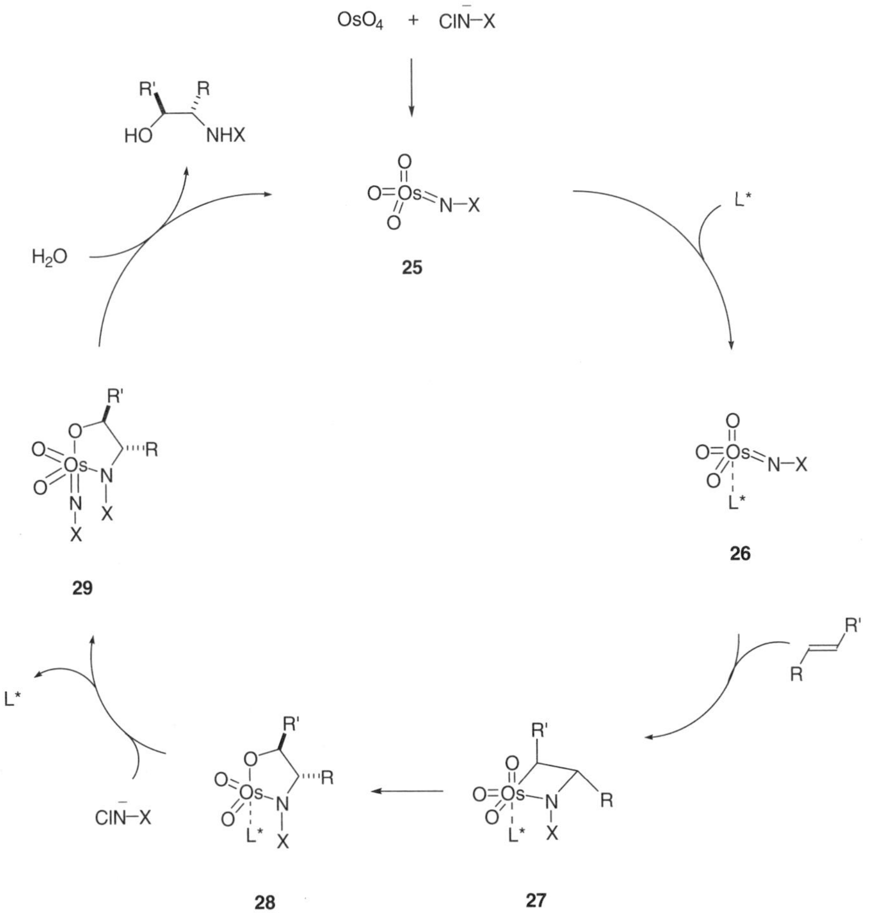

Scheme 6E.5.

the reaction in solvents containing 50% of water. Thus, the formerly rate-limiting hydrolysis of **29** is accelerated and the reaction rate of the overall process enhanced.

Despite the mechanistic similarities between the AA and the AD, the former faces additional challenges with respect to regio- and chemoselectivity. Particularly worth mentioning is the competition between aminohydroxylation and dihydroxylation in the AA via imidotrioxoosmium(VIII) complex **25** and OsO$_4$, respectively [72].

As in the AD, cinchona alkaloids are utilized as ligands in the AA. Because the face selectivities in these two processes are identical, it was concluded that the stereochemistry-determining steps should be identical or at least closely related with each other. Predictions of the absolute configurations of the amino alcohols obtained by AA can, therefore, be made by using the same mnemonic device as given for the AD (see Section 6D.1).

Generally, sulfonamides, carbamates, or amides are employed as nitrogen source giving access to different reaction courses. The first example of the sulfonamide variant was reported in 1996, and it relied on the use of Chloramine-T (**32**) [73]. To achieve reasonable catalytic

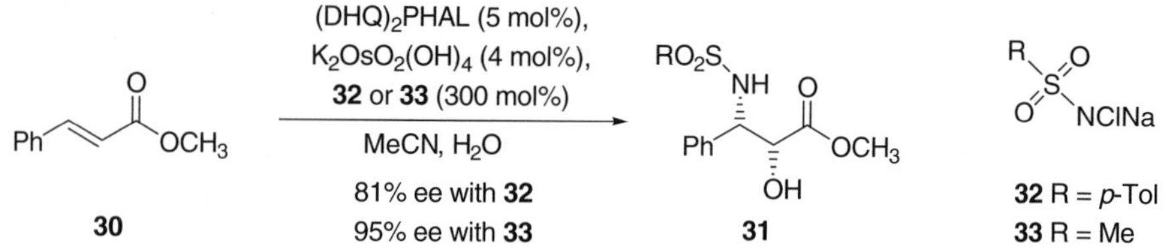

Scheme 6E.6.

conditions, a large excess (≥3 equivalents) of this reagent has to be employed. In accordance with the electronic factors discussed above, aminohydroxylation of methyl cinnamate **30** leads to product **31** with the amino group in the β-position to the ester function.

Studies on the structural effects of the nitrogen source revealed that the enantiopurities could be increased from 81 to 95% ee by changing from Chloramine-T (**32**) to Chloramine-M (**33**) [74]. Furthermore, reactions with the latter were ligand-accelerated, whereas transformations with the former were not.

Due to the sulfonamide group, the products are usually solids and the enantiomeric excesses can be increased by simple recrystallization. The difficulties in removing the sulfonyl group represent the drawback of this procedure.

Most remarkably, catalytic aminohydroxylation of cinnamate amides with sulfonamides in the absence of cinchona alkaloid ligands showed a remarkable chemoselectivity in favor of amino alcohol formation [75].

Another AA variant relies on the use of N-chlorocarbamates such as **34–36**. Again, a large excess of these in-situ reagents needs to be used and carbamates with smaller substituents give

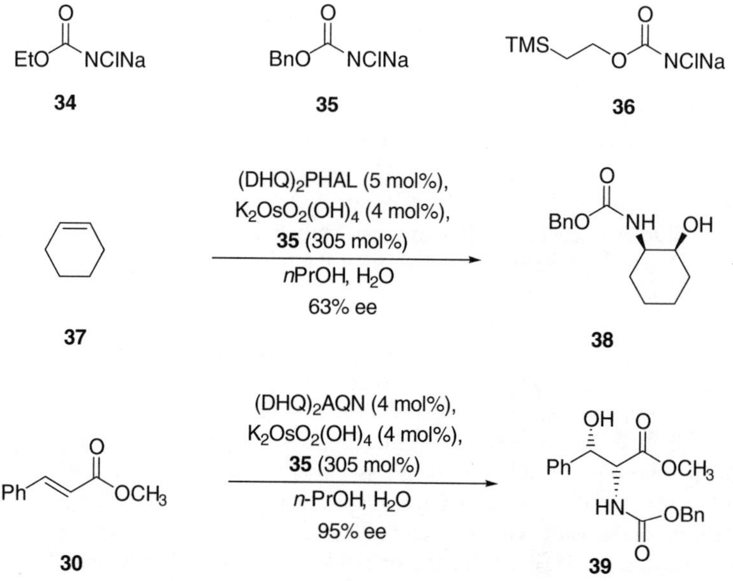

Scheme 6E.7.

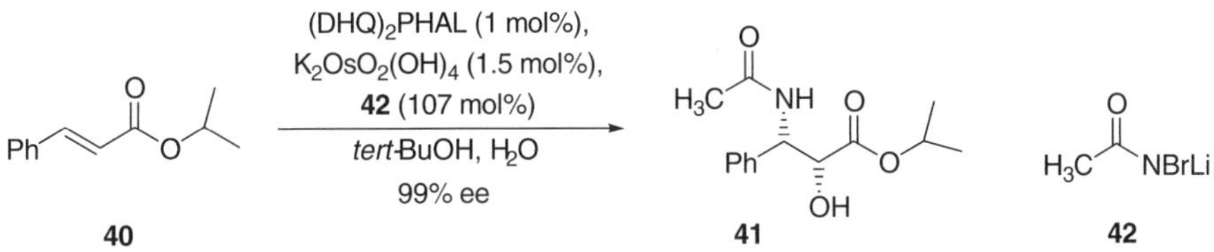

Scheme 6E.8.

superior results [76]. The reactions are ligand-accelerated, and some products crystallize from the reaction mixture, which allows their isolation by simple filtration. Furthermore, recrystallization increases the enantiopurities of the products. For example, AA of cyclohexene (**37**) gives the corresponding amino alcohol **38** with a moderate enantiopurity (63% ee). However, the subsequent recystallization of **38** from ethyl acetate/n-hexane raises its enantiopurity to 99% ee.

This procedure also allows to invert the regioselectivity. Replacement of the usual phthalazine ligand by an AQN-based one leads to an AA in which the formation of the "reversed regioisomer" **39** is preferred [77]. Apart from the obvious conclusion that the substrate orientation within the binding pockets of the different ligands is altered, the mechanistic reason for this reversal remains unknown. From a synthetic point of view this observation is important because it allows the conversion of cinnamate esters into the corresponding phenyl serine derivatives.

Sharpless also described the use of 2-trimethylsilylethyl carbamate **36** in the AA. This reagent leads to a product with a protective group that is easily removed by fluoride. Enantioselectivities in aminohydroxylation with **36** remain comparable with those obtained with benzyl carbamate **35** [78].

An alternative AA protocol is based on the use of amide precursors such as N-bromo acetamide (**42**). Here, it is of major advantage that the use of 1.1 equivalents of nitrenoid is sufficient for a high-yielding reaction. For example, conversion of isopropyl cinnamate (**40**) leads to the corresponding hydroxy ester **41** with a 20:1 regioselectivity and 99% ee. Most noteworthy is the fact that the ligand acceleration now allows to reduce the osmium and ligand amount to 1.5 and 1 mol %, respectively [79]. Again, the regioselectivity can be reversed by using AQN-based ligands. Terminal olefins such as styrenes or even acyclic olefins such as ethyl acrylate are among the best substrates.

Finally, Sharpless recently described the successful use of adenine-based nitrenoids for the AA. Although the yields were good and the regioselectivities were high, the reaction only proceeded in solvent systems containing alcohol-water mixtures and no enantioselectivity was observed [80].

6E.2.2. Applications of the Asymmetric Aminohydroxylation

An impressive application of the AA was performed recently by Sharpless, who reported the synthesis of enantiopure α-arylglycines like **46** from various styrenes [81]. With benzylcarbamate (**35**) the control of the regioselectivity was achieved by choosing an appropriate combination of ligand and solvent. Thus, with PHAL ligands in isopropanol the AA proceeded with moderate-to-excellent regioselectivity favoring the expected benzylic amine **44**. Enantioselectivities for all given 18 examples were in the range of 74–98% ee. Use of AQN ligands led to the predominant formation of regioisomeric products **45**.

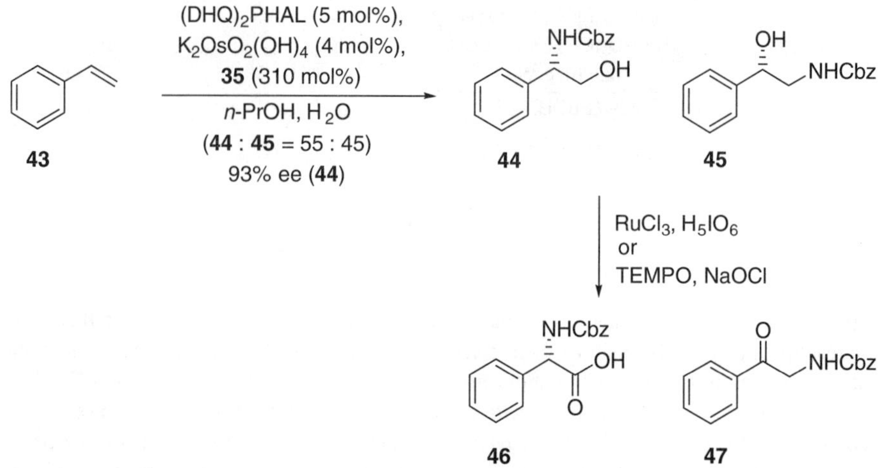

Scheme 6E.9.

Oxidation of the crude product mixtures containing both regioisomers **44** and **45** gave Cbz-protected amino acid **46** and amino ketone **47**, which could then be separated by simple acid-base-extraction. As oxidizing reagents, periodic acid and a catalytic amount of ruthenium trichloride were found to be suitable. For some cases in which chemical yields were low, a TEMPO–catalyzed oxidation with bleach as oxidant was more effective.

Another important application of the AA has been reported by Janda, who described a valuable strategy towards 1,2-diaminobutanoic acids [82]. For example, AA on *tert*-butyl crotonate (**48**) in the presence of $(DHDQ)_2PHAL$ using the carbamate variant provided a 9:1 mixture of regioisomeric amino alcohols favoring compound **49**. Its initial enantiopurity of 90% ee could be raised to 99% ee by a single recrystallization. Conversion of the free hydroxyl group into the corresponding mesylate afforded **50**, which could be transformed into *trans*-diamino acid **51** by nucleophilic substitution with sodium azide followed by catalytic hydrogenation and acidic hydrolysis. For the synthesis of *cis*-isomer **52** a reaction sequence involving an azide opening of an intermediate aziridine obtained from **50** was developed. The respective diamino

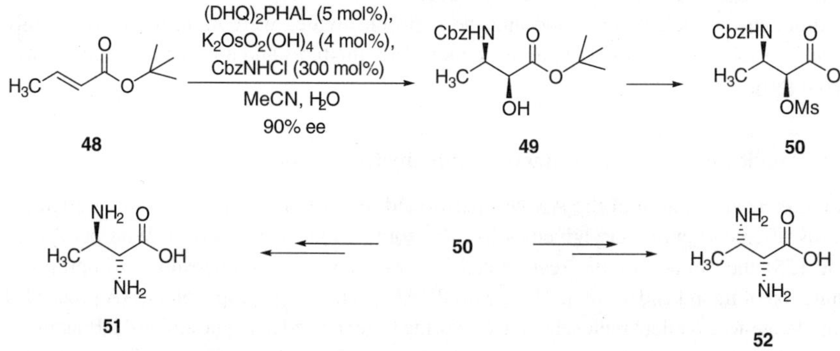

Scheme 6E.10.

acids with opposite absolute configurations could be prepared by AA with (DHQ)$_2$PHAL as ligand.

In a related contribution O'Brien described the AA on styrenes and converted the amino alcohols into enantiopure diamines by using a reaction strategy similar to Janda's [83]. Further synthetic applications of the AA include a new access to Evans' chiral oxazolidinones [84], the enantioselective synthesis of α-amino ketones from silyl enol ethers [85], the stereoselective synthesis of cyclohexylnorstatine [86], and a route towards amino cyclitols by aminohydroxylation of dienylsilanes [87].

Most recently, silica gel-supported bis-cinchona alkaloid **20** was successfully employed in the AA of *trans*-cinnamate derivatives [62b]. The resulting products had excellent enantiopurities (>99% ee). Recovered samples of **20** contained osmium and could be used in the AA again, though with a slight loss of activity. Therefore, recovered catalyst was regenerated upon addition of osmium salts.

6E.2.3 Application of Asymmetric Dihydroxylation and Aminohydroxylation in the Synthesis of Complex Natural Products

Due to its broad applicability, the Sharpless dihydroxylation of olefins has already found widespread use in the synthesis of natural products and valuable building blocks. This chapter highlights a few recent examples from the steadily growing number of applications of the AD in syntheses of complex natural products. It is interesting that some sequences also include the more recently developed catalytic aminohydroxylation of olefins already.

(−)-Ovalicin In 1994, Corey reported the enantioselective synthesis of the potent angiogenesis inhibitor (−)-ovalicin (**53**) [88,89]. An AD protocol was used for the preparation of the key intermediate **56**.

Applying standard AD-conditions with (DHQD)$_2$PHAL as alkaloid ligand afforded diol **55** with 99% ee in 93% chemical yield, which was then converted into epoxy enone **56** in a short reaction sequence.

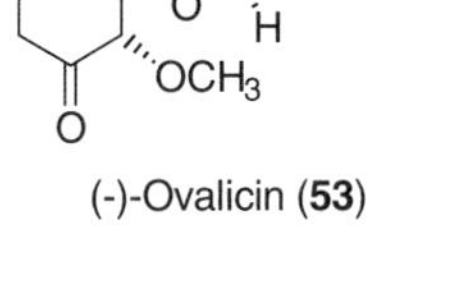

(-)-Ovalicin (**53**)

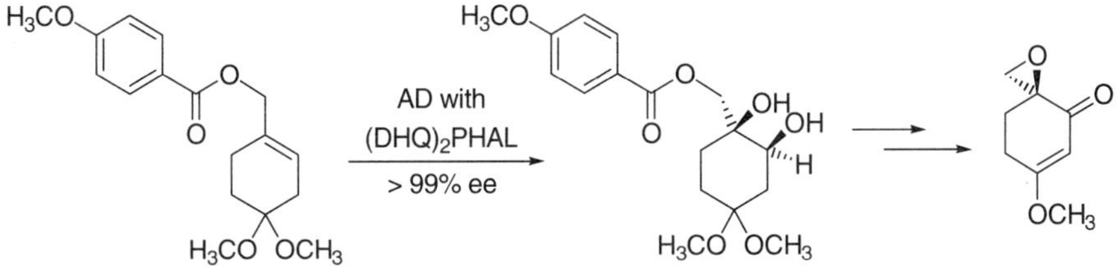

Scheme 6E.11.

Vancomycin Vancomycin (**57**) is an important antibiotic in the treatment of severe bacterial infections by a variety of pathogens and has been one of the most attractive targets for synthetic chemists over the past decade. In 1998, the completion of the total synthesis of the vancomycin aglycon was almost simultaneously reported by Evans [90] and Nicolaou [91].

For the preparation of intermediates **58**, **60**, and **62** Nicolaou used a combined AD/AA strategy. Standard AD-mixes afforded diols **58** and **62** with high enantiomeric excesses. Products **58** and **62a** were then further transformed into intermediates for the preparation of the A- and E-ring of vancomycin, respectively. Diol **62b** served as core precursor (D-ring). To obtain desired α-amino acid ester **60** by AA of **59**, use of an AQN ligand was essential. As expected from previous results (see Section 6E.2.2.), AA with (DHQD)$_2$PHAL afforded the regioisomeric phenylisoserine ester as the major product. Thus, with (DHQD)$_2$AQN the "correct" isomer **60** could be obtained with 87% ee [92]. Further transformations of **60** led to a C-ring precursor of vancomycin.

Sch 38516 (Fluvirucin B$_1$) One of the most remarkable total syntheses in recent years was reported in 1997 by Hoveyda [93]. A variety of efficient asymmetric metal-catalyzed transfor-

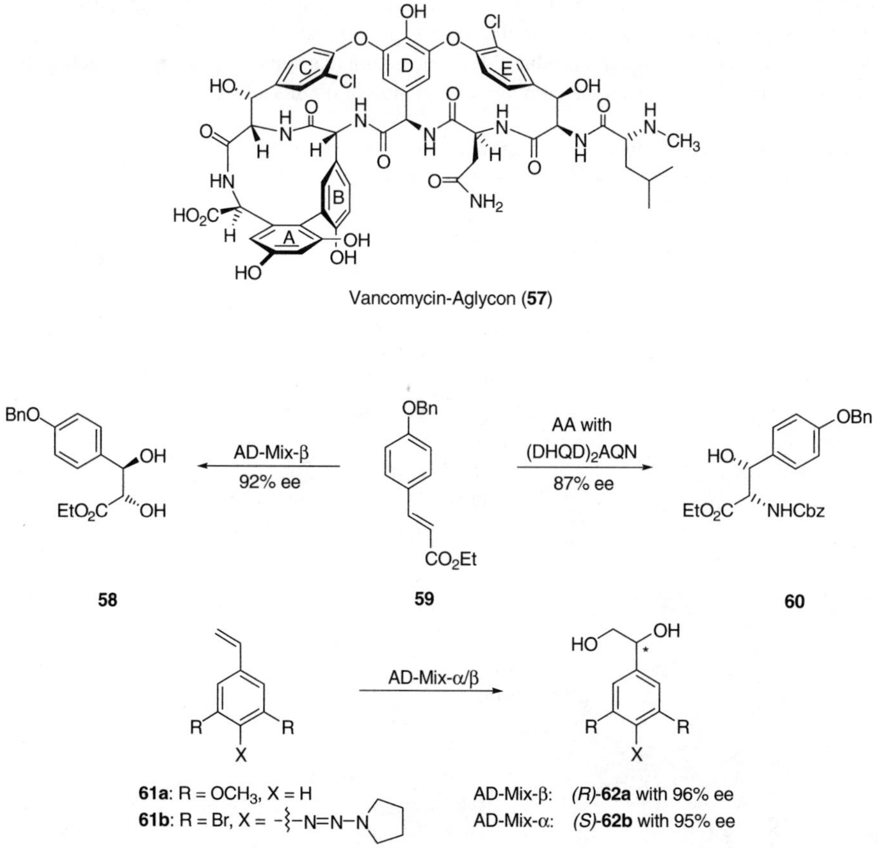

Vancomycin-Aglycon (**57**)

Scheme 6E.12.

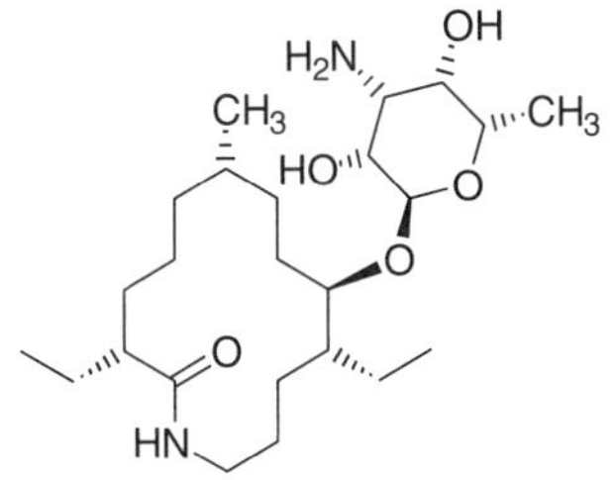

Sch 38516 (Fluvirucin B₁)
(**63**)

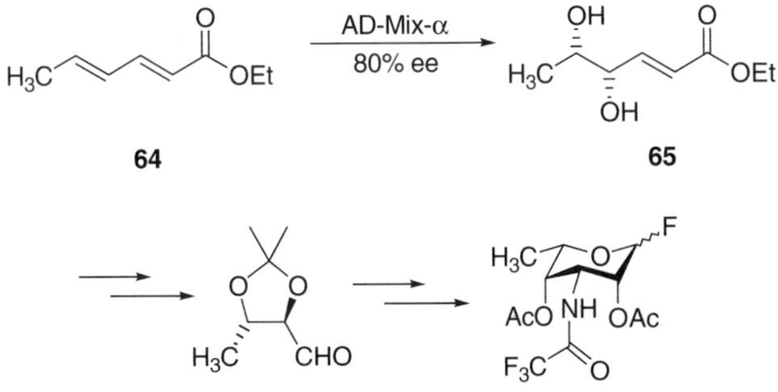

Scheme 6E.13.

mations was used for the synthesis of Sch 38516 (Fluvirucin B$_1$, **63**). While its aglycon was obtained by Sharpless' kinetic resolution of allylic alcohols and Hoveyda's asymmetric carbomagnesation, the construction of the fluvirucin amino sugar unit (3-amino-3,6-L-talopyranose) relied on an AD strategy.

Starting from commercially available ethyl sorbate (**64**) dihydroxylation with standard AD-mixes afforded diol **65** with 80% ee in 61% chemical yield [94]. Protection as the acetonide followed by ozonolysis led to the formation of key intermediate **66** in a four step sequence from diol **65** in 33% overall yield. Further transformations gave fluoroglycoside **67**, which was then used in a stereoselective glycosylation.

Acetogenins Annonaceous acetogenins are a class of potent antitumor, immunosuppressive, and pesticidal agents. (+)-Parviflorin (**68**), (+)-asimicin (**69**), and (+)-bullatacin (**70**) are among the most active members of the acetogenin family [95]. Common characteristic features of these compounds are the butenolide subunit, the bis(hydroxy)bistetrahydrofuran fragment, and the hydroxy group at C-4.

Trost used the dihydroxylation methodology extensively in the total synthesis of (+)-parviflorin (**68**) [96]. Both bistetrahydrofuran elements were prepared from δ, ω-dienyl alcohol **71**. To avoid overoxidation the reaction was stopped before complete conversion, giving triol **72** with 94% ee in 71% yield. Protection as the acetonide afforded **73** and subsequent hydrogenation furnished saturated alcohol **74**. The construction of **75** was accomplished by Julia olefination using derivatives of **73** and **74**. Diene **75** was then regioselectively dihydroxylated at the internal

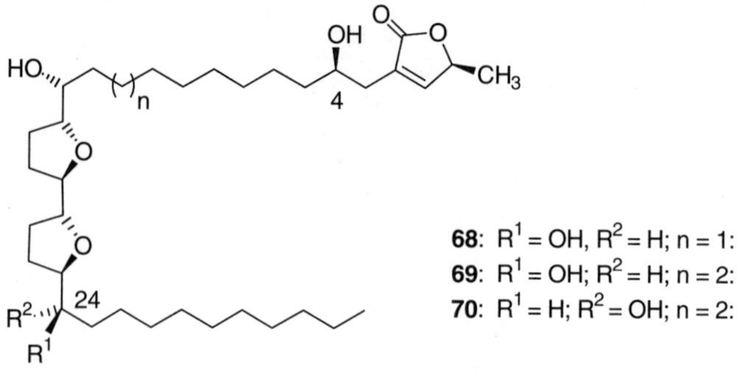

68: $R^1 = OH$, $R^2 = H$; $n = 1$:
69: $R^1 = OH$; $R^2 = H$; $n = 2$:
70: $R^1 = H$; $R^2 = OH$; $n = 2$:

Scheme 6E.14.

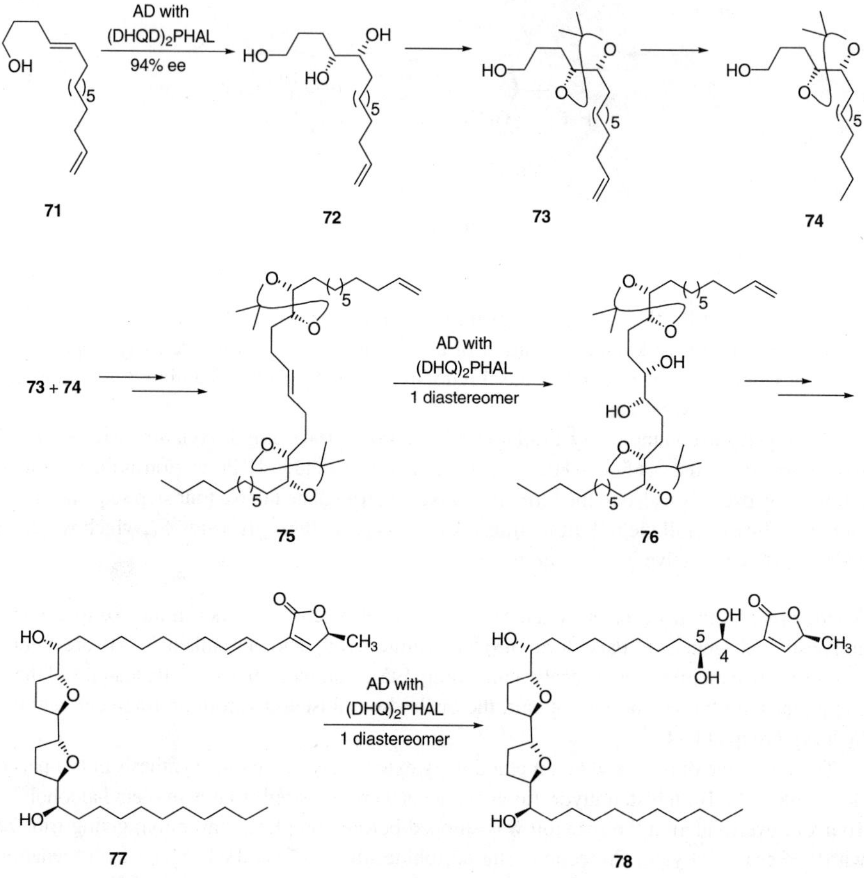

Scheme 6E.15.

double bond giving **76** as a single diastereomer, which could be further converted into butenolide **77**. To complete the synthesis of (+)-parviflorin, the double bond at C-4/C-5 was dihydroxylated under standard conditions delivering **78** in 78% chemical yield. Deoxygenation at C-5 via the corresponding 5-bromo-4-acetoxy intermediate and radical debromination furnished (+)-parviflorin (**68**).

In 1995, Hoye published total syntheses of (+)-asimicin (**69**) and (+)-bullatacin (**70**), which differ only in their configurations at C-24 [97]. Here, the AD was used for the synthesis of two key fragments, **81** and **82**.

Thus, 1,7-octadiene (**79**), which was subjected to monohydroboration followed by asymmetric dihydroxylation of the remaining double bond to give triol **80** with approximately 80% ee. Further transformations then afforded the desired butenolide **81**. Double asymmetric dihydroxylation of diene **83** and subsequent protection gave hydroxy lactone **84** [98], which was then converted into acetylenic bis(hydroxy)bistetrahydrofuran **82** as the required intermediate for the (+)-asimicin synthesis. Mitsunobu inversion at C-24 gave rise to the diastereomeric (+)-bullatacin precursor.

Paclitaxel (Taxol®)   Paclitaxel (**85**) is one of the most promising compounds against several types of cancer, and a number of elegant total syntheses have been reported in recent years [99]. It is produced by the slow growing pacific yew tree, and therefore its availability is limited. By partial synthesis, it is best prepared from 10-deacetylbaccatin III, a derivative lacking the C-10 acetoxy fragment as well as the C-13 side chain. Therefore, many efforts have been focused on

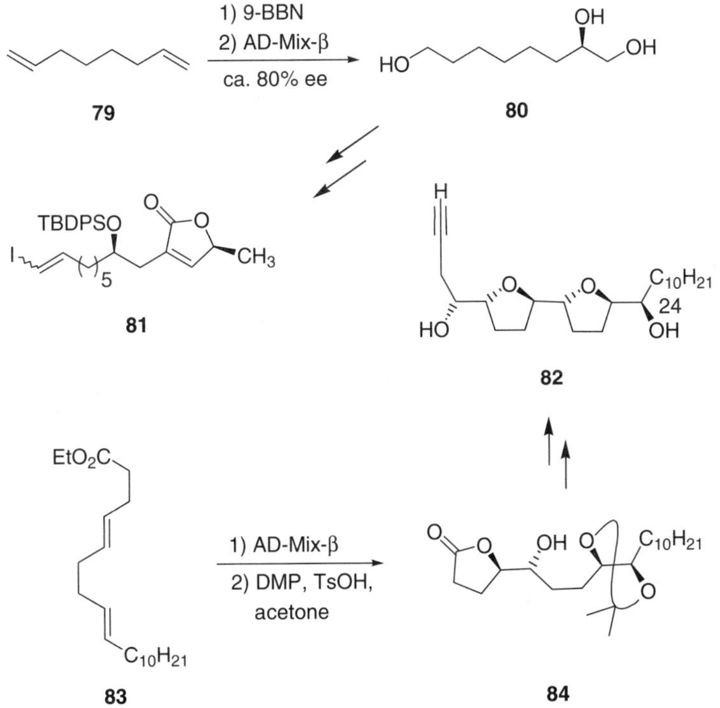

Scheme 6E.16.

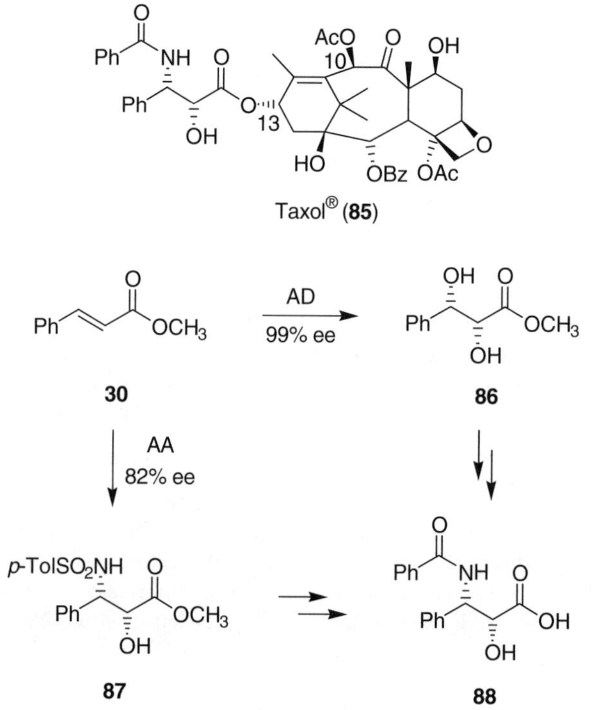

Scheme 6E.17.

the preparation of the latter. Catalytic approaches include syntheses by means of Sharpless epoxidation of allylic alcohols [100] (see Chapter 6A) and Jacobsen-Katsuki epoxidation of unfunctionalized olefins [101] (see Chapter 6B). Sharpless demonstrated that both the AD [102] and the AA strategy [71a, 103] could be used for the efficient construction of the C-13 side chain of paclitaxel.

Methyl cinnamate (**30**) served as starting material for both routes. The AD was carried out with only 0.5 mol % of the (DHQ)$_2$PHAL ligand and 0.2 mol % of potassium osmate(VI) dihydrate, which renders this process extremely cost-effective. Enantiomerically pure bis-hydroxyester **86** was obtained in 72% yield after one recrystallization. Conversion of **86** into *N*-benzoyl-(2*R*,3*S*)-3-phenylisoserine (**88**) was accomplished in only three steps giving the paclitaxel C-13 side chain in 23% overall yield.

Application of the AA [103] resulted in an even shorter synthesis of **88**. The aminohydroxylation was carried out with a slightly higher catalyst loading [1 mol % of Os, 2.5 mol % of (DHQ)$_2$PHAL] compared with the dihydroxylation sequence and furnished tosyl protected phenylisoserine ester **87** with 82% ee in 69% yield. The enantiomeric excess of **87** was further raised to 92% ee by trituration with ethyl acetate. Deprotection and conversion into benzamide **88** afforded the C-13 side chain in a total of three steps.

Zaragozic Acids The zaragozic acids (squalestatins) are a class of competitive inhibitors of mouse, rat, and HepG2 squalene synthases, which also display antifungal activity [104]. The characteristic structural feature of this family of natural products is its highly oxygenated core.

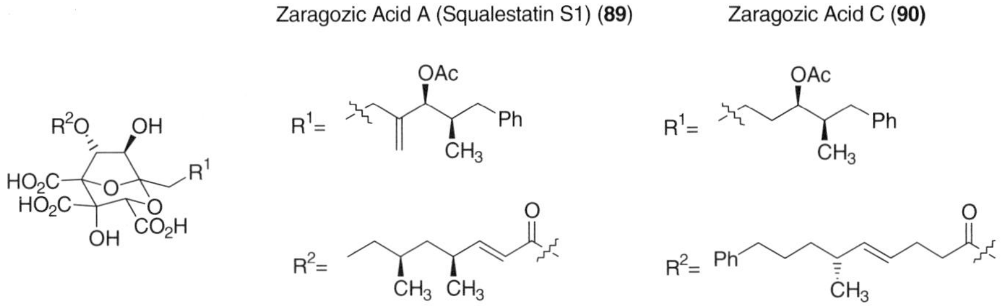

Zaragozic Acid A (Squalestatin S1) (**89**)   Zaragozic Acid C (**90**)

Scheme 6E.18.

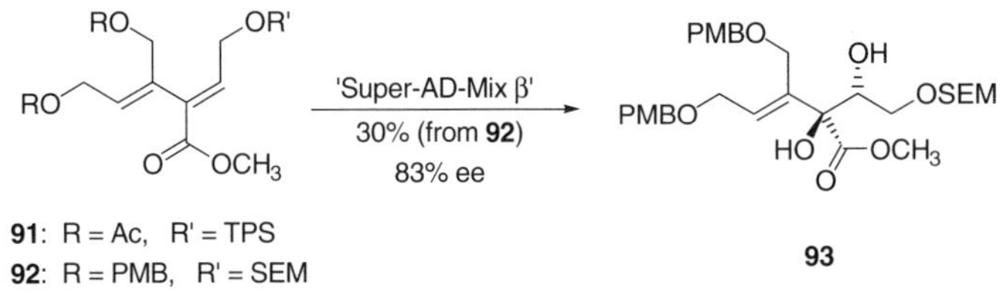

91: R = Ac, R' = TPS
92: R = PMB, R' = SEM

Scheme 6E.19.

Nicolaou used the asymmetric dihydroxlyation as key step toward the construction of zaragozic acid A (**89**) [105]. Initial studies with dienes such as **91** remained unsuccessful mainly because the dihydroxylation just did not proceed at all. After a "screening process" using a series of different protecting groups for the hydroxyl functionalities it was found that diene **92** could be transferred to mono-dihydroxylated product **93**. Although the enantiomeric excess of **93** was satisfying (83% ee), the chemical yield was only 30% together with 44% of recovered starting material. To achieve this dihydroxylation the authors used what they called a "super AD-mix

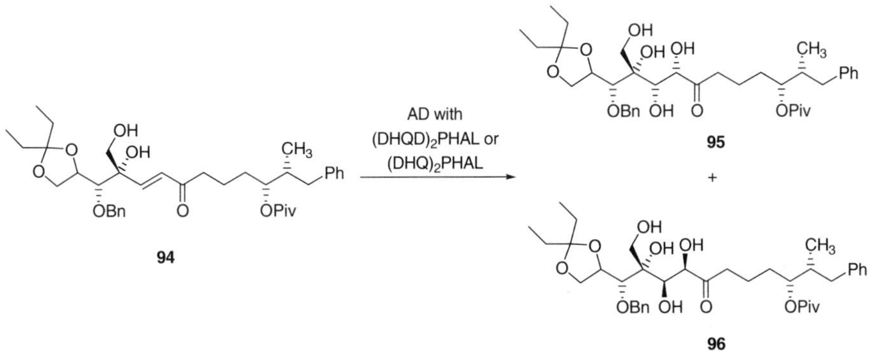

Scheme 6E.20.

β" consisting of 10 mol % of (DHQD)$_2$PHAL, 1 mol % of potassium osmate(VI) hydrate and 3 equiv. of methanesulfonamide [106].

The AD was also used in the synthesis of (+)-zaragozic acid C (**90**) by Carreira, who prepared the dioxabicyclooctane core by dihydroxylation of the late precursor **94** [107].

Here, the authors were faced with the problem that the AD of **94** proceeded in a sluggish and non-diastereoselective manner. Almost equal amounts of diastereomers **95** and **96** were obtained. After optimization both diastereoisomers were produced in 97% combined yield and in a 64:34 ratio in favor of the desired diastereomer **96**. It is interesting that the use of either (DHQ)$_2$PHAL or (DHQD)$_2$PHAL afforded this diastereoisomer predominantly. Attempts to increase the diastereomeric ratio by protection of the primary hydroxy group as the 4-methoxy benzoate failed.

ACKNOWLEDGMENT

We thank the Fonds der Chemischen Industrie for supporting our research.

REFERENCES

1. (a) Kolb, H. C.; Sharpless, K. B. In *Transition Metals for Organic Synthesis*; Beller, M.; Bolm, C. (Eds.); Wiley-VCH: Weinheim, 1998; Vol. 2, pp. 219. (b) Poli, G.; Scolastico, C. In *Stereoselective Synthesis*; Helmchen, G.; Hoffmann, R. W.; Mulzer, J.; Schaumann, E. (Eds.); Thieme: Stuttgart 1996, pp. 4547. (c) Markó, I. E.; Svendsen, J. S. In *Comprehensive Asymmetric Catalysis*; Jacobsen, E. N.; Pfaltz, A.; Yamamoto, H. (Eds.); Springer: Berlin, 1999, Vol. II, pp. 713.

2. (a) Becker, H.; Sharpless, K. B. *Angew. Chem., Int. Ed.* **1996**, *35*, 448. (b) For a recent application of the AQN ligands see: Hodgkinson, T. J.; Shipman, M. *Synthesis* **1998**, 1141.

3. For early AD results with (DHQD)$_2$PHAL see: (a) Crispino, G. A.; Sharpless, K. B. *Tetrahedron Lett.* **1992**, *33*, 4273. (b) Crispino, G. A.; Ho, P.-T.; Sharpless, K. B. *Science* **1993**, *259*, 64.

4. Corey, E. J.; Noe, M. C.; Shieh, W.-C. *Tetrahedron Lett.* **1993**, *34*, 5995.

5. Corey, E. J.; Noe, M. C.; Lin, S. *Tetrahedron Lett.* **1995**, *36*, 8741.

6. Corey, E. J.; Luo, G.; Lin, L. S. *J. Am. Chem. Soc.* **1997**, *119*, 9927.

7. Corey, E. J.; Noe, M. C.; Grogan, M. J. *Tetrahedron Lett.* **1994**, *35*, 6427.

8. Corey, E. J.; Guzman-Perez, A.; Noe, M. C. *J. Am. Chem. Soc.* **1995**, *117*, 10805.

9. Corey, E. J.; Guzman-Perez, A.; Noe, M. C. *Tetrahedron Lett.* **1995**, *36*, 3481.

10. Hawkins, J. M.; Meyer, A. *Science* **1993**, *260*, 1918.

11. Lohray, B. B.; Bhushan, V. *Tetrahedron Lett.* **1993**, *34*, 3911.

12. VanNieuwenhze, M. S.; Sharpless, K. B. *J. Am. Chem. Soc.* **1993**, *115*, 7864.

13. Corey, E. J.; Noe, M. C.; Guzman-Perez, A. *J. Am. Chem. Soc.* **1995**, *117*, 10817.

14. Ward, R. A.; Procter, G. *Tetrahedron Lett.* **1992**, *34*, 3363.

15. Cernerud, M.; Reina, J. A.; Tegenfeldt, J.; Moberg, C. *Tetrahedron: Asymmetry* **1996**, *7*, 2863.

16. Han, H.; Janda, K. D. *Angew. Chem., Int. Ed.* **1997**, *36*, 1731.

17. Riedl, R.; Tappe, R.; Berkessel, A. *J. Am. Chem. Soc.* **1998**, *120*, 8994.

18. Bräse, S.; Enders, D.; Köbberling, J.; Avemaria, F. *Angew. Chem., Int. Ed.* **1998**, *37*, 3413.

19. Chang, H.-T.; Chen, C.-T.; Kondo, T.; Siuzdak, G.; Sharpless, K. B. *Angew. Chem., Int. Ed.* **1996**, *35*, 182.

20. Rouhi, A. M. *Chemical and Engineering News* **1997**, November 3, p. 23.

21. (a) Böeseken, J.; van Giffen, J. *Rec. Trav. Chim. Pays-Bas* **1920**, *39*, 183. (b) Böeseken, J. *Rec. Trav. Chim. Pays-Bas* **1922**, *41*, 199.

22. (a) Criegee, R. *Justus Liebigs Ann. Chem.* **1936**, *522*, 75. (b) Criegee, R. *Angew. Chem.* **1937**, *50*, 153. (c) Criegee, R. *Angew. Chem.* **1938**, *51*, 519. (d) Criegee, R.; Marchand, B.; Wannowius, H. *Justus Liebigs Ann. Chem.* **1942**, *550*, 99, and references cited therein.

23. Sharpless, K. B.; Teranishi, A. Y.; Bäckvall, J.-E. *J. Am. Chem. Soc.* **1977**, *99*, 3120.

24. The term "[2+2]" mechanism is only a formal description of the two-step sequence leading to the osmaoxetane intermediate. For an excellent overview on the theoretical basics of oxidations see: Jørgensen, K. A. In *Transition Metals for Organic Synthesis*; Beller, M.; Bolm, C. (Ed.); Wiley-VCH:Weinheim, 1998; Vol. 2, pp. 157.

25. For a review on metallaoxetanes see: Jørgensen, K. A.; Schiøtt, B. *Chem. Rev.* **1990**, *90*, 1483.

26. (a) Norrby, P.-O.; Becker, H.; Sharpless, K. B. *J. Am. Chem. Soc.* **1996**, *118*, 35. (b) Norrby, P.-O.; Kolb, H. C.; Sharpless, K. B. *J. Am. Chem. Soc.* **1994**, *116*, 8470. (c) Becker, H.; Ho, P. T.; Kolb, H. C.; Loren, S.; Norrby, P.-O.; Sharpless, K. B. *Tetrahedron Lett.* **1994**, *35*, 7315.

27. Kolb, H. C.; Andersson, P. G.; Sharpless, K. B. *J. Am. Chem. Soc.* **1994**, *116*, 1278.

28. Göbel, Z.; Sharpless, K. B. *Angew. Chem., Int. Ed.* **1993**, *32*, 1329.

29. (a) Buschmann, H.; Scharf, H.-D.; Hoffmann, N.; Esser, P. *Angew. Chem., Int. Ed.* **1991**, *30*, 477. (b) Gypser, A.; Norrby, P.-O. *J. Chem. Soc. Perkin Trans. 2*, **1997**, 939. (c) Norrby, P.-O.; Gable, K. P. *J. Chem. Soc. Perkin Trans. 2*, **1996**, 171.

30. (a) Gable, K. P.; Phan, T. N. *J. Am. Chem. Soc.* **1994**, *116*, 833. (b) Gable, K. P.; Juliette, J.J.J. *J. Am. Chem. Soc.* **1995**, *117*, 955. (c) Gable, K. P.; Juliette, J. J. J. *J. Am. Chem. Soc.* **1996**, *118*, 2625.

31. (a) Corey, E. J.; Noe, M. C.; Sarshar, S. *J. Am. Chem. Soc.* **1993**, *115*, 3828. (b) Corey, E. J.; Lotto, G. I. *Tetrahedron Lett.* **1990**, *31*, 2665.

32. For a comparison of various ligands see: (a) Crispino, G. A., Makita, A.; Wang, Z.-M.; Sharpless, K. B. *Tetrahedron Lett.* **1994**, *35*, 543. (b) Corey, E. J.; Noe, M. C. *J. Am. Chem. Soc.* **1996**, *118*, 11038.

33. Inferring from kinetic studies, Sharpless had excluded a dihydroxylation pathway *via* a μ-oxo-bridged bis-osmium intermediate before: Kolb, H. C.; Andersson, P. G.; Bennani, Y. L.; Crispino, G. A.; Jeong, K.-S.; Kwong, H.-L.; Sharpless, K. B. *J. Am. Chem. Soc.* **1993**, *115*, 12226.

34. Corey, E. J.; Noe, M. C. *J. Am. Chem. Soc.* **1993**, *115*, 12579.

35. Corey, E. J.; Noe, M. C.; Sarshar, S. *Tetrahedron Lett.* **1994**, *35*, 2861.

36. Corey, E. J.; Sarshar, S.; Azimioara, M. D.; Newbold, R. C.; Noe, M. C. *J. Am. Chem. Soc.* **1996**, *118*, 7851.

37. Corey, E. J.; Noe, M. C. *J. Am. Chem. Soc.* **1996**, *118*, 319.

38. Nelson, D. W.; Gypser, A.; Ho, P. T.; Kolb, H. C.; Kondo, T.; Kwong, H.-L.; McGrath, D. V.; Rubin, A. E.; Norrby, P.-O.; Gable, K. P.; Sharpless, K. B. *J. Am. Chem. Soc.* **1997**, *119*, 1840.

39. (a) DelMonte, A. J.; Haller, J.; Houk, K. N.; Sharpless, K. B.; Singleton, D. A.; Strassner, T.; Thomas, A. A. *J. Am. Chem. Soc.* **1997**, *119*, 9907. (b) Ujaque, G.; Maseras, F.; Lledós, A. *J. Am. Chem. Soc.* **1999**, *121*, 1317. (c) Norrby, P.-O.; Rasmussen, T.; Haller, J.; Strassner, T.; Houk, K. N. *J. Am. Chem. Soc.* **1999**, *121*, 10186.

40. For further theoretical studies supporting the [3+2] mechanisms see: (a) Dapprich, S.; Ujaque, G.; Maseras, F.; Lledos, A.; Musaev, D. G.; Morokuma, K. *J. Am. Chem. Soc.* **1996**, *118*, 11660. (b) Pidun, U.; Boehme, C.; Frenking, G. *Angew. Chem., Int. Ed.* **1996**, *35*, 2817. (c) Torrent, M.; Deng, L.; Duran, M.; Sola, M.; Ziegler, T. *Organometallics* **1997**, *16*, 13.

41. (a) Tomioka, K.; Nakajima, M.; Koga, K. *J. Am. Chem. Soc.* **1987**, *109*, 6213. (b) Corey, E. J.; DaSilva Jardine, P.; Virgil, S.; Yuen, P.-W.; Connell, R. D. *J. Am. Chem. Soc.* **1989**, *111*, 9243. (c) Hanessian, S.; Meffre, P.; Girard, M.; Beaudoin, S.; Sancéau, J.-Y.; Bennani, Y. *J. Org. Chem.* **1993**, *58*, 1991.

42. Veldkamp, A.; Frenking, G. *J. Am. Chem. Soc.* **1994**, *116*, 4937.

43. Berrisford, D. J.; Bolm, C.; Sharpless, K. B. *Angew. Chem., Int. Ed.* **1995**, *34*, 1059.

44. Reviews: (a) Girard, C.; Kagan, H. B. *Angew. Chem., Int. Ed.* **1998**, *37*, 2922. (b) Bolm, C. In *Advanced Asymmetric Synthesis*; Stephenson, G. R. (Ed.); Chapman and Hall: London, 1996; pp. 9.

45. Zhang, S. Y.; Girard, C.; Kagan, H. B. *Tetrahedron: Asymmetry* **1995**, *6*, 2637.

46. Reviews: (a) Bolm, C.; Gerlach, A. *Eur. J. Org. Chem.* **1998**, 21. (b) Salvadori, P.; Pini, D.; Petri, A. *Synlett* **1999**, 1181.

47. For recent general reviews on polymer-supported ligands see: (a) Shuttleworth, S. J.; Allin, S. M.; Sharma, P. K. *Synthesis* **1997**, 1217. (b) Sherrington, D. C. *Chem. Commun.* **1998**, 2275. (c) Pu, L. *Tetrahedron: Asymmetry* **1998**, *9*, 1457.

48. Kim, B. M.; Sharpless, K. B. *Tetrahedron Lett.* **1990**, *31*, 3003.

49. (a) For the use of achiral polymer-bound OsO$_4$ see: Cainelli, G.; Contento, M.; Manescalchi, F.; Plessi, L. *Synthesis* **1989**, 45. (b) For the use of microencapsulated OsO$_4$ see: Nagayama, S.; Endo, M.; Kobayashi, S. *J. Org. Chem.* **1998**, *63*, 6094. (c) This process was recently shown to be applicable for the asymmetric dihydroxylation, too: Kobayashi, S.; Endo, M.; Nagayama, S. *J. Am. Chem. Soc.* **1999**, *121*, 11229.

50. Pini, D.; Petri, A.; Nardi, A.; Rosini, C.; Salvadori, P. *Tetrahedron Lett.* **1991**, *32*, 5175.

51. Lohray, B. B.; Thomas, A.; Chittari, P.; Ahuja, J. R.; Dhal, P. K. *Tetrahedron Lett.* **1992**, *33*, 5453.

52. Song, C. E.; Roh, E. J.; Lee, S.-g.; Kim, I. O. *Tetrahedron: Asymmetry* **1995**, *6*, 2687.

53. Canali, L.; Song, C. E.; Sherrington, D. C. *Tetrahedron: Asymmetry* **1998**, *9*, 1029.

54. Pini, D.; Petri, A.; Salvadori, P. *Tetrahedron: Asymmetry* **1993**, *4*, 2351.

55. (a) Pini, D.; Petri, A.; Salvadori, P. *Tetrahedron* **1994**, *50*, 11321. (b) Pini, D.; Petri, A.; Salvadori, P. *Chirality* **1995**, *7*, 580.

56. Lohray, B. B.; Nandanan, E.; Bhushan, V. *Tetrahedron Lett.* **1994**, *35*, 6559.

57. Petri, A.; Pini, D.; Salvadori, P. *Tetrahedron Lett.* **1995**, *36*, 1549.

58. Salvadori, P.; Pini, D.; Petri, A. *J. Am. Chem. Soc.* **1997**, *119*, 6929.

59. Song, C. E.; Yang, J. W.; Ha, H. J.; Lee, S.-G. *Tetrahedron: Asymmetry* **1996**, *7*, 645.

60. Nandanan, E.; Sudalai, A.; Ravindranathan, T. *Tetrahedron Lett.* **1997**, *38*, 2577.

61. Lohray, B. B.; Nandanan, E.; Bhushan, V. *Tetrahedron: Asymmetry* **1996**, *7*, 2805.

62. (a) Song, C. E.; Yang, J. W.; Ha, H.-J. *Tetrahedron: Asymmetry* **1997**, *8*, 841. (b) Recently, this ligand was also used in the AA. Song, C. E.; Oh, C. R.; Lee, S. W.; Lee, S.-G.; Canali, L.; Sherrington, D. C. *Chem. Commun.* **1998**, 2435.

63. (a) Bolm, C.; Maischak, A.; Gerlach, A. *Chem. Commun.* **1997**, 2353. (b) Bolm, C.; Maischak, A., unpublished results.

64. (a) Bayer, E.; Schurig, V. *Angew. Chem., Int. Ed.* **1975**, *14*, 493. (b) Schurig, V. *CHEMTECH* **1976**, *6*, 212.

65. Han, H.; Janda, K. D. *J. Am. Chem. Soc.* **1996**, *118*, 7632.

66. Han, H.; Janda, K. D. *Tetrahedron Lett.* **1997**, *38*, 1527.

67. For a review on organic synthesis on soluble polymer supports see: Gravert, D. J.; Janda, K. D. *Chem. Rev.* **1997**, *97*, 489.

68. Bolm, C.; Gerlach, A. *Angew. Chem., Int. Ed.* **1997**, *36*, 773.

69. Gerlach, A., PhD thesis, RWTH Aachen, **1997**.

70. (a) Sharpless, K. B.; Chong, A. O.; Oshima, K. *J. Org. Chem.* **1976**, *41*, 177. (b) Sharpless, K. B.; Herranz, E. *J. Org. Chem.* **1978**, *43*, 2544. (c) Herranz, E.; Biller, S. A.; Sharpless, K. B. *J. Am. Chem. Soc.* **1978**, *100*, 3596. (d) Herranz, E.; Sharpless, K. B. *J. Org. Chem.* **1980**, *45*, 2710. (e) Herranz, E.; Sharpless, K. B. *Org. Synth.* **1981**, *61*, 85. (f) Herranz, E.; Sharpless, K. B. *Org. Synth.* **1981**, *61*, 93.

71. For reviews and short overviews see: (a) Kolb, H. C.; Sharpless, K. B. In *Transition Metals for Organic Synthesis*; Beller, M.; Bolm, C. (Eds.); Wiley-VCH: Weinheim, 1998; Vol. 2, pp. 243. (b) Schlingloff, G.; Sharpless, K. B. in *Asymmetric Oxidation Reactions: A Practical Approach*; Katsuki, T. (Ed.),

Oxford University Press: London, in press. (c) Reiser, O. *Angew. Chem., Int. Ed.* **1996**, *35*, 1308. (d) Casiraghi, G.; Rassu, G.; Zanardi, F. *Chemtracts-Organic Chemistry* **1997**, *10*, 318. (e) O'Brien, P. *Angew. Chem. Int. Ed.* **1999**, *38*, 326.

72. General recommendations for an optimized AA protocol can be found in ref. 71a and b.

73. (a) Li, G.; Chang, H.-T.; Sharpless, K. B. *Angew. Chem., Int. Ed.* **1996**, *35*, 451. (b) For a recent application of this system in the synthesis of Vigabatrin® see: Chandrasekhar, S.; Mohapatra, S. *Tetrahedron Lett.* **1998**, *39*, 6415.

74. Rudolph, J.; Sennhenn, P. C.; Vlaar, C. P.; Sharpless, K. B. *Angew. Chem., Int. Ed.* **1996**, *35*, 2810.

75. Rubin, A. E.; Sharpless, K. B. *Angew. Chem., Int. Ed.* **1997**, *36*, 2637.

76. Li, G.; Angert, H. H.; Sharpless, K. B. *Angew. Chem., Int. Ed.* **1996**, *35*, 2813.

77. Tao, B.; Schlingloff, G.; Sharpless, K. B. *Tetrahedron Lett.* **1998**, *39*, 2507.

78. (a) Reddy, K. L.; Dress, K. R.; Sharpless, K. B. *Tetrahedron Lett.* **1998**, *39*, 3667. (b) Recently, Sharpless reported the use of *tert.*-butylsulfonamide as nitrogen source: Gontscharov, A. V.; Liu, Sharpless, K. B. *Org. Lett.* **1999**, *1*, 783.

79. Bruncko, M.; Schlingloff, G.; Sharpless, K. B. *Angew. Chem., Int. Ed.* **1997**, *36*, 1483.

80. (a) Dress, K. R.; Gooßen, L. J.; Liu, H.; Jerina, D. M.; Sharpless, K. B. *Tetrahedron Lett.* **1998**, *39*, 7669. (b) After the submission of this chapter Sharpless and co-workers reported on the asymmetric amonohydroxylation using amino-substituted, heterocyclic nitrene sources. Here, a large number of olefins were converted into the corresponding amino alcohols with enantioselectivities of up 99%: Gooßen, L. J.; Liu, H.; Dress, K. R.; Sharpless, K. B *Angew. Chem. Int. Ed.* **1999**, *38*, 1080.

81. Reddy, K. L.; Sharpless, K. B. *J. Am. Chem. Soc.* **1998**, *120*, 1207.

82. Han, H.; Yoon, J.; Janda, K. D. *J. Org. Chem.* **1998**, *63*, 2045.

83. (a) O'Brien, P.; Osborne, S. A.; Parker, D. D. *Tetrahedron Lett.* **1998**, *39*, 4099. (b) O'Brien, P.; Osborne, S. A.; Parker, D. D. *J. Chem. Soc., Perkin Trans I* **1998**, 2519.

84. Li, G.; Lenington, R.; Willis, S. Kim, S. H. *J. Chem. Soc., Perkin Trans I* **1998**, 1753.

85. Phukan, P.; Sudalai, A. *Tetrahedron: Asymmetry* **1998**, *9*, 1001.

86. Upadhya, T. T.; Sudalai, A. *Tetrahedron: Asymmetry* **1997**, *8*, 3685.

87. Angelaud, R.; Landais, Y.; Schenk, K. *Tetrahedron Lett.* **1997**, *38*, 1407.

88. Corey, E. J.; Guzman-Perez, A.; Noe, M. C. *J. Am. Chem. Soc.* **1994**, *116*, 12109.

89. For the synthesis of racemic **53** see: Corey, E. J.; Dittami, J. P. *J. Am. Chem. Soc.* **1985**, *107*, 256.

90. (a) Evans, D. A.; Wood, M. R.; Trotter, B. W.; Richardson, T. I.; Barrow, J. C.; Katz, J. L. *Angew. Chem., Int. Ed.* **1998**, *37*, 2700. (b) Evans, D. A.; Dinsmore, C. J.; Watson, P. S.; Wood, M. R.; Richardson, T. I.; Trotter, B. W.; Katz, J. L. *Angew. Chem., Int. Ed.* **1998**, *37*, 2704.

91. (a) Nicolaou, K. C.; Natarajan, S.; Li, H.; Jain, N. F.; Hughes, R.; Solomon, M. E.; Ramanjulu, J. M.; Boddy, C. N. C.; Takayanagi, M. *Angew. Chem., Int. Ed.* **1998**, *37*, 2708. (b) Nicolaou, K. C.; Jain, N. F.; Natarajan, S.; Hughes, R.; Solomon, M. E.; Li, H.; Ramanjulu, J. M.; Takayanagi, M.; Koumbis, A. E.; Bando, T. *Angew. Chem., Int. Ed.* **1998**, *37*, 2714. (c) Nicolaou, K. C.; Takayanagi, N.; Jain, N. F.; Natarajan, S.; Koumbis, A. E.; Bando, T.; Ramanjulu, J. M. *Angew. Chem., Int. Ed.* **1998**, *37*, 2717.

92. The yield of hydroxy α-amino acid ester **60** was only 45%. No information was given about the formation of regioisomeric products.

93. (a) Houri, A. F.; Xu, Z.; Cogan, D. A.; Hoveyda, A. H. *J. Am. Chem. Soc.* **1995**, *117*, 2943. (b) Xu, Z.; Johannes, C. W.; Salman S. S.; Hoveyda, A. H. *J. Am. Chem. Soc.* **1996**, *118*, 10926. (c) Xu, Z.; Johannes, C. W.; Houri, A. F.; La, D. S.; Cogan, D. A.; Hofilena, G. E.; Hoveyda, A. H. *J. Am. Chem. Soc.* **1997**, *119*, 10302. (d) Xu, Z.; Johannes, C. W.; La, D. S.; Hofilena, G. E.; Hoveyda, A. H. *Tetrahedron* **1997**, *53*, 16377.

94. For the synthesis of (+)-desoxynojirimycin starting with an AD of a dienic acid ester see: Lindstroem, U. M.; Somfai, P. *Tetrahedron Lett.* **1998**, *39*, 7173.

95. Reviews on the synthesis of acetogenins: (a) Hoppe, R.; Scharf, H.-D. *Synthesis* **1995**, 1447. (b) Figadère, B. *Acc. Chem. Res.* **1995**, *28*, 359. (c) Koert, U. *Synthesis* **1995**, 115.

96. Trost, B. M.; Calkins, T. L.; Bochet, C. G. *Angew. Chem., Int. Ed.* **1997**, *36*, 2632.

97. Hoye, T. R.; Tan, L. *Tetrahedron Lett.* **1995**, *36*, 1981.

98. No information was given about the enantioselectivity of this double dihydroxylation. For the synthesis of a closely related derivative *via* double dihydroxylation (>99% ee after recrystallization) see: Sinha, S. C.; Keinan, E. *J. Am. Chem. Soc.* **1993**, *115*, 4891.

99. (a) *Taxane Anticancer Agents: Basic Science and Current Status*; Georg, G. I.; Chen, T. T.; Ojima, I., Vyas, D. M., Eds.; ACS Symposium Series 583; American Chemical Society: Washington, DC, 1995. (b) *Taxol: Science and Applications*; Suffness, M. (Ed.); CRC: Boca Raton, FL, 1995. (c) Nicolaou, K. C.; Guy, R. K. *Angew. Chem., Int. Ed.* **1995**, *34*, 2079. (d) Cragg, G. M. *Med. Res. Rev.* **1998**, *18*, 315. (e) Mukaiyama, T.; Shiina, I.; Iwadare, H.; Saitoh, M.; Nishimura, T.; Ohkawa, N.; Sakoh, H.; Nishimura, K.; Tani, Y.-i.; Hasegawa, M.; Yamada, K.; Saitoh, K. *Chem. Eur. J.* **1999**, *5*, 121 and references therein.

100. (a) Denis, J.-N.; Greene, A. E.; Serra, A. A.; Luche, M.-J. *J. Org. Chem.* **1986**, *51*, 46. (b) Denis, J.-N.; Correa, A.; Greene, A. E. *J. Org. Chem.* **1990**, *55*, 1957.

101. Deng, L.; Jacobsen, E. N. *J. Org. Chem.* **1992**, *57*, 4320.

102. Wang, Z.-M.; Kolb, H. C.; Sharpless, K. B. *J. Org. Chem.* **1994**, *59*, 5104.

103. Li, G.; Sharpless, K. B. *Acta Chem. Scand.* **1996**, *50*, 649.

104. Review: Nadin, A.; Nicolaou, K. C. *Angew. Chem., Int. Ed.* **1996**, *35*, 1622.

105. Nicolaou, K. C.; Yue, E. W.; La Greca, S.; Nadin, A.; Yang, Z.; Leresche, J. E.; Tsuri, T.; Naniwa, Y.; De Riccardis, F. *Chem. Eur. J.* **1995**, *1*, 467.

106. In **92** the conjugated double bond reacted. For the AD of β,γ-unsaturated esters in the synthesis of chiral γ-lactones and butenolides see: (a) Harcken, C.; Brückner, R.; Rank, E. *Chem. Eur. J.* **1998**, *4*, 2342. (b) Harcken, C.; Brückner, R. *Angew. Chem., Int. Ed.* **1997**, *36*, 2750.

107. (a) Carreira, E. M.; Du Bois, J. *J. Am. Chem. Soc.* **1994**, *116*, 10825. (b) Carreira, E. M.; Du Bois, J. *J. Am. Chem. Soc.* **1995**, *117*, 8106. Note added in proof: After submission of this chapter several groups reported on the use of alternative oxygen sources for the AD reaction: (a) Bergstad, K.; Jonsson, S. Y.; Bäckvall, J.-E. *J. Am. Chem. Soc.* **1999**, *121*, 10424. (b) Döbler, C.; Mehltretter, G.; Beller, M. *Angew. Chem.* **1999**, *111*, 3211. *Angew. Chem. Int. Ed.* **1999**, *38*, 3026. (c) Krief, A.; Colaux-Castillo *Tetrahedron Lett.* **1999**, *40*, 4189. (d) See also: Wirth, T. *Angew. Chem. Int. Ed.* **2000**, *39*, 334.

7

ASYMMETRIC CARBONYLATIONS

Kyoko Nozaki
Department of Material Chemistry, Graduate School of Engineering, Kyoto University, Yoshida, Sakyo-ku, 606-8501, Japan

Iwao Ojima
Department of Chemistry, State University of New York at Stony Brook, NY 11794-3400, U.S.A.

7.1. INTRODUCTION

Transition metal–catalyzed carbonylations are very important reactions for the formation of a variety of carbonyl compounds in organic syntheses as well as commercial production. For example, hydroformylation of propene, producing butanal is the largest chemical process with homogeneous catalysis to date. In catalytic asymmetric synthesis, hydroformylation and other hydrocarbonylations of olefins have been extensively studied. These processes can be promoted only with man-made catalysts and are attracting much attention because of the fact that the products—optically active aldehydes, carboxylic acids, esters, and their derivatives—are very important as key intermediates for various pharmaceuticals, agrochemicals, and other fine chemicals. As discussed in the chapter on the asymmetric carbonylation written by Consiglio in the original edition of this book, the efficiency of catalytic asymmetric carbonylation reactions did not reach very practical level by the end of 1992. However, a remarkable improvement has been made, especially in the asymmetric hydroformylation of olefins, in the past several years so that this process has now reached highly practical level for organic syntheses in research laboratories as well as industries. This chapter describes recent advances in the asymmetric hydroformylation, hydrocarbonylations, copolymerization of carbon monoxide with olefins, and other related reactions.

7.2. ASYMMETRIC HYDROFORMYLATION

7.2.1. Development of Chiral Catalysts in Perspective

Asymmetric hydroformylation of prochiral olefins has been investigated both for the elucidation of reaction mechanism and for development of a potentially useful method for asymmetric

Catalytic Asymmetric Synthesis, Second Edition, Edited by Iwao Ojima
ISBN 0-471-29805-0 Copyright © 2000 Wiley-VCH, Inc.

organic synthesis. Rhodium and platinum complexes have been extensively studied and, to a lesser extent, cobalt complexes. When an appropriate chiral ligand is introduced to a catalyst, the differentiation of two enantiofaces of a prochiral olefin is conceptually possible in the hydroformylation reaction. There are three classes of alkenes from which enantiomerically enriched aldehydes can be obtained (Eqs. 7.1–3). The asymmetric hydroformylation of 1-alkenes to give the corresponding branched aldehydes regioselectively and enantioselectively (Eq. 7.1) is the most general process, but it is more complicated than other two ways (Eqs. 7.2 and 7.3) because of the formation of achiral linear aldehydes in substantial amounts as undesirable products.

$$R \diagup \xrightarrow[\text{H}_2,\ \text{CO}]{\text{Chiral Catalyst}} \underset{\text{CHO}}{R \diagup \overset{*}{\diagup} \text{Me}} \ + \ \left[R \diagdown \diagup \diagdown_{\underset{\text{achiral}}{\text{CHO}}} \right] \tag{7.1}$$

$$\underset{R^2}{R^1 \diagup} \diagdown \xrightarrow[\text{H}_2,\ \text{CO}]{\text{Chiral Catalyst}} \underset{R^2}{R^1 \diagdown} \overset{*}{\diagdown} \diagup^{\text{CHO}} \tag{7.2}$$

$$R \diagup \diagdown_R \ \text{or} \ R \diagup \overset{R}{\diagup} \xrightarrow[\text{H}_2,\ \text{CO}]{\text{Chiral Catalyst}} \underset{}{R \diagdown} \overset{R}{\underset{*}{\diagdown}} \diagup^{\text{CHO}} \tag{7.3}$$

A variety of enantiopure or enantiomerically enriched phosphines, diphosphines, phosphites, diphosphites, phosphinephosphites, thiols, dithiols, P,N-ligands, and P,S-ligands have been developed as chiral modifiers of rhodium and platinum catalysts [1–7]. Representative chiral ligands discussed in this chapter are shown in Figure 7.1.

In spite of extensive studies on the asymmetric hydroformylation of olefins using chiral rhodium and platinum complexes as catalysts in early days, enantioselectivity had not exceeded 60% ee until the reaction of styrene catalyzed by PtCl$_2$[DBP-DIOP (**1**)]/SnCl$_2$ was reported to attain 95% ee in 1982 [8]. Although the value was corrected to 73% ee in 1983 [9], this result spurred further studies of the reaction in connection to possible commercial synthesis of antiinflammatory drugs such as (*S*)-ibuprofen and (*S*)-naproxen. The catalyst PtCl$_2$[BPPM (**2**)]/SnCl$_2$ was effective for asymmetric hydroformylation of styrene and its derivatives, yielding the corresponding 2-arylpropanal (Eq. 7.4) [10]. For example, this catalyst gave the branched aldehyde with 70–80% ee for styrene, 80% ee for *p*-isobutylstyrene, and 81% ee for 2-ethenyl-6-methoxynaphthalene. Although the branched/linear ratios were low (0.5–0.7) in these cases, the enantioselectivities achieved were considerably higher than those realized by any other chiral catalyst system at that time. A chiral platinum catalyst, PtCl$_2$[BCO-DBP (**3**)]/SnCl$_2$, also achieved 86% ee for the reaction of styrene with much better branched/linear ratio (4:1) [11]. Higher enantioselectivity (91% ee) for the hydroformylation of styrene was achieved with the platinum/bisphosphite **4**/SnCl$_2$ system although regioselectivity did not exceed a branched/linear ratio of 60:40 [12].

$$\underset{}{\text{Ar} \diagdown} \diagdown \xrightarrow[\text{H}_2,\ \text{CO}]{\text{PtCl}_2(\text{DBP-DIOP})\text{-SnCl}_2} \underset{\text{CHO}}{\text{Ar} \diagdown \overset{*}{\diagdown} \text{Me}} \tag{7.4}$$

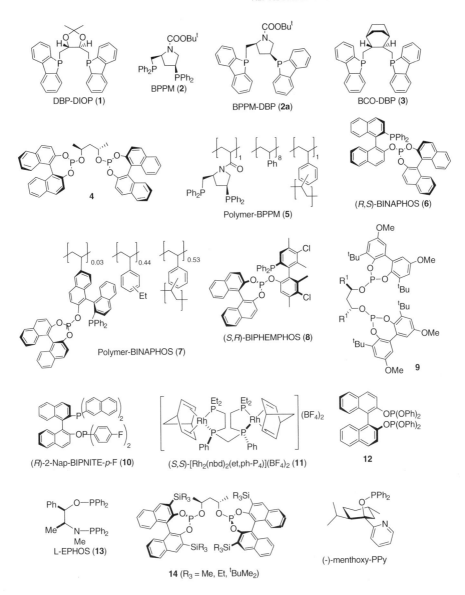

Figure 7.1. Chiral diphosphine, diphosphite, phosphine–phosphite, phosphine–phosphinite, amine–phosphinite, and polymer-anchored ligands used for asymmetric hydroformylation.

One of the difficulties in achieving high enantioselectivity in asymmetric hydroformylation is the propensity of chiral 2-arylpropanal to racemize under the reaction conditions. Accordingly, if the chiral aldehyde can be converted to a less-labile derivative in situ, higher enantioselectivity might be anticipated. In fact, when the asymmetric hydroformylation of styrene and its derivatives catalyzed by PtCl$_2$(BPPM)/SnCl$_2$ was carried out in triethyl orthoformate, the diethylacetal of chiral 2-arylpropanal was obtained with >96% ee (Eq. 7.5) [10,13,14]. It was reported that a chiral platinum-phosphine catalyst anchored on cross-linked beads bearing BPPM (2) as the pendant group [Polymer-BPPM (5)] gave virtually the same enantioselectivity (>98% ee) as that attained by the homogeneous catalyst system for the reaction of styrene [14]. Hydroformylations in triethyl orthoformate are very slow although enantioselectivity achieved is excellent. Thus, this protocol does not appear to be practical, unfortunately, because of the low reaction rate and low regioselectivity. Further studies toward the improvement of chiral platinum catalysts appear to be continuing [1,15–18].

$$Ar\diagdown\joinrel= \quad \xrightarrow[\substack{H_2,\ CO \\ CH(OEt)_3}]{PtCl_2(BPPM)/SnCl_2} \quad \underset{Ar\diagup\diagdown Me}{\overset{H\ CH(OEt)_2}{}} \quad \xrightarrow{PPTS} \quad \underset{Ar\diagup\diagdown Me}{\overset{H\ CHO}{}} \quad (7.5)$$

A breakthrough in asymmetric hydroformylation has been realized in 1993 by using rhodium complexes with a novel phosphine-phosphite ligand, (R,S)-BINAPHOS (6) [19]. The Rh(acac)(CO)$_2$–BINAPHOS catalyst can achieve excellent enantioselectivities (85–95% ee) in the hydroformylations of a variety of prochiral olefins such as vinyl acetate, N-vinylphthalimide, styrene and its derivatives, and 1,3-dienes [19–21] with high branched/linear ratios (84:16–92:8) and good reaction rates at 60–80°C. Polymer-supported (R,S)-BINAPHOS-Rh catalyst (7) has also achieved high enantioselectivities similar to those obtained in the homogeneous system, showing promise for the development of practical heterogenized reusable chiral catalysts (Eq. 7.6) [22].

$$R\diagup\diagdown\joinrel\diagdown \quad \xrightarrow[\substack{H_2,\ CO \\ >97\%}]{\substack{Rh(acac)(CO)_2 \\ BINAPHOS\ (6)}} \quad \underset{\substack{R\diagup\diagdown Me \\ 85\text{-}95\%\ ee \\ b/l = 85/15\text{~}95/5}}{\overset{CHO}{\vdots}} \quad (7.6)$$

R = AcO, phthalimide, phenyl, tolyl, 4-isobutylphenyl, etc.

A similar phosphine-phosphite ligand, (S,R)-BIPHENPHOS (8), has also been developed, and its Rh complex can achieve the same high-level enantioselectivity as that realized by (R,S)-BINAPHOS-Rh catalyst in the reactions of a variety of alkenes [23,24]. The Rh complex with the chiral diphosphite ligand 9 derived from (R,R)-pentane-2,4-diol has shown enantioselectivity up to 90% ee with 98% branched aldehyde selectivity in the reaction of styrene [25,26]. However, the enantioselectivity observed in the reaction of vinyl acetate is only 50% ee, indicating rather narrow scope for its applicability. Further studies toward the development of new and better chiral diphosphite and diphosphinite ligands are actively underway [27–32]. Phosphine-phosphinite ligands such as (R)-2-Nap-BIPNITE-p-F (10) show high asymmetric induction with moderate regioselectivity in the Rh-catalyzed hydroformylation of chiral 4-vinyl-β-lactams [33]. A unique bimetallic Rh-complex with a chiral tetraphosphine ligand (11) has attained 85% ee in the hydroformylation of vinyl acetate [34].

If we categorize the development of chiral catalysts for asymmetric hydroformylation in the past three decades described above, three types of chiral catalysts can be recognized. The "first-generation" catalysts are Rh(I)-complexes with chiral mono- or diphosphine ligands in the 1970s and early 1980s. The "second-generation" catalysts are Pt(II)-complexes with chiral diphosphine ligands in combination with SnCl$_2$ in the 1980s and early 1990s. Thus, there was a shift from Rh-based catalysts to Pt-based catalysts in this period. Then, the "third-generation" catalysts are back to Rh(I)-complexes, but with diphosphite, phosphine-phosphite, and phosphine-phosphinite ligands.

Since the invention of BINAPHOS, a phosphine-phosphite ligand, was a true breakthrough in this catalytic asymmetric process, it is worthwhile mentioning some background for this invention. In 1992, Takaya et al. reported the use of a chiral diphosphite ligand **12** derived from binaphthol for the asymmetric hydroformylation of vinyl acetate [31]. Although the enantiose-lectivity attained in this system was around 50% ee, an interesting difference was observed between the reaction with the chiral diphosphite **12** and that with chiral diphosphines. In the case of chiral diphosphine ligands, at least 3–6 equiv. (to Rh) of the ligand were required to achieve the best enantioselectivity obtainable with the chiral ligand. This requirement is due to the fact that the dissociation of the chiral diphosphine ligand produces "unmodified Rh species," which is active for hydroformylation to produce racemic aldehydes. Thus, to avoid this undesirable formation of achiral Rh species, the use of excess amounts of chiral ligand is necessary. However, the use of only 1.1 equiv. of the chiral diphosphite ligand **12** was sufficient to obtain the product aldehyde with 45% ee, the highest level of enantioselectivity with this ligand [31]. This observation prompted Takaya, Nozaki, and their co-workers to synthesize chiral phosphine-phosphite ligands, (*R,S*)-BINAPHOS (**6**), (*R,R*)-BINAPHOS (RR-**6**), and their enantiomers, which were designed to combine the high enantioselectivity achieved by disphosphine ligands, for example, BINAP in asymmetric hydrogenation, with excellent catalyst stability exhibited by diphosphite ligands apparently through strong coordination to Rh metal [19]. Then, it was found, serendipitously, that a Rh(I) complex with one of the two diastereomers, (*R,S*)-BINAPHOS (**6**), brought about much higher enantioselectivity, mostly >90% ee for a wide variety of substrates, than those achieved by Rh-complexes with chiral diphosphine or diphosphite ligands. The regioselectivity for the formation of branched aldehyde achieved by this ligand was much better than that attained by other ligands. It has been shown that in most cases the use of 2.0–2.5 equiv. of **6** to Rh metal is sufficient enough to achieve the highest enantioselectivity attainable by the system. Another characteristic feature of the phosphine-phosphite ligand is its unsymmetrical structure. Since the invention of DIPAMP, DIOP, and BINAP as excellent ligands for asymmetric hydrogenation, the ligand design had been dominated by the principle of C_2 symmetry [35] until BINAPHOS was invented. The discovery of BINAPHOS, an unsymmetrical bidentate ligand, as the most effective chiral ligand for the catalytic asymmetric hydroformylation has certainly changed the trends in chiral ligand design. Spurred by the exceptional success of BINAPHOS, other unsymmetrical ligands have recently been prepared and applied to asymmetric hydroformylation [1,36–42]. This trend in design may well be carried over to the next decade.

7.2.2 Mechanism of Asymmetric Hydroformylation

7.2.2.1. *Rhodium Complex Catalyzed Reactions*

Catalyst cycle of Rh(I)-phosphine system. Most mechanistic studies on ligand-modified rhodium catalysts have been performed with HRh(CO)(PPh$_3$)$_3$. Extensive mechanistic studies have revealed that HRh(CO)$_2$(PPh$_3$)$_2$ (18-electron species) is a key active catalyst species, which readily reacts with ethylene at 25°C [43]. Two mechanisms, an associative pathway and a dissociative pathway, were proposed [43–46], depending on the concentration of the catalyst.

According to the proposed mechanisms, the associative pathway dominates at catalyst concentrations $>6 \times 10^{-3}$ mol/L, whereas the dissociative pathway that includes the generation of a more active catalyst species, $HRh(CO)_2(PPh_3)$ (16-electron species), through loss of another triphenylphosphine ligand, became predominant at concentrations lower than 6×10^{-3} mol/L [43]. The fact that carbon monoxide and excess triphenylphosphine inhibit the reaction [43] strongly suggests the formation of pentacoordinated acyl-$Rh(CO)_2(PPh_3)_2$ (18-electron species), which does not react with molecular hydrogen because of the lack of a vacant coordination site for molecular hydrogen to undergo oxidative addition [47,48]. Under the industrial reaction conditions producing n-butanal from propene with high selectivity, the use of a large excess of triphenylphosphine is required, which certainly favors the associative pathway. In the associative pathway, it was originally assumed that the coordination of an olefin to $HRh(CO)_2(PPh_3)_2$ would take place to generate a hexacoordinated 20-electron π-olefin-Rh species [47,49]. However, this assumption appears to be unlikely based on the generally accepted 18-electron rule [48,50]. Thus, a modified mechanism that can accommodate these points has been proposed as shown in Scheme 7.1 (triphenylphosphine is replaced by L) [48]. This mechanism is applicable for rhodium complexes with chelating diphosphines and diphosphites, and thus widely accepted as the catalyst cycle of rhodium complex–catalyzed hydroformylations, providing the basis for mechanistic studies in asymmetric hydroformylation. In this mechanism, (i) $HRh(CO)_2L_2$ (18-electron species) is formed as key intermediate; (ii) dissociation of CO from this complex generates a coordinatively unsaturated 16-electron species $HRh(CO)L_2$ that is the active catalyst; (iii) coordination of an olefin, followed by olefin insertion into the Rh–H bond takes place to form alkyl-$Rh(CO)L_2$ complex; (iv) coordination of carbon monoxide is followed by migratory insertion of the alkyl group to one of the coordinated carbon monoxides; and (v) oxidative addition of molecular hydrogen, followed by the reductive elimination of the product aldehyde and regeneration of the active catalyst $HRh(CO)L_2$, which completes the catalytic cycle. It has been shown that the hydrometalation of olefin [step (iii)] proceeds through complete cis addition, and the subsequent migratory insertion of carbon monoxide [step (iv)] takes place with retention of configuration [47,48,51–53].

For reductive cleavage of the acyl-Rh complex, it is generally accepted that molecular hydrogen is the hydrogen donor as shown in Scheme 7.1. However, this step could be effected alternatively by another hydrido-Rh complex, that is, $RC(=O)MLn + LmMH \rightarrow RC(=O)H + LmM-MLn$. It is strongly indicated that this bimolecular reductive cleavage involving two rhodium species is operative under certain reaction conditions and catalyst systems [54–58].

Ab initio molecular orbital studies on the whole catalytic cycle of hydroformylation of ethylene catalyzed by $HRh(CO)_2(PH_3)_2$ has been performed [59,60], which points out the significance of the coordinating solvent—ethylene in this case—and identifies the oxidative addition of molecular hydrogen to the pentacoordinate acyl-Rh complex as the rate-determining step. In fact, this step is the only endothermic process in the catalytic cycle.

Kinetic studies on the effect of partial pressures of hydrogen and carbon monoxide on the reaction rate indicate that the oxidative addition of molecular hydrogen to the Rh complexes with phosphine [44,48] or diphosphite [28] ligands is the slowest step in the whole process. Further mechanistic studies, however, have revealed that the rate-determining step of the reaction depends on the nature of the ligand used and in some cases the alkene insertion into the Rh–H bond becomes the slowest step as observed for the Rh-BINAPHOS-catalyzed reactions [61]. It has also been shown on the basis of deuterium-labeling experiments that the alkene insertion to the Rh–H bond is irreversible for the formation of linear alkyl-Rh species in general, whereas this step may become reversible for the formation of branched alkyl-Rh species depending on the reaction conditions and the nature of the alkene [61].

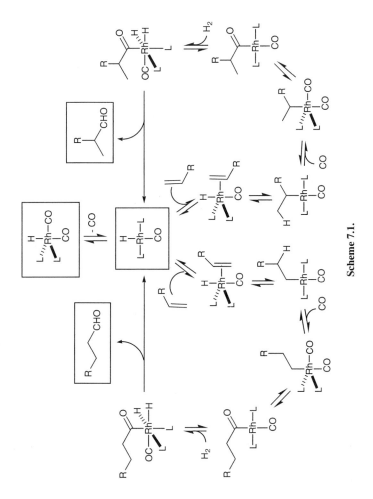

Scheme 7.1.

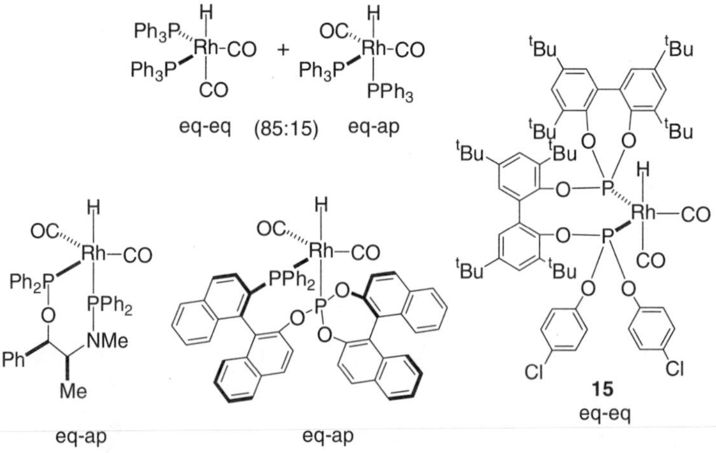

Figure 7.2.   Configurations of various rhodium complexes with chiral ligands.

Structural information about active catalyst species.   To achieve excellent selectivity, it is essential to generate the best catalytically active species exclusively because multiple species may give different products or selectivities through different reaction pathways. It has been shown that RhH(CO)$_2$(PPh$_3$)$_2$ exists as an 85:15 mixture of two isomers at –55°C, in which the two phosphines are placed at equatorial-equatorial and equatorial–apical positions, respectively, and these two isomers are in a rapid equilibrium at higher temperatures (Figure 7.2) [62]. Thus, it is highly desirable to control this type of equilibrium to generate the best catalyst structure for regio- and enantioselection in the asymmetric hydroformylation using chelating bidentate ligands.

The structure of HRh(CO)$_2$[L-EPHOS (**13**)]$_2$ bearing an aminophosphine-phosphinite ligand was determined to be trigonal bipyramidal, in which the aminophosphine moiety and the hydride occupy the two apical positions [63]. It has been shown that RhH(CO)$_2$[(R,S)-BINAPHOS (**6**)] bearing phosphine-phosphite ligand exists as a single species in which the phosphine moiety occupies an equatorial position and the phosphite moiety an apical position that is trans to the hydride [64,65]. The X-ray structure of RhH(CO)$_2$L$_2$ (**15**) bearing a bulky diphosphite was determined, in which two phosphorus atoms occupy the equatorial positions and the hydride as well as one of the two carbonyls, the apical position [30,66,67]. Systematic study on the correlation between the structure of chiral diphosphite ligands **14** and the enantioselectivity obtained in the Rh-catalyzed asymmetric hydroformylation of styrene has disclosed that the highest enantioselectivity is obtained when the chiral ligand coordinates to Rh metal exclusively in an equatorial-equatorial fashion, and the co-existence of an equatorial–apical isomer reduces enantioselectivity [27–30,67]. Although the enantioselectivity-determining step in the catalytic cycle is not directly related to the structure of RhH(CO)$_2$L$_2$, it should be noted that chiral ligands that form rigid RhH(CO)$_2$L$_2$ complex structures can achieve high enantioselectivity in asymmetric hydroformylation.

7.2.2.2.   Platinum Complex Catalyzed Reactions.   The catalyst cycle of the platinum complex–catalyzed hydroformylation in the presence of SnCl$_2$ includes a cationic [HPt^{II}(CO)L$_2$]$^+$ species as the active catalyst, which is generated by the removal of a chloride ligand from its precursor, PtCl$_2$L$_2$, via a mixed bimetallic complex, PtClL$_2$(SnCl$_3$) (Scheme 7.2)

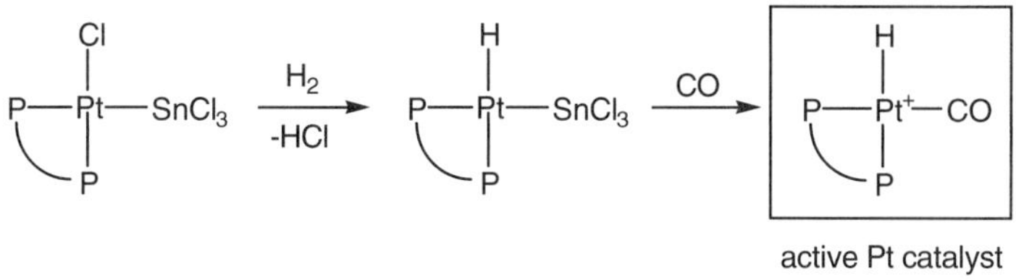

Scheme 7.2.

[68,69]. The rest of the catalyst cycle is identical to that illustrated for the rhodium complex–catalyzed reactions in Scheme 1. It has been proposed that the asymmetric induction occurs during the formation of alkyl-Pt(CO)L$_2$ intermediate through olefin insertion into the Pt–H bond [13].

7.2.3. Proposed Models for Enantioface Selection

Extensive mechanistic studies have been performed on the reactions catalyzed by rhodium and platinum complexes bearing enantiopure C$_2$-symmetric diphosphine ligands [70–72]. As discussed above, (i) the formation of the π-olefin-M(H) complex (M = Rh or Pt), (ii) stereospecific cis addition of the hydridorhodium to the coordinated olefin to form the alkyl-M complex, and (iii) the migratory insertion of a carbonyl ligand giving the acyl-M complex with retention of configuration, have been established in the hydroformylation of 1-alkenes or substituted ethenes. Thus, the enantioselectivity of the reaction giving a branched aldehyde can be determined at the diastereomeric (i) π-olefin-M complex formation step, (ii) alkyl-M complex formation step, or (iii) acyl-M complex formation step. Enantioselectivity-determining step may vary depending on the structure and nature of the chiral ligand as well as the olefin substrate used, and reaction conditions.

  An empirical rule was proposed for the prediction of the absolute configuration of the major aldehyde and regioselectivity in the asymmetric hydroformylation of prochiral olefins catalyzed by rhodium and platinum complexes with C$_2$-symmetrical chiral diphosphine ligands [70–72]. This simple quadrants model assumes the trigonal bipyramidal configuration of the π-olefin-M complex in the transition state of the alkyl-M complex formation (early transition-state model) and defines the large (L) and the small (S) ligands based on experimental results in a consistent manner (Figure 7.3). This model was successfully applied to the reactions of simple aliphatic olefins such as butenes, 2-methyl-1-butene, 2,3-dimethyl-1-butene, and norbornene. However,

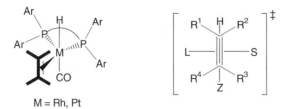

M = Rh, Pt

Figure 7.3. A quadrant model proposed to accommodate the enantioface-selection by rhodium complexes with chiral diphosphine ligands.

the model failed to give meaningful explanation and prediction for the reactions of unsaturated esters and vinylarenes such as styrene and 2-phenylpropene.

Recently, a similar model was proposed on the basis of a theoretical study of the rhodium catalyst with chiral phosphine-phosphite (R,S)-BINAPHOS (**6**) [73]. This model is assuming that the olefin insertion into the Rh–H bond is irreversible (vide supra), and the enantioface selection of a prochiral olefin takes place in this step, that is, the transition state of the alkyl-Rh complex formation step. Thus, the structure of the chiral Rh complex at the transition state becomes important. Two structures shown in Figure 7.4 as TS I and TS II indicate the possible transition states of olefin insertion into HRh(CO)[(R,S)-BINAPHOS (**6**)]. Density functional theory calculations on model rhodium complexes bearing PH_3 ligands were carried out first, followed by force-field calculations on the "real" catalyst system. These calculations indicate that both transition states, TS I and TS II, show lower energies for the *re*-face selection than the *si*-face selection in the case of styrene, which is consistent with the experimental results (Figure 7.4) The high enantioselectivity observed in the reaction of (Z)-2-butene, giving (S)-aldehyde, is also nicely explained with this model. Similar calculations predict that HRh(CO)[(R,R)-

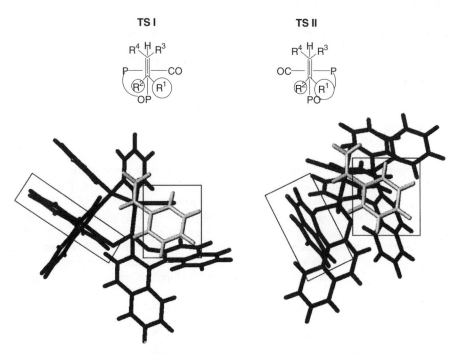

Figure 7.4. Upper: Quadrants model for two possible transition states, TS I (left) and TS II (right) for an olefin ($R^1R^2C{=}CR^3R^4$) insertion into the Rh–H bond of RhH(CO)[(R,S)-BINAPHOS]. Lower: Sybyl (Cache) representation of the molecular mechanics calculations of transition states, TS I (left) and TS II (right), for styrene (grey) insertion into the Rh–H bond of RhH(CO)[(R,S)-BINAPHOS] (black). *Re*-face binding of styrene to Rh is calculated to be lower energy than that of *si*-face binding to both TS I (by 7.1 kcal/mol) and in TS II (by 1.8 kcal/mol). In both structures, the enantioface selection seems to arise from the steric repulsion between the phenyl group of styrene and one of the naphthyls of (R,S)-BINAPHOS (marked with rectangles).

BINAPHOS] would favor the *si*-face selection by TS I and the *re*-face selection by TS II. This mixed result can explain the much lower enantioselectivities (<25% ee) observed when using Rh-(*R,R*)-BINAPHOS catalyst, that is a diastereomer of HRh(CO)[(*R,S*)-BINAPHOS (**6**)].

7.2.4. Scope and Limitations by Substrate Types

7.2.4.1. Vinylarenes Styrene and its derivatives have been most intensively studied as substrates for asymmetric hydroformylation because of their high reactivity and high selectivity for yielding branched aldehydes. Furthermore, aldehydes derived from these olefins can be converted to various pharmaceuticals such as non-steroidal antiinflammatory agents (vide infra) [74]. Representative results for the reaction of styrene (**16a**) are summarized in Scheme 7.3 and Table 7.1.

A good regioselectivity (branched/linear = 80:20) and a high enantioselectivity (85% ee) for the formation of (*S*)-1-phenylpropanal (**17a**) are achieved by using PtCl$_2$/SnCl$_2$/(*R,R*)-BCO-DBP (**3**), but the selectivity to aldehydes, **17a** and **17La**, is rather low (67%) and a substantial amount of hydrogenated compound, ethylbenzene, was formed as byproduct (entry 1) [11]. The reaction catalyzed by PtCl$_2$/SnCl$_2$/(*S,S*)-BPPM (**2**) gives **17a** with lower regioselectivity (branched/linear = 31:69) and enantiopurity (70% ee) (entry 2) [10,13]. Racemization of **17a** was observed due to the acidic conditions in this system. To avoid this racemization, CH(OEt)$_3$ was used as solvent so that the **17a** formed was converted in situ to the corresponding diethyl acetal. This protocol has improved the enantioselectivity dramatically (>96% ee), but the regioselectivity remains low (branched/linear = 33:67) and the reaction rate is very low (entry 3) [10]. The BPPM-Pt catalyst immobilized on a polymer support, PtCl$_2$/SnCl$_2$/Polymer-BPPM (**5**), is reported to exhibit virtually the same regio- as well as enantioselectivity as that achieved by its homogeneous counterpart (entries 4, 5) [75,76]. Platinum catalyst bearing BPPM-DBP (**2a**) with the orthoformate protocol is applied to the reactions of various vinylarenes with excellent enantioselectivity (>96% ee), and much improved regioselectivity (branched/linear 76:24~80:20) (entries 6, 17) [13]. The reaction of styrene (**16a**) catalyzed by Pt(PhCN)$_2$/SnCl$_2$/chiral diphosphite **4** gives **17a** with 91% ee, but the regioselectivity is modest and the aldehyde selectivity is low (46%) (entry 7) [12].

Chiral diphosphite **9** is successfully used in the Rh(I) complex-catalyzed asymmetric hydroformylation of styrene (**16a**), achieving excellent regioselectivity (branched/linear = 98:2) as well as enantioselectivity (90% ee) at 25°C (entry 8) [25]. Rhodium catalyst with chiral diphosphite **14** (R = Me) has also attained 87% ee and excellent regioselectivity at 25°C, although the conversion is low (26%) (entry 9) [30]. Rhodium complex-catalyzed reaction of styrene (**16a**) with chiral phosphine-phosphite ligands, (*R,S*)-BINAPHOS (**6**) or (*S,R*)-BIPHEMPHOS (**8**), gives excellent enantioselectivity (94% ee) with high branched aldehyde selectivity (entries 10, 12) [19,23,64,65]. For the Rh-BINAPHOS (**6**)-catalyzed reaction, a turnover frequency of 100 (mol of aldehydes)·(mol of Rh)$^{-1}$(h)$^{-1}$ is achieved at 40°C under a total pressure of 10 atm (H$_2$/CO = 1/1) (entry 11) [61]. A Rh-BINAPHOS complex supported on a highly cross-linked polymer matrix, Rh[Polymer-BINAPHOS (**7**)], also shows high regio- and enantioselectivities in the reaction of styrene (**16a**) (entry 13) [22].

(*S*)-2-(4-Isobutylphenyl)propanal (**17b**) with 92% ee is obtained from *p*-isobutylstyrene (**16b**) by using the Rh-BINAPHOS catalyst, which is the precursor of antiinflammatory drug (*S*)-ibuprofen (entry 15) [19,64,65]. In a similar manner, the precursor of (*S*)-naproxen is obtained with 85% ee and excellent regioselectivity in the reaction of **16c** catalyzed by Rh-(diphosphite **9**) complex (entry 16) [25]. Pentafluorostyrene (**16e**) is converted to the corresponding branched aldehyde **17e** by the catalysis of the Rh-BINASPHOS complex with excellent regio- as well as enantioselectivity (entry 18) [19,64,65]. The same catalyst is applied to the reactions of 1-propenylbenzenes (**16f** and **16g**) (entries 19, 20), which gives the corresponding (*R*)-aldehyde **17e** (=**17f**) that is known as an intermediate for the synthesis of the spasmolytic butetamate [74]. Reactions of cyclic vinylarenes, indene (**16h**) and 1,2-dihydronaphthalene (**16i**) with the same catalyst, give 1-formylindane (**17h**) and 1-formyl-1,2,3,4-tetrahydro-naphthalene (**17i**), respectively, with high regio- and enantioselectivities (entries 21, 22) [19,64,65]. The former aldehyde **17h** can be converted to the corresponding amines with hypotensive activity in a single step by reductive amination with a Ni or Pt catalyst, and the latter aldehyde **17i** is a synthetic intermediate for the synthesis of a vasoconstrictor tetrahydro-zoline [74].

As a unique medium for asymmetric hydroformylation, supercritical carbon dioxide has recently been examined, which can be carried out in an extremely low catalyst concentration. The reactions of styrene (**16a**) and pentafluorostyrene (**16e**) catalyzed by Rh-BINAPHOS appear to give mixed results that are highly dependent on the reaction conditions [77,78]. Enantioselectivity up to 92–95% ee for **16a** or 85% ee for **16e** has been observed [78]. A biphasic reaction system has also been examined for the reaction using $Rh(acac)(CO)_2$ with a sulfonated diphosphine ligand BINAS [79]. The reaction proceeds smoothly at 40°C and 100 atm in high conversion with excellent branched aldehyde selectivity (95%), but enantioselectivity is very low (18% ee). The use of these newer reaction conditions is still in the very early stage and further development is expected in the next decade.

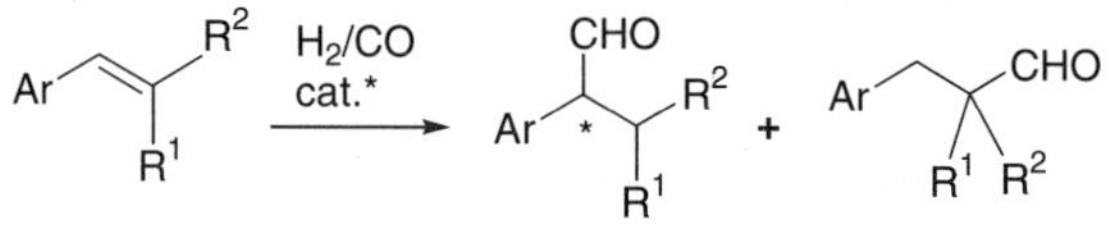

16a-g   **17a-g**   **17La-g**

a: Ar = Ph; R^1 = H; R^2 = H

b: Ar = 4-^{i}Bu-C_6H_4; R^1 = H; R^2 = H

c: Ar = 6-MeO-2-Naphthyl; R^1 = H; R^2 = H

d: Ar = 4-(Thiophene-2-carbonyl)phenyl; R^1 = H; R^2 = H

e: Ar = C_6F_5; R^1 = H; R^2 = H

f: Ar = Ph; R^1 = H; R^2 = Me

g: Ar = Ph; R^1 = Me; R^2 = H

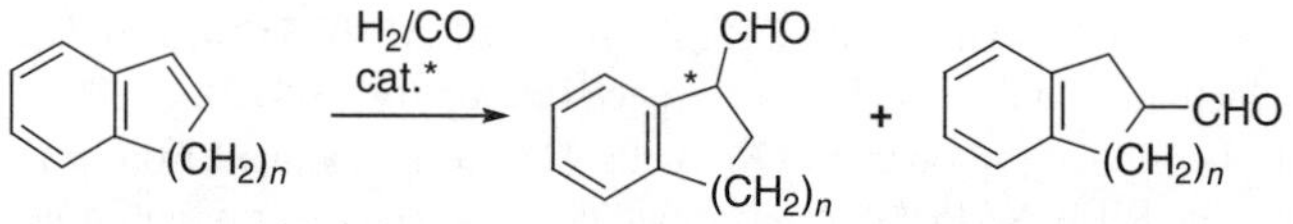

16h: $n = 1$ **17h:** $n = 1$ **17Lh:** $n = 1$
16i: $n = 2$ **17i:** $n = 2$ **17Li:** $n = 2$

Scheme 7.3.

TABLE 7.1. Hydroformylation of Styrene and Its Derivatives Catalyzed by Chiral Pt(II) or Rh(I) Complexes

Entry	Substrate	Metal (additive)	Chiral Ligand	Pressure atm/atm	Temp. °C	Conv. % (time, h)	TOF[a] ×10⁻³	Branched/Linear ratio 17/17L	% ee of 17 (config.)	Reference
1	16a	Pt (SnCl₂)	(R,R)-BCO-DBP (3)	70/140	60	95 (2.3)[b]	41	80/20	85 (S)	11
2	16a	Pt (SnCl₂)	(S,S)-BPPM (2)	81/81	60	40 (4)	4.0	31/69	77 (S)	13
3	16a	Pt (SnCl₂)	(S,S)-BPPM (2)[c]	81/81	60	100 (150)	0.27	33/67	≥96 (S)	10, 13
4	16a	Pt (SnCl₂)	Polymer-BPPM (5)	75/75	60	40 (90)	3.6	37/63	73 (S)	76
5	16a	Pt (SnCl₂)	Polymer-BPPM (5)[c]	92/92	60	22 (2400)	0.04	33/67	≥98 (S)	10
6	16a	Pt (SnCl₂)	(S,S)-BPPM-BDP (2a)[c]	81/81	60	56 (95)	2.4	76/24	≥96 (S)	13
7	16a	Pt (SnCl₂)	4	50/50	17	90 (70)[d]	30	60/40	91 (R)	12
8	16a	Rh	(R,R)-9	19/19	25	nr[e]	f	98/2	90 (S)	25
9	16a	Rh	14 (R = Me)	10/10	15	12 (5)	11	94/6	86 (S)	30
10	16a	Rh	(R,S)-BINAPHOS (6)	50/50	60	>99 (40)	>50	88/12	94 (R)	64
11	16a	Rh	(R,S)-BINAPHOS (6)	5/5	40	52 (5)	100	90/10	94 (R)	61
12	16a	Rh	(S,R)-BIPHEMPHOS	50/50	60	99 (42)	>2.4	90/10	94 (S)	23
13	16a	Rh	Polymer-BINAPHOS	10/10	60	99 (12)	>165	84/16	89 (S)	22
14	16b	Rh	(R,R)-9	15/5	nr[d]	nr[e]	g	99/1	82 (S)	25
15	16b	Rh	(S,R)-BINAPHOS	50/50	60	99 (66)	1.5	88/12	92 (S)	64
16	16c	Rh	(R,R)-9	82/82	60	nr[e]	h	99/1	85 (R)	25
17	16d	Pt (SnCl₂)	(S,S)-BPPM-BDP (2a)[c]	15/3	nr[d]	15 (143)	0.13	77/23	≥96 (S)	13
18	16e	Rh	(R,S)-BINAPHOS (6)	50/50	30	85 (44)	1.9	97/3	96 (R)	64
19	16f	Rh	(R,S)-BINAPHOS (6)	50/50	80	48 (61)	0.79	98/2	80 (R)	64
20	16f+16g (1:1)	Rh	(R,S)-BINAPHOS (6)	50/50	80	53 (38)	1.4	78/22	79 (R)	64
21	16h	Rh	(S,R)-BIPHEMPHOS	50/50	60	62 (20)	0.78	92/8	88 (+)	23
22	16i	Rh	(R,S)-BINAPHOS (6)	50/50	60	79 (20)	12	96/4	96 (−)	64

[a]TOF: (mol of product)·(mol of metal)⁻¹·h⁻¹. [b]Selectivity to aldehyde was 67%. Ethylbenzene is obtained as by product. [c]Triethyl orthoformate is used as a solvent. Product aldehydes are obtained as diethylacetals. [d]Not reported. [e]0.11 g-mole·l⁻¹·h⁻¹ at [Rh] = 250 ppm. [f]0.1 g-mole·l⁻¹·h⁻¹ at [Rh] = 250 ppm. [g]0.1 g-mole·l⁻¹·h⁻¹ at [Rh] = 250 ppm.

441

7.2.4.2. Aliphatic Alkenes and Dienes   In contrast to vinylarenes, the asymmetric hydroformylation of aliphatic alkenes, especially 1-alkenes, is still very challenging although the Rh-BINAPHOS (**6**) catalyst has made significant improvement in enantioselectivity as compared with that achieved by previous catalysts. Representative results are summarized in Scheme 7.4 and Table 7.2. Because symmetrical cis-olefins such as (Z)-2-butene and (Z)-3-hexene do not have an enantioface to be distinguished, the enantioselectivity should arise from the regioselection of the two sp^2-carbons of the olefin substrate coordinated to the metal center by the chiral catalyst. The reaction of (Z)-2-butene catalyzed by Rh(acac)(CO)$_2$/(R,S)-BINAPHOS (**6**) gives (S)-2-methylbutanal as single product with 82% ee (entry 1) [21,64,65]. The absence of pentanal formation in this reaction unambiguously excludes the possibility of isomerization of 2-butene to 1-butene under the reaction conditions. This fact is crucial for achieving high enantioselectivity because it has been shown that 1-butene and 2-butene are converted to 2-methylpropanal of opposite absolute configurations by the same catalyst (entries 1, 7). In contrast to this, the formation of 1-pentanal is observed (2-methylbutanal/1-pentanal = 87:13) in the reactions catalyzed by PtCl$_2$/SnCl$_2$/BCO-DBP (**3**) (entries 2, 4). As compared with the asymmetric hydroformylation of (Z)-2-butene, the reaction of its (E)-isomer proceeds more slowly, resulting in lower enantioselectivity (entry 3). A similar tendency on the relative reactivity between Z- and E-isomers is observed with 3-hexenes (entries 5, 6).

The asymmetric hydroformylation of 1-alkenes suffers from low regioselectivity for the formation of branched aldehydes, although the enantioselectivity has exceeded 80% ee (entries 7,9,11–15) [80]. Introduction of bulky substituent(s) at the C-3 position increases the regioselectivity for branched aldehydes to some extent, and enantioselectivity has reached up to >99% ee (entries 13, 15) [80]. This reaction needs much improvement in regioselectivity to be useful as synthetic method.

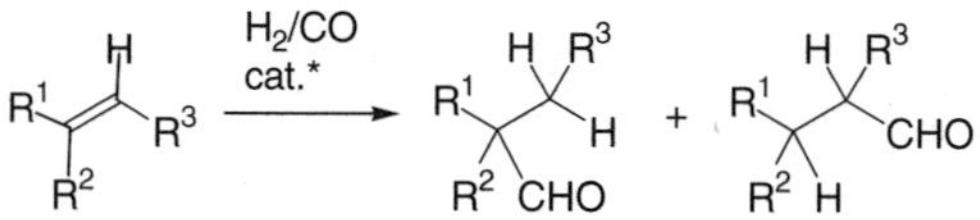

	R^1	R^2	R^3		R^1	R^2	R^3
a:	H	Me	Me	**g:**	^{i}Pr	H	H
b:	Me	H	Me	**h:**	^{t}Bu	H	H
c:	H	Et	Et	**i:**	^{t}BuCH$_2$	H	H
d:	Et	H	Et	**j:**	^{t}BuCH$_2$CH$_2$	H	H
e:	Bu	H	H	**k:**	Ph$_3$C	H	H
f:	Et	H	H				

Scheme 7.4.

Asymmetric hydroformylation of conjugated dienes has been almost unexplored. Recently, however, promising results were reported when Rh(acac)(CO)$_2$/(R,S)-BINAPHOS (**6**) is used as the catalyst. Reactions give the corresponding β,γ-unsaturated aldehydes with high enantiopurity (≤96% ee) and regioselectivity (78~95% branched) (Scheme 7.5) [20,81].

7.2.4.3. Functionalized Alkenes   Asymmetric hydroformylation of functionalized alkenes can serve as a useful method for the syntheses of polyfunctionalized intermediates to biologically

TABLE 7.2. Asymmetric Hydroformylation of Aliphatic Alkenes

Entry	Substrate	Metal (additive)	Chiral Ligand	Temp. °C	CO/H_2 atm/atm	Conv. % (time, h)	TOF^a $\times 13^{-3}$	Branched/Linear Ratio **19/20**	% ee of **19** (config)	Reference
1	**18a**	Rh	(R,S)-BINAPHOS (**6**)	30	50/50	nr[b] (44)	10	—	82 (S)	64, 65
2	**18a**	Pt (SnCl₂)	(R,R)-BCO-DBP (**3**)	80	70/150	67 (21)	31	—[d,e]	30 (R)	11
3	**18b**	Rh	(R,S)-BINAPHOS (**6**)	60	50/50	nr[b] (45)	0.5	—	48 (S)	64, 65
4	**18b**	Pt (SnCl₂)	(R,R)-BCO-DBP (**3**)	80	70/150	65 (30)	21	—[d,e]	29 (R)	11
5	**18c**	Rh	(R,S)-BINAPHOS (**6**)	30	50/50	32 (42)	0.77	—	79 (S)	64, 65
6	**18d**	Rh	(R,S)-BINAPHOS (**6**)	30	50/50	17 (42)	0.41	—	69 (R)	64, 65
7	**18f**	Rh	(R,S)-BINAPHOS (**6**)	30	35/35	nr[b]	23	21/79	83 (R)	64, 65
8	**18f**	Pt (SnCl₂)	(R,R)-BCO-DBP (**3**)	80	70/150	34 (4)	73	14/86	67 (S)	11
9	**18e**	Rh	(R,S)-BINAPHOS (**6**)	30	50/50	24	5.4	24/76	82 (R)	64, 65
10	**18e**		(R,R)-**9**	nr	22/22	nr[b]	c	67/33	20 (S)	25
11	**18g**	Rh	(R,S)-BINAPHOS (**6**)	30	50/50	48	18.7	8/92	83 (R)	64, 65
12	**18h**	Rh	(R,S)-BINAPHOS (**6**)	50	50/50	49	1.4	0/100	—	64, 65
13	**18i**	Rh	(R,S)-BINAPHOS (**6**)	50	50/50	87	1.1	43/57	92 (−)	64, 65
14	**18j**	Rh	(R,S)-BINAPHOS (**6**)	50	50/50	68	1.3	26/74	77(−)	64, 65
15	**18k**	Rh	(R,S)-BINAPHOS (**6**)	50	50/50	20	>5.0	60/40	>99 (+)	64, 65

[a]TOF: (mol of product)·(mol of metal)$^{-1}$·h^{-1}. [b]Not reported. [c]0.15 g-mole·l^{-1}·h^{-1} at [Rh] = 250 ppm. [d]1-pentanal was obtained as by-product. [e]2-methylbutanal/1-pentanal = 87/13.

443

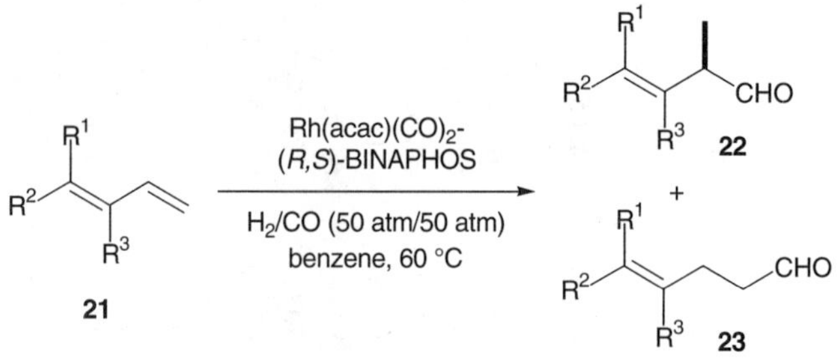

Scheme 7.5.

	R^1	R^2	R^3	Yield (%)	**22/23**	% ee of **22**
a:	H	-(CH$_2$)$_4$-		94	87/13	96
b:	Me	Me	H	92	78/22	72
c:	H	Ph	H	95	95/5	81

active compounds and materials. Representative results are listed in Scheme 7.6 and Table 7.3. Because the reactions catalyzed by platinum catalysts appear to give linear aldehydes (entries 4, 6) as the major products, the use of these catalysts does not seem to be synthetically useful, except for the cases of 1,1-disubstituted ethenes such as dimethyl itaconate (entry 11). Reaction of vinyl acetate (**24a**) catalyzed by Rh(acac)(CO)$_2$/(R,S)-BINAPHOS (**6**) gives (S)-2-ace-toxypropanal (**25a**) with 92% ee accompanied by a small amount of 3-acetoxypropanal (14%) in quantitative yield (entry 1) [64,65]. 2-Acetoxypropanal (**25a**) and its congeners can be readily converted to lactic acid derivatives [74] that have been attracting interest as a monomer of biodegradable or bioabsorbable polymers [82]. Threonine and its derivatives can also be synthesized from these aldehydes. The same reaction catalyzed by Rh(acac)(CO)$_2$/diphosphite **9** gives **25a** exclusively, but enantioselectivity is only 50% ee (entry 2) [25]. Chiral bimetallic catalyst Rh$_2$(allyl)$_2$(et,ph-P4)/HBF$_4$ (**11**) exhibits excellent catalytic activity (TOF = 125 h^{-1}) under low pressure (6 atm) to give **25a** with 85% ee (entry 3) [83].

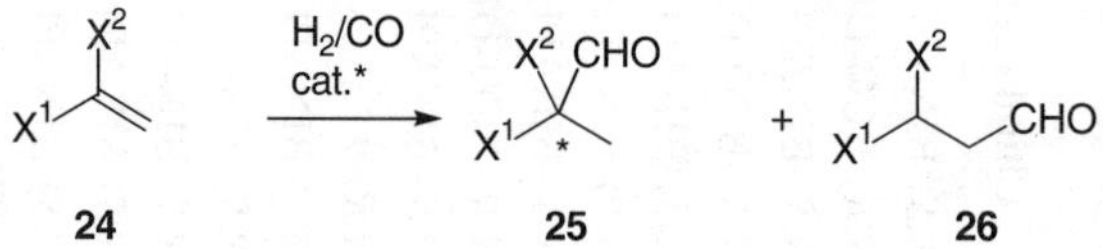

	X^1	X^2
a:	CH$_3$COO	H
b:	C(O)(C$_6$H$_4$)C(O)N	H
c:	p-MeC$_6$H$_4$S	H
d:	CF$_3$	H
e:	CO$_2$Me	H
f:	CO$_2$Me	NHCOMe
g:	CO$_2$Me	CH$_2$CO$_2$Me

Scheme 7.6.

TABLE 7.3. Hydroformylation of Functionalized Alkenes

Entry	Substrate	Metal (additive)	Chiral Ligand	Temp, °C	CO/H$_2$, atm/atm	Conv. % (time, h)	TOF^a h-1	25/26	% ee of 25 (config)	Reference
1	24a	Rh	(R,S)-BINAPHOS (6)	60	50/50	>99 (36)	11	86/14	92 (S)	64, 65
2	24a	Rh	(R,R)-9	50	5/5		c	100/0	50 (S)	25
3	24a	Rh$_2$(nbd)$_2$	[(S,S)-et,ph-P4]$^{2+}$ (11)	90	3/3	nr^b	125	80/20	85 (nr^b)	83
4	24a	Pt (SnCl$_2$)	(S,S)-BPPM (2)	60	90/90	70 (40)	7.6	30/70	80 (S)	10
5	24b	Rh	(R,S)-BINAPHOS (6)	60	50/50	98 (90)	3.3	89/11	85 (R)	64, 65
6	24b	Pt (SnCl$_2$)	(S,S)-BPPM (2)	60	90/90	52 (46)	4.2	33/67	73 (R)	10
7	24c	Rh	(R,S)-BINAPHOS (6)	40	50/50	96 (20)	4.8	96/4	74 (S)	64, 65
8	24d	Rh	(R,S)-BINAPHOS (6)	40	50/50	—(46)	8.3	95/5	93 (S)	64, 65
9	24e	Rh	(—)-menthoxy-PPy (15)	60	30/30	95 (16)	30	97/3	92 (R)	87
10	24f	Rh	(R,R)-DIOP	80	8/82	100 (70)	1.4	100/0	59 (R)	88, 89
11	24g	Pt (SnCl$_2$)	(R,R)-DIOP	100	40/40	80 (45)	3.6	0/100^d	82 (R)^e	90

^{a}TOF: (mol of product)·(mol of metal)$^{-1}$·h^{-1}. ^{b}Not reported. ^{c}0.12-g-mole·1^{-1}·h^{-1} at [Rh] = 250 ppm. ^{d}The selectivity to aldehyde was 35% and hydrogenated product was formed with 65% selectivity, 53% ee (R). ^{e}Aldehyde 25f was not produced and the enantiopurity of 26f is shown.

Reaction of *N*-vinylphthalimide (**24b**) catalyzed by the Rh-BINAPHOS complex gives *N*-(1-formylethyl)phthalimide (**25b**) that can be transformed to alanine and its derivatives (entry 6). In a similar manner, a vinyl sulfide **24c** is converted to the corresponding aldehyde **25c** with high regio- and enantioselectivities in spite of some concern that sulfides often behave as catalyst poison (entry 7) [64,84]. Reaction of trifluoropropene (**24d**) with the same catalyst gives 2-trifluoromethylpropanal with 93% ee (entry 8) [64,65], which is known to be a versatile intermediate to CF$_3$-containing amino acids and peptides of biological interest. This reaction provides a convenient way to synthesize such compounds [85,86]. The reaction of methyl acrylate (**24e**) catalyzed by a cationic rhodium complex with a P-N ligand, [Rh(CO)(PPh$_3$){(−)-menthoxy-PPy (**15**)}]ClO$_4$ gives 2-formylpropanoate (**25e**) with excellent enantioselectivity and regioselectivity (entry 9) [87]. Dehydro amino acid **24f** is converted to the corresponding tertiary aldehyde **25f** with 100% selectivity and 59% ee by using RhH(CO)(PPh$_3$)$_3$/(*R,R*)-DIOP (entry 10) [88,89]. With the same catalyst, however, the reaction of dimethyl itaconate (**24g**) yields the product with only 9% ee (95% branched) [90,91]. The same reaction catalyzed by PtCl$_2$/SnCl$_2$/(*R,R*)-DIOP gives the opposite regio isomer **26g** with 82% ee, but the selectivity to aldehyde is only 35% (entry 11) [90].

Asymmetric hydroformylation of heterocyclic olefins provides potentially useful synthetic building blocks for the syntheses of biologically active compounds. The reactions of 2,5-dihydrofuran **27a** (Eq. 7), dihydropyrroles (**27b**, **27c**, and **29**) (Eqs. 7.7, 7.8), and dioxepin **32** (Eq. 7.9) give the corresponding aldehydes with 47~88% ee [92]. Reaction of α-methylene-γ-butyrolactone (**34**) using cationic Rh-(*R*)-BINAP complex as the catalyst affords the formyllactone **35**, bearing a quaternary chiral center, with ≤37% ee (Eq. 7.10) [93].

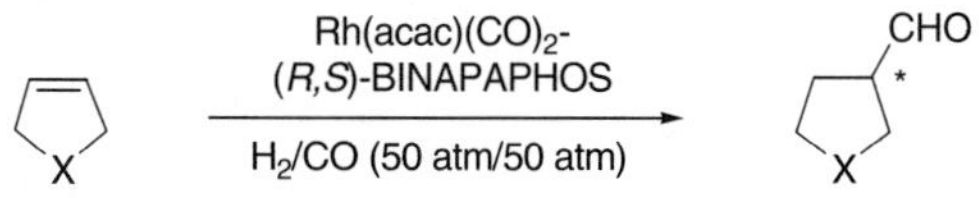

$$(7.7)$$

a: X = O	63%	62% ee (*R*)
b: X = NBoc	98%	47% ee (*R*)
c: X = NAc	92%	66% ee (-)

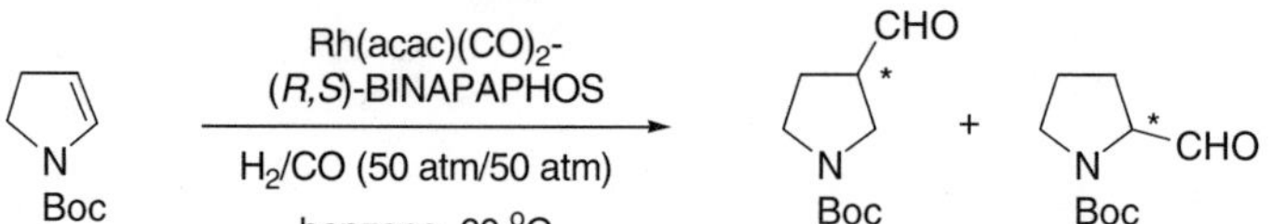

33% (71% ee (*S*))   67% (97% ee (*S*))

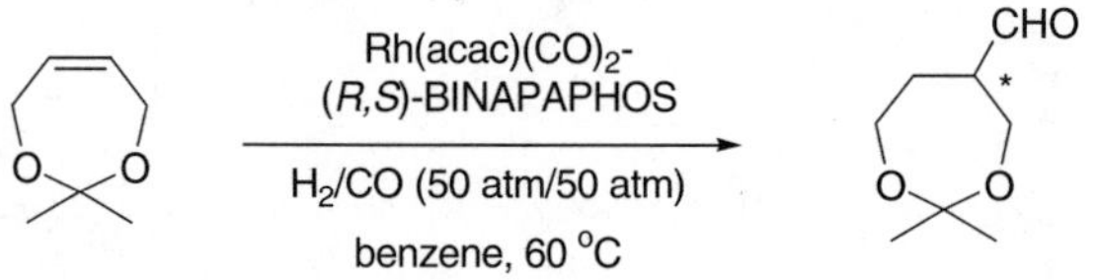

$$(7.8)$$

$$(7.9)$$

77% (73% ee (*R*))

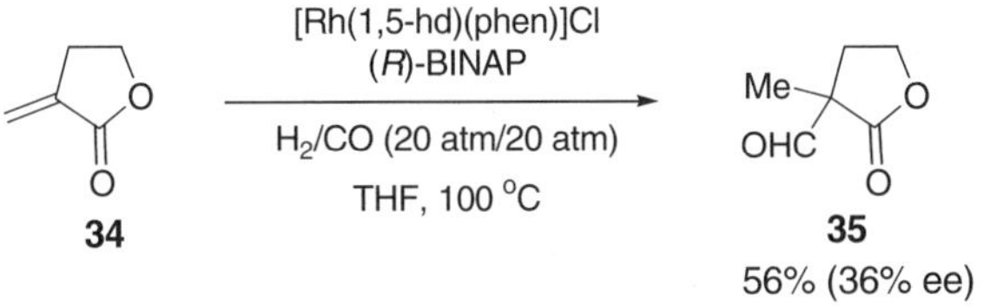

$$(7.10)$$

Reaction of cinnamyl alcohol (**36**) catalyzed by Rh-BINAPHOS gives the product as lactol **37** (1:1 mixture of diastereomers at the anomeric carbon) with high enantioselectivity (88% ee) [94] (Scheme 7.7). The enantiopurity of lactol **37** is determined by oxidizing the lactol to the corresponding lactone **38**. In the same manner, homoallyl alcohol (**39**) is converted to the corresponding α-methyl-γ-butyrolactone (**42**) with 73% ee via lactol **40** [94] (Scheme 7.7). However, the regioselectivity of the reaction is not favorable to the formation of **40**, forming achiral δ-lactol **41** as the major product.

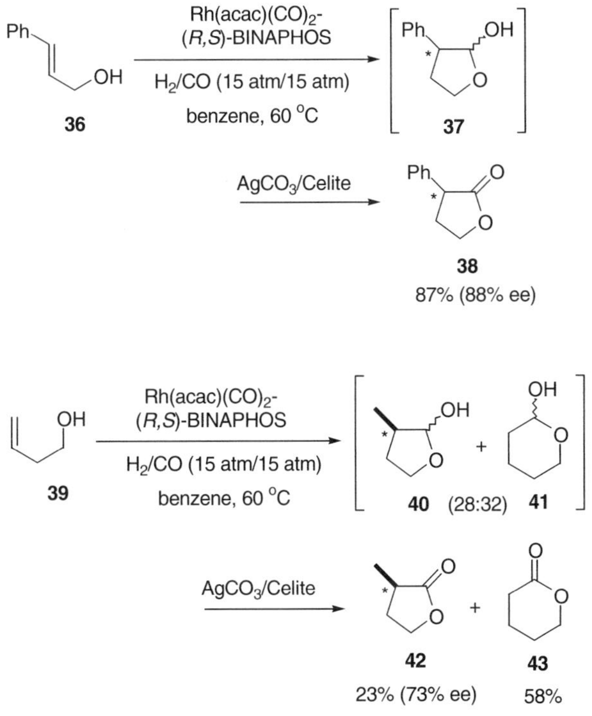

Scheme 7.7.

7.2.4.4. Application to the Synthesis of Fine Chemicals

One of the goals of homogeneous asymmetric catalysis is its applications to the production of fine chemicals such as pharmaceuticals, agrochemicals, flavors, and fragrances. To this end, chiral aldehydes with high enantiomeric purities obtained through asymmetric hydroformylation can serve as useful intermediates for pharmaceutical drugs [74]. For example, (S)-2-arylpropanals can be oxidized to the corresponding (S)-2-arylpropanoic acids **44**, which are antiinflammatory drugs such as

(*S*)-ibuprofen (**44b:** Ar = 4-isobutylphenyl), (*S*)-naproxen (**44c:** Ar = 6-MeO-naphthyl), and (*S*)-suprofen (**44d:** Ar = 4-(2-thienylcarbonyl)phenyl) (Eq. 7.11) [10]. Thus, the asymmetric hydroformylation of vinylarenes discussed above provides potentially efficient route to these drugs.

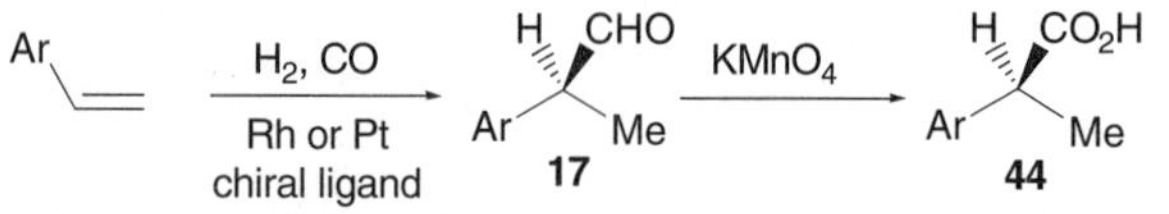

$$Ar\diagdown= \xrightarrow[\text{chiral ligand}]{\substack{H_2, CO \\ \text{Rh or Pt}}} \underset{\textbf{17}}{Ar\diagdown{Me}}^{H\diagup{CHO}} \xrightarrow{KMnO_4} \underset{\textbf{44}}{Ar\diagdown{Me}}^{H\diagup{CO_2H}} \qquad (7.11)$$

Application of catalytic asymmetric hydroformylation often includes the transformation of substrate olefins possessing one or more chiral centers. For example, the hydroformylation of (3*S*,4*R*)-3-{(*S*)-1-(*tert*-butyldimethylsilyloxy)ethyl}-4-vinyl-β-lactam (**45**) with (*R*,*S*)-BINAPHOS-Rh catalyst affords the desired product **46β**, its epimer **46α**, and their linear isomer **47** in a ratio of 51:4:45 (Scheme 7.8). In addition to the phosphine-phosphite ligand (*R*,*S*)-BINAPHOS, another class of chiral ligands, phosphine-phosphinites such as (*R*)-2-Nap-BIPNITE-*p*-F (**10**), is also effective, giving a better result (92% de; **46β**:**46α**:**47** = 71:3:26) [33]. The aldehydes **46** and **47** can be oxidized to the corresponding carboxylic acids without epimerization. Asymmetric synthesis of 1β-methylcarbapenem is of much interest due to its high potential as antibacterial antibiotic drug, and **46β** is an attractive intermediate for this family of compounds. Although the regioselectivity is apparently an issue, a commercial application of this method appears to be currently under investigation [95,96].

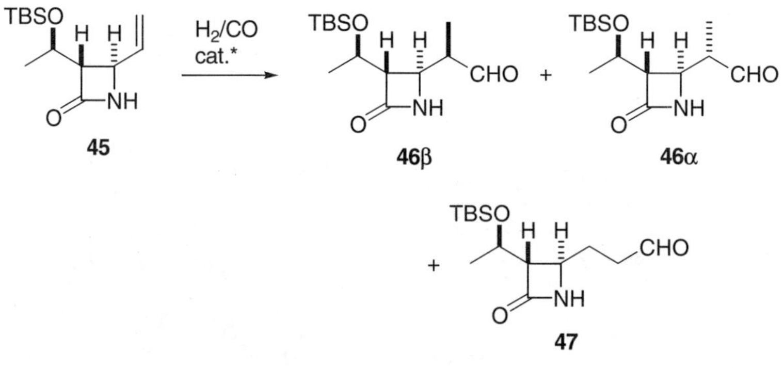

	yield (%)	**46β** : **46α** : **47**
(*R*,*S*)-BINAPHOS	95	51 : 4 : 45
(*R*)-2-Nap-BIPNITE-*p*-F	95	71 : 3 : 26

Scheme 7.8.

7.3. ASYMMETRIC HYDROCARBOXYLATION AND RELATED REACTIONS

Asymmetric hydrocarbalkoxylation of alkenes has been studied since the early 1970s, but the number of papers published on this subject is much less than that of asymmetric hydroformylation. This difference is mainly due to the fact that these reactions catalyzed by palladium complexes with chiral phosphine ligands usually require a very high pressure of carbon monoxide, which is rather difficult for academic laboratories to deal with. However, new processes that do not require high pressure have been developed, which make these potentially useful reactions in organic synthesis more attractive. Typical chiral ligands used in these reactions are shown below.

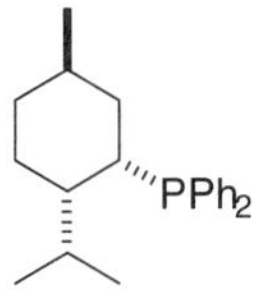

NMDPP (**48**) 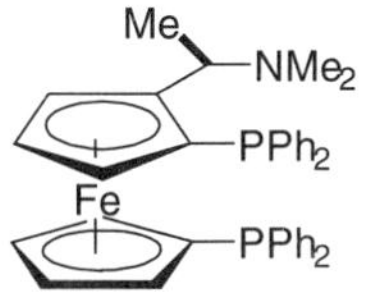   BPPFA (**49**) 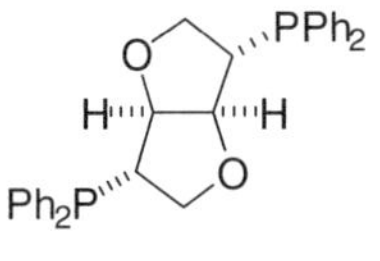   **50** 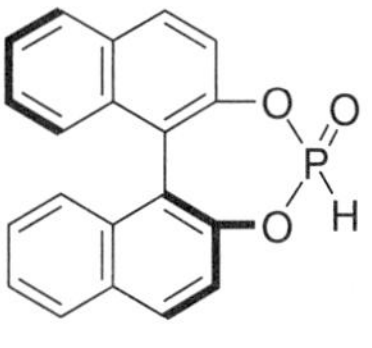   (*R*)-(–)-BNPPA (**51**)

Two possible mechanisms have been proposed for hydrocarboxylation as illustrated in Scheme 7.9 [71,97–100]. The originally proposed mechanism is similar to that of hydroformylation in which the catalyst cycle starts with a hydridometal complex (cycle A) [71,97,99]. In the catalyst cycle A, an olefin inserts into the Pd–H bond, followed by migratory insertion of CO to the resulting alkyl–Pd bond to form an acyl-Pd complex. Alcoholysis of the acyl-Pd complex yields the product ester and regenerates the catalytically active Pd-H species. In the catalyst cycle B, the catalytically active species is the carbalkoxy-Pd complex [98,100]. This mechanism includes the olefin insertion to the ROCO-Pd species and the subsequent alcoholysis of the resulting alkyl-Pd species, which forms the product ester and the alkoxy-Pd complex. Insertion of CO to the alkoxy-Pd species regenerates the carbalkoxy-Pd complex. It is likely that either mechanism is operative depending on the nature of the Pd catalyst and reaction conditions, especially in the presence of an acid or a base. It has been suggested that the enantioselectivity-determining step is the olefin complexation to a chiral Pd complex regardless of the catalyst cycle [71].

Representative results on the asymmetric hydrocarbalkoxylation and hydrocarbohydroxylation are summarized in Table 7.4. Apparently, it has been difficult to achieve high enantioselectivity in these reactions [71]. For example, until 1997 the best enantioselectivity attained in the asymmetric hydrocarbalkoxylation was 69% ee (at 8% conversion) in the reaction of 2-phenylpropene (**52a**) with *tert*-butanol catalyzed by a palladium complex with DBP-DIOP (**1**) (100°C and 238 atm of CO), giving *tert*-butyl 3-phenylbutanoate (**53a**) (entry 4) (Eq. 7.12) [101]. A closely related catalyst system, PdCl$_2$/DIOP, catalyzes the reactions of **52a** and methyl methacrylate (**52b**) at 100~120°C and 380~400 atm of carbon monoxide to give **53a** (59% ee) and methylsuccinic acid monomethyl ester (**53b**, 49% ee) (Eq. 7.12), respectively (entries 6, 7) [102,103]. Although an improved process for asymmetric hydrocarbalkoxylation under mild conditions (at 50°C and 1 atm of CO) was developed in 1982 by using Pd(dba)$_2$, neomenthyldiphenylphosphine (NMDPP **48**), and trifluoroacetic acid in methanol (entry 1) [104], further development has not been made, for some reason, in this catalyst system. However, quite recently a similar system, Pd(OAc)$_2$/BPPFA (**49**)/*p*-toluenesulfonic acid, has achieved 86% ee in the reaction of styrene (entry 2) [105]. Very recently, a PdCl$_2$/CuCl$_2$/diphosphine catalyst system with a unique chiral diphosphine **50** was reported to achieve 99% ee (entry 3) [106], which is very encouraging and warrants further investigation into the scope and limitation of this catalyst system.

For asymmetric hydrocarbohydroxylation, an efficient catalyst system consisting of PdCl$_2$, CuCl$_2$, and (*R*)-1,1′-binaphthyl-2,2′-diyl hydrogen phosphate, (*R*)-(–)-BNPPA (**51**) was intro-

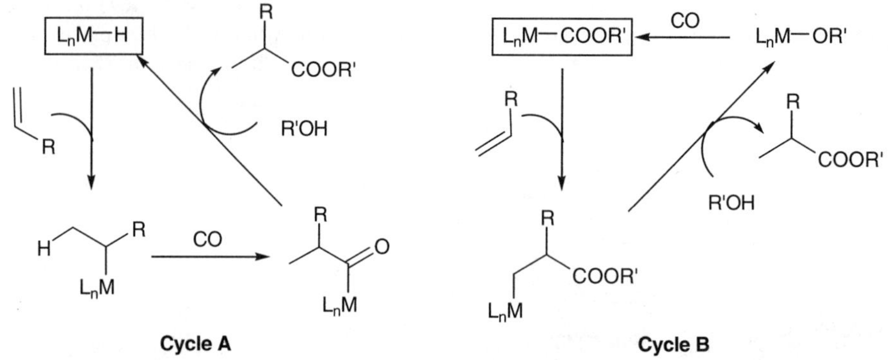

Cycle A   **Cycle B**

Scheme 7.9.

duced in 1990, which can promote the reaction at ambient temperature and pressure (entries 4, 5) [107]. The reactions of 4-isobutylstyrene (**16b**) and 6-methoxy-2-naphthylethene (**16c**) promoted by the palladium catalyst system PdCl$_2$/CuCl$_2$/(R)-(−)-BNPPA (**51**) give (S)-ibuprofen (**44b**) with 83–84% ee and (S)-naproxen (**44c**) with 91% ee (Scheme 7.10 and Eq. 7.13), respectively [108]. Although 10–25 mol % of the chiral palladium catalyst is required to promote the reaction efficiently, this process has high potential because of mild reaction conditions and high enantioselectivity achieved.

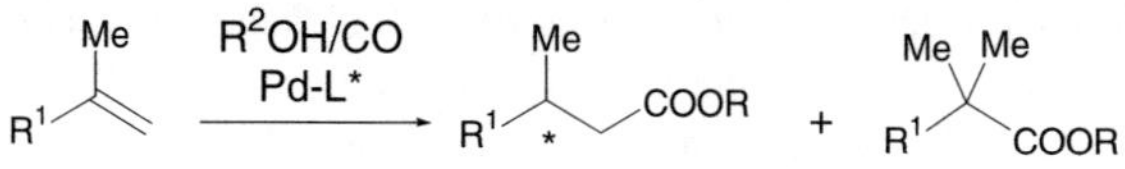

52   **53**

a: R^1 = Ph, R^2 = ^{t}Bu
b: R^1 = CO$_2$Me, R^2 = Me

(7.12)

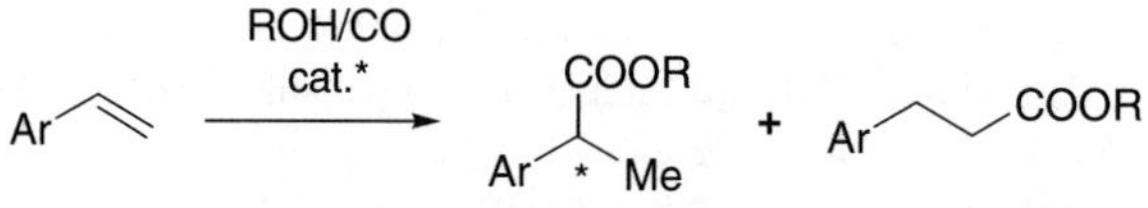

16   **44** (R = H)   **44L** (R = H)
              52 (R = Me)   **52L** (R =Me)

a: Ar = Ph
b: Ar = 4-^{i}Bu-C$_6$H$_4$
c: Ar = 6-MeO-2-naphthyl

Scheme 7.10.

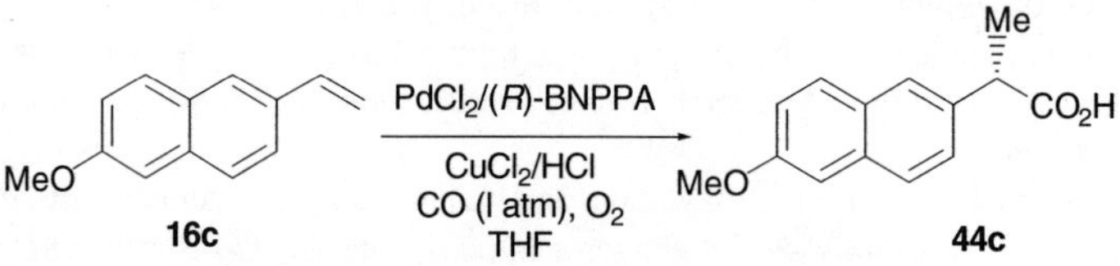

16c   **44c**   (7.13)

TABLE 7.4. Hydrocarboxylation of Olefins Catalyzed by Chiral Pd(II) Complexes

Entry	Substrate	ROH	Catalyst	CO atm	Temp. °C	Yield % (time, h)	Branched/Linear	Product	% ee (config.)	Reference
1	**16a**	MeOH	Pd(dba)$_2$/NMDPP (**48**)	1	50	94 (4)	94/6	**44a**	52 (nr[b])	104
2	**16a**	MeOH	Pd(OAc)$_2$/TsOH/ (S)-(R)-BPPFA (**49**)	20	r.t.	17 (20)	44/56	**44a**	86 (S)	105
3	**16a**	MeOH	PdCl$_2$/CuCl$_2$/**50**	50	80	97 (24)	96/4	**44a**	99 (S)	106
4	**16b**	H$_2$O	PdCl$_2$/(S)-BNPPA (**51**)/ HCl/H$_2$O/CuCl$_2$/O$_2$	<1	r.t.	89 (18)	100/0	**44b**	83 (S)[a]	107
5	**16c**	H$_2$O	PdCl$_2$/(S)-BNPPA (**51**)/ HCl/H$_2$O/CuCl$_2$/O$_2$	<1	r.t.	71 (18)	100/0	**44c**	85 (S)[a]	107
6	**52a**	t-BuOH	PdCl$_2$(PhCN)$_2$/ (R,R)-DBP-DIOP (**1**)	238	100	8 (94)	100/0	**53a**	69 (S)[a]	101
7	**52b**	MeOH	PdCl$_2$[(R,R)-DIOP]	380	120	100 (nr[c])	100/0	**53b**	59 (S)[a]	103

[a]Optical purity. [b]Not reported.

451

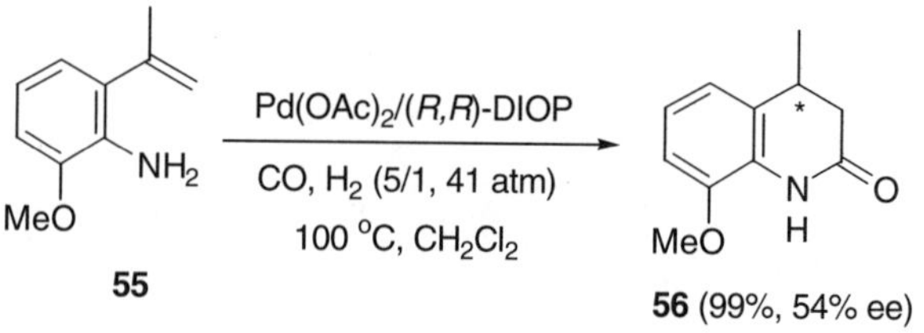

$$(7.14)$$

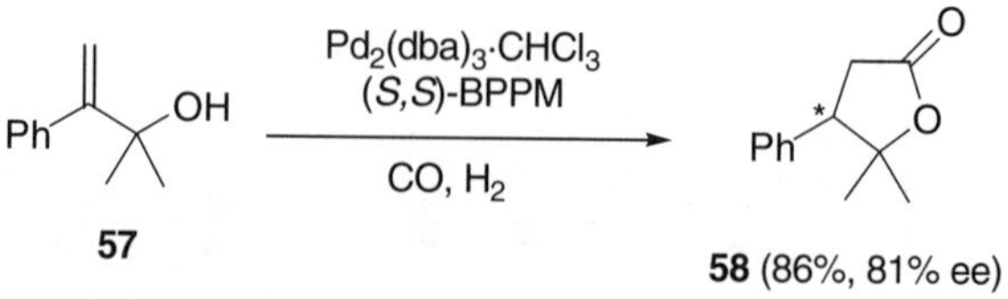

55

56 (99%, 54% ee)

$$(7.15)$$

57

58 (86%, 81% ee)

Asymmetric cyclohydrocarbonylation of 2-(1-methylvinyl)aniline **55** catalyzed by Pd(OAc)$_2$/(R,R)-DIOP at 100 °C and 41 atm (CO/H$_2$ = 5/1) gives the corresponding lactam **56** with 54% ee (Eq. 7.14) [109]. In a similar manner, the reaction of an allylic alcohol **57** Pd$_2$(dba)$_3$/(S,S)-BPPM affords γ-butyrolactone **58** with 81% ee (Eq. 7.15) [110,111]. These processes have excellent potential as useful synthetic methods in organic synthesis.

7.4. ASYMMETRIC ALTERNATING COPOLYMERIZATION OF OLEFINS WITH CO

Alternating copolymerization of carbon monoxide with ethene was discovered in the early 1950s [112] and has been attracting much interest lately. Among the several metals such as, Ni(II), Pd(II), and Rh(I) that can promote this reaction [113], Pd(II) complexes with a cis-chelating diphosphine ligand and a non-coordinating counter anion were selected as the catalysts of choice and have been studied extensively [114,115].

When a 1-alkene, for example, propene and styrene, is used in place of ethene, three more factors need to be controlled to obtain a stereoregular alternating copolymer (Figure 5): (i) *Regioselectivity*: Depending on the regioselectivity of the alkene incorporation into the polymer chain, three different units are possible that is, (a) head-to-tail, (b) head-to-head, and (c) tail-to-tail modes. If the polymer chain grows with the same regioselection, either 1,2- or 2,1-insertion, the polymer should solely consist of the head-to-tail units. (ii) *Tacticity*: The head-to-tail copolymer possesses chiral centers in the main chain. In a syndiotactic copolymer, the absolute configurations of the chiral centers are arranged in an alternating manner, that is, *RSRS*---, while an isotactic copolymer consists of the chiral centers of the same absolute configuration, either *RRRR*--- or *SSSS*---. (iii) *Enantioselectivity*: Isotactic copolymer is chiral and two enantiomers, *RRRR*--- and *SSSS*---, can be formed.

This section summarizes recent advances in the asymmetric copolymerization of 1-alkene and carbon monoxide, giving the corresponding head-to-tail isotactic copolymer with high regio- and enantioselectivity as well as tacticity.

7.4.1. Mechanism of Copolymerization of Alkenes with CO

The mechanism for the alternating copolymerization of ethene with CO has been studied both experimentally and theoretically to explain the completely alternating nature of this process.

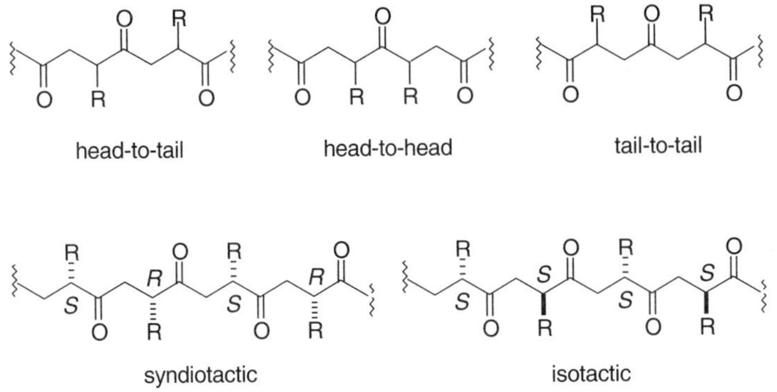

Figure 7.5. Possible modes of regiochemistry and main chain chirality in the stereoregular alternating alkene-CO copolymers.

The copolymer should be formed through the alternating migratory insertion of CO to alkyl-Pd species as well as alkene to acyl-Pd species, but not through the insertion of CO to acyl-Pd species or that of the alkene to alkyl-Pd species (Scheme 7.11). It has been shown that double carbonylation is thermodynamically very unfavorable in Pd-phosphine complexes [116,117]. Recently, the activation energy for ethene and CO insertion to an alkyl-Pd complex with a phenanthroline ligand was measured, which concluded that the CO insertion should be kinetically favored over the ethene insertion [118]. Theoretical studies with model ligands, (Z)-$H_2NCH=CHNH_2$ and (Z)-$H_2PCH=CHPH_2$, are consistent with these experimental results [119–121].

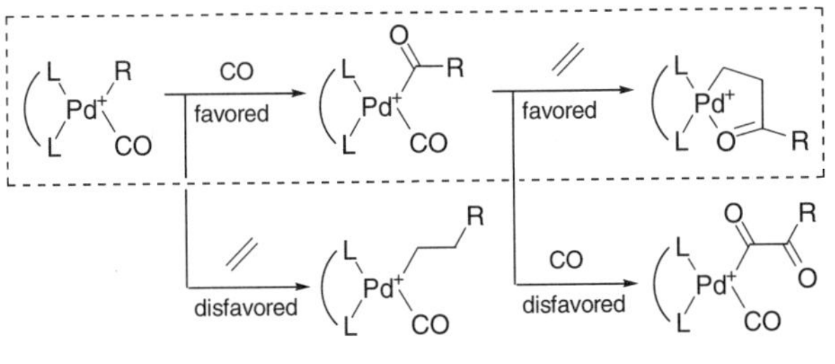

Scheme 7.11.

7.4.2. Scope and Limitation

7.4.2.1. Propene and Aliphatic 1-Alkenes The first example of alternating copolymerization of propene with carbon monoxide was claimed in 1985 in a patent and used L_2PdX_2 wherein L_2 represents a bidentate phosphine ligand and X stands for a weakly or non-coordinating anion [122]. By 1992, the head-to-tail selectivity was improved to almost 100% by using 1,3-(diisopropylphosphino)propane (dippp) in place of its 1,3-diphenylphosphino congener (dppp), and the isotacticity of the resulting copolymer triad was ca. 65% [123].

The first successful example of asymmetric copolymerization of propene with carbon monoxide was reported in 1992 and used a Pd complex with a chiral electron-rich bisphosphine

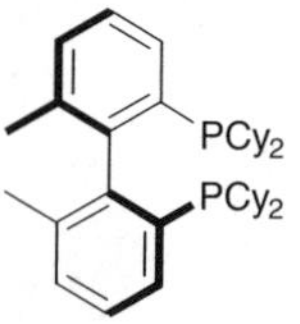

(S)-BICHEP (**59**)
Cy = cyclohexyl

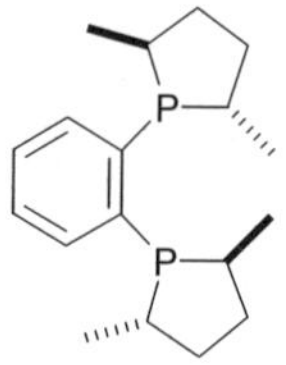

(S,S)-Me-DUPHOS (**60**)

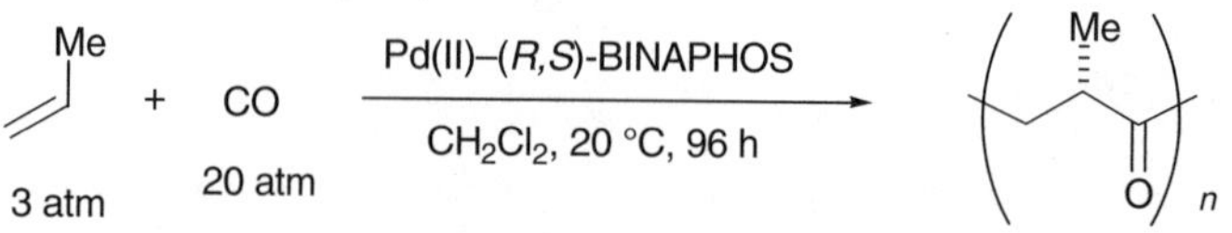

$[\Phi]_D^{24}$ + 40 (c 0.51, $(CF_3)_2CHOH$)
Mn = 65,000, Mw/Mn = 1.6

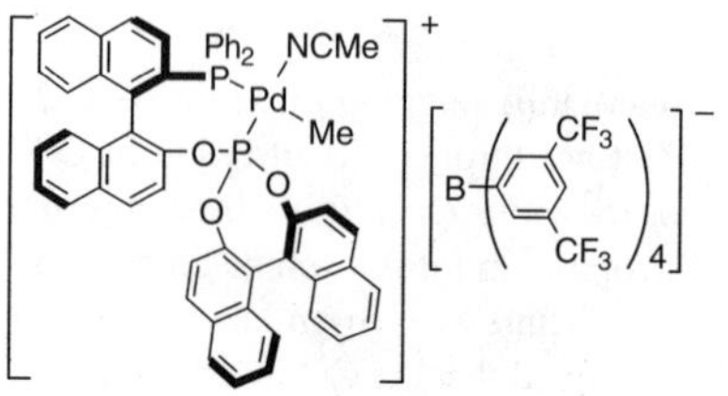

Pd(II)–(R,S)-BINAPHOS

Scheme 7.12.

ligand, BICHEP (**59**) [124], yielding an isotactic copolymer (isotacticity ca. 72%) [125]. The isotacticity in this reaction has been improved to almost 100% later [126]. Similar completely isotactic alternating copolymerization was achieved by using Pd complexes with Me-DUPHOS (**60**) [127] and (R,S)-BINAPHOS (**6**) (Scheme 7.12) [128].

Absolute configuration of the chiral centers in the isotactic copolymer main chain can be determined by comparing the CD spectrum of the copolymer that is a polyketone with that of (S)-3-methylpentan-2-one [126]. A recent model study has confirmed this assignment, showing that the copolymer with (S)-configuration in the main chain exhibits plus optical rotation in $(CF_3)_2CHOH$ and minus in $CHCl_3$ [128–130]. The study has also revealed that the enantioselectivity for the propene insertion is at least 95% ee.

Under the standard conditions, alternating copolymerization is initiated by the insertion of propene to the Pd-carbon bond of Pd(COOMe) species. Addition of a large amount of an oxidant results in shortening the copolymer chain, that is, inhibiting propagation. Thus, with this protocol, dimers were prepared to estimate the enantiopurity of the chiral centers in the polymer main chain [131]. For example, a head-to-tail and anti-dimer with >98% ee was obtained in the reaction catalyzed by a cationic Pd-[MeO-BICHEP] complex with 85% selectivity over other diastereomers (Scheme 7.13).

It has been shown that the propene/CO isotactic copolymer has a γ-polyketone structure in chloroform or $(CF_3)_2CHOH$ solution, whereas the copolymer partially forms polyspiroketal structure in the solid state on the basis of solution and solid state NMR studies [132]. The polyketone-polyspiroketal interconversion readily takes place in the presence of a Brønsted acid (Scheme 7.14).

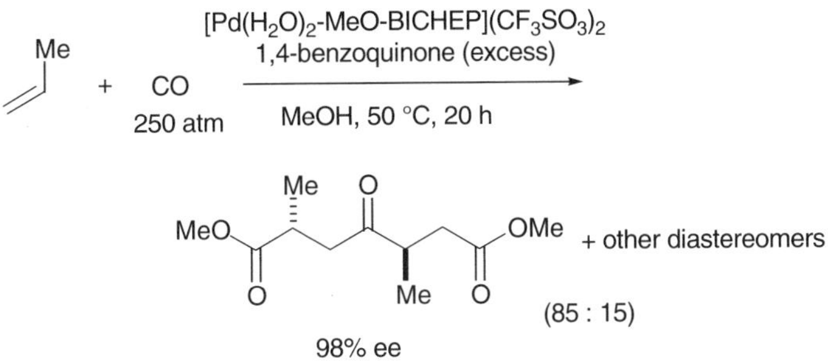

Scheme 7.13.

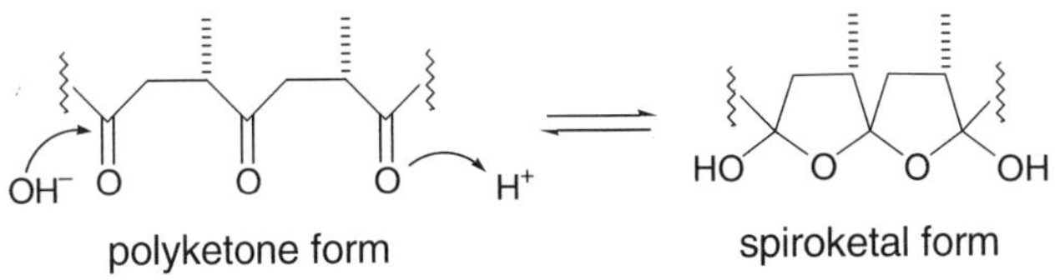

polyketone form   spiroketal form

Scheme 7.14.

Higher aliphatic 1-alkenes exhibit a slightly lower reactivity as compared with propene, but the head-to-tail as well as isotactic selectivity is as high as that observed for propene [132,133]. The copolymers arising from higher 1-alkenes tend to possess the polyspiroketal form more extensively as compared with the propene/CO copolymer. For example, poly(1-hexene-*alt*-CO) exists as a mixture of polyketone and polyspiroketal, even in a CDCl$_3$ solution based on ^{13}C NMR analysis [133].

7.4.2.2. Styrene Vinylarenes such as styrene and its derivatives, show different behaviors from those of aliphatic 1-alkenes described above. Completely alternating head-tail copolymer of styrene and CO can be obtained by using a Pd catalyst with an achiral bidentate nitrogen ligand, 1,10-phenanthroline or 2,2′-bipyridyl, but this copolymer is >90% syndiotactic (Eq. 7.16)

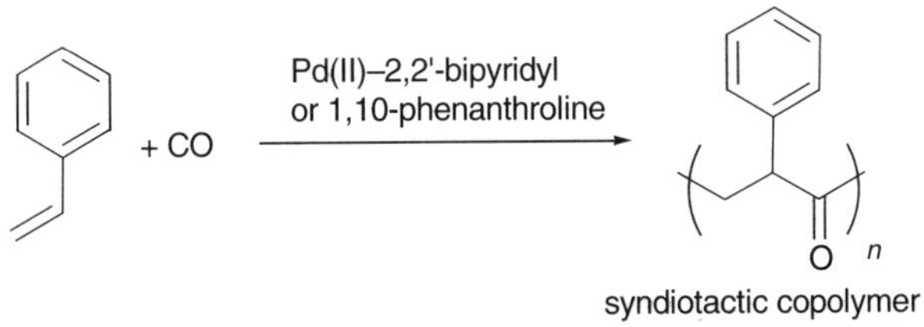

(7.16)

syndiotactic copolymer

[71,134,135]. This fact strongly suggests that the enantioface of the incoming styrene is distinguished by the chiral center arising from the last styrene unit that is incorporated to the copolymer chain, that is, chain-end control [136]. It should be noted that only low-molecular-weight oligoketone can be obtained when using Pd catalysts with phosphine ligands. In addition, the styrene/CO copolymer has a higher tendency than the propene/CO copolymer to be terminated by β-hydride elimination.

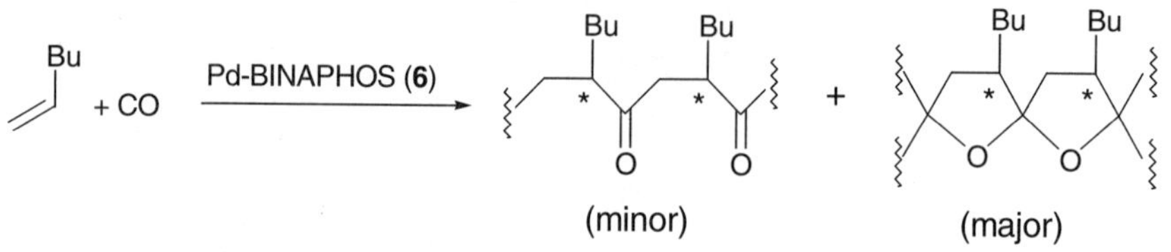

Scheme 7.15.

The first example of the asymmetric copolymerization of styrene with CO was realized by using a Pd catalyst with a chiral bisoxazoline ligand **61a** [137a] (Eq. 7.17). In this reaction, the enantioface selection was made by the chiral catalyst and not by the chain-end, giving completely isotactic copolymer. As an interesting extension of this reaction, a stereoblock polyketone has been synthesized through an alternating copolymerization of 4-*tert*-butylstyrene with carbon monoxide using a combination of Pd catalysts with achiral and chiral ligands (Eq. 7.18) [137]. The copolymerization is initiated with a chiral catalyst with a chiral bisoxazoline ligand **61b** affording the isotactic polyketone block [137b]. Addition of 2,2′-bipyridyl into the reaction mixture results in the elongation of the syndiotactic polyketone block. The efficient displacement of the bisoxazoline ligand by 2,2′-bipyridyl ligand is the key to the successful stereoblock copolymer formation.

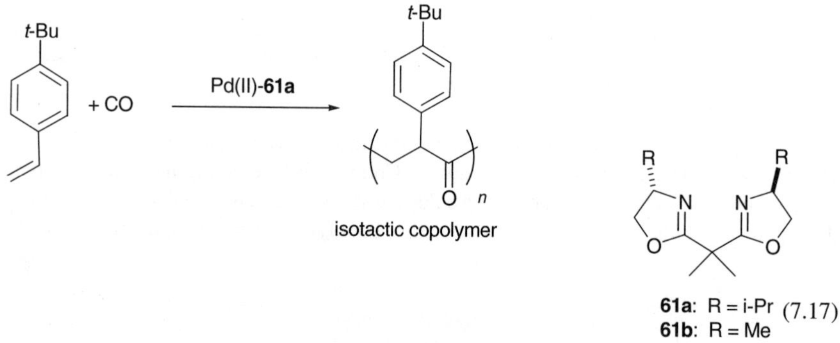

61a: R = i-Pr (7.17)
61b: R = Me

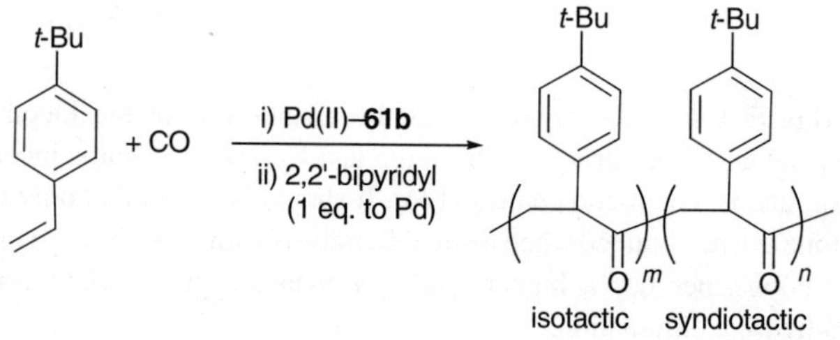

(7.18)

$$(7.19)$$

$$(7.20)$$

Ar = 4-*tert*-Bu-C$_6$H$_4$

unit A unit B

Asymmetric alternating copolymerization of styrene and carbon monoxide has also been successfully promoted by Pd catalysts with (R,S)-BINAPHOS (**6**) [129,130] and a chiral phosphine-imine ligand **62** [138]. It appears that these unsymmetrical bidentate ligands, **6** and **62**, circumvent the electronic problem of diphosphine ligands, that is, promotion of β-hydride elimination leading to termination of propagation. It is worth mentioning that the asymmetric terpolymerization of ethene, propene, styrene or its derivative, and carbon monoxide has been achieved by using the Pd catalyst with P,N-ligand **62** (Eq. 7.19) [139] or BINAPHOS (**6**) (Eq. 7.20) [140]. In the latter case, the units A and B are incorporated into a polymer chain in a random manner rather than long stereoblocks. The regiochemistry for the unit B has not been determined.

7.5. KINETIC RESOLUTION USING ASYMMETRIC CARBONYLATION

Kinetic resolution of chiral racemic olefins has been attempted without success [71]. Substantial rate difference ($S/R = 5.1$) between two enantiomers has been observed in the enantiomer discriminating carbonylation of racemic 1-phenylethyl bromide to 2-phenyl-propanoic acid (**44a**) under phase-transfer conditions catalyzed by a Pd-complex with a chiral oxaphospholane **63** (Eq. 7.21) [141]. Highly efficient kinetic resolution has been achieved in the reaction of racemic aziridines [142]. For example, 1-*tert*-butyl-2-phenylaz-iridine (**64**) is converted to the corresponding (S)-β-lactam **65** with 99.5% ee in 25% yield in the reaction catalyzed by [Rh(COD)Cl]$_2$ in the presence of *l*-menthol (3 equiv. to **64**) (Eq. 7.22). Enantiomerically enriched aziridine **64** (85% ee, *R*) is recovered in 54% yield. Asymmetric desymmetrization, that is, enantiotopic group discrimination, has been applied to cyclocarbonylations for the syntheses of lactones by using chiral Pd-diphosphine catalysts. For example, the reactions of cyclopentane-1,3-diol and its TMS-derivative bearing vinyl halide moieties, **66** and **68**, give the corresponding bicyclic α-methylenelac-tones, **67** and **69**, respectively, with moderate enantiopurities (Eqs. 7.23, 7.24) [143]. This type of reaction has a high potential as a useful synthetic method in natural product syntheses as well as in organic synthesis in general.

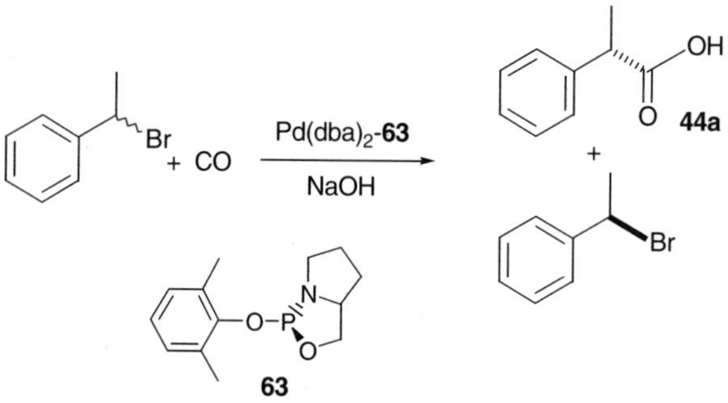

(7.21)

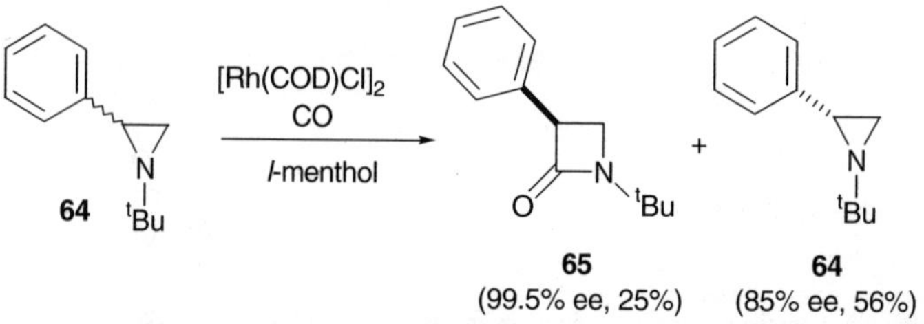

(7.22)

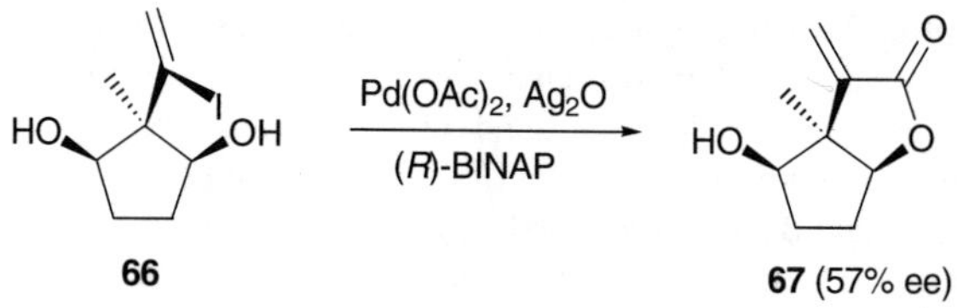

(7.23)

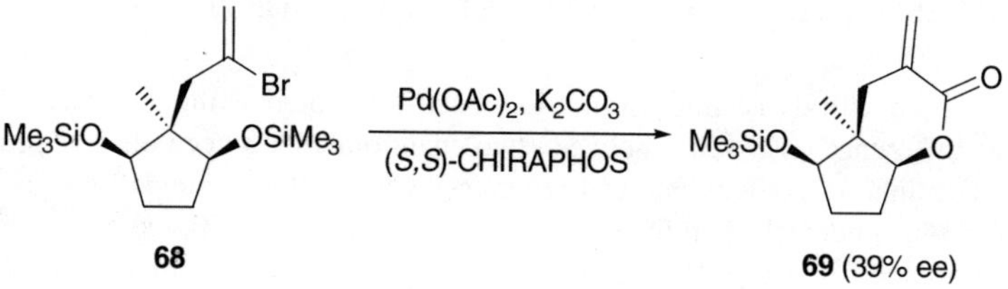

(7.24)

7.6. CONCLUSION AND OUTLOOK

The discovery and development of highly efficient Rh-catalysts with chiral diphosphites and phosphine-phosphites have dramatically improved the enantioselectivity of asymmetric hydroformylation to >90% ee in the first half of the 1990s. It appears that the success of the chiral Rh-catalyzed process has replaced the chiral Pt-catalyzed process used extensively in 1980s, which often suffer from side reactions, such as hydrogenation and isomerization, as well as low selectivity to branched aldehydes. It can be said that the enantioselectivity of asymmetric hydroformylation has reached the equivalent level to that of asymmetric hydrogenation and other processes that have been developed more extensively, although the catalyst-loading still needs substantial improvement to be commercially feasible. Accordingly, further development of more efficient chiral ligands, catalyst systems, and reaction conditions should be continued. From an economical as well as an environmental point of view, recycling of the chiral Rh-catalysts will become an important issue in the future.

Diastereoselective hydroformylation will find many applications in the total synthesis of complex natural products as well as the syntheses of biologically active compounds of medicinal and agrochemical interests in the near future. The diastereoselectivity of such reactions can be improved by the use of a matched-pair combination of a chiral ligand and chirality in the substrate.

Advance in asymmetric hydrocarboxylation has been much slower than that of asymmetric hydroformylation in spite of its high potential in the syntheses of fine chemicals. However, some very encouraging results have recently been reported, and thus much improvements in this reaction can be expected in the next decade.

The completely alternating, highly isotactic, and highly enantioselective copolymerization of 1-alkenes with carbon monoxide provides a new and efficient method for the syntheses of stereoregular polymers. The successful development of this process demonstrates that the concept of catalytic asymmetric synthesis of small molecules can be nicely applied to polymer synthesis. Although 1-alkene/CO isotactic alternating copolymers, that is, polyketones, have not been commercialized yet, these copolymers with extremely high stereoregularity and enantiopurity will find useful applications in material sciences in the future.

ACKNOWLEDGMENTS

Authors thank Professor W. A. Herrmann and D. Gleich for kindly sending us coordinates for Figure 7.4.

REFERENCES

1. Gladiali, S.; Bayon, J. C.; Claver, C. *Tetrahedron: Asymmetry* **1995**, *6*, 1453–1474.

2. Paumard, E.; Mutez, S.; Mortreux, A.; Petit, F. In *Eur. Pat. Appl.*; (Norsolor S. A.): 1989; pp. EP 335765.

3. Petit, M.; Mortreux, A.; Petit, F.; Bruno, G.; Peiffer, G. In *Fr. Dimande*; (Soc. Chim. Charbonnages): 1985; pp. FR 2550201.

4. Dessau, R. M. In *U.S. Pat.*; (Mobil Oil Corp.): 1985; pp. 4554262.

5. Botteghi, C.; Gladiali, S. G.; Marchetti, M.; Faedda, G. A. In *Eur. Pat. Appl.*; (Consiglio Nazionale Ricerche): 1983; pp. EP 81149.

6. Tinker, H. B.; Solodar, A. J. In *U.S. Pat.*; (Monsanto Co.): 1981; pp. 4268688.

7. Brunner, H.; Pieronczyk, W. In *Ger. Offen.*; (BASF A.-G.): 1980; pp. DE 2908358.

8. Pittman, C. U., Jr.; Kawabata, Y.; Flowers, L. I. *J. Chem. Soc., Chem. Commun.* **1982**, 473–474.

9. Consiglio, G.; Pino, P.; Flowers, L. I.; Pittman, C. U., Jr. *J. Chem. Soc., Chem. Commun.* **1983**, 612–613.

10. Parrinello, G.; Stille, J. K. *J. Am. Chem. Soc.* **1987**, *109*, 7122–7127.

11. Consiglio, G.; Nefkens, S. C. A.; Borer, A. *Organometallics* **1991**, *10*, 2046–2051.

12. Cserépi-Szücs, S.; Bakos, J. *Chem. Commun.* **1997**, 635–636.

13. Stille, J. K.; Su, H.; Brechot, P.; Parrinello, G.; Hegedus, L. S. *Organometallics* **1991**, *10*, 1183–1189.

14. Stille, J. K.; Parrinello, G. *PCT Int. Appl.* **1988**, WO 8808835; *Chem. Abstr.* **1988**, *111*, 96656.

15. Agbossou, F.; Carpentier, J.-F.; Mortreux, A. *Chem. Rev.* **1995**, *95*, 2485–2506.

16. Gladiali, S.; Fabbri, D.; Kollár, L. *J. Organometal. Chem.* **1995**, *491*, 91–96.

17. Scrivanti, A.; Beghetto, V.; Bastianini, A.; Matteoli, U.; Menchi, G. *Organometallics* **1996**, *15*, 4687–4694.

18. Tóth, I.; Elsevier, C. J.; de Vries, J. G.; Bakos, J.; Smeets, W. J. J.; Spek, A. L. *J. Organomet. Chem.* **1997**, *540*, 15–25.

19. Sakai, N.; Mano, S.; Nozaki, K.; Takaya, H. *J. Am. Chem. Soc.* **1993**, *115*, 7033–7034.

20. Horiuchi, T.; Ohta, T.; Nozaki, K.; Takaya, H. *Chem. Commun.* **1996**, 155–156.

21. Sakai, N.; Nozaki, K.; Takaya, H. *J. Chem. Soc., Chem. Commun.* **1994**, 395–396.

22. Nozaki, K.; Itoi, Y.; Shibahara, F.; Shirakawa, E.; Ohta, T.; Takaya, H.; Hiyama, T. *J. Am. Chem. Soc.* **1998**, *120*, 4051–4052; *idem*, *Bull. Chem. Soc. Jpn* **1999**, *72*, 1911–1918.

23. Higashizima, T.; Sakai, N.; Nozaki, K.; Takaya, H. *Tetrahedron Lett.* **1994**, *35*, 2023.

24. Takaya, H.; Sakai, N.; Nozaki, K. *Shokubai* **1994**, *36*, 259.

25. Babin, J. E.; Whiteker, G. T. *U.S. Pat.* **1994**, 5,360,938; *Chem Abstr.* **1994**, *122*, 186609.

26. Babin, J. E.; Whiteker, G. T. In *PCT Int. Appl.*; (Union Carbide Chem. Plastics Corp.): 1993; pp. WO 9303839.

27. Buisman, G. J. H.; Kamer, P. C. J.; van Leeuwen, P. W. N. M. *Tetrahedron: Asymmetry* **1993**, *4*, 1625–1634.

28. Buisman, G. J. H.; Martin, M. E.; Vos, E. J.; Klootwijk, A.; Kamer, P. C. J.; van Leeuwen, P. W. N. M. *Tetrahedron: Asymmetry* **1995**, *6*, 719–738.

29. Buisman, G. J. H.; Vos, E. J.; Kamer, P. C. J.; van Leeuwen, P. W. N. M. *J. Chem. Soc. Dalton Trans.* **1995**, 409–417.

30. Buisman, G. J. H.; van der Veen, L. A.; Klootwijk, A.; de Lang, W. G. J.; Kamer, P. C. J.; van Leeuwen, P. W. N. M.; Vogt, D. *Organometallics* **1997**, *16*, 2929–2939.

31. Sakai, N.; Nozaki, K.; Mashima, K.; Takaya, H. *Tetrahedron: Asymmetry* **1992**, *3*, 583–586.

32. RajanBabu, T. V.; Ayers, T. A. *Tetrahedron Lett.* **1994**, *35*, 4295–4298.

33. Nozaki, K.; Li, W.; Horiuchi, T.; Takaya, H. *J. Org. Chem.* **1996**, *61*, 7658–7659.

34. Stanley, G. G. *Adv. Catal. Processes* **1997**, *2*, 221–243.

35. Noyori, R. *Asymmetric Catalysis in Organic Synthesis*; John Wiley & Sons: New York, 1993.

36. Grazia, C.; Nicolo, F.; Drommi, D.; Bruno, G.; Faraone, F. *J. Chem. Soc., Chem. Commun.* **1994**, 2251–2252.

37. Gladiali, S.; Dore, A.; Fabbri, D. *Tetrahedron: Asymmetry* **1994**, *5*, 1143–1146.

38. Naieli, S.; Carpentier, J.-F.; Agbossou, F.; Mortreux, A.; Nowogrocki, G.; Wignacourt, J.-P. *Organometallics* **1995**, *14*, 401–406.

39. Kless, A.; Holz, J.; Heller, D.; Kadyrov, R.; Selke, R.; Fischer, C.; Börner, A. *Tetrahedron: Asymmetry* **1996**, *7*, 33–36.

40. Basoli, C.; Botteghi, C.; Cabras, M. A.; Chelucci, G.; Marchetti, M. *J. Organometal. Chem.* **1995**, *488*, C20–C22.

41. Chelucci, G.; Cabras, M. A.; Botteghi, C.; Basoli, C.; Marchetti, M. *Tetrahedron: Asymmetry* **1996**, *7*, 885–895.

42. Chelucci, G.; Marchetti, M.; Sechi, B. *J. Mol. Cat. A: Chem.* **1997**, *122*, 111–114.

43. Brown, C. K.; Wilkinson, G. *J. Chem. Soc. (A)* **1970**, 2753 and references cited therein.

44. Cornils, B. In *New Syntheses with Carbon Monoxide*; Falbe, J. (Ed.); Springer-Verlag: Berlin, 1980; pp. 1–225.

45. Pino, P.; Piacenti, F.; Bianchi, M. In *Organic Syntheses via Metal Carbonyls*; Wender, I.; Pino, P. (Ed.); John Wiley & Sons: New York, 1977; pp. 43–231.

46. Markó, L. *Aspects of Homogeneous Catal.* **1974**, *2*, 3.

47. Tkatchenko, I. In *Comprehensive Organometallic Chemistry*; Wilkinson, G. (Ed.); Pergamon: Oxford, 1982; Vol. 8; pp. Ch. 50.3; 101–223.

48. Collman, J. P.; Hegedus, L. S.; Norton, J. R.; Finke, R. G. *Principles and Applications of Organotransition Metal Chemistry*; University Science Books: Mill Valley, CA, 1987, pp. 621–630.

49. Yagupsky, G.; Brown, C. K.; Wilkinson, G. *J. Chem. Soc. (A)* **1970**, 1392.

50. Tolman, C. A.; Faller, J. W. In *Homogeneous Catalysis with Metal Phosphine Complexes*; Pignolet, L. H. (Ed.); Plenum Press: New York, 1983.

51. Stefani, A.; Consiglio, G.; Botteghi, C.; Pino, P. *J. Am. Chem. Soc.* **1973**, *95*, 6504.

52. Davidson, P. J.; Hignett, R. R.; Thomson, D. T. In *Catalysis*; Kemball, C. (Ed.); The Chemical Society: London, 1977; Vol. 1; pp. 369.

53. Stefani, A.; Tatone, D. *Helv. Chim. Acta* **1977**, *60*, 518.

54. Jongsma, T.; Challa, G.; van Leeuwen, P. W. N. M. *J. Organomet. Chem.* **1991**, *421*, 121–128.

55. Moasser, B.; Gladfelter, W. L.; Roe, D. C. *Organometallics* **1995**, *14*, 3832–3838.

56. Matthews, R. C.; Howell, D. K.; Peng, W.-J.; Train, S. G.; Treleaven, W. D.; Stanley, G. G. *Angew. Chem., Int. Ed.* **1996**, *35*, 2253–2256.

57. Broussard, M. E.; Juma, B.; Train, S. G.; Peng, W.-J.; Laneman, S. A.; Stanley, G. G. *Science* **1993**, *260*, 1784–1788.

58. Süss-Fink, G. *Angew. Chem., Int. Ed.* **1994**, *33*, 67–69.

59. Musaev, D. G.; Matsubara, T.; Mebel, A. M.; Koga, N.; Morokuma, K. *Oure & Appl. Chem.* **1995**, *67*, 257–263.

60. Matsubara, T.; Koga, N.; Ding, Y.; Musaev, D. G.; Morokuma, K. *Organometallics* **1997**, *16*, 1065–1078.

61. Horiuchi, T.; Shirakawa, E.; Nozaki, K.; Takaya, H. *Organometallics* **1997**, *16*, 2981–2986.

62. Brown, J. M.; Kent, A. G. *J. Chem. Soc., Perkin Trans* **1987**, *2*, 1597.

63. Pottier, Y.; Mortreux, A.; Petit, F. *J. Organomet. Chem.* **1989**, *370*, 333–342.

64. Nozaki, K.; Sakai, N.; Nanno, T.; Higashijima, T.; Mano, S.; Horiuchi, T.; Takaya, H. *J. Am. Chem. Soc.* **1997**, *119*, 4413–4423.

65. Nozaki, K.; Takaya, H.; Hiyama, T. *Top. Catal.* **1998**, *4*, 175–185.

66. van Rooy, A.; Kamer, P. C. J.; van Leeuwen, P. W. N. M.; Goubitz, K.; Fraanje, J.; Veldman, N.; Spek, A. L. *Organometallics* **1996**, *15*, 835–847.

67. Buisman, G. J. H.; van der Veen, L. A.; Kamer, P. C. J.; van Leeuwen, P. W. N. M. *Organometallics* **1997**, *16*, 5681–5687.

68. Pregosin, P. S.; Sze, S. N. *Helv. Chim. Acta* **1978**, *61*, 1848.

69. Rocha, W. R.; de Almeida, W. B. *Organometallics* **1998**, *17*, 1961.

70. Ojima, I.; Hirai, K. In *Asymmetric Synthesis, Vol. 5 Chiral Catalysis*; Morrison, J. D. (Ed.); Academic Press:San Diego, 1985; pp. 103–146.

71. Consiglio, G. In *Catalytic Asymmetric Synthesis*; Ojima, I. (Ed.); VCH Publishers: New York, 1993; pp. 273–302 and references cited therein.

72. Consiglio, G.; Pino, P. *Top. Curr. Chem.* **1982**, *105*, 77–123.

73. Gleich, D.; Schmid, R.; Herrmann, W. A. *Organometallics* **1998**, *17*, 2141–2143.

74. Botteghi, C.; Paganelli, S.; Schionato, A.; Marchetti, M. *Chirality* **1991**, 355–369.

75. Parrinello, G.; Stille, J. K. *Polym. Prepr.* **1986**, *27*, 9–10.

76. Stille, J. K.; Parrinello, G. *J. Mol. Catal.* **1983**, *21*, 203–210.

77. Kainz, S.; Leitner, W. *Catal. Lett.* **1998**, *55*, 223.

78. (a) Tzamarioudaki, M.; Chuang, C.-Y.; Inoue, O.; Ojima, I. *216th American Chemical Society National Meeting, Boston, MA, August 23–27, 1998; Abstrats* **1998**, ORGN 676 and unpublished results. (b) Francio, G.; Leitner, W. *Chem. Commun.* **1999**, 1663.

79. Eckl, R. W.; Priermeier, T.; Herrmann, W. A. *J. Organomet. Chem.* **1997**, *532*, 243–249.

80. Nozaki, K.; Nanno, T.; Takaya, H. *J. Organomet. Chem.* **1997**, *527*, 103–108.

81. Horiuchi, T.; Ohta, T.; Shirakawa, E.; Nozaki, K.; Takaya, H. *Tetrahedron* **1997**, *53*, 7795–7804.

82. Ikeda, Y. In *Polymers and Biomaterials*; Feng, H. H.; Huang, L. (Ed.); Elsevier: New York, 1991; pp. 273 and references therein.

83. Stanley, G. G. In *Catalysis of Organic Reactions*; Scaros, M. G.; Prunier, M. L. (Ed.); Marcel Dekker: New York, 1995; pp. 363.

84. Nanno, T.; Sakai, N.; Nozaki, K.; Takaya, H. *Tetrahedron: Asymmetry* **1995**, *6*, 2583–2591.

85. Ojima, I.; Dong, Q. In *Fluorine-containing Amino Acids: Synthesis and Properties*; Kukhar, V. P. V. P.; Soloshonok, V. A. (Ed.); John Wiley & Sons: Chichester, 1995; pp. 113–137.

86. Ojima, I. *Chem. Rev.* **1988**, *88*, 1011–1030.

87. Arena, C. G.; Nicolò, F.; Drommi, D.; Bruno, G.; Faraone, F. *J. Chem. Soc., Chem. Commun.* **1994**, 2251–2252.

88. Gladiali, S.; Pinna, L. *Tetrahedron: Asymmetry* **1990**, *1*, 693–696.

89. Gladiali, S.; Pinna, L. *Tetrahedron: Asymmetry* **1991**, *2*, 623–632.

90. Kollár, L.; Consiglio, G.; Pino, P. *Chimia* **1986**, *40*, 428–429.

91. Kollár, L.; Floris, B.; Pino, P. *Chimia* **1986**, *40*, 201–202.

92. Horiuchi, T.; Ohta, T.; Shirakawa, E.; Nozaki, K.; Takaya, H. *J. Org. Chem.* **1997**, *62*, 4285–4292.

93. Lee, C. W.; Alper, H. *J. Org. Chem.* **1995**, *60*, 499–503.

94. Nozaki, K.; Li, W.; Horiuchi, T.; Takaya, H. *Tetrahedron Lett.* **1997**, *38*, 4611–4614.

95. Takaya, H.; Sakai, N.; Tamao (Nozaki), K.; Mano, S.; Kumobayashi, H.; Tomita, T.; Saito, T.; Matsumura, K.; Kato, Y.; Sayo, N. *U.S. Patent* **1996**, US 5,530,150; *Chem. Abs.* **1996**, *125*, 143006.

96. Takaya, H.; Sakai, N.; Tamao, K.; Mano, S.; Kumobayashi, H.; Tomita, T. *Eur. Pat. Appl.* **1994**, EP 614870 A2; *Chem. Abstr.* **1994**, *123*, 198277.

97. Pino, P.; Piacenti, F.; Bianchi, M. In *Organic Synthesis via Metal Carbonyls*; Pino, P.; Wender, I. (Ed.); John Wiley: New York, 1977; Vol. 2; pp. 233–296.

98. Milstein, D. *Acc. Chem. Res.* **1988**, *21*, 428.

99. Cavinato, G.; Toniolo, L. *J. Organometal. Chem.* **1990**, *398*, 187.

100. Kawada, M.; Nakamura, S.; Watanabe, E.; Urata, H. *J. Organometal. Chem.* **1997**, *542*, 185.

101. Hayashi, T.; Tanaka, M.; Ogata, I. *Tetrahedron Lett.* **1978**, 3925.

102. Consiglio, G. *Adv. Chem. Ser.* **1982**, *196*, 371–388.

103. Consiglio, G.; Pino, P. *Chimia* **1976**, *30*, 193.

104. Commetti, G.; Chiusoli, G. P. *J. Organometal. Chem.* **1982**, *236*, C31.

105. Oi, S.; Nomura, M.; Aiko, T.; Inoue, Y. *J. Mol. Catal.* **1997**, *115*, 289.

106. Zhou, H.; Hou, J.; Chen, J.; Lu, S.; Fu, H.; Wang, H. *J. Organometal. Chem.* **1997**, *543*, 227.

107. Alper, H.; Hamel, N. *J. Am. Chem. Soc.* **1990**, *112*, 2803.

108. Alper, H.; Hamel, N. *J. Chem. Soc., Chem. Commun.* **1990**, 135.

109. Okuro, K.; Kai, H.; Alper, H. *Tetrahedron: Asymmetry* **1997**, *8*, 2307–2309.

110. El Ali, B.; Alper, H. In *Transition Metal Catalyzed Reactions; IUPAC Monograph "Chemistry for the 21st Century"*; Davies, S. G.; Murahashi, S. (Ed.); Blackwell Science: Oxford, 1999.

111. El Ali, B.; Alper, H. *Synlett.* **2000**, 161.

112. Reppe, W.; Magin, A. *U.S. Pat.* **1951**, 2,577,208; *Chem. Abstr.* **1952**, *46*, 16143.

113. Gough, A. *Brit. Pat.* **1967**, 1,081,304; *Chem. Abstr.* **1967**, *67*, 100569.

114. Drent, E.; Budzelaar, P. H. M. *Chem. Rev.* **1996**, *96*, 663.

115. Sen, A. *Acc. Chem. Res.* **1993**, *26*, 303.

116. Lai, T.-W.; Sen, A. *Organometallics* **1984**, *3*, 866.

117. Chen, J.-T.; Sen, A. *J. Am. Chem. Soc.* **1984**, *106*, 1506.

118. Rix, F. C.; Brookhart, M.; White, P. S. *J. Am. Chem. Soc.* **1996**, *118*, 4746.

119. Svensson, M.; Matsubara, T.; Morokuma, K. *Organometallics* **1996**, *15*, 5568.

120. Margl, P.; Ziegler, T. *J. Am. Chem. Soc.* **1996**, *118*, 7337.

121. Margl, P.; Ziegler, T. *Organometallics* **1996**, *15*, 5519.

122. Drent, E.; Wife, R. L. *Eur. Pat. Appl.* **1985**, 181014; *Chem. Abstr.* **1985**, *108*, 6617.

123. Batistini, A.; Consiglio, G.; Suster, U. W. *Angew. Chem,. Int. Ed.* **1992**, *104*, 306.

124. Miyashita, M.; Karino, H.; Shimamura, J.-I.; Chiba, T.; Nagano, K.; Nohira, H.; Takaya, H. *Chem. Lett.* **1989**, 1849.

125. Barsacchi, M.; Batistini, A.; Consiglio, G.; Suster, U. W. *Macromolecules* **1992**, *25*, 3604.

126. Bronco, S.; Consiglio, G.; Hutter, R.; Batistini, A.; Shuster, U. W. *Macromolecules* **1994**, *27*, 4436.

127. Jiang, Z.; Adams, S. E.; Sen, A. *Macromolecules* **1994**, *27*, 2694.

128. Nozaki, K.; Sato, N.; Takaya, H. *J. Am. Chem. Soc.* **1995**, *117*, 9911.

129. Nozaki, K.; Sato, N.; Tonomura, Y.; Yasutomi, M.; Takaya, H.; Hiyama, T.; Matsubara, T.; Koga, N. *J. Am. Chem. Soc.* **1997**, *119*, 12779.

130. Nozaki, K.; Yasutomi, M.; Nakamoto, K.; Hiyama, T. *Polyhedron* **1998**, *17*, 1159.

131. Sperrle, M.; Consiglio, G. *J. Am. Chem. Soc.* **1995**, *117*, 12130.

132. Jiang, Z.; Sen, A. *J. Am. Chem. Soc.* **1995**, *117*, 4455.

133. Nozaki, K.; Hiyama, T. *J. Organomet. Chem.* **1999**, *576*, 248–253; Nozaki, K.: Kawashima, Y.; Nakamoto, K.; Hiyama, T. *Polymer J.* **1999**, *31*, 1057 and unpublished results.

134. Barsacchi, M.; Consiglio, G.; Medici, L.; Petrucci, G.; Suster, U. W. *Angew. Chem., Int. Ed.* **1991**, *103*, 992.

135. Drent, E. *Eur. Pat. Appl.* **1986**, 229408; *Chem. Abstr.* **1988**, *108*, 6617.

136. Brookhart, M.; Rix, F. C.; DeSimone, J. M.; Barborak, J. C. *J. Am. Chem. Soc.* **1992**, *114*, 5894.

137. (a) Brookhart, M.; Wagner, M. I.; Balavoine, G. G. A.; Haddou, H. A. *J. Am. Chem. Soc.* **1994**, *116*, 3641. (b) Brookhart, M.; Wagner, M. I. *ibid* **1996**, *118*, 7219.

138. Aperrle, M.; Aeby, A.; Consiglio, G.; Pfaltz, A. *Helv. Chim. Acta* **1996**, *79*, 1387.

139. Aeby, A.; Consiglio, G. *Helv. Chim. Acta* **1998**, *81*, 35.

140. Nozaki, K.; Kawashima, Y.; Nakamoto, K.; Hiyama, T. *Macromolecules* **1999**, *32*, 5168.

141. Azoumanian, H.; Buono, G.; Choukrad, M.; Petrigani, J.-F. *Organometallics* **1988**, *7*, 59.

142. Calet, S.; Urso, F.; Alper, H. *J. Am. Chem. Soc.* **1989**, *111*, 931.

143. Suzuki, T.; Uozumi, Y.; Shibasaki, M. *J. Chem. Soc., Chem. Commun.* **1991**, 1593.

8
ASYMMETRIC CARBON–CARBON BOND-FORMING REACTIONS

8A

ASYMMETRIC CYCLOADDITION REACTIONS

Keiji Maruoka

Department of Chemistry, Graduate School of Science, Hokkaido University Sapporo 060-0810

8A.1. INTRODUCTION

Lewis acids such as BF_3, $AlCl_3$, MgX_2, $ZnCl_2$, $SnCl_4$, $TiCl_4$ and $Sc(OTf)_3$ have been used for a variety of carbon–carbon bond-forming reactions in organic synthesis. Most often stoichiometric amounts of Lewis acids are employed. Undoubtedly, a most efficient way to create new, carbon–carbon bonds enantioselectively would be to use catalytic amounts of a chiral Lewis acid in these reactions. In all cases, the complexation of a Lewis acid with a carbonyl or imine functionality would activate the system. Therefore, a particular chiral Lewis acid has the potential to be a common catalyst for different carbon–carbon bond-forming reactions [1].

A comprehensive review of asymmetric reactions with chiral Lewis acid catalysts appeared in the first edition of this book, covering a variety of reactions through the end of 1991 [2]. This review aims to discuss the more recent important developments in this area from 1992 with particular emphasis on the asymmetric cycloadditions promoted by chiral Lewis acid catalysts [3].

From several pioneering works described in the original book [2], there is no doubt that carefully designed chiral Lewis acids have vast potential for the asymmetric synthesis of carbon skeletons and others. Choice of a proper metal and design of a suitable chiral ligand may be the most important elements in obtaining successful results.

8A.2. ASYMMETRIC DIELS–ALDER REACTIONS

8A.2.1. Asymmetric Diels–Alder Reactions with Unsaturated Aldehydes

The remarkable reactivity of borane toward carboxylic acids over esters is one of the conspicuous characteristics of this element, which is rarely seen in any other hydride reagent. An acyloxyborane is recognized as an initial intermediate, hence it became of interest to evaluate the asymmetric induction ability of appropriate chiral auxiliaries by introducing them into such

Catalytic Asymmetric Synthesis, Second Edition, Edited by Iwao Ojima
ISBN 0-471-29805-0   Copyright © 2000 Wiley-VCH, Inc.

an intermediate. Readily available sulfonamides of various amino acids seem to be the chiral auxiliaries of choice. Preparation of the starting sulfonamide is quite simple, that is, an amino acid in aqueous sodium hydroxide is reacted with sulfonyl chloride to afford the desired sulfonamide as white crystalline product. The sulfonamide was treated with an equimolar amount of borane-THF complex. The catalyst thus obtained was successfully utilized in asymmetric Diels–Alner reactions [4,5] (Eqs. 8A.1 and 2).

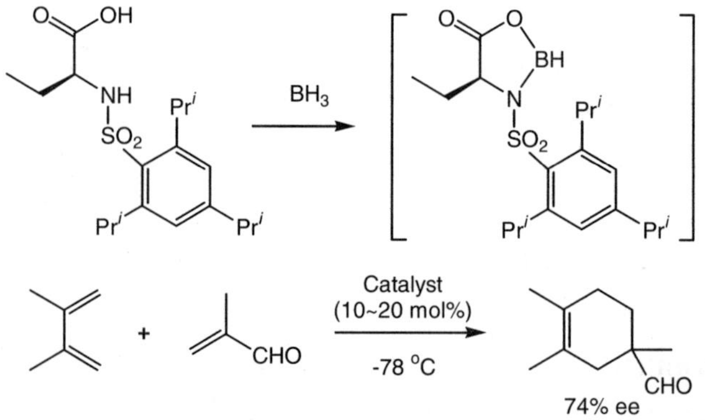

(8A.1)

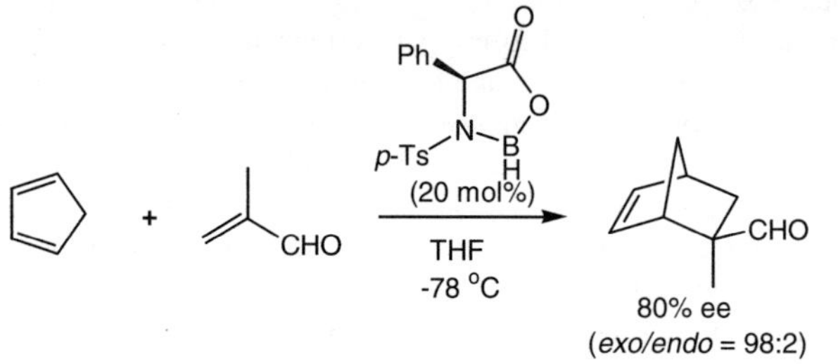

74% ee

(8A.2)

80% ee
(*exo/endo* = 98:2)

Although the enantiomeric excess (ee) of the cycloadducts from these reactions was not particularly high at the initial stage of the development of this chemistry, several new catalysts have revealed a broader range of applicability. The complexation phenomena with carbonyl substrates have been studied by means of *ab initio* molecular orbital methods [6]. A polymer-supported catalyst of this type has also been developed [7]. Corey screened sulfonamides of various amino acids for designing suitable chiral auxiliary, and found that N-tosyl ($\alpha S,\beta R$)-β-methyltryptophan is the best ligand [8]. Indeed, an oxazaborolidine derived from N-tosyl ($\alpha S,\beta R$)-β-methyltryptophan catalyzed the Diels–Alder reaction of 2-bromoacrolein with furan with 92% enantioselectivity, leading to an efficient synthesis of a variety of chiral 7-oxabicyclo[2.2.1]heptene derivatives (Eq. 8A.3). The power of this synthetic method has been demonstrated by its efficient application to the syntheses of Cassiol and Gibberellic acid [9].

The characteristic feature of the aforementioned oxazaborolidine catalyst system consists of α-sulfonamide carboxylic acid ligand for boron reagent, where the five-membered ring system seems to be the major structural feature for the active catalyst. Accordingly, tartaric acid-derived chiral (acyloxy)borane (CAB) complexes can also catalyze the asymmetric Diels–Alder reaction of α,β-unsaturated aldehydes with a high level of asymmetric induction [10] (Eq. 8A.4). Similarly, a chiral tartrate-derived dioxaborolidine has been introduced as a catalyst for enantioselective Diels–Alder reaction of 2-bromoacrolein [11] (Eq. 8A.5).

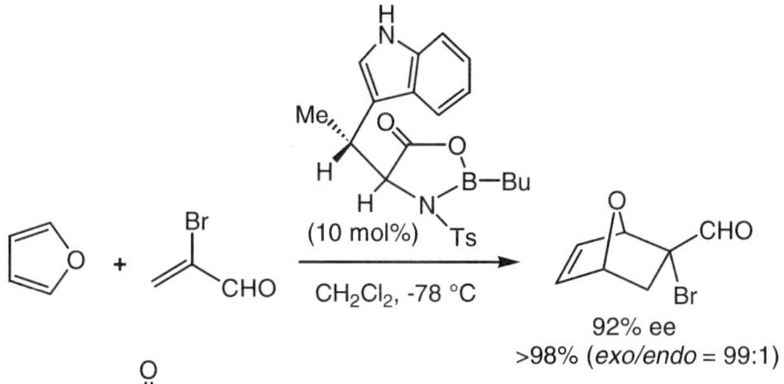
(10 mol%)
CH2Cl2, -78 °C
92% ee
>98% (exo/endo = 99:1)

(8A.3)

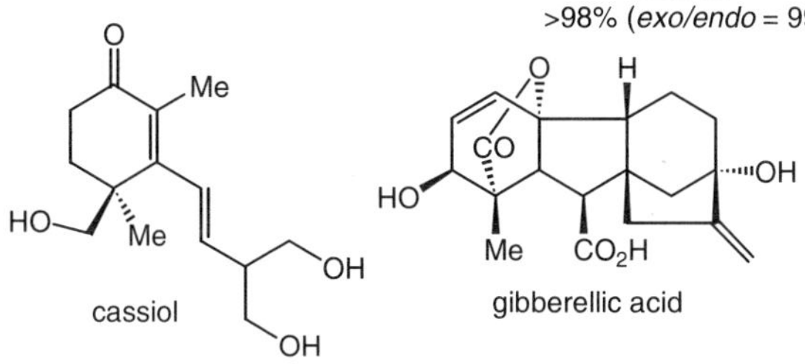
cassiol
gibberellic acid

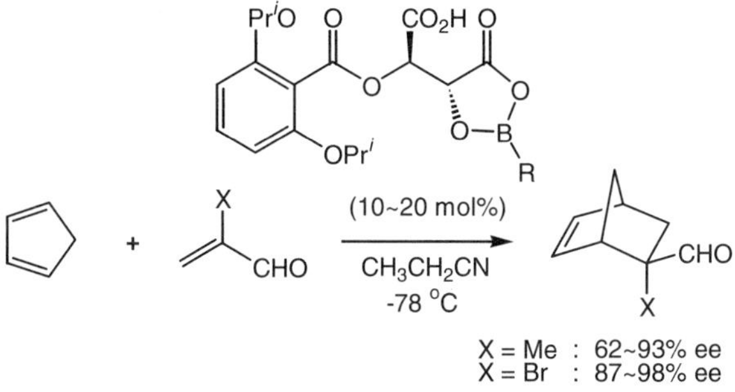
(10~20 mol%)
CH3CH2CN
-78 °C
X = Me : 62~93% ee
X = Br : 87~98% ee

(8A.4)

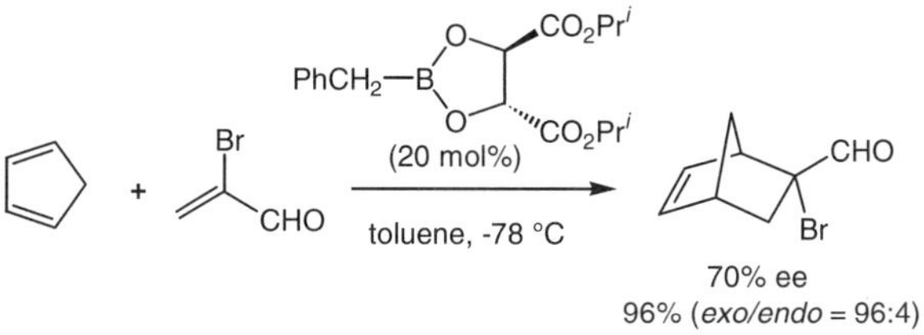
(20 mol%)
toluene, -78 °C
70% ee
96% (exo/endo = 96:4)

(8A.5)

Cationic chiral Rh and Ru complexes were prepared by reaction of $[(\eta\text{-}C_5H_5)RhCl_2]_2$ and $[RuCl_2(\eta^6\text{-mes})]_2$ with chiral bidentate or monodentate oxazoline ligands, respectively. Treatment of these monocationic metal complexes, with $AgSbF_6$ produced dicationic complexes, which were also found to be highly effective for the enantioselective Diels–Alder reaction of methacrolein [12,13] (Eq. 8A.6). On the basis of spectroscopic and structural studies, a full catalytic cycle of a chiral Ru complex was proposed for the Diels–Alder reaction of cyclopentadiene with methacrolein [14].

A new type of chiral helical titanium reagent has been designed based on $Ti(OPr^i)_4$ and a chiral ligand derived from enantiopure binaphthol. These Ti reagents have been successfully utilized as efficient chiral templates for the conformational fixation of α,β-unsaturated aldehydes, thereby allowing excellent enantioselective recognition of the substrates to achieve

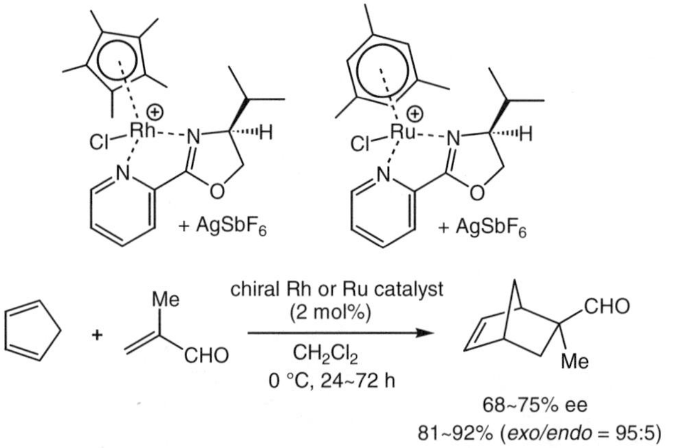

$$\text{(8A.6)}$$

68~75% ee
81~92% (*exo/endo* = 95:5)

uniformly high asymmetric induction in asymmetric Diels–Alder reactions of dienes in a wide range of reaction temperature [15] (Eq. 8A.7).

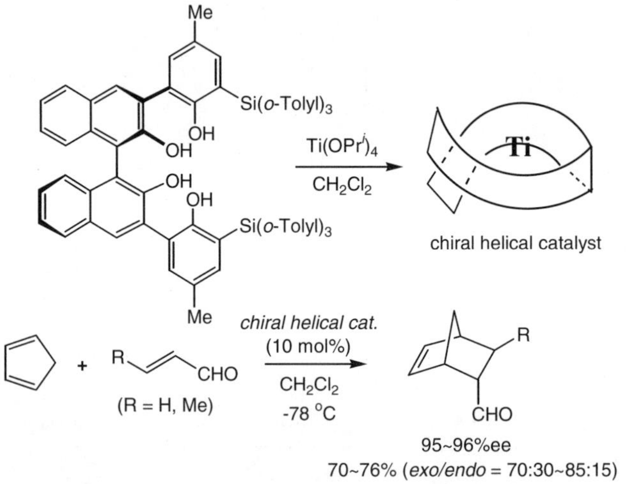

$$\text{(8A.7)}$$

95~96%ee
70~76% (*exo/endo* = 70:30~85:15)

This type of chiral tetraphenol ligand is quite suitable for preparing Brønsted acid-assisted chiral Lewis acid catalyst, which is successfully applied for the enantioselective Diels–Alder reaction of various α,β-unsaturated aldehydes with high asymmetric induction [16] (Eq. 8A.8). Another type of Brønsted acid-assisted chiral Lewis acid catalyst has been designed by using a chiral binaphthol-based triphenol and arylboronic acid, wherein an extremely large acceleration of reaction rate is observed in the presence of the Brønsted acid [17] (Eq. 8A.9). These catalysts are also highly effective to achieve excellent asymmetric induction in the asymmetric Diels–Alder reaction of acetylenic aldehydes, a transformation hitherto difficult to accomplish [18].

Novel bidentate chiral Lewis acids derived from 1,8-naphthalenediylbis(dichloroborane) and modified amino acids as chiral auxiliary have been successfully utilized as effective catalysts for the asymmetric Diels–Alder reaction of α,β-unsaturated aldehydes. The enantioselectivity is highly sensitive to the kind of chiral amino acids. Moderate enantioselectivity was obtained with the tryptophan-derived ligand for the *endo* adduct, but amino acids without aromatic groups

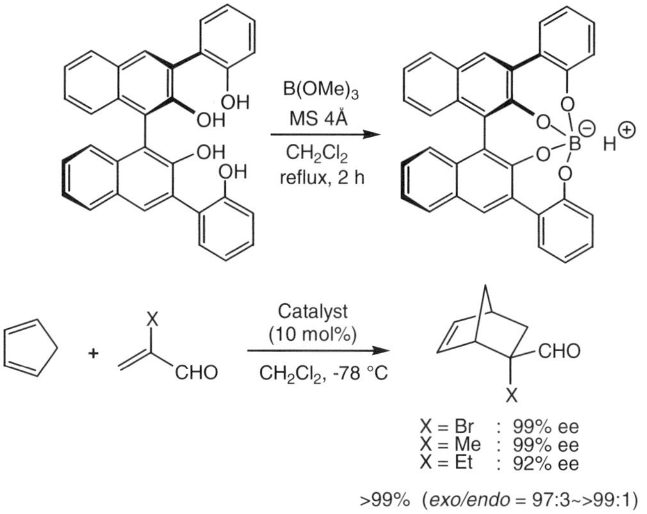

$$X = Br \quad : \ 99\% \ ee$$
$$X = Me \quad : \ 99\% \ ee$$
$$X = Et \quad : \ 92\% \ ee$$

$$>99\% \ (exo/endo = 97:3 \sim >99:1)$$

(8A.8)

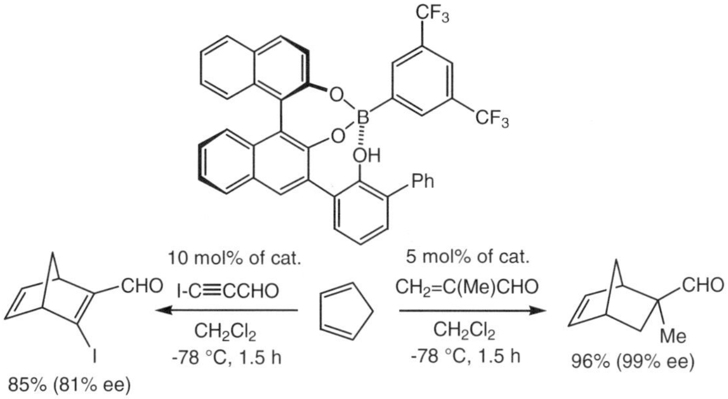

85% (81% ee)   96% (99% ee)   (8A.9)

exhibited very low enantioselectivity, suggesting the importance of π-π interaction in the transition state [19] (Eq. 8A.10).

A titanium complex derived from chiral N-arenesulfonyl-2-amino-1-indanol [20], a cationic chiral iron complex [21], and a chiral oxo(salen)manganese(V) complex [22] have been developed for the asymmetric Diels–Alder reaction of α,β-unsaturated aldehydes with high asymmetric induction (Eq. 8A.11). In addition, a stable, chiral diaquo titanocene complex is utilized for the enantioselective Diels–Alder reaction of cyclopentadiene and a series of α,β-unsaturated aldehydes at low temperature, where catalysis occurs at the metal center rather than through activation of the dienophile by protonation. The high endo/exo selectivity is observed for α-substituted aldehydes, but the asymmetric induction is only moderate [23] (Eq. 8A.12).

Currently available chiral Diels–Alder catalysts have major limitations with regard to the range of dienes to which they can be applied successfully. Indeed, most of the reported catalytic enantioselective Diels–Alder reactions involve reactive dienes such as cyclopentadiene, but 1,3-butadiene and 1,3-cyclohexadiene have not been successfully utilized without reactive 2-bromoacrolein. To solve this problem, a new class of "super-reactive" chiral Lewis acid catalysts has been developed from chiral tertiary amino alcohols and BBr_3 [24] (Eq. 8A.13). This type of chiral super Lewis acid works well for α,β-acetylenic aldehydes [25].

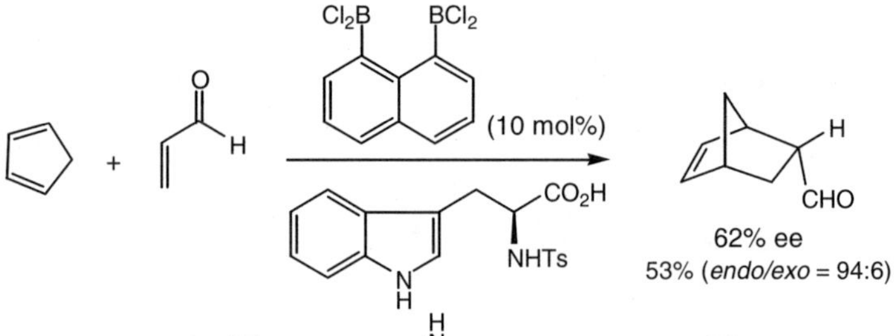
(10 mol%)
62% ee
53% (endo/exo = 94:6)

(8A.10)

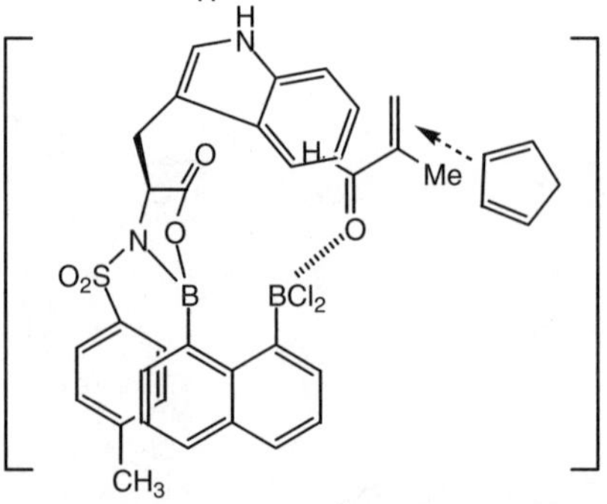

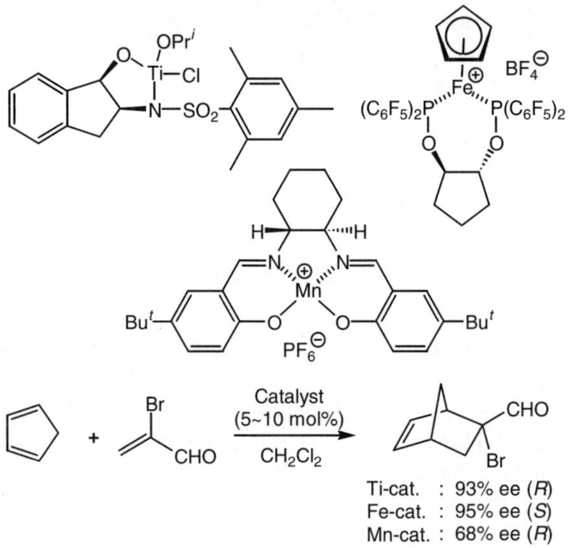
Catalyst
(5~10 mol%)
CH2Cl2
Ti-cat. : 93% ee (R)
Fe-cat. : 95% ee (S)
Mn-cat. : 68% ee (R)

(8A.11)

8A.2.2. Asymmetric Diels–Alder Reactions with Acryloyl Oxazolidinones

With Chiral Bis(oxazoline)/metal Complexes Several research groups have developed chiral Lewis acids by using chiral 1,3-bis(oxazoline) ligands for asymmetric Diels–Alder reactions. Evans designed C_2-symmetric bis(oxazoline)/Cu(II) complexes derived from chiral bis(oxazoline)and $Cu(OTf)_2$, and applied them to asymmetric cycloadditions of acryloyl oxazolidinones and thiazolidine-2-thione analogues. Attractive features of this catalyst system include a clearly interpretable geometry for the catalyst-dienophile complex, which rationalizes the sense of asymmetric induction for the cycloaddition process [26] (Eq. 8A.14).

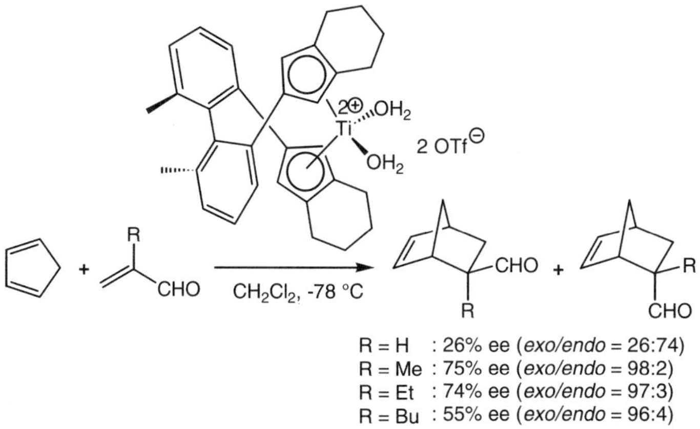
2 OTf
CH2Cl2, -78 °C
R = H : 26% ee (exo/endo = 26:74)
R = Me : 75% ee (exo/endo = 98:2)
R = Et : 74% ee (exo/endo = 97:3)
R = Bu : 55% ee (exo/endo = 96:4)

(8A.12)

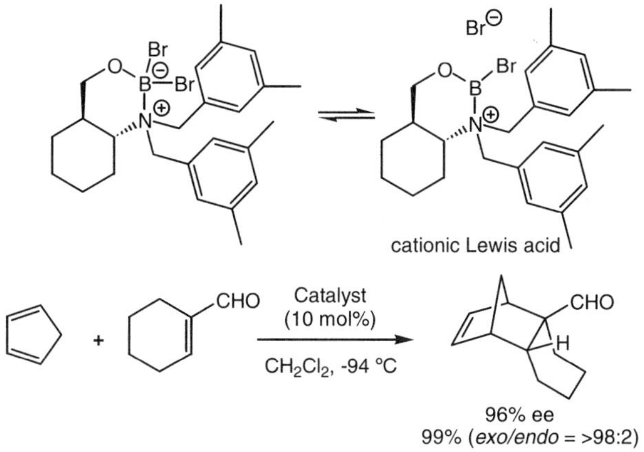
cationic Lewis acid
Catalyst
(10 mol%)
CH2Cl2, -94 °C
96% ee
99% (exo/endo = >98:2)

(8A.13)

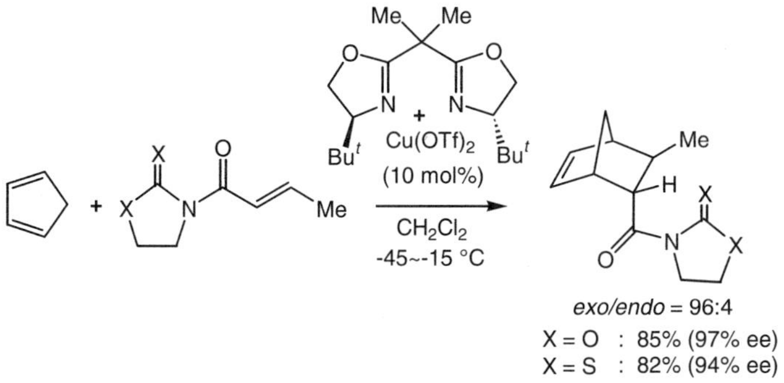
Cu(OTf)2
(10 mol%)
CH2Cl2
-45~-15 °C
exo/endo = 96:4
X = O : 85% (97% ee)
X = S : 82% (94% ee)

(8A.14)

Evans, and later Jørgensen, studied the counterion effect of these C_2-symmetric bis(oxazoline)/Cu(II) complexes, and found that the counterion structure dramatically affected the catalytic efficiency, and SbF_6^- was the best among the anions examined (SbF_6>PF_6>OTf>BF_4) [27,28] (Eq. 8A.15). This cationic bis(oxazoline)/Cu(II) catalyst has been successfully applied to asymmetric synthesis of *ent*-Δ^1-tetrahydrocannabinol [29] and *ent*-shikimic acid [30].

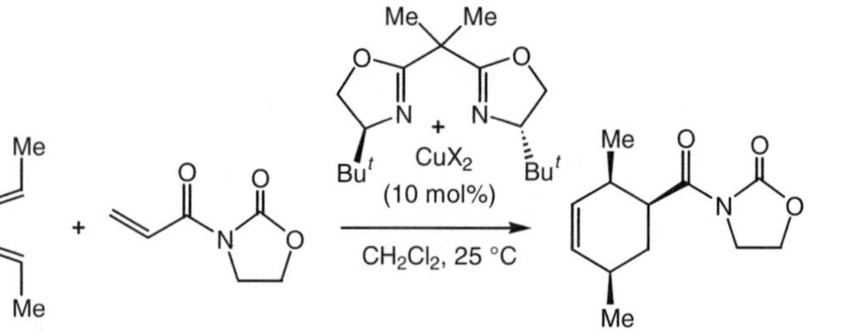

(8A.15)

exo/endo = 77~78:22~23
X = OTf : 66% (84% ee)
X = SbF$_6$: 59% (93% ee)

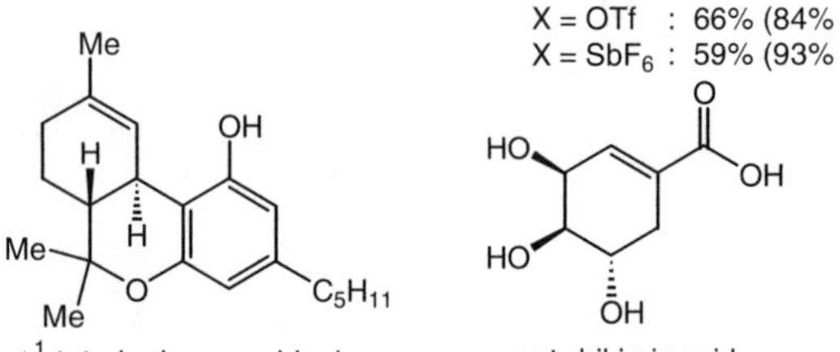

Δ^1-tetrahydrocannabinol

ent-shikimic acid

Using a similar concept, a square-planar bis(imine)/Cu(OTf)$_2$ complex has been developed, and it has been shown that the 2,6-dichlorophenyl-substituted bis(imine) ligand is, in general, the most effective for the asymmetric Diels–Alder reaction of N-acyl oxazolidinones (83~94% ee) [31] (Eq. 16).

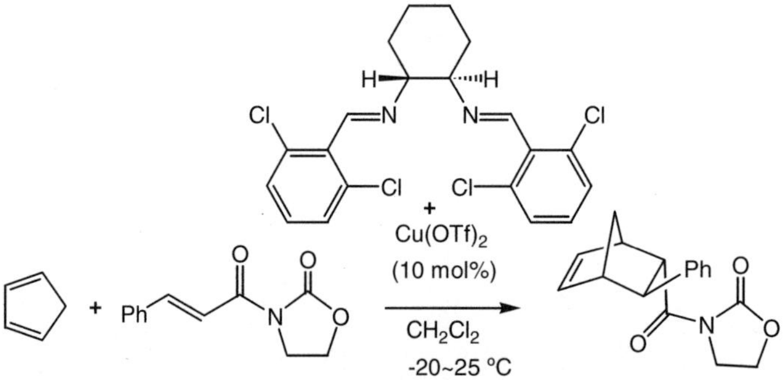

(8A.16)

X = O : 85% ee; 83% (endo/exo = 60:40)
X = S : 92% ee; 84% (endo/exo = 92:8)

A chiral bis(oxazoline) ligand forms effective catalysts in combination with magnesium salts for enantioselective Diels–Alder addition of acryloyl oxazolidinone [32] (Eq. 8A.17). Recently, Desimoni found that both Diels–Alder adduct enantiomers could be obtained by using the same bis(oxazoline)/MgClO$_4$ catalyst, that is, in the presence of water (2 equiv.), the other enantiomer was obtained selectively, which can be rationalized by taking into account the coordination of two water molecules to the metal, changing its geometry to octahedral [33].

A series of chiral bis(oxazoline) ligands, **A**, **B**, and **C**, differing in the length of the chain connecting the chiral oxazoline subunits and in the nature of the substituent at the chiral center, were examined in the M(OTf)$_2$ (M = Zn, Mg, Cu) catalyzed reaction of N-crotonyl oxazolidinone with cyclopentadiene [34] (Eq. 8A.18). A 1,4-bis(oxazoline) ligand **C** proved best for

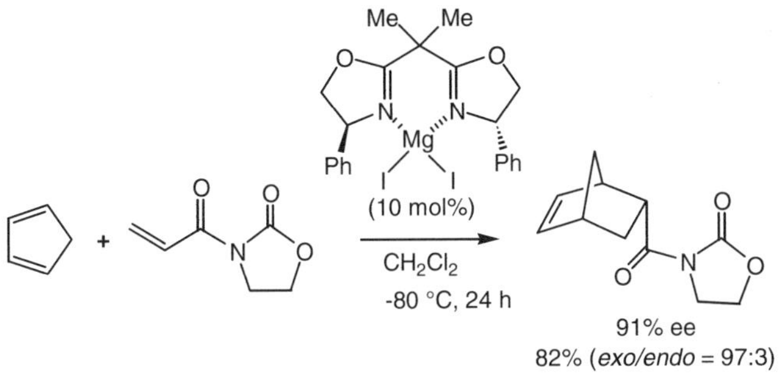

$$(8A.17)$$

91% ee
82% (*exo/endo* = 97:3)

Zn(OTf) and afforded product consistent with reaction via an octahedral model, whereas 1,3-bis(oxazoline) ligands **B** were superior to **A** and **C** for Mg(OTf)$_2$ and Cu(OTf)$_2$. In addition, among several 1,4-bis(oxazoline) ligands, a non-C$_2$-symmetric bis(oxazoline) **D** bearing a meso backbone was also found to be highly efficient [35].

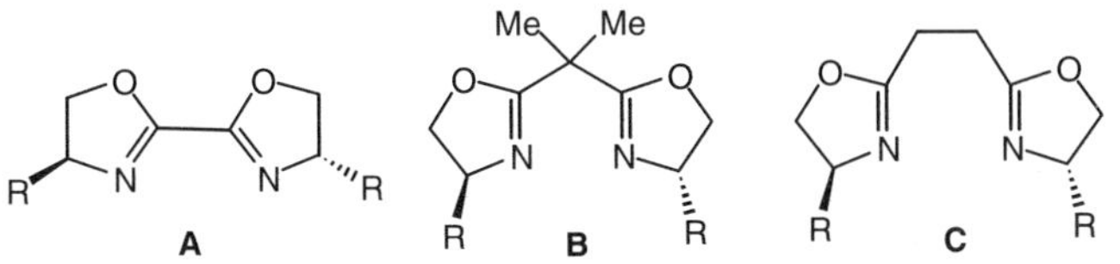

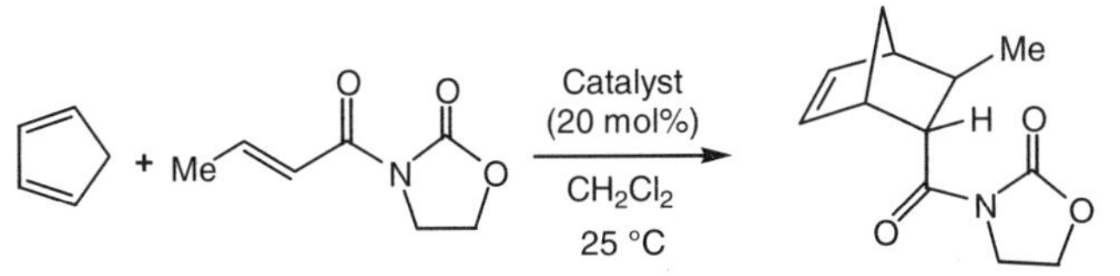

$$(8A.18)$$

A (R = Ph, Bu^t)/M(OTf)$_2$: ineffective
B (R = Ph)/Mg(OTf)$_2$: 95% (87% ee)
B (R = Bu^t)/Cu(OTf)$_2$: 90% (76% ee)
C (R = Ph)/Zn(OTf)$_2$: 99% (78% ee)

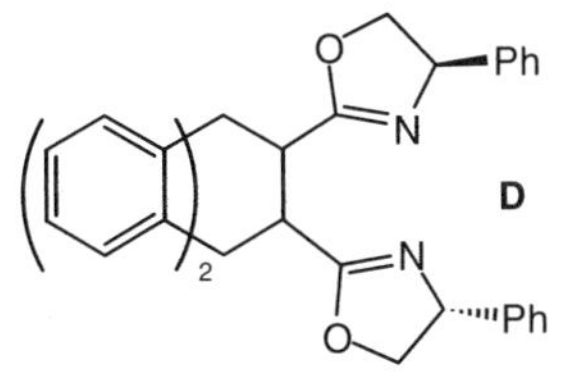

Highly efficient indane-derived C$_2$-symmetric bis(oxazoline)/Cu(OTf)$_2$ complex [36] and transition-metal (Ni, Fe, Co, Cu, Zn, etc.) hydrate complexes of 4,6-dibenzofurandiyl-2,2′-bis(oxazoline) [37] have been developed for enantioselective Diels–Alder addition of acryloyl oxazolidinone derivatives (Eq. 8A.19).

With Chiral Mono(oxazoline)/metal Complexes 2-(Tosylamino)phenyloxazoline can also be utilized as a chiral ligand for the enantioselective Diels–Alder addition of acryloyl oxazolidi-nones. The requisite chiral magnesium complex is derived from methylmagnesium iodide and this chiral ligand [38] (Eq. 8A.20).

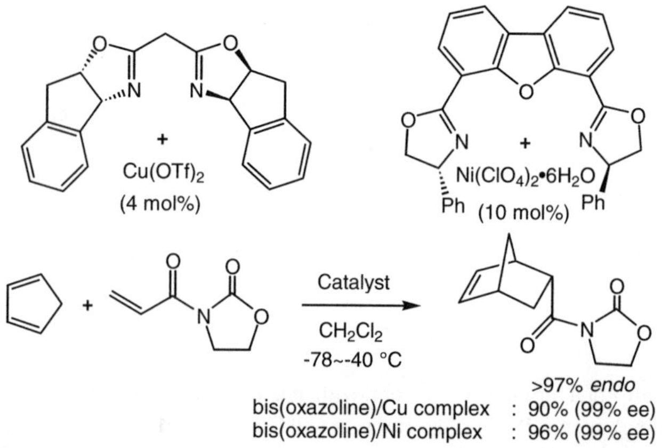

$$(8A.19)$$

>97% *endo*
bis(oxazoline)/Cu complex : 90% (99% ee)
bis(oxazoline)/Ni complex : 96% (99% ee)

(Phosphino-oxazoline)/Cu(II) complexes with bulky aryl groups at phosphorus were found to be excellent for a similar asymmetric cycloaddition reaction [39] (Eq. 8A.20). Enantioselectivity up to 79% ee was achieved in the cycloaddition of cyclohexadiene and acryloyl oxazolidinone.

tosylamino-oxazoline/Mg complex : 69~81% (84~92% ee) (*R*)
phosphino-oxazoline/Cu complex : 92% (97% ee) (*S*)

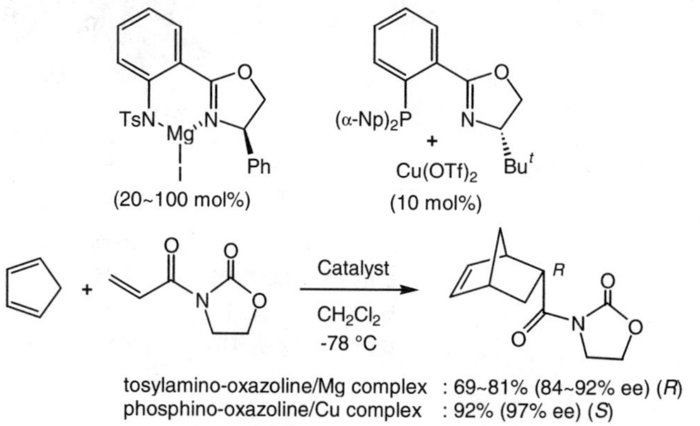

$$(8A.20)$$

With Tartrate-Derived Chiral 1,4-Diol/Ti Complexes A catalytic asymmetric Diels-Alder reaction is promoted by the use of a chiral titanium catalyst prepared in situ from $(Pr^iO)_2TiCl_2$ and a tartrate-derived $(2R,3R)$-1,1,4,4-tetraphenyl-2,3-O-(1-phenylethylidene)-1,2,3,4-butanetetrol. This chiral titanium catalyst, developed by Narasaka, has been successfully executed with oxazolidinone derivatives of 3-borylpropenoic acids as β-hydroxy acrylic acid equivalents [40] (Eq. 8A.21). The resulting chiral adduct can be utilized for the first asymmetric total synthesis of a highly oxygenated sesquiterpene, (+)-Paniculide.

Various α,α,α′,α′-tetraaryl-1,3-dioxolane-4,5-dimethanols have been prepared from (R,R)-tartrate, which are called "TADDOLs" by Seebach et al. They studied the influence of the Ti catalyst preparation methods, the presence of molecular sieves, and the TADDOL structure in the enantioselective Diels–Alder reaction of acryloyl oxazolidinones [41] (Eq. 8A.22). Seebach also prepared polymer- and dendrimer-bound Ti-TADDOLates and used in catalytic asymmetric cycloadditions [42].

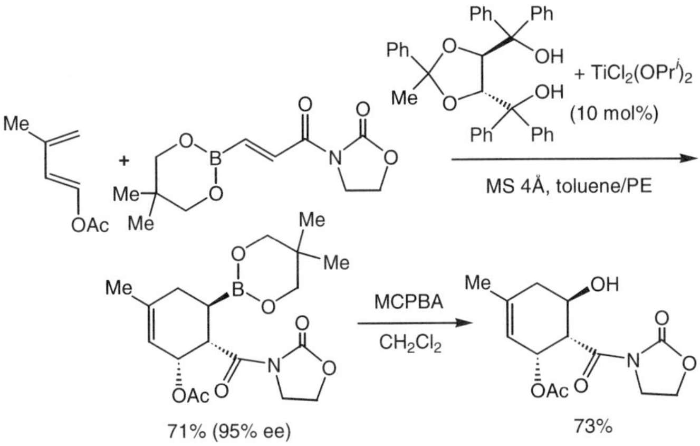

(8A.21)

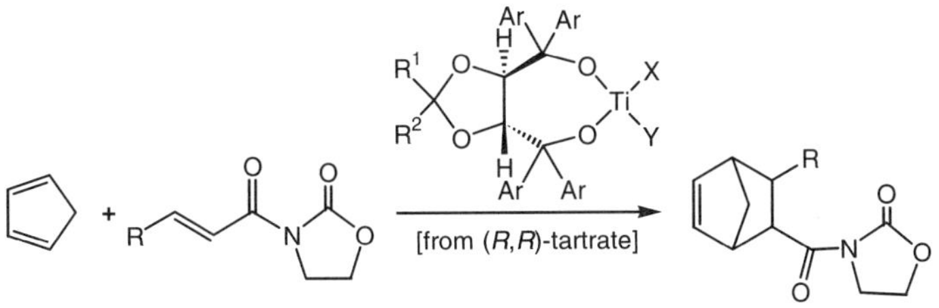

71% (95% ee)   73%

(8A.22)

Jørgensen successfully determined the structure of a Ti-TADDOLate/dienophile complex by X-ray crystallography and proposed a possible reaction mechanism for the enantioselective Diels–Alder addition of cyclopentadiene to *N*-acyloxazolidinones using the Ti-TADDOLate catalyst [43].

With Binaphthol/M(OTf)₃ Complexes (M = Yb, Sc) A chiral ytterbium triflate, derived from $Yb(OTf)_3$, (*R*)-binaphthol, and a tertiary amine, has been applied to the enantioselective Diels–Alder reaction of cyclopentadiene with crotonoyl oxazolidinones. Among various tertiary amines, *cis*-1,2,6-trimethylpiperidine was found to be highly effective [44] (Eq. 8A.23). The unique structure of such chiral Yb catalysts is characterized by hydrogen bonding between the phenolic hydrogens of (*R*)-binaphthol and the nitrogens of tertiary amines.

Recently, Kobayashi found that both Diels–Alder adduct enantiomers were accessible with $Yb(OTf)_3$ by using a single chiral source, (*R*)-binaphthol, and a choice of achiral ligands [45] (Eq. 8A.24). In the presence of 3-phenylacetylacetone (PAA) as the achiral ligand, one of the two active coordination sites in the chiral Yb catalyst is occupied, hence a dienophile coordinates to the remaining active sites, giving the other enantiomer selectively.

With Chiral Diazaaluminolidine A chiral (*S,S*)-diazaaluminolidine, developed by Corey, has been applied for enantioselective Diels–Alder addition of 5-(benzyloxymethyl)-1,3-cyclopentadiene to acryloyl oxazolidinone [46] (Eq. 8A.25). The mechanistic details of this useful process have been studied with regard to the control of enantioselectivity by the catalyst-substrate assembly by an X-ray crystallographic and NMR analyses.

With Chiral Sulfoxide/metal Complexes Certain chiral hydroxysulfoxide and bis-sulfoxide are found useful as chiral ligands for the enantioselective Diels–Alder addition of acryloyl

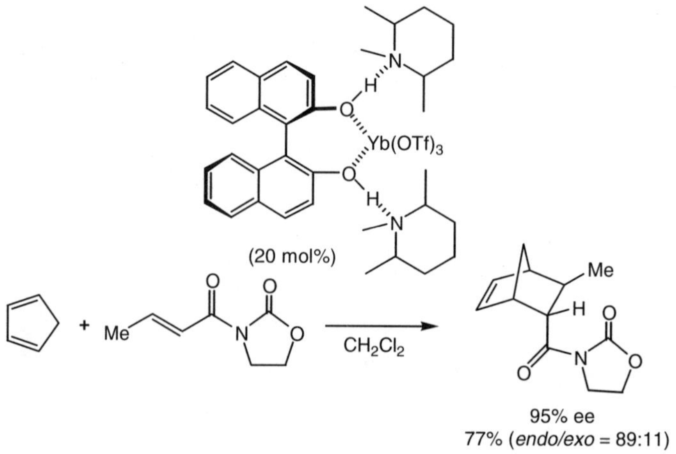

95% ee
77% (*endo/exo* = 89:11)

(8A.23)

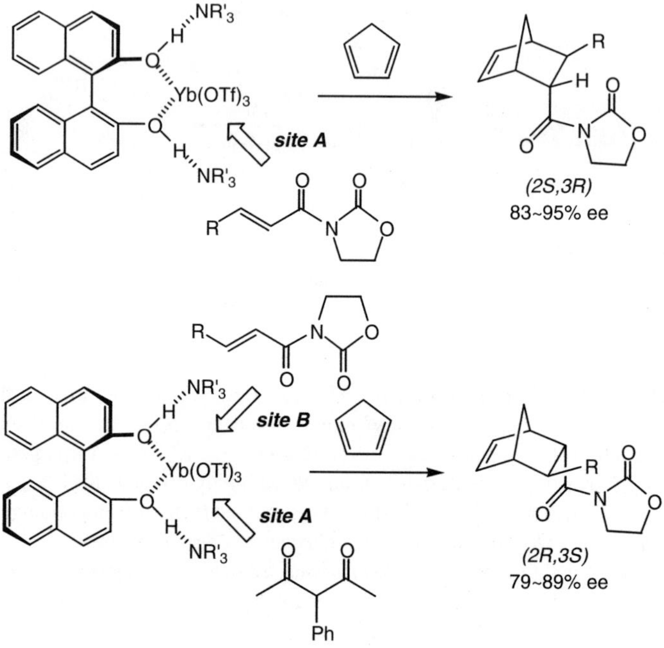

(2S,3R)
83~95% ee

(2R,3S)
79~89% ee

(8A.24)

93% (94% ee)

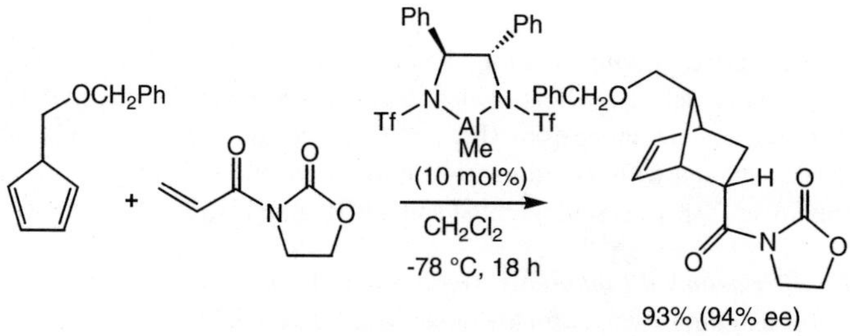

(8A.25)

oxazolidinone. Chiral hydroxysulfoxides are readily prepared from ketones and the corresponding anion of chiral methyl 1-naphthyl sulfoxide, and then converted to chiral magnesium complexes with MgI_2 [47] (Eq. 8A.26).

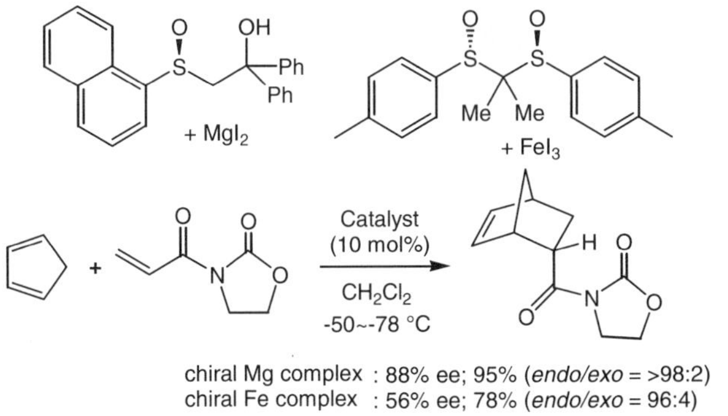

chiral Mg complex : 88% ee; 95% (*endo/exo* = >98:2)
chiral Fe complex : 56% ee; 78% (*endo/exo* = 96:4)

(8A.26)

Chiral bis-sulfoxides with a C_2 symmetry axis can be readily prepared from the known (*R*)-methyl *p*-tolyl sulfoxide and commercially available methyl (*S*)-*p*-toluenesulfinate. Such chiral ligands are very attractive because of their easy synthesis and their ready availability in both enantiomers from inexpensive starting materials. Their complexes with FeI_3 are shown to be good chiral catalysts for asymmetric Diels–Alder reactions [48] (Eq. 8A.26).

8A.2.3. Asymmetric Diels–Alder Reactions with Unsaturated Esters

With Chiral Bis(oxazoline)/Mg Complexes A chiral bis(oxazoline)/MgI_2 complex has been utilized for the enantioselective Diels–Alder reaction of cyclopentadiene and ethyl 2-benzoylacrylate that acts as a two-point binding dienophile [49] (Eq. 8A.27). This reaction proceeds with virtually complete *endo/exo* selectivity (>99:1).

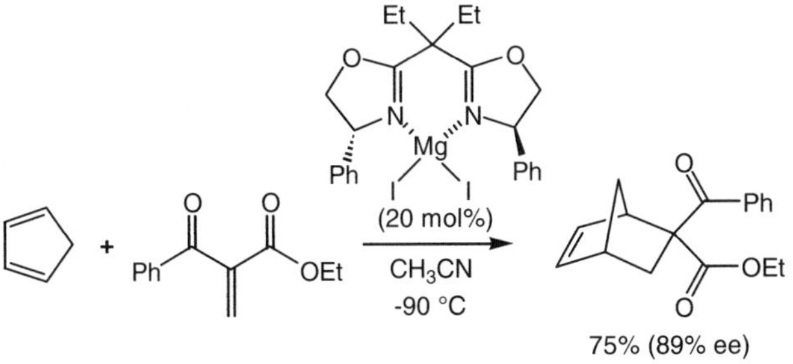

75% (89% ee)

(8A.27)

With Chiral Al Complexes Chiral bis(silyl)binaphthol-modified aluminum catalyst, which is originally developed for asymmetric hetero-Diels–Alder reaction [50], is successfully applied to asymmetric Diels–Alder reaction of cyclopentadiene with methyl acrylate or methyl propiolate [51] (Eq. 8A.28). The latter is a rather rare example in the literatures.

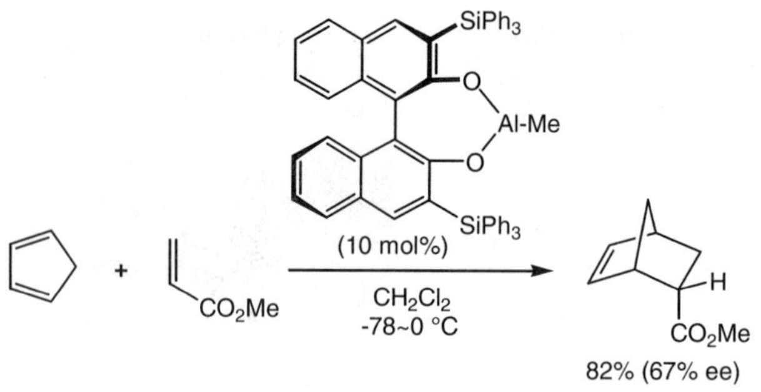

$$(8A.28)$$

82% (67% ee)

With Chiral Diol/Ti Complexes Harada investigated the influence of the torsional angles of biaryl rings on the enantioselectivity in the asymmetric Diels–Alder reaction of cyclopentadiene with methyl acrylate catalyzed by a series of Ti complexes possessing chiral 2,2'-biaryldiols. The best results were obtained by the use of chiral binaphthol or 6,6'-hexylenedioxy-2,2'-biphenyldiol [52] (Eq. 8A.29).

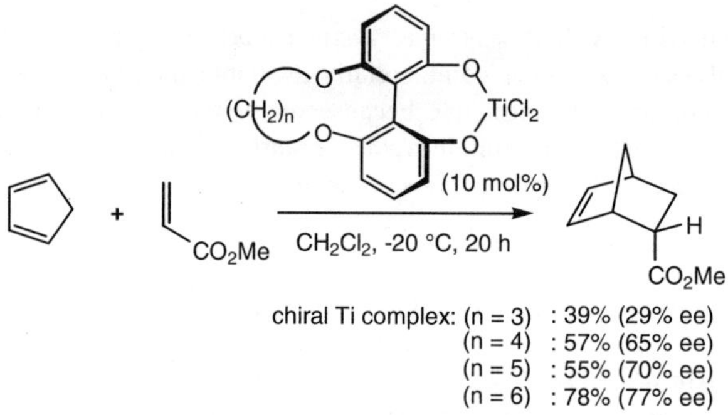

chiral Ti complex: (n = 3) : 39% (29% ee)
(n = 4) : 57% (65% ee)
(n = 5) : 55% (70% ee)
(n = 6) : 78% (77% ee)

$$(8A.29)$$

Chiral Ti complex, derived from hydrobenzoin dilithium salt and $TiCl_4$, can be used for the asymmetric Diels–Alder reaction of several dienes with fumarate [53] (Eq. 8A.30). However, attempted use of acrylate as dienophile resulted in low enantioselectivity.

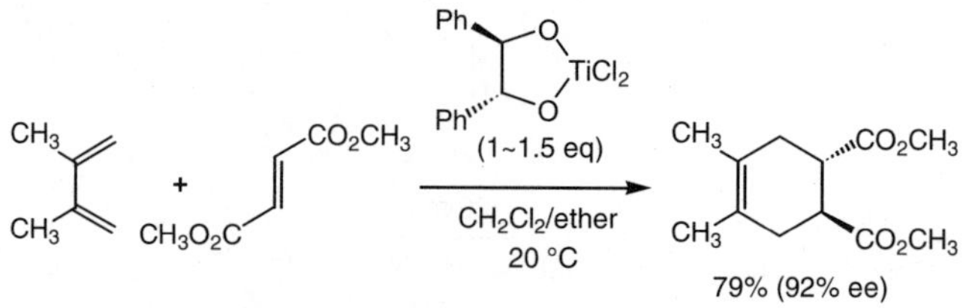

79% (92% ee)

$$(8A.30)$$

8A.2.4. Asymmetric Diels–Alder Reactions with Unsaturated Ketones

The asymmetric Diels–Alder reaction of diene and cyclopentenone derivatives can be promoted by a chiral titanium catalyst prepared in situ from $(Pr^iO)_2TiCl_2$ and a tartrate-derived $\alpha,\alpha,\alpha',\alpha'$-tetraalkyl-1,3-dioxolane-4,5-dimethanol [54] (Eq. 8A.31). The resulting adducts can easily be tranformed to estrogens and progestogens.

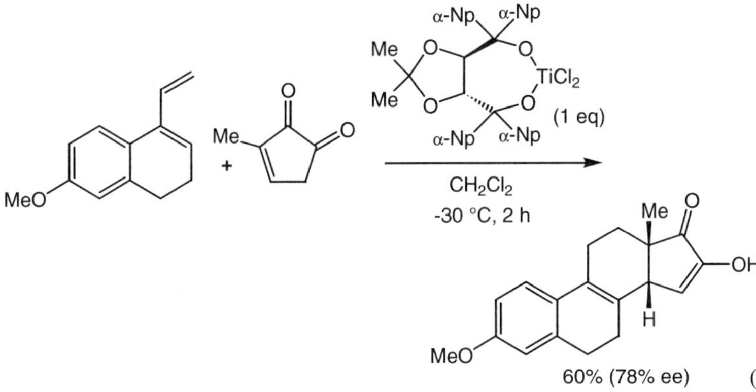

60% (78% ee)   (8A.31)

(*E*)-1-Phenylsulfonyl-3-alken-2-one as a new type of chelating dienophile is utilized for asymmetric Diels–Alder cycloaddition of cyclopentadiene catalyzed by chiral Ti complex with (4*R*,5*R*)-α,α,α′,α′-tetraaryl-2,2-dimethyl-1,3-dioxolane-4,5-dimethanol in the presence or absence of molecular sieves 4Å [55] (Eq. 8A.32). The enantioselectivity of this reaction depends upon the size of aryl substituents involved in the chiral ligands.

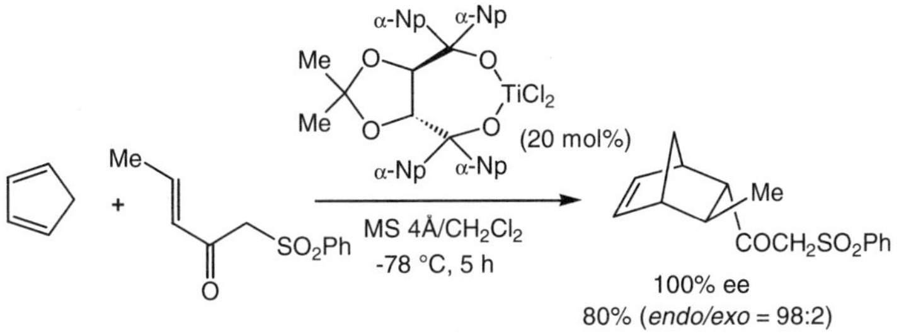

100% ee
80% (*endo/exo* = 98:2)

(8A.32)

Chiral bis(oxazoline)/Copper(II) complexes are evaluated for asymmetric Diels–Alder reaction of naphthoquinone derivatives, and moderate levels of enantiomeric excess are observed in certain cases [56] (Eq. 8A.33).

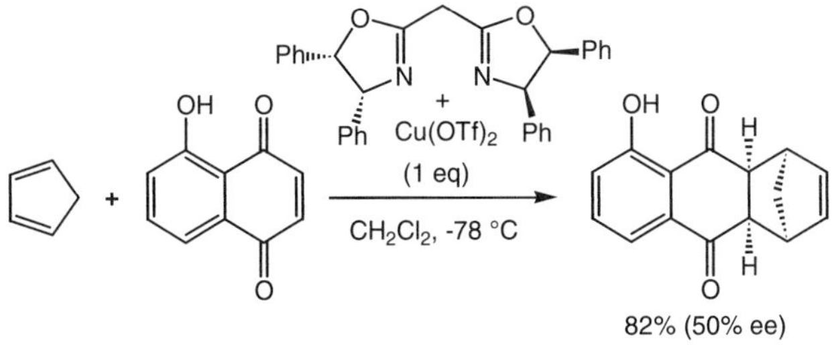

82% (50% ee)   (8A.33)

8A.2.5. Asymmetric Diels–Alder Reactions with Maleimides

Several research groups studied the asymmetric Diels–Alder reaction of maleimides, and applied those methods to natural product synthesis.

Chiral (S,S)-diazaaluminolidine catalyst brought about the first highly enantioselective catalytic Diels–Alder reaction of an achiral C_{2v}-symmetric dienophile with an achiral diene. Addition of 2-methoxybutadiene to N-o-tolylmaleimide in the presence of 20 mol % (S,S)-diazaaluminolidine gave rise to the cycloadduct in 98% yield and 93% ee; one recrystallization from i-PrOH-hexane furnished the enantiomerically pure compound [57] (Eq. 8A.34). The Diels–Alder reaction of 2-((trimethylsilyl)methyl)butadiene and N-arylmaleimide promoted by this catalyst has been successfully applied to the enantioselective total synthesis of Gracilins B and C [58].

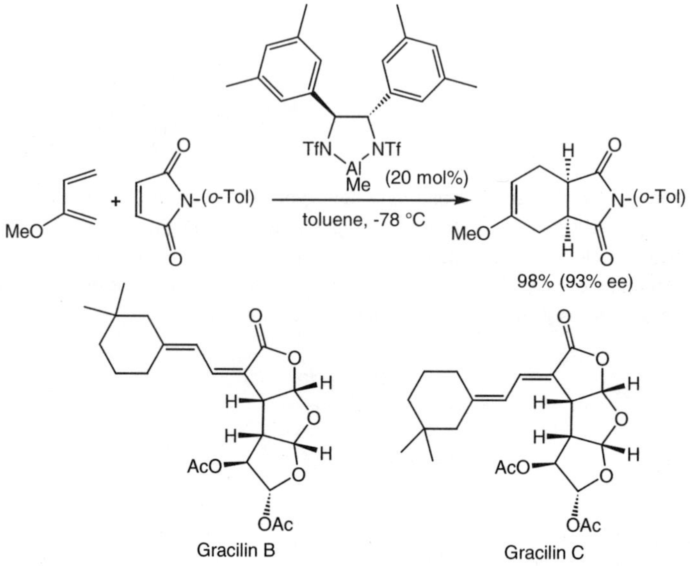

98% (93% ee)

Gracilin B   Gracilin C

(8A.34)

More recently, a Ti-TADDOLate catalyst was found useful for a similar asymmetric transformation, and this reaction has been successfully applied to the asymmetric synthesis of the potent, non-peptidic, NK1-selective, substance P antagonist RPR 107880 [59] (Eq. 8A.35).

Itsuno developed a chiral binaphthylaluminum chloride to promote the asymmetric Diels–Alder polymerization of a bis(diene) with a diimide, which furnished the corresponding optically active functionalized polymers in high yield [60] (Eq. 8A.36).

8A.2.6.   Asymmetric Diels–Alder Reactions with Other Dienophiles

A new and easy access to the enantiomerically pure rigid bidentate ligand, $(-)$-[5-(dipheny-larsino)-2,3-dimethyl-7-phenyl-7-(S)-phosphabicyclo[2.2.1]hept-2-ene is established via an asymmetric cycloaddition between diphenylvinylarsine and 1-phenyl-3,4-dimethylphosphole using chiral organopalladium(II) complex containing *ortho*-metalated dimethyl[1-(2-naphthyl)ethyl]amine as the reaction promoter [61] (Eq. 8A.37).

8A.3.   ASYMMETRIC HETERO DIELS–ALDER REACTIONS

Several chiral Lewis acid catalysts have been used for asymmetric hetero Diels–Alder reactions. Jørgensen applied chiral bis(oxazoline)/Cu(II) catalyst to asymmetric hetero Diels–Alder

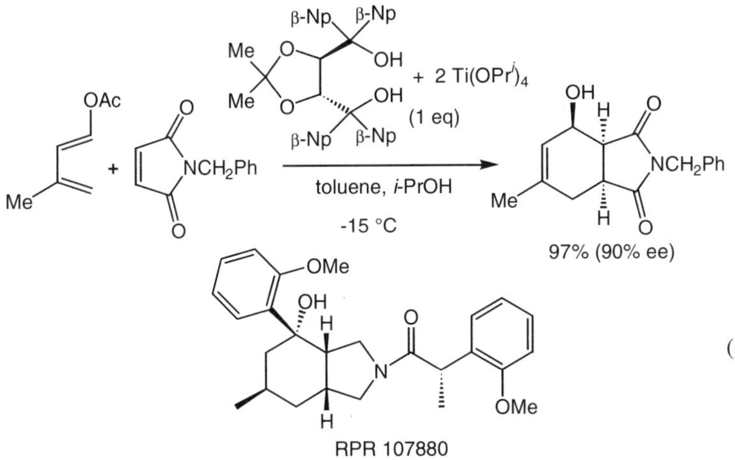

$$(8A.35)$$

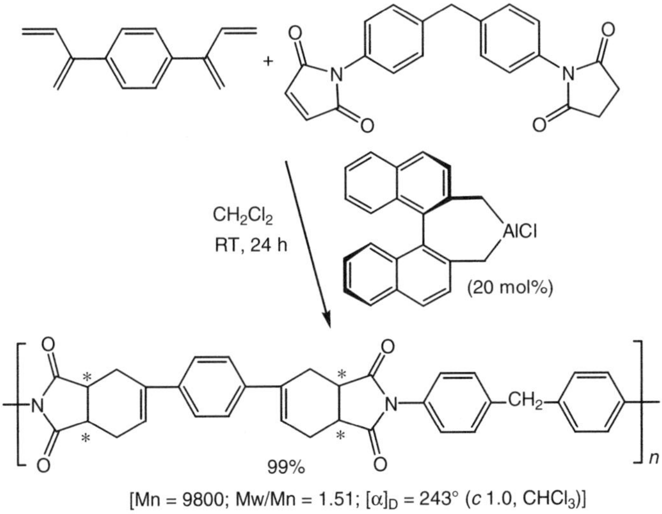

RPR 107880

$$(8A.36)$$

$$[Mn = 9800; Mw/Mn = 1.51; [\alpha]_D = 243° (c\ 1.0, CHCl_3)]$$

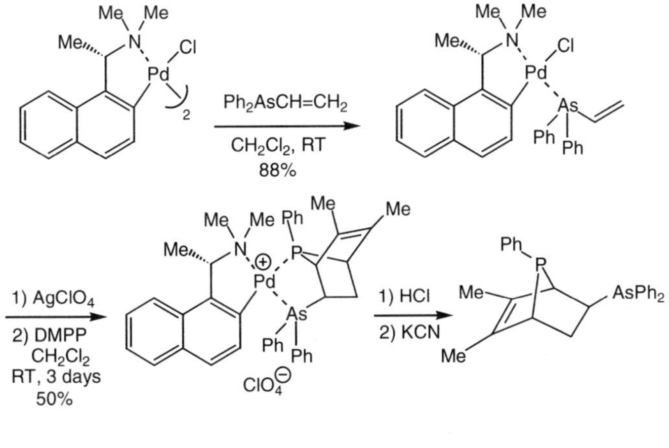

$$(8A.37)$$

reaction of glyoxylates with dienes [62] (Eq. 8A.38). In this reaction, the use of polar solvents such as nitromethane and 2-nitropropane led to a significant improvement of the catalytic efficiency. The resulting chiral adduct is easily transformed to a versatile synthon for sesquiterpene lactones.

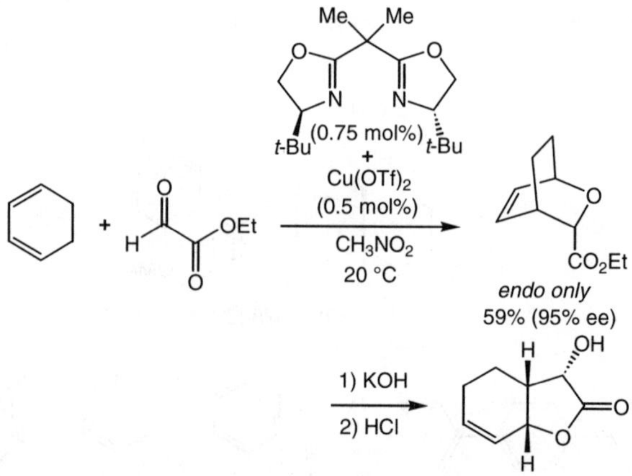

(8A.38)

A similar chiral bis(oxazoline)/Cu(II) catalyst is useful for the asymmetric hetero Diels–Alder reaction of Danishefsky's diene and glyoxylates [63] (Eq. 8A.39). Other bis(oxazoline)/M(OTf)$_2$ (M = Sn, Mg) complexes are not effective. This method provides new routes to asymmetric aldol synthesis upon hydrolysis of the resulting adducts.

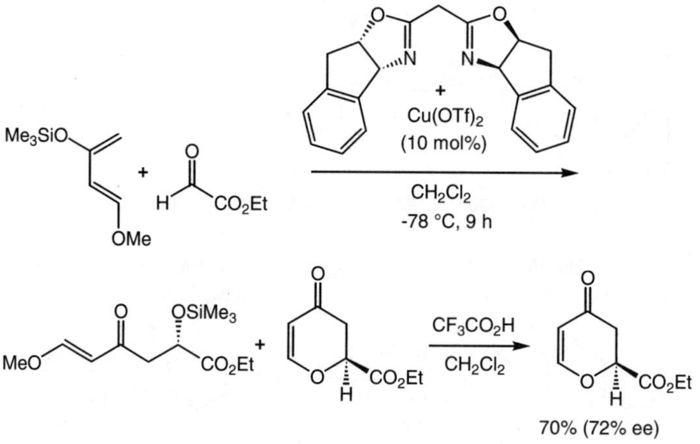

(8A.39)

The chiral bis(oxazoline)/Cu(II) complex with $^-$OTf or $^-$SbF$_6$ as a counter anion effectively promotes the enantioselective hetero Diels–Alder reaction of enol ethers with acyl phosphonates to give chiral enol phosphonates as synthetically useful chiral building blocks [64] (Eq. 8A.40).

Jacobsen studied the efficiency of a series of chiral (salen)Cr(III)X complex (X = Cl, N$_3$, F, BF$_4$) for the asymmetric hetero Diels–Alder reaction of Danishefsky's diene with aldehydes. The best result was obtained with the terafluoroborate catalyst [65] (Eq. 8A.41).

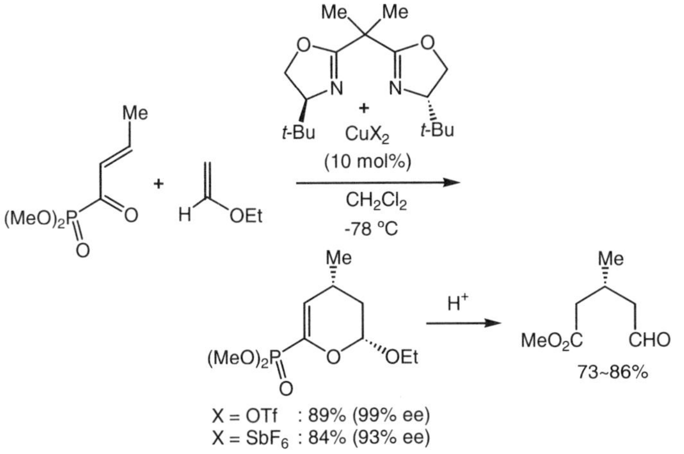

X = OTf : 89% (99% ee)
X = SbF$_6$: 84% (93% ee)

(8A.40)

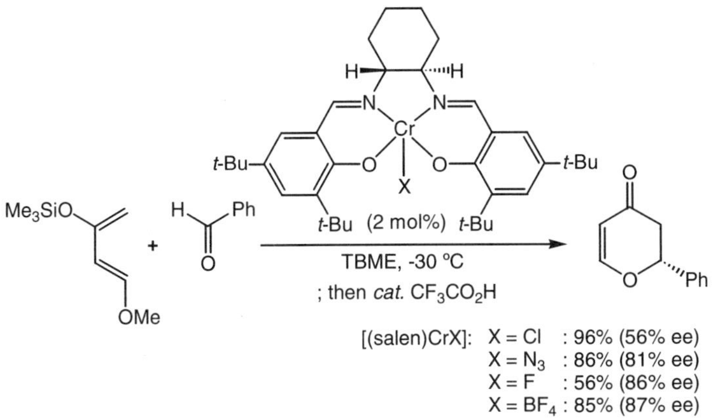

[(salen)CrX]: X = Cl : 96% (56% ee)
X = N$_3$: 86% (81% ee)
X = F   : 56% (86% ee)
X = BF$_4$: 85% (87% ee)

(8A.41)

A series of trivalent lanthanoid complexes, scandium, and ytterbium tris-(R)-(–)-1,1'-binaphthyl-2,2'-diyl phosphonate, have been introduced as new chiral and stable Lewis acids for the asymmetric hetero Diels–Alder reaction of Danishefsky's diene and aldehydes. 2,6-Lutidine was found to be an effective additive to improve the enantioselectivity up to 89% ee [66] (Eq. 8A.42).

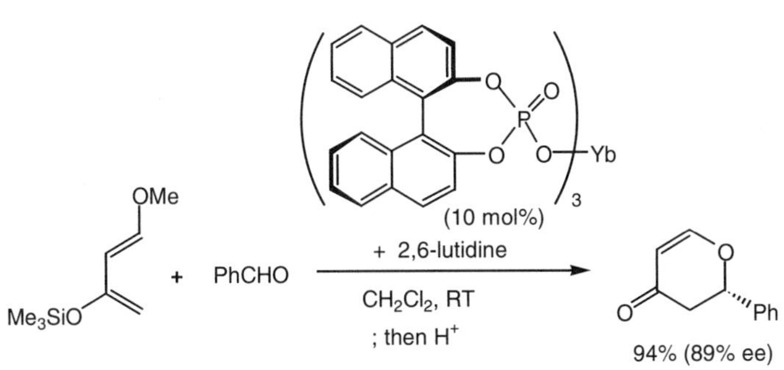

(8A.42)

94% (89% ee)

The chiral boron complex prepared in situ from chiral binaphthol and $B(OPh)_3$ is utilized for the asymmetric aza-Diels–Alder reaction of Danishefsky's diene and imines [67] (Eq. 8A.43). Although the asymmetric reaction of prochiral imine affords products with up to 90% ee, the double asymmetric induction with chiral imine by using α-benzylamine as a chiral auxiliary has achieved almost complete diastereoselectivity for both aliphatic and aromatic aldimines. This method has been successfully applied to the efficient asymmetric synthesis of anabasine and coniine of piperidine alkaloides.

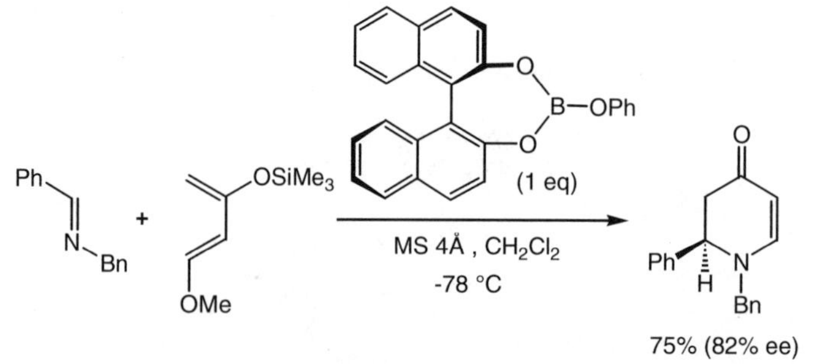

$$\text{(8A.43)}$$

75% (82% ee)

8A.4. ASYMMETRIC [2+2] CYCLOADDITIONS

Narasaka's chiral titanium catalyst, prepared from $(Pr^iO)_2TiCl_2$ and a tartrate-derived $(2R,3R)$-1,1,4,4-tetraphenyl-2,3-O-(1-phenylethylidene)-1,2,3,4-butanetetrol, is utilized for the asymmetric [2+2] cycloaddition of N-acyl oxazolidinones to 1,2-propadienyl sulfides possessing α-substituents, which afford methylenecyclobutane derivatives with high enantiomeric purity. These chiral adducts are readily transformed to seven- and eight-membered carbocycles with chiral side chains by the ring-cleavage reaction and subsequent cationic cyclization of the chiral cyclobutane derivative [68] (Eq. 8A.44).

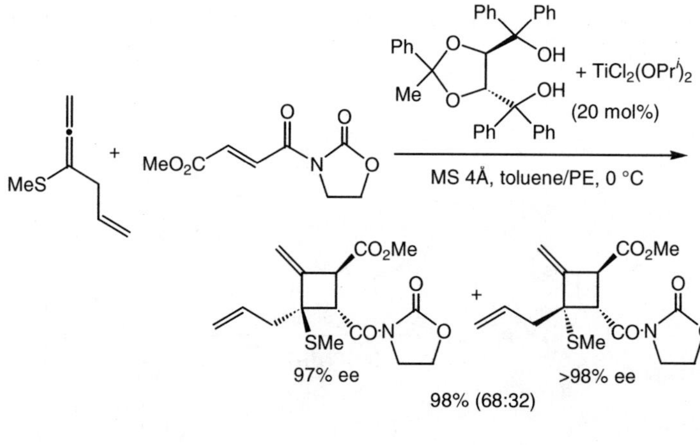

$$\text{(8A.44)}$$

97% ee   >98% ee

98% (68:32)

A chiral bis(silyl)binaphthol-modified aluminum catalyst has been successfully applied to the asymmetric [2+2] cycloaddition of ketenes to aldehydes, giving optically active 4-substituted oxetan-2-ones with moderate enantioselectivity [69] (Eq. 8A.45).

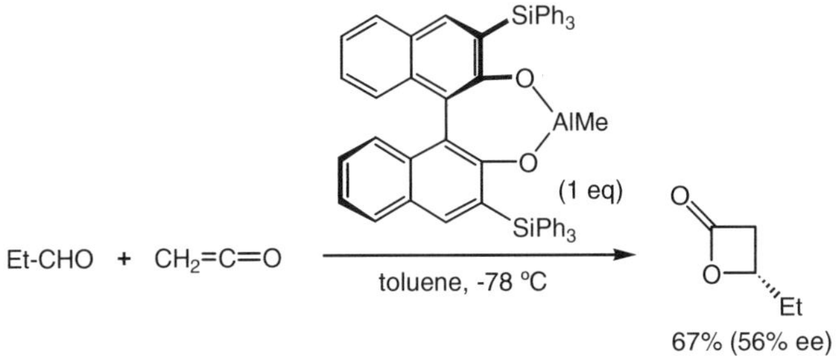

$$Et\text{-}CHO \ + \ CH_2{=}C{=}O \quad \xrightarrow[\text{toluene, -78 °C}]{} \qquad (8A.45)$$

67% (56% ee)

8A.5. ASYMMETRIC DIPOLAR CYCLOADDITIONS

Asymmetric 1,3-dipolar cycloaddition of nitrones to ketene acetals is effectively catalyzed by chiral oxazaborolidines derived from N-tosyl-L-α-amino acids to afford 5,5-dialkoxyisoxazolidines with high regio- and stereoselectivity [70] (Eq. 8A.46). Hydrolysis of the N-O bond of the resulting chiral adducts under mild conditions yields the corresponding β-amino esters quantitatively.

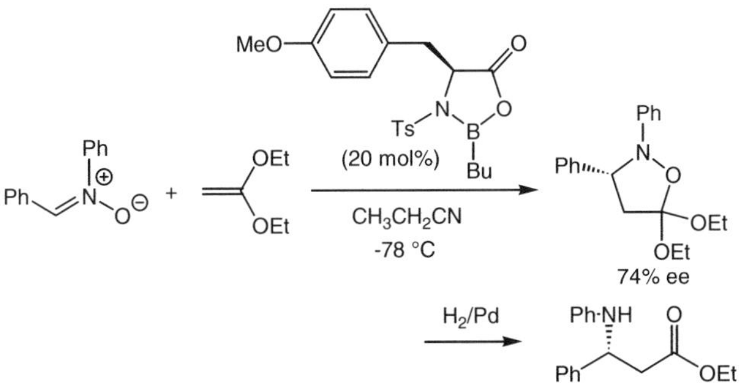

$$ \qquad (8A.46)$$

74% ee

A chiral zinc(II) complex derived from Et_2Zn and diisopropyl (R,R)-tartrate as a chiral auxiliary is applied to the asymmetric 1,3-dipolar cycloaddition of nitrile oxides to an achiral allylic alcohol, giving the corresponding (R)-2-isoxazolines with high enantioselectivity. Addition of a small amount of ethereal compounds such as DME and 1,4-dioxane is crucial for achieving the high asymmetric induction in a reproducible manner [71] (Eq. 8A.47).

8A.6. OTHER ASYMMETRIC CYCLOADDITIONS

Asymmetric [4+1] cycloaddition of vinylallenes and carbon monoxide is promoted by a cationic Rh complex formed in situ from $[Rh(cod)_2]PF_6$ and chiral diphosphine ligand, (R,R)-Me-DuPHOS to afford 2-alkylidene-3-cyclopentenones with high asymmetric induction [72] (Eq. 8A.48).

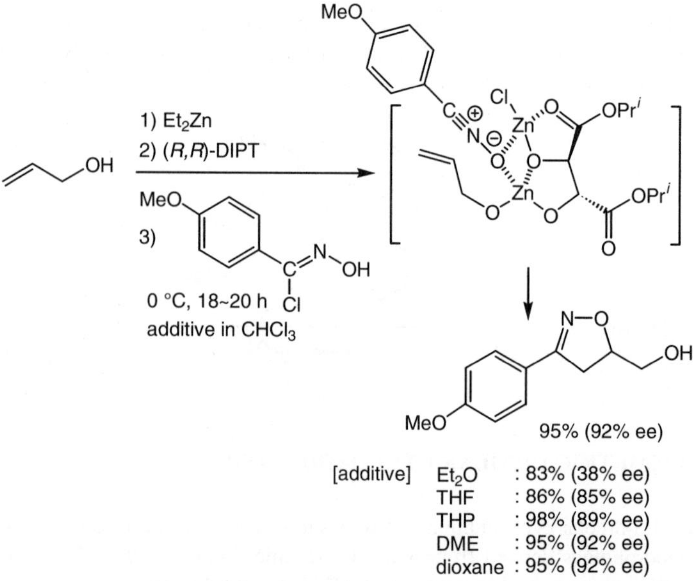

(8A.47)

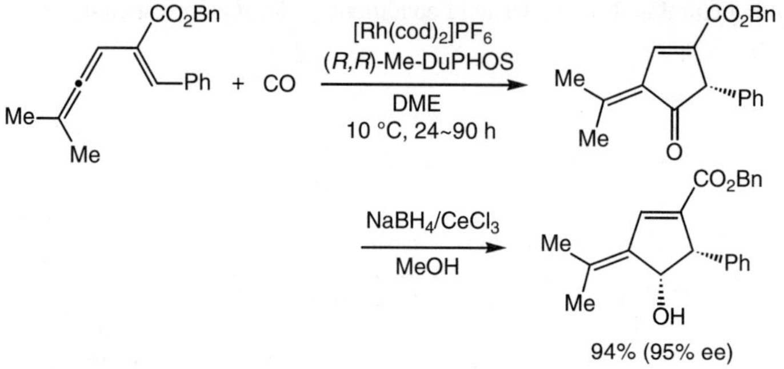

[additive] Et$_2$O   : 83% (38% ee)
            THF   : 86% (85% ee)
            THP   : 98% (89% ee)
            DME   : 95% (92% ee)
            dioxane : 95% (92% ee)

(8A.48)

Enantioselective tandem carbonyl ylide formation-cycloaddition of α-diazo-β-keto esters is achieved in hexane with [Rh$_2$(S-DOSP)$_4$] (1 mol %) at room temperature to give the corresponding cycloadducts with moderate enantioselectivity [73] (Eq. 8A.49).

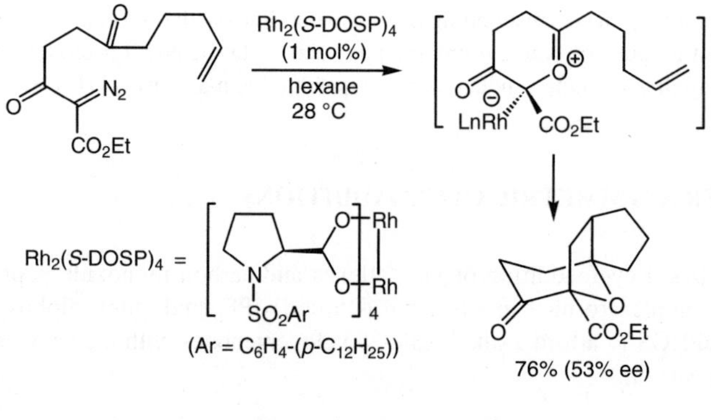

(8A.49)

Asymmetric cycloaddition of 2-vinyloxiranes to carbodiimides proceeds in the presence of $Pd_2(dba)_3 \cdot CHCl_3$ and TolBINAP as the chiral ligand in THF at room temperature to yield 4-vinyl-1,3-oxazolidin-2-imines with up to 95% ee [74] (Eq. 8A.50). The enantio-determining step is assumed to be the nucleophilic attack of a nitrogen nucleophile on a π-allylpalladium intermediate. Reactions of 2-vinyloxiranes with isocyanates using the same catalyst system afford 4-vinyl-1,3-oxazolidin-2-ones with low enantioselectivity.

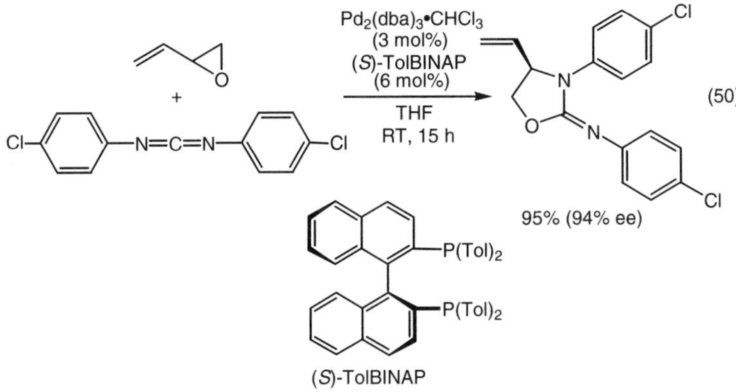

$$(8A.50)$$

8A.7. CONCLUSION

As discussed above, the vast synthetic potential of chiral Lewis acid catalysts in asymmetric cycloadditions has obviously been demonstrated. There is no doubt that further research and development of asymmetric cycloadditions promoted by chiral Lewis acid catalysts will continue to provide exciting results in the coming years.

REFERENCES

1. Narasaka, K. *Synthesis,* **1991**, 1.

2. Maruoka, K.; Yamamoto, H. In *Catalytic Asymmetric Synthesis,* Ojima, I. (Ed.); VCH: New York, 1993, Ch. 9, pp. 413.

3. (a) Kagan, H. B.; Riant, O. *Chem. Rev.* **1992**, *92*, 1007. (b) Dias, L. C. *J. Braz. Chem. Soc.* **1997**, *8*, 289.

4. (a) Takasu, M.; Yamamoto, H. *Synlett* **1990**, 194. (b) Sartor, D.; Saffrich, J.; Helmchen, G. *Synlett* **1990**, 197. (c) Sartor, D.; Saffrich, J.; Helmchen, G.; Richards, C. J.; Lambert, H. *Tetrahedron: Asymmetry,* **1991**, *2*, 639.

5. Seerden, J. G.; Scheeren, H. W. Tetrahedron Lett., **1993**, *34*, 2669.

6. (a) Nevalainen, V. *Tetrahedron: Asymmetry* **1993**, *4*, 1565. (b) Nevalainen, V. *Tetrahedron: Asymmetry* **1994**, *5*, 767. (c) Nevalainen, V. *Tetrahedron: Asymmetry* **1996**, *7*, 1449.

7. Kamahori, K.; Ito, K.; Itsuno, S. *J. Org. Chem.* **1996**, *61*, 8321.

8. Corey, E. J.; Loh, T.-P. *Tetrahedron Lett.* **1993**, *34*, 3979.

9. Corey, E. J.; Guzman-Perez, A.; Loh, T.-P. *J. Am. Chem. Soc.* **1994**, *116*, 3611.

10. (a) Ishihara, K.; Gao, Q.; Yamamoto, H. *J. Am. Chem. Soc.* **1993**, *115*, 10412. (b) Ishihara, K.; Gao, Q.; Yamamoto, H. *J. Org. Chem.* **1993**, *58*, 6917.

11. Loh, T.-P.; Wang, R.-B.; Sim, K.-Y. *Tetrahedron Lett.* **1996**, *37*, 2989.

12. Davenport, A. J.; Davies, D. L.; Fawcett, J.; Garratt, S. A.; Lad, L.; Russell, D. R. *Chem. Commun.* **1997**, 2347.

13. Davies, D. L.; Fawcett, J.; Garratt, S. A.; Russell, D. R. *Chem. Commun.* **1997**, 1351.

14. Carmona, D.; Cativiela, C.; Elipe, S.; Lahoz, F. J.; Lamata, M. P.; Víu, M.P.L.-R.; Oro, L. A.; Vega, C.; Viguri, F. *Chem. Commun.* **1997**, 2351.

15. Maruoka, K.; Murase, N.; Yamamoto, H. *J. Org. Chem.* **1993**, *58*, 2938.

16. Ishihara, K.; Yamamoto, H. *J. Am. Chem. Soc.* **1994**, *116*, 1561.

17. Ishihara, K.; Kurihara, H.; Yamamoto, H. *J. Am. Chem. Soc.* **1996**, *118*, 3049.

18. Ishihara, K.; Kondo, S.; Kurihara, H.; Yamamoto, H. *J. Org. Chem.* **1997**, *62*, 3026.

19. Reilly, M.; Oh, T. *Tetrahedron Lett.* **1995**, *36*, 221.

20. Corey, E. J.; Roper, T. D.; Ishihara, K.; Sarakinos, G. *Tetrahedron Lett.* **1993**, *34*, 8399.

21. Kündig, E. P.; Bourdin, B.; Bernardinelli, G. *Angew. Chem., Int. Ed.* **1994**, *33*, 1856.

22. Yamashita, Y.; Katsuki, T. *Synlett* **1995**, 829.

23. Odenkirk, W.; Bosnich, B. *Chem. Commun.* **1995**, 1181.

24. Hayashi, Y.; Rohde, J. J.; Corey, E. J. *J. Am. Chem. Soc.* **1996**, *118*, 5502.

25. Corey, E. J.; Lee, T. W. *Tetrahedron Lett.* **1997**, *38*, 5755.

26. Evans, D. A.; Miller, S. J.; Lectka, T. *J. Am. Chem. Soc.* **1993**, *115*, 6460.

27. Evans, D. A.; Murry, J. A.; Matt, P.; Norcross, R. D.; Miller, S. J. *Angew. Chem., Int. Ed.* **1995**, *34*, 798.

28. Johannsen, M.; Jørgensen, K. A. *J. Chem. Soc., Perkin Trans. 2* **1997**, 1183.

29. Evans, D. A.; Shaughnessy, E. A.; Barnes, D. M. *Tetrahedron Lett.* **1997**, *38*, 3193.

30. Evans, D. A.; Barnes, D. M. *Tetrahedron Lett.* **1997**, *38*, 57.

31. Evans, D. A.; Lectka, T.; Miller, S. J. *Tetrahedron Lett.* **1993**, *34*, 7027.

32. Corey, E. J.; Ishihara, K. *Tetrahedron Lett.* **1992**, *33*, 6807.

33. (a) Desimoni, G.; Faita, G.; Righetti, P. P. *Tetrahedron Lett.* **1996**, *37*, 3027. (b) Desimoni, G.; Faita, G.; Invernizzi, A. G.; Righetti, P. *Tetrahedron* **1997**, *53*, 7671.

34. Takacs, J. M.; Lawson, E. C.; Reno, M. J.; Youngman, M. A.; Quincy, D. A. *Tetrahedron: Asymmetry* **1997**, *8*, 3073.

35. Takacs, J. M.; Quincy, D. A.; Shay, W.; Jones, B. E.; Ross, II, C. R. *Tetrahedron: Asymmetry* **1997**, *8*, 3079.

36. (a) Ghosh, A. K.; Mathivanan, P.; Cappiello, J. *Tetrahedron Lett.* **1996**, *37*, 3815. (b) Davies, I. W.; Senanayake, C. H.; Larsen, R. D.; Verhoeven, T. R.; Reider, P. J. *Tetrahedron Lett.* **1996**, *37*, 1725.

37. Kanemasa, S.; Oderaotoshi, Y.; Sakaguchi, S.; Yamamoto, H.; Tanaka, J.; Wada, E.; Curran, D. P. *J. Am. Chem. Soc.* **1998**, *120*, 3074.

38. (a) Fujisawa, T.; Ichiyanagi, T.; Shimizu, M. *Tetrahedron Lett.* **1995**, *36*, 5031. (b) Ichiyanagi, T.; Shimizu, M.; Fujisawa, T. *J. Org. Chem.* **1997**, *62*, 7937.

39. Sagasser, I.; Helmchen, G. *Tetrahedron Lett.* **1998**, *39*, 261.

40. Narasaki, K.; Yamamoto, I. *Tetrahedron*, **1992**, *48*, 5743.

41. Seebach, D.; Dahinden, R.; Marti, R. E.; Beck, A. K.; Platter, D. A.; Kühnle, F. N. K. *J. Org. Chem.* **1995**, *60*, 1788.

42. Seebach, D.; Marti, R. E.; Hintermann, T. *Hely. Chim. Acta* **1996**, *79*, 1710.

43. (a) Gothelf, K.V.; Hazell, R. G.; Jørgensen, K. A. *J. Am. Chem. Soc.* **1995**, *117*, 4435. (b) Gothelf, K. V.; Jørgensen, K. A. *J. Org. Chem.* **1995**, *60*, 6847.

44. (a) Kobayasi, S.; Hachiya, I.; Ishitani, H.; Arai, M. *Tetrahedron Lett.* **1993**, *34*, 4535. (b) Kobayasi, S.; Ishitani, H.; Araki, M.; Hachiya, I. *Tetrahedron Lett.* **1994**, *35*, 6325.

45. (a) Kobayashi, S.; Ishitani, H. *J. Am. Chem. Soc.* **1994**, *116*, 4083. (b) Kobayashi, S.; Ishitani, H.; Hachiya, I.; Araki, M. *Tetrahedron* **1994**, *50*, 11623.

46. Corey, E. J.; Sarshar, S.; Bordner, J. *J. Am. Chem. Soc.* **1992**, *114*, 7938.

47. Ordoñez, M.; Rosa, V. G.; Labastida, V.; Llera, J. M. *Tetrahedron: Asymmetry* **1996**, *7*, 2675.

48. Khiar, N.; Fernández, I.; Alcudia, F. *Tetrahedron Lett.* **1993**, *34*, 123.

49. Honda, Y.; Date, T.; Hiramatsu, H.; Yamauchi, M. *Chem. Commun.* **1997**, 1411.

50. Maruoka, K.; Itoh, T.; Shirasaka, T.; Yamamoto, H. *J. Am. Chem. Soc.* **1988**, *110*, 310.

51. Maruoka, K.; Concepcion, A. B.; Yamamoto, H. *Bull. Chem. Soc. Jpn.* **1992**, *65*, 3501.

52. Harada, T.; Takeuchi, M.; Hatsuda, M.; Ueda, H.; Oku, A. *Tetrahedron: Asymmetry* **1996**, *7*, 2479.

53. Devine, P. N.; Oh, T. *J. Org. Chem.* **1992**, *57*, 396.

54. Quinkert, G.; Grosso, M.; Bucher, A.; Bauch, M.; Döring, W.; Bats, J. W.; Dürner, G. *Tetrahedron Lett.* **1992**, *33*, 3617.

55. Wada, E.; Pei, W.; Kanemasa, S. *Chem. Lett.* **1994**, 2345.

56. Brimble, M. A.; McEwan, J. F. *Tetrahedron: Asymmetry* **1997**, *8*, 4069.

57. Corey, E. J.; Sarshar, S.; Lee, D.-H. *J. Am. Chem. Soc.* **1994**, *116*, 12089.

58. Corey, E. J.; Letavic, M. A. *J. Am. Chem. Soc.* **1995**, *117*, 9616.

59. Bienaymé, H. *Angew. Chem., Int. Ed.* **1997**, *36*, 2670.

60. Itsuno, S.; Tada, S.; Ito, K. *Chem. Commun.* **1997**, 933.

61. Aw, B.-H.; Leung, P.-H.; White, A.J.P.; Williams, D. J. *Organometallics* **1996**, *15*, 3640.

62. Johannsen, M.; Jørgensen, A. *Tetrahedron* **1996**, *52*, 7321.

63. Ghosh, A. K.; Mathivanan, P.; Cappiello, J.; Krishnan, K. *Tetrahedron: Asymmetry* **1996**, *7*, 2165.

64. Evans, D. A.; Johnson, J. S. *J. Am. Chem. Soc.* **1998**, *120*, 4895.

65. Schaus, S. E.; Branalt, J.; Jacobsen, E. N. *J. Org. Chem.* **1998**, *63*, 403.

66. Hanamoto, T.; Furuno, H.; Sugimoto, Y.; Inanaga, J. *Synlett* **1997**, 79.

67. Hattori, K.; Yamamoto, H. *Tetrahedron* **1993**, *49*, 1749.

68. Narasaka, K.; Hayashi, K.; Hayashi, Y. *Tetrahedron* **1994**, *50*, 4529.

69. Tamai, Y.; Someya, M.; Fukumoto, J.; Miyano, S. *J. Chem. Soc., Perkin Trans. 1* **1994**, 1549.

70. Seerden, J.-P. G.; Reimer, A. W. A. S.; Scheeren, H. W. *Tetrahedron Lett.* **1994**, *35*, 4419.

71. Shimizu, M.; Ukaji, Y.; Inomata, K. *Chem. Lett.* **1996**, 455.

72. Murakami, M.; Itami, K.; Ito, Y. *J. Am. Chem. Soc.* **1997**, *119*, 2950.

73. Hodgson, D. M.; Stupple, P. A.; Johnstone, C. *Tetrahedron Lett.* **1997**, *38*, 6471.

74. Larksarp, C.; Alper, H. *J. Am. Chem. Soc.* **1997**, *119*, 3709.

8B1

ASYMMETRIC ALDOL REACTIONS— DISCOVERY AND DEVELOPMENT

MASAYA SAWAMURA AND YOSHIHIKO ITO

Department of Synthetic Chemistry and Biological Chemistry, Graduate School of Engineering, Kyoto University, Kyoto, Japan

8B1.1. INTRODUCTION

The aldol reaction is one of the most important organic reactions because it is a carbon–carbon bond-forming reaction, which produces highly functionalized compounds with a pair of newly generated chiral centers. Therefore, the pursuit of high stereoselectivity in the aldol reaction has been of much interest in recent years. In 1981 Masamune et al. [1] and Evans et al. [2], independently, reported extremely stereoselective asymmetric aldol reactions by means of chiral boron enolates. These discoveries are milestones in the history of asymmetric synthesis. Especially Evans's boron enolates, which can easily be prepared from optically active amino acids, have been widely applied to natural product syntheses. The next advance was made in 1982 by Iwasawa and Mukaiyama [3], who reported highly enantioselective aldol reactions of achiral tin(II) enolates with achiral aldehydes in the presence of a stoichiometric amount of chiral diamine ligands. This process does not require tedious procedures for attaching the auxiliary or removing it from the product. The next challenging goal in this field was the catalytic asymmetric aldol reaction.

8B1.2. GOLD-CATALYZED ASYMMETRIC ALDOL REACTION OF α-ISOCYANOCARBOXYLATES

In 1986 Ito, Sawamura, and Hayashi [4] reported that gold(I) complexes prepared from cationic gold complex **1** and chiral ferrocenylphosphine ligands (**2**) bearing a tertiary amino group at the terminal position of a pendant chain are effective catalysts for asymmetric aldol reaction of

Catalytic Asymmetric Synthesis, Second Edition, Edited by Iwao Ojima
ISBN 0-471-29805-0   Copyright © 2000 Wiley-VCH, Inc.

isocyanoacetate **3a** with aldehydes, giving optically active 5-alkyl-2-oxazoline-4-carboxylates **4** (Scheme 8B1.1) [4–11]. The reaction shows high selectivity for the formation of trans isomers with enantiomeric excesses (ee) ranging as high as 97% (Table 8B1.1). Various substituents on aromatic aldehydes are acceptable for the highly stereoselective aldol reaction with the isocyanoacetate (Table 8B1.1: entries 5–10). Secondary and tertiary alkyl aldehydes give the corresponding *trans*-oxazolines almost exclusively with high enantioselectivity (entries 18, 19). The gold catalysts are effective for the reaction of α,β-unsaturated aldehydes as well (entries 20, 21). The *trans*-oxazolines thus obtained can be readily hydrolyzed to β-hydroxy-α-amino acids and their derivatives. The turnover efficiency of the gold catalyst is shown in entry 6, where the [substrate]/[catalyst] ratio can be raised to 10,000:1 without significant loss of the enantiomeric purity of the *trans*-oxazoline, indicating that the gold-catalyzed aldol reaction may provide a practical process of great promise [12].

Both the enantio- and diastereoselectivity of this reaction are significantly dependent on the structure of the terminal amino group of the ligand (Table 8B1.1, entries 1–4, 11–16), indicating that the amino group plays a key role in the stereoselection [5,7]. In general, six-membered ring amino groups such as piperidino and morpholino groups are effective. Results listed in Table 8B1.2 [4] indicate that the existence of a terminal amino group in the proper position is crucial for high stereoselectivity. Thus, the use of ligand **5**, which is analogous to **2a** but with a longer tether between the terminal amino group and the ferrocene moiety, or ligand **6**, which has a hydroxyl group instead of the terminal amino group, causes a drastic decrease in stereoselectivity. Gold catalysts containing ferrocenylphosphines BPPFA (**7**) and **8**, which lack the terminal

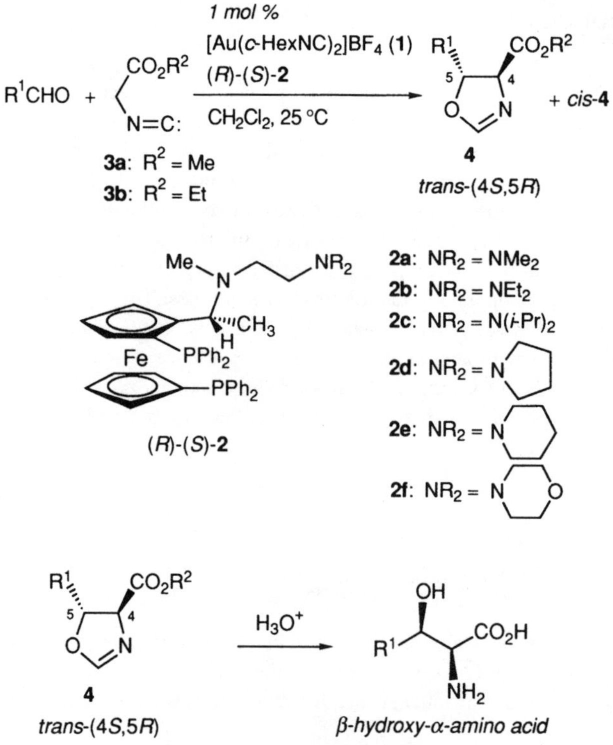

Scheme 8B1.1.

TABLE 8B1.1. Gold-Catalyzed Asymmetric Aldol Reaction of Isocyanoacetate 3a with Aldehydes (Scheme 8B1.1)[a]

Entry	Aldehyde	L*	Yield (%) of **4a**	trans/cis	% ee of trans-**4a**
1	PhCHO	**2a**	91	90/10	91
2	PhCHO	**2b**	98	89/11	93
3	PhCHO	**2e**	94	94/6	95
4	PhCHO	**2f**	93	95/5	95
5	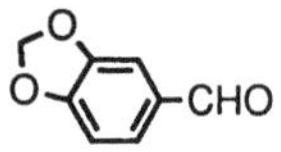 (benzodioxole aldehyde)	**2f**	86	95/5	96
6[b]	(benzodioxole-CHO)	**2f**	c	91/9	91
7	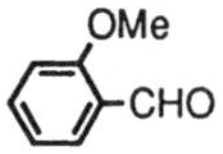 OMe-substituted benzaldehyde	**2f**	98	92/8	92
8	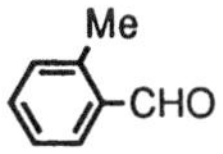 Me-substituted benzaldehyde	**2f**	98	96/4	95
9	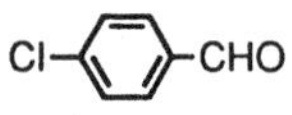 Cl—⟨⟩—CHO	**2f**	97	94/6	94
10	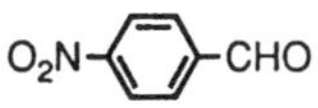 O₂N—⟨⟩—CHO	**2f**	80	83/17	86
11	MeCHO	**2a**	94	78/22	37
12	MeCHO	**2b**	100	84/16	72
13	MeCHO	**2c**	99	70/30	55
14	MeCHO	**2d**	83	87/13	74
15	MeCHO	**2e**	100	85/15	85
16	MeCHO	**2f**	99	89/11	89
17	i-BuCHO	**2f**	99	96/4	87
18	i-PrCHO	**2e**	99	99/1	94
19	t-BuCHO	**2f**	94	100/0	97
20	(E)-n-PrCH=CHCHO	**2f**	85	87/13	92
21	(E)-MeCH=CMeCHO	**2a**	89	91/9	93

[a]Reaction time: 20–40 h. [b]Substrates/catalyst = 1×10^4/L. Reaction at 40°C for 9 days. [c]Not isolated. Conversion: 100%.

amino group, as well as CHIRAPHOS, DIOP, and p-Tol-BINAP, are catalytically much less active and give almost racemic oxazolines.

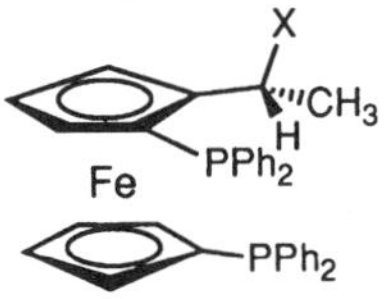

5: X = NMe(CH₂)₃NMe₂

6: X = NMeCH₂CH₂OH

7: X = NMe₂

8: X = OMe

The high efficiency of the gold catalysts has been explained by a hypothetical transition state as shown in Figure 8B1.1: The chiral ligand chelates to gold with the two phosphorus atoms,

TABLE 8B1.2. Gold-Catalyzed Asymmetric Aldol Reaction of Isocyanoacetate 3a with PhCHO Using Various Chiral Ligands (Scheme 8B.1)

Entry	L*	Time (h)	Yield of **4a**	*trans/cis*	% ee of *trans*-**4a**
1	**5**	40	99	89/11	23
2	**6**	22	91	69/31	37
3	**7**	51	80	68/32	*racemic*
4	**8**	90	87	86/14	*racemic*
5	(*S,S*)-CHIRAPHOS	100	90	75/25	*racemic*
6	(−)-DIOP	300	57	76/24	*racemic*
7	(+)-*p*-TolBINAP	200	69	69/31	8

leaving the two nitrogen atoms in the pendant chain uncoordinated; the terminal amino group abstracts one of the α-methylene protons of the isocyanoacetate coordinated to the gold, forming an ion pair between enolate anion and ammonium cation. The attractive interaction (*secondary ligand–substrate interaction*) permits a favorable arrangement of the enolate and the aldehyde on the gold in the stereodifferentiating transition state.

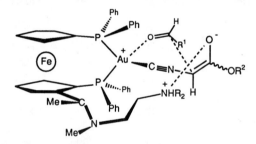

Figure 8B1.1. Proposed transition-state model of the gold-catalyzed asymmetric aldol reaction.

The usefulness of the gold-catalyzed aldol reaction was demonstrated by application of the method to the asymmetric synthesis of the important membrane components D-*erythro* and *threo*-sphingosines, and their stereoisomers (Scheme 8B1.2) [13], and MeBmt, an unusual amino acid in the immunosuppressive undecapeptide cyclosporine (Scheme 8B1.3) [14].

As pointed out by Togni and Pastor, enantioselectivities in the gold-catalyzed aldol reaction of aldehydes containing an α-heteroatom are significantly different from those of simple aldehydes (Table 8B1.3) [15,16]. Low enantioselectivities for *trans*-oxazolines are observed in the aldol reactions of 2-thiophene-, 2-furan-, and 2-pyridinecarboxaldehyde (entries 2, 4, 7). In the reactions of the 2-furan- and 2-pyridinecarboxaldehyde, *cis*-oxazolines with fairly high enantiomeric purities are formed as the minor product but in a rather low trans/cis ratio. A similar α-heteroatom effect is also observed in the aldol reaction of 2,3-*O*-isopropylidene-D-glyceraldehyde.

The catalytic asymmetric aldol reaction is tolerant of α-isocyanocarboxylates bearing an α-alkyl substituent (**9a–d**) as shown in Scheme 4 and Table 8B1.4 [17,18]. The enantioselectivities in the reaction with paraformaldehyde are generally moderate, with the exception of the

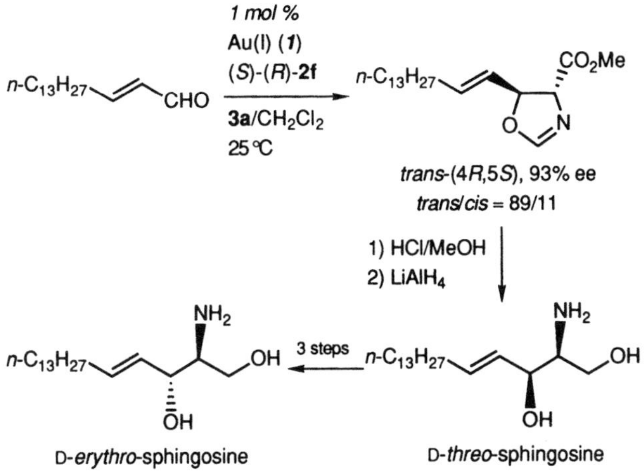

Scheme 8B1.2.

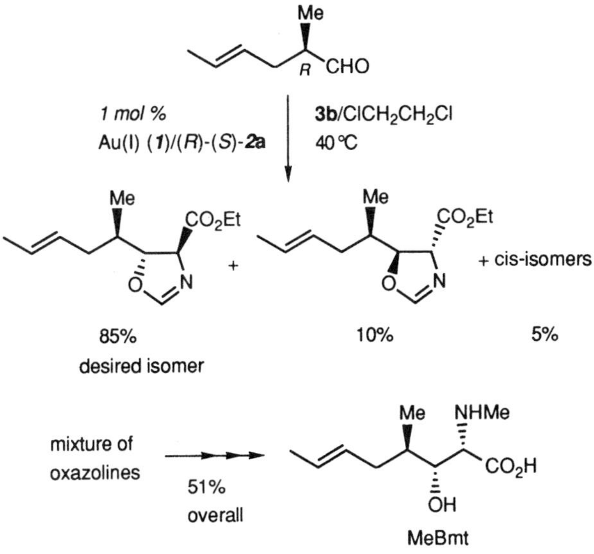

Scheme 8B1.3.

reaction of **9c** (81% ee), which has an α-isopropyl substituent (Table 8B1.4, entry 3). The oxazolines can be converted to optically active α-alkylserine derivative, a class of biologically interesting compounds.

N,N-Dimethyl-α-isocyanoacetamide **10** is the substrate of choice in the reaction with acetaldehyde (98.6% ee) or primary aldehydes such as propionaldehyde (96.3% ee) or isovaleraldehyde (97.3% ee) (Scheme 8B1.5, Table 8B1.5) [19]. The oxazolinecarboxamides **11** thus prepared can be converted to β-hydroxy-α-amino acids by acidic hydrolysis. The aldol reaction

TABLE 8B1.3. Asymmetric Aldol Reaction of Isocyanoacetate 3b with Functionalized Aldehydes in the Presence of Gold(I) Catalyst Containing Ligand 2a (Scheme 8B1.1)[a]

Entry	Aldehyde R^1	Yield (%) of **4b**	trans/cis	% ee of trans-**4b**	% ee of cis-**4b**
1	3-thienyl	78	90/10	78	1 (4S,5S)
2	2-thienyl	90	95/5	33	17 (4S,5S)
3	3-furyl	80	86/14	87	7 (4S,5S)
4	2-furyl	62	68/32	32	83 (4S,5S)
5	4-pyridyl	38	88/12	75	78 (4S,5S)
6	3-pyridyl	55	88/12	79	62 (4S,5S)
7	2-pyridyl	45	75/25	6	84 (4S,5S)

[a]Reaction at 50°C in dichloromethane.

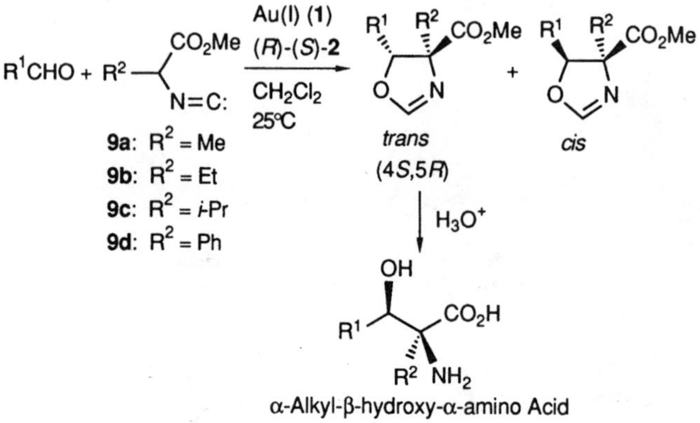

Scheme 8B1.4.

TABLE 8B1.4. Gold-Catalyzed Asymmetric Aldol Reaction of α-Isocyanocarboxylates 9 with Aldehydes (Scheme 8B1.4)

Entry	Aldehyde	**9** R^2	Ligand	Yield (%)	trans/cis	% ee trans	% ee cis
1	$(CH_2O)_n$	Me	**2a**	100	-	64	-
2	$(CH_2O)_n$	Et	**2a**	89	-	70	-
3	$(CH_2O)_n$	i-Pr	**2e**	96	-	81	-
4	$(CH_2O)_n$	Ph	**2a**	75	-	67[a]	-
5	MeCHO	Me	**2f**	86	56/44	86	54 (4S,5S)
6	MeCHO	Et	**2f**	92	54/46	87	66 (4S,5S)
7	MeCHO	i-Pr	**2f**	100	24/76	26	51 (4S,5S)
8	PhCHO	Me	**2f**	97	93/7	94	53 (4S,5S)
9	PhCHO	i-Pr	**2e**	86	54/46	92	28 (4R,5R)

[a]The configuration has not been determined.

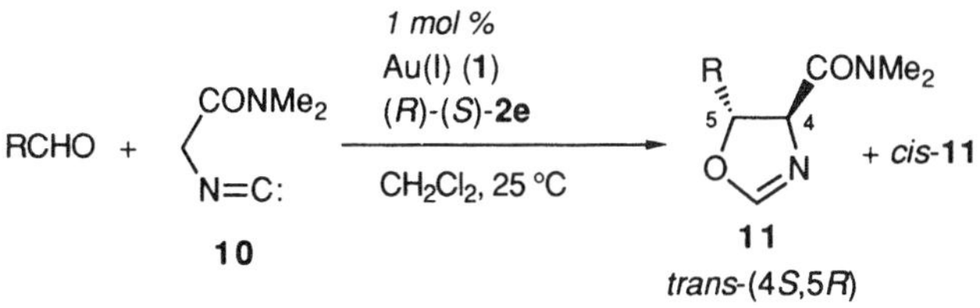

Scheme 8B1.5.

TABLE 8B1.5. Gold-Catalyzed Asymmetric Aldol Reaction of Isocyanoacetamide 10 with Aldehydes (Scheme 8B1.5)

Entry	Aldehyde	Time (h)	Yield (%) of **11**	*trans/cis*	% ee of *trans*-**11**
1	MeCHO	40	85	91/9	98.6
2	EtCHO	40	84	95/5	96.3
3	*i*-BuCHO	6	92	94/6	97.3
4	*p*-BnOC$_6$H$_4$CH$_2$CHO	80	84	>95/5	94.5
5	PhCHO	25	74	94/6	94.1

of α-keto esters with isocyanoacetamide **10** proceeds with moderate-to-high enantioselectivity to give the corresponding oxazolines with up to 90% ee [20].

The methodology of the catalytic asymmetric aldol reaction has been further extended to the aldol-type condensation of (isocyanomethyl)phosphonates (**12**) with aldehydes, providing a useful method for the synthesis of optically active (1-aminoalkyl)phosphonic acids, which are a class of biologically interesting phosphorous analogs of α-amino acids (Scheme 8B1.6) [21,22]. Higher enantioselectivity and reactivity are obtained with diphenyl ester **12b** than with diethyl ester **12a** (Table 8B1.6).

Pastor and Togni pointed out that the central chirality and the planar chirality in the ferrocenylphosphine ligand **2** are cooperative for stereoselection (the concept of *internal cooperativity of chirality*) [16,23,24]. As Table 8B1.7 shows, the change of chirality of the stereogenic carbon atom from *R* to *S* results in the formation of the other *trans*-oxazoline enantiomer with moderate enantiomeric excess.

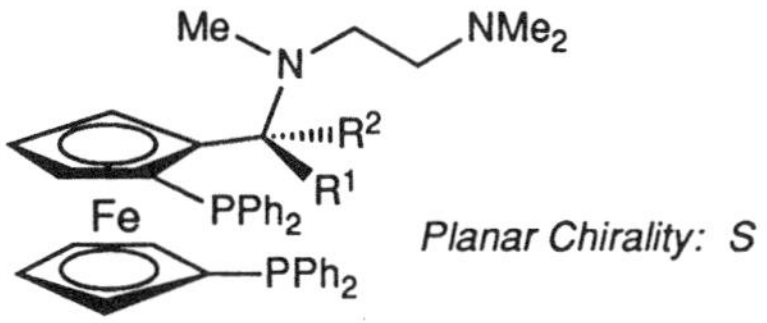

(*R*)-(*S*)-**2a**: R^1 = H, R^2 = Me
(*S*)-(*S*)-**2a**: R^1 = Me, R^2 = H

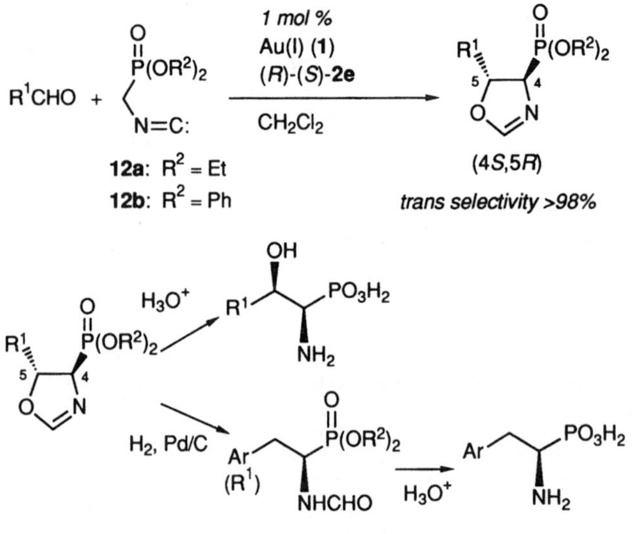

Scheme 8B1.6.

TABLE 8B1.6. Gold-Catalyzed Aldol-Type Condensation of (Isocyanomethyl)phosphonate 12 with Aldehydes (Scheme 8B1.6)

Entry	Isocyanide (R^2)	Aldehyde	Temp (°C)	Time (h)	Yield (%)	% ee
1	**12a** (Et)	PhCHO	40	90	78	92
2	**12a** (Et)	*i*-PrCHO	60	99	85	88
3	**12b** (Ph)	PhCHO	25	156	83	96
4	**12b** (Ph)	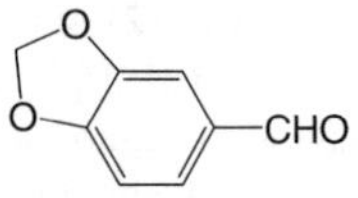	40	98	92	95
5	**12b** (Ph)	*i*-PrCHO	40	75	88	95

Recently it has been found that high stereoselectivity in the asymmetric aldol reaction of an isocyanoacetate is also obtainable with the silver(I) catalyst containing ferrocenylphosphine ligands **2e**, by keeping the isocyanoacetate concentration low throughout the reaction by the slow addition of **3a** over a period of 1 h (Scheme 8B1.7, Table 8B1.8) [25].

Mechanistic studies on the structure of gold(I) and silver(I) complexes coordinated with ferrocenylphosphine ligand **2** in the presence of an isocyanoacetate revealed that the most significant difference between these metal catalysts is in the number of the isocyanoacetates coordinating to the metal (Scheme 8B1.8). Thus, the gold complex adopts tricoordinated structure **13** with one isocyanoacetate even in the presence of a large excess of isocyanoacetate, while the silver complex is in equilibrium between tricoordinated complex **14** and tetracoordinated complex **15** bearing two isocyanoacetates [6,25]. The slow addition protocol in the

TABLE 8B1.7. Gold-Catalyzed Asymmetric Aldol Reaction of Isocyanoacetate 3a with PhCHO by the Use of (R)-(S)-2a or (S)-(S)-2a (Scheme 8B1.1)

Entry	L^*	trans/cis	trans-**4a** % ee	cis-**4a** % ee
1	(R)-(S)-**2a**	90/10	91 (4S,5R)	7 (4S,5S)
2	(S)-(S)-**2a**	84/16	41 (4R,5S)	20 (4S,5S)

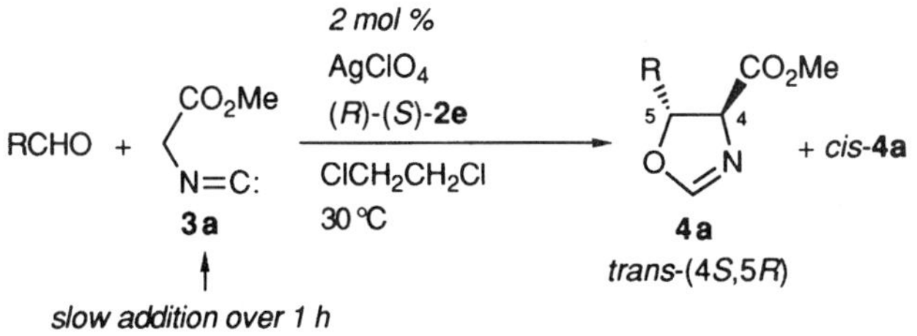

Scheme 8B1.7.

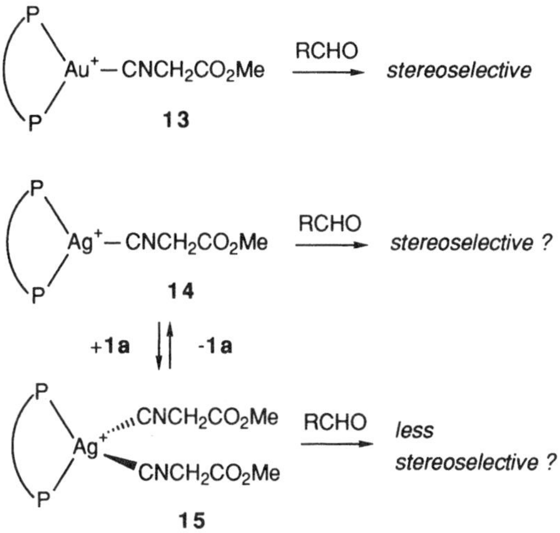

Scheme 8B1.8.

TABLE 8B1.8. Silver-Catalyzed Asymmetric Aldol Reaction of Isocyanoacetate 3a with Aldehydes (Scheme 8B1.7)

Entry	Aldehyde	Yield (%) of **4**	trans/cis	% ee of trans-**4a**
1	PhCHO	90	96/4	80
2	i-PrCHO	90	99/1	90
3	t-BuCHO	91	>99/1	88
4	CH_2=CMeCHO	90	97/3	87

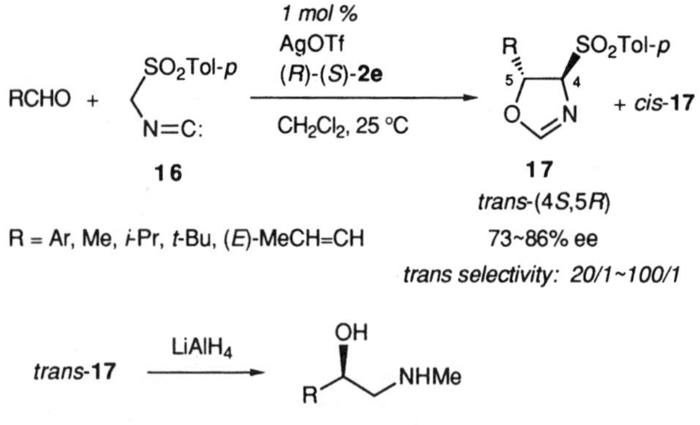

RCHO + **16** (SO₂Tol-p, N=C:)

1 mol %
AgOTf
(R)-(S)-**2e**

CH₂Cl₂, 25 °C

→ **17** trans-(4S,5R) + cis-**17**

R = Ar, Me, i-Pr, t-Bu, (E)-MeCH=CH

73~86% ee

trans selectivity: 20/1~100/1

trans-**17** —LiAlH₄→ R–CH(OH)–CH₂–NHMe

Scheme 8B1.9.

silver-catalyzed aldol reaction should keep the isocyanoacetate concentration low in the reaction system, diminishing the unfavorable tetracoordinate species. The high stereoselectivity of the tricoordinated catalyst may result from the presence of one vacant coordination site on the metal, where an aldehyde can coordinate. The aldol reaction of an aldehyde in the chiral coordination site may well be more stereoselective (transition state as shown in Fig. 8B1.1) than that of the aldehyde without coordination to the metal catalyst.

It is interesting that aldol-type condensation of tosylmethyl isocyanide (**16**) with aldehydes is catalyzed by the silver catalyst more stereoselectively than that catalyzed by the gold catalyst under the standard reaction conditions (Scheme 8B1.9) [26]. Elucidation of the mechanistic differences between the gold and silver catalysts in the asymmetric aldol reaction of **16** needs further study. Oxazoline **17** can be converted to optically active α-alkyl-β-(N-methyl-amino)ethanols.

8B1.3.   ASYMMETRIC NITROALDOL REACTION

Shibasaki et al. have reported an asymmetric nitroaldol reaction catalyzed by chiral lanthanum alkoxide **18** to produce an optically active 2-hydroxy-1-nitroalkane with moderate-to-high enantiomeric excesses (Scheme 8B1.10) [27]. Apparently this novel catalyst acts as Lewis base. The proposed reaction mechanism is shown in Scheme 8B1.11, where the first step of the reaction is the ligand exchange between binaphthol and nitromethane. This reaction is probably the first successful example of the catalytic asymmetric reaction promoted by a Lewis base metal catalyst. Future application of this methodology is quite promising.

8B1.4.   LEWIS ACID CATALYZED ASYMMETRIC ALDOL REACTION

The aldol reaction of p-nitrobenzaldehyde with acetone is catalyzed by Zn^{2+} complexes of α-amino acid esters in MeOH, giving the optically active aldol adduct (Scheme 8B1.12) [28]. Although the enantiomeric excesses of the product have not been determined, the extent of asymmetric induction is dependent on the structure of α-substituents in the amino acids. The Zn^{2+} complexes of amino acid esters bearing an aromatic substituent such as esters of phenyl-alanine, tyrosine, and tryptophane (Trp) are more effective in terms of both catalytic activity and asymmetric induction. The highest asymmetric induction is observed with Trp–OEt ligand.

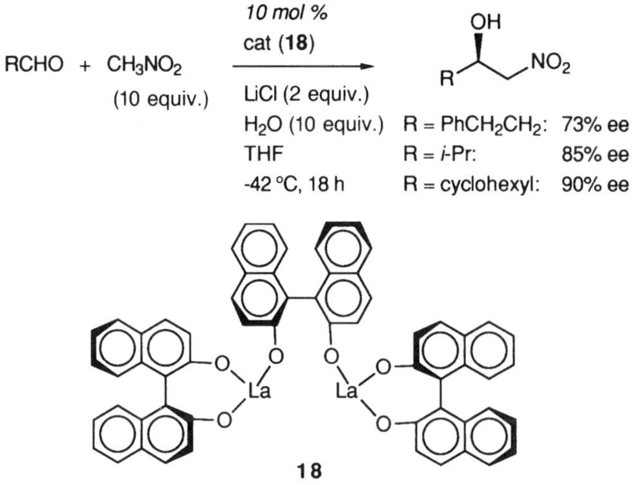
10 mol %
cat (18)
RCHO + CH3NO2
(10 equiv.)
LiCl (2 equiv.)
H2O (10 equiv.)
THF
-42 °C, 18 h
OH
R
NO2
R = PhCH2CH2: 73% ee
R = i-Pr: 85% ee
R = cyclohexyl: 90% ee
O
La
18

Scheme 8B1.10.

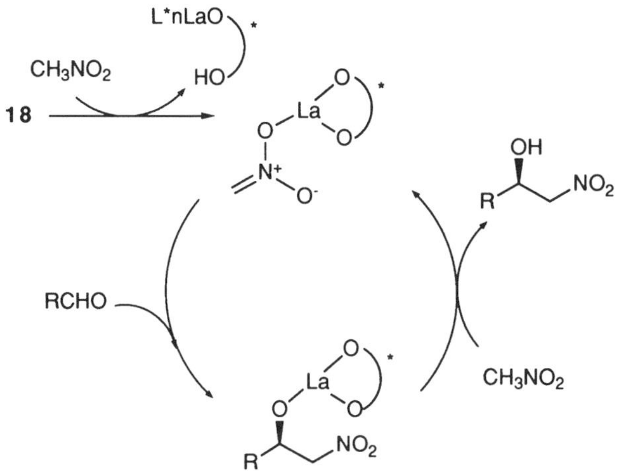
L*nLaO
CH3NO2
HO
18
O
La
N+
O-
OH
R
NO2
RCHO
CH3NO2

Scheme 8B1.11.

It has been reported that the chiral NMR shift reagent Eu(DPPM)$_3$, represented by structure **19**, catalyzes the Mukaiyama-type aldol condensation of a ketene silyl acetal with enantioselectivity of up to 48% ee (Scheme 8B1.13) [29–32]. The chiral alkoxyaluminum complex **20** [33] and the rhodium–phosphine complex **21** [34] under hydrogen atmosphere are also used in the asymmetric aldol reaction of ketene silyl acetals (Scheme 8B1.14), although the catalyst TON is quite low for the former complex.

Recently, highly enantioselective Mukaiyama-type aldol reactions using a substoichiometric amount of chiral Lewis acid have been reported from three research groups. Mukaiyama et al.

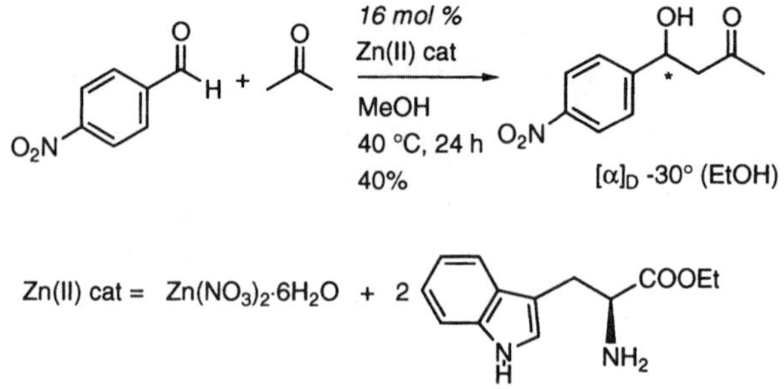

Scheme 8B1.12.

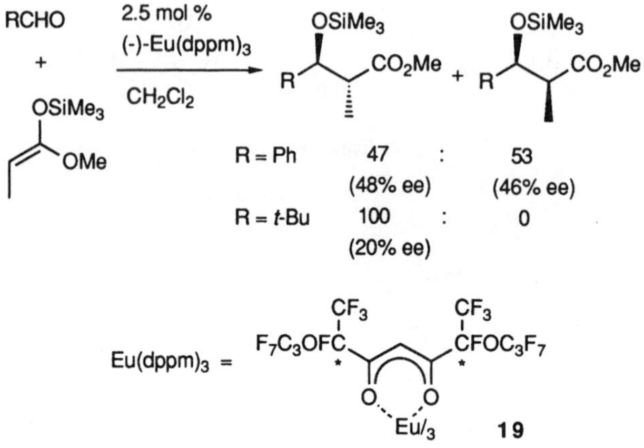

Scheme 8B1.13.

have achieved a high enantioselectivity in the aldol reaction of silyl enol ethers of thioesters (**22**) with aldehydes in the presence of a substoichiometric amount of chiral promoters (20 mol %), which consist of tin(II) triflate and chiral diamine **23**, under reaction conditions in which a solution of an aldehyde and **22** is slowly added to a solution of the chiral promoters (Scheme 8B1.15) [35–40]. As shown in Table 8B1.9, various achiral aldehydes are applicable to the aldol reaction. The reaction of the silyl ether of propanethiolate (**22a**) shows selectivity for the formation of the *syn*-aldol adduct with high enantioselectivity (entries 1–7). α-Unsubstituted aldol adducts are produced with high enantioselectivities in the reactions of the silyl ether of ethanethiolate (**22b**) (entries 8–11) [38]. The slow addition protocol may keep the trimethylsilyl triflate concentration low to suppress the undesirable TMSOTf-mediated aldol reaction. The proposed transition state is shown in Figure 8B1.2, where the *re* face of the aldehyde is shielded by the naphthyl (or tetrahydronaphthyl) group of the chiral diamine ligand. The stereochemical outcome of the aldol reaction of chiral α-siloxyaldehydes **24** is almost completely controlled by the chiral catalyst regardless of the inherent diastereofacial preference of the chiral aldehydes (Scheme 8B1.16) [40].

The same research group has reported that the chiral oxotitanium(IV) complex **26** is an efficient catalyst for the asymmetric aldol reaction of the silyl ether of thioester **27** with

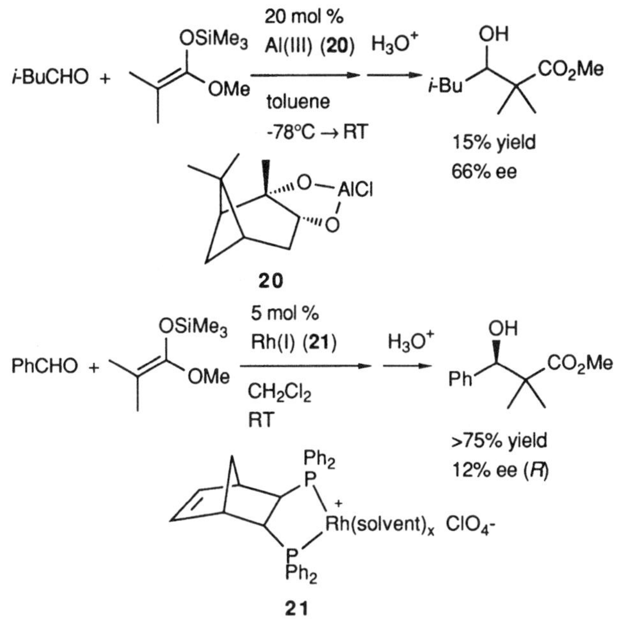

Scheme 8B1.14.

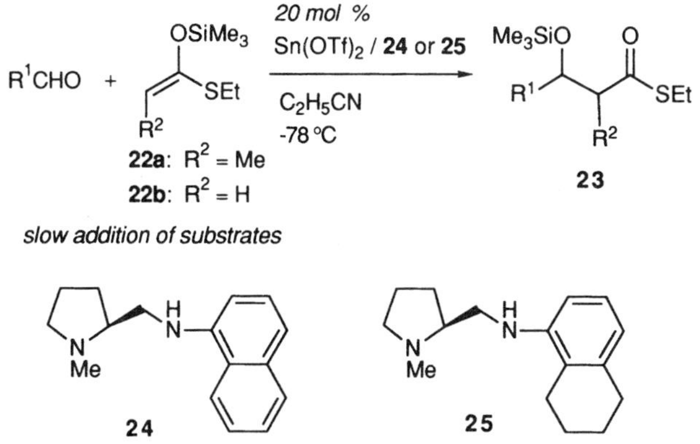

Scheme 8B1.15.

aldehydes to produce the corresponding aldol adducts with moderate-to-high enantiomeric excess (Scheme 8B1.17) [41]. The bulkiness of the substituents on both silicon and sulfur atoms of **27** is important for enantioselection.

Mukaiyama's group also reported the catalytic asymmetric aldol reaction of ketene silyl acetals (**28**) promoted by chiral zinc complexes. These complexes are prepared in situ from

TABLE 8B1.9. Asymmetric Aldol Reaction of 22 with Aldehydes Catalyzed by Chiral Tin(II) Catalysts (Scheme 8B1.15)

Entry	Aldehyde R^1	**22** R^2	L^*	Addition Time (h)	Yield (%)	syn/anti	% ee of syn-**23**
1	Ph	Me	**24**	3	77	93/7	90
2	(E)-n-PrCH=CH	Me	**24**	3	93	97/3	93
3	n-C_7H_{15}	Me	**24**	3	80	100/0	>98
4	c-C_6H_{11}	Me	**24**	3	71	100/0	>98
5[a]	$Me_3SiC\equiv C$	Me	**24**	4.5	73	95/5	91
6[a]	$PhC\equiv C$	Me	**24**	4.5	82	90/10	86
7[a]	n-$BuC\equiv C$	Me	**24**	4.5	67	93/7	91
8[a]	n-C_7H_{15}	H	**25**	4	79	-	93
9	c-C_6H_{11}	H	**25**	4.5	81	-	92
10	(E)-n-PrC=C	H	**24**	20	65	-	72
11[a]	n-$Buc\equiv C$	H	**25**	5	68	-	88

[a]solvent: CH_2Cl_2

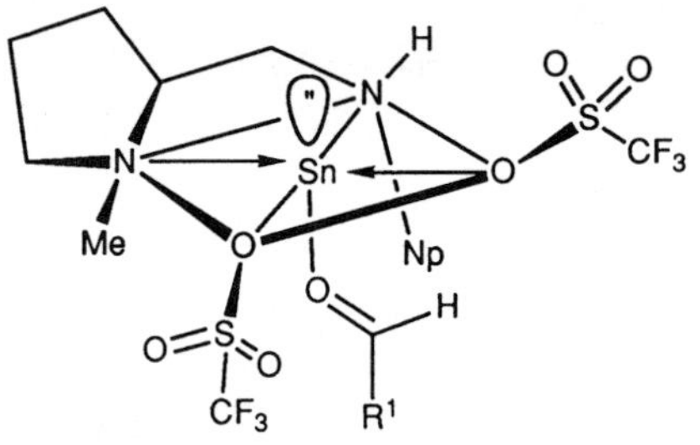

Figure 8B1.2. Proposed structure for a chiral Lewis acid–aldehyde complex.

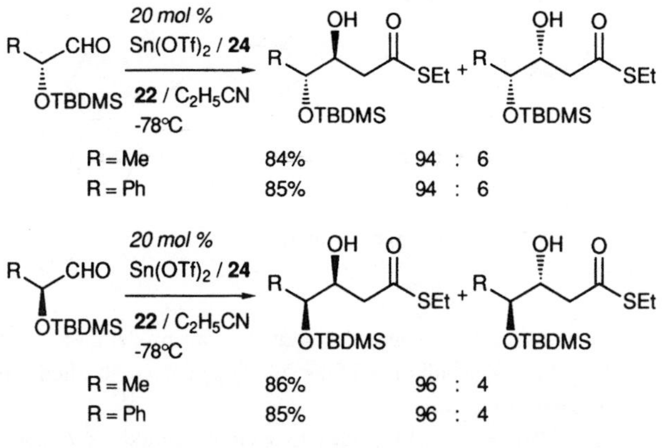

Scheme 8B1.16.

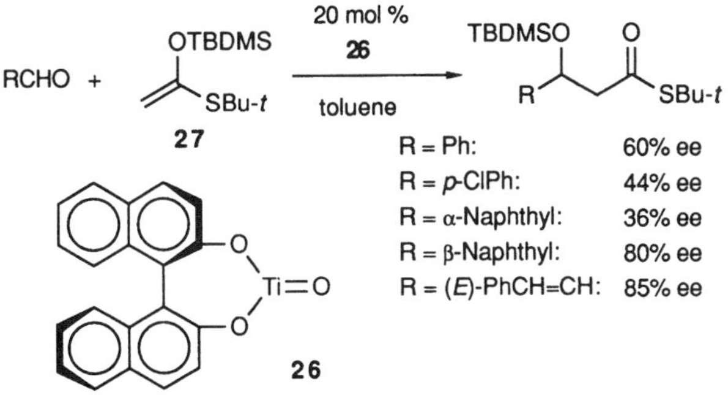

Scheme 8B1.17.

diethylzinc and sulfonamide ligands **29**, derived from *optically amino acids*, in a ratio of 1:2 (Scheme 8B1.18) [42]. Although the enantioselectivities in the reaction with simple aldehydes such as benzaldehyde and cyclohexanecarboxaldehyde are low to moderate, high enantioselectivities of around 90% ee are obtainable with chloral and bromal (Table 8B1.10). A possible catalyst precursor generated in situ is presented as **30**.

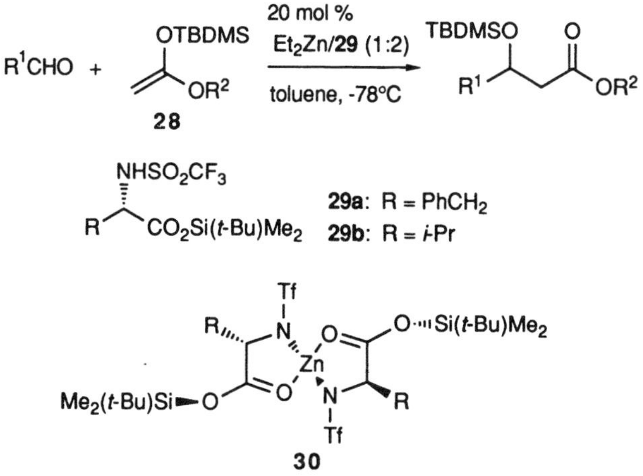

Scheme 8B1.18.

TABLE 8B1.10. Asymmetric Aldol Reaction of Ketene Silyl Acetals 28 with Aldehydes Promoted by Chiral Zinc(II) Catalysts (Scheme 8B1.18)

Entry	Aldehyde R^1	R^2	29	Yield (%)	% ee
1	Ph	Et	29a	79	60
2	c-C_6H_{11}	Bn	29a	76	23
3	CCl_3	Bn	29a	70^a	72
4	CCl_3	Bn	29b	61^a	88
5	CBr_3	Bn	29b	66^a	93

^{a}Isolated as alcohol form.

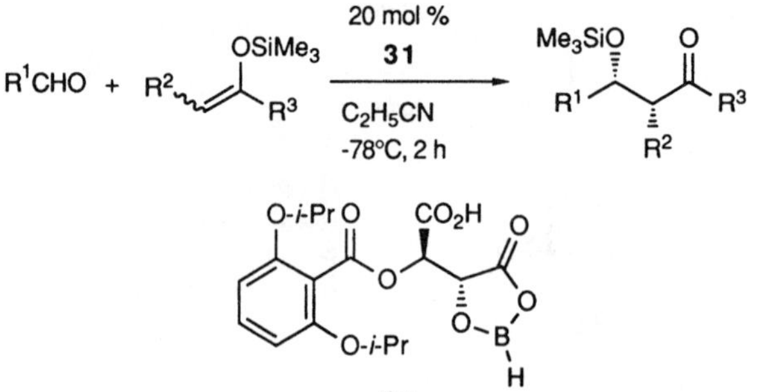

Scheme 8B1.19.

Recently, Yamamoto et al. have shown that the chiral acyloxyborane complex **31** is an excellent catalyst for the asymmetric Mukaiyama condensation of simple silyl enol ethers (Scheme 8B1.19; Table 8B1.11: entries 1–7) [43]. The *syn*-aldol adducts are formed preferentially with high enantiomeric excess regardless of the stereochemistry (*E/Z*) of the silyl enol ethers, suggesting an extended transition state (entries 4, 7). This methodology has been

TABLE 8B1.11. Asymmetric Aldol Reaction Catalyzed by Chiral Acyloxyborane Complex 31 (Scheme 8B1.19)

Entry		Aldehyde R^1	Yield (%)	*syn/anti*	*syn* % ee
1	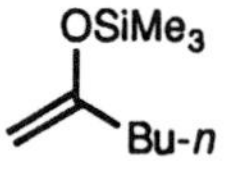 OSiMe$_3$ / Bu-*n*	Ph	81	-	85
2		*n*-Bu	70	-	80
3		(*E*)-PhCH=CH	88	-	83
4	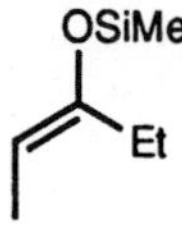 OSiMe$_3$ / Ph / OSiMe$_3$ / Et	Ph	96	94/6	96 (3*R*)
5		*n*-Pr	61	80/20	88 (3*S*)
6		(*E*)-MeCH=CH	79	>94/6	93 (3*R*)
7		Ph	97	93/7	94 (3*R*)
8	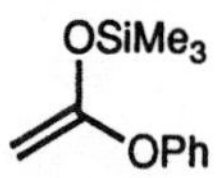 OSiMe$_3$ / Et / OSiMe$_3$ / OPh	Ph	63	-	84 (3*R*)
9		*n*-Pr	49	-	76 (3*S*)
10		Ph	83	79/21	92 (3*R*)
11	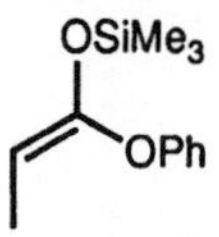 OSiMe$_3$ / OPh	*n*-Pr	57	65/35	88
12		*i*-Pr	45	64/36	79
13		(*E*)-*n*-PrCH=CH	97	96/4	97
14		(*E*)-MeCH=CMe	86	>95/5	94

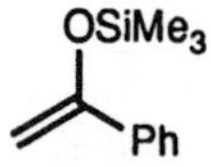

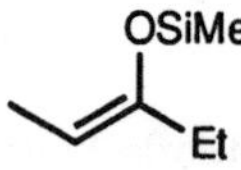

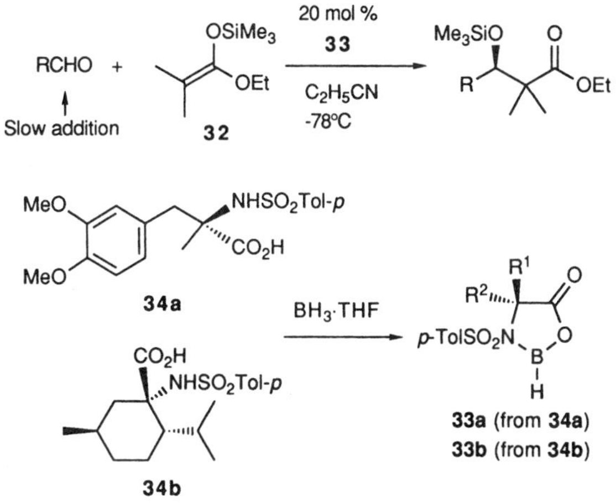

Scheme 8B1.20.

extended to the aldol reaction of ketene silyl acetals (entries 8–14) [44]. Ketene silyl acetals derived from phenyl esters reacted with higher diastereo- and enantioselectivities than those from the corresponding ethyl ester.

Another clever approach by Masamune et al. is an asymmetric aldol reaction of the ketene silyl acetal **32**. This reaction is effectively promoted by 20 mol % of chiral borane complexes **33**, prepared from BH3·THF and α,α-disubstituted glycine tosylamide **34**, under reaction conditions in which aldehydes are slowly added to the reaction mixture (Scheme 8B1.20, Table 8B1.12). The catalysts and the reaction conditions have been designed according to the proposed catalytic cycle shown in Scheme 8B1.21. Thus, the use of a geminally disubstituted catalyst accelerates the ring closure of intermediate **35** as expected from the Thorpe–Ingold effect, and the slow addition of the aldehyde reduces the accumulation of **35**, which might catalyze the aldol reaction with low enantioselectivity. Extremely

TABLE 8B1.12. Asymmetric Aldol Reaction of Ketene Silyl Acetal 32 with Aldehydes Catalyzed by Borane Complex 33 (Scheme 8B1.20)

Entry	Aldehyde R^1	Catalyst **33**	Yield (%)	% ee
1	Ph	**33a**	80	84 (R)
2	Ph	**33b**	83	91 (R)
3	$c\text{-}C_6H_{11}$	**33a**	68	91 (R)
4	$c\text{-}C_6H_{11}$	**33b**	59	96 (R)
5	$n\text{-}Pr$	**33a**	81	>98
6	$n\text{-}Pr$	**33b**	82	>98
7	$Ph(CH_2)_2$	**33a**	83	>98
8	$Ph(CH_2)_2$	**33b**	83	>98
9	$BnO(CH_2)_2$	**33a**	86	99

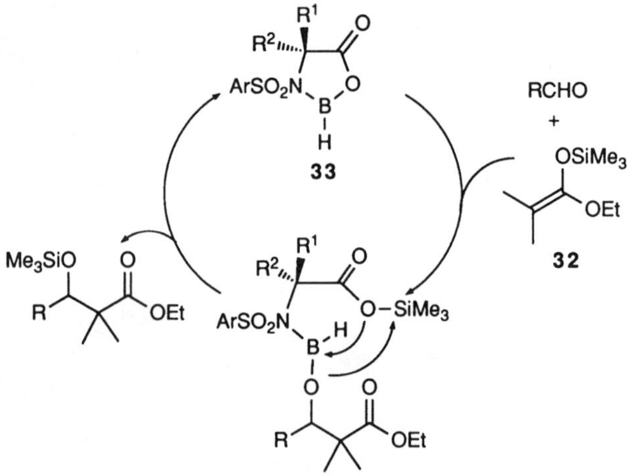

Scheme 8B1.21.

high enantioselectivities are obtainable for the reactions with primary alkyl aldehydes. It has recently been shown that unsubstituted and monosubstituted ketene silyl acetals can also achieve high enantioselectivity [46].

Kiyooka et al. have shown that the asymmetric aldol reaction of ketene silyl acetals is promoted by 20 mol % of oxazaborolidine catalyst derived from (S)-valine with enantioselectivity employing nitromethane as the solvent [47].

8B1.5. CONCLUSION

Recent advancements of the catalytic asymmetric aldol reaction are reviewed. The gold-catalyzed asymmetric aldol reaction simultaneously provides high stereoselectivity and catalyst turnover efficiency, although the reaction is limited to α-isocyanocarboxylates or their analogs. Success in the gold-catalyzed aldol reaction has shown that a synthetic approach with multiple recognition of the reacting substrates is quite promising for the future development of catalytic asymmetric synthesis. Chiral Lewis base catalysts, one of which was first used in the nitroaldol reaction, may provide a generally useful synthetic methodology, because most aldol reactions are catalyzed by base. Although it is now possible to achieve high enantioselectivities in some Mukaiyama-type aldol reactions by using substoichiometric amounts of chiral Lewis acid catalysts, low catalyst turnover is still a major hurdle to be overcome. A major breakthrough in the catalytic asymmetric synthesis of aldols remains to be achieved.

REFERENCES

1. Masamune, S.; Choy, W.; Kerdesky, F. A. J.; Imperiali, B. *J. Am. Chem. Soc.* **1981**, *103*, 1566.
2. Evans, D. A.; Bartroli, J.; Shih, T. L. *J. Am. Chem. Soc.* **1981**, *103*, 2127.

3. Iwasawa, N.; Mukaiyama, T. *Chem. Lett.* **1982**, 1441.

4. Ito, Y.; Sawamura, M.; Hayashi, T. *J. Am. Chem. Soc.* **1986**, *108*, 6405.

5. Ito, Y.; Sawamura, M.; Hayashi, T. *Tetrahedron Lett.* **1987**, *28*, 6215.

6. Sawamura, M.; Ito, Y.; Hayashi, T. *Tetrahedron Lett.* **1990**, *31*, 2723.

7. Hayashi, T.; Sawamura, M.; Ito, Y. *Tetrahedron* **1992**, *48*, 1999.

8. Pastor, S. D. *Tetrahedron* **1988**, *44*, 2883.

9. Togni, A.; Häusel, R. *Synlett* **1990**, 633.

10. Pastor, S. D.; Togni, A. *Tetrahedron Lett.* **1990**, *31*, 839.

11. Togni, A.; Pastor, S. D. *J. Organomet. Chem.* **1990**, *381*, C21.

12. Ito, Y., et al. Unpublished result.

13. Ito, Y.; Sawamura, M.; Hayashi, T. *Tetrahedron Lett.* **1988**, *29*, 239.

14. Togni, A.; Pastor, S. D.; Rihs, G. *Helv. Chim. Acta* **1989**, *72*, 1471.

15. Togni, A.; Pastor, S. D. *Helv. Chim. Acta* **1989**, *72*, 1038.

16. Togni, A.; Pastor, S. D. *J. Org. Chem.* **1990**, *55*, 1649.

17. Ito, Y.; Sawamura, M.; Shirakawa, E.; Hayashizaki, K.; Hayashi, T. *Tetrahedron Lett.* **1988**, *29*, 235.

18. Ito, Y.; Sawamura, M.; Shirakawa, E.; Hayashizaki, K.; Hayashi, T. *Tetrahedron* **1988**, *44*, 5253.

19. Ito, Y.; Sawamura, M.; Kobayashi, M.; Hayashi, T. *Tetrahedron Lett.* **1988**, *29*, 6321.

20. Ito, Y.; Sawamura, M.; Hamashima, H.; Emura, T.; Hayashi, T. *Tetrahedron Lett.* **1989**, *30*, 4681.

21. Sawamura, M.; Ito, Y.; Hayashi, T. *Tetrahedron Lett.* **1989**, *30*, 2247.

22. Togni, A.; Pastor, S. D. *Tetrahedron Lett.* **1989**, *30*, 1071.

23. Pastor, S. D.; Togni, A. *J. Am. Chem. Soc.* **1989**, *111*, 2333.

24. Pastor, S. D.; Togni, A. *Helv. Chim. Acta* **1991**, *74*, 905.

25. Hayashi, T.; Uozumi, Y.; Yamazaki, A.; Sawamura, M.; Hamashima, H.; Ito, Y. *Tetrahedron Lett.* **1991**, *32*, 2799.

26. Sawamura, M.; Hamashima, H.; Ito, Y. *J. Org. Chem.* **1990**, *55*, 5935.

27. Sasai, H.; Suzuki, T.; Arai, S.; Arai, T.; Shibasaki, M. *J. Am. Chem. Soc.* **1992**, *114*, 4418.

28. Nakagawa, M.; Nakao, H.; Watanabe, K. *Chem. Lett.* **1985**, 391.

29. Mikami, K.; Terada, M.; Nakai, T. Paper 3Y29, presented at the 52nd Annual Meeting of the Chemical Society of Japan, Kyoto, April 1–4, 1986.

30. Mikami, K.; Terada, M.; Nakai, T. *J. Org. Chem.* **1991**, *56*, 5456.

31. Mikami, K.; Terada, M.; Nakai, T. *Tetrahedron: Asymmetry* **1991**, *2*, 993.

32. Terada, M.; Gu, J.-H.; Deka, D. C.; Mikami, K.; Nakai, T. *Chem. Lett.* **1992**, 29.

33. Reetz, M. T.; Kyung, S.-H.; Bolm, C.; Zierke, T. *Chem. Ind.* **1986**, 824.

34. Reetz, M. T.; Vougioukas, A. E. *Tetrahedron Lett.* **1987**, *28*, 793.

35. Mukaiyama, T.; Kobayashi, S.; Uchiro, H.; Shiina, I. *Chem. Lett.* **1990**, 129.

36. Kobayashi, S.; Fujishita, Y.; Mukaiyama, T. *Chem. Lett.* **1990**, 1455.

37. Mukaiyama, T.; Furuya, M.; Ohtsubo, A.; Kobayashi, S. *Chem. Lett.* **1991**, 989.

38. Kobayashi, S.; Furuya, M.; Ohtsubo, A.; Mukaiyama, T. *Tetrahedron: Asymmetry* **1991**, *2*, 635.

39. Mukaiyama, T.; Uchiro, H.; Kobayashi, S. *Chem. Lett.* **1990**, 1147.

40. Kobayashi, S.; Ohtsubo, A.; Mukaiyama, T. *Chem. Lett.* **1991**, 831.

41. Mukaiyama, T.; Inubushi, A.; Suda, S.; Hara, R.; Kobayashi, S. *Chem. Lett.* **1990**, 1015.

42. Mukaiyama, T.; Takashima, T.; Kusaka, H.; Shimpuku, T. *Chem. Lett.* **1990**, 1777.

43. Furuta, K.; Maruyama, T.; Yamamoto, H. *J. Am. Chem. Soc.* **1991**, *113*, 1041.

44. Furuta, K.; Maruyama, T.; Yamamoto, H. *Synlett* **1991**, 439.

45. Parmee, E. R.; Tempkin, O.; Masamune, S. *J. Am. Chem. Soc.* **1991**, *113*, 9365.

46. Parmee, E. R.; Hong, Y.; Tempkin, O.; Masamune, S. *Tetrahedron Lett.* **1992**, *33*, 1729.

47. Kiyooka, S.; Kaneko, Y.; Kume, K. *Tetrahedron Lett.* **1992**, *33*, 4927.

8B2

RECENT ADVANCES IN ASYMMETRIC ALDOL ADDITION REACTIONS

ERICK M. CARREIRA

Laboratorium für Organische Chemie, ETH-Zentrum, Universitätstrasse 16, CH-8092 Zürich, Switzerland

8B2.1. INTRODUCTION

The venerable aldol addition has been one of the most widely used synthetic methods for the construction of stereochemically complex natural and nonnatural products. Its versatility is evident in the innumerable syntheses in which the aldol transform has served indispensably as the focus of convergent syntheses. It has been utilized at a variety of strategic points in syntheses such as in the construction of basic building blocks early in a synthetic plan or in the constitutive assembly of large, stereochemically complex fragments late in the synthetic plan [1,2].

Since the first edition of *Catalytic Asymmetric Synthesis* in 1993 the pace of innovative discoveries in the area of asymmetric catalysis of the aldol reaction has been breathtaking [3,4]. The fast-paced evolution is evident in the significant improvements in substrate generality, experimental simplicity, catalytic loading, and the enantiopurity of the adducts isolated. In parallel with these advances in preparative chemistry of catalytic aldol addition processes, there has been increased sophistication in our understanding of the mechanistic aspects of the wide-ranging transforms included in the aldol rubric.

The advances that have taken place over the past five years in catalytic, enantioselective aldol addition reactions is evident in a number of important respects. The types of transition metals and their complexes that function competently as catalysts have been expanded considerably. Thus, in addition to B(III), Ag(I), Au(I), Sn(II), and La(III), chiral catalysts prepared from Cu(II), Ti(IV), Ln(III), Si(IV), Pt(II) and Pd(II) have been introduced. The expansion in the use of transition metals has taken place hand-in-hand with the design and synthesis of new bidentate and tridentate organic ligands based on nitrogen, oxygen, and phosphorus donors. Additionally, whereas the older methods primarily relied on the use of Lewis-acids for the activation of

Catalytic Asymmetric Synthesis, Second Edition, Edited by Iwao Ojima
ISBN 0-471-29805-0   Copyright © 2000 Wiley-VCH, Inc.

aldehydes towards addition, new methods have been reported proceeding by activation of the enolate component via metalloenolate intermediates.

This chapter aims to highlight the new catalytic methods that have been reported since 1992. An organization based on aldol type has been selected to facilitate chemists in easily identifying and selecting the transform and accompanying method towards the preparation of a specific target. As a testimony to the advances in this area, it is interesting to note how little overlap exists with those notable methods discussed in the first edition of *Catalytic Asymmetric Synthesis* [3a].

8B2.2.   KETONE ALDOLS

Oxazaborolidenes.   There are noteworthy advances in the design, synthesis, and study of amino acid-derived oxazaborolidene complexes as catalysts for the Mukaiyama aldol addition. Corey has documented the use of complex **1** prepared from *N*-tosyl (*S*)-tryptophan in enantioselective Mukaiyama aldol addition reactions [5]. The addition of aryl or alkyl methyl ketones **2a–b** proceeded with aromatic as well as aliphatic aldehydes, giving adducts in 56–100% yields and up to 93% ee (Scheme 8B2.1, Table 8B2.1). The use of 1-trimethylsilyloxycyclopentene **3** as well as dienolsilane **4** has been examined. Thus, for example, the cyclopentanone adduct with benzaldehyde **5** (R = Ph) was isolated as a 94:6 mixture of diastereomers favoring the syn diastereomer, which was formed with 92% ee. Dienolate adducts **6** were isolated with up to 82% ee; it is important that these were shown to afford the corresponding dihydropyrones upon treatment with trifluoroacetic acid. Thus this process not only allows access to aldol addition adducts, but also the products of hetero Diels–Alder cycloaddition reactions.

The (*S*)-tryptophan-derived oxazaborolidenes utilized in this aldol study have been previously examined by Corey as effective catalysts for enantioselective Diels–Alder cycloaddition reactions [6]. Corey has documented unique physical properties of the complex and has proposed that the electron-rich indole participates in stabilizing a donor-acceptor interaction with the metal-bound polarized aldehyde. More recently, Corey has formulated a model exemplified by **7** in which binding by the aldehyde to the metal is rigidified through the formation of a hydrogen-bond between the polarized formyl C-H and an oxyanionic ligand [7]. The model illustrates the sophisticated design elements that can be incorporated into the preparation of transition-metal complexes that lead to exquisite control in aldehyde enantiofacial differentiation.

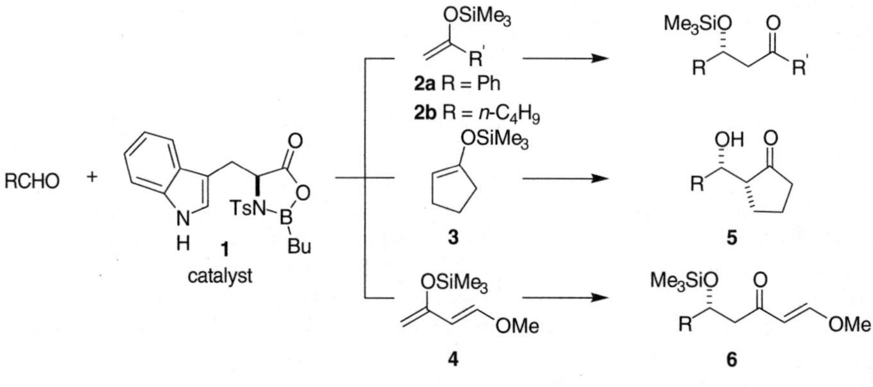

Scheme 8B2.1.

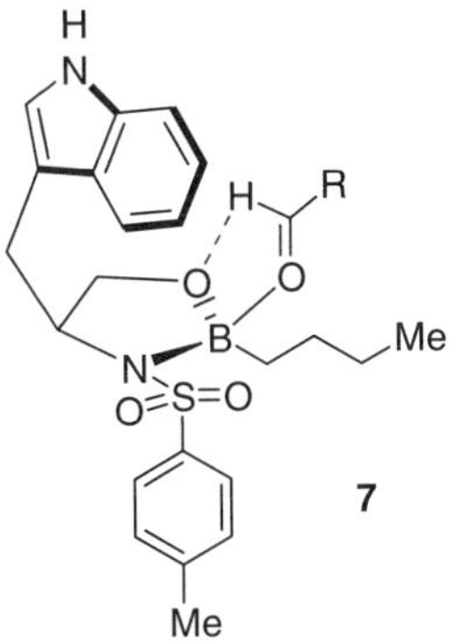

7

Cu(II) and Sn(II) Bisoxazoline Complexes. Evans has prepared and studied a family of Cu(II) complexes prepared from bisoxazoline ligands [8]. Utilizing these complexes a number of different addition reactions can be successfully conducted on pyruvate, benzyloxyacetaldehyde, and glyoxylates. Whereas the focus of the work in the context of aldol addition reactions has been on the use of silyl ketene acetals (*vide infra*), the addition of ketone-derived enoxy silanes **8a–b** with methyl pyruvate has been examined (Eq. 8B2.1). The additions of **8a–b** proceed in the presence of 10 mol % Cu(II) catalyst at –78°C in CH_2Cl_2, affording adducts of acetophenone **9a** and acetone **9b** with 99% and 93% ee, respectively.

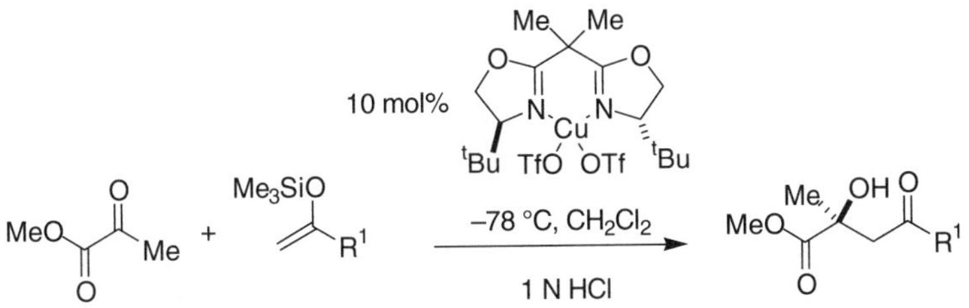

8a R = Ph
8b R = Me

9a R = Ph 99% ee
9b R = Me 93% ee

(8B2.1)

Complexes of Lanthanides with BINOL. In a series of innovative studies, Shibasaki has developed and studied a family of aldol addition reactions that proceed by the direct deprotonation of ketone precursors (Eq. 8B2.2) [9,10]. The addition reactions are conducted by using heterobimetallic complexes **10** that are suggested to function in a dual role of aldehyde activation and binding the enolate. In the presence of 20 mol % **10**, excess acetophenone, acetonaphthone, acetone, or 2-butanone undergo addition to a range of aldehydes to give adducts **11** in good yields and up to 94% ee (Table 8B2.2). One of the intriguing features of this remarkable process is that certain enolizable aldehydes such as cyclohexane carboxaldehyde may be utilized as electrophilic partners in the reaction. In this regard, the addition of acetophenone enolate with cyclohexane carboxaldehyde gave aldol adduct in 72% yield and 44% ee. The complexes derived from the various lanthanides, and bis alkali metal binaphthoxides have been studied in great detail utilizing laser desorption time-of-flight mass spectrometry, nuclear magnetic resonance spectroscopy, and X-ray crystallography. On the basis of ^1H NMR shift data, Shibasaki has suggested that activation of the aldehyde proceeds by coordination of C=O to the electrophilic lanthanide metal center. The nucleophilic ketone enolate partner undergoes

TABLE 8B2.1. Aldol addition reactions catalyzed by 1 (Scheme 8B2.1)[a]

Entry	Aldehyde	Enolsilane	% ee
1	PhCHO	**2a**	89
2	EtCHO	**2b**	90
		4	82
3	c-C$_6$H$_{11}$CHO	**2a**	89
		4	73
		2a	93
		3b	86
4	CHO	**4**	76
	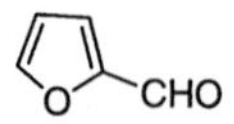	**2a**	92
		4	67
5	 Ph⌒CHO	**4**	69

[a]Reaction Conditions: 20–40 mol % **1**, EtCN, –78°C, 14 h, then H$_2$O, H$_3$O$^+$; Yields: 56-quantitative.

activation and enolization at a separate metal site; the synergistic operation of these two elements leads to enantioselective product formation.

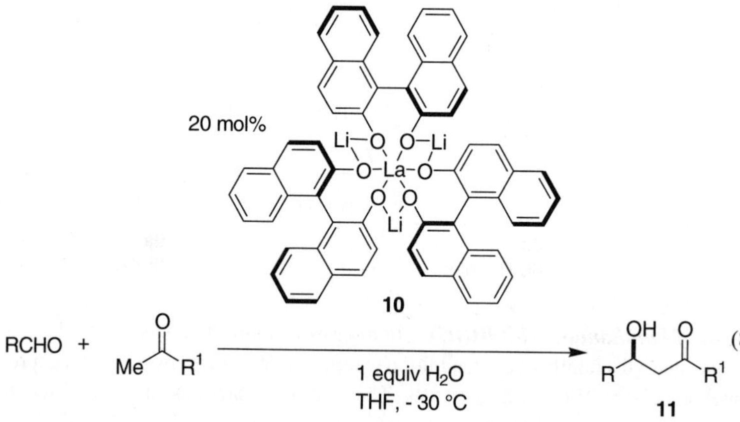

$$RCHO + \underset{Me}{\overset{O}{\|}}{\diagdown}R^1 \xrightarrow[\text{THF, - 30 °C}]{\text{1 equiv H}_2\text{O}} \underset{R}{\overset{OH}{\diagup}}\underset{}{\overset{O}{\|}}R^1 \quad (8B2.2)$$

11

Shibasaki and co-workers have also generated and studied the complex that assembles in situ between the mono *O*-methyl ether of BINOL **12** and Ba(O^{i}Pr)$_2$ in dimethoxyethane (Eq. 8B2.3) [11]. This complex is suggested to catalyze the addition of acetophenone to aldehydes, giving adducts in 77–99% yield and up to 70% ee (Eq. 8B2.4). An important benefit of this system over the corresponding Ln-based process is evident by the fact that fewer equivalents of ketone are utilized (2 equiv. ketone with **13** instead of the typical 5–8 equiv. with **10**); additionally, the reaction times are considerably shorter when compared with those prescribed for **10**. In contrast to the models that have been proposed for the heterobimetallic systems, in these systems the barium is suggested to function singularly in a dual capacity of activating the aldehyde and ketone components.

TABLE 8B2.2. Aldol addition reactions of methyl ketones catalyzed by 10 (Eq. 8B2.2)[a]

Aldehyde (R)	Ketone (R^1)	% ee
Tert-butyl	Ph	91
	Me	73
	1-Naphthyl	76
	CH$_3$CH$_2$	94
c-C$_6$H$_{11}$CHO	Ph	44
CHMe$_2$	Ph	54

[a]Typical Reaction Conditions: 20 mol % **10**, THF, −30°C, 72–277 h.

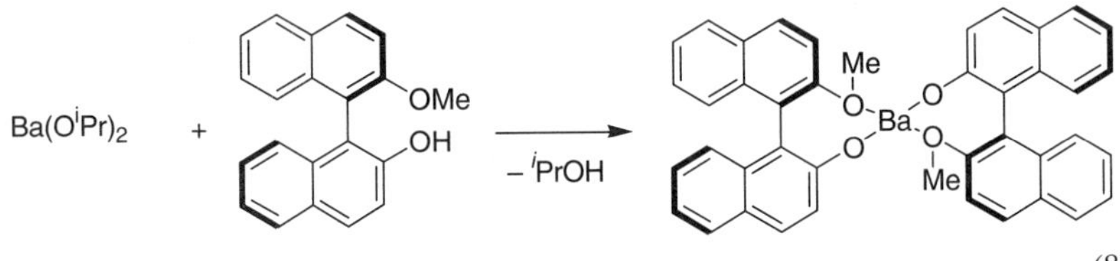

$$(8B2.3)$$

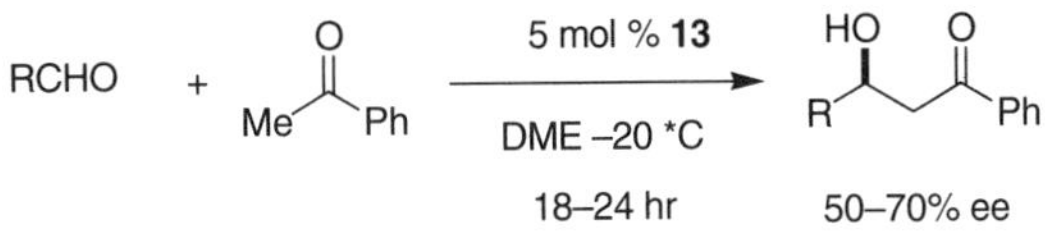

$$(8B2.4)$$

Complexes of Pd(II) and Ag(I) with BINAP. Since the pioneering work of Hayashi and Ito on the Ag(I)- and Au(I)-catalyzed aldol addition reactions of isocyanoacetates to aldehydes [12], only a few reports had appeared until recently describing the use of soft-metal complexes as catalysts for the aldol addition reaction. Most of the developments that have appeared over the past two decades have focused on the design and synthesis of Lewis acids derived from hard oxophilic metals such as titanium, boron, and tin. Among the new developments in the area of enantioselective, catalytic aldol addition methods are the increasing number of processes that utilize soft-metals, that is, Cu(I), Cu(II), Pd(II), Pt(II), and Ag(I) coordination complexes. The ensuing discussion is limited to the use of such complexes in the addition reactions of ketone-derived enol silanes to aldehydes. Subsequently additional examples will be detailed involving the addition reaction of ester and lactone derived silyl enolates.

Shibasaki has examined catalysis of a complex, prepared in situ from PdCl$_2$, AgOTf, (*R*)- or (*S*)-BINAP, 4 Å molecular sieves, and H$_2$O, in the aldol addition reaction of enolsilanes by (Eq. 8B2.5) [13]. Under these conditions, aryl methyl ketone-derived trimethylsilyl enolates add to benzaldehyde and hydrocinnamaldehyde, affording adducts with up to 73% ee.

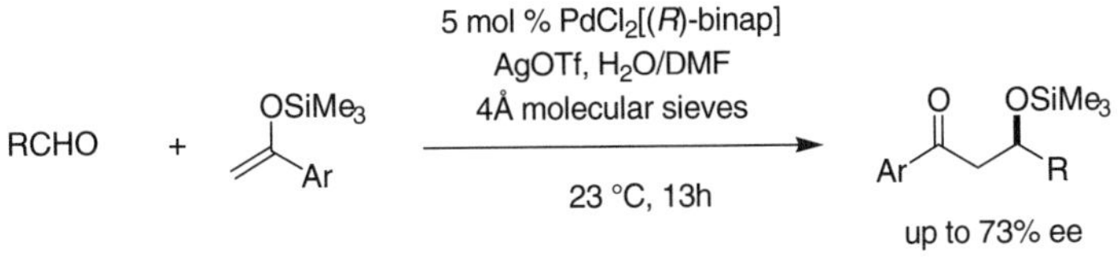

$$(8B2.5)$$

On the basis of ^{1}H NMR spectroscopic studies, Shibasaki has suggested that the catalytic process involves Pd(II) complex **14** and the derived enolate **16** (Scheme 8B2.2). The added water is proposed to lead to the generation of **16** by desilylation of **15**. This species serves as an active Pd-enolate that plays a critical role in the catalytic cycle. Following addition to aldehyde, aldolate **17** is protonated off the metal to yield adduct **18** and regenerate **14**.

Yamamoto has recently described a novel catalytic, asymmetric aldol addition reaction of enol stannanes **19** and **21** with aldehydes (Eqs. 8B2.6 and 8B2.7) [14]. The stannyl ketones are prepared solvent-free by treatment of the corresponding enol acetates with tributyltin methoxide. Although, in general, these enolates are known to exist as mixtures of C- and O-bound tautomers, it is reported that the mixture may be utilized in the catalytic process. The complexes Yamamoto utilized in this unprecedented process are noteworthy in their novelty as catalysts for catalytic C–C bond-forming reactions. The active complex is generated upon treatment of Ag(OTf) with (R)-BINAP in THF. Under optimal conditions, 10 mol % catalyst **20** effects the addition of enol stannanes with benzaldehyde, hydrocinnamaldehyde, or cinnamaldehyde to give the adducts of acetone, *tert*-butyl methyl ketone (pinacolone), and acetophenone in good yields and 41–95% ee (Table 8B2.3).

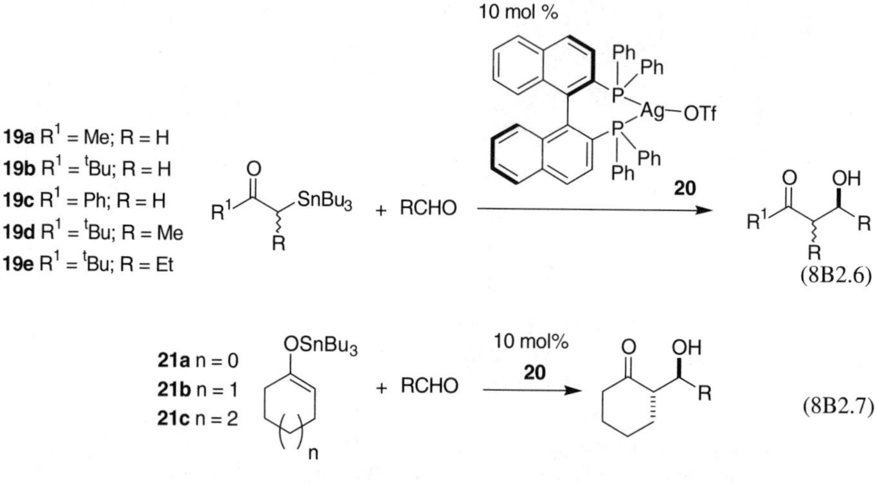

19a R^1 = Me; R = H
19b R^1 = ^{t}Bu; R = H
19c R^1 = Ph; R = H
19d R^1 = ^{t}Bu; R = Me
19e R^1 = ^{t}Bu; R = Et

$$R^1 \overset{O}{\underset{R}{\diagup}} SnBu_3 \ + \ RCHO \ \xrightarrow{\ \ 20\ \ } \ R^1 \overset{O}{\underset{R}{\diagup}} \overset{OH}{\diagup} R \qquad (8B2.6)$$

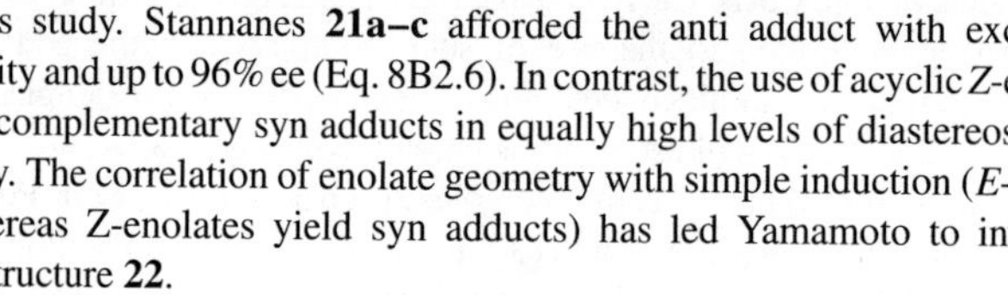

21a n = 0
21b n = 1
21c n = 2

$$\text{(OSnBu}_3\text{ cyclohexene)} \ + \ RCHO \ \xrightarrow{\ 10\text{ mol\%}\ 20\ } \ \text{(cyclohexanone with }OH, R) \qquad (8B2.7)$$

The addition of acyclic substituted enolates **19d–e** and cyclic enolates **21a–c** were also examined in this study. Stannanes **21a–c** afforded the anti adduct with excellent simple diastereoselectivity and up to 96% ee (Eq. 8B2.6). In contrast, the use of acyclic Z-enol stannanes **21** provided the complementary syn adducts in equally high levels of diastereoselectivity and enantioselectivity. The correlation of enolate geometry with simple induction (*E*-enolates yield antiadducts, whereas Z-enolates yield syn adducts) has led Yamamoto to invoke a cyclic transition-state structure **22**.

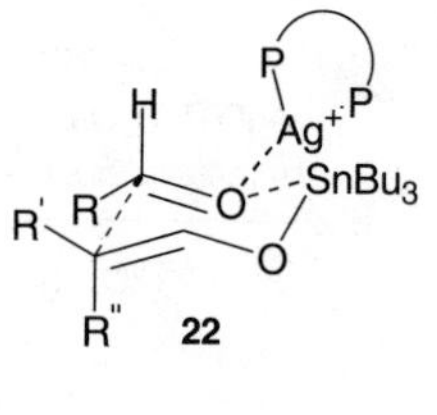

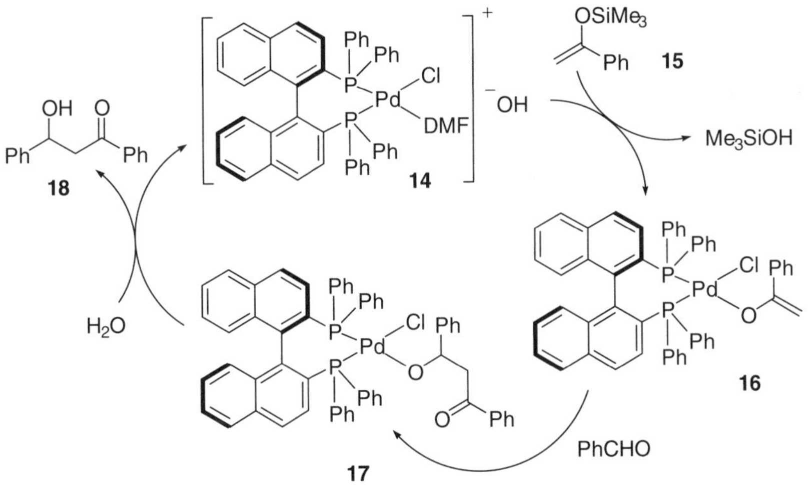

Scheme 8B2.2.

Phosphoramide Catalysis. In a series of elegant studies, Denmark has described a process utilizing trichlorosilyl enolates **27** and **28** as the nucleophilic component in catalytic, enantioselective aldol addition reactions (Scheme 8B2.3) [15]. These enoxysilanes are prepared by treatment of stannyl enolates with $SiCl_4$. Although trichlorosilyl enolates are sufficiently reactive entities to undergo addition to aldehydes rapidly at $-78°C$, Denmark documented that

TABLE 8B2.3. Addition reactions of 19a–e and 21a–c with aldehydes[a]

Aldehyde	Stannane	% ee
PhCHO	**19a**	77
	19b	95
	19c	71
	19d	95 (syn/anti 99:1)
	19e	91% (syn/anti 99:1)
PhCHO	**21a**	92 (syn/anti 11:89)
	21b	93 (syn/anti 8:92)
	21c	96 (syn/anti 15:85)
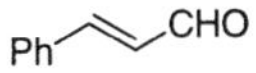 Ph~~CHO	**19a**	53
	19b	86
	19c	41
 Ph~~CHO	**19a**	59
	19b	94
	19c	50
	19d	95 (syn/anti 99: 1)

[a]Typical Reaction Conditions: 10 mol % **20**, molecular sieves, DMF/H_2O 23°C, 13 h.

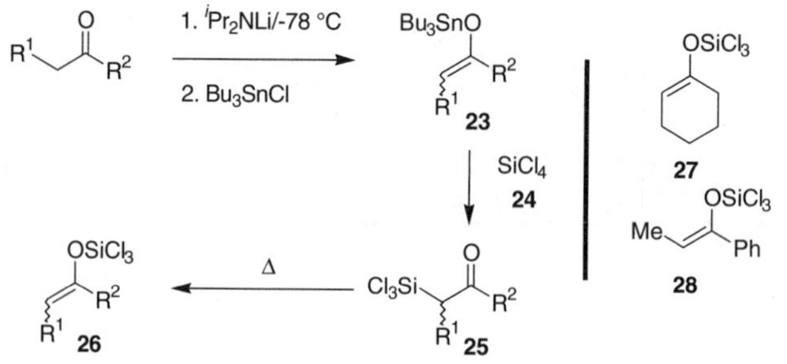

Scheme 8B2.3.

it is possible to accelerate nucleophilic addition to aldehydes in the presence of chiral phosphonamides **29–31** as Lewis base catalysts (Chart 8B2.1). Although the reported selectivities are modest for trichlorosilyl ketene acetals, the addition of the less reactive ketone-derived trichlorosilyl enolates are impressive (Eqs. 8B2.8 and 8B2.9, Table 8B2.4). Thus treatment of the cyclohexanone or propiophenone-derived trichloroenolsilanes **27** and **28** with a variety of aldehydes afforded adducts displaying high levels of simple diastereoselectivity and enantioselectivity (up to 97% ee). The observation of high levels of simple anti diastereoselectivity has led Denmark to propose an ordered chair-like transition-state structure wherein the enolate and aldehyde are highly organized around a hexacoordinate siliconate center.

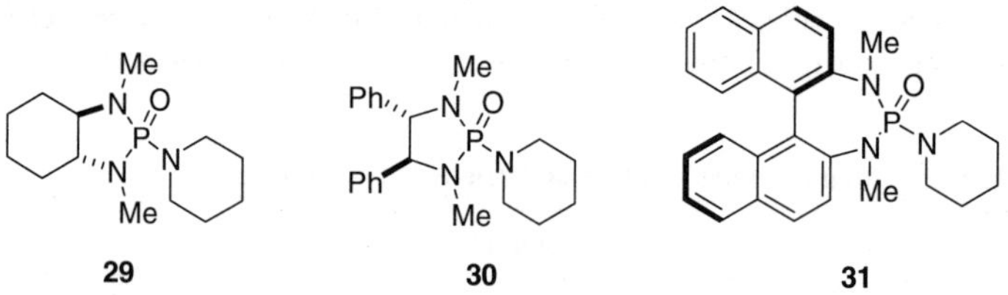

Chart 8B2.1.

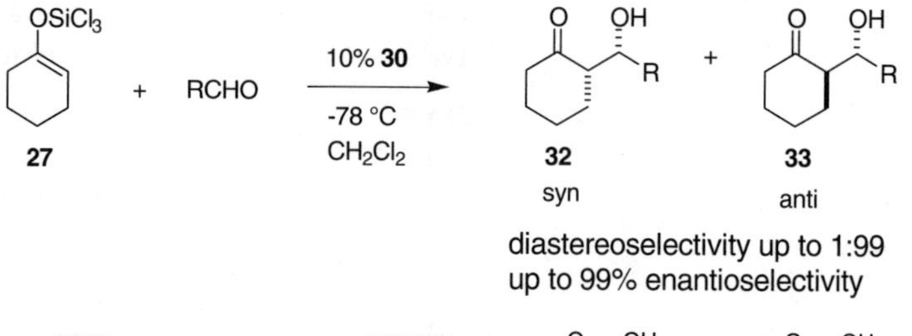

diastereoselectivity up to 1:99
up to 99% enantioselectivity

(8B2.8)

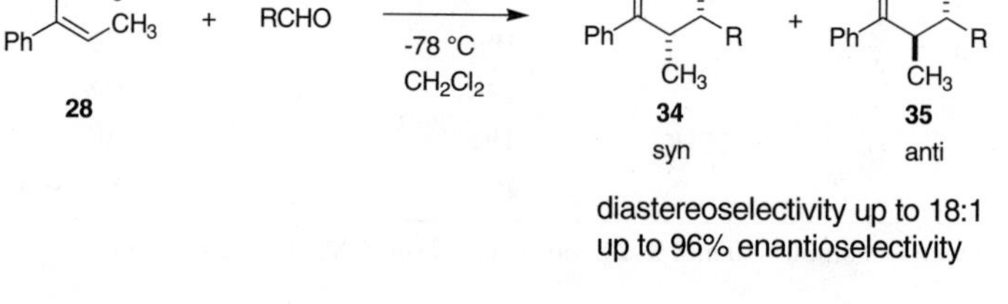

diastereoselectivity up to 18:1
up to 96% enantioselectivity

(8B2.9)

TABLE 8B2.4. Lewis base-catalyzed aldol addition reactions (Eq. 8B2.9)[a]

Aldehyde	Enol silane	*syn/anti*	% ee
PhCHO	27	1:61	93
	28	18:1	95
α-NpCHO	27	<1:99	97
	28	3:1	84
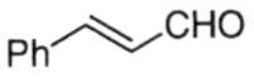 Ph⌒CHO	27	<1:99	88
	28	9.4:1	92
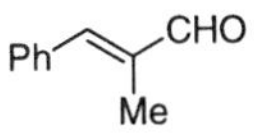 Ph⌒CHO / Me	27	<1:99	92
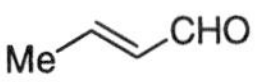 Me⌒CHO	28	7:1	91
	27	1:5.3	82
PhC≡CCHO	28	1:3.5	58

[a]Typical Reaction Conditions: 10–15 mol % **30**, CH$_2$Cl$_2$, −78°C, 2 h.

Ti(IV) Catalysis. In a related series of reactions, Carreira has reported the enantioselective addition reaction of 2-methoxypropene **36** to aldehydes using the catalyst generated upon mixing ligand **37** and Ti(O^{i}Pr)$_4$ with subsequent removal of the released isopropanol (Eq. 8B2.10 and Table 8B2.5) [16]. Although the process is best considered mechanistically as an ene reaction, the products that are isolated from an acidic workup correspond to the acetone aldol addition adduct **39** (Scheme 8B2.4). The use of 2-methoxypropene has added advantages over related reactions involving enol silanes. 2-Methoxypropene is an item of commerce; thus, its use obviates the necessary preparation of the enol silanes that is required for all but a few of the catalytic enantioselective aldol-like addition reactions that have been reported to date. The unique features of the Ti(IV) catalyst derived from **37** result from the fact that the complex is sufficiently electrophilic to activate a bound aldehyde towards addition by 2-methoxypropene without leading to extensive polymerization of the 2-methoxypropene. An additional attractive feature of the adducts is that, in lieu of an acidic workup, selection of various oxidative work-up procedures can provide access to the corresponding methyl acetate **40** or the adduct of hydroxyacetone **41** adducts.

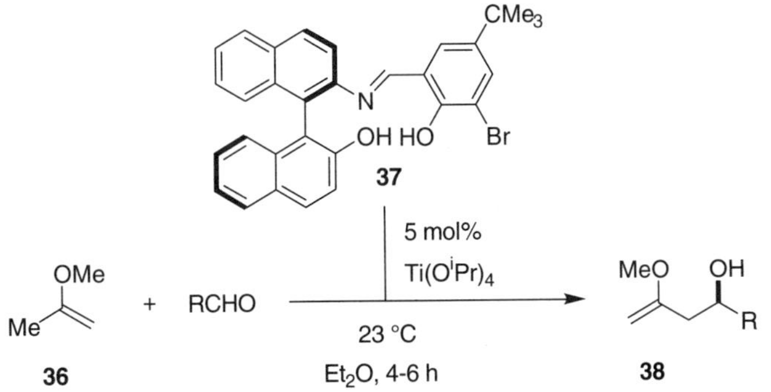

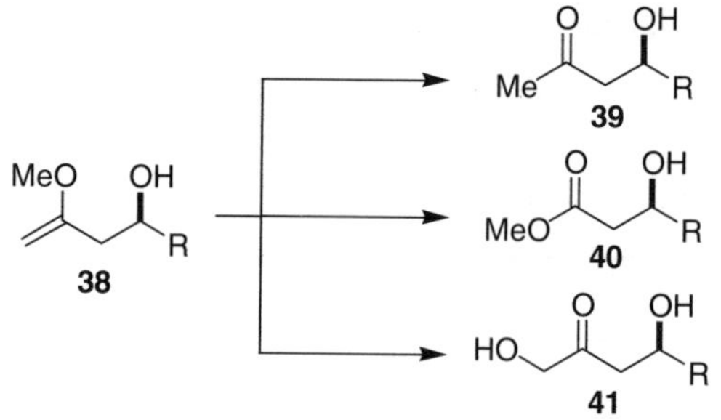

Scheme 8B2.4.

TABLE 8B2.5. Additions of methoxypropene to aldehydes (Eq. 8B2.10)[a]

Entry	Aldehyde	% ee
1	Ph(CH$_2$)$_3$—≡—CHO	98
2	TBSOCH$_2$ —≡— CHO	93
3	PhC ≡ CCHO	91
4	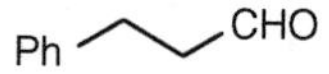 Ph CHO	90
5	c-C$_6$H$_{11}$CHO	75
6	PhCHO	66

[a]Typical Reaction Conditions: 5 mol % catalyst, 2,6-lutidine, 0 °C, 6–8 h.

Mikami has also reported a related ene-like process involving glyoxylates and ketone-derived enolsilanes **42** (Eq. 8B2.11) [17]. The enol ether adducts **43** yield the corresponding β-hydroxy ketone upon treatment with mild acid. On the basis of an analysis of the stereo- and regiochemical outcome of the addition reaction Mikami has invoked a monodentate complex between aldehyde and metal, in contrast to the typical transition-state structures involving glyoxylates that are suggested to involve metal/aldehyde chelates.

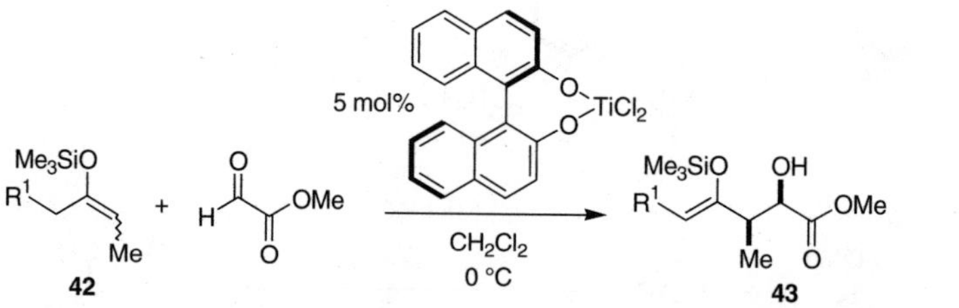

syn:anti up to 99:1   (8B2.11)
up to 99% enantioselectivity

Antibody Catalysis. Recent advances in biocatalysis have led to the generation of catalytic antibodies exhibiting aldolase activity by Lerner and Barbas. The antibody-catalyzed aldol addition reactions display remarkable enantioselectivity and substrate scope [18]. The requisite antibodies were produced through the process of reactive immunization wherein antibodies were raised against a β-diketone hapten. During the selection process, the presence of a suitably oriented lysine leads to the condensation of the ε-amine with the hapten. The formation of enaminone at the active site results in a molecular imprint that leads to the production of antibodies that function as aldol catalysts via a lysine-dependent class I aldolase mechanism (Eq. 8B2.12).

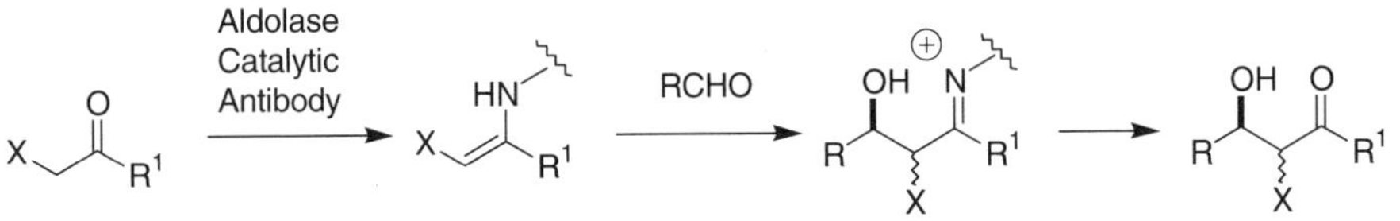

$$(8B2.12)$$

A wide range of donor ketones, including acetone, butanone, 2-pentanone, cyclopentanone, cyclohexanone, hydroxyacetone, and fluoroacetone with an equally wide range of acceptor aromatic and aliphatic aldehydes were shown to serve as substrates for the antibody-catalyzed aldol addition reactions (Chart 2, Table 8B2.6). It is interesting to note that the aldol addition reactions of functionalized ketones such as hydroxyacetone occurs regioselectively at the site of functionaliztion to give α-substituted-β–hydroxy ketones. The nature of the electrophilic and nucleophilic substrates utilized in this process as well as the reaction conditions complement those that are used in transition-metal and enzymatic catalysis.

TABLE 8B2.6. Antibody-catalyzed aldol addition reactions (Eq. 8B2.11)[a]

Substrate	Donor	Antibody	Selectivity
PhCHO	**44**	38C2	>99% ee
		33F12	>99% ee
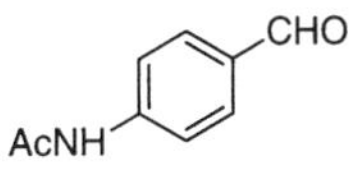	**44**	38C2	98% ee
		33F12	99% ee
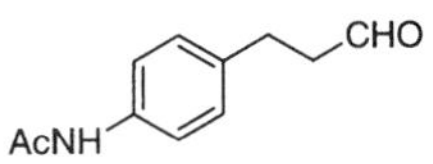	**44**	38C2	58% ee
		33F12	69% ee
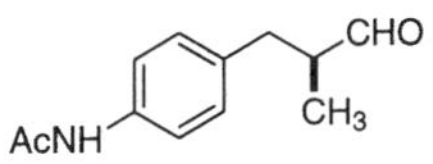	**44**	38C2	>95% de
		33F12	>95% de
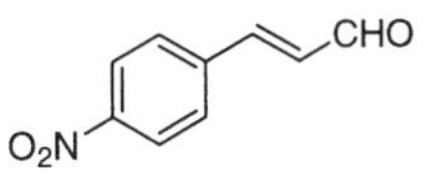	**44**	38C2	99% ee
		33F12	98% ee
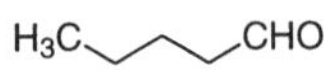	**49**	38C2	>98% ee
		33F12	89% ee

[a]Typical Reaction Conditions: 0.4 mol % antibody, buffer, H$_2$O/DMF, 23°C, 7 days.

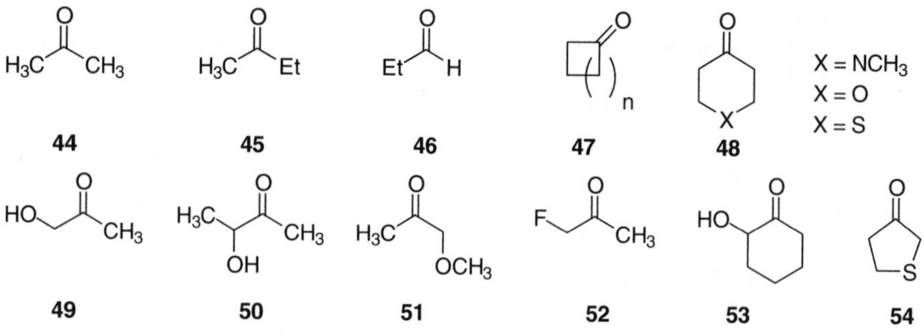

Chart 8B2.2.

Enzymatic Catalysis. Another area of rapid development in biocatalysis is the use of enzymatic aldolases. Wong has demonstrated that recombinant and isolated aldolases catalyze the condensation of acetaldehyde, acetone, fluoroacetone, and propionaldehyde to a wide range of aldehydes, including substrates that incorporate highly polar functionality in the form of free hydroxyls, azides, and phosphates (Scheme 8B2.5) [19]. Wong has further examined the ability of these enzymatic processes to perform sequential aldol addition reactions. Thus, the adduct of acetaldehyde with chloroacetaldehyde **57** (Eq. 8B2.13) is a subsequent substrate for a second aldol addition reaction in situ giving polyketide **58**. In a recent report that further extends the utility of the methodology, Wong documents the use of recombinant D- and L-threonine aldolases for the enzymatic synthesis of β-hydroxy–α-amino acids (Eq. 8B2.14). For example, treatment of glycine with acetaldehyde gives the corresponding aldol adduct in 40% yield and 91:9 diastereoselectivity with preferential formation of the anti stereoisomer [20]. As with the previous aldolase-mediated reactions, the range of substrates that are tolerated is impressive, providing access to a broad range of amino alcohols derived from a large range of aldehydes. It is interesting to note the complementary nature of the substrates used in biocatalysis when compared with those typically utilized with transition-metal based catalysts. Thus, biocatalytic processes are tolerant of highly polar substrates possessing unmasked polar functionality.

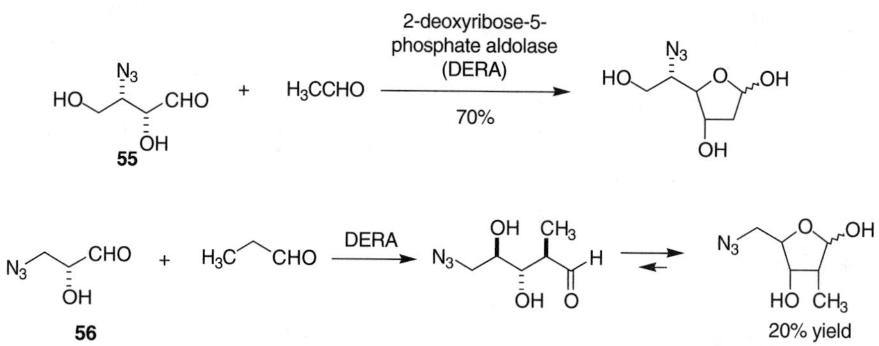

Scheme 8B2.5.

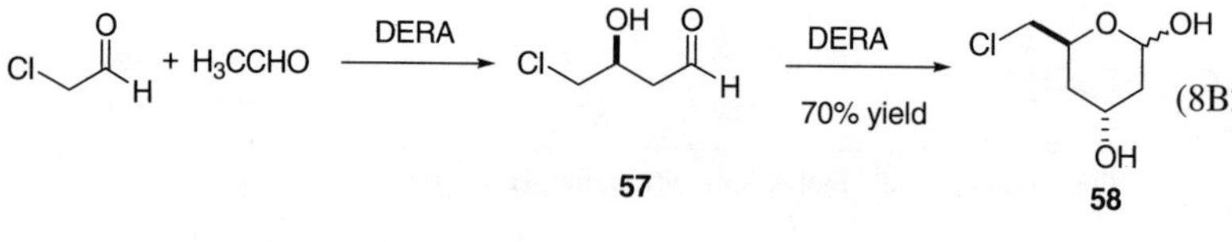

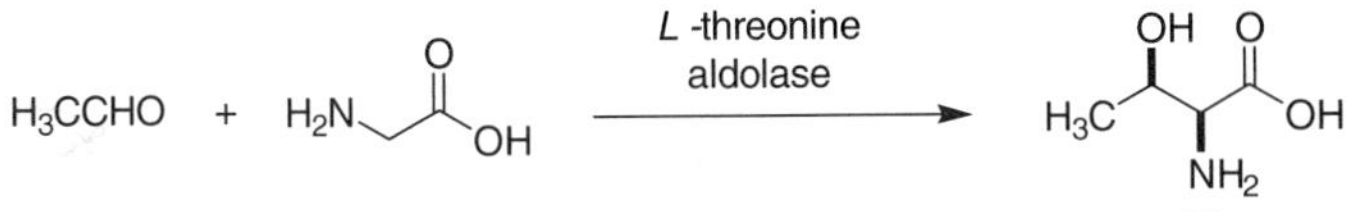

$$H_3CCHO \ + \ H_2N\diagdown\!\!\!\diagdown_{OH}^{O} \xrightarrow[\text{aldolase}]{L\text{-threonine}} \ H_3C\diagdown\!\!\!\diagup_{\underset{NH_2}{OH}}^{OH \ O}\diagup OH$$

63

(8B2.14)

8B2.3. ACETATE ADDITION REACTIONS

Some of the most impressive advances in the area of catalytic, enantioselective aldol addition reactions have taken place in the development of catalytic methods for enantioselective acetate aldol additions, a reaction type that has long been recalcitrant. Thus, although prior to 1992 a number of chiral-auxiliary based and catalytic methods were available for diastereo- and enantiocontrol in propionate aldol addition reactions, there was a paucity of analogous methods for effective stereocontrol in the addition of the simpler acetate-derived enol silanes. However, recent developments in this area have led to the availability of several useful catalytic processes. Thus, in contrast to the state of the art in 1992, it is possible to prepare acetate-derived aldol fragments utilizing asymmetric catalysis with a variety of transition-metal based complexes of Ti(IV), Cu(II), Sn(II), and Ag(I).

Catalysis by Complexes of Ti(IV). Mikami has documented the aldol addition reactions of thioacetate-derived silyl ketene acetals **59** and **60** to various highly functionalized aldehydes (Eq. 8B2.15 and Table 8B2.7) [21]. The active catalyst is generated upon mixing $TiCl_2(O^iPr)_2$ and BINOL. Utilizing as little as 5 mol % of the catalyst the addition reactions furnished adducts in excellent yields and up to 96% ee. The fact that the process tolerates aldehyde substrates containing polar substituents is noteworthy as few other transition-metal catalyzed aldol addition reactions exhibit such tolerance of functionality. The adducts of chloroacetaldehyde and *N*-Boc aminoacetaldehyde can serve as useful starting materials for the preparation of pharmacologically important compounds such as carnitine and GABOB, respectively.

Mikami has carried out a number of investigations aimed at elucidating mechanistic aspects of this Si-atom transfer process. In particular, when the aldol addition reaction was conducted with a 1:1 mixture of enoxysilanes **60** and **62**, differentiated by the nature of the *O*-alkyl and *O*-silyl moieties, only the adducts of intramolecular silyl-group transfer **63** and **64** are obtained (Scheme 8B2.6). This observation in addition to results obtained with substituted enol silanes have led Mikami to postulate a silatropic ene-like mechanism involving a cyclic, closed transition-state structure organized around the silyl group (Scheme 8B2.6).

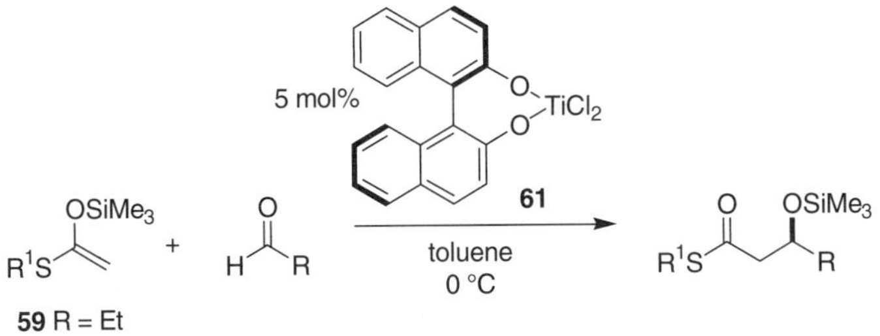

59 R = Et
60 R = ^tBu

(8B2.15)

TABLE 8B2.7. Thioacetate aldol additions catalyzed by 61 (Eq. 8B2.15)^a

Entry	Aldehyde	Enolate	% ee
1	Ph ⌒O⌒CHO	**59**	94
		60	96
2	$ClCH_2CHO$	**60**	91
3	$BocNHCH_2CHO$	**59**	88
4	$C_8H_{17}CHO$	**59**	86
		60	91
5	Me_2CHCHO	**59**	85
6	Me⌒⌒CHO	**59**	81
7	$n\text{-}BuO_2CCHO$	**59**	95

^{a}Typical Reaction Conditions: 5 mol % **61**, toluene or CH_2Cl_2, 0°C, 2 h.

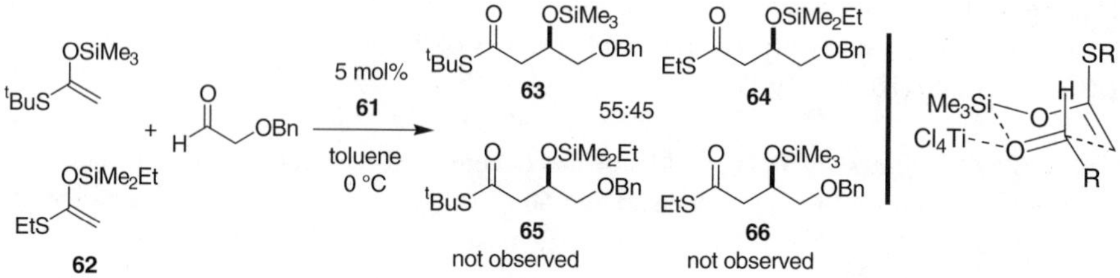

Scheme 8B2.6.

Carreira and co-workers reported novel Ti(IV) complexes **69** derived from $Ti(O^iPr)_4$, tridentate ligands **67**, and salicylic acids such as **68**. The complexes serve as competent catalysts for the addition of the methyl acetate-derived silyl ketene acetal to a large range of aldehydes (Eqs. 8B2.16 and 8B2.17) [22]. The salient features of this system include: the wide range of functionalized aliphatic and aromatic aldehydes that may be used; the ability to carry out the reaction with 0.2–5 mol % catalyst-loading; and experimental ease with which the process is executed (Table 8B2.8). Thus the reaction can be carried out at −10 to 0°C, in a variety of solvents, without recourse to slow addition of reagents. The adducts from the catalytic reaction were isolated with excellent enantiopurities up to 99% ee. The original catalyst-preparation

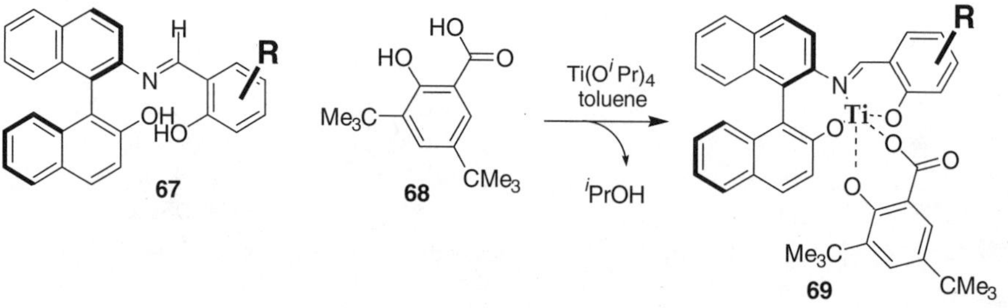

(8B2.16)

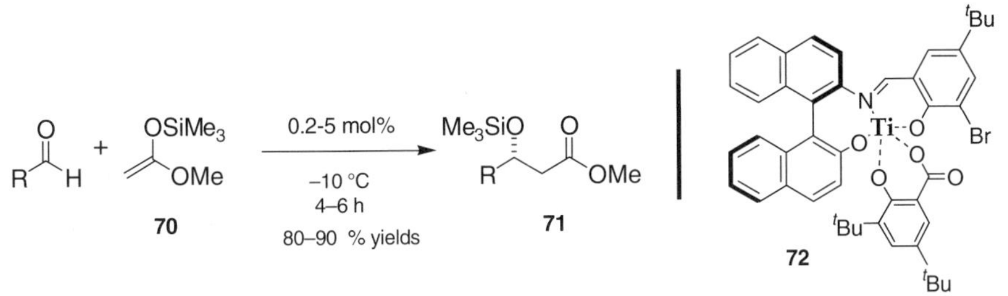

(8B2.17)

TABLE 8B2.8. Catalytic, Enantioselective methyl acetate aldol addition reactions (Eq. 8B2.17)^a

Entry	Aldehyde	% ee	Entry	Aldehyde	% ee
1	PhCHO	96	9	^{t}BuMe$_2$SiOCH$_2$—≡—CHO	96
2	Ph(CH$_2$)$_2$CHO	94	10	Ph(CH$_2$)$_3$—≡—CHO	96
3	PhC≡CCHO	94	11	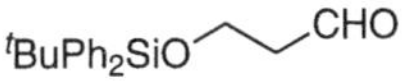 ^{t}BuPh$_2$SiO⌒⌒CHO	91
4	^{i}Pr$_3$SiC≡CCHO	97	12	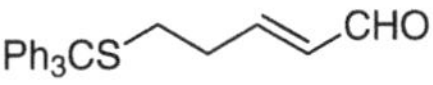 Ph$_3$CS⌒⌒⌒CHO	98
5	C$_6$H$_{11}$CHO	95	13	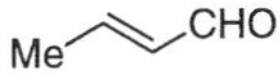 Me⌒⌒CHO	98
6	Me(CH$_2$)$_2$CHO	95	14	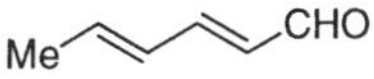 Me⌒⌒⌒CHO	98
7	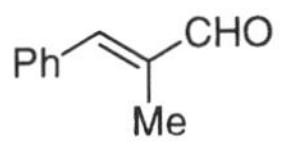 Ph⌒(Me)CHO	98	15	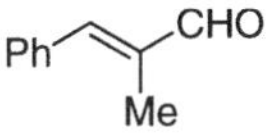 Ph⌒(Me)CHO	95
8	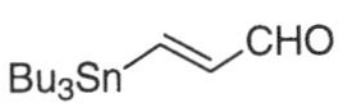 Bu$_3$Sn⌒⌒CHO	92	16	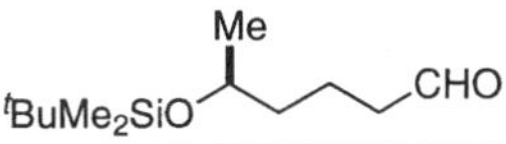 ^{t}BuMe$_2$SiO⌒(Me)⌒⌒CHO	95

^{a}Typical Reaction Conditions: 0.2–5 mol % **72**, 2,6-lutidine, Et$_2$O, –10°C, 4–8 h.

protocol prescribed mixing the tridentate ligand, Ti(O^{i}Pr)$_4$, salicylic acid, and lutidine in toluene with subsequent evaporation of the solvent to remove the released isopropanol. Since then, however, a modification of the experimental procedure has been developed, which greatly facilitates the preparation of the active catalyst mixture [23]. Thus, the more recently described procedure prescribes that the ligands and metal are mixed in the presence of Me$_3$SiCl and Et$_3$N. The released isopropanol is sequestered as the corresponding silyl ether, obviating the requirement for its evaporative removal.

The acetate aldol addition reactions using catalyst **72** have found application in a number of syntheses. Simon and co-workers utilized aldol adduct **74** as a key building block in the total synthesis of depsipeptide antibiotic **75** (Eq. 8B2.18) [24]. Rychnovsky and co-workers have conducted a diastereoselective methyl acetate aldol addition reaction with aldehyde **76** (Eq. 8B2.19), an advanced intermediate in an elegant convergent total synthesis of roflamycoin [25]. This addition reaction represents an interesting application of a catalyst developed for absolute stereocontrol (enantioselective synthesis) in the context of a diastereoselective reaction process.

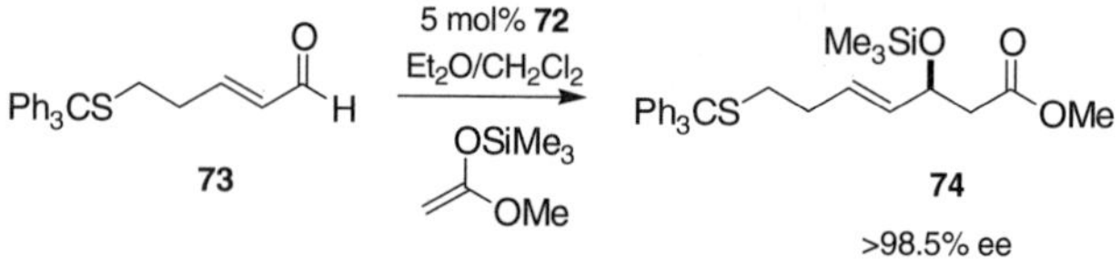

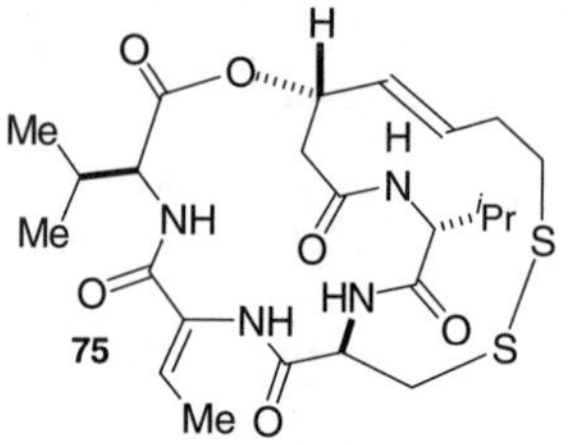

(8B2.18)

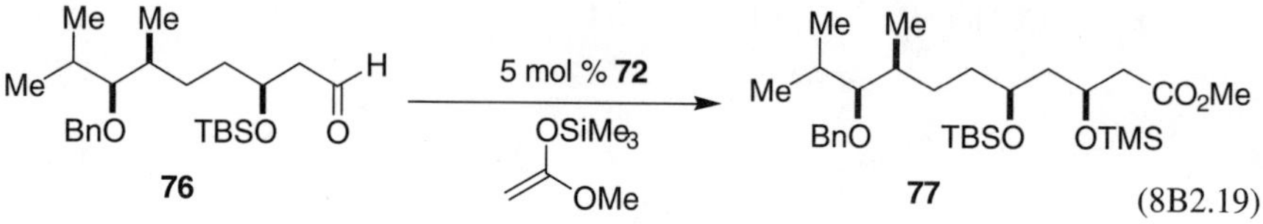

(8B2.19)

Keck has reported a catalytic enantioselective acetate aldol addition reaction that utilizes a Ti(IV) catalyst **79** that is readily prepared in situ (Eq. 8B2.20) [26]. The reaction protocol is noteworthy as a consequence of its simplicity of execution; thus BINOL, $TiCl_2(O^iPr)_2$, and 4 Å molecular sieves are mixed in Et_2O at $-20°C$; after an aging period, the catalyst mixture is ready for use (Table 8B2.9). The addition reactions are best conducted with 10 mol % catalyst in ether at $-20°C$ with the *tert*-butyl thioacetate-derived silyl ketene acetal **78**, giving adducts in 90% yield and up to 98% ee for a range of aldehyde substrates (Table 8B2.9). Following a meticulous examination of reaction parameters (solvent, temperature, catalytic loading, concentration) Keck has suggested that the active catalyst likely consists of an aggregate; consistent with this hypothesis is the fact that the reaction enantioselectivity is sensitive to catalyst concentration and loading.

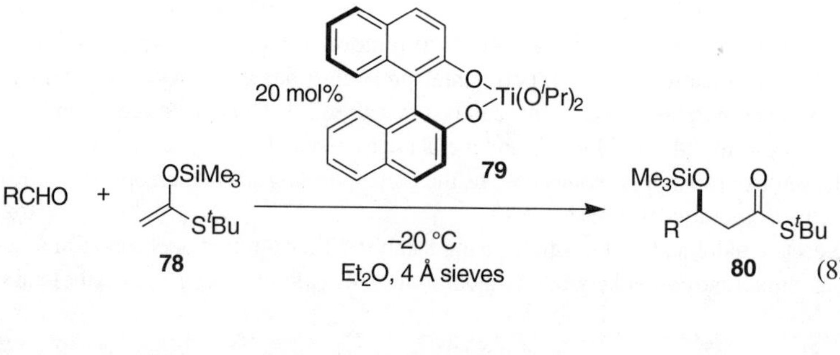

(8B2.20)

TABLE 8B2.9. Thioacetate aldol addition reactions catalyzed by 79 (Eq. 8B2.20)[a]

Aldehyde	Yield	% ee
PhCHO	90%	97
Ph(CH$_2$)$_2$CHO	80%	97
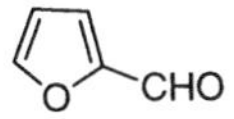	88%	>98
c-C$_6$H$_{11}$CHO	79%	89
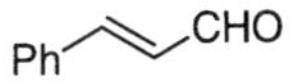	76%	89
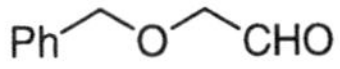	82%	>98

[a]Typical Reaction Conditions: 20 mol % **79**, Et$_2$O, –20°C, 12 h.

Catalysis with Complexes of Sn(II) and Cu(II). Recent investigations by Evans in the design of C$_2$-symmetric bisoxazoline ligands and their derived Sn(II) and Cu(II)-complexes (Chart 8B2.3) have led to the development of catalytic, enantioselective aldol addition reactions of benzyloxyacetaldehyde **86**, glyoxal **87**, and pyruvates **88** (Scheme 8B2.7) [8,27]. The reaction of benzyloxyacetaldehyde with *tert*-butyl thioacetate-derived enoxy silane catalyzed by as little as 0.5 mol % **83** in CH$_2$Cl$_2$ at –78°C afforded the aldol adduct with up to 99% ee (Table 8B2.10). This catalytic process is noteworthy not only as a result of the high stereoselectivity observed, but also as a consequence of the structural insight that has been revealed about the complexes. On the basis of experimental and X-ray crystallographic data, the C$_2$-symmetric complexes are suggested to form chelates with 1,2-donors such as **86**, **87**, and **88** to afford activated electrophilic species that undergo addition by the silyl ketene acetals.

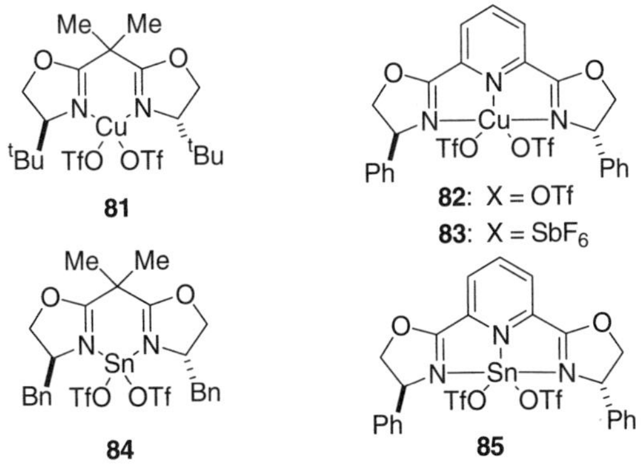

Chart 8B2.3.

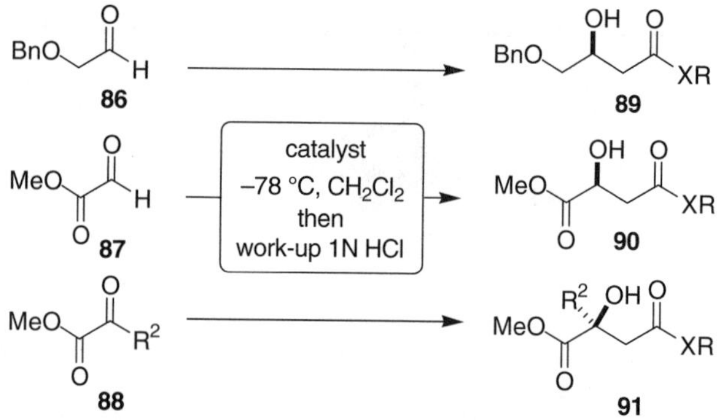

Scheme 8B2.7.

TABLE 8B2.10. Thioacetate aldol additions (Scheme 8B2.7)[a]

Electrophile	Enol Silane		Catalyst	% ee
BnOCH$_2$CHO	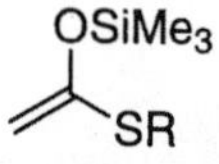 OSiMe$_3$ SR	R = S^{t}Bu	83	99
		R = SEt	83	98
		R = OEt	83	98
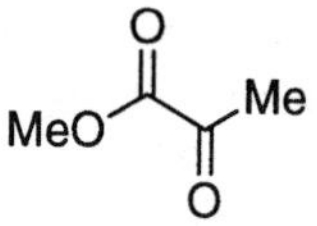 MeO (with Me, O, O)	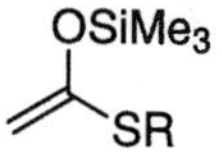 OSiMe$_3$ SR	R = S^{t}Bu	81	99
		R = SEt	81	97
		R = SPh	84	98
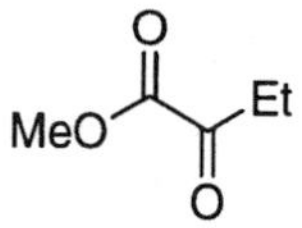 MeO (with Et, O, O)	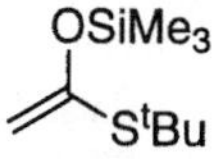 OSiMe$_3$ S^{t}Bu		81	94
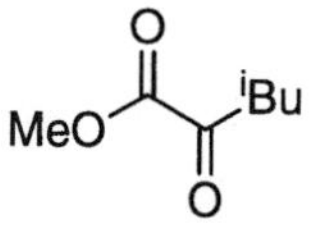 MeO (with ^{i}Bu, O, O)	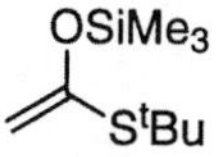 OSiMe$_3$ S^{t}Bu		81	94
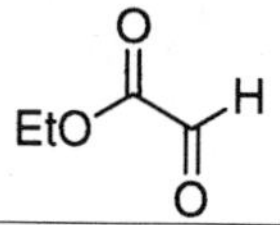 EtO (with H, O, O)	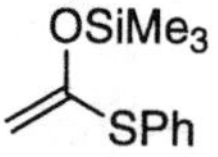 OSiMe$_3$ SPh		84	98

[a]Typical Reaction Conditions: 0.5–10 mol % catalyst, CH$_2$Cl$_2$, –78°C, 5 min.

8B2.4. PROPIONATE ADDITION REACTIONS

Impressive advances in catalytic, enantioselective propionate aldol addition reactions have also been documented since 1992. Mikami has described a Ti(IV) catalyst readily prepared from BINOL and TiCl$_2$(O^{i}Pr)$_2$. A propionate aldol addition process by Evans utilizes complexes prepared with bisoxazoline ligands and Sn(II) and Cu(II). In analogy to the acetate aldol additions, the structural requirement for the aldehyde substrates is that they putatively can form five membered-ring chelates with the metal complex.

Catalysis with BINOL·Ti(IV) Complex. Mikami has reported the Mukaiyama aldol addition reaction of thiopropionate-derived enolsilanes **92** and **93** with benzyloxyacetaldehyde and *n*-butyl glyoxaldehyde (Eq. 8B2.21 and Table 8B2.11) [21]. The active catalyst mixture is generated upon treatment of $TiCl_2(O^iPr)_2$ with BINOL in toluene. The Z-enol silane **92** gave adduct of benzyloxyacetaldehyde as an 92:8 mixture of diastereomers in which the major was *anti* stereoisomer **94b** with 90% ee. However, additions of *E*- or *Z-S*-ethyl thiopropionate-derived silyl ketene acetals **93** to benzyloxyacetaldehyde afforded adducts with high enantiose-lectivity (up to 98% ee), albeit with diminished levels of simple diastereoselectivity 72:28-48:52, *syn/anti*). In contrast, additions of *E*-or *Z*-**92** with *n*-butyl glyoxylate gave adducts with lower diastereo- and enantio-selectivity. However, the addition of *E*-**93** to this same aldehyde led to the preferential formation of syn diastereomer **94a** (92:8 *syn/anti*) with impressive 98% ee. Thus, appropriate matching of aldehyde and silyl ketene acetal can lead to divergent diastereo-chemical outcomes in the addition reaction, providing ready access to either *syn* or *anti* aldol adducts possessing high enantiopurity.

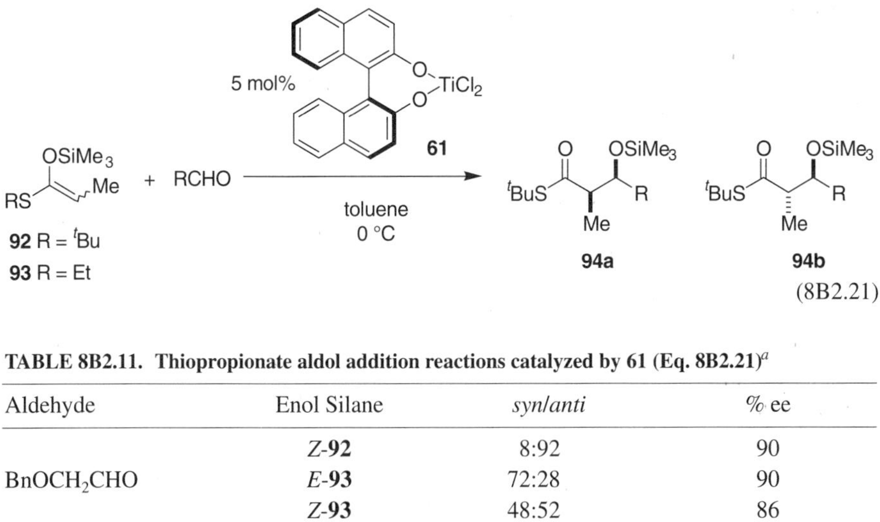

(8B2.21)

TABLE 8B2.11. Thiopropionate aldol addition reactions catalyzed by 61 (Eq. 8B2.21)[a]

Aldehyde	Enol Silane	*syn/anti*	% ee
	Z-**92**	8:92	90
BnOCH₂CHO	E-**93**	72:28	90
	Z-**93**	48:52	86
	Z-**92**	20:80	86
BuO₂CCHO	E-**92**	57:43	88
	E-**93**	92:8	98

[a]Typical Reaction Conditions: 5 mol % **61**, toluene, 0°C, 2h.

Catalysis with Bisoxazoline Complexes of Sn(II) and Cu(II). The bisoxazoline Cu(II) and Sn(II) complexes **81–85** that have proven successful in the acetate additions with aldehydes **86, 87, 88** also function as catalysts for the corresponding asymmetric propionate Mukaiyama aldol addition reactions (Scheme 8B2.8) [27]. It is worth noting that either syn or anti simple diastereoselectivity may be obtained by appropriate selection of either Sn(II) or Cu(II) complexes (Table 8B2.12).

When catalyzed by the Cu(II) complexes **81–83**, the addition of either *E*- or *Z*-thiopropion-ate-derived silyl ketene acetals afford adducts displaying 86:14–97:3 simple diastereoselectiv-

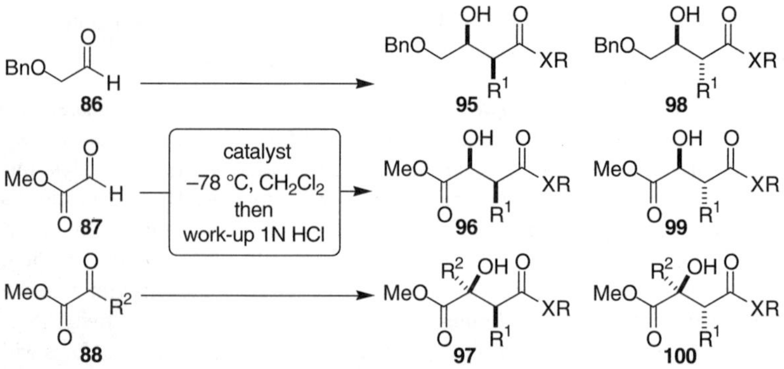

Scheme 8B2.8.

TABLE 8B2.12. Thiopropionate aldol addition reactions catalyzed by 81–85 (Scheme 8B2.8)[a]

Substrate	Enolsilane		Catalyst 10 mol %	syn/anti	% ee major
BnOCH₂CHO	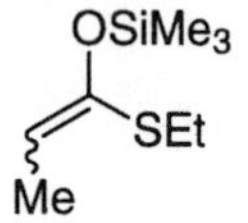 OSiMe₃ / SEt / Me	Z	83	97:3	97
		E		86:14	85
	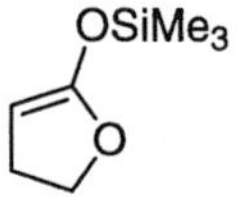 OSiMe₃ (furanyl)		83	96:4	95
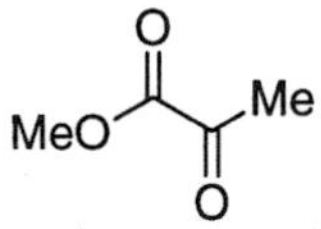 MeO₂C–CO–Me	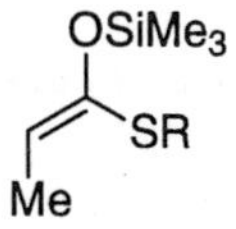 OSiMe₃ / SR / Me	R = ᵗBu	81	94:6	96
		R = Et	81	94:6	93
		R = ᵗBu	85	1:99	99
		R = Et	85	5:95	92
	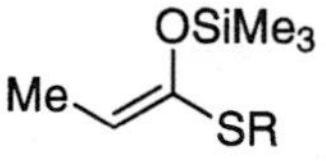 Me / OSiMe₃ / SR	R = ᵗBu	81	95:5	98
		R = Et	81	98:2	98
		R = ᵗBu	85	1:99	96
	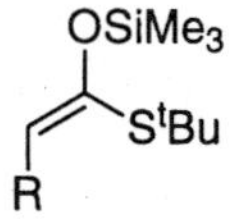 OSiMe₃ / SᵗBu / R	R = Et	85	2:98	97
		R = ⁱBu	85	1:99	99
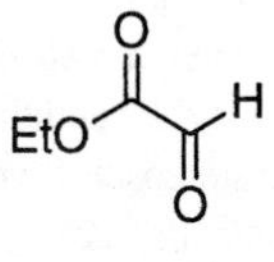 EtO₂C–CO–H	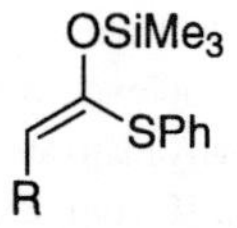 OSiMe₃ / SPh / R	R = Me	84	10:90	95
		R = Et		8:92	95
		R = ⁱPr		7:93	95
		R = ⁱBu		8:92	98

[a]Typical Reaction Conditions: 10 mol % catalyst, CH₂Cl₂, −78°C, 5 min.

ity, with the *syn* stereoisomers **95–97** consistently dominating. For the aldehydes examined, the enantiopurity of the major *syn* aldol adducts isolated from the Cu(II)-catalyzed additions of both Z- and E-enol silanes were excellent (85–99% ee). The Cu(II)-catalyzed addition of butyrolactone enol silane also proved to be highly selective for the *syn* diastereomer (96:4 *syn/anti*), which was formed with 95% ee.

The stereochemical outcome of the aldol addition reactions to pyruvate and glyoxylates catalyzed by Sn(II) were complementary to the Cu(II)-catalyzed process, giving the *anti* stereoisomers. Thus, aldol adducts were isolated as mixtures of 10:90–1:99 *syn/anti* diastereomers in 92–99% ee.

The Evans Cu(II)- and Sn(II)-catalyzed processes are unique in their ability to mediate aldol additions to pyruvate. Thus, the process provides convenient access to tertiary α-hydroxy esters, a class of chiral compounds not otherwise readily accessed with known methods in asymmetric catalysis. The process has been extended further to include α-diketone **101** (Eqs. 8B2.22 and 8B2.23). It is remarkable that the Cu(II) and Sn(II) complexes display enzyme-like group selectivity, as the complexes can differentiate between ethyl and methyl groups in the addition of thiopropionate-derived Z-silyl ketene acetal to **101**. As discussed above, either *syn* or *anti* diastereomers may be prepared by selection of the Cu(II) or Sn(II) catalyst, respectively.

$$(8B2.22)$$

$$(8B2.23)$$

8B2.5. ALDOL ADDITIONS OF DIENOLATES AND ISOBUTYRATES

Significant efforts have extended the scope of catalytic enantioselective Mukaiyama aldol addition reactions beyond the acetate and propionate enoxysilanes and have been used traditionally. Recent reports describe novel addition reactions of silyl dienolates along with isobutyrate-derived enol silanes.

Catalysis with Ti(IV) Complexes and Boronates. Carreira has documented the addition of dienolsilane **105** to a broad range of aldehydes [28]. Enolization of the commercially available acetone-ketene adduct **104** with LDA, followed by quenching with chlorotrimethyl silane, gave **105** in 78% yield as a clear colorless liquid that can be conveniently purified by distillation (Eq. 8B2.24). The addition reactions are conducted at 23°C utilizing 5 mol % **72** to give adducts with up to 94% ee (Eq. 8B2.25, Table 8B2.13). The aldol adducts **106** were isolated fully protected as the corresponding O-silyl ethers with the β-keto ester masked in the form of a dioxinone.

TABLE 8B2.13. Dienolate addition reactions (Eq. 8B2.25)[a]

Entry	Aldehyde	% ee
1	^{i}Pr$_3$Si—≡—CHO	91
2	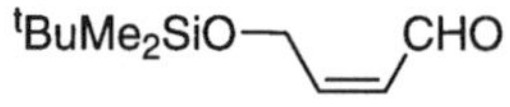 ^{t}BuMe$_2$SiO⁀＼＿CHO	94
3	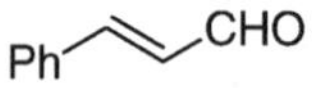 Ph⁀⁀CHO	92(99)
4	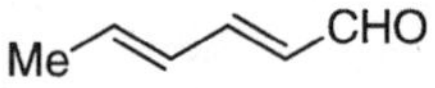 Me⁀⁀⁀CHO	92
5	PhCHO	84(96)
6	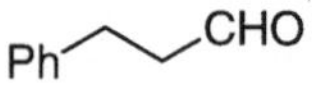 Ph⁀⁀CHO	80
7	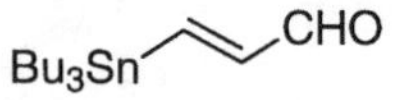 Bu$_3$Sn⁀⁀CHO	92

[a]Typical Reaction Conditions: 1–3 mol %, **72**, Et$_2$O, 0°C, 2h.

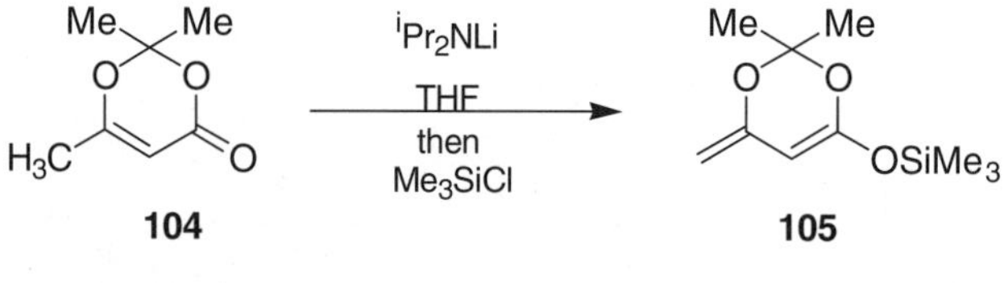

$$(8B2.24)$$

$$(8B2.25)$$

106 R$'$ = Me$_3$Si
107 R$'$ = H

Carreira and co-workers have also extended the scope of aldehydes that may be utilized in catalytic addition reactions to include stannylpropenal **108** as a substrate (Table 8B2.12, Entry 7). The adduct produced from the aldol addition of **105** is isolated with 92% ee and serves as a useful building block, as it is amenable for further synthetic elaboration (Scheme 8B2.9). Thus, vinylstannane **109** is a substrate for Stille cross-coupling reactions to give a diverse family of protected acetoacetate adducts **110**. Following deprotection of the masked keto ester, the corresponding hydroxy keto ester **111** may be converted to either the *syn* or *anti* skipped polyols **112** or **113**. A recent total synthesis of macrolactin A by Carreira and co-workers utilizes aldol

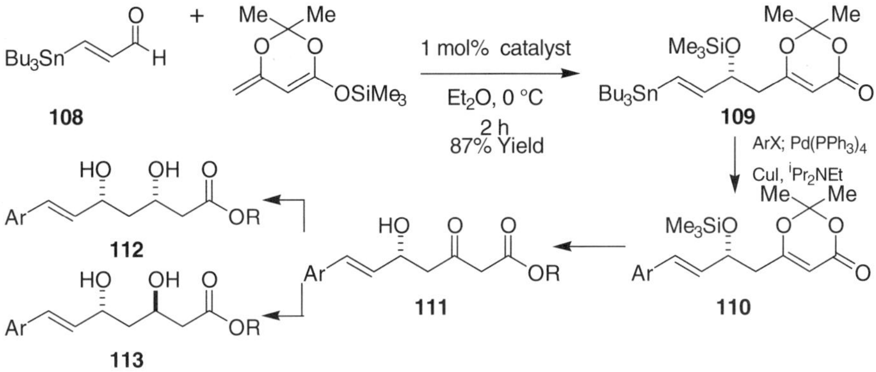

Scheme 8B2.9.

adduct of **109** as a versatile building block. In the convergent strategy, access to aldol adduct **109** allows for rapid assembly of two key optically active subunits that in turn are joined via Stille cross-coupling reactions [29]. The full synthetic potential of the dual electrophilic and nucleophilic reactive sites in **109** has also been exploited in a synthesis of the C_1–C_3 polyol portion of amphotericin [30].

Sato and co-workers have also documented related additions utilizing **105** (Scheme 8B2.10) [31]. The addition reaction of **105** to benzaldehyde and pentanal with 20 mol % **61** afforded adducts **114** and **115** in 38% yield/88% ee and 55% yield/92% ee, respectively. When Yamamoto's oxazaborolidene catalyst **116** was used, the same adducts **114** and **115** were isolated in 69% yield/67% ee and 52% yield/70% ee, respectively.

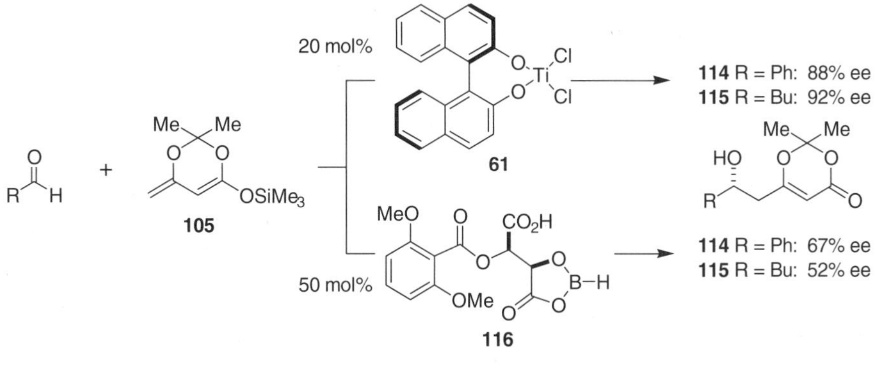

Scheme 8B2.10.

Catalysis with Lewis-Acidic Complexes of Cu(II) and Sn(II). Evans has reported the addition of dienolate **105** and dienolate **119** to benzyloxyacetaldehyde utilizing Cu(II) catalyst **81** Eqs. 8B2.26 and 8B2.27) [8]. The addition of **105** in the presence of 5 mol% **81** gave **117** in

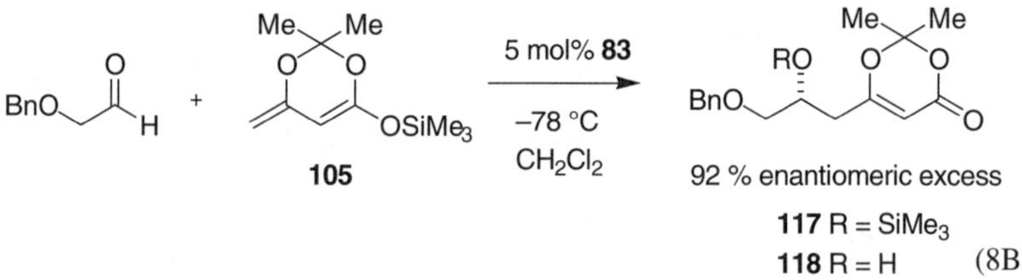

$$(8B2.26)$$

$$(8B2.27)$$

94% yield and 92% ee. With **119**, the addition reaction could be conducted with as little as 0.5 mol % catalyst to give directly a hydroxy keto ester that was analyzed in the following stereoselective reduction to be *anti* diol **120**.

Catalysis with BINAP·CuF₂. Carreira and co-workers have recently reported a novel aldol addition reaction using a putative Cu(I) fluoride complex as catalyst [32]. The complex is readily assembled in situ upon dissolving BINAP and $Cu(OTf)_2$, in THF followed by addition of $Bu_4NPh_3SiF_2$, as an anhydrous fluoride source. Dienolate **105** undergoes addition to a wide range of aldehyde substrates at −78°C with 5 mol % catalyst, giving the protected acetoacetate adducts with up to 94% ee. The reaction has been mechanistically examined in some detail, leading to the postulation that the addition process includes a metalloenolate in the catalytic cycle as relevant reactive intermediate (Scheme 8B2.11) [33]. This perspective contrasts the more traditional role of transition-metal catalysis of the Mukaiyama aldol addition reaction wherein the metal functions as a Lewis acid and activates the electrophilic aldehyde partner.

Carreira has reported that the aldol addition reaction to furan 2-carboxaldehyde may be performed on preparative scale (0.5 mol) with as little as 0.5 mol % catalyst (Scheme 8B2.12) [30]. The adduct is isolated as a crystalline solid with 94% ee; a single recrystallization allows access to **122** with >99% ee. Ozonolytic cleavage of the furan unmasks a carboxyl function, providing acid **124**. This acid has been used in a convergent synthesis of the amphotericin C_1-C_3 polyol portion. Moreover, **124** has been utilized in the synthesis of a number of pharmacologically important structures such as HMG CoA reductase inhibitors for the treatment of hyper-cholesterolemia [34].

Catalysis with Pt(II) Complexes. Fujimura has recently reported the enantioselective addition reaction of isobutyrate-derived dimethyl substituted silyl ketene acetal **125** to aldehydes (Eq. 8B2.29 and Table 8B2.15) [35]. The process utilizes a novel Pt(II) complex first reported by Pregrosin and has several salient features. In this regard, the complex is readily assembled and

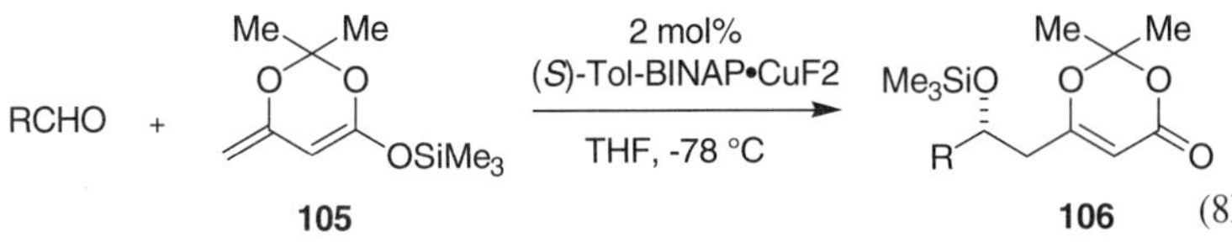

$$\text{RCHO} \quad + \quad \underset{\textbf{105}}{\text{105}} \quad \xrightarrow[\text{THF, -78 °C}]{\begin{array}{c} 2 \text{ mol\%} \\ (S)\text{-Tol-BINAP•CuF2} \end{array}} \quad \underset{\textbf{106}}{\text{106}} \qquad (8B2.28)$$

TABLE 8B2.14. Dienolate aldol addition reactions (Eq. 8B2.28)[a]

Entry	Aldehyde	Yield	% ee
1	PhCHO	92%	94
2	2-NapCHO	86%	93
3	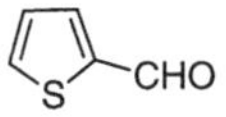 ⟨thienyl⟩-CHO	98%	95
4	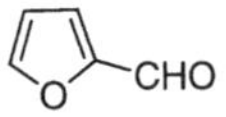 ⟨furyl⟩-CHO	91%	94
5	4-MeOC6H4CHO	93%	94
6	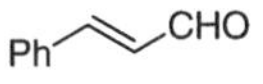 Ph⌒CHO	83%	85
7	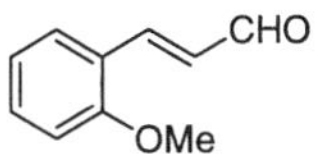 (2-OMe-C6H4)⌒CHO	82%	90
8	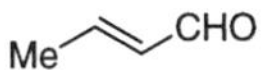 Me⌒CHO	48%	91
9	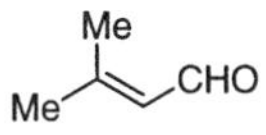 Me₂C=CH-CHO	81%	83
10	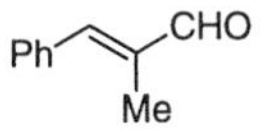 Ph⌒C(Me)=CH-CHO	74%	65

[a]Typical Reaction Conditions: 2 mol % catalyst, THF, −78 °C, 6–8 h.

can be easily handled in the laboratory [36]. In the presence of 5 mol % each of **126**, triflic acid, and lutidine in CH_2Cl_2 **125** undergoes additions to aldehydes at −25°C to afford adducts with up to 95% ee. It is interesting to note that, although adducts are isolated from the reaction as a mixture of **127** and **128**, both forms possess identical enantiopurity. A meticulous investigation of the reaction parameters and the effect of additives on the product selectivities has revealed that the addition of oxygen and water leads to significant improvement in the enantiopurity of the product. A series of spectroscopic studies by ^{31}P NMR and IR has led Fujimura to postulate that the reaction involves a C-bound platinum enolate intermediate in the catalytic cycle.

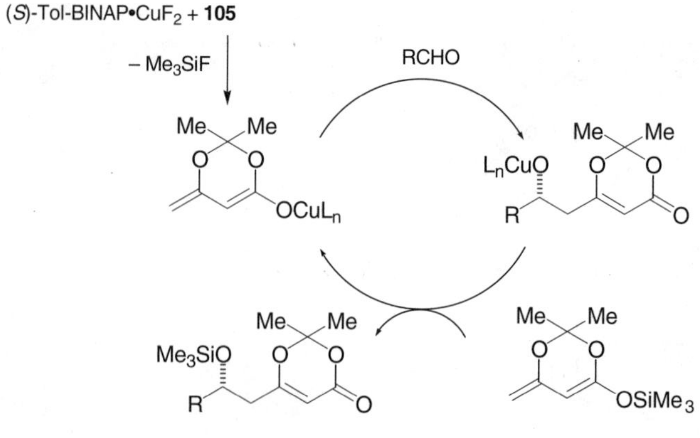

(S)-Tol-BINAP•CuF$_2$ + **105**

Scheme 8B2.11.

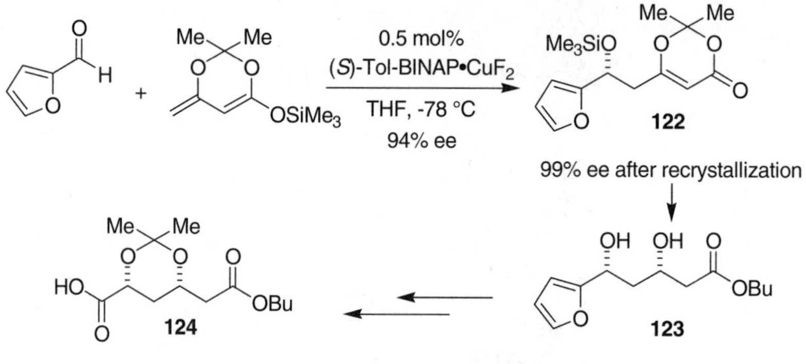

Scheme 8B2.12.

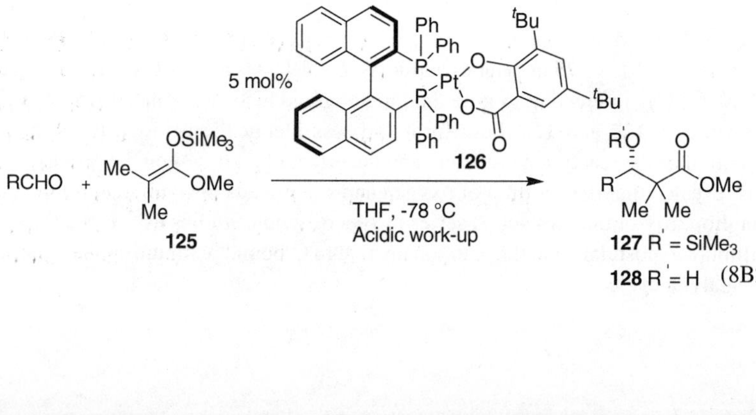

127 R$'$ = SiMe$_3$
128 R$'$ = H   (8B2.29)

TABLE 8B2.15. Aldol additions of 125 to aldehydes (Eq. 8B2.29)^a

Entry	Aldehyde	Catalyst	**120/121**	% ee
1	PhCHO	**118**	65:35	59
		119	82:18	56
2	Ph⌒⌒CHO	**118**	65:35	46
		119	84:16	46
3	Ph⌒⌒CHO	**118**	51:49	95
		119	84:16	91
4	Me$_2$CHCH$_2$CHO	**118**	82:18	88
		119	94:6	80

^{a}Typical Reaction Conditions: 5 mol % **126**, CH$_2$Cl$_2$, −25°C, 17–170 h.

8B2.6. CONCLUSION

The survey of catalytic, enantioselective aldol addition reactions presented herein attests to the phenomenal advances that have been made in the field recently. Thus, in contrast to the conclusion to the preceding chapter 8B.1 that covers achievements by mid-1992 [3a], it can be stated that the past few years have witnessed major breakthroughs in the catalytic asymmetric synthesis of aldol adducts. In this regard, the range of substrates that can be reliably utilized in catalytic aldol additions have been expanded considerably and the enantiopurity of the adducts isolated from various procedures are uniformly higher than was attainable with prior art. Moreover, in general, the accessibility of catalysts as well as the experimental execution of the prescribed procedures have improved. The realization of these advances has been possible as a consequence of novel catalysts operating in concert with new mechanistic insight. However, despite the rapid evolution of the catalytic aldol reaction, when set against the advances that have taken place in catalytic asymmetric hydrogenation, epoxidation, and dihydroxylation, it is clear that much still remains to be done. These gold-standards of asymmetric catalysis include methods remarkable in their ease of execution and prescribe catalytic loadings, volumetric efficiencies, and turnover numbers yet to be matched in the aldol field. Building upon the discoveries highlighted in this chapter, the next decade holds the promise to yield further revolutionary advances for the asymmetric catalysis of aldol addition reactions.

REFERENCES

1. (a) Gennari, C. In *Comprehensive Organic Synthesis: Additions to C-X π-Bonds*; Trost, B. M., Fleming, I.; Heathcock, C. H. (Ed.); Pergamon Press:New York, 1991; Ch. 2.4, pp. 629. (b) Ito, Y.; Braun, M.; Sacha, H. *J. Prakt. Chem.* **1993**, *335*, 653. (c) Franklin, A. S.; Paterson, I. *Contemporary Organic Synthesis* **1994**, 317. Bach, T. *Angew. Chem., Int. Ed.* **1994**, *33*, 417. (d) Santelli, M.; Pons, J. M. *Lewis Acids and Selectivity in Organic Synthesis;* CRC:Boca Raton, 1995.

2. For entry into this vast literature, see: (a) Evans D. A. *Aldrichimica Acta* **1982**, *15*, 23. (b) Evans, D. A.; Kim, A. S.; Metternich, R.; Novak, V. J. *J. Am. Chem. Soc.* **1998**, *120*, 5921. (c) Patterson. I.; Norcross, R. D.; Ward. R. A.; Romea. P.; Lister. M. A. *J. Am. Chem. Soc.* **1994**, *116*, 11287.

3. (a) Sawamura, M.; Ito, Y. In *Catalytic Asymmetric Synthesis;* Ojima, I. (Ed.); VCH:New York, 1993; Ch. 7.2, pp. 367; (see Chapter 8B.1 in this book). (b) Yamamoto, H.; Maruoka, K. In *Catalytic*

Asymmetric Synthesis; Ojima, I. (Ed.); VCH: New York, 1993, Ch. 9, pp. 413. (c) Noyori, R. *Asymmetric Catalysis in Organic Synthesis;* Wiley:New York, 1994.

4. For a recent review, see: Nelson, S. G. *Tetrahedron: Asymmetry* **1998**, *9*, 357.

5. Corey, E. J.; Cywin, C. L.; Roper, T. D. *Tetrahedron Lett.* **1992**, *33*, 6907.

6. Corey, E. J.; Loh, T.-P.; Roper, T. D.; Azimioara, M. D.; Noel M. C. *J. Am. Chem. Soc.* **1992**, *114*, 8290.

7. Corey, E. J.; Rohde, J. J.; Fischer, A.; Azimioara, M. D. *Tetrahedron Lett.* **1997**, *38*, 33. (b) Corey, E. J.; Rohde J. J. *Tetrahedron Lett.* **1997**, *38*, 37. (c) Corey, E. J.; Barnes-Seeman, D.; Lee, T. W. *Tetrahedron Lett.* **1997**, *38*, 1699. (d) Corey, E. J.; Barnes-Seeman, D.; Lee, T. W. *Tetrahedron Lett.* **1997**, *38*, 4351. (e) Corey, E. J.; Barnes-Seeman, D.; Lee, T. W.; Goodman, S. N. *Tetrahedron Lett.* **1997**, *38*, 6513.

8. Evans, D. A.; Murry, J.A.; Kozlowski, M. C. *J. Am. Chem. Soc.* **1996**, *118*, 5814.

9. (a) Yamada, Y. M.; Yoshikawa, N.; Sasai, H.; Shibasaki, M. *Angew. Chem., Int. Ed.* **1997**, *36*, 1871. (b) Yamada, Y. M.; Shibasaki, M. *Tetrahedron Lett.* **1998**, *39*, 5561.

10. For a comprehensive review of these bimetallic catalysts, see: Shibasaki, M.; Sasai, H.; Arai, T. *Angew. Chem., Int. Ed.* **1997**, *36*, 1236.

11. Yamada, Y. M. A.; Shibasaki, M. *Tetrahedron Lett.* **1998**, *39*, 5561.

12. (a) Ito, Y.; Sawamura, M.; Hayashi, T. *J. Am. Chem. Soc.* **1986**, *108*, 6405. (b) For a recent report of a Pd(II) catalyzed reaction of isonitriles, see: Nesper, R.; Pregrosin, P. S.; Püntener, K.; Wörle, M. *Helv. Chim. Acta.* **1993**, *76*, 2239.

13. Sodeoka, M.; Ohrai, K.; Shibasaki, M. *J. Org. Chem.* **1995**, *60*, 2648.

14. Yanagisawa, A.; Matsumoto, Y.; Nakashima, H.; Asakawa, K.; Yamamoto, H. *J. Am. Chem. Soc.* **1997**, *119*, 9319.

15. (a) Denmark, S. E.; Wong, K.-T.; Stavenger, R. A. *J. Am. Chem. Soc.* **1997**, *119*, 2333. (b) Denmark, S. E.; Winter, S.B.D.; Su, X.; Wong, K.-T. *J. Am. Chem. Soc.* **1996**, *118*, 7404.

16. Carreira, E. M.; Lee, W.; Singer, R. A. *J. Am. Chem. Soc.* **1995**, *117*, 3649.

17. Mikami, K.; Matsukawa, S. *J. Am. Chem. Soc.* **1993**, *115*, 7039.

18. (a) Hoffman, T.; Zhong, G.; List, B.; Shabat, D.; Anderson, J.; Gramatikova, S.; Lerner, R. A.; Barbas, C. F., III *J. Am. Chem. Soc.* **1998**, *120*, 2768. (b) Barbas III, C. F.; Heine, A.; Zhong, G.; Hoffman, T.; Gramatikova, S.; Björnestedt, R.; List, B.; Anderson, J.; Stura, E. A.; Wilson, E. A.; Lerenre, R. A. *Science* **1997**, *278*, 2085. (c) Wagner, J.; Lerner, R. A.; Barbas III, C. F. *Science* **1995**, *270*, 1797.

19. (a) Wong, C.-H.; Garcia-Junceda, E.; Chen, L.; Blanco, O.; Gijsen, H.J.M.; Steensma, D. H. *J. Am. Chem. Soc.* **1995**, *117*, 3333. (b) Chen, L.; Dumas, D. P.; Wong, C.-H. *J. Am. Chem. Soc.* **1992**, *114*, 3741. (c) Wong, C.-H.; Whitesides, G. M. In *Enzymes in Synthetic Organic Chemistry,* Baldwin, J. E., Magnus, P. D. (Eds.) Pergamon Press:Tarrytown, New York, 1994; Vol. 12, pp. 195.

20. Kimura, T.; Vassilev, V. P.; Shen, G.-J.; Wong, C,-H. *J. Am. Chem. Soc.* **1997**, *119*, 11734.

21. Mikami, K.; Matsukawa, S. *J. Am. Chem. Soc.* **1994**, *116*, 4077.

22. (a) Carreira, E. M.; Singer, R. A.; Lee, W. *J. Am. Chem. Soc.* **1994**, *116*, 8837. (b) Carreira, E. M.; Singer, R. A. *DDT* **1996**, *1*, 145.

23. Singer, R. A.; Carreira, E. M. *Tetrahedron Lett.* **1997**, *38*, 927.

24. Li, K. W.; Wu, J.; Xing, W.; Simon, J. A. *J. Am. Chem. Soc.* **1996**, *118*, 7237.

25. Rychnovsky, S. D.; Khire, U. R.; Yang, G. *J. Am. Chem. Soc.* **1997**, *119*, 2058.

26. Keck, G. E.; Krishnamurthy, D. *J. Am. Chem. Soc.* **1995**, *117*, 2363.

27. (a) Evans, D. A.; MacMillan, D. W. C.; Campos, K. R. *J. Am. Chem. Soc.* **1997**, *119*, 10859. (b) Evans, D. A.; Kozlowski, M. C.; Burgey, C. S.; MacMillan, D. W. C. *J. Am. Chem. Soc.* **1997**, *119*, 7893.

28. Singer, R. A.; Carreira, E. M. *J. Am. Chem. Soc.* **1995**, *117*, 12360.

29. Kim, Y.; Singer, R. A.; Carreira, E. M. *Angew. Chem., Int. Ed.* **1998**.

30. Krüger, J.; Carreira, E. M. *Tetrahedron Lett.* **1998**, *39*, 7013.

31. (a) Sato, M.; Sunami, S.; Sugita, Y.; Kaneko, C. *Heterocycles* **1995**, *41*, 1435. (b) Sato, M.; Sunami, S.; Sugita, Y.; Kaneko, C. *Chem. Pharm. Bull.* **1994**, *42*, 839.

32. Krüger, J.; Carreira E. M. *J. Am. Chem. Soc.* **1998**, *120*, 837.

33. Pagenkopf, B.; Krüger, J.; Stojanovic, A.; Carreira, E. M. *Angew. Chem., Int. Ed.* **1998**, 3124.

34. (a) Boberg, M.; Angerbauer, R.; Fey, P.; Kanhai, W.; Karl, W.; Kern, A.; Ploschke, J.; Radkte, M. *Drug Metab. Dispos.* **1997**, *25*, 321. (b) Konioke, T.; Araki, Y. *J. Org. Chem.* **1994**, *59*, 7849. (b) Takahashi, K. *Tetrahedron Lett.* **1993**, *34*, 8263.

35. Fujimura, O. *J. Am. Chem. Soc.* **1998**, *120*, 10032.

36. Anklin, C.; Pregosin, P. S.; Bachechi, F.; Mura, P.; Zambonelli, L. *J. Organomet. Chem.* **1981**, *222*, 175.

8C

ASYMMETRIC ENE REACTIONS

KOICHI MIKAMI AND TAKESHI NAKAI

Department of Chemical Technology, Tokyo Institute of Technology, Meguro-ku, Tokyo 152, Japan

C–H Bond activation [1] and C–C bond formation are two of the key issues in organic synthesis. In principle, the ene reaction is one of the simplest ways for C–C bond formation, which converts readily available olefins into more functionalized products with activation of an allylic C–H bond and allylic transposition of the C=C bond. The ene reaction encompasses a vast number of variants in terms of the enophile used [2].

8C.1. CARBONYL-ENE REACTION

The class of ene reactions which involves a carbonyl compound as the enophile and what we refer to as the carbonyl-ene reaction [2c], constitutes a useful synthetic method for the stereocontrolled construction of carbon skeleton using either a stoichiometric or catalytic amount of various Lewis acids [3,4] (Scheme 8C.1). From the synthetic point of view, the carbonyl-ene reaction should—in principle—constitute a more-efficient alternative to the allylmetal carbonyl addition reaction, which has currently been one of the most useful methods for stereocontrol [5].

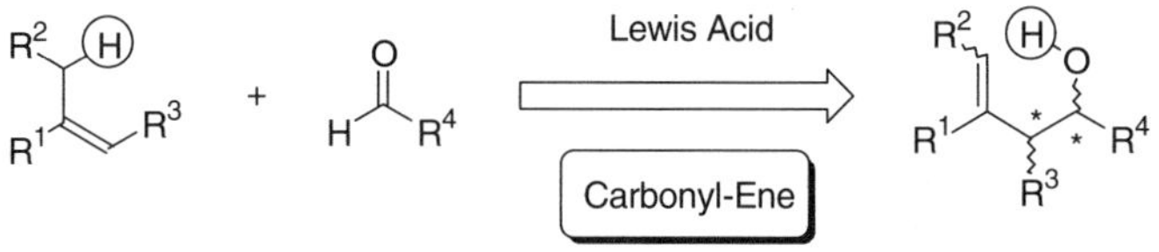

Scheme 8C.1. Carbonyl-ene reaction catalyzed by chiral Lewis acids.

Catalytic Asymmetric Synthesis, Second Edition, Edited by Iwao Ojima
ISBN 0-471-29805-0   Copyright © 2000 Wiley-VCH, Inc.

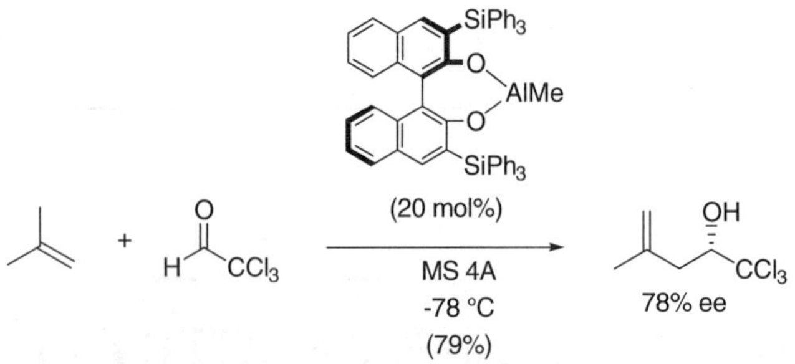

Scheme 8C.2. Asymmetric carbonyl-ene reaction catalyzed by chiral Al complex.

Yamamoto et al. have reported the asymmetric catalysis of a chiral Lewis acid in a carbonyl-ene reaction, which uses chloral as the enophile and an aluminum catalyst with enantiopure 3,3′-bissilylated binaphthol (BINOL) to give the corresponding homoallylic alcohol with 78% ee in 79% yield (Scheme 8C.2) [6]. It should be noted that 3,3′-diphenyl–BINOL-derived aluminum catalyst provides the racemic product in low yield.

Mikami and Nakai et al. have developed a chiral titanium catalyst for the glyoxylate-ene reaction, which provides the corresponding α-hydroxy esters of biological and synthetic importance [7] in an enantioselective fashion (Scheme 8C.3) [8,9]. Various chiral titanium catalysts were screened [10]. The best result was obtained with the titanium catalyst (**1**) prepared in situ in the presence of MS 4A from diisopropoxytitanium dihalides ($X_2Ti(OPr^i)_2$: X=Br [11] or Cl [12]) and enantiopure BINOL or 6-Br-BINOL [13]. The remarkable levels of enantioselectivity and rate acceleration observed with these BINOL-Ti catalysts (**1**) [14] stem from the

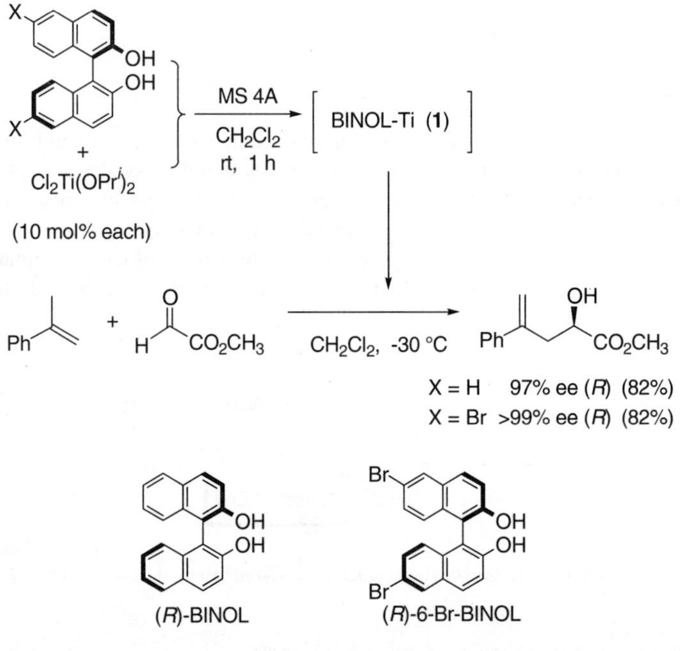

Scheme 8C.3. Asymmetric carbonyl-ene reaction catalyzed by BINOL-Ti complex.

TABLE 8C.1. Asymmetric catalytic glyoxylate-ene reaction with various olefins

Run	olefin	$X_2Ti(OPr^i)_2$ (X)	catalyst (mol %)	products	% yield	% ee
A		Cl	10	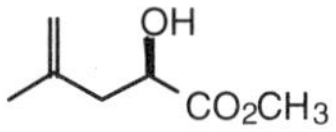	72	95
		Cl	1.0	CO_2CH_3, OH	78	93
		Br	10		87	94
B	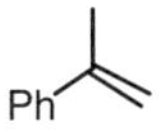 Ph	Cl	1.0	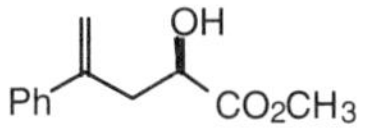 Ph, OH, CO_2CH_3	97	97
		Br	1.0		98	95
C	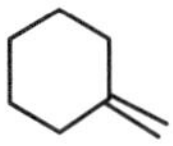	Cl	10	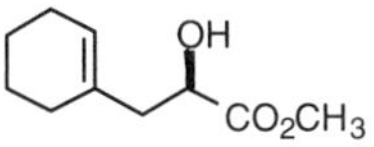 OH, CO_2CH_3	82	97
		Br	5		89	98
D	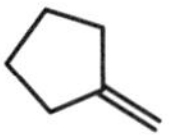	Cl	10	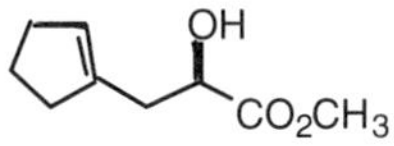 OH, CO_2CH_3	87	88
		Br	5		92	89

favorable influence of the inherent C_2 symmetry and the higher acidity of BINOLs compared with those of aliphatic diols. The reaction is applicable with a variety of 1,1-disubstituted olefins to provide the ene products with extremely high enantiopurity. Representative results are shown in Table 8C.1.

In the reactions with mono- and 1,2-disubstituted olefins, however, no ene product was obtained. This limitation has been overcome by the use of vinylic sulfides and selenides instead of mono- and 1,2-disubstituted olefins. With these substrates, the ene products are obtained with comparably high enantioselectivity and high diastereoselectivity [15]. The synthetic utility of the vinylic sulfide and selenide is shown in the synthesis of enantiopure (R)-$(-)$-ipsdienol, an insect-aggregation pheromone (Scheme 8C.4) [16].

The lanthanide complex-catalyzed asymmetric glyoxylate-ene reaction has also been reported (Scheme 8C.5) [17,18]. Although the reaction of glyoxylate and α-methylstyrene

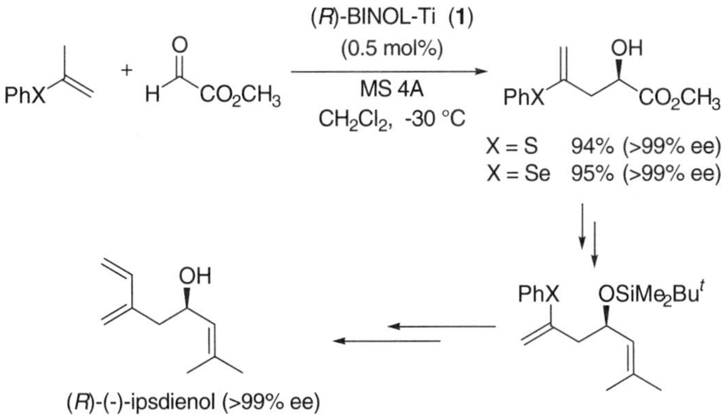

$$PhX + H\text{-}CO_2CH_3 \xrightarrow[\substack{MS\ 4A \\ CH_2Cl_2,\ -30\ °C}]{\substack{(R)\text{-BINOL-Ti (1)} \\ (0.5\ mol\%)}} PhX\text{-}CO_2CH_3,\ OH$$

X = S 94% (>99% ee)
X = Se 95% (>99% ee)

(R)-(-)-ipsdienol (>99% ee) ← PhX, OSiMe$_2$Bu^t

Scheme 8C.4. Asymmetric carbonyl-ene reaction of vinylic sulfides and selenides.

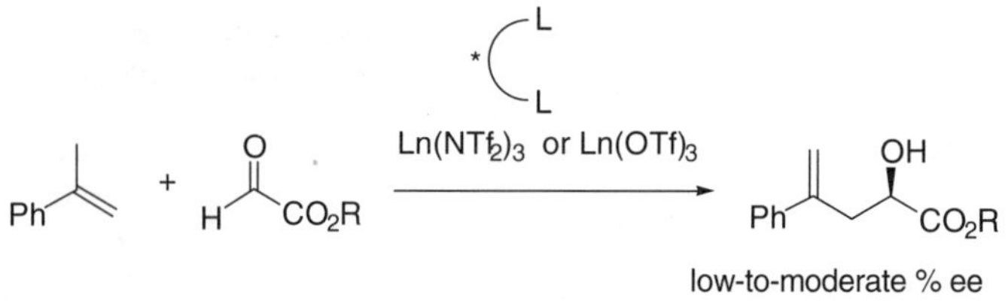

Scheme 8C.5. Asymmetric carbonyl-ene reaction catalyzed by chiral lanthanide complexes.

proceeds catalytically under the influence of chiral lanthanide complexes, the enantioselectivity is low to moderate [19].

The synthetic potential of the asymmetric catalytic carbonyl-ene reaction depends markedly on the reactivity of carbonyl enophile. Indeed, the types of enophile that can be used in the asymmetric catalytic ene reaction have previously been limited to reactive aldehydes such as glyoxylate [9,10,14] and chloral [6,20]. Thus, it is highly desirable to develop other types of carbonyl enophiles to afford enantio-enriched molecules with a broad variety of functionalities. Mikami et al. have developed an asymmetric catalytic fluoral-ene reaction [21], which provides an efficient approach to the asymmetric synthesis of some fluorine-containing compounds [22]. The reaction of fluoral with 1,1-disubstituted and trisubstituted olefins proceeds quite smoothly under the catalysis by the BINOL-Ti complex (1) to give the corresponding homoallylic alcohol with extremely high enantioselectivity (>95% ee) and *syn*-diastereoselectivity (>90%) (Scheme 8C.6). The sense of asymmetric induction in the fluoral-ene reaction is the same as that observed for the glyoxylate-ene reaction, that is, (R)-BINOL-Ti (1) provides the (R)-α-CF$_3$ alcohol. The *syn*-diastereomer of α-trifluoromethyl-β-methyl-substituted liquid crystalline molecule thus synthesized with *two stereogenic centers* shows anti-ferroelectric properties more favorable than the *anti* diastereomer [23].

BINOL-Ti catalysis is also applicable to carbonyl-ene reaction with formaldehyde or vinylogous and alkynylogous analogs of glyoxylates in the catalytic desymmetrization (vide infra) approach to the asymmetric synthesis of isocarbacycline analogs (Scheme 8C.7) [24].

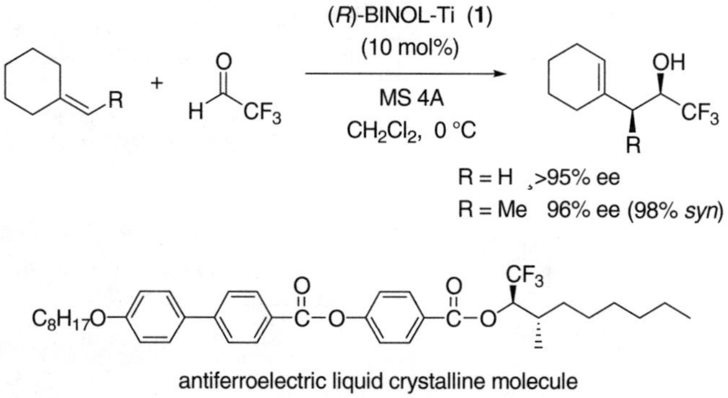

antiferroelectric liquid crystalline molecule

Scheme 8C.6. Asymmetric carbonyl-ene reaction of fluoral.

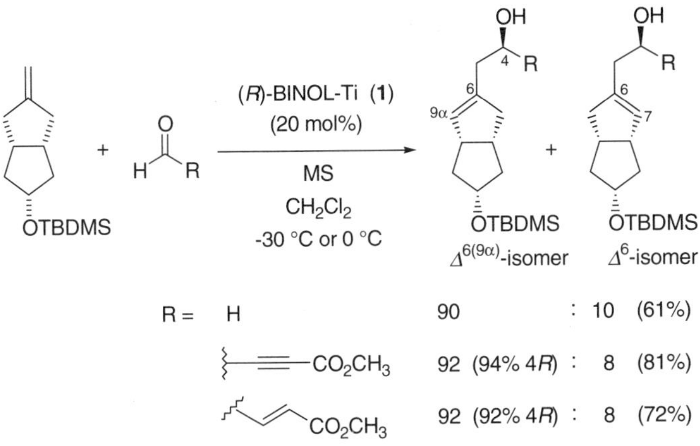

Scheme 8C.7. Asymmetric carbonyl-ene reaction of formaldehyde or vinyl and alkynyl analogues of glyoxylate.

8C.2. ENE VERSUS HETERO DIELS–ALDER REACTION

In the carbonyl-ene reaction with a conjugated diene having an allylic hydrogen, such as isoprene, there is a problem of so-called "periselectivity", arising from the competitive formation of the ene product versus the hetero Diels–Alder (HDA) product. In the reaction of glyoxylate with isoprene under the catalysis of BINOL-Ti complex (**1**), the ratio of ene/HDA products depends not only on the solvent used but also on the chiral ligand of titanium complex and further on the bulkiness of alkyl group (R) in glyoxylate (Scheme 8C.8, Table 8C.2, Figure 8C.1) [13c].

Dramatic changeover is observed not only in the ene/HDA product ratio, but also in the absolute stereochemistry upon changing the central metal from Ti to Al. Thus, Jørgensen et al. reported the HDA-selective reaction of ethyl glyoxylate with 2,3-dimethyl-1,3-butadiene catalyzed by a BINOL-derived Al complex [25], where the HAD product was obtained with up to 89% periselectivity and high enantiopurity (Scheme 8C.9). The absolute configuration was opposite to that observed by using BINOL-Ti catalyst.

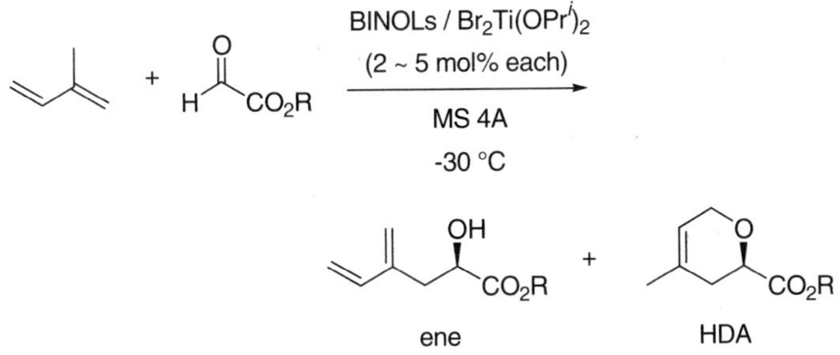

Scheme 8C.8. Ene *vs.* hetero Diels–Alder reaction catalyzed by BINOL-Ti complex.

TABLE 8C.2. The reaction of glyoxylate with isoprene

Run	R	BINOLs	(mol %)	% yield	ene (% ee) / HDA (% ee)
1	Me	BINOL	(2)	94	79 (97) : 21 (97)
2[a]	Me	BINOL	(2)	85	74 (98) : 26 (–)
3	Me	6-Br-BINOL	(2)	95	83 (99) : 17 (–)
4	Bu^n	6-Br-BINOL	(2)	86	85 (>99) : 15 (–)
5	Pr^i	6-Br-BINOL	(5)	61	90 (92) : 10 (–)
6	CH(CF$_3$)$_2$	6-Br-BINOL	(2)	84	92 (>99) : 8 (–)

[a]The reaction was carried out in toluene.

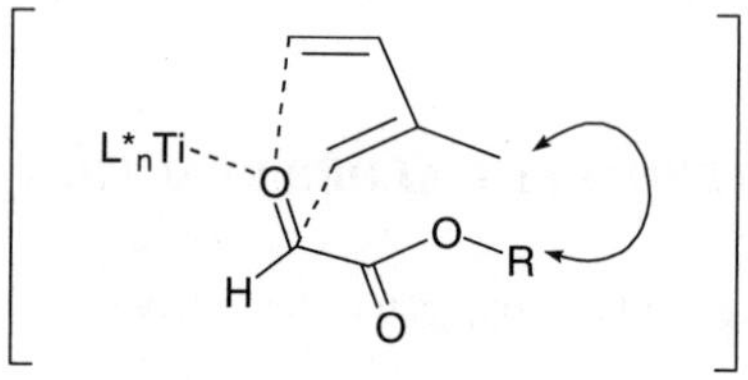

A

Figure 8C.1. *Endo* transition state of hetero Diels–Alder reaction.

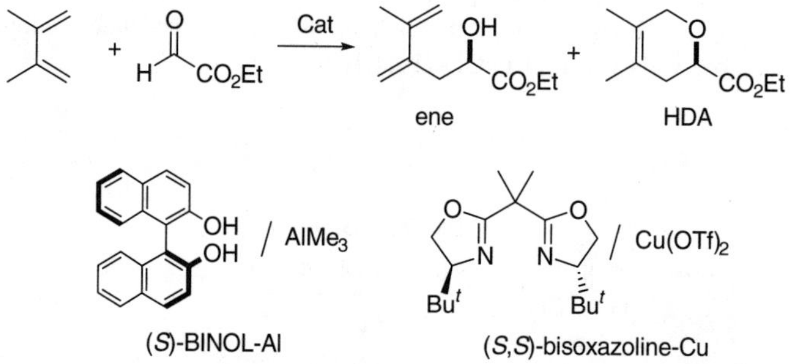

Cat	solvent	% yield	ene	HDA
(S)-BINOL-Al	CH$_2$Cl$_2$	82	11 (88% ee S)	: 89 (97% ee S)
(S,S)-bisoxazoline-Cu	CH$_2$Cl$_2$	56	64 (83% ee R)	: 36 (85% ee R)
(S,S)-bisoxazoline-Cu	CH$_3$NO$_2$	68	44 (78% ee R)	: 56 (90% ee R)

Scheme 8C.9. Ene *vs.* hetero Diels–Alder reaction catalyzed by BINOL-Al or bisoxazoline-Cu complex.

Solvent effects on the reaction of ethyl glyoxylate and 2,3-dimethyl-1,3-butadiene catalyzed by the cationic bisoxazoline-Cu complex have also been reported (Scheme 8C.9) [26]. In a less polar solvent, CH_2Cl_2, the ene product is obtained predominantly, whereas the reaction in a more polar solvent, CH_3NO_2, leads to the preferential formation of the HDA product.

8C.3. ASYMMETRIC DESYMMETRIZATION

Desymmetrization of an achiral, symmetrical molecule through a catalytic process is a potentially powerful but relatively unexplored concept for asymmetric synthesis. Whereas the ability of enzymes to differentiate enantiotopic functional groups is well-known [27], little has been explored on a similar ability of non-enzymatic catalysts, particularly for C–C bond-forming processes. The asymmetric desymmetrization through the catalytic glyoxylate-ene reaction of prochiral ene substrates with planar symmetry provides an efficient access to remote [28] and internal [29] asymmetric induction (Scheme 8C.10) [30]. The (2R,5S)-syn-product is obtained with >99% ee and >99% diastereoselectivity. The diene thus obtained can be transformed to a more functionalized compound in a regioselective and diastereoselective manner.

Ziegler et al. have reported the asymmetric desymmetrization approach to the synthesis of tricothecene, anguidine, via an ene cyclization (Scheme 8C.11) [31]. The (2,4) ene cyclization (vide infra) of the prochiral aldehyde on silica gel gives a 1:1 diastereomeric mixture. Cyclization with purified $Eu(fod)_3$ as Lewis acid catalyst gives an 8:1 mixture. The major isomer is a potential intermediate for the synthesis of anguidine. However, use of $(+)$-$Eu(hfc)_3$, $(+)$-$Eu(dppm)_3$, or (S)-BINOL-$TiCl_2$ complex as chiral Lewis acid affords the major product with only 20~38% ee.

8C.4. KINETIC OPTICAL RESOLUTION

The kinetic optical resolution [32] of a racemic substrate might be considered as an intermolecular version of desymmetrization. In principle, the kinetic resolution of a racemic allylic ether

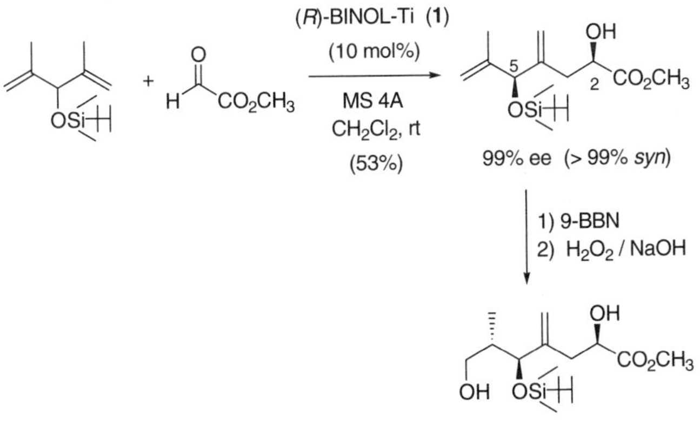

Scheme 8C.10. Asymmetric desymmetrization in carbonyl-ene reaction catalyzed by BINOL-Ti complex.

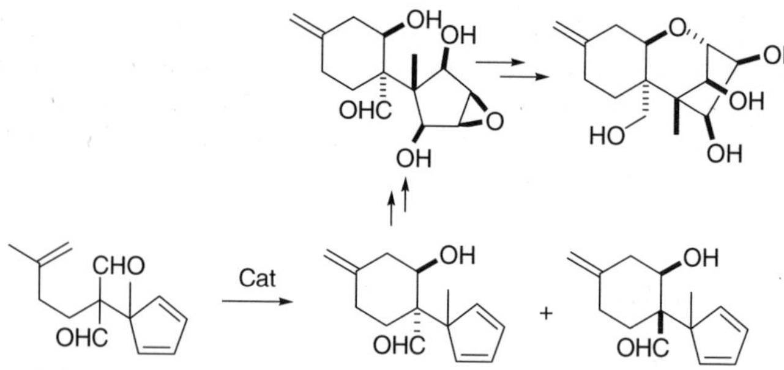

(+)-Eu(hfc)$_3$	5 (20% ee)	:	1
(+)-Eu(dppm)$_3$	4.5 (31% ee)	:	1
(*S*)-BINOL-TiCl$_2$	4.5 (38% ee)	:	1

Scheme 8C.11. Asymmetric desymmetrization in asymmetric ene cyclization.

through glyoxylate-ene reaction might provide an efficient access to remote but relative [29] asymmetric induction. The reaction of allylic ethers catalyzed by the (*R*)-BINOL-derived complex (**1**) provides (2*R*,5*S*)-*syn*-products with >99% diastereoselectivity and >95% ee (Scheme 8C.12). The relative rate of the reactions for the two enantiomeric substrates, calculated by using the equation, $\ln[(1\text{-}c)(1\text{-}ee_{recov})] * \{\ln[(1\text{-}c)(1+ee_{recov})]\}^{-1}$ (wherein c is the fraction of consumption $c = (ee_{recov}) * (ee_{recov}+ee_{prod})^{-1}$, $c > 0$, $ee < 1$), were *ca.* 700 for R as *i*-Pr and 65 for R as Me. As expected, the double asymmetric induction [33,34] in the reaction of the (*R*)-ene component using the catalyst (*S*)-**1** ("matched" catalyst) leads to the complete (>99%) 2,5-*syn*-diastereoselectivity in high chemical yield, whereas the reaction of (*R*)-ene using (*R*)-**1** ("mismatched" catalyst) produces a diastereomeric mixture in quite low yield (Scheme 8C.13).

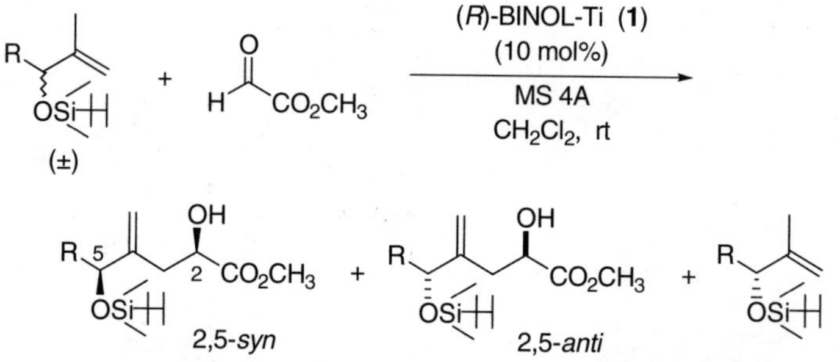

R	ene-product	recovered ene	relative rate (k_R/k_S)
i-Pr	99.6% ee (>99% *syn*)	37.8% ee	720
Me	96.2% ee (>99% *syn*)	22.0% ee	64

Scheme 8C.12. Kinetic optical resolution in asymmetric carbonyl-ene reaction catalyzed by BINOL-Ti complex.

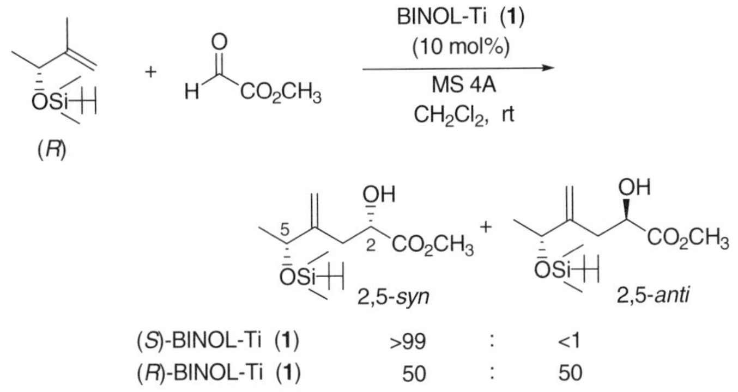

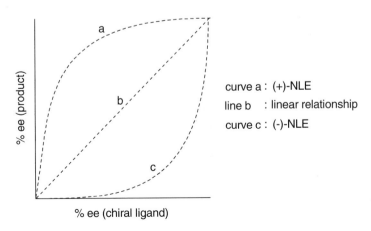

Scheme 8C.13. Double asymmetric induction in carbonyl-ene reaction catalyzed by BINOL-Ti com plex.

8C.5. POSITIVE NON-LINEAR EFFECT OF NON-RACEMIC CATALYSTS

A chiral catalyst is not necessarily in enantiopure form. Deviation from the linear relationship, namely "non-linear effect", is sometimes observed between the enantiomeric purity of the chiral catalysts and the optical yields of the products (Figure 8C.2). The convex deviation—which Kagan [35a] and Mikami [36] independently refer to as "positive non-linear effect" (abbreviated as (+)-NLE) and what Oguni refers to as [35c] "asymmetric amplification"—is currently attracting much attention for achieving a higher level of asymmetric induction that exceeds the enantiopurity of the non-racemic (partially resolved) catalysts. In turn, (–)-NLE stands for the opposite phenomenon of concave deviation, namely negative non-linear effect.

A remarkable level of (+)-NLE has been observed in the catalytic ene reaction. For instance, the glyoxylate-ene reaction with 2-phenylpropene using a catalyst prepared from BINOL with 33.0% ee provides the corresponding ene product with 91.4% ee in 92% chemical yield (Scheme

Figure 8C.2. Relationship between the enantiomeric purity of chiral ligands and the optical yield of products.

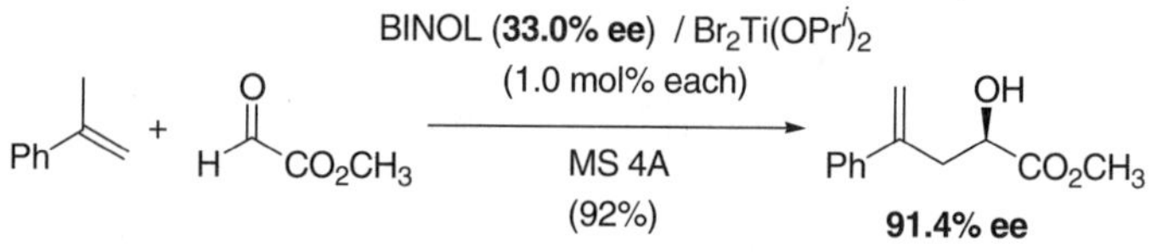

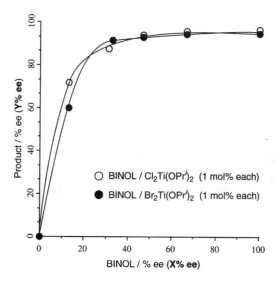

Scheme 8C.14. Positive non-linear effect in asymmetric carbonyl-ene reaction.

Figure 8C.3. Positive non-linear effect in aymmetric glyoxylate-ene reaction.

8C.14) [36]. The enantioselectivity thus attained is very close to the value (94.6% ee) obtained by using enantiopure BINOL (Figure 8C.3).

In an effort to develop new chiral BINOL-Ti complexes, chemical modifications of the chiral complex (R)-BINOL-Ti(OPr^i)$_2$ (R-2) that can easily be prepared by simply mixing (^{i}PrO)$_4$Ti and (R)-BINOL *in the absence of MS4A* have been studied [37c–e]. A dimeric form has been reported for the single-crystal X-ray structure of complex R-2 [38]. "(R)-BINOL-Ti-μ$_3$-oxo" complex, prepared via hydrolysis of complex R-2 has been shown to serve as an efficient and moisture-tolerant asymmetric catalyst [37d,e]. It is noteworthy that the (R)-BINOL-Ti-μ$_3$-oxo catalyst [37e] shows a remarkable level of (+)-NLE (asymmetric amplification), thereby attaining the maximum enantioselectivity for this system by using (R)-BINOL with only 55–60% ee as the chiral source (consult Scheme 8C.14).

8C.6. ASYMMETRIC ACTIVATION OF RACEMIC CATALYSTS

Whereas non-racemic catalysts can generate non-racemic products with or without NLE, racemic catalysts inherently produce only racemic products. A strategy whereby a racemic

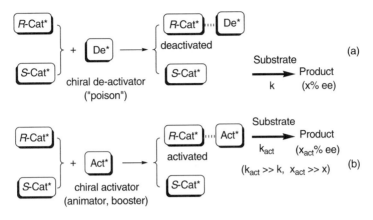

Figure 8C.4. Asymmetric activation *vs.* de-activation.

catalyst is enantiomer-selectively, deactivated by a chiral molecule, has shown to yield non-racemic products [39,40]. However, the level of asymmetric induction does not exceed the level attained by the enantiopure catalyst (Figure 8C.4a). The term "chiral poisoning" [41] has been used for such a deactivation strategy. In contrast to this, Mikami and Matsukawa have reported an alternative and conceptually opposite strategy for asymmetric catalysis by racemic catalysts, wherein a chiral activator selectively activates one enantiomer of a racemic catalyst. In principle, enantioselectivity higher than that achieved by an enantiopure catalyst (% $ee_{act} \gg \%$ $ee_{enantiopure}$) might be attained in addition to a higher-level catalytic efficiency ($k_{act} \gg k_{enantio-pure}$) by using this strategy (Figure 8C.4b).

Catalysis of racemic BINOL-Ti(OPr^i)$_2$ ($\pm$**2**) in the glyoxylate-ene reaction with 2-phenyl-propene achieves extremely high enantioselectivity by adding another diol for enantiomer-selective activation (Scheme 8C.15, Table 8C.3) [42]. For example, excellent enantioselectivity (90% ee, R) was achieved by adding just a half-molar equiv. (5 mol %) of (R)-BINOL activator to racemic ($\pm$)-BINOL-Ti(OPr^i)$_2$ complex ($\pm$**2**) (10 mol %) in this reaction (Table 8C.3, entry 4).

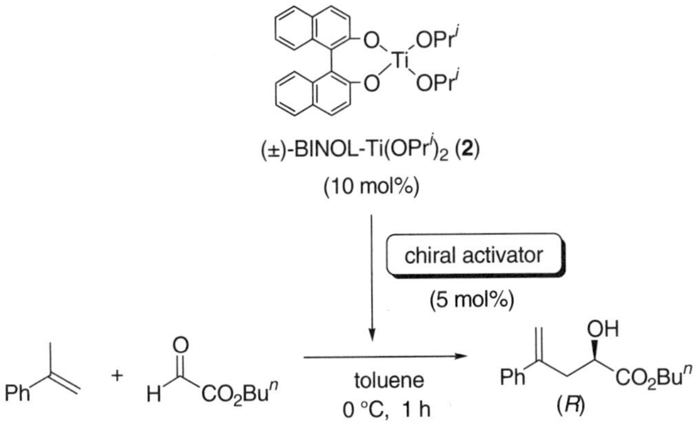

($\pm$)-BINOL-Ti(OPr^i)$_2$ (**2**)

(10 mol%)

chiral activator

(5 mol%)

Ph + H–CO$_2$Bu^n → toluene
0 °C, 1 h → Ph ... OH ... CO$_2$Bu^n
(R)

Scheme 8C.15. Enantiomer selective activation of racemic BINOL-Ti(OPr^i)$_2$.

TABLE 8C.3. Enantiomer selective activation of racemic BINOL-Ti(OPr^i)$_2$ (2)

Run	chiral activator	% yield	% ee
1	none	1.6	0
2	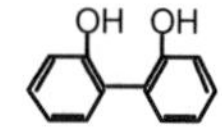	20	0
3	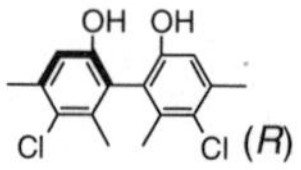	38	81
4	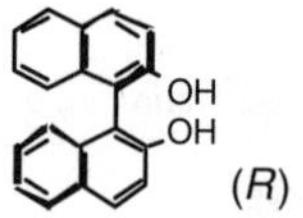	52	90
5^a		35	80

^{a}2.5 mol % of (R)-BINOL was used as a chiral activator.

For the same ene reaction, a possible activation of the enantiopure (R)-BINOL-Ti(OPr^i)$_2$ [43] catalyst (R-2) by adding (R)-BINOL was investigated (Scheme 8C.16, Table 8C.4) [42a]. In fact, the reaction in the presence of additional (R)-BINOL proceeded quite smoothly to provide the carbonyl-ene product in higher chemical yield (82%) and enantioselectivity (97% ee) than those without additional (R)-BINOL (20% yield, 95% ee). Based on the comparison of these results of enantiomer-selective activation of the racemic catalyst (90% ee, R) with those of the enantiopure catalyst with (97% ee, R) or without activator (95% ee, R), the reaction catalyzed by (R)-BINOL-Ti(OPr^i)$_2$/(R)-BINOL complex (2') is calculated to be 27 times as fast as that catalyzed by (S)-BINOL-Ti(OPr^i)$_2$ (S-2) in the reaction using racemic (±)-BINOL-Ti(OPr^i)$_2$ (±2) (Figure 8C.5a) [42a]. Indeed, kinetic studies indicated that the reaction catalyzed by the complex 2' is 26 times as fast as that catalyzed by the complex S-2 [42a]. These results

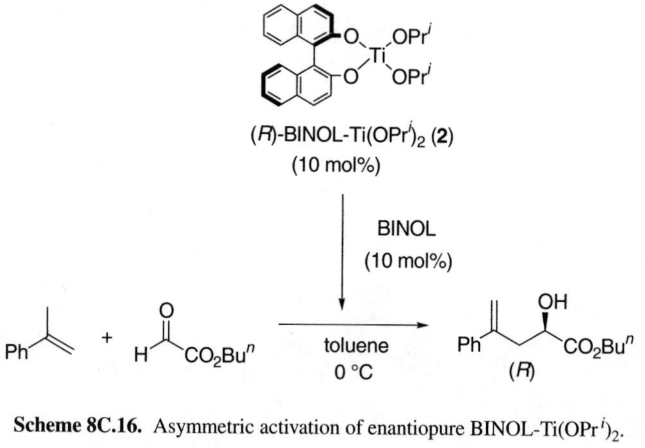

Scheme 8C.16. Asymmetric activation of enantiopure BINOL-Ti(OPr^i)$_2$.

TABLE 8C.4. Asymmetric activation of enantiopure (R)-BINOL-Ti(OPr^i)$_2$ (2)

Run	BINOL	time (min)	% yield	% ee
1	none	60	20	95
2		1	1.6	95
3	(R)-BINOL	60	82	97
4		1	41	97
5		0.5	24	97
6	(S)-BINOL	60	48	86
7		0.5	2.6	86
8	(±)-BINOL	60	69	96

imply that the racemic complex (±2) and a half-molar amount of (R)-BINOL assemble preferentially to form (R)-BINOL-Ti(OPr^i)$_2$/(R)-BINOL complex (2′) and unchanged S-2. In contrast, when the other enantiomer, (S)-BINOL, is added, (R)-BINOL-Ti(OPr^i)$_2$ (2) is activated to a smaller degree, thus providing a lower enantioselectivity (86% ee, R) and chemical yield (48%) than (R)-BINOL does.

The advantage of asymmetric activation of the racemic BINOL-Ti(OPr^i)$_2$ complex (±2) is highlighted in a catalytic version (Table 8C.3, entry 5) wherein high enantioselectivity (80.0% ee) is obtained by adding less than the stoichiometric amount (0.25 molar amount) of (R)-BI-NOL [42a]. A similar phenomenon has been observed in the aldol [42c] and (hetero) Diels–Alder [44] reactions catalyzed by the racemic BINOL-Ti(OPr^i)$_2$ catalyst (±2).

Another possibility has been explored by using racemic BINOL as an activator in the same ene reaction [42a]. Racemic BINOL was added to (R)-BINOL-Ti(OPr^i)$_2$ (R-2), giving higher yield and enantioselectivity (69% yield; 96% ee, R) than those (20% yield; 95% ee, R) obtained

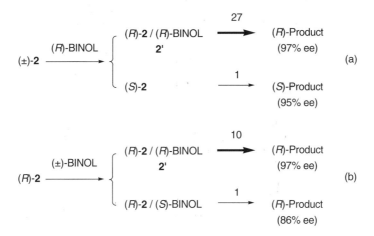

Figure 8C.5. Kinetic feature of asymmetric activation of BINOL-Ti (OPr^i)$_2$.

by the original catalyst (R)-BINOL-Ti(OPr^i)$_2$ (R-**2**) without additional BINOL. Based on the comparison of this result (96% ee, R) by using the racemic activator with those attained by using the enantiopure catalyst, (R)-BINOL-Ti(OPr^i)$_2$/(R)-BINOL (**2'**) (97% ee, R) or (R)-BINOL-Ti(OPr^i)$_2$/(S)-BINOL (86% ee, R), it is calculated that the reaction catalyzed by the (R)-BINOL-Ti(OPr^i)$_2$ catalyst/(R)-BINOL complex (**2'**) is 10 times as fast as that catalyzed by the (R)-BINOL-Ti(OPr^i)$_2$/(S)-BINOL (Figure 8C.5b) [42a]. Kinetic studies have also shown that the reaction catalyzed by the (R)-BINOL-Ti(OPr^i)$_2$/(R)-BINOL complex (**2'**) is 9.2 times as fast as that catalyzed by (R)-BINOL-Ti(OPr^i)$_2$/(S)-BINOL [42a].

8C.7. ENE CYCLIZATION

Conceptually, intramolecular ene reactions [45] (ene cyclizations) can be classified into six different modes (Figure 8C.6) [8a,46]. In the ene cyclizations, the carbon numbers where the tether connects the [1,5]-hydrogen shift system, are indicated as (m,n). A numerical prefix stands for the forming ring size, for example, 6-(2,4) stands for a (2,4) ene cyclization to give a six-membered ring.

Catalytic asymmetric ene reactions were initially explored in the intramolecular cases because these processes are more facile than their intermolecular counterparts. The first reported

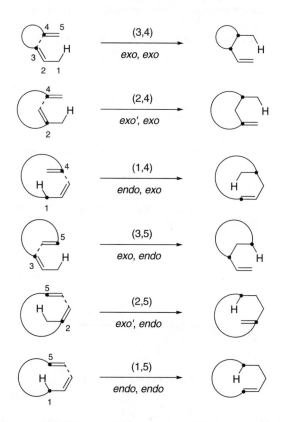

Figure 8C.6. Classification of ene cyclization.

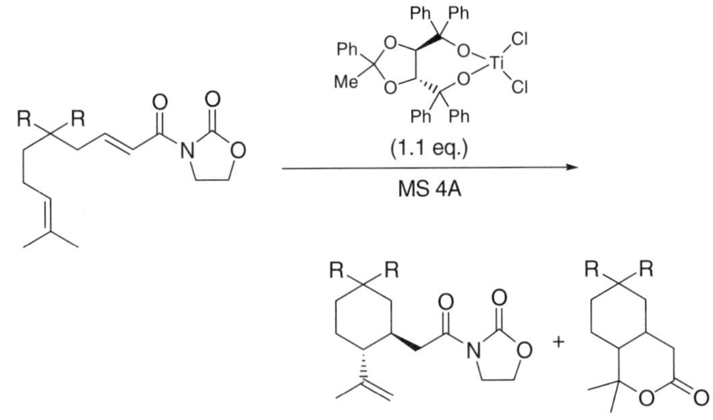

toluene	R = H	20 days	17%	37%
	R = Me	4 days	39% (82% ee)	36% (92% ee)
mesitylene	R = Me	4 days	32% (86% ee)	37% (>98% ee)
$CFCl_2CF_2Cl$	R = Me	4 days	63% (>98% ee)	25% (-)

Scheme 8C.17. Asymmetric ene cyclization catalyzed by TADDOL-Ti complex.

example of an enantioselective 6-(3,4) carbonyl-ene cyclization was with a BINOL-derived zinc reagent [47]. However, this reaction was successful only when an excess of the zinc reagent (at least 3 equiv.) was used. Recently, an enantioselective 6-(3,4) olefin-ene cyclization was developed with a stoichiometric amount of a TADDOL-derived chiral titanium complex (Scheme 8C.17) [48]. In this case, the HDA product was also obtained in which the product ratio depended critically on the solvent used. In both cases, the geminal-disubstitution is required and the stoichiometric use of the chiral complexes is needed to obtain high enantioselectivity.

Mikami et al. reported the first examples of catalytic asymmetric intramolecular carbonyl-ene reactions of types (3,4) and (2,4), using the BINOL-derived titanium complex (**1**) [46,49]. The catalytic 7-(2,4) carbonyl-ene cyclization gives the corresponding oxepane with high enantiopurity, and the *gem*-dimethyl groups are not required (Scheme 8C.18). In a similar catalytic 6-(3,4) ene cyclization, *trans*-tetrahydropyran is preferentially obtained with high enantiopurity (Scheme 8C.19). The sense of asymmetric induction is the same as that observed for the glyoxylate-ene reaction, that is, (*R*)-BINOL-Ti catalyst provides (*R*)-alcohol. Therefore, the

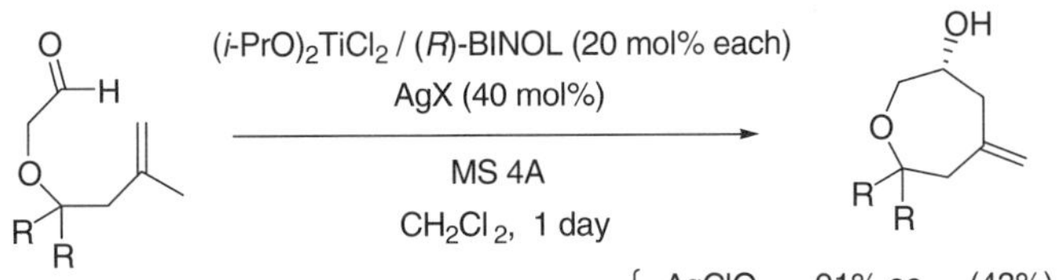

R = H
R = Me

R = H,	$AgClO_4$,	91% ee	(43%)
	AgOTf	92% ee	(40%)
R = Me	$AgClO_4$,	82% ee,	(40%)

Scheme 8C.18. Asymmetric 7-(2,4) carbonyl-ene cyclization catalyzed by BINOL-Ti complex.

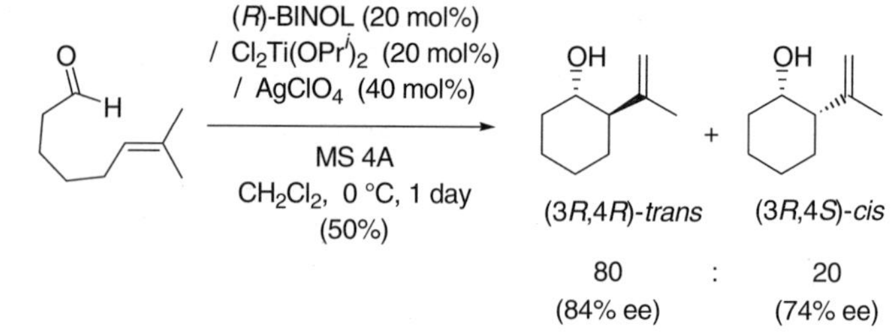

Scheme 8C.19. Asymmetric 6-(3,4) carbonyl-ene cyclization catalyzed by BINOL-Ti complex.

chiral BINOL-Ti catalyst works efficiently for both the chiral recognition of the enantioface of the aldehyde and the discrimination of the diastereotopic protons of the ene component.

Recently, much effort has been directed to the synthesis of analogs of the biologically active form of vitamin D_3, 1α,25-dihydroxyvitamin D_3 [1α,25$(OH)_2D_3$] as an important new field in medicinal chemistry [50]. As a new and efficient route to this class of compounds, the "symmetry"-assisted enantiospecific synthesis of the A-ring of the vitamin D-hybrid analogs, 19-nor-22-oxa-1α,25$(OH)_2D_3$ was reported (Scheme 8C.20) [51]. It should be noted that any *gem*-dialkyl substituents is not necessary to attain a high level of enantioselectivity.

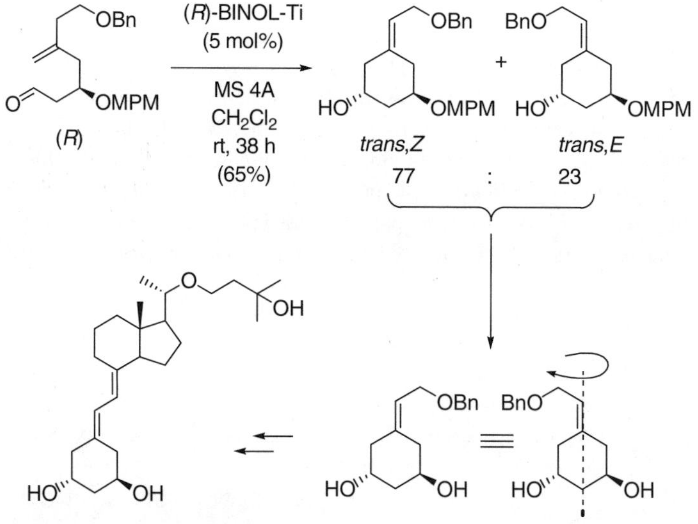

19-nor-22-oxa-1α,25$(OH)_2D_3$

Scheme 8C.20. "Symmetry"-assisted enantiospecific synthesis of A-ring of Vitamin D analogues based on asymmetric ene cyclization.

8C.8. ALLYLMETAL CARBONYL ADDITION REACTION

BINOL-Ti complexes (**1**) has been shown to serve as efficient asymmetric catalysts for the carbonyl addition reaction of allylic stannanes and silanes [52,53]. The addition reactions to glyoxylates of (E)-2-butenylsilane and –stannane proceed smoothly to afford the corresponding *syn*-product with high enantiomeric excess (Scheme 8C.21) [52].

It has also been shown that the Sakurai-Hosomi reaction of methallylsilanes with glyoxylates is catalyzed by the BINOL-Ti complex (**1**) to give the products, surprisingly, in an allylic silane (ene product) form with high enantioselectivity (Scheme 8C.22) [53].

Asymmetric catalysis of BINOL-Ti complexes in the reaction of aliphatic and aromatic aldehydes with an allylstannane has also been reported independently by Umani-Ronchi [54] and Keck [55]. The former group has suggested that a new complex generated by the reaction of the BINOL-Ti complex with allylstannane is the catalytic species that provides remarkably high enantioselectivity (Scheme 8C.23). It is interesting that no reaction occurs if dry MS 4A

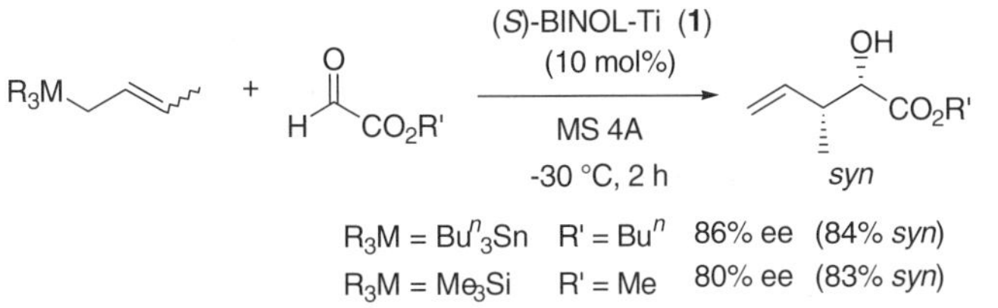

$R_3M = Bu^n_3Sn$ $R' = Bu^n$ 86% ee   (84% *syn*)
$R_3M = Me_3Si$ $R' = Me$ 80% ee   (83% *syn*)

Scheme 8C.21. Asymmetric carbonyl addition reaction catalyzed by BINOL-Ti.

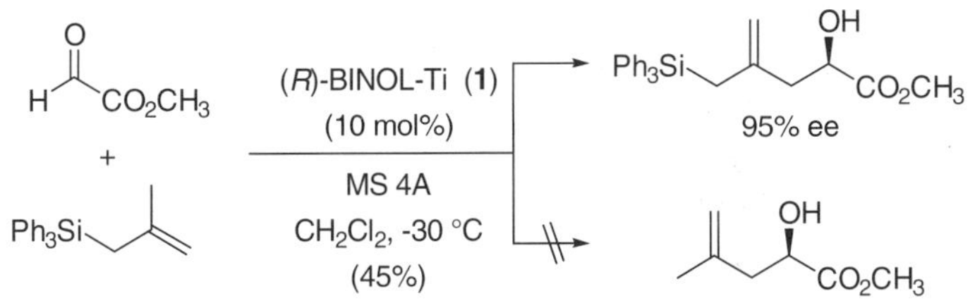

Scheme 8C.22. Asymmetric ene-type reaction of methallylsilanes.

$R = n\text{-}C_5H_{11}$ 98% ee   (75%)
PhCH=CH   94% ee   (38%)
Ph   82% ee   (96%)

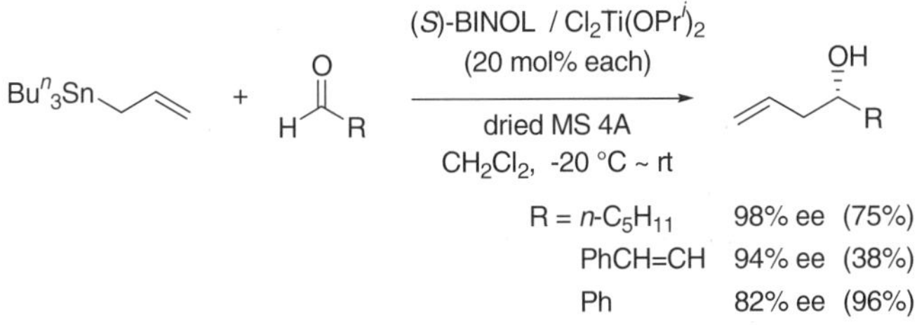

Scheme 8C.23. Asymmetric carbonyl addition reaction of aliphatic and aromatic aldehydes.

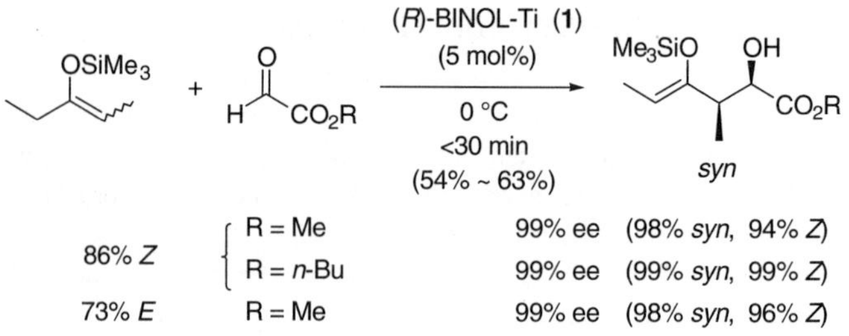

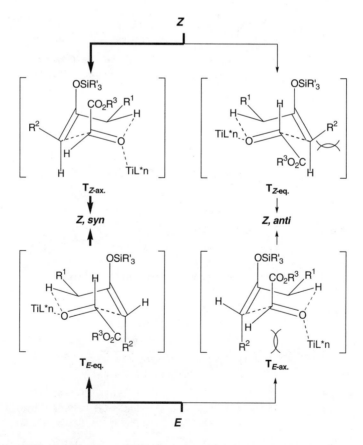

Scheme 8C.24. Asymmetric aldol (prototropic ene-type) reaction.

is not present at the preparation stage of the chiral catalyst, and MS 4A affects the subsequent allylation reaction. MS 4A dried for 12 h at 250°C and 0.1 Torr was recommended. Keck has reported that addition of CF_3CO_2H or CF_3SO_3H strongly accelerates the reactions catalyzed by $BINOL\text{-}Ti(OPr^i)_2$ complex (**2**) [55].

Figure 8C.7. Transition states of Mukaiyama-type aldol reaction of ketone silyl enol ethers.

8C.9. ALDOL-TYPE REACTION

The aldol reaction constitutes one of the most fundamental bond-construction processes in organic synthesis [56]. Therefore, much attention has been focused on the development of asymmetric catalysts for the Mukaiyama aldol reaction in recent years.

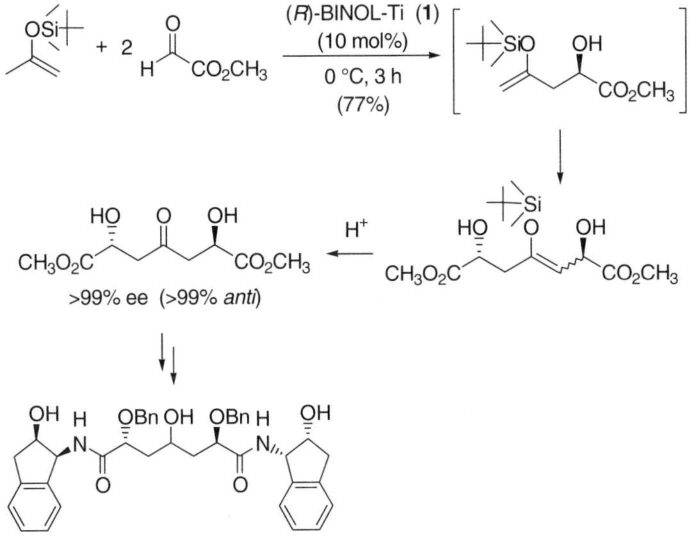

Scheme 8C.25. Tandem and two-directional aldol-type reaction.

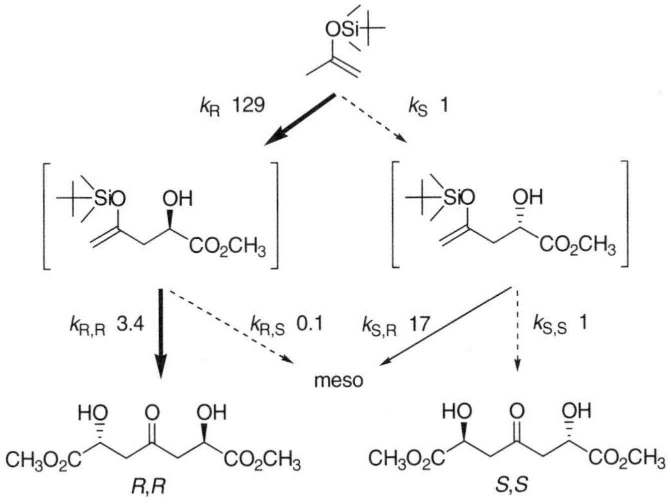

Figure 8C.8. Kinetic feature of two-directional aldol reaction.

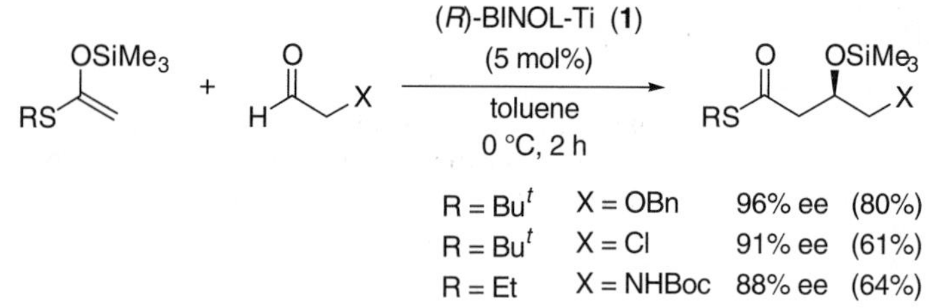

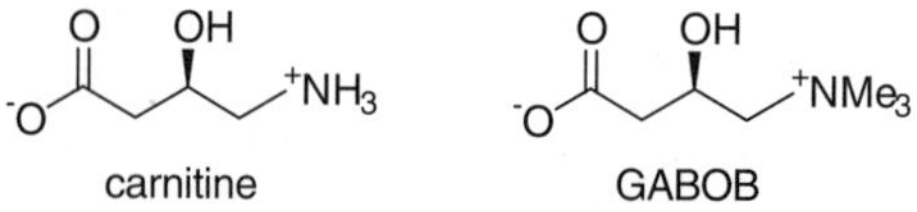

Scheme 8C.26. Asymmetric aldol (metallotropic ene-type) reaction.

BINOL-derived titanium complex was found to serve as an efficient catalyst for the Mukaiyama-type aldol reaction of ketone silyl enol ethers with good control of both absolute and relative stereochemistry (Scheme 8C.24) [57]. It is surprising, however, that the aldol products were obtained in the silyl enol ether (ene product) form, with high *syn*-diastereoselectivity from either geometrical isomer of the starting silyl enol ethers.

It appears likely that the reaction proceeds through the ene reaction pathway, although such an ene reaction pathway has not been previously recognized as a possible mechanism in the Mukaiyama aldol reaction. In general, an acyclic antiperiplanar transition-state model has been used to explain the formation of the *syn*-diastereomer from either (E)- or (Z)-silyl enol ethers [58]. However, the cyclic ene mechanism now provides another rationale for the *syn*-diastereoselection regardless of the enol silyl ether geometry (Figure 8C.7).

The asymmetric catalytic aldol reaction of a silyl enol ether can be performed in a double and two-directional fashion to give the 1:2 adduct in the silyl enol ether form with >99% ee and 99% de in 77% isolated yield (Scheme 8C.25) [59]. The present catalytic asymmetric aldol reaction is characterized by a kinetic amplification phenomenon of the product chirality, going from the one-directional aldol intermediate to the two-directional product (Figure 8C.8). Further transformation of the *pseudo* C_2 symmetric product, while still being protected as the silyl enol ether, leads to a potent analog of an HIV protease inhibitor.

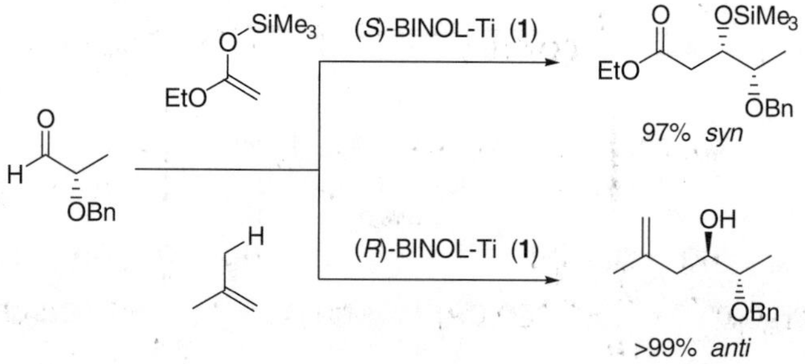

Scheme 8C.27. Asymmetric carbonyl-ene *vs.* aldol-type reaction of α-benzyloxy aldehyde.

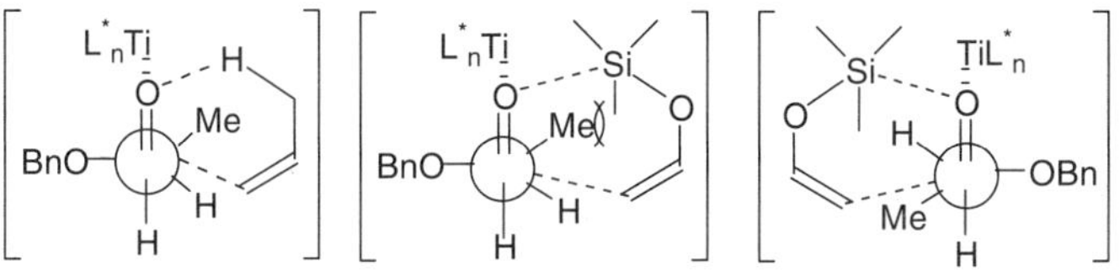

Figure 8C.9. Transition states of carbonyl-ene and aldol-type reaction of α-alkoxy aldehyde.

The silatropic ene pathway, that is, direct silyl transfer from an silyl enol ether to an aldehyde, may be involved as a possible mechanism in the Mukaiyama aldol-type reaction. Indeed, *ab initio* calculations show that the silatropic ene pathway involving the cyclic (boat and chair) transition states for the BH_3-promoted aldol reaction of the trihydrosilyl enol ether derived from acetaldehyde with formaldehyde is favored [60]. Recently, we have reported the possible intervention of a silatropic ene pathway in the catalytic asymmetric aldol-type reaction of silyl enol ethers of thioesters [61]. Chlorine- and amine-containing products thus obtained are useful intermediates for the synthesis of carnitine and GABOB (Scheme 8C.26) [62].

In similar reactions of α-alkoxy aldehydes, there is a dichotomy in terms of *syn-* versus *anti*-diastereofacial preference, which might be dictated by the bulkiness of the migrating group [60]. The sterically demanding silyl group shows *syn*-preference, but the less-demanding proton leads to *anti*-preference (Scheme 8C.27). The *anti*-diastereoselectivity in carbonyl-ene reactions can be explained by the Felkin-Anh-like cyclic transition-state model (T_1) (Figure 8C.9). In the aldol reaction, however, the *syn*-selectivity is explained best by the *anti*-Felkin-like cyclic transition-state model (T_2'). This is due to the fact that the inside-crowded transition state (T_1') is less favorable than T_2' because of the steric repulsion between the trimethylsilyl group and the inside methyl group of aldehyde (T_1').

Keck [63] and Carreira [64] have independently reported catalytic asymmetric Mukaiyama aldol reactions. Keck et al. also reported the aldol reaction of an α-benzyloxy aldehyde with a Danishefsky's diene. The aldol product was transformed to the corresponding HDA-type product through acid-catalyzed cyclization. In these reactions, the catalyst that is claimed to

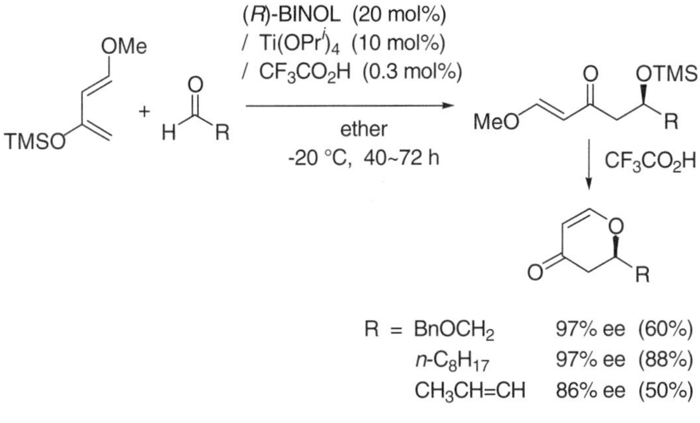

R =	BnOCH$_2$	97% ee	(60%)
	n-C$_8$H$_{17}$	97% ee	(88%)
	CH$_3$CH=CH	86% ee	(50%)

Scheme 8C.28. Asymmetric aldol-type reaction of Danishefky's diene.

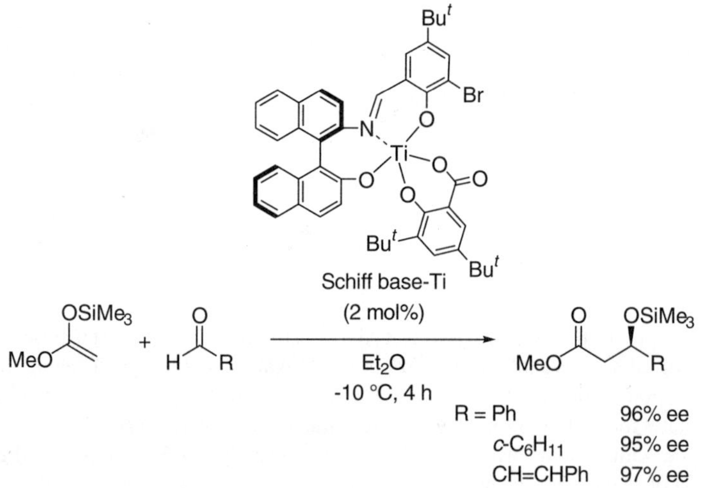

Scheme 8C.29. Asymmetric aldol-type reaction catalyzed by chiral Schiff base-Ti.

have BINOL-Ti(OPr^i)$_2$ structure was prepared with a 1:1 or 2:1 ratio of BINOL with Ti(OPr^i)$_4$ in the presence of oven-dried MS under reflux (Scheme 8C.28) [65].

Carreira et al. used a chiral BINOL-derived Schiff base-titanium complex as the catalyst for the aldol reactions of acetate-derived ketene silyl acetals (Scheme 8C.29) [64a]. The catalyst was prepared in toluene in the presence of salicylic acid, which was reported to be crucial to attain a high enantioselectivity. A similar Schiff base-titanium complex is also applicable to the carbonyl-ene type reaction with 2-methoxypropene [64b]. Although the reaction, when con-

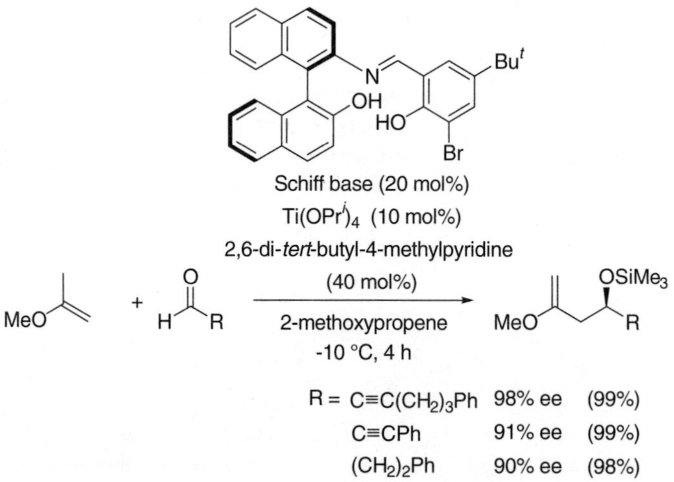

Scheme 8C.30. Asymmetric ene-type reaction of 2-methoxypropene.

ducted in toluene or ether, did not provide any addition product, excellent chemical yield and enantioselectivity were achieved by using 2-methoxypropene as the solvent (Scheme 8C.30).

8C.10. CONCLUSION

The class of ene reactions, carbonyl-ene reaction in particular, has gained a wide scope of synthetic applications and mechanistic basis for the stereo-controlled construction of carbon skeletons by using a catalytic amount of various chiral Lewis acids. In a practical application, the development of more efficient catalysts is important, for which molecular design of asymmetric catalysts is the key in view of the structure-catalytic activity relationship. Quite recently, Evans and Vederas reported that Cu(II) complexes with bidentate bis(oxazolinyl) (box), [Cu((S,S)-t-Bu-box)](SbF$_6$)$_2$ and [Cu((S,S)-Ph-box)]OTf$_2$ complex, respectively, serve as efficient catalysts for the glyoxylate-ene reaction even with a less nucleophilic mono-substituted olefin to give good chemical yield and high enantioselectivity [66]. Any progress along this line in the future is highly desirable and worth the effort.

REFERENCES

1. Reviews: Shilov, A. E.; Shul'pin, G. B. *Chem. Rev.* **1997**, *97*, 2879; Hall, C.; Perutz, R. N. *Chem. Rev.* **1997**, *97*, 3125; Schneider, J. J. *Angew. Chem., Int. Ed.* **1996**, *35*, 1068; Arndtsen, B. A.; Bergman, R. G.; Mobley, T. A.; Peterson, T. H. *Acc. Chem. Res.* **1995**, *28*, 154; Crabtree, R. H. *Angew. Chem., Int. Ed.* **1993**, *32*, 789; Davies, J. A.; Watson, P. L.; Liebman, J. F.; Greenberg, A. *Selective Hydrocarbon Activation*; Wiley-VCH: New York, 1990.

2. Comprehensive reviews on ene reactions: (a) Mikami, K.; Shimizu, M. *Chem. Rev.* **1992**, *92*, 1021; (b) Snider, B. B. In *Comprehensive Organic Synthesis*; Trost, B. M.; Fleming, I. (Ed.); Pergamon: London, 1991; Vol. 2, p. 527 and Vol. 5, pp. 1; (c) Mikami, K.; Terada, M.; Shimizu, M.; Nakai, T. *J. Synth. Org. Chem. Jpn.* **1990**, *48*, 292; (d) Hoffmann, H. M. R. *Angew. Chem., Int. Ed.* **1969**, *8*, 556.

3. Reviews: (a) Mikami, K. In *Advances in Asymmetric Synthesis*; Hassner, A. (Ed.); JAI Press: Greenwich, Connecticut, 1995; Vol. 1, pp. 1; (b) Whitesell, J. K. *Acc. Chem. Res.* **1985**, *18*, 280; (b) Snider, B. B. *Acc. Chem. Res.* **1980**, *13*, 426.

4. Mikami, K.; Loh, T.-P.; Nakai, T. *Tetrahedron Lett.* **1988**, *29*, 6305; Mikami, K.; Loh, T.-P.; Nakai, T. *J. Chem. Soc.; Chem. Commun.* **1988**, 1430; Mikami, K.; Loh, T.-P.; Nakai, T. *J. Am. Chem. Soc.* **1990**, *112*, 6737.

5. Reviews: (a) Hoffmann, R. W. *Angew. Chem., Int. Ed.* **1983**, *22*, 489; (b) Weidmann, B.; Seebach, D. *Angew. Chem., Int. Ed.* **1983**, *22*, 31; (c) Yamamoto, Y. *Acc. Chem. Res.* **1987**, *20*, 243; (d) Hoffmann, R. W. *Angew. Chem., Int. Ed.* **1987**, *26*, 489; (e) Roush, W. R. In *Comprehensive Organic Synthesis*, Trost, B. M.; Fleming, I. (Ed.); Pergamon: London, 1991; Vol. 2, pp. 1; (f) Marshall, J. A. *Chemtracts* **1992**, *5*, 75; (g) Yamamoto, Y.; Asao, N. *Chem. Rev.* **1993**, *93*, 2207.

6. Maruoka, K.; Hoshino, Y.; Shirasaka, T.; Yamamoto, H. Annual Meeting of the Chemical Society of Japan, Tokyo, 1988, April 1–4, Abstract No. 1XIIB27; *Tetrahedron Lett.* **1988**, *29*, 3967.

7. Hanessian, S. *Total Synthesis of Natural Products: The "Chiron" Approach*; Pergamon: Oxford, 1983; Mori, K. *The Total Synthesis of Natural Products*; Wiley: New York, 1981; Vol. 4; Seebach, D.; Hungerbuhler, E. In *Modern Synthetic Methods*; Scheffold, R. (Ed.); Otto Salle Verlag: Frankfurt am Main, 1980; Vol. 2, pp. 91.

8. Mikami, K. *Pure. Appl. Chem.* **1996**, *68*, 639; Mikami, K.; Terada, M.; Nakai, T. In *Advances in Catalytic Processes*; Doyle, M. P. (Ed.); JAI Press: London, 1995, Vol. 1, pp. 123; Mikami, K.; Terada, M.; Narisawa, S.; Nakai, T. *Synlett* **1992**, 255.

9. (a) Mikami, K.; Terada, M.; Nakai, T. Annual Meeting of the Chemical Society of Japan, Tokyo, 1988, April 1–4, Abstract No. 1XIB43; *J. Am. Chem. Soc.* **1989**, *111*, 1940; (b) Mikami, K.; Terada, M.; Nakai, T. *J. Am. Chem. Soc.* **1990**, *112*, 3949.

10. Mikami, K.; Terada, M.; Nakai, T. *Chem. Express* **1989**, *4*, 589.

11. Mikami, K.; Terada, M.; Narisawa, S.; Nakai, T. *Org. Synth.* **1992**, *71*, 14.

12. Dijkgraff, C.; Rousseau, J. P. G. *Spectrochim. Acta* **1968**, *2*, 1213.

13. 6-Br-BINOL-Ti catalyst: (a) Mikami, K.; Motoyama, Y.; Terada, M. *Inorg. Chim. Acta* **1994**, *222*, 71; (b) Terada, M.; Motoyama, Y.; Mikami, K. *Tetrahedron Lett.* **1994**, *35*, 6693; (c) Terada, M.; Mikami, K. *J. Chem. Soc., Chem. Commun.* **1995**, 2391; (d) Motoyama, Y.; Terada, M.; Mikami, K. *Synlett* **1995**; 967.

14. Mikami, K. In *Encyclopedia of Reagents for Organic Synthesis*; Paquette, L. A. (Ed.); Wiley: Chichester, 1995; Vol. 1, pp. 403.

15. Terada, M.; Matsukawa, S.; Mikami, K. *J. Chem. Soc., Chem. Commun.* **1993**, 327.

16. Syntheses of enantio-enriched ipsdienol: (a) Mori, K.; Takigawa, H. *Tetrahedron* **1991**, *47*, 2163 (>96% ee); (b) Brown, H. C.; Randad, R. S. *Tetrahedron* **1990**, *46*, 4463 (96% ee); (c) Ohloff, G.; Giersch, W. *Helv. Chim. Acta* **1977**, *60*, 1496 (91% ee).

17. Mikami, K.; Kotera, O.; Motoyama, Y.; Tanaka, M.; Maruta, M.; Sakaguchi, H. Annual Meeting of the Chemical Society of Japan, Tokyo, 1997, March 27–30, Abstract No. 1F108.

18. Qian, C.; Huang, T. *Tetrahedron Lett.* **1997**, *38*, 6721.

19. Kobayashi, S.; Hachiya, I.; Ishitani, H.; Araki, M. *Synlett* **1993**, 472.

20. (a) Faller, J. W.; Liu, X. *Tetrahedron Lett.* **1996**, *37*, 3449; (b) Akhmedov, I. M.; Tanyeli, C.; Akhmedov, M. A.; Mohammadi, M.; Demir, A. S. *Synth. Commun.* **1994**, *24*, 137.

21. Mikami, K.; Yajima, T.; Terada, M.; Uchimaru, T. *Tetrahedron Lett.* **1993**, *34*, 7591; Mikami, K.; Yajima, T.; Terada, M.; Kato, E.; Maruta, M. *Tetrahedron Asymm.* **1994**, *5*, 1087; Mikami, K.; Yajima, T.; Takasaki, T.; Matsukawa, S.; Terada, M.; Uchimaru, T.; Maruta, M. *Tetrahedron* **1996**, *52*, 85.

22. Review: Welch, J. T.; Eswarakrishnan, S. *Flurine in Bioorganic Chemistry*; Wiley: New York, 1990.

23. (a) Mikami, K.; Siree, N.; Yajima, T.; Terada, M.; Suzuki, Y. Annual Meeting of the Chemical Society of Japan, Tokyo, 1995, March 28–31, Abstract No. 3H218; (b) Mikami, K.; Yajima, T.; Siree, N.; Terada, M.; Suzuki, Y.; Kobayashi, I. *Synlett* **1996**, 837; (c) Mikami, K.; Yajima, T.; Terada, M.; Kawauchi, S.; Suzuki, Y.; Kobayashi, I. *Chem. Lett.* **1996**, 861; (d) Mikami, K.; Yajima, T.; Terada, M.; Suzuki, Y.; Kobayashi, I. *Chem. Commun.* **1997**, 57.

24. (a) Mikami, K.; Yoshida, A. *Synlett* **1995**, 29; (b) Mikami, K.; Yoshida, A.; Matsumoto, Y. *Tetrahedron Lett.* **1996**, *37*, 8515.

25. Graven, A.; Johannsen, M.; Jørgensen, K. A. *Chem. Commun.* **1996**, 2373.

26. Johannsen, M.; Jørgensen, K. A. *Tetrahedron* **1996**, *52*, 7321; *J. Org. Chem.* **1995**, *60*, 5757.

27. Ward, R. S. *Chem. Soc. Rev.* **1990**, *19*, 1; Baba, S. E.; Sartor, K.; Poulin, J.-C.; Kagan, H. B. *Bull. Soc. Chim. Fr.* **1994**, *131*, 525; Rautenstrauch, V. *Bull. Soc. Chim. Fr.* **1994**, *131*, 515; (a) Ward, D. E.; How, D.; Liu, Y. *J. Am. Chem. Soc.* **1997**, *119*, 1884; (b) Schreiber, S. L.; Schreiber, T. S.; Smith, D. B. *J. Am. Chem. Soc.* **1987**, *109*, 1525.

28. Review: Mikami, K.; Shimizu, M. *J. Synth. Org. Chem. Jpn.* **1993**, *51*, 21.

29. Bartlett, P. A. *Tetrahedron* **1980**, *36*, 2.

30. Mikami, K.; Narisawa, S.; Shimizu, M.; Terada, M. *J. Am. Chem. Soc.* **1992**, *114*, 6566; **1992**, *114*, 9242.

31. Ziegler, F. E.; Sobolov, S. B. *J. Am. Chem. Soc.* **1990**, *112*, 2749.

32. (a) Kagan, H. B.; Fiaud, J. C. *Topics in Stereochemistry*; Interscience: New York, 1988; Vol. 18. (b) Brown, J. M. *Chem. Ind. (London)* **1988**, 612.

33. (a) Masamune, S.; Choy, W.; Peterson, J.; Sita, L. R. *Angew. Chem., Int. Ed.* **1985**, *24*, 1; (b) Heathcock, C. H. In *Asymmetric Synthesis*; Morrison, J. D. (Ed.); Academic Press: New York, 1985; Vol. 3, pp. 111.

34. Double asymmetric induction with terpenes as chiral ene components: Terada, M.; Sayo, N.; Mikami, K. *Synlett* **1995**, 411.

35. (a) Guillaneux, D.; Zhao, S.-H.; Samuel, O.; Rainford, D.; Kagan, H. B. *J. Am. Chem. Soc.* **1995**, *116*, 9430; Puchot, C.; Samuel, O.; Dunach, E.; Zhao, S.; Agami, C.; Kagan, H. B. *J. Am. Chem. Soc.* **1986**, *108*, 2353; (b) Noyori, R.; Kitamura, M. *Angew. Chem., Int. Ed.* **1991**, *30*, 49; Kitamura, M.; Suga, S.; Niwa, M.; Noyori, R. *J. Am. Chem. Soc.* **1995**, *117*, 4832; (c) Oguni, N.; Matsuda, Y.; Kaneko, T. *J. Am. Chem. Soc.* **1988**, *110*, 7877.

36. Mikami, K.; Terada, M. *Tetrahedron* **1992**, *48*, 5671; Mikami, K.; Motoyama, Y.; Terada, M. *J. Am. Chem. Soc.* **1994**, *116*, 2812; Terada, M.; Mikami, K.; Nakai, T. *J. Chem. Soc., Chem. Commun.* **1990**, 1623.

37. (a) Mukaiyama, T.; Inubushi, A.; Suda, S.; Hara, R.; Kobayashi, S. *Chem. Lett.* **1990**, 1015; (b) Terada, M.; Mikami, K. *J. Chem. Soc., Chem. Commun.* **1994**, 883; (c) Matsukawa, S.; Mikami, K. *Tetrahedron Asymm.* **1995**, *6*, 2571; (d) Kitamoto, D.; Imma, H.; Nakai, T. *Tetrahedron. Lett.* **1995**, *36*, 1861; (e) Terada, M.; Matsumoto, Y.; Mikami, K. *J. Chem. Soc., Chem. Commun.* **1997**, 281.

38. Martin, C. A. Ph.D. Thesis, MIT, 1988.

39. Alcock, N. W.; Brown, J. M.; Maddox, P. J. *J. Chem. Soc., Chem. Commun.* **1986**, 1532.

40. Maruoka, K.; Yamamoto, H. *J. Am. Chem. Soc.* **1988**, *111*, 789.

41. Faller, J. W.; Parr, J. *J. Am. Chem. Soc.* **1993**, *115*, 804.

42. (a) Mikami, K.; Matsukawa, S. *Nature* **1997**, *385*, 613; (b) Matsukawa, S.; Mikami, K. *Enantiomer* **1996**, *1*, 69.

43. Mikami, K. In *Encyclopedia of Reagents for Organic Synthesis*; Paquette, L. A. (Ed.); Wiley: Chichester; Vol. 1, pp. 407.

44. Matsukawa, S.; Mikami, K. *Tetrahedron Asymm.* **1997**, *8*, 815.

45. (a) Taber, D. F. *Intramolecular Diels–Alder and Alder Ene Reactions*; Springer Verlag: Berlin, 1984; (b) Fujita, Y.; Suzuki, S.; Kanehira, K. *J. Synth. Org. Chem. Jpn.* **1983**, *41*, 1152; (c) Oppolzer, W.; Snieckus, V. *Angew. Chem., Int. Ed.* **1978**, *17*, 476; (d) Conia, J. M.; Le Perchec, P. *Synthesis* **1975**, 1.

46. Mikami, K.; Sawa, E.; Terada, M. *Tetrahedron Asymm.* **1991**, *2*, 1403.

47. Sakane, S.; Maruoka, K.; Yamamoto, H. *Tetrahedron* **1986**, *42*, 2203; *Tetrahedron Lett.* **1985**, *26*, 5535.

48. Narasaka, K.; Hayashi, Y.; Shimada, S. *Chem. Lett.* **1988**, 1609; Narasaka, K.; Hayashi, Y.; Shimada, S.; Yamada, J. *Isr. J. Chem.* **1991**, *31*, 261.

49. Mikami, K.; Terada; M.; Sawa, E.; Nakai, T. *Tetrahedron Lett.* **1991**, *32*, 6571.

50. Reviews on the synthesis and structure-function relationships of vitamin D analogues: Bouillon, R.; Okamura, W. H.; Norman, A. W. *Endocrine Reviews* **1995**, *16*, 200; Dai, H.; Posner, G. H. *Synthesis* **1994**, 1383.

51. Mikami, K.; Osawa, A.; Isaka, A.; Sawa, E.; Shimizu, M.; Terada, M.; Kubodera, N.; Nakagawa, K.; Tsugawa, N.; Okano, T. *Tetrahedron Lett.* **1998**, *39*, 3359. Okano, T.; Nakagawa, K.; Ozono, K.; Kubodera, N.; Isaka, A.; Osawa, A.; Terada, M.; Mikami, K. *Chem. Biol.* **2000**, *7*, 173.

52. Aoki, S.; Mikami, K.; Terada, M.; Nakai, T. *Tetrahedron* **1993**, *49*, 1783.

53. Mikami, K.; Matsukawa, S. *Tetrahedron Lett.* **1994**, *35*, 3133.

54. Costa, A. L.; Piazza, M. G.; Tagliavini, E.; Trombini, C.; Umani-Ronchi, A. *J. Am. Chem. Soc.* **1993**, *115*, 7001.

55. Keck, G. E.; Tarbet, K. H.; Geraci, L. S. *J. Am. Chem. Soc.* **1993**, *115*, 8467; Keck, G. E.; Krishnamurthy, D.; Grier, M. C. *J. Org. Chem.* **1993**, *58*, 6543. Also see: Weigand, S.; Brückner, R. *Chem. Eur.* **1996**, *2*, 1077.

56. (a) Evans, D. A.; Nelson, J. V.; Taber, T. R. *Topics in Stereochemistry*; Interscience: New York, 1982; Vol. 13; (b) Mukaiyama, T. *Org. React.* **1982**, *28*, 203.

57. Mikami, K.; Matsukawa, S. *J. Am. Chem. Soc.* **1993**, *115*, 7039.

58. (a) Murata, S.; Suzuki, M.; Noyori, R. *J. Am. Chem. Soc.* **1980**, *102*, 3248; (b) Yamamoto, Y.; Maruyama, K. *Tetrahedron Lett.* **1980**, *21*, 4607.

59. Mikami, K.; Matsukawa, S.; Nagashima, M.; Funabashi, H.; Morishima, H. *Tetrahedron Lett.* **1997**, *38*, 579.

60. Mikami, K.; Matsukawa, S.; Sawa, E.; Harada, A.; Koga, N. *Tetrahedron Lett.* **1997**, *38*, 1951.

61. Mikami, K.; Matsukawa, S. *J. Am. Chem. Soc.* **1994**, *116*, 4077.

62. (a) Kolb, H. C.; Bennari, Y. L.; Sharpless, K. B. *Tetrahedron Asymm.* **1993**, *4*, 133; (b) Larcheveque, M.; Henrot, S. *Tetrahedron* **1990**, *46*, 4277.

63. Keck, G. E.; Krishnamurthy, D. *J. Am. Chem. Soc.* **1995**, *117*, 2363.

64. (a) Carreira, E. M.; Singer, R. A.; Lee, W. *J. Am. Chem. Soc.* **1994**, *116*, 8837; (b) Carreira, E. M.; Lee, W.; Singer, R. A. *J. Am. Chem. Soc.* **1995**, *117*, 3649.

65. Keck, G. E.; Li, X.-Y.; Krishnamurthy, D. *J. Org. Chem.* **1995**, *60*, 5998.

66. Evans, D. A.; Burgey, C. S.; Paras, N. A.; Vojkovsky, T.; Tregay, S. W. *J. Am. Chem. Soc.* **1998**, *120*, 5824; Gao, Y.; Lane-Bell, P.; Vederas, J. C. *J. Org. Chem.* **1998**, *63*, 2133.

8D

ASYMMETRIC MICHAEL REACTIONS

MOTOMU KANAI AND MASAKATSU SHIBASAKI
Graduate School of Pharmaceutical Sciences, The University of Tokyo, Tokyo, Japan

8D.1. INTRODUCTION

Michael addition to α,β-unsaturated systems is a fundamental process in organic chemistry, because it forms a carbon–carbon, carbon–nitrogen, carbon–sulfur, or carbon–oxygen bond at the β-position of the carbonyl, and produces highly functionalized compounds with one or two chiral centers. The initial adduct is an enolate, and this important reactive species can perform a variety of nucleophilic reactions such as aldol reaction, Michael reaction, and alkylation reaction, which again can generate new chiral centers. In this chapter, recent advances in catalytic asymmetric Michael reactions are divided into four categories in terms of nucleophiles: (1) organometallics, (2) malonates, (3) radicals, and (4) heteroatoms.

8D.2. CATALYTIC ASYMMETRIC MICHAEL ADDITION OF ORGANOMETALLICS

Since the pioneering work of Kharasch [1], it has been well-established that the regioselectivity of the addition of organometallic reagents, such as organolithium and Grignard reagents, to α,β-unsaturated carbonyl compounds can be controlled to afford 1,4-adducts by the use of a catalytic amount of copper salts [2]. Although the detailed mechanism is not fully understood, the currently accepted concept postulates the preliminary and reversible formation of the d,π*-complex, in which the lithium (or magnesium) atom interacts with the carbonyl group, and the copper interacts with the π-olefin moiety [3]. Therefore, it seems logical to design a chiral catalyst that recognizes both metals. In 1995, Kanai and Tomioka reported a catalytic asymmetric conjugate addition of Grignard reagents to cyclic enones (Scheme 8D.1) [4,5]. The selective coordination of the urea oxygen and the phosphorus atoms of **1** to magnesium and copper, respectively, was the key concept for designing this catalyst. However, a large catalyst loading

Catalytic Asymmetric Synthesis, Second Edition, Edited by Iwao Ojima
ISBN 0-471-29805-0   Copyright © 2000 Wiley-VCH, Inc.

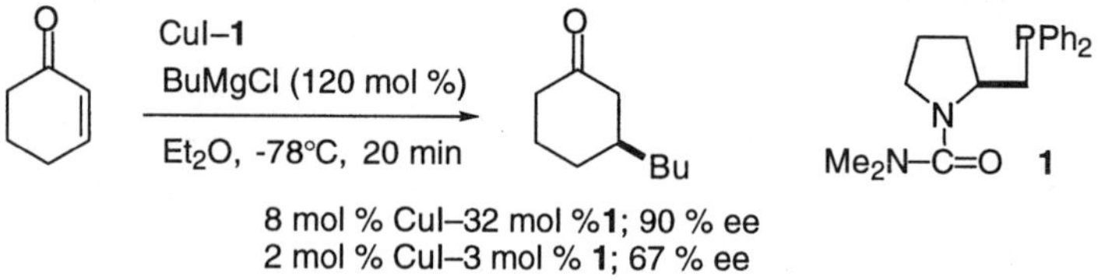

Scheme 8D.1.

(32 mol %) was required to obtain products with excellent enantiomeric excess (% ee), probably due to the dissociation of **1** from the copper atom in the presence of highly nucleophilic Grignard reagents.

Organozinc compounds are another group of versatile organometallic reagents, which can be readily prepared from the corresponding alkenes through hydroboration followed by metal exchange, or by using the corresponding Grignard reagents [6]. Organozinc reagents are not as reactive as organolithium or Grignard reagents and do not react with α,β-unsaturated ketones by themselves. However, diorganozinc compounds smoothly react with enones in the presence of catalytic amounts of copper or nickel salts as well as additives such as HMPA and TMSCl [7].

Soai [8], Bolm [9], Feringa [10], and Alexakis [11] have reported nickel- or copper-catalyzed asymmetric conjugate addition of diethylzinc, with chiral amino alcohols or a chiral trivalent phosphorus compound (Scheme 8D.2).

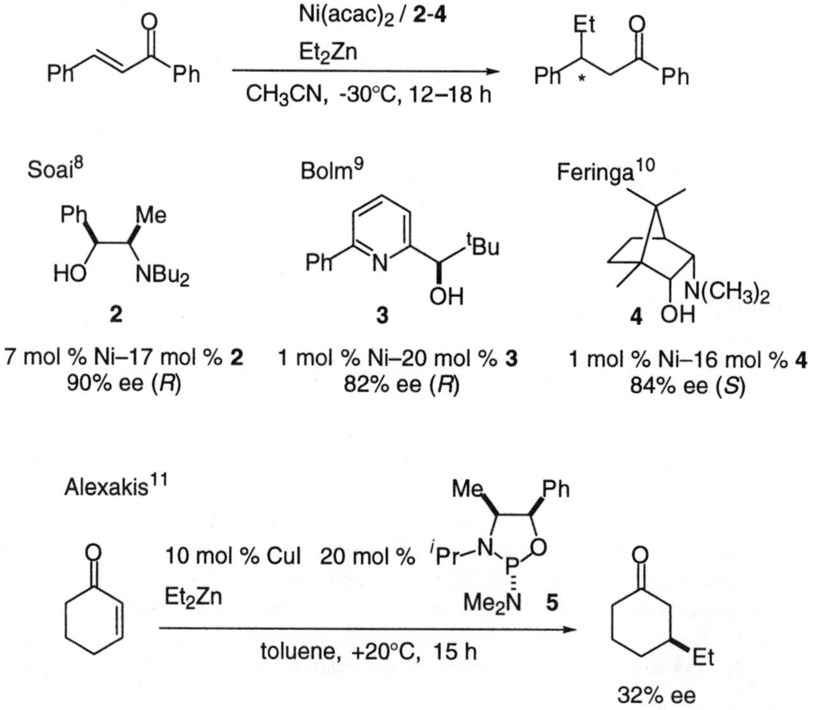

Scheme 8D.2.

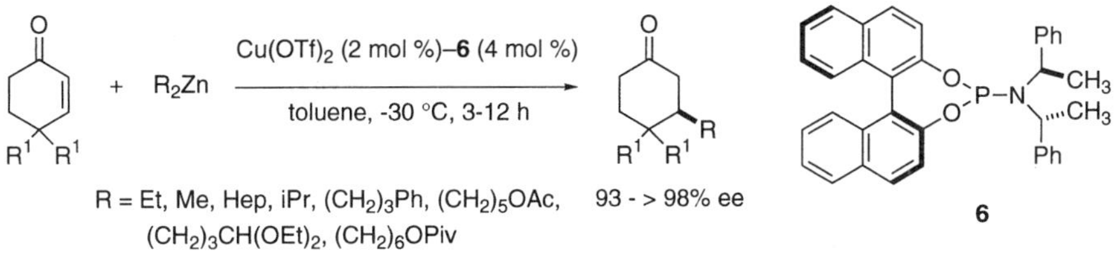

R = Et, Me, Hep, iPr, (CH$_2$)$_3$Ph, (CH$_2$)$_5$OAc,   93 - > 98% ee
(CH$_2$)$_3$CH(OEt)$_2$, (CH$_2$)$_6$OPiv

Scheme 8D.3.

In 1997, Feringa et al. reported the most efficient conjugate addition, to date, of dialkylzinc reagents catalyzed by Cu-chiral phosphorus amidite **6** complexes [12]. The reaction is mediated by 2 mol % of Cu(OTf)$_2$ and 4 mol % of chiral ligand **6**, at −30°C for 3 h, and the enantiopurity of product is generally greater than 93% ee for cyclohexenones (Scheme 8D.3). Various alkyl groups, specifically those with ester or acetal functionalities, can be introduced in excellent enantioselectivities. The reaction rate is greatly accelerated by the presence of ligand **6**, whereas without the ligand the reaction is very slow and many side products are formed. The combination of (R,R)-Bis(1-phenylethyl)amine and (S)-2,2′-binaphthol for **6** was found to be a matched pair, achieving 93~98% ee, whereas the mismatched pair (S,S,S)-**6** afforded the product (R = Et) with only 75% ee.

The regio- and enantioselective catalytic three-component coupling was also achieved by trapping the zinc enolate intermediate with various aldehydes (Scheme 8D.4). For example, when the zinc enolate, formed by the conjugate addition of dialkylzinc to cyclohexenone in the presence of Cu(OTf)$_2$ (1.2 mol %) and ligand **6** (2.4 mol %), was treated with various aldehydes, a mixture of *trans-erythro-* and *trans-threo-*coupling products was obtained, which was converted to the corresponding 2-acylcyclohexanones with greater than 91% ee.

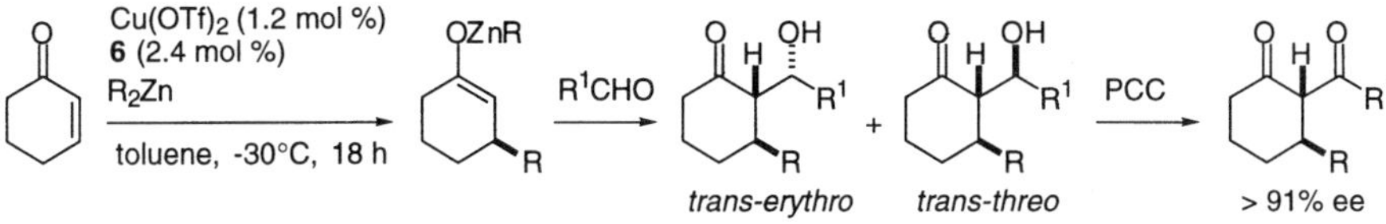

Scheme 8D.4.

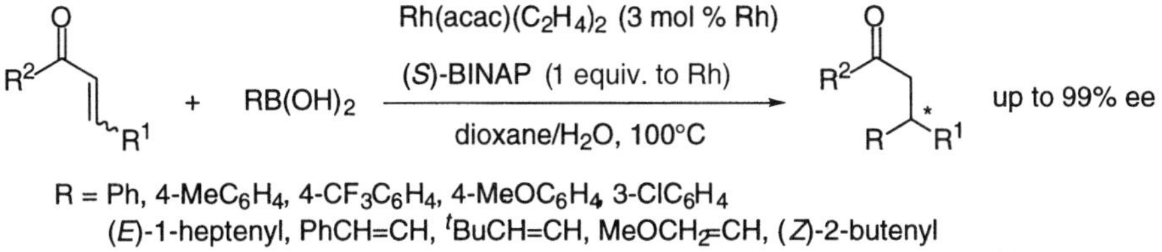

R = Ph, 4-MeC$_6$H$_4$, 4-CF$_3$C$_6$H$_4$, 4-MeOC$_6$H$_4$ 3-ClC$_6$H$_4$
(E)-1-heptenyl, PhCH=CH, ^{t}BuCH=CH, MeOCH$_2$CH, (Z)-2-butenyl

Scheme 8D.5.

Hayashi et al. and Miyaura et al. have reported that far less nucleophilic aryl- and alkenyl-boronic acids can react with a variety of enones in the presence of a BINAP-rhodium catalyst to give adducts with high enantiopurity in general (Scheme 8D.5) [13]. The one pot procedure, involving the hydroboration of alkynes as the first step (R = alkenyl), was achieved in the presence of amines without affecting the enantioselectivity [13].

8D.3. CATALYTIC ASYMMETRIC MICHAEL ADDITION OF STABILIZED CARBON NUCLEOPHILES

In this section, Michael reactions of carbon nucleophiles stabilized by electron-withdrawing groups such as carbonyl groups or nitro groups will be discussed. This category can be classified into two groups: (1) Michael addition of silyl ketene acetals and (2) direct Michael addition of malonates or nitroalkanes. Kobayashi and Mukaiyama developed a chiral titanium catalyst **7** that gave Michael adducts with up to 90% ee for the selected examples shown below (Scheme 8D.6) [14]. However, this section will focus on the group (2) reactions as this approach has been more successful to date [15].

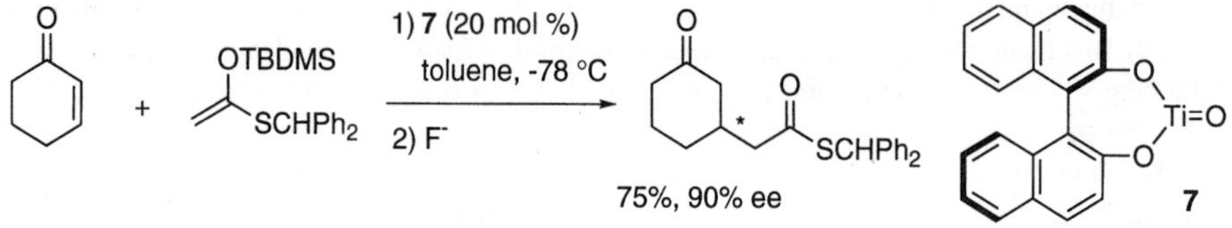

Scheme 8D.6.

For the group (2) reactions, it is essential that the Brønsted basicity of catalyst is high enough to allow the deprotonation of the acidic proton of the pre-nucleophile (malonate or nitroalkane) to occur, thus generating a reactive nucleophile. A postulated catalytic cycle is shown in Figure 8D.1. Deprotonation of malonate (**II**) by a catalyst (**I**, B*–M, M; metal) affords a metal malonate–chiral ligand complex (**III**). A substrate (**IV**) is then activated by the Lewis acid M through coordination of the carbonyl oxygen to M. From this ternary complex (**V**) of malonate, substrate, and chiral ligand B*H, Michael addition proceeds to give enolate **VII**. Finally, proton transfer to the enolate gives product **VIII** and re-generates catalyst **I**.

Chiral metal alkoxides and naphthoxides have been used as catalysts for asymmetric Michael reaction. An early successful example was reported by Cram et al., who used 4 mol % of KO^{t}Bu-chiral crown ether **8** complex as the catalyst to afford the Michael adduct with up to 99% ee (Scheme 8D.7) [16]. In this case KO^{t}Bu complexed with chiral crown ether **8** plays two

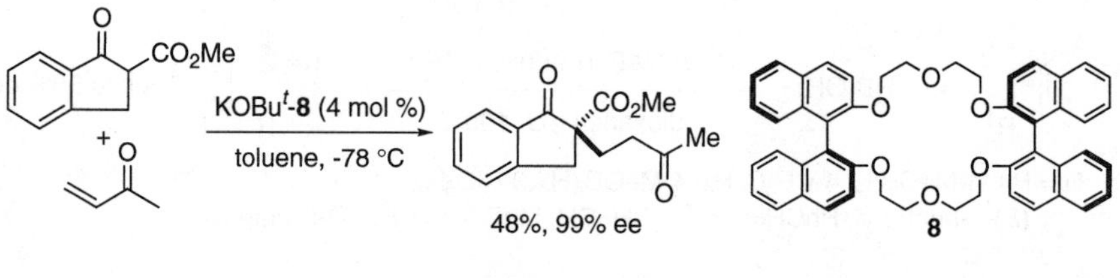

Scheme 8D.7.

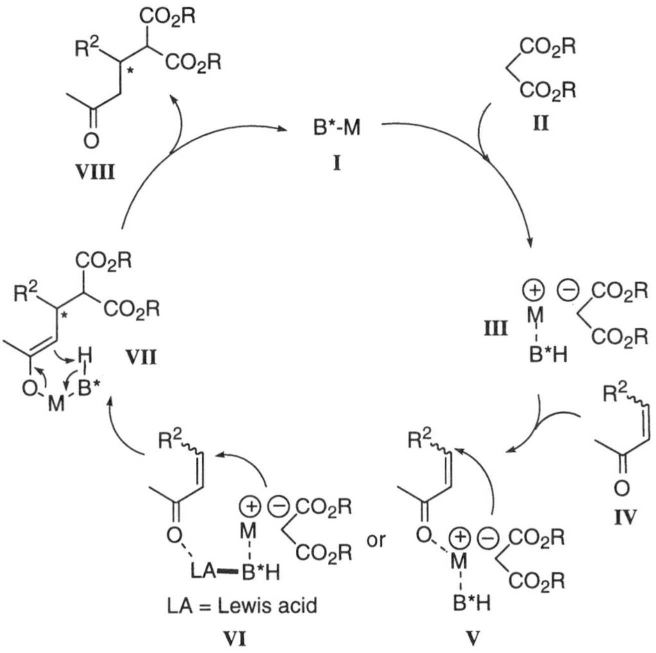

Figure 8D.1.

roles, that is, KO^{t}Bu-**8** functions as (i) a Brønsted base to abstract a proton and (ii) a Lewis acid to activate the enone. Potassium *tert*-butoxide may be a suitable Brønsted base for generation of an enolate, but it may not be the best Lewis acid for activating enones. If these two roles of a catalyst, that is, acting as Brønsted base and a Lewis acid, are separated to have both working ideally at a defined position in one catalyst, the catalyst may achieve excellent enantioselectivity as well as reactivity for a broad range of substrates (Figure 8D.1, **VI**).

This idea was realized very successfully by Shibasaki and Sasai in their heterobimetallic chiral catalysts [17]. Two representative well-defined catalysts, LSB **9** (Lanthanum/Sodium/BINOL complex) and ALB **10** (Aluminum/Lithium/BINOL complex), are shown in Figure 8D.2, whose structures were confirmed by X-ray crystallography. In these catalysts, the alkali metal (Na, Li, or K)-naphthoxide works as a Brønsted base and lanthanum or aluminum works as a Lewis acid.

Complex LSB **9** is readily prepared either by the reaction of La(O^{i}Pr)$_3$ with 3 equiv. of BINOL followed by the addition of NaO^{t}Bu (3 equiv.) or by the reaction of LaCl$_3$•nH$_2$O with sodium binaphthoxide. The complex **9** is stable to oxygen and moisture and has been proven to be effective in the catalytic Michael reaction of various enones with either malonates or β-keto esters. The Michael adducts with up to 92% ee were obtained in almost quantitative yield. Typical results with malonates are summarized in Table 8D.1 (Ln = lanthanide) [18]. In general, the use of THF as solvent gave the best results except for the case of the LSB-catalyzed reaction of *trans*-chalcone with dimethyl malonate, wherein the use of toluene was essential to give the adduct with good enantiomeric excess. The effects of the central metal (La, Pr, and Gd) on asymmetric induction were also examined in the same reaction, and LSB was found to be the best catalyst.

What is the origin of the catalytic activity and the mode of enantioselection of the LSB-catalyzed Michael reactions? To clarify the interaction between the enone and the chiral

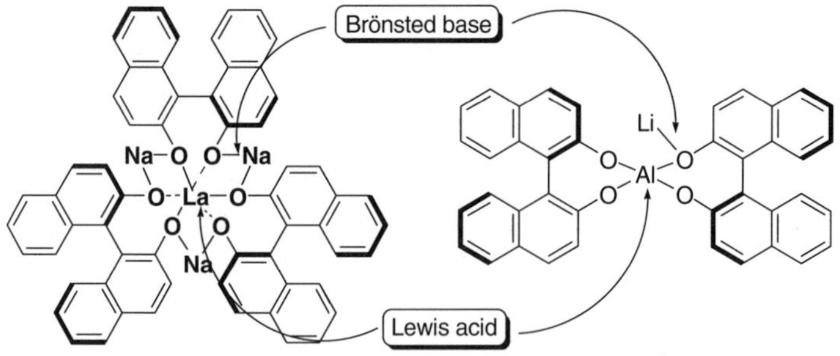

Figure 8D.2. Heterobimetallic Multifunctional Asymmetric Catalysts.

TABLE 8D.1. Catalytic Asymmetric Michael Reaction with LnSB (10 mol %)

enone	Michael donor	cat.	solvent	temp (°C)	time (h)	yield (%)	% ee
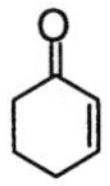	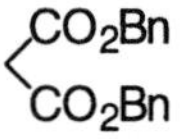	LSB	THF	0	24	97	88
	CO_2Bn	LSB	THF	rt	12	98	85
	CO_2Bn	LSB	toluene	rt	12	96	82
		LLB	THF	rt	12	78	2
		LPB	THF	rt	12	99	48
	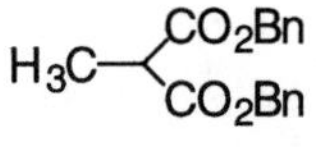 H_3C CO_2Bn CO_2Bn	LSB	THF	0	24	91	92
		LSB	THF	rt	12	96	90
	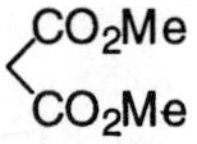 CO_2Me CO_2Me	LSB	THF	rt	12	98	83
	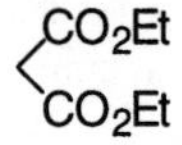 CO_2Et CO_2Et	LSB	THF	rt	12	97	81
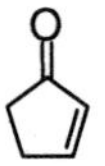	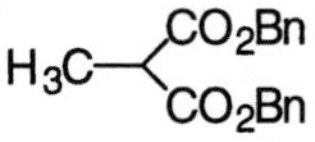 H_3C CO_2Bn CO_2Bn	LSB	THF	−40	36	89	72
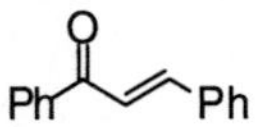	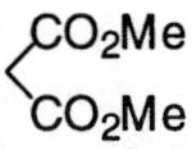	LSB	THF	−50	36	62	0
	CO_2Me	LSB	toluene	−50	24	93	77
Ph Ph	CO_2Me	PrSB	toluene	−50	24	96	56
		GdSB	toluene	−50	24	54	6

LLB = LaLi$_3$tris(binaphthoxide). LPB = LaK$_3$tris(binaphthoxide). PrSB = PrNa$_3$tris(binaphthoxide). GdSB = GdNa$_3$tris(binaphthoxide).

catalyst, the nature of complexation was studied by mixing cyclohexenone and the chiral bimetallic complexes to observe the ^{1}H-NMR chemical shift of the α-proton of the enone as compared with that of the parent (Figure 8D.3). It is well-known, in general, that praseodymium complexes induce up-field shifts whereas europium complexes induce down-field shifts [19]. In addition, ordinary Lewis acids such as La(OTf)$_3$ and Et$_2$AlCl induce down-field shifts. It was observed that complexation to LSB induced a small down-field shift of the α-proton of cyclohexenone, whereas complexation to PrNa$_3$tris(binaphthoxide) complex (PrSB), a modest chiral catalyst for Michael reactions, induced a large up-field shift. It is worth mentioning that in the case of EuNa$_3$tris(binaphthoxide) complex (EuSB) or LaLi$_3$tris(binaphthoxide) complex (LLB), which gave almost racemic Michael adducts, the ^{1}H-NMR spectra of the mixture of cyclohexene and these complexes showed no change in the chemical shift of the α-proton of cyclohexenone. These NMR studies indicate that the carbonyl group of the enone coordinates to the lanthanum and praseodymium metal in the LnSB molecule, whereas such complexation does not take place with LLB and EuSB. These changes of chemical shift were observed even in the presence of dimethyl malonate. The chemical phenomena described above could be explained by considering the difference in the twisted angles of the BINOL moiety in each heterobimetallic catalyst.

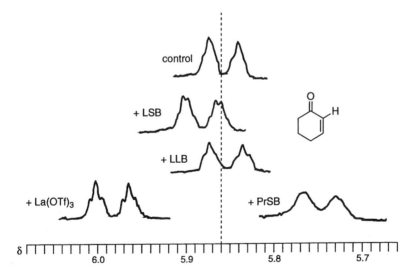

Figure 8D.3. Effect of the Complexation on the Chemical Shift of the α-Proton of Cyclohexenone.

Based on the X-ray structure of LnSB, computational simulations of enantioselection were performed by the aid of Rappé's universal force field (UFF) [20]. As shown in Figure 8D.4, when cyclohexenone is coordinated to the lanthanum metal, the plane of the cyclohexenone ring should be occupied in parallel to one of the naphthyl rings of LSB. Then, the enone should be attacked by the sodium-enolate of dimethyl malonate to give the Michael adduct. The UFF calculation and conformational search of models for pro-(R) and pro-(S) adducts show that the pro-(R) adduct of cyclohexenone with (R)-LSB complex is substantially more favorable than the pro-(S) adduct (ΔE = 4.9 kcal/mol). Namely, the LSB complex can generate the intermediate with high diastereomeric purity such as the one shown at the top of Figure 8D.4. Furthermore, resulting sodium enolates of optically active Michael adducts appear to abstract a proton from

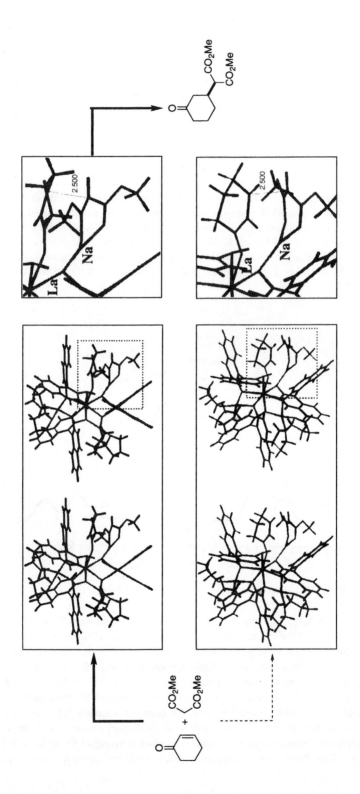

Figure 8D.4. UFF Computational Simulation of Michael Reaction Catalyzed by (*R*)-LSB.

an acidic OH to regenerate the LSB catalyst. The proposed catalytic cycle is shown in Figure 8D.5. Thus, the basic LSB complex also acts as a Lewis acid to control the orientation of the carbonyl function as well as activate the enone. We believe that the multiple functions of the LSB catalyst make the formation of Michael adducts with high enantiomeric purity possible, even at room temperature.

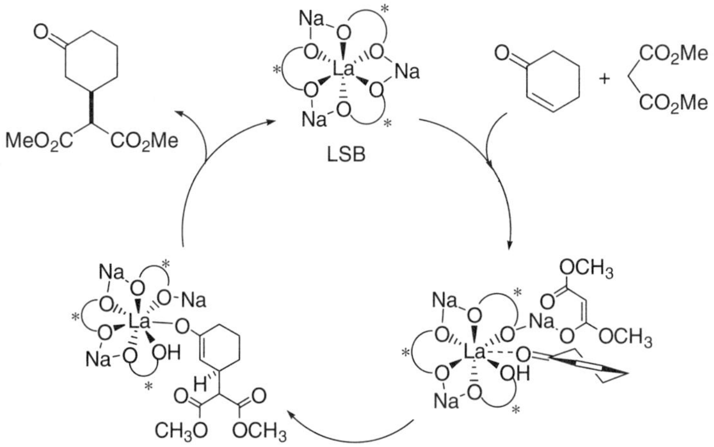

Figure 8D.5. Proposed Catalytic Cycle of the Asymmetric Michael Reaction Promoted by LSB.

The LSB catalyst was next applied to another type of catalytic asymmetric Michael reaction, which induces a new chiral center on the Michael donor side [21]. In a preliminary study, it was found that the reaction of ketoester **11** with methyl vinyl ketone (MVK), in THF with 10 mol % of LSB, gave the adduct with 23% ee, whereas the reaction in toluene afforded the adduct with 75% ee. However, when the amount of LSB was reduced to 5 mol % in toluene, the enantiomeric purity of the adduct declined to 25% ee. In an attempt to offset this decline while maintaining the lower level of catalyst loading, the effects of slow addition of ketoester **11** were examined. As expected, the use of a syringe pump gave the adduct with high enantiomeric excess. In marked contrast to this result, when CH_2Cl_2 was used instead of toluene, the reaction catalyzed by 5 mol % of LSB gave the adduct with 91% ee in 89% yield without using slow addition. Moreover, it was found that the catalytic asymmetric Michael reaction of **11** was not affected so much by the choice of rare-earth metal. As shown in Table 8D.2, the slow addition of ketoester and the use of CH_2Cl_2 as the solvent are generally quite effective in keeping high enantiomeric purity of various Michael adducts. However, malonates give the adducts with high % ee regardless of the solvent used. These results can be rationalized by the comparison of the pKa of a ketoester with that of a malonate, that is, the former is significantly more acidic than the latter. Thus, the concentration of the resulting Na-enolate can be expected to be greater in the case of ketoester, and furthermore the Na-enolate of ketoester will react with an enone more slowly than the Na-enolate derived from a malonate. It is suggested that a combination of more rapid formation and longer lifetime of the Na-enolate of ketoester increases the likelihood of dissociation of free Na-enolate from the enolate/LSB/acceptor complex, giving a product with lower % ee. On the other hand, in less polar CH_2Cl_2 the Na-enolate would remain part of the complex, thereby affording the product with high % ee (Figure 8D.6). The slow addition of the

TABLE 8D.2. Catalytic Asymmetric Michael Reactions Promoted by LSB in CH₂Cl₂

Michael donor	Michael acceptor	product	catalyst amount (mol %)	temp (°C)	time (h)	yield (%)	% ee
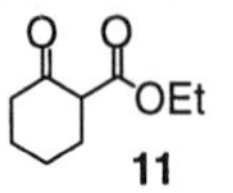 (2-oxocyclohexanecarboxylic acid ethyl ester) **11**	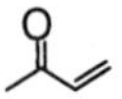 (methyl vinyl ketone)	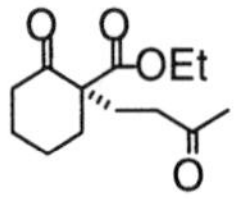	5	−50	19	89	91
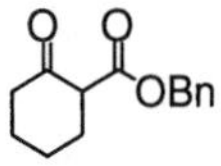 (2-oxocyclohexanecarboxylic acid benzyl ester, OBn)		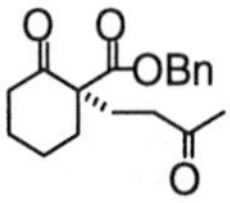	10	−50	12	73	91
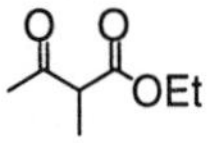 (ethyl 2-methylacetoacetate, OEt)		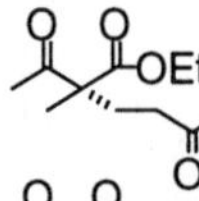	5	−50	20	94	74
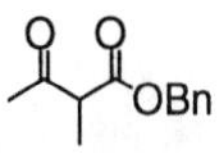 (benzyl 2-methylacetoacetate, OBn)		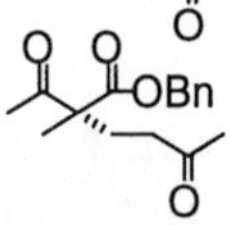	5	−50	16	93	83
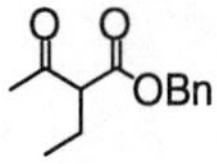 (benzyl 2-ethylacetoacetate, OBn)		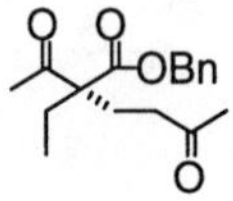	5	−50	16	98	89
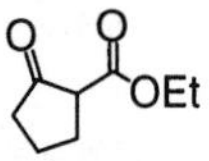 (2-oxocyclopentanecarboxylic acid ethyl ester, OEt)		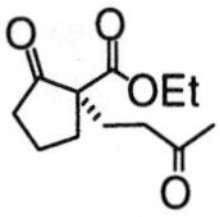	10	−50	18	95	62
			10	−50	17	95	75[a]
			20	−50	18	97	84
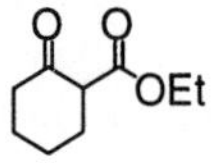 (2-oxocyclohexanecarboxylic acid ethyl ester, OEt)	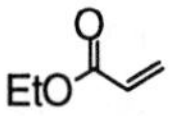 EtO (ethyl acrylate)	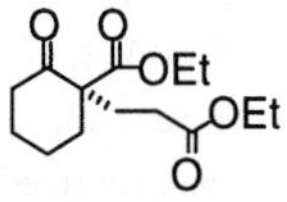	10	0 → rt	17	60	76
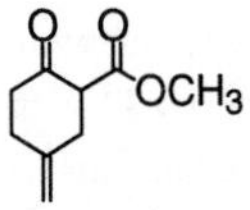 (OCH₃)	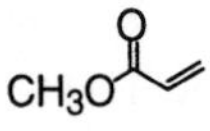 CH₃O (methyl acrylate)	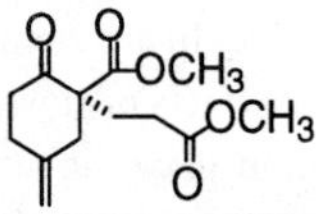	20	0 → rt	93	69	89

[a]Ketoester was added slowly (8 h).

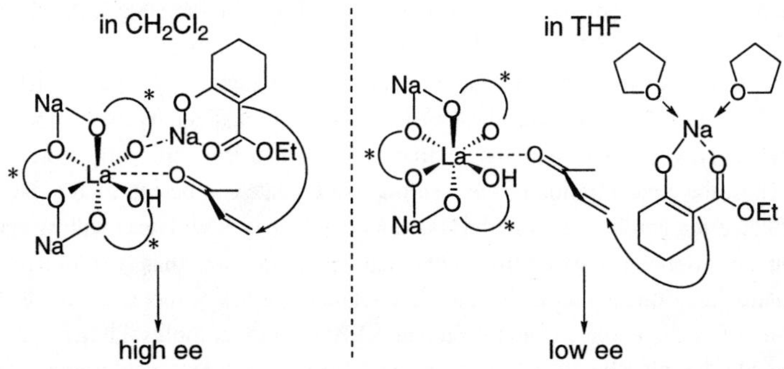

Figure 8D.6.

ketoester is believed to limit undesired ligand exchange between BINOL moieties and the Michael donor.

Other nucleophiles such as nitromethane can also be used for this reaction. Thus, by the catalysis of (R)-LPB (LaK$_3$tris((R)-binaphthoxide) (20 mol %), in which La works as a Lewis acid and K-naphthoxide works as a Brønsted base, nitromethane reacted with chalcone to give the Michael adduct in 85% yield and 93% ee (Scheme 8D.8) [22]. Addition of ^{t}BuOH (120 mol %) gave a beneficial effect on the reactivity as well as the enantioselectivity of this reaction.

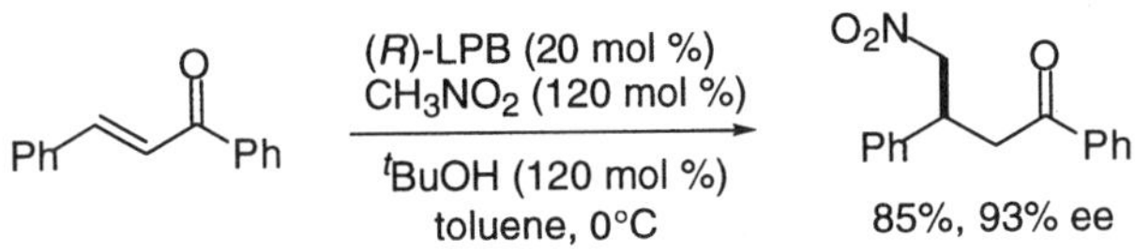

Scheme 8D.8.

Another highly useful heterobimetallic catalyst is the aluminum-lithium-BINOL complex (ALB) prepared from LiAlH$_4$ and 2 equiv. of (R)-BINOL. The ALB catalyst (10 mol %) is also effective in the Michael reaction of enones with various malonates, giving Michael products generally with excellent enantioselectivity (91–99% ee) and in excellent yields [23]. These results are summarized in Table 8D.3. Although LLB and LSB complement each other in their ability to catalyze asymmetric nitroaldol and Michael reactions, respectively, the Al-M-(R)-BINOL complexes (M = Li, Na, K, and Ba) are commonly useful for the catalytic asymmetric Michael reaction.

The structure of AlLibis((R)-binaphthoxide) complex (ALB) was determined unequivocally by X-ray crystallographic analysis shown in Figure 8D.2 [23]. The complex has a tetrahedral geometry around aluminum with an average Al-O distance of 1.75 Å}. The long Li-O distance of 2.00 Å} is indicative of the ionic character between Li$^+$ and Al-2BINOL$^-$.

The small electronegativity value of lithium (1.0) as compared with that of aluminum (1.5) suggests that a lithium enolate should be generated from a malonate derivative. However, the role of aluminum in this reaction is not well-understood. To observe the interaction between the enone and aluminum, an ^{27}Al-NMR study of an ALB-enone complex was performed. First, one broad signal was observed at δ 75 for ALB itself (S$_0$). Upon addition of 3 equiv. of cyclohexenone to ALB, two additional relatively sharp signals were obtained at δ 40 (S$_1$) (the structure giving this signal is not known) and δ 23 (S$_2$). The latter signal (S$_2$) became the major peak under the catalytic Michael reaction conditions. This observation provides strong evidence for the octahedral arrangement of six magnetically equivalent ligands around the aluminum cation in THF solution, based on Feringa's results [24,25]. This NMR study clearly indicates that the carbonyl groups of the enones coordinate to the aluminum. Consequently, ALB is believed to act as a heterobimetallic multifunctional chiral catalyst, making efficient Michael reactions possible to occur even at room temperature.

Feringa et al. applied ALB to a reaction that induces a chiral center on the Michael donor side (Scheme 8D.9) [25]. Thus, ALB (5 mol %) was used to catalyze the addition of nitroester **12** to methyl vinyl ketone, giving the Michael adduct **13** in 81% yield and 80% ee. This paper reported that ALB prepared from 1 equiv. of LiAlH$_4$ and 2.45 equiv. of BINOL afforded a better result in terms of chemical yield. However, this deviation from the original composition (LiAlH$_4$/BINOL = 1:2) reported by Shibasaki et al. is very likely to stem from an impurity of LiAlH$_4$ used. Feringa reported that ^{27}Al NMR of both ALB (LiAlH$_4$/BINOL = 1:2 and 1:2.45) showed a mixture of three peaks in almost the same

TABLE 8D.3. Catalytic Asymmetric Michael Reactions Promoted by Al-M-bis((R)-binaphthoxide) complex (AMB)

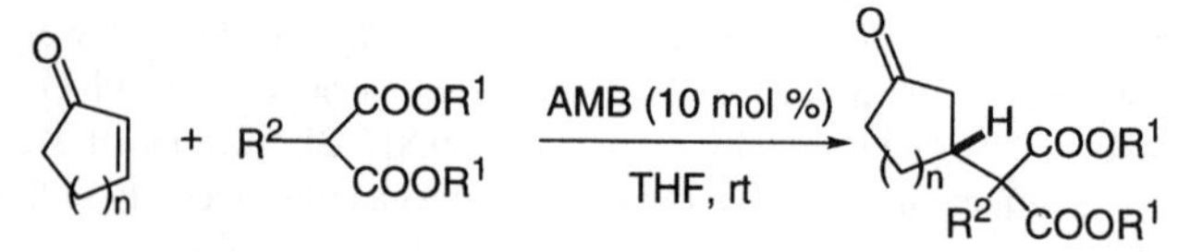

enone	Michael donor	M	time (h)	yield (%)	% ee
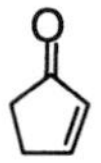 (cyclopentenone)	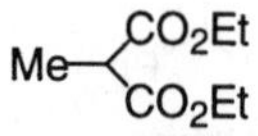 Me—C(CO₂Et)(CO₂Et)	Li	72	84	91
	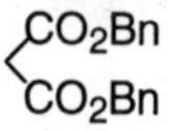 C(CO₂Bn)(CO₂Bn)	Li	60	93	91
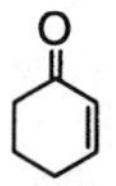 (cyclohexenone)	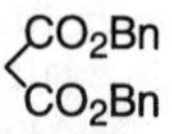 C(CO₂Bn)(CO₂Bn)	Li	72	88	99
		Na	72	50	98
		K	72	43	87
		Ba	6	100	84
	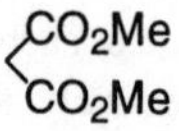 C(CO₂Me)(CO₂Me)	Li	72	90	93
	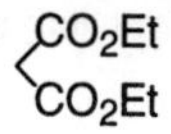 C(CO₂Et)(CO₂Et)	Li	72	87	95

81% (80% ee)

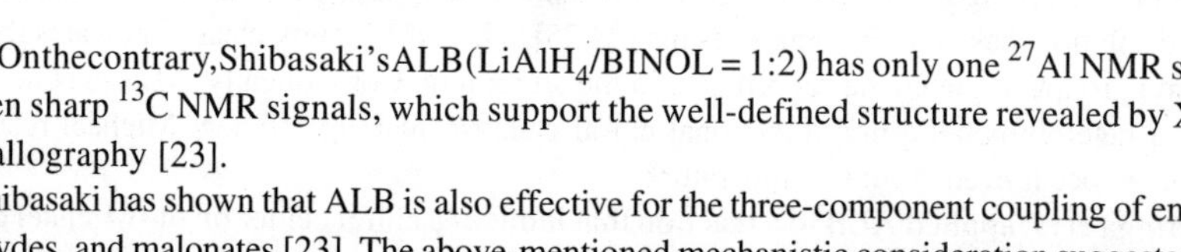

Scheme 8D.9.

ratio. On the contrary, Shibasaki's ALB (LiAlH$_4$/BINOL = 1:2) has only one ^{27}Al NMR signal and ten sharp ^{13}C NMR signals, which support the well-defined structure revealed by X-ray crystallography [23].

Shibasaki has shown that ALB is also effective for the three-component coupling of enones, aldehydes, and malonates [23]. The above-mentioned mechanistic consideration suggested that the reaction of a lithium enolate derived from a malonate derivative with an enone would lead to the formation of an intermediary aluminum enolate. Thus, further studies were carried out to obtain direct evidence for the formation of an aluminum enolate. The larger electronegativity of aluminum (1.5) as compared with that of lithium, sodium, or lanthanoid suggests that the protonation of the aluminum enolate should be slower than that of the corresponding lithium, sodium, and/or lanthanoid enolates. Then, is it possible to trap such an Al-enolate by an electrophile such as an aldehyde? As anticipated, the reaction of cyclopentenone, diethyl methylmalonate, and hydrocinnamaldehyde in the presence of 10 mol % of ALB indeed gave the three-component coupling product **14** as a single diastereomer with 91% ee in 64% yield (Scheme 8D.10). It should be noted that the use of LLB, LSB, and/or lithium-free lanthanum-BINOL [26] complex gave very unsatisfactory results in the three-component coupling reaction. This result is probably due to the fact that the lanthanum enolate is more reactive to acidic proton. The use of benzaldehyde instead of hydrocinnamaldehyde gave the three-component coupling product **15** in 82% yield (Scheme 8D.10). Although this product **15** was a mixture of diastereomers, the oxidation of **15** gave the corresponding diketone derivative **16** as a single diastereomer with 89% ee. These examples of catalytic asymmetric tandem Michael-aldol reaction are the first.

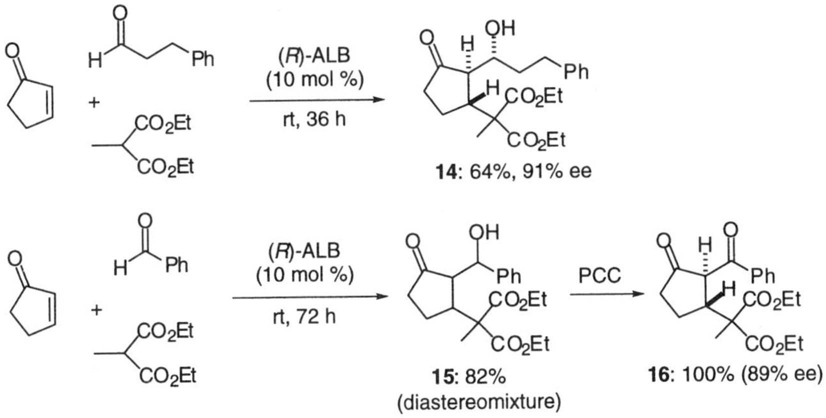

Scheme 8D.10.

The proposed mechanism for the three-component coupling reaction is illustrated in Figure 8D.7. The proposed reaction pathway includes the following steps. First, the reaction of diethyl methylmalonate with ALB gives the corresponding lithium enolate **I**. Enolate **I** then reacts with cyclopentenone pre-coordinated to the aluminum metal of ALB, giving the aluminum enolate **II** in an enantioselective manner. Further reaction of enolate **II** with an aldehyde leads to alkoxide **III**. Although it is unclear whether alkoxide **III** is an Al-alkoxide or Li-alkoxide, the resulting alkoxide **III** then abstracts a proton from an acidic OH to give the three-component coupling product and regenerate ALB to complete the catalytic cycle.

Significant improvement in the catalytic activity of ALB was realized without any loss of enantioselectivity by using the "second-generation ALB" [27] generated by the self-assembled complex formation of ALB with alkali metal-malonate or alkoxide. This protocol allowed the catalyst loading to be reduced to 0.3 mol %, for example, the Michael addition of methyl malonate to cyclohexenone catalyzed by the self-assembled complex of (R)-ALB (0.3 mol %) and KO^{t}Bu (0.27 mol %) in the presence of MS 4A gave the adduct in 94% yield and 99% ee [28]. This reaction has been successfully carried out on a 100-g scale wherein the product was purified by recrystallization. The kinetic studies of the reactions catalyzed by ALB and ALB/Na-malonate have revealed that the reactions are second-order to these catalysts (the rate constant: $k_{ALB} = 0.273$ M^{-1}h^{-1}; $k_{ALB/Na\text{-}malonate} = 1.66$ M^{-1}h^{-1}) [27]. This reaction was used as the first key step for the catalytic asymmetric total synthesis of tubifolidine (Scheme 8D.11) [28].

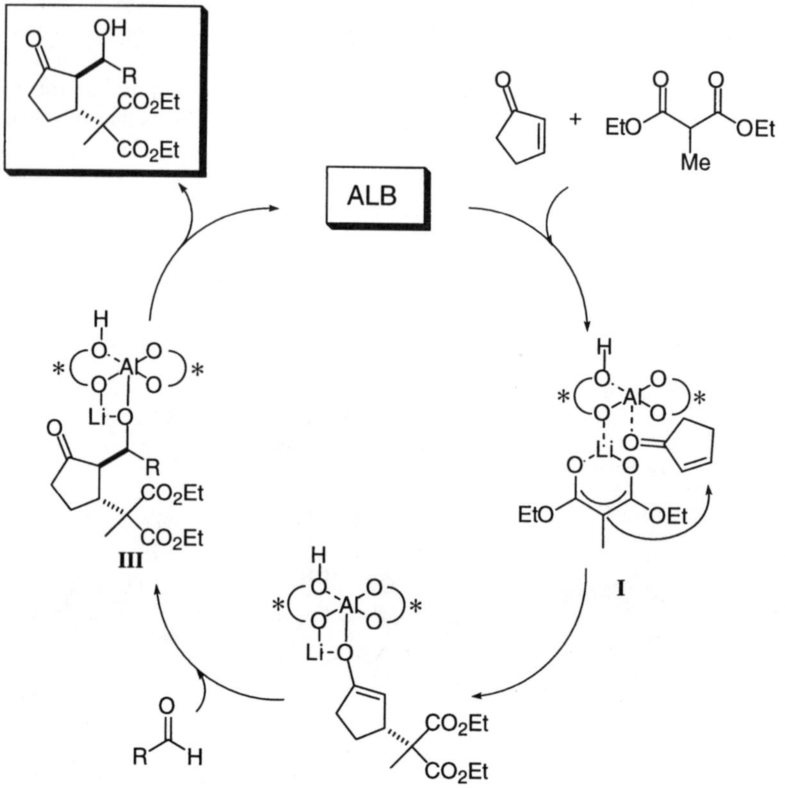

Figure 8D.7. Proposed Mechanism for the Catalytic Asymmetric Tandem Michael-Aldol Reaction Promoted by ALB.

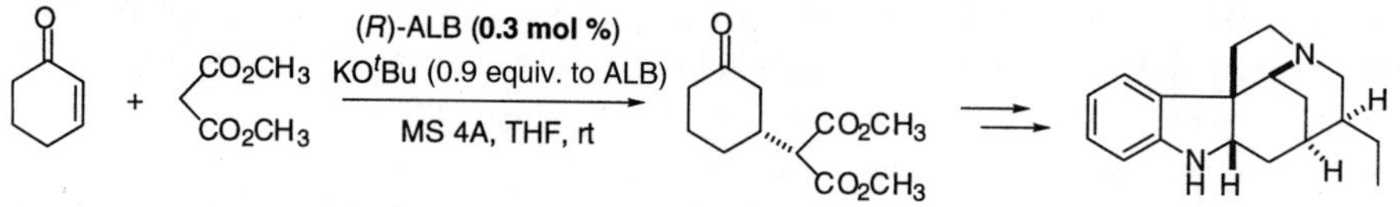

Scheme 8D.11.

Both regioselectivity and enantioselectivity are efficiently controlled by ALB in the Michael addition of Horner-Wadsworth-Emmons reagents **17** with enones (Scheme 8D.12) [29]. Although the reaction catalyzed by ALB itself did not afford the product, the use of a combination of ALB (10 mol %) and NaO^tBu (0.9 equiv. to ALB) gave the Michael adduct of **17** to cyclohexenone in 64% yield and 99% ee. The reaction of cyclohexenone with **17** promoted by standard bases such as NaO^tBu and BuLi gave the 1,2-adduct in only 8–9% yield. The adduct

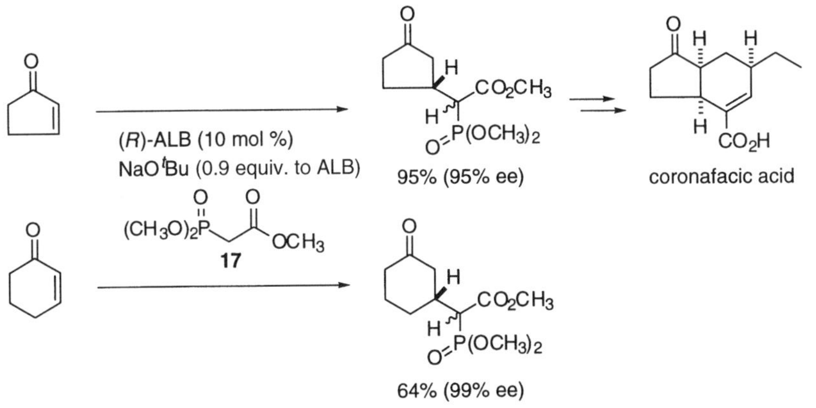

Scheme 8D.12.

obtained from the reaction of cyclopentenone with **17** was successfully transformed into coronafacic acid by using a reported procedure [30].

An interesting insight into the reactive species was obtained from this study. Treatment of ALB with the Li-enolate of **17** (1.0 equiv. to ALB) in THF was found to give stable and crystalline complex **21**, whose structure was determined by X-ray crystallography. The mechanism for the formation of **21** is proposed below (Scheme 8D.13). This species proved not to be the actual reactive species, that is, the reaction of cyclopentenone with Horner-Wadsworth-Emmons reagent **17**, in the presence of **21** in THF at room temperature for 56 h, gave the adduct in 78% yield, but with only 12% ee. To investigate the reactivity of another possible reactive species **19** (Nu = enolate), which can be generated from ALB and Li-enolate of **17**, the basic precursor of **19** (Nu = CH$_3$) was prepared from (CH$_3$)$_3$Al and BINOL (1 equiv.) in THF and the structure of **19** (Nu = CH$_3$) was determined by the X-ray analysis. The species **19** was found not to promote the reaction at all. However, it was found that a mixture of **21** (10 mol %) and **19** (5 mol %) promoted the reaction at 50°C, giving the Michael adduct in 92% ee and 83% yield after 92 h. These results clearly indicate that there is an equilibrium between the ALB and the hexacoordinated aluminum complex **21** in the reaction medium, and the actual catalyst is

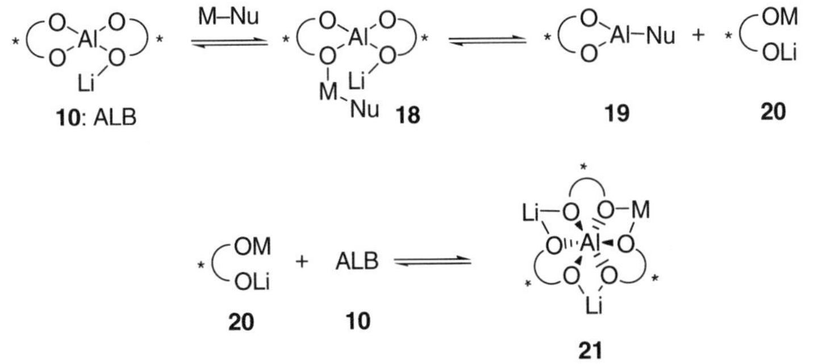

Scheme 8D.13.

probably the self-assembled complex **18** derived from ALB and a standard base or an alkali metal enolate.

The self-complexed catalyst of ALB and a base was also successfully applied to the above-mentioned three-component coupling reaction of an enone, an aldehyde, and a malonate. A catalytic asymmetric synthesis of 11-deoxy-PGF$_{1\alpha}$ was achieved by utilizing this reaction as the first key step (Scheme 8D.14) [31]. Furthermore, a potential key intermediate **25** for the synthesis of PGF$_{1\alpha}$ was synthesized in relatively high yield with extremely high enantioselectivity (97%) by the kinetic resolution of the three-component coupling between racemic enone **24**, aldehyde **22**, and methyl dibenzylmalonate **23** [31].

Third heterobimetallic asymmetric catalyst reported by Shibasaki et al., gallium-sodium-BINOL complex (GaSB) **26** and indium-potassium-BINOL complex (InPB), are also rather effective catalysts for asymmetric Michael reactions, and GaSB was better than InPB in terms of enantioselectivity. The GaSB catalyst was prepared from GaCl$_3$, NaO^{t}Bu (4 mol equiv. to

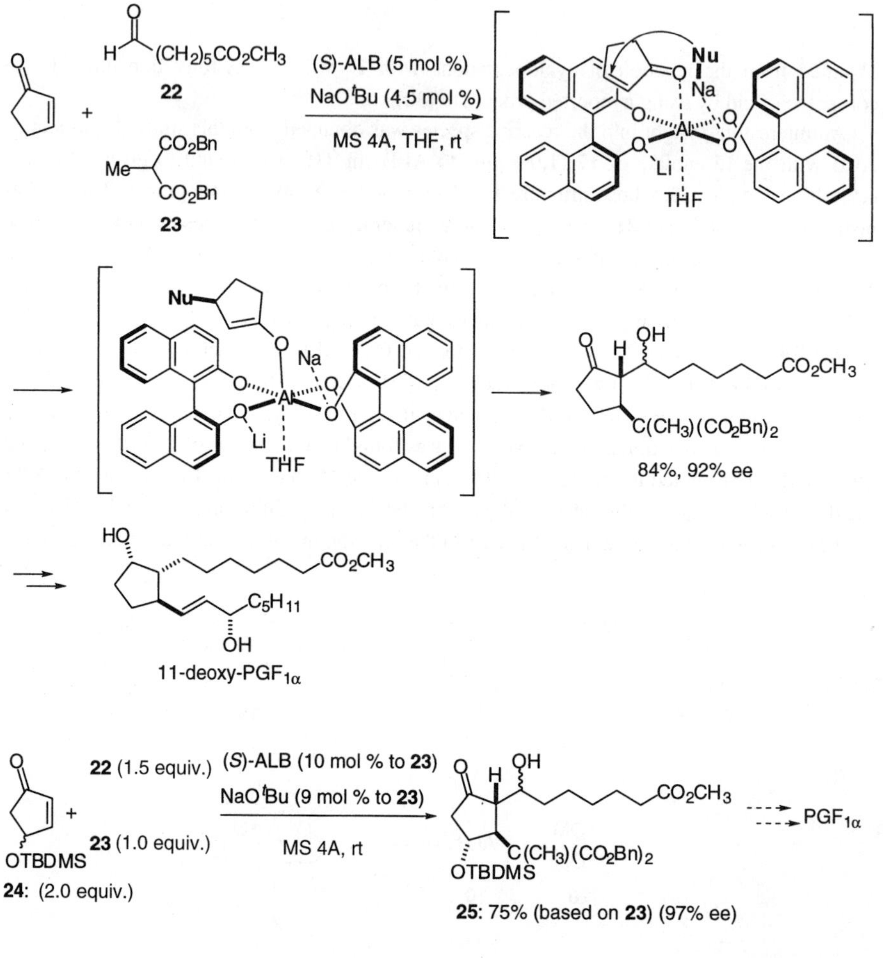

Scheme 8D.14.

GaCl$_3$), and (*R*)-BINOL (2 mol equiv. to GaCl$_3$) in THF-ether, and the structure of GaSB, consisting of one Ga atom, one Na atom and two molecules of BINOL, was determined by ^{13}C-NMR and LDI-TOF mass spectrum. The self-assembled GaSB complex, in a manner similar to that of ALB, has an enhanced catalytic activity without reducing the enantiopurity of the Michael adducts. As shown in Table 8D.4, the asymmetric Michael reaction of cyclo-hexenone with dibenzyl malonate, catalyzed by 10 mol % GaSB, required 143 h at room temperature to give the product with 98% ee in 45% yield. However, the same reaction with 10 mol % GaSB and 9 mol % NaO^{t}Bu at room temperature gave the adduct with 98% ee in 87% yield after only 21 h. Addition of the sodium salt of dibenzyl malonate (9 mol %) instead of NaO^{t}Bu gave almost the same result (96% ee, quantitative yield). The asymmetric Michael reaction of cyclopentenone with dibenzyl malonate in the presence of 10 mol % GaSB and 9 mol % NaO^{t}Bu also proceeded smoothly to give the adduct with 98% ee in 96% yield (room temperature, 22 h). This Michael product was used as a key intermediate for the asymmetric synthesis of (+)-coronafacic acid by Toshima, Ichihara et al. [30b]. A highly efficient catalytic asymmetric Michael addition to cycloheptenone has also been realized for the first time (>99% ee, 79% yield).

Another interesting example was reported by Yamaguchi et al. by using a simple L-proline rubidium salt **27** (Scheme 8D.15) [32]. A reversible iminium salt formation, involving the amine moiety of the catalyst and the carbonyl group of an enone, was proposed as the key intermediate.

A chiral quarternary ammonium salt works as a chiral-phase transfer catalyst, and this chemistry has been applied to asymmetric Michael reaction by Corey et al. (Scheme 8D.16) [33]. It has been shown that the cinchonidine salt **28**, which has been designed by rigidifying

TABLE 8D.4

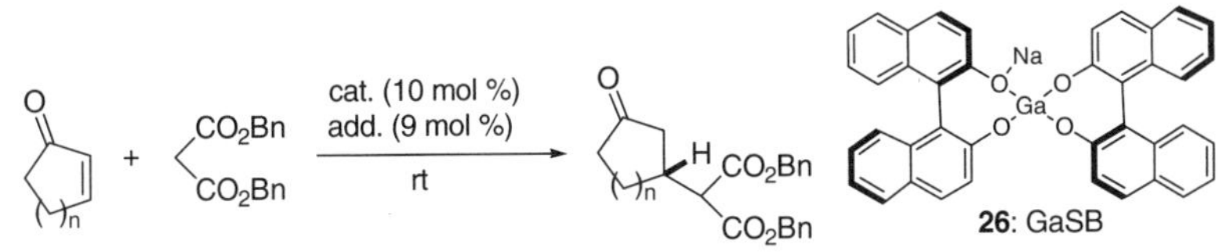

entry	enone	catalyst	additive	time (h)	yield (%)	ee (%)
1		GaSB	–	143	45	98
2		GaSB	A	21	87	98
3	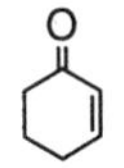	GaSB	A^a	6	81	84
4		GaSB	A^b	6	91	60
5		GaSB	B	21	quant	96
6		GaSB	–	72	32	89
7		GaSB	A	22	96	98
8	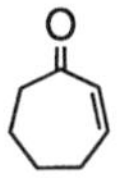	GaSB	–	73	trace	–
9		GaSB	A	73	79	>99

A: NaO-*t*-Bu, B: Na-malonate
^{a}2 mol equiv of NaO-*t*-Bu was used to GaSB.
^{b}3 mol equiv of NaO-*t*-Bu was used to GaSB.

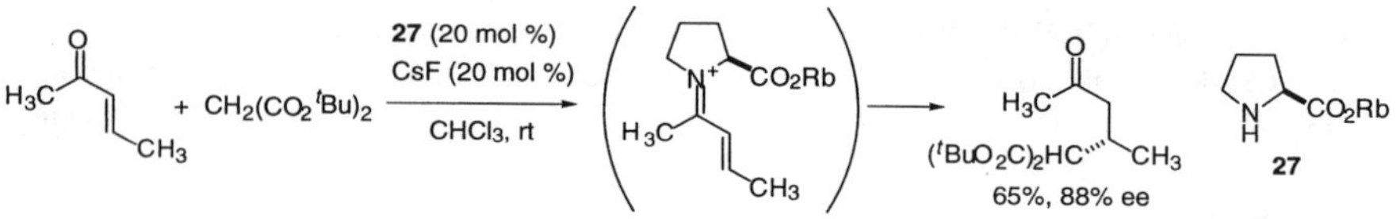

Scheme 8D.15.

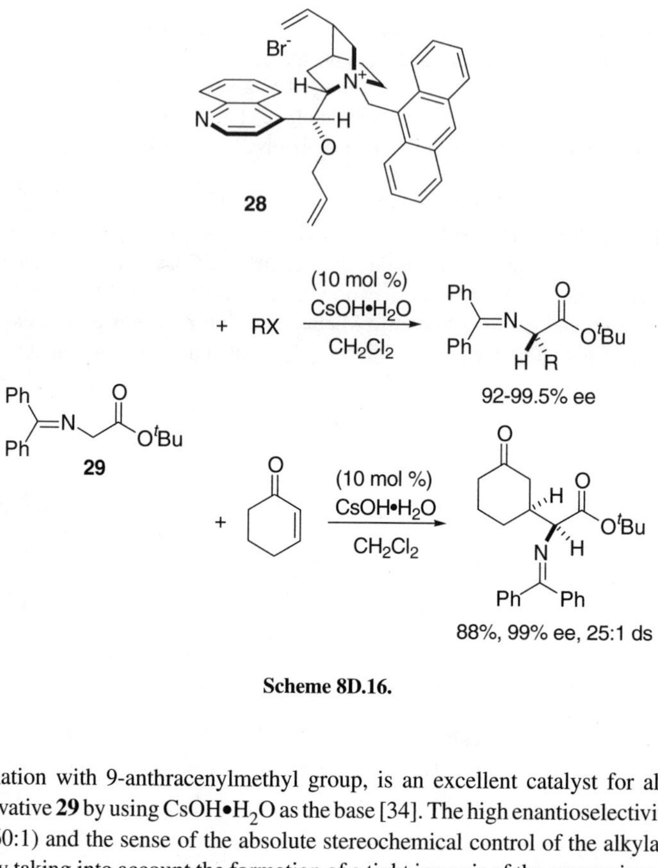

28

92-99.5% ee

88%, 99% ee, 25:1 ds

Scheme 8D.16.

its conformation with 9-anthracenylmethyl group, is an excellent catalyst for alkylation of glycine derivative **29** by using $CsOH \cdot H_2O$ as the base [34]. The high enantioselectivity (between 400:1 and 60:1) and the sense of the absolute stereochemical control of the alkylation can be explained by taking into account the formation of a tight ion pair of the ammonium cation with the enolate, whose structure was postulated on the basis of the X-ray crystal structure of **28** [34]. The enolate oxygen of the ion pair positions itself at the location of the bromide ion in the X-ray crystal structure and the remainder of the complex is arranged so as to allow maximum van der Waals contact. This energetically favorable three-dimensional arrangement blocks one face of the nucleophilic carbon center of the enolate but leaves the other face free for the attack of an alkylating agent. As an extension of this extremely enantioselective reaction, α,β-unsaturated carbonyl compounds have been used as electrophile [33]. For example, 2-cyclohexenone undergoes Michael reaction with ester **29** under the phase-transfer conditions, in the presence of catalyst **28** and $CsOH \cdot H_2O$, affording the adduct with 99% ee and 25:1 diastereoselectivity.

The absolute configuration of the Michael adduct formed can be deduced from the postulated structure of the ion pair of the enolate of **29** and catalyst **28**.

8D.4. CATALYTIC ASYMMETRIC MICHAEL ADDITION OF ALKYL RADICALS

Carbon–carbon bond formation with radical intermediates constitutes an important category in organic synthesis [35]. An enantioselective version of this reaction has recently begun to emerge and the first example of chiral Lewis acid–catalyzed conjugate radical addition was reported by Sibi et al. [36]. The introduction of an isopropyl group to N-cinnamoyloxazolidinone, by ^{i}PrI and Bu$_3$SnH with Et$_3$B/O$_2$ as the radical initiator, was studied in the presence of chiral

TABLE 8D.5. Catalytic Asymmetric Michael Addition of ^{i}Propyl Radical to N-Cinnamoyloxazolidinone

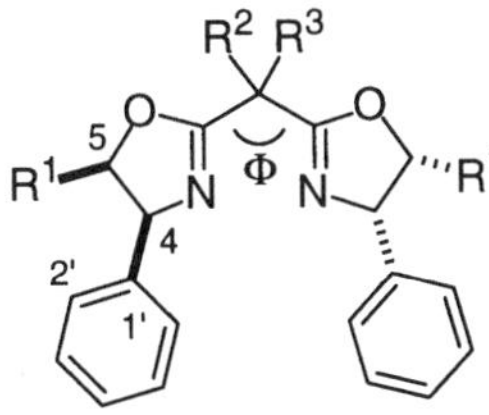

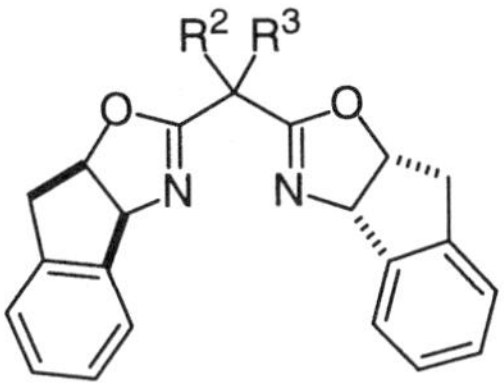

30a $R_1 = H$, $R_2 = R_3 = CH_3$
30b $R_1 = H$, $R_2, R_3 = (CH_2CH_2)$
30c $R_1 = CH_3$, $R_2 = R_3 = CH_3$
30d $R_1 = CH_3$, $R_2, R_3 = (CH_2CH_2)$

30e $R_1 = R_2 = CH_3$
30f $R_1, R_2 = (CH_2CH_2)$
30g $R_1, R_2 = (CH_2CH_2CH_2)$
30h $R_1, R_2 = (CH_2CH_2CH_2CH_2)$

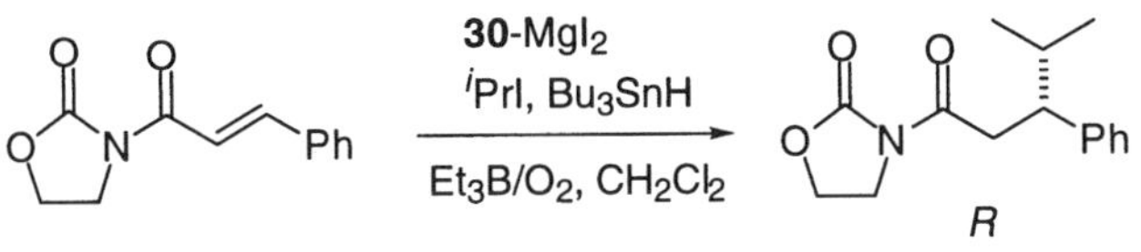

entry	ligand	(mol %)	Φ	θ	yield (%)	% ee	S/R
1	**30a**	(100)	–	–	88	47	S
2	**30b**	(100)	–	70	87	37	S
3	**30c**	(100)	–	–	79	31	S
4	**30d**	(100)	–	81	88	36	S
5	**30e**	(100)	104.7	–	88	89	R
6	**30f**	(100)	110.6	10	88	93	R
7	**30g**	(100)	108.0	–	90	82	R
8	**30h**	(100)	105.8	–	92	82	R
9	**30f**	(30)	110.6	10	91	97	R
10	**30f**	(5)	110.6	10	92	90	R
11	**30f**	(1)	110.6	10	29	63	R

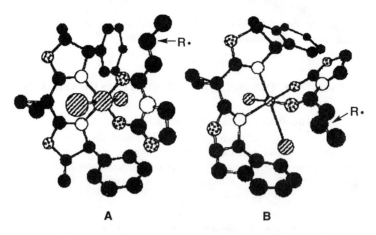

A   B

Figure 8D.8.

bisoxazoline **30**–MgI_2 complexes (Table 8D.5). Chiral ligands were modified in terms of the bite angle (represented by the angle Φ) [37] and the dihedral angle of C_4 substituent (the C_5-C_4-C_1-C_2, dihedral angle θ). The best enantioselectivity (93% ee) was obtained with the use or ligand **30f** that has the largest Φ and fully constrained C4-phenyl group. It is interesting that the absolute configuration of the products with ligands **30a–d** is opposite to that with ligands **30e–h**. A plausible explanation for the observed difference in enantioselection is illustrated in Figure 8D.8. In the reactions with ligands in which the C4-Ph bond is flexible (**30a–d**), the substrate-MgI_2-ligand complex adopts an octahedral geometry with the iodides in trans-geometry. The observed low-level enantioselectivity with ligands **30a–d** also indicates that the flexible C4-Ph substituent does not provide optimal face shielding. Attack of a radical on the least-hindered *si*-face of the substrate accounts for the fact that the absolute stereochemistry of the product is S (Figure 8D.8, A). In the case of aminoin-danol-derived ligands **30e–h**, the ligand-MgI_2-substrate complex adopts an octahedral geometry where the two iodides take cis geometry and the more Lewis basic carbonyl oxygen of the cinnamoyl moiety of the substrate is located *trans* to the iodide. The ring constraint and the larger bite angle in the **30f**-MgI_2-substrate complex provides optimal face shielding for the radical addition, which accounts for the observed high enantioselec-tivity. Attack of the radical on the least-hindered side also explains the observed absolute stereochemistry of the product being R (Figure 8D.8, B).

8D.5. CATALYTIC ASYMMETRIC MICHAEL ADDITION OF HETEROATOM NUCLEOPHILES

The last topic in this chapter is the Michael addition of heteroatoms such as nitrogen and sulfur that affords β-aminocarbonyl and β-thiocarbonyl compounds.

Encouraged by the excellent enantioselectivity in the conjugate radical addition, Sibi et al. further investigated the Michael addition of *O*-benzylhydroxylamine to pyrazol-derived croto-namides catalyzed by chiral bisoxazoline **30**-$MgBr_2$ complex (Scheme 8D.17) [38]. Good-to-high enantioselectivity (up to 95% ee) was achieved with 10 mol % of the catalyst. The trajectory

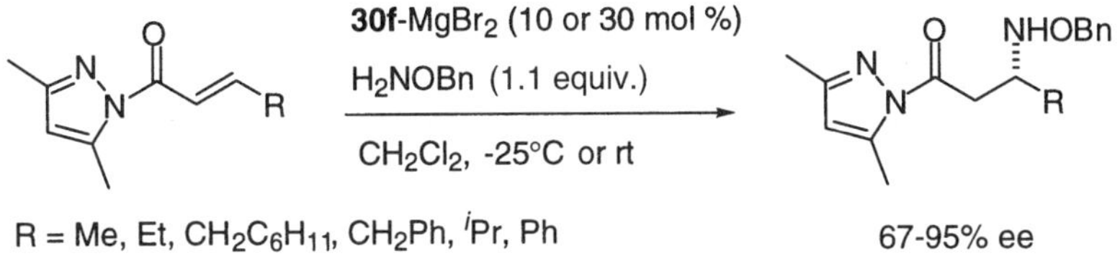

Scheme 8D.17.

of the attack of the nucleophile on the substrate complex from the less-hindered side is the same as that of the radical reaction discussed above, and the model proposed in Figure 8D.8 is applicable for this reaction as well.

Tomioka et al. reported the asymmetric Michael addition of lithium thiolates catalyzed by chiral aminoether **31** (Scheme 8D.18) [39]. Thus, in the presence of catalytic amounts of **31** (10 mol %) and lithium 2-(trimethylsilyl)thiophenolate **32**-Li (8 mol %), thiol **32** (3 equiv.) reacted with α,β-unsaturated esters at $-78°C$ in toluene-hexane solvent to give the Michael adduct with up to 97% ee. In the absence of **31**, the reaction of thiophenol proceeded in only 0.5% yield at room temperature. A monomeric complex consisting of **31** and lithium is proposed as the key reactive species in this asymmetric reaction. The trimethylsilyl group at the *ortho*-position of the thiol moiety in **32** contributes to the formation of the stereochemically defined monomeric chelated structure, wherein the lithium cation is coordinated with the three heteroatoms of the tridentate ligand **31**. The reactions of acyclic *trans*-α,β-unsaturated esters (R^1 = Me, Et, Pr, Bu, ^iBu, $PhCH_2$; R^2 = H) proceeds with high enantioselectivity in general. Cyclic cis-α,β-unsaturated esters (R^1, R^2 = $(CH_2)_3$, $(CH_2)_4$) gave cis products selectively with high enantiopurity.

The heterobimetallic multifunctional complexes LnSB developed by Shibasaki and Sasai described above are excellent catalysts for the Michael addition of thiols [40]. Thus, phenylmethanethiol reacted with cycloalkenones in the presence of (R)-LSB (LaNa$_3$tris(binaphthoxide)) (10 mol %) in toluene-THF (60:1) at $-40°C$, to give the adduct with up to 90% ee. A proposed catalytic cycle for this reaction is shown in Figure 8D.9. Because the multifunctional catalyst still has the internal naphthol proton after deprotonation of the thiol (bold-H in **I** and **II**), this acidic proton in the chiral environment can serve as the source of asymmetric protonation of the intermediary enolate, which is coordinated to the catalyst **II**. In fact, the Michael addition of 4-*tert*-butylbenzenethiol to ethyl thiomethacrylate afforded the product with up to 93% ee using (R)-SmSB as catalyst. The catalyst loading could be reduced to 2 mol % without affecting enantioselectivity of the reaction.

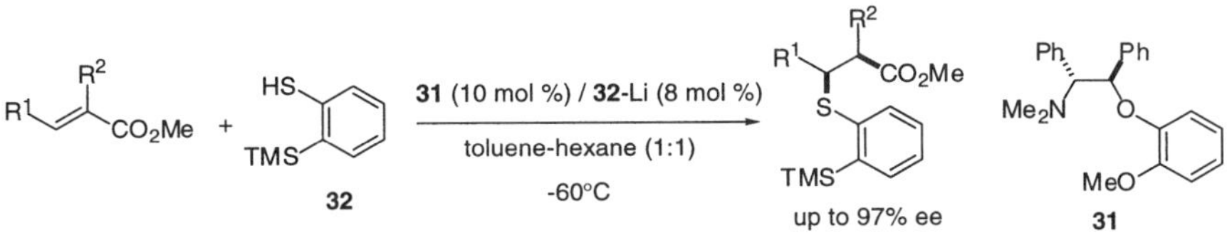

Scheme 8D.18.

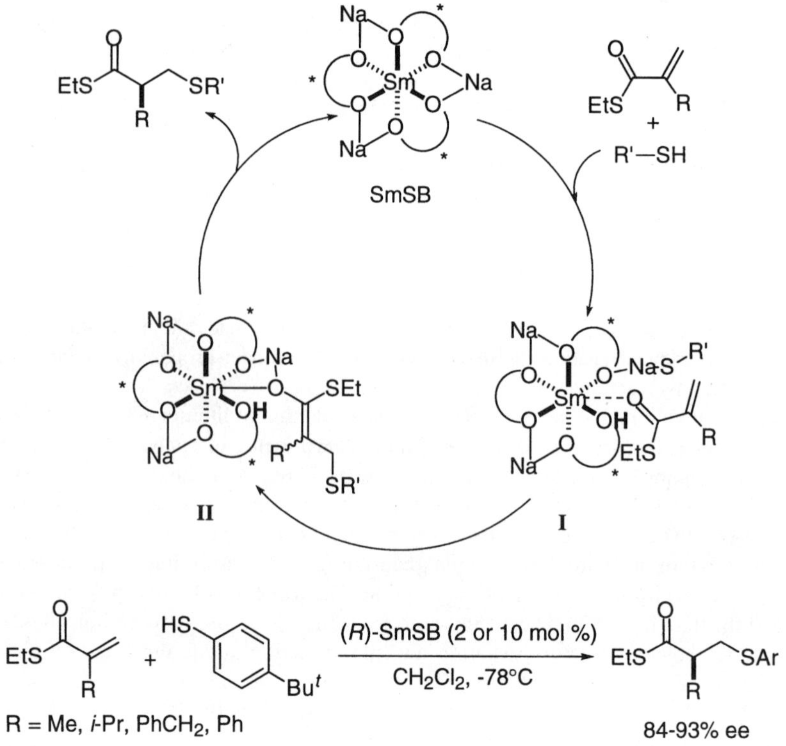

Figure 8D.9.

The addition of oxygen nucleophiles (peroxides) to α,β-unsaturated ketones is also catalyzed by the lanthanoid catalysts, leading to the formation of the corresponding epoxides with up to 96% ee (Scheme 8D.19) [41]. This reaction shall be reviewed in another chapter.

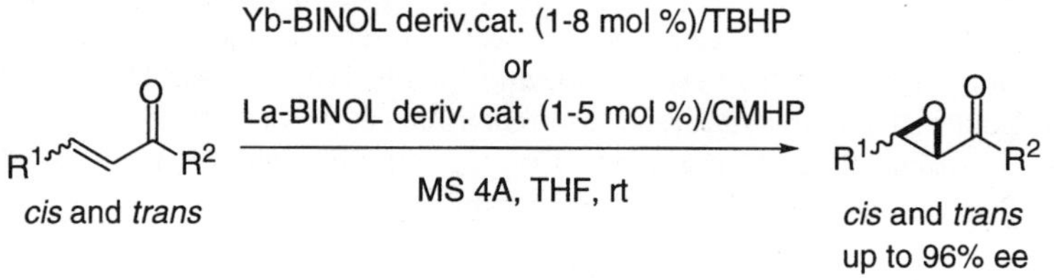

Scheme 8D.19.

8D.6. CONCLUSION AND FUTURE PROSPECTS

Recent advances in the catalytic asymmetric Michael reactions have made it possible to achieve enantioselectivity higher than 90% ee in these processes rather routinely. These reactions have just begun to be used as key steps in the syntheses of natural products or other useful compounds. In many cases, however, more than 10 mol % of catalysts are necessary to achieve high chemical yield and enantioselectivity at present. In consideration of the limited natural resources on

Mother Earth and recent environmental problems, the development of catalysts exerting much higher catalytic activity and higher enantioselectivity with broad applicability is still an urgent demand and a challenge for synthetic chemists.

REFERENCES

1. Kharasch, M. S.; Tawney, P. O. *J. Am. Chem. Soc.* **1941**, *63*, 2308–2316.

2. Reviews: (a) Posner, G. H. *Org. React.* **1972**, *19*, 1–113. (b) Lipshutz, B. H.; Sengupta, S. *Org. React.* **1992**, *41*, 135–631.

3. (a) Nakamura, E.; Mori, S.; Morokuma, K. *J. Am. Chem. Soc.* **1997**, *119*, 4900–4910. (b) Krause, M.; Wagner, R.; Gerold, A. *J. Am. Chem. Soc.* **1994**, *116*, 381–382. (c) Christenson, B.; Olsson, T.; Ullenius, C. *Tetrahedron* **1989**, *45*, 523–534. (d) Bertz, S. H.; Smith, A. J. *J. Am. Chem. Soc.* **1989**, *111*, 8276–8277. (e) Corey, E. J.; Boaz, N. W. *Tetrahedron Lett.* **1985**, *26*, 6015–6018. (f) Corey, E. J.; Boaz, N. W. *Tetrahedron Lett.* **1984**, *25*, 3063–3066.

4. Kanai, M.; Tomioka, K. *Tetrahedron Lett.* **1995**, *36*, 4275–4278.

5. Other examples: (a) Stangeland, E. L.; Sammakia, T. *Tetrahedron* **1997**, *53*, 16503–16510. (b) van Klaveren, M.; Lambert, F.; Eijkelkamp, D. J. F. M.; Grove, D. M.; van Koten, G. *Tetrahedron Lett.* **1994**, *35*, 6135–6138. (c) Zhou, Q.-L.; Pfaltz, A. *Tetrahedron* **1994**, *50*, 4467–4478. (d) Spescha, M.; Rihs, G. *Helv. Chim. Acta* **1993**, *76*, 1219–1230. (e) Ahn, K.-H.; Klassen, R. B.; Lippard, S. J. *Organometallics* **1990**, *9*, 3178–3181. (f) Review: Rossiter, B. E.; Swingle, N. M. *Chem. Rev.* **1992**, *92*, 771–806.

6. (a) Knochel, P.; Singer, R. D. *Chem. Rev.* **1993**, *93*, 2117–2188. (b) Nakamura, M.; Nakamura, E. *J. Synth. Org. Chem. Jpn.* **1998**, *56*, 632–644.

7. (a) Lipshutz, B. H.; Wood, M. R.; Tirado, R. *J. Am. Chem. Soc.* **1995**, *117*, 6126–6127. (b) Tamaru, Y.; Tanigawa, H.; Yamamoto, T.; Yoshida, Z. *Angew. Chem., Int. Ed.* **1989**, *28*, 351–353. (c) Nakamura, E.; Kuwajima, I. *J. Am. Chem. Soc.* **1984**, *106*, 3368–3370.

8. Soai, K.; Hayasaka, T.; Ugajin, S. *Chem. Commun.*, **1989**, 516–517.

9. Bolm, C.; Ewald, M.; Felder, M. *Chem. Ber.* **1992**, *125*, 1205–1215.

10. de Vries, A. H. M.; Jansen, J. F. G. A.; Feringa, B. L. *Tetrahedron* **1994**, *50*, 4479–4491.

11. Alexakis, A.; Frutos, J.; Mangeney, P. *Tetrahedron: Asymmetry* **1993**, *4*, 2427–2430.

12. Feringa, B. L.; Pineschi, M.; Arnold, L. A.; Imbos, R.; de Vries, A. H. M. *Angew. Chem., Int. Ed.* **1997**, *36*, 2620–2623.

13. (a) Takaya, Y.; Ogasawara, M.; Hayashi, T.; Sakai, M.; Miyaura, N. *J. Am. Chem. Soc.* **1998**, *120*, 5579–5580. (b) Takaya, Y.; Ogasawara, M.; Hayashi, T. *Tetrahedron Lett.* **1998**, *39*, 8479–8482.

14. Kobayashi, S.; Suda, S.; Yamada, M.; Mukaiyama, T. *Chem. Lett.* **1994**, 97–100. During the preparation of this manuscript, a more general example was reported: Evans, D. A.; Rovis, T. ; Kozlowski, M. C.; Tedrow, J. S. *J. Am. Chem. Soc.* **1999**, *121*, 1994–1995.

15. Catalytic asymmetric Michael reactions other than those in this chapter: (a) Helder, R.; Wynberg, H. *Tetrahedron Lett.* **1975**, 4057. (b) Hermann, K.; Wynberg, H. *J. Org. Chem.* **1979**, *44*, 2238. (c) Brunner, H.; Hammer, B. *Angew. Chem., Int. Ed.* **1984**, *23*, 312. (d) Sera, A.; Takagi, K.; Katayama, H.; Yamada, H.; Matsumoto, K. *J. Org. Chem.* **1988**, *53*, 1157. (e) Tamai, Y.; Kamifuku, A.; Koshiishi, E.; Miyano, S. *Chem. Lett.* **1995**, 957. (f) Desimoni, G.; Dusi, G.; Faita, G.; Quadrelli, P.; Righetti, P. *Tetrahedron* **1995**, *51*, 4131. (g) Aoki, S.; Sasaki, S.; Koga, K. *Tetrahedron Lett.* **1989**, *30*, 7229. (h) Sawamura, M.; Hamashima, H.; Shinoto, H.; Ito, Y. *Tetrahedron Lett.* **1995**, *36*, 6479 and references cited therein. (i) Takasu, M.; Wakabayashi, H.; Furuta, K.; Yamamoto, H. *Tetrahedron Lett.* **1988**, *29*, 6943. (j) Aoki, S.; Sasaki, S.; Koga, K. *Heterocycles* **1992**, *33*, 493. (k) Kawara, A.; Taguchi, T. *Tetrahedron Lett.* **1994**, *35*, 8805. (l) Minickam, G.; Sundararajan, G. *Tetrahedron Asymmetry* **1997**, *8*, 2271–2278. (m) Bakó, P.; Vizvárdi, K.; Bajor, Z.; Tōke, L. *Chem. Commun.* **1998**, 1193–1194. (n) Conn, R. S. E.; Lovell, A. V.; Karady, S.; Weinstock, L. M. *J. Org. Chem.* **1986**,

51, 4710–4711. (o) Chen, Z.; Zhu, G.; Jiang, Q.; Xiao, D.; Cao, P.; Zhang, X. *J. Org. Chem.* **1998**, *63*, 5631–5635. (p) End, N.; Macko, L.; Zehnder, M.; Pfaltz, A. *Chem. Eur. J.* **1998**, *4*, 818–824. (q) Alexakis, A.; Vastra, J.; Burton, J.; Benhaim, C.; Mangeney, P. *Tetrahedron Lett.* **1998**, *39*, 7869–7872.

16. Cram, D. J.; Sogah, G. D. Y. *Chem. Commun.* **1981**, 625–627.

17. (a) Shibasaki, M.; Sasai, H.; Arai, T. *Angew. Chem., Int. Ed.* **1997**, *36*, 1236–1256. (b) Shibasaki, M.; Sasai, H.; Arai, T.; Iida, T. *Pure & Appl. Chem.* **1998**, *70*, 1027–1034. (c) Shibasaki, M.; Iida, T.; Yamada, Y. M. A. *J. Synth. Org. Chem. Jpn.* **1998**, *56*, 344–356.

18. Sasai, H.; Arai, T.; Satow, Y.; Houk, K. N.; Shibasaki, M. *J. Am. Chem. Soc.* **1995**, *117*, 6194–6198.

19. Cockerill, A. F.; Davis, L. O.; Harden, R. C.; Rackham, D. M. *Chem. Rev.* **1973**, *73*, 553–588.

20. Rappé, A. K.; Casewit, C. J.; Colwell, K. S.; Goddard III, W. A.; Skiff, W. M. *J. Am. Chem. Soc.* **1992**, *114*, 10024; Casewit, C. J.; Colwell, K. S.; Rappé, A. K. *J. Am. Chem. Soc.* **1992**, *114*, 10035; Casewit, C. J.; Colwell, K. S.; Rappé, A. K. *J. Am. Chem. Soc.* **1992**, *114*, 10046; Rappé, A. K.; Colwell, K. S.; Casewit, C. J. *Inorg. Chem.* **1993**, *32*, 3438.

21. Sasai, H.; Emori, E.; Arai, T.; Shibasaki, M. *Tetrahedron Lett.* **1996**, *37*, 5561–5564.

22. Funabashi, K.; Saida, Y.; Kanai, M.; Arai, T.; Sasai, H.; Shibasaki, M. *Tetrahedron Lett.* **1998**, *39*, 7557–7558.

23. Arai, T.; Sasai, H.; Aoe, K.; Okamura, K.; Date, T.; Shibasaki, M. *Angew. Chem., Int. Ed.* **1996**, *35*, 104–106.

24. Canet, D.; Delpuech, J. J.; Khaddar, M. R.; Rubini, P. R. *J. Mag. Res.* **1973**, *9*, 329; Delpuech, J. J.; Khaddar, M. R.; Peguy, A. A.; Rubini, P. R. *J. Chem. Soc., Chem. Commun.* **1974**, 154; *Idem, J. Am. Chem. Soc.* **1975**, *97*, 3373.

25. Keller, E.; Veldman, N.; Spek, A. L.; Feringa, B. L. *Tetrahedron: Asymmetry* **1997**, *8*, 3403–3413.

26. Sasai, H.; Arai, T.; Shibasaki, M. *J. Am. Chem. Soc.* **1994**, *116*, 1571–1572.

27. Arai, T.; Yamada, Y. M. A.; Yamamoto, N.; Sasai, H.; Shibasaki, M. *Chem. Eur. J.* **1996**, *2*, 1368–1372.

28. Shimizu, S.; Ohori, K.; Arai, T.; Sasai, H.; Shibasaki, M. *J. Org. Chem.* **1998**, *63*, 7547–7551.

29. Arai, T.; Sasai, H.; Yamaguchi, K.; Shibasaki, M. *J. Am. Chem. Soc.* **1998**, *120*, 441–442.

30. (a) Nara, S.; Toshima, H.; Ichihara, A. *Tetrahedron Lett.* **1996**, *37*, 6745–6748. (b) Nara, S.; Toshima, H.; Ichihara, A. *Tetrahedron* **1997**, *53*, 9509–9524.

31. Yamada, K.-I.; Arai, T.; Sasai, H.; Shibasaki, M. *J. Org. Chem.* **1998**, *63*, 3666–3672.

32. (a) Yamaguchi, M.; Shiraishi, T.; Hirama, M. *Angew. Chem., Int. Ed.* **1993**, *32*, 1176–1177. (b) Yamaguchi, M.; Shiraishi, T.; Igarashi, Y.; Hirama, M. *Tetrahedron Lett.* **1994**, *44*, 8233–8236. (c) Yamaguchi, M.; Shiraishi, T.; Hirama, M. *J. Org. Chem.* **1996**, *61*, 3520–3530.

33. Corey, E. J.; Noe, M. C.; Xu, F. *Tetrahedron Lett.* **1998**, *39*, 5347–5350.

34. Corey, E. J.; Xu, F.; Noe, M. C. *J. Am. Chem. Soc.* **1997**, *119*, 12414–12415.

35. (a) Giese, B. *Radical in Organic Synthesis. Formation of Carbon–Carbon Bond*; Pergamon: Oxford, 1986. (b) Giese, B. *Angew. Chem., Int. Ed.* **1989**, *28*, 969.

36. Sibi, M. P.; Ji, J. *J. Org. Chem.* **1997**, *62*, 3800–3801.

37. Davies, I. W.; Gerena, L.; Castonguay, L.; Senanayake, C. H.; Larsen, R. D.; Verhoeven, T. R.; Reider, P. J. *Chem. Commun.* **1996**, 1753–1754.

38. Sibi, M. P.; Shay, J. J.; Liu, M.; Jasperse, C. P. *J. Am. Chem. Soc.* **1998**, *120*, 6615–6616. During the preparation of this manuscript, highly efficient asymmetric conjugate addition of HN_3 by a salen-Al complex was reported: Myers, J. K.; Jacobsen, E. N. *J. Am. Chem. Soc.* **1999**, *121*, 8959–8960.

39. Nishimura, K.; Ono, M.; Nagaoka, Y.; Tomioka, K. *J. Am. Chem. Soc.* **1997**, *119*, 12974–12975.

40. Emori, E.; Arai, T.; Sasai, H.; Shibasaki, M. *J. Am. Chem. Soc.* **1998**, *120*, 4043–4044.

41. (a) Bougauchi, M.; Watanabe, S.; Arai, T.; Sasai, H.; Shibasaki, M. *J. Am. Chem. Soc.* **1997**, *119*, 2329–2330. (b) Watanabe, S.; Arai, T.; Sasai, H.; Bougauchi, M.; Shibasaki, M. *J. Org. Chem.* **1998**, *63*, 8090–8091. (c) Watanabe, S.; Kobayashi, Y.; Arai, T.; Sasai, H.; Bougauchi, M.; Shibasaki, M. *Tetrahedron Lett.* **1998**, *39*, 7353–7356.

8E

ASYMMETRIC ALLYLIC ALKYLATION REACTIONS

BARRY M. TROST AND CHULBOM LEE
Department of Chemistry, Stanford University, Stanford, CA 94305-5080

8E.1. INTRODUCTION

Inducing chirality in chemical synthesis by catalytic methods represents one of the most important frontiers in modern organic chemistry [1]. Among various approaches to address this challenge, the use of chiral metal catalysts is a particularly appealing strategy because enantiocontrol may often be achieved by simple use of a chiral ligand for the reaction. Although the most thoroughly studied asymmetric transition-metal catalyzed reactions involve transfers of hydrogen [2] and oxygen (epoxidation [3] and dihydroxylation [4]), the alkylation reaction presents an advantage to form many different types of carbon–carbon and carbon–heteroatom bonds. Transition-metal catalyzed allylic alkylation reactions have emerged as one of the more powerful tools for the controlled introduction of various chemical bonds to organic compounds (Eq. 8E.1) [5–8]. The reaction involves a (π-allyl)metal complex as a key intermediate, which can be exploited for various transformations with high chemo-, regio- and stereoselectivities [9–11]. Although the ambiphilic nature of a metal-bound allyl ligand allows the (π-allyl)metal species to behave as both electrophile and nucleophile, this review focuses on those reactions in which the allyl unit undergoes a nucleophilic displacement. This process is catalyzed by a variety of transition-metal complexes derived from nickel, palladium, platinum, cobalt, rhodium, iridium, iron, ruthenium, molybdenum, and tungsten.

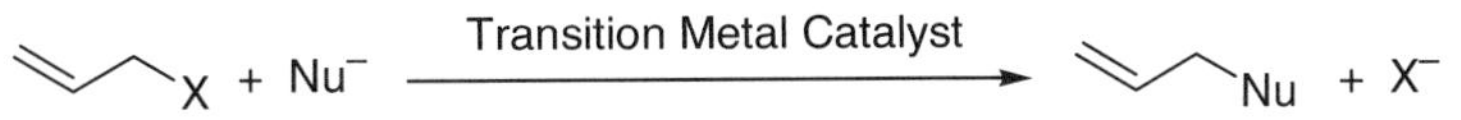

$$\text{X} + \text{Nu}^- \xrightarrow{\text{Transition Metal Catalyst}} \text{Nu} + \text{X}^- \qquad (8E.1)$$

Since the first example of inducing asymmetry at the allylic fragment with a palladium catalyst in 1977 [12], the asymmetric allylic alkylation (AAA) reaction has undergone a revolutionary development in recent years to establish its synthetic viability. New ligands of

Catalytic Asymmetric Synthesis, Second Edition, Edited by Iwao Ojima
ISBN 0-471-29805-0 Copyright © 2000 Wiley-VCH, Inc.

various types have been synthesized and tested in reactions involving a wide range of substrates and nucleophiles. Consequently, new reactions have been developed to give enantiomerically pure products whose enantiopurities are routinely greater than 90% ee. Several reviews have already covered the chemistry of enantioselective transition-metal catalyzed allylic alkylations up to the mid-1990s [13–16]. This chapter presents an overview of the various approaches for asymmetric allylic alkylation, with particular emphasis on their synthetic applications, which began to appear in the literature only within the past few years. Examples presented herein are organized primarily according to the mode of enantioselection because understanding the origin of enantiodiscrimination in the individual mechanistic step or bond formation is indispensable in designing such reactions as well as extending to synthetic applications.

8E.2. FUNDAMENTAL ELEMENTS OF ENANTIOSELECTIVE ALLYLIC ALKYLATION

8E.2.1. Catalytic Cycle and Mechanism

The mechanism of metal-catalyzed allylic alkylation is generally believed to involve the four fundamental steps illustrated in Scheme 8E.1 [17–20]. The key feature of the catalytic cycle is the intermediacy of a (π-allyl)metal complex. Its generation and subsequent reaction represent, respectively, the bond-breaking and -making events in which the source of chiral induction can be derived. Depending on the structure of the substrate, every step provides an opportunity for enantioselection, except for decomplexation, which occurs after bond formation. Because the rates of these catalytic steps are delicately balanced, both the rate-limiting step and the selectivity-determining step may vary depending on the reaction conditions.

Although ionization usually proceeds with inversion of stereochemistry regardless of the nucleophile, the addition of a nucleophile involves two pathways in which the nature of the nucleophile leads to different stereochemical consequences (Scheme 8E.2). The "soft" (stabilized) nucleophiles, defined as those derived from conjugate acids whose $pK_a < 25$, usually add to the allyl ligand from the side opposite to the metal, giving rise to product with overall retention of stereochemistry [21–23]. On the other hand, with "hard" (unstabilized) nucleophiles ($pK_a > 25$), the addition normally occurs via transmetallation where the nucleophile attacks the metal center of the π-allyl intermediate and the resultant adduct collapses to product by reductive

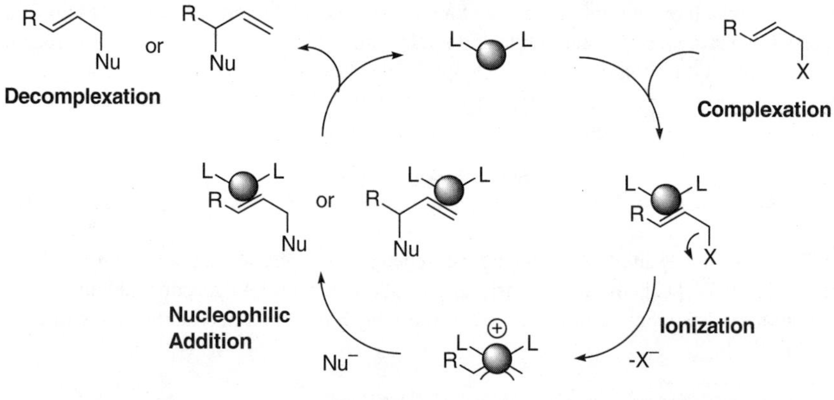

Scheme 8E.1. Catalytic cycle of transition metal catalyzed allylic alkylation.

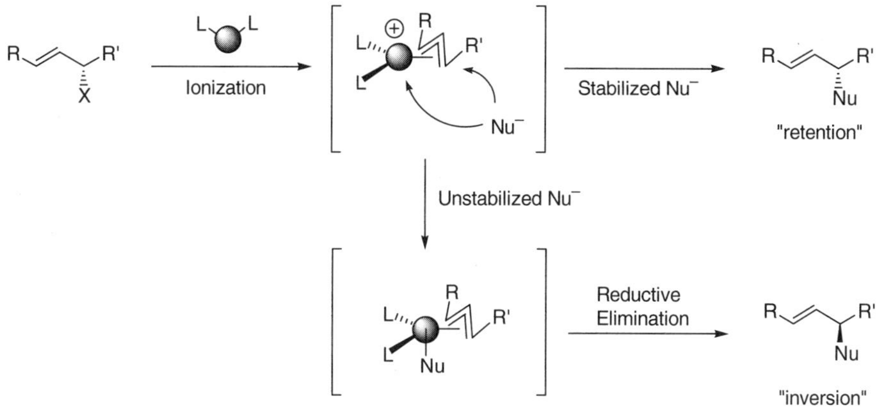

Scheme 8E.2. Addition of stabilized and unstabilized nucleophiles.

elimination, resulting in inversion of stereochemistry [24–30]. Exceptions to these generalities have been claimed.

In the alkylation with most commonly used stabilized nucleophiles, one quite obvious problem associated with this process is the physical separation between the chiral ligand and the reaction site. Both bond-breaking and -forming occur on the π-allyl face opposite the metal and its attendant ligand. Thus, efficient transfer of the chirality over the allyl barrier is key to a successful strategy for asymmetric allylic alkylation.

8E.2.2. Stereodynamics of Transition-Metal–Allyl Complexes

The cationic π-allyl complex is often isolable and has been the subject of considerable study. X-ray structures of the π-allyl complexes with chiral ligands have been a primary source of structural information from which the design and predictive model of chiral catalysts derive. Chiral-metal–olefin complexes, which constitute another important class of intermediates have also been isolated, albeit few in number [31]. These static studies have been complemented by a growing number of NMR studies taking advantage of modern heteronuclear correlation and NOE techniques, which offer opportunities to monitor solution structures of the catalytic species [32–34].

The (π-allyl)metal complexes exist in dynamic equilibrium under typical alkylation reaction conditions during which their conformation and geometry may change by intramolecular processes as well as by reversible ligand-dissociation and reassociation processes. In general, the equilibration occurs at a faster or comparable rate compared with that of the nucleophilic addition. Depending on the mechanism from which enantioselectivity derives, such behavior of (π-allyl)metal complexes can be detrimental or mandatory. Thus, the understanding of this fundamental phenomenon in the (π-allyl)metal chemistry is necessary to correctly interpret otherwise confusing results.

A metal-bound allyl ligand can undergo geometric interconversion by changing its hapticity. The well known π–σ–π (η^3–η^1–η^3) isomerization, which involves the rotation around the carbon–carbon bond in the allyl unit, leads to the *syn/anti* interconversion as illustrated in Scheme 8E.3. This process does not switch the relative position of the allyl carbons and ligands in the square-planar structure; the carbon-bearing R_1 group, which changed its configuration,

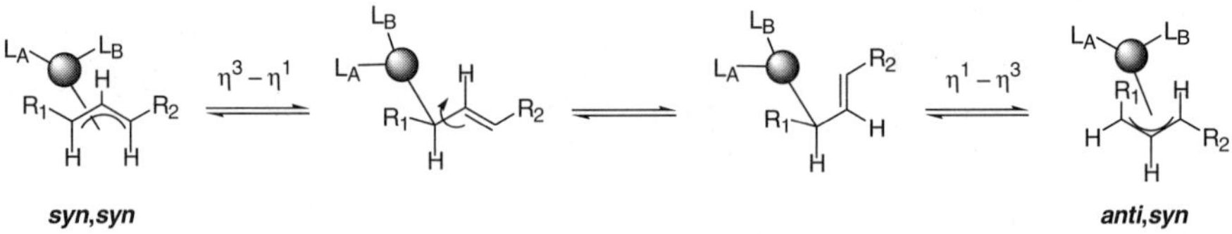

Scheme 8E.3. *Syn/anti* isomerization by η^3-η^1-η^3 mechanism.

still remains *trans* to the ligand L_B. If the metal is bound to a non-C_2 symmetric ligand, the exchange process represents an important *endo/exo* isomerization process: the "W"-shaped allyl unit becomes "M"-shaped.

Another type of isomerization is apparent "allyl rotation" which results in an *endo/exo* isomerization without change of *syn/anti* geometry. There is no evidence for a simple mechanism involving rotation around the metal-allyl axis. The reorganization of the relative orientation between the ligand and the allyl unit is more likely to follow the mechanisms depicted in Scheme 8E.4. Note that the *cis/trans* descriptor is used instead of the "W" versus "M" relationship in this case. First, changing the hapticity followed by rotating the metal–carbon bond can effect isomerization in a similar fashion to the previous *syn/anti* exchange process [35,36]. These dissociative processes involving trisubstituted square planar d^8 complexes are likely to depend on the donor/acceptor properties of the ligands [37]. Alternatively, NMR experiments using dinitrogen ligands suggest that dissociation of one of the ligands can induce allyl rotation via isomerization, followed by reassociation [38,39]. External ligands such as halide ions [38,40–42] and solvents [43] are also known to accelerate the allyl rotation process by an associative mechanism. The apparent allyl rotation may be completed by pseudorotation of the pentacoordinated allyl complex, or the external ligand may induce allyl dissociation, which then enters into the mechanistically indistinguishable pathways involving the change of hapticity.

Isomerization of a π-allyl species can arise from a bimolecular reaction as shown in Scheme 8E.5 [44]. In the case of unsymmetrically substituted allyl systems, this process constitutes

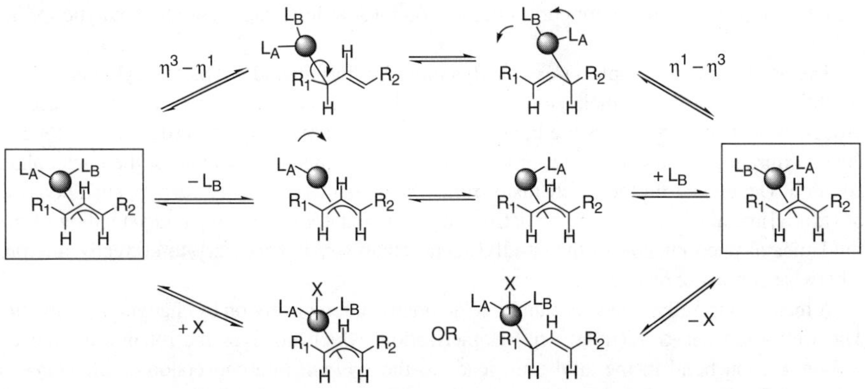

Scheme 8E.4. Mechanisms for apparent allyl rotation.

Scheme 8E.5. Mechanism for racemization.

enantioface exchange; the stereochemistry of all three allyl carbons is simultaneously inverted. This path can also be accomplished by ligand exchange of the type seen in Scheme 8E.4. Although the precise mechanism for this apparent S_N2 type reaction is not yet clear, several factors facilitating or inhibiting this intermolecular ligand exchange are known [20,45].

8E.2.3. Strategies for Enantioselection and Ligand Design

A unique feature of the transition-metal–catalyzed allylic alkylation is its ability to convert starting materials of various symmetry types, such as racemic, *meso-* and achiral compounds, into optically pure material. Strategies to effect such transformations derive from recognition of the stereochemical courses in each step of the catalytic cycle and analysis of symmetry elements in the substrate or intermediate. Figure 8E.1 summarizes potential sources of enantiodiscrimination in transition-metal–catalyzed allylic alkylation.

The first strategy involves discrimination between enantiotopic leaving groups (Type A). In the second approach, two enantiomers of a racemic substrate converge into a *meso-π*-allyl complex wherein preferential attack of the nucleophile at one of either allylic termini leads to asymmetric induction, a process that may be referred to as a dynamic kinetic enantioselective transformation (Type B). The third requires differentiation between two enantiotopic transition

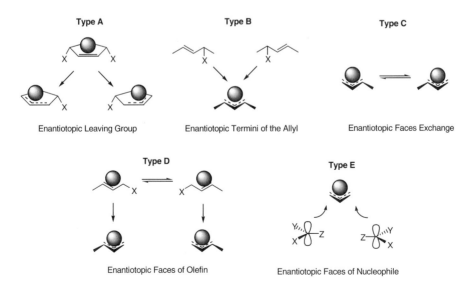

Figure 8E.1. Various sources of enantiodiscrimination in asymmetric allylation.

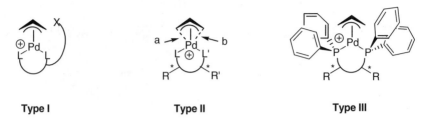

Type I   **Type II**   **Type III**

Figure 8E.2. Concepts used in ligand design.

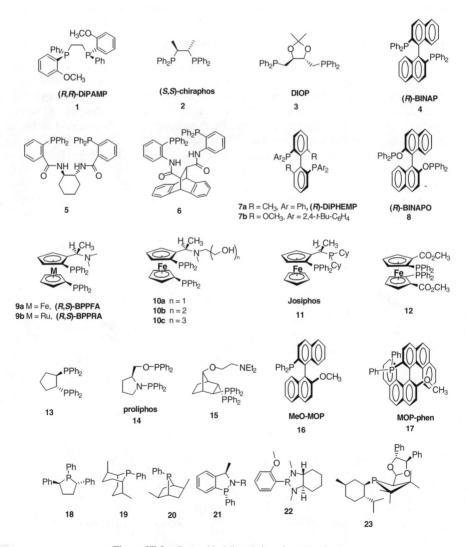

Figure 8E.3. Some chiral ligands based on phosphorus.

states in the nucleophilic addition to rapidly interconverting enantiomeric π-allyl intermediates, also a dynamic kinetic enantioselective transformation (Type C). Alternatively, an enantiomeric intermediate is generated from enantiofacial complexation of an alkene (Type D). When the nucleophile is a prochiral compound or a rapidly equilibrating racemic mixture, asymmetric induction can be also possible at the nucleophile (Type E).

The problem in metal-catalyzed allylic alkylation has naturally led to design of chiral catalysts in such a way that asymmetry is effectively transmitted to the reaction event occurring distal to the ligand. As depicted in Figure 8E.2, three concepts have been proposed: 1) attaching a functional tether to the ligand to induce interaction with an incoming nucleophile [46,47]; 2) imposing electronic desymmetrization on the donor atoms of the ligand whereby different bond lengths **a** and **b** promote differential reactivity at each allylic terminus [48]; and 3) providing the substrate with chiral space, which may derive, for example, by a propeller-like array of aryl groups whose chirality is induced by a conformational bias of edge-face interactions which, in turn, ultimately originates from primary stereogenic centers [49]. In all of these cases, non-C_2 symmetric as well as symmetric ligands may be envisioned.

8E.2.4. Chiral Ligands

As many endeavors in transition-metal catalysis, the design, synthesis and screening of chiral ligands have played a pivotal role in the development of the asymmetric allylic alkylation reaction. A series of C_2-symmetric diphosphines such as DiPAMP, chiraphos, DIOP, and BINAP,

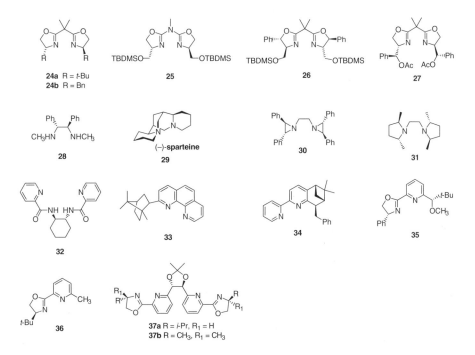

Figure 8E.4.   Chiral dinitrogen ligands.

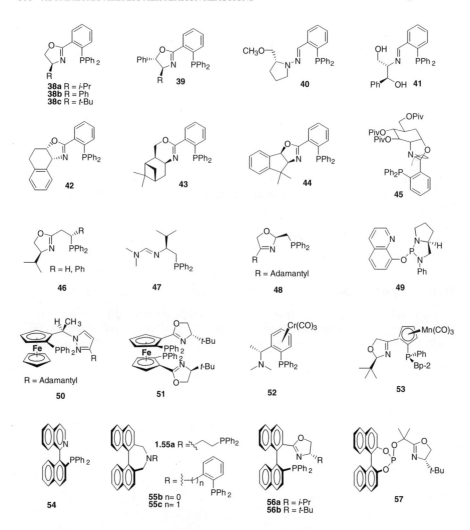

Figure 8E.5. Chiral P,N-chelate ligands.

which gave high enantioselectivities in asymmetric hydrogenations, were used in early work. However, these "borrowed ligands" exhibited only modest performance in this area. Soon after, many new ligands have been developed by modifying these relatively simple structured ligands. Virtually all of the elements constituting a chiral ligand such as the type of symmetry, donor atoms, mode of denticity (e.g., mono- vs. bidentation), and combination of these have been explored. Consequently, impressive progress has been made in the design of ligands, which are primarily targeted for AAA.

Figures 8E.3–8E.6 show a selection of some ligands that have been successfully used in the palladium-catalyzed AAA reaction. They are classified into four categories: 1) chiral phosphines, 2) dinitrogen ligands, 3) P,N-chelate ligands, and 4) mixed chelate ligands. Although this

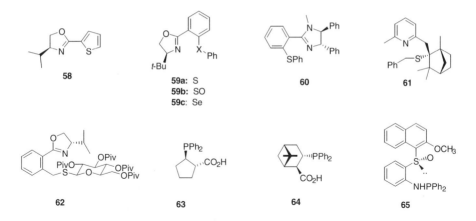

Figure 8E.6. Chiral ligands with mixed donor atoms.

division is primarily made based on the type of donor atoms, each category is not mutually exclusive.

The chirality at the phosphorus atom such as in DiPAMP (**1**), which was originally thought a necessary element, is not required to achieve high enantioselectivity. Neither is C_2-symmetry of the ligand although it provides a simple way of reducing the number of possible transition states by a factor of two [50]. A growing number of new ligands, which lack C_2-symmetry, have given successful results in many different reactions. In addition, it has been shown that high enantioselectivities can also be obtained from the use of several monophosphines. Another notable feature of the ligand morphology is the diverse source of the ligand chirality. Starting from the point chirality of phosphorus and carbon atoms, various types of stereogenicity such as axial and planar chirality and combination of these have been utilized.

Although most of the early studies focused on the use of phosphine ligands, recently, nitrogen ligands have been used in enantioselective allylic alkylations with increasing frequency. Both sp^2 (imine) and sp^3 (amine) hybridized nitrogens have proved effective as donor atoms. As shown through the examples, a remarkably high level of enantioselectivity has been recorded with the dinitrogen ligands listed in Figure 8E.4. However, up to the present, such strong donor-type ligands have exhibited limited scope in the catalytic processes.

The distinct electronic (i.e., electronic desymmetrization) and steric properties of different donor atoms influence the reactivity of allyl substrates through the metal. Much interest has recently been drawn to this type of ligand, resulting in the synthesis of a variety of ligands involving metal coordination by both N and P as illustrated in Figure 8E.5. Bidentate ligands with different ligating atoms such as phosphinooxazolines can lead to high enantioselectivity. It is thought that the oxazoline moiety provides the source of asymmetry and the phosphorus atom enhances reactivity. The stereochemical outcome is most consistent with the mechanism involving nucleophilic addition *trans* to the phosphorus atom as supported by X-ray and NMR studies.

Variations on this theme have led to the design of mixed chelate-type ligands as shown in Figure 8E.6. Bidentate ligands derived from various combinations of phosphorus, nitrogen, oxygen, sulfur, and selenium are able to provide high enantioselectivities. Similar to the P,N-chelate ligands, the origin of enantioselectivity has been proposed to be the preferential nucleophilic attack trans to the better π-acceptor. However, the poor binding ability of some donor atoms raises doubts as to the bidentate binding of these ligands. Therefore, the results obtained with these ligands should be interpreted with caution.

8E.3. PALLADIUM-CATALYZED ENANTIOSELECTIVE ALLYLATION

8E.3.1. Enantioselective Ionization: Desymmetrization of *meso*-Diol Derivatives

8E.3.1.1. Desymmetrization of Cyclic meso-2-ene-1,4-diesters Asymmetric induction in the ionization step depends on the ability of the chiral catalyst to promote differential ionization of enantiotopic leaving groups. Because this event occurs independently of the nucleophile, interpretation of the resulting enantioselectivity becomes relatively uncomplicated. Desymmetrization of *meso*-2-ene-1,4-diols as shown in Scheme 8E.6 is particularly interesting not only because the alkylation products have high synthetic potential as demonstrated by the syntheses of mannostatin [51] and allosamizoline [52] but also the simplicity of the process provides a good testing ground for evaluating ligand efficacy [53,54].

Although modest, yet encouraging, enantioselectivities (55% ee) were obtained from an early effort with BINAPO (**8**) as ligand, palladium-catalyzed cyclization of bis-urethane **67** to oxazolidinone **68** or **69** has led to development of various ligands by a modular synthesis [55,56]. As illustrated in Scheme 8E.7, primary chirality of a chiral scaffold induces conformational chirality through a linker which, in turn, creates chiral space. Thus, the substrate binds to palladium and undergoes ionization in the "chiral pocket" provided by propeller-like array of phenyl groups on the phosphorus atom. Using readily available 2-(diphenylphosphino)benzoic acid (DPPBA) and 2-(diphenylphosphino)aniline (DPPA) as platforms, a series of C_2 symmetric ligands have been easily constructed with various chiral diols, diamines and dicarboxylic acids as shown in Figure 8E.7 [56,57].

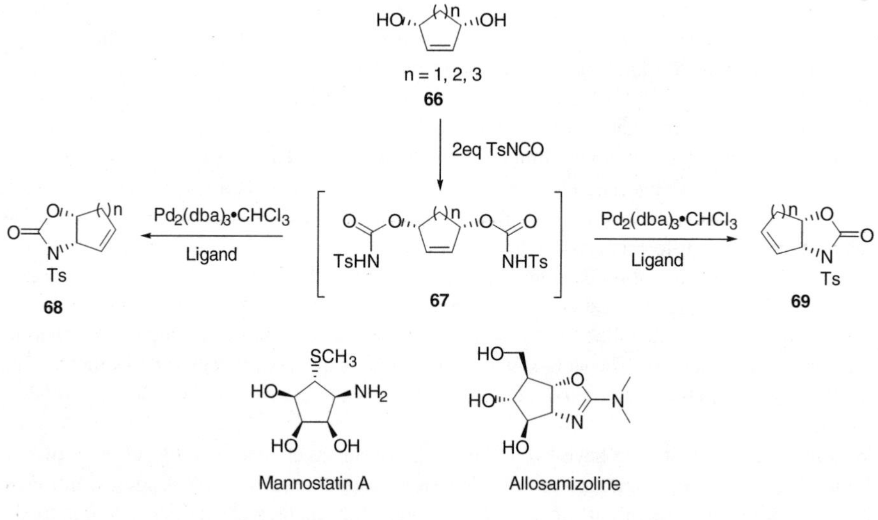

Scheme 8E.6. Desymmetrization of *meso*-2-ene-1,4-diols by intramolecular alkylation.

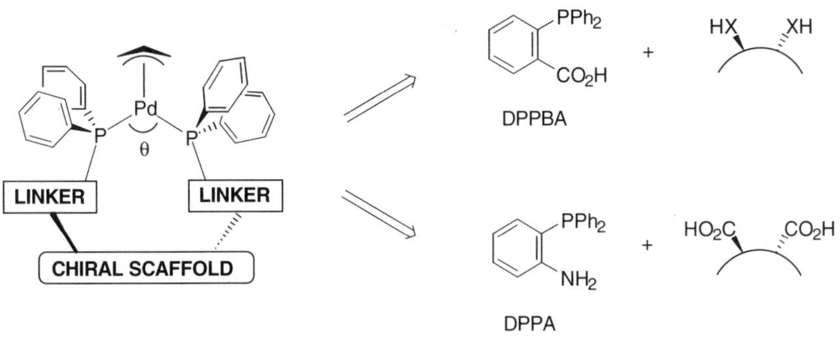

Scheme 8E.7. Modular design and synthesis of ligands.

The reactions with these ligands have revealed several interesting aspects as summarized in Table 8E.1. In the DPPBA series (Figure 8E.7), the ligands possessing the conformationally tighter amide linkage always give higher enantioselectivities than those based on the corresponding ester linkage. It is surprising that the invertomer ligand **75**, which maintains the same sense of chirality of the scaffold as the normal ligand **70** but has an inverted orientation of the amide

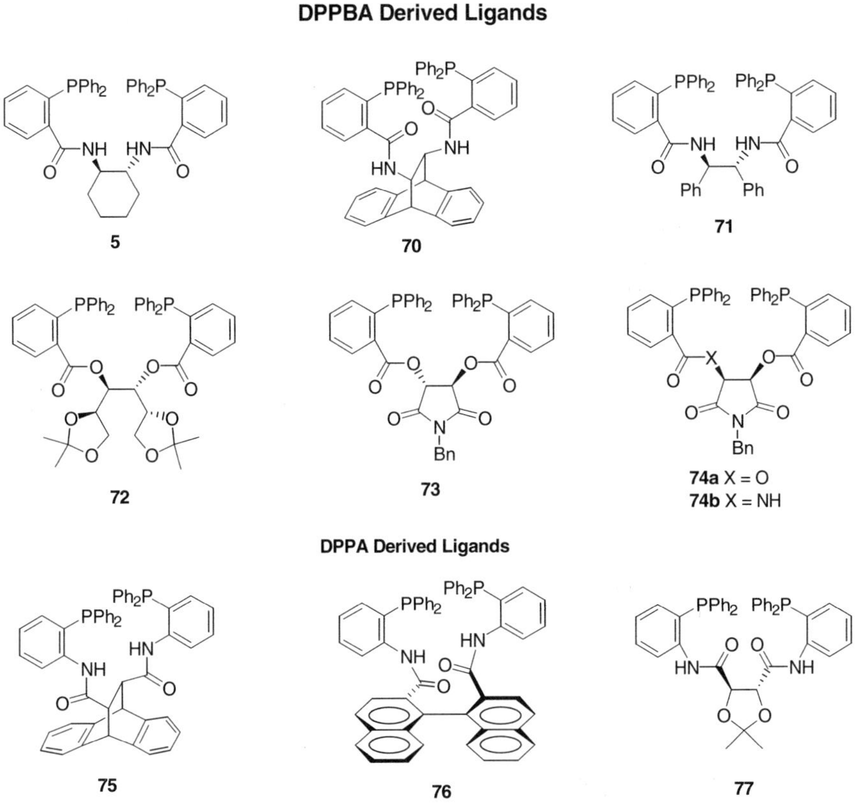

DPPBA Derived Ligands

DPPA Derived Ligands

Figure 8E.7. Modular ligands based on DPPBA and DPPA.

TABLE 8E.1. Enantioselective Formation of Oxazolidinone **68** and **69**

Entry	Ligand	Ring size	% Yield	% ee	Major enantiomer	Ref.
1	**72**	5	100	64	(+)-**68**	55
2	**73**	5	68	75	(−)-**69**	55
3	**74b**	5	nd^a	78	(−)-**69**	58
4	**71**	5	97	80	(+)-**68**	55
5	**70**	5	94	88	(−)-**69**	55
6	**75**	5	99	88	(+)-**68**	57
7	**75**	6	82	97	(+)-**68**	57
8	**75**	7	82	95	(+)-**68**	57
9	**5**	5	84	>99	(+)-**68**	56,59
10	**5**	6	85	99	(+)-**68**	56,59
11	**5**	7	70	99	(+)-**68**	56,59

^{a}Not determined.

function, reverses the sense of chirality in the product. Higher enantioselectivities are obtained as the dihedral angle θ between the two DPPBA or DPPA on the scaffold becomes larger. The beneficial effect of a larger dihedral angle is believed to derive from the increased P-Pd-P angle (i.e., bite angle) in the 13-membered chelated palladium complexes. This opening of the bite angle may provide a deeper chiral pocket wherein the alkene and allyl unit can experience enhanced chiral recognition. A considerably larger bite angle of 110.05° observed in an X-ray crystal structure of a π-allylpalladium complex derived from **75**, as compared with typical 90° of common square-planar complexes, lends support to this notion [57].

A most-intriguing observation is that the pseudo-*meso* ligand **74b** gives 78% ee, which is higher than the normal chiral tartrate ligand **73** (entries 2 vs. 3). This result indicates that the orientation of the chiral backbone does not directly determine the enantioselectivity. MM2 calculations on the olefin palladium complex of **74b** generated two low-energy conformations of the phenyl rings in which their propeller-like twists adopted opposite orientations. If the backbone is achiral or *meso* such as **74a**, the two conformers are equal in energy, thereby giving a racemic product. The large energy difference (5.4 kcal/mol) between the two chiral spaces derived from **74b** is more than enough to account for the 78% ee. Thus, the origin of enantioselectivity for this ligand is the interaction of the substrate with the propeller-like array of the phenyl rings whose chirality is induced by the chiral scaffold as is shown in Figure 8E.2 and Scheme 8E.7.

Another interesting aspect observed in the reactions with DPPBA-based ligands is that stereochemistry of the ligand correlates with enantiotopicity of ionization. As illustrated in Figure 8E.8, when viewed along its backbone, which is perpendicular to the principal C_2-axis, the two phosphine posts relate to each other either in clockwise or counterclockwise fashion. With the bis-urethane substrate oriented with the leaving groups pointing below and the palladium-diphosphine complex above the plane of the page, a clockwise-type ligand preferentially promotes ionization via a clockwise motion, or vice versa in the case of a counterclockwise-type ligand. This mnemonic provides convenience with which one can predict the sense of chirality of the product simply based on the stereochemistry of the ligand. Notably, the mnemonic works for all of the known "normal"-type ligands based on DPPBA. Furthermore,

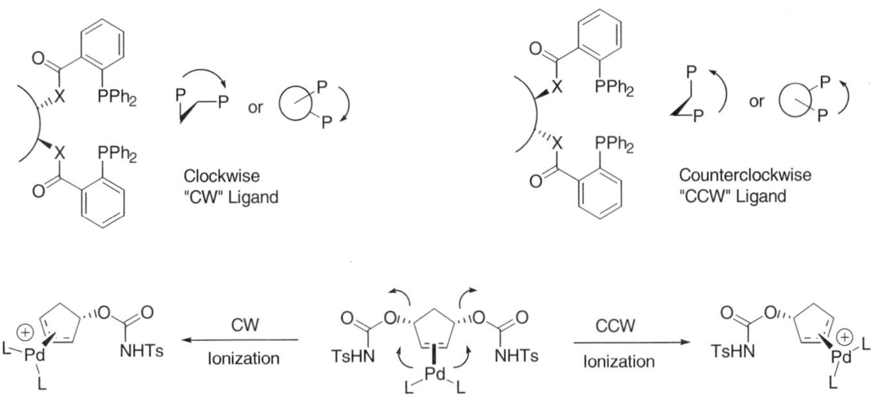

Figure 8E.8. Mnemonic for DPPBA-based ligand in the ionization.

it has been successfully applied in a number of reactions to predict the stereochemistry of products even when the enantioselectivity is determined by a nucleophilic addition step.

Initially, good enantioselectivities have been obtained by using preferably the "invertomer" ligand **75** rather than the "normal" ligand (entries 6–8). Recently, it has been found that the simple addition of triethylamine to the reactions with ligand **5** remarkably increases the enantioselectivities to 99% ee in five-, six- and seven-membered ring-forming reactions [59]. The lower enantioselectivities (78, 85, and 90% ee's, respectively for the reactions as in entries 9–11) obtained from earlier experiments, which were performed in the absence of triethylamine, seem to derive from the reversibility of the ionization process, not from the poor enantiodiscriminating ability of the ligand. Because the enantioselectivity erodes in accord with the reversal of the incipient π-allylpalladium intermediate to the bis-urethane, the higher enantioselectivities can be achieved by kinetic capture of the intermediate. The added base increases the

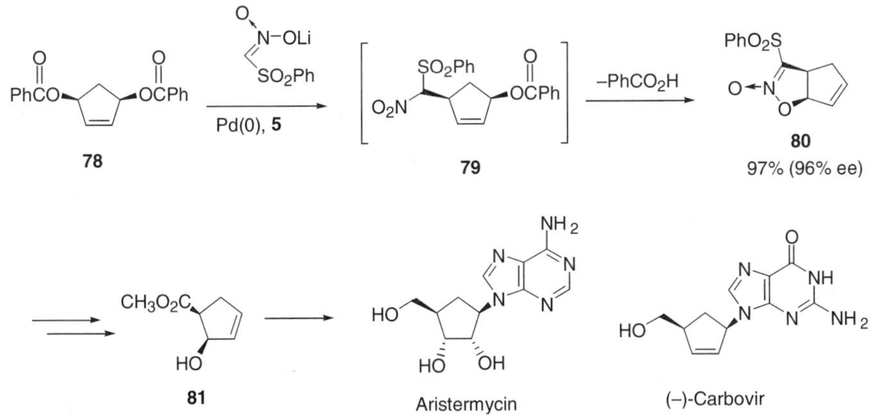

Scheme 8E.8. Asymmetric synthesis of isoxazoline *N*-oxides.

effective concentration of the nucleophile and thus accelerates the rate of nucleophilic addition, which, in this case, is a cyclization event.

This strategy of desymmetrizing *meso*-compounds worked well for intermolecular alkylations with the use of the lithium salt of benzenesulfonylnitromethane as nucleophile (Scheme 8E.8). The initially generated product **79** undergoes a further intramolecular cyclization to give the isoxazoline *N*-oxide **80** in 96% ee [60]. The stereochemical outcome of this reaction can be predicted by the same mnemonic as the intramolecular reactions. The introduced nitrosulfonylmethyl group was easily converted to the carbomethoxy group as in **81**, which becomes the hydroxymethyl unit of antiviral agents carbovir and aristeromycin.

To a first approximation, the chiral discrimination should be independent of the nucleophile. The palladium-catalyzed desymmetrization protocol utilizing a heterocyclic nucleophile provides enantio- and diastereoselective entries to diverse carbo-nucleosides. As shown in Scheme 8E.9, introduction of purine bases rather than the hydroxymethyl synthon also affords high enantioselectivities [61]. A variety of natural and unnatural nucleosides can be flexibly prepared because the simple change of ligand chirality or, alternatively, switching the alkylation sequence leads to opposite enantiomers. The palladium-catalyzed approach sharply contrasts with the chiral-pool method, whose enantiodivergency is limited by the availability of the starting material.

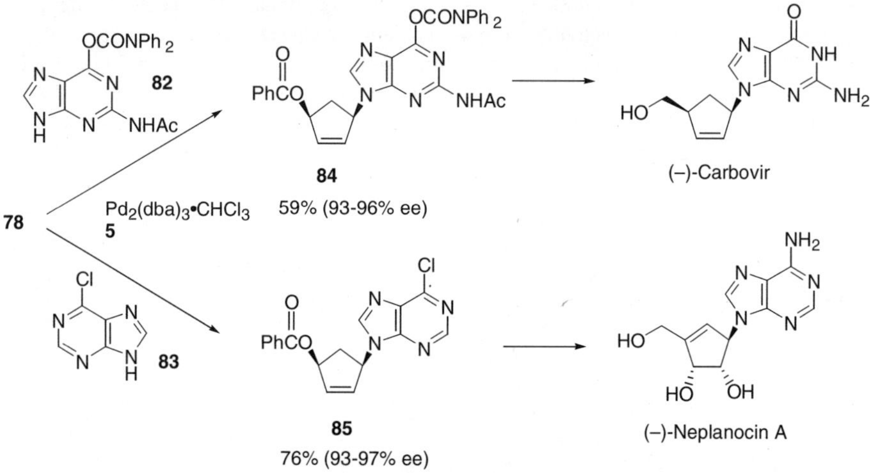

Scheme 8E.9. Desymmetrization with purine derivatives.

With the same protocol, a heterocyclic dibenzoate **86** derived from furan in one step has been efficiently desymmetrized to provide facile entry to either D or L nucleosides (see Scheme 8E.10). As depicted in Scheme 8E.10, the catalyst derived from ligand **71** gave rise to high enantioselectivities in the alkylation with both a purine **83** and a pyrimidine **87** [62]. Subsequent allylic alkylations with an achiral ligand introduced the tartronate and aminomalonate moieties to furnish enantiomerically pure *cis*-2,5-disubstituted-2,5-dihydrofurans **89** and **91**, respectively. Only six steps from furan were required to synthesize the *allo* and *talo* isomers of the nucleoside skeleton of the polyoxin-nikkomycin complexes. It should be noted that the corresponding enzymatic desymmetrization of substrate **86** is impossible because the product is labile.

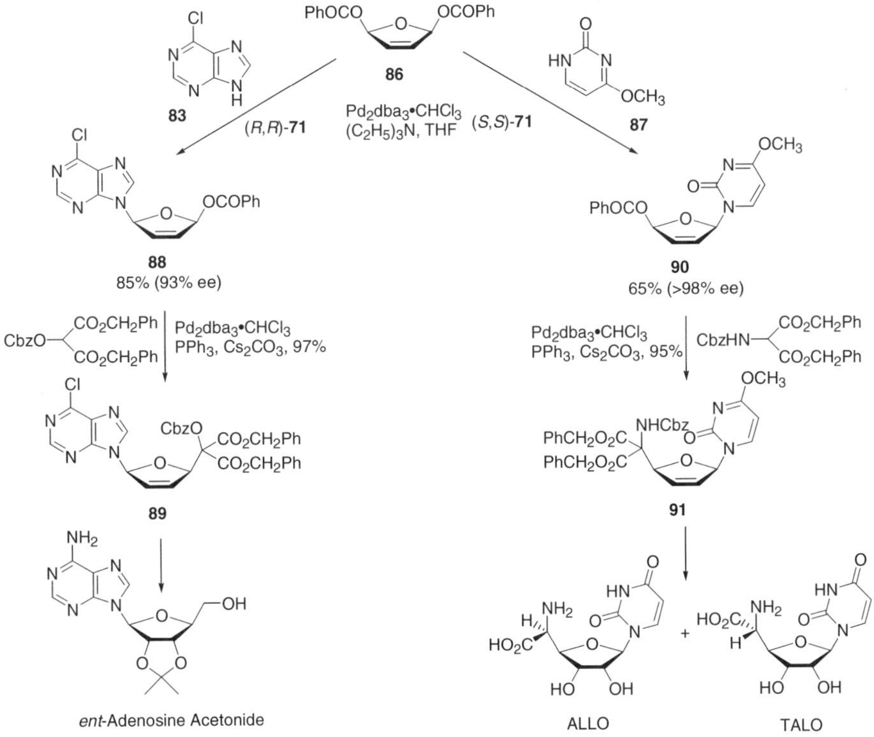

Scheme 8E.10.   Asymmetric synthesis of nucleosides from Furan.

Desymmetrization of cyclic *meso*-diols by this strategy is often advantageous over biological protocols because chiral induction occurs as part of the synthetic sequence and not as an additional step, and high enantioselectivities can be obtained regardless of the ring size. In particular, the metal-catalyzed method becomes more valuable in the desymmetrization of diols derived from a six-membered ring because the enzymatic procedure proved ineffective with these substrates. An application of the asymmetric isoxazoline *N*-oxide synthesis with six-membered ring substrate **92** resulted in efficient asymmetric synthesis of valienamine (Scheme 8E.11) [63]. In contrast to the five-membered ring case, the subsequent palladium-catalyzed cyclization step proceeded with an extremely slow rate due to the mismatched nature of this event with the chiral catalyst. Thus, after the first alkylation, an achiral catalyst had to be added for the subsequent cyclization to give the enantiopure bicyclic product **94**.

Cyclopropane and lactone functions can be prepared by the same double alkylation protocol as shown in Scheme 8E.12 [64]. Both dibenzyl malonate and Meldrum's acid served well as nucleophiles to give excellent enantioselectivities. While subjecting the malonate adduct **95** to the second alkylation generated cyclopropane **96**, the alkylation of the Meldrum's adduct **97** under similar conditions produced lactone **98** via an unusual *O*-alkylation.

With BINAPO as ligand, the alkylation of dibenzoate **92** with β-ketoester **101** provides the mono-alkylated product **102** in modest ee (Scheme 8E.13) [65]. Subsequently, a simultaneous allylation-Heck annulation reaction provided the pentacycle **103**, which was further function-

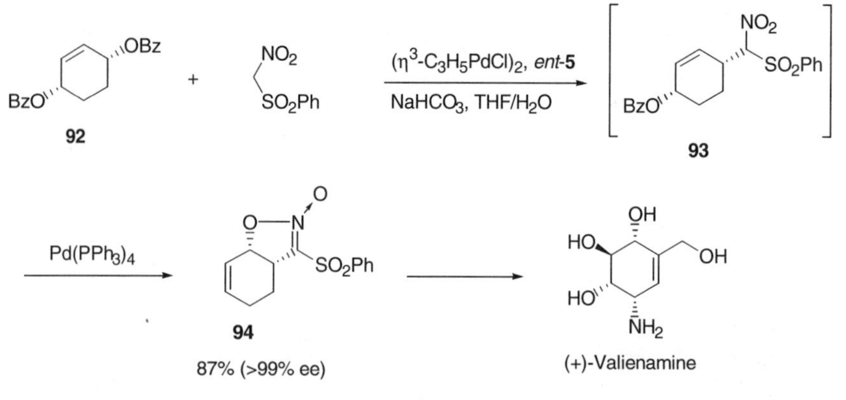

Scheme 8E.11. Synthesis of valienamine by double alkylation.

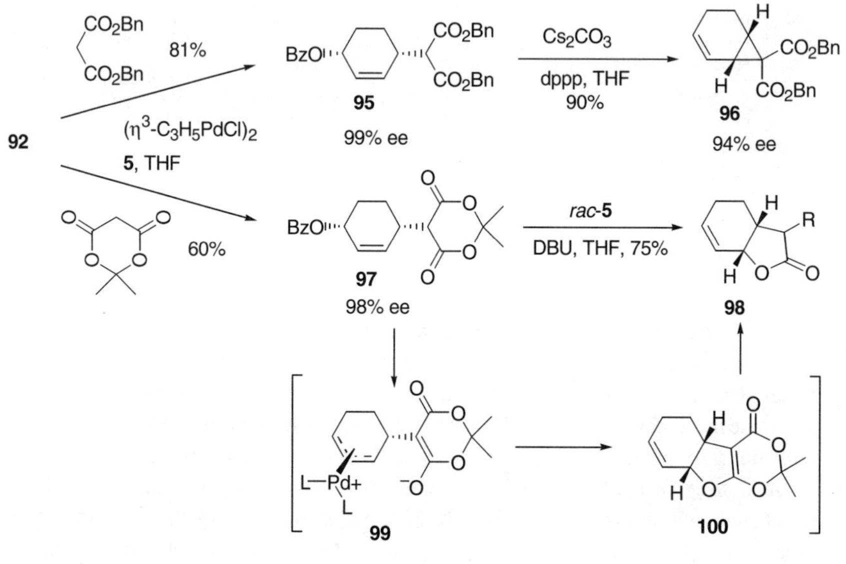

Scheme 8E.12. Synthesis of cyclopropane and lactone by tandem alkylation.

alized to the alkaloid (+)-γ-lycorane. It is interesting that a similar reaction with an analogous five-membered ring substrate gave the enantiomer as the major product.

Azide serves as an excellent nucleophile in the alkylation of *cis*-2-cyclohexen-1,4-diol derivatives to give enantioselectivities greater than 95% ee as shown in Scheme 8E.14. It is noteworthy that the mildness of the reaction conditions effectively suppressed the strong propensity of the allylic azide to undergo a [3,3]-sigmatropic rearrangement. Thus, the obtained regio- and enantiopure allylic azides **104** and **105** have been utilized as key building blocks for the synthesis of the anticancer agent, pancratistatin, and non-opioid analgesic, epibatidine, respectively [66,67].

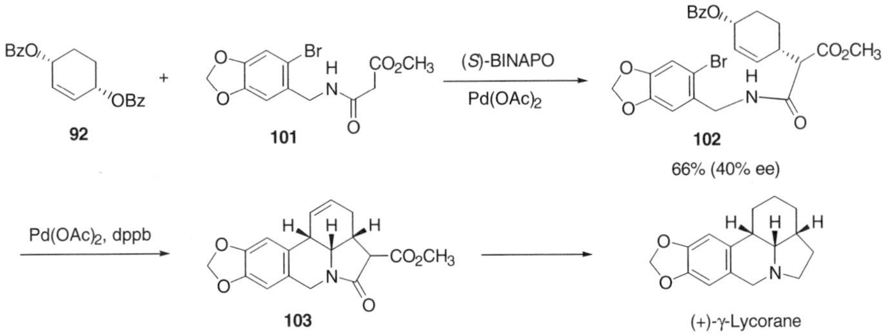

Scheme 8E.13. Synthesis of lycorane.

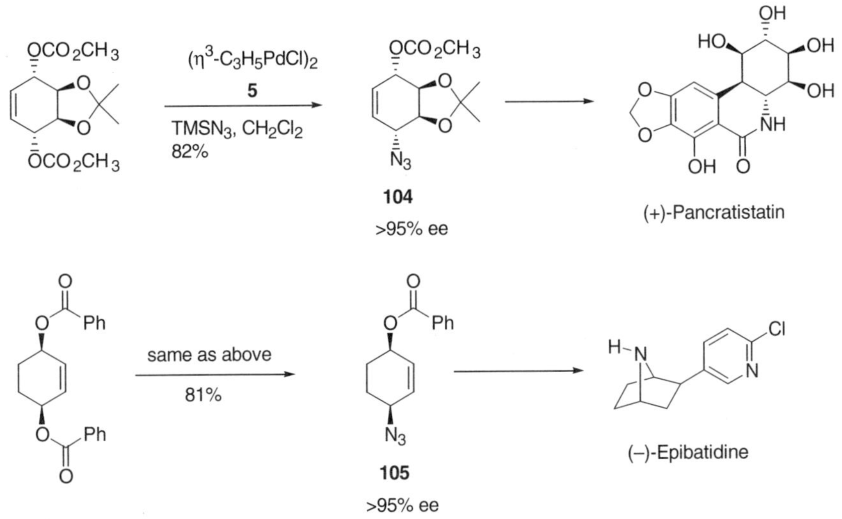

Scheme 8E.14. Allylic azidation and synthesis of pancratistatin and epibatidine.

Desymmetrization of a seven-membered cyclic *meso* compound, followed by a metallo-ene reaction, provides ready access to the trans-hydroazulene skeleton (Scheme 8E.15) [69]. By using BINAP in preference to BINAPO ligand, which gave a lower enantioselectivity (70% ee), two alkylation products were obtained in differential enantioselectivities. The high enantiopurity of the major product **106** may be the consequence of a subsequent kinetic resolution because the second ionization, followed by proton loss from *ent*-**106**, involved a matched event.

The application of the desymmetrization method can be extended to a bicyclic system containing two primary acetate leaving groups (Scheme 8E.16). By using the chiral pyridine-phosphine ligand **49**, the diacetate could be converted into the allylic amine in 93% yield and 89% ee [70].

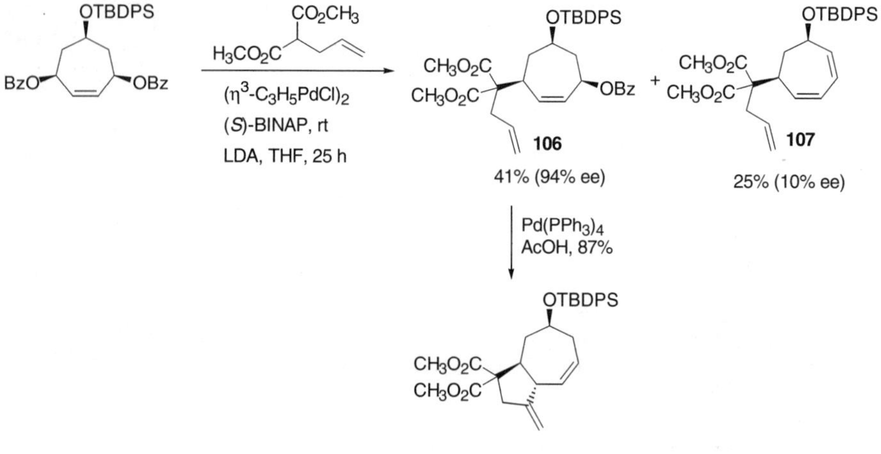

Scheme 8E.15. Desymmetrization of seven-membered *meso*-cyclic compound.

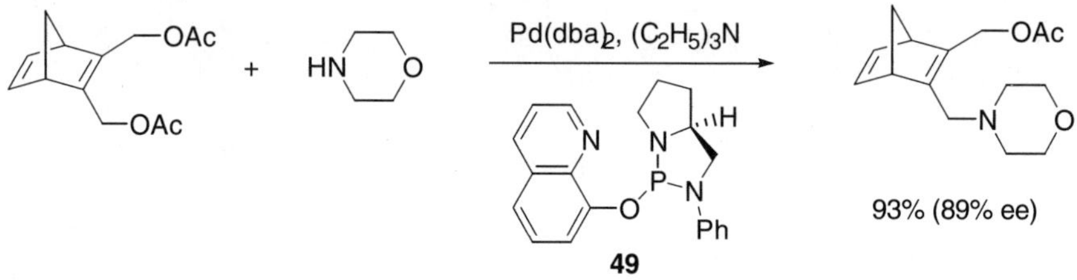

Scheme 8E.16. Desymmetrization of bicyclic diacetate.

8E.3.1.2. Desymmetrization of gem-Dicarboxylates An equivalent of asymmetric carbonyl addition can be achieved by the alkylation of *gem*-dicarboxylates (Scheme 8E.17). The alkylation of *gem*-dicarboxylates, which are easily prepared by the Lewis acid-catalyzed addition of acid anhydrides to an aldehyde, converts the problem of differentiating the two enantiotopic π-faces of a carbonyl group into that of asymmetric substitution of either enantiotopic C–O bond of the *gem*-dicarboxylate. Although asymmetric induction may be derived from enantio-discrimination in the ionization step or in the alkene coordination step, the fast and reversible nature of alkene coordination suggests that the ionization step is more likely to be the source of enantio-discrimination.

Excellent enantioselectivities are observed in the alkylation of various *gem*-dicarboxylates with both carbon and heteroatom nucleophiles as summarized in Table 8E.2 [71]. Screening various chiral ligands from the DPPBA module revealed that 1,2-diaminocyclohexane derived ligand **5** gives the best results, and the mnemonic of Figure 8E.8 correctly predicts the enantiomer obtained in this reaction. Notably, the corresponding desymmetrization by enzymatic hydrolysis or acylation is not feasible with these 1,1-diol derivatives.

When the size of the terminal alkyl-substituent became smaller, the formation of regioisomer was noted (Scheme 8E.18) [72]. On the other hand, the regioselectivity was increased from 2.5:1 to 5.7:1 by using a bidentate ligand, and dramatically to 10:1 by using a chiral ligand,

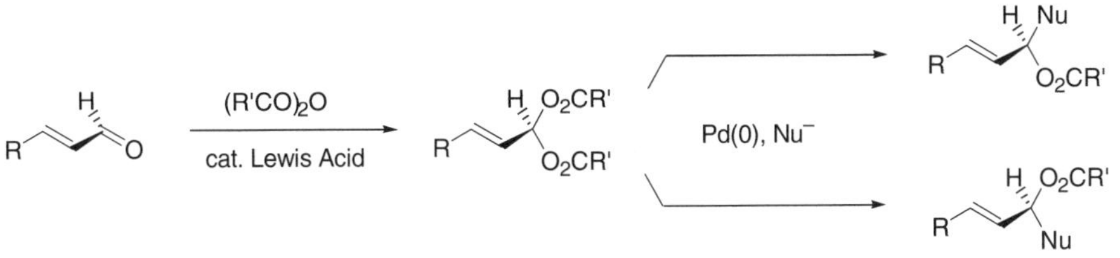

Scheme 8E.17. Desymmetrization of *gem*-dicarboxylate.

TABLE 8E.2. Desymmetrization of *gem*-Dicarboxylates (Scheme 8E.17)

Entry	R	R′	Nu	% Yield	% ee
1	Ph	CH$_3$	CH$_3$C(CO$_2$CH$_3$)$_2$	92	>95
2	Ph	CH$_3$	CH$_3$C(CO$_2$CH$_2$Ph)$_2$	75	>95
3	Ph	CH$_3$	CH$_3$C(SO$_2$Ph)$_2$	54	95
4	Ph	CH$_3$	SO$_2$Ph	92	94
5[a]	CH$_3$	CH$_3$	CH$_3$C(CO$_2$CH$_3$)$_2$	86	92
6	i-C$_3$H$_7$	CH$_3$	CH$_3$C(CO$_2$CH$_3$)$_2$	75	95
7	TMS	CH$_3$	SO$_2$Ph	68	96
8	TBDPSOCH$_2$	CH$_3$	PhCH$_2$C(CN)$_2$	80	90
9	TBDPSOCH$_2$	CH$_3$	SO$_2$Ph	94	92
10	TBDPSOCH$_2$	CH$_3$	Phthalimide	80	95
11	TBDPSOCH$_2$	CH$_3$	CH$_3$C(CO$_2$CH$_3$)$_2$	85	93

[a]A mixture of regioisomers was obtained in 11:1 ratio. See text.

favoring the formation of the allylic acetate with 92% ee. The product of this alkylation retains an allylic acetate, which may be useful for further transformations. For example, chirality transfer by allylic transposition and amination with tosylamine can be effected by subsequent palladium-catalyzed processes with high stereospecificity [72].

8E.3.2. Desymmetrization of *meso*-π-Allyl Complexes

8E.3.2.1. Deracemization of Acyclic Substrates When chiral allylic substrates generate *meso* π-allyl intermediates, the two allylic termini of such intermediates are enantiotopic. Thus, enantioselectivity is derived from the regiochemistry of the nucleophilic addition (**a** vs. **b**), and the alkylation process corresponds to a deracemization event (Eq. 8E.2). One of the prototypical reactions, which has been studied in extensive detail, is the 1,3-diphenylallyl system. Since its introduction as a different mechanistic motif in contrast with the 1,1,3-triphenylallyl system, this reaction has become a "benchmark" for design and comparison of a variety of different ligands in recent years [73].

Due to direct involvement of the nucleophile in the enantiodiscriminating step, the enantioselectivity of this reaction is greatly affected by the structure of the nucleophile. In terms of the nucleophile structure, both the covalent constitution and the nature of the ion pair influence

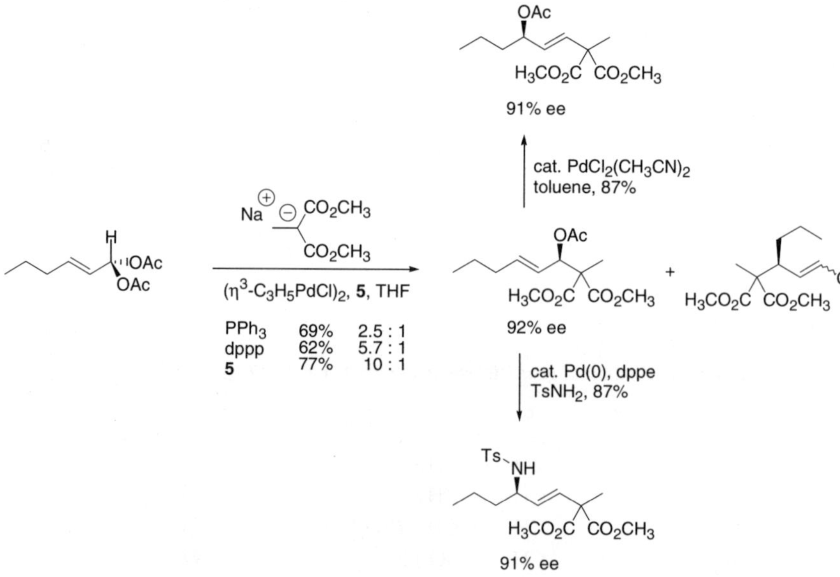

Scheme 8E.18. Enantio- and regiocontrol and utilization of alkylation product.

the enantioselectivity. One of the early investigations with BINAP as ligand showed that the alkylation of 1,3-diphenylallyl acetate with sodium salts of malonate, methylmalonate, and acetamidomalonate gave the corresponding alkylation products with 30, 39 and 94% ee's, respectively [74]. With these sodium salts, which are most frequently used, differential formation of aggregates can lead to significantly different enantioselectivities depending on the solvent used. For example, the alkylation with dimethyl sodiomalonate in Equation 8E.2 with ligand **54** gives the product with 67% ee in THF, 78% ee in acetonitrile, but addition of 15-crown-5 to the latter reaction in acetonitrile dramatically increased the enantioselectivity to 95% ee [43]. In a number of cases, better enantioselectivities have been obtained when the nucleophile is generated by using N,O-bis(trimethylsilyl)acetamide (BSA) as base, often in the presence of acetate ion (KOAc or LiOAc) [73].

Nearly 100 different ligands have been synthesized and their efficacy has been tested in this reaction. Table 8E.3 summarizes the ligands that give high enantioselectivities (>90% ee) in the alkylation of the 1,3-diphenylallyl system with dimethyl malonate. As is readily seen from the

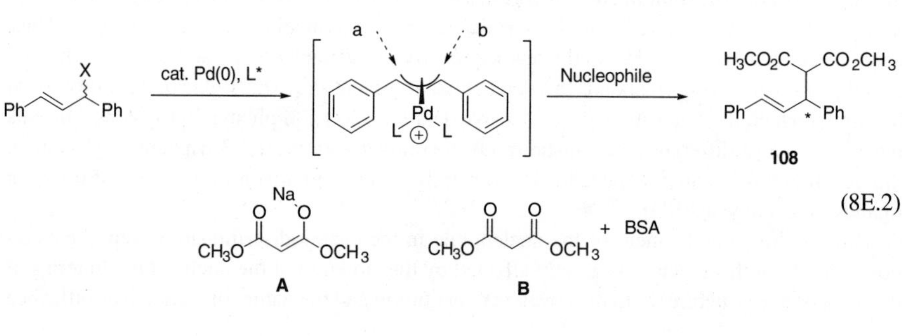

(8E.2)

TABLE 8E.3. Alkylation of 1,3-Diphenylprop-2-enyl System with Carbon Nucleophiles

Entry	Ligand	X	Nu	% Yield	% ee	Ref.
1	**55b**	OAc	A	95	96	75,76
2	**29**	OAc	A	81	95	77,78
3	**2**	OAc	A	86	90	74
4	**10b**	OAc	A	40	92	79
5	**10c**	OAc	A	85	96	46
6	**30**	OAc	B	89	99	42
7	**38b**	OAc	B	99	99	80,81
8	**38a**	OAc	B	98	98	80-82
9	**27**	OAc	B	97	97	48
10	**59a**	OAc	B	92	96	83
11	**55a**	OAc	B	96	96	84
12	**25**	OAc	B	97	95	85
13	**28**	OAc	B	83	95	86
14	**59c**	OAc	B	50–84	95	35
15	**11**	OAc	B	nd	93	36,87
16	**59b**	OAc	B	100	92	88
17	**21**	OAc	B	56	92	89
18	**31**	OAc	B	98	91	90
19	*ent*-**4**	OAc	B	85	90	43
20	**54**	OAc	A, 15-c-5	nd	90	43
21	**38b**	OAc	A, 15-c-5	98	97	80
22	**48**	OAc	B	78	94	91
23	**55c**	OAc	B	99	96	76,92
24	**52**	OAc	A	41	94	93
25	**61**	OAc	B	96	98	94
26	**62**	OAc	B	40	97	95
27	**40**	OAc	B, LiOAc	99	92	96
28	**12**	OAc	B	95	92	97
29	**47**	OPiv	B, LiOAc	97	95	98
30	**43**	OAc	A	99	95	99
31	**46b**	OAc	B, THAB[a]	99	97	100
32	**15**	OPiv	B, LiOAc	90	99	101
33	**27**	OAc	A	100	90	102
34	**51**	OAc	A	90	99	103,104
35	**37a**	OAc	B	91	98	105
36	**36**	OAc	B	92	91	106
37	**35**	OAc	B	97	99	107
38	**60**	OPiv	B, LiOAc	78	96	108
39	**22**	OAc	A	90	91	109
40	**19**	OPiv	B, LiOAc	100	94	110
41	**20**	OAc	B	99	97	111
42	**56a**	OAc	A	99	91	112
43	**56b**	OAc	B	90	96	113
44	**45**	OAc	B	94	98	114
45	**65**	OAc	A	49(90)	97	115
46	**76**	OAc	B	70	91	116
47	**33**	OAc	B	80	92	117

[a]THAB = Tetrahexylammonium bromide.

table, good results have been obtained from a wide variety of ligands whose electronic and structural motifs are vastly different. Various non-C_2-symmetric ligands and the ligands possessing multiple types of chirality give excellent results. The most significant and major advances for this reaction using a sterically bulky acyclic substrate have come from the development of phosphinooxazoline ligands **38** [80–82]. In addition to the excellent enantioselectivity, the availability of various ligands of this type through synthetic sequences from readily accessible chiral amino alcohols as in Equation 8E.3 adds more value to the utility of these ligands.

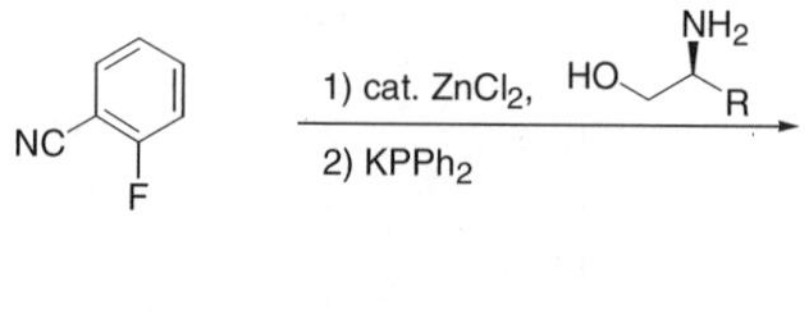

$$\text{(8E.3)}$$

Structural studies on the 1,3-diphenylallylpalladium complex of **38a** have provided insights into the origin of the high selectivity [35]. ^{1}H and ^{31}P NMR studies at room temperature reveal that the complex exists as an 8:1 mixture of the *exo* and *endo* isomers, which can equilibrate 50 times faster than the nucleophilic attack. Somewhat surprisingly, the more stable complex is the *exo* isomer, and the X-ray structure also adopts this conformation. In the crystal structure, the palladium-carbon bond *trans* to phosphorus was longer than that *cis* to nitrogen (2.25Å vs. 2.13Å), reflecting the differing σ-donor/π-acceptor properties of nitrogen and phosphorus. Based on these data and the observed *S*-configuration of the product, it is generally believed that the major enantiomer is derived from the nucleophilic attack *trans* to the phosphorus center of the exo-π-allyl complex **109** (see Scheme 8E.19). A more recent NOE study performed on the olefin-Pd(0) complex **110** also lends further support to this mechanism [33].

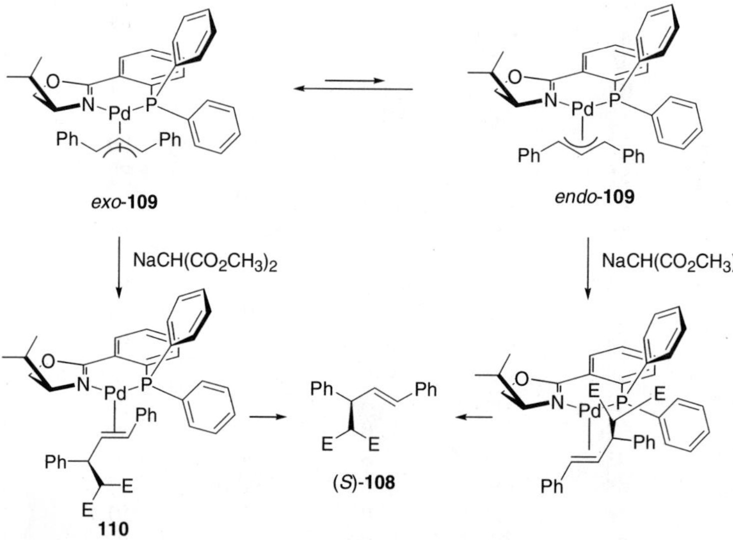

Scheme 8E.19. Possible reaction pathways leading to the major enantiomer.

Good results have also been found in the alkylations with heteroatom nucleophiles as summarized in Table 8E.4. In addition to the nitrogen nucleophiles, both sulfinate and sulfide have proved to give high enantioselectivities. Most interesting among these results is the remarkable anion effect on the enantioselectivity, which was noted in the alkylation using ferrocenyl ligand **50**. Although excellent enantioselectivities (>99.5%) are obtained in the presence of small hard anions such as F^- and BH_4^-, non-coordinating anions (e.g., PF_6^-) have a detrimental effect and give very low enantioselectivities (<10%) [118]. Because this catalyst system does not induce kinetic resolution, the effect is most likely to derive from the efficiency of the added anion to facilitate exchange between the enantiomeric π-allylpalladium intermediates. The halide ion effect observed in other reactions presumably involves a similar allyl isomerization process via a pentacoordinated palladium complex [38,40–42]. Another interesting aspect is that a comparably high enantioselectivity could be obtained by the use of bimetallic ligand **111** (entry 14). X-ray analysis of the π-allylpalladium complex of **111** indicated that both reaction sites are catalytically functioning [119]. In addition, it was shown that simple precipitation allowed for easy recovery of this catalyst, which exhibited almost equal catalytic activity, suggesting the possibility of developing dendrimeric or polymeric catalysts with potential advantages.

Although the reactions of the 1,3-diphenylallyl system have provided a valuable means of comparing various ligands, several drawbacks of this system as a standard reaction should be considered to correctly assess ligand efficacy. First, the factors leading to high enantioselectivities in this reaction frequently do not translate readily into other substrates or reactions, many of which have more synthetic utility. Second, although the high reactivity of the substrate generally gives a good reaction, no reaction or poor yields are frequently observed in other systems, especially with the ligands that weakly ligate palladium. Finally, many examples have shown a particular sensitivity of this system to combinations of ligand, substrate, or nucleophile. Therefore, proper optimization of reaction conditions should precede the evaluation of ligand potency.

Despite the success in the 1,3-diphenylallyl system, use of many of these ligands in the alkylation of 1,3-dialkylallyl system as Equation 8E.4 has produced mixed results, as summarized in Table 8E.5. With the phosphinooxazoline-type ligands, good selectivities (>90% ee) are still obtained from the reactions of substrates possessing bulky allylic substituents such as isopropyl groups (entries 8–10), but smaller substrates such as 1,3-dimethylallyl derivatives give only a modest level of enantioselectivities (entries 1–7). The disparity between these results appears to be sterically derived as the enhanced preference of *syn* versus *anti* orientation in the π-allyl structure by the bulky phenyl or isopropyl groups may not be present with the smaller substrates.

$$ \tag{8E.4} $$

A contrasting pattern of reactivity is noted with the DPPBA ligand derived from a chiral module (entries 12–18). For instance, the use of ligand **5** gives rise to 92% ee in the alkylation of 1,3-dimethylallyl acetate with dimethyl malonate (entry 12). A high level of enantioselectivities has been similarly achieved in the alkylations of a series of linear substrates ranging from dimethyl- to dipentylallyl acetate with heteroatom nucleophiles. However, branching at the allylic position results in a significant diminution in the enantioselectivity (entries 17 and 18). Thus, there exists a complementarity between the DPPBA-derived ligand and those that

TABLE 8E.4. Alkylation of 1,3-Diphenylprop-2-enyl System with Heteroatom Nucleophiles

Entry	Ligand	X	Nu	% Yield	% ee	Ref.
1	**10b**	$CO_2C_2H_5$	$BnNH_2$	93	97	120
2	**10b**	$CO_2C_2H_5$	Veratrylamine	87	95	120
3	**38b**	CO_2CH_3	$BnNH_2$	98	94	121
4	**38b**	Ac	t-Boc$_2$NNa	98	86	121
5	**38c**	Ac	TsNHNa	98	97	121
6	**38c**	Ac	$NaPhCONNH_2$	95	97	121
7	**44**	Ac	$BnNH_2$	89	97	122
8	**44**	Ac	PhthNK	88	99	122
9	**55c**	Ac	$BnNH_2$	94	84	92
10	**55c**	Ac	TsNHNa′	92	84	92
11	**55c**	Ac	Boc$_2$NNa	92	92	92
12	**49**	Ac	$BnNH_2$	95	93	123
13	**50**	$CO_2C_2H_5$	$BnNH_2$	90–95	>99	118,124
14	**111**	$CO_2C_2H_5$	$BnNH_2$	nd	99	119
15	**38b**	$CO_2C_2H_5$	t-BuSO$_2$Li	69	93	125
16	**38b**	$CO_2C_2H_5$	t-BuSTMS	62	92	125

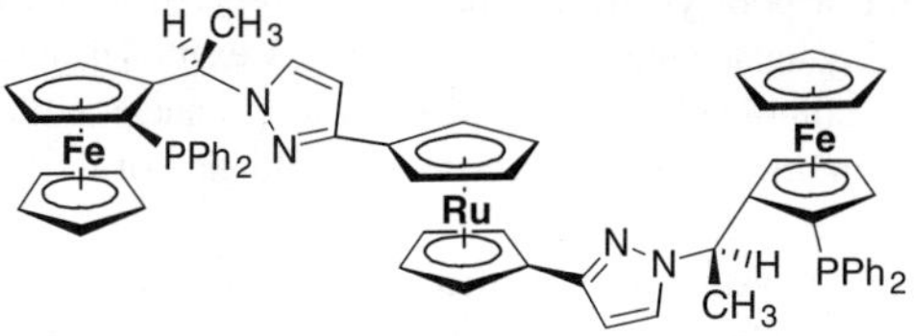

111

give good results with bulky acyclic substrates. On the other hand, a recent study with the modified phosphinooxazoline ligand **42** shows that the diminished steric interaction between the ligand and the smaller allyl substrate may be compensated by incorporating more steric hindrance into the ligand (entry 11). By using a tricyclic phosphinooxazoline ligand, enantiose-lectivities approaching 90% ee have been obtained from the alkylation of 1,3-dimethylallyl acetate with dimethyl malonate [127].

An application of the deracemization strategy has provided efficient entry to a novel amino acid substituent of the antifungal agents, polyoxins and nikkomycins, as shown in Scheme 8E.20. The versatile five-carbon building block was obtained from phthalimidation of the hydroxymethyl-substituted epoxide in 87% yield and 82% ee. Straightforward synthesis of polyoxamic acid was then accomplished by subsequent dihydroxylation and selective oxidation of the alkylation product.

The use of a chiral nucleophile in a deracemization reaction creates the issue of double stereodifferentiation. The ability of a chiral ligand to dictate the regiochemistry with respect to the π-allylpalladium intermediate (which corresponds to the diastereoselectivity in this case) is affected by the stereochemistry of the nucleophile either in a matched or mismatched fashion.

TABLE 8E.5. Alkylation of 1,3-Dialkylallyl Substrates

Entry	Ligand	R	X	Nu[a]	% Yield	% ee	Ref.
1	**38a**	CH_3	OAc	CH_2E_2/NaH	52	62	83
2	**38b**	CH_3	OAc	$PhSO_2Na$	83	55	126
3	**38b**	CH_3	Cl	$PhSO_2Na$	55	52	126
4	**38c**	CH_3	OAc	$BnNH_2$	87	57	121
5	**38c**	CH_3	OAc	$t\text{-Boc}_2NNa$	44	75	121
6	**38c**	CH_3	OAc	TsNHNa	61	66	121
7	**38c**	CH_3	OAc	CH_2E_2/BSA	96	71	80
8	**38c**	$i\text{-}C_3H_7$	$OPO(C_2H_5)_2$	TsNHNa	57	90	121
9	**38c**	$i\text{-}C_3H_7$	$OPO(C_2H_5)_2$	$t\text{-Boc}_2NNa$	29	97	121
10	**38c**	$i\text{-}C_3H_7$	OAc	TsNHNa	88	96	80
11	**42**	CH_3	OAc	CH_2E_2/NaH	97	85	127
12	**5**	CH_3	$OCO_2C_2H_5$	CH_2E_2/Cs_2CO_3	98	92	128
13	**5**	C_2H_5	$OCO_2C_2H_5$	PhthNK	78	94	128
14	**5**	C_2H_5	$OCO_2C_2H_5$	$PhSO_2NHex_4$	92	91	128
15	**5**	$n\text{-}C_3H_7$	$OCO_2C_2H_5$	$PhSO_2NHex_4$	99	97	128
16	**5**	$n\text{-}C_5H_{11}$	$OCO_2C_2H_5$	$PhSO_2NHex_4$	78	95	128
17	**5**	$c\text{-}C_6H_{11}$	$OCO_2C_2H_5$	$PhSO_2NHex_4$	31	68	128
18	**5**	Ph	OAc	CH_2E_2/Cs_2CO_3	9	52	128

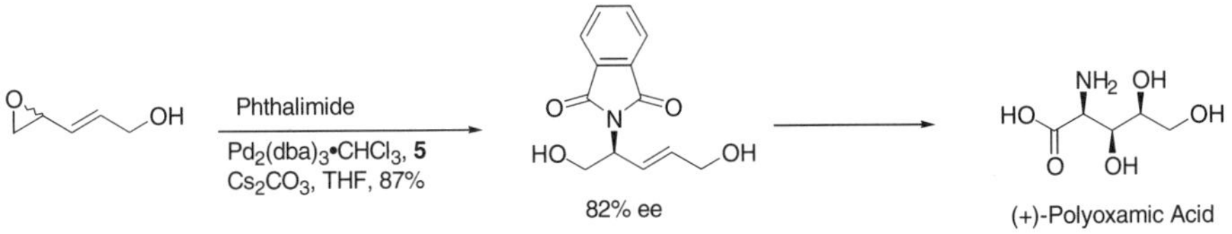

Scheme 8E.20. Synthesis of polyoxamic acid.

With an α-amino ester as nucleophile, the example shown in Equation 8E.5 demonstrates that the catalyst rather than the substrates can dominate in determining the diastereoselectivity. The 3:1 selectivity exhibited in the alkylation with the achiral catalyst was increased to 19:1 with (R,R)-**5** whereas the (S,S)-**5**, under identical conditions, gave a 1:3 ratio.

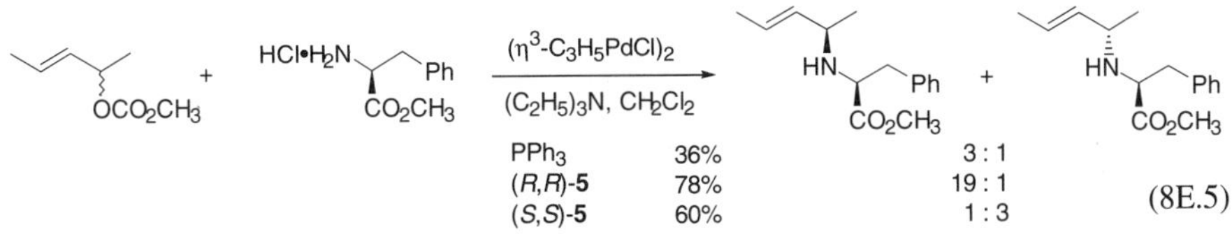

8E.3.2.2. Deracemization of Cyclic Substrates Although the ambiguity of *syn/anti* geometry can be removed by the rigidity of the cyclic framework, the interaction between the ligand and a cyclic allyl substrate becomes quite different from that of acyclic systems because relatively small *syn,syn* allylic systems are generated. Only mediocre enantioselectvities have been recorded with the ligands, which perform extremely well in the 1,3-diphenylallyl system. Although many of these ligands exhibit somewhat similar reactivity profiles as in the alkylation with small acyclic substrates, excellent results have come from the use of the DPPBA-based diphosphine ligand **5** (Eq. 8E.6 and Table 8E.6). With this ligand, the alkylation of five-, six-, and seven-membered allylic acetates with dimethyl malonate could be carried out with high enantioselectivities (entries 1–3). Best enantioselectivites were consistently obtained with tetrahexylammonium ion as the counterion of the nucleophile.

Recently, good results have been obtained from using two non-C_2-symmetric ligands, which possess a P,O- and a P,N-chelate type. Compared with the poor performance of the phosphinooxazoline ligands **38**, the success of **64** was attributed to the small size of the carboxylic acid moiety, which emphasizes the interactions of the cyclic substrate with the diphenylphosphino group. Alternatively, the steric fortification on the phosphorus part of the ligand gave rise to a better selectivity by controlling the *endo/exo* orientations of the π-allylpalladium complexes. X-ray and ^{31}P NMR studies of (η^3-cyclohexenyl)palladium complex of **53** concluded that high enantioselectivity is due to the exclusive formation of an "exo" π-allylpalladium intermediate induced by the interaction between the 2-biphenylyl group and the Mn(CO)$_3$ moiety.

$$(8E.6)$$

TABLE 8E.6. Alkylation of Cyclic Allylic Acetates with Malonates

Entry	Ligand	n	R	% Yield	% ee	Ref.
1	5	5	CH$_3$	>98	>98	129
2	5	6	CH$_3$	98	98	129
3	5	7	CH$_3$	93	93	129
4	64	5	t-C$_4$H$_9$	74	85	130
5	64	6	t-C$_4$H$_9$	89	98	130
6	64	7	t-C$_4$H$_9$	73	>99	130
7	53	5	CH$_3$	73	96	131,132
8	53	6	CH$_3$	62	93	131,132
9	53	7	CH$_3$	86	98	131,132

Systematic studies on the alkylation of cyclic allylic systems revealed that the nature of the ion pair has a large effect on the enantioselectivity (Table 8E.7). In the alkylations with the modular ligand **5**, bigger counterions generally gave rise to higher enantioselectivities; an optimal result (98% ee) was reached with tetrahexylammonium ion (entries 1–5) [129]. This effect of cation was even more pronounced when a series of alkali metal ions were used as counterion (entries 6–11). Tightening the ion pair by changing the solvent from THF to methylene chloride also had an equally dramatic influence on the enantioselectivity (entries 3 vs. 4 and entries 10 vs. 11). On the other hand, when the ligand **112** possessing a cation-binding

TABLE 8E.7. Dependence of Enantioselectivity on the Nature of the Ion Pair

Entry	Ligand	n	M	Solvent	% Yield	% ee
1	**5**	5	$(CH_3)_4N^+$	THF	88	41
2	**5**	5	$(C_4H_9)_4N^+$	THF	74	57
3	**5**	5	$(C_6H_{13})_4N^+$	THF	92	68
4	**5**	5	$(C_6H_{13})_4N^+$	CH_2Cl_2	81	>98
5	**5**	5	$(C_8H_{17})_4N^+$	THF	74	66
6	**5**	5	Li^+	THF	75	(−)63
7	**5**	5	Na^+	THF	77	38
8	**112**	5	Na^+	CH_2Cl_2	90	96
9	**5**	5	K^+	THF	90	51
10	**5**	5	Cs^+	THF	76	76
11	**5**	5	Cs^+	CH_2Cl_2	93	>99
12	**64**	6	Li^+	THF	89	98
13	**64**	6	Na^+	THF	92	96
14	**64**	6	K^+	THF	nd	62

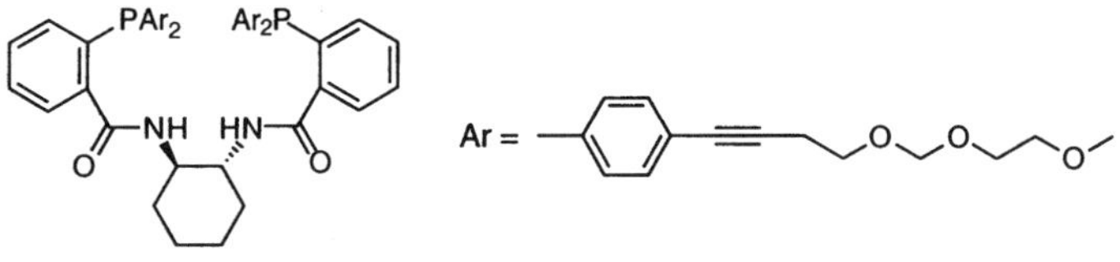

112

site was used, the enantioselectivity became cation-independent (entry 8) [133]. The use of sodium as counterion gave a high enantioselectivity with a large rate acceleration. By contrast, when the myrtenate-derived P,O-chelate ligand was used, the best enantioselectivities were obtained with lithium cation (entries 12–14) [130]. In these cases, the *t*-butyl malonate esters give a slightly better ee than the corresponding methyl esters.

An allylic halide has been used to give a better result than the corresponding allylic acetate (Scheme 8E.21) [134]. Notably, only 0.05 mol% of catalyst was sufficient to produce the enantiopure product in 96% yield. To achieve high enantioselectivity, the reactivity of the substrate had to be modulated by slow addition of the nucleophile. This deracemization strategy offers an efficient alternative method for the preparation of hydroxylactone, which has served as a synthetically useful building block for various natural product syntheses [135,136].

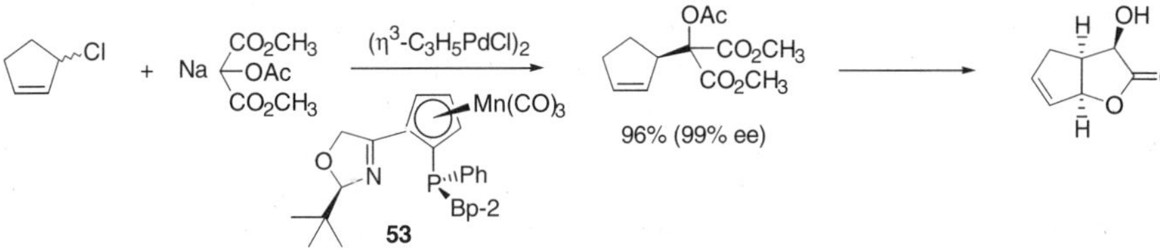

Scheme 8E.21. Deracemization of 2-cyclopentyl chloride.

With tetrahexylammonium as the cation of choice and **5** as ligand, the established conditions for the alkylation with carbon nucleophiles has been successfully applied to asymmetric introduction of heteroatoms such as nitrogen, oxygen, and sulfur (Eq. 8E.7). As shown in Table 8E.8, excellent enantioselectivities are observed independent of the counterion of the heteroatom nucleophiles and the ring size.

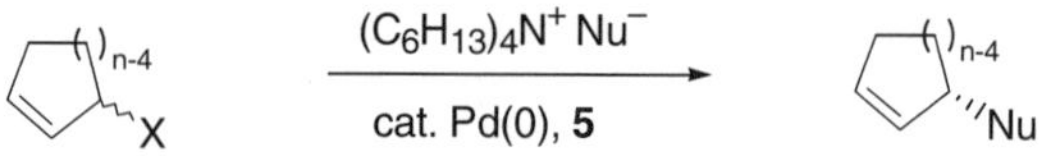

$$\text{(C}_6\text{H}_{13})_4\text{N}^+ \text{ Nu}^-$$
$$\xrightarrow{\hspace{2cm}}$$
$$\text{cat. Pd(0), } \mathbf{5}$$

(8E.7)

TABLE 8E.8. Deracemization of Cyclic Allyl Substrates with Heteroatom Nucleophiles

Entry	R	n	Heteroatom	Nu⁻	% Yield	% ee	Ref.
1	Ac	5	N	Phth⁻	87	94	129
2	Ac	6	N	Phth⁻	95	97	129
3	Ac	7	N	Phth⁻	84	98	129
4	CO_2CH_3	5	O	$(CH_3)_3CO_2^-$	91	97	137
5	CO_2CH_3	6	O	$(CH_3)_3CO_2^-$	94	92	137
6	CO_2CH_3	7	O	$(CH_3)_3CO_2^-$	98	98	137
7	CO_2CH_3	5	S	$PhSO_2^-$	99	98	138
8	CO_2CH_3	6	S	$PhSO_2^-$	95	98	138
9	CO_2CH_3	7	S	$PhSO_2^-$	95	98	138

The kinetic resolution of C_2-symmetric racemic substrate as shown in Scheme 8E.22 requires differentiation between the starting material and the monoalkylated product **113** in addition to the enantiodiscrimination in the ionization step. With a bulky carboxylic acid as nucleophile, the racemic tetraacetate of conduritol B was efficiently resolved under phase-transfer conditions [68]. Further elaboration of the hydrolytically desymmetrized alcohol resulted in the synthesis of the important glycosidase inhibitor, (+)-cyclophellitol.

The alkylation products are synthetically useful because simple subsequent transformations furnishes precursors of important natural products as illustrated in Scheme 8E.23. Simple oxidative cleavage of allylic phthalimide **45** generates protected (S)-2-aminopimelic acid, whose dipeptide derivatives have shown antibiotic activity. The esterification via deracemization protocol is not limited to the use of bulky pivalic acid. The alkylation with sterically less hindered propionic acid also occurs with high enantioselectivity to give allylic ester **116**, which has been utilized as an intermediate towards the antitumor agent phyllanthocin and the insect sex excitant periplanone. Dihydroxylation of the enantiopure allylic sulfone gives diol **117** with complete diastereoselectivity. Upon further transformation, the structurally versatile γ-hydroxy-α,β-unsaturated sulfone **118** is readily obtained enantiomerically pure.

Introduction of an oxygen function can also be achieved by using phenol as nucleophile. Excellent enantioselectivities have been obtained from the alkylation of cyclohexenyl carbonate with various phenols [139]. In particular, this asymmetric O-alkylation of phenol becomes an efficient C-alkylation protocol when combined with subsequent Claisen rearrangement. As shown in Scheme 8E.24, the aryl ether undergoes a [3,3]-sigmatropic rearrangement on treatment with a europium catalyst at 50°C to give the ortho-alkylated product with high chirality

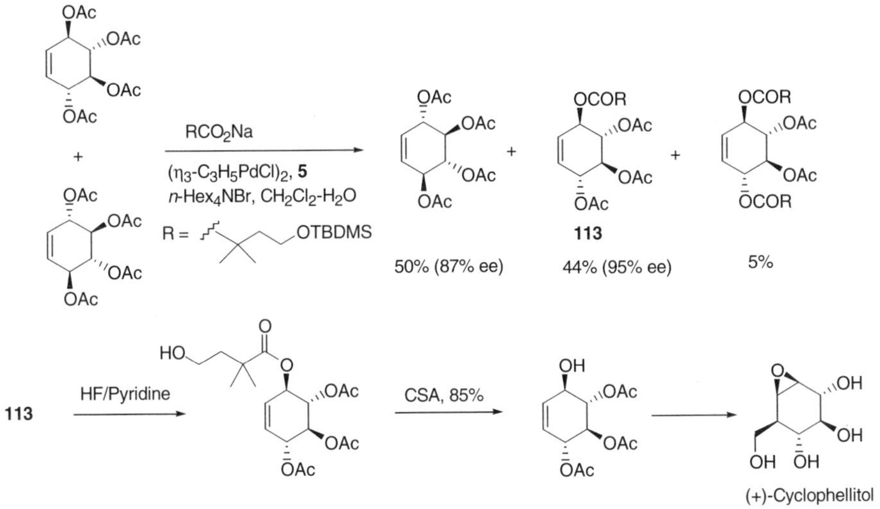

Scheme 8E.22. Kinetic resolution of conduritol B and synthesis of cyclophellitol.

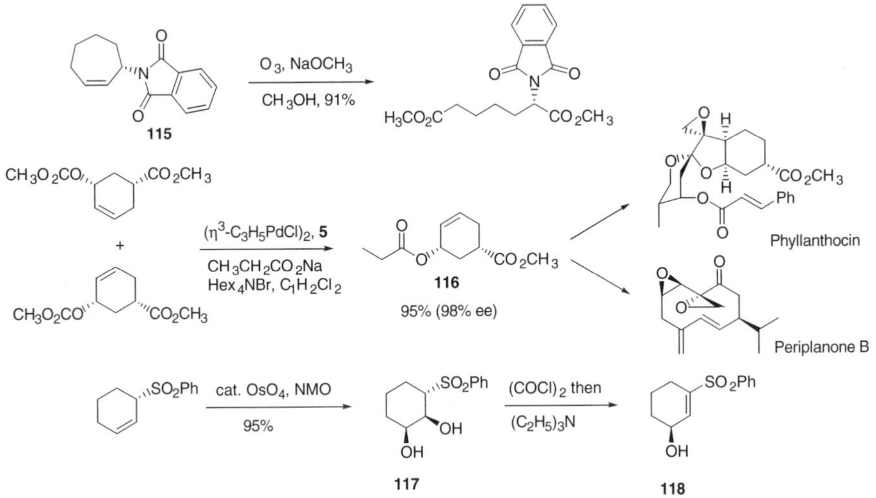

Scheme 8E.23. Synthetic utility of heteroatom containing deracemized products.

transfer. The *O*-alkylation of benzyl alcohol using the same ligand has been reported to give 98% ee, albeit in poor yield [140].

The deracemization strategy has been applied to a 2-substituted allylic system as shown in Scheme 8E.25 [141]. Amination of 2-arylcyclohexenyl carbonate with *N*-allyltosylamine was performed by using *ent*-**8** as ligand to give 86% ee, while the use of BINAP gave a poor result (12% yield, 9% ee). After the enantiopurity was enriched to 99% ee by recrystallization, the obtained bisallylic amine was further elaborated to mesembrine natural products.

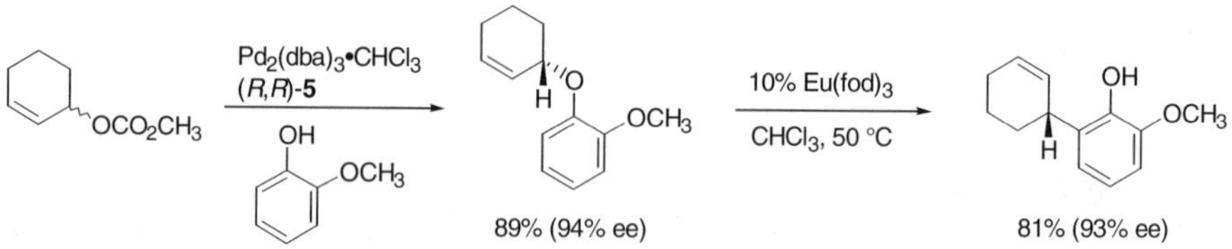

Scheme 8E.24. Asymmetric *O*- and *C*-alkylation of phenol.

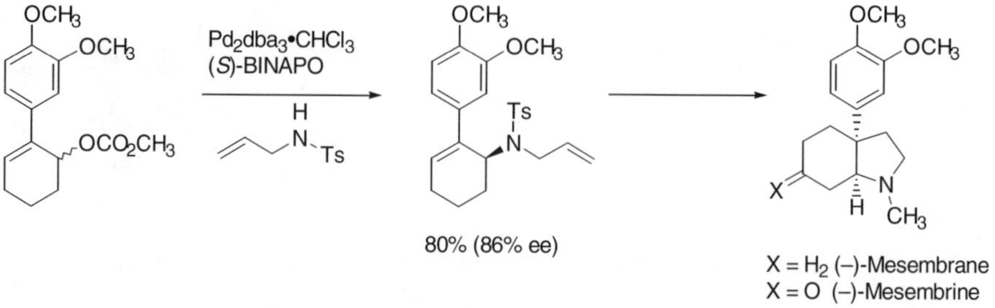

Scheme 8E.25. Synthesis of mesembrane and mesembrine.

8E.3.3. Enantioface Exchange

One of the desirable features of metal-catalyzed allylic alkylations is the ability to convert racemic starting material into optically pure material even when both the starting olefin and π-allyl intermediate lack a symmetry element. Because racemic starting material gives rise to equal amounts of enantiomeric intermediates, which ultimately leads to enantiomeric products, a mechanism must exist for enantioconversion whereby the simple kinetic resolution that gives 50% theoretical yield is avoided. The dynamic η^3-η^1-η^3 process of a metal-bound π-allyl complex provides such a pathway through which the intermediate loses the stereochemical information of the starting material.

If the substrate contains two identical substituents at one terminus of the allylic position such as shown in Scheme 8E.26, the π-allyl intermediate can undergo enantioface exchange via the formation of a σ-palladium species at that terminus. This process should occur faster than the nucleophilic addition, which is the enantio-determining step ($k_1 > k_2[\text{Nu}^-]$ and $k'_2[\text{Nu}^-]$). Thus, enantioselection can be derived from the relative rate of the nucleophilic addition to each diastereomer; the relative stabilities of the two diastereomeric complexes need not have a direct effect on the enantioselectivity (Curtin-Hammett conditions). Although the achiral allylic isomer **120** is expected to follow the same kinetic pathway as the racemic substrate **119**, the difference between the results from the two systems often gives an indication as to the origin of enantioselection—complexation or ionization versus nucleophilic addition.

The 1,1,3-triphenylallyl system (R = Ph) was one of the early examples that was thoroughly studied and gave high asymmetric induction [44,142]. The alkylation exclusively occurs at the less substituted carbon to give the more conjugated regioisomer as shown in Scheme 8E.26. Excellent enantioselectivities have been obtained from both the racemic **119** and achiral substrate **120** by using P,P-, N,N- and P,N-chelate type ligands as summarized in Table 8E.9.

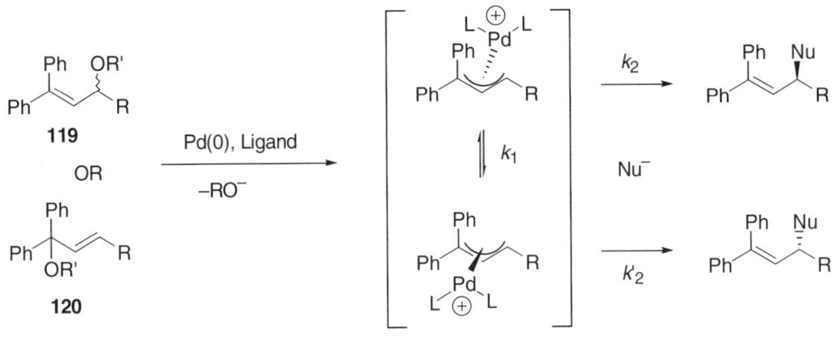

Scheme 8E.26. Asymmetric induction involving enantioface exchange.

TABLE 8E.9. Alkylation of 1,1,3-Trisubstituted Allyl System

Entry	Ligand	Substrate, R	Nu	% Yield	% ee	Ref.
1	2	**119**, Ph	CH(CO$_2$CH$_3$)$_2$	100	86	142
2	2	**120**, Ph	CH(CO$_2$CH$_3$)$_2$	100	84	142
3	29	**120**, Ph	CH(CO$_2$CH$_3$)$_2$	61	85	77
4	38a	**119**, Ph	CH(CO$_2$CH$_3$)$_2$	97	99	143,144
5	38a	**120**, Ph	CH(CO$_2$CH$_3$)$_2$	95	97	143,144
6	38a	**119**, CH$_3$	CH(CO$_2$CH$_3$)$_2$	95	95	143,144
7	44	**119**, CH$_3$	PhCH$_2$NH$_2$	82	98	122
8	44	**119**, n-C$_4$H$_9$	PhCH$_2$NH$_2$	30	94	122

The obligatory enantiofacial exchange for efficient deracemization also occurs in a very congested allylic system giving high enantioselectivity (86% ee) as shown in Eq. 8E.8 [145]. The silyl groups in the alkylation product could be removed without loss of stereochemical integrity to effect a net-benzylic substitution.

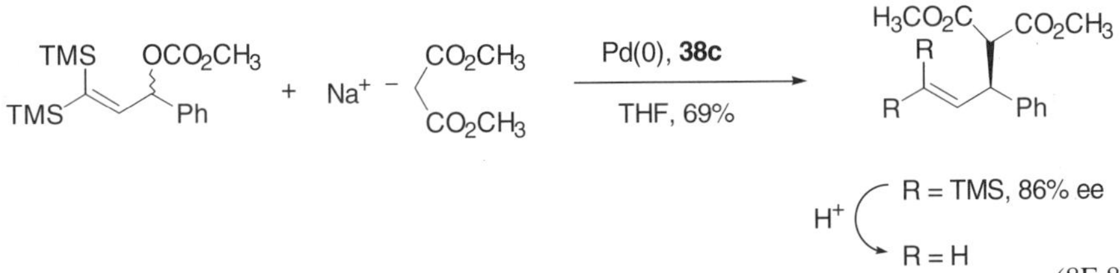

$$(8E.8)$$

A functionalized amino acid derivative has been synthesized by alkylation of a heteroatom-substituted π-allyl intermediate (Scheme 8E.27) [146]. The alkylation of a Schiff base, which forms a 2-aza-π-allyl intermediate, furnishes the malonate adduct with 85% ee despite the epimerizable nature of the newly generated stereogenic center.

An additional issue of regiocontrol arises in asymmetric induction when the π-allyl complex possesses a primary terminus. Although steric factors favor the formation of the achiral linear product, alkylations with reactive nucleophiles often benefit from electronic effects leading to the branched product [147,148]. Of particular interest is the reaction of the crotyl system because

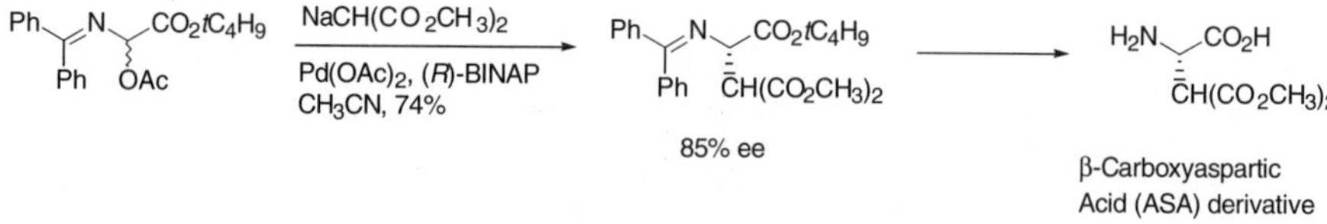

Scheme 8E.27. Deracemization by alkylation of (2-Aza-π-allyl)palladium complex.

alkylations with heteroatom nucleophiles tend to give the major product that is derived from attack at the more substituted carbon. An early study showed that 88% ee could be achieved in the sulfinylation of crotyl chloride by using DIOP as ligand [149]. However, DIOP as well as conventional phosphine ligands such as chiraphos and BINAP exhibited rather poor enantiose-lectivities (7, 9 and 41% ee, respectively) in the amination of crotyl acetate with benzylamine [150]. In the alkylation of this system, multiple modes of enantioselection, for example, type A, C and D, are generally operative. As shown in Eq. 8E.9, the amination reactions gave differential enantioselectivities depending on the structure of the starting material. Nonidentical enantioselectivities obtained from these reactions suggest that the nucleophilic addition occurs competitively with enantioface exchange [150].

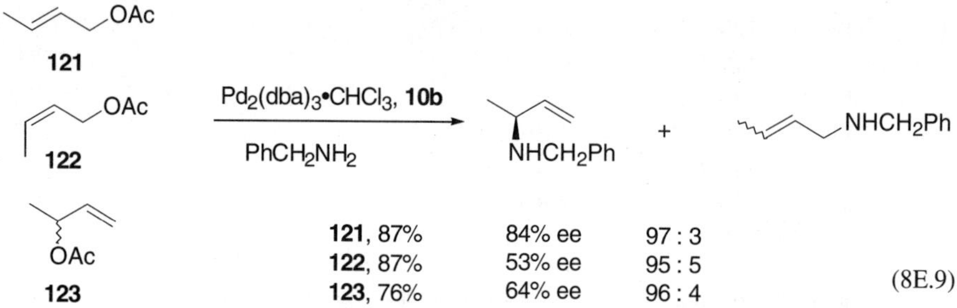

121, 87%	84% ee	97 : 3	
122, 87%	53% ee	95 : 5	
123, 76%	64% ee	96 : 4	(8E.9)

Despite some success achieved in the crotyl system, alkylation at the more substituted allylic position with a carbon nucleophile is a more challenging problem. Promising results have mainly come from the use of other metal (Mo, W, Rh and Ir) catalyst systems (see section 8E.4). Recently, the control of both regio- and enantioselectivity in such transformations has been realized by palladium-based catalysts in the alkylation of a cinnamyl system (Eq. 8E.10 and Table 8E.10).

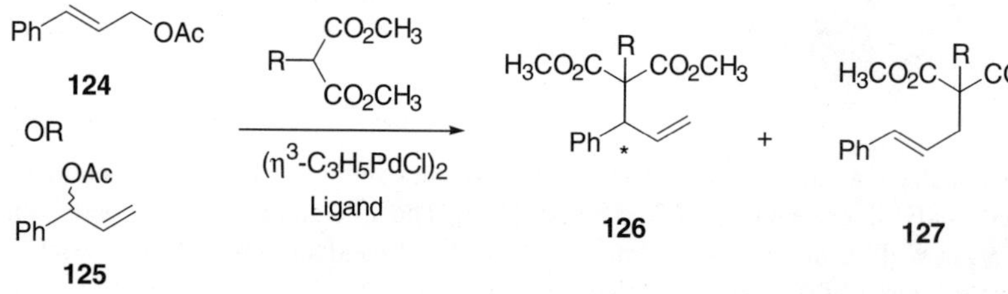

Modifying the parent phosphinooxazoline **38** to ligand **57**, exciting improvement in the regio- and enantioselectivity has been realized (entries 1 and 2) [151,152]. Similar enantio- and regioselectivities obtained from the reaction of the racemic substrate **125** suggest a facile

TABLE 8E.10. Asymmetric Alkylation of Cinnamyl System

Entry	Ligand	Substrate	R	Ratio **126:127**	% Yield	% ee of **126**	Ref.
1	**38c**	**124**	H	4:96	90	78	151,152
2	**57**	**124**	H	76:24	86	90	151,152
3	**57**	**125**	H	66:34	82	88	151,152
4	**16**	**124**	CH$_3$	21:79	99	nd	153
5	**16**	**125**	CH$_3$	82:18	97	86	154

enantioface exchange in this alkylation reaction (entry 3). The use of monophosphine ligand **16** also gives comparable regio- and enantioselectivities in the alkylation of racemic substrate, but quite different results when the starting substrate was the achiral olefin **124** (entries 4 and 5) [154]. Based on this unusually strong memory effect, which leads to retention of regiochemistry [155], a different mechanism invoking a trans effect has been advanced [153].

Reactions of vinyl epoxides present a more serious problem for asymmetric induction due to the strong propensity of these substrates to undergo alkylation at the primary carbon [156]. With a nitrogen nucleophile, which could be delivered to the 2-position by coordination, the alkylation of butadiene monoepoxide with ligand **5** afforded a gratifying 10:1 regioselectivity favoring the 1,2-isomer with 76% ee [157]. Significant improvement in the selectivities has been achieved by using a modified ligand in which the benzoic acid linker in the standard ligand is switched to a 1-naphthoic acid derivative, thereby restricting rotation around the carboxamide moiety. Under the same conditions, the use of **128** increased the enantioselectivity to 98% ee and regioselectivity to better than 75:1 (Scheme 8E.28). Simple transformations of the major regioisomer furnish synthetically versatile vinylglycinol and clinically important agents, vigabatrin for epilepsy and ethambutol for tuberculosis [158].

When natural amino acids are used as the nucleophile, the alkylation of isoprene monoepoxide exhibits a matched and mismatched phenomenon (Scheme 8E.29). Good control of diastereoselectivity by the catalyst was observed with the naphthyl ligand **128** [159]. The complementary selectivity could be increased to 12:1 and 14:1 when the ligands used were **71** and *ent*-**71** respectively, albeit with somewhat lower yields (45% and 50%, respectively).

Less satisfactory results have been frequently encountered in the related asymmetric cycloaddition of a vinyl epoxide to an isocyanate as in Scheme 8E.30 [160]. The modest enantioselectivities of this process are indicative of the competitive intramolecular nucleophilic addition with enantioface exchange. When the oxazolidinone was generated from an achiral substrate, somewhat higher enantioselectiviites were obtained presumably due to superposition of the enantioselection obtained in the ionization step.

A tandem palladium-catalyzed reaction can effect a similar transformation to produce 2-vinyl-substituted heterocyclic systems as in Eq. 8E.11. By varying the amino acid moiety of the ligand, 83% ee could be obtained from the use of the glycine-derived ligand **129** [161]. A maximum enantioselectivity of 65% ee has been recorded for this type of reaction in an earlier study with BINAP as ligand [162]. Because both (*E*)- and (*Z*)-isomers gave the same enantioselectivity, attack on the rapidly interconverting π-allyl intermediates seems to determine the selectivity. Modest enantioselectivities have been reported for the related asymmetric preparation of 2-vinylpiperazine and 1,4-benzodioxane derivatives [163,164].

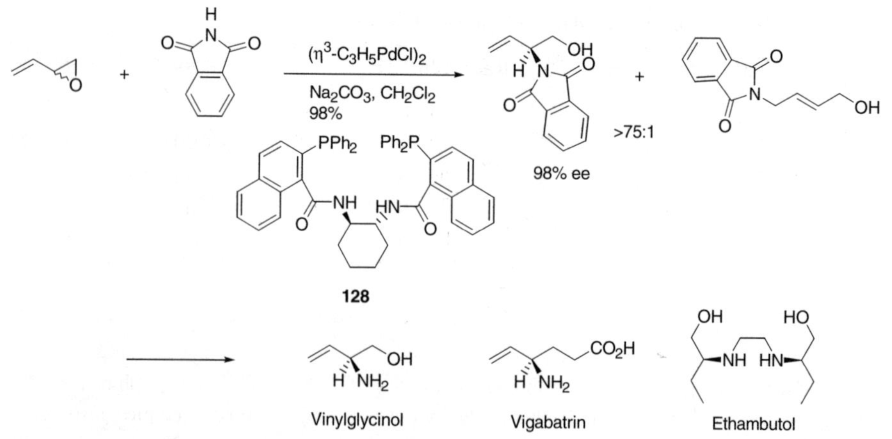

Scheme 8E.28. Asymmetric alkylation of vinyl epoxide with nitrogen nucleophile.

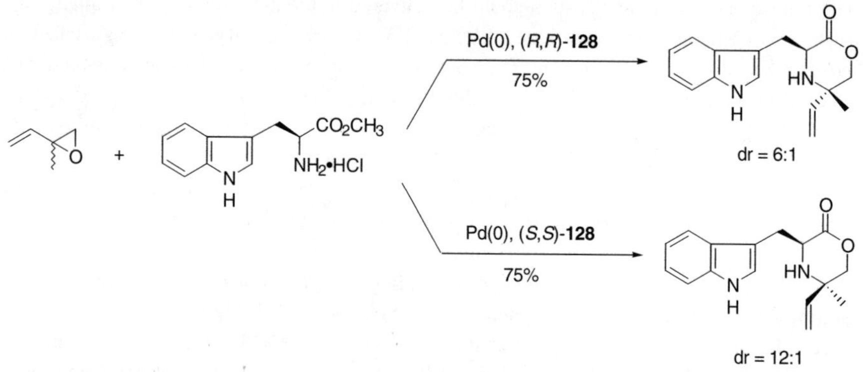

Scheme 8E.29. Deracemization of isoprene monoepoxide with α-amino ester.

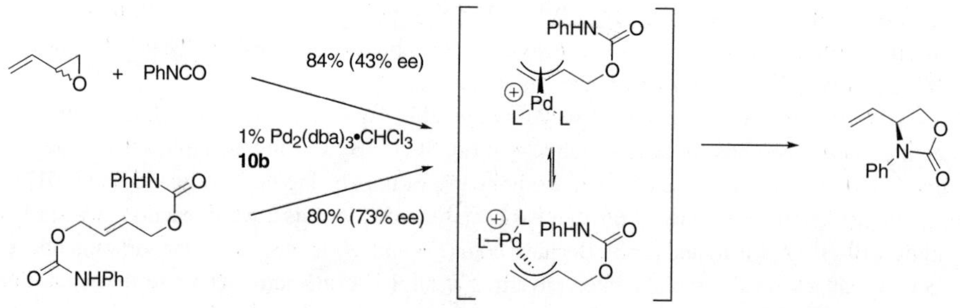

Scheme 8E.30. Formation of 4-vinyl-2-oxazoline.

The asymmetric intermolecular *O*-alkylation with aliphatic alcohols has been difficult due to the poor reactivity of these nucleophiles. Recently, this elusive goal has been achieved by a

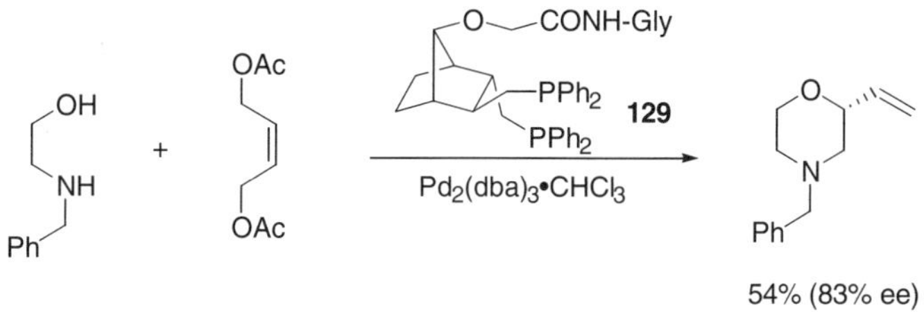

$$(8E.11)$$

54% (83% ee)

novel two-component catalyst system in which the nucleophile and electrophile are activated by palladium and borane catalysts, respectively [165]. Although elucidation of the detailed mechanism requires further studies, it appears that the oxygen nucleophile adds to the π-allyl-palladium complex in an intramolecular fashion via the "ate" complex as drawn in Scheme 8E.31 and summarized in Table 8E.11. The successful participation of various functionalized alcohols in this reaction is particularly noteworthy because the alkylation products constitute differentiated diols wherein the more hindered hydroxyl group is selectively protected. Thus, this method provides valuable and flexible entry to the versatile vinylglycidol building blocks.

The palladium complex derived from TolBINAP catalyzes cycloaddition of vinyl epoxide to a carbodiimide with high enantioselectivity (Eq. 8E.12) [166]. The stereochemical course of the cycloaddition of the epoxide with the heterocumulene was noticeably influenced by the structure of

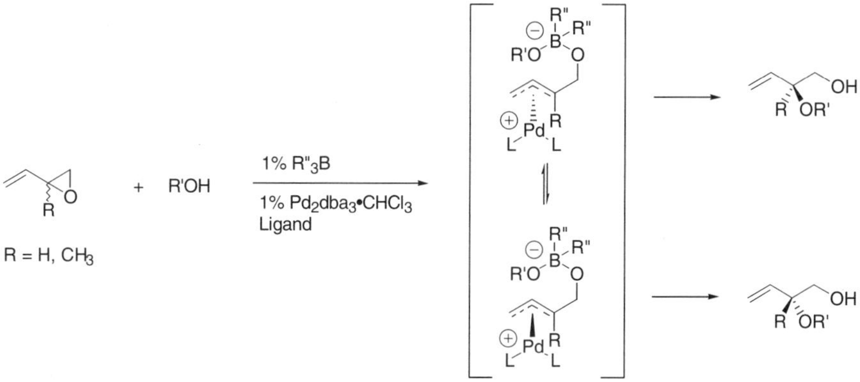

Scheme 8E.31. Enantioselective alkylation of vinyl epoxide with aliphatic alcohols.

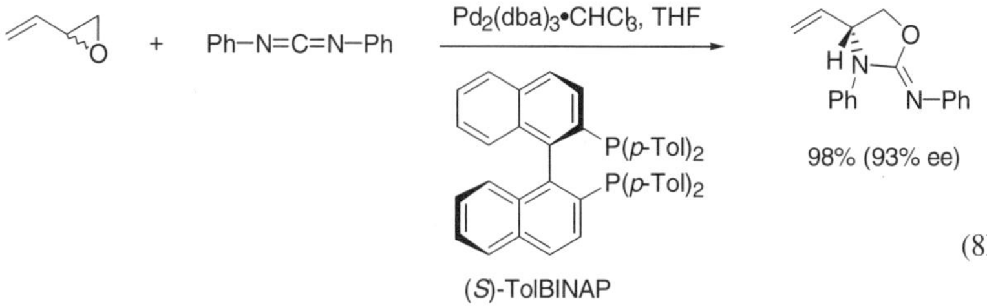

(*S*)-TolBINAP

$$(8E.12)$$

98% (93% ee)

TABLE 8E.11. Palladium and Boron Co-catalyzed Alkylation of Alcohols to Vinylepoxides

Entry	Ligand, R	R'	R''$_3$B	% Yield	% ee
1	**5**, CH$_3$	CH$_3$	(C$_2$H$_5$)$_3$B	88	94
2	**5**, CH$_3$	H$_2$C=CHCH$_2$	(C$_2$H$_5$)$_3$B	83	95
3	**5**, CH$_3$	p-CH$_3$OC$_6$H$_4$CH$_2$	(C$_2$H$_5$)$_3$B	91	94
4	**5**, CH$_3$	NCCH$_2$CH$_2$	(C$_2$H$_5$)$_3$B	82	81
5	**5**, CH$_3$	NCCH$_2$CH$_2$	(s-C$_4$H$_9$)$_3$B	81	90
6	**5**, CH$_3$	CH$_3$CH(OH)(CH$_2$)$_2$	(s-C$_4$H$_9$)$_3$B	43	98
7	**128**, H	CH$_3$	(C$_2$H$_5$)$_3$B	70	84
8	**128**, H	CH$_3$	(s-C$_4$H$_9$)$_3$B	82	89
9	**128**, H	HC≡CHCH$_2$	(C$_2$H$_5$)$_3$B	78	88
10	**128**, H	(CH$_3$)$_3$Si(CH$_2$)$_2$	(C$_2$H$_5$)$_3$B	85	94

the heterocumulene substrates. Similar to the previous studies (cf. Scheme 8E.30), the corresponding cycloaddition reactions with isocyanate to form 2-oxazolidinone gave lower enantioselectivities (43–49% ee).

Intramolecular cyclization by allylic alkylations has been a straightforward and attractive strategy for the construction of carbocycles. However, only few asymmetric versions of this reaction have been reported in the literature. The cyclization of a β-ketoester to a six-membered ring with (R,S)-BPPFA (**9a**) as ligand gave a modest enantioselectivity (48% ee) [167]. Recently, much higher enantioselectivities have been achieved in the analogous intramolecular reactions leading to cyclopentane derivatives (Eq. 8E.13) [168]. Both the enantioselectivity and the yield were shown to depend strongly on a number of reaction parameters. After optimization with the fluorinated BSA as base, 87% ee could be obtained from the reaction of the (E)-isomer (Eq. 8E.13). The (Z)-isomer gives very similar enantioselectivity and the same absolute stereochemistry, which is most consistent with a reaction path involving rapid equilibration of the π-allyl intermediates.

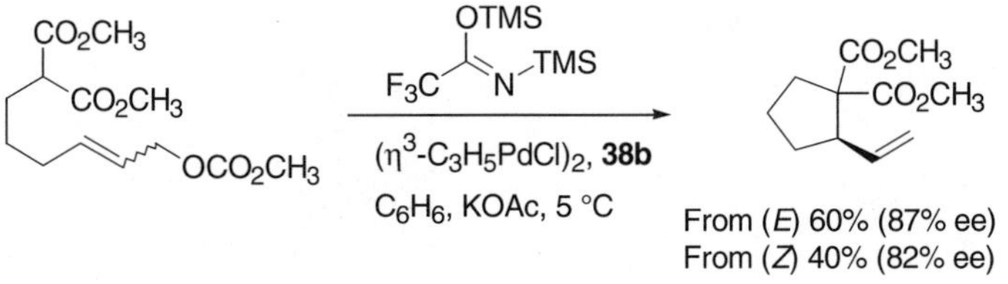

From (E) 60% (87% ee)
From (Z) 40% (82% ee)   (8E.13)

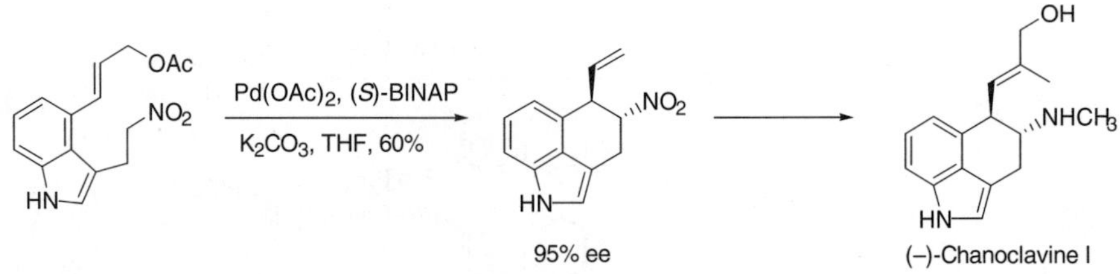

Scheme 8E.32. Synthesis of chanoclavine I.

A nitroalkane has also served as nucleophile in the cyclization as shown in Scheme 8E.32 to give the ergoline ring system in high diastereo- and enantioselectivity. The initially obtained modest enanitoselectivity (66% ee) by using a catalyst derived from $Pd_2(dba)_3 \bullet CHCl_3$ and (*S,S*)-chiraphos (**2**) was optimized to 95% ee by using a complex of (*S*)-BINAP (*ent*-**4**) and $Pd(OAc)_2$ [169,170]. The obtained tricyclic alkylation product provided an expeditious access to (−)-chanoclavine I.

8E.3.4. Enantiotopic Alkene Coordination

Although metal-olefin complexation can be a source of enantioselection, reactions exploiting this mechanistic motif have not been developed much. Due to the facile enantioface interconversion process, the origin of the enantioselection often reverts back to Type C alkylation (Figure 8E.1). To transfer chiral recognition of the coordination process to the ee of the product, kinetic trapping of the incipient π-allyl complex is required prior to any isomerization process. For this reason, few successful examples have come from the use of more reactive heteroatom nucleophiles (N, O and S) and/or intramolecular reactions.

In contradistinction to the alkylation result in the crotyl system described in the previous section (cf. Eq. 8E.9), a rare example depicted in Scheme 8E.33 demonstrates that chiral recognition in the first step of the catalytic cycle can be a source of enantioselectivity [171]. The alkylations of (*E*)- and (*Z*)-crotyl carbonate with a sulfur nucleophile generate the branched product in 92% and 29% ee's, respectively. On the other hand, the same reaction using the chiral substrate gives a nearly racemic product with a similar 5:1 regioselectivity. These results clearly indicate that the nucleophilic addition occurs more rapidly than the enantioface exchange process.

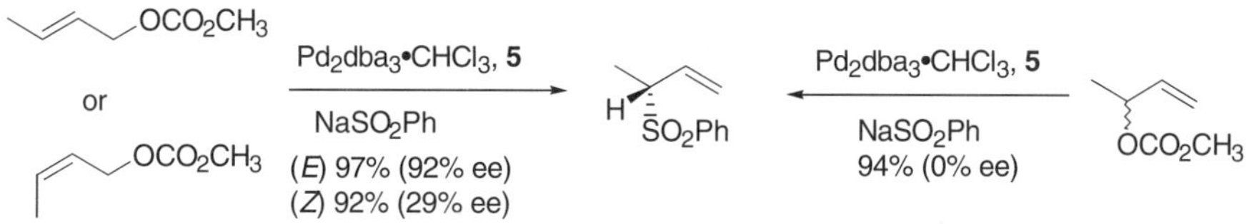

Scheme 8E.33. Enantioselective complexation.

An axially dissymmetric product has also been produced by using this mode of enantioselection. By using dimethyl sodiomalonate as nucleophile and acetate (X = OAc) as leaving group in the alkylation shown in Eq. 8E.14, modest enantioselectivities (5–40% ee) have been obtained with several popular diphosphines, among which BINAP gives the best enantioselectivity [172]. Whereas the chemical and optical yields are slightly dependent on the reaction conditions, the nature of the leaving group, which is undoubtedly involved in the enantiodiscriminating event, has the greatest effect on the enantioselectivity of the reaction rising to 90% ee with the methoxybenzoate leaving group [173].

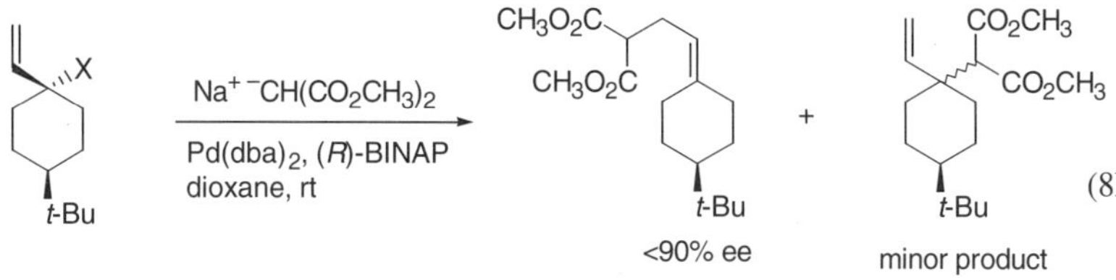

(8E.14)

Distinction between enantiodiscrimination by complexation and by alkylation of equilibrating intermediates is less clear in a number of related cases. It is likely that more than one type of chiral discrimination may be involved. For example, when a conformationally flexible four-membered ring substrate is used for the same reaction, the enantioselectivity was only 56% ee (Eq. 8E.15) [175]. In this case, it has been proposed that equilibration via a tertiary σ-palladium species may be possible, switching the origin of enantio-discrimination to the alkylation step. A more contrasting example involves the formation of an asymmetric diene via selective β-elimination of similar diastereomeric π-allyl intermediates (Eq. 8E.16). Evidence suggests that the enantio-determining elimination process occurs after the equilibration of the π-allyl intermediates [176].

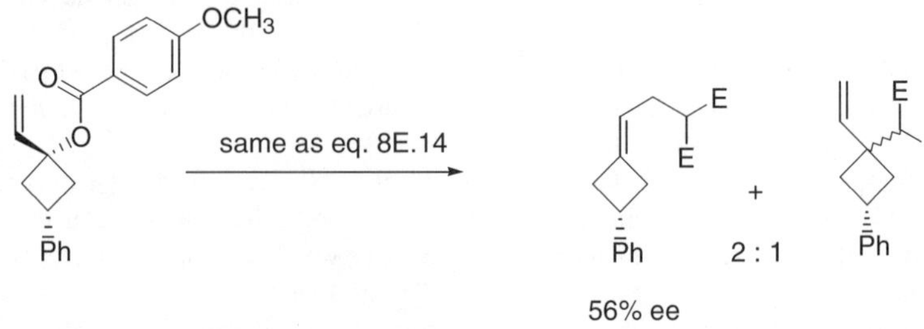

A recent study of intramolecular cyclizations with a nitrogen nucleophile has shown that the enantioselective coordination phenomenon can be exploited to provide high enantioselectivities (Scheme 8E.34) [171]. Using ligand **5** and the achiral olefin substrates **130**, cyclization reactions furnished the pyrrolidine and piperidine in 91% yield and 88% ee, respectively. However, when the chiral isomers **131** were subjected to the same conditions, near-racemic products were obtained. A completely different kinetic picture exists for the seven-membered ring forming reactions. High enantioselectivity could be obtained with the racemic substrate **131** (n = 3), whereas the reaction of achiral substrate **130** (n = 3) gave a modest enantioselectivity, favoring the formation of *ent*-**132** (n = 3). The relatively slow rate of seven-membered ring formation may have changed the selectivity-determining step. Ozonolysis, followed by reduction of the cyclized products, provides the cyclic amino alcohol, which is a useful building block for the synthesis of indolizidines and quinolizidines.

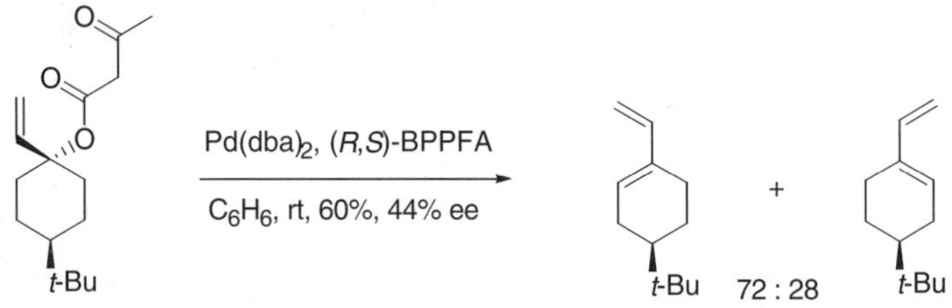

Utilization of an oxygen nucleophile gives similar results (Scheme 8E.35). Whereas modest enantioselectivities (7–54% ee) have been recorded with various ligands [177], the use of **5** results in the efficient cyclization of phenol to furnish the nucleus of tocopherol (vitamin E) with 86% ee [178]. Extension of this methodology to intermolecular reactions requires control of regiochemistry, a problem that is not present in the corresponding intramolecular

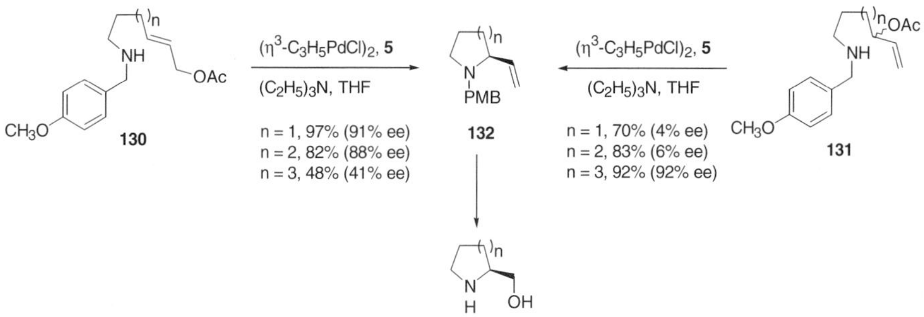

Scheme 8E.34. Intramolecular cyclization using nitrogen nucleophiles.

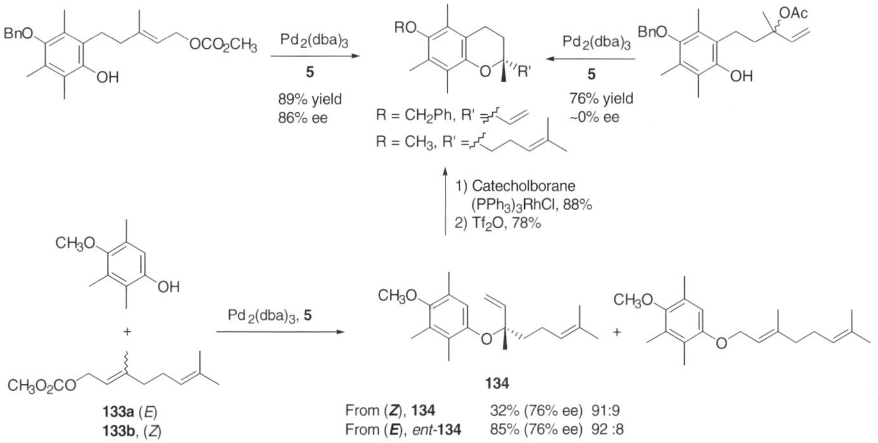

Scheme 8E.35. Synthesis of vitamin e nucleus by alkylation of phenols.

ring-forming process. While electronic rather than steric factors play a beneficial role, especially with the oxygen nucleophile, the alkylation at the more substituted position may be achieved under the influence of the chiral pocket of the ligand. With the geranyl **133a** and neryl **133b** carbonates, the formation of the desired tertiary ether has been realized to give excellent regioselectivity and reasonable enantioselectivity [179]. Note that the change of olefin geometry from (E) to (Z) leads to the formation of enantiomeric products.

Even with the sterically demanding coumarin derivative as nucleophile, the secondary ether could be formed in excellent regio- (92:8) and enantioselectivity (98% ee) by using the bicyclic ligand **135** (Scheme 8E.36). The alkylation product has efficiently served as a key intermediate in the synthesis of (−)-calanolides A and B, which have the most potent HIV-1-specific reverse-transcriptase inhibitory activity among the chromanol family.

An intramolecular reaction was used to establish a quaternary stereogenic center in a cis-decalin system by differentiation of two enantiotopic double bonds (Eq. 8E.17). Despite the quite good enantioselectivity (83% ee), the major product was unfortunately the achiral bicyclic [4.2.2] compound.

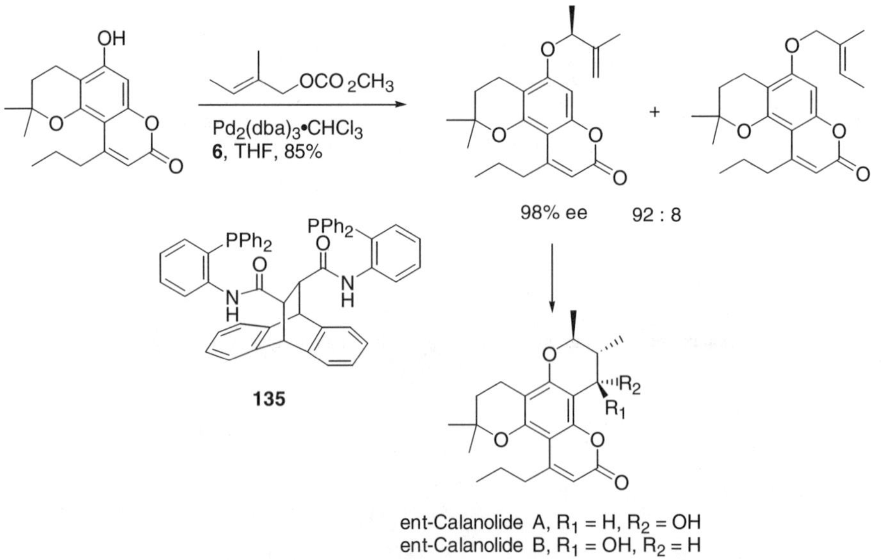

Scheme 8E.36. Synthesis of calanolide by *O*-alkylation of phenol.

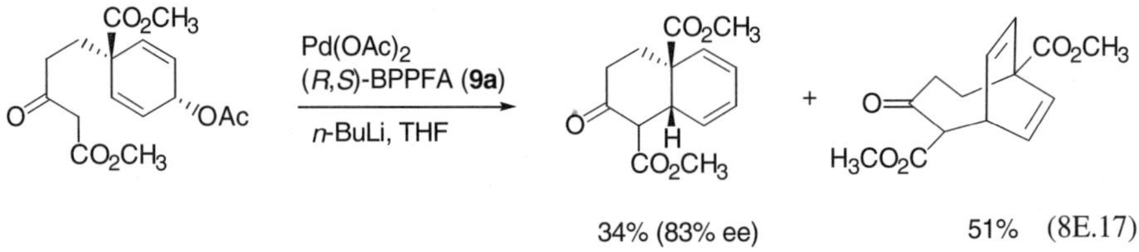

34% (83% ee)   51%   (8E.17)

8E.3.5. Enantioselective Allylation of Prochiral Nucleophiles

The insulation of stereochemical information between the chiral ligand and the reaction site poses the most difficult situation when one wants to induce asymmetry at the incoming nucleophile. Although early efforts to create stereogenic centers at prochiral nucleophiles resulted in low selectivity [180], methods for chiral induction at the nucleophile have improved dramatically in the past decade. Modest success has come from the use of ligands whose functional arm can reach beyond the allyl barrier to direct the orientation of the enolate [181]. Ferrocenylamine ligands possessing a hydroxyethyl group (**10a**) or a crown ether (**136**) provide a site for interaction with the nucleophile or the counterion of the nucleophile in the reaction mixture. Indeed, such ligands increased the enantioselectivity of the reaction as in Eq. 8E.18, in which (*R*)-BINAP and (*S,S*)-DIOP gave poor results (<10% ee). Because the crown ether ligand **136** gives the *R* configuration for the product as opposed to **10a**, which generates *S*-configuration for the major enantiomer, the interaction of the two ligands with the nucleophile may be fundamentally different [182].

The "secondary interaction" between the chiral ligand and the nucleophile via complexation of the counterion led to 80% ee in the preparation of nitroesters, which can serve as precursor of α-alkylated amino acids (Scheme 8E.37) [182]. Enantioselectivity increased in accord with

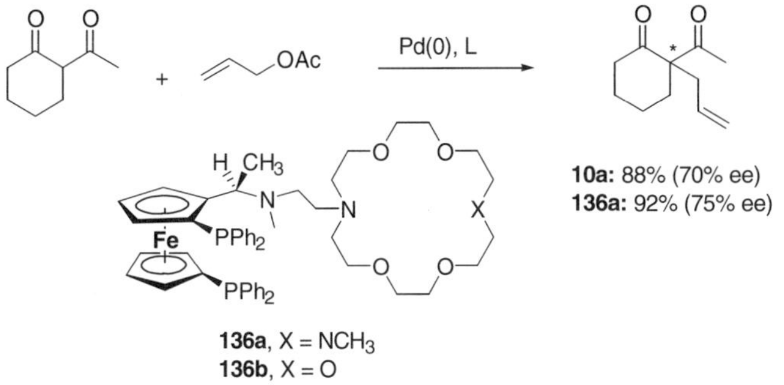

$$(8E.18)$$

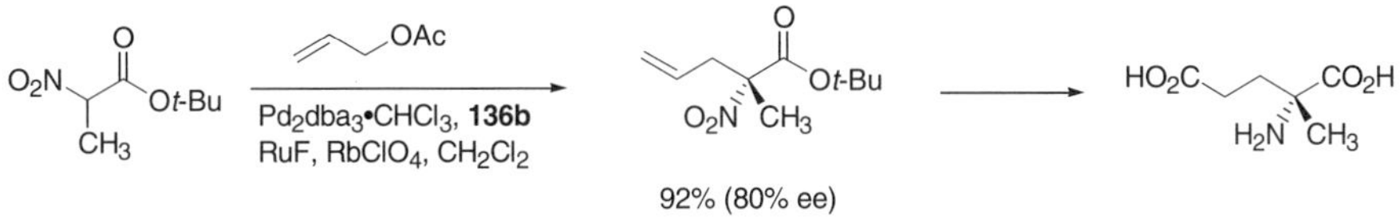

136a, X = NCH$_3$
136b, X = O

92% (80% ee)

Scheme 8E.37. Asymmetric allylation of α-nitroester.

increasing steric hindrance of the ester alkyl group; furthermore, rubidium proved to be the most effective cation within the alkali metal series.

A double alkylation reaction of a tetralone produced the key tricyclic intermediate for the synthesis of (−)-huperzine (Scheme 8E.38). A better enantioselectivity was obtained with the BPPFA-derived ligand **137b**, possessing a more extended side arm (n = 3) [183–185].

The use of nucleophiles, which can coordinate to a transition metal, provides an opportunity to control the orientation of a prochiral enolate (Eq. 8E.19). With α-cyanoesters as substrate,

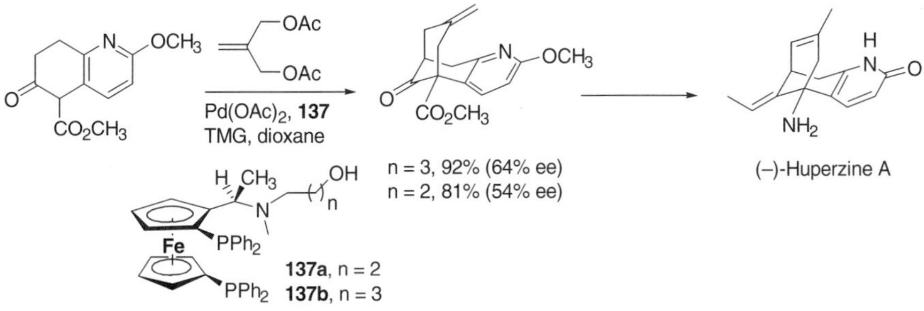

n = 3, 92% (64% ee)
n = 2, 81% (54% ee)

(−)-Huperzine A

137a, n = 2
137b, n = 3

Scheme 8E.38. Synthesis of huperzine A.

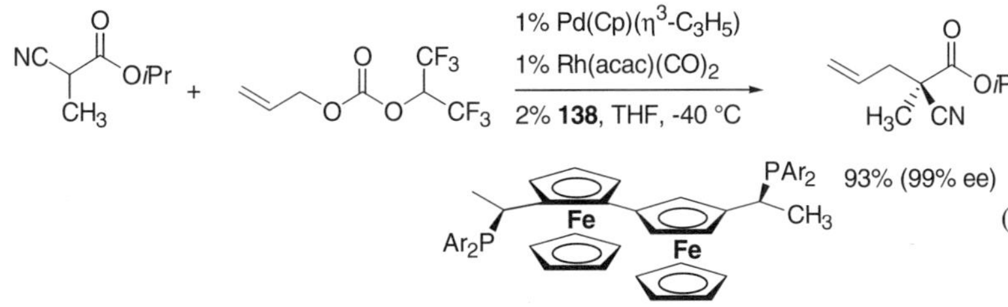

93% (99% ee)

$$(8E.19)$$

138, Ar = p-OMe-C$_6$H$_4$

asymmetric induction at the nucleophile has been achieved with high enantioselectivities by the two-component catalyst system in which the rhodium and palladium catalysts activate the nucleophile and electrophile, respectively. Using one equivalent of bisferrocenyl ligand **138** with respect to the palladium and rhodium combination, the alkylation of the isopropyl ester with allyl carbonate gave the allylated product with 99% ee [186].

When the alkylation was performed with ethyl allyl carbonate as the precursor of the π-allyl intermediate, only 32% ee was obtained, indicative of a subtle proton-transfer process involved in the catalytic process such as in Scheme 8E.39. The chiral rhodium catalyst was shown to be the primary source of the asymmetric induction because the same reaction in the absence of the rhodium catalyst generated a racemic product in 91% yield. It is interesting that the use of only half an equivalent of the chiral ligand together with half an equivalent of achiral ligand (dppb) with respect to [Pd + Rh] was sufficient to give a high enantioselectivity (93% ee).

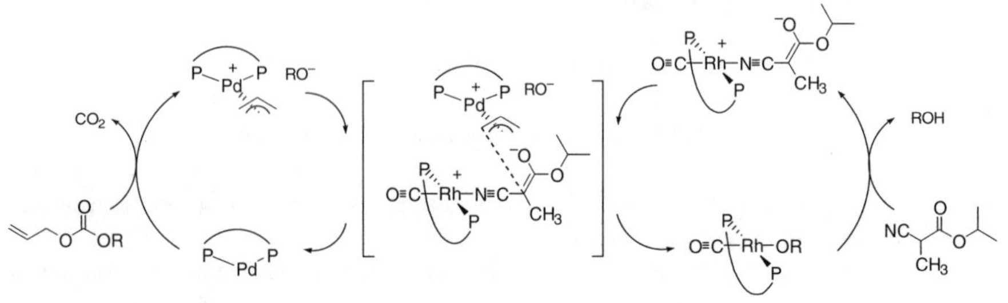

Scheme 8E.39.   Proposed catalytic cycle of Pd-Rh catalyst system.

The chiral pocket approach to attack the problem of physical separation between the chiral ligand and the incoming nucleophile proved highly effective. By using the modular ligand **5**, 2-carboethoxycyclohexanone was quantitatively allylated with 86% ee [187]. The creation of a quaternary center bearing three different functionalities (ketone, ester, and allyl) can be useful for further structural elaborations as shown in Scheme 8E.40. Hydroboration of the alkene effected simultaneous diastereoselective reduction of the ketone to give a single diol, which was elaborated to (−)-nitramine in only six total steps.

When a prochiral nucleophile is reacted with 1,3-disubstituted allylic systems, the issue of diastereo- as well as enantioselectivity arises. In the alkylation of a tetralone, both the acyclic

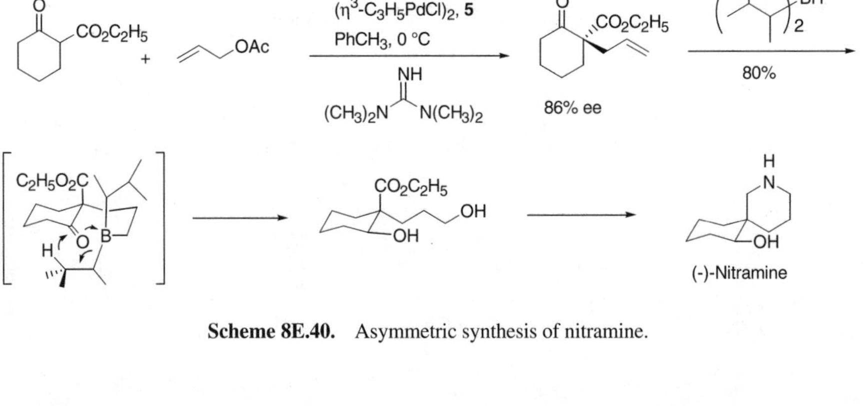

Scheme 8E.40.   Asymmetric synthesis of nitramine.

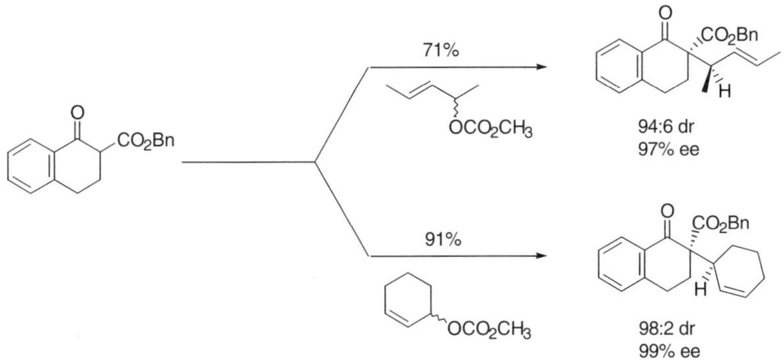

Scheme 8E.41. Asymmetric allylation of tetralone.

and cyclic allylic substrates participated well to give excellent enantio- and diastereoselectivities (Scheme 8E.41).

The double stereodifferentiation is also found in the reactions of acyclic nucleophiles (Scheme 8E.42). Whereas the alkylation of N-imino(methylphosphonates) affords the simple allylated product in poor enantioselectivity (11% ee) using ligand **38a**, excellent enantioselectivity and good diastereoselectivity can be obtained from the alkylation with the 1,3-diphenyl-allyl system [188].

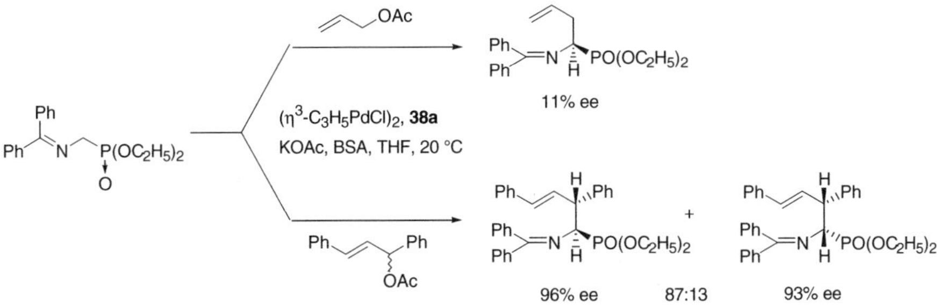

Scheme 8E.42. Asymmetric allylation of N-imino(methylphosphonate).

Whereas preparation of α-amino acid derivatives by asymmetric allylation of an acyclic iminoglycinate gave a modest enantioselectivity (62% ee) in an early investigation [189], the use of conformationally constrained nucleophiles in an analogous alkylation resulted in high selectivities (Scheme 8E.43) [190]. With 2-cyclohexenyl acetate, the alkylation of azlactones occurred with good diastereomeric ratios as well as excellent enantioselectivities. This method provides very facile access to a variety of α-alkylamino acids, which are difficult to synthesize by other methods. When a series of azlactones were alkylated with a prochiral *gem*-diacetate, excellent enantioselectivities were uniformly obtained for both the major and minor diastereomers (Eq. 8E.20 and Table 8E.12).

Constructing a quaternary center by the alkylation of azlactones has led to development of a new strategy for the synthesis of sphingosine analogs. The azlactone derived from alanine was alkylated with a *gem*-diacetate to give a 10.5:1 mixture of diastereomers both with 89% ee [191].

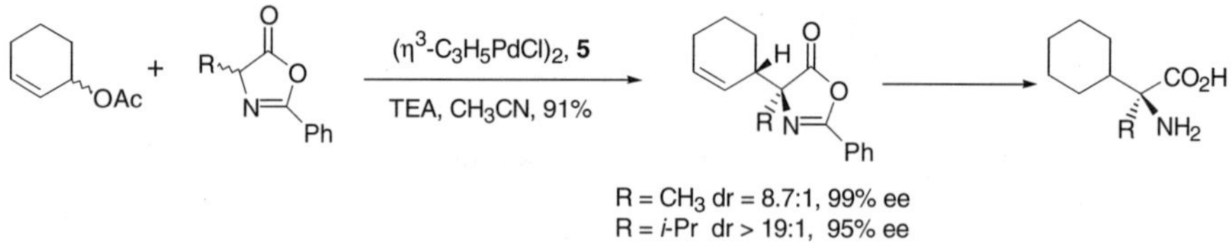

Scheme 8E.43. Asymmetric alkylation of azlactones.

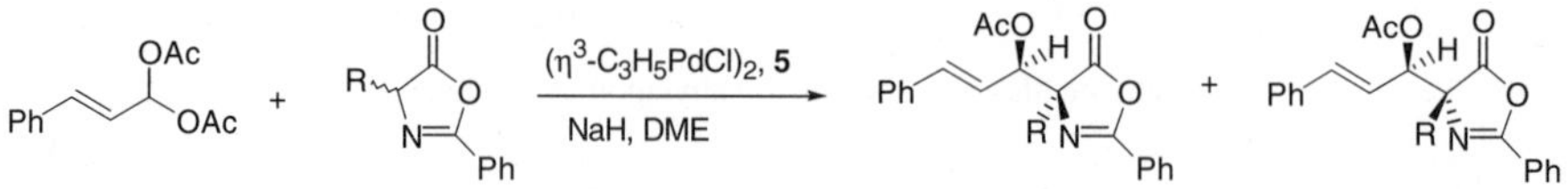

R = CH$_3$ dr = 8.7:1, 99% ee
R = i-Pr dr > 19:1, 95% ee

(8E.20)

TABLE 8E.12. Alkylations of Azlactones with Cinnamyl 1,1-Diacetate

Entry	R	Temp.	Ratio (139:140)	% Yield (% ee) 139	% Yield (% ee) 140
1	CH$_3$	rt	6.6:1	60(99)	9(96)
2	CH$_2$Ph	0~5 °C	9.7:1	75(99)	6(96)
3	CH$_2$CH(CH$_3$)$_2$	0~5 °C	15:1	91(99)	6(95)
4	CH(CH$_3$)$_2$	0~5 °C	>19:1	88(99)	4(–)

In addition to providing good enantioselectivites, the chiral ligand was shown to overcome the intrinsic diastereoselective bias of the system as the same reaction with triphenylphosphine as ligand gave a reversed 1:1.6 diastereomeric ratio. The major diastereomer was ultimately elaborated to an antifungal agent, sphingofungin F (see Scheme 8E.44).

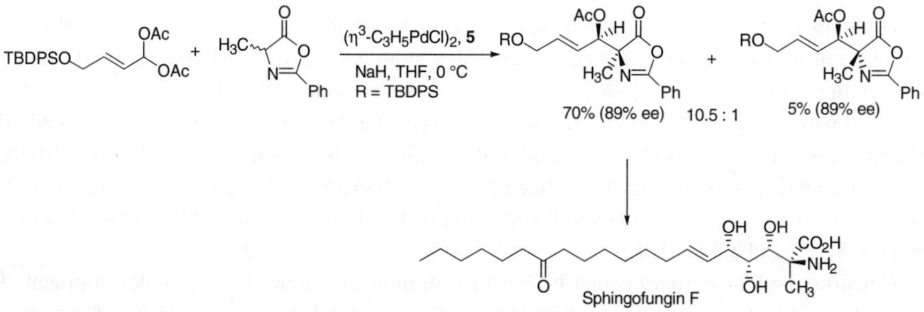

Scheme 8E.44. Synthesis of sphingofungin F.

8E.3.6. Asymmetric Induction with Unstabilized Nucleophiles

Considering the reductive elimination mechanism that takes place within the coordination sphere of the palladium, one might expect the nucleophilic addition of unstabilized nucleophiles to be more enantioselective than that of stabilized nucleophiles because the nucleophile can directly interact with the chiral ligand. However, there are only a few examples in the literature that give high enantioselectivity. In the case of the alkylation with unstabilized carbanions, nickel catalysts have been more frequently used (see next section).

The efforts to effect asymmetric allylic cross-couplings have involved the use of organozinc and organomagnesium reagents as nucleophiles (Scheme 8E.45). Using monodentate ligand **18**, palladium-catalyzed alkylation of phenylzinc chloride with cyclohex-2-enyl acetate affords the adduct in 60% yield with 10% ee [192]. A similar alkylation with a Grignard reagent in the presence of proliphos (**14**) as ligand gives vinylsilane in 53% yield and 33% ee [193].

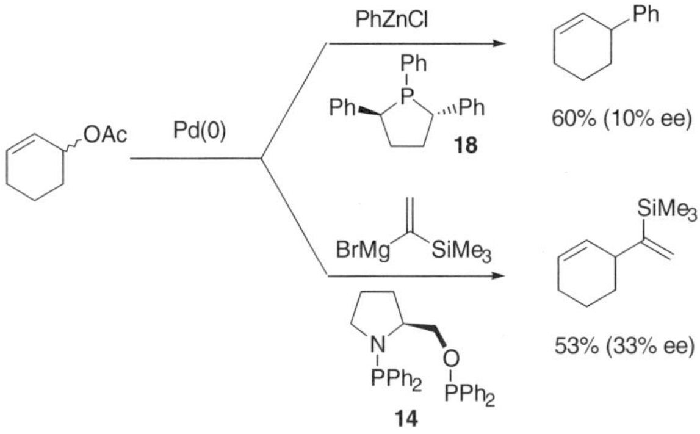

Scheme 8E.45. Allylation with unstabilized nucleophiles.

Promising results were obtained from the silyl transfer reaction, which utilized an allylic chloride as electrophile (Eq. 8E.21). Although the allylsilane can be prepared with 92% ee, the reaction suffered from low regioselectivity [194]. The ruthenocene ligand, which has a larger P-Pd-P bite angle, gives better enantio- and regioselectivity than the ferrocenyl analog.

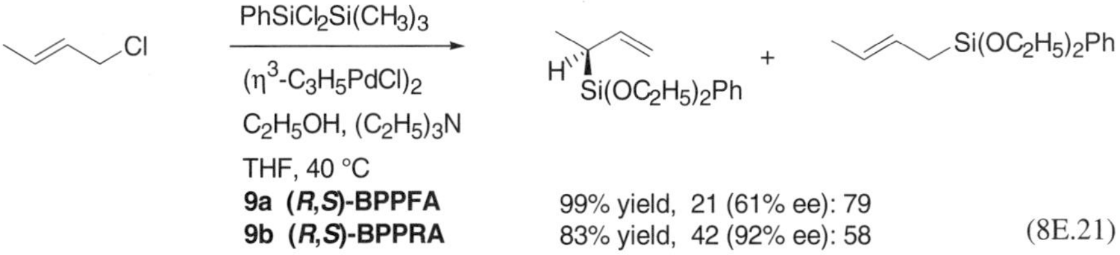

Significantly better results in addition of non-stabilized nucleophiles have come from hydrogenolysis reactions using formate as a hydride donor as shown in Scheme 8E.46. The racemic cyclic acetate and prochiral linear carbonates were reduced in good enantioselectivities by monophosphine ligands (*R*)-MOP (**16**) and (*R*)-MOP-phen (**17**), respectively [195]. The chirality of the allylsilane can be efficiently transferred to the carbinol center of the homoallylic alcohol by the subsequent Lewis acid catalyzed carbonyl addition reaction [196]. The analogous

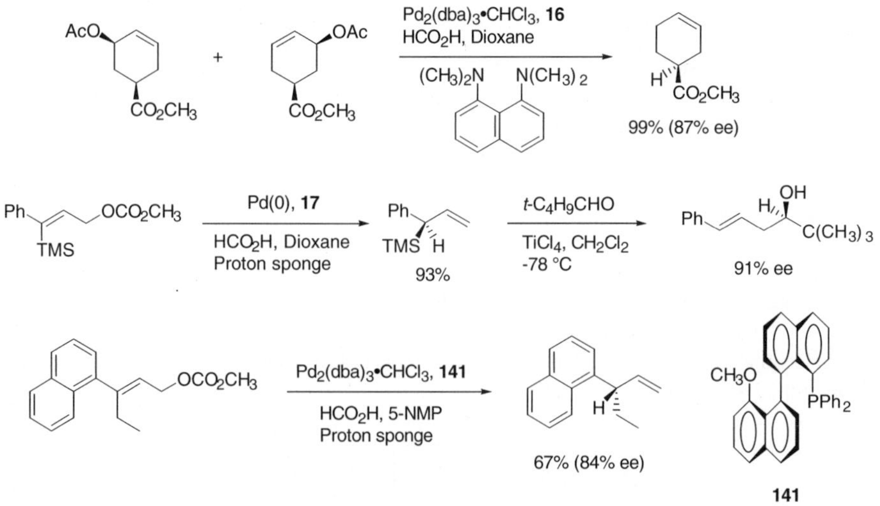

Scheme 8E.46.   Asymmetric hydrogenolysis.

ligand **141** based on 8,8′-disubstituted binaphthyl system was also shown to be an effective ligand for this process [197].

In these hydrogenolyses, the geometry of olefin correlates with the stereochemistry of the alkene as shown in Eq. 8E.22. The reduction of geranyl carbonate gives *S*-**142** with 85% ee, whereas the neryl substrate produces *R*-**142** with 82% ee under the same conditions [195]. With two regioisomeric carbonates that can generate the same π-allyl intermediate mixture, the facile enantioface exchange leads to the formation of the same enantiomer as the major product (Eq. 8E.23) [198]. These results suggest that under these conditions the *syn/anti* isomerization of the π-allyl intermediate, which would lead to low enantioselectivity, is much slower than hydride transfer.

Carbon monoxide, which has a strong propensity for coordination to a metal, can react with a π-allylpalladium intermediate. Although most efforts in the asymmetric carbonylation have

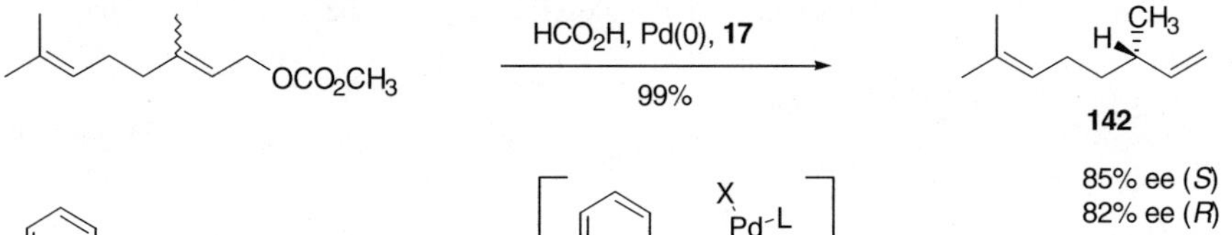

(8E.22)

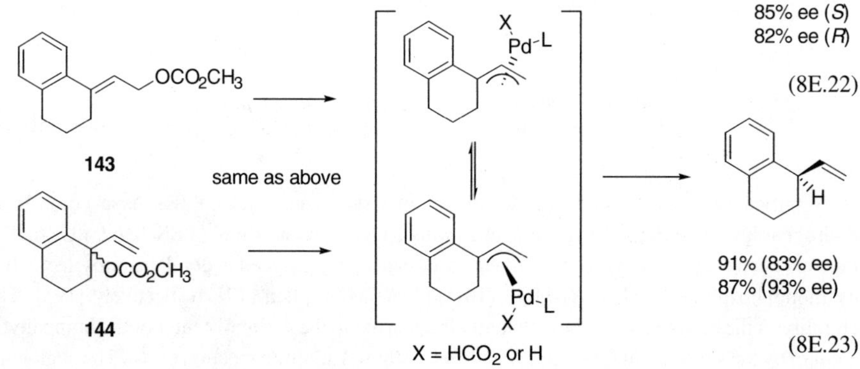

(8E.23)

centered on the use of an olefin as substrate, an example has been recently reported in which an allylic phosphate was converted to a β,γ-unsaturated ester in modest enantioselectivity (Eq. 8E.24) [199].

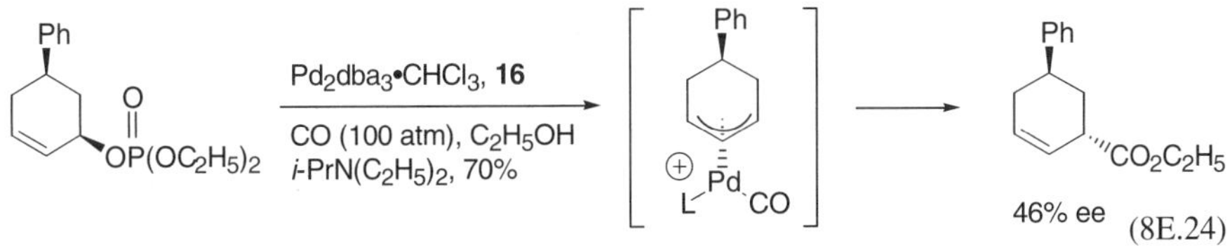

$$\text{(8E.24)}$$

8E.4. ENANTIOSELECTIVE ALLYLATION BY OTHER METALS

8E.4.1. Nickel and Platinum

Although the use of nickel catalysts in asymmetric allylations with a soft nucleophile has been recorded in the literature [200], the greatest use of nickel has been as a catalyst for allylic alkylation reactions with unstabilized nucleophiles. Similar to the corresponding palladium-catalyzed process involving "hard" nucleophiles, the reactions are thought to proceed via an inner-sphere reductive elimination mechanism, which results in overall inversion at the allylic stereogenic center. Because highly reactive nucleophiles (e.g., Grignard reagents) are used in these reactions, less reactive allylic ethers rather than allylic esters have been frequently used as the electrophile. In these cross-coupling reactions, nickel-catalyzed alkylations generally provide higher yields and enantioselectivities than the corresponding palladium-catalyzed processes. Various chiral ligands and Grignard reagents have been tested in the nickel-catalyzed allylation of 2-cyclopentenyl phenyl ether to give moderate-to-good enantioselectivities (Eq. 8E.25).

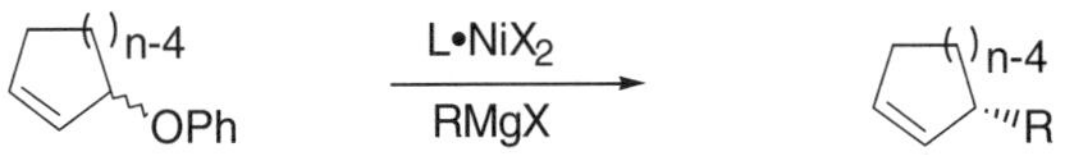

$$\text{(8E.25)}$$

As is readily noted from the results summarized in Table 8E.13, enantioselectivity is very sensitive to a variety of factors such as the nucleophile and the nature of the allylic system as well as the ligand used. As expected, the enantioselectivity varied greatly with the structure of the nucleophile. Higher enantioselectivities were consistently obtained from the reactions of 2-cyclopentenyl phenyl ether than from the corresponding reactions of 2-cyclohexenyl ether. The biphenyl-derived DiPHEMP (**7a**) proved to be more effective than closely related BINAP (**4**) for this reaction.

Prochiral acyclic olefins have been used in the reaction with an aryl Grignard reagent. With the complex of nickel(II) chloride and chiraphos (**2**), a high enantioselectivity could be obtained, albeit with poor regioselectivity (Eq. 8E.26) [204]. When the sterically more bulky 1,3-diphenylallyl

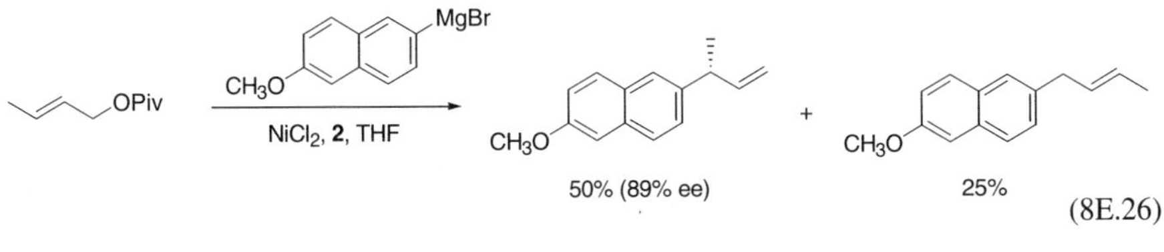

$$\text{(8E.26)}$$

TABLE 8E.13. Nickel Catalyzed Allylic Alkylation

Entry	Ligand	n	R	% Yield	% ee	Ref.
1	(S,S)-chiraphos (2)	5	C_2H_5	60	90	201
2	(S)-BINAP (ent-4)	5	C_2H_5	67	82	202
3	30	5	C_2H_5	91	83	203
4	(R)-DiPHEMP (7a)	5	C_2H_5	90	94	203
5	(R)-DiPHEMP (7a)	5	CH_3	53	51	203
6	(R)-DiPHEMP (7a)	5	n-C_3H_7	10	61	203
7	(R)-DiPHEMP (7a)	5	i-C_3H_7	8	36	203
8	(S,S)-chiraphos (2)	6	C_2H_5	80	51	201
9	(S)-BINAP (ent-4)	6	C_2H_5	11	65	202
10	30	6	C_2H_5	99	74	203
11	(R)-DiPHEMP (7a)	6	C_2H_5	84	84	203

derivative was subjected to similar conditions, a quite high degree of kinetic resolution was observed, indicating that enantiodiscrimination has occurred in both the ionization and reductive elimination steps (Eq. 8E.27) [205].

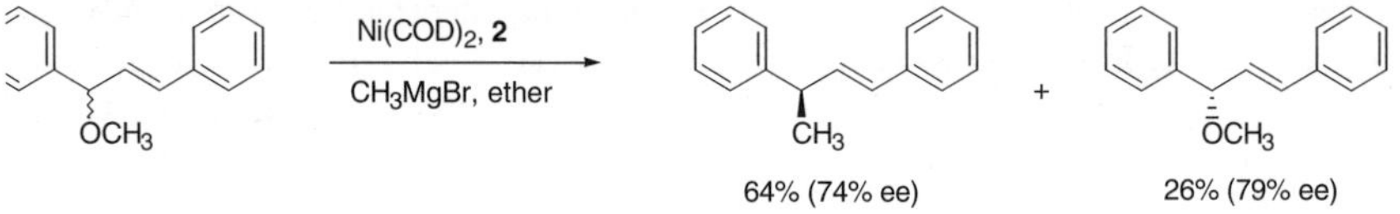

$$(8E.27)$$

An interesting use of the nickel-catalyzed allylic alkylation has prochiral allylic ketals as substrate (Scheme 8E.47) [206]. In contrast to the previous kinetic-resolution process, the enantioselectivity achieved in the ionization step is directly reflected in the stereochemical outcome of the reaction. Thus, the commonly observed variation of the enantioselectivity with respect to the structure of the nucleophile is avoided in this type of reaction. Depending on the method of isolation, the regio- and enantioselective substitution gives an asymmetric Michael adduct or an enol ether in quite good enantioselectivity to provide further synthetic flexibility.

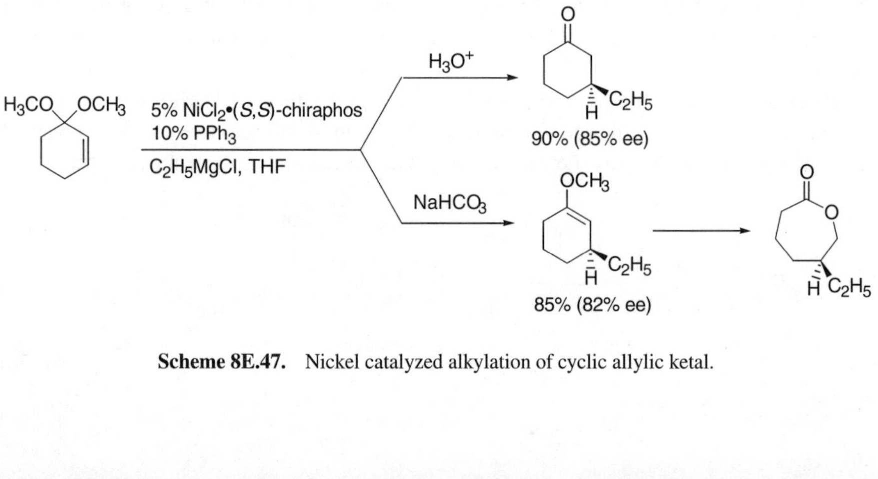

Scheme 8E.47. Nickel catalyzed alkylation of cyclic allylic ketal.

With a racemic mixture of the secondary Grignard reagent, asymmetric cross-coupling with chiral catalysts creates a stereogenic center on the nucleophile. Using (*S,S*)-chiraphos as ligand, the facile interconversion between the two enantiomers of α-phenethylmagnesium bromide allows the formation of the allylated product in 87% yield and 58% ee by a dynamic kinetic asymmetric transformation (Eq. 8E.28) [207].

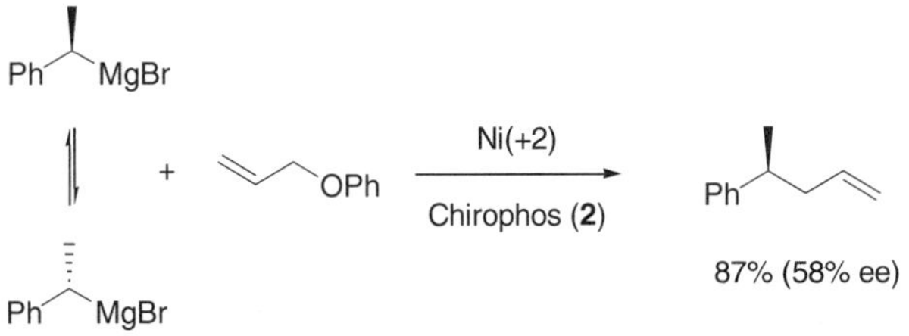

(8E.28)

87% (58% ee)

There have been no significant advances in the area of platinum-catalyzed allylic alkylations since the first work in 1985 [208]. With the platinum complex derived from DIOP, low ee (11%) and modest regioselectivity (4:1) were obtained for the alkylation of (*E*)-crotyl acetate with dimethyl malonate.

8E.4.2. Molybdenum and Tungsten

Despite the progress in certain palladium catalysts, the issue of regiocontrol in the reactions of monosubstituted allylic substrates has led to a search for alternative transition-metal catalysts. In the group VI series, molybdenum and tungsten catalysts have been used in allylation reactions to provide the regioselectivity that is often complementary to that observed in the palladium catalyzed reaction. An asymmetric version of this process has been realized first with a tungsten catalyst system based on the phosphinoaryloxazoline ligand [209]. The complex **143**, which was prepared from **38a** and W(CO)$_3$(CH$_3$CN)$_3$ or W(cycloheptatriene)(CO)$_3$ and characterized by X-ray crystallography, proved to be effective as a catalyst in the alkylation of mono-substituted allylic substrates. With this catalyst, the alkylation of 3-aryl-2-propenyl phosphate with dimethyl sodiomalonate occurred at the more substituted benzylic position to generate the desired chiral regioisomer as the major product with good regioselectivity and high enantiose-lectivity (Eq. 8E.29). Due to the decreased reactivity of the tungsten catalyst, the corresponding allyl carbonate substrate failed to react.

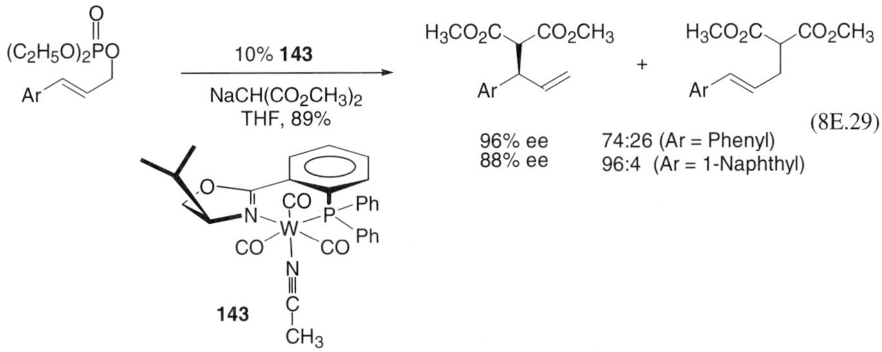

(8E.29)

96% ee
88% ee

74:26 (Ar = Phenyl)
96:4 (Ar = 1-Naphthyl)

An excellent level of regio- and enantiocontrol in the alkylation of cinnamyl-like systems has been achieved by using the dipyridyl ligand **32** (Eq. 8E.30). The in situ generated complex

(8E.30)

of molybdenum and the ligand catalyzed the allylation of malonate nucleophiles with a variety of aromatic allylic esters [210]. Table 8E.14 reveals that remarkably high ee and regioselectivity can be obtained by using the molybdenum-catalyst system. In addition to the phenyl and 1-naphthyl substituted substrates, both electron-poor and -rich heterocyclic substrates were alkylated with unsubstituted ($R' = H$) and substituted ($R' = CH_3$, allyl) nucleophiles in high enantioselectivity (Entries 6–10). Unlike the previous tungsten catalyst, the substrates possessing a less reactive acetate-leaving group could be alkylated with this molybdenum catalyst (entries 9 and 10). Notably, both the achiral linear and racemic chiral isomers (entries 3, 9, and 10) participate well to give good results, indicating that the obtained enantioselectivity is the consequence of the nucleophilic attack on equilibrating π-allylmolybdenum complexes. In contrast to the phosphinoaryloxazoline ligand, whose molybdenum catalyst analogous to **143** was found less effective and gave poor conversion and much lower enantioselectivity, the tungsten complex of the pyridyl ligand also gave a good result (entry 4).

TABLE 8E.14. Molybdenum Catalyzed Allylic Alkylation[a]

Entry	Substrate (Ar)	R	R′	**146:147**	% Yield	% ee
1	**144** (Ph)	CO_2CH_3	H	49:1	70	99
2	**144** (Ph)	CO_2CH_3	CH_3	24:1	67	98
3	**145** (Ph)	CO_2CH_3	H	32:1	61	97
4[a]	**144** (Ph)	CO_2CH_3	H	49:1	55	98
5	**144** (1-Naphthyl)	CO_2CH_3	H	99:1	82	87
6	**144** (2-Thienyl)	CO_2CH_3	H	19:1	78	88
7	**144** (2-Pyridyl)	CO_2CH_3	CH_3	8:1	69	96
8	**144** (2-Furyl)	CO_2CH_3	CH_3	32:1	71	97
9	**145** (2-Furyl)	CH_3	CH_3	99:1	54	95
10	**145** (2-Furyl)	CH_3	Allyl	99:1	50	98

[a] The catalyst was generated from **32** and $(C_2H_5CN)_3W(CO)_3$.

8E.4.3. Rhodium and Iridium

The formation of a branched chiral product from the alkylation of monosubstituted substrates is not limited to the catalysis of metals described thus far. Allylic alkylation reactions catalyzed with rhodium [211] and iridium [212] complexes have been shown to occur at the more substituted terminus of an allylic unit. Despite the significant progress in the structural [213] and mechanistic studies [214,215], the asymmetric rhodium-catalyzed process has not been fully explored since the early effort, which involved decarboxylative alkylation of phenol as shown in Eq. 8E.31 [216].

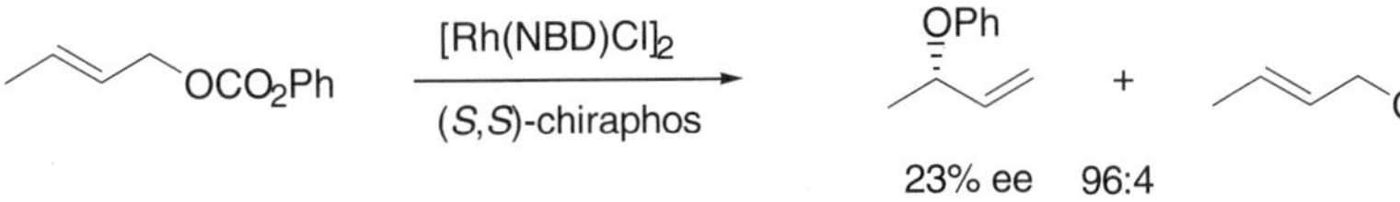

An exciting development has come from work with an iridium catalyst. The use of the complex derived from phosphinoaryloxazoline ligand **149** leads to an efficient alkylation of *E*-cinnamyl acetate (Eq. 8E.32) [217]. It is of note that electron-withdrawing substituents on the phosphorus atom, which are known to be required to give a good regioselectivity in general [218], also increased the enantioselectivity dramatically.

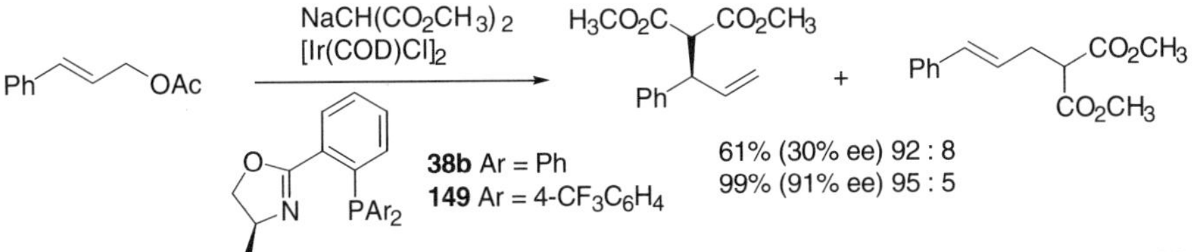

8E.4.4. Other Metals

A regio- and stereocontrolled copper catalyzed allylic alkylation has been recently reported [219]. With the thiophenolate copper catalyst **150**, the Grignard-coupling reaction provided only the γ-allylated product in quantitative yield, albeit with modest enantioselectivity (Eq. 8E.33).

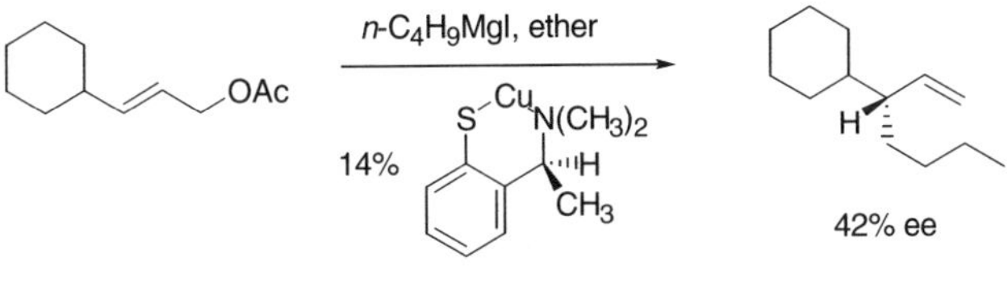

Although mechanistically different, a successful kinetic resolution of cyclic allyl ethers has recently been achieved by zirconium catalysis [220]. Other metals such as cobalt [221], ruthenium [222], and iron [223] have been shown to catalyze allylic alkylation reactions via metal-allyl complexes. However, their catalytic systems have not been thoroughly investigated, and the corresponding asymmetric catalytic processes have not been forthcoming. Nevertheless, increasing interest in the use of alternative metals for asymmetric alkylation will undoubtedly promote further research in this area.

8E.5. CONCLUSIONS

The past several years have witnessed the explosive growth of interest and impressive progress in the transition-metal–catalyzed allylic alkylation reactions. Various approaches towards the development of new and efficient ligands and substrates for this process have led to better understanding of the reaction, and good enantioselectivities have been obtained. Although most of the work in this area has focused on the use of palladium catalysts, utilization of other transition metals is emerging to complement the patterns of reactivity. Many reactions have been developed and have reached a level where they can emulate other well-developed asymmetric catalytic methods. In contrast to hydrogenations or oxygenations, the allylic alkylation reaction can form many different types of bonds to carbon, not limited to C–H or C–O bonds, in a highly stereoselective fashion. In addition, the stereodynamic processes involved in the catalytic cycle provide ample opportunities for chiral induction based on different mechanisms. Both reacting partners, the nucleophile and the allyl unit, can be the site for asymmetric induction. A number of parameters that affect the selectivity may look discouraging at first glance, but on the contrary, this feature is conducive to optimization. The exceptional chemoselectivity and synthetic versatility of this process have resulted in a number of new strategies for the synthesis of natural products and have found elegant applications. As is evident from this review, the potential of the asymmetric allylation reaction will lead to continuing efforts to discover efficient catalysts and reaction systems. Thus, it should not be surprising if research in this area undergoes more extensive development over the years to come.

REFERENCES

1. Ojima, I. *Catalytic Asymmetric Synthesis*; VCH Publishers Inc.: New York, 1993.
2. Noyori, R. *Asymmetric Catalysis in Organic Synthesis*; John Wiley & Sons, Inc.: New York, 1994.
3. Jacobsen, E. N.; Zhang, W.; Muci, A. R.; Ecker, J. R.; Deng, L. *J. Am. Chem. Soc.* **1991**, *113*, 7063.
4. Kolb, H. C.; VanNieuwenhze, M. S.; Sharpless, K. B. *Chem. Rev.* **1994**, *94*, 2483.
5. Trost, B. M. *Acc. Chem. Res.* **1980**, *13*, 385.
6. Trost, B. M.; Verhoeven, T. R. In *Comprehensive Organometallic Chemistry*; Wilkinson, G., Ed.; Pergamon Press: Oxford, 1982; Vol. 8, Ch. 57.
7. Tsuji, J. *Pure Appl. Chem.* **1982**, *54*, 197.
8. Tsuji, J. *Tetrahedron* **1986**, *42*, 4361.
9. Trost, B. M. *Angew. Chem., Int. Ed.* **1989**, *28*, 1173.
10. Godleski, S. A. In *Comprehensive Organic Synthesis*; Trost, B. M., Fleming, I. (Ed.); Pergamon Press: New York, 1991; Vol. 4, Ch. 3.3.
11. Tsuji, J. *Palladium Reagents and Catalysts*; John Wiley & Sons, Inc.: New York, 1996.
12. Trost, B. M.; Strege, P. E. *J. Am. Chem. Soc.* **1977**, *99*, 1650.
13. Consiglio, G.; Waymouth, R. M. *Chem. Rev.* **1989**, *89*, 257.
14. Frost, C. G.; Howarth, J.; Williams, J. M. J. *Tetrahedron: Asymmetry* **1992**, *3*, 1089.
15. Hayashi, T. In *Catalytic Asymmetric Synthesis*; Ojima, I. (Ed.); VCH Publishers, Inc.: New York, 1993.
16. Trost, B. M.; Van Vranken, D. L. *Chem. Rev.* **1996**, *96*, 395.
17. Trost, B. M.; Verhoeven, T. R. *J. Am. Chem. Soc.* **1980**, *102*, 4730.
18. Yamamoto, K.; Deguchi, R.; Ogimura, Y.; Tsuji, J. *Chemistry Lett.* **1982**, 1657.

19. Hayashi, T.; Yamamoto, A.; Hagihara, T. *J. Org. Chem.* **1986**, *51*, 723.

20. Granberg, K. L.; Backvall, J. E. *J. Am. Chem. Soc.* **1992**, *114*, 6858.

21. Trost, B. M.; Strege, P. E. *J. Am. Chem. Soc.* **1975**, *97*, 2534.

22. Trost, B. M.; Verhoeven, T. R. *J. Org. Chem.* **1976**, *41*, 3215.

23. Hayashi, T.; Hagihara, T.; Konishi, M.; Kumada, M. *J. Am. Chem. Soc.* **1983**, *105*, 7767.

24. Temple, J. S.; Schwartz, J. *J. Am. Chem. Soc.* **1980**, *102*, 7381.

25. Matsushida, H.; Negishi, E. *J. Chem. Soc., Chem. Commun.* **1982**, 160.

26. Temple, J. S.; Riediker, M.; Schwartz, J. *J. Am. Chem. Soc.* **1982**, *104*, 1310.

27. Labadie, J. W.; Stille, J. K. *J. Am. Chem. Soc.* **1983**, *105*, 6129.

28. Goliaszewski, A.; Schwartz, J. *J. Am. Chem. Soc.* **1984**, *106*, 5028.

29. Keinan, E.; Roth, Z. *J. Org. Chem.* **1983**, *48*, 1769.

30. Fiaud, J.-C.; Legros, J.-Y. *J. Org. Chem.* **1987**, *52*, 1907.

31. Tschoerner, M.; Trabesinger, G.; Albinati, A.; Pregosin, P. S. *Organometallics* **1997**, *16*, 3447 and references cited therein.

32. *Advanced Application of NMR to Organometallic Chemistry*; Gielen, M.; Willem, R.; Wrackmeyer, B. (Ed.); John Wiley & Sons: Chichester, 1996.

33. Steinhagen, H.; Reggelin, M.; Helmchen, G. *Angew. Chem., Int. Ed.* **1997**, *36*, 2108.

34. Pregosin, P. S.; Trabesinger, G. *J. Chem. Soc., Dalton Trans.* **1998**, 727.

35. Sprinz, J.; Kiefer, M.; Helmchen, G.; Huttner, G.; Walter, O.; Zsolnai, L.; Reggelin, M. *Tetrahedron Lett.* **1994**, *35*, 1523.

36. Pregosin, P. S.; Salzmann, R.; Togni, A. *Organometallics* **1995**, *14*, 842.

37. Tatsumi, K.; Hoffmann, R.; Yamamoto, A.; Stille, J. K. *Bull. Chem. Soc. Jpn.* **1981**, *54*, 1857.

38. Gogoll, A.; Ornebro, J.; Grennberg, H.; Backvall, J. E. *J. Am. Chem. Soc.* **1994**, *116*, 3631.

39. Albinati, A.; Kunz, R. W.; Ammann, C. J.; Pregosin, P. S. *Organometallics* **1991**, *10*, 1800.

40. Hansson, S.; Norrby, P. O.; Sjogren, M. P. T.; Akermark, B.; Cucciolito, M. E.; Giordano, F.; Vitagliano, A. *Organometallics* **1993**, *12*, 4940.

41. Bovens, M.; Togni, A.; Venanzi, L. M. *J. Organomet. Chem.* **1993**, *451*, C28.

42. Andersson, P. G.; Harden, A.; Tanner, D.; Norrby, P. O. *Chem. Eur. J.* **1995**, *1*, 12.

43. Brown, J. M.; Hulmes, D. I.; Guiry, P. J. *Tetrahedron* **1994**, *50*, 4493.

44. Mackenzie, P. B.; Whelan, J.; Bosnich, B. *J. Am. Chem. Soc.* **1985**, *107*, 2046.

45. Trost, B. M.; Keinan, E. *J. Am. Chem. Soc.* **1978**, *100*, 7779.

46. Hayashi, T. *Pure Appl. Chem.* **1988**, *60*, 7.

47. Sawamura, M.; Ito, Y. *Chem. Rev.* **1992**, *92*, 857.

48. Pfaltz, A. *Acc. Chem. Res.* **1993**, *26*, 339.

49. Trost, B. M. *Acc. Chem. Res.* **1996**, *29*, 355.

50. Whitesell, J. K. *Chem. Rev.* **1989**, *89*, 1581.

51. Trost, B. M.; Van Vranken, D. L. *J. Am. Chem. Soc.* **1991**, *113*, 6317.

52. Trost, B. M.; Van Vranken, D. L. *J. Am. Chem. Soc.* **1991**, *112*, 1261.

53. Hayashi, T.; Yamamoto, A.; Ito, Y. *Tetrahedron Lett.* **1987**, *28*, 4837.

54. Trost, B. M.; Van Vranken, D. L. *J. Am. Chem. Soc.* **1993**, *115*, 444.

55. Trost, B. M.; Van Vranken, D. L. *Angew. Chem., Int. Ed.* **1992**, *31*, 228.

56. Trost, B. M.; Van Vranken, D. L.; Bingel, C. *J. Am. Chem. Soc.* **1992**, *114*, 9327.

57. Trost, B. M.; Breit, B.; Peukert, S.; Zambrano, J.; Ziller, J. W. *Angew. Chem., Int. Ed.* **1995**, *34*, 2386.

58. Trost, B. M.; Breit, B.; Organ, M. G. *Tetrahedron Lett.* **1994**, *35*, 5817.

59. Trost, B. M.; Patterson, D. E. *J. Org. Chem.* **1998**, *63*, 1339.

60. Trost, B. M.; Li, L.; Guile, S. D. *J. Am. Chem. Soc.* **1992**, *114*, 8745.

61. Trost, B. M.; Madsen, R.; Guile, S. D. *Tetrahedron Lett.* **1997**, *38*, 1707.

62. Trost, B. M.; Shi, Z. *J. Am. Chem. Soc.* **1996**, *118*, 3039.

63. Trost, B. M.; Chupak, L. S.; Lubbers, T. *J. Am. Chem. Soc.* **1998**, *120*, 1732.

64. Trost, B. M.; Tanimori, S.; Dunn, P. T. *J. Am. Chem. Soc.* **1997**, *119*, 2735.

65. Yoshizaki, H.; Satoh, H.; Sato, Y.; Nukui, S.; Shibasaki, M.; Mori, M. *J. Org. Chem.* **1995**, *60*, 2016.

66. Trost, B. M.; Pulley, S. R. *J. Am. Chem. Soc.* **1995**, *117*, 10143.

67. Trost, B. M.; Cook, G. R. *Tetrahedron Lett.* **1996**, *37*, 7485.

68. Trost, B. M.; Hembre, E. J. *Tetrahedron Lett.* **1999**, *40*, 219.

69. Yoshizaki, H.; Yoshioka, K.; Sato, Y.; Mori, M. *Tetrahedron* **1997**, *53*, 5433.

70. Muchow, G.; Brunel, J. M.; Maffei, M.; Pardigon, O.; Buono, G. *Tetrahedron* **1998**, *54*, 10435.

71. Trost, B. M.; Lee, C. B.; Weiss, J. M. *J. Am. Chem. Soc.* **1995**, *117*, 7247.

72. Trost, B. M.; Lee, C. B., Unpublished results.

73. Trost, B. M.; Murphy, D. J. *Organometallics* **1985**, *4*, 1143.

74. Yamaguchi, M.; Shima, T.; Yamagishi, T.; Hida, M. *Tetrahedron: Asymmetry* **1991**, *2*, 663.

75. Wimmer, P.; Widhalm, M. *Tetrahedron: Asymmetry* **1995**, *6*, 657.

76. Bourghida, M.; Widhalm, M. *Tetrahedron: Asymmetry* **1998**, *9*, 1073.

77. Togni, A. *Tetrahedron: Asymmetry* **1991**, *2*, 683.

78. Kang, J.; Cho, W. O.; Cho, H. G. *Tetrahedron: Asymmetry* **1994**, *5*, 1347.

79. Hayashi, T.; Yamamoto, A.; Hagihara, T.; Ito, Y. *Tetrahedron Lett.* **1986**, *27*, 191.

80. von Matt, P.; Pfaltz, A. *Angew. Chem., Int. Ed.* **1993**, *32*, 566.

81. Dawson, G. J.; Frost, C. G.; Williams, J.M.J. *Tetrahedron Lett.* **1993**, *34*, 3149.

82. Sprinz, J.; Helmchen, G. *Tetrahedron Lett.* **1993**, *34*, 1769.

83. Allen, J. V.; Coote, S. J.; Dawson, G. J.; Frost, C. G.; Martin, C. J.; Williams, J. M. J. *J. Chem. Soc., Perkin Trans. 1* **1994**, 2065.

84. Kubota, H.; Koga, K. *Tetrahedron Lett.* **1994**, *35*, 6689.

85. Leutenegger, U.; Umbricht, G.; Fahrni, C.; von Matt, P.; Pfaltz, A. *Tetrahedron* **1992**, *48*, 2143.

86. Gamez, P.; Dunjic, B.; Fache, F.; Lemaire, M. *J. Chem. Soc., Chem. Commun.* **1994**, 1417.

87. Abbenhuis, H. C. L.; Burckhardt, U.; Gramlich, V.; Kollner, C.; Pregosin, P. S.; Salzmann, R.; Togni, A. *Organometallics* **1995**, *14*, 759.

88. Allen, J. V.; Bower, J. F.; Williams, J. M. J. *Tetrahedron: Asymmetry* **1994**, *5*, 1895.

89. Brenchley, G.; Merifield, E.; Wills, M.; Fedouloff, M. *Tetrahedron Lett.* **1994**, *35*, 2791.

90. Kubota, H.; Nakajima, M.; Koga, K. *Tetrahedron Lett.* **1993**, *34*, 8135.

91. Porte, A. M.; Reibenspies, J.; Burgess, K. *J. Am. Chem. Soc.* **1998**, *120*, 9180.

92. Kubota, H.; Koga, K. *Heterocycles* **1996**, *42*, 543.

93. Hayashi, Y.; Sakai, H.; Kaneta, N.; Uemura, M. *J. Organomet. Chem.* **1995**, *503*, 143.

94. Koning, B.; Meetsma, A.; Kellogg, R. M. *J. Org. Chem.* **1998**, *63*, 5533.

95. Boog-Wick, K.; Pregosin, P. S.; Trabesinger, G. *Organometallics* **1998**, *17*, 3254.

96. Mino, T.; Imiya, W.; Yamashita, M. *Synlett* **1997**, 583.

97. Zhang, W. B.; Kida, T.; Nakatsuji, Y.; Ikeda, I. *Tetrahedron Lett.* **1996**, *37*, 7995.

98. Saitoh, A.; Morimoto, T.; Achiwa, K. *Tetrahedron: Asymmetry* **1997**, *8*, 3567.

99. Evans, P. A.; Brandt, T. A. *Tetrahedron Lett.* **1996**, *37*, 9143.

100. Gilbertson, S. R.; Chang, C. W. T. *Chem. Commun.* **1997**, 975.

101. Achiwa, I.; Yamazaki, A.; Achiwa, K. *Synlett* **1998**, 45.

102. Hoarau, O.; Aithaddou, H.; Castro, M.; Balavoine, G. G. A. *Tetrahedron: Asymmetry* **1997**, *8*, 3755.

103. Zhang, W. B.; Hirao, T.; Ikeda, I. *Tetrahedron Lett.* **1996**, *37*, 4545.

104. Ahn, K. H.; Cho, C. W.; Park, J. W.; Lee, S. W. *Tetrahedron: Asymmetry* **1997**, *8*, 1179.

105. Chelucci, G. *Tetrahedron: Asymmetry* **1997**, *8*, 2667.

106. Chelucci, G.; Medici, S.; Saba, A. *Tetrahedron: Asymmetry* **1997**, *8*, 3183.

107. Nordstrom, K.; Macedo, E.; Moberg, C. *J. Org. Chem.* **1997**, *62*, 1604.

108. Morimoto, T.; Tachibana, K.; Achiwa, K. *Synlett* **1997**, 783.

109. Marinetti, A.; Kruger, V.; Ricard, L. *J. Organomet. Chem.* **1997**, *529*, 465.

110. Hamada, Y.; Matsuura, F.; Oku, M.; Hatano, I.; Shioiri, T. *Tetrahedron Lett.* **1997**, *38*, 8961.

111. Chen, Z. G.; Jiang, Q. Z.; Zhu, G. X.; Xiao, D. M.; Cao, P.; Guo, C.; Zhang, X. M. *J. Org. Chem.* **1997**, *62*, 4521.

112. Ogasawara, M.; Yoshida, K.; Kamei, H.; Kato, K.; Uozumi, Y.; Hayashi, T. *Tetrahedron: Asymmetry* **1998**, *9*, 1779.

113. Imai, Y.; Zhang, W. B.; Kida, T.; Nakatsuji, Y.; Ikeda, I. *Tetrahedron Lett.* **1998**, *39*, 4343.

114. Glaser, B.; Kunz, H. *Synlett* **1998**, 53.

115. Hiroi, K.; Suzuki, Y. *Tetrahedron Lett.* **1998**, *39*, 6499.

116. Trabesinger, G.; Albinati, A.; Feiken, N.; Kunz, R. W.; Pregosin, P. S.; Tschoerner, M. *J. Am. Chem. Soc.* **1997**, *119*, 6315.

117. Pena-Cabrera, E.; Norrby, P.-O.; Sjogren, M.; Vitagliano, A.; Defelice, V.; Oslob, J.; Ishii, S.; O'Neill, D.; Åkermark, B.; Helquist, P. *J. Am. Chem. Soc.* **1996**, *118*, 4299.

118. Burckhardt, U.; Baumann, M.; Togni, A. *Tetrahedron: Asymmetry* **1997**, *8*, 155.

119. Burckhardt, U.; Baumann, M.; Trabesinger, G.; Gramlich, V.; Togni, A. *Organometallics* **1997**, *16*, 5252.

120. Hayashi, T.; Yamamoto, A.; Ito, Y.; Nishioka, E.; Miura, H.; Yanagi, K. *J. Am. Chem. Soc.* **1989**, *111*, 6301.

121. von Matt, P.; Loiseleur, O.; Koch, G.; Pfaltz, A.; Lefeber, C.; Feucht, T.; Helmchen, G. *Tetrahedron: Asymmetry* **1994**, *5*, 573.

122. Sudo, A.; Saigo, K. *J. Org. Chem.* **1997**, *62*, 5508.

123. Constantieux, T.; Brunel, J. M.; Labande, A.; Buono, G. *Synlett* **1998**, 49.

124. Togni, A.; Burckhardt, U.; Gramlich, V.; Pregosin, P. S.; Salzmann, R. *J. Am. Chem. Soc.* **1996**, *118*, 1031.

125. Gais, H. J.; Eichelmann, H.; Spalthoff, N.; Gerhards, F.; Frank, M.; Raabe, G. *Tetrahedron: Asymmetry* **1998**, *9*, 235.

126. Eichelmann, H.; Gais, H. J. *Tetrahedron: Asymmetry* **1995**, *6*, 643.

127. Wiese, B.; Helmchen, G. *Tetrahedron Lett.* **1998**, *39*, 5727.

128. Trost, B. M.; Krueger, A. C.; Bunt, R. C.; Zambrano, J. *J. Am. Chem. Soc.* **1996**, *118*, 6520.

129. Trost, B. M.; Bunt, R. C. *J. Am. Chem. Soc.* **1994**, *116*, 4089.

130. Knuhl, G.; Sennhenn, P.; Helmchen, G. *J. Chem. Soc., Chem. Commun.* **1995**, 1845.

131. Kudis, S.; Helmchen, G. *Angew. Chem., Int. Ed.* **1998**, *37*, 3047.

132. Helmchen, G.; Kudis, S.; Sennhenn, P.; Steinhagen, H. *Pure Appl. Chem.* **1997**, *69*, 513.

133. Trost, B. M.; Radinov, R. *J. Am. Chem. Soc.* **1997**, *119*, 5962.

134. Kudis, S.; Helmchen, G. *Tetrahedron* **1998**, *54*, 10449.

135. Lubineau, A.; Auge, J.; Lubin, N. *Tetrahedron Lett.* **1991**, *32*, 7529.

136. Burlina, F.; Clivio, P.; Fourrey, J.-L.; Riche, C.; Thomas, M. *Tetrahedron Lett.* **1994**, *35*, 8151.

137. Trost, B. M.; Organ, M. G. *J. Am. Chem. Soc.* **1994**, *116*, 10320.

138. Trost, B. M.; Organ, M. G.; O'Doherty, G. A. *J. Am. Chem. Soc.* **1995**, *117*, 9662.

139. Trost, B. M.; Toste, F. D. *J. Am. Chem. Soc.* **1998**, *120*, 815.

140. Iourtchenko, A.; Sinou, D. *J. Mol. Catal. A Chem.* **1997**, *122*, 91.

141. Mori, M.; Kuroda, S.; Zhang, C.-S.; Sato, Y. *J. Org. Chem.* **1997**, *62*, 3263.

142. Auburn, P. R.; Mackenzie, P. B.; Bosnich, B. *J. Am. Chem. Soc.* **1985**, *107*, 2033.

143. Dawson, G. J.; Williams, J. M. J.; Coote, S. J. *Tetrahedron Lett.* **1995**, *36*, 461.

144. Dawson, G. J.; Williams, J. M. J.; Coote, S. J. *Tetrahedron: Asymmetry* **1995**, *6*, 2535.

145. Romero, D. L.; Fritzen, E. L. *Tetrahedron Lett.* **1997**, *38*, 8659.

146. O'Donnell, M. J.; Chen, N.; Zhou, C. Y.; Murray, A.; Kubiak, C. P.; Yang, F.; Stanley, G. G. *J. Org. Chem.* **1997**, *62*, 3962.

147. Trost, B. M.; Huang, M.-H. *J. Am. Chem. Soc.* **1984**, *106*, 6837.

148. Trost, B. M.; Bunt, R. C. *Tetrahedron Lett.* **1993**, *34*, 7513.

149. Hiroi, K.; Makino, K. *Chemistry Lett.* **1986**, 617.

150. Hayashi, T.; Kishi, K.; Yamamoto, A.; Ito, Y. *Tetrahedron Lett.* **1990**, *31*, 1743.

151. Pretot, R.; Lloyd-Jones, G. C.; Pfaltz, A. *Pure Appl. Chem.* **1998**, *70*, 1035.

152. Pretot, R.; Pfaltz, A. *Angew. Chem., Int. Ed.* **1998**, *37*, 323.

153. Hayashi, T.; Kawatsura, M.; Uozumi, Y. *J. Am. Chem. Soc.* **1998**, *120*, 1681.

154. Hayashi, T.; Kawatsura, M.; Uozumi, Y. *Chem. Commun.* **1997**, 561.

155. Trost, B. M.; Bunt, R. C. *J. Am. Chem. Soc.* **1996**, *118*, 235.

156. Trost, B. M.; Tenaglia, A. *Tetrahedron Lett.* **1988**, *29*, 2931.

157. Trost, B. M.; Bunt, R. C. *Angew. Chem., Int. Ed.* **1996**, *35*, 99.

158. Trost, B. M.; Lemoine, R. C. *Tetrahedron Lett.* **1996**, *37*, 9161.

159. Trost, B. M.; Calkins, T. L.; Oertelt, C.; Zambrano, J. *Tetrahedron Lett.* **1998**, *39*, 1713.

160. Hayashi, T.; Yamamoto, A.; Ito, Y. *Tetrahedron Lett.* **1988**, *29*, 99.

161. Yamazaki, A.; Achiwa, K. *Tetrahedron: Asymmetry* **1995**, *6*, 1021.

162. Uozumi, Y.; Tanahashi, A.; Hayashi, T. *J. Org. Chem.* **1993**, *58*, 6826.

163. Massacret, M.; Goux, C.; Lhoste, P.; Sinou, D. *Tetrahedron Lett.* **1994**, *35*, 6093.

164. Yamazaki, A.; Achiwa, I.; Achiwa, K. *Tetrahedron: Asymmetry* **1996**, *7*, 403.

165. Trost, B. M.; McEachern, E. J.; Toste, F. D. *J. Am. Chem. Soc.* **1999**, *120*, 9074.

166. Larksarp, C.; Alper, H. *J. Am. Chem. Soc.* **1997**, *119*, 3709.

167. Yamamoto, K.; Tsuji, J. *Tetrahedron Lett.* **1982**, *23*, 3089.

168. Koch, G.; Pfaltz, A. *Tetrahedron: Asymmetry* **1996**, *7*, 2213.

169. Genet, J. P.; Grisoni, S. *Tetrahedron Lett.* **1988**, *29*, 4543.

170. Kardos, N.; Genet, J. P. *Tetrahedron: Asymmetry* **1994**, *5*, 1525.

171. Trost, B. M.; Krische, M. J.; Radinov, R.; Zanoni, G. *J. Am. Chem. Soc.* **1996**, *118*, 6297.

172. Fiaud, J. C.; Legros, J. Y. *Tetrahedron Lett.* **1988**, *29*, 2959.

173. Fiaud, J. C.; Legros, J. Y. *J. Org. Chem.* **1990**, *55*, 4840.

174. Legros, J. Y.; Fiaud, J.-C. *Tetrahedron* **1994**, *50*, 465.

175. Gil, R.; Fiaud, J. C. *Bull. Soc. Chim. Fr.* **1994**, *131*, 584.

176. Hayashi, T.; Kishi, K.; Uozumi, Y. *Tetrahedron: Asymmetry* **1991**, *2*, 195.

177. Mizuguchi, E.; Achiwa, K. *Chem. Pharm. Bull.* **1997**, *45*, 1209.

178. Trost, B. M.; Asakawa, N. *Synthesis* **1999**, 1491.

179. Trost, B. M.; Toste, F. D. *J. Am. Chem. Soc.* **1998**, *120*, 9074.

180. Fiaud, J. C.; Degournay, A. H.; Larcheveque, M.; Kagan, H. B. *J. Organomet. Chem.* **1978**, *154*, 175.

181. Hayashi, T.; Kanehira, K.; Hagihara, T.; Kumada, M. *J. Org. Chem.* **1988**, *53*, 113.

182. Sawamura, M.; Nagata, H.; Sakamoto, H.; Ito, Y. *J. Am. Chem. Soc.* **1992**, *114*, 2586.

183. Kaneko, S.; Yoshino, T.; Katoh, T.; Terashima, S. *Tetrahedron: Asymmetry* **1997**, *8*, 829.

184. Kaneko, S.; Yoshino, T.; Katoh, T.; Terashima, S. *Tetrahedron* **1998**, *54*, 5471.

185. He, X. C.; Wang, B.; Bai, D. L. *Tetrahedron Lett.* **1998**, *39*, 411.

186. Sawamura, M.; Sudo, M.; Ito, Y. *J. Am. Chem. Soc.* **1996**, *118*, 3309.

187. Trost, B. M.; Radinov, R.; Grenzer, E. M. *J. Am. Chem. Soc.* **1997**, *119*, 7879.

188. Baldwin, I. C.; Williams, J. M. J.; Beckett, R. P. *Tetrahedron: Asymmetry* **1995**, *6*, 679.

189. Genet, J. P.; Juge, S.; Montes, J. R.; Gaudin, J.-M. *J. Chem. Soc., Chem. Commun.* **1988**, 718.

190. Trost, B. M.; Ariza, X. *Angew. Chem., Int. Ed.* **1997**, *36*, 2635.

191. Trost, B. M.; Lee, C. B. *J. Am. Chem. Soc.* **1998**, *120*, 6818.

192. Fiaud, J. C.; Aribi-Zouioueche, L. *J. Organomet. Chem.* **1985**, *295*, 383.

193. Fotiadu, F.; Cros, P.; Faure, B.; Buono, G. *Tetrahedron Lett.* **1990**, *31*, 77.

194. Hayashi, T.; Ohno, A.; Lu, S. J.; Matsumoto, Y.; Fukuyo, E.; Yanagi, K. *J. Am. Chem. Soc.* **1994**, *116*, 4221.

195. Hayashi, T.; Iwamura, H.; Naito, M.; Matsumoto, Y.; Uozumi, Y.; Miki, M.; Yanagi, K. *J. Am. Chem. Soc.* **1994**, *116*, 775.

196. Hayashi, T.; Iwamura, H.; Uozumi, Y. *Tetrahedron Lett.* **1994**, *35*, 4813.

197. Fuji, K.; Sakurai, M.; Kinoshita, T.; Kawabata, T. *Tetrahedron Lett.* **1998**, *39*, 6323.

198. Hayashi, T.; Kawatsura, M.; Iwamura, H.; Yamaura, Y.; Uozumi, Y. *Chem. Commun.* **1996**, 1767.

199. Imada, Y.; Fujii, M.; Kubota, Y.; Murahashi, S.-I. *Tetrahedron Lett.* **1997**, *38*, 8227.

200. Bricout, H.; Carpentier, J. F.; Mortreux, A. *Tetrahedron Lett.* **1996**, *37*, 6105.

201. Consiglio, G.; Piccolo, O.; Roncetti, L.; Morandini, F. *Tetrahedron* **1986**, *42*, 2043.

202. Consiglio, G.; Indolese, A. *Organometallics* **1991**, *10*, 3425.

203. Indolese, A. F.; Consiglio, G. *Organometallics* **1994**, *13*, 2230.

204. Hiyama, T.; Wakasa, N. *Tetrahedron Lett.* **1985**, *27*, 3259.

205. Nomura, N.; RajanBabu, T. V. *Tetrahedron Lett.* **1997**, *38*, 1713.

206. Gomez-Bengoa, E.; Heron, N. M.; Didiuk, M. T.; Luchaco, C. A.; Hoveyda, A. H. *J. Am. Chem. Soc.* **1998**, *120*, 7649.

207. Consiglio, G.; Indolese, A. *J. Organomet. Chem.* **1991**, *417*, C36.

208. Brown, J. M.; Macintyre, J. E. *J. Chem. Soc. Perkin Trans. 2* **1985**, 961.

209. Lloyd-Jones, G. C.; Pfaltz, A. *Angew. Chem., Int. Ed.* **1995**, *34*, 462.

210. Trost, B. M.; Hachiya, I. *J. Am. Chem. Soc.* **1998**, *120*, 1104.

211. Tsuji, J.; Minami, I.; Shimizu, I. *Tetrahedron Lett.* **1984**, *25*, 5157.

212. Takeuchi, R.; Kashio, M. *Angew. Chem., Int. Ed.* **1997** *36*, 263.

213. Lange, S.; Wittmann, K.; Gabor, B.; Mynott, R.; Leitner, W. *Tetrahedron: Asymmetry* **1998**, *9*, 475.

214. Evans, P. A.; Nelson, J. D. *Tetrahedron Lett.* **1998**, *39*, 1725.

215. Evans, P. A.; Nelson, J. D. *J. Am. Chem. Soc.* **1998**, *120*, 5581.

216. Consiglio, G.; Scalone, M.; Rama, F. *J. Mol. Catal.* **1988**, *50*, L11.

217. Janssen, J. P.; Helmchen, G. *Tetrahedron Lett.* **1997**, *38*, 8025.

218. Takeuchi, R.; Kashio, M. *J. Am. Chem. Soc.* **1998**, *120*, 8647.

219. van Klaveren, M.; Persson, E. S. M.; del Villar, A.; Grove, D. M.; Backvall, J.-E.; van Koten, G. *Tetrahedron Lett.* **1995**, *36*, 3059.

220. Visser, M. S.; Harrity, J. P. A.; Hoveyda, A. H. *J. Am. Chem. Soc.* **1996**, *118*, 3779.

221. Bhatia, B.; Reddy, M. M.; Iqbal, J. *Tetrahedron Lett.* **1993**, *34*, 6301.

222. Kondo, T.; Ono, H.; Satake, N.; Mitsudo, T.-A.; Watanabe, Y. *Organometallics* **1995**, *14*, 1945.

223. Xu, Y. Y.; Bo, Z. *J. Org. Chem.* **1987**, *52*, 974.

8F

ASYMMETRIC CROSS-COUPLING REACTIONS

MASAMICHI OGASAWARA AND TAMIO HAYASHI
Department of Chemistry, Graduate School of Science, Kyoto University, Kyoto 606-8502, Japan

8F.1. INTRODUCTION

Nickel- or palladium-catalyzed cross-coupling reaction of a main-group organometallic reagent (R-m) with an alkenyl or an aryl halide (R′-X) is one of the most useful carbon–carbon bond-forming reactions in organic transformations catalyzed by transition-metal complexes [1–4]. As main-group metals (or metalloids) in R-m, a variety of elements such as Mg, Zn, Al, Zr, Sn, B, or Si are applicable to the coupling reaction. A wide range of simple or substituted aryl and alkenyl halides (and very rarely alkynyl halides) are reactive substrates to R-m, but simple alkyl halides are usually inert. The catalytic cycle of the reaction is generally explained as shown in Scheme 8F.1. The catalytically active species, σ-aryl or σ-alkenyl complexes $L_nM(II)R'X$, are generated by the oxidative addition of R′-X to low-valent transition-metal species $L_nM(0)$, occasionally which are produced in situ from divalent precursors L_nMX_2 by reduction with R-m. Transfer of an alkyl group from R-m to the intermediate by transmetallation produces the diorganometal complex L_nMRR'. The reductive elimination of the two organic substituents from this key intermediate gives the product R-R′ and regenerates $L_nM(0)$ species. Note that the described mechanism must be too simple to explain all of the reaction features. For instance, it does not account for the difference of reactivity between Grignard reagents and organolithium reagents in the palladium-catalyzed cross-coupling, because it has been known that both of the organometallic reagents undergo transmetallation to $L_nPd(II)R'X$. One of the advantageous characteristics of the process is that most of the catalysts used successfully have tertiary phosphines as ligands. Without doubt, chiral tertiary phosphines are the chiral ligands most extensively studied and developed for transition-metal–catalyzed asymmetric reactions. Accordingly, the catalysts have been conveniently modified by enantiomerically pure phosphine ligands to make the metal complexes function as chiral catalysts. As organometallic reagents

Catalytic Asymmetric Synthesis, Second Edition, Edited by Iwao Ojima
ISBN 0-471-29805-0   Copyright © 2000 Wiley-VCH, Inc.

$$R\text{-}m \ + \ R'\text{-}X \xrightarrow{\text{[M] (catalyst)}} R\text{-}R' \ + \ mX$$

M = Ni, Pd
m = Mg, Zn, Al, Zr, Sn, B, Si, etc.
R' = aryl, alkenyl, alkynyl
X = Cl, Br, I, OSO_2CF_3, $OPO(OR)_2$, etc.

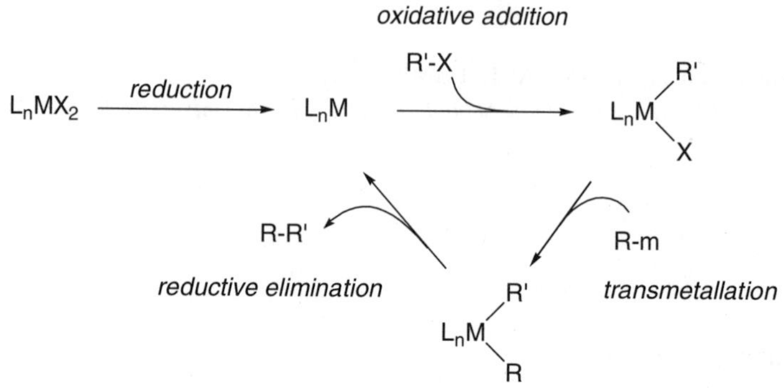

oxidative addition

$$L_nMX_2 \xrightarrow{\text{reduction}} L_nM \xrightarrow{\quad R'\text{-}X \quad} L_nM\begin{smallmatrix}R'\\X\end{smallmatrix}$$

R-R'   R-m

reductive elimination $L_nM\begin{smallmatrix}R'\\R\end{smallmatrix}$ *transmetallation*

Scheme 8F.1.

R-m, relatively reactive organomagnesium and -zinc reagents have been often used for the asymmetric cross-coupling. Aryl and alkenyl pseudo halides, such as triflates, are also applicable to the reaction as the organic electrophiles R'-X. In the cross-coupling product, the new carbon–carbon bond is formed at the sp^2 (or sp) carbon center of the electrophile, thus introduction of chirality into the product is not always easy. For the asymmetrization of this cross-coupling process, special systems have been designed and some of them have shown fairly successful results [5]. In the following sections, the examples will be explained in detail.

8F.2. STRATEGIES FOR ASYMMETRIZATION OF CROSS-COUPLING REACTIONS

The reaction patterns of successful examples of the *asymmetric* cross-coupling can be categorized into the following three groups. The first one is the reaction of secondary alkylmagnesium (or -zinc) reagents. Because there is a chiral carbon center in the Grignard reagent applied for the asymmetric cross-coupling, a kinetic resolution of the racemic reagent will be a way to an enantiomerically enriched product. The second example is a synthesis of a sterically constrained biaryl compound, which possesses axial chirality due to restricted rotation around the carbon–carbon bond between the two aryls. The last representative is an enantioposition-selective cross-coupling. In a substrate for this reaction, there are two identical leaving groups that are enantiotopic each other. Substitution of one of the two leaving groups by the cross-coupling produces chirality in the product. In accordance with importance of carbon central chirality compared with axial or planar chirality in organic chemistry, the examples reported so far in the last two categories are still very limited and the majority of investigations on the asymmetric cross-coupling are related to the first classification, the asymmetric cross-coupling of secondary alkyl organometallic reagents.

8F.2.1. Asymmetric Cross-Coupling of Secondary Alkyl Organometallic Reagents

As described in the introduction section, *alkyl* halides are not available as the organic electrophiles R'-X for the cross-coupling reaction, hence the carbon-central chirality can be introduced only to the R moieties, which are from the organometallic reagents R-m. Asymmetric synthesis by the catalytic cross-coupling reaction has been most extensively studied with secondary alkyl Grignard reagents, such as 1-phenylethylmagnesium halide. The asymmetric cross-coupling with a chiral catalyst allows transformation of a racemic mixture of the secondary alkyl Grignard reagent into an optically active product by a kinetic resolution of the Grignard reagent. That is, one of the two enantiomers of the Grignard reagent reacts with the chiral nickel- (or palladium-) alkyl intermediate faster than the other enantiomer, inducing the enantiomeric enrichment in the final product. It should be noted that the reaction is an example of so-called *dynamic kinetic resolution*. Because the secondary alkyl Grignard reagent usually undergoes racemization at a rate comparable with or faster than that of the cross-coupling, enantiomerically enriched coupling product is formed even if the conversion of the Grignard reagent is nearly 100% (Scheme 8F.2).

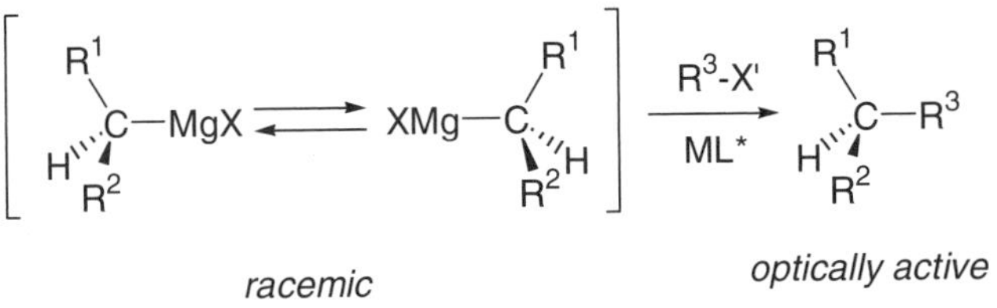

racemic *optically active*

Scheme 8F.2.

8F.2.2. Asymmetric Cross-Coupling of Aryl Grignard Reagents Forming Axially Chiral Biaryls

Despite growing importance of axially chiral biaryls as chiral auxiliaries in asymmetric synthesis, direct synthetic methods accessing to the enantiomerically enriched biaryls from achiral precursors are still very rare. Application of asymmetric cross-coupling to construction of the chiral biaryls is one of the most exciting strategies to this goal. The reported application

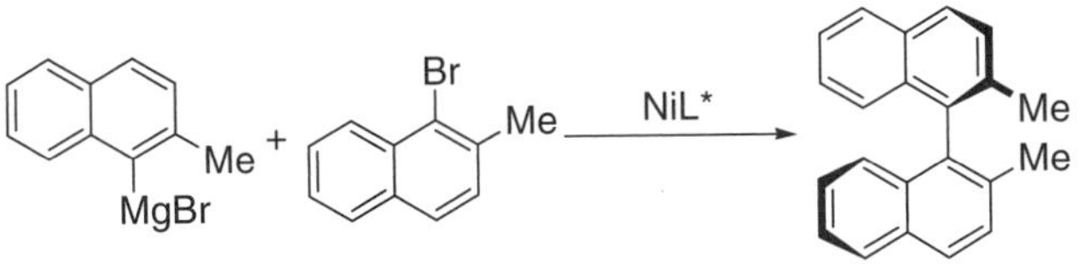

Scheme 8F.3.

of this methodology has been limited to synthesis of atropisomeric substituted binaphthyls at this moment. A representative example is the preparation of 2,2′-dimethyl-1,1′-binaphthyl. The cross-coupling of 2-methyl-1-naphthylmagnesium bromide with 1-bromo-2-methyl-naphthalene gives the chiral 2,2′-dimethyl-1,1′-binaphthyl in the presence of a chiral nickel catalyst [6] (Scheme 8F.3).

8F.2.3. Enantioposition-Selective Asymmetric Cross-Coupling

Three types of reaction systems have been designed and applied for the enantioposition-selective asymmetric cross-coupling reactions so far. First example is asymmetric induction of planar chirality on chromium-arene complexes [7,8]. Two chloro-substituents in a tricarbonyl(η^6-o-dichlorobenzene)chromium are prochiral with respect to the planar chirality of the π-arene-metal moiety, thus an enantioposition-selective substitution at one of the two chloro substituents takes place to give a planar chiral monosubstitution product with a minor amount of the disubstitution product. A similar methodology of monosubstitution can be applicable to the synthesis of axially chiral biaryl molecules from an achiral ditriflate in which the two tri-fluoromethanesulfonyloxy groups are enantiotopic [9–11]. The last example is intramolecular alkylation of alkenyl triflate with one of the enantiotopic alkylboranes, which leads to a chiral cyclic system [12]. The structures of the three representative substrates are illustrated in Figure 8F.1.

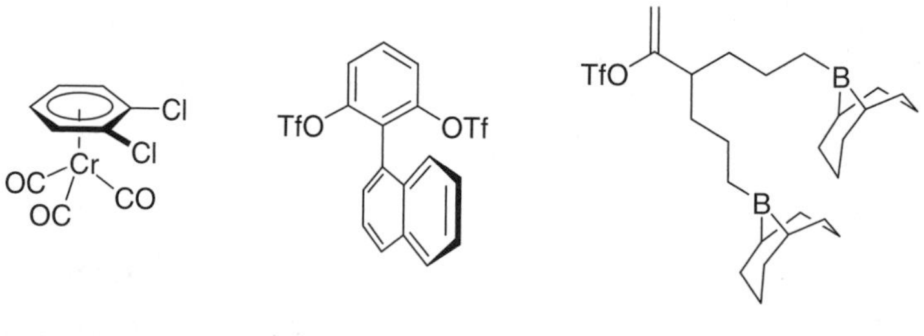

Figure 8F.1. Substrates for enantioposition-selective asymmetric cross-coupling.

8F.3. ASYMMETRIC CROSS-COUPLING REACTIONS ACCORDING TO CHIRAL LIGANDS

In this section, examples of asymmetric cross-coupling will be surveyed according to ligand types. A major part of successful chiral ligands to the asymmetric cross-coupling possesses a similar structural feature, that is, both phosphorus- and nitrogen-donors are in a single molecule. An important roll of the latter, the nitrogen moiety, has been proposed to coordinate with the magnesium (or zinc) atom in the Grignard reagent at the transmetallation step in the catalytic cycle, where the coordination occurs selectively with one of the enantiomers of the racemic Grignard reagent to bring about high selectivity, though the coordination has not been supported by NMR studies of a palladium complex [13]. Another feature of these ligands is lack of symmetry, though this is closely related to the first feature. This feature makes a sharp contrast to C_2-symmetry in most of successful asymmetric bidentate phosphine ligands such as diop, chiraphos, and binap. Because these structural characteristics of the ligands, which show high stereoselectivity in the asymmetric cross-coupling, are very unique, some of the ligands were designed and prepared for targeting application to the asymmetric cross-coupling.

8F.3.1. Early Works: Historical Background

The very first examples of the asymmetric induction in the cross-coupling were achieved in the early 1970s for 1-phenylethyl- (**2**) and 2-butyl- (**3**) magnesium halide with vinyl chloride (**4a**) or phenyl halide (**5**) by using a nickel complex of (−)-diop (**1**) as a catalyst [14,15] (Scheme 8F.4). The enantiomeric purity of the corresponding coupling products, (*R*)-3-phenyl-1-butene (**6**) and (*R*)-2-phenylbutane (**7**), was dependent slightly on the halide atoms of both the Grignard reagents and organic halides, the highest is 13% ee for **6** and 17% ee for **7**. Although these results were promising, the reported enantioselectivity was far from practical applications.

After these findings, asymmetric cross-coupling of the secondary alkyl Grignard reagents has been attempted by using nickel- or palladium-catalysts with various kinds of chiral ligands. The reaction most extensively studied so far is that of 1-phenylethylmagnesium chloride (**2a**) with vinyl bromide (**4b**) or bromostyrene (**8**) forming 3-phenyl-1-butene (**6**) or 1,3-diphenyl-1-butene (**9**), respectively (Scheme 8F.5). These reactions set standards for examining enantio-inducing ability of chiral ligands for the asymmetric cross-coupling reactions and many examples have been reported for the reactions so far. Some of the representative results are summarized in Table 8F.1. The cross-coupling proceeds generally in high yields in ether at or below 0°C with not more than 1 mol % of nickel-phosphine precatalyst $NiCl_2L_n^*$ or catalyst generated in situ from NiX_2 (X = Cl or Br) and a chiral phosphine ligand L^*. Preformed palladium complex $PdCl_2L_n^*$ also catalyzes the asymmetric cross-coupling.

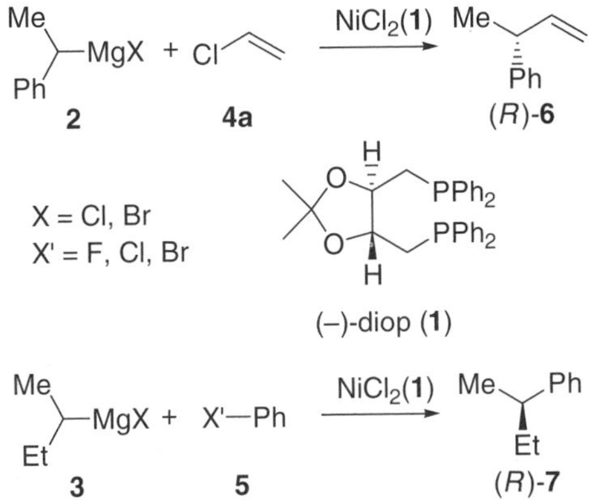

Scheme 8F.4.

8F.3.2. Phosphines Based on Chiral Ferrocenes

Among chiral phosphine ligands used for the nickel- or palladium-catalyzed asymmetric cross-coupling, a series of the ferrocenylphosphines has shown a great success in terms of enantioselectivity and a variety of modified chiral ferrocenylphosphines has been applied to the reaction (**10–16**). The first example of asymmetric cross-coupling catalyzed by

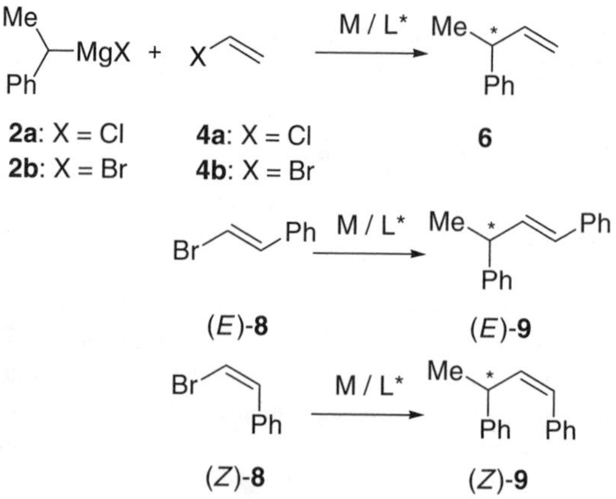

Scheme 8F.5.

nickel-ferrocenylphosphine complexes appeared in 1976 and demonstrated the first practical level enantioselectivity for the cross-coupling of **2a** with **4b** [68% ee with (S)-(R)-PPFA (**10a**), 65% ee with **12**] [16–18]. The ferrocenylphosphines have two distinctive structural characteristics: (1) two independent chiral elements, the ferrocenyl planar chirality and the carbon central chirality on the side chain, and (2) dialkylamino groups on the side chain. Between the two independent chiral elements, the planar chirality plays an important role in the enantiocontrol at the asymmetric cross-coupling rather than the carbon central chirality in the ferrocene side chain, which was confirmed by comparison of the enantioselectivity of the Ni/**10a** catalyst with that observed for a Ni catalyst of its diastereoisomer (R)-(R)-PPFA (**11**) or **12**, which lacks the central chirality on the side arm [16]. The design of the (dialkylamino)alkyl side chain is of primary importance for the high enantioselectivity. The ferrocenylphosphine **13**, whose side chain has no dialkylamino substituent and lacks a carbon central chirality, gave the coupling product **6** with only 5% ee [16]. A bisphosphine ligand (S)-(R)-BPPFA (**14**), which is analogous to PPFA **10a** but has a second diphenylphosphino moiety on the second Cp ring, showed comparable enantioselectivity with **10** for the nickel-catalyzed asymmetric cross-coupling of **2a** with **4b** to give (R)-**6** of 65% ee. The influence of the extent of conversion on enantioselectivity has been studied in the reaction of the Grignard reagent **2b** with **4a** catalyzed by nickel complex of **14** [19]. A ferrocenylphosphine **16**, which is analogous to PPFA but has a tetrahydroindenyl moiety, was more enantioselective than **10a** for the palladium-catalyzed cross-coupling of **2a** with **4b** to give (R)-**6** with 79% ee [20] (Table 8F.1, entry 15).

An example of practical application of the asymmetric cross-coupling, enantioselective synthesis of α-curcumene (**19**), is shown in Scheme 8F.6 [21]. The compound was synthesized with 66% ee through Ni/(S)-(R)-PPFA (**10a**)-catalyzed cross-coupling of 1-(p-tolyl)ethylmagnesium chloride (**17**) and vinyl bromide (**4b**).

The use of 1-phenylethylzinc reagents in place of the corresponding magnesium reagents sometimes shows interesting selectivity (Scheme 8F.7). The results are summarized in Table 8F.2, which also contains the data obtained by using the Grignard reagents for comparison. The reaction of zinc reagents **20** prepared from the magnesium reagents **2a** and zinc halide in THF in the presence of a palladium catalyst coordinated with a chiral ferrocenylphosphine [(R)-(S)-PPFA (**10a**)] proceeded with 85–86% enantioselectivity [22] (entries 1 and 2). The selectivity is higher than that observed for the reaction with 1-phenylethyl Grignard reagents (entry 3, see

TABLE 8F.1. Asymmetric Cross-Coupling of 1-Phenylethylmagnesium Halides 2 with Alkenyl Halides 4 or 8

Entry	Halide 4 or 8	PhCH(Me)M X (equiv. to halide)	Catalyst (mol %)	solvent	temp (°C)	time (h)	Yield %	% ee (config)	Reference
				Reaction Conditions					
1	4a	2a	(−)-diop (1)/Ni (0.4)	Et$_2$O	−80	20	74 (6)	7 (R)	14
2	4a	2a (3.9)	(−)-diop (1)/Ni (0.1)	Et$_2$O	0	1	81 (6)	13 (R)	15
3	4b	2a (4)	(S)-(R)-PPFA (10a)/Ni (0.5)	Et$_2$O	−20	24	>95 (6)	68 (R)	16, 17
4	4b	2a (3)	(S)-(R)-10b/Ni (0.5)	Et$_2$O	0	24	>95 (6)	62 (R)	17
5	4b	2a (3)	(S)-(R)-10c/Ni (0.5)	Et$_2$O	0	24	43 (6)	42 (S)	17
6	4b	2a (2)	(R)-(R)-PPFA (11)/Ni (0.5)	Et$_2$O	−20	24	>95 (6)	54 (R)	17
7	4b	2a (4)	(S)-12/Ni (0.5)	Et$_2$O	0	24	>95 (6)	65 (S)	16, 17
8	4b	2a (3)	(R)-13/Ni (0.5)	Et$_2$O	0	24	86 (6)	5 (S)	16, 17
9	4b	2a (3)	(S)-(R)-PPFA (10a)/Pd (0.5)	Et$_2$O	25	70	82 (6)	61 (R)	17
10	4b	2a (4)	(S)-(R)-BPPFA (14)/Pd (0.5)	Et$_2$O	0	24	73 (6)	65 (R)	17
11	4b	2a (2)	(R)-(S)-15/Ni (0.5)	Et$_2$O	0	40	77 (6)	17 (R)	18
12	(E)-8	2a (2)	(S)-(R)-PPFA (10a)/Ni (0.5)	Et$_2$O	0	24	62 ((E)-9)	52 (R)	17
13	(E)-8	2a (2)	(S)-(R)-PPFA (10a)/Pd (2)	Et$_2$O	25	20	−((E)-9)	73	13
14	4a	2b (1.2)	(S)-(R)-BPPFA (14)/Ni (0.5)	Et$_2$O	25	–	~100 (6)	60–70 (R)	19
15	4b	2a (2)	16/Pd (0.4)	Et$_2$O	0	24	95 (6)	79 (R)	20
16	4b	2a (2)	(S)-alaphos (49a)/Ni (0.5)	Et$_2$O	0	48	>95 (6)	38 (S)	33, 34
17	4b	2a (2)	(S)-phephos (49b)/Ni (0.5)	Et$_2$O	0	48	>95 (6)	71 (S)	33, 34
18	4b	2a (2)	(S)-valphos (49c)/Ni (0.5)	Et$_2$O	0	48	>95 (6)	81 (S)	33, 34
19	4b	2a (2)	(S)-ilephos (49d)/Ni (0.5)	Et$_2$O	0	48	>95 (6)	81 (S)	34
20	4b	2a (2)	(R)-phglyphos (49e)/Ni (0.5)	Et$_2$O	0	48	>95 (6)	70 (R)	33, 34
21	4b	2a (2)	(R)-t-leuphos (49f)/Ni (0.5)	Et$_2$O	0	48	>95 (6)	83 (R)	33, 34
22	4b	2a (2)	50/Ni (0.5)	Et$_2$O	0	40	73 (6)	49 (S)	35
23	4b	2a (2)	(R)-51a/Ni (0.8)	Et$_2$O	−5	16	(6)	88 (R)	36
24	4b	2a (2)	(S)-51b/Ni (0.5)	Et$_2$O	−5	14	>90 (6)	38 (S)	37
25	4b	2a (2)	(S)-51c/Ni (0.5)	Et$_2$O	−5	14	>90 (6)	65 (S)	37

(continued)

TABLE 8F.1. Continued

Entry	Halide 4 or 8	PhCH(Me)M X (equiv. to halide)	Catalyst (mol %)	Reaction Conditions			Yield %	% ee (config)	Reference
				solvent	temp (°C)	time (h)			
26	**4a**	**2a** (1.2)	(S)-**67a**/Ni (1)	Et₂O	−78 → rt	20	68 (**6**)	89 (R)	41, 42
27	**4b**	**2a** (1.2)	(S)-**67a**/Ni (1)	Et₂O	−78 → rt	20	96 (**6**)	73 (R)	41, 42
28	**4a**	**2a** (1.2)	(S)-**67b**/Ni (1)	Et₂O	−78 → rt	20	64 (**6**)	75 (R)	41
29	**4a**	**2a** (1.2)	(S)-**67c**/Ni (1)	Et₂O	−78 → rt	20	82 (**6**)	72 (R)	41, 42
30	**4b**	**2a** (2)	(1R,2S)-**68**/Ni	Et₂O	0	48	91 (**6**)	66	43
31	(E)-**8**	**2a** (2)	(1R,2S)-**68**/Ni	Et₂O	0	48	95 ((E)-**9**)	94	43
32	**4b**	**2a**	(S-**69**)/Ni (0.5)	Et₂O	0	20	67 (**6**)	46 (R)	44, 45
33	**4b**	**2a** (1.7)	(S)-**70**/Ni (0.5)	Et₂O	0	20	54 (**6**)	16 (R)	46
34	(E)-**8**	**2a** (1.5)	**71**/Pd (0.5)	Et₂O	−45	20	65 ((E)-**9**)	11 (R)	47
35	(E)-**8**	**2a** (2)	**72**/Pd (0.5)	Et₂O	−45 → 0	20	95 ((E)-**9**)	40 (R)	48
36	(E)-**8**	**2a** (1.5)	(S)-(S)-**73**/Pd (1)	Et₂O	0	6	74 ((E)-**9**)	45 (S)	49
37	(Z)-**8**	**2a** (2.5)	(S)-**74**/Ni (1.5)	Et₂O	−19		>95 ((Z)-**9**)	45 (S)	50
38	**4b**	**2a** (2)	**75**/Ni (0.5)	Et₂O	−40 → rt	2	95 (**6**)	67 (S)	51
39	**4a**	**2a** (2)	**76**/Ni (0.3)	Et₂O			45 (**6**)	47 (S)	19, 52
40	**4b**	**2a**	**33**/Ni	Et₂O	0	24	(**6**)	11 (S)	30
41	**4b**	**2a** (2)	**77**/Ni (0.7)	Et₂O	−78 → rt	12	52 (**6**)	17 (S)	53
42	(E)-**8**	**2a** (2)	**78**/Pd (0.5)	Et₂O	−45 → **20**	20	90 ((E)-**9**)	13 (S)	54
43	**4b**	**2a**	**79**/Ni	Et₂O	0	18	66 (**6**)	32 (R)	55
44	**4b**	**2a** (1.5–2)	**80**/Ni (1)	Et₂O	−78 → 25		<53 (**6**)	22 (S)	56
45	**4b**	**2a** (2)	**86**/Ni (0.5)	Et₂O	−10 → 0	17	50 (**6**)	46 (R)	61, 62
46	**4b**	**2a**	**87**/Ni	Et₂O	−10		46 (**6**)	11 (R)	63

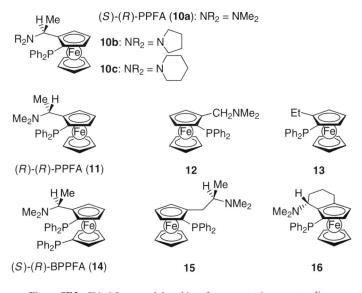

Figure 8F.2. Chiral ferrocenylphosphines for asymmetric cross-coupling.

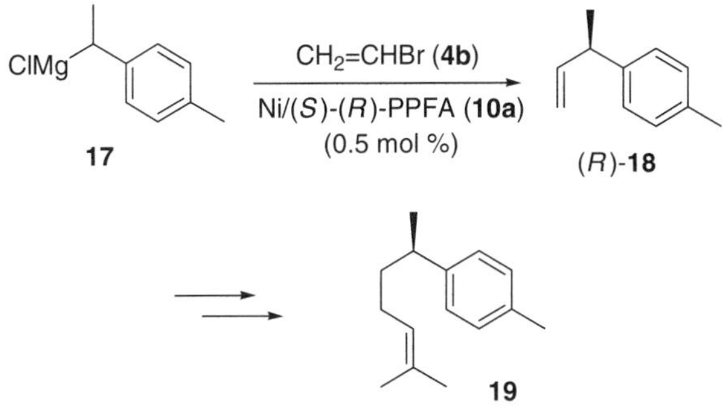

Scheme 8F.6.

also entry 3 in the Table 8F.1). The highest reported enantioselectivity (93% ee) in forming (*R*)-**6** was obtained with C_2-symmetric ferrocenylphosphine ligand **21** that has two identical cyclopentadienyl moieties, each of which possesses diphenylphosphino- and 1-dimethylaminoethyl substituents [23,24] (entry 5). An aminoalkylphosphine ligand **22**, which is analogous to PPFA (**10a**) but has (η^6-benzene)-chromium structure in place of ferrocenyl skeleton, showed a little lower selectivity (61% ee) in the reaction of 1-phenylethylzinc reagents [25] (entries 6 and 7).

The asymmetric cross-coupling of zinc reagent **20** with (*E*)- and (*Z*)-1-bromo-2-(phenylthio)ethenes (**23**) catalyzed by Pd/(*S*)-(*R*)-PPFA (**10a**) gave optically active alkenyl sulfides **24**, which could be a potential substrate for the second cross-coupling, that is, the sulfide is replaced by the Grignard reagent in the presence of a nickel catalyst [26] (Scheme 8F.8).

The asymmetric cross-coupling was successfully applied to the synthesis of optically active allylsilanes [27,28] (Scheme 8F.9). The reaction of α-(trimethylsilyl)benzylmagnesium bro-

TABLE 8F.2. Asymmetric Cross-Coupling of 1-Phenylethylzinc Halides 20 with Alkenyl Halides 4 or 8

Entry	Halide 4 or 8	PhCH(Me)MX (equiv. to halide)	Catalyst (mol %)	Reaction Conditions			Yield %	% ee (config)	Reference
				solvent	temp (°C)	time (h)			
1	4a	2a + ZnI₂ (3)	(R)-(S)-PPFA (10a)/Pd (0.5)	THF/Et₂O	0	21	>95 (6)	86 (S)	22
2	4a	2a + ZnCl₂ (3)	(R)-(S)-PPFA (10a)/Pd (0.5)	THF/Et₂O	0	40	>95 (6)	85 (S)	22
3	4a	2a (3)	(R)-(S)-PPFA (10a)/Pd (0.5)	Et₂O	0	21	>95 (6)	65 (S)	22
4	(E)-8	2a + ZnCl₂ (3)	(R)-(S)-PPFA (10a)/Pd (0.5)	THF/Et₂O	0	22	88 ((E)-9)	60 (S)	22
5	4a	2a + ZnCl₂ (3)	21/Pd (0.5)	THF/Et₂O	0	20	>95 (6)	93 (R)	23
6	4a	2a + ZnCl₂ (3)	22/Pd (0.5)	THF/Et₂O	0	18	67 (6)	61 (S)	25
7	4a	2a (3)	22/Pd (0.5)	THF/Et₂O	0	18	56 (6)	13 (S)	25
8	4a	2a + ZnBr₂ (2)	(S)-51a/Ni (0.1)	Et₂O	−34	21	73 (6)	70 (R)	38, 39
9	4a	2a (2)	(S)-51a/Ni (0.1)	Et₂O	0	20	95 (6)	61 (S)	38, 39
10	4a	2a + ZnBr₂ (2)	(S)-51b/Ni (0.1)	Et₂O	2	16	88 (6)	52 (R)	38, 39
11	4a	2a (2)	(S)-51b/Ni (0.1)	Et₂O	4	96	>95 (6)	60 (S)	38, 39

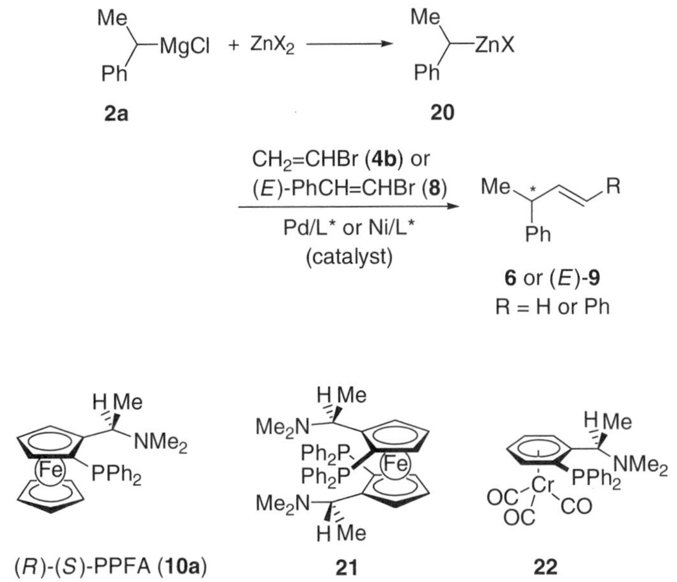

Scheme 8F.7.

mide (**25**) with vinyl bromide (**4b**), (*E*)-bromopropene ((*E*)-**26**), and (*E*)-bromostyrene ((*E*)-**8**) in the presence of 0.5 mol % of palladium complex coordinated with chiral ferrocenylphosphine, (*R*)-(*S*)-PPFA (**10a**), gave the corresponding (*R*)-allylsilane (**27**) with 95%, 85%, and 95% ee, respectively, which were substituted with phenyl groups at the chiral carbon center bonded to the silicon atom. These allylsilanes were used for the S_E' reactions forming optically active homoallyl alcohols and π-allylpalladium complexes. Lower stereoselectivity was observed with (*Z*)-alkenyl bromides (*Z*)-**26** and (*Z*)-**8**. The palladium/**10a** catalyst was also effective for the reaction of 1-(trialkylsilyl)ethylmagnesium chlorides **28** with (*E*)-bromostyrene ((*E*)-**8**). The enantioselectivity was dependent on the trialkylsilyl group, and triethylsilyl is best to produce (*S*)-1-phenyl-3-silyl-1-butene (**29c**) with 93% ee. Dienylsilane (*S*)-**31** with 45% ee was also prepared by the asymmetric cross-coupling of **28b** with dienyl bromide (*E*)-**30**. The palladium-catalyzed asymmetric cross-coupling of α-(trimethylsilyl)benzylmagnesium bromide (**25**) was also applied to the synthesis of optically active propargylsilane **32** (18% ee) by using 1-bromo-2-phenylacetylene as a coupling partner [29] (Scheme 8F.10).

The ferrocenylphosphine-nickel catalysts are also applied to asymmetric synthesis of axially chiral biaryl compounds through the cross-coupling reaction. Although initial attempts to this

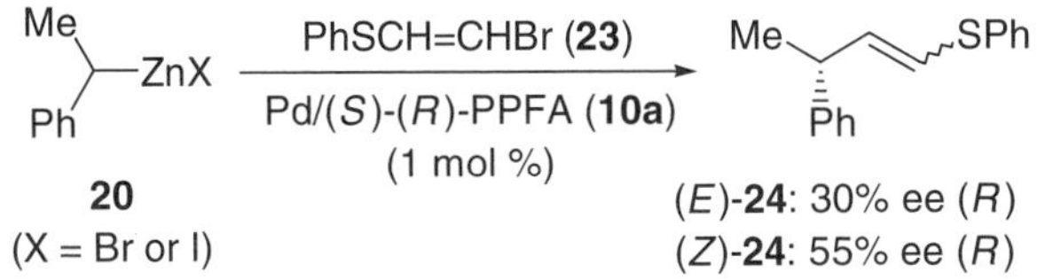

Scheme 8F.8.

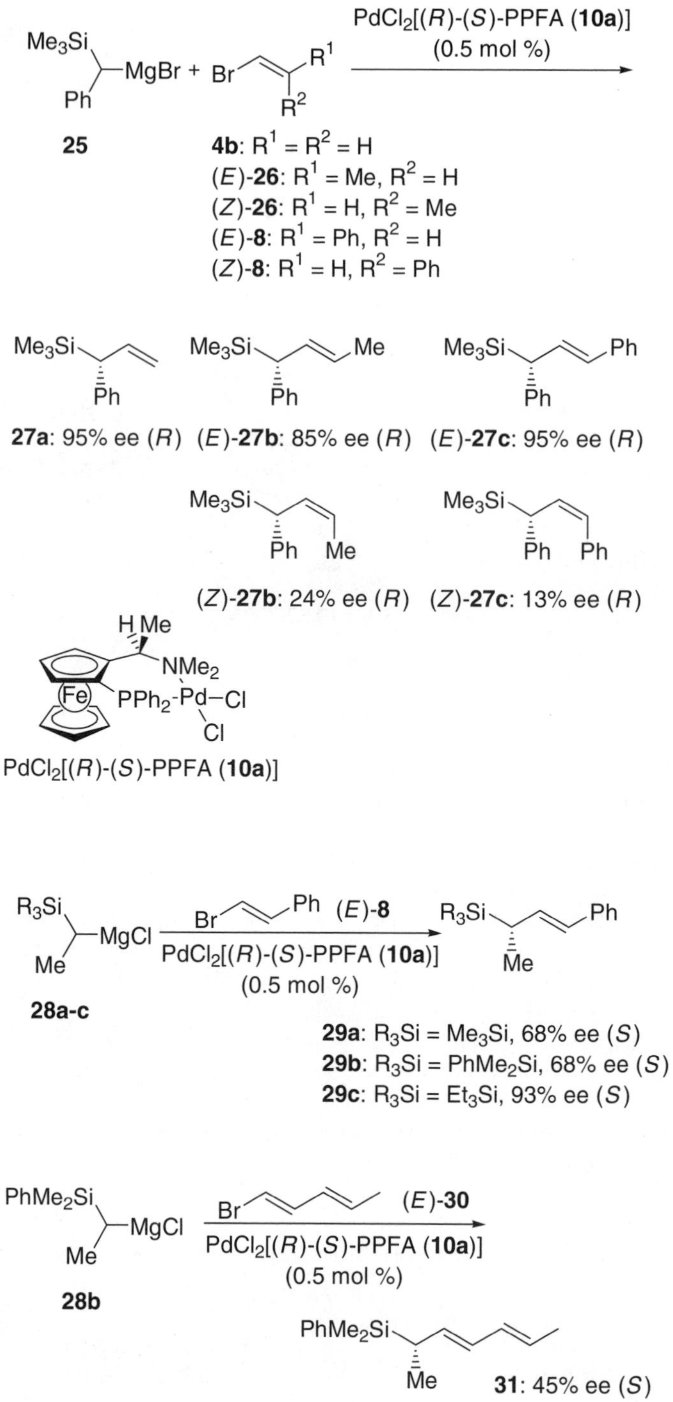
Me3Si
MgBr + Br
R1
R2
Ph
PdCl2[(R)-(S)-PPFA (10a)]
(0.5 mol %)
25
4b: R1 = R2 = H
(E)-26: R1 = Me, R2 = H
(Z)-26: R1 = H, R2 = Me
(E)-8: R1 = Ph, R2 = H
(Z)-8: R1 = H, R2 = Ph
Me3Si
Ph
Me
27a: 95% ee (R)
(E)-27b: 85% ee (R)
(E)-27c: 95% ee (R)
(Z)-27b: 24% ee (R)
(Z)-27c: 13% ee (R)
H Me
NMe2
Fe
PPh2-Pd–Cl
Cl
PdCl2[(R)-(S)-PPFA (10a)]
R3Si
MgCl
Me
28a-c
Br
Ph
(E)-8
PdCl2[(R)-(S)-PPFA (10a)]
(0.5 mol %)
29a: R3Si = Me3Si, 68% ee (S)
29b: R3Si = PhMe2Si, 68% ee (S)
29c: R3Si = Et3Si, 93% ee (S)
PhMe2Si
MgCl
Me
28b
Br
(E)-30
PdCl2[(R)-(S)-PPFA (10a)]
(0.5 mol %)
PhMe2Si
Me
31: 45% ee (S)

Scheme 8F.9.

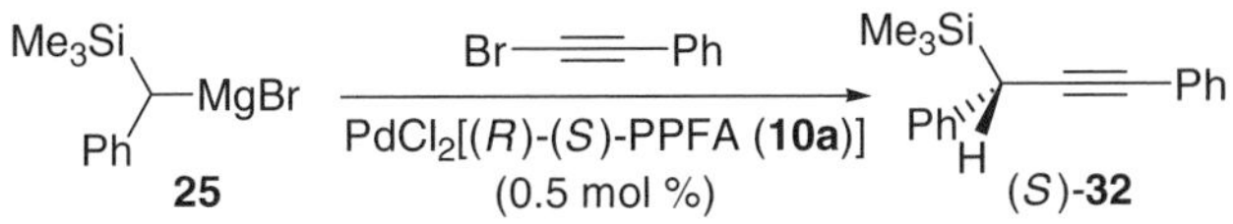

Scheme 8F.10.

type of reaction were reported for the coupling of 2-methyl-1-naphthylmagnesium bromide (**34a**) with 1-bromo-2-methylnaphthalene (**35a**) to form 2,2'-dimethyl-1,1'-binaphthyl (**36a**) using nickel catalysts coordinated with (−)-diop (**1**), (*S*)-(*R*)-BPPFA (**14**), and (*S*)-**33**, the enantioselectivity induced was rather poor (2%, 5%, and 13% ee, respectively) [30,31]. Use of the ferrocenylphosphine ligand (*S*)-(*R*)-**37**, which is a chiral monophosphine ligand containing a methoxy group on the side chain, dramatically increased the stereoselectivity to produce a high yield of (*R*)-**36a** with 95% ee [6]. High enantioselectivity was also attained in the reaction of **34a** with 1-bromonaphthalene (**35b**), which gave (*R*)-2-methyl-1,1'-binaphthyl (**36b**) with 83% ee. Binaphthyl (*R*)-**36b** with a much lower enantiomeric purity was produced in the reaction of the other combination, that is, cross-coupling of 1-naphthylmagnesium bromide (**34b**) with 1-bromo-2-methylnaphthalene (**35a**). The 2-ethyl-1-naphthyl Grignard reagent **34c** was also successfully applied to the reaction with **35b**, which gave **36c** with 77% ee.

A similar methodology was successfully extended to the asymmetric synthesis of ter-naphthalenes [32] (Scheme 8F.12). Reaction of 1,5-dibromonaphthalene (**38**) with 2 equiv. of **34a** in the presence of nickel catalyst coordinated with (*S*)-(*R*)-**37** gave a high yield of diastereomeric mixture of ternaphthalene **39** consisting of chiral and *meso* isomers in a ratio of 84:16. The chiral isomer turned out to be 98.7% ee with (*R*,*R*) configuration. The very high enantiomeric purity can be rationalized by the double asymmetric induction at the first and the second cross-coupling process. The reaction of **34a** with 1,4-dibromonaphthalene (**40**) gave ternaphthalene (*R*,*R*)-**41** with 95.3% ee together with a small amount of meso-**41**.

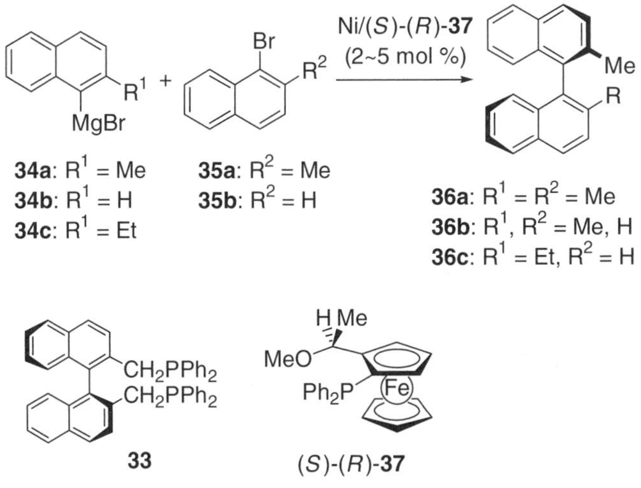

Scheme 8F.11.

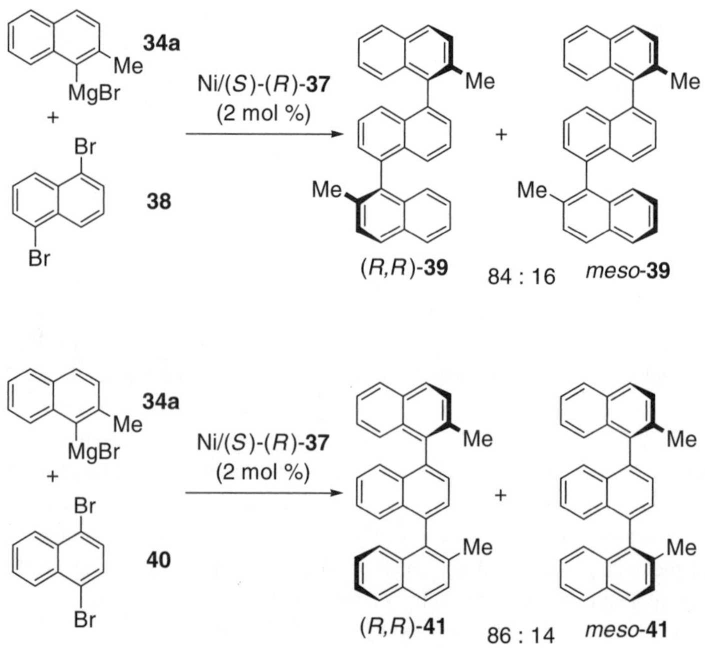

Scheme 8F.12.

The first examples of the enantioposition-selective asymmetric cross-coupling are reported for asymmetric induction of planar chirality on chromium-arene complexes using the palladium-(S)-(R)-PPFA (**10a**) complex [7,8] (Scheme 8F.13). Tricarbonyl(η^6-o-dichloroben-zene)chromium (**42**) reacts with alkenyl- or arylmetal reagents in the presence of 10 mol % of a palladium catalyst generated in situ from [PdCl(η^3-C_3H_5)]$_2$ and ferrocenylmonophosphine (S)-(R)-PPFA (**10a**). An enantioposition-selective substitution at one of the two chloro-substituents takes place to give planar chiral monosubstitution products **43** together with a minor amount of disubstitution products **44**, which are achiral. The highest enantiopurity of the monosubstitution products is 69% ee, which was achieved for the phenylation of **42** with phenylboronic acid, giving (1S,2R)-**43a**. Alkenylation with ethenylboronic acid or propen-2-ylboronic acid also

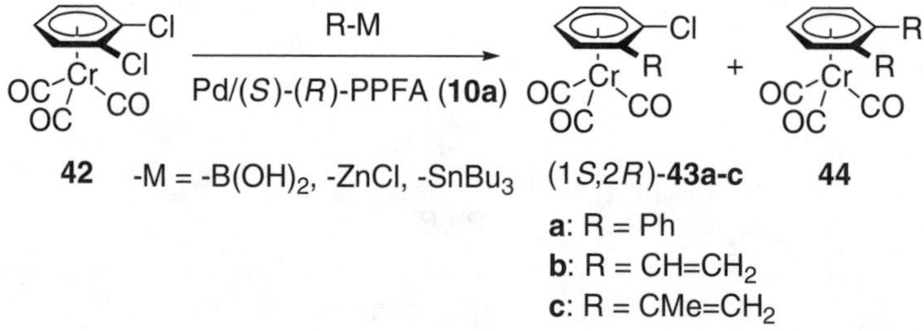

Scheme 8F.13.

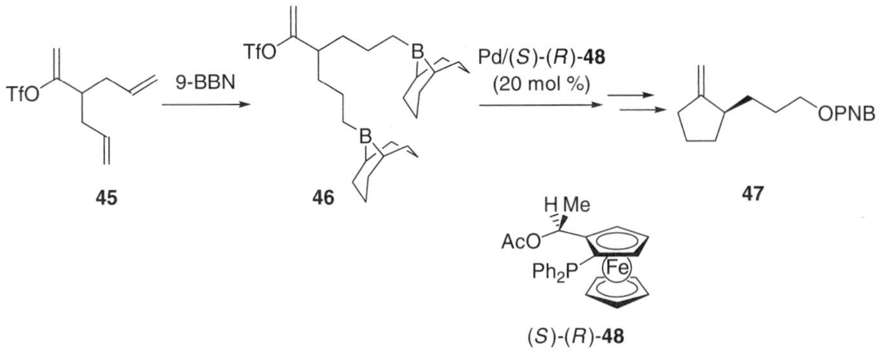

Scheme 8F.14.

proceeded enantioselectively to give the corresponding monoalkenylation product **43b** with 38% ee or **43c** with 44% ee. It is interesting that the use of ethenyltributyltin as a vinylation reagent in place of the vinylboronic acid resulted in the formation of racemic product **43b**, whereas vinylzinc chloride gave **43b** with 42% ee.

Asymmetric induction of central chirality at a carbon atom was achieved by an intramolecular enantioposition-selective asymmetric cross-coupling [12]. Treatment of the prochiral bisborane **46**, which was prepared from the alkenyl triflate **45** and 2 equiv. of 9-BBN with 20 mol % of Pd/(S)-(R)-BPPFOAc (**48**) catalyst generated in situ in THF, brings about intramolecular Suzuki coupling. The following oxidative workup and p-nitrobenzoylation affords the chiral cyclopentane derivative (R)-**47** in 58% yield and 28% ee (Scheme 8F.14).

8F.3.3. Phosphines Derived from Amino Acids

Because importance of the (dialkylamino)alkyl side chain has been recognized in the ferrocenylphospines, a series of β-(dialkylamino)alkylphosphines was prepared (Figure 8F.3) and applied to the asymmetric cross-coupling of **2** with **4**. First generations of these P–N ligands (**49**) are derivatives of chiral amino acids that are obtainable from a natural chiral pool. Generally, these amino acid base ligands, which have bulky alkyl substituents at the chiral carbon centers, are more effective in both enantioselectivity and reactivity than the ferrocenylphosphine ligands for the asymmetric cross-coupling of secondary alkyl Grignard reagents (see Table 8F.1). Valphos (**49c**), ilephos (**49d**), and t-leuphos (**49f**), which were prepared from valine, isoleucine, and tert-leucine, respectively, gave the product **6** with more than 81% ee [33,34] (entries 16–21). An interesting example is a polymer-supported amino-phosphine ligand (**50**) whose structure is similar to that of valphos (**49c**), although its stereoselectivity is considerably lower [35]. Some chiral β-(dialkylamino)alkylphosphines with sulfide groups in the alkyl side chains (**51**) were prepared from sulfur-containing amino acids such as methionine and homomethionine [36,37]. The sulfide substituents were expected to interact with magnesium of Grignard reagents, which would lead to higher enantioselectivity in asymmetric cross-coupling reaction. However, most of observed enantioselectivity was comparable with those reported for the analogous aminophosphine ligands without sulfide substituents (entries 24 and 25), but the highest value (88% ee, entry 23) obtained for **51a**/Ni-catalyzed cross-coupling of 1-phenylethylmagnesium chloride (**2a**) was better than the enantioselectivity attained by the phosphines with simple alkyl side chains.

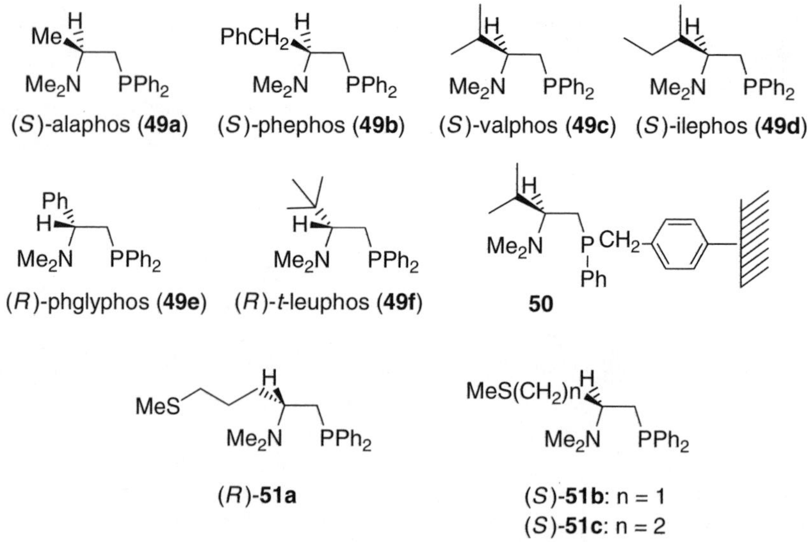

(S)-alaphos (**49a**) (S)-phephos (**49b**) (S)-valphos (**49c**) (S)-ilephos (**49d**)

(R)-phglyphos (**49e**) (R)-t-leuphos (**49f**) **50**

(R)-**51a**

(S)-**51b**: n = 1
(S)-**51c**: n = 2

Figure 8F.3. Chiral phosphine ligands derived from amino acids.

With nickel catalyst coordinated with (S)-valphos (**49c**), arylethylmagnesium chlorides **52a** and **52b** were coupled with **4b** to give the coupling products **53a** and **53b**, respectively. Subsequent oxidation of the coupling products gave optically active 2-(4-isobutyl-phenyl)propionic acid (ibuprofen, **54a**, 80% ee) and its biphenyl analogue (**54b**, 82% ee), both of which are antiinflammatory agents [34].

Reversal of the absolute configuration of the product **6** by addition of a zinc salt was observed in the cross-coupling of **2a** with **4a** catalyzed by **51**/Ni [38,39] (Scheme 8F.7). Thus, the cross-coupling of the Grignard reagents **2a** with **4a** in the presence of nickel catalyst coordinated with aminoalkylphosphine (S)-**51a** gave (S)-**6** with 61% ee, whereas the reaction of the

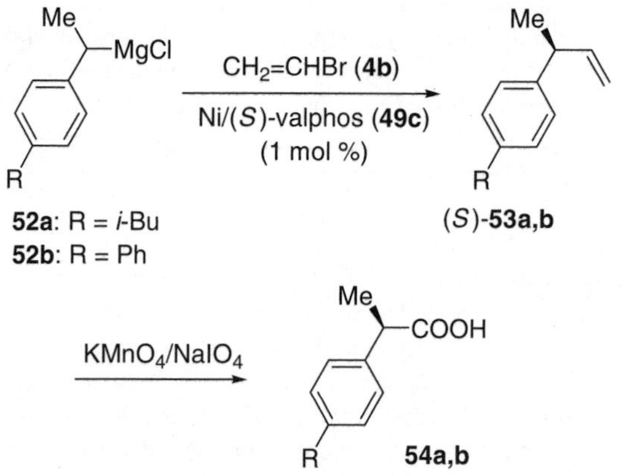

52a: R = i-Bu
52b: R = Ph

(S)-**53a,b**

54a,b

Scheme 8F.15.

organozinc reagent generated from **2a** and zinc bromide gave its enantiomer (R)-**6** with 70% ee (see Table 8F.2, entries 8–11).

Examples of the asymmetric cross-coupling of organometallic reagents other than 1-phenylethyl Grignard **2** or its Zn analogue **20** are still very limited and most of them did not show sufficient stereoselectivity. Kinetic resolution of chiral Grignard reagents, which do not undergo the racemization, have been examined for 2-phenylpropylmagnesium chloride (**55**) and norbornyl Grignard reagent (**57**) in the presence of less then 1 equiv. of vinyl bromide, though the efficiency of the resolution is not high [40] (Scheme 8F.16). One of the enantiomers of the racemic **57** underwent the cross-coupling in the presence of Ni/(S)-valphos (**49c**) catalyst 2.4 times faster than the other enantiomer. The enantiomeric purity of the coupling product **58** was 37% ee at 19% conversion. It is interesting that all of the coupling product **58** has vinyl group at *exo* position.

The second example of enantio-position selective asymmetric cross-coupling is achieved for a synthesis of axially chiral biaryl molecules by using a Pd-(S)-phephos (**49b**) catalyst [9,11] (Scheme 8F.17). Reaction of achiral ditriflate **59** with excess (2 equiv.) phenylmagnesium bromide in the presence of lithium bromide and 5 mol % of PdCl$_2$[(S)-phephos (**49b**)] at -30°C for 48 h gave monophenylated product (S)-**60** in 87% yield, which is 93% ee and an achiral diphenylated product **61** in 13% yield. The enantiomeric purity of the monophenylated product (S)-**60** is dependent on the yield of diphenylated product **61**. A kinetic resolution is demonstrated to take place at the second cross-coupling, forming **61**. Minor enantiomer from the first cross-coupling, that is (R)-**60**, is consumed approximately five times faster than the major enantiomer (S)-**60** at the second cross-coupling, which causes an increase in the enantiomeric purity of (S)-**60** as the amount of **61** increases. Addition of lithium bromide to the reaction mixture is essential for both high-enantioselectivity and good reactivity, though roles of the inorganic additive remain still uncertain. High enantioselectivity was also reported in the reaction of *o*-tolyl analogue **62**, which gave monophenylated product **63** with 84% ee. The reaction has been successfully extended to the enantioposition-selective alkynylation [10]. For the alkynylation, (S)-alaphos (**49a**) ligand is much more effective than (S)-phephos (**49b**) in terms of enantioselectivity. For examples, the reaction of achiral ditriflates **59** and **65** with (triphenylsilyl)ethynylmagnesium bromide in the presence of palladium catalyst coordinated with **49a** gave the corresponding monoalkynylated products (S)-**64** (92% ee) and **66** (99% ee), respectively. The great utility of this method is the presence of the second leaving groups in the products after induction of the chirality. Since the very reactive organometallic species, such as

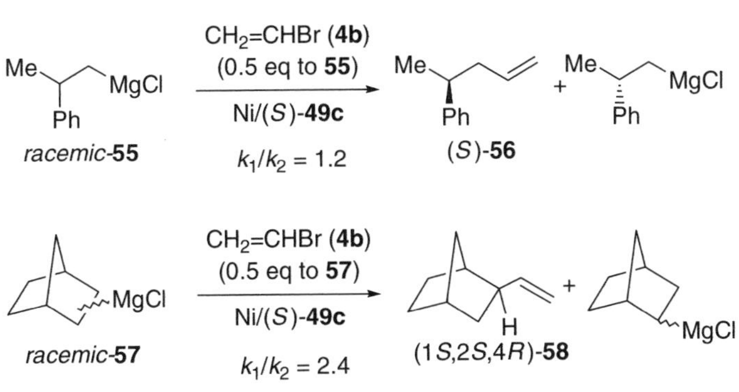

Scheme 8F.16.

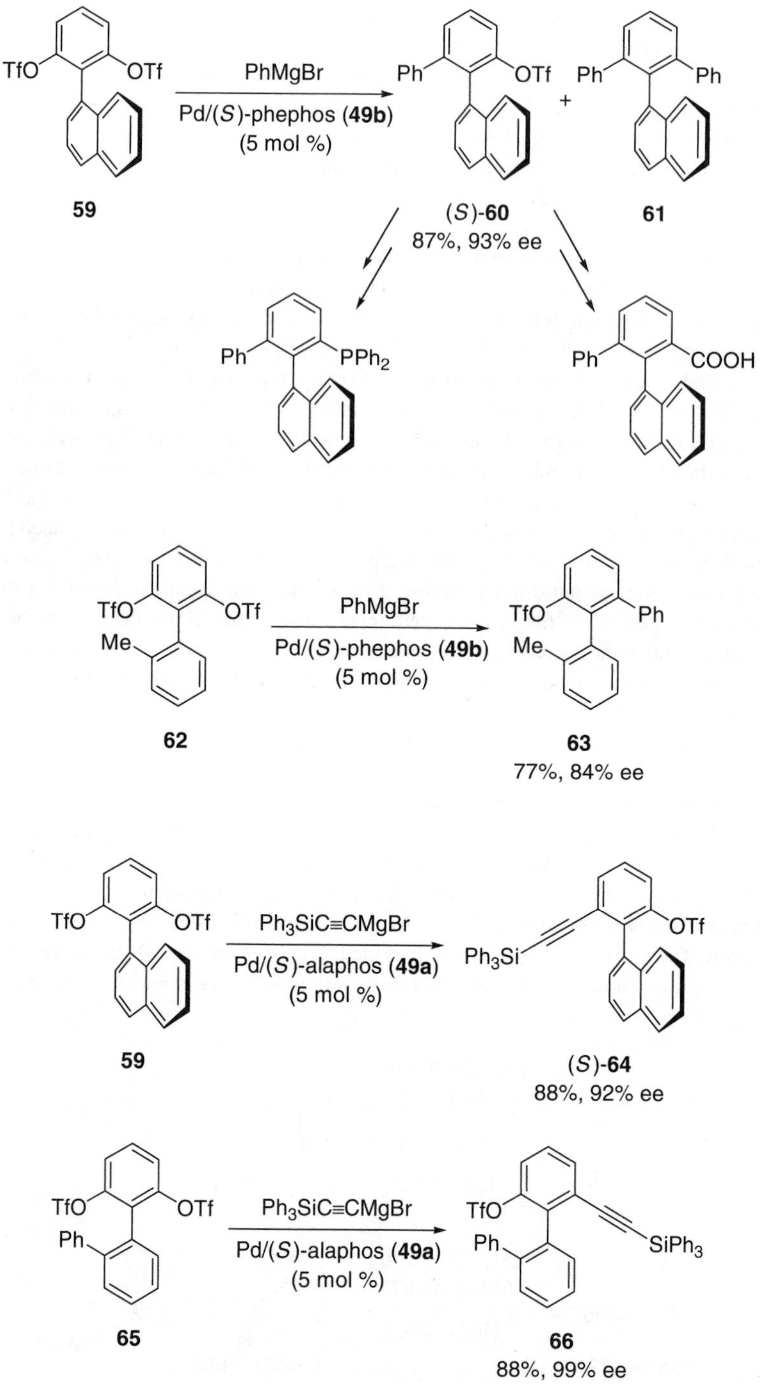
TfO
OTf
PhMgBr
Pd/(S)-phephos (49b)
(5 mol %)
Ph
OTf
Ph
Ph
59
(S)-60
87%, 93% ee
61
Ph
PPh2
Ph
COOH
TfO
OTf
Me
PhMgBr
Pd/(S)-phephos (49b)
(5 mol %)
TfO
Ph
Me
62
63
77%, 84% ee
TfO
OTf
Ph3SiC≡CMgBr
Pd/(S)-alaphos (49a)
(5 mol %)
Ph3Si
OTf
59
(S)-64
88%, 92% ee
TfO
OTf
Ph
Ph3SiC≡CMgBr
Pd/(S)-alaphos (49a)
(5 mol %)
TfO
Ph
SiPh3
65
66
88%, 99% ee

Scheme 8F.17.

Grignard reagents or organozinc halides, are required for the nickel- or palladium-catalyzed cross-coupling reactions, compounds that include substituents reactive to the organometallic reagents, such as –COOR, –OH, and –CN, cannot be used as substrates for the cross-coupling reaction. However, a variety of functionalization in the monosubstituted products, which are chiral, is now possible by utilizing the remaining second leaving groups. Enantiomerically pure monotriflate (S)-**60** was converted into a chiral phosphine or a chiral carboxylic acid in high yield via palladium-catalyzed phosphinylation or carbonylation, respectively [9].

8F.3.4. Other Phosphine Ligands

Several 3-diphenylphosphinopyrrolidine-type ligands **67** were prepared and used for the nickel-catalyzed Grignard cross-coupling of **2a** [41,42] (Table 8F.1, entries 26–29). N-Benzyl derivative **67a** is the most effective, giving (R)-**6** with 89% ee in the reaction with vinyl chloride (**4a**). An asymmetric amplification was observed to some extent in the asymmetric cross-coupling with ligands **67**. High enantioselectivity (94% ee) was reported in the cross-coupling of **2a** with (E)-**8** in the presence of nickel or palladium catalyst coordinated with a new chiral (β-amino-alkyl)phosphine ligand (1R,2S)-**68**, which was derived from erythro-2-amino-1,2-diphenylethanol [43] (Table 8F.1, entries 30–31). Some other (aminoalkyl)phosphines, based on axially chiral 1,1'-binaphthyl skeleton, **69** [44,45] and **70** [46], dimenthylphosphine **71** [47], and 1-phenylethylamine derivative **72** [48] have also been examined, although the enantioselectivity was not always high (Table 8F.1, entries 32–35). Another category of chiral phosphine ligands used for the asymmetric cross-coupling is chiral phosphino-oxazolines. A diastereomeric pair of phosphinoferrocenyloxazolines was applied for the palladium-catalyzed asymmetric cross-coupling reactions, and it was found that (S)-(S) isomer **73** was more stereoselective than the other diastereomer, giving (E)-**9** with 45% ee from **2a** and (E)-**8** [49] (entry 36). A nickel complex coordinated with phosphinophenyloxazoline (S)-**74** (phox), which has been an efficient ligand for palladium-catalyzed asymmetric allylic substitutions, was studied on its structure and its use as a catalyst for the asymmetric cross-coupling with (Z)-β-bromostyrene (Z)-**8** [50] (entry 37).

Nickel catalyst complexed with unfunctionalized chelating bisphosphine ligands, (R,R)-norphos (**75**) [51] and **76** [19,52], also induced a high selectivity in the reaction shown in Scheme 8F.5 (Table 8F.1, entries 38–39). The results reported with other phosphine ligands **33**, **77–80** [30,53–56] are summarized in the Table 8F.1 (entries 40–44).

Asymmetric cross-coupling of secondary alkyl Grignard reagents that contain no aryl group, such as phenyl, on the chiral carbon center has been relatively unsuccessful in terms of enantioselectivity as that of the 1-arylethyl Grignard reagent. The reaction of the 2-butyl Grignard reagents **3** with phenyl halides **5** was mainly studied with nickel catalysts complexed with chiral homologues of 1,2-bis(diphenylphosphino)ethane [52,57,58] (Scheme 8F.18). Highest enantiomeric purity (55% ee) of the product, 2-phenylbutane, was reported in the reaction of **3b** (X = Br) with **5b** (X' = Br) in the presence of a nickel complex with 1,2-bis(diphenylphosphino)cyclopentane (**76**). Use of (S,S)-chiraphos (**81**) as a chiral ligand produced (S)-**7** with 43% ee. Detailed studies on the reaction of **3** (X = Cl, Br, I) with **5** (X' = Cl, Br, I) in the presence of nickel/(R)-prophos (**82**) catalyst revealed that the absolute configuration of the coupling product as well as the enantioselectivity is dependent on the halogen atoms in both the Grignard reagents and phenyl halides. For examples, reaction of **3b** (X = Br) with **5b** (X' = Br) gave (R)-2-phenylbutane (**7**) with 40% ee, whereas that of **3a** (X = Cl) with **5c** (X = I) gave (S)-**7** with 15% ee. In the reaction of substituted phenyl halides, the enantioselectivity was found to be strongly influenced by steric factors but only slightly by electronic factors.

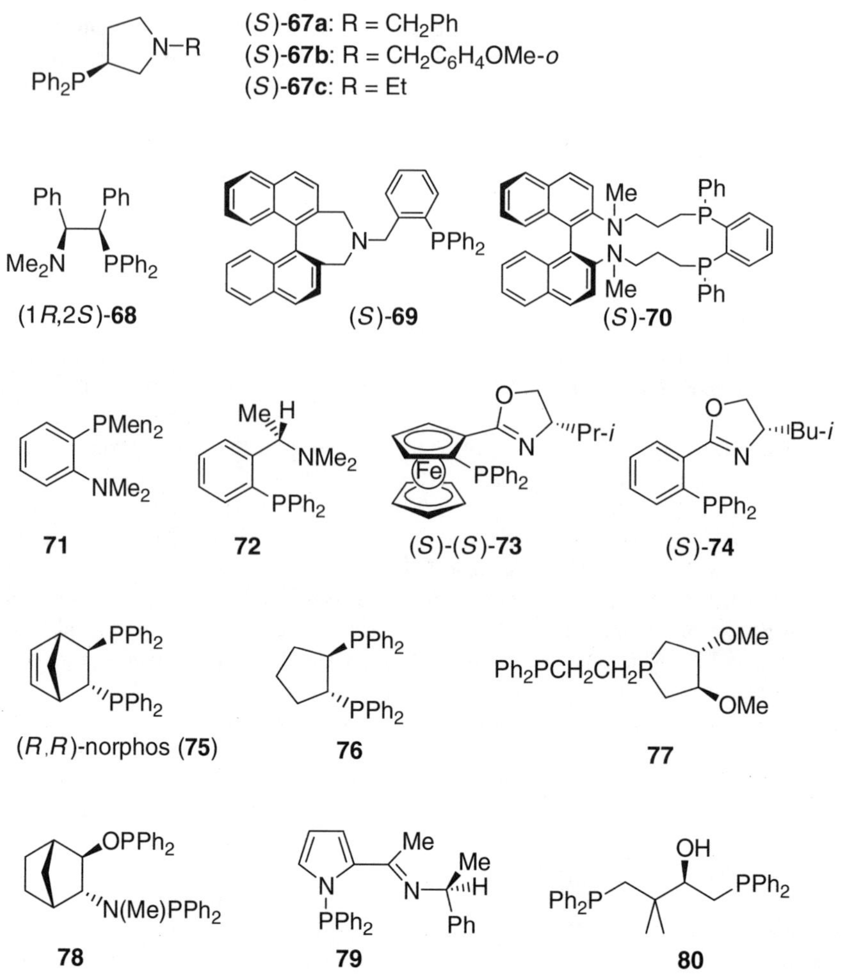

(S)-**67a**: R = CH$_2$Ph
(S)-**67b**: R = CH$_2$C$_6$H$_4$OMe-o
(S)-**67c**: R = Et

(1R,2S)-**68**

(S)-**69**

(S)-**70**

71

72

(S)-(S)-**73**

(S)-**74**

(R,R)-norphos (**75**)

76

77

78

79

80

Figure 8F.4. Miscellaneous phosphine ligands for asymmetric cross-coupling.

The kinetic resolution of racemic 1-phenylethyl Grignard reagent **2a** was also observed in the cross-coupling with allylic substrates, such as allyl phenyl ether, in the presence of NiCl$_2$[(S,S)-chiraphos (**81**)] which gave (R)-4-phenyl-1-butene (**56**) with 58% ee [59] (Scheme 8F.19).

A 1,3-substituted allene, which has axial chirality instead of carbon central chirality, has been prepared by a palladium-catalyzed cross-coupling of 4,4-dimethylpenta-1,2-dienylzinc chloride (**83**) with phenyl iodide (**5c**) or by that of 1-bromo-4,4-dimethylpenta-1,2-diene (**84**) with phenylzinc chloride [60] (Scheme 8F.20). The highest enantiomeric purity (25% ee) of the allene (S)-**85** was obtained in the former combination with (R,R)-diop (**1**) as chiral ligand. It is interesting that the enantiomeric purity was independent of the ratio of the reagents though the reaction seems to involve a kinetic resolution of the racemic **83**.

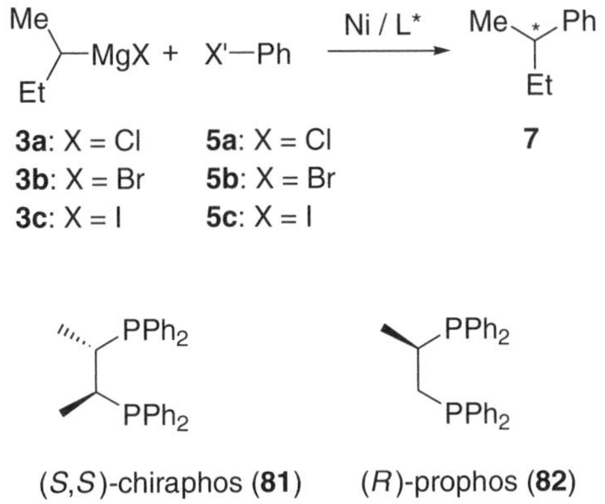

$$\underset{\substack{\text{Me}\\\text{Et}}}{\diagup}\!\!-\text{MgX} \;+\; \text{X'}\!-\!\text{Ph} \xrightarrow{\text{Ni / L*}} \underset{\substack{\text{Me}\\\text{Et}}}{\diagup}\!\!\overset{*}{\diagdown}\text{Ph}$$

3a: X = Cl   **5a**: X = Cl   **7**
3b: X = Br   **5b**: X = Br
3c: X = I   **5c**: X = I

(S,S)-chiraphos (**81**)   (R)-prophos (**82**)

Scheme 8F.18.

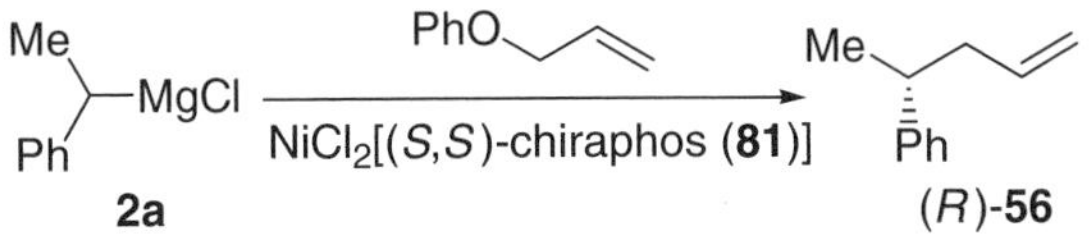

$$\underset{\substack{\text{Me}\\\text{Ph}}}{\diagup}\!\!-\text{MgCl} \xrightarrow[\text{NiCl}_2[(S,S)\text{-chiraphos (81)}]]{\text{PhO}\diagup\!\!\diagup} \underset{\substack{\text{Me}\\\text{Ph}}}{\diagup}\!\!\diagup\!\!\diagdown$$

2a   (R)-**56**

Scheme 8F.19.

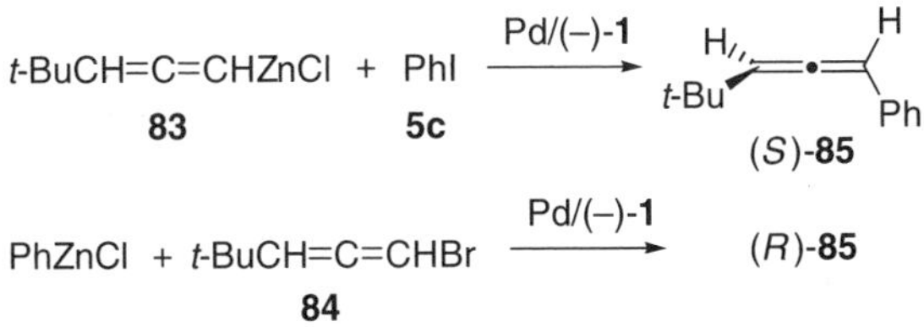

$$t\text{-BuCH=C=CHZnCl} \;+\; \text{PhI} \xrightarrow{\text{Pd/(-)-1}} \underset{(S)\text{-85}}{\overset{\text{H}}{\underset{t\text{-Bu}}{\diagup}}\!\!=\!\!\bullet\!\!=\!\!\overset{\text{H}}{\underset{\text{Ph}}{\diagdown}}}$$

83   **5c**

$$\text{PhZnCl} \;+\; t\text{-BuCH=C=CHBr} \xrightarrow{\text{Pd/(-)-1}} (R)\text{-85}$$

           84

Scheme 8F.20.

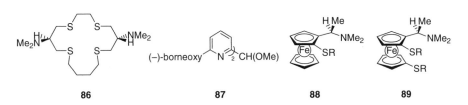

    86   **87**   **88**   **89**

Figure 8F.5. Non-phosphorus chiral ligands for asymmetric cross-coupling.

8F.3.5. Miscellaneous Ligands

Some examples of non-phosphorus chiral ligands have also been reported (Figure 8F.5). The chiral macrocyclic sulfide **86** [61,62] and the pyridine derivative **87** [63] have been prepared and examined for the nickel-catalyzed cross-coupling of **2a** with **4b**, though the enantioselectivity of this process was modest to low (46% ee with **86** and 11% ee with **87**; see Table 8F.1, entry 45 and 46). Several chiral ferrocenylsulfides **88** and **89**, of which skeletons are analogous to PPFA (**10a**) or BPPFA (**14**) but have sulfide groups instead of diphenylphosphino groups, were used for the reactions of allylmagnesium chloride with 1-phenylethyl chloride in the presence of nickel or palladium catalysts to give 4-phenyl-1-pentene with up to 28% ee [64,65].

8F.4. CONCLUSION

This chapter has discussed the enantioselective carbon–carbon bond formation through Grignard cross-coupling reactions catalyzed by palladium and nickel complexes with various chiral ligands. High enantioselectivities (>90% ee) have been achieved in certain systems, including allylsilanes and biaryls, but detailed mechanism for enantioselection is not well-understood. More efficient catalyst systems will be developed when a detailed mechanistic understanding becomes available. These reactions will find a plenty of applications in asymmetric organic syntheses.

REFERENCES

1. Farina V. In *Comprehensive Organometallic Chemistry II*, Hegedus L. S. (Ed.); Pergamon: Oxford 1995, Vol. 12, pp. 161.

2. (a) Tsuji, J. *Palladium Reagents and Catalysts*; John Wiley and Sons: Chichester, 1995. (b) Heck, R. F., *Palladium Reagents in Organic Synthesis*; Academic Press: New York, 1985.

3. Trost, B. M.; Verhoeven, T. R. In *Comprehensive Organometallic Chemistry*; Wilkinson G.; Stone, F.G.A.; Abel, E. W. (Ed.); Pergamon: Oxford, 1982; Vol. 8, pp. 799.

4. Jolly, P. W. In *Comprehensive Organometallic Chemistry*, Wilkinson G.; Stone, F. G. A.; Abel, E. W. (Ed.); Pergamon: Oxford, 1982; Vol. 8, pp. 713.

5. Hayashi, T. In *Catalytic Asymmetric Synthesis*, Ojima, I. (Ed.); VCH: New York, 1983; pp. 325.

6. Hayashi, T.; Hayashizaki, K.; Kiyoi, T.; Ito, Y. *J. Am. Chem. Soc.* **1988**, *110*, 8153.

7. Uemura, M.; Nishimura, H.; Hayashi, T. *Tetrahedron Lett.* **1993**, *34*, 107.

8. Uemura, M.; Nishimura, H.; Hayashi, T. *J. Organomet. Chem.* **1994**, *473*, 129.

9. Hayashi, T.; Niizuma, S.; Kamikawa, T.; Suzuki, N.; Uozumi, Y. *J. Am. Chem. Soc.* **1995**, *117*, 9101.

10. Kamikawa, T.; Uozumi, Y.; Hayashi, T. *Tetrahedron Lett.* **1996**, *37*, 3161.

11. Kamikawa, T.; Hayashi, T. *Tetrahedron* **1999**, *55*, 3455.

12. Cho, S. Y.; Shibasaki, M. *Tetrahedron: Asymmetry* **1998**, *9*, 3751.

13. Baker, K. V.; Brown, J. M.; Cooley, N. A.; Hughes, G. D.; Taylor, R. J. *J. Organomet. Chem.* **1989**, *370*, 397.

14. Consiglio, G.; Botteghi, C. *Helv. Chim. Acta* **1973**, *56*, 460.

15. Kiso, Y.; Tamao, K.; Miyake, N.; Yamamoto, K.; Kumada, M. *Tetrahedron Lett.* **1974**, 3.

16. Hayashi, T.; Tajika, M.; Tamao, K.; Kumada, M. *J. Am. Chem. Soc.* **1976**, *98*, 3718.

17. Hayashi, T.; Konishi, M.; Fukushima, M.; Mise, T.; Kagotani, M.; Tajika, M.; Kumada, M. *J. Am. Chem. Soc.* **1982**, *104*, 180.

18. Hayashi, T.; Konishi, M.; Hioki, T.; Kumada, M.; Ratajczak, A.; Hiedbala, H. *Bull. Chem. Soc. Jpn.* **1981**, *54*, 3615.

19. Indolese, A.; Consiglio, G. *J. Organomet. Chem.* **1993**, *463*, 23.

20. Jedlicka, B.; Kratky, C.; Weissensteiner, W.; Widhalm, M. *J. Chem. Soc., Chem. Commun.* **1993**, 1329.

21. Tamao, K.; Hayashi, T.; Matsumoto, H.; Yamamoto, H.; Kumada, M. *Tetrahedron Lett.* **1979**, *23*, 2155.

22. Hayashi, T.; Hagihara, T.; Katsuro, Y.; Kumada, M. *Bull. Chem. Soc. Jpn.* **1983**, *56*, 363.

23. Hayashi, T.; Yamamoto, A.; Hojo, M.; Ito, Y. *J. Chem. Soc., Chem. Commun.* **1989**, 495.

24. Hayashi, T.; Yamamoto, A.; Hojo, M.; Kishi, K.; Ito, Y.; Nishioka, E.; Miura, H.; Yanagi, K. *J. Organomet. Chem.* **1989**, *370*, 129.

25. Uemura, M.; Miyake, R.; Nishimura, H.; Matsumoto, Y.; Hayashi, T. *Tetrahedron: Asymmetry* **1992**, *3*, 213.

26. Cardellicchio, C.; Fiandanese, V.; Naso, F. *Gazz. Chim. Ital.* **1991**, *121*, 11.

27. Hayashi, T.; Konishi, M.; Ito, Y.; Kumada, M. *J. Am. Chem. Soc.* **1982**, *104*, 4962.

28. Hayashi, T.; Konishi, M.; Okamoto, Y.; Kabeta, K.; Kumada, M. *J. Org. Chem.* **1986**, *51*, 3772.

29. Hayashi, T.; Okamoto, Y.; Kumada, M. *Tetrahedron Lett.* **1983**, *24*, 807.

30. Tamao, K.; Yamamoto, H.; Matsumoto, H.; Miyake, N.; Hayashi, T.; Kumada, M. *Tetrahedron Lett.* **1977**, 1389.

31. Tamao, K.; Minato, A.; Miyake, N.; Natsuda, T.; Kiso, Y.; Kumada, M. *Chem. Lett.* **1975**, 133.

32. Hayashi, T.; Hayashizaki, K.; Ito, Y. *Tetrahedron Lett.* **1989**, *30*, 215.

33. Hayashi, T.; Fukushima, M.; Konishi, M.; Kumada, M. *Tetrahedron Lett.* **1980**, *21*, 79.

34. Hayashi, T.; Konishi, M.; Fukushima, M.; Kanehira, K.; Hioki, T.; Kumada, M. *J. Org. Chem.* **1983**, *48*, 2195.

35. Hayashi, T.; Nagashima, N.; Kumada, M. *Tetrahedron Lett.* **1980**, *21*, 4623.

36. Vriesema, B. K.; Kellogg, R. M. *Tetrahedron Lett.* **1986**, *27*, 2049.

37. Griffin, J. H.; Kellogg, R. M. *J. Org. Chem.* **1985**, *50*, 3261.

38. Cross, G.; Kellogg, R. M. *J. Chem. Soc., Chem. Commun.* **1987**, 1746.

39. Cross, G.; Vriesema, B. K.; Boven, G.; Kellogg, R. M. *J. Organomet. Chem.* **1989**, *370*, 357.

40. Hayashi, T.; Kanehira, K.; Hioki, T.; Kumada, M. *Tetrahedron Lett.* **1981**, *21*, 137.

41. Nagel, U.; Nedden, H. G. *Chem. Ber./Recl.* **1997**, *130*, 535.

42. Nagel, U.; Nedden, H. G. *Chem. Ber./Recl.* **1997**, *130*, 989.

43. Hayashi, M.; Takaoki, K.; Hashimoto, Y.; Saigo, K. *Enantiomer* **1997**, *2*, 293.

44. Wimmer, P.; Widhalm, M. *Tetrahedron: Asymmetry* **1995**, *6*, 657.

45. Wimmer, P.; Widhalm, M. *Monatsh. Chem.* **1996**, *127*, 668.

46. Widhalm, M.; Wimmer, P.; Klintschar, G. *J. Organomet. Chem.* **1996**, *523*, 167.

47. Döbler, C.; Kinting, A. *J. Organomet. Chem.* **1991**, *401*, C23.

48. Kreuzfeld, H.-J.; Döbler, C.; Abicht, H.-P. *J. Organomet. Chem.* **1987**, *336*, 287.

49. Richards, C. J.; Hibbs, D. E.; Hursthouse, M. B. *Tetrahedron Lett.* **1995**, *36*, 3745.

50. Lloyd-Jones, G. C.; Butts, C. P. *Tetrahedron* **1998**, *54*, 901.

51. Brunner, H.; Pröbster, M. *J. Organomet. Chem.* **1981**, *209*, C1.

52. Consiglio, G.; Indolese, A. *J. Organomet. Chem.* **1991**, *417*, C36.

53. Brunner, H.; Limmer, S. *J. Organomet. Chem.* **1991**, *413*, 55.

54. Döbler, C.; Kreuzfeld, H.-J. *J. Organomet. Chem.* **1988**, *344*, 249.

55. Brunner, H.; Li, W.; Weber, H. *J. Organomet. Chem.* **1985**, *288*, 359.

56. Brunner, H.; Lautenschlager, H.-J.; König, W. A.; Krebber, R. *Chem. Ber.* **1990**, *123*, 847.

57. Consiglio, G.; Piccolo, O.; Morandini, F. *J. Organomet. Chem.* **1979**, *177*, C13.

58. Consiglio, G.; Morandini, F.; Piccolo, O. *Tetrahedron* **1983**, *39*, 2699.

59. Consiglio, G.; Piccolo, O.; Roncetti, L.; Morandini, F. *Tetrahedron* **1986**, *42*, 2043.

60. de Graaf, W.; Boersma, J.; van Koten, G.; Elsevier, C. J. *J. Organomet. Chem.* **1989**, *378*, 115.

61. Lemaire, M.; Vriesema, B. K.; Kellogg, R. M. *Tetrahedron Lett.* **1985**, *26*, 3499.

62. Vriesema, B. K.; Lemaire, M.; Buter, J.; Kellogg, R. M. *J. Org. Chem.* **1986**, *51*, 5169.

63. Wright, M. E.; Jin, M.-J. *J. Organomet. Chem.* **1990**, *387*, 373.

64. Okoroafor, M. O.; Ward, D. L.; Brubaker, Jr., C. H. *Organometallics* **1988**, *7*, 1504.

65. Naiini, A. A.; Lai, C.-K.; Ward, D. L. Brubaker, Jr., C. H. *J. Organomet. Chem.* **1990**, *390*, 73.

8G

ASYMMETRIC INTRAMOLECULAR HECK REACTIONS

YARIV DONDE AND LARRY E. OVERMAN
Department of Chemistry, University of California, Irvine, Irvine, CA 92697-2025

8G.1. INTRODUCTION

Although the coupling of aryl halides with alkenes (commonly referred to as the Heck reaction) was first reported more than 25 years ago [1], only in the past decade has its enormous synthetic potential been realized [2]. Within that time, the reaction has been extended to many substrates, including vinyl iodides and bromides and enol triflates. Moreover, the intramolecular variant has become one of the more important reactions for the formation of carbon–carbon bonds and has emerged as a premier method for the construction of quaternary carbon centers. The ability of intramolecular Heck reactions to reliably fashion carbon–carbon bonds in polyfunctional molecules has led to wide application of this reaction at the strategy level for the synthesis of complex natural products [2g].

In addition to extending the scope of Heck couplings, notable progress has been made in achieving catalytic asymmetric Heck reactions through the use of chiral bisphosphine ligands [3,4]. Due to early reports that bisphosphines were unsuitable ligands for intermolecular Heck reactions [2a], the first examples of asymmetric Heck reactions were disclosed only in 1989, independently, by Shibasaki and Overman [5,6]. Although only modest enantiocontrol was achieved in these early studies, many asymmetric Heck cyclizations have now been demonstrated that proceed with >90% ee.

Recent development of the Heck reaction has also led to greater understanding of its mechanistic details. The general outlines of the mechanism of the Heck reaction have been appreciated since the 1970s and are discussed in numerous reviews [2,3]. More recently, two distinct pathways, termed the neutral and cationic pathways, have been recognized [2c–g,3,7,8,9]. The neutral pathway is followed for unsaturated halide substrates and is outlined in Scheme 8G.1 for the Heck cyclization of an aryl halide. Thus, oxidative addition of the aryl halide **1.2** to a (bisphosphine)Pd(0) (**1.1**) catalyst generates intermediate **1.3**. Coordination of

Catalytic Asymmetric Synthesis, Second Edition, Edited by Iwao Ojima
ISBN 0-471-29805-0   Copyright © 2000 Wiley-VCH, Inc.

Neutral pathway (X = halide):

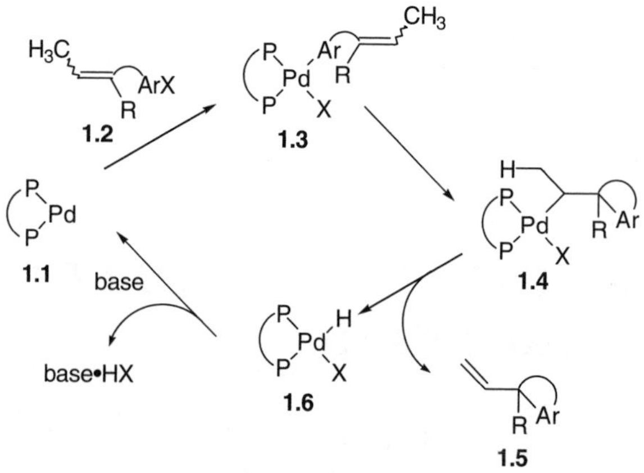

Scheme 8G.1.

an alkene followed by migratory insertion generates intermediate **1.4**, which undergoes syn β-hydride elimination to give the product **1.5** and the hydridopalladium complex **1.6**. The catalyst is then regenerated by reaction of **1.6** with base.

The cationic pathway is followed for unsaturated triflate substrates, or for unsaturated halide substrates in the presence of halide scavengers such as Ag(I) or Th(I) salts [2,3], and is shown in Scheme 8G.2. The individual steps are similar to the neutral pathway, but the difference in the two mechanisms lies in the nature of the Pd(II) intermediates **2.1–2.3**, which are now cationic. As will be discussed in more detail later in this review, this difference has a marked effect on both reactivity and enantioselectivity.

An important consideration common to both pathways is the regioselectivity of β-hydride elimination. To preserve a newly created stereogenic center, β-hydride elimination must be directed away from the newly formed center in intermediates such as **1.4** or **2.2**. This is not an

cationic pathway (X = triflate or other noncoordinating ligand):

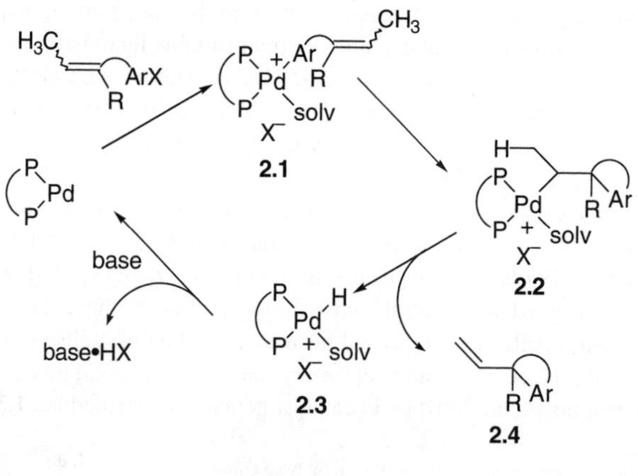

Scheme 8G.2.

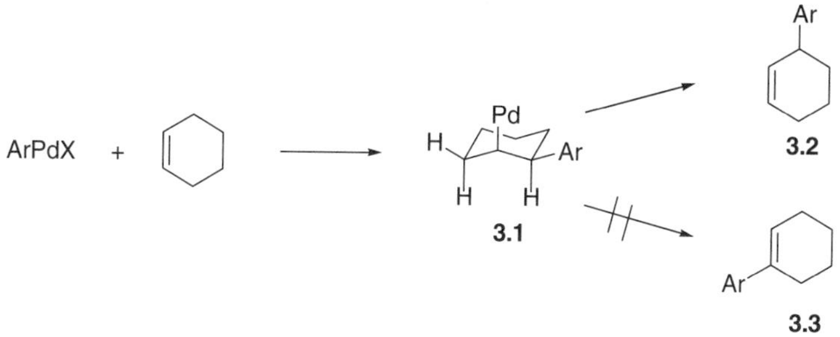

Scheme 8G.3.

issue for reactions that form quaternary centers (R ≠ H, Scheme 8G.1 and 8G.2) and is also easily controlled for cyclic alkenes. However, in other cases, an inherent structural bias must be present. In the case of cyclic alkenes, regiocontrol is implicit in the stereospecific nature of the individual steps as shown in Scheme 8G.3. *Syn* addition of ArPdX to cyclohexene generates σ-alkyl complex **3.1**, which can only undergo *syn* β-hydride elimination away from the newly formed stereogenic center.

Another potential complication is palladium-promoted double-bond isomerization of the alkene product, which can also destroy a newly formed stereogenic center [10]. This process involves re-addition of palladium hydride to the alkene in the initially generated (alkene)palladium hydride complex (the reverse of β-hydride elimination). Double-bond isomerization results when re-addition occurs with different regioselectivity than the original β-hydride elimination.

It is the intent of this chapter to outline the major synthetic developments of the intramolecular asymmetric Heck reaction as well as to examine the mechanistic factors that have been revealed to date that effect enantioselectivity.

8G.2. SYNTHETIC DEVELOPMENTS

8G.2.1. Group-Selective Reactions

The first intramolecular asymmetric Heck reactions reported by Shibasaki in 1989 involved cyclization of prochiral vinyl iodides **4.1a–c** to give *cis*-decalins **4.2a–c** (Scheme 8G.4) [5]. This reaction is unusual in that two stereocenters are created in one step. In the initial report, Heck products **4.2a–c** were obtained with only modest enantioselectivity. As is often the case, the optimum ligand was BINAP [2,2′-bis(diphenylphosphino)-1,1′-binapthyl] [11], with a number of other ligands such as (S,S)-BPPM (**4.3**) and (R,S)-BPPFA (**4.4**) giving much lower enantioselectivitiy. Optimization of the vinyl iodide cyclization conditions eventually led to considerable improvement in enantioselectivity [12]. Thus, treatment of **4.1b** with PdCl$_2$[(R)-BINAP], Ag$_3$PO$_4$, and CaCO$_3$ in N-methyl-2-pyrrolidinone (NMP) improved the enantiopurity of **4.2b** to 80% ee (67% yield). Cationic conditions, that is, the presence of a Ag(I) salt, were found to be critical as reaction in the absence of a Ag(I) salt (neutral conditions) resulted in a very slow reaction with no enantioselectivity. The optimized conditions developed in this study have been widely used for cyclization of unsaturated iodide substrates; most often, a polar aprotic solvent is used with Ag$_3$PO$_4$ often the optimum silver source. If a silver salt with a

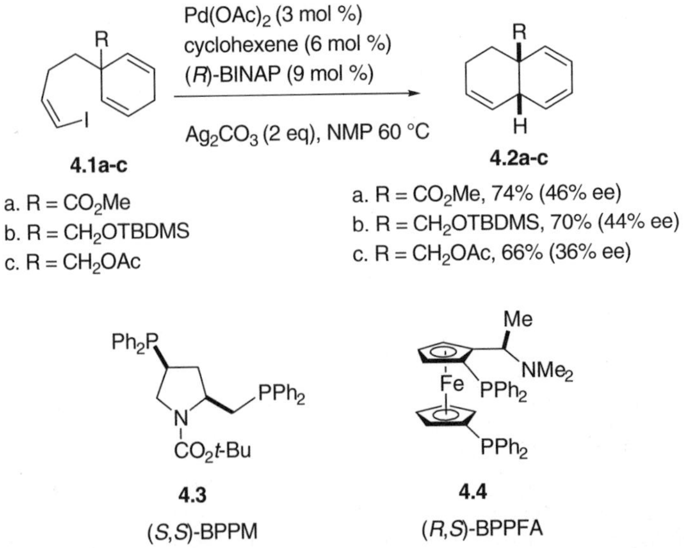

$$\begin{array}{c}\text{Pd(OAc)}_2\ (3\ \text{mol \%})\\ \text{cyclohexene}\ (6\ \text{mol \%})\\ (R)\text{-BINAP}\ (9\ \text{mol \%})\\ \hline \text{Ag}_2\text{CO}_3\ (2\ \text{eq}),\ \text{NMP}\ 60\ ^\circ\text{C}\end{array}$$

4.1a-c

a. R = CO$_2$Me
b. R = CH$_2$OTBDMS
c. R = CH$_2$OAc

4.2a-c

a. R = CO$_2$Me, 74% (46% ee)
b. R = CH$_2$OTBDMS, 70% (44% ee)
c. R = CH$_2$OAc, 66% (36% ee)

4.3

(S,S)-BPPM

4.4

(R,S)-BPPFA

Scheme 8G.4.

nonbasic counter ion (e.g., AgBF$_4$) is used, an inorganic or organic base must be added to regenerate the catalyst. A variety of palladium sources have been used, and Pd(OAc)$_2$ and Pd$_2$(dba)$_3$•CHCl$_3$ are most common. In general, a particular asymmetric Heck cyclization is sensitive to each of these variables and considerable optimization is required to achieve high enantioselectivities and yields.

For this particular *cis*-decalin synthesis, it was eventually found that best results were obtained with the corresponding vinyl triflates and Pd(BINAP) (Scheme 8G.5) [13]. As illustrated in these examples, the reaction of unsaturated triflates under cationic conditions typically is done in a nonpolar solvent, in the presence of an inorganic base such as K$_2$CO$_3$ or a tertiary amine; Pd(OAc)$_2$ and Pd$_2$(dba)$_3$•CHCl$_3$ have been the most widely used palladium sources.

The usefulness of this *cis*-decalin synthesis was demonstrated by an enantioselective synthesis of the sesquiterpene (+)-vernolepin **6.4** (Scheme 8G.6). Thus, Heck cyclization of the

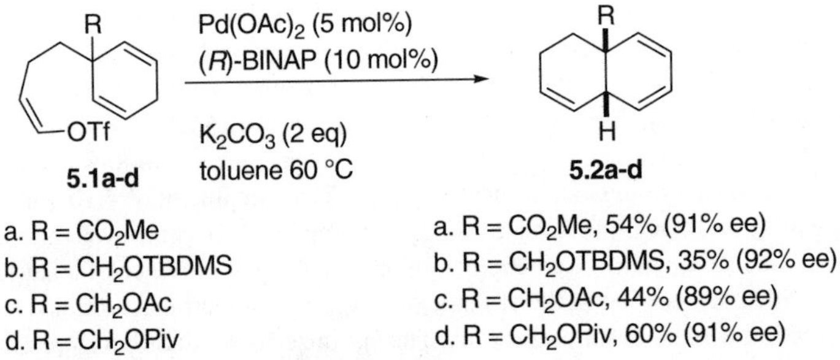

$$\begin{array}{c}\text{Pd(OAc)}_2\ (5\ \text{mol\%})\\ (R)\text{-BINAP}\ (10\ \text{mol\%})\\ \hline \text{K}_2\text{CO}_3\ (2\ \text{eq})\\ \text{toluene}\ 60\ ^\circ\text{C}\end{array}$$

5.1a-d

a. R = CO$_2$Me
b. R = CH$_2$OTBDMS
c. R = CH$_2$OAc
d. R = CH$_2$OPiv

5.2a-d

a. R = CO$_2$Me, 54% (91% ee)
b. R = CH$_2$OTBDMS, 35% (92% ee)
c. R = CH$_2$OAc, 44% (89% ee)
d. R = CH$_2$OPiv, 60% (91% ee)

Scheme 8G.5.

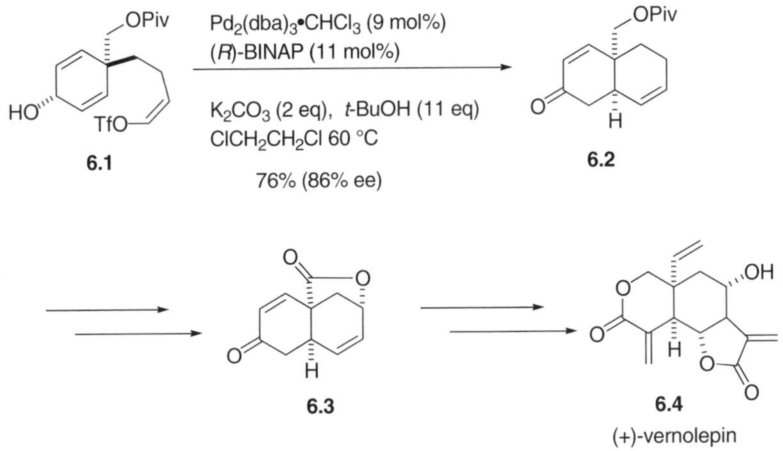

Scheme 8G.6.

functionalized prochiral triflate **6.1** led to decalin **6.2** in 76% yield and in 86% ee [14–16]. For this reaction, the optimum conditions were found to require dichloroethane as the solvent and *tert*-butanol as an additive. More typical solvents such as toluene or THF afforded much lower enantioselectivities (<50% ee). It was demonstrated that the *tert*-butanol additive improves both the rate and yield of the cyclization by preventing oxidation of the Pd(0) catalyst by the solvent [15]. Use of dichloroethane as the solvent with a tertiary alcohol additive has also been successfully used in other asymmetric Heck cyclizations [15,16]. In particular, the cyclization of **5.1a** was improved by use of dichloroethane as solvent and pinacol as the additive, which gave **5.2a** in an improved 78% yield with 95% ee [15]. Both **5.2a** and **6.2** could be converted to intermediate **6.3**, which was subsequently elaborated to (+)-vernolepin **6.4** following the Danishefsky route to racemic vernolepin [17].

The enantiotopic group discrimination strategy has also been used for the synthesis of other ring systems. The *cis*-hydrindane ring system was accessed by cyclization of either vinyl iodides **7.1a–c** or vinyl triflate **7.3** (Scheme 8G.7) [18]. In these examples, the vinyl iodide provided the *cis*-hydrindane products **7.2a–c** with higher enantioselectivity than the corresponding vinyl triflate, when both reactions were run under typical conditions. The related cyclization of Z vinyl iodide **7.4** under similar conditions resulted in nearly racemic product, illustrating the sensitivity of enantioselectivity to subtle structural changes in the substrate. Hydrindanes **7.2a–c** were used in the formal total syntheses of (−)-oppositol (**7.5**) and (−)-prepinnaterpene (**7.6**) [19].

An extension of this strategy to the synthesis of the bicyclo[3.3.0]octane ring system illustrates the formation and trapping of π-allylpalladium intermediates (Scheme 8G.8) [20,21]. Asymmetric Heck cyclization of vinyl triflate **8.1** proceeded with moderately good enantioselectivity to generate π-allyl intermediate **8.2**, which could be trapped with nucleophiles such as acetate (forming **8.3**) and benzylamine (forming **8.4**) [20]. The optimum solvent was DMSO, whereas more typical nonpolar solvents such as toluene gave much reduced enantioselectivity. Cyclization did not proceed in the absence of the nucleophile. This π-allyl trapping sequence was also extended to stabilized carbon nucleophiles, which furnished products **8.5–8.7** with high enantioselectivity [21]. In the latter cases, the presence of NaBr as an additive was found to improve enantioselectivity by 10–15%. It is somewhat surprising that counterion exchange

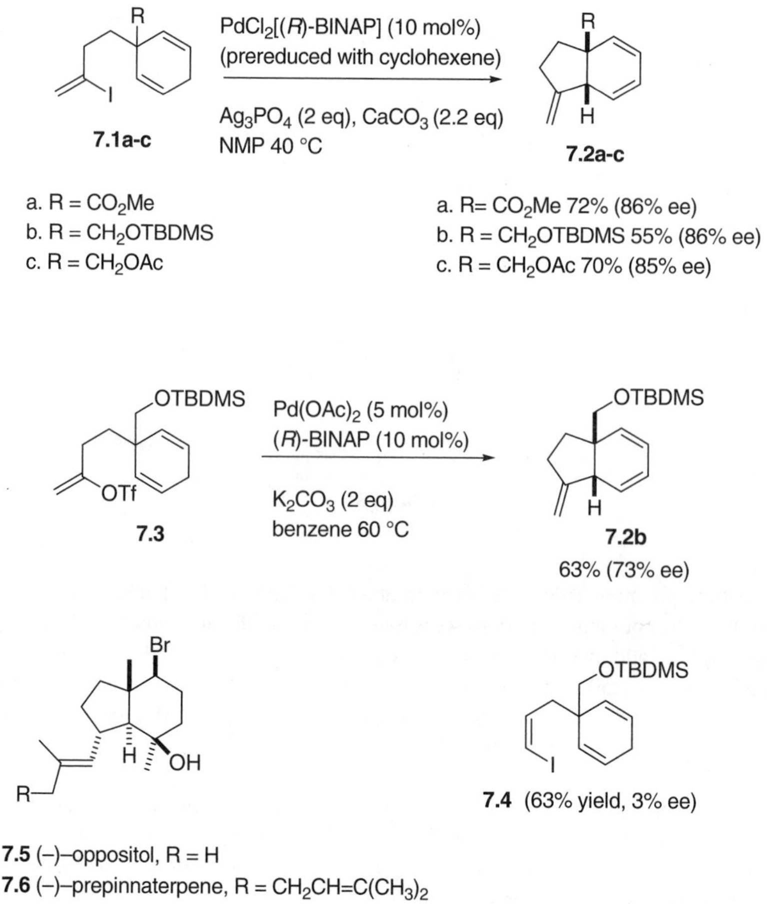

7.1a-c

PdCl$_2$[(R)-BINAP] (10 mol%)
(prereduced with cyclohexene)

Ag$_3$PO$_4$ (2 eq), CaCO$_3$ (2.2 eq)
NMP 40 °C

7.2a-c

a. R = CO$_2$Me
b. R = CH$_2$OTBDMS
c. R = CH$_2$OAc

a. R= CO$_2$Me 72% (86% ee)
b. R = CH$_2$OTBDMS 55% (86% ee)
c. R = CH$_2$OAc 70% (85% ee)

7.3

Pd(OAc)$_2$ (5 mol%)
(R)-BINAP (10 mol%)

K$_2$CO$_3$ (2 eq)
benzene 60 °C

7.2b

63% (73% ee)

7.4 (63% yield, 3% ee)

7.5 (–)–oppositol, R = H
7.6 (–)–prepinnaterpene, R = CH$_2$CH=C(CH$_3$)$_2$

Scheme 8G.7.

does not occur between the triflate and bromide anions, as this would direct the reaction toward the neutral manifold and potentially reduce enantioselectivity. Heck product **8.7** was subsequently elaborated to realize the first catalytic asymmetric synthesis of the angular triquinane $\Delta^{9(12)}$-capnellene **8.8** [21].

8G.2.2. Six- and Seven-Membered Ring Formation

One of the most common applications of the asymmetric Heck cyclization has been in the formation of six-membered rings. All cases reported to date involve 6-exo cyclization under cationic Heck conditions. This strategy has been used as the key step in the asymmetric synthesis of several natural products.

In a series of studies, Shibasaki examined the formation of chiral benzylic quaternary centers by using asymmetric intramolecular Heck reactions. The effect of double-bond stereochemistry was examined in the cyclization of aryl triflates **9.1** and **9.3** (Scheme 8G.9) [22]. As is commonly

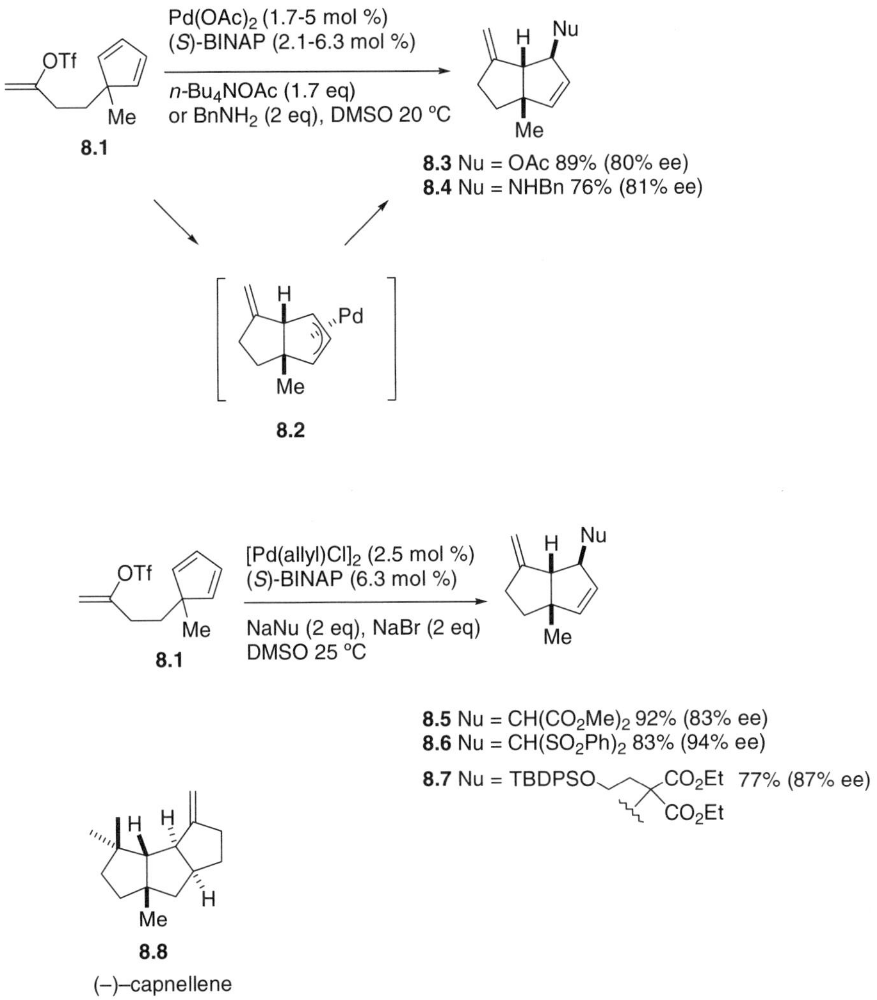

Pd(OAc)₂ (1.7-5 mol %)
(S)-BINAP (2.1-6.3 mol %)

n-Bu₄NOAc (1.7 eq)
or BnNH₂ (2 eq), DMSO 20 °C

8.1

8.3 Nu = OAc 89% (80% ee)
8.4 Nu = NHBn 76% (81% ee)

8.2

[Pd(allyl)Cl]₂ (2.5 mol %)
(S)-BINAP (6.3 mol %)

NaNu (2 eq), NaBr (2 eq)
DMSO 25 °C

8.1

8.5 Nu = CH(CO₂Me)₂ 92% (83% ee)
8.6 Nu = CH(SO₂Ph)₂ 83% (94% ee)
8.7 Nu = TBDPSO⟋⟍CO₂Et 77% (87% ee)
                      ⟍CO₂Et

8.8
(–)–capnellene

Scheme 8G.8.

observed in asymmetric Heck cyclizations involving acyclic alkene components, the Z isomer cyclized with higher enantioselection under standard cationic conditions. The products from each reaction were of opposite absolute configuration, which is also usually the case in asymmetric Heck cyclizations carried out under cationic conditions. Further optimization of the reaction by the use of Pd(OAc)₂ instead of Pd₂(dba)₃•CHCl₃ as the palladium source gave **9.4** with improved enantioselectivity (87% ee), whereas reduction of the temperature to 50°C achieved a further improvement in enantioselectivity to 91% ee. Again, BINAP was found to be the ligand of choice.

In an application of this strategy to the synthesis of (–)-eptazocine, substrate **9.5** was cyclized under optimized conditions to furnish Heck product **9.6** in excellent enantioselectivity (Scheme 8G.9). Tetralin **9.6** was subsequently converted in several steps to (–)-eptazocine (**9.7**) [22].

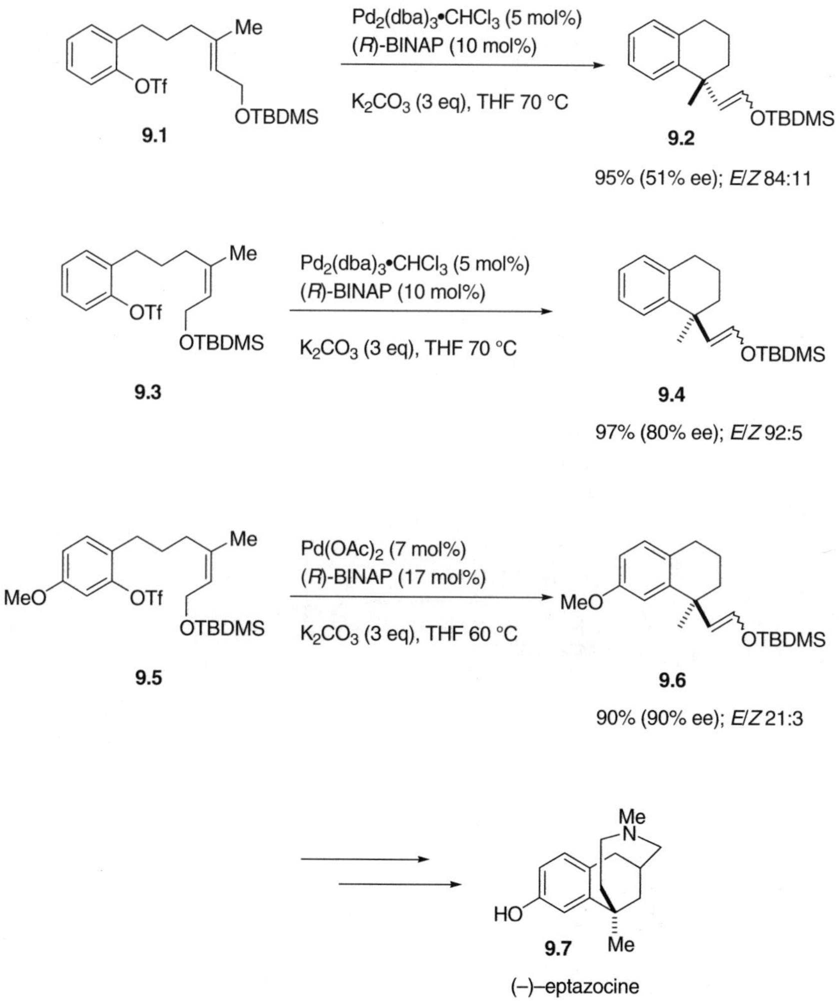

9.1 → Pd$_2$(dba)$_3$•CHCl$_3$ (5 mol%) / (R)-BINAP (10 mol%) / K$_2$CO$_3$ (3 eq), THF 70 °C → 9.2

95% (51% ee); E/Z 84:11

9.3 → Pd$_2$(dba)$_3$•CHCl$_3$ (5 mol%) / (R)-BINAP (10 mol%) / K$_2$CO$_3$ (3 eq), THF 70 °C → 9.4

97% (80% ee); E/Z 92:5

9.5 → Pd(OAc)$_2$ (7 mol%) / (R)-BINAP (17 mol%) / K$_2$CO$_3$ (3 eq), THF 60 °C → 9.6

90% (90% ee); E/Z 21:3

9.7

(–)–eptazocine

Scheme 8G.9.

In a related study, the Shibasaki group examined cyclization of naphthyl triflate **10.1** (Scheme 8G.10) [23]. Cyclization of **10.1** under standard cationic conditions gave Heck product **10.2** in 78% yield and 87% ee. Evidently, the reaction is fairly tolerant of the nature of the aryl group, because both **10.1** and **9.3** behaved similarly. An interesting variation of this reaction was also demonstrated in which Suzuki coupling and asymmetric Heck cyclization were performed in a one-pot operation. Thus, treatment of ditriflate **10.3** with borane **10.4** under standard Heck conditions provided **10.2** in similar enantioselectivity to the stepwise procedure, albeit in quite low yield. Heck product **10.2** was converted in several steps to the natural products, halenaquinone (**10.5**) and halenaquinol (**10.6**).

An approach to the related natural product, (+)-xestoquinone (**11.3**), has been reported by Keay and co-workers in which an asymmetric palladium-catalyzed polyene cyclization is used

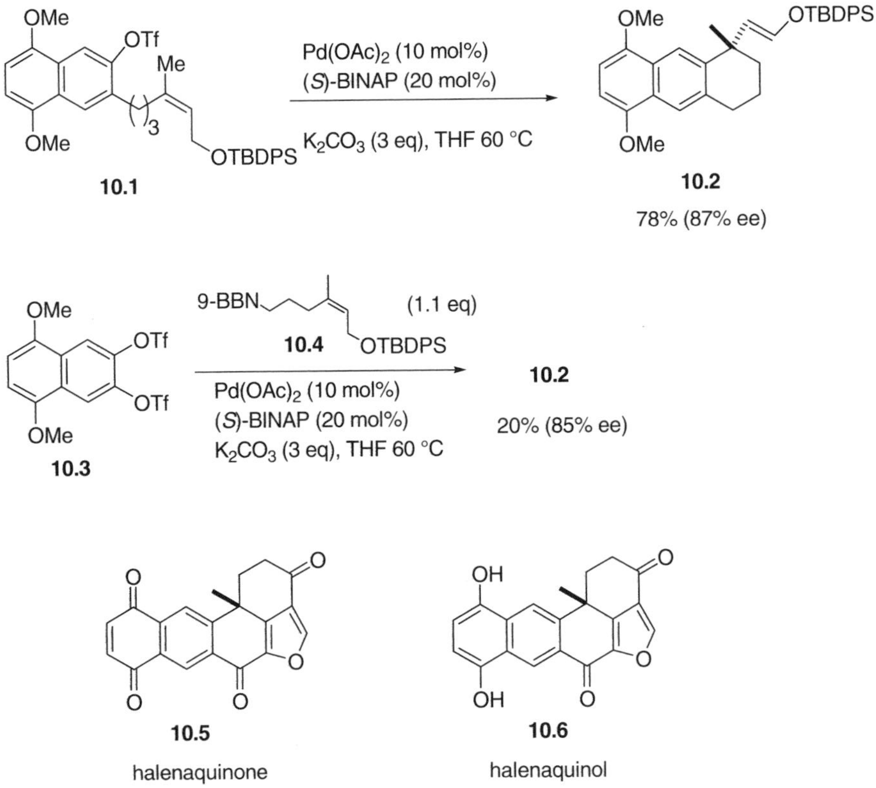

Scheme 8G.10.

as the key step (Scheme 8G.11) [24]. Thus, cyclization of triflate **11.1** under usual cationic conditions provided pentacycle **11.2** in 82% yield with 68% ee. The corresponding aryl bromide derivative resulted in low enantioselectivity (<15% ee) even in the presence of Ag(I) salts. This polycyclization sequence involves an initial 6-exo cyclization, which forms the quaternary asymmetric center, and this is followed by a rare 6-endo cyclization, which forms the last ring. Pentacycle **11.2** was converted to (+)-xestoquinone by reduction of the double bond and subsequent oxidation of the aromatic unit to the quinone.

Shibasaki and co-workers have described a regioselective Heck cyclization of aryl triflate **12.1**, which ultimately provides tricyclic enone **12.4**, a key intermediate in a number of diterpene syntheses (Scheme 8G.12) [25]. Treatment of **12.1** under typical cationic conditions resulted in preferential 6-exo closure to give **12.2** and **12.3** as a 3:1 mixture in 62% overall yield and with 95% ee for both products. The complete selectivity for 6-exo cyclization is noteworthy because 6-endo, 5-exo, and 7-endo cyclization modes were also possible. An analysis of the steric interactions involved in the various cyclization modes was presented and was used to rationalize the observed selectivity. Non-conjugated diene **12.2** could be isomerized to the fully conjugated diene **12.3** in quantitative yield by using catalytic naphthalene•Cr(CO)$_3$. Both Heck products could be converted to the enone **12.4**,

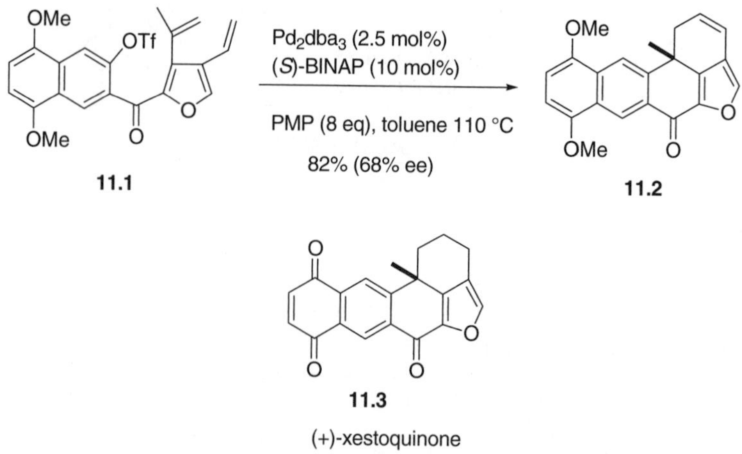

82% (68% ee)

11.1

11.2

11.3

(+)-xestoquinone

Scheme 8G.11.

which has been used as the key intermediate in syntheses of the diterpenes kaurene (**12.5**) and abietic acid (**12.6**) [26].

An example of a 6-exo asymmetric Heck cyclization where an enamide is the alkene component was recently reported by Ripa and Hallberg (Eq. 8G.1) [27]. Cyclization of aryl triflate **e1.1**, under optimum conditions, provided the spirocyclic products **e1.3** (87% ee) and **e1.4** (>99% ee) in a 6:1 ratio and 71% overall yield. Because the two products differ in enantiopurity, a kinetic resolution is probably occurring during the double-bond migration step [10]. It is noteworthy that the optimum ligand in this case was not BINAP, but the oxazoline derivative **e1.2**. This ligand was first described by Pfaltz and has been successfully used in intermolecular asymmetric Heck reactions [28]. The recent successful applications of ligand **e1.2** in asymmetric Heck reactions demonstrates that C_2 symmetry is not required and bodes well for future development of additional useful ligands for asymmetric Heck reactions.

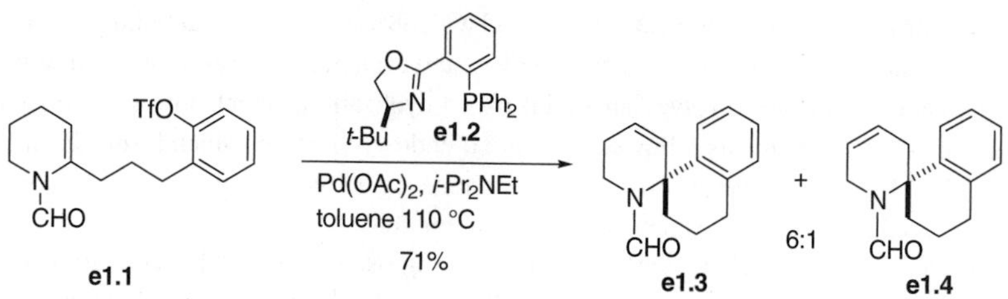

(8G.1)

Most of the studies discussed thus far have avoided the issue of regioselection in β-hydride elimination by forming quaternary centers or by using cyclic alkenes. A potentially more general solution to the problem of controlling the β-hydride elimination step has been described by Tietze in which an allylsilane is used as the alkene component (Scheme 8G.13) [29]. Iodide **13.1** was cyclized under standard cationic asymmetric Heck conditions to give **13.2** in good yield and with moderate enantioselectivity. The minor vinylsilane product **13.3** was found to have low enantiopurity, presumably as a result of reversible double-bond migration. The related aryl iodide **13.4** with a hydrocarbon tether cyclized in good yield, and with enhanced

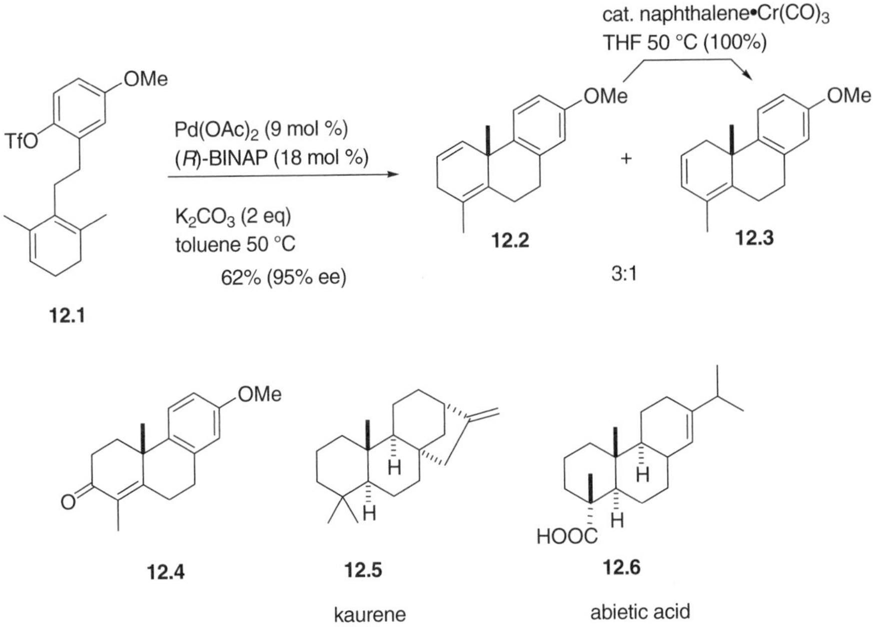

Scheme 8G.12.

enantioselectivity, to give **13.5** in 92% ee [30,31]. This product was subsequently used in an asymmetric synthesis of the simple norsesquiterpene 7-desmethyl-2-methoxycalamenene (**13.6**) [30,31]. In a similar manner, **13.7** was cyclized to benzoazepine **13.8** with moderate enantioselectivity; again, the minor vinylsilane product **13.9** was produced in low ee [29]. Cyclization precursors related to **13.1** with an unfunctionalized alkene instead of the allylsilane group provided mixtures of ethylidene and vinyl products as a result of nonregioselective β-hydride elimination steps [29].

8G.2.3. Five-Membered Ring Formation

Another common application of asymmetric Heck reactions has been in the synthesis of five-membered rings. Shibasaki has described the cyclization of iodide **14.1** to give indolizidinones **14.2** and **14.3** (Scheme 8G.14) [32,33]. Cationic conditions were found to be optimum, and with (*R,S*)-BPPFOH (**14.4**) as the ligand [34], products **14.2** and **14.3** (1:1.4 ratio) were formed in 86% ee and 94% combined yield. Both products had the same enantiopurity, so kinetic resolution is not occurring during the alkene isomerization step in this case. Product **14.2** could be isomerized to **14.3** in quantitative yield by treatment with catalytic Pd/C in MeOH. The use of Ag-zeolite was important; in the presence of the more typical halide scavenger Ag₃PO₄, no reaction was observed at 0°C and at higher temperatures **14.2** and **14.3** were formed with reduced enantiopurity. It is noteworthy that the best ligand was found to be (*R,S*)-BPPFOH, which provides a catalyst that is much more reactive than Pd(BINAP). For example, cyclization of **14.1** by using BINAP as ligand was slow even at 90°C and gave **14.2** of only 34% ee. Compound

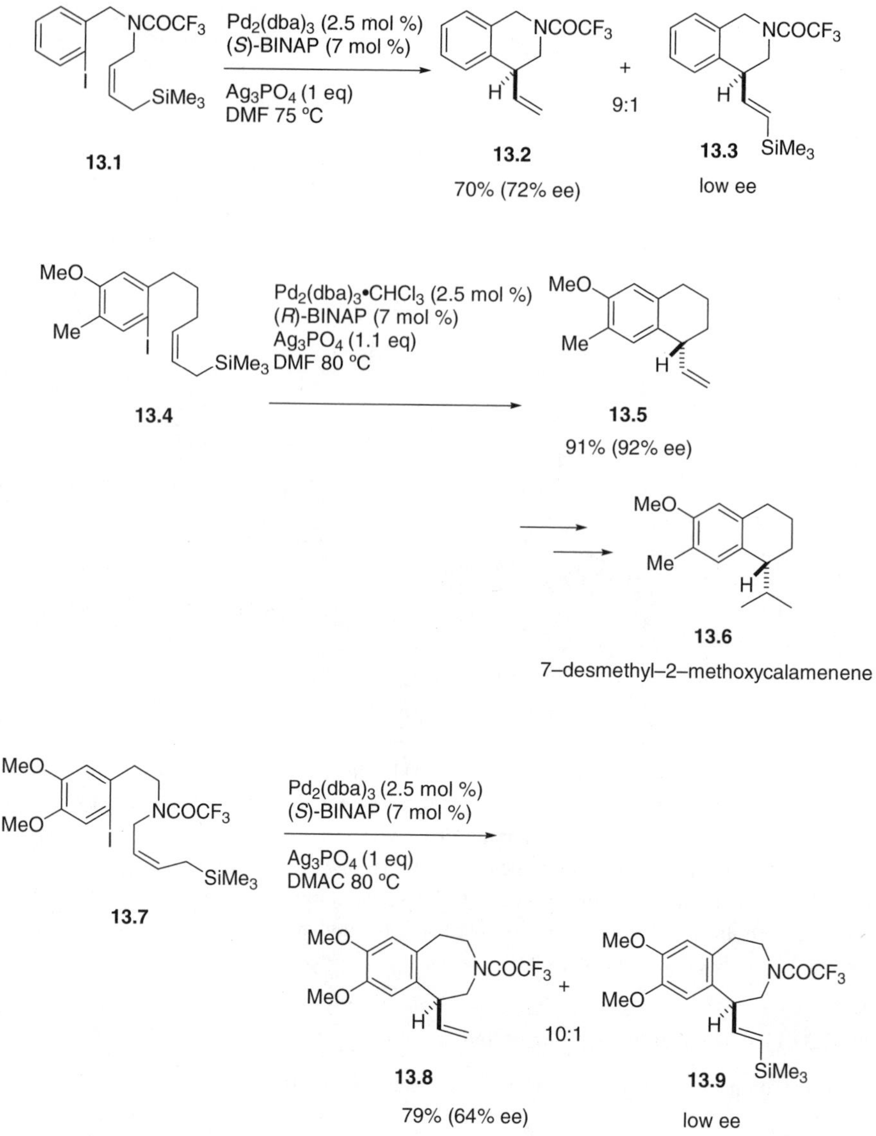

13.1

Pd$_2$(dba)$_3$ (2.5 mol %)
(*S*)-BINAP (7 mol %)

Ag$_3$PO$_4$ (1 eq)
DMF 75 °C

13.2

70% (72% ee)

+

9:1

13.3

low ee

13.4

Pd$_2$(dba)$_3$•CHCl$_3$ (2.5 mol %)
(*R*)-BINAP (7 mol %)
Ag$_3$PO$_4$ (1.1 eq)
DMF 80 °C

13.5

91% (92% ee)

13.6

7–desmethyl–2–methoxycalamenene

13.7

Pd$_2$(dba)$_3$ (2.5 mol %)
(*S*)-BINAP (7 mol %)

Ag$_3$PO$_4$ (1 eq)
DMAC 80 °C

13.8

79% (64% ee)

+

10:1

13.9

low ee

Scheme 8G.13.

14.3 was used in enantioselective total syntheses of the indolizidine alkaloids 1,2-diepilentigi-nosine (**14.5**) and gephyrotoxin 209D (**14.6**) [35].

A strategy involving a palladium-catalyzed cyclization/hydride capture process has been used in the synthesis of chiral benzofuran derivatives (Scheme 8G.15) [36]. Aryl iodides **15.1** and **15.3** were cyclized under cationic conditions in the presence of HCO$_2$Na to give benzofurans **15.2** and **15.4**, respectfully, both with good enantioselection. Products resulting from a 6-endo-

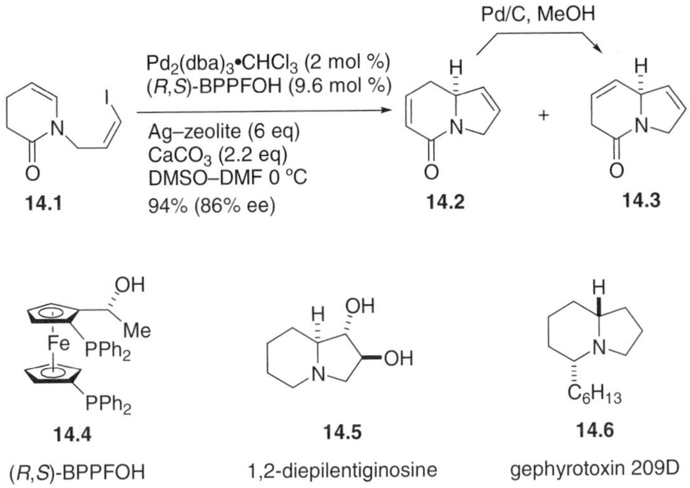

Scheme 8G.14.

cyclization mode were not observed under these conditions. In these reactions, the σ-bonded palladium species, generated by 5-exo cyclization, cannot undergo the usual β-hydride elimination because no β-hydrogens are available; however, in the presence of a hydride source, in this case HCO_2Na, this species is converted to the reduced product with concomitant recycling of the Pd(0) catalyst. The benzofuran products, **15.2** and **15.4**, were used as conformationally restricted retinoids.

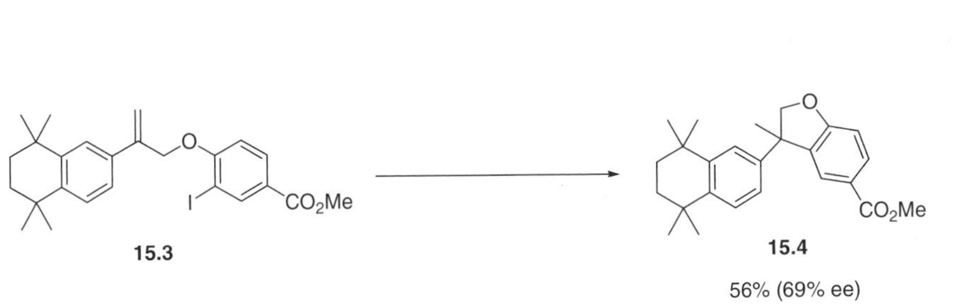

Scheme 8G.15.

The synthesis of oxindoles bearing quaternary asymmetric centers has been studied extensively by Overman and co-workers, from both a synthetic and mechanistic standpoint. These workers were the first to demonstrate that high enantioselection could be obtained under neutral Heck conditions [37,38]. From extensive optimization studies, BINAP was found to be the best ligand and two sets of reaction conditions for the synthesis of chiral 3,3-disubstituted oxindoles emerged. Optimum conditions for the cationic pathway involve use of Pd$_2$(dba)$_3$•CHCl$_3$, BINAP and Ag$_3$PO$_4$ in the polar aprotic solvent DMA (dimethylacetamide), conditions that are fairly standard for asymmetric Heck cyclizations of aryl iodides. Optimum conditions for the neutral pathway are similar to the cationic conditions but the key difference is the absence of a silver salt and the presence of the highly basic tertiary amine 1,2,2,6,6-pentamethylpiperdine (PMP).

The first examples of this asymmetric oxindole synthesis, and the first example of a successful Heck cyclization under neutral conditions, was described by Overman and co-workers in 1992 [37]. Heck cyclization of **16.1** furnished oxindoles **16.2** or *ent*-**16.2** in good yield under the two sets of conditions described above, with comparable enantioselectivity realized under both cationic and neutral conditions (Scheme 8G.16). Significantly, the oxindoles obtained under the two sets of conditions were enantiomeric even though a single enantiomer of the BINAP ligand was used. A number of cyclic alkenes were examined in this series and representative results are shown in Scheme 8G.16. Thus, in the case of forming cycloheptene analog **16.3**, no enantioselectivity was realized under cationic conditions, whereas high enantioselectivity was obtained under neutral conditions. The high enantiopurity of **16.3** obtained under neutral conditions undoubtedly reflects some enhancement by a kinetic resolution during the double-bond isomerization [10], because two other double-bond isomers are isolated as minor products in this reaction. The cyclopentene product **16.4** was obtained in very low enantiopurity under cationic conditions, while a fair enantiopurity was realized under neutral

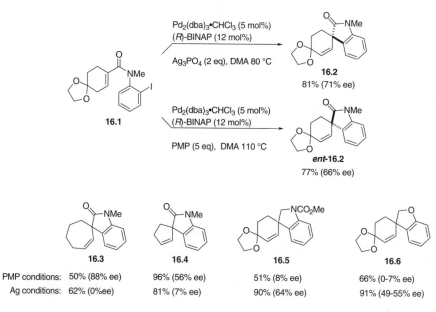

	16.3	**16.4**	**16.5**	**16.6**

PMP conditions: 50% (88% ee)   96% (56% ee)   51% (8% ee)   66% (0-7% ee)

Ag conditions: 62% (0%ee)   81% (7% ee)   90% (64% ee)   91% (49-55% ee)

Scheme 8G.16.

conditions. Formation of products **16.5** and **16.6**, where the nature of the tether connecting the aryl iodide with the trisubstituted double bond is changed, was quite different. In these cases the Heck products were formed with very low enantiopurities under neutral conditions, whereas fair enantioselectivities were realized under cationic conditions. For all but 1 of the 15 substrates examined in this study, products of opposite configuration were obtained from the two pathways [37]. From the data in Scheme 8G.16 and the many more examples reported, it is evident that the structure of the substrate can affect enantioselection in dramatic and subtle ways.

The related formation of 3,3-disubstituted oxindoles from acyclic E alkenes is summarized in Scheme 8G.17 [37]. Substrates **17.1a–c** cyclized under the standard set of conditions to provide oxindoles **17.2a–c** in good yields. These enol ether products **17.2a–c** were subsequently converted to the corresponding alcohols **17.3a–c** by hydrolysis and reduction. When the α-substituent R^1 was Me, poor enantioselectivity was realized under both cationic and neutral conditions. However, when the α substituent was bulkier such as t-Bu (**17.3b**) or Ph (**17.3c**), moderate enantioselectivities were obtained under cationic conditions, although low enantioselectivities were again realized under neutral conditions. The effect of the terminal substituent R^2 was less pronounced, with the β-methyl congener of **17.1a** ($R^2 = $ H) exhibiting similar behavior to **17.1a** [37b]. It is interesting that, in the t-Bu series, the R enantiomer was obtained under both conditions, whereas all other substrates of this type, similar to their cyclic analogs, produced products of opposite absolute configuration when cyclized under cationic or neutral conditions.

The utility of the neutral pathway is highlighted in the synthesis of chiral 3,3-disubstituted oxindoles from the corresponding acyclic Z iodide substrates, which can be realized with enantioselectivities as high as 97% ee under neutral conditions (Scheme 8G.18, Table 8G.1) [38]. Again, enantioselectivity is sensitive to the α substituent R^1 with Me or *prim*-alkyl (entries 1–6 and 11) giving very high enantioselectivities under neutral conditions. Most notable, cyclization of **18.1c** under neutral conditions in THF (rather than DMA) yielded oxindole **17.3a** of 97% ee. Substrates with the bulky α substituents (**18.1d,e**) gave markedly lower enantioselectivities under both sets of conditions (entries 7–10). Enantioselectivity is less sensitive to the β substituent R^2 as can be seen by comparison of cyclizations of substrates **18.1a–c** (entries

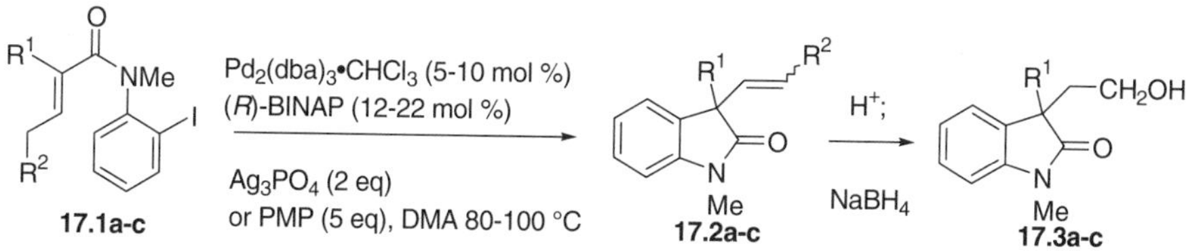

	17.3a	17.3b	17.3c
PMP conditions:	85% (38% ee, R)	90% (27% ee, R)	74% (35% ee, R^*)
Ag conditions:	80% (45% ee, S)	41% (72% ee, R)	93% (73% ee, S^*)

Scheme 8G.17.

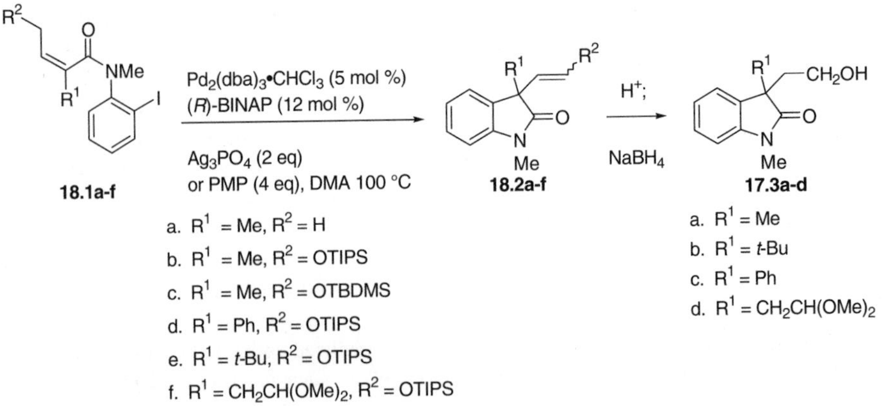

Scheme 8G.18.

1–6). In contrast to the related E substrates, cationic and neutral conditions produce oxindole products of the same absolute configuration.

The analogous Z aryl triflate **19.1** reacts under the cationic manifold to give, ultimately, oxindole (R)-**17.3a** in 72% yield and 43–48% ee (Scheme 8G.19) [38]. An important synthetic advance is the observation that Heck cyclization of this substrate could be diverted to the more selective neutral pathway by addition of halide salts. For example, Heck cyclization of triflate **19.1** in the presence of 1 equiv. of n-Bu$_4$NI gave (R)-**17.3a** in 62% yield and 90% ee, which is similar to the enantioselectivity obtained for cyclization of the corresponding iodide **18.1c** under neutral conditions (see entry 6, Table 8G.1). Conversely, cyclization of iodide **18.1c** in the

TABLE 8G.1. Scope of Asymmetric Heck Cyclization of (Z)-α, β-Unsaturated 2-Iodoanilides with Pd(R)-BINAP[a]

Entry	Substrate	HX scavenger	Yield[b]	% ee	Abs. config.
1	**18.1a**	Ag$_3$PO$_4$	85	69	R
2	**18.1a**	PMP	89	85	R
3	**18.1b**	Ag$_3$PO$_4$	73	80	R
4	**18.1b**	PMP	87	90	R
5	**18.1c**	Ag$_3$PO$_4$	53	78	R
6	**18.1c**	PMP	80	92	R
7	**18.1d**	Ag$_3$PO$_4$	86	65	c
8	**18.1d**	PMP	nd	19	c
9	**18.1e**	Ag$_3$PO$_4$	55	33	S
10	**18.1e**	PMP	76	0	—
11	**18.1f**	PMP	93	91	S

[a]Conditions: 5 mol % Pd$_2$(dba)$_3$•CHCl$_3$, 12 mol % (R)-BINAP, 4 equiv. PMP or 2 equiv. Ag$_3$PO$_4$, DMA, 100°C. [b]Yield of alcohols **17.3a–d** except for entries 1 and 2 where yield is of oxindole **18.2a**. [c]The PMP and Ag$_3$PO$_4$ promoted cyclization produced the same enantiomer.

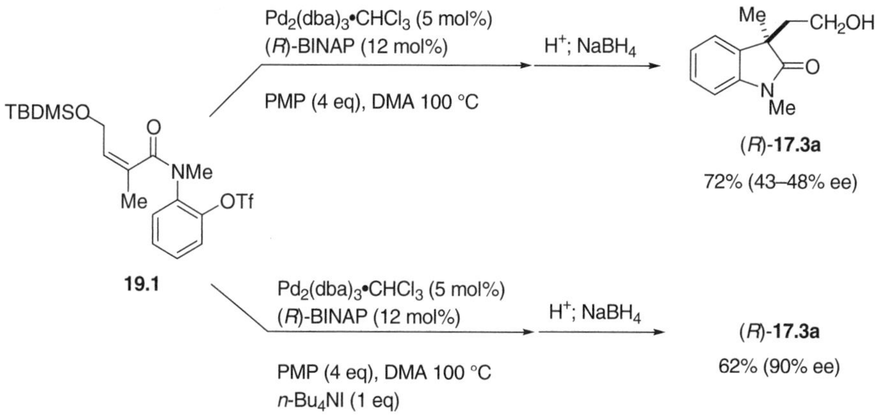

Scheme 8G.19.

presence of AgOTf gave oxindole (*R*)-**17.3a** in 43% ee, which is identical to the results obtained with triflate **19.1**. These results demonstrate that the cationic or neutral pathway can be accessed with either an iodide or triflate substrate. This observation is potentially important for synthetic applications since it introduces some flexibility in the choice of cyclization precursors.

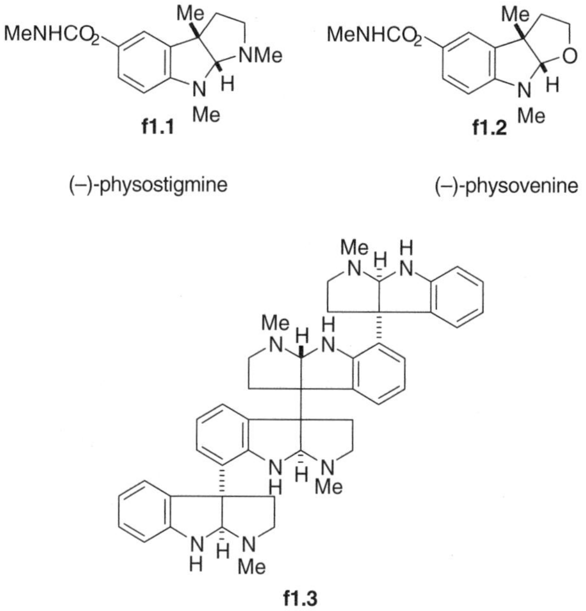

Figure 8G.1. Pyrroloindoline alkaloids and a dodecacyclic polyindoline that are the targets of the application of asymmetric Heck cyclization.

Several generalizations can be drawn from the cyclizations of the E and Z α,β-unsaturated 2-iodoanilide substrates. Thus, when the 2-substituent is Me or *prim*-alkyl, highest selectivity is obtained in cyclizations of the Z substrates under neutral conditions. When the 2-substituent is large, highest enantioselectivity is obtained with the E substrates under cationic conditions. Contributions from the β-substituent are minor. Finally, the E and Z isomers always give products of opposite absolute configuration under cationic conditions, whereas the sense of stereoinduction is independent of alkene geometry under neutral conditions. From a synthetic point of view, the ability to vary both the alkene geometry and the reaction manifold provides maximum opportunity for optimization of a particular cyclization.

Asymmetric Heck cyclizations to form chiral 3,3-disubstituted oxindoles were key strategic steps in enantioselective syntheses of the pyrroloindoline alkaloids (–)-physostigmine **f1.1** [39], (–)-physovenine (**f1.2**) [39], and complex dodecacyclic polyindolines such as **f1.3** (Figure 8G.1) [40].

8G.3. MECHANISTIC STUDIES

As discussed earlier, the generally accepted mechanism for the Heck reaction involves the steps of oxidative addition, coordination of the alkene, migratory insertion, and β-hydride elimination [2,3]. With the intramolecular Heck reaction emerging as an important synthetic reaction over the past decade, the individual steps of this mechanism have come under closer scrutiny, and attention is beginning to be directed at determining the identity of the enantioselective step [41].

Extensive studies by Amatore, Jutand, and co-workers have shed light on the structure and oxidative addition chemistry of a number of synthetically important palladium complexes [42]. In particular, these workers have shown that the major species in a solution of Pd(dba)$_2$ and BINAP is Pd(dba)BINAP and that oxidative addition of PhI to this complex generates (BINAP)Pd(Ph)I [42d,43]. In addition, it has been demonstrated that palladium halide complexes such as (Ph$_3$P)$_2$(aryl)PdCl do not dissociate the halide ligand in DMF solution [44], whereas the corresponding triflate complex is completely dissociated [44,45]. As noted earlier, the nature of the oxidative addition intermediates defines two mechanistic pathways for the Heck reaction: the neutral pathway for unsaturated halide substrates and the cationic pathway for unsaturated triflate substrates [2c–g,3,7–9]. Further, it is possible for halide substrates to be diverted to the cationic pathway by addition of Ag(I) or Th(I) salts [3], and it is possible to divert some triflate substrates to the neutral pathway by addition of halide additives [38]. Individual steps of these two pathways have recently received some scrutiny.

Salient features of the cationic pathway, which was introduced independently by Cabri [8] and Hayashi [9], are presented in Scheme 8G.20. Thus, subsequent to oxidative addition, a vacant coordination site is generated either by triflate dissociation or by halide abstraction by the Ag(I) salt in intermediate **20.4**. This vacant coordination site facilitates double-bond coordination to form cationic intermediate **20.5**, which ultimately forms the Heck product.

The first detailed study of the individual steps of the cationic pathway of the intramolecular Heck reaction was recently described by Brown (Scheme 8G.21) [46]. Oxidative addition of aryl iodide **21.1** to [1,1′-bis(diphenylphosphino)ferrocene](cyclooctatetraene)palladium generated **21.2**. Complex **21.2** was stable at room temperature and was characterized by X-ray crystallography; no interaction between the palladium center and the tethered alkene was observed in this intermediate. Treatment of **21.2** with AgOTf at –78°C removed iodide from the palladium coordination sphere, which facilitated a rapid alkene coordination and subsequent

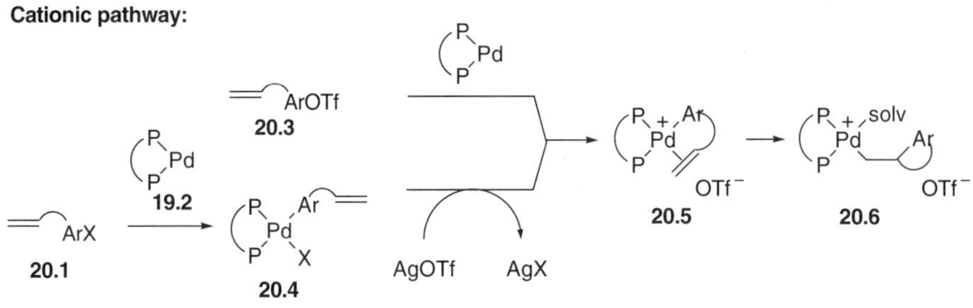

Cationic pathway:

Scheme 8G.20.

insertion (at ~ −40°C) to form, as the next detectable complex, **21.4**. Presumably **21.3** intervenes in this final multistep sequence.

A key feature of the cationic mechanism is that removal (or dissociation) of an anionic ligand from the palladium coordination sphere allows alkene complexation to occur while maintaining coordination of both phosphines of a bisphosphine ligand. That both phosphines can be accommodated in a square-planar four-coordinate intermediate during the insertion step has provided a simple rationalization for the higher enantioselectivities often observed for the cationic pathway. Concrete information on the enantioselective step of asymmetric Heck reactions proceeding by the cationic pathway has not been reported to date. It is likely to be either coordination of the alkene to generate **20.5** or the insertion step (**20.5** → **20.6**, Scheme 8G.20).

The neutral pathway differs from the cationic pathway in the absence of a vacant coordination site in the square-planar four-coordinate palladium(II) intermediate prior to alkene coordination. The key question is then how does alkene coordination take place. Early studies pointed out that Heck reactions of aryl or vinyl halides promoted by (bisphosphine)palladium complexes could be sluggish, and this sluggishness was attributed to a reluctance of one of the phosphines of the

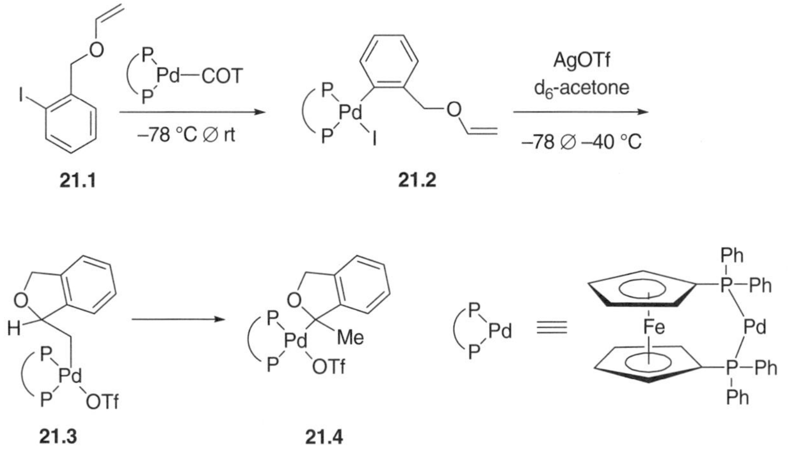

Scheme 8G.21.

bisphosphine ligand to dissociate from palladium [2,3,7]. An asymmetric Heck pathway where alkene coordination is accompanied by dissociation of one of the two phosphines of a chiral bisphosphine ligand from palladium would be expected to be less selective due to a loss in rigidity. In light of the high enantioselectivities that can be realized in some asymmetric Heck reactions that proceed by the neutral pathway, Overman and co-workers have investigated the mechanism of the neutral pathway with particular emphasis on identifying the enantioselective step. Until these recent studies, essentially no information was available on how a (bisphosphine)(aryl)palladium(halide) complex is converted to a Heck product.

Their most detailed investigations focused on the Heck cyclization of iodide **18.1c** to form oxindole **17.3a** (Scheme 8G.18) [38a,b]. A chiral-amplification study [47] established that the catalytically active species is a monomeric Pd-BINAP complex, a conclusion also corroborated by NMR studies by Amatore and co-workers [42d,43]. In addition, two possibilities for the enantioselective step of the neutral pathway were easily eliminated [38a]. Oxidative addition was precluded as the enantioselective step, because iodides cyclize with very different enan-tioselectivities in the presence of Ag(I) salts. A scenario where migratory insertion is reversible and β-hydride elimination is the enantioselective step was also ruled out, because this is not consistent with the dependence of enantioselectivity on the geometry of the double bond of the cyclization precursor.

Three possibilities for the C-C bond-forming step in the neutral pathway were considered and are represented in Scheme 8G.22 [38a,b]. To examine whether the neutral pathway involves phosphine dissociation [2,3,7], Heck cyclizations were conducted with analog **f2.1a–c** designed to mimic a monocoordinated BINAP (Figure 8G.2). Heck cyclization with these ligands resulted in very fast rates and enantioselectivity opposite to that realized in the corresponding Pd(BINAP) catalyzed reactions [38a,b]. These results suggest that both phosphines of the BINAP ligand are bonded to palladium during the enantioselective step.

Because substitution chemistry at square-planar palladium is dominated by associative processes [48], coordination of the alkene in **22.2** would undoubtedly initially generate penta-coordinate intermediate **22.6**. Complex **22.6** could then either evolve to square-planar complex **22.5** by a series of pseudorotations and eventual expulsion of the halide ligand or undergo

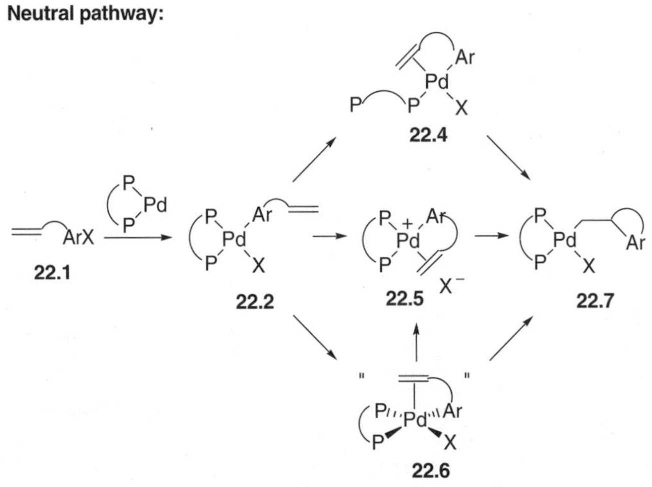

Neutral pathway:

Scheme 8G.22.

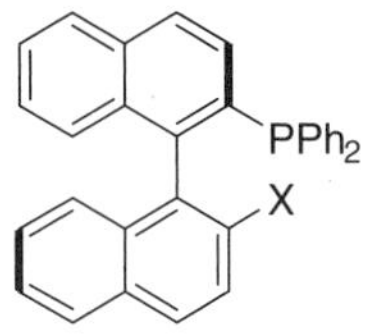

f2.1a-c

a. X = O-*i*-Pr
b. X = OTBDMS
c. X = CHPh$_2$

Figure 8G.2. Monocoordinated BINAP analogs used for asymmetric Heck cyclization.

migratory insertion to directly form **22.7**. Although migratory insertion from the pentacoordinate intermediate **22.6** could not be ruled out, the authors considered the pathway **22.2** → **22.6** → **22.5** to be more likely [38]. It was then reasoned that the enantioselective step in asymmetric Heck reactions proceeding by the neutral pathway is formation of **22.6**, or the conversion of **22.6** to **22.5**. The evolution of **22.6** to **22.5** is undoubtedly a complex process involving a number of discrete pentacoordinate intermediates; the enantioselective step could be formation or breakdown of any one of these intermediates. That a pentacoordinate complex is a viable intermediate is indicated by a number of reports of stable pentacoordinate Pd(II)-alkene complexes [49]. There is also some precedent for olefin displacement of halide in (bisphosphine)(aryl) Pd(halide) complexes [50]. The provocative suggestion that the enantioselective step in the neutral pathway occurs during the process in which halide is displaced by the tethered alkene is certain to provoke additional mechanistic work in this area.

8G.4. CONCLUSIONS

Many intramolecular asymmetric Heck cyclizations have now been demonstrated that proceed with high enantioselectivity, and this reaction has been used as the defining step in asymmetric total syntheses of a number of natural products. In addition, recognition of two synthetically useful reaction manifolds, the cationic and neutral pathways, has increased the opportunities for optimizing asymmetric intramolecular Heck reactions to achieve high levels of enantioselection. In spite of these advances and the growing understanding of mechanistic nuances, much has yet to be revealed. An important area for further studies will be the development of new ligands that not only result in high enantioselectivity but also in higher reactivity so that lower temperatures and lower catalyst loadings can be used [51]. In addition, the structural requirements in a substrate that lead to high enantioselectivity are subtle and are not understood at this point. A better understanding of this aspect of asymmetric intramolecular Heck reactions will be required before this reaction can be routinely used for enantioselective synthesis of broad classes of chiral cyclic products.

REFERENCES

1. (a) Mizoroki, T.; Mori, K.; Ozaki, A. *Bull. Chem. Soc. Jpn.* **1971**, *44*, 581. (b) Heck, R. F.; Nolley, J. P., Jr. *J. Org. Chem.* **1972**, *37*, 2320.

2. Selected reviews include: (a) Heck, R. F. *Organic Reactions* **1982**, *27*, 345. (b) Heck, R. F. In *Comprehensive Organic Synthesis*; Trost, B. M. (Ed.); Pergamon: London, 1991; Vol. 4, pp. 833–863. (c) Davies, G. D.; Hallberg, A. *Chem. Rev.* **1989**, *89*, 1433. (d) de Meijere, A.; Meyer, F. E. *Angew. Chem., Int. Ed.* **1994**, *33*, 2379. (e) Gibson, S. E.; Middleton, R. J. *Contemp. Org. Synth.* **1996**, *3*, 447. (f) Bräse, S.; de Meijere, A. In *Metal-Catalyzed Cross Coupling Reactions*; Stang, P. J. and Diederick, F. (Ed.); Wiley–VCH: Weinheim, 1998; Ch. 3. (g) Link, J. T.; Overman, L. E. In *Metal-Catalyzed Cross Coupling Reactions*; Stang, P. J. and Diederick, F. (Ed.); Wiley–VCH: Weinheim, 1998; Ch. 6.

3. For recent reviews on the asymmetric Heck reaction, see: (a) Shibasaki, M. In *Advances in Metal-Organic Chemistry*; Liebeskind, L. S. (Ed.); JAI: Greenwich, 1996; pp. 119–151. (b) Shibasaki, M.; Boden, C.D.J.; Kojima, A. *Tetrahedron* **1997**, *53*, 7371.

4. For a recent review on catalytic asymmetric synthesis of quaternary carbon centers, see: Corey, E. J.; Guzman-Perez, A. *Angew. Chem., Int. Ed.* **1998**, *37*, 388.

5. Sato, Y.; Sodeoka, M.; Shibasaki, M. *J. Org. Chem.* **1989**, *54*, 4738.

6. Carpenter, N. E.; Kucera, D. J.; Overman, L. E. *J. Org. Chem.* **1989**, *54*, 5846.

7. Cabri, W.; Candiani, I. *Acc. Chem. Res.* **1995**, *28*, 2.

8. Cabri, W.; Candiani, I.; DeBernardis, S.; Francalanci, F.; Penco, S. *J. Org. Chem.* **1991**, *56*, 5796.

9. Ozawa, F.; Kubo, A.; Hayashi, T. *J. Am. Chem. Soc.* **1991**, *113*, 1417.

10. Ozawa, F.; Kubo, A.; Matsumoto, Y.; Hayashi, T.; Nishioka, E.; Yanagi, K.; Moriguchi, K. *Organometallics* **1993**, *12*, 4188.

11. Takaya, H.; Mashima, K.; Koyano, K.; Yagi, M.; Kumobayashi, H.; Taketomi, T.; Akutagawa, S.; Noyori, R. *J. Org. Chem.* **1986**, *51*, 629.

12. Sato, Y.; Sodeoka, M.; Shibasaki, M. *Chem. Lett.* **1990**, 1953.

13. Sato, Y.; Watanabe, S.; Shibasaki, M. *Tetrahedron Lett.* **1992**, *35*, 2589.

14. Kondo, K.; Sodeoka, M.; Mori, M.; Shibasaki, M. *Tetrahedron Lett.* **1993**, *34*, 4219.

15. Ohrai, K.; Kondo, K.; Sodeoka, M.; Shibasaki, M. *J. Am. Chem. Soc.* **1994**, *116*, 11737.

16. Kondo, K.; Sodeoka, M.; Mori, M.; Shibasaki, M. *Synthesis* **1993**, 920.

17. Danishefsky, S.; Schuda, P. F.; Kitahara, T.; Etheridge, S.J. *J. Am. Chem. Soc.* **1977**, *99*, 6066.

18. Sato, Y.; Honda, T.; Shibasaki, M. *Tetrahedron Lett.* **1992**, *33*, 2593.

19. (a) Sato, Y.; Mori, M.; Shibasaki, M. *Tetrahedron: Asymmetry* **1995**, *6*, 757. (b) Fukuzawa, A.; Sato, H.; Masamune, T. *Tetrahedron Lett.* **1987**, *28*, 4303.

20. (a) Kagechika, K.; Shibasaki, M. *J. Org. Chem.* **1991**, *56*, 4093. (b) Kagechika, K.; Ohshima, T.; Shibasaki, M. *Tetrahedron* **1993**, *49*, 1773.

21. Ohshima, T.; Kagechika, K.; Adachi, M.; Sodeoka, M.; Shibasaki, M. *J. Am. Chem. Soc.* **1996**, *118*, 7108.

22. Takemoto, T.; Sodeoka, M.; Sasai, H.; Shibasaki, M. *J. Am. Chem. Soc.* **1993**, *115*, 8477.

23. (a) Kojima, A.; Takemoto, T.; Sodeoka, M.; Shibasaki, M. *J. Org. Chem.* **1996**, *61*, 4876. (b) Kojima, A.; Takemoto, T.; Sodeoka, M.; Shibasaki, M. *Synthesis* **1998**, 581. (c) Shibasaki, M.; Kojima, A.; Shimizu, S. *J. Heterocyclic Chem.* **1998**, *35*, 1057.

24. Maddaford, S. P.; Andersen, N. G.; Cristofoli, W. A.; Keay, B. A. *J. Am. Chem. Soc.* **1996**, *118*, 10766.

25. (a) Kondo, K.; Sodeoka, M.; Shibasaki, M. *J. Org. Chem.* **1995**, *60*, 4322. (b) Kondo, K.; Sodeoka, M.; Shibasaki, M. *Tetrahedron: Asymmetry* **1995**, *6*, 2453.

26. (a) Nakanishi, K.; Goto, T.; Ito, G.; Natori, S.; Nozoe, S. *Natural Products Chemistry*; Academic Press, Inc.: New York, 1974, 1983; Vols. 1,3. (b) Kuehne, M. E.; Nelson, J. A. *J. Org. Chem.* **1970**, *35*, 161. (c) Bell, R. A.; Ireland, R. E.; Partyka, R. A. *J. Org. Chem.* **1962**, *27*, 3741. (d) Chiu, C.K-F.; Govindan, S. V.; Fuchs, P. L. *J. Org. Chem.* **1994**, *59*, 311.

27. (a) Ripa, L.; Hallberg, A. *J. Org. Chem.* **1997**, *62*, 595. (b) Ripa, L.; Hallberg, A. *J. Org. Chem.* **1996**, *61*, 7147.

28. (a) Loiseleur, O.; Hayashi, M.; Schmees, N.; Pfaltz, A. *Synthesis* **1997**, 1338. (b) Loiseleur, O.; Meier, P.; Pfaltz, A. *Angew. Chem., Int. Ed.* **1996**, *35*, 200.

29. Tietze, L. F.; Schimpf, R. *Angew. Chem., Int. Ed.* **1994**, *33*, 1089.

30. Tietze, L. F.; Raschke, T. *Synlett* **1995**, 597.

31. Tietze, L. F.; Raschke, T. *Liebigs Ann. Chem.* **1996**, 1981.

32. Nukui, S.; Sodeoka, M.; Shibasaki, M. *Tetrahedron Lett.* **1993**, *34*, 4965.

33. Sato, Y.; Nukui, S.; Sodeoka, M.; Shibasaki, M. *Tetrahedron* **1994**, *50*, 371.

34. Hayashi, T.; Mise, T.; Kumada, M. *Tetrahedron Lett.* **1976**, *17*, 4351.

35. Nukui, S.; Sodeoka, M.; Sasai, H.; Shibasaki, M. *J. Org. Chem.* **1995**, *60*, 398.

36. Diaz, P.; Gendre, F.; Stella, L.; Charpentier, B. *Tetrahedron* **1998**, *54*, 4579.

37. (a) Ashimori, A.; Overman, L. E. *J. Org. Chem.* **1992**, *57*, 4571. (b) Ashimori, A.; Bachand, B.; Overman, L. E.; Poon, D. J. *J. Am. Chem. Soc.* **1998**, *120*, 6477.

38. (a) Ashimori, A.; Bachand, B.; Calter, M. A.; Govek, S. P.; Overman, L. E.; Poon, D. J. *J. Am. Chem. Soc.* **1998**, 120, 6488. (b) Overman, L. E.; Poon, D. J. *Angew. Chem., Int. Ed.* **1997**, *36*, 518. (c) Ashimori, A.; Matsuura, T.; Overman, L. E.; Poon, D. J. *J. Org. Chem.* **1993**, *58*, 6949.

39. Matsuura, T.; Overman, L. E.; Poon, D. J. *J. Am. Chem. Soc.* **1998**, *120*, 6500.

40. Overman, L. E. Abstracts from the 216th National Meeting of the American Chemical Society, Boston, MA, August 23–27, 1998; ORGN 47.

41. The enantioselective step is typically the first irreversible enantiodifferentiating step, see: Landis, C. R.; Halpern, J. *J. Am. Chem. Soc.* **1987**, *109*, 1746.

42. (a) Amatore, C.; Azzabi, M.; Jutand, A. *J. Am. Chem. Soc.* **1991**, *113*, 8375. (b) Amatore, C.; Jutand, A.; M'Barki, M. A. *Organometallics* **1995**, *14*, 1818. (c) Amatore, C.; Jutand, A.; Suarez, *J. Am. Chem. Soc.* **1993**, *115*, 9531. (d) Amatore, C.; Broeker, G.; Jutand, A.; Khalil, F. *J. Am. Chem. Soc.* **1997**, *119*, 5176.

43. Buchwald has recently isolated and characterized Pd[(*R*)-BINAP](dba): Wolfe, J. P.; Wagaw, S.; Buchwald, S. L. *J. Am. Chem. Soc.* **1996**, *118*, 7215.

44. Jutand, A.; Mosleh, A. *Organometallics* **1995**, *14*, 1810.

45. (a) Stang, P. J.; Kowalski, M. H.; Schiavilli, M. D.; Longford, D. *J. Am. Chem. Soc.* **1989**, *111*, 3347. (b) Stang, P. J.; Kowalski, M. H. *J. Am. Chem. Soc.* **1989**, *111*, 3356. (c) Hinkle, R. J.; Stang, P. J.; Kowalski, M. H. *J. Org. Chem.* **1990**, *55*, 5033. (d) Brown, J. M.; King, K. *Angew. Chem., Int. Ed. Eng.* **1996**, *35*, 657.

46. Brown, J. M.; Pérez-Torrente, J. J.; Alcock, N. W.; Clase, H. J. *Organometallics* **1995**, *14*, 207.

47. Puchot, C.; Samuel, O.; Duñach, E.; Zhao, S.; Agami, C.; Kagan, H. B. *J. Am. Chem. Soc.* **1986**, *108*, 2353.

48. For reviews, see: (a) Cross, R. J. *Adv. Inorg. Chem.* **1989**, *34*, 219. (b) Tobe, M. L. *In Comprehensive Coordination Chemistry*; Wilkinson, G.; Gillard, R. D.; McCleverty, J. A. (Eds.); Pergamon: Oxford, 1987; Vol. 1, pp. 281–329.

49. Albano, V. G.; Castellari, C.; Cucciolito, M. E.; Panunzi, A.; Vitagliano, A. *Organometallics* **1990**, *9*, 1269. (b) Albano, V. G.; Natile, G.; Panunzi, A. *Coord. Chem. Rev.* **1994**, *133*, 67.

50. Portnoy, M.; Ben-David, Y.; Rousso, I.; Milstein, D. *Organometallics* **1994**, *13*, 3465.

51. (a) Kojima, A.; Boden, C. D. J.; Shibasaki, M. *Tetrahedron Lett.* **1997**, *38*, 3459. (b) Young Cho, S.; Shibasaki, M. *Tetrahedron Lett.* **1998**, *39*, 1773.

9

ASYMMETRIC AMPLIFICATION AND AUTOCATALYSIS

Kenso Soai and Takanori Shibata
Department of Applied Chemistry, Faculty of Science, Science University of Tokyo, Kagurazaka, Shinjuku-ku, Tokyo, 162-8601 Japan

9.1. INTRODUCTION

In experimental studies on catalytic asymmetric synthesis, considerable attention has been paid to the interaction between chiral catalysts and the reactant(s) to achieve high enantioselectivity. However, very little attention has been paid to the interaction between the enantiomers of a chiral catalyst when this catalyst is not enantiomerically pure.

Recently, examples of catalytic asymmetric synthesis have been reported in which the enantiomeric purity of the product is much higher than that of the chiral catalyst. A positive nonlinear effect, that is, asymmetric amplification, is synthetically useful because a chiral catalyst of high enantiopurity is not needed to prepare a chiral product with high enantiomeric excess (% ee) (Scheme 9.1).

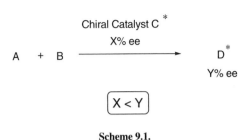

Scheme 9.1.

In asymmetric autocatalysis, the chiral catalyst and the product have the same structure; that is, the chiral product acts as a chiral catalyst for its own multiplication. Asymmetric autocatalysis

Catalytic Asymmetric Synthesis, Second Edition, Edited by Iwao Ojima
ISBN 0-471-29805-0   Copyright © 2000 Wiley-VCH, Inc.

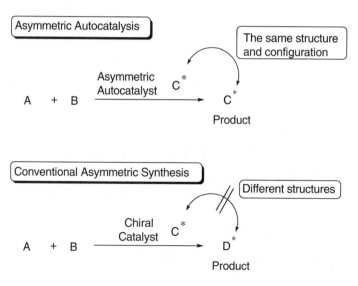

Scheme 9.2.

differs from conventional catalytic asymmetric syntheses where the chiral catalyst and the product have different structures (Scheme 9.2).

Highly enantioselective asymmetric autocatalysis has recently been reported. In such reactions, a trace amount of chiral molecule automultiplies without the assistance of another chiral molecule. Moreover, asymmetric autocatalysis with an amplification of enantiopurity has been reported, that is, the enantiopurity of the initial chiral molecule increases from very low to very high during automultiplication.

This chapter presents an overview of the recent developments in asymmetric amplification [1] and asymmetric autocatalysis [2].

9.2. ASYMMETRIC AMPLIFICATION

9.2.1. The First Asymmetric Amplification

The enantiomer recognition and interaction of chiral organic molecules in solution were reported by Horeau [3] and Wynberg [4]. However, such recognition and interaction of a chiral catalyst were hardly examined in asymmetric synthesis. The first reactions to show an obvious nonlinear effect (NLE) between the enantiopurity of a chiral catalyst and that of a product were reported by Kagan in 1986 (Scheme 9.3) (Figure 9.1) [5]. In the presence of (S)-proline of various enantiopurity, the asymmetric aldol reactions of triketone **1** (Robinson-type annulation) were performed, and a weak and negative NLE was observed between the enantiopurity of (S)-proline and the aldol adduct **2**. An NLE was also observed in asymmetric oxidation under the Sharpless conditions [chiral catalyst: $Ti(O-i-Pr)_4$-diethyl tartrate (DET), oxidant: $t-BuO_2H$]. In the Sharpless-Katsuki asymmetric epoxidation of geraniol (**3**), an obvious positive NLE (asymmetric amplification) was detected between the enantiopurity of (R,R)-DET and epoxide **4**. These results are consistent with the proposed mechanism of asymmetric epoxidation where a dimeric species is generated

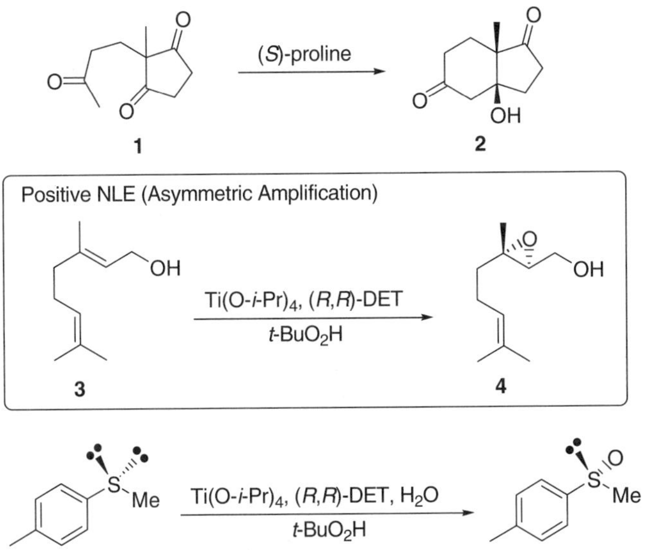

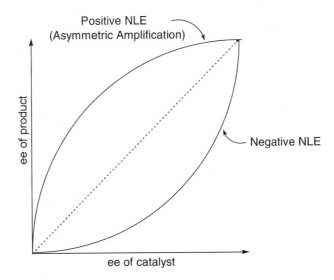

Scheme 9.3.

from two tartrates [6]. In asymmetric sulfoxidation, a negative NLE was observed when (*R,R*)-DET of up to 70% ee was used. These positive and negative NLEs observed in asymmetric synthesis were brought about by the formation of diastereomeric associations of chiral ligands outside and/or inside of the catalytic cycles, and Kagan illustrated these systems by computer simulation using simplified mathematical models [7].

Figure 9.1. Nonlinear relationship between the ee of catalyst and product.

9.2.2. Asymmetric Alkylation

An amino alcohol was found to accelerate the addition reaction of diethlylzinc to aldehyde [8], and then chiral amino alcohols were proved to be efficient chiral catalysts for asymmetric alkylation by using dialkylzinc reagents [9]. Oguni reported a remarkable asymmetric amplification in chiral amino alcohol-promoted alkylation (Scheme 9.4). In the presence of (–)-1-piperidino-3,3-dimethyl-2-butanol (**5**) of 11% ee, benzaldehyde is alkylated enantioselectively to give (*R*)-1-phenylpropanol with 82% ee [10]. Asymmetric amplification was also observed by Noyori using partially resolved (2*S*)-3-*exo*-(dimethylamino)isoborneol (**6**) [11].

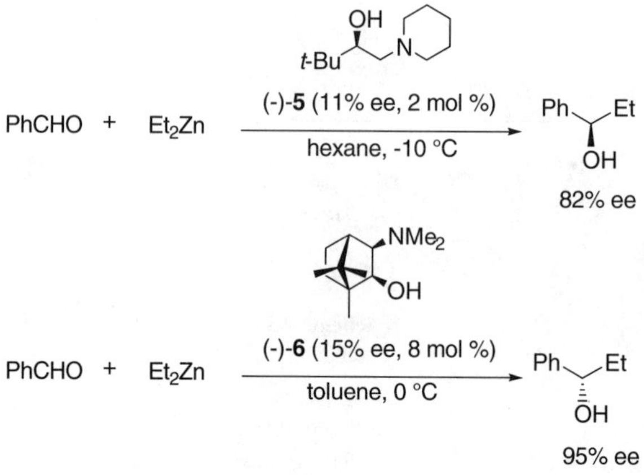

Scheme 9.4.

These deviations from linearity indicate the existence of an oligomeric distribution of chiral ligands. Noyori proposed a rationale as follows: Due to the different dissociability (stability) of homochiral and heterochiral dimer, the enantiopurity of the remaining reactive catalyst (monomer) is improved as compared with that of the submitted chiral ligand **6** (Scheme 9.5) [11]. Heterochiral dimer is thermodynamically more stable than homochiral dimer, which is consistent with Noyori's rationale mentioned above [12a]. An *ab initio* molecular orbital study was also reported in a simplified model reaction between formaldehyde and dimethylzinc catalyzed by achiral 2-aminoethanol [12b].

Fu reported the enantioselective addition of diphenylzinc to ketones catalyzed by chiral amino alcohol **6**, and observed a slight asymmetric amplification [13]. Bolm also reported asymmetric amplification in enantioselective alkylations using diethylzinc promoted by a chiral 2-pyridyl alkanol **7** and β-hydroxy sulfoximine **8** (Scheme 9.6) [14,15].

9.2.3. Asymmetric Conjugate Addition

Tanaka reported the synthesis of (*R*)-muscone (**10**) by an enantioselective conjugate addition of chiral alkoxydimethylcuprate, which was prepared from chiral *endo*-3-[(1-methylpyrrol-2-yl)methylamino]-1,7,7-trimethylbicyclo[2.2.1]heptan-2-ol (**9**), methyllithium, and copper iodide (Scheme 9.7) [16]. In this reaction, convex deviation from a linear correlation was observed when the chiral ligand had a higher enantiopurity. This positive NLE was probably induced by the formation of a reactive homochiral dinuclear copper complex to give (*R*)-muscone. Rossitter also observed asymmetric amplification in a copper-catalyzed conjugate addition of methyl-

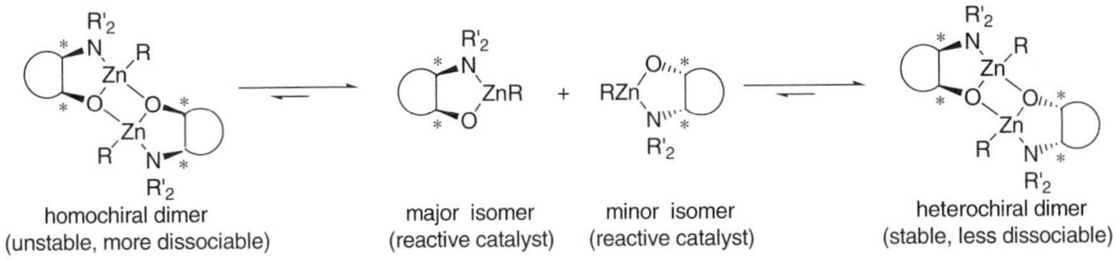

Scheme 9.5.

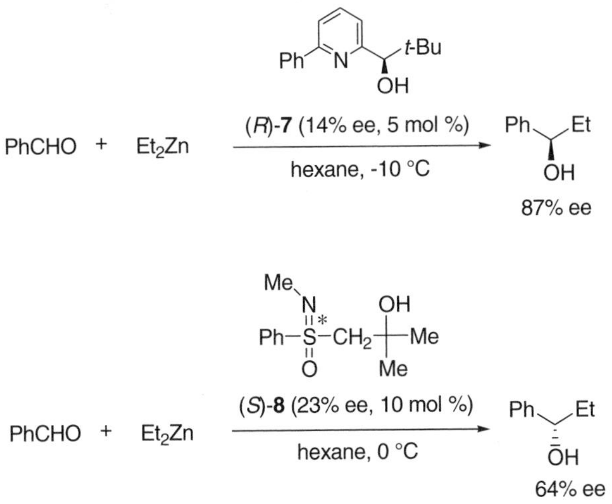

Scheme 9.6.

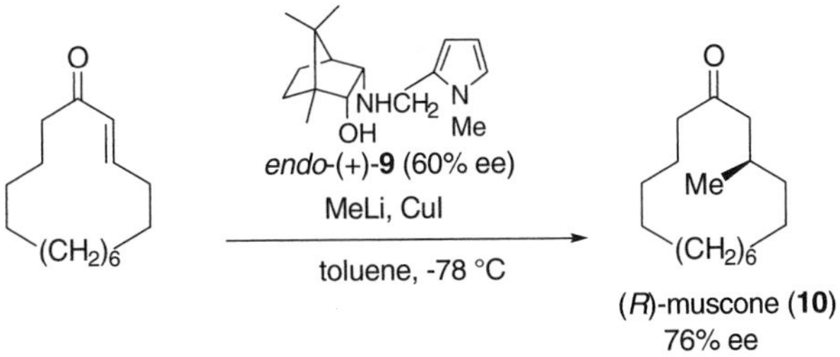

Scheme 9.7.

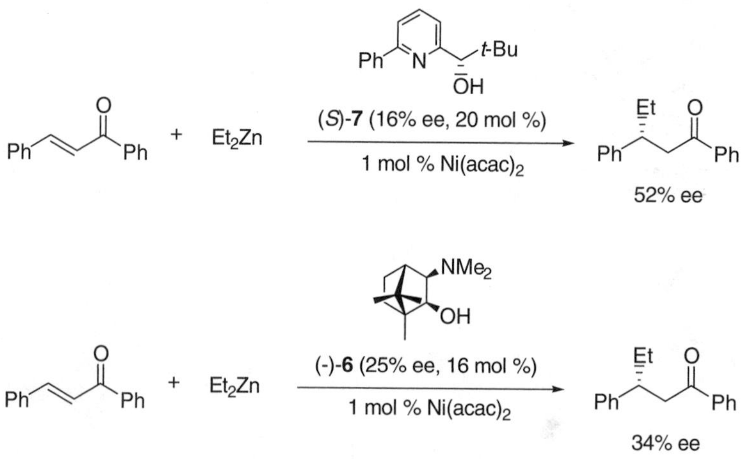

Scheme 9.8.

lithium to cyclic enones using chiral diamines [17]. However, all of the conjugate additions mentioned above are stoichiometric asymmetric reactions.

Whereas Pfaltz reported an example of catalytic copper-catalyzed 1,4-addition, a negative NLE was observed by using a chiral mercaptoaryl-oxazoline as ligand [18]. An enantioselective nickel(II)-catalyzed conjugate addition of diethylzinc to enones was achieved in the presence of a catalytic amount of chiral amino alcohols [19]. Bolm reported asymmetric amplification in a nickel-catalyzed conjugate addition of diethylzinc to chalcone using chiral 2-pyrimidyl alkanol **7** [20]. Feringa also reported the same reaction but with **6** as the chiral ligand wherein a positive NLE was observed with only (−)-**6** of low enantiopurity (Scheme 9.8) [21].

9.2.4. Asymmetric Ene, Allylation, and Aldol Reaction

Mikami reported a highly enantioselective carbonyl-ene reaction where a chiral titanium complex **11** prepared from enantiomerically pure binaphthol (BINOL) and Ti(O-i-Pr)$_2$Br$_2$ catalyzed a glyoxylate-ene reaction with α-methylstyrene to give chiral homoallyl alcohol **12** with 94.6% ee [22]. In this reaction, a remarkable asymmetric amplification was observed and almost the same enantioselectivity (94.4% ee) was achieved by using chiral catalyst prepared

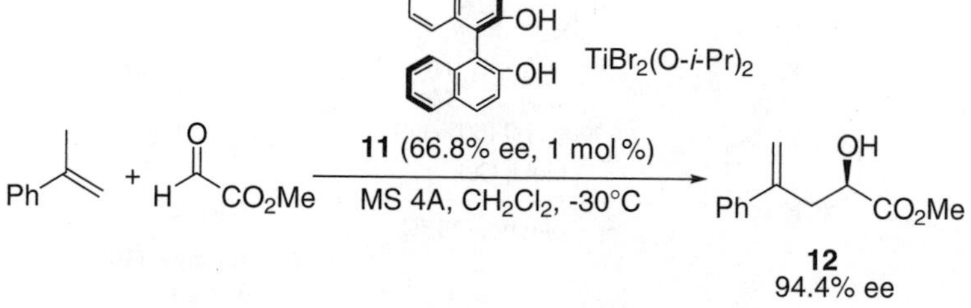

Scheme 9.9.

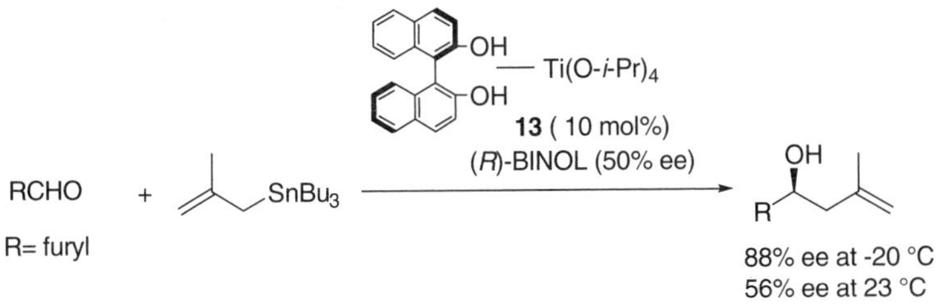

Scheme 9.10.

from BINOL of 66.8% ee (Scheme 9.9) [23]. A simple kinetic study revealed that the catalytic activity of the chiral titanium complex **11** prepared from enantiomerically pure BINOL is 35 times greater than that from racemic BINOL.

Keck reported an asymmetric allylation with a catalytic amount of chiral titanium catalyst [24]. The enantioselective addition of methallylstannane to aldehydes is promoted by a chiral catalyst **13** prepared from chiral BINOL and Ti(O-i-Pr)$_4$ (Scheme 9.10). An example of asymmetric amplification was reported by using (R)-BINOL of 50% ee, and the degree of asymmetric amplification was dependent on the reaction temperature. Tagliavini also observed an asymmetric amplification in the enantioselective allylation with a BINOL-Zr(O-i-Pr)$_2$ catalyst [25].

Asymmetric amplification has also been observed in lanthanum-catalyzed nitro-aldol reaction. Shibasaki used a chiral lanthanum complex **15** prepared from LaCl$_3$ and dilithium alkoxide of chiral BINOL for the enantioselective aldol reaction between naphthoxyacetaldehyde **14** and nitromethane (Scheme 9.11) [26]. When chiral catalyst **15** was prepared from BINOL with 56% ee, the corresponding aldol adduct **16** with 68% ee was obtained. This result indicates that the lanthanum **15** complex should exist as oligomer(s).

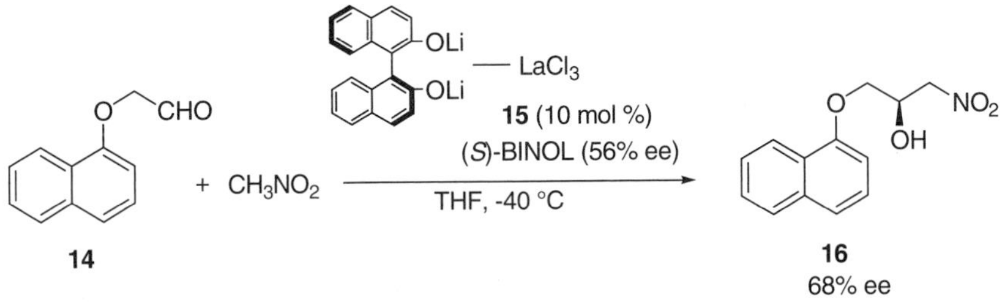

Scheme 9.11.

9.2.5. Asymmetric Diels–Alder Reaction

Narasaka reported a highly enantioselective Diels–Alder reaction between isoprene and a fumaric acid derivative **17** using a catalytic amount of a chiral titanium reagent prepared from tartrate-derived chiral 1,4-diol **18** and TiCl$_2$(O-i-Pr)$_2$ [27]. When a stoichiometric amount of the partially resolved chiral diol (25% ee) was used, the corresponding cycloadduct **19** with 83% ee was obtained (Scheme 9.12). In this reaction, white precipitate was formed, which proved to

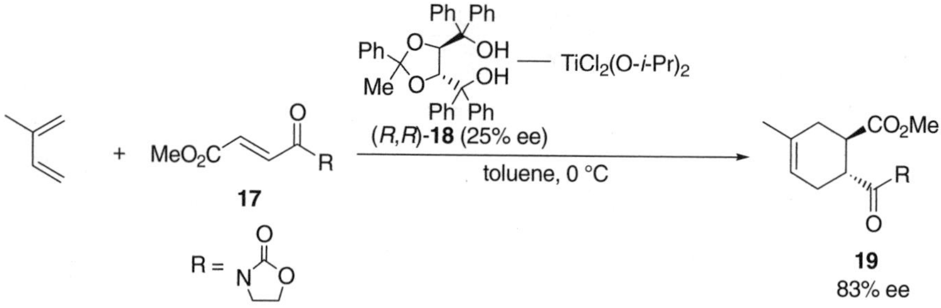

Scheme 9.12.

be racemic 1,4-diol **18** after hydrolysis. This finding means that the precipitated titanium species is a dimer (or oligomer) consisting of the (R,R)- and (S,S)-diols **18** in a 1:1 ratio [28].

Also in a titanium-BINOL system, Mikami reported asymmetric amplification in the Diels–Alder reaction of 1-acetoxy-1,3-butadiene (**20**) with methacrolein [29]. It is interesting that the deviation from linearity is dependent on the way of preparing a partially resolved chiral catalyst **22** (Table 9.1). When a catalyst with low enantiopurity was prepared by mixing the (R)-catalyst and the racemic catalyst, a positive NLE was observed. On the other hand, when the catalyst was prepared by mixing the (R)- and (S)-catalysts, a slightly negative NLE was observed. These results also support the existence of oligomeric (dimeric) complexes generated from racemates.

TABLE 9.1. Asymmetric Amplification in Diels–Alder Reaction

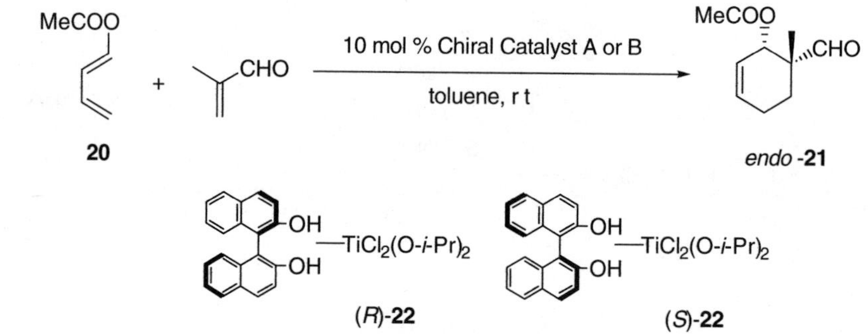

Chiral Catalyst	BINOL (% ee)	21 (% ee)
A	33	50
A	68	80
B	32	29
B	73	63

A: prepared by the mixture of (R)-**22** and racemic catalyst.
B: prepared by the mixture of (R)-**22** and (S)-**22**.

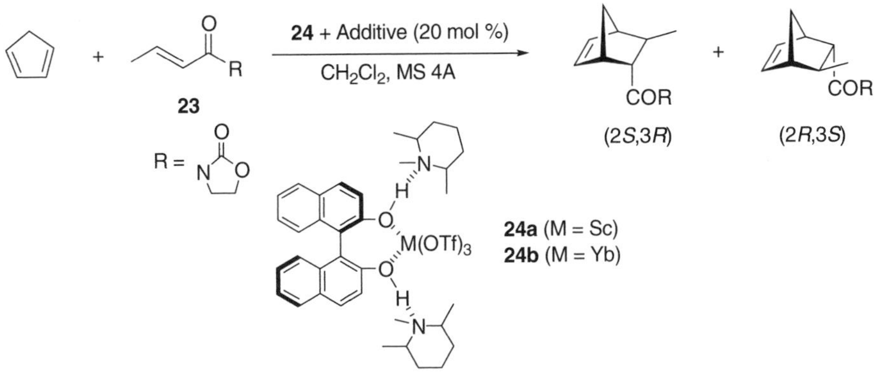

Scheme 9.13.

Kobayashi reported an asymmetric Diels–Alder reaction catalyzed by a chiral lanthanide(III) complex **24**, prepared from ytterbium or scandium triflate [Yb(OTf)$_3$ or Sc(OTf)$_3$], (R)-BINOL and tertiary amine (ex. 1,2,6-trimethylpiperidine) [30]. A highly enantioselective and endoselective Diels–Alder reaction of 3-(2-butenoyl)-1,3-oxazolidin-2-one (**23**) with cyclopentadiene (Scheme 9.13) takes place in the presence of **24**. When chiral Sc catalyst **24a** was used, asymmetric amplification was observed with regard to the enantiopurity of (R)-BINOL and that of the endoadduct [31]. On the other hand, in the case of chiral Yb catalyst **24b**, NLE was affected by additives, that is, when 3-acetyl-1,3-oxazolidin-2-one was added, almost no deviation was observed from linearity, whereas a negative NLE was observed with the addition of 3-phenylacetylacetone.

9.2.6. Asymmetric Oxidation and Reduction

Feringa reported an enantioselective allylic oxidation of cyclohexene to optically active 2-cyclohexenyl propionate **25** by using a chiral copper complex prepared from Cu(OAc)$_2$ and (S)-proline, as chiral catalyst (Scheme 9.14) [32]. In the absence of additives, a negative NLE was observed, whereas in the presence of a catalytic amount of anthraquinone, a positive NLE (asymmetric amplification) was observed. Moreover, higher enantioselectivity was attained when enantiopure (S)-proline was used. However, the role of the additive remains elusive.

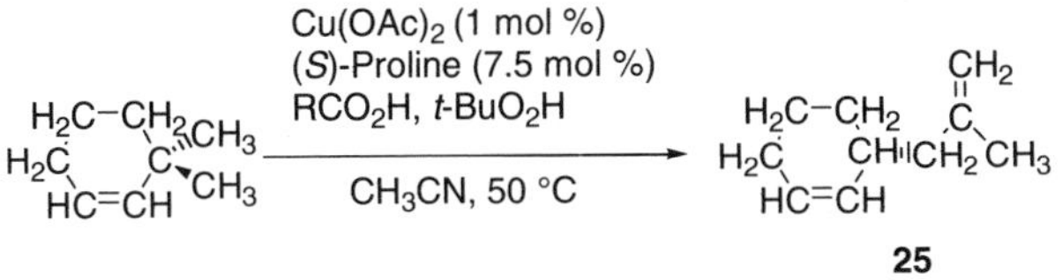

Scheme 9.14.

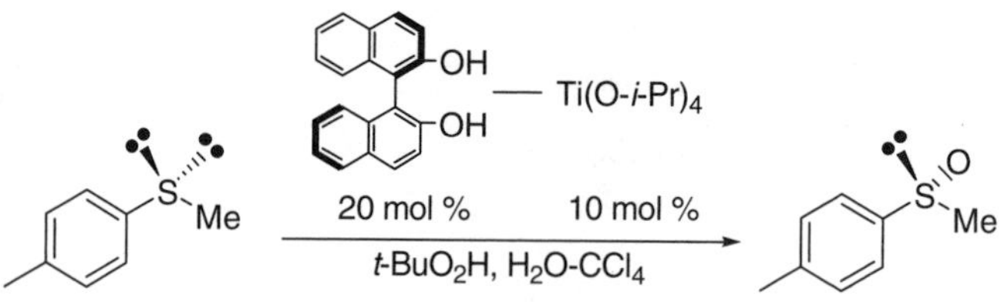

Scheme 9.15.

Uemura reported a highly enantioselective oxidation of sulfides to sulfoxides using a chiral titanium complex prepared from chiral BINOL and Ti(O-i-Pr)$_4$, and this reaction exhibits a remarkable asymmetric amplification (Scheme 9.15) [33].

Evans reported an enantioselective Meerwein–Ponndorf–Verley reduction using a catalytic amount of chiral samarium complex **26** prepared from samarium (III) iodide and a chiral amino diol (Scheme 9.16) [34]. Even when a partially resolved ligand (80% ee) was used, the enantiopurity of the resulting alcohol **27** reached 95% ee, which is the same value as that obtained when the enantiopure amino diol was used.

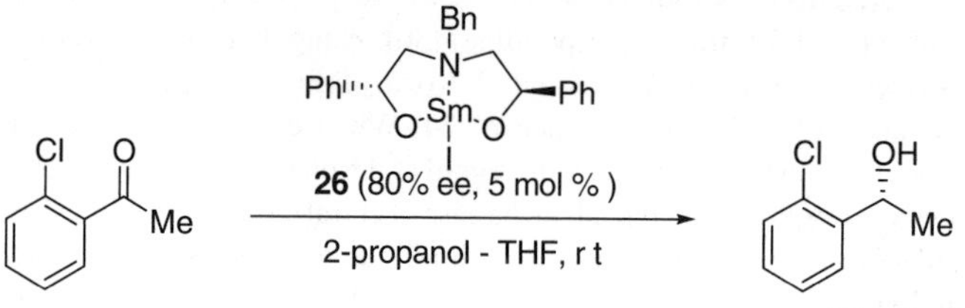

Scheme 9.16.

9.2.7. Activation or Deactivation of Racemic Catalysts by Chiral Additives

As described in the preceding sections, asymmetric amplification is generally a consequence of the formation of aggregates (i.e., dimers or oligomers that are homochiral or heterochiral) of a chiral catalyst. However, even a racemic catalyst can be used as a chiral catalyst with the aid of chiral additives (a simple model consisting of dimers is depicted in Scheme 9.17). If a chiral additive (*R*)-**B** is selectively associated with (*S*)-**A** in the racemic catalyst, the remaining (*R*)-**A** could operate as the chiral monomer catalyst (asymmetric deactivation). Conversely, the chiral additive (*R*)-**B** can be selectively associated with (*R*)-**A** in racemic catalyst to generate an active dimeric catalyst (asymmetric activation).

Based on the concept mentioned above, Brown realized the asymmetric deactivation of a racemic catalyst in asymmetric hydrogenation (Scheme 9.18) [35]. One enantiomer of (±)-CHIRAPHOS **28** was selectively converted into an inactive complex **30** with a chiral iridium complex **29**, whereas the remaining enantiomer of CHIRAPHOS forms a chiral rhodium complex **31** that acts as the chiral catalyst for the enantioselective hydrogenation of dehydroamino acid derivative **32** to give an enantio-enriched phenylalanine derivative **33**.

Scheme 9.17. Concept of asymmetric deactivation and activation of racemic catalyst by a chiral additive.

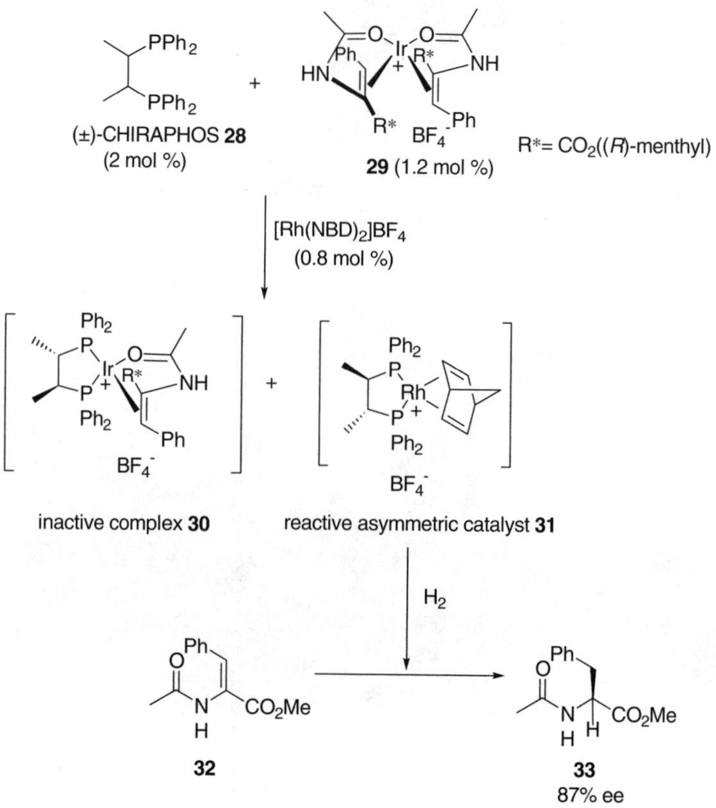

Scheme 9.18.

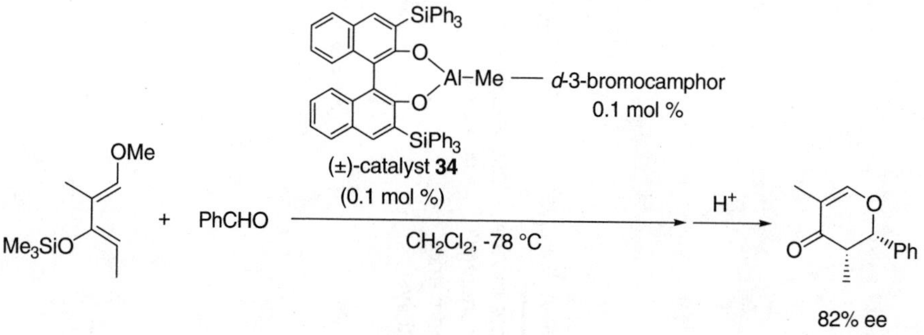

Scheme 9.19.

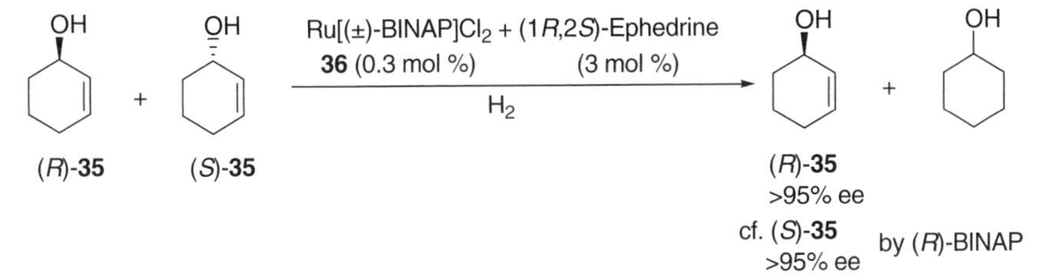

Scheme 9.20.

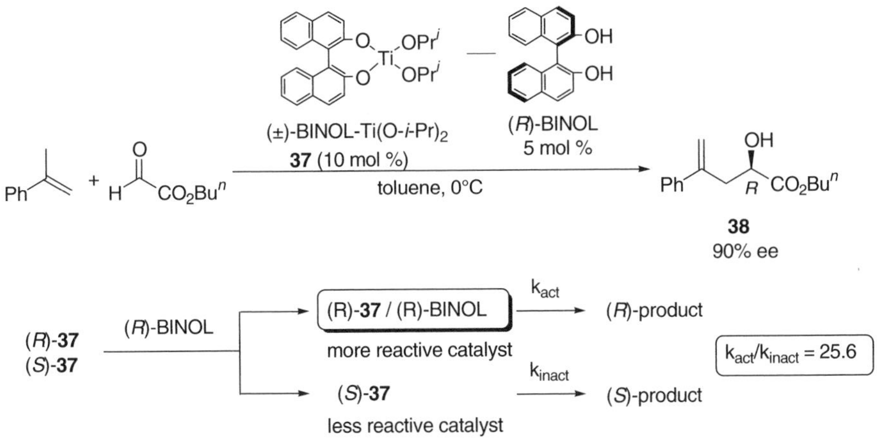

Scheme 9.21.

Yamamoto reported an aluminum complex-catalyzed asymmetric hetero Diels–Alder reaction (Scheme 9.19) [36]. Chiral ketone d-3-bromocamphor discriminates (R)-organoaluminum complex **34** from (S)-complex **34** by diastereoselective complexation, whereas the remaining (S)-isomer **34** catalyzes the enantioselective cycloaddition of an activated diene to benzaldehyde.

Faller also achieved asymmetric deactivation in asymmetric hydrogenation by the combination of racemic [(CHIRAPHOS)Rh]$^+$ and (S)-METHOPHOS [Ph$_2$POCH$_2$CH(NMe$_2$)CH$_2$CH$_2$-SMe] [37]. He used the term "chiral poison" to refer to the chiral additive (ex. (S)-METHOPHOS). The asymmetric deactivation (chiral poisoning) is very effective in the kinetic resolution of allylic alcohol **35** by using ruthenium complex **36** (Scheme 9.20) [38]. Simply by adding the inexpensive and readily available chiral compound ephedrine as the chiral poison, racemic BINAP can be used as the chiral ligand for **36**, and the reaction gives allyl alcohol **35** with >95% ee. At first, (1R,2S)-ephedrine selectively binds to (i.e., poisons) Ru-(R)-BINAP complex, and then Ru[(S)-BINAP]Cl$_2$ plays a role as the chiral catalyst.

The activation of a racemic catalyst by a chiral additive was achieved by Mikami in a chiral titanium complex-catalyzed asymmetric carbonyl-ene reaction (Scheme 9.21) [39]. The racemic catalyst ($\pm$)-BINOL-Ti-(O-i-Pr)$_2$ **37** (10 mol %) is activated by adding (R)-BINOL (5 mol %), and the ene product **38** with 90% ee is obtained. (R)-BINOL is selectively associated with (R)-BINOL-Ti-(O-i-Pr)$_2$ to give a dimeric catalyst whose activity is kinetically calculated to be 25.6 times greater than that of the remaining (S)-BINOL-Ti-(O-i-Pr)$_2$.

Mikami et al. also reported the activation of a racemic catalyst in Ru-BINAP-catalyzed asymmetric hydrogenation (Scheme 9.22) [40]. The combination of racemic RuCl$_2$[($\pm$)-**39**] and

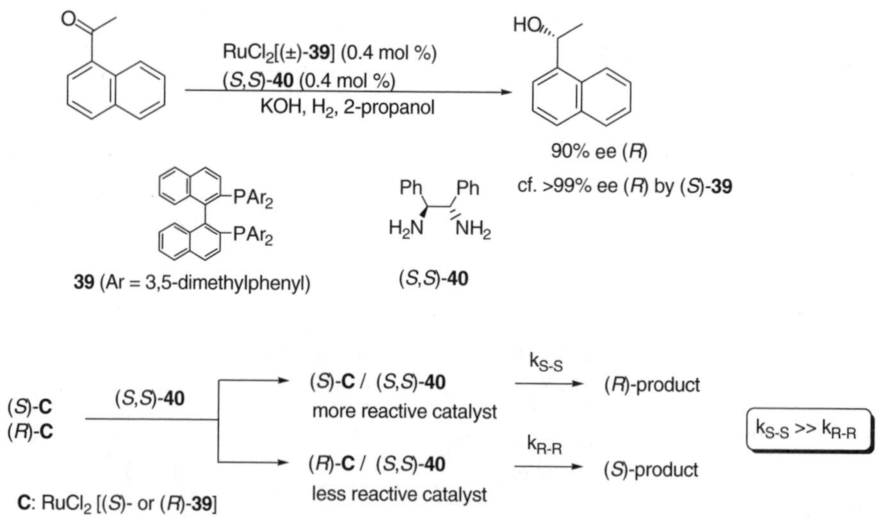

Scheme 9.22.

chiral diamine **40** enables highly enantioselective reduction. However, in contrast to the above ene reaction, the association between RuCl$_2$[(±)-**39**] and the chiral activator (*S,S*)-**40** is not selective, that is, (*S*)-**C**/(*S,S*)-**40**:(*R*)-**C**/(*S,S*)-**40** = 1:1. Therefore, the observed high enantioselectivity should have been induced by the difference in the catalytic activity of the two diastereomeric complexes (i.e., $k_{S-S} >> k_{R-R}$).

9.3. ASYMMETRIC AUTOCATALYSIS

9.3.1. Implications

In 1953, Frank proposed a reaction mechanism without showing any chemical structure for the molecules, in which a chiral product acts as a chiral catalyst for its own production (asymmetric autocatalysis) and prohibits the formation of its antipode [41]. In such a reaction, if it exists, the enantiomeric purity of the product would increase as the reaction progresses. Since then, asymmetric autocatalysis has attracted considerable conceptual interest [42]. However, it was not until 1990 that the first asymmetric autocatalysis in asymmetric synthesis was reported [43].

Unlike conventional catalytic asymmetric synthesis, asymmetric autocatalysis possesses the following novel features:

(1) No chiral catalyst with a structure different from that of the product is required.

(2) No separation of the product from the chiral catalyst is required.

(3) The chiral molecule automultiplies exponentially because newly formed chiral molecules act as a chiral catalyst.

Thus, asymmetric autocatalysis is one of the most energy- and resource-saving processes in asymmetric synthesis.

9.3.2. Background

Seebach recognized the importance of the effect of mixed aggregates of products (lithium enolates) on enantioselectivity [44]. Alberts and Wynberg reported an asymmetric autoinduction (Scheme 9.23) in which ethyllithium adds to benzaldehyde to give in situ lithium alkoxide of chiral 1-phenyl-1-propanol **42** with 17% ee in the presence of a stoichiometric amount of the lithium alkoxide of 1-phenyl-1-propanol-d_1 **41** of the same configuration [45]. They also described an enantioselective addition (32% ee) of diethylzinc to benzaldehyde using titanium (IV) tetraalkoxide of chiral 1-phenyl-1-propanol-l-d_1 **43**. In this reaction, the structures of the chiral catalyst **43** and the product **44** (zinc alkoxide before quenching the reaction) are different [45,46]. Danda reported that, in an asymmetric cyanohydrin-forming reaction with 2,5-diketopiperazine as a chiral catalyst, the presence of a chiral product enhances the enantioselectivity of the chiral catalyst [47].

In the course of the continuing study [9a,b] on the enantioselective addition of dialkylzincs to aldehydes by using chiral amino alcohols such as diphenyl(1-methyl-2-pyrrolidinyl)methanol (**45**) (DPMPM) [48] *N,N*-dibutylnorephedrine **46** (DBNE) [49], and 2-pyrrolidinyl-1-phenyl-1-propanol (**47**) [50] as chiral catalysts, Soai et al. reacted pyridine-3-carbaldehyde (**48**) with dialkylzincs using (1*S*,2*R*)-DBNE **46**, which gave the corresponding chiral pyridyl alkanols **49** with 74–86% ee (Scheme 9.24) [51]. The reaction with aldehyde **48** proceeded more rapidly (1 h) than that with benzaldehyde (16 h), which indicates that the product (zinc alkoxide of pyridyl alkanol) also catalyzes the reaction to produce itself. This observation led them to search for an asymmetric autocatalysis by using chiral pyridyl alkanol.

9.3.3. The First Asymmetric Autocatalysis

The first asymmetric autocatalysis was attained with chiral pyridyl alkanols **49**. When 3-pyridinecarbaldehyde **48** was reacted with dialkylzinc by using chiral pyridyl alkanol **49** (20 mol %) as asymmetric autocatalyst, enantio-enriched pyridyl alkanol **49** with the same structure and configuration as that of the catalyst was obtained [43]. Regarding the substituents (R) of pyridyl

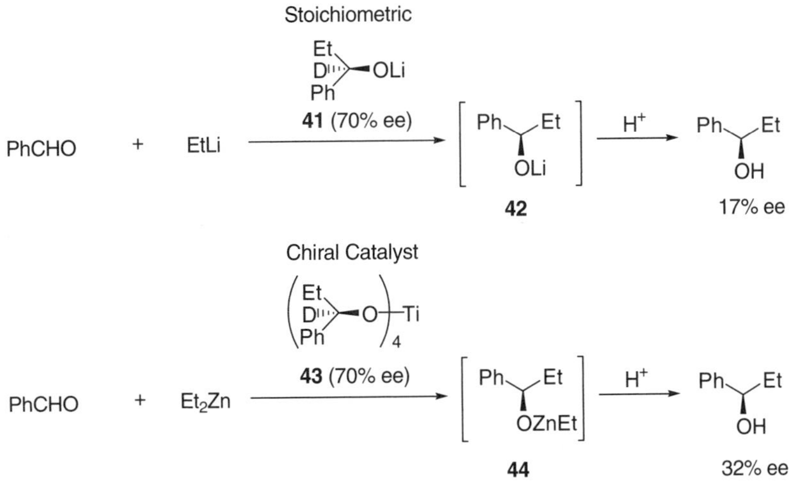

Scheme 9.23.

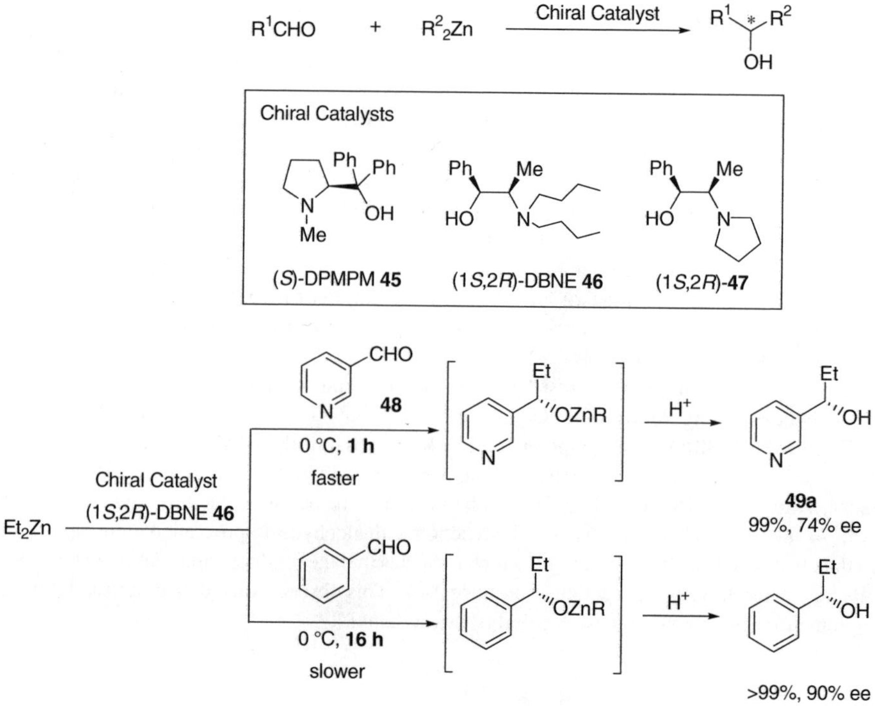

Scheme 9.24.

alkanol and dialkylzinc (Note: the substituents of alkanol and dialkylzinc must be identical in asymmetric autocatalysis), isopropyl (*i*-Pr) is more enantioselective than ethyl (Et). When (*S*)-2-methyl-1-(3-pyridyl)-1-propanol (**49b**) with 86% ee was used as asymmetric autocatalyst in the reaction of **48** with *i*-Pr$_2$Zn, (*S*)-**49b** with 35% ee was newly formed in 67% yield.

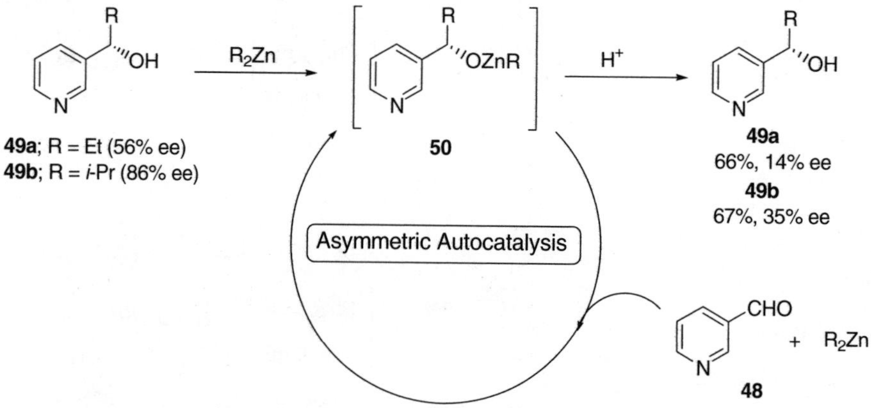

Scheme 9.25.

Chiral isopropylzinc alkoxide **50** of pyrimidyl alkanol formed in situ from the chiral pyrimidyl alkanol and i-Pr$_2$Zn is considered to be an initial actual asymmetric autocatalyst, which automultiplies itself by catalyzing the addition of i-Pr$_2$Zn to aldehyde **48** (Scheme 9.25).

Chiral diol **52** [52] and ferrocenyl alkanol **54** [53] also work as asymmetric autocatalysts in the enantioselective addition of dialkylzincs to the corresponding dialdehyde **51** and ferrocene-carbaldehyde **53**, respectively (Scheme 9.26).

Thus, chiral pyrimidyl alkanol, ferrocenyl alkanol and diol are asymmetric autocatalysts, although the enantiopurities of the newly formed products are moderate.

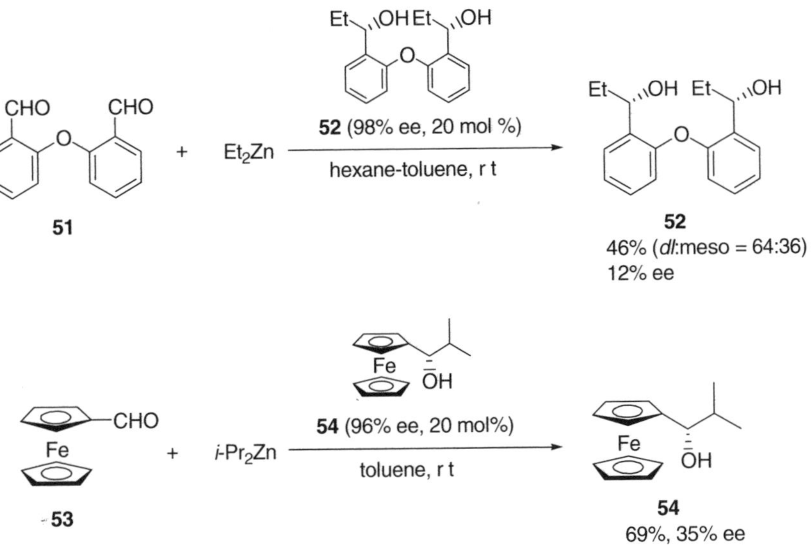

Scheme 9.26.

9.3.4. Highly Enantioselective Asymmetric Autocatalysis of Pyrimidyl Alkanol

The first highly enantioselective asymmetric autocatalytic reaction was achieved in the addition of i-Pr$_2$Zn to pyrimidine-5-carbaldehydes **55** by using chiral 5-pyrimidyl alkanols **56** as asymmetric autocatalysts. When chiral pyrimidyl alkanol **56b** with 95% ee was used, it was automultiplied without any loss of enantiopurity to give itself with 96% ee [54]. The enantiopurity of the newly formed pyrimidyl alkanol **56b** reached 98.2% ee when asymmetric autocatalyst **56b** with > 99.5% ee was used (Scheme 9.27).

By optimizing the structure of the substituent at the 2-position of the pyrimidine ring, practically perfect asymmetric autocatalysis was realized [55]. When (S)-2-methyl-1-(2-t-butylethynyl-5-pyrimidyl)-1-propanol (**56c**) with >99.5% ee was used as the asymmetric autocatalyst (20 mol %) in cumene, (S)-**56c** was obtained as the product with >99.5% ee in almost quantitative (>99%) yield (Scheme 9.28). The autocatalytic reaction was performed successively, with the products of one round serving as the asymmetric autocatalysts for the next. Even in 10 rounds, the enantiopurity and yield of the product **56c** were always almost perfect (>99%, >99.5% ee) (Table 9.2). Thus, the autocatalyst **56c** multiplied by a factor of ca. 10^7 during the 10 rounds, with no deterioration.

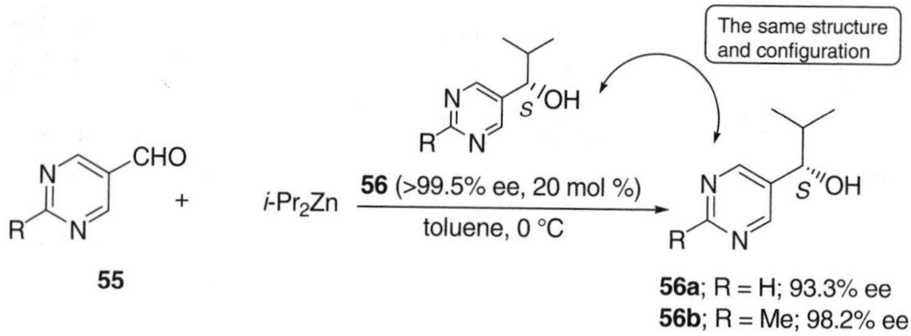

Scheme 9.27.

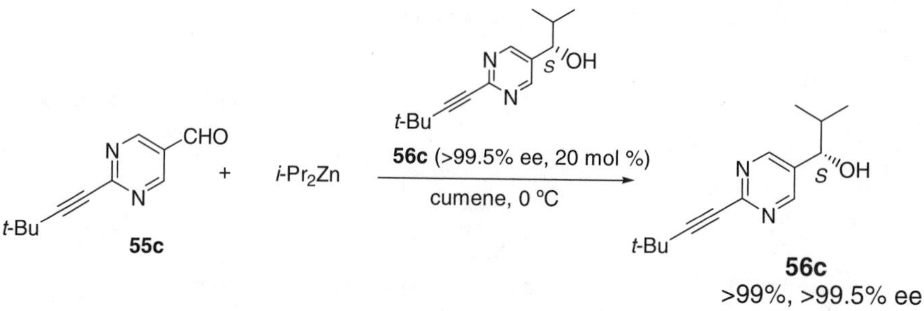

Scheme 9.28.

TABLE 9.2. Practically Perfect Asymmetric Autocatalysis of (S)-56c

Run[a]	Asym. Autocat.	Product		
	% ee	Yield (%)	% ee	Amplified factor[b]
1	>99.5(**56ca**)	>99	>99.5(**56cb**)	6
2	>99.5(**56cb**)	>99	>99.5(**56cc**)	6^2
3	>99.5(**56cc**)	>99	>99.5(**56cd**)	6^3
4	>99.5(**56cd**)	>99	>99.5(**56ce**)	6^4
5	>99.5(**56ce**)	>99	>99.5(**56cf**)	$6^5 \approx 8 \times 10^3$
6	>99.5(**56cf**)	>99	>99.5(**56cg**)	6^6
7	>99.5(**56cg**)	>99	>99.5(**56ch**)	6^7
8	>99.5(**56ch**)	>99	>99.5(**56ci**)	6^8
9	>99.5(**56ci**)	>99	>99.5(**56cj**)	6^9
10	>99.5(**56cj**)	>99	>99.5(**56ck**)	$6^{10} \approx 6 \times 10^7$

[a] Molar ratio. Aldehyde **55c**: i-Pr$_2$Zn (in cumene): catalyst **56c** = 1.0:1.7:0.2. [b] The factor by which the amount of **56c** has multiplied based on the amount of **56ca** which is used as asymmetric autocatalyst in Run 1.

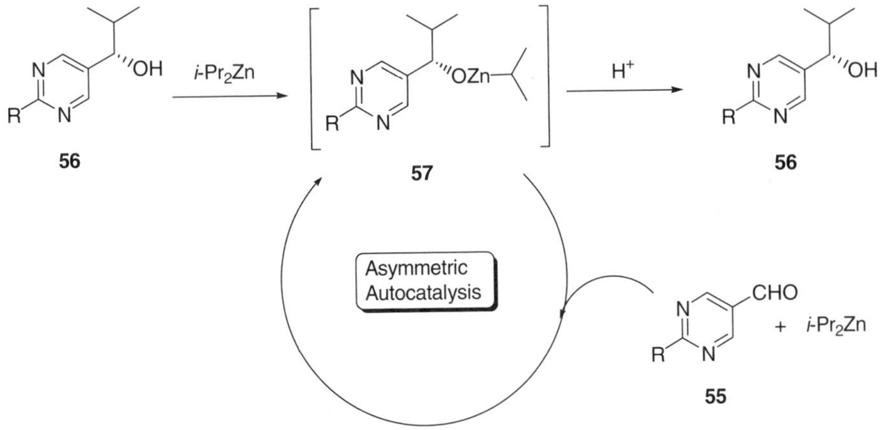

Scheme 9.29.

Chiral pyrimidyl alkanol reacts with i-Pr$_2$Zn to form chiral isopropylzinc alkoxide **57**, which serves as the true asymmetric autocatalyst to multiply itself with the same configuration in the addition reaction of i-Pr$_2$Zn to pyrimidine-5-carbaldehyde **55** (Scheme 9.29).

One of the reasons that the enantioselectivity of 5-pyrimidyl alkanol is higher than that of 3-pyridyl alkanol may be due to the presence of symmetry in the pyrimidine ring. When the bond between the asymmetric carbon and the pyridine ring rotates, conformational isomers of pyridyl alkanol will be formed (Figure 9.2). However, with pyrimidyl alkanol, the rotation does not form conformational isomers. Thus, the reduced number of possible conformational isomers may enable 5-pyrimidyl alkanol to serve as a highly enantioselective asymmetric autocatalyst.

X = H, Zn(i-Pr)

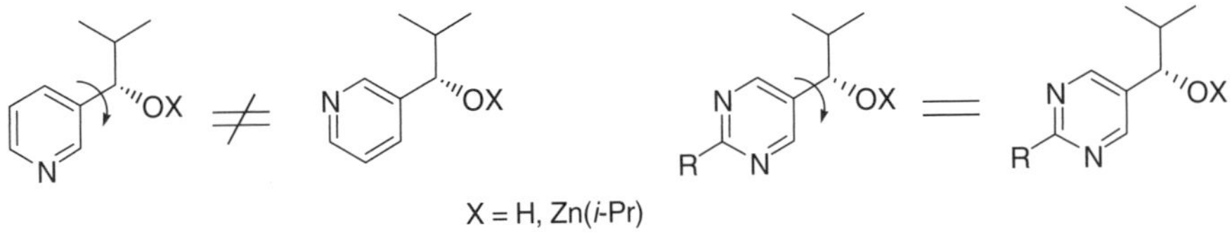

Figure 9.2. Difference between pyridiyl and pyrimidyl alkanols in the rotation of rings.

9.3.5. Asymmetric Autocatalysis of Pyrimidyl Alkanol with Amplification of Enantiopurity

Asymmetric autocatalysis with amplification of enantiopurity involves the catalytic automultiplication of a chiral molecule along with increase in its enantiopurity. This process is a direct method for synthesizing highly enantio-enriched compounds from the same compound with very low enantiopurity and without any other chiral catalysts.

The reaction with chiral pyrimidyl alkanol **56** with low enantiopurity as the asymmetric autocatalyst gives a compound with the same structure and configuration with higher enantiopurity (Scheme 9.30) [56]. Starting from (S)-pyrimidyl alkanol **56a** with only 2% ee, an asym-

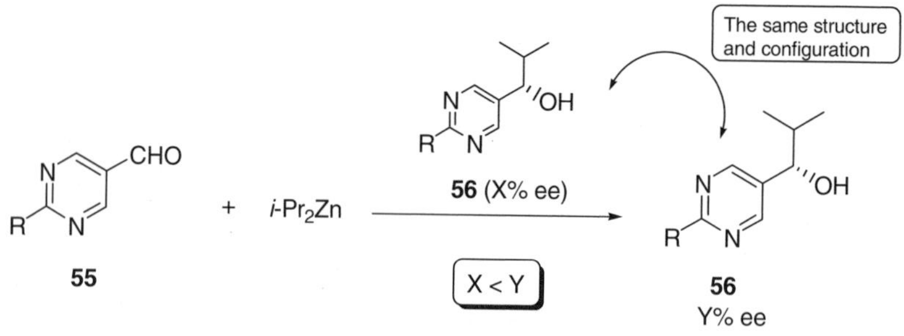

Scheme 9.30.

metric autocatalytic reaction gave (S)-pyrimidyl alkanol **56a** consisting of the initial and the newly formed alkanol with an increased enantiopurity of 10% ee. The reaction was performed successively, with the products of one round serving as the asymmetric autocatalyst for the next, resulting in a successive increase in enantiopurity of the products; that is, 57% ee (after third run), 81% ee (after fourth run) and 88% ee (after fifth run) (Table 9.3). This example of asymmetric autocatalysis with amplification of enantiopurity is the first. As shown in Figure 9.3, the initial (S)-**56a** was multiplied by a factor of 238 during the four successive asymmetric autocatalytic reactions, whereas (R)-**56a** was multiplied only by a factor of 16 (Figure 9.3).

TABLE 9.3. Consecutive Asymmetric Autocatalysis of (S)-56a with Amplification of Enantiopurity

| | Asym. autocat. **56a** | Catalyst and product **56a** | |
Run	% ee	Yield (%)	% ee
1	2	38	10
2	10	63	57
3	57	67	81
4	81	63	88
5	88	66	88

Asymmetric autocatalysis with amplification of enantiopurity differs from all other non-autocatalytic asymmetric amplifications in that even an extremely low enantiopurity can be amplified to a high enantiopurity by successive asymmetric autocatalysis by using the product with a better enantiopurity as the asymmetric autocatalyst for the next round.

Indeed, chiral (S)-pyrimidyl alkanol **56b** with only 0.28% ee was automultiplied by a factor of ca. 100 in quantity (92%) with significant amplification of enantiopurity (87.0% ee) in a one-pot reaction by three portion-wise additions of 2-methylpyrimidine-5-carbaldehyde (**56b**) and i-Pr$_2$Zn (Table 9.4, Run 1) [57]. As shown in Figure 9.4, the initially slightly predominant (S)-**56b** was multiplied by a factor of 189, whereas the initially slightly minor (R)-**56b** was multiplied by a factor of only 13 (Figure 9.4). Even when (R)-**56b** with enantiopurity as low as 0.18% ee was used as the asymmetric autocatalyst, (R)-**56b** with 83.9% ee was obtained in 84% yield through one-pot asymmetric autocatalysis (Run 2).

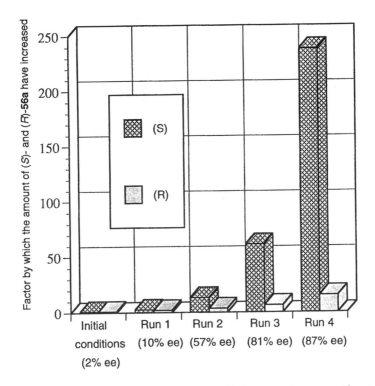

Figure 9.3. The increase in the amounts of (S)- and (R)-**56a** in consecutive asymmetric autocatalytic reaction.

Coordination of the nitrogen atoms of both pyrimidinecarbaldehyde and pyrimidyl alkanol to zinc atom(s) may participate in the transition state of the present asymmetric autocatalysis. Although elucidation of the mechanism of asymmetric autocatalysis with amplification of enantiopurity requires further investigation, one possible mechanism may be the disproportionation of enantiomeric autocatalyst through the formation of an aggregate of (+)- and (−)-enantiomeric autocatalysts, that is, alkylzinc alkoxide of pyrimidyl alkanol. The enantiopurity of the remaining monomeric autocatalyst may become higher than that of the autocatalyst in the aggregate. If the monomeric autocatalyst is more reactive than the aggregate, the enantiopurity of the newly formed product may become higher than that of the initial autocata-

TABLE 9.4. One-pot Asymmetric Autocatalysis of 56b with an Amplification of ee

| | Asym. autocat. **56b** | | Obtained alcohol **56b** | | |
Run	Amount (mg)	% ee	Amount (mg)	Yield (%)	% ee
1	3.2	0.28 (S)	323.5	92	87.0 (S)
2	3.3	0.18 (R)	297.4	84	83.9 (R)

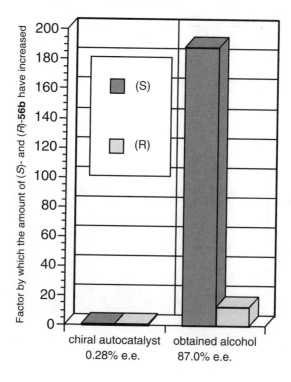

Figure 9.4. Change of the amount of (S)- and (R)-**56b** in one pot reaction.

lyst. Unlike the asymmetric amplification without autocatalysis, the newly formed chiral product with a higher enantiopurity promotes automultiplication as asymmetric autocatalyst. This mechanism may proceed continuously during asymmetric autocatalysis, allowing the initial autocatalyst with very low enantiopurity to automultiply with a significant increase in enantiopurity.

9.3.6. Asymmetric Autocatalysis of Quinolyl Alkanol with Amplification of Enantiopurity

Chiral 3-quinolyl alkanol acts as a highly enantioselective asymmetric autocatalyst. Chiral (S)-**59** with 94% ee catalyzes the enantioselective addition of i-Pr$_2$Zn to quinoline-3-carbaldehyde (**58**) to

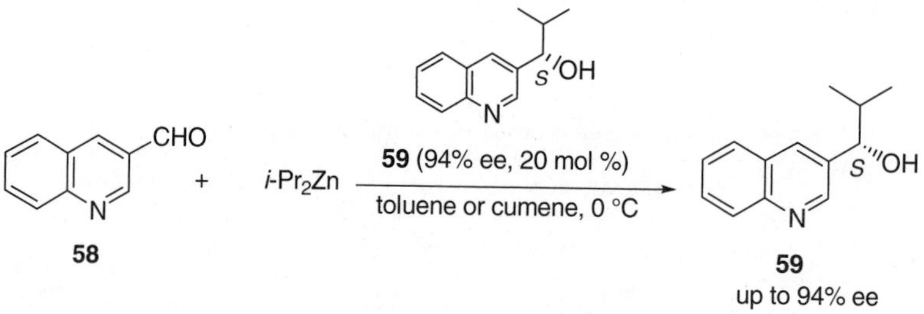

Scheme 9.31.

TABLE 9.5. Consecutive Asymmetric Autocatalysis of Quinolyl Alkanol (R)-59 with Amplification of Enantiopurity

Run	Asym. autocat. **59** % ee	Obtained alcohol **59** Yield (%)	Obtained alcohol **59** % ee
1	8.9	47	43.3
2	43.3	56	67.0
3	67.0	43	81.6
4	81.6	60	85.5
5	85.5	65	86.2
6	86.2	53	88.1

give (S)-**59** with high enantiopurity (up to 94% ee) and the same configuration (Scheme 9.31) [58]. With chiral 3-quinolyl alkanol **59**, asymmetric autocatalysis with amplification of enantiopurity is observed. Thus, the initial (R)-**59** with 8.9% ee is automultiplied to give (R)-**59** with 88.1% ee as a result of the six successive asymmetric autocatalytic reactions (Table 9.5) [59]. During these reactions, the (R)-**59** in the initial compound was multiplied by a factor of ca. 7600, whereas (S)-**59** was multiplied by a factor of only 586. The change in the amount of (R)- and (S)-**59** is shown in Figure 9.5.

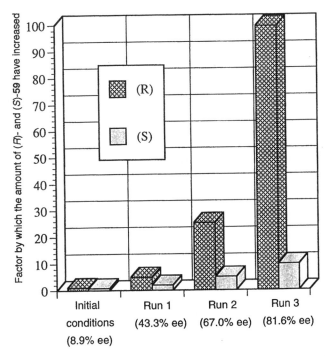

Figure 9.5. The increase in the amounts of (R)- and (S)-**59** in consecutive asymmetric autocatalytic reaction.

9.3.7. Asymmetric Autocatalysis of 5-Carbamoyl-3-pyridyl Alkanol

The introduction of a carbamoyl group to the 5-position of chiral 3-pyridyl alkanol enhances its efficacy as asymmetric autocatalyst. (S)-5-Carbamoyl-3-pyridyl alkanols **61** are automultiplied with up to 86% ee in the enantioselective addition of i-Pr$_2$Zn to 5-carbamoyl-3-pyridine-carbaldehyde **60** (Scheme 9.32) [60]. The enantioselectivity depends on the structure of the substituent on the nitrogen atom of the amide. A bulky i-Pr substituent is efficient for achieving high enantioselectivity. The amplification of the enantiopurity of **61** to a certain degree is also observed [61].

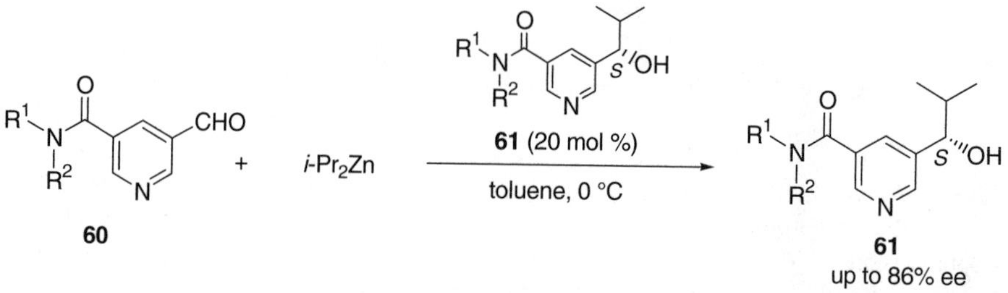

Scheme 9.32.

9.4. CONCLUSION

Due to the intensive studies by many groups, the number of examples of asymmetric amplification has been substantially increasing. Aggregation state of the enantiomers of a chiral catalyst can be estimated based on the observation of a nonlinear effect between the enantiopurity of the chiral catalyst and that of the product.

Asymmetric autocatalysis of chiral pyridyl, pyrimidyl and quinolyl alkanols in the enantioselective addition of i-Pr$_2$Zn to the corresponding aldehydes gives chiral pyridyl, pyrimidyl and quinolyl alkanols with the same structure and configuration as those of the respective asymmetric autocatalysts. Amplification of the enantioselectivity is observed in the consecutive and one-pot asymmetric autocatalysis of pyrimidyl and quinolyl alkanols. Starting from trace amounts of pyrimidyl and quinolyl alkanols with very low enantiopurity, increased amounts of chiral pyrimidyl and quinolyl alkanols with very high enantiopurity can be formed without the assistance of any other chiral molecules. Thus, asymmetric autocatalysis with amplification of enantioselectivity is one of the most straightforward methods for the multiplication of chiral molecules leading to high enantio-enrichment. Moreover, asymmetric autocatalysis is characterized by self-replication (automultiplication), which is a major feature of life [62].

Future research will reveal more examples and find many applications of asymmetric amplification [63] and asymmetric autocatalysis [64].

ACKNOWLEDGMENT

The authors are grateful to their enthusiastic co-workers whose names appear in their papers for the work on asymmetric autocatalysis. Financial support from the Ministry of Education,

Science, Sports, and Culture of Japan; New Energy and Industrial Technology Development Organization (NEDO); Novartis (Ciba-Geigy) Foundation (Japan) for the Promotion of Science; the Daicel Award in Synthetic Organic Chemistry, Japan; Kurata Foundation; and SUT (Special) Grant for the Promotion of Research is gratefully acknowledged.

REFERENCES

1. Reviews including asymmetric amplification: (a) Ojima, I.; Clos, N.; Bastos, C. *Tetrahedron* **1989**, *45*, 6901. (b) Brown, J. M.; Davies, S. G. *Nature* **1989**, *342*, 631. (c) Kagan, H. B.; Girard, C.; Guillaneux, D.; Rainford, D.; Samuel, O.; Zhang, S. Y.; Zhao, S. H. *Acta Chem. Scand.* **1996**, *50*, 345. (d) Avalos, M.; Babiano, R.; Cintas, P.; Jiménez, J. L.; Palacios, J. C. *Tetrahedron: Asymmetry* **1997**, *8*, 2997.

2. Short reviews: (a) Soai, K.; Shibata, T. *Yuki Gosei Kagaku Kyokaishi (J. Synth. Org. Chem. Jpn.)* **1997**, *55*, 994. (b) Bolm, C.; Bienewald, F.; Seger, A. *Angew. Chem., Int. Ed.* **1996**, *35*, 1657.

3. Horeau, A.; Guetté, J. P. *Tetrahedron* **1974**, *30*, 1923.

4. Wynberg, H.; Feringa, B. *Tetrahedron* **1976**, *32*, 2831.

5. Puchot, C.; Samuel, O.; Duñach, E.; Zhao, S.; Agami, C.; Kagan, H. B. *J. Am. Chem. Soc.* **1986**, *108*, 2353.

6. Finn, M. G.; Sharpless, K. B. *J. Am. Chem. Soc.* **1991**, *113*, 113.

7. Guillaneux, D.; Zhao, S.-H.; Samuel, O.; Rainford, D.; Kagan, H. B. *J. Am. Chem. Soc.* **1994**, *116*, 9430.

8. Mukaiyama, T.; Soai, K.; Sato, T.; Shimizu, H.; Suzuki, K. *J. Am. Chem. Soc.* **1979**, *101*, 1455.

9. Reviews, see: (a) Soai, K.; Niwa, S. *Chem. Rev.* **1992**, *92*, 833. (b) Soai, K.; Shibata, T. In *Comprehensive Asymmetric Catalysis*; Jacobsen, E. N.; Pfaltz, A.; Yamamoto, H. (Ed.); Springer: New York, 1999; Ch. 28.1. (c) Noyori, R.; Kitamura, M. *Angew. Chem., Int. Ed.* **1991**, *30*, 49.

10. Oguni, N.; Matsuda, Y.; Kaneko, T. *J. Am. Chem. Soc.* **1988**, *110*, 7877.

11. Kitamura, M.; Okada, S.; Suga, S.; Noyori, R. *J. Am. Chem. Soc.* **1989**, *111*, 4028.

12. (a) Kitamura, M.; Suga, S.; Niwa, M.; Noyori, R.; Zhai, Z.-X.; Suga, H. *J. Phys. Chem.* **1994**, *98*, 12776. (b) Yamakawa, M.; Noyori, R. *J. Am. Chem. Soc.* **1995**, *117*, 6327.

13. Dosa, P. I.; Fu, G. C. *J. Am. Chem. Soc.* **1998**, *120*, 445.

14. Bolm, C.; Schlingloff, G.; Harms, K. *Chem. Ber.* **1992**, *125*, 1191.

15. Bolm, C.; Müller, J.; Schlingloff, G.; Zehnder, M.; Neuburger, M. *J. Chem. Soc., Chem. Commun.* **1993**, 182.

16. (a) Tanaka, K.; Matsui, J.; Kawabata, Y.; Suzuki, H.; Watanabe, A. *J. Chem. Soc., Chem. Commun.* **1991**, 1632. (b) Tanaka, K.; Matsui, J.; Suzuki, H. *J. Chem. Soc., Perkin Trans. 1* **1993**, 153.

17. Rossiter, B. E.; Eguchi, M.; Hernández, A. E.; Vickers, D.; Medich, J.; Marr, J.; Heinis, D. *Tetrahedron Lett.* **1991**, *32*, 3973.

18. Zhou, Q.-L.; Pfaltz, A. *Tetrahedron* **1994**, *50*, 4467.

19. Soai, K.; Hayasaka, T.; Ugajin, S. *J. Chem. Soc., Chem. Commun.* **1989**, 516.

20. Bolm, C. *Tetrahedron: Asymmetry* **1991**, *2*, 701.

21. de Vries, A. H. M.; Jansen, J. F. G. A.; Feringa, B. L. *Tetrahedron* **1994**, *50*, 4479.

22. Mikami, K.; Terada, M.; Nakai, T. *J. Am. Chem. Soc.* **1990**, *112*, 3949.

23. Mikami, K.; Terada, M. *Tetrahedron* **1992**, *48*, 5671.

24. Keck, G. E.; Krishnamurthy, D.; Grier, M. C. *J. Org. Chem.* **1993**, *58*, 6543.

25. Bedeschi, P.; Casolari, S.; Costa, A. L.; Tagliavini, E.; Umani-Ronchi, A. *Tetrahedron Lett.* **1995**, *36*, 7897.

26. Sasai, H.; Suzuki, T.; Itoh, N.; Shibasaki, M. *Tetrahedron Lett.* **1993**, *34*, 851.

27. Narasaka, K.; Iwasawa, N.; Inoue, M.; Yamada, T.; Nakashima, M.; Sugimori, J. *J. Am. Chem. Soc.* **1989**, *111*, 5340.

28. Iwasawa, N.; Hayashi, Y.; Sakurai, H.; Narasaka, K. *Chem. Lett.* **1989**, 1581.

29. Mikami, K.; Motoyama, Y.; Terada, M. *J. Am. Chem. Soc.* **1994**, *116*, 2812.

30. Kobayashi, S.; Ishitani, H. *J. Am. Chem. Soc.* **1994**, *116*, 4083.

31. Kobayashi, S.; Ishitani, H.; Araki, M.; Hachiya, I. *Tetrahedron Lett.* **1994**, *35*, 6325.

32. Zondervan, C.; Feringa, B. L. *Tetrahedron: Asymmetry* **1996**, *7*, 1895.

33. Komatsu, N.; Hashizume, M.; Sugita, T.; Uemura, S. *J. Org. Chem.* **1993**, *58*, 4529.

34. Evans, D. A.; Nelson, S. G.; Gagné, M. R.; Muci, A. R. *J. Am. Chem. Soc.* **1993**, *115*, 9800.

35. (a) Alcock, N. W.; Brown, J. M.; Maddox, P. J. *J. Chem. Soc., Chem. Commun.* **1986**, 1532. (b) Brown, J. M.; Maddox, P. J. *Chirality* **1991**, *3*, 345.

36. Maruoka, K.; Yamamoto, H. *J. Am. Chem. Soc.* **1989**, *111*, 789.

37. Faller, J. W.; Parr, J. *J. Am. Chem. Soc.* **1993**, *115*, 804.

38. Faller, J. W.; Tokunaga, M. *Tetrahedron Lett.* **1993**, *34*, 7359.

39. Mikami, K.; Matsukawa, S. *Nature* **1997**, *385*, 613.

40. Ohkuma, T.; Doucet, H.; Pham, T.; Mikami, K.; Korenaga, T.; Terada, M.; Noyori, R. *J. Am. Chem. Soc.* **1998**, *120*, 1086.

41. Frank, F. C. *Biochim. Biophys.* **1953**, *11*, 459.

42. (a) M. Calvin, *Chemical Evolution*; Clarendon: London, 1969; Ch.7. (b) Wynberg, H. *J. Macromol. Sci. Chem. A* **1989**, *26*, 1033.

43. Soai, K.; Niwa, S.; Hori, H. *J. Chem. Soc., Chem. Commun.* **1990**, 982.

44. Seebach, D.; Amstutz, R.; Dunitz, J. D. *Helv. Chim. Acta* **1981**, *64*, 2622.

45. Alberts, A. H.; Wynberg, H. *J. Am. Chem. Soc.* **1989**, *111*, 7265.

46. For related reactions, (a) Soai, K.; Inoue, Y.; Takahashi, T.; Shibata, T. *Tetrahedron* **1996**, *52*, 13355. (b) Shibata, T.; Takahashi, T.; Konishi, T.; Soai, K. *Angew. Chem., Int. Ed.* **1997**, *36*, 2458.

47. Danda, H.; Nishikawa, H.; Otaka, K. *J. Org. Chem.* **1991**, *56*, 6740.

48. (a) Soai, K.; Ookawa, A.; Ogawa, K.; Kaba, T. *J. Chem. Soc., Chem. Commun.* **1987**, 467. (b) Soai, K.; Ookawa, A.; Kaba, T.; Ogawa, K. *J. Am. Chem. Soc.* **1987**, *109*, 7111.

49. (a) Soai, K.; Yokoyama, S.; Ebihara, K.; Hayasaka, T. *J. Chem. Soc., Chem. Commun.* **1987**, 1690. (b) Watanabe, M.; Soai, K. *J. Chem. Soc., Perkin Trans. 1* **1994**, 3125. (c) Soai, K.; Shimada, C.; Takeuchi, M.; Itabashi, M. *J. Chem. Soc., Chem. Commun.* **1994**, 567.

50. (a) Soai, K.; Yokoyama, S.; Hayasaka, T. *J. Org. Chem.* **1991**, *56*, 4264. (b) Soai, K.; Hayase, T.; Takai, K.; Sugiyama, T. *J. Org. Chem.* **1994**, *59*, 7908.

51. Soai, K.; Hori, S.; Niwa, S. *Heterocycles* **1989**, *29*, 2965.

52. Soai, K.; Hayase, T.; Shimada, C.; Isobe, K. *Tetrahedron: Asymmetry* **1994**, *5*, 789.

53. Soai, K.; Hayase, T.; Takai, K. *Tetrahedron: Asymmetry* **1995**, *6*, 637.

54. Shibata, T.; Morioka, H.; Hayase, T.; Choji, K.; Soai, K. *J. Am. Chem. Soc.* **1996**, *118*, 471.

55. Shibata, T.; Yonekubo, S.; Soai, K. *Angew. Chem., Int. Ed.* **1999**, *38*, 659.

56. Soai, K.; Shibata, T.; Morioka, H.; Choji, K. *Nature* **1995**, *378*, 767.

57. Shibata, T.; Hayase, T.; Yamamoto, J.; Soai, K. *Tetrahedron: Asymmetry* **1997**, *8*, 1717.

58. Shibata, T.; Choji, K.; Morioka, H.; Hayase, T.; Soai, K. *Chem. Commun.* **1996**, 751.

59. Shibata, T.; Choji, K.; Hayase, T.; Aizu, Y.; Soai, K. *Chem. Commun.* **1996**, 1235.

60. Shibata, T.; Morioka, H.; Tanji, S.; Hayase, T.; Kodaka, Y.; Soai, K. *Tetrahedron Lett.* **1996**, *37*, 8783.

61. Tanji, S.; Kodaka, Y.; Aoyama, A.; Shibata, T.; Soai, K. The 74th National Meeting of the Chemical Society of Japan, Tokyo, March, **1998**, 1D605.

62. Breslow, R. In *Chemistry Today and Tomorrow. The Central, Useful, and Creative Science*; Jones and Bartlett Publishers, Inc.: Sudbury, 1997; Chap. 6.

63. Zhang, S. Y.; Girard, C.; Kagan, H. B. *Tetrahedron: Asymmetry* **1995**, *6*, 2637.

64. Shibata, T.; Yamamoto, J.; Matsumoto, N.; Yonekubo, S.; Osanai, S.; Soai, K. *J. Am. Chem. Soc.* **1998**, *120*, 12157.

10

ASYMMETRIC PHASE-TRANSFER REACTIONS

Martin J. O'Donnell
Department of Chemistry, Indiana University-Purdue University at Indianapolis, Indianapolis, IN 46205, U.S.A.

10.1. INTRODUCTION

Phase-transfer catalysis (PTC) is a reaction method that typically involves a simple reaction procedure, mild conditions, inexpensive and safe reagents and solvents, and the ability to easily scale-up the reaction [1–4]. The use of PTC for the preparation of chiral, non-racemic compounds from prochiral substrates using chiral catalysts is becoming an important area in catalysis [5], as is demonstrated by a number of notable recent successes in the field. The purpose of this chapter is to review asymmetric phase-transfer reactions, especially those that give promising levels of induction, and give perspective to the field as a whole. Coverage has been limited to chiral quaternary ammonium compounds, crown ethers [6], and associated species as catalysts. In general, polymer-bound catalysts [7] and supramolecular systems [8] will not be discussed. Related areas involving free-alkaloid catalysis, asymmetric polymerizations, chiral stationary phases, resolutions, stoichiometric enzyme model studies, and micellar catalysis will also not be covered.

Reactions accomplished by enantioselective phase-transfer catalysis are summarized in Table 10.1 according to type of catalyst and synthetic transformation [9–81]. Highest reported enantioselectivities (% ee) or optical purities (% op) are listed to give perspective to the overall field [82]. General aspects of phase-transfer systems, including catalysts are then discussed, followed by particular reaction classes.

10.2. THE PHASE-TRANSFER SYSTEM

10.2.1. Mechanism

A mechanistic scheme (Scheme 10.1) for the monoalkylation of active methylene compounds [5p,12] is presented to illustrate the variables common to many of the systems that have been

Catalytic Asymmetric Synthesis, Second Edition, Edited by Iwao Ojima
ISBN 0-471-29805-0 Copyright © 2000 Wiley-VCH, Inc.

TABLE 10.1. Enantioselective PTC Reactions^{a,b}

PTC Reaction	Cinchona Quat	Ephedra Quat	Crown Ether	Other
C–C bond formation				
Alkylation	99.5% ee [14]	7% op [11e]	.	82% ee [23]
1,2-Carbonyl addition	72% ee [32a]	74% ee [30]	40% ee [33a]	
Michael addition	99% ee [21]	82% ee [44a]	99% ee [35]	0% [47g]
Cyclopropanation		1% op ?? [49a]		>2% ee ?? [49a]
Horner–Wadsworth–Emmons	57% ee [51]			
Darzens	81% ee [53]		64% ee [38b]	
Reduction				
Ketone	71% ee [55]	53% op [56]	8% ee [60g]	1% ee [11i]
Imine	11% op [57a]	22% op [57b]		
Other	25% ee [61]			
Oxidation				
Epoxidation	89% ee [65,68]	37% ee [67a]	12% ee [48b]	
α-Hydroxy ketone formation	79% ee [73]	7% ee [73]	72% ee [74]	8% ee [73]
Other	8% ee [76]	4% ee [76]		"Low" [75]
C–X bond formation				
C–N	47% ee [77c]	3% op [60d]		
Aziridination	45% ee [79]			
C–O	3% op [11e]	0.03% op [11e]		48% ee ?? [80a]
C–S	36% ee [47f]	3% op [47f]		
C–X	"Low" [33b]	"Low" [33b]		
Deracemization			40% ee [39]	

^{a}Enantiomeric excess (% ee) is the percentage of the major enantiomer minus that of the min or enantiomer; optical purity (% op) is the ratio, in percent, of the optical rotation of a mixture of en antiomers to that of the pure enantiomer.

^{b}Disputed reactions indicated by ?? (for others, see references 26a, 27a, 28a, 29a, 62a, and 88a).

studied and to alert the reader to key problems that need to be addressed in such reactions. Three main steps are required in this process [2f]: (1) deprotonation of the active methylene compound with base, which generally occurs at the interface between the two layers (liquid-liquid (L/L) or solid-liquid (S/L) PTC); (2) ion-exchange of the anion (A^-) with the cation of the chiral quaternary ammonium compound (quat) to form a lipophilic ion-pair (**D**), which then either reacts from the interface (step 3) or is extracted into the bulk organic phase; and (3) creation of the new chiral center in product **P*** by alkylation of the ion-pair (**D**) with concomitant regeneration of the catalyst.

A number of undesirable processes can occur in competition with the formation of the optically active product: (1) alkylation of the "wrong" ion-pair leading to the enantiomer of the desired product (step c); (2) side-reactions of either the starting substrate [e.g., ester saponifi-

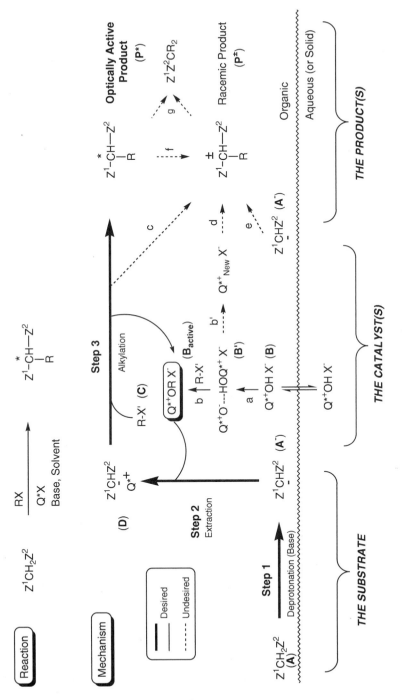

Scheme 10.1.

cation or imine hydrolysis of starting materials such as **35** (see Scheme 10.4)] or the reaction product [racemization (step f) and/or dialkylation (step g) following product formation as well as the hydrolyses reactions mentioned above for the starting material]; (3) interfacial alkylation (step e) of substrate anion (**A⁻**) in the absence of the chiral quat cation, which necessarily yields racemic product; and (4) reaction of the chiral quat (**B**) to form a new organic compound, which could function either as the reactive catalyst species (step b) or as a compound (step b′), which is either an ineffective catalyst or one that leads to racemic product. Recent studies of chiral PTC alkylations, which are detailed below, have shown that the active catalyst (**B**$_{active}$) is the O-alkylated derivative formed by in situ O-alkylation of the alkoxide from the N-alkyl quaternary salt.

Key variables for controlling these processes include solvent; temperature; concentration of the various reactants; stirring rate; type of PTC process (L/L or S/L); structures and nature of various protecting groups in the nucleophilic and electrophilic reactants; structure of the catalyst and, finally, identity of counterions in the catalyst, alkyl halide, and base.

10.2.2. Chiral Nonracemic Catalysts

Catalysts (**3** and **6**) derived from the *cinchona* alkaloids (Chart 10.1) [83] have been utilized extensively in chiral PTC because the parent alkaloids (**1–4**) are inexpensive, readily available in both pseudoenantiomeric forms [84], and can be easily quaternarized to a variety of different salts.

Typically one catalyst of the pair provides one enantiomer in excess, whereas the other catalyst gives the opposite enantiomer as the major product. Thus, by a simple choice of catalyst, it is possible to prepare either of the two enantiomeric products at will. Overall the *cinchona*

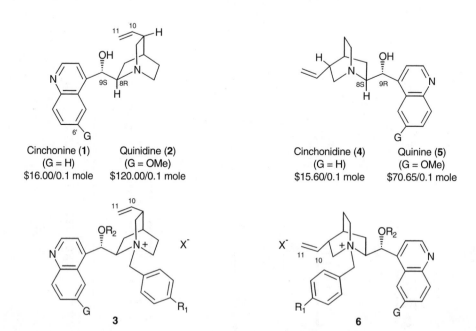

Cinchonine (**1**)   Quinidine (**2**)   Cinchonidine (**4**)   Quinine (**5**)
(G = H)   (G = OMe)   (G = H)   (G = OMe)
$16.00/0.1 mole $120.00/0.1 mole $15.60/0.1 mole $70.65/0.1 mole

Chart 10.1.

TABLE 10.2. Reactions with "First- and Second-Generation" *Cinchona*-**Derived Catalysts**

Entry	Q*X 3/6	X	G	R₁	R₂	Reaction	ee	Ref.
1	3a	Br	H	4-CF$_3$	H	Alkylation	92%	[9a]
2	6a	Br	H	H	Benzyl	Alkylation	81%	[5p][12c]
3	3b	Cl	H	3,4-Cl$_2$	H	Alkylation	78%	[10]
4	3c	Cl	H	H	Allyl	Alkylation	75%	[5p]
5	6b	Cl	H	H	H	Alkylation	64%	[12a]
6	3d	F	H	H	H	Aldol	72%	[32a]
7	3e	F	H	4-CF$_3$	H	Aldol	51%	[31]
8	6c	Br	H	4-CF$_3$	H	Michael	91%	[37]
9	6d	Br	H	4-CF$_3$	H	Michael	87%	[11d]
10	3a	Br	H	4-CF$_3$	H	Michael	80%	[36]
11	3a	Br	H	4-CF$_3$	H	Michael	70%	[43]
12	3f	Cl	H	H	H	HWE[b]	57%	[51]
13	6e	Cl	OMe	4-CF$_3$	H	Darzens	81%	[53]
14	3a	Br	H	4-CF$_3$	H	Darzens	69%	[52]
15	6f	F	OMe	H	H	Reduction	71%	[55]
16	6g	Cl	H	H	H	Epoxidation	89%	[68]
17	6h	Cl	OMe	H	H	Epoxidation	78%	[69]
18	6i	Cl	OMe	H	H	Epoxidation	90%	[72]
19	6j	Cl	OMe	H	H	Epoxidation	54%	[47b]
20	3a	Br	H	4-CF$_3$	H	α-Hydroxylation	79%	[73]
21	3f	Cl	H	H	H	C–N Bond	47%	[77c]
22	3a	Br	H	4-CF$_3$	H	Aziridination	45%	[79]

[a]10,11-Dihydro. [b]Horner–Wadsworth–Emmons.

TABLE 10.3. Reactions with "Third-Generation" *Cinchona*-Derived Catalysts

Entry	Q*X 7/8	X	R$_2$	Reaction	ee	Ref.
1	**8a**	Br	Allyl	Alkylation	99.5%	[14]
2	**7a**	Br	Allyl	Alkylation	87%	[15]
3	**8b**[a]	Cl	H	Alkylation	91%	[13]
4	**7b**	Cl	H	Alkylation	89%	[13]
5	**8a**	Br	Allyl	Michael	99%	[21]
6	**8c**[a]	OH	Bn	Epoxidation	89%	[65]
7	**7c**	OH	Bn	Epoxidation	86%	[65]

[a]10,11-Dihydro.

quats have given more impressive enantioselectivities for a range of reactions than other classes of phase-transfer catalysts (see Tables 2 and 3 and Charts 2–4, below).

Crown ethers have given impressive enantioselectivites in Michael additions (Chart 10.2). Purely synthetic chiral crowns are of limited use on large scale based on cost; although, in general, the crowns are less susceptible to catalyst degradation and, therefore, have higher catalyst turnover numbers than the chiral quaternary ammonium salts. Of interest are crowns with symmetry, aza-crowns, and those with sugars or other chiral-pool units as sources of chirality (Charts 2 and 4).

A number of other types of compounds have been used as chiral catalysts in phase-transfer reactions. Many of these compounds embody the key structural component, a β-hydroxyammonium salt-type structure, which has been shown to be crucial to the success of the above described *cinchona*-derived quats. Although they have not been as successful as the *cinchona* catalysts, the *ephedra*-alkaloid derived catalysts (see **20, 22, 23** and **25** in Charts 3 and 4) have been used effectively in several reactions. In general, quats with chirality derived only from a single chiral center, which cannot participate in a multipoint interaction with other reaction species, have not been effective catalysts [80].

Early work from the McIntosh group [11h,85] and extensive research from the Dehmlow group [24e–i,48b] concerning chiral catalyst design is noted. Recently, Lygo and co-workers have reported short enantio- and diastereoselective syntheses of the four stereoisomers of 2-(phenylhydroxymethyl)quinuclidine. The authors report that these compounds, which contain the basic core structure of the *cinchona* alkaloids, will be examined as possible chiral control elements in a variety of asymmetric transformations [86].

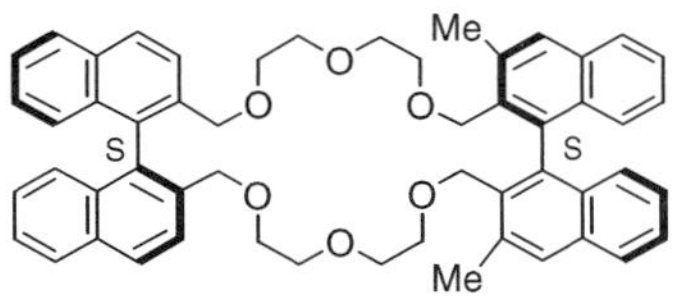

11 (Michael, 90% ee)
[38a]

14 (Michael, 79% ee)
[41a]

17 (Michael, 70% ee)
[42a]

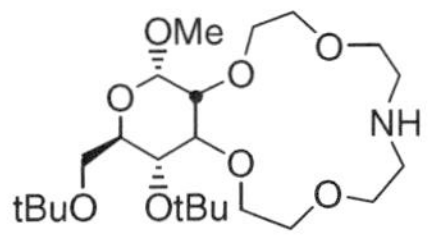

10 (Michael, 83% ee)
[35]

13 (Michael, 83% ee)
[40]

16 (Michael, 71% ee)
[41b]

9 (Michael, 99% ee)
[35]-Scheme 7

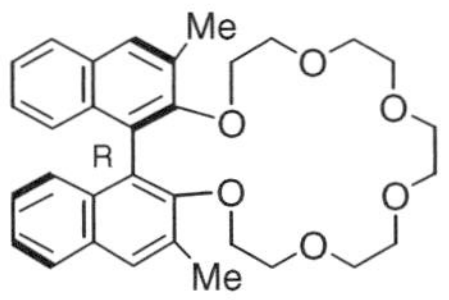

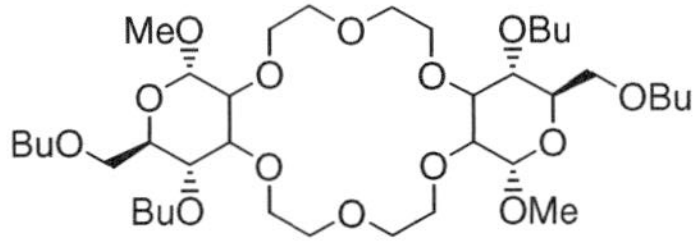

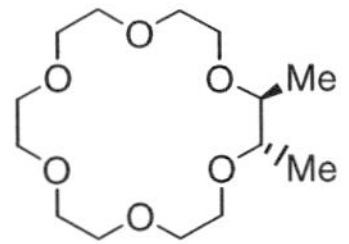

12 (Michael, 84% ee)
(Deracemization, 40% ee)
[39]

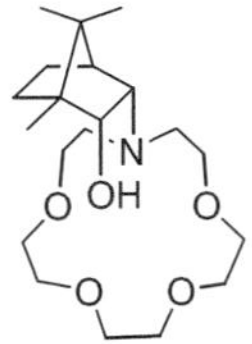

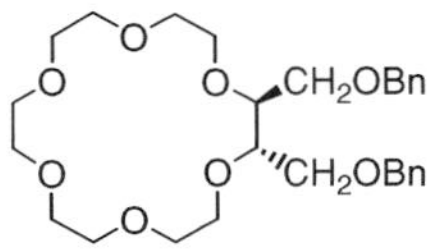

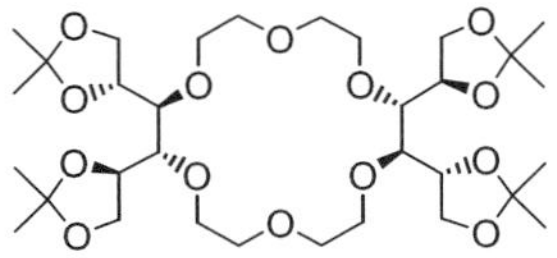

15 (Michael, 71% ee)
[24e]

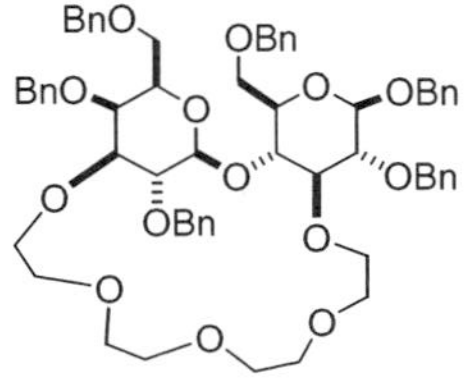

Chart 10.2.

733

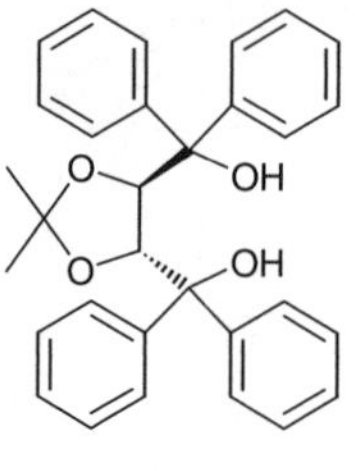

18 (Alkylation, 82% ee)
[23]
(Michael, 45% ee)
[45a]

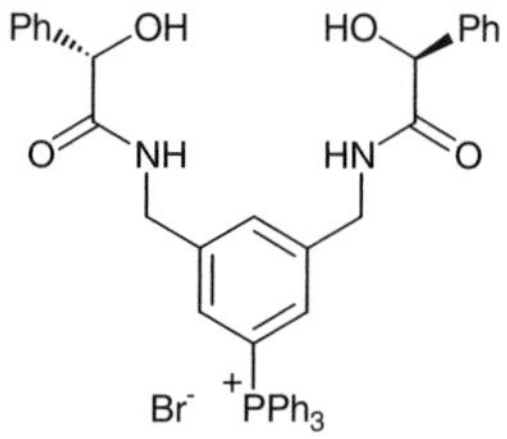

19 (Alkylation, 50% ee)
[11c]

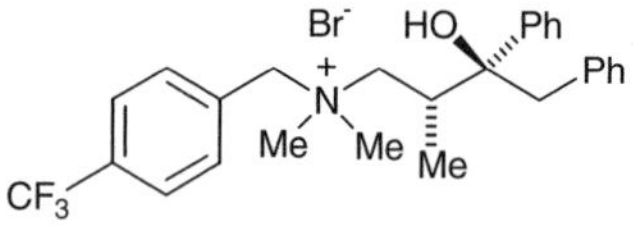

20 (Alkylation, 49% ee)
[24h]

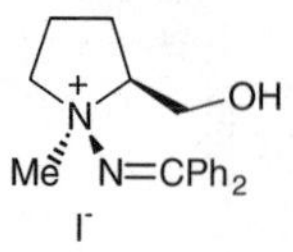

21 (Alkylation, 94% ee)
[29, Disputed]

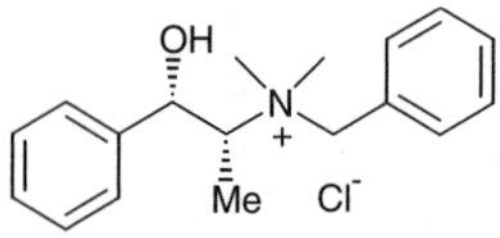

22 (Aldol, 74% ee)
[30]
(Michael, 68% ee)
[44b]

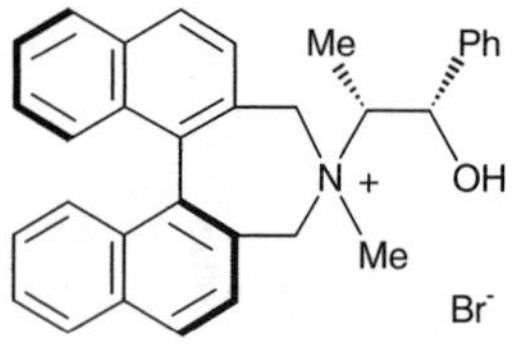

23 (Michael, 82% ee)
[44a]

Chart 10.3.

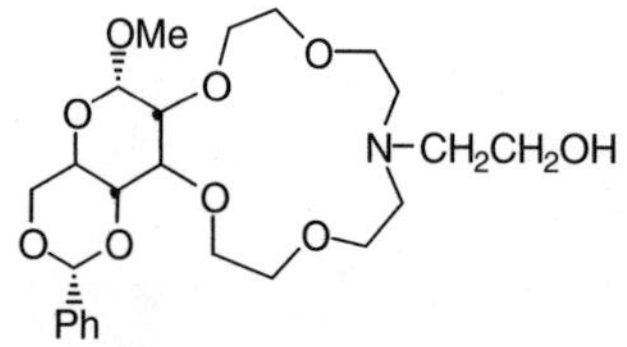

24 (Darzens, 64% ee)
[38b]

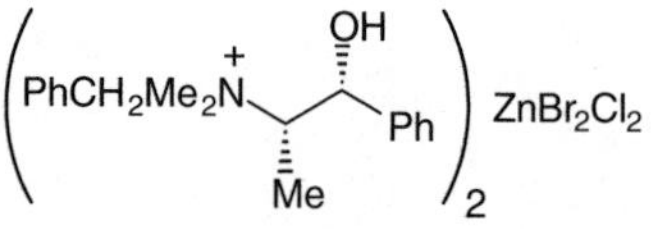

25 (Reduction, 53% op)
[56]

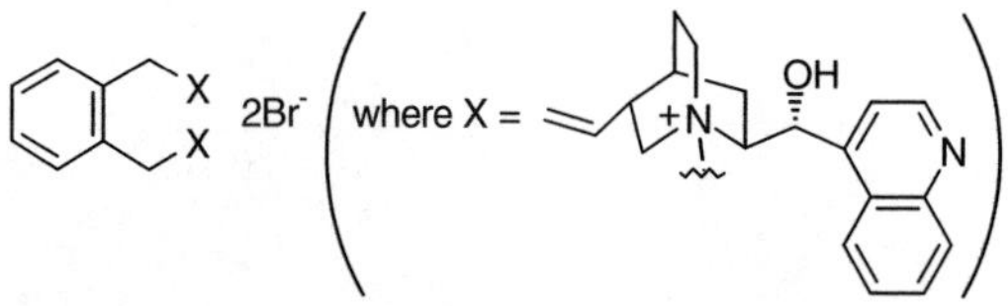

26 (Epoxidation, 63% ee)
[71a]

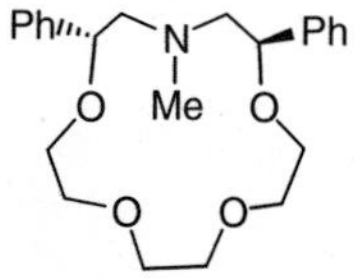

27 (α-Hydroxylation, 72% ee)
[74]

Chart 10.4.

10.2.3. Catalyst Decomposition and Product Analysis

β-Hydroxyammonium salts can react under the strongly basic reaction conditions present in many phase-transfer reactions and the newly formed products could, in principle, serve either as effective or ineffective catalysts (Scheme 10.1) [9c]. The development of a new class of chiral phase-transfer catalysts, the N-alkyl-O-alkyl *cinchona* quats (B_{active} in Scheme 10.1 and **30** in Scheme 10.2), resulted from detailed mechanistic studies of these systems [5p,12]. These catalysts are formed by in situ deprotonation of **28** to the alkoxide **29** followed by alkylation to form the active catalyst **30**. Such catalysts offer an important second site of variation (R_2 in **30**) for catalyst development, which has been rapidly utilized for the preparation of more effective catalysts.

Two routes of catalyst decomposition are also possible from alkoxide **29**, fragmentation to form an epoxide or *O*-alkylation and subsequent fragmentation to an enol ether. Both of these tertiary amines can then be N-alkylated to form new chiral, non-racemic quat salts. The quaternary ammonium catalyst can also be dequaternarized by nucleophiles to a tertiary amine, which can then undergo subsequent reactions [9c,11i,26b,87].

The optically active catalyst decomposition products can be carried through the reaction workup and contaminate the desired product, leading to false optical rotations. Numerous high inductions have been reported [26–29,49,62,80,88], which, on careful examination of the reaction products, have been questioned. Because of this problem, methods for direct analysis of product mixtures such as chiral HPLC or NMR with chiral shift reagents are preferable to optical rotations of products.

Other Studies. Various studies concerning the mechanism of the phase-transfer process have appeared [1a–c,1h–l,2a,2c,2f,2g,2k,2m,2r,5c–e,5m,5p,5r,9c,12b,12c,60e,89]. Catalyst [11i,12b,44a,90] and structural studies by crystal structure [9a,12b,14,85d,91], NMR [85c,85e,91b,91c,92], and calculations [39,40,91] are noted.

10.2.4. Isolation of Pure Enantiomers

Unless asymmetric induction is complete, it is necessary to remove the undesired enantiomer from the product mixture. Whereas in conventional diastereoselective asymmetric syntheses this removal can typically be readily accomplished by crystallization or chromatography, the separation of enantiomeric products can be problematic. Often, though, with enantio-enriched samples it is possible to recrystallize either the racemate from the pure enantiomer or, preferably, one enantiomer from the other [12a,16,17]. Another very effective method to produce enantiopure compounds is by enzymatic resolution of the enantio-enriched product from chiral PTC [16,18]. These methods are illustrated by examples in the alkylation section of this chapter (Chart 10.6).

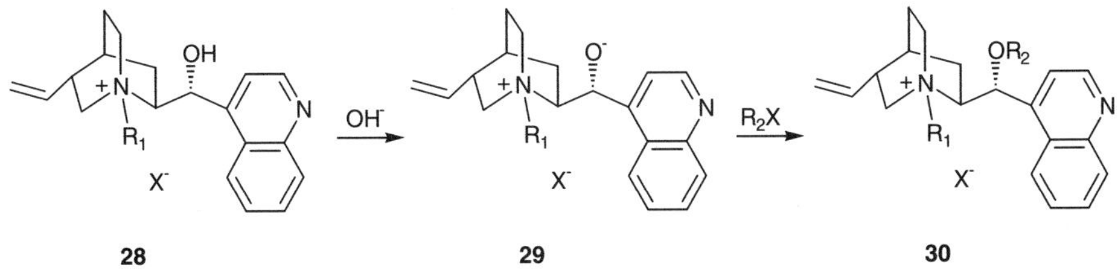

Scheme 10.2.

10.3. PHASE-TRANSFER REACTIONS

10.3.1. Carbon–Carbon Bond Formation

10.3.1.1. Alkylation Alkylation of the phenylindanone **31** with catalyst **3a** by the Merck group demonstrates the reward that can accompany a careful and systematic study of a particular phase-transfer reaction (Scheme 10.3) [5d,5f,9,36]. The numerous reaction variables were optimized and the kinetics and mechanism of the reaction were studied in detail. It has been proposed that the chiral induction step involves an ion-pair in which the enolate anion fits on top of the catalyst and is positioned by electrostatic and hydrogen-bonding effects as well as π–π stacking interactions between the aromatic rings in the catalyst and the enolate. The electrophile then preferentially approaches the ion-pair from the top (front) face, because the catalyst effectively shields the bottom-face approach. A crystal structure of the catalyst as well as calculations of the catalyst-enolate complex support this interpretation [9a,91]. Alkylations of related active methine compounds, such as **33** to **34** (Scheme 10.3), have also appeared [10,11].

α-Amino acids can be prepared by alkylation of the Schiff base ester **35** using chiral PTC with *cinchona* quats (Scheme 10.4) [5p,12–15]. The situation here is somewhat more complex than that described above, because the acyclic enolate can exist in either *E*- or *Z*-isomeric forms and also, the monoalkylated product is susceptible to deprotonation followed by either reprotonation (racemization) or a second alkylation (dialkylation). Early "first-generation catalysts" (**6b**) with optimized substrate **35** gave moderate enantioselectivities (≤75% chemical yield, ≤64% ee) with 50% aqueous NaOH and CH₂Cl₂ at room temperature for 9 h. A single recrystallization of the crude reaction product (64% ee) to remove racemic crystals, followed by deprotection, gave the product α-amino acid in high enantiopurity on a multigram scale (50% overall yield from **35**, >99% ee) [12a].

Subsequent generations of catalyst (Scheme 10.4 and Tables 2 and 3) have led to increased product enantioselectivity. In the "second generation," the *N*-benzyl-*O*-allyl cinchonidinium

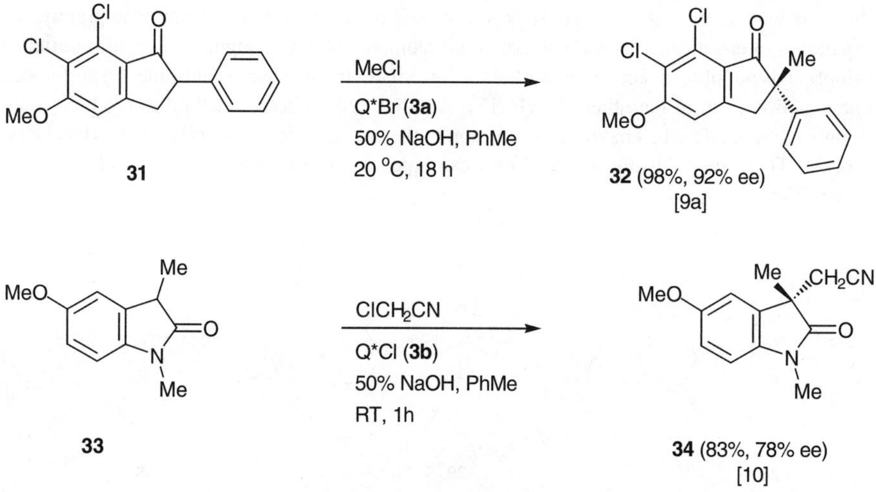

31

MeCl
Q*Br (**3a**)
50% NaOH, PhMe
20 °C, 18 h

32 (98%, 92% ee)
[9a]

33

ClCH₂CN
Q*Cl (**3b**)
50% NaOH, PhMe
RT, 1h

34 (83%, 78% ee)
[10]

Scheme 10.3.

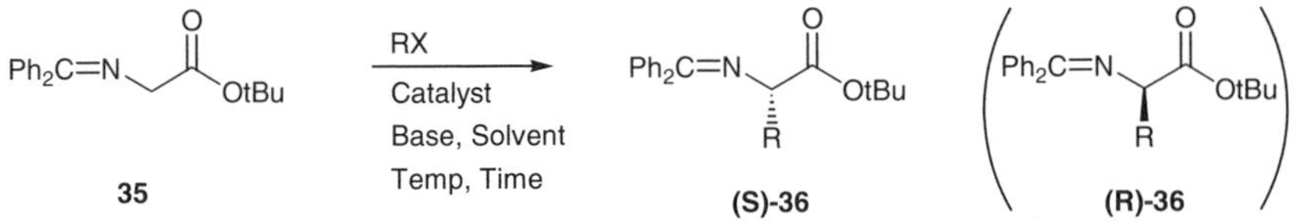

2nd Generation Catalyst

3rd Generation Catalysts

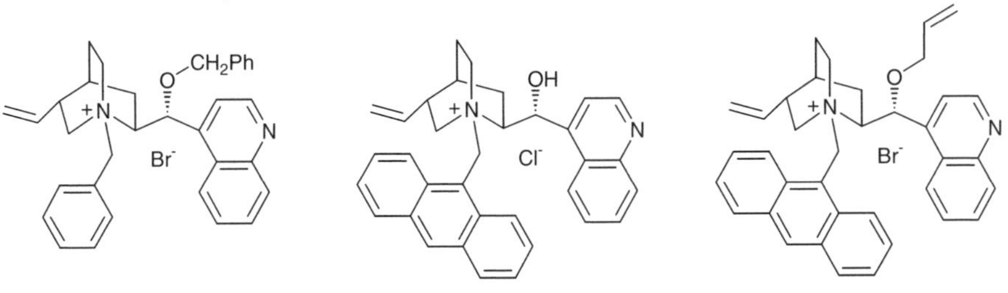

6a (81% ee)
[5p,12c]

8b (91% ee)
[13]

8a (99.5% ee)
[14]

Conditions for Benzylation to form (S)-36 (RBr = PhCH$_2$Br) from catalyst CC or CA:

[SA5,SA6]	Lygo [13]	Corey [14]	O'Donnell [15]
87%	68%*	87%	88%
81% ee (10:1)	91% ee (21:1)	94% ee (32:1)	91% ee (21:1)
50% NaOH, CH$_2$Cl$_2$/PhMe	50% KOH, PhMe	CsOH•H$_2$O, CH$_2$Cl$_2$	BEMP, CH$_2$Cl$_2$
+5 °C., 0.5 h	RT, 18 h	-78 °C., 23 h	-78 °C., 7 h

*% Yield to hydrolyzed
amino acid product.

Scheme 10.4.

salt **6a** gave 81% ee [5p,12b,12c], whereas in the "third generation," a further very significant increase in enantioselectivity (up to 91% ee with **8b** and 99.5% ee with **8a**) was achieved nearly simultaneously by the Lygo [13] and Corey [14] groups by introduction of the *N*-(9-anthracenyl-methyl) group in catalysts **8b** and **8a**, respectively. In addition to these modifications of the catalyst structure, different reaction conditions were also reported for these optimal catalysts (Scheme 10.4 and Table 10.4).

Thus, with catalyst **8b** (this catalyst is likely converted in situ to the *O*-alkyl derivative as discussed above in Scheme 10.2), a typical solid-liquid room-temperature PTC process took

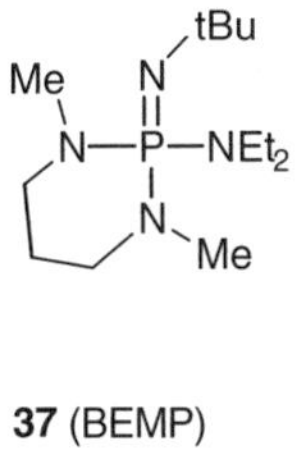

37 (BEMP)

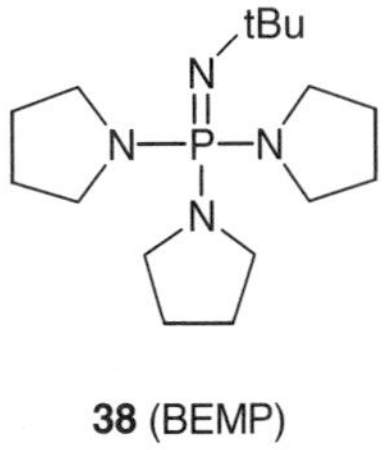

38 (BEMP)

Chart 10.5.

TABLE 10.4. Alkylation of Schiff Base Ester 35 with Catalysts 8b(7b) or 8a(7a) using Three Different Reaction Systems (see Scheme 10.4)

RX	Cat. 8b(7b)[a]-Lygo [13]		Cat. 8a-Corey [14]		Cat. 8a(7a)[a]-O'Donnell [15]	
	Time, Yield[b]	ee	Time, Yield	ee	Time, Yield	ee
n-Alkyl						
MeI	3h, 41% (40%)	89% ee (86% ee)	28h, 71%	97% ee	4h, 92%	94%ee
EtI	--	--	30h, 82%	98% ee	6h, 89% (90%)	89% ee (85%ee)
nButylI	18h, 42% (56%)	88% ee (87% ee)	--	--	3.5h, 88	91% ee
nHexylI	--	--	32h, 79%	99.5% ee	--	--
nOctylI	--	--	--	--	3.5h, 83%	93% ee
2° β-Center						
iBuI	--	--	--	--	24h, 93% (87%)	97% ee(87% ee)
c-C$_3$H$_5$CH$_2$Br	--	--	36h, 75%	99% ee	--	--
Allyl						
AllylBr	18h, 76% (62%)	88% ee (88% ee)	22h, 89%	97% ee	6h, 96% (90%)	90% ee(87% ee)
CH$_2$=C(Me)CH$_2$Br	--	--	20h, 91%	92% ee	4h, 91%	94% ee
tBuMe$_2$SiC≡CCH$_2$Br	--	--	18h, 68%	95% ee	--	--
Benzyl						
PhCH$_2$Br	18h, 68% (63%)	91% ee (89% ee)	23h, 87%	94% ee	7h, 88% (89%)	91% ee(83% ee)
4-NO$_2$C$_6$H$_4$CH$_2$Br	--	--	--	--	4h, 93%	89% ee
4-NCC$_6$H$_4$CH$_2$Br	--	--	--	--	4h, 89%	85% ee
4-CF$_3$C$_6$H$_4$CH$_2$Br	--	--	--	--	4h, 94%	88% ee
4-FC$_6$H$_4$CH$_2$Br	--	--	--	--	5h, 91%	84% ee
4-PhC$_6$H$_4$CH$_2$Br	--	--	--	--	5h, 90%	86% ee
4-MeC$_6$H$_4$CH$_2$Br	--	--	--	--	6h, 95%	91% ee
Ph$_2$CHBr	--	--	22h, 73%	99.5% ee	c	c
α-Halo Ester						
tBuO$_2$CCH$_2$I	4h, 84% (83%)	72%ee(67%ee)	--	--	4h, 91% (89%)[d]	85%ee(56%ee)[d]

[a]Yield and % ee in parentheses are using the pseudoenantiomeric catalysts **7b** or **7a**. [b]% Yield to hydrolyzed amino acid product. [c]No Product Obtained. [d]tBuO$_2$CCH$_2$Br.

738

from 3–18 h, depending on the nature of the alkylating agent. As observed previously [12c], whereas the stirring rate does not affect the level of enantioselectivity, higher stirring rates do decrease reaction times substantially. The other product enantiomer ((**R**)-**36**) is available by using the corresponding pseudoenantiomeric catalyst derived from cinchonine (**7b**) [13]. Reaction with catalyst **8a** also involves a solid-liquid PTC with $CsOH \cdot H_2O$ as base in CH_2Cl_2 at −60 to −78°C. for 18–36 h, a system that gave the highest overall enantioselectivities in these various studies. X-ray structures of catalyst **8a** and the ion-pair of **8a** and *p*-nitrophenoxide led to a proposed structure for the key ion-pair from the enolate of **35** and the quaternary cation of **8a** [14]. The final set of reaction conditions shown in Scheme 10.4 involves a *homogeneous* catalytic enantioselective synthesis [15]. In this case, the achiral organic soluble, non-ionic Schwesinger bases [BEMP (**37**) or BTPP (**38**), Chart 10.5] are used in conjunction with catalytic amounts of **8a** [or **7a** to form the enantiomeric products (**R**)-**36**] in CH_2Cl_2 at −50 or −78°C for 3.5–24 h to yield products with up to 97% ee. In general, these homogeneous reactions were faster than their heterogeneous counterparts [15]. Table 10.4 summarizes the results with catalysts **8b** (or **7b**) and **8a** (or **7b**) and the three different reaction systems outlined above and in Scheme 10.4.

Several other practical syntheses of enantiopure amino acid derivatives have been accomplished recently from substrate **35** (Chart 10.6). The Imperiali group has used two techniques following PTC alkylations that occurred with modest enantioselectivity (50–53% ee). The first involved fractional recrystallization followed by subsequent deprotection/reprotection to give **39** (>99% ee). In the second method, enzymatic hydrolysis of the amino acid methyl ester with alkaline protease and then nitrogen acylation gave **40** (99% ee) [16]. Several other publications that deal with related purification techniques have appeared [17–19].

A caution has been noted for chiral PTC alkylations involving alkyl halides that can be easily reduced. Attempted alkylation of **35** with (bromomethyl)cyclooctatetraene with a chiral *cinchona*-derived catalyst gave only racemic product [20].

The Corey group used catalyst **8a** in a selective monoalkylation of **35** with 1-chloro-4-iodobutane to give the 4-chlorobutyl product (99% ee) (Chart 10.6). The intermediate product was reduced and then converted, by an *N*-alkylation and subsequent steps, into the (*S*)-pipecolic acid derivative **41** (77% overall chemical yield from **35**, 99% ee, Chart 10.6) [21].

α,α-Dialkylamino acid derivatives can also be prepared by using PTC with chiral catalysts. Alkylation of the 4-chlorobenzaldehyde Schiff base t-butyl ester of alanine (**42**) with **43** in a solid-liquid PTC gave the α-methyltryptophan derivative **44** with 75% ee (Scheme 10.5) [5p,22]. Recently, an efficient (82% ee) room-temperature solid–liquid PTC alkylation with "TADDOL" (**18**) as catalyst was reported by Belokon and Kagan et al. for the synthesis of α-methylphenylalanine **46** (Scheme 10.5) [23].

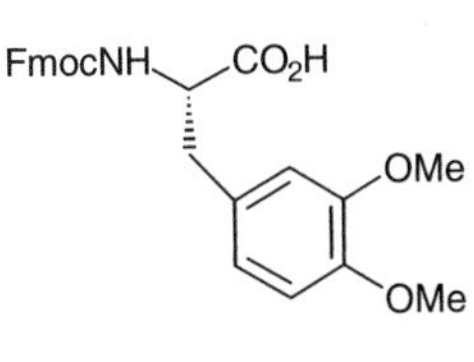

39 (>99% ee)
[16]

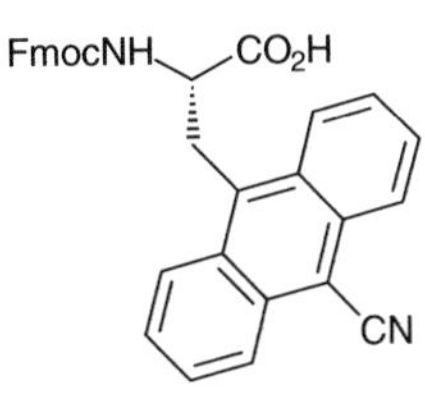

40 (99% ee)
[16]

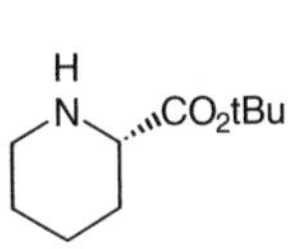

41 (99% ee)
[21]

Chart 10.6.

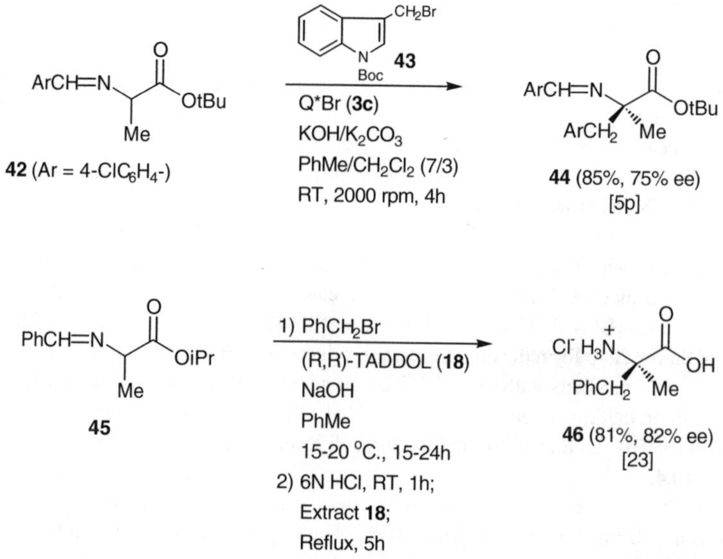

ArCH=N. ... OtBu
|
Me

42 (Ar = 4-ClC$_6$H$_4$-)

CH$_2$Br on indole N-Boc **43**

Q*Br (**3c**)
KOH/K$_2$CO$_3$
PhMe/CH$_2$Cl$_2$ (7/3)
RT, 2000 rpm, 4h
$\longrightarrow$

ArCH=N. ... OtBu
|
ArCH$_2$ Me

44 (85%, 75% ee)
[5p]

PhCH=N. ... OiPr
|
Me

45

1) PhCH$_2$Br
(R,R)-TADDOL (**18**)
NaOH
PhMe
15-20 °C., 15-24h
2) 6N HCl, RT, 1h;
Extract **18**;
Reflux, 5h
$\longrightarrow$

Cl$^-$ H$_3$N$^+$. ... OH
|
PhCH$_2$ Me

46 (81%, 82% ee)
[23]

Scheme 10.5.

Other chiral PTC alkylations of active methylene compounds leading to amino acid derivatives have been reported [24] as have other alkylations [25]. Several reported asymmetric PTC alkylations have been disputed [26–29].

10.3.1.2. 1,2-Carbonyl Addition Diethylzinc has been added to benzaldehyde at room temperature in the presence of an *ephedra*-derived chiral quat (**8**) to give optically active secondary alcohols, a case in which the chiral catalyst affords a much higher enantioselectivity in the solid state than in solution (**47** to **48**, Scheme 10.6) [30]. Asymmetric trifluoromethylation of aldehydes and ketones (**49** to **50**, Scheme 10.6 [31]) is accomplished with trifluoromethyl-trimethylsilane, catalyzed by a quaternary ammonium fluoride (**3d**). Catalyst **3d** was first used by the Shioiri group for catalytic asymmetric aldol reactions from silyl enol ethers **51** or **54** (Scheme 10.6) [32]. Various other 1,2-carbonyl additions [33] and aldol reactions [34] have been reported.

10.3.1.3. Michael Addition A spectacular asymmetric induction (ca. 99% ee) was achieved in 1981 by Cram and co-workers in the Michael addition of the cyclic substrate **56** to methyl vinyl ketone (MVK) in the presence of 4 mol % each of crown ether **9** and KO^{t}Bu (Scheme 10.7) [35]. As expected, lower reaction temperatures provided higher stereospecificity, for example, when the reaction was run at room temperature, product with 67% ee was obtained. With crown **10**, products with the opposite *S*-configuration were obtained. Reactions of acyclic substrates (PhCH$_2$CO$_2$Me and PhCH(Me)CO$_2$Me) with methyl acrylate at –78°C gave Michael adducts with up to 65% ee and 83% ee, respectively. Other cyclic substrates have been used in similar Michael additions with *cinchona*-derived catalysts to prepare products **59** (catalyst **3a**, 50% NaOH, PhMe, RT, 5m, [36]) and **60** (catalyst **6d**, 60% KOH, PhMe, –20°C., 2.5h, [11d]). The key stereogenic center in a total synthesis of (+)-triptoquinone A was introduced by using this

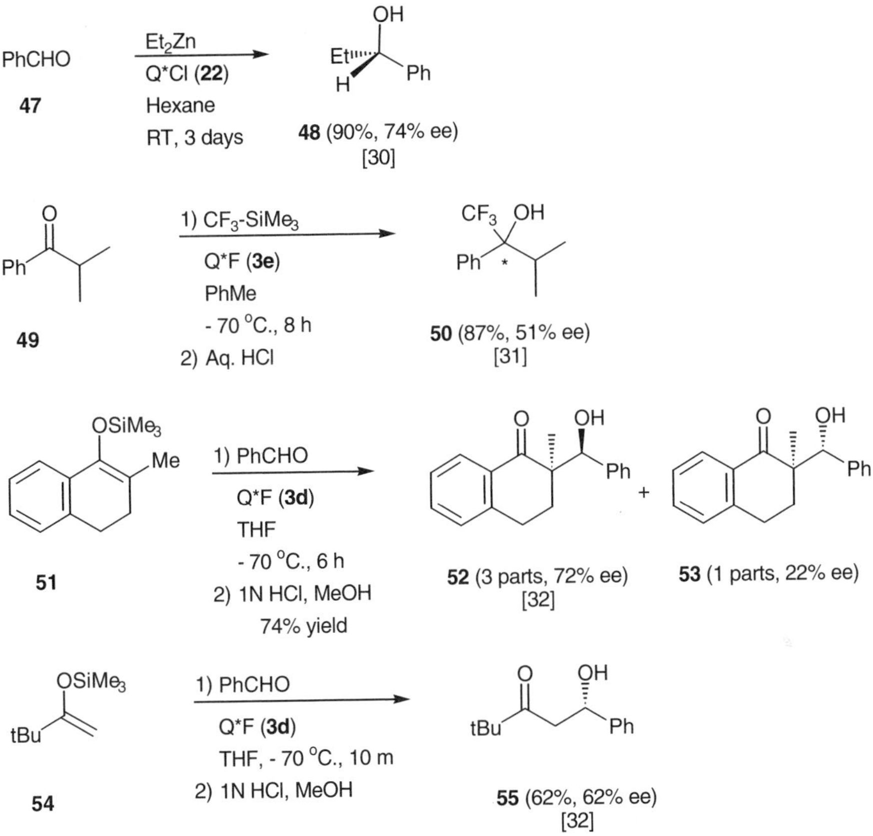

Scheme 10.6.

methodology, which involves a two-step Robinson annelation procedure (**61** to **62**, Scheme 10.7) [37].

Several other high inductions have been reported by using crown ethers as catalysts (Scheme 10.8). The Tőke group has used a chiral crown **11** (Chart 10.2), which incorporates a glucose unit, for the addition of 2-nitropropane to a chalcone (Scheme 10.8) [38]. Several other effective chiral crowns (**12–17**, Chart 10.2 and Scheme 10.8) are noted [24e,39–42,48b]. An interesting study of the Michael addition under both solvent-free (0% ee) and liquid–liquid conditions (up to 70% ee) was reported by Diez-Barra and co-workers, who also addressed the question of free –OH quats (**28**, 58% ee) verses O-benzyl quats (**30**, 46% ee) [43].

Protected glycine derivatives have been used as the nucleophilic partner in enantioselective syntheses of amino acid derivatives by chiral PTC (Scheme 10.9). Loupy and co-workers have reported the addition of diethyl acetylaminomalonate to chalcone without solvent with enantioselectivity up to 82% ee [44]. The recent report from the Corey group, with catalyst **8a** used in conjunction with the benzophenone imine of glycine t-butyl ester **35**, discussed earlier, results in highly enantioselective reactions (91–99% ee) with various Michael acceptors (2-cyclo-hexenone, methyl acrylate, and ethyl vinyl ketone) to yield products **71–73** [21]. Other Michael reactions resulting in amino acid products are noted [45].

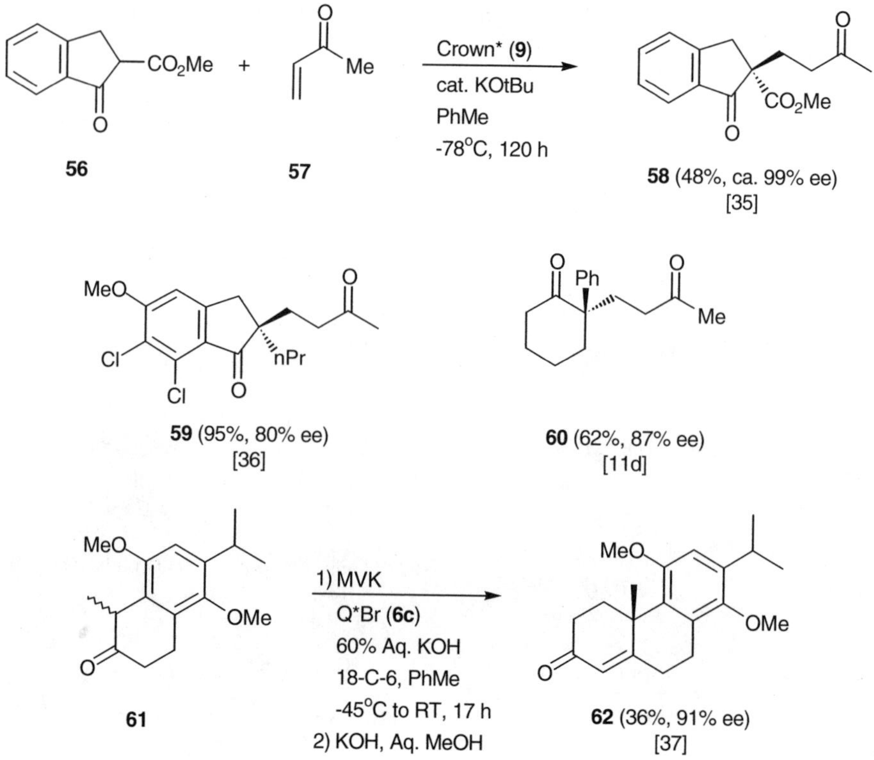

56 + **57** → **58** (48%, ca. 99% ee) [35]

Crown* (9)
cat. KOtBu
PhMe
-78°C, 120 h

59 (95%, 80% ee) [36]

60 (62%, 87% ee) [11d]

61

1) MVK
Q*Br (6c)
60% Aq. KOH
18-C-6, PhMe
-45°C to RT, 17 h
2) KOH, Aq. MeOH

62 (36%, 91% ee) [37]

Scheme 10.7.

Reports of other Michael additions that resulted in only modest selectivities (40–50% ee) [24f,24h,24i,46] and earlier studies—especially those from the Wynberg and Colonna groups—of mechanistic or catalyst interest [47,48] are noted. Michael additions have been reviewed [5g].

10.3.1.4. Cyclopropanation, Horner-Wadsworth Emmons Reaction, and Darzens Condensation Although induction in the cyclopropanation of alkenes was reported early, this work was disputed [49]. Other reports of cyclopropanations have yielded, at best, low asymmetric inductions [11h,50]. The first example of a catalytic asymmetric Horner–Wadsworth Emmons reaction, which is promoted by a chiral quaternary ammonium salt, was reported recently by the Shioiri group (Scheme 10.10) [51]. The reaction of the prochiral ketone **74** gives optically active α,β-unsaturated ester **76** with 57% ee.

Promising examples of the catalytic asymmetric Darzens condensation, which yields an epoxide product via carbon–carbon and carbon–oxygen bond formation, have been reported recently by two groups (Scheme 10.11). Töke and co-workers used crown ether **24** in the reaction to form the α,β-unsaturated ketone **78** [38b] with 64% ee, whereas the Shioiri group used the *cinchona*-derived salt **3a** [52], which resulted in **78** with 69% ee. The latter authors propose a catalytic cycle involving generation of a chiral enolate in situ from an achiral inorganic base

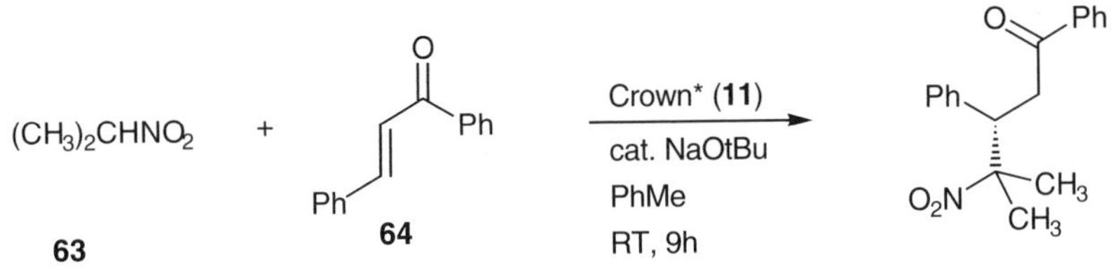

65 (82%, 90% ee)
[38a]

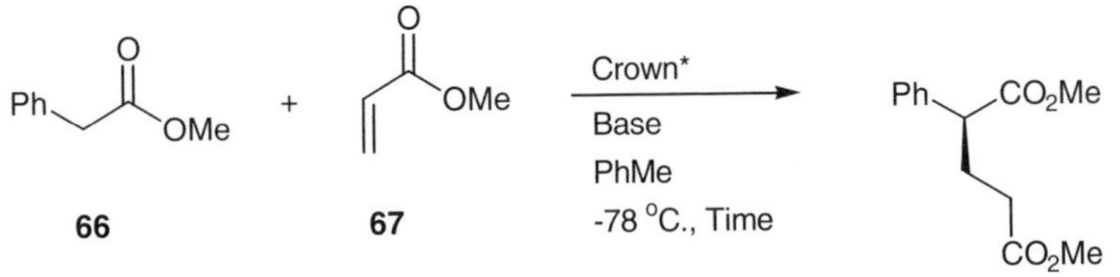

Crown*	Base	Time	Yield	%ee	Ref.
12	KOtBu	1m	82%	84%ee	[39]
13	NaOtBu	3h	65%	83%ee	[40]
14·KOtBu		3h	95%	79%ee	[41a]

Scheme 10.8.

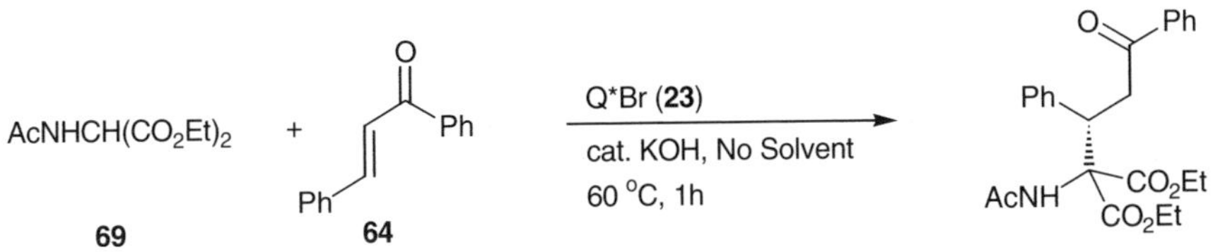

70 (51%, 82% ee)
[44a]

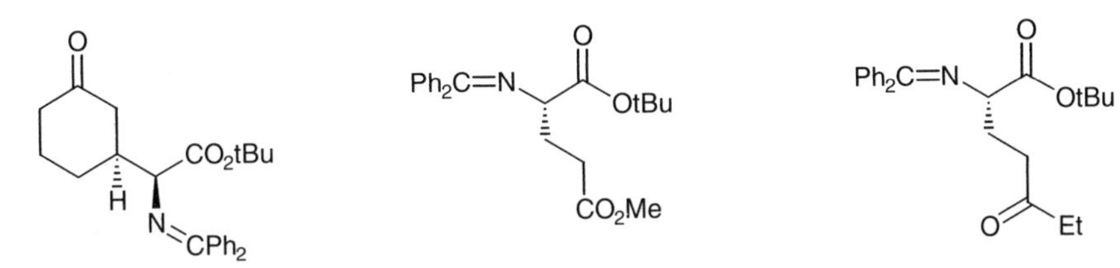

71 (88%, 99% ee, 25:1 ds)
[21]

72 (85%, 95% ee)
[21]

73 (85%, 91% ee)
[21]

Scheme 10.9.

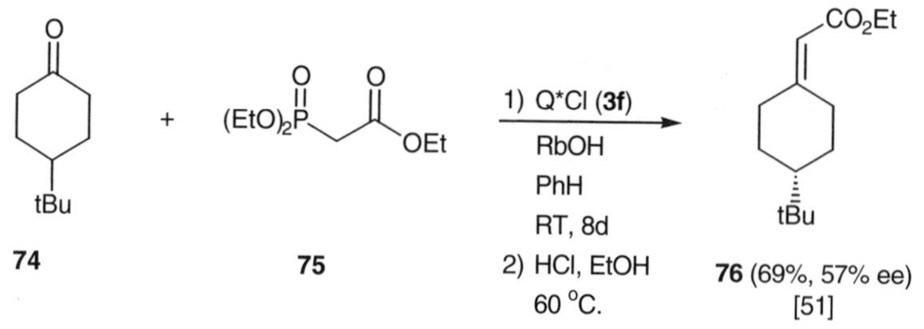

Scheme 10.10.

and the chiral quaternary ammonium halide. Subsequent reaction with the aldehyde followed by closure to the epoxide and liberation of the ammonium salt completes the catalytic cycle. α,β-Epoxysulfones (**83**) are prepared in up to 81% ee by a similar Darzen's condensation with chloromethyl phenylsulfone [53]. Other Darzens condensations have resulted in low asymmetric inductions [54].

10.3.1.5. Reduction The asymmetric reduction of a series of aryl alkyl ketones with quaternary ammonium fluorides and silanes was reported by Drew and Lawrence [55]. In these reactions, the best catalysts (e.g., **6f**) were from the quinine/quinidine series; in fact, a fluoride salt prepared from cinchonine gave no induction. The use of trimethoxysilane resulted in faster rates but lower enantioselectivites when compared with tris(trimethoxy)silane. It is interesting that, with the

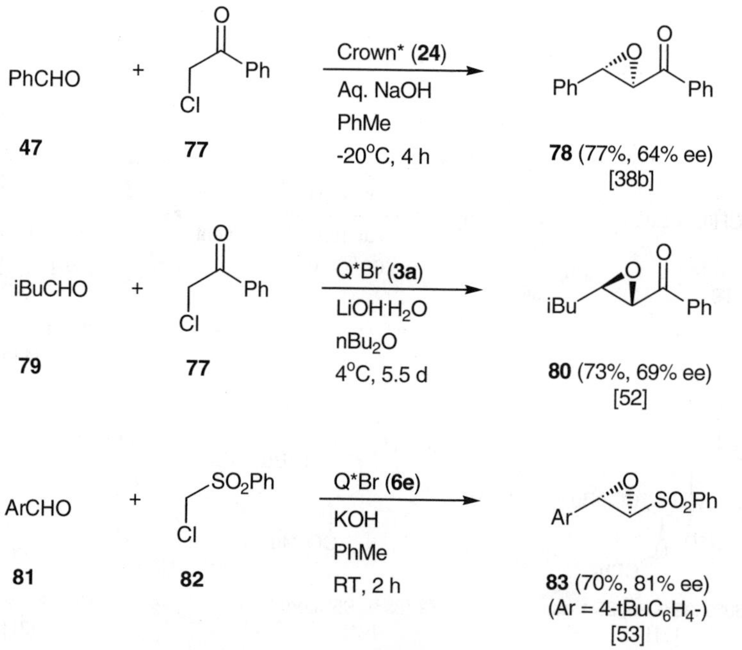

Scheme 10.11.

PMHS corresponding polymeric reagent (PMHS, polymethylhydrosiloxane), a substantial rate increase was observed over the monomeric model (complete reduction of acetophenone in less than 1 min with PMHS vs. only 60% conversion in 1 h with (EtO)$_2$SiHMe) [55]. The related hydrosilylation of **86** by chiral PTC uses an interesting *ephedra*-derived halometallated catalyst **25** (Scheme 10.12) [56].

Electroreduction [5b] (with chiral quat as the supporting electrolyte) has been compared with chemical reduction (NaBH$_4$) in the presence of chiral quats for ketone (up to 28% op) and imine (up to 22% op) reductions [57,58]. The reduction (NaBH$_4$) of a chiral α,β-enone prostaglandin intermediate in the presence of *ephedra*-derived catalysts led to the formation of the enol with 70% de [59]. Other reductions with lower asymmetric inductions are noted for ketone [11i,24h,24i,47e,60], imine [5b,57], and hydrodehalogenation of a cyclic α,α-dichloroamide [61].

An early report of a promising level of asymmetric induction in the reduction of ketones by chiral PTC [62a] was disputed [26b,60c]. Several reviews concerning hydrogenation by enantioselective catalysis have appeared [5j–l].

10.3.1.6. Oxidation

Epoxidation. Epoxidations of acyclic and cyclic electron-deficient alkenes, such as α,β-unsaturated carbonyl systems and quinones, have been studied in detail by several different groups (Scheme 10.13). Of special note is the early pioneering research of the Wynberg group in the mid- to late-1970s [5c,47b,54a,63,64]. One of the several advances made concerned the importance of solvent effects. For example, the oxidation of a chalcone with hydrogen peroxide using quinine-derived catalyst **6j** gave 54% ee in benzene but only 28% ee in methylene chloride [47b].

Lygo and Wainwright recently reported a detailed study of the asymmetric phase-transfer mediated epoxidation of a variety of acyclic α,β-unsaturated ketones of the chalcone type. The "third-generation" *cinchona*-derived quats (**8c** and **7c**), related to those discussed earlier in the alkylation section and Scheme 10.4, gave the best inductions (89% ee, **88** to **89**, Scheme 10.13 and 86% ee for the pseudoenantiomeric catalyst **7c** to give, as product, the enantiomer of **89**).

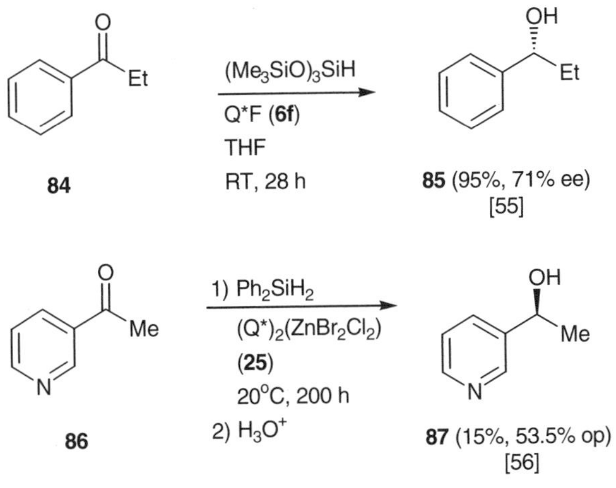

Scheme 10.12.

In this case, the *O*-benzyl quat hydroxide was used in conjunction with sodium hypochlorite in toluene [65]. Other reports of epoxidations of acyclic enones have appeared, showing 62% ee [66a] or lower enantioselectivity [24e,24h,48b,54a,54c–e,67].

An equally impressive chiral PTC epoxidation of a cyclic enone was accomplished by Taylor and co-workers in the syntheses of various epoxycyclohexanone natural products and analogs (89% ee, **90** to **91**, Scheme 10.13). In this reaction a full equivalent of the quaternary ammonium salt was used, that is, **6g** formally serves as a chiral reagent rather than a catalyst. A model study demonstrated an improvement in enantioselectivity in the stoichiometric case (77% ee) versus the catalytic one (69% ee) [68]. The Onda group reported an early success in the epoxidation of quinone **92** (Scheme 10.13) [69] (Scheme 10.13). Recently, the Shioiri group has used a quaternary ammonium catalyst derived from quinidine (**2**) by *N*-alkylation with 1-(chloromethyl)naphthalene for epoxidation of 2-phenyl-1,4-naphthoquinone (76% ee). This catalyst is an interesting derivative of the "third-generation" catalysts (Scheme 10.4) discussed earlier in the alkylation section of this chapter [70]. 2-Cyclohexenone was epoxidized in up to 63% ee by using 9-alkylfluorenyl peroxides and the novel bis-quat derivative **26** [71]. Other epoxidations of cyclic enones are noted [26b,64].

Jacobsen and co-workers have found that catalytic amounts of chiral quaternary ammonium salts, such as **6i**, promote a dramatic reversal in the *diastereoselectivity* of (salen)Mn-catalyzed epoxidation of cis-alkenes, resulting in a highly enantioselective catalytic route to trans-epoxides (Scheme 10.14) [72].

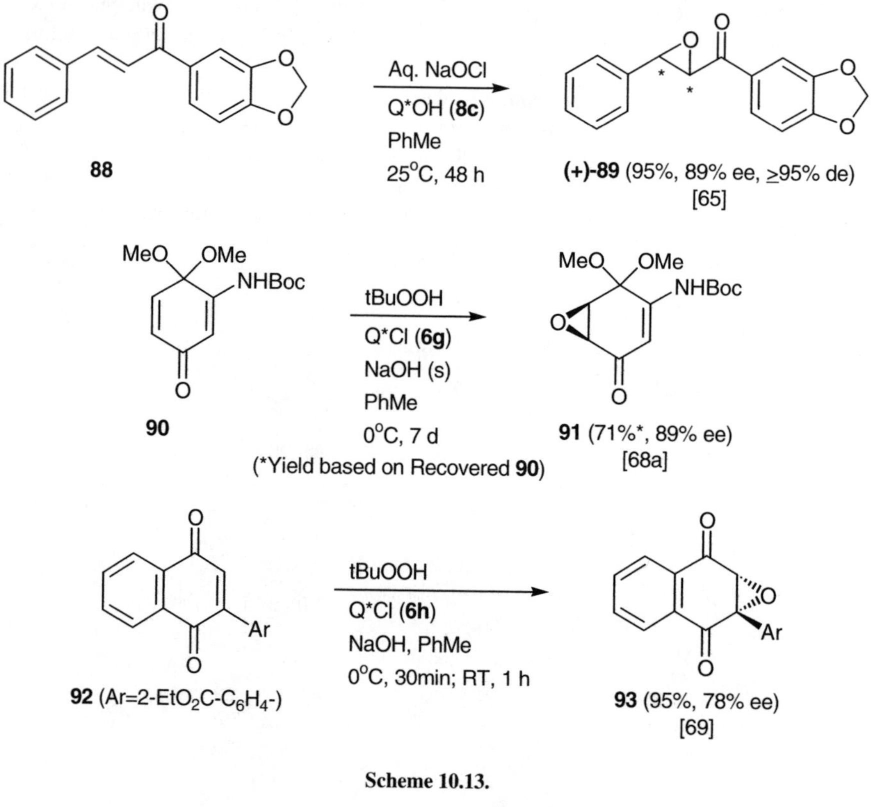

Scheme 10.13.

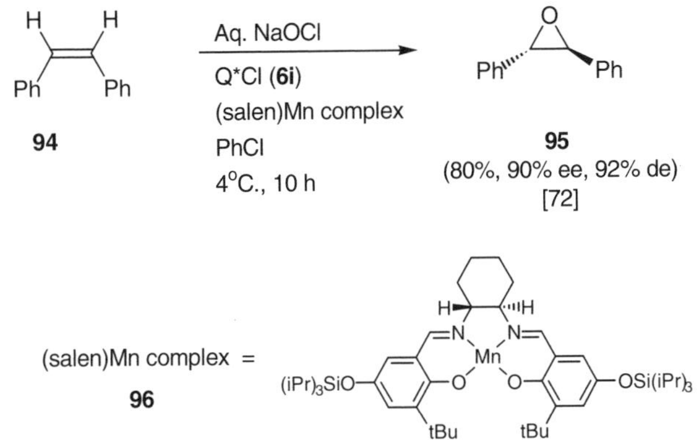

Scheme 10.14.

α-Hydroxylation of Ketones. Cyclic ketones have been converted to optically active α-hydroxy ketones with good induction by reaction with molecular oxygen in the presence of either *cinchona*-derived (**3a**) or crown ether (**27**) catalysts (Scheme 10.15) [73,74]. Other α-hydroxylations are noted as well [24e,24f,24h].

Other Oxidations. Glycol formation by oxidation of styrene [75], as well as oxidation of prochiral phosphines to the optically active phosphine oxides [76] by chiral PTC, gave only low asymmetric inductions.

10.3.2. Carbon-Hetero Bond Formation

10.3.2.1. Carbon-Nitrogen Bonds. Several groups have studied the synthesis of optically active α-amino acids from inexpensive and readily available α-haloesters by displacement with phthalimide in the presence of chiral *cinchona* catalysts [11e,24h,24i,47e,60d,77]. Early studies, with chiral, non-racemic starting material, showed that this reaction occurs with partial

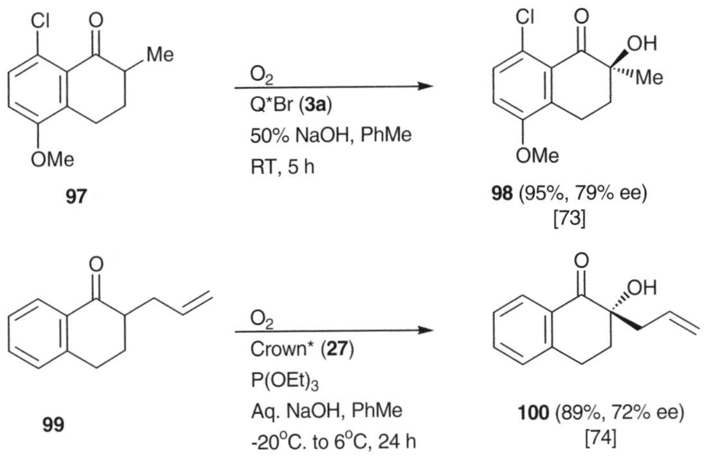

Scheme 10.15.

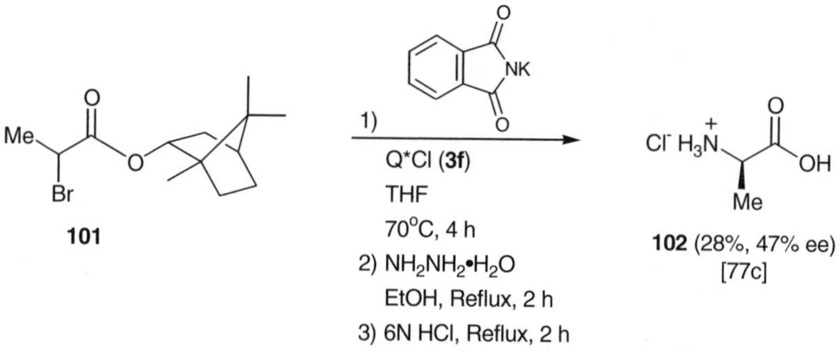

Scheme 10.16.

inversion of configuration and likely involves a kinetic resolution when racemic starting material is used [77a]. More recently, a double asymmetric induction, involving reaction of the bornyl ester **101** with catalyst **3f** gave a moderate enantioselectivity (Scheme 10.16) [77c]. Other C–N bond-forming reactions are noted [78].

10.3.2.2. Aziridination. A catalytic enantioselective synthesis of *N*-arylaziridines, with *N*-acyl-*N*-arylhydroxylamine **104** serving as the aziridinating agent for electron-deficient olefins, has been reported by Aires-de-Sousa and co-workers (Scheme 10.17). Modest enantioselectivities (absolute configurations were not assigned) were obtained with the cinchonine-derived catalyst **3a**. Although higher enantioselectivity was achieved with a milder base (9% NaOH), the chemical yield of the reaction was reduced substantially (12% yield, 61% ee). It is surprising that the use of the pseudo-enantiomeric catalysts gave the same major enantiomer under identical conditions (18% yield, 18% ee) [79].

10.3.2.3. Carbon–Oxygen and Carbon–Sulfur Bonds. A report of modest enantioselectivity up to 48% ee in the *O*-alkylation of racemic secondary alcohols (a kinetic resolution) in the presence of a chiral non-racemic non-functionalized quat, (*S*)-Et$_3$NCH$_2$CH(Me)Et Br, could not be repeated [80]. Such catalysts would not be capable of making the multipoint interaction between catalyst and reactants in the transition state, which are thought to govern the stereochemistry of these types of reactions. Other *O*-alkylations are noted [11e].

The addition of thiophenol to cyclohexenone in the presence of a *cinchona* quat gave the Michael adduct in 85% yield and 36% op [47f]. Other C–S bond formations are also noted [11e,26b,47d,47e,81].

10.3.2.4. Carbon–Halogen Bond. Early chlorination of various alkenes gave only low rotations in the isolated products with either *ephedra* or *cinchona* quats [33b].

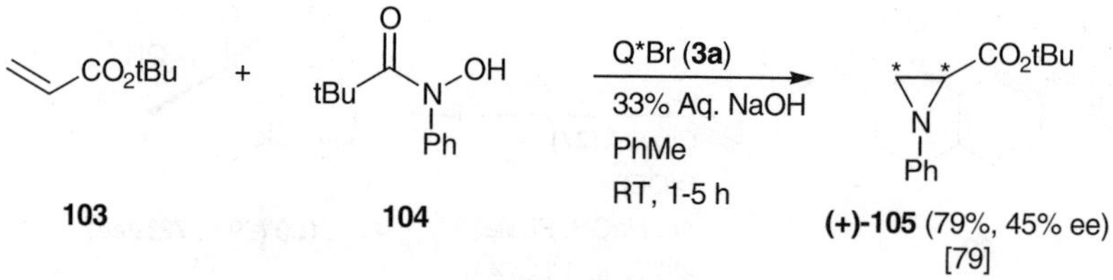

Scheme 10.17.

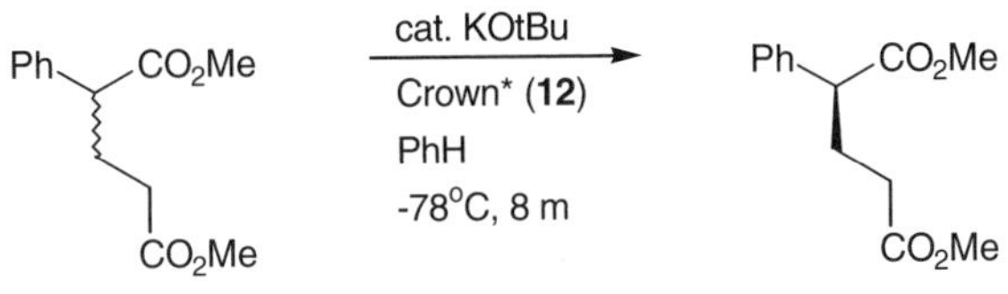

Scheme 10.18.

10.3.3. Other Reactions

10.3.3.1. Deracemization. Results from Michael additions described earlier (Scheme 10.8) led Tõke and co-workers to an interesting deracemization study. When racemic Michael adduct **106** was reacted with a catalytic amount of base in the presence of the chiral crown **12** for 8 min, the resulting product was optically active (40% ee). The authors propose that a deprotonation followed by reprotonation of the resulting chiral ion-pair accounts for the asymmetric induction [39].

10.4. CONCLUSION AND FUTURE PROSPECTS

Catalytic enantioselective synthesis by phase-transfer catalysis is a field of great potential that continues to grow. Important advances have occurred since the first edition of this book [5m], including alkylations, Michael additions, Horner–Wadsworth–Emmons reactions, Darzens condensations, ketone reductions, and epoxidations. As many of the examples cited in this review demonstrate, continued advancement will depend, in large part, on the careful and systematic study of these reactions to understand the details of the chiral induction step as well as the various other processes that occur during the reaction. With such an in-depth appreciation, it will be possible to design substrates, catalysts, and reactions systems that lead to both high chemical yield and high levels of asymmetric induction.

ACKNOWLEDGMENTS

I wish to express my appreciation to all past and current members of my research group. I also wish to thank the National Institutes of Health (GM 28193), the North Atlantic Treaty Organization, and Eli Lilly and Company for financial support.

REFERENCES

1. PTC books: (a) Brändström, A. *Preparative Ion Pair Extraction*; Apotakarsocieten/Hässle: Läke-medel, Sweden, 1974. (b) Weber, W. P.; Gokel, G. W. *Phase Transfer Catalysis in Organic Synthesis*; Springer-Verlag: Berlin, 1977. (c) Caubère, P. *Le transfert de phase et son utilisation en chimie organique*; Masson: Paris, 1982. (d) Keller, W. E. *Phase-Transfer Reactions. Fluka Compendium*; Georg Thieme Verlag: Stuttgart, 1986; Vol. 1. (e) Keller, W. E. *Phase-Transfer Reactions. Fluka*

Compendium; Georg Thieme Verlag: Stuttgart, 1987; Vol. 2. (f) Keller, W. E. *Phase-Transfer Reactions. Fluka Compendium*; Georg Thieme Verlag: Stuttgart, 1990; Vol. 3. (g) *Phase Transfer Catalysis (ACS Symposium Series: 326)*; Starks, C. M. (Ed.); American Chemical Society: Washington, D.C., 1987. (h) Goldberg, Y. *Phase Transfer Catalysis. Selected Problems and Applications*; Gordon and Breach: Amsterdam, 1992. (i) Dehmlow, E. V.; Dehmlow, S. S. *Phase Transfer Catalysis,* 3rd ed.; VCH: Weinheim, 1993. (j) Starks, C. M.; Liotta, C. L.; Halpern, M. *Phase-Transfer Catalysis. Fundamentals, Applications, and Industrial Perspectives*; Chapman & Hall: New York, 1994. (k) *Phase-Transfer Catalysis: Mechanisms and Synthesis (ACS Symposium Series: 659)*; Halpern, M. E. (Ed.); American Chemical Society: Washington, D.C., 1997. (l) *Handbook of Phase Transfer Catalysis*; Sasson, Y.; Neumann, R. (Ed.); Blackie Academic & Professional: London, 1997.

2. PTC reviews: (a) Brändström, A. *Adv. P. Org. Chem.* **1977**, *15*, 267–330. (b) Sjöberg, K. *Aldrichimica Acta* **1980**, *13*, 55–58. (c) Montanari, F.; Landini, D.; Rolla, F. *Top. Curr. Chem.* **1982**, *101*, 147–200. (d) Starks, C. M. *Israel J. Chem.* **1985**, *26*, 211–215. (e) Freedman, H. H. *Pure Appl. Chem.* **1986**, *58*, 857–868. (f) Rabinovitz, M.; Cohen, Y.; Halpern, M. *Angew. Chem., Int. Ed.* **1986**, *25*, 960–970. (g) Makosza, M.; Fedorynski, M. *Adv. Catal.* **1987**, *35*, 375–422. (h) Bram, G.; Galons, H.; Labidalle, S.; Loupy, A.; Miocque, M.; Petit, A.; Pigeon, P.; Sansoulet, J. *Bull. Soc. Chim. Fr.* **1989**, 247–251. (i) Landini, D.; Maia, A.; Podda, G. *Gazz. Chim. Ital.* **1995**, *125*, 583–587. (j) Kryshtal, G. V.; Serebryakov, E. P. *Russian Chem. Bull.* **1995**, *44*, 1785–1804; *Chem. Abstr.* **1996**, *124*, 259923. (k) Dehmlow, E. V. *Russian Chem. Bull.* **1995**, *44*, 1998–2005. (l) Luche, J.-L. *C. R. Acad. Sci., Ser. IIb* **1996**, *323*, 337–354. (m) Weng, H.-S.; Wang, D.-H. *J. Chin. Inst. Chem. Eng.* **1996**, *27*, 419–426; *Chem. Abstr.* **1997**, *126*, 119306. (n) Goldberg, Y.; Alper, H. In *Applied Homogeneous Catalysis with Organometallic Compounds*; Cornils, B; Herrmann, W. A. (Ed.); Wiley-VCH: Weinheim, 1996; pp. 844–865. (o) Makosza, M.; Fedorynski, M. *Polish J. Chem.* **1996**, *70*, 1093–1110. (p) Tavener, S. J.; Clark, J. H. *Chem. Ind. (London)* **1997**, 22–24. (q) Cook, M. M.; Halpern, M. E. *Chim. Oggi*, **1998**, *16*, 44–48; *Chem. Abstr.* **1998**, *128*, 262378. (r) Naik, S. D.; Doraiswamy, L. K. *AIChE Journal* **1998**, *44*, 612–646. (s) Loupy, A.; Petit, A.; Hamelin, J.; Texier-Boullet, F.; Jacquault, P.; Mathé, D. *Synthesis* **1998**, 1213–1234.

3. International PTC Conferences: (a) "Phase-Transfer Catalysis," O'Donnell, M. J.; Alper, H.; Shioiri, T. (Organizers), Pacifichem '95, Honolulu, Hawaii, December 19–21, 1995. (b) "PTC '97," Shioiri, T. (Organizer), The 1997 International Phase-Transfer Catalysis Conference, Nagoya, Japan, September 24–27, 1997.

4. PTC Newsletters: (a) *Phases*, a newsletter published by SACHEM, Inc. 821 E. Woodward, Austin, TX 78704. (b) *Phase-Transfer Catalysis Communications*, a newsletter published by PTC Communications, Inc., Suite 627, 1040 N. Kings Highway, Cherry Hill, NJ 08034.

5. Asymmetric PTC reviews: (a) Kong, F. *Huaxue Tongbao* **1985**, 35–42; *Chem. Abstr.* **1986**, *105*, 41955v. (b) Tallec, A. *Bull. Soc. Chim. Fr.* **1985**, 743–761. (c) Wynberg, H. In *Topics in Stereochemistry*; Eliel, E. L.; Wilen, S.; Allinger, N. L. (Ed.); Wiley: New York, 1986; Vol. 16; pp. 87–129. (d) Dolling, U.-H.; Hughes, D. L.; Bhattacharya, A.; Ryan, K. M.; Karady, S.; Weinstock, L. M.; Grabowski, E. J. J. In *Phase Transfer Catalysis (ACS Symposium Series: 326)*; Starks, C. M. (Ed.); American Chemical Society: Washington, D.C., 1987; pp. 67–81. (e) Zhang, J.; Wu, Y. *Huaxue Tongbao* **1987**, 18–23, 27; *Chem. Abstr.* **1988**, *108*, 203978w. (f) Dolling, U.-H.; Hughes, D. L.; Bhattacharya, A.; Ryan, K. M.; Karady, S.; Weinstock, L. M.; Grenda, V. J.; Grabowski, E. J. J. In *Catalysis of Organic Reactions (Chem. Ind.: 33)*; Rylander, P. N.; Greenfield, H.; Augustine, R. L. (Ed.); Dekker: New York, 1988; pp. 65–86. (g) Oare, D. A.; Heathcock, C. H. In *Topics in Stereochemistry*; Eliel, E. L.; Wilen, S. H. (Ed.); Wiley: New York, 1989; Vol. 19; pp. 227–403. (h) Chen, Z.; Zeng, Z. *Huaxue Shiji* **1989**, *11*, 243–247, 228; *Chem. Abstr.* **1989**, *111*, 231422r. (i) Baba, N. *Okayama Daigaku Nogakubu Gakujutsu Hokoku* **1990**, *75*, 31–45; *Chem. Abstr.* **1990**, *113*, 210919j. (j) Blaser, H.-U.; Müller, M. In *Studies in Surface Science and Catalysis*; Guisnet, M.; Barrault, J.; Bouchoule, C.; Duprez, D.; Perot, G.; Maurel, R.; Montassier, C. (Ed.); 1991; Vol. 59; pp. 73–92. (k) Blaser, H. U.; Jalett, H. P.; Monti, D. M.; Baiker, A.; Wehrli, J. T. In *Structure-Activity and Selectivity Relationships in Heterogeneous Catalysis*; Grasselli, R. K.; Sleight, A. W. (Ed.); Elsevier: Amsterdam, 1991; Vol. 67; pp. 147–155. (l) Blaser, H.-U. *Tetrahedron: Asymmetry.* **1991**, *2*, 843–866. (m) O'Donnell, M. J. In *Catalytic Asymmetric Synthesis*; Ojima, I. (Ed.); VCH: New

York, 1993, pp. 389–411. (n) Rao, J. M.; Rao, T. B. *J. Indian Inst. Sci.* **1994**, *74*, 373–400; *Chem. Abstr.* **1995**, *123*, 55043. (o) Mi, A.; Lou, R.; Jiang, Y. *Hecheng Huaxue* **1996**, *4*, 13–22; *Chem. Abstr.* **1996**, *125*, 166832. (p) O'Donnell, M. J.; Esikova, I. A.; Mi, A.; Shullenberger, D. F.; Wu, S. In *Phase-Transfer Catalysis: Mechanisms and Synthesis* (*ACS Symposium Series: 659*); Halpern, M. E. (Ed.); American Chemical Society: Washington, D.C., 1997; pp. 124–135. (q) Mi, A.; Lou, R. *Huaxue* **1997**, *55*, 105–115; *Chem. Abstr.* **1997**, *127*, 50080. (r) Shioiri, T. In *Handbook of Phase Transfer Catalysis*; Sasson, Y.; Neumann, R. (Ed.); Blackie Academic & Professional: London, 1997, pp. 462–479. (s) Ebrahim, S.; Wills, M. *Tetrahedron: Asymmetry* **1997**, *8*, 3163–3173. (t) O'Donnell, M. J. *Phases* **1998**, Issue 4, 5–8 (see reference 4a).

6. (a) Stoddart, J. F. In *Topics in Stereochemistry*; Eliel, E. L.; Wilen, S. H. (Ed.); Wiley: New York, 1987; Vol. 17; pp. 207–288. (b) Lehn, J.-M. *Angew. Chem., Int. Ed.* **1988**, *27*, 89–112. (c) Cram, D. J. *Angew. Chem., Int. Ed.* **1988**, *27*, 1009–1020. (d) Pedersen, C. J. *Angew. Chem., Int. Ed.* **1988**, *27*, 1021–1027. (e) Parker, D. In *Macrocycle Synthesis*; Parker, D. (Ed.); Oxford University Press: Oxford, 1996; pp. 25–47. (f) Parker, D. In *Macrocycle Synthesis*; Parker, D. (Ed.); Oxford University Press: Oxford, 1996, pp. 49–70. (g) Amabilino, D. B.; Preece, J. A.; Stoddart, J. F. In *Macrocycle Synthesis*; Parker, D. (Ed.); Oxford University Press: Oxford, 1996; pp. 71–91. (h) Gokel, G. W.; Abel, E. In *Comprehensive Supramolecular Chemistry*; Gokel, G. W. (Ed.); Elsevier: Oxford, 1996; pp. 511–535. (i) Bradshaw, J. S. *J. Inclusion Phenom. Mol. Recognit. Chem.* **1997**, *29*, 221–246.

7. Shuttleworth, S. J.; Allin; S. M.; Sharma, P. K. *Synthesis* **1997**, 1217–1239.

8. (a) Osa, T.; Suzuki, I. In *Comprehensive Supramolecular Chemistry*; Szejtli, J.; Osa, T. (Ed.); Elsevier: Oxford, 1996; Vol. 3, pp. 367–400. (b) Montanari, F.; Quici, S.; Banfi, S. In *Comprehensive Supramolecular Chemistry*; Reinhoudt, D. N. (Ed.); Elsevier: Oxford, 1996; Vol. 10, pp. 389–416. (c) Vögtle, F.; Hoss, R.; Händel, M. In *Applied Homogeneous Catalysis with Organometallic Compounds*; Cornils, B.; Herrmann, W. A. (Ed.); VCH: Weinheim, 1996; pp. 801–832. (d) Reetz, M. T. *Catalysis Today* **1998**, *42*, 399–411.

9. (a) Dolling, U.-H.; Davis, P.; Grabowski, E. J. J. *J. Am. Chem. Soc.* **1984**, *106*, 446–447. (b) Bhattacharya, A.; Dolling, U.-H.; Grabowski, E. J. J.; Karady, S.; Ryan, K. M.; Weinstock, L. M. *Angew. Chem., Int. Ed.* **1986**, *25*, 476–477. (c) Hughes, D. L.; Dolling, U.-H.; Ryan, K. M.; Schoenewaldt, E. F.; Grabowski, E. J. J. *J. Org. Chem.* **1987**, *52*, 4745–4752.

10. Lee, T. B. K.; Wong, G. S. K. *J. Org. Chem.* **1991**, *56*, 872–875.

11. (a) Chen, B.-H.; Ji, Q.-E. *Huaxue Xuebao* **1989**, *47*, 350–354; *Chem. Abstr.* **1989**, *111*, 194508a. (b) Li, Y.; Yu, Q. *Huaxue Shiji*, **1998**, *20*, 1–3; *Chem. Abstr.* **1998**, *128*, 257301. (c) Manabe, K. *Tetrahedron Lett.* **1998**, *39*, 5807–5810. (d) Nerinckx, W.; Vandewalle, M. *Tetrahedron: Asymmetry* **1990**, *1*, 265–276. (e) Juliá, S.; Ginebreda, A.; Guixer, J.; Tomás, A. *Tetrahedron Lett.* **1980**, *21*, 3709–3712. (f) Saigo, K.; Koda, H.; Nohira, H. *Bull. Chem. Soc. Jpn.* **1979**, *52*, 3119–3120. (g) Valli, V. L. K.; Sarma, G. V. M.; Choudary, B. M. *Ind. J. Chem.* **1990**, *29B*, 481–482. (h) McIntosh, J. M.; Acquaah, S. O. *Can. J. Chem.* **1988**, *66*, 1752–1756. (i) Esikova, I. A.; Serebryakov, É. P. *Izv. Akad. Nauk. SSSR, Ser. Khim.* **1989**, 1836–1843; *Chem. Abstr.* **1990**, *112*, 54595x.

12. (a) O'Donnell, M. J.; Bennett, W. D.; Wu, S. *J. Am. Chem. Soc.* **1989**, *111*, 2353–2355. (b) O'Donnell, M. J.; Wu, S.; Huffman, J. C. *Tetrahedron* **1994**, *50*, 4507–4518. (c) Esikova, I. A.; Nahreini, T. S.; O'Donnell, M. J. In *Phase-Transfer Catalysis: Mechanisms and Synthesis* (*ACS Symposium Series: 659*); Halpern, M. E. (Ed.); American Chemical Society: Washington, D.C., 1997; pp. 89–96.

13. Lygo, B.; Wainwright, P. G. *Tetrahedron Lett.* **1997**, *38*, 8595–8598.

14. Corey, E. J.; Xu, F.; Noe, M. C. *J. Am. Chem. Soc.* **1997**, *119*, 12414–12415.

15. O'Donnell, M. J.; Delgado, F.; Hostettler, C.; Schwesinger, R. *Tetrahedron Lett.* **1998**, *39*, 8775–8778.

16. Torrado, A.; Imperiali, B. *J. Org. Chem.* **1996**, *61*, 8940–8948.

17. (a) Tohdo, K.; Hamada, Y.; Shioiri, T. *Peptide Chem.* **1992**, *29*, 7–12. (b) Imperiali, B.; Fisher, S. L. *J. Org. Chem.* **1992**, *57*, 757–759. (c) Tohdo, K.; Hamada, Y.; Shioiri, T. *Synlett* **1994**, 247–249. (d) Imperiali, B.; Roy, R. S. *J. Org. Chem.* **1995**, *60*, 1891–1894.

18. Imperiali, B.; Prins, T. J.; Fisher, S. L. *J. Org. Chem.* **1993**, *58*, 1613–1616.

19. Kim, M. H.; Lai, J. H.; Hangauer, D. G. *Int. J. Pept. Protein Res.* **1994**, *44*, 457–465.

20. Pirrung, M. C.; Krishnamurthy, N. *J. Org. Chem.* **1993**, *58*, 954–956.

21. Corey, E. J.; Noe, M. C.; Xu, F. *Tetrahedron Lett.* **1998**, *39*, 5347–5350.

22. O'Donnell, M. J.; Wu, S. *Tetrahedron: Asymmetry.* **1992**, *3*, 591–594.

23. Belokon', Y. N.; Kochetkov, K. A.; Churkina, T. D.; Ikonnikov, N. S.; Chesnokov, A. A.; Larionov, O. V.; Parmár, V. S.; Kumar, R.; Kagan, H. B. *Tetrahedron: Asymmetry* **1998**, *9*, 851–857.

24. (a) Belokon', Y. N.; Maleev, V. I.; Savel'eva, T. F.; Garbalinskaya, N. S.; Saporovskaya, M. B.; Bakhmutov, V. I.; Belikov, V. M. *Izv. Akad. Nauk. SSSR, Ser. Khim.* **1989**, 631–635; *Chem. Abstr.* **1990**, *112*, 36400a. (b) Belokon', Y. N.; Maleev, V. I.; Videnskaya, S. O.; Saporovskaya, M. B.; Tsyryapkin, V. A.; Belikov, V. M. *Izv. Akad. Nauk SSSR, Ser. Khim.* **1991**, 126–134; *Chem. Abstr.* **1991**, *115*, 50229v. (c) Belokon', Y. N. *Izv. Akad. Nauk SSSR, Ser. Khim.* **1992**, 1106–1127; *Chem. Abstr.* **1993**, *118*, 234426. (d) Sun, P.; Zhang, Y. *Synth. Commun.* **1997**, *27*, 4173–4179. (e) Dehmlow, E. V.; Knufinke, V. *Liebigs Ann. Chem.* **1992**, 283–285. (f) Dehmlow, E. V.; Nachstedt, I. *J. Prakt. Chem.* **1993**, *335*, 371–374. (g) Schrader, S.; Dehmlow, E. V. *Zh. Org. Khim.* **1994**, *30*, 1431–1432. (h) Dehmlow, E. V.; Schrader, S. *Polish J. Chem.* **1994**, *68*, 2199–2208. (i) Dehmlow, E. V.; Klauck, R.; Neumann, B.; Stammler, H.-G. *J. Prakt. Chem.* **1998**, *340*, 572–575. (j) Uziel, J.; Genêt, J. P. *Russian J. Org. Chem.* **1997**, *33*, 1521–1542.

25. (a) An, X.-X.; Ding, M.-X. *Huaxue Xuebao* **1991**, *49*, 507–512; *Chem. Abstr.* **1991**, *115*, 158668b. (b) Lou, R. L.; Mi, A.; Zhou, Z. Y.; Zhang, J. M.; Jiang, Y. Z. *Chin. Chem. Lett.* **1997**, *8*, 1015–1016; *Chem. Abstr.* **1998**, *128*, 140278.

26. (a) Fiaud, J.-C. *Tetrahedron Lett.* **1975**, 3495–3496. Reference 26b questions this paper. (b) Dehmlow, E. V.; Singh, P.; Heider, J. *J. Chem. Res. (S)* **1981**, 292–293.

27. (a) Chen, J.-T.; Chen, Y.-H.; Sheng, H.-Y. *Youji Huaxue* **1988**, *8*, 164–166; *Chem. Abstr.* **1989**, *110*, 24259g. Note 27b questions this paper. (b) Reference 27a describes the chiral PTC benzylation of the benzaldehyde imine of glycine ethyl ester using catalyst Q*Cl (**6**, G=R=H) in a S/L PTC system of KOH/K$_2$CO$_3$ in CH$_2$Cl$_2$ at 30°C to give, following hydrolysis, D-phenylalanine in 61.6% yield and 89.9% op. We have not been able to obtain the reported level of induction when this reaction was repeated in our laboratory.

28. (a) Zhu, Z.; Yu, L. *Beijing Shifan Daxue Xuebao, Ziran Kexueban* **1989**, 63–67; *Chem. Abstr.* **1990**, *113*, 24466j. Note 28b questions this paper. (b) Reference 28 describes the chiral PTC benzylation (with 4-O$_2$N-C$_6$H$_4$CH$_2$Cl) of the benzophenone imine of glycine ethyl ester using catalyst Q*Cl (**3**, G=R=H) in a S/L PTC system of KOH/K$_2$CO$_3$ in CH$_2$Cl$_2$ at 20°C for 6 h to give the benzylated product in 51% yield and 96.6% op. We have not been able to obtain the reported level of induction when this reaction was repeated in our laboratory.

29. (a) Jamal Eddine, J.; Cherqaoui, M. *Tetrahedron: Asymmetry* **1995**, *6*, 1225–1228. Alkylation of Ph$_2$C=NCH$_2$Ph using catalyst **21** was reported to give up to 94% ee. Reference 29b questions this paper. (b) Dehmlow, E. V.; Klauck, R.; Duttmann, S.; Neumann, B.; Stammler, H.-G. *Tetrahedron: Asymmetry* **1998**, *9*, 2235–2244.

30. Soai, K.; Watanabe, M. *Chem. Commun.* **1990**, 43–44.

31. Iseki, K.; Nagai, T.; Kobayashi, Y. *Tetrahedron Lett.* **1994**, *35*, 3137–3138.

32. (a) Ando, A.; Miura, T.; Tatematsu, T.; Shioiri, T. *Tetrahedron Lett.* **1993**, *34*, 1507–1510. (b) Shioiri, T.; Bohsako, A.; Ando, A. *Heterocycles* **1996**, *42*, 93–97.

33. (a) Stoddart, J. F. *Biochem. Soc. Trans.* **1987**, *15*, 1188–1191. (b) Juliá, S.; Ginebreda, A. *Tetrahedron Lett.* **1979**, 2171–2174. (c) Juliá, S.; Ginebreda, A. *Afinidad* **1980**, *37*, 194–196. (d) Tong, Y.-J.; Ding, M.-X. *Youji Huaxue* **1990**, *10*, 464–470; *Chem. Abstr.* **1991**, *114*, 101294b.

34. (a) Sasai, H.; Itoh, N.; Suzuki, T.; Shibasaki, M. *Tetrahedron Lett.* **1993**, *34*, 855–858. (b) Gasparski, C. M.; Miller, M. J. *Tetrahedron* **1991**, *47*, 5367–5378.

35. Cram, D. J.; Sogah, G. D. Y. *Chem. Commun.* **1981**, 625–628.

36. Conn, R. S. E.; Lovell, A. V.; Karady, S.; Weinstock, L. M. *J. Org. Chem.* **1986**, *51*, 4710–4711.

37. Shishido, K.; Goto, K.; Miyoshi, S.; Takaishi, Y.; Shibuya, M. *J. Org. Chem.* **1994**, *59*, 406–414.

38. (a) Bakó, P.; Kiss, T.; Töke, L. *Tetrahedron Lett.* **1997**, *38*, 7259–7262. (b) Bakó, P.; Szöllösy, A.; Bombicz, P.; Töke, L. *Synlett* **1997**, 291–292. (c) Bakó, P.; Töke, L.; Szöllösy, A.; Bombicz, P. *Heteroatom Chem.* **1997**, *8*, 333–337.

39. Töke, L.; Bakó, P.; Keserü, G. M.; Albert, M.; Fenichel, L. *Tetrahedron* **1998**, *54*, 213–222.

40. Brunet, E.; Poveda, A. M.; Rabasco, D.; Oreja, E.; Font, L. M.; Batra, M. S.; Rodríguez-Ubis, J. C. *Tetrahedron: Asymmetry* **1994**, *5*, 935–948.

41. (a) Aoki, S.; Sasaki, S.; Koga, K. *Tetrahedron Lett.* **1989**, *30*, 7229–7230. (b) Aoki, S.; Sasaki, S.; Koga, K. *Heterocycles* **1992**, *33*, 493–495.

42. (a) Alonso-López, M.; Martín-Lomas, M.; Penadés, S. *Tetrahedron Lett.* **1986**, *27*, 3551–3554. (b) Alonso-López, M.; Jimenez-Barbero, J.; Martín-Lomas, M.; Penadés, S. *Tetrahedron* **1988**, *44*, 1535–1543. (c) Vicent, C.; Martín-Lomas, M.; Penadés, S. *Tetrahedron* **1989**, *45*, 3605–3612.

43. Díez-Barra, E.; de la Hoz, A.; Merino, S.; Rodríguez, A.; Sánchez-Verdú, P. *Tetrahedron* **1998**, *54*, 1835–1844.

44. (a) Loupy, A.; Zaparucha, A. *Tetrahedron Lett.* **1993**, *34*, 473–476. (b) Loupy, A.; Sansoulet, J.; Zaparucha, A.; Merienne, C. *Tetrahedron Lett.* **1989**, *30*, 333–336. (c) Delee, E.; Jullien, I.; Le Garrec, L.; Loupy, A.; Sansoulet, J.; Zaparucha, A. *J. Chromatogr.* **1988**, *450*, 183–189.

45. (a) Belokon', Y. N.; Kochetkov, K. A.; Churkina, T. D.; Chesnokov, A. A.; Smirnov, V. V.; Ikonnikov, N. S.; Orlova, S. A. *Russ. Chem. Bull.* **1998**, *47*, 74–81; *Chem. Abstr.* **1998**, *128*, 321883. (b) Belokon', Y. N.; Kochetkov, K. A.; Churkina, T. D.; Ikonnikov, N. S.; Orlova, S. A.; Smirnov, V. V.; Chesnokov, A. A. *Mendeleev Commun.* **1997**, 137–138; *Chem. Abstr.* **1997**, *127*, 293581. (c) Cui, J.; Yu, L. *Synth. Commun.* **1990**, *20*, 2895–2900. (d) Li, J.; Chen, Z.; Yu, L. *Beijing Shifan Daxue Xuebao, Ziran Kexueban* **1992**, *28*, 363–366; *Chem. Abstr.* **1994**, *121*, 180133.

46. (a) Takasu, M.; Wakabayashi, H.; Furuta, K.; Yamamoto, H. *Tetrahedron Lett.* **1988**, *29*, 6943–6946. (b) Crosby, J.; Stoddart, J. F.; Sun, X.; Venner, M. R. W. *Synthesis* **1993**, 141–145.

47. (a) Colonna, S.; Hiemstra, H.; Wynberg, H. *Chem. Commun.* **1978**, 238–239. (b) Wynberg, H.; Greijdanus, B. *Chem. Commun.* **1978**, 427–428. (c) Hermann, K.; Wynberg, H. *J. Org. Chem.* **1979**, *44*, 2238–2244. (d) Annunziata, R.; Cinquini, M.; Colonna, S. *Chem. Ind. (London)* **1980**, 238. (e) Colonna, S.; Annunziata, R. *Afinidad* **1981**, *38*, 501–502. (f) Colonna, S.; Re, A.; Wynberg, H. *J. Chem. Soc., Perkin Trans. 1* **1981**, 547–552. (g) Banfi, S.; Cinquini, M.; Colonna, S. *Bull. Chem. Soc. Jpn.* **1981**, *54*, 1841–1843.

48. (a) Raguse, B.; Ridley, D. D. *Aust. J. Chem.* **1984**, *37*, 2059–2071. (b) Dehmlow, E. V.; Sauerbier, C. *Liebigs Ann. Chem.* **1989**, 181–185. (c) Brunner, H.; Zintl, H. *Monatsh. Chem.* **1991**, *122*, 841–848. (d) Lavilla, R.; Gotsens, T.; Guerrero, M.; Masdeu, C.; Santano, M. C.; Minguillón, C.; Bosch, J. *Tetrahedron* **1997**, *53*, 13959–13968.

49. (a) Hiyama, T.; Sawada, H.; Tsukanaka, M.; Nozaki, H. *Tetrahedron Lett.* **1975**, 3013–3016. Reference 49b questions this paper. (b) Dehmlow, E. V.; Lissel, M.; Heider, J. *Tetrahedron* **1977**, *33*, 363–366.

50. Goldberg, Y.; Abele, E.; Bremanis, G.; Trapenciers, P.; Gaukhman, A.; Popelis, J.; Gomtsyan, A.; Kalvins, I.; Shymanska, M.; Lukevics, E. *Tetrahedron* **1990**, *46*, 1911–1922.

51. Arai, S.; Hamaguchi, S.; Shioiri, T. *Tetrahedron Lett.* **1998**, *39*, 2997–3000.

52. Arai, S.; Shioiri, T. *Tetrahedron Lett.* **1998**, *39*, 2145–2148.

53. Arai, S.; Ishida, T.; Shioiri, T. *Tetrahedron Lett.* **1998**, *39*, 8299–8302.

54. (a) Hummelen, J. C.; Wynberg, H. *Tetrahedron Lett.* **1978**, 1089–1092. (b) Annunziata, R. *Synth. Commun.* **1979**, *9*, 171–178. (c) Shi, M.; Masaki, Y. *J. Chem. Res. (S)* **1994**, 250–251. (d) Shi, M.; Kazuta, K.; Satoh, Y.; Masaki, Y. *Chem. Pharm. Bull.* **1994**, *42*, 2625–2628. (e) Shi, M.; Itoh, N.; Masaki, Y. *J. Chem. Res. (S)* **1995**, 46–47.

55. Drew, M. D.; Lawrence, N. J.; Watson, W.; Bowles, S. A. *Tetrahedron Lett.* **1997**, *38*, 5857–5860.

56. Rubina, K. I.; Goldberg, Y. S.; Shymanska, M. V.; Lukevics, E. *Appl. Organomet. Chem.* **1987**, *1*, 435–439; *Chem. Abstr.* **1989**, *110*, 39049n.

57. (a) Horner, L.; Skaletz, D. H. *Liebigs Ann. Chem.* **1977**, 1365–1409. (b) Horner, L.; Brich, W. *Liebigs Ann. Chem.* **1978**, 710–716.

58. Kwee, S.; Lund, H. *Bioelectrochem. Bioenerg.* **1980**, *7*, 693–698; *Chem. Abstr.* **1981**, *94*, 204217b.

59. Passarotti, C.; Bandi, G. L.; Fossati, A.; Valenti, M.; Dal Bo, L. *Boll. Chim. Farm.* **1990**, *129*, 195–198; *Chem. Abstr.* **1991**, *114*, 228584b.

60. (a) Colonna, S.; Fornasier, R. *Synthesis* **1975**, 531–532. (b) Balcells, J.; Colonna, S.; Fornasier, R. *Synthesis* **1976**, 266–267. (c) Colonna, S.; Fornasier, R. *J. Chem. Soc., Perkin Trans. 1* **1978**, 371–373. (d) Juliá, S.; Ginebreda, A.; Guixer, J.; Masana, J.; Tomás, A.; Colonna, S. *J. Chem. Soc., Perkin Trans. 1* **1981**, 574–577. (e) Innis, C.; Lamaty, G. *Nouv. J. Chim.* **1977**, *1*, 503–509. (f) Kinishi, R.; Nakajima, Y.; Oda, J.; Inouye, Y. *Agric. Biol. Chem.* **1978**, *42*, 869–872. (g) Shida, Y.; Ando, N.; Yamamoto, Y.; Oda, J.; Inouye, Y. *Agric. Biol. Chem.* **1979**, *43*, 1797–1799. (h) Kinishi, R.; Uchida, N.; Yamamoto, Y.; Oda, J.; Inouye, Y. *Agric. Biol. Chem.* **1980**, *44*, 643–648. (i) Li, S.; Meng, S.; Jian, H. *Beijing Gongye Daxue Xuebao* **1987**, *13*, 117–120; *Chem. Abstr.* **1988**, *109*, 148995t. (j) Sarkar, A.; Rao, B. R. *Tetrahedron Lett.* **1991**, *32*, 1247–1250. (k) Takeshita, M.; Yanagihara, H.; Terada, K.; Akutsu, N. *Annu. Rep. Tohoku Coll. Pharm.* **1992**, *39*, 247–250; *Chem. Abstr.* **1994**, *120*, 244600. (l) Gomtsyan, A.; Gevorgyan, V.; Kalvins, I.; Lukevics, E. *Latv. Kim. Z.* **1992**, 361–365; *Chem. Abstr.* **1993**, *118*, 192137z. (m) Zhdanov, Y. A.; Alekseev, Y. E.; Korol, E. L.; Sudareva, T. P.; Alekseeva, V. G. *Zh. Obshch. Khim.* **1993**, *63*, 1495–1501.

61. Blaser, H.-U.; Boyer, S. K.; Pittelkow, U. *Tetrahedron: Asymmetry* **1991**, *2*, 721–732.

62. (a) Massé, J. P.; Parayre, E. R. *Chem. Commun.* **1976**, 438–439. References 26b and 60c question this paper. (b) Massé, J. P.; Parayre, E. *Bull. Soc. Chim. Fr. II* **1978**, 395–400.

63. (a) Helder, R.; Hummelen, J. C.; Laane, R. W. P. M.; Wiering, J. S.; Wynberg, H. *Tetrahedron Lett.* **1976**, 1831–1834. (b) Marsman, B.; Wynberg, H. *J. Org. Chem.* **1979**, *44*, 2312–2314. (c) Wynberg, H. *Chimia* **1976**, *30*, 445–451.

64. (a) Wynberg, H.; Marsman, B. *J. Org. Chem.* **1980**, *45*, 158–161. (b) Pluim, H.; Wynberg, H. *J. Org. Chem.* **1980**, *45*, 2498–2502. (c) Snatzke, G.; Wynberg, H.; Feringa, B.; Marsman, B. G.; Greydanus, B.; Pluim, H. *J. Org. Chem.* **1980**, *45*, 4094–4096.

65. Lygo, B.; Wainwright, P. G. *Tetrahedron Lett.* **1998**, *39*, 1599–1602.

66. (a) Onda, M.; Li, S.; Li, X.; Harigaya, Y.; Takahashi, H.; Kawase, H.; Kagawa, H. *J. Nat. Prod.* **1989**, *52*, 1100–1106. (b) Takahashi, H.; Kubota, Y.; Miyazaki, H.; Onda, M. *Chem. Pharm. Bull.* **1984**, *32*, 4852–4857.

67. (a) Mazaleyrat, J. P. *Tetrahedron Lett.* **1983**, *24*, 1243–1246. (b) Domagala, J. M.; Bach, R. D. *J. Org. Chem.* **1979**, *44*, 3168–3174. (c) Banfi, S.; Colonna, S.; Julia, S. *Synth. Commun.* **1983**, *13*, 1049–1052.

68. (a) Macdonald, G.; Alcaraz, L.; Lewis, N. J.; Taylor, R. J. K. *Tetrahedron Lett.* **1998**, *39*, 5433–5436. (b) Alcaraz, L.; Macdonald, G.; Ragot, J. P.; Lewis, N.; Taylor, R. J. K. *J. Org. Chem.* **1998**, *63*, 3526–3527.

69. (a) Harigaya, Y.; Yamaguchi, H.; Onda, M. *Heterocycles* **1981**, *15*, 183–185. (b) Harigaya, Y.; Yamaguchi, H.; Onda, M. *Chem. Pharm. Bull.* **1981**, *29*, 1321–1327.

70. Arai, S.; Oku, M.; Miura, M.; Shioiri, T. *Synlett* **1998**, 1201–1202.

71. (a) Baba, N.; Oda, J.; Kawaguchi, M. *Agric. Biol. Chem.* **1986**, *50*, 3113–3117. (b) Baba, N.; Kawahara, S.; Hamada, M.; Oda, J. *Bull. Inst. Chem. Res., Kyoto Univ.* **1987**, *65*, 144–146; *Chem. Abstr.* **1988**, *109*, 170140g. (c) Baba, N.; Oda, J.; Kawahara, S.; Hamada, M. *Bull. Inst. Chem. Res., Kyoto Univ.* **1989**, *67*, 121–127; *Chem. Abstr.* **1990**, *113*, 23546y.

72. Chang, S.; Galvin, J. M.; Jacobsen, E. N. *J. Am. Chem. Soc.* **1994**, *116*, 6937–6938.

73. Masui, M.; Ando, A.; Shioiri, T. *Tetrahedron Lett.* **1988**, *29*, 2835–2838.

74. de Vries, E. F. J.; Ploeg, L.; Colao, M.; Brussee, J.; van der Gen, A. *Tetrahedron: Asymmetry* **1995**, *6*, 1123–1132.

75. Inoue, K.; Noguchi, H.; Hidai, M.; Uchida, Y. *Yukagaku* **1980**, *29*, 397–401; *Chem. Abstr.* **1981**, *94*, 102469t.

76. Bourson, J.; Goguillon, T.; Jugé, S. *Phos. Sulfur* **1983**, *14*, 347–356.

77. (a) Juliá, S.; Ginebreda, A.; Guixer, J. *Chem. Commun.* **1978**, 742–743. (b) Yu, L.; Wang, F. *Gaodeng Xuexiao Huaxue Xuebao* **1987**, *8*, 336–340; *Chem. Abstr.* **1988**, *108*, 112909h. (c) Guifa, S.; Lingchong, Y. *Synth. Commun.* **1993**, *23*, 1229–1234.

78. Wakabayashi, T.; Watanabe, K. *Chem. Lett.* **1978**, 1407–1410.

79. Aires-de-Sousa, J.; Lobo, A. M.; Prabhakar, S. *Tetrahedron Lett.* **1996**, *37*, 3183–3186.

80. (a) Verbicky, J. W., Jr.; O'Neil, E. A. *J. Org. Chem.* **1985**, *50*, 1786–1787. Reference 80b questions this paper. (b) Dehmlow, E. V.; Sleegers, A. *J. Org. Chem.* **1988**, *53*, 3875–3877.

81. (a) Annunziata, R.; Cinquini, M.; Colonna, S. *J. Chem. Soc., Perkin Trans. 1* **1980**, 2422–2424. (b) Ahuja, R. R.; Bhole, S. I.; Bhongle, N. N.; Gogte, V. N.; Natu, A. A. *Indian J. Chem.* **1982**, *21B*, 299–303.

82. Enantiomeric excess (% ee) is the percent of the major enantiomer minus that of the minor enantiomer, whereas optical purity (% op) is the ratio, expressed as a percent, of the optical rotation of a mixture of enantiomers to that of the pure enantiomer.

83. Prices listed in Chart 1 are for 0.1 mole of parent alkaloids from Aldrich Chemical Company Catalog Handbook of Fine Chemicals, 1998–1999.

84. The pseudoenantiomeric catalysts derived from the alkaloid pairs cinchonine/cinchonidine and quinidine/quinine (**1/4** or **2/5**, respectively) are related as diastereomers but are enantiomeric with respect to the front "working-face" defined by carbons 8 and 9.

85. (a) McIntosh, J. M. *Tetrahedron Lett.* **1979**, 403–404. (b) McIntosh, J. M. *Can. J. Chem.* **1980**, *58*, 2604–2609. (c) McIntosh, J. M. *Tetrahedron* **1982**, *38*, 261–266. (d) McIntosh, J. M.; Khan, M. A.; Delbaere, L. T. J. *Can. J. Chem.* **1982**, *60*, 1073–1077. (e) McIntosh, J. M. *J. Org. Chem.* **1982**, *47*, 3777–3779.

86. Lygo, B.; Crosby, J.; Lowdon, T. R.; Wainwright, P. G. *Tetrahedron Lett.* **1997**, *38*, 2343–2346.

87. (a) Mikhlina, E. E.; Yakhontov, L. N. *Khim. Geterotsikl. Soedin.* **1984**, 147–161; *Chem. Abstr.* **1984**, *100*, 209536g. (b) Singh, P.; Arora, G. *Indian J. Chem.* **1986**, *25B*, 1034–1037. (c) Dehmlow, E. V.; Knufinke, V. *J. Chem. Res. (S)* **1989**, 224–225, 400.

88. (a) Hiyama, T.; Mishima, T.; Sawada, H.; Nozaki, H. *J. Am. Chem. Soc.* **1975**, *97*, 1626–1627; **1976**, *98*, 641. Reference 69 questions this paper. (b) Hernandez, O.; Bhatia, A. V.; Walker, M. *J. Liq. Chromatogr.* **1983**, *6*, 1475–1489.

89. Liotta, C. L.; Burgess, E. M.; Ray, C. C.; Black, E. D.; Fair, B. E. In *Phase Transfer Catalysis (ACS Symposium Series: 326)*; Starks, C. M. (Ed.); American Chemical Society: Washington, D.C., 1987; pp. 15–23.

90. (a) Mason, D.; Magdassi, S.; Sasson, Y. *J. Org. Chem.* **1990**, *55*, 2714–2717. (b) Moberg, R.; Bökman, F.; Bohman, O.; Siegbahn, H.O.G. *J. Am. Chem. Soc.* **1991**, *113*, 3663–3667.

91. (a) Lipkowitz, K. B.; Cavanaugh, M. W.; Baker, B.; O'Donnell, M. J. *J. Org. Chem.* **1991**, *56*, 5181–5192. (b) Vicent, C.; Jiménez-Barbero, J.; Martín-Lomas, M.; Penadés, S.; Cano, F. H.; Foces-Foces, C. *J. Chem. Soc., Perkin Trans. 2* **1991**, 905–912. (c) Vicent, C.; Bosso, C.; Cano, F. H.; de Paz, J. L. G.; Foces-Foces, C.; Jiménez-Barbero, J.; Martín-Lomas, M.; Penadés, S. *J. Org. Chem.* **1991**, *56*, 3614–3618.

92. (a) Struchkova, M. I.; Él'perina, E. A.; Abylgaziev, R. I.; Serebryakov, E. P. *Izv. Akad. Nauk. SSSR, Ser. Khim.* **1989**, 2492–2500; *Chem. Abstr.* **1990**, *112*, 217371n. (b) Pochapsky, T. C.; Stone, P. M.; Pochapsky, S. S. *J. Am. Chem. Soc.* **1991**, *113*, 1460–1462. (c) Mo, H.; Pochapsky, T. C. *Prog. NMR Spec.* **1997**, *30*, 1–38.

11

ASYMMETRIC POLYMERIZATION

YOSHIO OKAMOTO AND TAMAKI NAKANO
Department of Applied Chemistry, Graduate School of Engineering, Nagoya University, Chikusa-ku, Nagoya 464-8603, JAPAN

11.1. INTRODUCTION

Most naturally occurring macromolecules (polymers) such as proteins, polysaccharides, and nucleic acids are optically active. The chiral nature of these polymers appears essential for them to exhibit sophisticated functions. Therefore, the synthesis of optically active polymers is not only interesting from an academic view point for the chiral structure, but it is also important as a practical method to produce functional polymeric materials with wide potential applications [1–9]. Optically active polymers are synthesized either by the polymerization of monomers having an optically active moiety without chiral induction to the polymer chain or by polymerization in which chirality is introduced into the polymer structure through the polymerization reaction. In this chapter, the latter type of polymerization (asymmetric polymerization) is reviewed, dividing it into three major categories, namely, asymmetric synthesis polymerization, helix-sense-selective polymerization, and enantiomer-selective polymerization. Chiral induction can be achieved either by catalytic reaction by using achiral or prochiral monomers or by non-catalytic reaction systems typically based on chiral monomer structure. Although non-catalytic syntheses may not exactly be within the scope of this book, some examples of them are included in this chapter because they still occupy an important position in the field of asymmetric polymerization. Also, in the following text, "catalysts" include the initiators for polymerization, which attach a chiral group to the chain end of a polymer although they are not catalysts in a strict sense.

11.2. ASYMMETRIC SYNTHESIS POLYMERIZATION

Asymmetric synthesis polymerization is the reaction that produces the polymers with configurational chirality of the main chain from an optically inactive, prochiral monomer or a prochiral

Catalytic Asymmetric Synthesis, Second Edition, Edited by Iwao Ojima
ISBN 0-471-29805-0   Copyright © 2000 Wiley-VCH, Inc.

monomer bearing an optically active auxiliary. Chiral induction is achieved by selective attack of the growing species on a monomer on one enantioface of unsaturated bonding. Various olefinic monomers have been used in this type of polymerization. However, the extent of chiral induction was determined in only limited cases. The relation between the optical activity of polymers and the configuration of the main chain has been discussed in several reports [7–13].

11.2.1. Catalytic Syntheses

11.2.1.1. Vinyl Polymers Stereoregular (e.g., isotactic) polymers of vinyl monomers (1-substituted and 1,1-disubstituted olefins) cannot be optically active even if effective chiral induction takes place through polymerization reaction because the polymer chain has a plane of symmetry and the main-chain chiral centers are pseudoasymmetric [7–19] as exemplified by polystyrene [15,16], polypropylene [17], polymethacrylates [18,19], and polyacrylonitrile [18]. The polymers do not show optical activity except for that based on the chiral centers in the vicinity of the chain ends. However, if some higher-order tacticities are realized for a vinyl polymer with effective chiral induction, the polymer can be optically active. Chirality of a polymer chain can be tested by representing a stereotactic infinite chain as a cyclic compound and by considering the symmetry of the compound (Figures 11.1 and 11.2 [1a]) [1a,7,8,13]. Whereas all stereosequences of diad, triad, tetrad, and pentad have a plane of symmetry, and, therefore, they are not chiral, a hexad sequence having triad tacticity of mm/mr/rr = 0/67/33 is chiral [Figure 11.1 (T) and (U)]. The number of chiral sequences increases as the stereosequence becomes longer. Chirality of copolymers can be tested in the same manner (Figure 11.2). Among the three A-A-B sequences, the one shown as (D) is chiral. This method of chirality test can be applied to other vinyl polymers, including 1,2-disubstituted olefins.

The synthesis of optically active PMMA has been realized through the anionic polymerization of 9-phenylfluoren-9-yl methacrylate (**PFMA**) by using the complexes of *N,N'*-diphenylethylenediamine monolithium amide (**DPEDA-Li**) with sparteine (**Sp**), (+)-1-(2-pyrrolidinylmethyl)pyrrolidine (**PMP**), and (+)-2,3-dimethoxy-1,4-bis(dimethylamino)butane (**DDB**) (Table 11.1) [20]. Although **PFMA** has a bulky structure similar to triphenylmethyl methacrylate (**TrMA**), helix formation was not confirmed in its polymerization (see next section). However, the PMMAs derived from the poly(**PFMA**) exhibit significant optical activity. The optical activity of PMMA derived from the poly(**PFMA**) obtained with **DPEDA-Li-PMP** decreases as the degree of polymerization (DP) increases and reaches an almost constant value ($[\alpha]_{365}$ −11~−15°) in the DP range of ca. 120–150, which suggests that the chiroptical property was partly due to the chiral centers in the vicinity of chain terminals, but the configurational chirality of in-chain asymmetric centers makes a significant contribution to the optical activity.

PFMA   DPEDA-Li   Sp   (+)-DDB   (+)-PMP

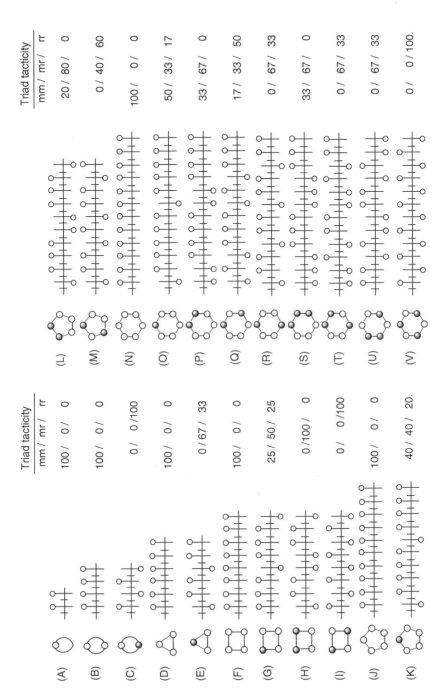

Figure 11.1 Chirality of vinyl polymer chain. [Reprinted with permission from ref. 1a. [Copyright 1994 American Chemical Society.]

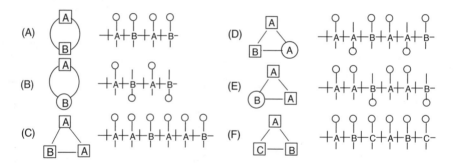

Figure 11.2 Chirality of copolymer chain. [Reprinted with permission from ref. 1a. Copyright 1994 American Chemical Society.]

TABLE 11.1. Asymmetric Anionic Polymerization of PFMA with DPEDA-Li Complexes in Toluene at−78°C[a]

Monomer	Ligand	Yield[b] (%)	$[\alpha]_{365}^{25}$ [c] (deg)	DP[d]	Mw/Mn[d]	Tacticity[e] (%)			$[\alpha]_{365}^{25}$ of PMMA[c] (deg)
						mm	mr	rr	
PFMA	(+)-**PMP**	95	+30	41	1.20	54	25	21	−44
PFMA	(+)-**DDB**	96	−16	39	1.17	46	32	22	−2
PFMA	(−)-**Sp**	50	−2	29	1.18	48	35	17	+38
TrMA	(+)-**PMP**	>99	+1350[f]	39[f]	1.12[f]	>99[f]	<1[f]	<1[f]	+3

[a]Conditions: [M]/[I] = 20 =, time 48 hr. Data cited from ref. 20.
[b]MeOH-insoluble part.
[c]In THF.
[d]Determined by GPC analysis of the PMMA derived from the original polymer.
[e]Determined by ^{1}H NMR analysis of the PMMA derived from the original polymer.
[f]Properties of benzene-hexane (1/1)-insoluble part of the products.

Enantioselective processes in olefin oligomerizations have been investigated for the systems involving propylene, 1-pentene, and 4-methyl-1-pentene using optically active Zr catalysts (**1** and **2**) with methylaluminoxane (**MAO**) in the presence of H_2 [21,22]. Chiral induction is observed in the oligomerization of 1-butene by using $(RO)_2TiCl_2/MgCl_2$ catalysts whose R is based on chiral monosaccharides (**3–5**) in the presence of organoaluminum cocatalysts [23]. It was found in the oligomerization that only trimer or higher oligomers were optically active. The isotactic trimer (**6**, n = 2) and tetramer (**6**, n = 3) obtained after workup by using O_2/aq.NaOH had 29–71% ee depending on the ligand and reaction conditions.

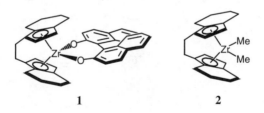

1   2

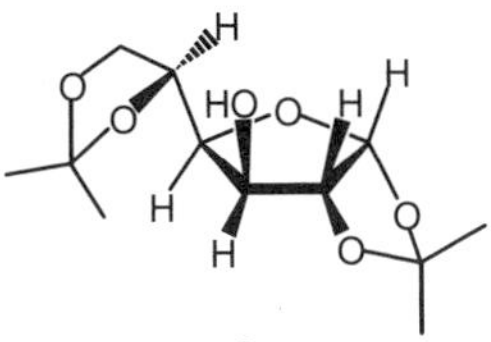

3 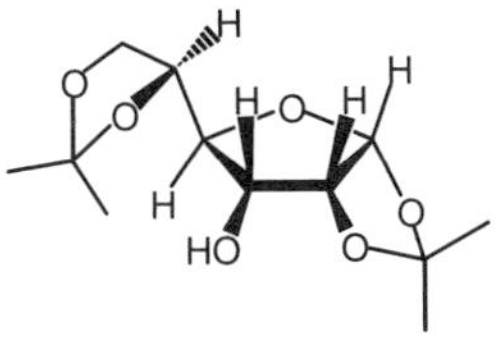   4 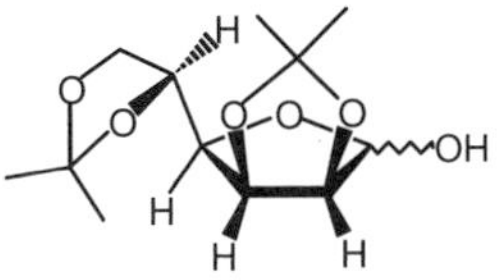   5

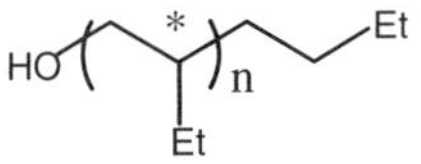

6

11.2.1.2. Copolymers of Olefin and Carbon Monoxide

Palladium catalysts are known to give an alternating copolymer (poly(1,4-ketone)) from propylene and carbon monoxide [24–31]. When 1,3-bis(diisopropylphosphino)propane (dippp) is used as a ligand, a polymer with an almost completely head-to-tail structure and an isotactic configuration is obtained. This type of a polymer does not have a plane of symmetry in the main chain, and, therefore, application of chiral ligands for the polymerization can lead to optically active polymers. Ligands (R,S)-**BINAPHOS** [24–26], (S,S)-**Me-DUPHOS** [27,28], (S)-**BICHEP** [29,30], and $(R)(S_p)$-**JOSIPHOS** [31] were successfully used for the asymmetric synthesis polymerization.

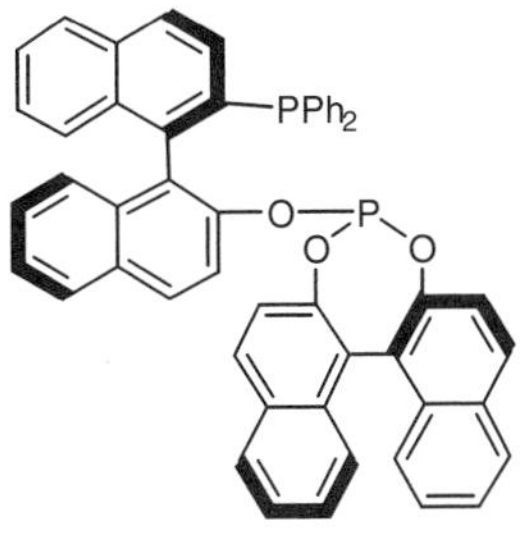

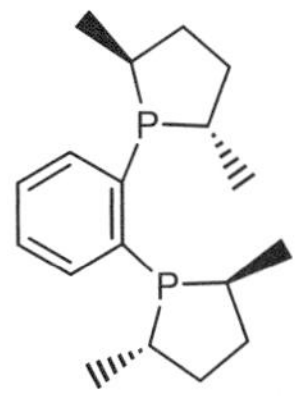

(S,S)-**Me-DUPHOS**

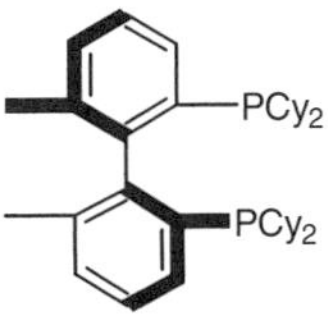

(S)-**BICHEP**
Cy = cyclohexyl

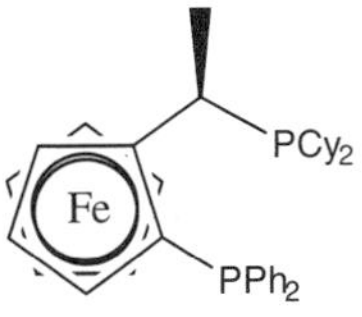

$(R)(S_p)$-**JOSIPHOS**
Cy = cyclohexyl

(R,S)-**BINAPHOS**

The polymer prepared with $[Pd(CH_3)(CH_3CN)\{(R,S)$-**BINAPHOS**$\}]$ $[B\{3,5\text{-}(CF_3)_2\text{-}C_6H_3\}_4]$ had a high molecular weight (Mn 65,000) and exhibited optical activity of $[\Phi]_D$ +40° (in $(CF_3)_2CHOH$) (Scheme 11.1) [24,25]. The extent of chiral induction in the polymerization with **BINAPHOS** was studied by the oligomerization reaction leading to **7**. The ee of **7** was

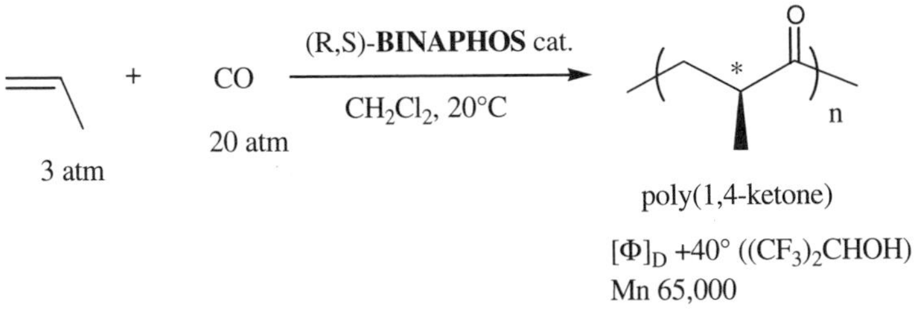

3 atm

20 atm

(R,S)-**BINAPHOS** cat.

CH_2Cl_2, 20°C

poly(1,4-ketone)

$[\Phi]_D$ +40° $((CF_3)_2CHOH)$
Mn 65,000

Scheme 11.1

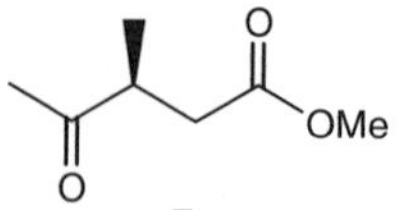

over 95%, which suggests a high extent of chiral induction during the polymerization reaction. Reaction intermediates have been identified spectroscopically for this polymerization. Though the catalyst failed to copolymerize 2-butene (internal olefin) with carbon monoxide, it gave optically active polymers in the copolymerization of α,ω-diene and carbon monoxide. The copolymers obtained from 1,4-pentadiene and 1,5-hexadiene had structures **8** and **9**, respectively, and the latter polymer showed $[\Phi]_D$ 13.8°.

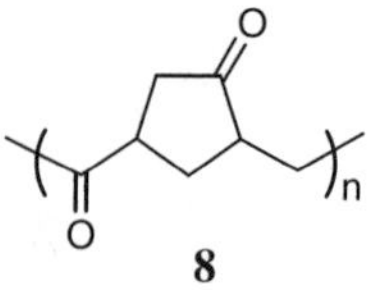

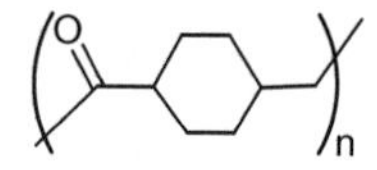

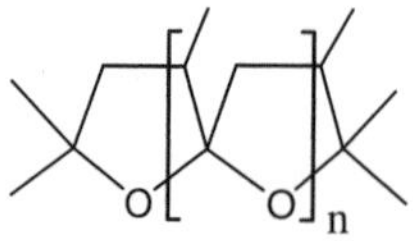

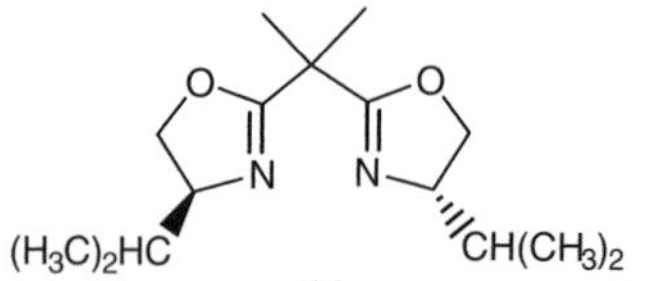

The catalyst [Pd(**Me-DUPHOS**)(MeCN)$_2$](BF$_4$)$_2$ was also effective in the alternating asymmetric copolymerization of aliphatic α-olefins with carbon monoxide [27,28]. The polymer synthesized in a CH$_3$NO$_2$-CH$_3$OH mixture has both 1,4-ketone and spiroketal (**10**) units in the main chain. The propylene-CO copolymer consisting only of a 1,4-ketone structure shows $[\Phi]_D$ +22° (in (CF$_3$)$_2$CHOH), and the optical purity of the main chain chiral centers is over 90% as estimated by NMR analysis using a chiral Eu shift reagent.

In addition to α-olefins, p-t-butylstyrene can be used as a comonomer for the isotactic specific copolymerization with carbon monoxide with chiral induction [32]. Whereas bipyridyl as ligand gives a syndiotactic polymer, the Pd-catalyzed polymerization with ligand **11** leads to a polymer with isotacticity of over 98% and high optical activity ($[\Phi]_D$ −536° (in CH$_2$Cl$_2$)) [33, 34]. The **BINAPHOS** catalyst described above is also effective in this type of copolymerization [25].

11.2.1.3. 1,3-Diene Polymers

Polymerization of a 1,3-diene yields a polymer having true asymmetric centers in the main chain and ozonolysis of the polymer gives a chiral diacid compound (**12**) whose analysis of optical purity discloses the extent of chiral induction in the polymerization (Scheme 11.2) [12,35–39]. The polymerization of methyl and butyl sorbates; methyl and butyl styrylacrylates; and methyl, ethyl, butyl, and t-butyl 1,3-butadiene-1-carboxylates using (+)-2-methylbutyllithium, butyllithium/(−)-menthyl ethyl ether, butyllithium/menthoxy-Na, butyllithium/borneoxy-Na, butyllithium/Ti((−)-menthoxy)$_4$, and butyllithium/bornyl ethyl ether initiators [35–37] and that of 1,3-pentadiene in the presence of

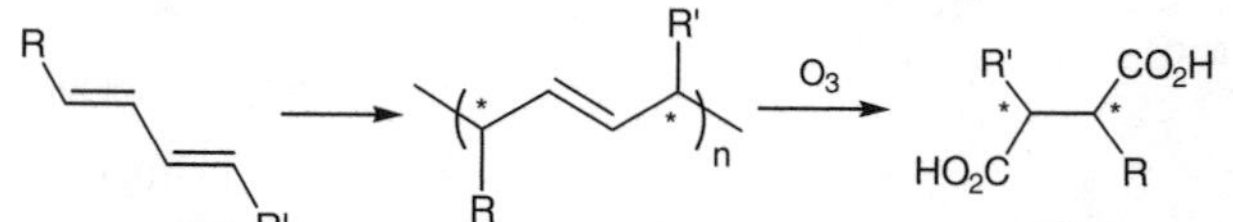

Scheme 11.2

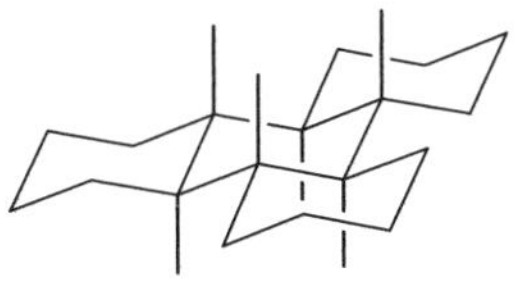

13 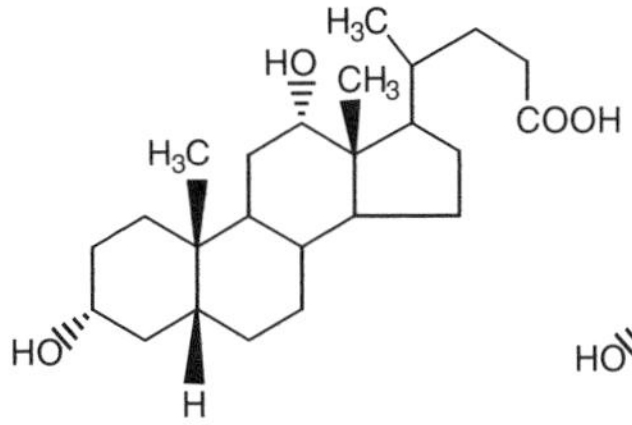 **14** 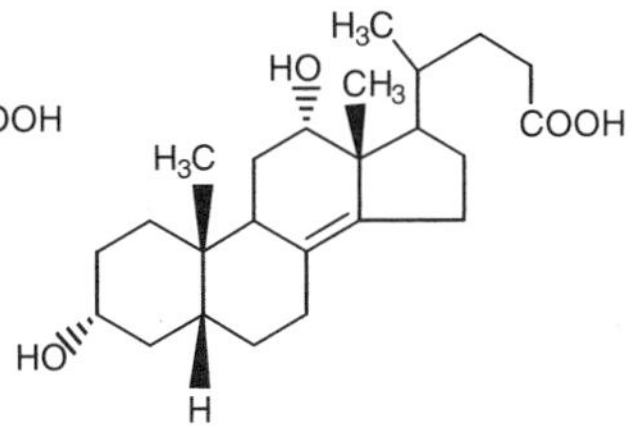 **15**

AlEt$_3$/Ti((−)-menthoxy)$_4$, Ti(OBu)$_4$/Al((+)-2-methylbutyl)$_3$, and VCl$_3$/Al((+)-2-methyl-butyl)$_3$-Et$_2$O catalysts [38,39] are known to give optically active polymers. Optical purity of **12** obtained from the polymers of the sorbates and the styrylacrylates has been reported to be 0.4–6%.

Chiral solid matrices are used for asymmetric synthesis polymerization of 1,3-dienes (inclusion polymerization), although the matrix reaction is not exactly a catalytic synthesis [40,41]. (*R*)-*trans-anti-trans-anti-trans*-Perhydrotriphenylene (**13**) [42,43], deoxyapocholic acid (**14**) [44,45], and apocholic acid (**15**) [46,47] are known as effective matrices for the 1,3-diene polymerization. Monomer molecules are included in chiral channels in the matrix crystals, and the polymerization takes place in chiral environment. The γ-ray irradiation polymerization of *trans*-1,3-pentadiene included in **13** gives an optically active isotactic polymer with a *trans*-structure. The polymerization of (*Z*)-2-methyl-1,3-butadiene using **15** as a matrix leads to a polymer having an optical purity of the main-chain chiral centers of 36% [47].

11.2.1.4. Cyclic Olefin Polymers Benzofuran (**16**) gives an optically active polymer by cationic polymerization with AlEtCl$_2$ or AlCl$_3$ in the presence of an optically active cocatalyst such as β-phenylalanine and 10-camphorsulfonic acid [12,48–50]. The optically active polymer is considered to have an erythro- or threodiisotactic structure with no plane of symmetry. Initiator systems of AlCl$_3$/(−)-menthoxytriethyltin, -germanium, and -silicon also give an optically active polymer [51,52].

Maleimides (**17**) with a substituent on the N atom have a C$_2$ axis of symmetry. Therefore, a polymaleimide can be optically active only when the main chain has one of the two enantiomeric *trans*-structures (Scheme 11.3) [53,54], whereas all asymmetric centers in polybenzofuran are

16 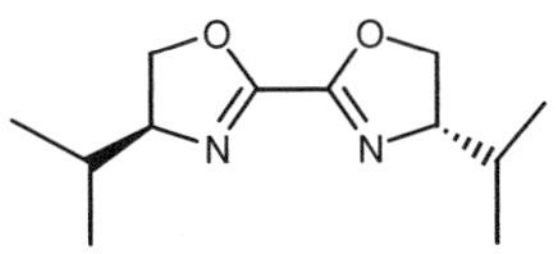 **18** 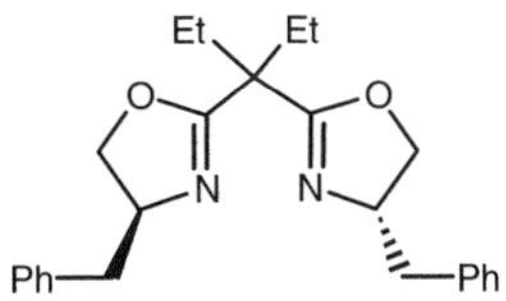 **19**

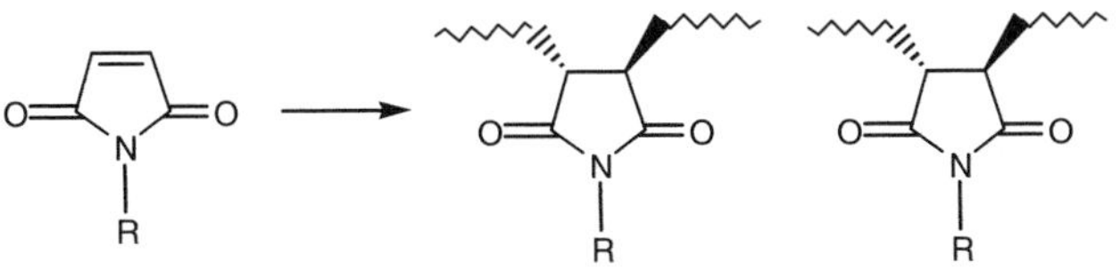

17

Scheme 11.3

true chiral centers. Optically active poly-**17**s have been synthesized by anionic polymerization using the n-BuLi/**Sp** complex (N-phenylmaleimide, $[\alpha]_{435}$ −16.9° [53]; N-cyclohexylmaleimide, $[\alpha]_D$ −39.2° [54]); the n-BuLi-CuI/**Sp** complex (N-phenylmaleimide, $[\alpha]_{435}$ −23.7° [53]); the sodium bis(2-methoxyethoxy)aluminum hydride/**Sp** complex; the n-BuLi/(−)-6-ethylsparteine (**EtSp**) complex; and the n-BuLi/(+)-**DDB** complex [53]. Ligands based on an oxazoline structure are also effective in the asymmetric synthesis polymerization of **17**, and polymerization with the n-BuLi/**18** complex [55] (N-phenylmaleimide, $[\alpha]_{435}$ −5.8°; N-cyclohexylmaleimide, $[\alpha]_{435}$ −7.4°); the n-BuLi/**19** complex [56] (N-phenylmaleimide, $[\alpha]_{435}$ +0.9°; N-cyclohexylmaleimide, $[\alpha]_{435}$ +111.4°); and the Et$_2$Zn/**19** complex [56] (N-phenylmaleimide, $[\alpha]_{435}$ +94.6°; N-cyclohexylmaleimide, $[\alpha]_{435}$ +117.5°) has been reported. In addition, polymerization with Li-(−)-menthoxide and lithium amide of (+)-PMP gives optically active poly(N-triphenylmethylmaleimide) [57].

11.2.1.5. α,ω-Diolefin Polymers (cyclopolymerization) Cyclopolymerization of 1,5-hexadiene using metallocene catalyst (S,S)-**1** in the presence of MAO gives an optically active polymer **20** (Scheme 11.4) [58,59]. The polymerization proceeded exclusively via the cyclization mechanism, and the obtained polymer **20** was rich in *trans*-units (73%) having no plane of symmetry and it showed molecular rotation of $[\Phi]_{405}$ −51.2°.

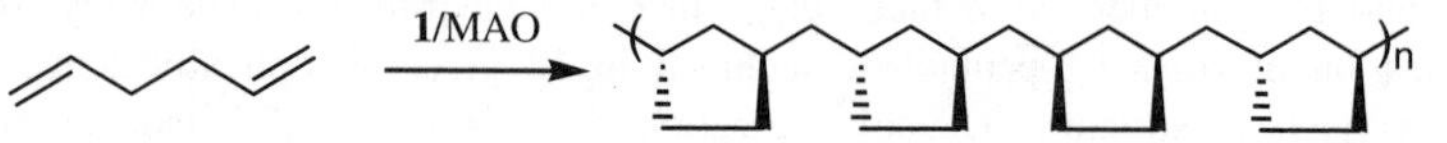

20

Scheme 11.4

Optically active catalyst **1** can be obtained either by enantiomer-selective reaction of rac.-**2** with optically active lithium (1,1′-binaphthyl)-2,2′-diolate or by direct resolution by chiral HPLC. Optically active **21** and **22** in addition to **1** were successfully obtained by HPLC resolution and used for the polymerization of 1,5-hexadiene [60–62]. Both catalysts gave an optically active polymer through cyclopolymerization. The optical activity and the content of *trans*-structure in the main chain of the polymers obtained with **21** and **22** were comparable with those of the polymers synthesized with **1** [61,62].

N-Phenyl-N-allylmethacrylamide (**23**) gives a polymer with a five-membered ring structure with true asymmetric centers in the main chain by free-radical cyclopolymerization [63]. When the polymerization is carried out in the presence of SnCl$_4$/(−)-menthol, the resulting polymer was optically active ($[\alpha]_D$ −5.6°). Chiral induction was also observed in the copolymerization of **23** with MMA. Cationic cyclopolymerization of **24** using a ZnCl$_2$/10-camphorsulfonic acid (**25**) initiator system gives an optically active polymer having a 1,3-dioxane structure in the main chain ($[\alpha]_{435}$ −17°) [64].

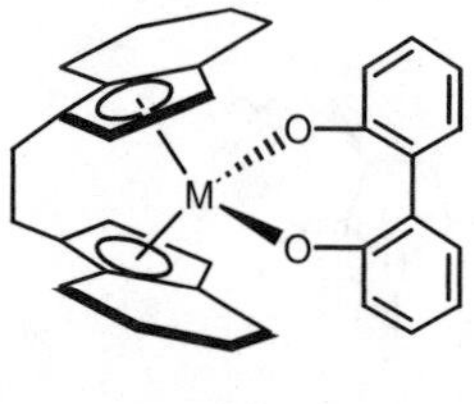

21: M = Zr
22: M = Hf

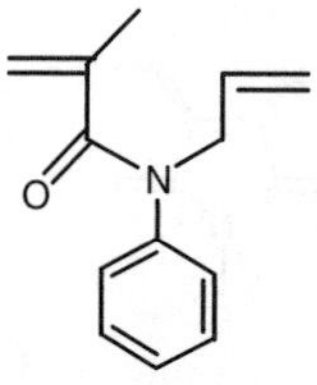

23

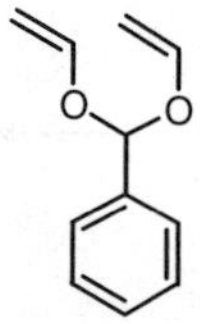

24

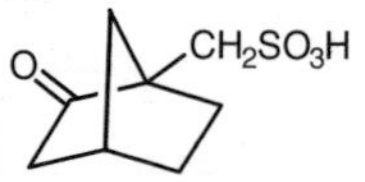

25

11.2.1.6. Ring-Opening Polymerization An asymmetric synthesis process has been found in the ring-opening polymerization of three-membered cyclic monomers having a hetero atom (O or S) using optically active initiators. In the polymerization of monomers shown in Scheme 11.5, chirality of the polymers depends on that of the monomer. Because inversion of one of the asymmetric centers takes place in the polymerization, *cis*-monomers can give a chiral main-chain structure whose configuration is determined by on which side of the monomer the C-X bond breaks (α or β in Scheme 11.5 A), whereas *trans*-monomers lead to a chain structure with a plane of symmetry (Scheme 11.5 B).

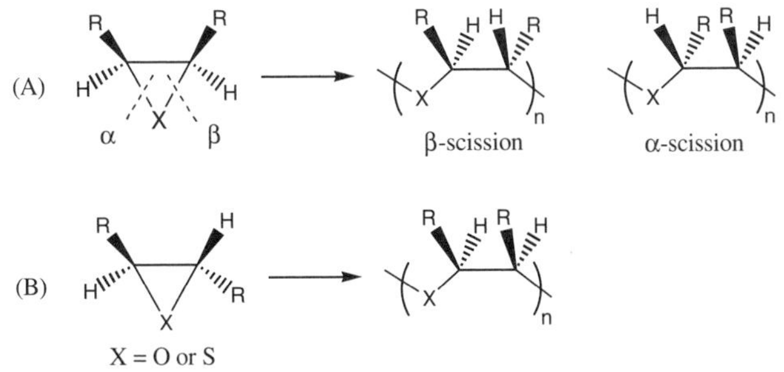

Scheme 11.5

Polymerization of *cis*-dimethyloxirane, cyclohexene oxide, cis-dimethylthiirane, and cyclohexene sulfide was carried out using $ZnEt_2$/(R)-3,3-dimethyl-1,2-butanediol (**26**) (1/1) [65,66], $CdMe_2$/**26** (1/1) [65,66], $ZnEt_2$/(R,R)-1,2-diphenyl-1,2-ethanediol (**27**) (1/1) [66], $ZnEt_2$/(S)-binaphthol (**28**) (1/1) [65], and $ZnEt_2$/(1S,2R)-ephedrine (**29**) [67], and, for example, a polymer obtained from *cis*-dimethylthiirane with the $ZnEt_2$/**28** initiator system in toluene containing tetrahydrothiophene showed levorotation of $[\alpha]_D$ $-150°$. This polymer was shown to have a ratio of (S,S)-unit to (R,R)-unit (or the reverse) of 11.0 by NMR analysis [65].

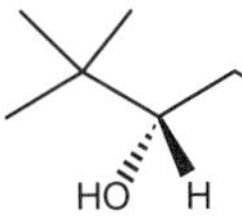

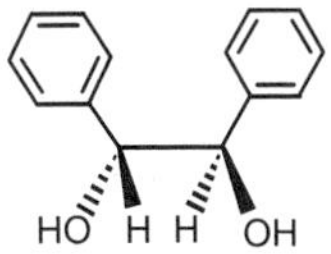

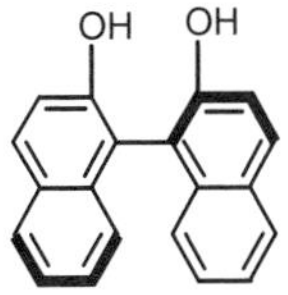

11.2.1.7. Diels–Alder Polymerization Diels–Alder type polymerization of a bisdienophile monomer (**30**) and bisdiene monomers (**31, 32a,b**) by using chiral Lewis acidic catalysts (**33–35**) affords optically active polymers [68]. For instance, the polymerization of **30** in CH_2Cl_2 with **32b** by using **33** as a catalyst at $-30°C$ gives a polymer with structure **36** showing molecular rotation of $[\Phi]_D$ $+243°$. When $CHCl_3$ or tetrahydrofuran is used as solvent, the polymers with only low optical activity are produced.

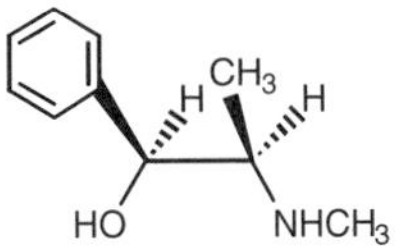

11.2.1.8. Topochemical Polymerization The chiral crystalline environment of a monomer itself can be a source of asymmetric induction in solid-state polymerization [69–72]. Prochiral monomers such as **37** give enantiomorphic crystals, one of which can be preferentially formed by recrystallization with a trace amount of optically active compounds. Photoir-

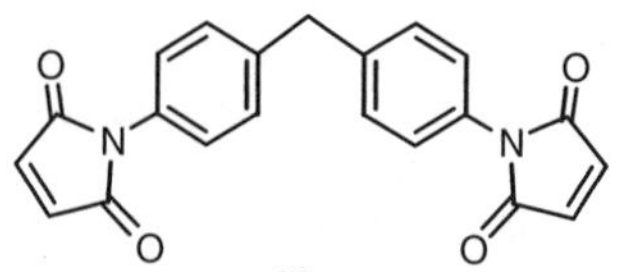

30

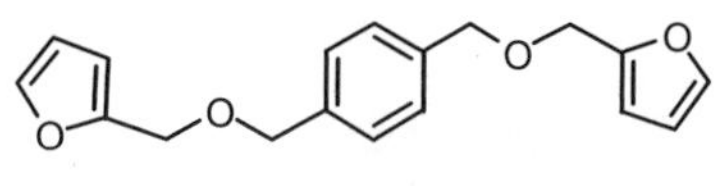

31

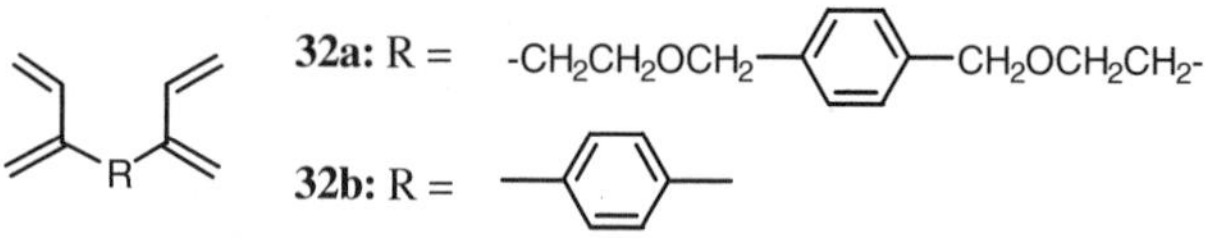

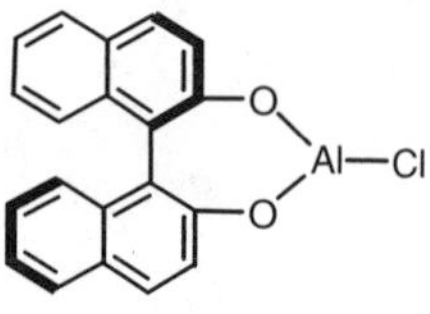

32a: R = -CH₂CH₂OCH₂— ⬡ —CH₂OCH₂CH₂-

32b: R = — ⬡ —

33

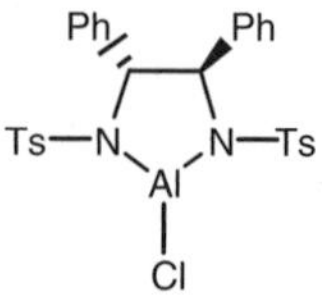

34

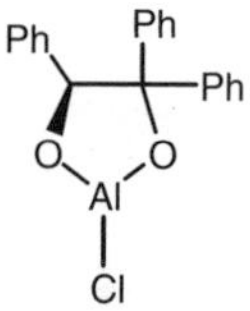

35

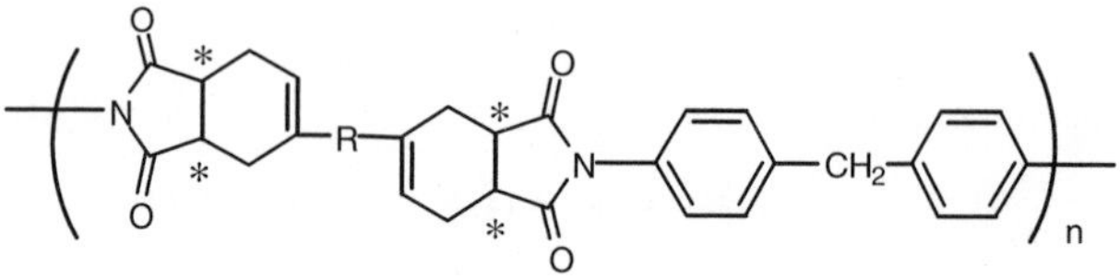

36

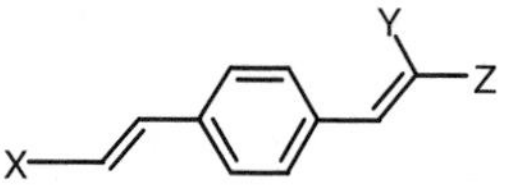

X = 4-pydiryl, Y = -H, Z = -COOEt;
X = -COOR', Y = -CN, Z = COOR"

37

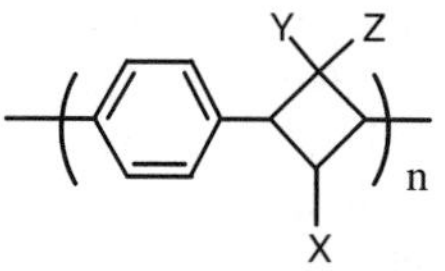

38

radiation of the crystal leads to [2+2] topochemical asymmetric synthesis polymerization in the solid state giving polymers having a cyclobutane structure in the main chain (**38**).

11.2.2. Non-catalytic Syntheses

Chirality induction can be achieved in homo- and copolymerization of vinyl monomers based on chiral monomer structure [1,3,8,9]. The first example of this type of polymerization was the copolymerization of (S)-α-methylbenzyl methacrylate with maleic anhydride: the polymerization product showed $[\alpha]_D$ +23° after removal of the chiral side group [73]. For another example, the copolymerization of an optically active styrene derivative (**39**) with N-phenylmaleimide (**17**, R = –Ph) followed by removal of the optically active side group and deboronation gave an optically active N-phenylmaleimide-styrene copolymer [74].

Asymmetric free-radical oligomerization has been performed by using acrylamides bearing an optically active side group (**40**) [75–77]. The in-chain asymmetric centers of the resulting oligomers had preferential single-handed chirality due to the effects of the chiral auxiliary group. When the polymerization of **40b–d** was carried out by free-radical catalysis, isotactic polymers (mm ~92%) were obtained [78]. Asymmetric oligomerization of a chiral monomer has also been realized using vinyl phenyl sulfoxide [79].

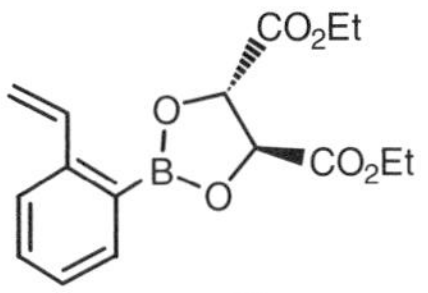

39

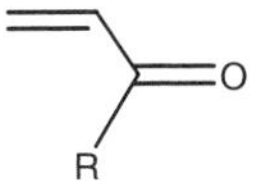

40

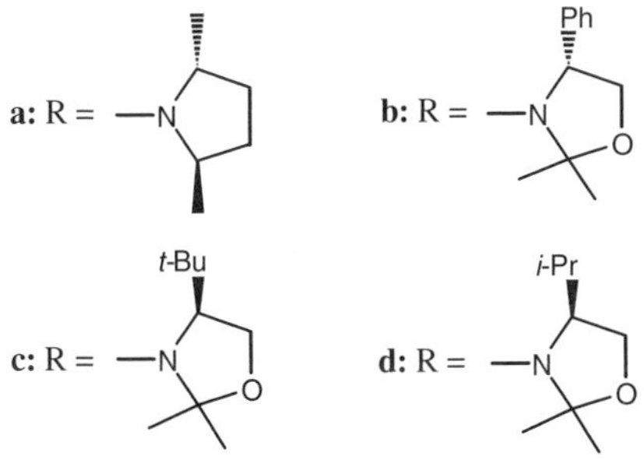

a: R = (pyrrolidine) b: R = (oxazolidinone, Ph)

c: R = (oxazolidinone, t-Bu) d: R = (oxazolidinone, i-Pr)

Asymmetric induction to main-chain chiral centers can also be achieved by the radical polymerization of sorbates having chiral ester groups [80,81] and 1,3-butadiene-1-carboxylic acid complexed with optically active amines [82,83].

Cyclopolymerization of bifunctional monomers is an effective method of chirality induction. Optically active vinyl homopolymers and copolymers have been synthesized by using optically active distyrenic monomers (**41**) based on a readily removable chiral template moiety. Free-radical copolymerization of **41a** with styrene and removal of the chiral template moiety from the obtained copolymer led to polystyrene, which showed optical activity ($[\alpha]_{365}$ -0.5~$3.5°$) (Scheme 11.6) [84]. The optical activity was explained in terms of chiral (S,S)-diad units generated in the polymer chain through cyclopolymerization of **41a** [85]. Several different bifunctional monomers have been synthesized and used for this type of copolymerization [86–90].

Chiral induction was observed in the cyclopolymerization of optically active di-methacrylate monomer **42** [88]. Free-radical polymerization of **42** proceeds via a cycliza-tion mechanism, and the resulting polymer can be converted to PMMA. The PMMA exhibits optical activity ($[\alpha]_{405}$ $-4.3°$) and the tacticity of the polymer (mm/mr/rr = 12 / 49 / 39) is different from that of free-radical polymerization products of MMA. Free-radical polymerization of vinyl ethers with a chiral binaphthyl structure also involved chiral induction [91,92]. Optically active PMMA was also synthesized through the polymeriza-tion of methacrylic acid complexed with chitosan and conversion of the resulting polymer into methyl ester [93,94].

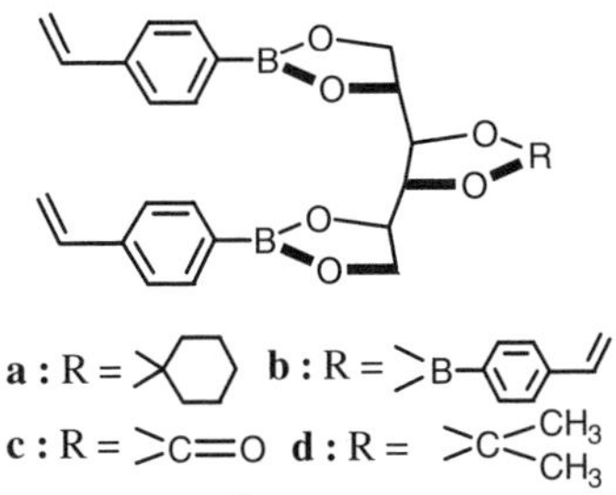

a : R = (cyclohexyl) b : R = >B-(phenyl with vinyl)

c : R = >C=O d : R = >C<$^{CH_3}_{CH_3}$

41

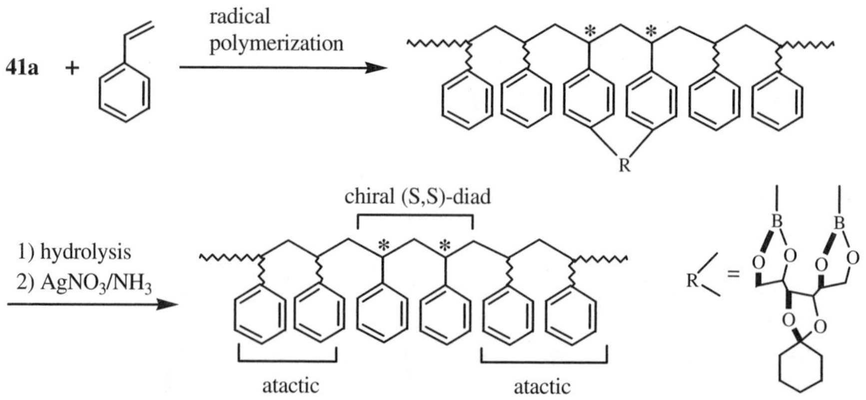

Scheme 11.6

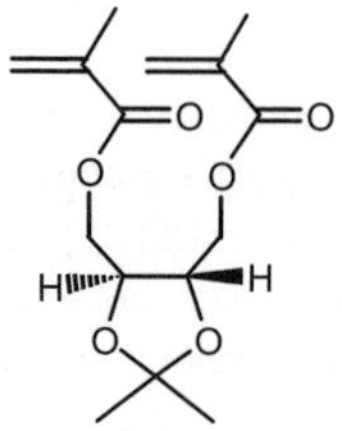

42

There is an interesting theoretical report concluding that the polymerization giving an atactic polymer can be asymmetric synthesis polymerization without using chiral catalysts [95]. According to the report, when DP of an atactic polymer exceeds 70, the polymer sample cannot be a racemic mixture because the number of possible diastereomers is far larger than that of polymer chains and only one antipode of enantiomer can exist for some of the diastereomeric chains. However, the polymer sample does not show optical activity due to compensation of optical rotations contributed from different diastereomers (crypto-chirality).

11.3. HELIX-SENSE-SELECTIVE POLYMERIZATION

Helix-sense-selective polymerization is the reaction that yields polymers with a helical confor-mation with excess screw sense. Right- and left-handed helices are mirror images (atropisom-ers), and if one of the two is preferentially synthesized for a polymer chain, the polymer can be optically active [1–9].

Although many stereoregular polymers have a helical conformation in the solid state [5,96], the conformation is lost in solution in most cases, except in the case of some polyolefins with optically active side groups [12], because the dynamics of the polymer chain are extremely fast in solution. Therefore, isotactic polystyrene [15,16] and polypropylene [17] prepared with an optically active catalyst do not show optical activity due to a helical conformation. However, a helical conformation can be maintained in solution for some polymers having a rigid main chain or bulky side groups that prevent mutation to random conformation, and the conformation may have excess single-handed helicity when the polymers are synthesized using optically active catalysts or when they consist of chiral monomeric units.

11.3.1. Catalytic Syntheses

11.3.1.1. *Polymers of Methacrylates, Acrylates, Acrylamides and Related Monomers*

11.3.1.1.1. Anionic Catalysis Several bulky methacrylates afford highly isotactic, optically active polymers having a single-handed helical structure by asymmetric polymerization. The effective polymerization mechanism is mainly anionic but free-radical catalysis can also lead to helix-sense-selective polymerization. The anionic initiator systems can also be applied for the polymerization of bulky acrylates and acrylamides. The one-handed helical polymethacrylates show an excellent chiral recognition ability when used as a chiral stationary phase for high-performance liquid chromatography (HPLC) [97,98].

The methacrylate monomers listed below have been successfully used for helix-sense-selective polymerization [99–134]. Among these monomers, triphenylmethyl methacrylate (**TrMA**) is the first example of a vinyl monomer that gives an optically active polymer whose chirality arises exclusively from a single-handed helical conformation of the main chain. Helical poly(TrMA) is obtained by anionic polymerization by using complexes of achiral organolithiums with optically active ligands or optically active organolithiums. (−)-**Sp** [99,100,102,103,106], (*S,S*)-(+)- and (*R,R*)-(−)-**DDB** [101,103], (+)-**PMP** [103], and the ligands bearing a biphenyl or binaphthyl moiety **58a,b**, and **59–61** [104,105] are effective as chiral ligands in inducing the single-handed helical structure of poly(**TrMA**). The polymerization of TrMA with optically active lithium amide (**62**) and lithium enolate (**63**) also yields optically active polymers. The results of the polymerization of **TrMA** using (−)-**Sp**-, (+)-**DDB**-, and (+)-**PMP**-9-fluorenyllithium (**FlLi**) complexes are shown in Table 11.2 [103]. The optically active complexes gave almost perfectly isotactic polymers having a single-handed helical conformation, which show high optical activity and intense CD absorption bands. The helical conformation is supported by steric repulsion of the bulky side groups. Therefore, when the side groups are removed by hydrolysis from the polymer chain, the helical conformation is lost and the chiroptical properties of the polymer are not observed. The stereochemical mechanism of the helix-sense-selective polymerization of **TrMA** has been investigated in detail [103]. Stable helical conformation starts at DP ~9, and one turn of the helix consists of three or four monomeric units. The isotactic chain of

TrMA	**D2PyMA**	**43**	**44**	**45**	**46**
[99-110]	[111-116]	[117]	[111,113]	[118]	[119]

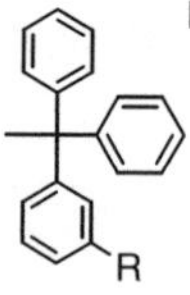

R = -F, -Cl   R = -CH₃, -Cl
47   **48**   **49**   **50**   **51**
[120]   [120]   [121]   [122a]   [122b]

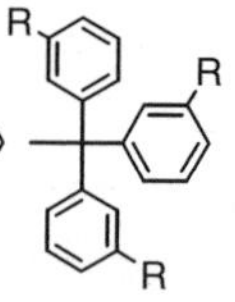

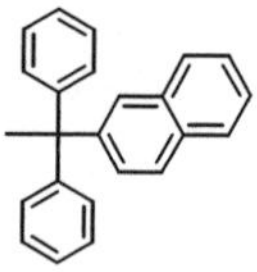

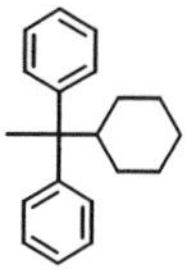

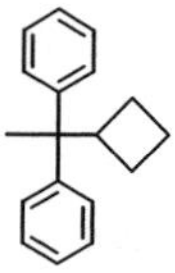

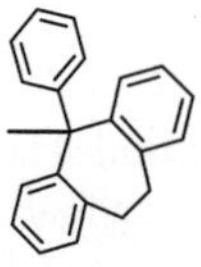

PDBSMA   **52**   **53**   **54**
[123-126]   [123]   [127]   [128]

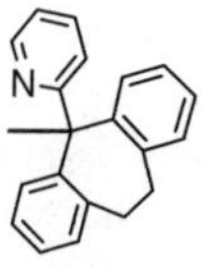

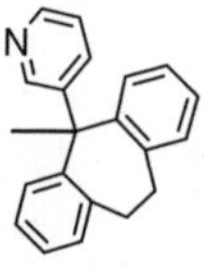

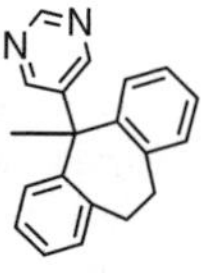

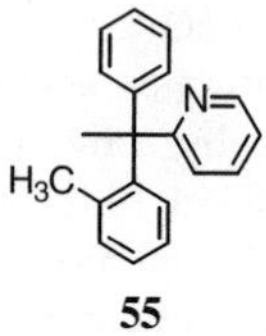

55   **56**   **57**
[129-132]   [133]   [134]

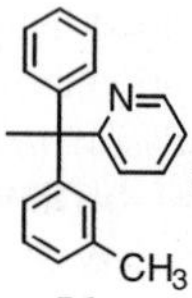

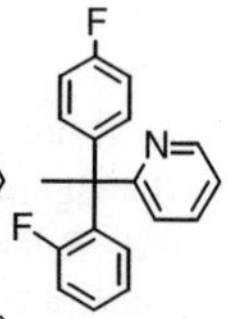

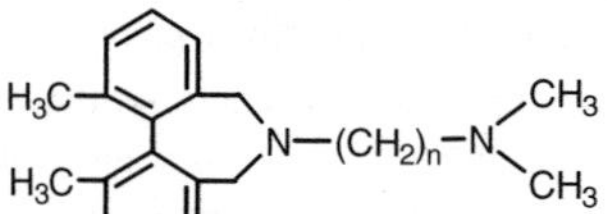

58   **59**

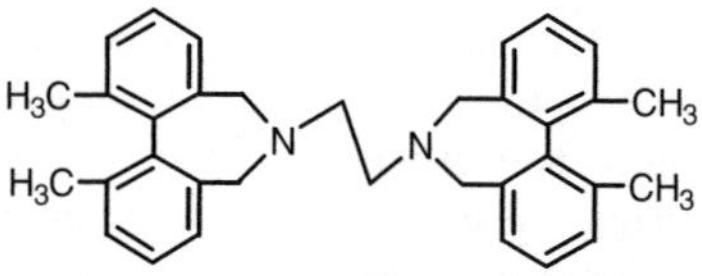

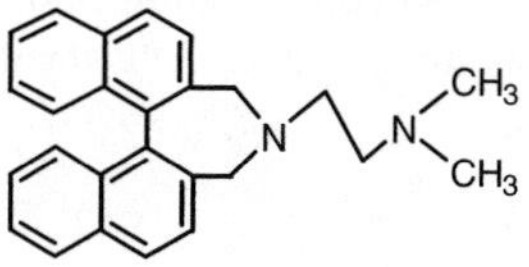

60   **61**

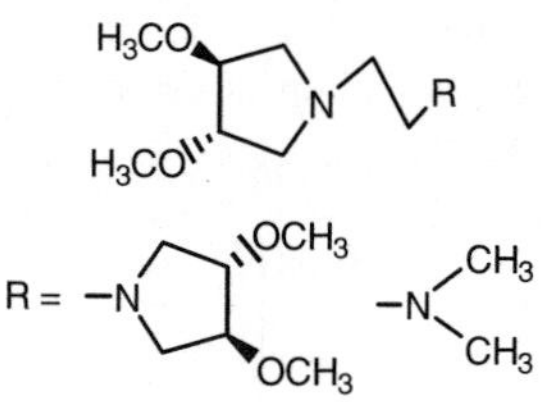

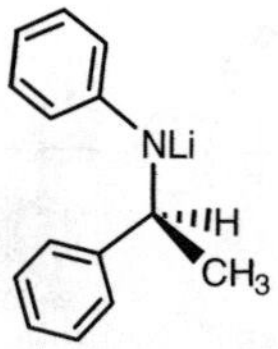

62   **63**

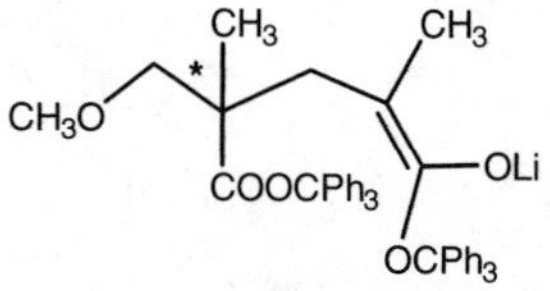

TABLE 11.2. Helix-Sense-Selective Polymerization of TrMA with (−)-Sp-, (+)-DDB, and (+)-PMP-FlLi Complexes in Toluene at −78°C[a]

Initiator	Benzene-Hexane (1/1, v/v)-Insoluble Part of Products[b]				
	Yield (%)	$[\alpha]^{25}$ D[c] (deg)	$[\theta]^d \times 10^{-4}$ (235 nm)	$[\theta]^d \times 10^{-5}$ (210 nm)	DP[e]
(−)-**Sp**-FlLi	82	+383	+9.42	+2.32	60
(+)-**DDB**-FlLi	93	+344	+8.45	+1.86	47
(+)-**PMP**-FlLi	94	+334	+7.78	+1.76	39

[a]Conditions: [TrMA]/[Li] = 20. Product yield was nearly qantitative. Data cited from ref. 103.
[b]Crude products contained oligomers which were removed by washing with the benzene-hexane mixture.
[c]In tetrahydrofuran.
[d]Measured in tetrahydrofuran at cat 25°C. Units: cm²dmol⁻¹.
[e]Determined by GPC of poly(MMA) derived from poly(TrMA).

the polymers obtained by using the (−)-**Sp** initiator system has ---*RRR*--- absolute configuration and that of the polymers obtained using the (+)-**DDB** and (+)-**PMP** initiator systems has ---*SSS*--- absolute configuration, although all the polymers have the same screw sense of the helix.

In the polymerization of the monomers bearing a pyridyl group, stereocontrol by a chiral ligand may not be effective because the pyridyl group of the monomer can competitively coordinate to the Li cation at the active end of the growing species, interrupting the effective coordination of the chiral ligand. In fact, the polymerizations of **D2PyMA** with **Sp** and **DDB** as chiral ligands give the polymers with much lower optical activity ($[\alpha]_{365}$ −153° (using **Sp**-FlLi); $[\alpha]_{365}$ +132° (using **DDB**-FlLi)) than the single-handed helical poly(**TrMA**) ($[\alpha]_{365}$ 1400°) [116]. The poly(**D2PyMA**)s are mixtures of right- and left-handed helices. The chiral ligands, 64–82, have been used as complexes with FlLi or DPEDA-Li for the helix-sense-selective polymerization of **D2PyMA**; however, the specific rotation of the obtained polymers was rather low (Table 11.3) [116]. In contrast to these chiral ligands, **PMP** can lead to a single-handed helical polymer in **D2PyMA** polymerization (Table 11.4) [115,116]. The polymers obtained using **PMP** exhibit high optical rotation and have highly isotactic main-chain configuration and narrow molecular weight distribution. The polymerization behavior of **D2PyMA** using **PMP** is quite in contrast to that using **80**, which differs from **PMP** only in the substituent on the N atom. The excellence of **PMP** as a chiral ligand may be explained in terms of the proposed coordination structure **83** at the growing end [116]. The smaller substituent on the N atom in **PMP** (hydrogen) may result in tighter and closer ligand coordination to a Li cation, which can be reinforced by hydrogen bonding of the hydrogen with the enolate oxygen. The tight and close coordination may be the reason for the strict stereocontrol using on **PMP**. The ligand **58a** also gives a single-handed helical

$(C_2H_5)_2N$ ⋯ $N(C_2H_5)_2$

Me   Me
64

Me   Me
65

Me   Me
66

Me   Me
67

68

69

70

71

72

73

74

75

76

77

78

79

80

81

82

83

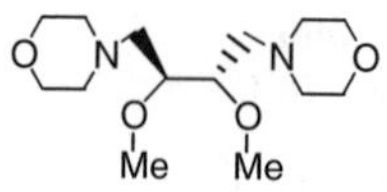

poly(**D2PyMA**) [105]. **PMP** was also effective in controlling the stereochemistry of anionic polymerization of other pyridyl (pyrimidyl) group-containing monomers, including **43**, **46**, **52**, **53**, and **54**.

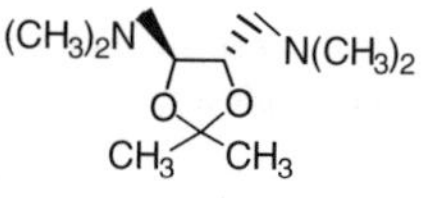

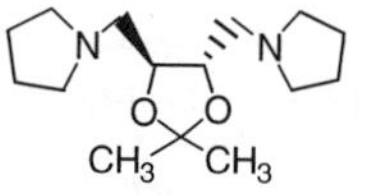

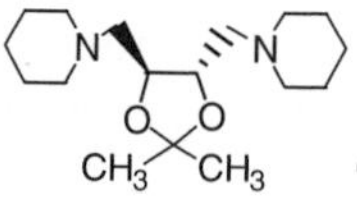

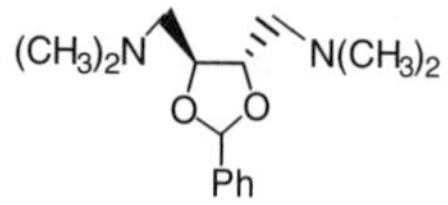

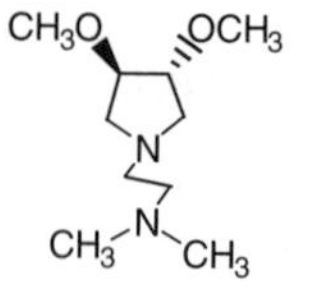

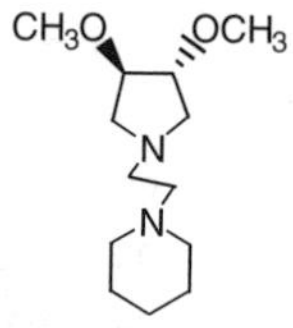

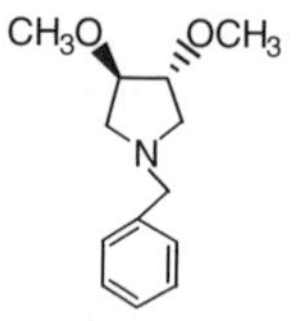

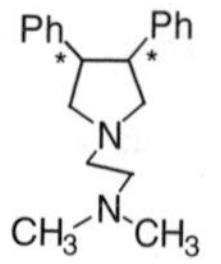

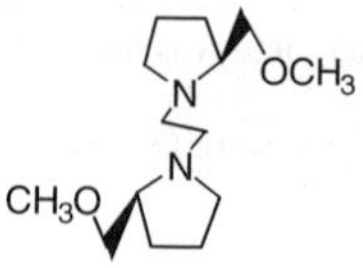

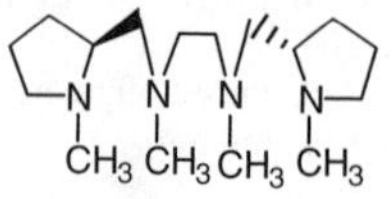

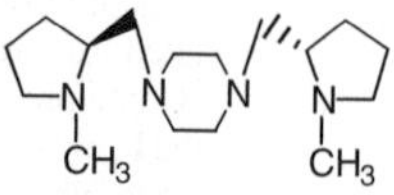

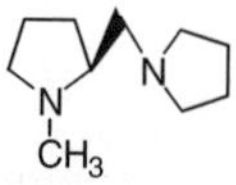

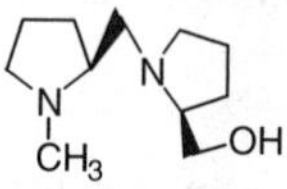

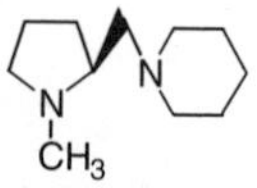

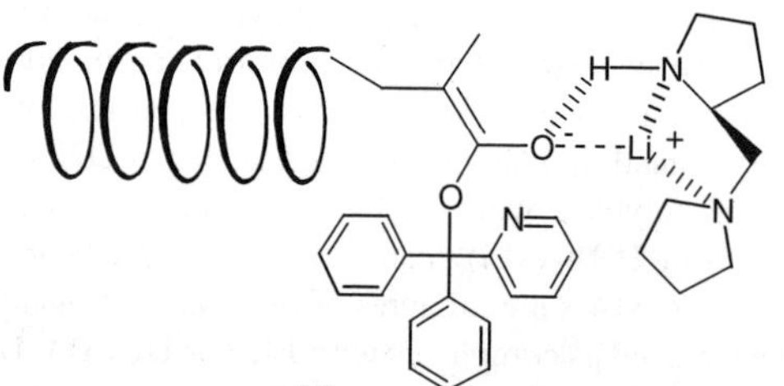

Optically active poly(**D2PyMA**) [135], poly-**43** [117], poly-**46** [119], and poly-**50** [122] are known to undergo helix-helix transition of the main chain in solution through which a single-handed helix racemizes or stereomutates from a kinetically controlled form to a thermodynamically stable form.

In contrast to the above-mentioned bulky methacrylates, monomers **84** [120] and **85** [136] do not give a polymer, and **86** [4], **87** [136], and **88** [122a] give atactic polymers by polymerization using **Sp** or **DDB** as chiral ligand. The esters groups of **84** and **85** seem to be too bulky and those of **86–88** too small to give an isotactic, helical polymer. Monomer **89** [4] gives a syndiotactic polymer, suggesting that the ester group is too flexible to form a rigid helix of the main chain. **90** having a triptycenyl group gives a polymer showing only low optical activity ($[\alpha]_D$ −5°) by the polymerization of (−)-**Sp**- or (+)-**DDB**-*n*-BuLi complex [137]. Tacticity of the polymer is unknown.

84

85

86

87

88

89

90

91

92

93

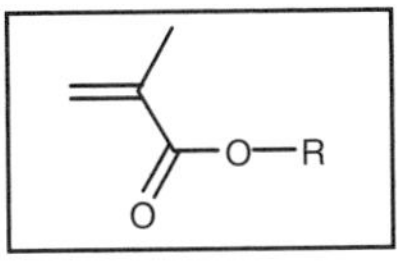

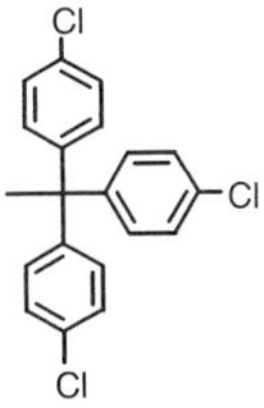

TABLE 11.3. Optical Activity of Poly(D2PyMA)s Obtained Using the Complexes of 64–82 with DPEDA-Li in Toluene at −78°C[a]

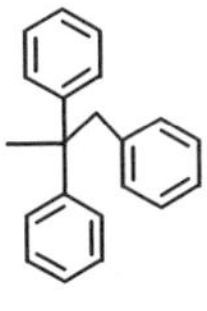

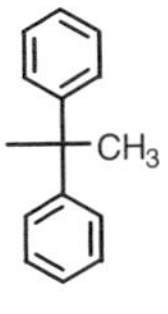

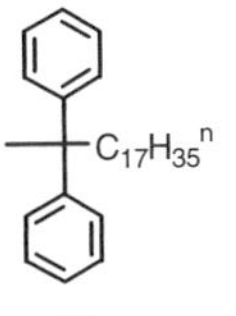

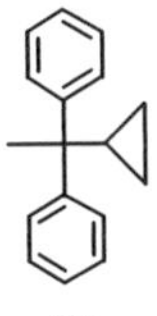

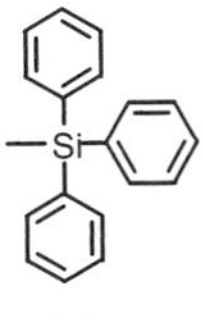

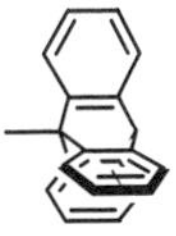

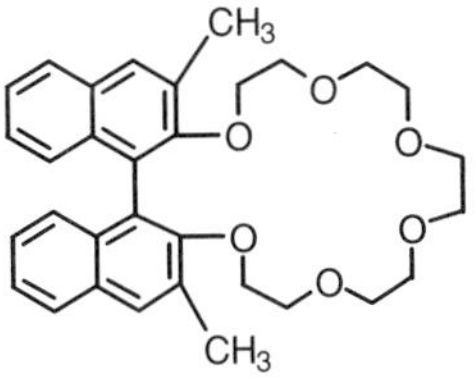

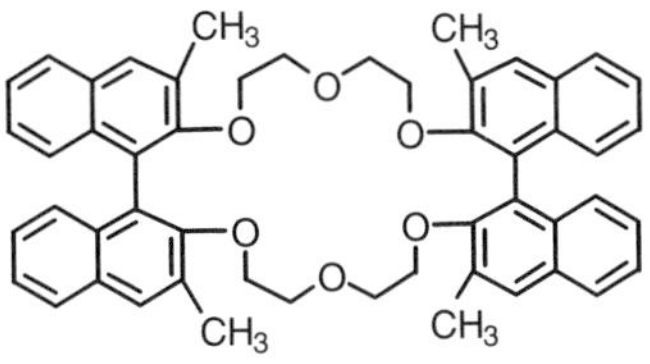

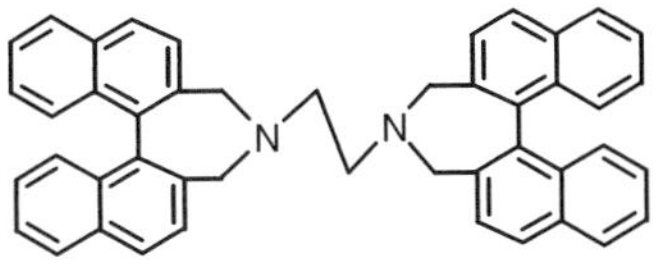

Ligand	$[\alpha]^{25}_{365}$ (deg)	Ligand	$[\alpha]^{25}_{365}$ (deg)
64	−661	**74**	−100
65	−457	**75**	−100
66	−1108	**76**	+53
67	−1170	**77**	+228
68	−32	**78**	−657
69	+202	**79**	−221
70	−105	**80**	−212
71	−21	**81**[b,c]	+14
72	+473	**82**	−324
73	+1012		

[a]Conditions: [D2PyMA]/[Li] = 20. Polymer yield was >99~86% except for the reaction with **81**. Data cited from ref. 116.
[b]Used as *O*-lithiated form.
[c]Polymer yield was 3.5%.

TABLE 11.4. Asymmetric Polymerization of D2PyMA with the Complexes of (+)-PMP with DPEDA-Li and FlLi in Toluene at −78°C[a]

RLi	[D2PyMA]/[Li]	$[\alpha]_{365}^{25,b}$ (deg)	Dp[c]	Mw/Mn[c]
DPEDA-Li	15	+1325	27	1.04
DPEDA-Li	20	+1406	30	1.08
DPEDA-Li	30	+1651	45	1.13
DPEDA-Li	50	+1675[d]	81	1.12
FlLi	15	+1530	30	1.14
FlLi	30	+1641	60	1.10

[a]Conditions: D2PyMA 1.0 g; toluene 20 mL. Polymer yield was almost quantitative in all cases. Data cited from ref. 116.
[b]Measured in a mixture of $CHCl_3$ and 2,2,2-trifluoro-ethanol (9/1, v/v).
[c]Determined by GPC of poly(MMA) derived from poly(D2PyMA).
[d]$[\alpha]^{25}$ D +414°.

Helix-sense-selective polymerization of methyl, benzyl, and *t*-butyl methacrylates was attempted by using the complexes of chiral crown ethers, **91** and **92**, and that of a chiral diamine **93** with *n*-BuLi; however, these esters seem to be too small to form and maintain helical conformation [138,139]. The complexes of BuLi with **58a** and **59** failed in producing an optically active, helical polymer in the polymerization of methyl and benzyl methacrylates [104b].

Catalytic helix-sense-selective polymerization of acrylates (**94**, **95**) [140,141] and acrylamides (**96–105**) [142–147] have been investigated by using (+)-**PMP**, (−)-**Sp**, and (+)-**DDB** as chiral ligands. Stereocontrol in the polymerzation of the acrylates and acrylamides was more difficult compared with that in the methacrylate polymerization. Specific rotations ($[\alpha]_{365}^{25}$) of poly-**94** and poly-**95** obtained by the asymmetric polymerization were much smaller than those of the corresponding polymethacrylates prepared under similar conditions and were up to +102° (ligand: **PMP**, diad isotacticity 70%) and −94° (ligand: **DDB**, diad isotacticity 61%), respectively. The isotactic part of the polymers is considered to have a helical conformation with excess helicity. For the polymerization of the bulky acrylamides, (−)-**Sp** has been mainly used as the chiral ligand. **Sp** was shown to be a better ligand compared with **DDB** and **PMP** in the polymerization of **105d**. The highest isotacticity (m = 87%) and optical activity ($[\alpha]_{365}^{25}$ −657°)

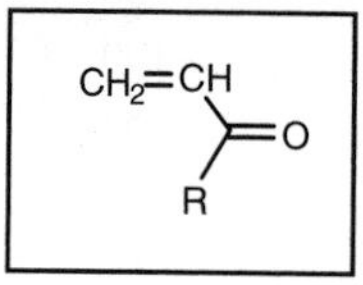

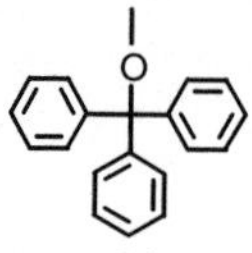

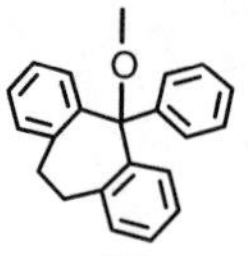

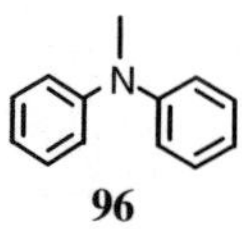

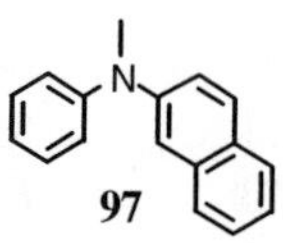

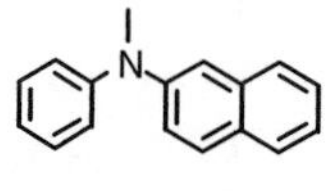

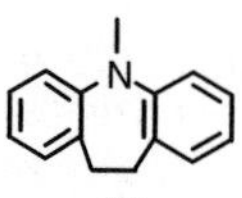

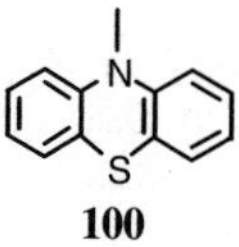

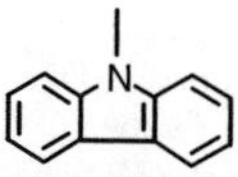

102   **103**   **104**

105

a: R_1 = -CH₃, R_2 = R_3 = R_4 = -H
b: R_1 = R_3 = R_4 = H, R_2 = -CH₃
c: R_1 = R_3 = R_4 = H, R_2 = -Cl
d: R_1 = R_3 = R_4 = H, R_2 = -OCH₃
e: R_1 = R_3 = R_4 = H, R_2 = -OSi(CH₃)₂C(CH₃)₃
f: R_1 = R_3 = R_4 = H, R_2 = -OCC(CH₃)₃
g: R_1 = R_3 = R_4 = H, R_2 = -OCOCH(Ph)C₂H₅
h: R_1 = R_2 = R_4 = -H, R_3 = -CH₃
i: R_1 = R_2 = R_4 = -H, R_3 = hexyl
j: R_1 = R_2 = -H, R_3 = R_4 = hexyl
k: R_1 = R_2 = -H, R_3 = R_4 = butyl

in the asymmetric polymerization of acrylamide was achieved in the polymerization of **105h** by using the (−)-**Sp-FlLi** complex as an initiator at −98°C [145]. The stereostructure of the poly-**105h** depended on molecular weight, and the high-molecular weight fractions separated by GPC fractionation exhibited large levorotation $[\alpha]_{365}^{25}$ −1122° (m = 94%) which is comparable to the optical activity of single-handed helical polymethacrylates [145].

Helix-sense-selective polymerization of some other several bulky monomers has been attempted by using chiral initiators. A bulky acrylonitrile derivative, **106a**, gave an optically active polymer ($[\alpha]_D$ +115°) by polymerzation with the (+)-**DDB-FlLi** complex [148]. The poly-**106a** was insoluble in solvents and the specific rotation was measured in suspension. Monomers **106b–d** [149] and **107a** [150] do not give high polymers probably because of the highly crowded monomer structures. A bulky crotonate **107b** [151,152] affords optically active, helical polymers by polymerization using **DDB-FlLi** and **PMP-FlLi** complexes.

a: R = -CN
b: R = -CO₂Me
c: R = -CO₂Et
d: R = -Ph

106

a: R_1 = -H, R_2 = -Et
b: R_1 = -Me, R_2 = -H

107

11.3.1.1.2. Free-Radical Catalysis **PDBSMA** gives an almost completely isotactic polymer not only by anionic polymerization but also by free-radical polymerization [124–126]. Poly(**PDBSMA**) obtained by free-radical polymerization under achiral conditions is an equimolar mixture of right- and left-handed helices. By using optically active initiators (**108**, **109**), chain-transfer agents (**110**, **111**), solvents (**112**, **113**) [126], and a metallic radical species (**114**) [153], optically active poly(**PDBSMA**)s having excess single-handed helicity were obtained. Table 11.5 shows the results of polymerization using (*i*-PrOCOO)₂ in the presence of (+)- and

(−)–**110**. The obtained polymers were mostly insoluble in common solvents, but the THF-soluble part exhibited optical activity depending on the amount of **110** used for the polymerization. The levorotatory polymer obtained using 40 mol% of (+)-**110** consisted of a higher-molecular-weight (−)-fraction and a lower-molecular-weight (+)-fraction, and the (−)-fraction separated by GPC fractionation exhibited $[\alpha]_{365}$ −750° corresponding to a mixture of (+)- and (−)-helices in a ratio of 3:7 (40% ee). Based on detailed analyses of the polymer structure, it was proposed that chiral induction (helix-sense selection) takes place in the termination process, that is, the hydrogen transfer from **110** to the polymer radical is helix-sense selective.

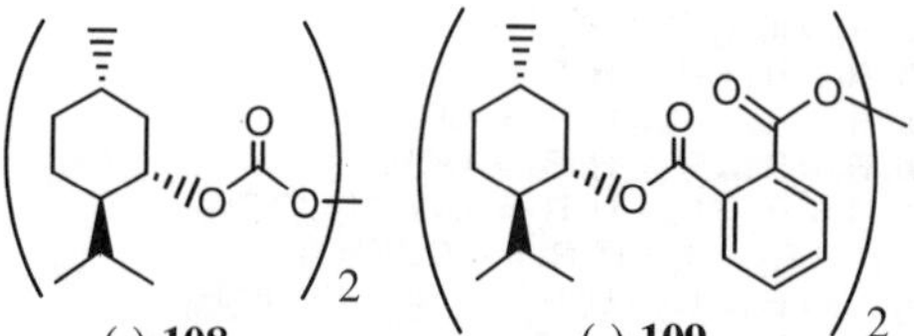

(-)-**108**   (-)-**109**

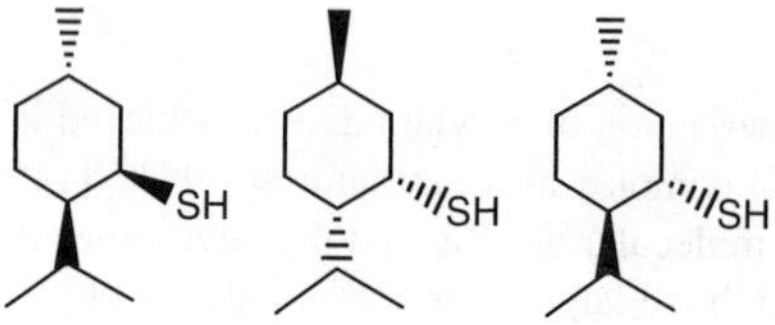

(+)-**110**   (-)-**110**   (-)-**111**

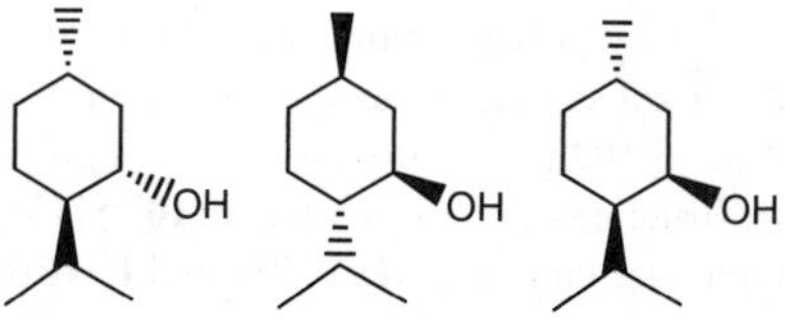

(-)-**112**   (+)-**112**   (+)-**113**

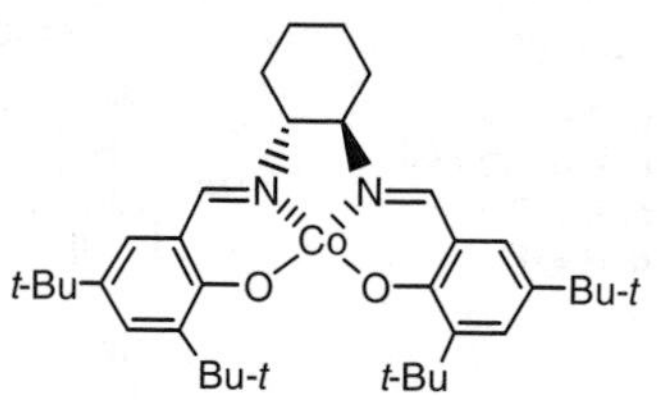

114

11.3.1.2. Isocyanide Polymers Bulky isocyanides give polymers having a 4:1 helical conformation (**115**) [154]. An optically active polyisocyanide was first obtained by chromatographic resolution of poly(*t*-butyl isocyanide) (poly-**116**) using optically active poly((*S*)-*sec*-butyl isocyanide) as a stationary phase and the polymer showing positive rotation was found to possess an M-helical conformation on the basis of CD spectral analysis [155,156]. Polymerization of bulky isocyanides with chiral catalysts also leads to optically active polymers.

TABLE 11.5. Free-Radical Polymerization of PDBSMA using (i-PrOCOO)$_2$ in the Presence of (+)- and (−)-110^a

		THF-sol. part			
		B/H^b-insol. polymer			B/H^b sol. oligomers
Additive	Yield^c (%)	Yield (%)	$[\alpha]_{365}^{25, e}$ (deg)	DP^d	Yield (%)
(+)-**110**(5 mol%^{f})	82	2	−80°	42	7
(+)-**110**(10 mol%^{f})	80	3	−130°	41	7
(+)-**110**(40 mol%^{f})	18	11	−140°	40	7
(−)-**110**(40 mol%^{f})	19	10	+110°	51	9

^{a}Conditions: [monomer]/[(i-PrOCOO)$_2$] = 50; temp. 40°C; time 24 h. Data cited from ref. 126.
^{b}A mixture of benzene and hexane (1/1).
^{c}Hexane-insoluble products.
^{d}Determined by GPC of poly(MMA) derived from the original polymer using polystyrene standard samples.
^{e}In THF.
^{f}With respect to [PDBSMA].

As effective catalysts (initiators), Ni(CNR)$_4$(ClO$_4$)/optically active amine systems [157], the Ni(II) complexes, **117**–**119** [158], and the dinuclear complex containing Pd and Pt, which have a single-handed oligomeric isocyanide chain (**120**) [159] are known. By the polymerization of **115** by using Ni(CN-Bu^t)$_4$(ClO$_4$)/(R)-(+)-C$_6$H$_5$CH(CH$_3$)NH$_2$, an M-helical polymer with ee of 62% can be synthesized and the complex **119** leads **115** to a levorotatory polymer with 69% ee [157]. The complex **120** can smoothly polymerize bulky monomers, **121** and **122**, in a helix-sense-selective manner; for an instance, the polymerization of **122** with **120** (n = 10, Mn = 3720, $[\alpha]_D$ = +22°) affords a polymer with Mw = 13.5 × 10^3 and $[\alpha]_D$ +126° [159].

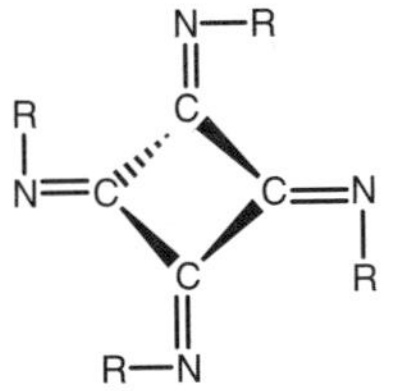

115

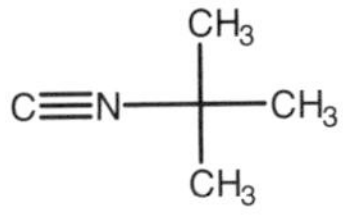

116

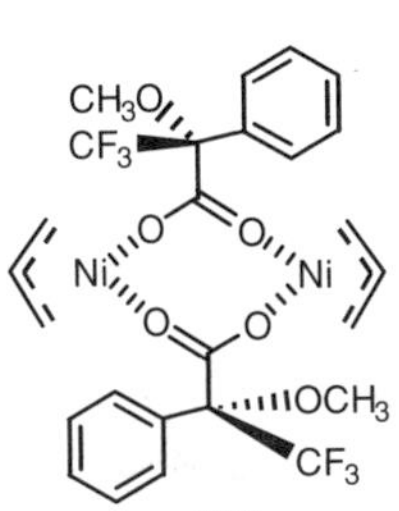

117

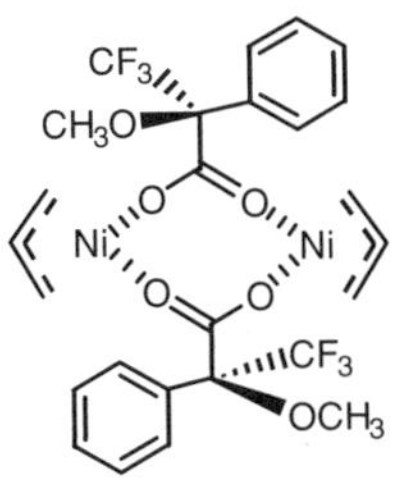

118

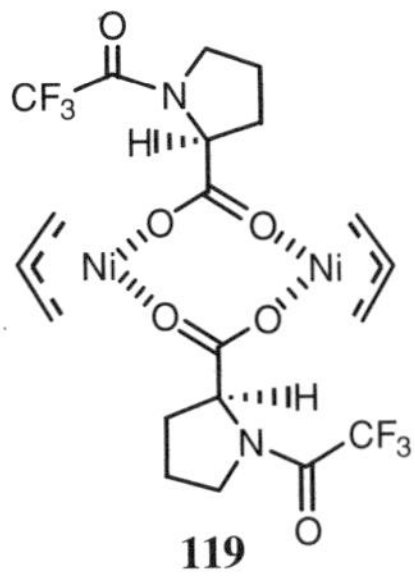

119

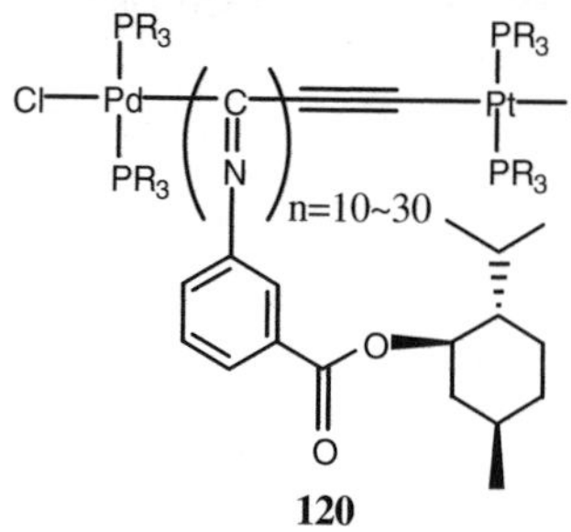

120

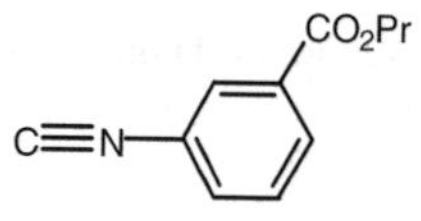

121

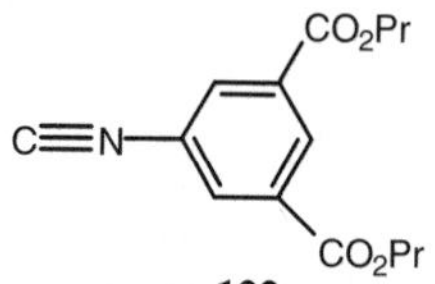

122

By the polymerization with Pd and Ni complexes, 1,2-diisocyanobenzene derivatives give helical polymers via cyclopolymerization mechanism. Optically active polymers were first obtained by the method illustrated in Scheme 11.7 [160–164]. Monomer **123** was reacted with an optically active Pd complex to form diastereomeric pentamers **124** which were separated into (+)- and (−)-forms by HPLC. When the separated pentamer was used to initiate the polymerization of **125**, a one-handed helical polymer was obtained [160]. The polymerization of **123** using the initiators based on chiral binaphthyl groups, **126** and **127**, also gives optically active polymers [163]. In the polymerization with **126**, helix-sense-selectivity was affected by the polymerization procedure. The polymer obtained by direct polymerization with **126** had a much lower helix-sense excess compared with the polymer prepared by using a pentamer synthesized with **126** and purified into a single-handed helical form. The latter polymer seemed to have a single-handed helical structure. In contrast to **126**, **127**, without purification of intermediate oligomeric species, gives poly-**123** with high helix-sense selectivity. Helix-sense selectivity in the polymerization of **123** using **128** as initiator was estimated to be over 95% [164].

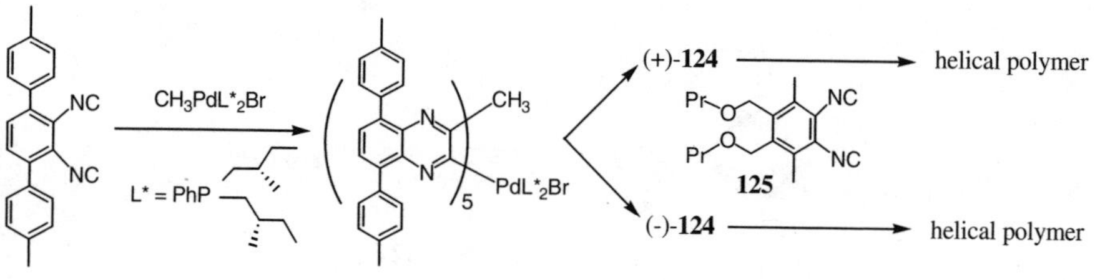

123

124

Scheme 11.7

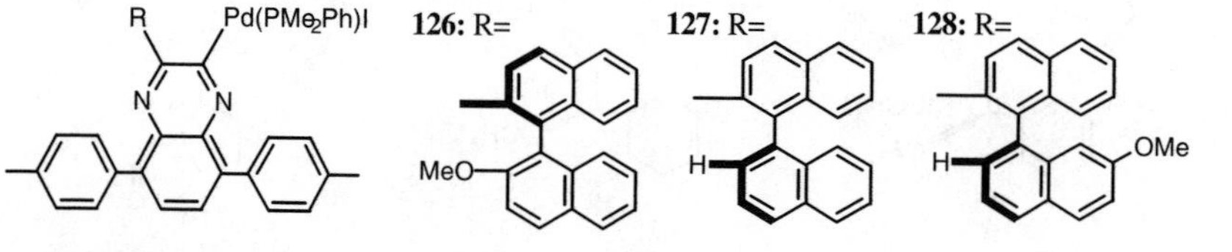

126: R=   **127:** R=   **128:** R=

11.3.1.3. Isocyanate Polymers Anionic polymerization of isocyanates gives polymers with the structure of 1-nylon (**129**), which possess a dynamic helical conformation [165,166]. Though optically active polyisocyanates having excess right- or left-handed helicity were first made by polymerization of optically active monomers (see section **3.2.4**), they can also be prepared by the anionic polymerization of butyl isocyanate and other achiral monomers (**130**) using optically active anionic initiators (**131–136**) [167–169]. The poly(butyl isocyanate) (Mn 9000) obtained by using **131a** exhibits $[\alpha]_{435} = +416°$. The optical activity of the polymers arises from the helical part extending from the chain terminal bearing the chiral group originated from the initiator to a certain length (persistence length) that has a single-screw sense due to the influence of the terminal chiral group.

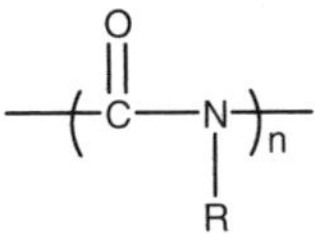

129

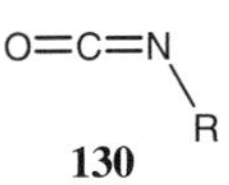

130

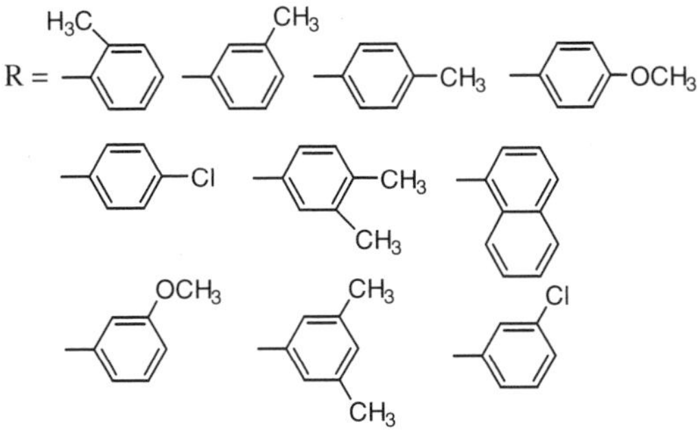

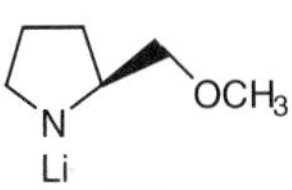

131a 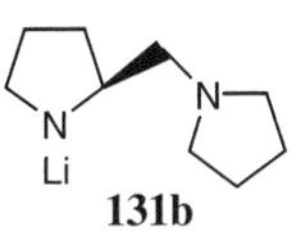 **131b** 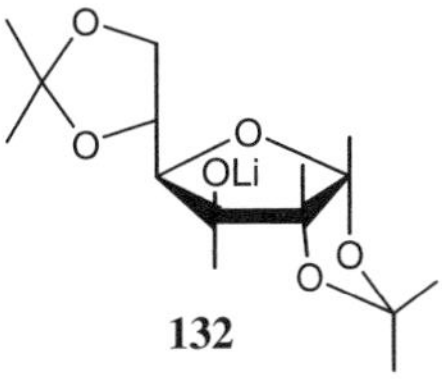 **132**

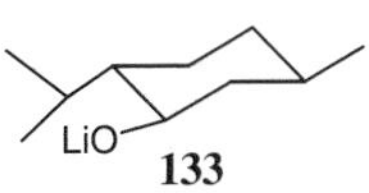

133 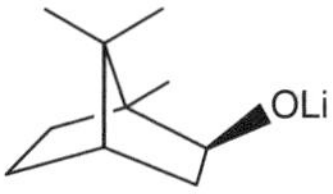 **134**

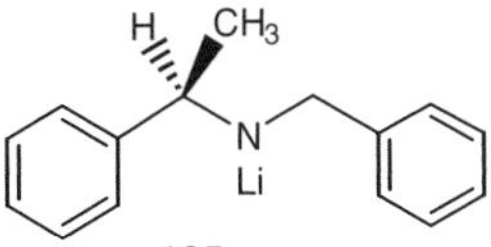

135 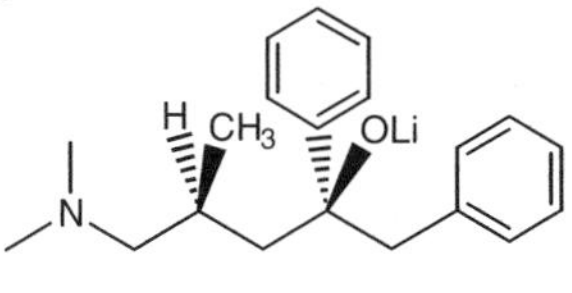 **136**

11.3.1.4. Aldehyde Polymers Asymmetric anionic polymerization can lead trichloroacetalde-hyde (chloral) to a one-handed helical, isotactic polymer having a 4/1-helical conformation with high optical activity ($[\alpha]_D$ 4000° in film) (**137**) [170–174]. Anionic initiators such as **138** [171], **139** [171], and **140** [173] and Li salts of optically active carboxylic acids or alcohols have been used for the polymerization. Although the polymers are insoluble in solvents and their conformation in solution cannot be directly elucidated, a helical structure has been verified by NMR and crystallographic analyses of the oligomers [174].

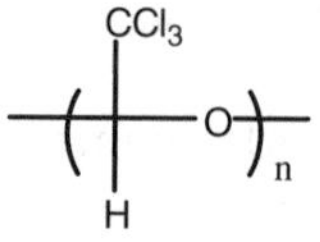

137

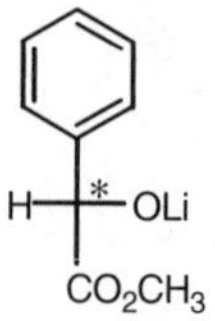

138 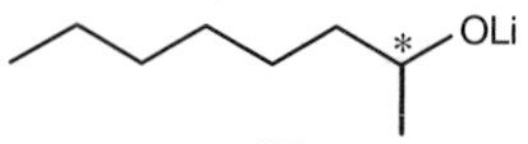　　　**139** 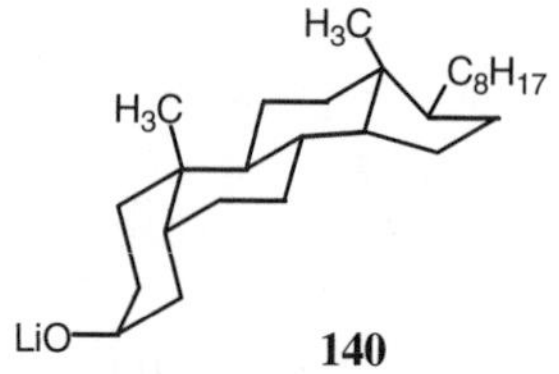　　　**140**

Anionic polymerization of 3-phenylpropanal (**141**) using the complexes of **Sp** with ethylmagnesium bromide (EtMgBr) and *n*-octylmagnesium bromide (OctMgBr) gives optically active polymers showing levorotation ($[\alpha]_{365}^{25}$ −33°~−56°), which may be based on a predominant single-handed helical conformation of the main chain [175]. Reaction with the initiator and **141** gives an ester (**142**) and the (3-phenylpropoxy)magnesium bromide-**Sp** complex (**143**) through a Tishchenko reaction (Scheme 11.8 A). **143** rather than EtMgBr is the actual initiator of the polymerization, and the termination reaction takes place through a Tishchenko reaction leading to the polymer structure, **144**, having an ester moiety at the chain terminal (Scheme 11.8 B). The major diastereomer of dimer (**144**, n = 2) (diastereomeric stereostructure not identified) prepared by oligomerization using the EtMgBr-**Sp** complex was found to be rich in the (+)-isomer with 70% ee. This finding suggests that a certain oligomer anion with a configu-

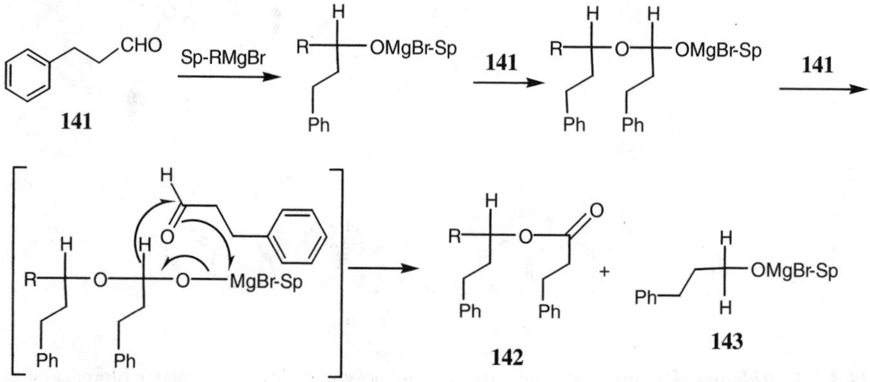

Scheme 11.8A

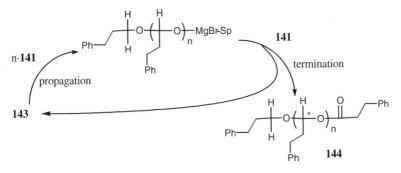

Scheme 11.8

ration, for instance, (S,S) or (R,R), may propagate preferentially to higher-molecular-weight oligomers and polymers.

11.3.1.5. Olefin Polymers Isotactic polymers of propylene and 1-butene obtained by optically active metallocene catalyst (**145**) have been reported to show large specific rotation in suspension ($[\alpha]_D$ −123°, −250° for polypropylene; $[\alpha]_D$ +130° for polybutene), which was lost when the polymers were completely dissolved or heated [176,177]. The optical activity was ascribed to a helical conformation of the polymer chain with preferential screw sense.

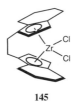

145

11.3.2. Non-Catalytic Syntheses

11.3.2.1. Olefin Polymers Stereoregular polymers obtained from optically active olefins assume a helical conformation with excess single-handed screw sense in solution based on the steric effects of the chiral side groups [12,178–183]. For example, the isotactic polymers obtained by using Ziegler-Natta type catalysts from a series of olefins (**146a**) showed larger

n = 0~3

146a

n = 0~3
R = -H or -CH$_3$

146b

optical activity than the monomers or the low-molecular-weight model compounds of the monomeric units (**146b**) based on excess helicity of the main chain. The discrepancy between the optical activity of the polymers and that of the model compounds was larger when the asymmetric carbon in **146a** is closer to the double bond, and the polymers with higher stereoregularity exhibited larger optical rotation. Chiroptical properties based on helical conformation have also been found for syndiotactic poly{(S)-4-methyl-1-hexene} prepared with a homogeneous Zr-based catalyst [180]. Isotactic polymers obtained by anionic polymerization from optically active vinyl ketones also assume helical conformation in solution [181–183].

11.3.2.2. Polymers of Methacrylates, Acrylates, Acrylamides, and Related Monomers

Helix-sense selection has been observed in the anionic or free-radical polymerization of the optically active methacrylates, **55** [131,132], **56** [133], **57** [134], **147** [184], and **148** [185]. The selection is based on the chirality of the monomers. **147** ($[\alpha]_{365}$ −82°) gives dextrorotatory, isotactic polymers ($[\alpha]_{365}$ +786~+939°) by anionic polymerization using **PMP-**, **Sp-**, **DDB-**, and **TMEDA-DPEDA-Li** complexes regardless of ligand chirality [184]. The optical activity seems to be based on the helical conformation with excess single-handed helical sense. The poly-**147** exhibits reversible helix–helix transition in which the helical content is determined by solvent properties. Optically active (+)-**55** ($[\alpha]_D$ +88°) gives levorotatory polymers ($[\alpha]_D$ −315~−357°) by the polymerization using (+)-**DDB**- and (−)-**DDB-DPEDA**-Li complexes in toluene or n-BuLi in THF [131]. Helix-sense selection is also achieved in the radical polymerization of optically active **55** [132]. In contrast to the anionic polymerization of **55**, the polymer obtained from (+)-**56** ($[\alpha]_{365}$ +14°, $[\alpha]_D$ +4.5°) by using **TMEDA-DPEDA-Li** initially shows much smaller optical activity ($[\alpha]_{365}$ +262°) compared with the poly-(+)-**56** ($[\alpha]_{365}$ +1456°) prepared by using (+)-**PMP-DPEDA-Li**. However, the former polymer exhibits the mutarotation in solution at 60°C, which leads to much larger optical activity ($[\alpha]_{365}$ +1481°), possibly based on a helix–helix transition. The free-radical polymerization as well as the anionic polymerization of optically pure **148** gives a highly isotactic polymer whose chiroptical properties appear to arise from excess single-handed helicity of the main chain. Free-radical and anionic polymerizations of **149** proceed exclusively via a cyclization mechanism, and the obtained polymer seems to have a helical conformation with excess helicity [186,187]. Similar monomers have been synthesized and polymerized [188].

147   **148**   **149**

11.3.2.3. Isocyanide Polymers

Optically active polyisocyanides with excess helix sense are obtained from optically active monomers by polymerization with $NiCl_2$. The polymers obtained from (R)-$(CH_3)_2CHCH(CH_3)NC$, (R)-$(CH_3)_2CHCH_2CH(CH_3)NC$, and (R)-n-$C_6H_{13}CH(CH_3)NC$ have M helical sense with screw-sense excesses of 62%, 56%, and 20%, respectively [189]. The copolymerization of achiral phenyl isocyanide with optically active

(S)-$(CH_3)_2$CHCH(COOCH$_3$)NC affords an optically active polymer with ca. 30% of optically active monomer incorporated into the chain [190,191]. The prevailing helix sense of the copolymer was P, whereas the homopolymerization of the optically active monomer gives an M-helix. This is because the polymerizability of the optically active monomer is much lower than that of the achiral monomer. In the copolymerization, phenyl isocyanide preferentially polymerizes in the early stages of the polymerization to lead to a mixture of P- and M-helical growing chains, and the optically active monomer is incorporated into the M-helical growing chain. As a consequence, the propagation rate of the M-chain becomes smaller than that of the P-chain, giving a P-helical chain consisting mainly of phenyl isocyanide as the major component of the product.

11.3.2.4. Isocyanate Polymers The helix of a polyisocyanate chain can have excess helicity based on the effects of a chiral side chain [192–202]. For example, optically active monomer (R)-**150a**, whose chirality is based only on the difference between -H and -D ($[\alpha]_D < 1°$), gives a polymer showing $[\alpha]_D$ −367° by anionic polymerization with NaCN [194]. It is interesting that **150a** and **150b** with the same absolute configuration and that very similar structures result in opposite helical sense of the polymers [198]. An optically active aromatic isocyanate, (S)-**151** [200,201], was synthesized and polymerized. Poly-(S)-**151** showed a very large levorotation ($[\alpha]_{365}$ −1969°), which was not affected by temperature. This result is in contrast to the fact that the optical activity of polyisocyanates with chiral side chains is often greatly dependent on temperature and may suggest that the poly-(S)-**151** has a perfectly single-handed helical conformation.

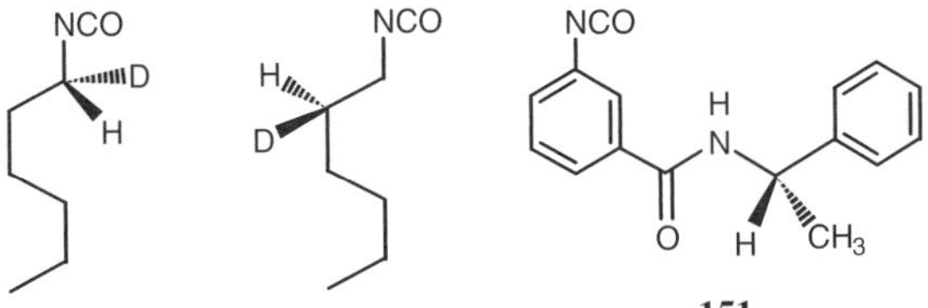

150a   **150b**   **151**

11.3.2.5. Carbodiimide Polymers An optically active carbodiimide, (R)-**152** ($[\alpha]_{365}$ +7.6°), gives a polymer by polymerization using a titanium (IV) isopropoxide catalyst (Scheme 11.9) [203]. The polymer showed optical activity essentially identical to the monomer; however, on heating, the polymer indicated mutarotation and specific rotation reached a plateau value of $[\alpha]_{365}$ −157.5°, which is considered to be based on excess helical sense of the main chain. The mutarotation has been ascribed to a conformational transition from a kinetically controlled one to a thermodynamically controlled one. Excess single-handed helical conformation can be induced for poly(di-n-hexylcarbodiimide) by protonating the polymer with chiral camphorsulfonic acid.

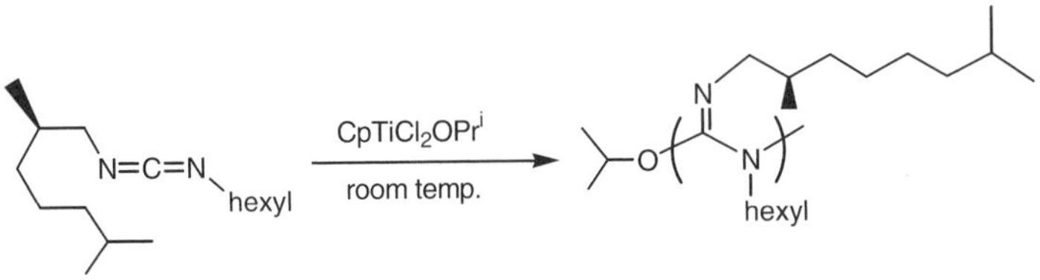

152

Scheme 11.9

11.3.2.6. Acetylene Polymers Homopolymers of optically active acetylenes, including (*R*)-**153** synthesized by [RhCl(norbornadiene)]$_2$ catalyst, show intense CD bands in the UV-visible region, probably based on a predominant helical sense of the main chain [204]. Excess single-handed helicity of the main chain can be induced for polymers of achiral acetylenes (**154** and **155**) by adding chiral molecules. The chiral induction is based on acid-base interaction or complex formation between the polymer and the additives [205–208].

153　　　　**154** CO_2H　**155** $B(OH)_2$

11.3.2.7. Other Polymers (+)- or (–)-*m*-Tolyl vinyl sulfoxide ([α]$_D$ +486°, –486°) gives optically active polymers by polymerization with BuLi. When the polymers were reduced to poly(tolyl vinyl sulfone) with an achiral side chain, they still showed optical activity ([α]$_D$ +42°~–41°). The optical activity might be based on a helical structure of the main chain [209]. Chiral polymers having a single-handed helical or propeller conformation have been prepared based on an optically active binaphthyl moiety [210–212]. By anionic polymerization with t-BuOK, an optically active carbonate monomer (**156**) gives the polymer (**157**), which is expected to have a single-handed 4$_1$-helical conformation [211]. Single-handed helical conformation has been observed for poly(methyl{(*S*)-2-methylbutyl)silane} [213] .

156　　　　**157**

11.4. ENANTIOMER-SELECTIVE POLYMERIZATION

In enantiomer-selective polymerization, one antipode of racemic chiral monomers is preferentially polymerized to give an optically active polymer. This type of polymerization has been attained for α-olefins, methacrylates, and cyclic monomers such as propylene oxide and sulfide, lactones, and α-amino acid *N*-carboxy anhydrides (NCAs). Enantiomer-selective polymerization has a close connection with stereoselective polymerization. In stereoselective polymerization, racemic chiral monomers are polymerized to give a mixture of a polymer preferentially consisting of one antipode and that consisting of the opposite enantiomer. There are comprehensive review articles about these types of polymerizations [4,214]. Enantiomer selection has been mainly performed by using chiral catalysts (including chiral initiators) for racemic, optically inactive monomers, but there are a few examples in which optically active monomers having ee of less than 100% were polymerized, and selection was confirmed by comparing the ee's of the starting monomer and the resulting polymer.

11.4.1. α-Olefins and Vinyl Ethers

The polymerizations of olefins **158–161** using Ziegler–Natta catalysts are stereoselective, and a mixture of polymers consisting of preferentially (R)-monomer and (S)-enantiomer is obtained (stereoselective polymerization) [215–225]. When the polymerization is carried out with $TiCl_4/Zn[(S)$-2-$CH_2CH(CH_3)C_2H_5]_2$, enantiomer selection is observed [219–221]. The selectivity decreases as the distance between the chiral center and the vinyl group in a monomer increases, and no enantiomer selection is attained for 5-methyl-1-heptene (**161**) [12,220,221]. $MgCl_2$-supported $TiCl_3$ catalysts modified with bis{(S)-2-methyl-butyl} phthalate were effective in enantiomer-selective polymerization of **158–160** [223]. Enantiomer selection was also observed in copolymerization of optically active monomers with racemic monomers [224,225].

158   **159**   **160**   **161**

Enantiomer selection is also found in vinyl ether polymerization [226,227]. The polymerization of *cis*- and *trans*-1-methylpropyl propenyl ethers using (−)-menthoxyaluminum dichloride [227] and the copolymerization of *rac*-1-methylpropyl vinyl ether with optically active monomers [226] are enantiomer selective.

11.4.2. Methacrylates

High enantiomer selectivity has been realized in the anionic polymerization of α-methylbenzyl methacrylate (**MBMA**) and its analogs by using the complexes of Grignard reagents and (−)-**Sp** [228–236]. Before the finding of this polymerization system, there were examples of attempted enantiomer-selective polymerization of menthyl methacrylate by using (−)-amylmagnesium bromide [237], that of **MBMA** using [(+)-2-methylbutyl]lithium or a complex of butyllithium with (−)-menthoxylithium [238], and those of **MBMA** and *sec*-butyl methacrylate with the $AlEt_3$-(−)-**Sp** complex [239]; however, selectivity was very low.

In the **MBMA** polymerization with cyclohexylmagnesium-(−)-**Sp** complex in toluene at −78°C, the (S)-monomer is preferentially polymerized over the (R)-isomer and the enantiomeric excess of the remaining monomer reaches greater than 90% at 50–70% monomer conversion ratio [228]. The polymerization is highly isotactic specific, indicating that the chiral **Sp**-complex simultaneously differentiates enantiomers and enantiofaces (diastereotopic faces) of the double bond of the monomer. The polymerization with Grignard reagent alone gives an atactic polymer. The Grignard reagent-(−)-**Sp** complexes are also effective in enantiomer-selective polymerization of **MBMA** analogues with the structure **162** [233–236].

MBMA   **162**

R = -H, R' = -CH₃
R = -C₂H₅, R' = -CH₃
R = -CH₃, R' = -C₂H₅
R = -CH₃, R' = -C₃H₇-*i*
R = -CH₃, R' = -C₄H₉-*t*
R = -CH₃, R' = -CH₂Ph
R = -CH₃, R' = -CPh₃

The enantiomer selectivity and polymerization activity is greatly dependent on the structure of initiator complexes [232,240] and monomer conformation [236,241]. The complexes with **163**, **164**, and **165** were nearly inactive in **MBMA** polymerization, whereas the complexes with **166** and **167** can polymerize **MBMA** with much lower enantiomer selectivity compared with the **Sp**-complex [4]. **Sp** and **163–167** have different shapes and sizes of the cavity of the ligand, which play an important role in chiral discrimination and influence the activity of the complex.

163

164

165

166

167

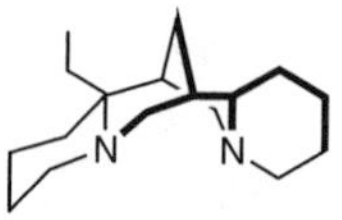

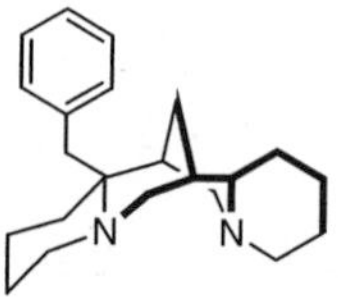

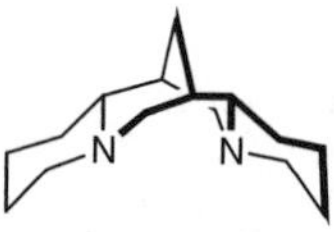

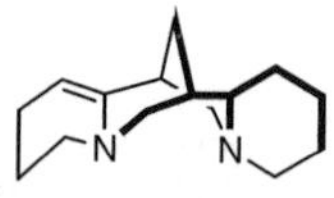

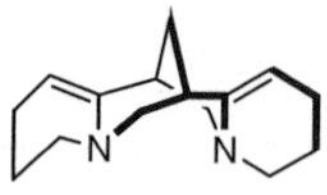

Enantiomer-selective polymerization of **MBMA** has also been attained by using the reaction products of chiral amine compounds, **168** and **169**, with cyclohexylmagnesium bromide as initiator [242,243] and by using the aluminum porphyrin complex **170** in the presence of optically active aluminum alkoxide compounds 171**a–e** [244]. In the latter systems, the enantiomer selection is based on enantiomer-selective coordination of the chiral aluminum compounds to **MBMA** as revealed by ^{1}H NMR analysis. With **171e** as a catalyst, the ee of the unreacted monomer is 40% at 75% monomer conversion ratio in the polymerization at –70°C.

R = R' = -NH$_2$
R = R' = -NHCH$_3$
R = -N(CH$_3$)$_2$, R' = -NH$_2$

168

169

170

a: R = -SiMe$_2$Bu-*t*, **b:** R = -CH$_2$Bu-*t*
c: R = -SiPh$_2$Bu-*t*, **d:** R = -COBu-*t*

e: R =

171

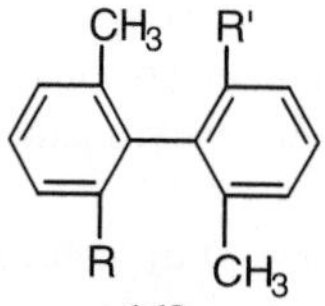

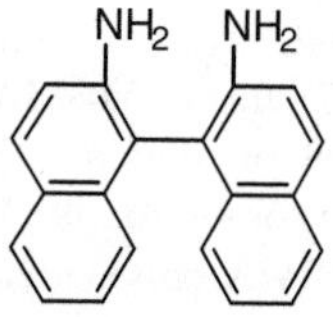

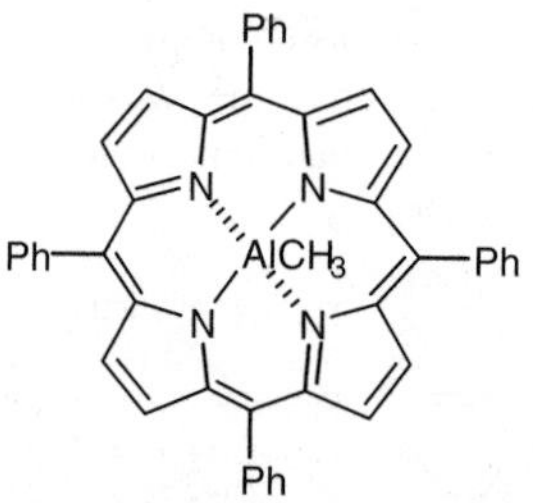

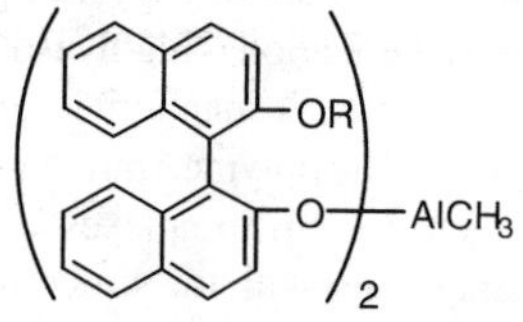

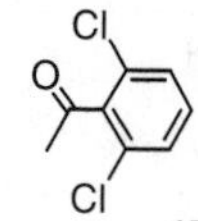

TABLE 11.6. Enantiomer-Selective Polymerization of (+)-55 Using Preformed Living Poly- or Oligo[(+)-55] Anion in Toluene at −78°C[a]

Poly- or Oligo[(+)-55] Anion		ee (%) (Polymer)	ee (%) (Remaining Monomer)
Initiator	DP	Conversion (%)	
FlLi-(+)-DDB	30	16.1	100	20.0 (−)
FlLi-(+)-DDB	1–5	30.1	37.8	16.3 (−)
FlLi-(−)-DDB	30	22	43.1	12.2 (−)
FlLi-(−)-DDB	1–5	20	14.0	3.6 (−)
FlLi-TMEDA	40	24.9	40.0	13.3 (−)

[a]Data cited from ref. 130.

The anionic polymerization of racemic **55** by using chiral initiator complexes is enantiomer selective [129,130]. In the polymerization using **FlLi-(−)-DDB**, enantiomer selectivity is low in the early stages of polymerization and increases as the reaction proceeds [130]. This is because the helical conformation of the growing anion rather than the chiral ligand (**DDB**) has an important role in enantiomer selection as evidenced by the polymerization using the polymer anion (Table 11.6) [130]. Poly[(+)-**55**] anions having a DP of 30 were prepared by the polymerization of (+)–**55** with **FlLi-(−)-DDB** and **FlLi-(+)-DDB** complexes. The anions are considered to have a single-handed helical conformation with the same screw sense due to the effects of the chiral side chains and were used as initiators for the polymerization of racemic **55** in toluene at −78°C. Both anions preferentially polymerized the (+)-monomer regardless of the chirality of **DDB**. Furthermore, the poly[(+)-**55**] anion prepared **FlLi-TMEDA** complex with a DP of 40 preferentially polymerized the (+)-monomer. In addition, oligo[(+)-**55**] anions with a DP of 1–5 synthesized using **FlLi-(+)-DDB** and **FlLi-(−)-DDB**, which are too short to maintain a stable helical conformation, exhibited different enantiomer selectivities depending on the chirality of **DDB**. These findings clearly indicate that the enantiomer selection is mainly governed by the helicity of the polymer anion.

Enantiomer-selection was also confirmed in the free-radical polymerizations of **55** [132] and **148** [185] having various enantioselectivities, though the selectivity was much lower than that observed in the anionic polymerization.

11.4.3. Cyclic Monomers

The first enantiomer-selective polymerization was performed with propylene oxide (**172**) as a monomer [245]. The polymerization was carried out with a $ZnEt_2/(+)$-borneol or $ZnEt_2/(−)$-menthol initiator system. The obtained polymer was optically active and the unreacted monomer was rich in (S)-isomer. Various examples are known concerning the polymerization and copolymerization of **172** [246–251]. A Schiff base complex **173** has been shown to be an effective catalyst: In the polymerization at 60°C, the enantiopurity of the remaining monomer was 9% ee at 50% monomer conversion [250].

Propylene sulfide (**174**) can also be polymerized enantiomer-selectively [252–262]. In the polymerization with a $ZnEt_2$-(−)-binaphthol initiator system at room temperature, the

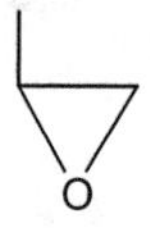

172

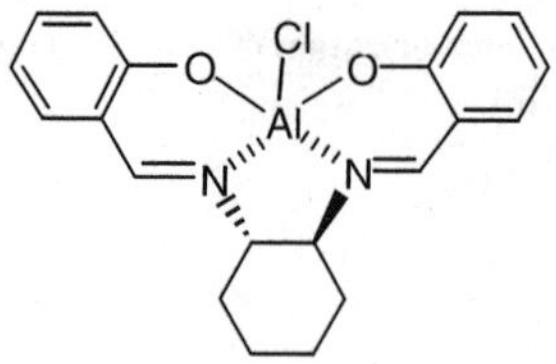

173

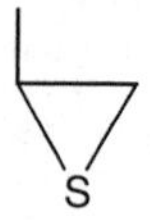

174

enantiopurity of the remaining monomer was 92% ee at 67% monomer conversion and the ratio of (*S*)-monomer consumption rate to (*R*)-monomer consumption rate (k_S/k_R) was 20 [257,258].

Enantiomer-selective polymerization has also been achieved for lactones (**175**) [250,255,260–265] and lactide (**176**) [266]. Polymerization using $ZnEt_2$-(*R*)-**26** gave k_R/k_S values of 1.7, 1.25, and 1.02–1.07 for **175a** [264], **175b** [261], and **175c** [262], respectively. In the polymerization of **176** with **177** at 70°C, the enantiopurity of unreacted monomer at 50% monomer conversion was around 80% ee and the k_D/k_L value was estimated to be 20 [266]. **175d** is polymerized enantiomer-selectively by using the lipase from Pseudomonas fluorescens (P530) [265]. By the polymerization in toluene at 35°C, the enantiopurity of the unreacted monomer was 45% ee at 47% conversion. There are several other examples of enantiomer-selective polymerization using enzymes [267–269].

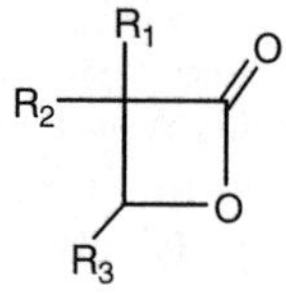

a: R_1 = -H; R_2 = -H; R_3 = -CH$_3$
b: R_1 = -CH$_3$; R_2 = -C$_3$H$_7$-*n*; R_3 = -H
c: R_1 = -CH$_3$; R_2 = -C$_2$H$_5$; R_3 = -H
d: R_1 = -H; R_2 = -CH$_3$; R_3 = -H

175

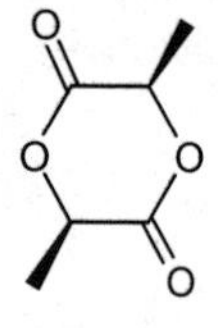

176

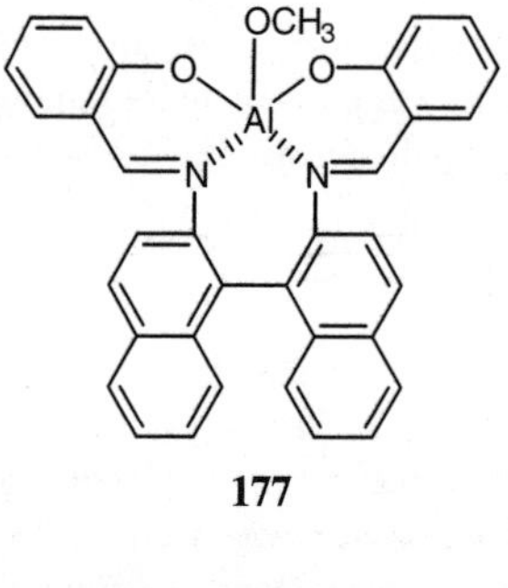

177

Enantiomer-selective polymerization of α-amino acid NCAs has been reported [270–274]. In the polymerization of γ-benzylglutamate NCA using $[(+)-C_2H_5(CH_3)CHCOO]_2NiPBu_3^n$ in dioxane at 30°C, the enantiopurity of monomeric units in the polymer chain was estimated to be (−) 34% ee at 28% polymer yield [272].

11.5. CONCLUSION

Asymmetric polymerization is an expanding field of polymer synthesis: various novel methods of synthesis are currently being developed and even wider varieties of polymer structures and polymerization methods are expected to be devised in the future. Advance in this field would take place most effectively when we take advantage of the progress in the field of asymmetric organic synthesis.

REFERENCES

1. (a) Okamoto, Y.; Nakano, T. *Chem. Rev.* **1994**, *94*, 349. (b) Nakano, T.; Okamoto, Y. In *Catalysis in Precision Polymerization*; Kobayashi, S (Ed.); John Wiley & Sons: Chishester, Sussex, 1997; pp. 271–309. (c) Nakano, T.; Okamoto, Y. In *The Polymeric Materials Encyclopedia*; Salamone, J. C. (Ed.); CRC Press: Florida, 1996; pp. 417–423.

2. Wulff, G. *Angew. Chem., Int. Ed.* **1995**, *34*, 1812.

3. *Liquid Crystal Polymers: From Structures to Applications*; Colloyer, A. A. (Ed.); Elsevier: London, 1992.

4. Okamoto, Y.; Yashima, E. *Prog. Polym. Sci.* **1990**, *15*, 263.

5. Vogl, O.; Jaycox, G. D. *Polymer* **1987**, *28*, 2179.

6. Wulff, G. *CHEMTECH* **1991**, 364.

7. Wulff, G. *Angew. Chem. Int. Ed.* **1989**, *28*, 21.

8. Farina, M. *Top. Stereochem.* **1987**, *17*, 1.

9. (a) *Optically Active Polymers*; Selegny, E. (Ed.); Reidel: Dordrecht, 1979. (b) Ciardelli, F. In *Encyclopedia of Polymer Science and Engineering*, Vol 10; Wiley: New York, 1987; pp. 463–493.

10. (a) Arcus, C. L. *J. Chem. Soc.* **1955**, 2801. (b) Arcus, C. L. *J. Chem. Soc.* **1957**, 1189. (c) Arcus, C. L. *Prog. Stereochem.* **1962**, *3*, 264.

11. Schulz, R. C.; Kaiser, E. *Adv. Polym. Sci.* **1965**, *4*, 236.

12. Pino, P. *Adv. Polym. Sci.* **1965**, *4*, 393.

13. Farina, M.; Peraldo, M.; Natta, G. *Angew. Chem., Int. Ed.* **1965**, *4*, 107.

14. Frisch, H. L.; Schuerch, C.; Szwarc, M. *J. Polym. Sci.* **1953**, *11*, 559.

15. Murahashi, S.; Nozakura, S.; Takeuchi, S. *Bull. Chem. Soc. Jpn.* **1960**, *33*, 658.

16. Braun, D.; Kern, W. *J. Polym. Sci., Part C* **1964**, *4*, 197.

17. Fray, G. I.; Robinson, R. *Tetrahedron* **1962**, *18*, 261.

18. Marvel, C. S.; Frank, R. L.; Prill, E. *J. Am. Chem. Soc.* **1943**, *65*, 1647.

19. Arcus, C. L.; West, D. W. *J. Chem. Soc.* **1959**, 2699.

20. Nakano, T.; Hidaka, Y.; Okamoto, Y. *Polym. J.* **1998**, *30*, 596.

21. Pino, P.; Cioni, P.; Wei, J. *J. Am. Chem. Soc.* **1987**, *109*, 6189.

22. Pino, P.; Galimberti, M.; Prada, P.; Consiglio, G. *Makromol. Chem.* **1990**, *191*, 1677.

23. Fuhrman, H.; Kortus, K.; Fuhrman, C. *Macromol. Chem. Phys.* **1996**, *197*, 3869.

24. Nozaki, K.; Sato, N.; Takaya, H. *J. Am. Chem. Soc.* **1995**, *117*, 9911.

25. Nozaki, K.; Sato, N.; Tonomura, Y.; Yasutomi, M.; Takaya, H.; Hiyama, T.; Matsubara, T.; Koga, N. *J. Am. Chem. Soc.* **1997**, *119*, 12779.

26. Nozaki, K.; Sato, N.; Nakamoto, K.; Takaya, H. *Bull. Chem. Soc. Jpn.* **1997**, *70*, 659.

27. Jiang, Z.; Sen, A. *J. Am. Chem. Soc.* **1995**, *117*, 4455.

28. Kacker, S.; Jiang, Z.; Sen, A. *Macromolecules* **1996**, *29*, 5852.

29. Sperrle, M.; Consiglio, G. *J. Am. Chem. Soc.* **1995**, *117*, 12130.

30. Bronco, S.; Consiglio, G.; Hutter, R.; Batistini, A.; Suter, U. W. *Macromolecules* **1994**, *27*, 4436.

31. Bronco, S.; Consiglio, C.; Di Benedetto, S.; Fehr, M. *Helv. Chim. Acta* **1995**, *78*, 883.

32. Brookhart, M.; Rix, R. C.; DeSimone, J. M.; Barborak, J. C. *J. Am. Chem. Soc.* **1992**, *114*, 5894.

33. Brookhart, M.; Wagner, M. I.; Balavoine, G. G. A.; Haddou, H. A. *J. Am. Chem. Soc.* **1994**, *116*, 3641.

34. Brookhart, M.; Wagner, M. I. *J. Am. Chem. Soc.* **1996**, *118*, 7219.

35. Farina, M. *Chim. Ind. (Milan)* **1964**, *46*, 761 (*Chem. Absr.* **1964**, *61*, 8409a)

36. Farina, M.; Modena, M.; Ghizzoni, W. Rend. *Acc. Naz. Lincei* **1962**, *32*, 91 (*Chem. Abstr.* **1962**, *57*, 15320h).

37. Natta, G.; Farina, M.; Donati, M. *Makromol. Chem.* **1961**, *43*, 251.

38. Natta, G.; Farina, M.; Carbonaro, A.; Lugli, G. *Chim. Ind. (Milan)* **1961**, *43*, 529.

39. Natta, G.; Porri, L.; Valenti, S. *Makromol. Chem.* **1963**, *67*, 225.

40. Takemoto, K.; Miyata, M. *J. Macromol. Sci. Rev. Macromol. Chem.* **1980**, *C18(1)*, 83.

41. Farina, M. In *Inclusion Compounds, Vol. 3*; Atwood, J. L., Davies, J. D. E., MacNicol, D. D. (Ed.); Academic Press: London, 1984; pp. 297.

42. Farina, M.; Audisio, G.; Natta, G. *J. Am. Chem. Soc.* **1967**, *89*, 5071.

43. Farina, M. In *Proceedings of the International Symposium on Macromolecules, Rio de Janeiro*; Mano, E. B. (Ed.); Elsevier: Amsterdam, 1975; pp. 21.

44. Audisio, G.; Silvani, A. *J. Chem. Soc., Chem. Commun.* **1976**, 481.

45. Miyata, M.; Kitahara, Y.; Takamoto, K. *Polym. J.* **1981**, *13*, 111.

46. Miyata, M.; Kitahara, Y.; Takemoto, K. *Polym. Bull.* **1980**, *2*, 671.

47. Miyata, M.; Kitahara, Y.; Takemoto, K. *Makromol. Chem.* **1983**, *184*, 1771.

48. Natta, G.; Farina, M.; Peraldo, M.; Bressan, G. *Makromol. Chem.* **1961**, *43*, 68.

49. Farina, M.; Bressan, G. *Makromol. Chem.* **1963**, *61*, 79.

50. Natta, G.; Farina, M. *Tetrahedron. Lett.* **1963**, 703.

51. Takeda, Y.; Hayakawa, Y.; Fueno, T.; Furukawa, J. *Makoromol Chem.* **1965**, *83*, 234.

52. Hayakawa, Y.; Fueno, T.; Furukawa, J. *J. Polym. Sci. Part A–1* **1967**, *5*, 2099.

53. Okamoto, Y.; Nakano, T.; Kobayashi, H.; Hatada, K. *Polym. Bull.* **1991**, *25*, 5.

54. Oishi, T.; Yamasaki, H.; Fujimoto, M. *Polym. J.* **1991**, *23*, 795.

55. Oninumra, K.; Tsutsumi, H.; Oishi, T. *Polym. Bull.* **1997**, *39*, 437.

56. Onimura, K.; Tsutsumi, H.; Oishi, T. *Macromolecules* **1998**, *31*, 5971.

57. Liu, W.; Chen, C.; Chen, Y.; Xi, F. *Polym. Bull.* **1997**, *38*, 509.

58. Coates, G. W.; Waymouth, R. M. *J. Am. Chem. Soc.* **1991**, *113*, 6207.

59. Coates, G. W.; Waymouth, R. M. *J. Am. Chem. Soc.* **1993**, *115*, 91.

60. Habaue, S.; Sakamoto, H.; Okamoto, Y. *Chem. Lett.,* **1996**, 383.

61. Habaue, S.; Sakamoto, H.; Okamoto, Y. *Polym. J.* **1997**, *29*, 384.

62. S. Habaue, S.; Sakamoto, H.; Baraki, H.; Okamoto, Y. *Macromol. Chem. Rapid Commun.* **1997**, *18*, 707.

63. Seno, M.; Kawamura, Y.; Sato, T. *Macromolecules* **1997**, *30*, 6417.

64. Haba, O.; Kakuchi, T.; Yokota, K. *Macromolecules* **1993**, *26*, 1782.

65. Spassky, N.; Momtaz, A.; Kassamaly, A.; Sepulchre, M. *Chirality* **1992**, *4*, 295.

66. Momtaz, A.; Spassky, N.; Sigwalt, P. *Polym. Bull.* **1979**, *1*, 267.

67. Sepulchre, M.; Kassamaly, A.; Spassky, N. *Makromol. Chem., Macromol. Symp.* **1991**, *42/43*, 489.

68. Itsuno, S.; Tada, S.; Ito, K. *Chem. Commun.* **1997**, 933.

69. Addadi, L.; Cohen, M. D.; Lahav, M. *J. Chem. Soc. Chem. Commun.* **1975**, 471.

70. Addadi, L.; Lahav, M. *J. Am. Chem. Soc.* **1979**, *101*, 2152.

71. Addadi, L.; van Mil. J.; Lahav, M. *J. Am. Chem. Soc.* **1982**, *104*, 3422.

72. Chung, C.-M.; Hasegawa, M. *J. Am. Chem. Soc.* **1991**, *113*, 7311.

73. Beredjick, N.; Schuerch, C. *J. Am. Chem. Soc.* **1958**, *80*, 1933.

74. De, B. B.; Sivaram, S.; Dhal, P. K. *Macromolecules* **1996**, *29*, 468.

75. Porter, N. A.; Swann, E.; Nally, J.; McPhail. A. T. *J. Am. Chem. Soc.* **1990**, *112*, 6740.

76. Porter, N. A.; Breyer, R.; Swann, T.; Nally, J.; Pradhan, J.; Allen. T.; McPhail A. T. *J. Am. Chem. Soc.* **1991**, *113*, 7002.

77. Porter, N. A.; Bruhnke, J. D.; Wu, W.-X.; Rosenstein, I. J.; Breyer, R. A. *J. Am. Chem. Soc.* **1991**, *113*, 7788.

78. Porter, N. A.; Allen, T. R.; Breyer, R. A. *J. Am. Chem. Soc.* **1992**, 114, 7676.

79. (a) Buese, M. A.; Hogen-Esch, T. E. *J. Am. Chem. Soc* **1985**, *107*, 4509. (b) Buese, M. A.; Hogen-Esch, T. E. *Macromolecules* **1984**, *17*, 119.

80. Doiuchi, T.; Dodoh, T.; Yamaguchi, H. *Makromol. Chem.* **1990**, *191*, 1253.

81. Doiuchi, T.; Dodoh, T.; Yamaguchi, H. *Makromol. Chem.* **1992**, *193*, 221.

82. Bando, Y.; Minoura, Y. *Eur. Polym. J.* **1979**, *15*, 333.

83. Yamaguchi, H.; Iwama, T.; Hayashi, T.; Doiuchi, T. *Makromol. Chem.* **1990**, *191*, 1243.

84. Wulff, G.; Dhal, P. K. *Angew., Chem. Int. Ed.* **1989**, *28*, 196.

85. Wulff, G.; Kemmerer, R.; Vogt, B. *J. Am. Chem. Soc.* **1987**, *109*, 7449.

86. Wulff, G.; Hohn, J. *Macromolecules* **1982**, *15*, 1255.

87. Wulff, G.; Dhal, P. K. *Macromolecules* **1990**, *23*, 4525.

88. Kakuchi, T.; Kawai, H.; Katoh, S.; Haba, O.; Yokota, K. *Macromolecules* **1992**, *25*, 5545.

89. Obata, M.; Kakuchi, T.; Yokota, K. *Macromolecules* **1997**, *30*, 348.

90. Kakuchi, T.; Haba, O.; Uesaka, T.; Obata, M.; Morimoto, Y.; Yokota, K. *Macromolecules* **1996,** *29*, 3812.

91. (a) Yokota, K.; Kakuchi, T.; Sasaki, H.; Ohmori, H. *Makromol. Chem.* **1989**, *190*, 1269. (b) Yokota, K.; Kakuchi, T.; Yamamoto, T.; Hasegawa, T.; Haba, O. *Makromol. Chem.* **1992**, *193*, 1805.

92. Yokota, K.; Kakuchi, T.; Sakurai, K.-I.; Iwata, Y.; Kawai, H. *Makromol. Chem. Rapid Commun.* **1992**, *13*, 343.

93. Kataoka, S.; Ando, T. *Kobunshi Ronbunshu (Japan)* **1980**, *37*, 185 (*Chem. Abstr.* **1980**, *92*, 198833).

94. Kataoka, S.; Ando, T. *Polym. Commun.* **1984**, *25*, 24.

95. Green, M. M.; Garetz, B. A. *Tetrahedron Lett.* **1984**, *25*, 2831.

96. Tadokoro, H. *Structure of Crystalline Polymers*; Wiley: New York, 1979.

97. Okamoto, Y. *CHEMTECH* **1987**, 144.

98. Okamoto, Y.; Hatada, K. In *Chromatographic Chiral Separations*; Zief, M., Crane, L. J. (Ed.); Marcel Dekker: New York, 1988; pp. 199.

99. Okamoto, Y.; Suzuki, K.; Ohta, K.; Hatada, K.; Yuki, H. *J. Am. Chem. Soc.* **1979**, *101*, 4763.

100. Okamoto, Y.; Suzuki, K.; Yuki, H. *J. Polym. Sci., Polym. Chem. Ed.* **1980**, *18*, 3043.

101. Okamoto, Y.; Shohi, H.; Yuki, H. *J. Polym. Sci., Polym. Lett. Ed.* **1983**, *21*, 601.

102. Okamoto, Y.; Yashima. E.; Nakano, T.; Hatada, K. *Chem. Lett.* **1987**, 759.

103. Nakano, T.; Okamoto, Y.; Hatada, K. *J. Am. Chem. Soc.* **1992**, *114*, 1318.

104. (a) Kanoh, S.; Suda, H.; Kawaguchi, N.; Motoi, M. *Makromol. Chem.* **1986**, *187*, 53. (b) Kanoh, S.; Kawaguchi, N.; Sumino, T.; Hongo, Y.; Suda, H. *J. Polym. Sci., Part A, Polym. Chem.* **1987**, *25*, 1603.

105. Kanoh, S.; Sumino, T.; Kawguchi, N.; Motoi, M.; Suda, H. *Polym. J.* **1988**, *20*, 539.

106. Wulff, G.; Sczepan, R.; Steigel, A. *Tetrahedron Lett.* **1986**, *27*, 1991.

107. Wulff, G.; Vogt, B.; Petzold, J. *ACS Polym. Mat. Sci. Eng.* **1988**, *58*, 859.

108. Okamoto, Y.; Okamoto, I.; Yuki, H. *J. Polym. Sci. Polym. Lett. Ed.* **1981**, *19*, 451.

109. Cavallo, M.; Corradini, P.; Vacatello, M. *Polym. Commun.* **1989**, *30*, 236.

110. Nakano, T.; Ute, K.; Okamoto, Y.; Matsuura, Y.; Hatada, K. *Polym. J.* **1989**, 935.

111. Okamoto, Y.; Ishikura, M.; Hatada, K.; Yuki, H. *Polym. J.* **1983**, *15*, 851.

112. Okamoto, Y.; Mohri, H.; Ishikura, M.; Hatada, K.; Yuki, H. *J. Polym. Sci., Polym. Symp.* **1986**, *74*, 125.

113. Okamoto, Y.; Mohri, H.; Ishikura, M.; Hatada, K. In *Current Topics in Polymer Science, Vol. I*; Ottenbrite, R. M., Utracki, L. A., Inoue, S., (Ed.); Hanser: New York, 1987; pp. 31.

114. Okamoto, Y.; Mohri, H.; Hatada, K. *Polym. Bull.* **1980**, *20*, 25.

115. Okamoto, Y.; Mohri, H.; Hatada, K. *Chem. Lett.* **1988**, 1879.

116. Okamoto, Y.; Mohri, H.; Nakano, T.; Hatada, K. *Chirality* **1991**, *3*, 277.

117. Nakano, T.; Taniguchi, K.; Okamoto, Y. *Polym. J.* **1997**, *29*, 540.

118. Mohri, H.; Okamoto, Y.; Hatada, K. *Polym. J.* **1989**, *21*, 719.

119. Ren, C.; Chen, F.; Xi, F.; Nakano, T.; Okamoto, Y. *J. Polym. Chem., Part A., Polym. Chem.* **1993**, *31*, 2721.

120. Okamoto, Y.; Yashima, E.; Ishikura, M.; Hatada, K. *Polym. J.* **1987**, *19*, 1183.

121. Okamoto, Y.; Shohi, H.; Ishikura, M.; Hatada, K. *Proc. IUPAC 28th Macromolec. Symp. (Amherst)* **1982**, 232.

122. (a)Okamoto, Y.; Nakano,. T.; Fukuoka, T.; Hatada, K. *Polym. Bull.* **1991**, *26*, 259. (b) Okamoto, Y.; Nakano, T.; Hasegawa, T. *Polym. Prepr. Jpn. (English Edition)* **1991**, *40*, E826.

123. Nakano, T.; Matsuda, A.; Mori, M.; Okamoto, Y. *Polym. J.* **1996**, *28*, 300.

124. Nakano, T.; Mori, M.; Okamoto, Y. *Macromolecules* **1993**, *26*, 867.

125. Nakano, T.; Matsuda, A.; Okamoto, Y. *Polym. J.* **1996**, *28*, 56.

126. (a) Nakano, T.; Shikisai, Y.; Okamoto, Y. *Polym. J.* **1996**, *28*, 51. (b) Nakano, T.; Shikisai, Y.; Okamoto, Y. *Proc. Japan Acad.* **1995**, *71*, Ser. B, 251.

127. Nakano, T.; Satoh, Y.; Okamoto, Y. *Polym. J.* **1998**, *30*, 635.

128. Satoh, Y.; Nakano, T.; Okamoto, Y. *Polym. Prepr. Jpn.* **1997**, *46*, 1165.

129. Yashima, E.; Okamoto, Y ; Hatada, K. *Polym. J.* **1987**, *19*, 897.

130. Yashima, E.; Okamoto, Y.; Hatada, K. *Macromolecules* **1988**, *21*, 854.

131. Okamoto, Y.; Yashima, E.; Hatada, K. *J. Polym. Sci., Part C, Polym. Lett.* **1987**, *25*, 297.

132. Okamoto, Y.; Nishikawa, M.; Nakano, T.; Yashima, E.; Hatada, K. *Macromolecules* **1995**, *28*, 5135.

133. (a) Okamoto, Y.; Nakano, T.; Asakura, T.; Mohri, H.; Hatada, K. *Polym. Prepr.* **1992**, *30(2)*, 437. (b) Okamoto, Y.; Nakano, T.; Asakura, T.; Mohri, H.; Hatada, K. *J. Polym. Sci., Part A, Polym. Chem.* **1991**, *29*, 287.

134. Wu, J.; Nakano, T.; Okamoto, Y. *J. Polym. Sci., Part A, Polym. Chem.* **1998**, *36*, 2013.

135. Okamoto, Y.; Mohri, H.; Nakano, T.; Hatada, K. *J. Am. Chem. Soc.* **1989**, *111*, 5952.

136. Okamoto, Y. et al. unpublished results.

137. Tamai, Y.; Kinouchi, S.; Matsuda, T. *Polym. Prepr. Jpn.* **1986**, 35, 220.

138. Cram, D. J.; Sogah, D. Y. *J. Am. Chem. Soc.* **1985**, *107*, 8301.

139. Okamoto, Y.; Nakano, T.; Hatada, K. *Polym. J.* **1989**, *21*, 199.

140. Habaue, S.; Tanaka, T.; Okamoto, Y. *Macromolecules* **1995**, *28*, 5973.

141. Tanaka, T.; Habaue, S.; Okamoto, Y. *Polym. J.* **1995**, *27*, 1202.

142. Okamoto, Y.; Adachi, M.; Shohi, H.; Yuki, H. *Polym. J.* **1981**, *13*, 175.

143. Okamoto, Y.; Hayashida, H.; Hatada, K. *Polym. J.* **1989**, *21*, 543.

144. Shiohara, K.; Habaue, S.; Okamoto, Y. *Polym. J.* **1996**, *28*, 682.

145. Shiohara, K.; Habaue, S.; Okamoto, Y. *Polym. J.* **1998**, *30*, 249.

146. Habaue, S.; Shiohara, K.; Uno, T.; Okamoto, Y. *Enantiomer* **1996**, 1, 55.

147. Uno, T.; Shiohara, K.; Habaue, S.; Okamoto, Y. *Polym. J.* **1998**, *30*, 352.

148. Wulff, G.; Wu, Y. *Makromol. Chem.* **1990**, *191*, 2993.

149. Wulff, G.; Wu, Y. *Makromol. Chem.* **1990**, *191*, 3005.

150. Okamoto Y. et. al. unpublished data.

151. Ute, K.; Asada, T.; Nabeshima, Y.; Hatada, K. *Acta Polymer* **1995**, *46*, 458.

152. Ute, K.; Asada, T.; Nabeshima, Y.; Hatada, K. *Macromolecules* **1993**, *26*, 7086.

153. Nakano, T; Okamoto, Y. *Macromolecules* **1999**, *32*, 2391.

154. (a) Millich, F. *Chem. Rev.* **1972**, *72*, 101. (b) Millich, F. *Adv. Polym. Sci.* **1975**, *19*, 117. (c) Millich, F. *Macromol. Rev.* **1980**, *15*, 207.

155. Nolte, R. J. M.; van Beijnen, A. J. M.; Drenth, W. *J. Am. Chem. Soc.* **1974**, *96*, 5932.

156. van Beijnen, A. J. M.; Nolte, R. J. M.; Drenth, W.; Hezemans A.M.F. *Tetrahedron* **1976**, *32*, 2017.

157. Kamer, P. C. J.; Nolte, R. J. M.; Drenth, W. *J. Am. Chem. Soc.* **1988**, *110*, 6818.

158. Deming, T. J.; Novak, B. M. *J. Am Chem. Soc.* **1992**, *114*, 7926.

159. (a) Takei, F.; Yanai, K.; Onitsuka, K.; Takahashi, S. *Angew. Chem., Int. Ed.* **1996**, *35*, 1554. (b) Takahashi, S.; Onitsuka, K.; Takei, F. *Proc. Jpn. Acad., Ser B* **1998**, *74B*, 25.

160. Ito, Y.; Ihara, E.; Murakami, M. *Angew. Chem., Int. Ed.* **1992**, *31*, 1509.

161. Ito, Y.; Ihara, E.; Murakami, Y.; Sisido, M. *Macromolecules* **1992**, *25*, 6810.

162. Ito, Y.; Kojima, Y.; Murakami, M. *Tetrahedron Lett.* **1993**, *34*, 8279.

163. Ito, Y.; Ohara, T.; Shima, R.; Suginome, M. *J. Am. Chem. Soc.* **1996**, *118*, 9188.

164. Ito, Y.; Miyake, T.; Ohara, T.; Suginome, M. *Macromolecules* **1998**, *31*, 1697.

165. Bur, A.; Fetters, L. J. *Chem. Rev.* **1976**, *76*, 727.

166. Shashoua, V. E.; Sweeny, W.; Tietz, R. F. *J. Am. Chem. Soc.* **1960**, *82*, 866.

167. Okamoto, Y.; Matsuda, M.; Nakano, T.; Yashima, E. *Polym. J.* **1993**, *25*, 391.

168. Okamoto, Y.; Matsuda, M.; Nakano, T.; Yashima, E. *J. Polym. Sci., Part A., Polym. Chem.* **1994**, *32*, 309.

169. Maeda, K.; Matsuda, M.; Nakano, T.; Okamoto, Y. *Polym. J.* **1995**, *27*, 141.

170. (a) Vogl, O.; Miller, H. C.; Sharkey, W. H. *Macromolecules* **1972**, *5*, 658. (b) Corley, L. S.; Vogl, O. Polym. Bull. 1980, 3, 211.

171. Vogl, O.; Corley, L. S.; Harris, W. J.; Jaycox, G. D.; Zhang, J. *Makromol. Chem. Suppl.* **1985**, *13*, 1.

172. Zhang, J.; Jaycox, G. D.; Vogl, O. *Polym. J.* **1987**, *19*, 603.

173. Jaycox, G. D.; Vogl, O. *Makromol. Chem., Rap. Commun.* **1990**, *11*, 61.

174. (a) Ute, K.; Hirose, K.; Kashimoto, H.; Hatada, K.; Vogl, O. *J. Am. Chem. Soc.* **1991**, *113*, 6305. (b) Ute, K.; Hirose, K.; Kashimoto, H.; Nakayama, K.; Hatada, K.; Vogl, O. *Polym. J.* **1993**, *25*, 1175. (c) Vogl, O.; Xi, F.; Vass, F.; Ute, K.; Nishimura, T.; Hatada, K. *Macromolecules* **1989**, *22*, 4660. (d) Ute, K.; Oka, K.; Okamoto, Y.; Hatada, K.; Xi, F.; Vogl, O. *Polym. J.* **1991**, *23*, 142.

175. Choi, S.-H.; Yashima, E.; Okamoto, Y. *Macromolecules* **1996**, *29*, 1880.

176. Kaminsky, W. *Angew. Makromol. Chem.* **1986**, *145/146*, 149.

177. Kaminsky, W.; Kupler, K.; Niedoba, S. *Makromol Chem., Makromol. Symp.* **1986**, *3*, 377.

178. Pino, P.; Ciardelli, F.; Lorenzi, G. P.; Montagnoli, G. *Makromol. Chem.* **1963**, *61*, 207.

179. Pino, P *Polym. Prepr.* **1989**, *30(2)*, 433.

180. Zambelli, A.; Grassi, A.; Galimberti, M.; Perego, Gabriele, P. *Makromol. Chem., Rapid. Commun.* **1992**, *13*, 269.

181. Pieroni, O.; Ciardelli, F.; Botteghi, C.; Lardicci, L.; Salvadori, P.; Pino, P. *J. Polym. Sci., Part C* **1969**, *22*, 993.

182. Thien, N.-T.; Suter, U. W.; Pino, P. *Makromol. Chem.* **1983**, *184*, 2335.

183. Alli, A.; Pino, P. *Helv. Chim. Acta* **1974**, *57*, 616.

184. Okamoto, Y.; Nakano, T.; Ono, E.; Hatada, K. *Chem. Lett.* **1991**, 525.

185. Nakano, T.; Kinjo, N.; Hidaka, Y.; Okamoto, Y. *Polym. J.* **1999**, *31*, 464.

186. Nakano, T.; Okamoto, Y., Sogah, D. Y.; Zheng, S. *Macromolecules* **1995**, 28, 8705.

187. Wulff, G; Gladow, S.; Kühneweg, B.; Krieger, S. *Macromol. Symp.* **1996**, *101*, 335.

188. Zheng, S.; Sogah, D. Y. *Tetrahedron* **1997**, *53*, 15469.

189. van Beijnen, A. J. M.; Nolte, R. J. M.; Drenth, W.; Hezemans, A. M. F.; van de Coolwijk, P. J. F. M. *Macromolecules* **1980**, *13*, 1386.

190. Harada, T.; Cleij, M. C.; Nolte, R.J.M.; Hezemans, A.M.F.; Drenth, W. *J. Chem. Soc., Chem. Commun.* **1984**, 726.

191. Kamer, P. C. J.; Cleij, M. C.; Nolte, R. J. M.; Harada, T.; Hezemans, A. M. F.; Drenth, W. *J. Am. Chem. Soc.* **1988**, *110*, 1581.

192. Goodman, M.; Chen, S.-C. *Macromolecules* **1970**, *3*, 398.

193. Goodman, M.; Chen, S.-C. *Macromolecules* **1971**, *4*, 625.

194. Green, M. M.; Gross, R. A.; Crosby III, C.; Schilling, R. C. *Macromolecules* **1987**, *20*, 992.

195. Green, M. M.; Gross, R. A.; Cook, R.; Schilling, R. C. *Macromolecules* **1987**, *20*, 2638.

196. Green, M. M.; Andreola, C.; Munoz, B.; Reidy, M. P.; Zero, K. *J. Am. Chem. Soc.* **1988**, *110*, 4063.

197. Green, M. M.; Reidy, M. P.; Johnson, R. J.; Darling, G.; O'Leary, D. J.; Willson, G. *J. Am. Chem. Soc.* **1989**, *111*, 6452.

198. Green, M. M.; Peterson, N. C.; Sato, T.; Teramoto, A.; Cook, R.; Lifson, S. *Science* **1995**, *268*, 1860.

199. Green, M. M.; Garetz, B. A.; Munoz, B.; Chang, S.; Hoke S.; Cooks, R. G. *J. Am. Chem. Soc.* **1995**, *117*, 4182.

200. Maeda, K.; Okamoto, Y. *Macromolecules* **1998**, *31*, 1046.

201. Maeda, K.; Okamoto, Y. *Macromolecules* **1998**, *31*, 5164.

202. Maeda, K.; Matsunaga, M.; Yamada, H.; Okamoto, Y. *Polym. J.* **1997**, 29, 333.

203. Schlitzer, D. S.; Novak, B. M. *J. Am. Chem. Soc.* **1998**, *120*, 2196.

204. Yashima, E.; Huang, S; Matsushima, T.; Okamoto, Y. *Maromolecules* **1995**, *28*, 4184.

205. Yashima, E.; Matsushima, T.; Okamoto, Y. *J. Am. Chem. Soc.* **1995**, *117*, 11596.

206. Yashima, E.; Maeda, Y.; Okamoto, Y. *Chem. Lett.* **1996**, 955.

207. Yashima, E.; Nimura, T.; Matushima, T.; Okamoto, Y. *J. Am. Chem. Soc.* **1996**, 118, 9800.

208. Yashima, E.; Matushima, T.; Okamoto, Y. *J. Am. Chem. Soc.* **1997**, *119*, 6345.

209. Toda, F.; Mori, K. *J. Chem. Soc., Chem. Commun.* **1986**, 1059.

210. Ma, L.; Hu, Q.-S.; Vitharana, D.; Wu, C.; Kwan, C. M. S.; Pu, L. *Macromolecules* **1997**, *30*, 204.

211. Takata, T.; Furusho, Y.; Murakawa, K.-I.; Endo, T.; Matsuoka, H.; Hirasa, T.; Matsuo, J.; Sisido, M. *J. Am. Chem. Soc.* **1998**, *120*, 4530.

212. Puts, R. D.; Sogah, D. Y. *Macromolecules* **1997**, *30*, 6826.

213. Fujiki, M. *J. Am. Chem. Soc.* **1994**, *116*, 11976.

214. Tsuruta, T. *J. Polym. Sci., Part D* **1972**, *6*, 179.

215. Pino, P.; Lorenzi, G. P.; Lardicci, L. *Chim. Ind. (Milan)* **1960**, *42*, 712.

216. Pino, P.; Ciardelli, F.; Lorenzi, G. P.; Natta, G. *J. Am. Chem. Soc.* **1962**, *84*, 1487.

217. Pino, P.; Mantagnoli, G.; Ciardelli, F.; Benedetti, E. *Makromol Chem.* **1966**, *93*, 158.

218. Pino, P.; Ciardelli, F.; Montagnoli, G. *J. Polym. Sci, Part C* **1968**, *16*, 3265.

219. Pino, P.; Ciardelli, F.; Lorenzi, G. P. *J. Am. Chem. Soc.* **1963**, *85*, 3888.

220. Pino, P.; Ciardelli, F.; Paolo, L. *J. Polym. Sci., Part C* **1963**, *4*, 21.

221. Pino, P.; Ciardelli, F.; Lorenzi, G P. *Makromol. Chem.* **1964**, *70*, 182.

222. Ciardelli, F.; Carlini, C.; Montaudo, G.; Lardicci, L.; Pino, P. *Chim. Ind. (Milan)* **1968**, *50*, 860 (*Chem. Abstr.* **1968**, *69*, 97242r)

223. Vizzini, J.; Ciardelli, F.; Chien, C. W. *Macromolecules* **1992**, *25*, 108.

224. Carlini, C.; Ciardelli, F.; Pino, P. *Makromol. Chem.* **1968**, *119*, 244.

225. Ciardelli, F.; Carlini, C.; Montagnoli, G. *Macromolecules* **1969**, *2*, 296.

226. Chiellini, E. *Macromolecules* **1970**, *3*, 527.

227. Higashimura, T.; Hirokawa, Y. *J. Polym. Sci., Polym. Chem. Ed.* **1977**, *15*, 1137.

228. Okamoto, Y.; Ohta, K.; Yuki, H. *Chem. Lett.* **1977**, 617.

229. Okamoto, Y.; Urakawa, K.; Ohta, K.; Yuki, H. *Macromolecules* **1978**, *11*, 719.

230. Okamoto, Y.; Ohta, K.; Yuki, H. *Macromolecules* **1978**, *11*, 724.

231. Okamoto, Y.; Suzuki, K.; Ohta, K.; Yuki, H. *J. Polym. Sci., Polym. Lett. Ed.* **1979**, *17*, 293.

232. Okamoto, Y.; Suzuki, K.; Kitayama, T.; Yuki, H.; Kageyama, H.; Miki, K.; Tanaka, N.; Kasai, N. *J. Am. Chem. Soc.* **1982**, *104*, 4618.

233. Okamoto, Y.; Gamaike, H.; Yuki, H. *Makromol. Chem.* **1981**, *182*, 2732.

234. Okamoto, Y.; Uarkawa, K.; Yuki, H. *J. Polym. Sci., Polym. Chem. Ed.* **1981**, *19*, 1385.

235. Okamoto, Y.; Yashima, E.; Hatada, K.; Kageyama, Miki, K.; Kasai, N. *J. Polym. Sci., Polym. Chem. Ed.* **1984**, *22*, 1831.

236. Yashima, E.; Okamoto, Y.; Hatada, K.; Kageyama, H.; Kasai, N. *Bull. Chem. Soc. Jpn.* **1988**, *61*, 2071.

237. Matsuzaki, K.; Tateno, N. *J. Polym. Sci., Part C* **1968**, *23*, 733.

238. Solomantia, I. P.; Aliev, A. D.; Krentzel, B. A. *Vysokomolek. Soedin., Ser. A* **1969**, *11*, 871 (*Chem. Abstr.* **1969**, *71*, 3731r).

239. Ikeda, M.; Hirano, T.; Nakayama, S.; Tsuruta, T. *Makromol. Chem.* **1974**, *175*, 2775.

240. Kageyama, H.; Miki, K.; Kai, Y.; Kasai, N.; Okamoto, Y.; Yuki, H. *Bull. Chem. Soc. Jpn.* **1984**, *57*, 1189.

241. Kageyama, H.; Miki, K.; Kasai, N.; Okamoto, Y.; Yashima, E.; Hatada, K.; Yuki, H. *Makromol. Chem.* **1984**, *185*, 913.

242. Suda, H.; Kanoh, S.; Murose, N.; Goka, S.; Motoi, M. *Polym. Bull.* **1983**, *10*, 162.

243. Kanoh, S.; Kawaguchi, N.; Suda, H. *Makromol. Chem.* **1987**, *188*, 463.

244. Watanabe, Y.; Kinugawa, M.; Aida, T.; Inoue, S. *Polym. Prepr. Jpn. (English Edition)* **1992**, *41*, E871.

245. (a) Inoue, S.; Tsuruta, T.; Furukawa, J. *Makromol. Chem.* **1962**, *53*, 215. (b) Tsuruta, T.; Inoue, S.; Yoshida, M.; Furukawa, J. *Makromol. Chem.* **1962**, *55*, 230.

246. Furukawa, J.; Kumata, Y.; Yamada, K.; Fueno, T. *J. Polym. Sci., Part C* **1968**, *23*, 711.

247. Nakaniwa, M.; Kameoka, M.; Ozaki, K.; Furukawa, J. *Makromol. Chem.* **1970**, *138*, 209.

248. Coulon, C.; Spassky, N.; Sigwalt, P. *Polymer* **1976**, *17*, 821.

249. Haubenstck, H.; Panchalingam, V.; Odian, G. *Makromol. Chem.* **1987**, *188*, 2789.

250. (a) Vincens, V.; Le Borgne, A.; Spassky, N. *Makromol. Chem., Makromol. Symp.* **1991**, *47*, 285. (b) Vincens, V.; Le Borgne, A.; Spassky, N. *Makromol. Chem., Rap. Commun.* **1989**, *10*, 623.

251. Yamaguchi, H.; Nagasawa, M.; Minoura, Y. *J. Polym. Sci., Part A-1* **1972**, *10*, 1207.

252. Furukawa, J.; Kawabata, N.; Kato, A. *J. Polym. Sci., Polym. Lett.* **1967**, *5*, 1073.

253. Sepulchre, M.; Spassky, N.; Sigwalt, P. *Macromolecules* **1972**, *5*, 92.

254. Aliev, A. D.; Solomantia, I. P.; Krentsel, B. A. *Macromolecules* **1973**, *6*, 797.

255. Spassky, N.; Leborgne, A.; Momtaz, A. *J. Polym. Sci., Polym. Chem. Ed.* **1980**, *18*, 3089.

256. Dumas, P.; Sigwalt, P.; Guerin, P. *Makromol. Chem.* **1981**, *182*, 2225.

257. Sepulchre, M.; Spassky, N. *Makromol. Chem., Rap. Commun.* **1981**, *2*, 261.

258. Sepulchre, M. *Makromol. Chem.* **1987**, *188*, 1583.

259. Dumas, P.; Sigwalt, P. *Chirality* **1991**, *3*, 484.

260. Spassky, N.; Le Borgne, A.; Reix, M.; Prud'homme, R. E.; Bigdeli, E.; Lenz, R. W. *Macromolecules* **1978**, *11*, 716.

261. Le Borgne, A.; Spassky, N.; Sigwalt, P. *Polym. Bull.* **1979**, *1*, 825.

262. Le Borgne, A.; Greiner, D.; Prud'homme, R. E.; Spassky, N. *Eur. Polym. J.* **1981**, *17*, 1103.

263. Takeichi, T.; Hieda, Y.; Takayama, Y. *Polym. J.* **1988**, *20*, 159.

264. Le Borgne, A.; Spassky, N. *Polymer* **1989**, *30*, 2312.

265. Svirkin, Y. Y.; Xu, J.; Gross, R. A.; Kaplan, D. L; Swift, G. *Macromolecules* **1996**, *29*, 4591.

266. Spassky, H.; Wisniewski, M.; Pluta, C.; Le Borgne, A. *Macromol. Chem. Phys.* **1996**, *197*, 2627.

267. Margolin, A. L.; Crenne, J.-Y.; Klivanov, A. M. *Tetrahedron Lett.* **1987**, *28*, 1607.

268. Wallace, J. S.; Morrow, C. J. *J. Polym. Sci.: Part A: Polym. Chem.* **1989**, *27*, 2553.

269. Khani, D., Kohn, D. H. *J. Polym. Sci.: Part A: Polym. Chem.* **1993**, *31*, 2887.

270. Lundberg, R. D.; Doty, P. *J. Am. Chem. Soc.* **1957**, *79*, 3961.

271. Inoue, S.; Matsuura, K.; Tsuruta, T. *J. Polym. Sci., Part C* **1968**, *23*, 721.

272. Oguni, N.; Kuboyama, H.; Nakamura, A. *J. Polym. Sci., Polym. Chem. Ed.* **1983**, *21*, 1559.

273. Hashimoto, Y.; Imanishi, Y. *Biopolymers* **1981**, *20*, 489.

274. Yamashita, S.; Yamawaki, N.; Tani, H. *Macromolecules* **1974**, *7*, 724.

EPILOGUE

After reading through all of the chapters, all readers have surely acquired an updated understanding of *Catalytic Asymmetric Synthesis* achieved with man-made chiral catalysts. It is evident that these man-made chiral catalysts, bearing much simpler and smaller ligands than the proteins of naturally occurring enzymes, can efficiently create molecules with extremely high enantiomeric purities. The chiral catalysts have a beautiful C_2 symmetry in some cases and in other cases a fascinating dissymmetry. It is breath-taking to realize that such simple and beautiful small molecules can compete, practically and efficiently, with highly sophisticated enzymes that nature has created. This situation is a great encouragement for synthetic organic chemists to continue and to expand their efforts for the design and development of highly efficient chiral catalysts. FDA's decision on "Chiral Drugs" has spawned a new technology called "chirotechnology," and the chirotechnology industries have emerged in a manner similar to the biotechnology industries and are growing strongly. As correctly predicted in the first edition of this book, *Catalytic Asymmetric Synthesis* has been, indeed, taking a center role in these developments.

Also, I should mention that numbers of catalytic asymmetric processes have now been incorporated into one of the key steps in the productions of pharmaceutical drugs, that is these new processes have been replacing traditional stoichiometric reactions that were used in the process research and manufacturing in industries. This clearly indicates that the catalytic asymmetric reactions are recognized as feasible and practical processes by industrial process chemists in general. In this regard, the impacts of Sharpless oxidation, Noyori-Takaya's second-generation as well as Noyori's third-generation asymmetric hydrogenation, Sharpless dihydroxylation, and Jacobsen–Katsuki's epoxidation have been tremendous. Now, asymmetric allylic substitution reactions, aldol reactions, and other carbon–carbon bond forming processes as well as asymmetric polymerizations appear to be emerging as the next forefront technology.

Catalytic asymmetric hydrogenation processes have been at the forefront of practical applications. Following the classical Monsanto's L-DOPA production using DiPAMP-Rh catalyst, BINAP-Ru catalysts have been used in the industrial synthesis of a β-lactam key intermediate to carbapenem antibiotics (Takasago Int. Corp.), 1,2-propanediol (50 tons/year), a key intermediate to antibacterial agent levofloxacin (Takasago Int. Corp./Dai-ichi Pharmaceutical Co.) and others. Pantolactone has been synthesized in a 200-Kg batch scale using the m-CH$_3$POPPM-Rh catalyst (Hoffmann-La Roche). Metolachlor, a herbicide, is produced (>10,000 tons/year) through asymmetric hydrogenation of the imine of 1-methoxy-2-propanone (Novartis) using the XYLIPHOS-Ir catalyst. DuPHOS-Rh catalysts have been used for the commercial synthesis of a key intermediate of an HIV protease inhibitor (Chiroscience). These are just several examples that are available to public at this moment and surely many more to come in the near future. Sharpless dihydroxylation and Jacobsen's epoxidation were licensed to Sepracor, one of the chirotechnology industries, forming the basis of a supply of new enantiomeric intermediates. Sharpless catalyst kits for asymmetric dihydroxylation and aminohydroxylation are commercially available.

I should emphasize that any of the catalytic asymmetric synthesis discussed in this book could become a future industrial process not only in chirotechnology industries, but also in pharmaceutical, chemical and agricultural industries when its development reaches a certain efficiency. Asymmetric isomerization was successfully applied to the large scale commercial synthesis of l-menthol by using the "Takasago Process," and other applications of this type of asymmetric catalytic process will emerge in the future. Many promising results have accumulated for asymmetric cyclopropanation, cyclopropenation, and C–H insertion reactions, from which many useful applications can be envisioned. This is especially true after the successful commercial asymmetric synthesis of a silastatin component using an Aratani catalyst ("Sumitomo Process"). As mentioned above, asymmetric carbon–carbon bond-forming reactions have already been widely used in research laboratories and it is only a matter of time before some of these reactions become important commercial processes, especially in chirotechnology industries. Asymmetric syntheses catalyzed by chiral Lewis acids have rapidly gained significant attention in the synthetic community, and these processes have a very high potential for commercial applications. Asymmetric hydrosilylation has a somewhat limited scope in commercial applications, but there are highly enantioselective reactions which cannot be achieved by other methods and more importantly this "chirotechnology" may well be applicable for the production of silicon-based chiral "new materials." There was a clear breakthrough in asymmetric hydroformylation by Takaya's invention of the BINAPHOS-Rh catalyst in early 1990s, and this process is reaching a highly practical level, which will have a significant impact on chirotechnology. Other carbonylations are still a great challenge for synthetic chemists, but there are encouraging results that surely provide hints to develop practical processes in the future. Asymmetric phase-transfer reactions are in an early stage of development, but these reactions obviously have great potential as commercial processes. I would like to call readers' attention to the fact that most of the inventions in *Catalytic Asymmetric Synthesis* have been made in academic institutions and those inventions have been immediately secured by industries for development as commercial processes. Accordingly, it is evident that creative basic research in universities is crucial for the birth and growth of innovative chirotechnology.

It is noteworthy that *Catalytic Asymmetric Synthesis* has made significant advances in terms of substrate-catalyst interactions and the chiral recognition of substrate structures. Knowles' epoch-making "Monsanto Process" established in the early 1970s for the asymmetric synthesis of L-DOPA, was based on the interactions between the multi-functionalized substrate, N-acetyldehydroamino acid, and a chiral diphosphine-rhodium catalyst. However, the second and third generation asymmetric hydrogenation based on chiral diphosphine-ruthenium catalysts can now be appled to simple acrylic acids with exceptionally high enantioselectivity, as demonstrated for the asymmetric synthesis of (S)-Naproxen, a potent inflammatory drug. Now, the third- generation asymmetric hydrogenation can reduce ketones and imines with extremely high efficiency. Sharpless oxidation of allylic alcohols, extensively developed in the 1980s, required hydroxyl functionality to anchor the substrates to a chiral titanium catalyst. In sharp contrast with this, the Sharpless dihydroxylation and Jacobsen-Katsuki's epoxidation work extremely well with *unfunctionalized* olefins. Asymmetric carbometallation of simple alkenes with high enantioselectivity is another remarkable advance in recent years. Synthetic organic chemists should now have confidence in the rational design of a chiral catalyst, which brings about extremely high enantioselectivity without the assistance of a huge protein backbone. It is worthy of note that the TOF (turn over frequency) of the Ir-XYLIPHOS catalyst for asymmetric hydrogenation has reached 500/s and the substrate/catalyst ratio for the process has reached 1,000,000. It is not rare to find the operating substrate/catalyst ratio above 100,000 in asymmetric hydrogenations. Thus, not only the enantioselectivity, but also the efficiency of man-made chiral catalysts, are getting very close to those of enzymes.

I would like to emphasize the importance of detailed mechanistic studies for accurate understanding of asymmetric induction steps as well as key catalytic species, because these studies have been and will be essential for breakthroughs in *Catalytic Asymmetric Synthesis*. "Asymmetric amplification" and "Autocatalysis" are very intriguing. Now, enantiopure catalysts are not always necessary to achieve perfect asymmetric induction. Another very intriguing and useful protocol is the activation or deactivation of racemic catalysts by chiral additives, which deserve considerable attention.

When only low enantioselectivities are observed on using man-made chiral catalysts, synthetic organic chemists tend to think that their design is insufficient and the system is too simple. However, detailed mechanistic study has shown that in many cases it is not the efficacy of the designed catalyst system, but other coexisting catalyst species in the system, that are responsible for the poor results. Once the well-designed catalyst species is generated selectively or coaxed to work selectively, excellent results are obtained. This is a very important message to synthetic organic chemists.

Although I put emphasis on the remarkable efficacy of man-made chiral catalysts, it is not my intention to undermine the use of enzymatic and biological methods to produce enantiopure compounds. Enzymes are surely excellent chiral catalysts for asymmetric organic transformations. The catalytic asymmetric organic reactions compiled in this book and those promoted by enzymes or microorganisms will complement each other and both are essential for the development of chirotechnology.

I sincerely hope that this book attracts the interests of a broad range of synthetic organic and medicinal chemists, especially among the younger generation in both academia and industry, so that many talented young chemists will introduce creative ideas into this fascinating area of research, ones that will promote significant future advances in *Catalytic Asymmetric Synthesis*.

Finally, I would like to thank Dr. Subrata Charavarty, Dr. Michael L. Miller, Dr. Songnian Lin, Dr. Dominique Bonafoux, Xudong Geng, and Dr. An T. Vu in my laboratories at Department of Chemistry, State University of New York at Stony Brook for their help in proofreading all chapters.

Iwao Ojima
August, 1999

APPENDIX

LIST OF CHIRAL LIGANDS

The list of chiral ligands is provided in this Appendix. Typical chiral ligands are summarized for each chapter and references given are taken from the corresponding References Section of the chapter.

Catalytic Asymmetric Synthesis, Second Edition, Edited by Iwao Ojima
ISBN 0-471-29805-0   Copyright © 2000 Wiley-VCH, Inc.

CHAPTER 1

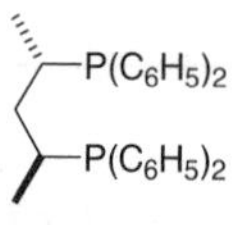

(S,S)-**BDPP** [17, 145]
(SKEWPHOS)

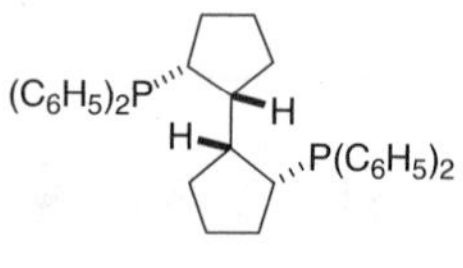

(R,R)-**BICP** [22]

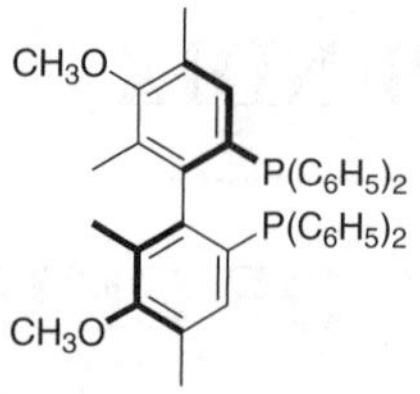

(R)-**BIMOP** [184]

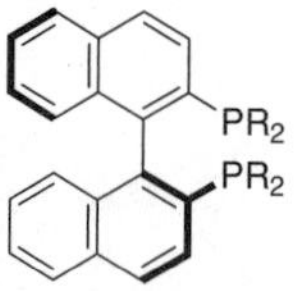

(S)-**BINAP**
BINAP: R = C$_6$H$_5$ [56]
TolBINAP: R = 4-CH$_3$C$_6$H$_4$ [168]
XylBINAP: R = 3,5-(CH$_3$)$_2$C$_6$H$_3$ [168]
DTBBINAP: R = 3,5-(*t*-C$_4$H$_9$)$_2$C$_6$H$_3$ [168]
(absolute configuration unlnown)
p-**MeO-BINAP**: R = 4-CH$_3$OC$_6$H$_4$ [168]
BINAP-SO$_3$Na: R = 4-NaOSO$_2$C$_6$H$_4$ [114]
Cy-BINAP: R = *cyclo*-C$_6$H$_{11}$ [135]

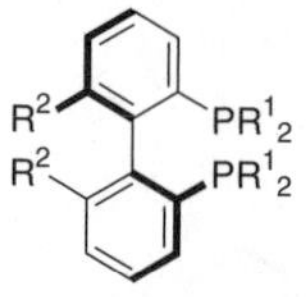

(S)-**BIPHEMP**
BIPHEMP: R^1 = C$_6$H$_5$; R^2 = CH$_3$ [83]
MeO-BIPHEP: R^1 = C$_6$H$_5$; R^2 = CH$_3$O [86, 171]
p-**Tol-MeO-BIPHEP**: R^1 = 4-CH$_3$C$_6$H$_4$; R^2 = CH$_3$O [86]
di-*t*-Bu-MeO-BIPHEP: R^1 = 3,5-(*t*-C$_4$H$_9$)$_2$C$_6$H$_3$; R^2 = CH$_3$O [86]
Fr-MeO-BIPHEP: R^1 = 2-furyl; R^2 = CH$_3$O [86]
BICPEP: R^1 = *cyclo*-C$_5$H$_9$; R^2 = CH$_3$ [124]
BICHEP: R^1 = *cyclo*-C$_6$H$_{11}$; R^2 = CH$_3$ [21]

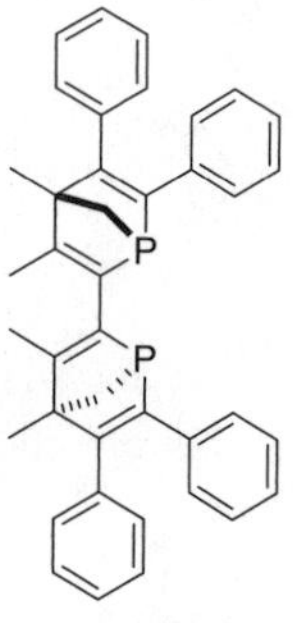

(R,R)-**BisP*** [30]
R = 1-adamantyl

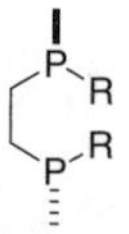

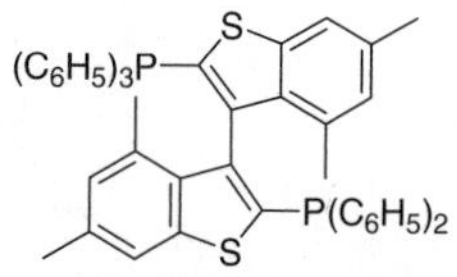

tetraMe-BITIANP [109]
(absolute configuration unknown)

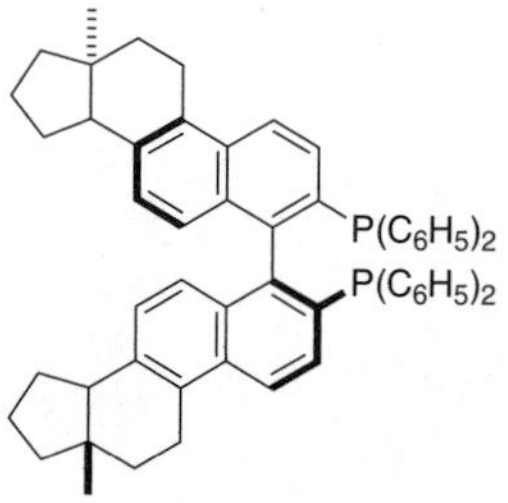

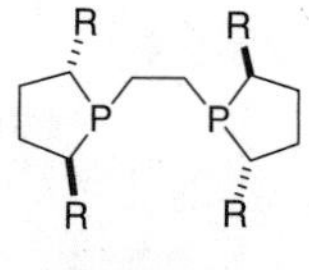

(S)-bis-steroidal
phosphine [185]

(R,R)-**BIPNOR** [26]

BPE [19, 63]
Me-BPE: R = CH$_3$
Et-BPE: R = C$_2$H$_5$
i-**Pr-BPE**: R = (CH$_3$)$_2$CH

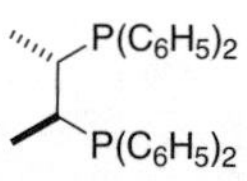

(S,S)-**CHIRAPHOS** [13]

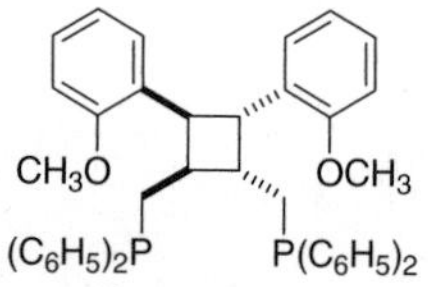

(R,R)-**CDP** [72]

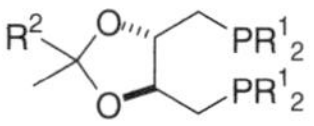

(S,S)-DIOP

DIOP: $R^1 = C_6H_5$; $R^2 = CH_3$ [7]

DIOP-OH: $R^1 = C_6H_5$; $R^2 = HOCH_2$ [47]

MOD-DIOP: $R^1 = 3,5-(CH_3)_2-4-(CH_3O)C_6H_2$; $R^2 = CH_3$ [122]

CyDIOP: $R^1 = cyclo-C_6H_{11}$; $R^2 = CH_3$ [198]

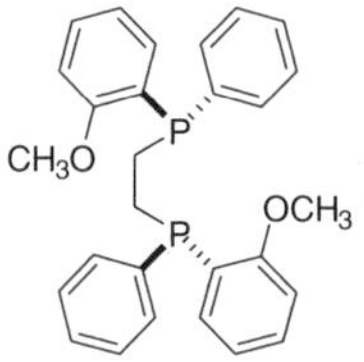

(S,S)-DIPAMP [9]

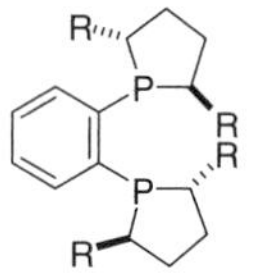

DuPHOS [19, 63]

Me-DuPHOS: R = CH_3

Et-DuPHOS: R = C_2H_5

i-Pr-DuPHOS: R = $(CH_3)_2CH$

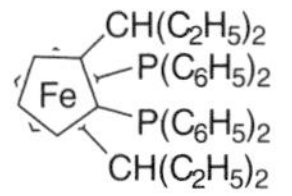

(S,S)-FerroPHOS [31]

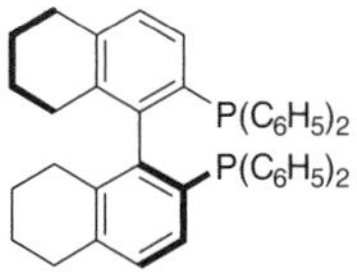

(S)-H₈-BINAP [111]

(S,S)-NORPHOS [14]

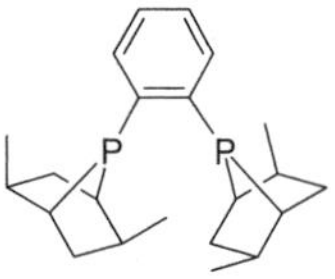

(R,S,R,S)-Me-PennPhos [261]

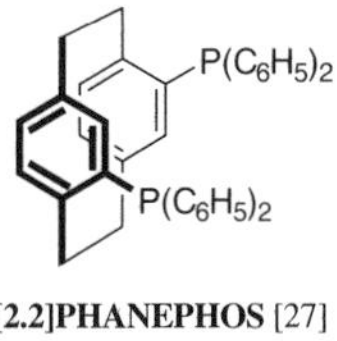

(S)-[2.2]PHANEPHOS [27]

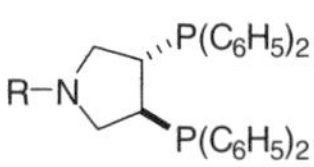

(S,S)-PYRPHOS [18]
(DEGUPHOS)

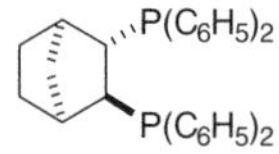

(S,S)-RENORPHOS [10d]

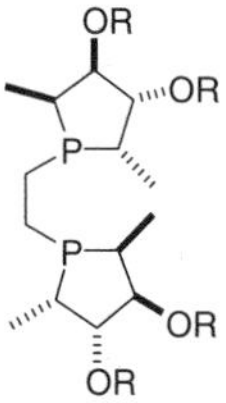

(S,S,S,S)-RoPHOS [32]

R = $CH_2C_6H_5$ or $t-C_4H_9$

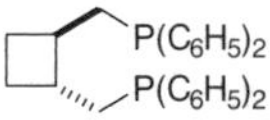

(R,R)-TBPC [132]

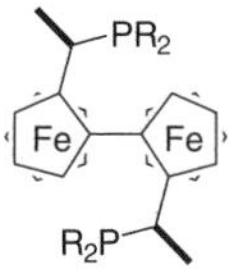

(R,R)-(S,S)-TRAP [24]

EtTRAP: R = C_2H_5

i-BuTRAP: R = C_2H_5

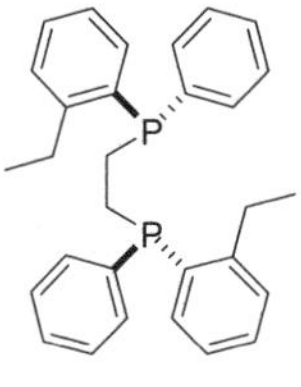

(S,S)-1 [25]

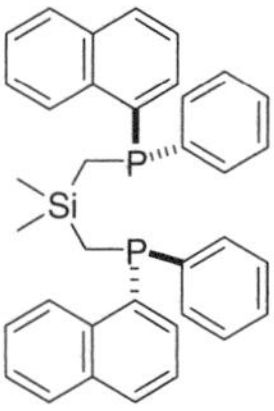

(S,S)-2 [29]

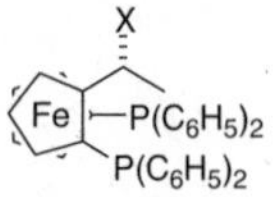

(R)-(S)-**BPPFA**

BPPFA: X = (CH₃)₂N [11]

3: X = $\bigcirc$ N(CH₂)₂(CH₃)N [117]

BPPFOH: X = HO [220]

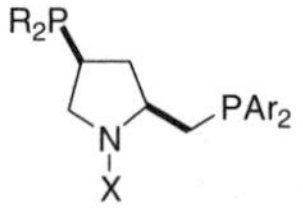

(2S,4S)-**BPPM**

BPPM: Ar = R = C₆H₅; X = (CH₃)₃COCO [12, 172]

R'-CAPP: Ar = R = C₆H₅; X = R'NHCO [119]

BCPM: Ar = C₆H₅; R = cyclo-C₆H₁₁; X = (CH₃)₃COCO [173]

MCCPM: Ar = C₆H₅; R = cyclo-C₆H₁₁; X = CH₃NHCO [163]

m-CH₃POPPM: Ar = R = 3-CH₃C₆H₄; X = (C₆H₅)₂PO [174]

MCCXM: Ar = 3,5-(CH₃)₂C₆H₃; R = cyclo-C₆H₁₁;
    X = CH₃NHCO [195]

MOD-BCPM: Ar = 3,5-(CH₃)₂-4-(CH₃O)C₆H₂;
    R = cyclo-C₆H₁₁; X = (CH₃)₃COCO [121]

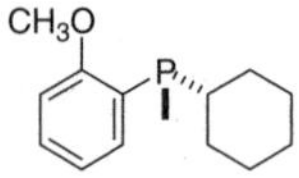

(S)-**CAMP** [8]

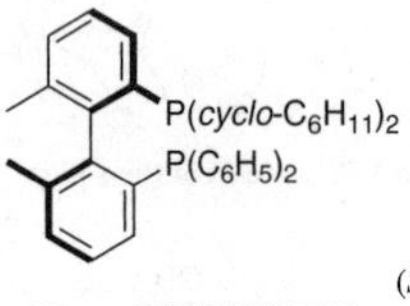

(R)-cy₂-**BIPHEMP** [68]

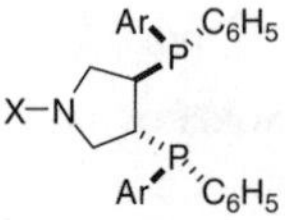

(S,R,R,R)-**TMO-DEGUPHOS**
    [267]
Ar = 2,4,6-(CH₃O)₃C₆H₂;
X = (CH₃)₃COCO

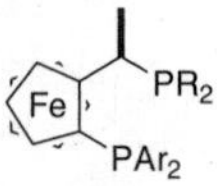

(R)-(S)-JOSIPHOS

JOSIPHOS: R = cyclo-C₆H₁₁; Ar = C₆H₅ [188]

XYLIPHOS: R = 3,5-(CH₃)₂C₆H₃; Ar = C₆H₅ [68]

xyl₂PF-Pxyl₂: R = 3,5-(CH₃)₂C₆H₃; Ar = 3,5-(CH₃)₂C₆H₃
    [335f]

MOD-XYLIPHOS: R = 3,5-(CH₃)₂-4-((CH₃)₂CH)₂NC₆H₂;
    Ar = C₆H₅ [335f]

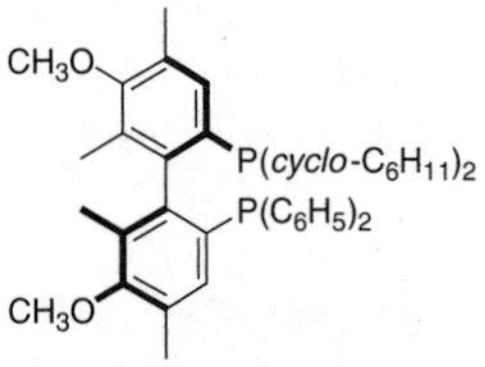

(R)-**MOC-BIMOP** [217]

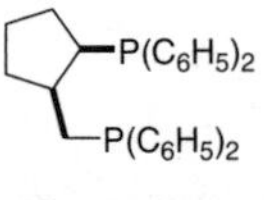

(1R,2R)-**PPCP** [20]

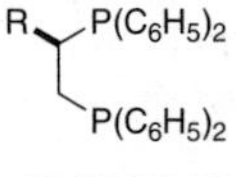

(R)-PROPHOS

PROPHOS: R = CH₃ [91]

BENZPHOS: R = C₆H₅CH₂ [337]

CyCPHOS: R = cyclo-C₆H₁₁ [16]

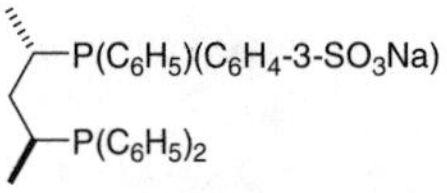

(S,S)-**SulfBDPP** [340]
(diastereo mixture)

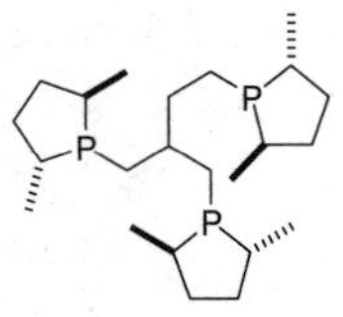

(R,R)-**4** [23]

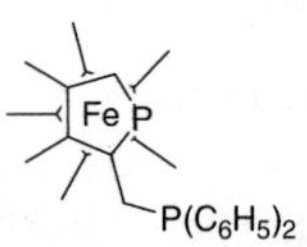

(1S,2S,3R,4S,5S)-**5** [28]

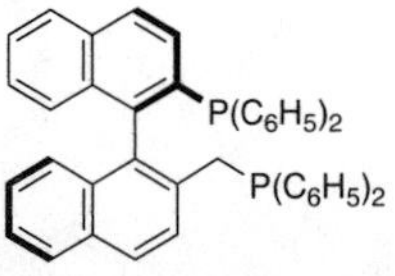

(R)-**6** [146]

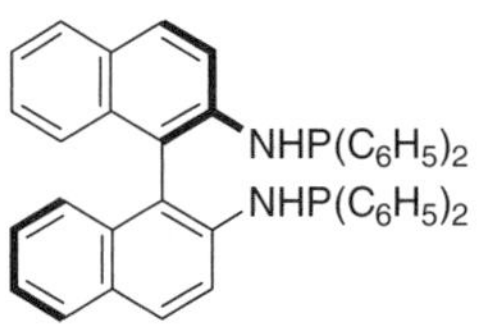

(R)-**BDPAB** [70]

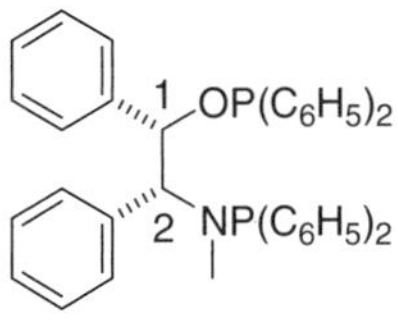

(1S,2R)-**DPAMPP** [40]

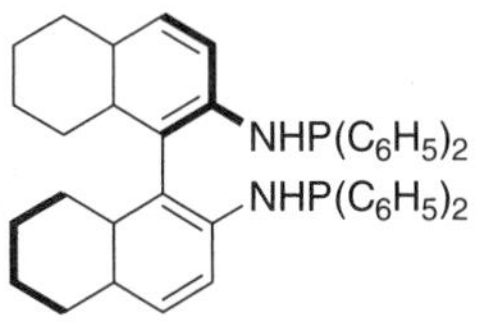

(R)-H$_8$-**BDPAB** [70]

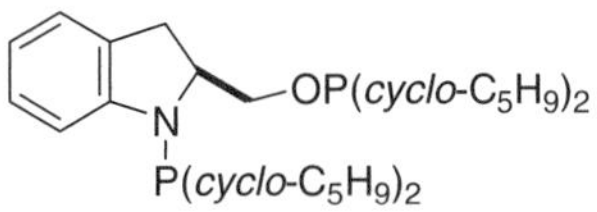

(S)-**Cp,Cp-IndoNOP** [165]

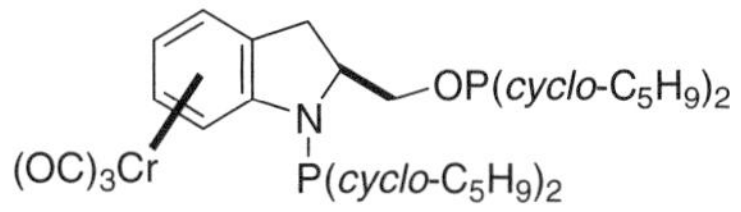

(S,2S)-**Cr(CO)$_3$-Cp,Cp-IndoNOP** [165]

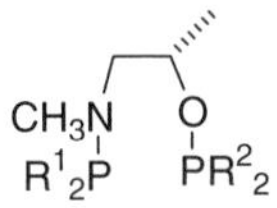

(S)-**isoAlaNOP**

Ph,Cp-isoAlaNOP: R^1 = C$_6$H$_5$; R^2 = cyclo-C$_5$H$_9$ [166]

Cp,Cp-isoAlaNOP: R^1 = R^2 = cyclo-C$_5$H$_9$ [176a]

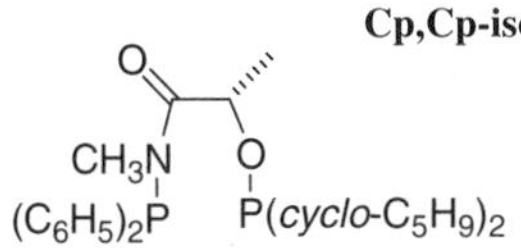

(S)-**Ph,Cp-methyllactamide**
[176a]

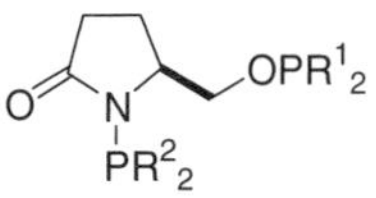

(S)-**oxoProNOP** [176b]

(R)-**PINDOPHOS**

Ph,Ph-oxoProNOP: R^1 = R^2 = C$_6$H$_5$

Cp,Cp-oxoProNOP: R^1 = R^2 = cyclo-C$_5$H$_9$

Cy,Cy-oxoProNOP: R^1 = R^2 = cyclo-C$_6$H$_{11}$

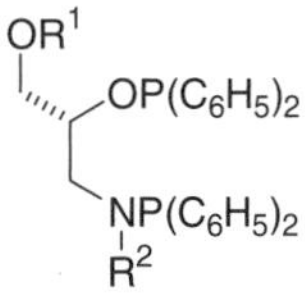

PINDOPHOS: R^1 = 7-indolyl; R^2 = (CH$_3$)$_2$CH [39]

11: R^1 = 1-naphthyl; R^2 = cyclo-C$_5$H$_9$ [41]

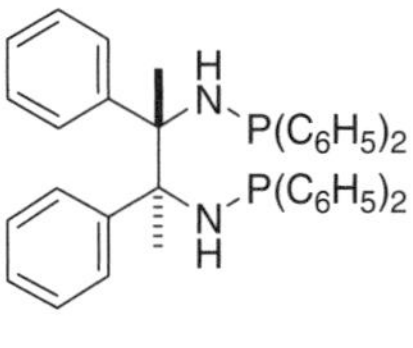

(S,S)-**PNNP** [42]

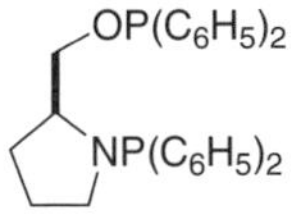

(S)-**PROLOPHOS** [38]

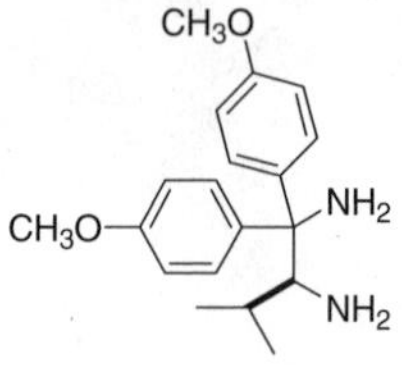

(S)-**DAIPEN** [257]

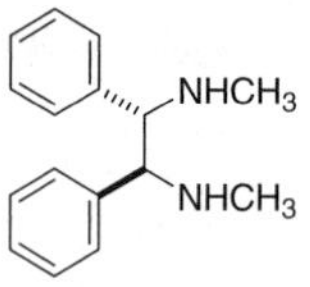

(S,S)-**DMDPEN** [313]

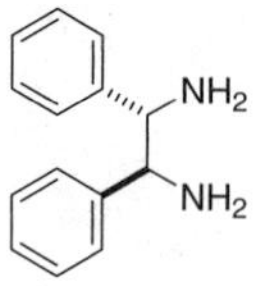

(S,S)-**DPEN** [257]

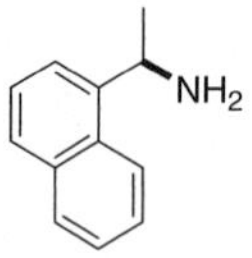

(R)-1-**NEA** [208]

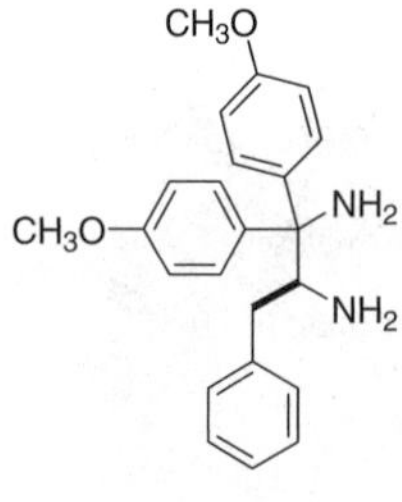

(S)-**12** [308]

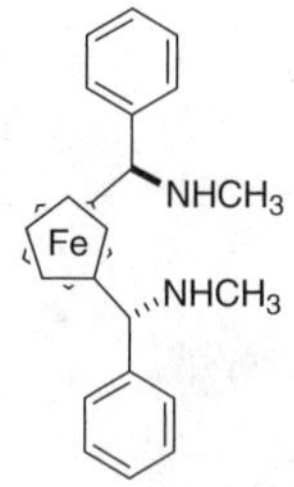

(R,R)-**13** [294]

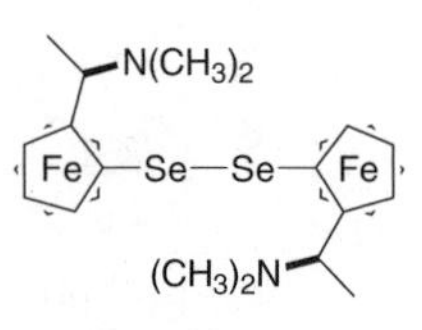

(R)-(S)-**14** [315]

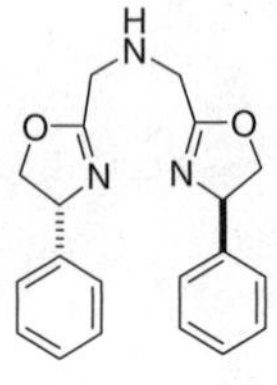

(R)-**AMBOX** [296]

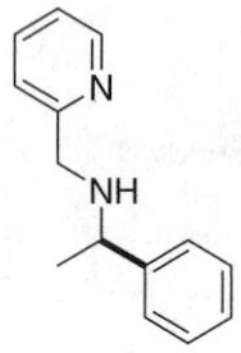

(R)-**DHPPEI** [2c]

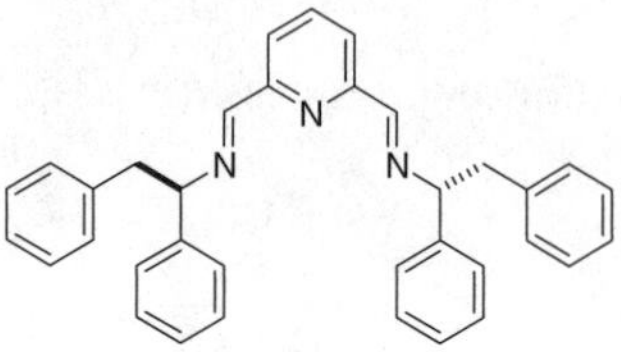

(R,R)-**PDPBI** [326]

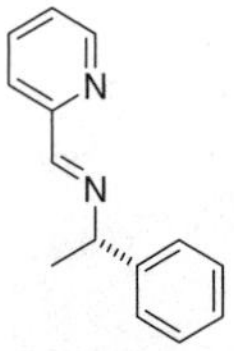

(S)-**PPEI** [309]

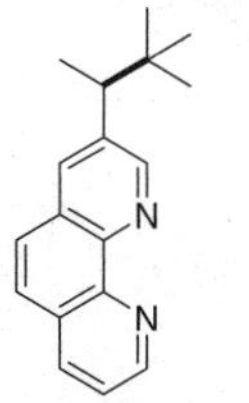

(S)-**15** [310]

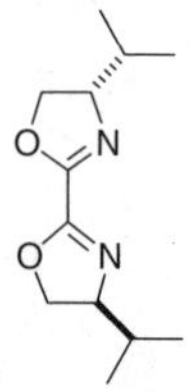

(S,S)-**16** [306]

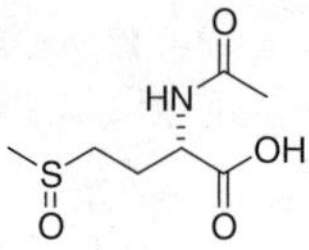

(S)-**AMSO** [314]

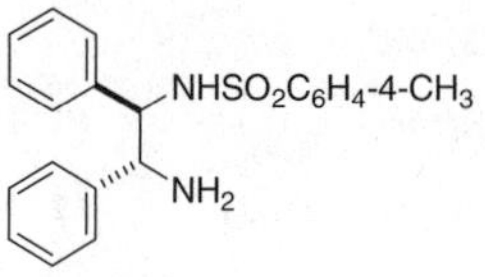

(R,R)-**TsDPEN** [292, 293]

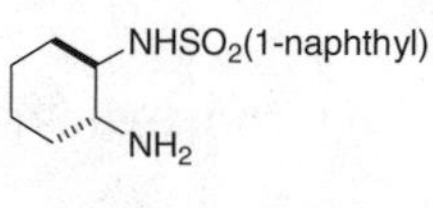

(R,R)-**17** [294]

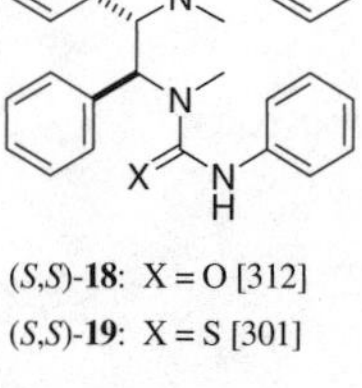

(S,S)-**18**: X = O [312]
(S,S)-**19**: X = S [301]

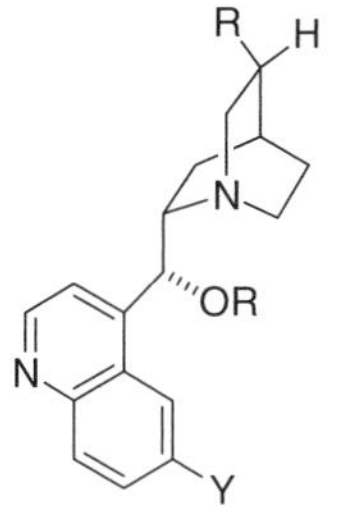

cinchona alkaloids

cinchonidine: R = CH$_2$=CH; R = H; Y = H [202]

quinine: R = CH$_2$=CH; R = H; Y = CH$_3$O [202]

MeOHCd: R = C$_2$H$_5$; R =CH$_3$; Y = H [206]

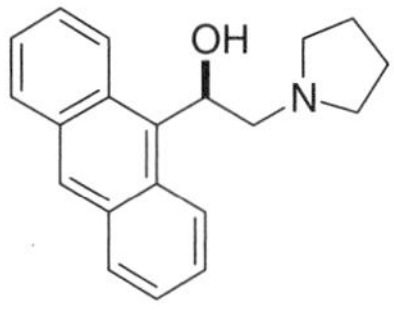

(R)-**20** [209]

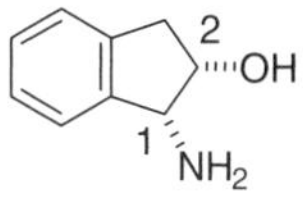

(1R,2S)-**21** [299a]

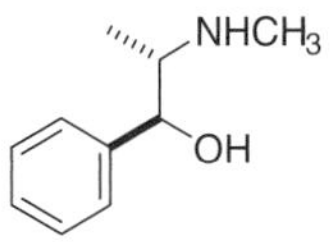

(S,S)-**22** [298]

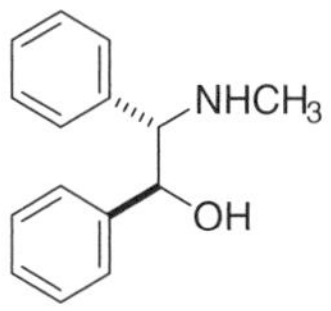

(S,S)-**23** [298]

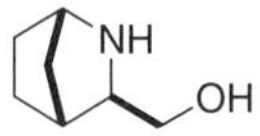

(S,R,R)-**24** [299b]

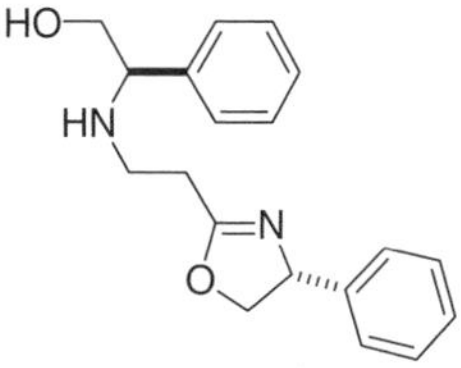

(R,R)-**25** [300]

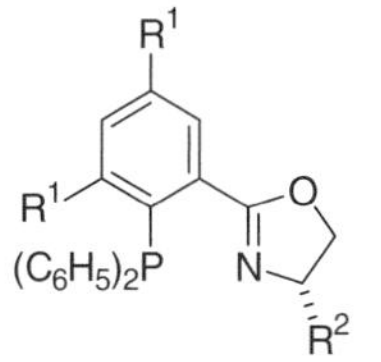

(S)-**26**: R^1 = H; R^2 = (CH$_3$)$_3$C [141]
(S)-**27**: R^1 = CH$_3$; R^2 = (CH$_3$)$_2$CH [344]

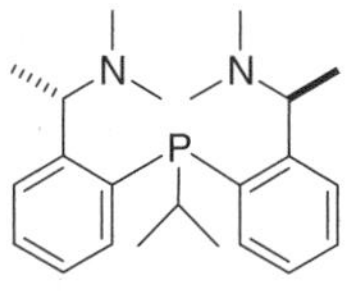

(S,S)-**28** [116]

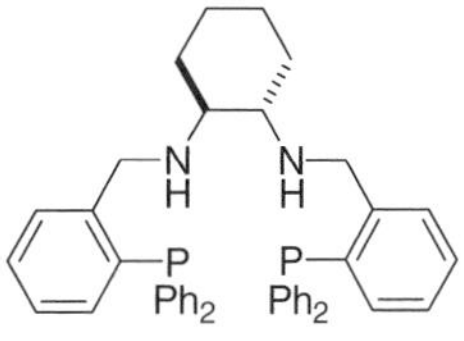

(S,S)-**29** [305]

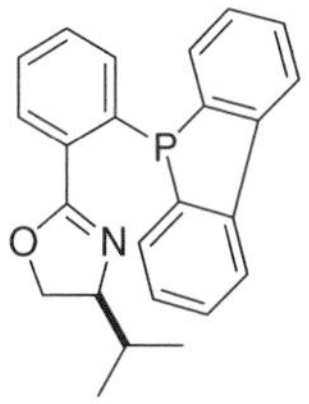

(S)-**30** [302]

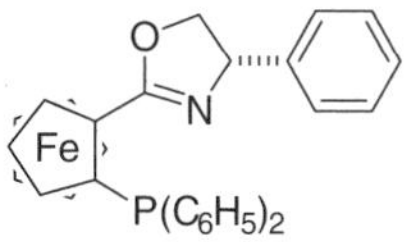

(S)-**31** [304]

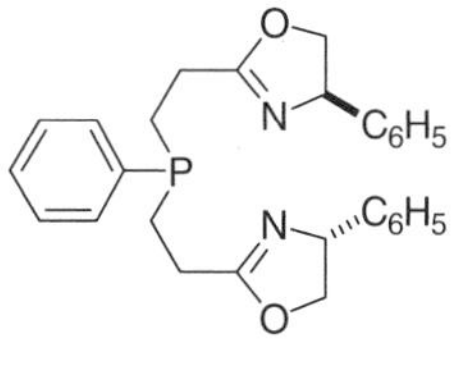

(R,R)-**32** [316]

CHAPTER 2

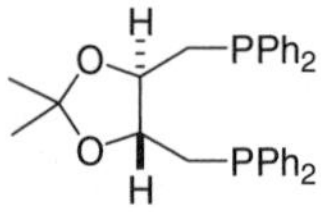

(*R*,*R*)-(-)-**DIOP** (**5**) [5]

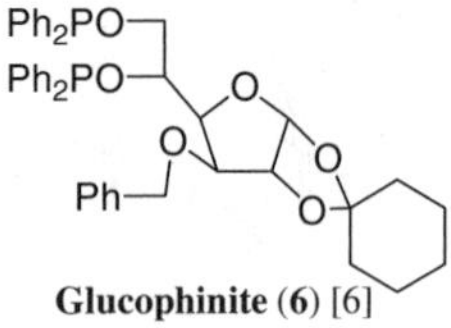

Glucophinite (**6**) [6]

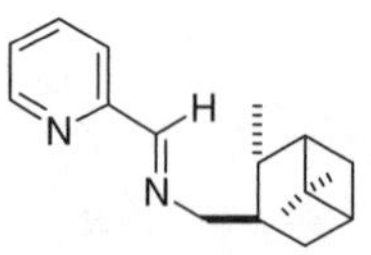

Iminopyridine (**7**) [7]

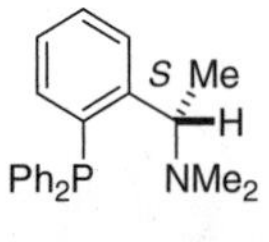

Amphos-(*S*) (**8**) [8]

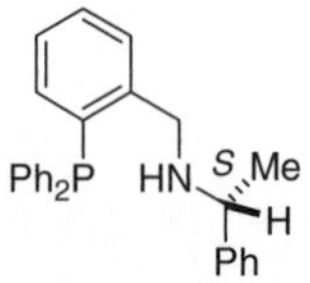

Aminphos (**9**) [9]

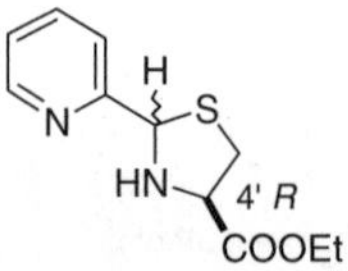

Pythia (**10**) [10]

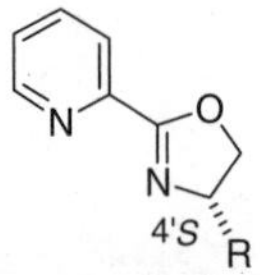

11a R = Et **Pymox-Et** [12,13]
11b R = *i*-Pr **Pymox-*i*-Pr** [12,13]
11c R = *t*-Bu **Pymox-*t*-Bu** [12,13]

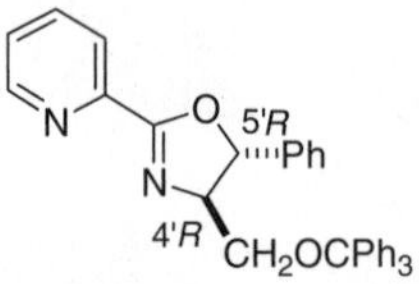

11d [14]

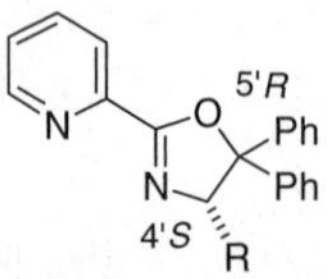

11e R = Bz [15]
11f R = *i*-Pr [15]

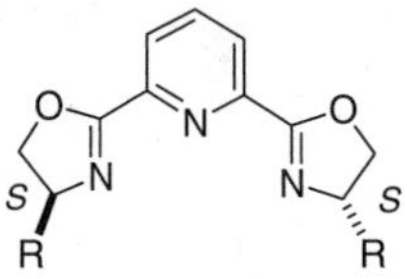

12a R = *i*-Pr **Pybox-*i*-Pr** [13,16,17]
12b R = *t*-Bu **Pybox-*t*-Bu** [13,16,17]

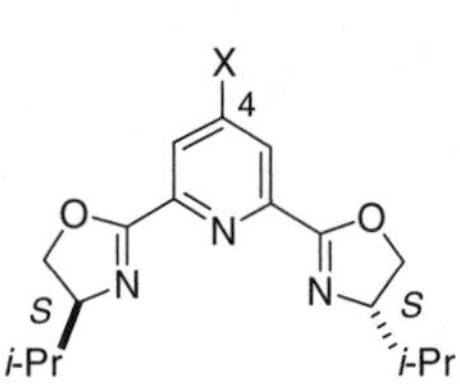

12c X = CO_2Me [18,19]
12d X = NMe_2 [18]

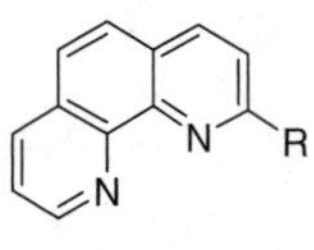

13a R =
[20]

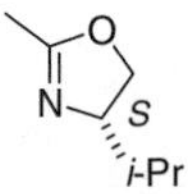

13b R =
[20]

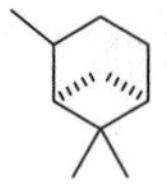

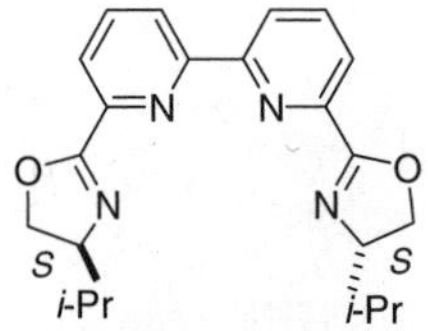

Bipymox-*i*-Pr (**14**) [21]

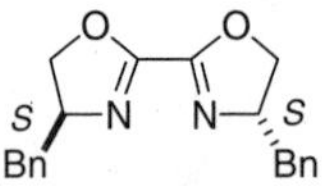

15 [23]

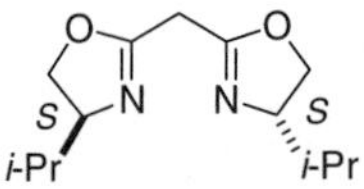

16 [23]

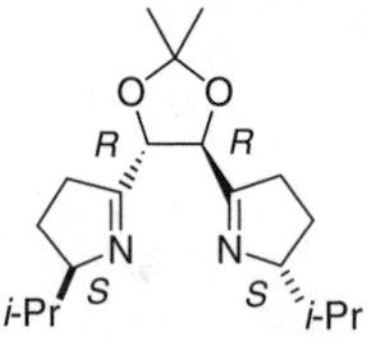

17a [24]

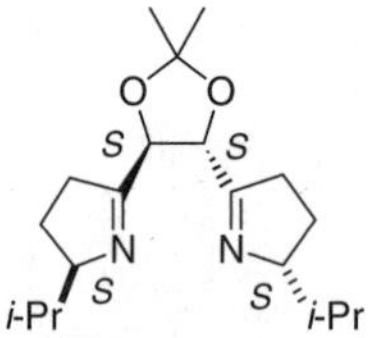

17b [24]

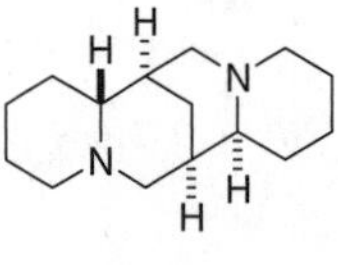

(-)-**Sparteine** (**18**) [25]

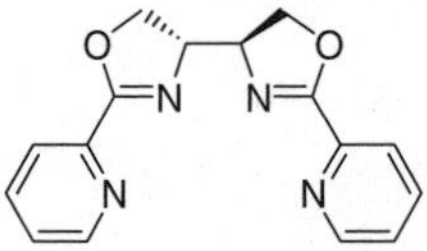

19 [26]

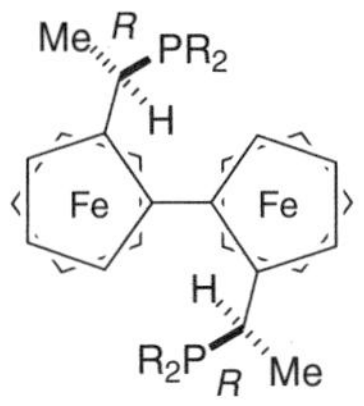

20a R = *n*-Bu (*R,R*)-(*S,S*)-**TRAP-*n*-Bu** [27]
20b R = Et (*R,R*)-(*S,S*)-**TRAP-Et** [27]

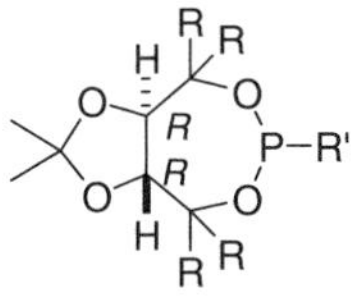

(*R,R*)-**TADDOL-Phosphonites**:

21a R = Ph, R' = 4-Me-C$_6$H$_4$ [28]
21b R = 2-Np, R' = Ph [28]
21c R = 2-Np, R' = 2-Np [28]

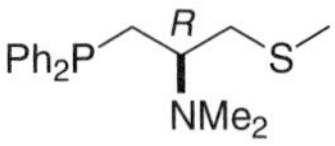

22 (*R*)-**Cystphos** [29]

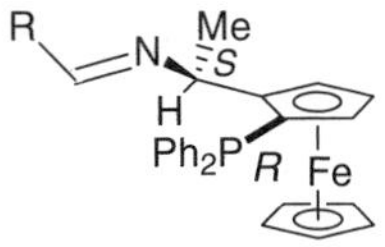

Ferrocenylphosphine-Imines [30]
23a-(*R,S*) R = Ph
23b-(*R,S*) R = CF$_3$-C$_6$H$_4$
23c-(*R,S*) R = 4-CF$_3$-C$_6$H$_4$

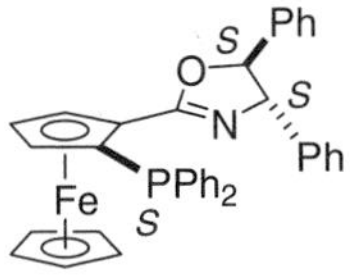

Ferrocenylphosphine-oxazoline [32,33]
24 R = Ph (*S,S,S*)-DIPOF

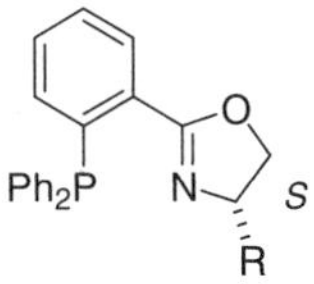

25a R = *i*-Pr [34]
25b R = *t*-Bu [34]

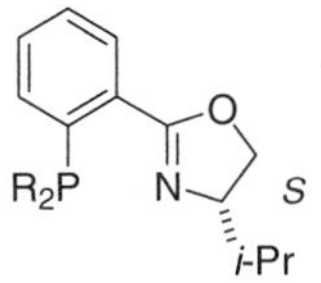

25c R = 3,5-(CF$_3$)$_2$C$_6$H$_3$ [34]

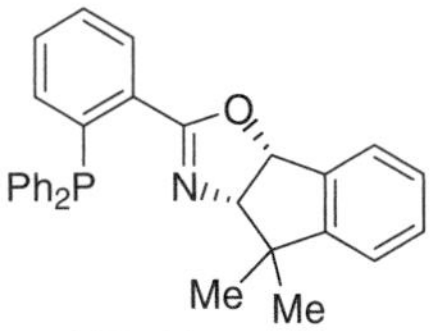

25d [34]

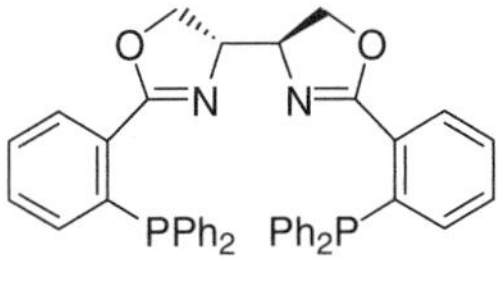

Phos-Biox (26) [35]

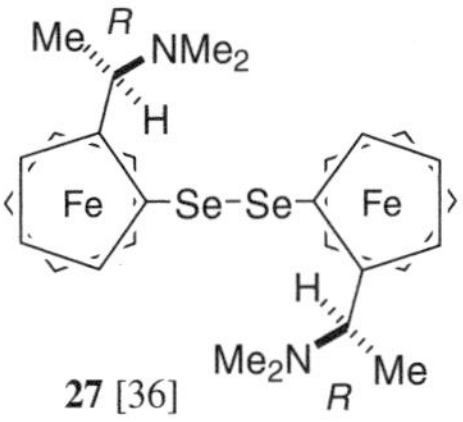

27 [36]

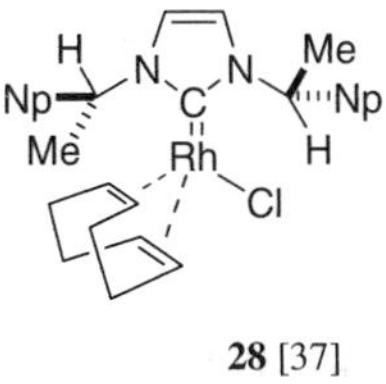

28 [37]

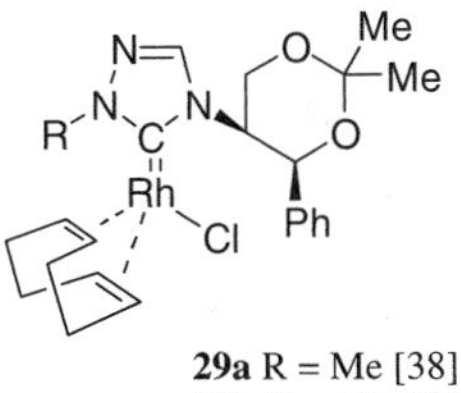

29a R = Me [38]
29b R = t-Bu [38]

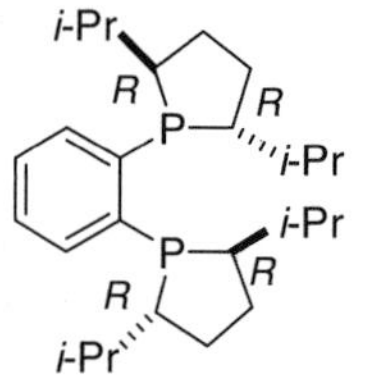

R = *i*-Pr, **DuPHOS-*i*-Pr** (**56**) [42]

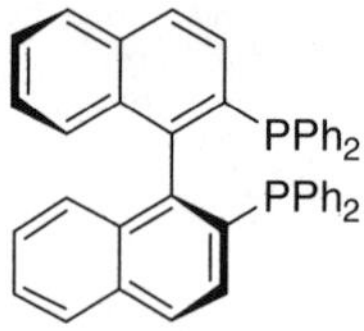

(*S*)-(-)-**BINAP** (**58**) [42]

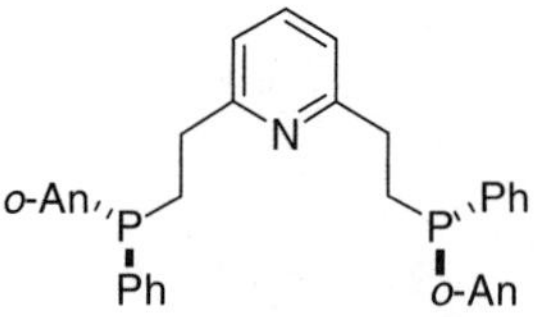

63 *o*-An = *ortho*-Anisyl [44]

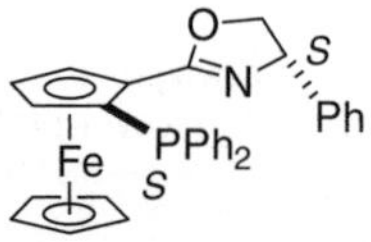

Ferrocenylphosphine-oxazoline (**64**) [45]

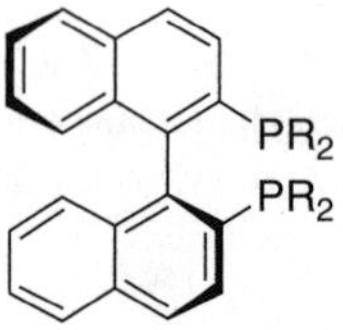

R = Ph-4-Me

(*S*)-(-)-**Tol-BINAP** (**70**) [46]

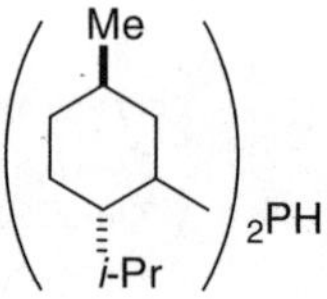

74 [47]

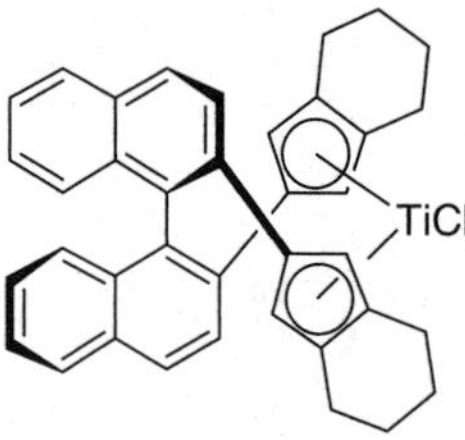

75 [49]

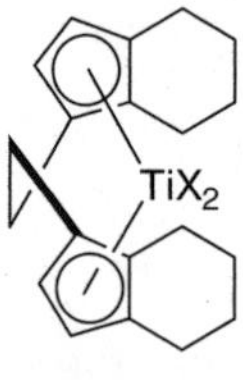

76-(*R*,*R*)
X$_2$ = 1,1'-binaphth-2,2'-diolate
[50-52]

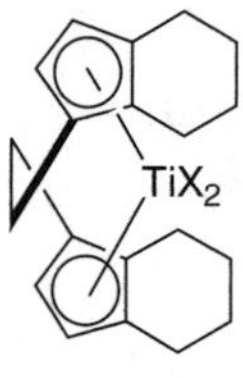

77-(*S*,*S*) X = Cl [53]
78-(*S*,*S*) X = F [53]

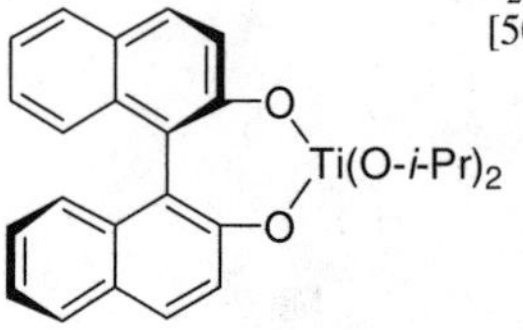

79-(*R*) [54]

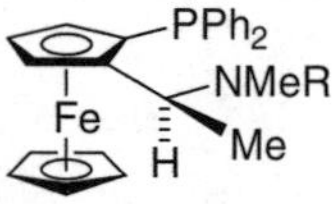

91 R = Me, **PPFA-(*S*)-(*R*)-(+)** [59]
96 R = CH$_2$C$_8$F$_{17}$ [60]

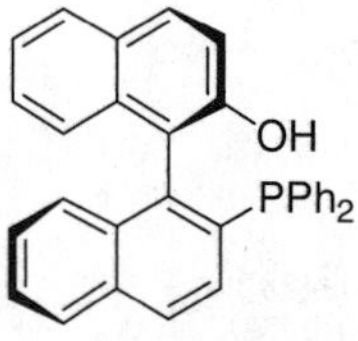

98 [61]

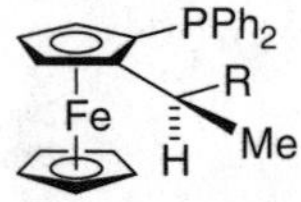

101 R = OH, **PPF-OH** [62]
102 R = OAc, **PPF-OAc** [62]

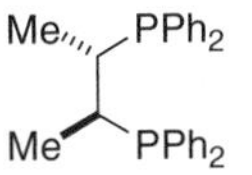

Chiraphos-(*S,S*) (**106**) [64]

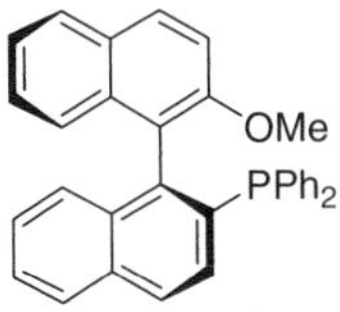

(*S*)-(-)-**MOP** (**116**) [67,68]

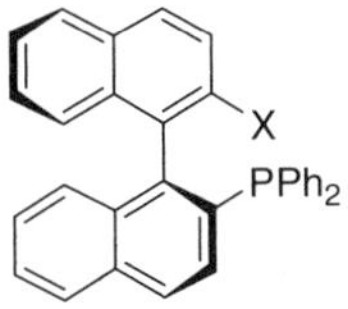

X = H **125** (*S*)-**MOP-H** [67,68]
= OH, OAc, Et etc.

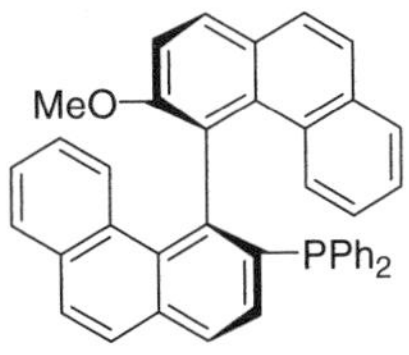

(*R*)-**MOP-phen** (**131**) [9]

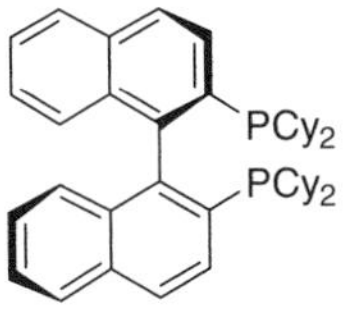

(*R*)-**Cybinap** (**136**) [75]

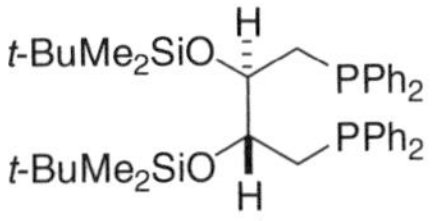

SILOP-TBDM-(*R,R*) (**141**) [77]

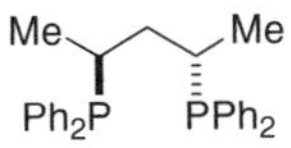

145 [81b]

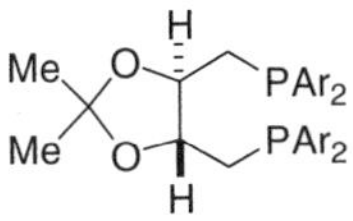

146 Ar = 2-MeOPh [81b]

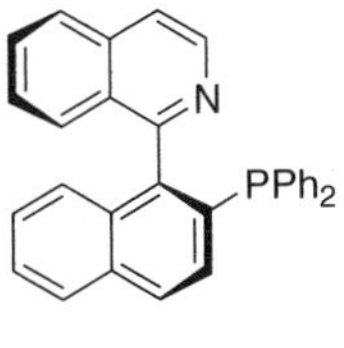

148 [83]

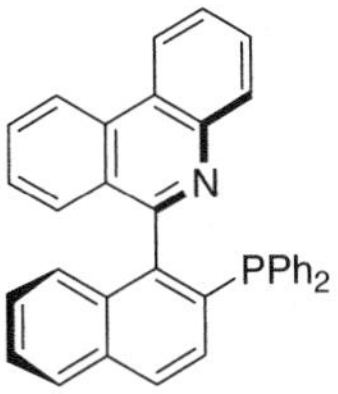

151 [83~85]

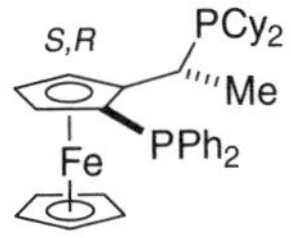

153 [87]

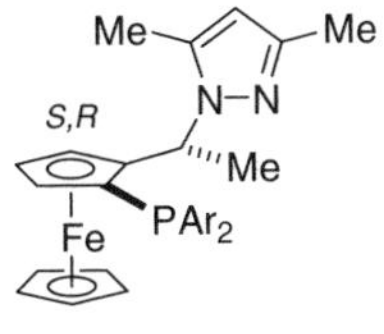

154a Ar = Ph [88]
154b Ar = 4-CF₃Ph [88]

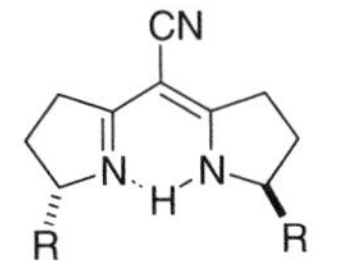

161 R = CH₂OSiMe₂-*t*-Bu
[93]

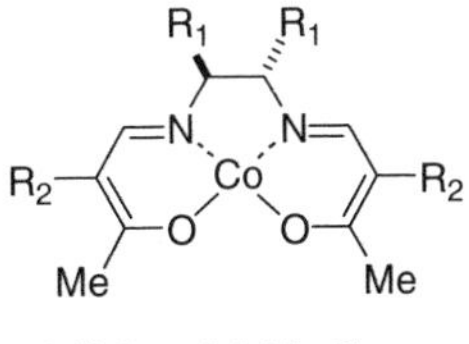

164 R₁ = 3,5-Me₂Ph,
R₂ = 2,4,6-Me₃PhCO [94]

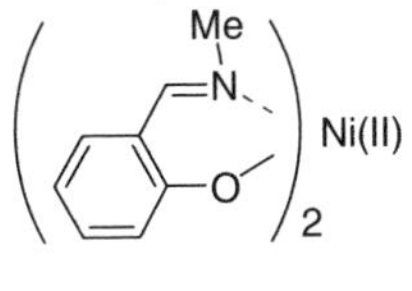

173 Ni(Mesal)₂ [98]

CHAPTER 3

For the chiral ligands in this Chapter, see the Ligand List for Chapter 1.

CHAPTER 4

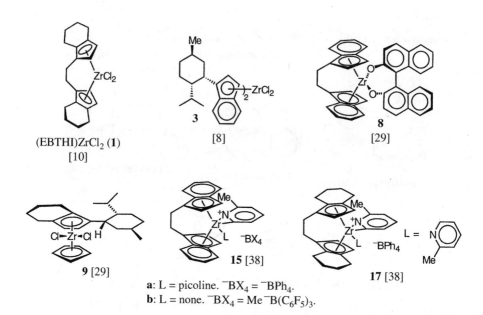

(EBTHI)ZrCl$_2$ (**1**)
[10]

3

[8]

8

[29]

9 [29]

15 [38]

17 [38]

a: L = picoline. $^-$BX$_4$ = $^-$BPh$_4$.
b: L = none. $^-$BX$_4$ = Me$^-$B(C$_6$F$_5$)$_3$.

CHAPTER 5

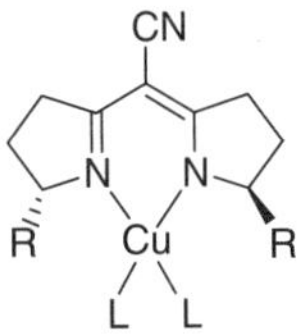

4 [36]

a R = CMe₂OH
b R = CH₂OSiMe₂ᵗBu
c R = COOMe

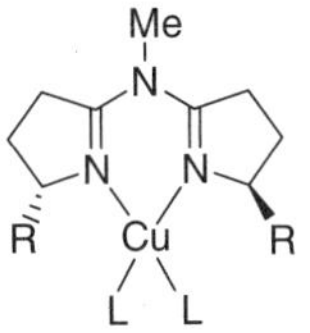

5 [35]

a R = CH₂OSiMe₂ᵗBu
b R = CMe₂OSiMe₃
c R = CMe₂OSiMe₂ᵗBu

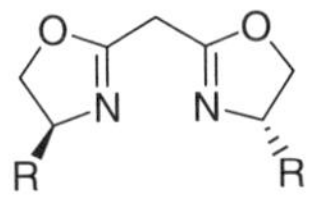

6 [39,40,42]

a R = Ph
b R = Me₂CH
c R = (S)-EtCH(Me)
d R = Et
e R = PhCH₂
f R = ᵗBu
g R = CH₂OH
h R = Me₂C(OH)
i R = CH₂OSiMe₂ᵗBu

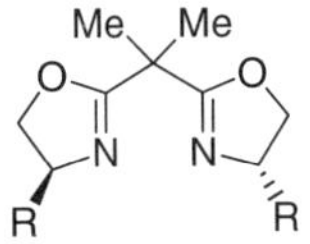

7 [40]

a R = ⁱPr
b R = ᵗBu

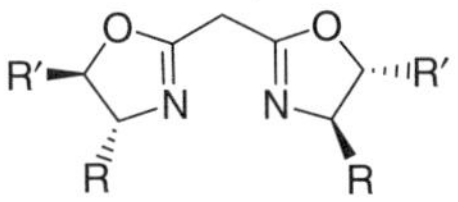

8 [41]

a R = Me, R′ = Ph
b R = CH₂OH, R′ = Ph

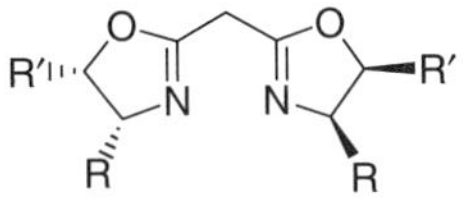

9 [41]

a R = Me, R′ = PH
b R = Et, R′ = ᵗBu
c R = R′ = Ph

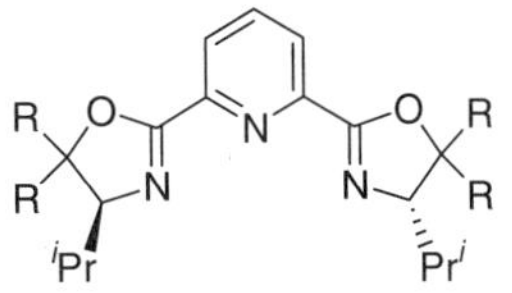

10 [45]

a R = Et
b R = Ph

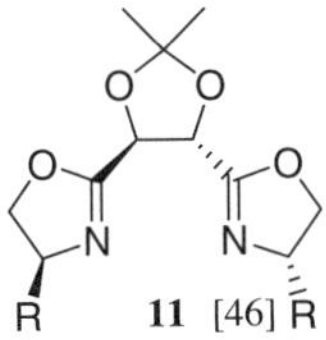

11 [46]

a R = PhCH₂
b R = Me₂CH
c R = ᵗBu
d R = Ph

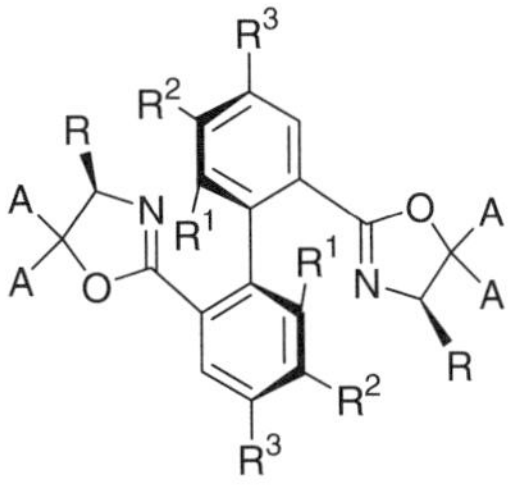

12 [47]

a R¹ = R² = R³ = OMe, A = H, R = ⁱPr
b R¹ = R² = R³ = OMe, A = H, R = ᵗBu

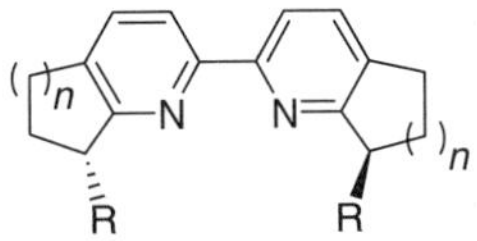

13 [48]

a n = 1, R = ᵗBu
b n = 2, R = ᵗBu
c n = 2, R = CMe₂(OMe)
d n = 2, R = TMS

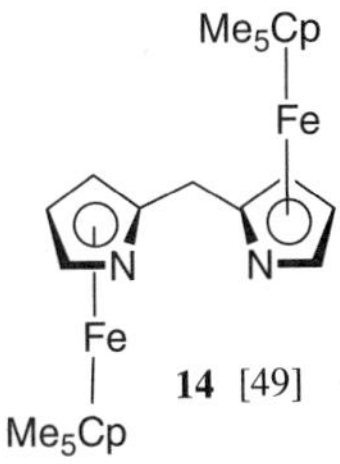

14 [49]

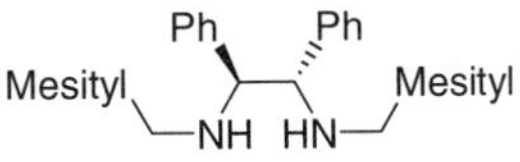

15 [50]

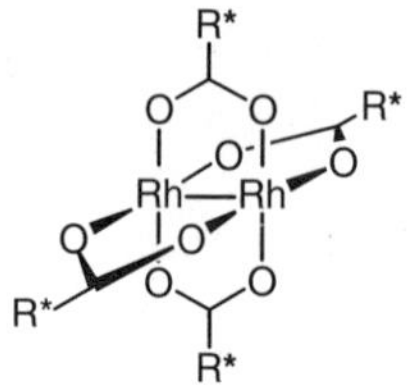

16 [52]

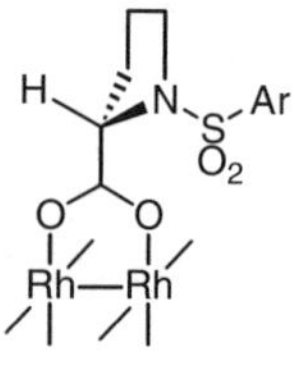

17 [53,61]

17a Ar = Ph
17b Ar = Nap
17c Ar = C₆H₄X

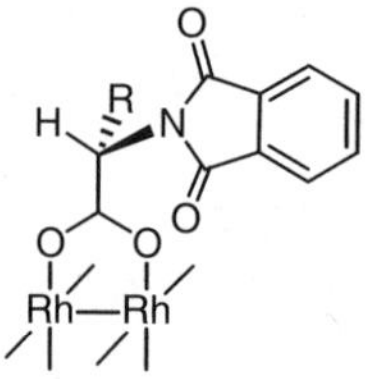

18a R = PhCH₂ [54]
18b R = ᵗBu [55]
18c R = Me [54]

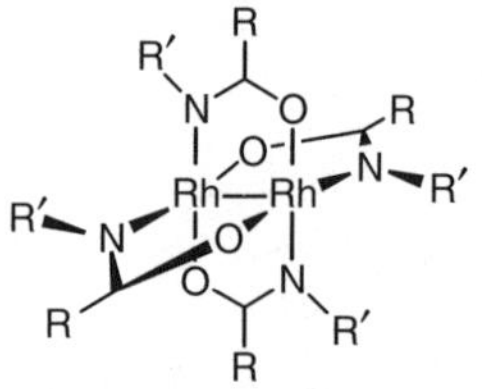

19 [7]

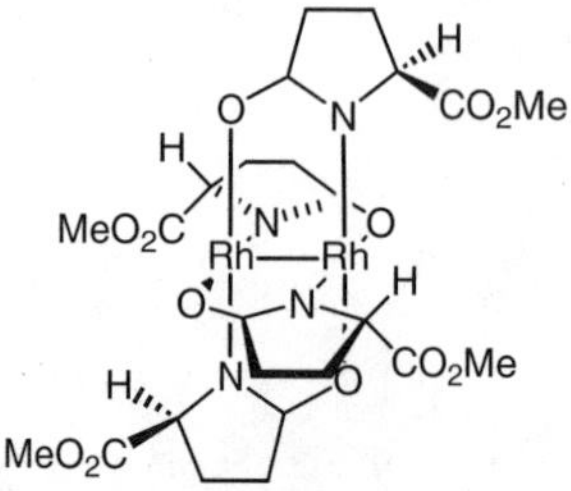

20: Rh₂(5*S*-MEPY)₄ [71]

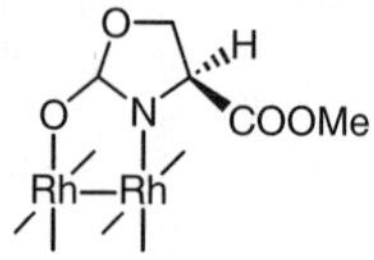

21 [7]
Rh₂(4*S*-MEOX)₄

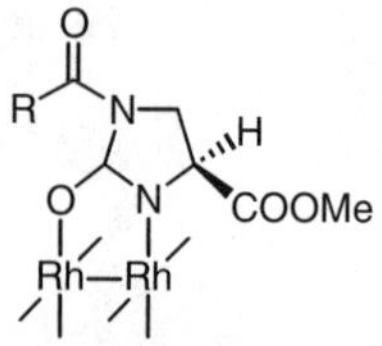

22 [65]

a. R = Me : Rh₂(4*S*-MACIM)₄
b. R = Ph : Rh₂(4*S*-MBOIM)₄
c. R = PhCH₂CH₂ : Rh₂(4*S*-MPPIM)₄

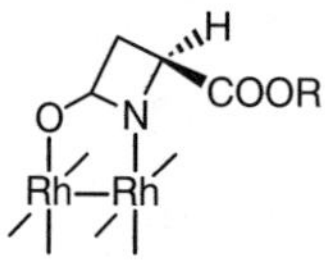

23 [7,10]

a. R = ᵗBu : Rh₂(4*S*-IBAZ)₄
b. R = Bn : Rh₂(4*S*-BNAZ)₄

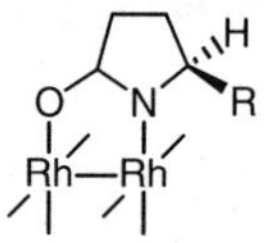

25 [68,71]

a R = Bn : Rh₂(5*R*-BNOX)₄
b R = ᶦPr : Rh₂(5*R*-IPOX)₄
c R = Ph : Rh₂(5*R*-PHOX)₄

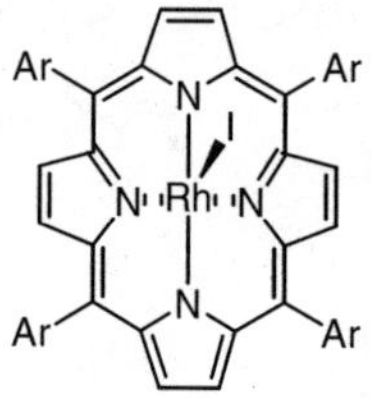

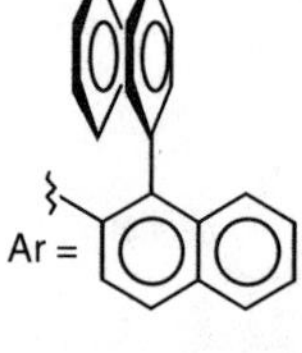

27 [74]

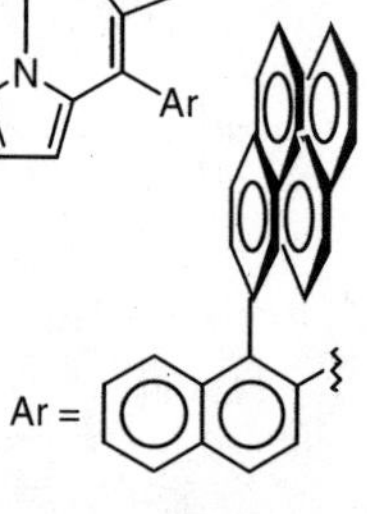

Ar =

28 [75]

Ar =

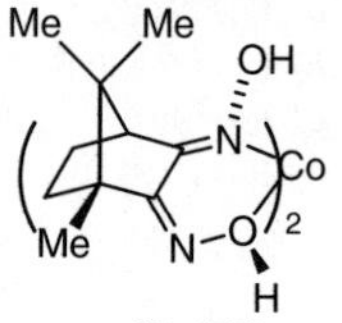

29 [77]

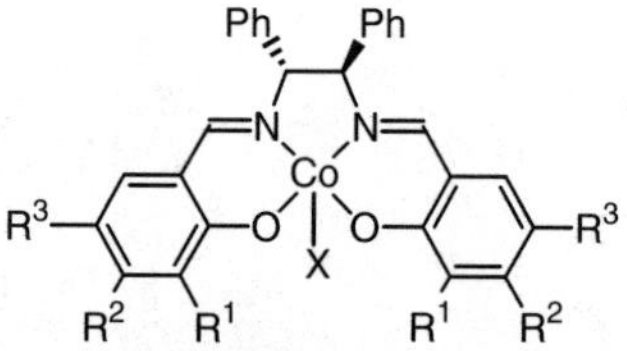

30 [78]

a. X = Br, R¹ = R³ = H, R² = ᵗBu
b. X = Br, R¹ = R² = H, R³ = OMe

CHAPTER 6A

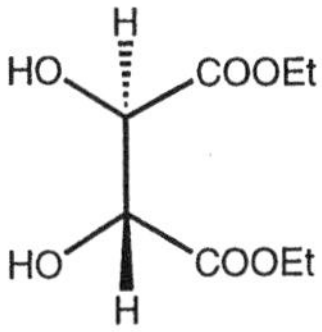

(+) DET

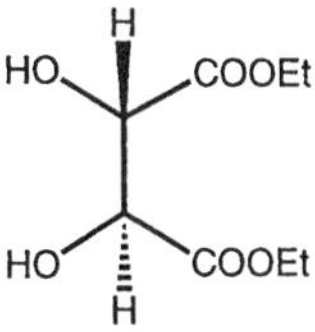

(-) DET

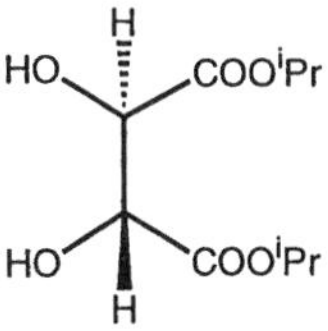

(+)-DIPT

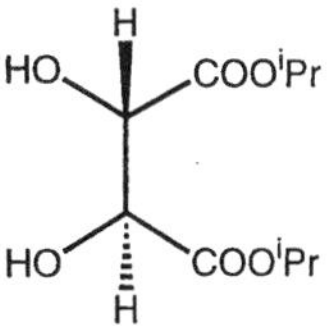

(-) DIPT

CHAPTER 6B

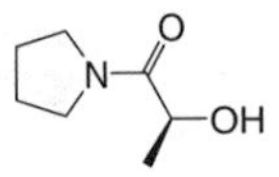

[16]

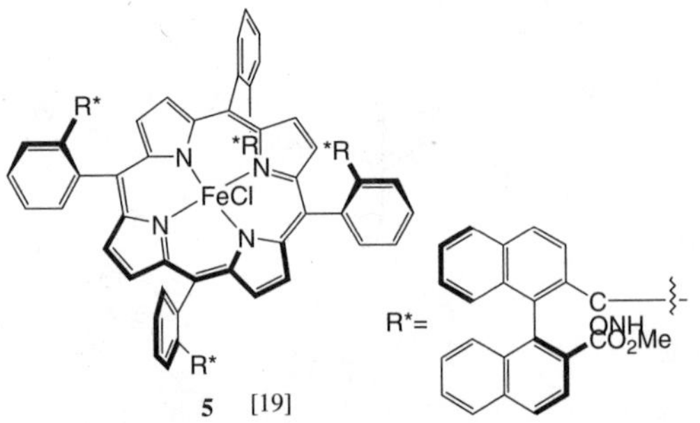

5   [19]

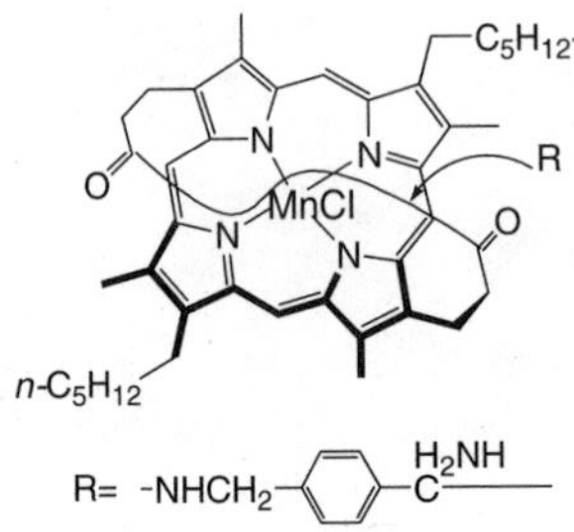

R*=

R= -NHCH₂—⟨ ⟩—CH(H₂NH)

R= $-NHCH_2$

6   [20f]

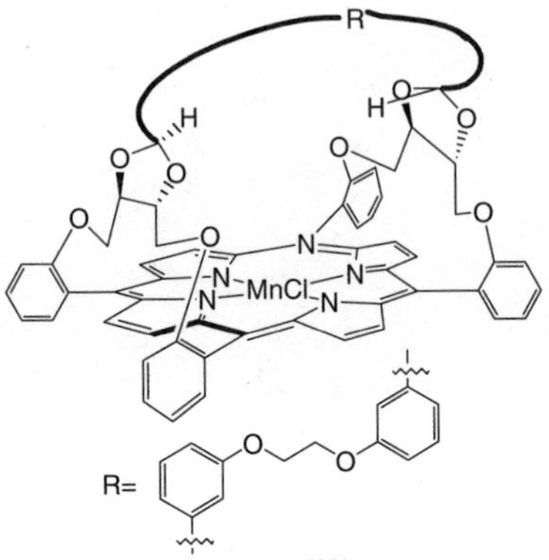

7   [22]

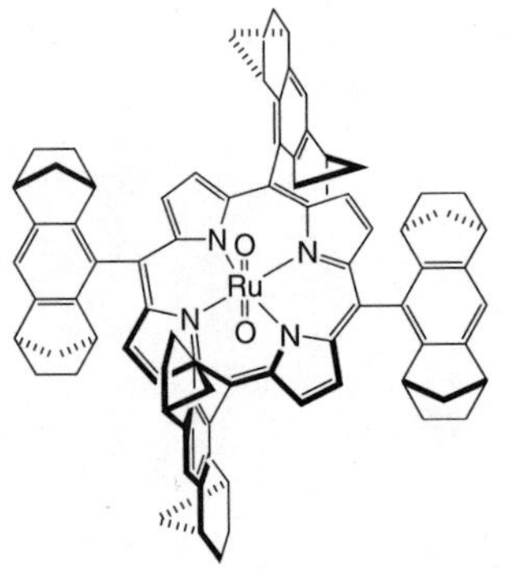

8   [24]

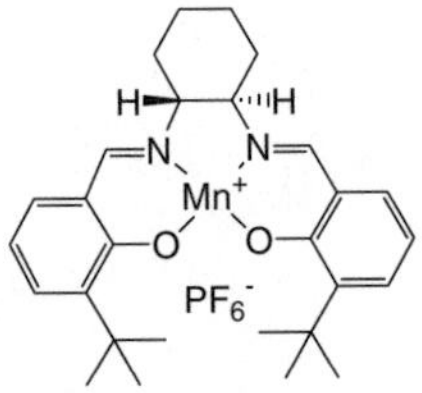

9   [29a]

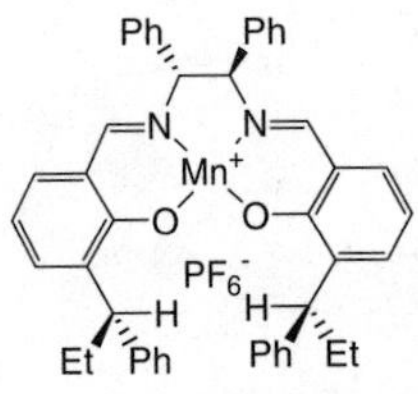

10   [29b]

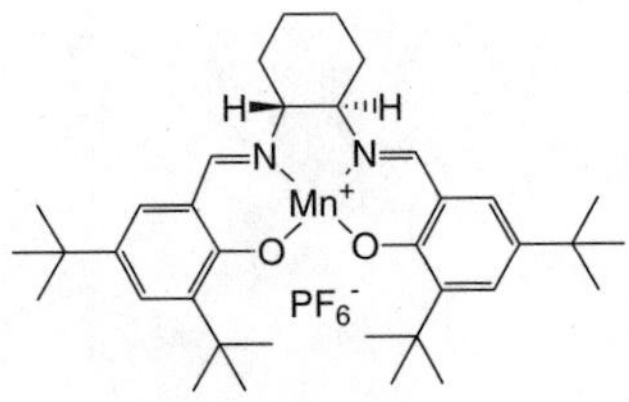

11   [30]

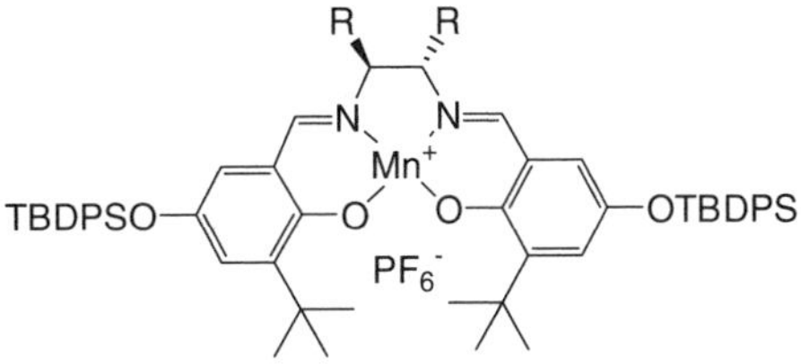

12 a: R= Ph   [31]
    b: R,R= -(CH₂)₄-

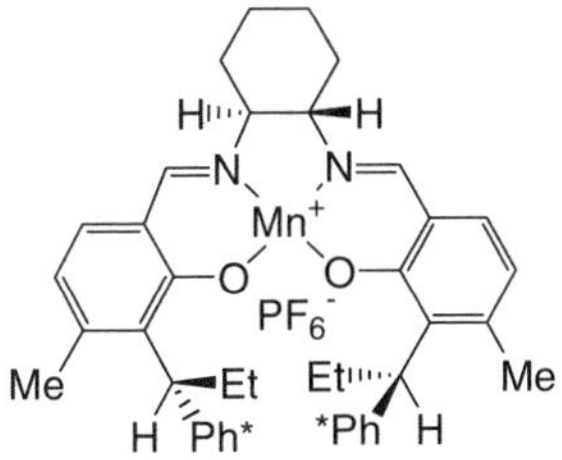

13   Ph*= 4-*t*-BuC₆H₄
    [32]

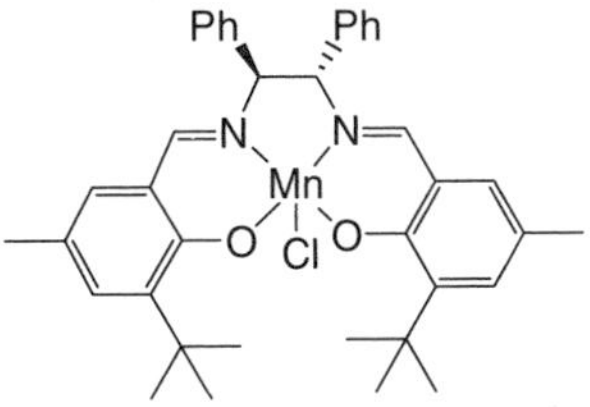

14   [37]

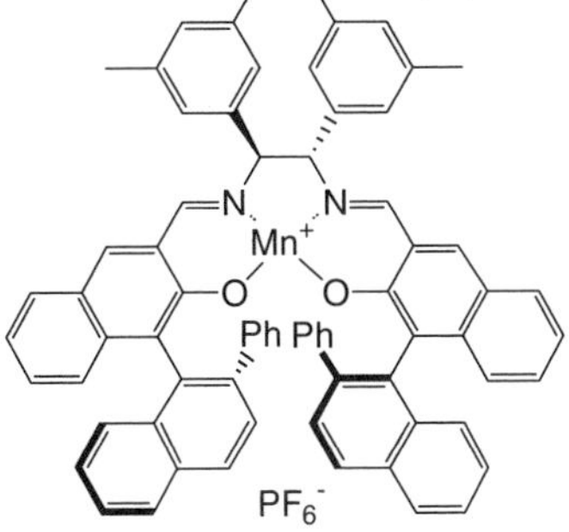

15   [45c]

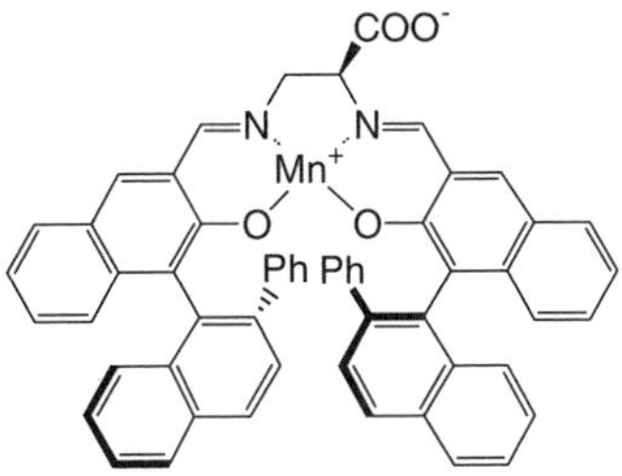

18   [51]

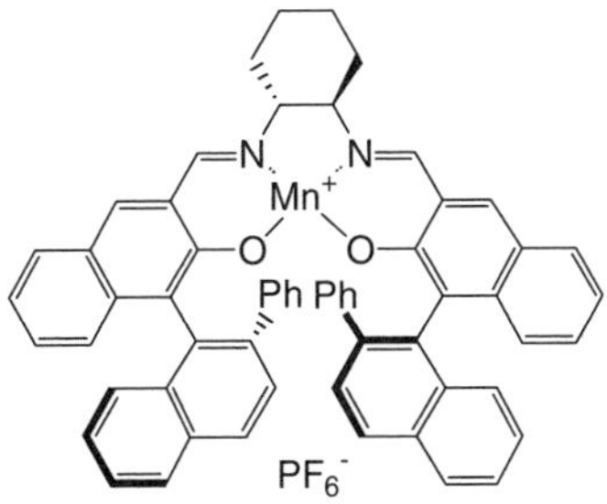

(R,R)-**19**   [58c]

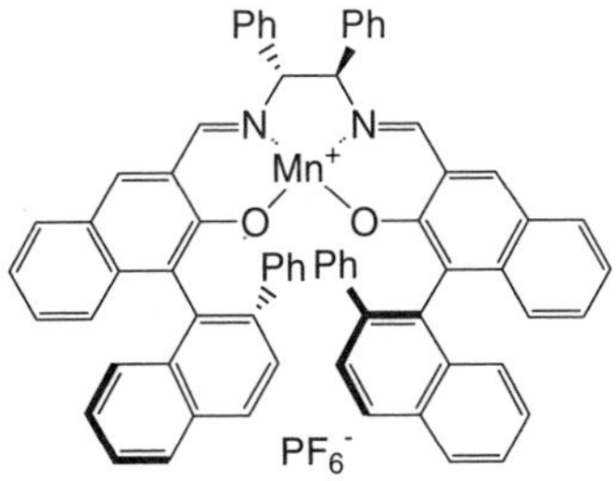

(R,R)-**20**   [58e]

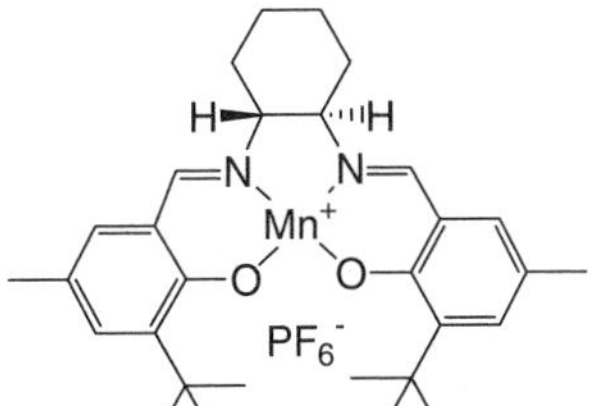

22   [64]

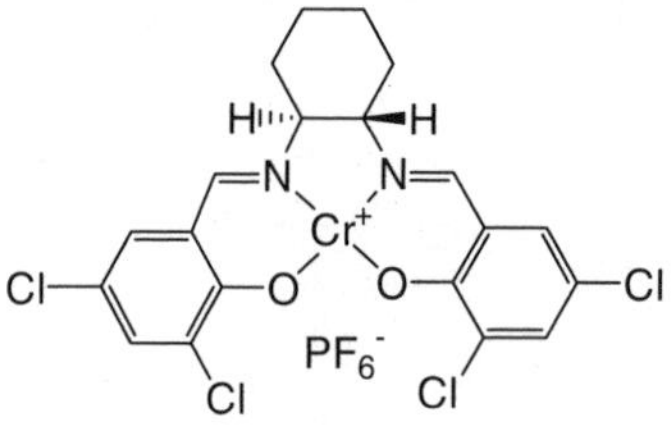

23 [65]

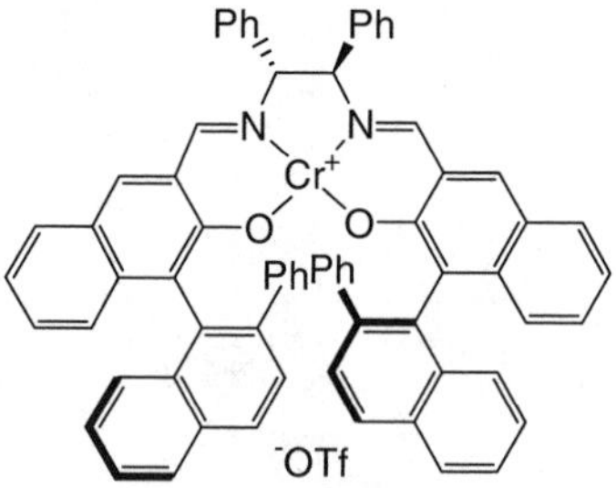

24 [67]

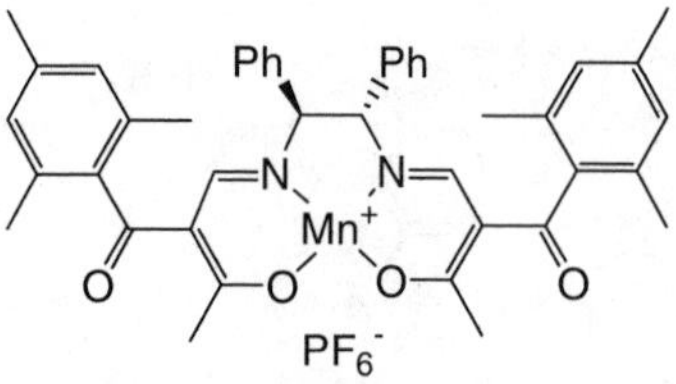

25 [69]

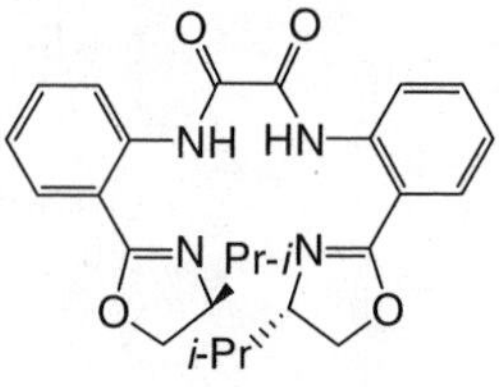

26 [70]

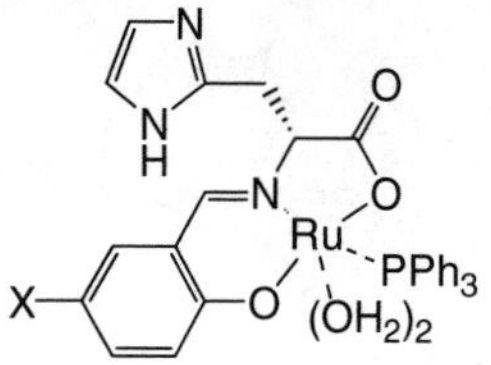

27 **a**: X= MeO [71]
b: X= H
c: X= Cl

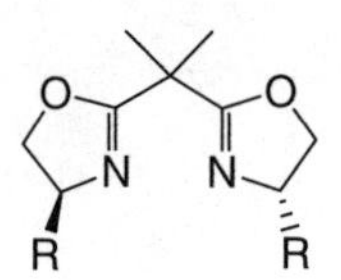

28a, R= C6H5
28b, R= CMe3

[74b]

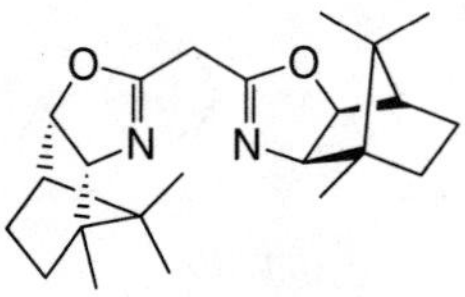

29 [75]

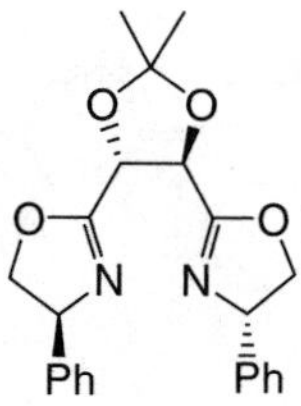

30 [77]

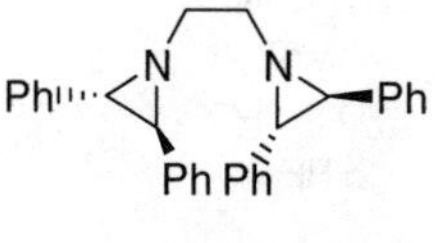

31 [78]

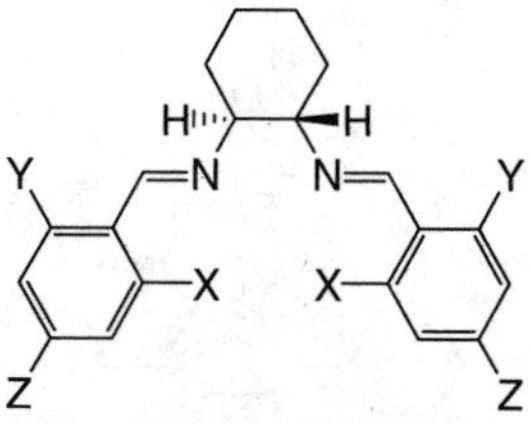

32a: X=Y= Cl, Z= H [79]
32b: X=Y=Z= H
32c: X=Y=Z= CH3

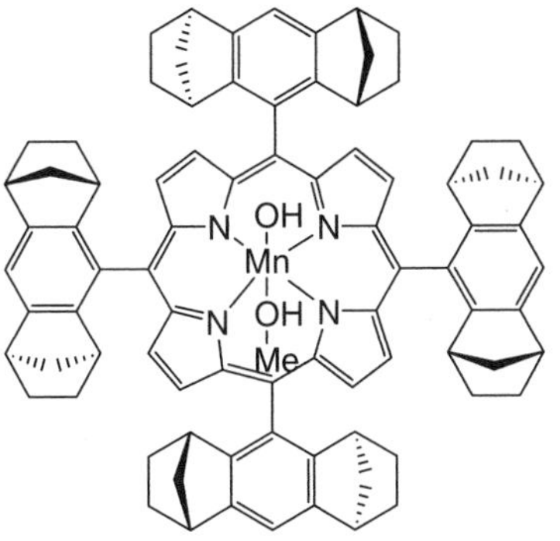

33   [80]

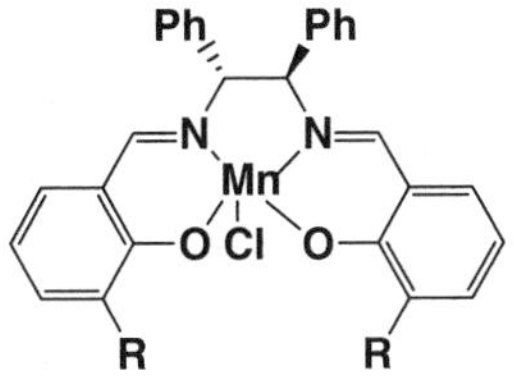

35 a: R= Me$_3$Si   [81]
    b: R= t-BuMe$_2$Si

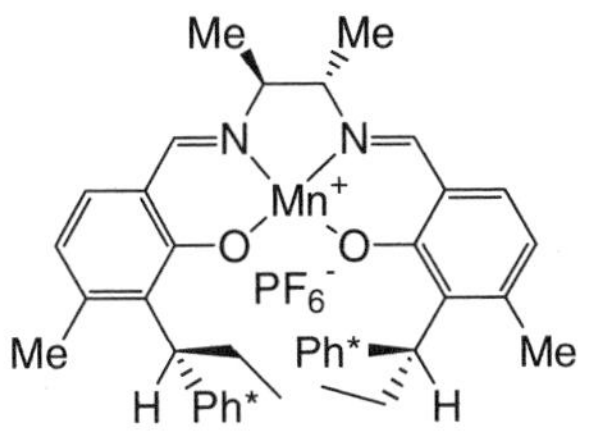

Ph* = 4-(t-Bu)C$_6$H$_4$

38   [82]

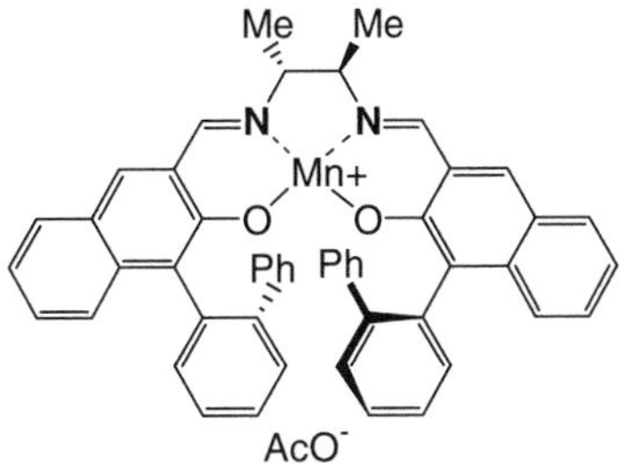

39   [84]

CHAPTER 6C

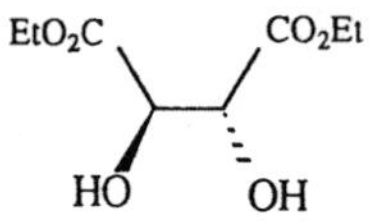

EtO₂C, CO₂Et structure

HO OH

(+)-(R,R)-DET
commercially avaliable

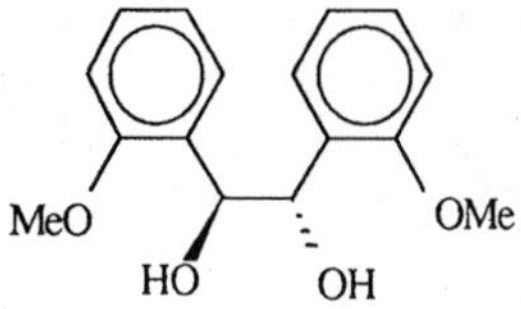

MeO structure OMe

HO OH

2

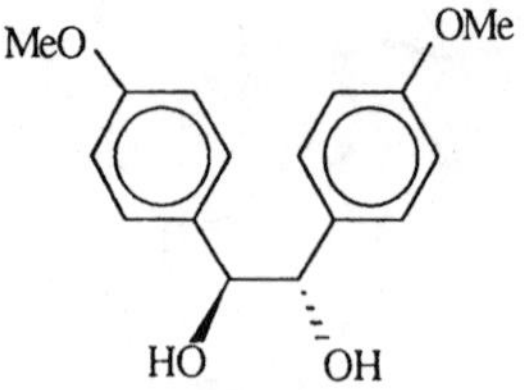

MeO structure OMe

HO OH

3

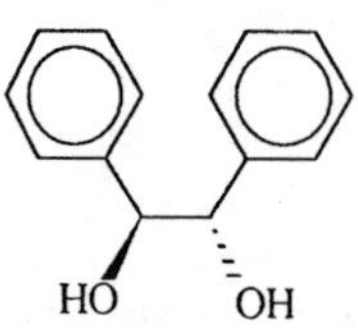

structure

HO OH

4

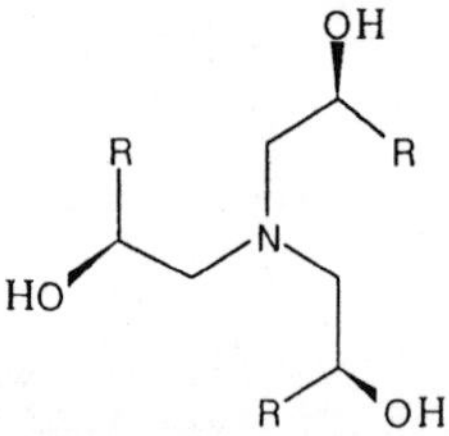

OH

R R

HO N

R OH

(S,S,S)-5

a: R = Me
b: R = t-Bu
c: R = Ph

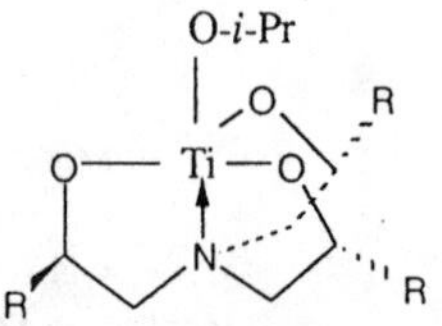

O-i-Pr

Ti structure

R R

(S,S,S,)-6

a: R = Me
b: R = t-Bu
c: R = Ph

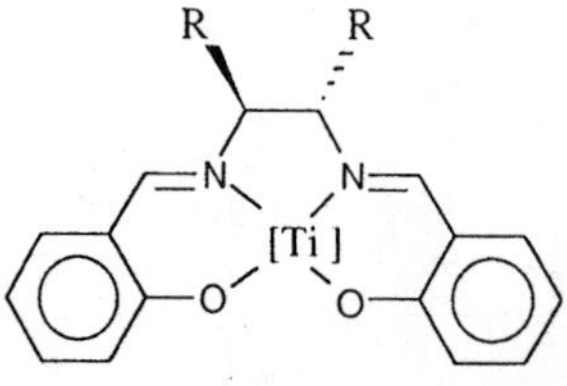

R R

N N

[Ti]

O O

15 R = Me or Ph
[67]

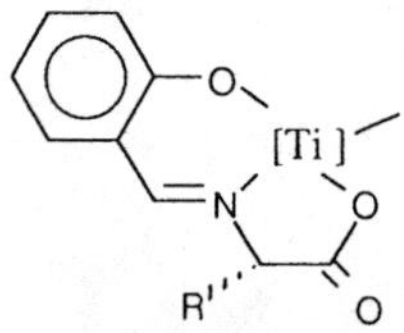

structure [Ti]

R'

16 **a** R = i-Pr   **c** R = CH₂— (imidazole)
        b R = t-Bu
        [68]

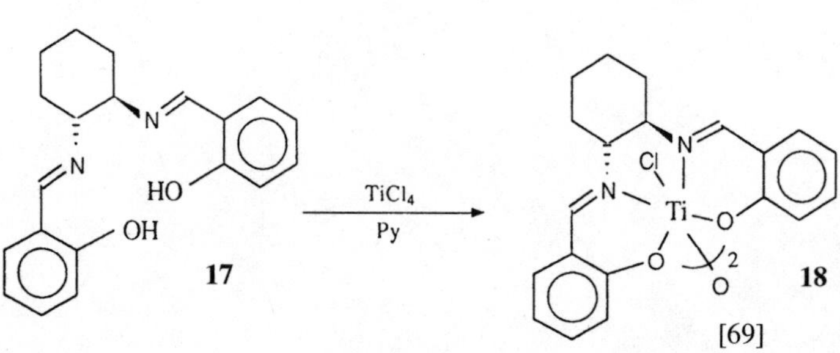

structure

N

HO

N

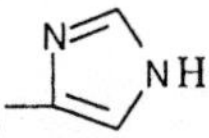

OH

17

$\xrightarrow[\text{Py}]{\text{TiCl}_4}$

structure

Cl N

Ti

O

O

O ⟩₂

18

[69]

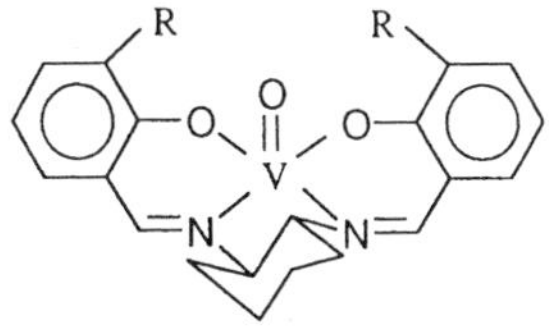

19 R = OMe, OEt or *t*-Bu [70]

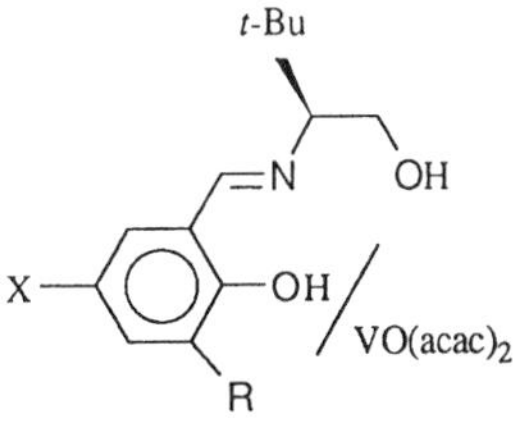

20 **a** R=*t*-Bu, X=NO_2 **b** R=X=*t*-Bu [91]

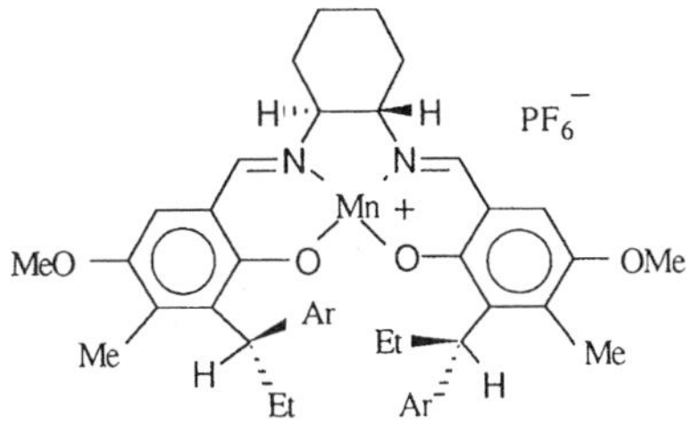

23

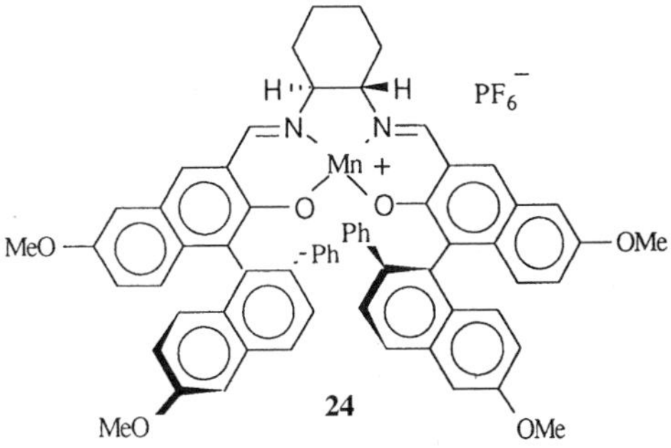

24

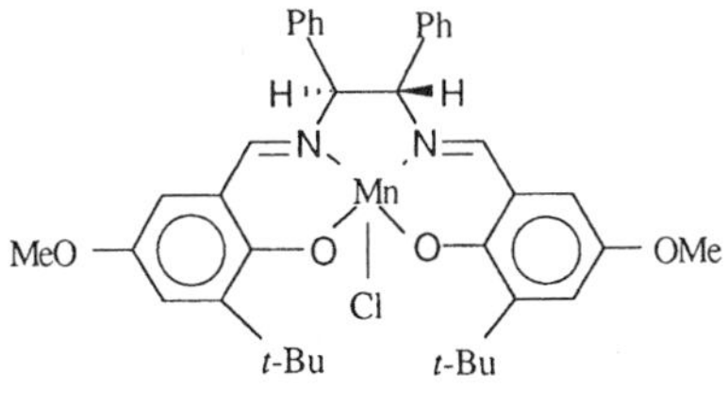

25

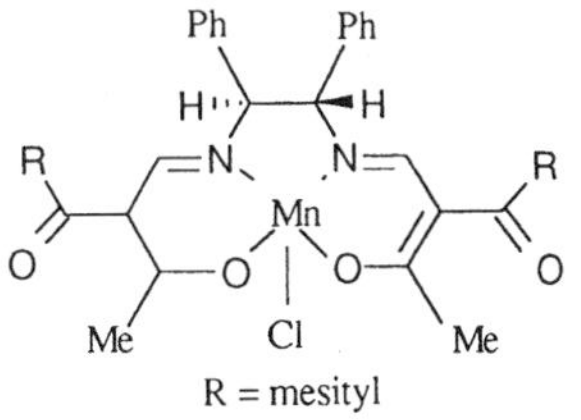

R = mesityl

26

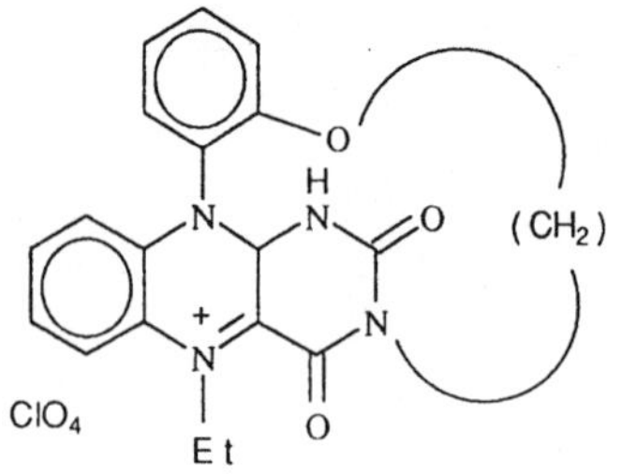

30 [87]

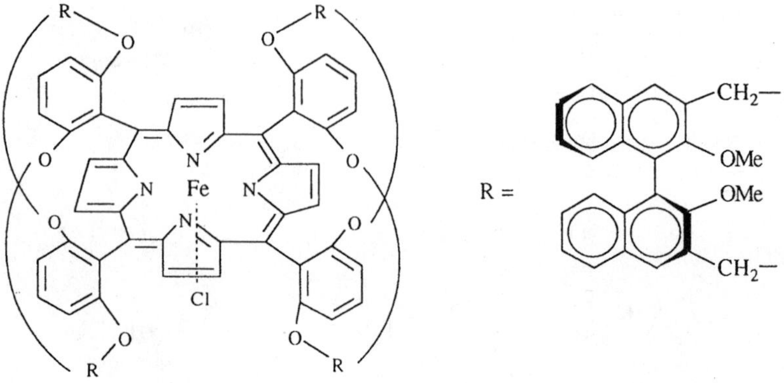

R =

27 [79, 80]

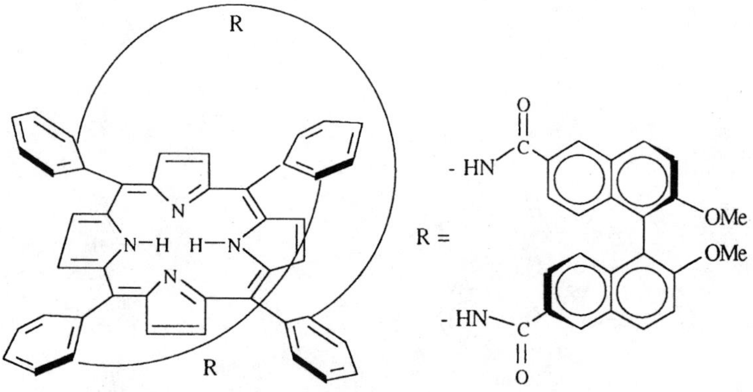

R =

28 [81]

R =

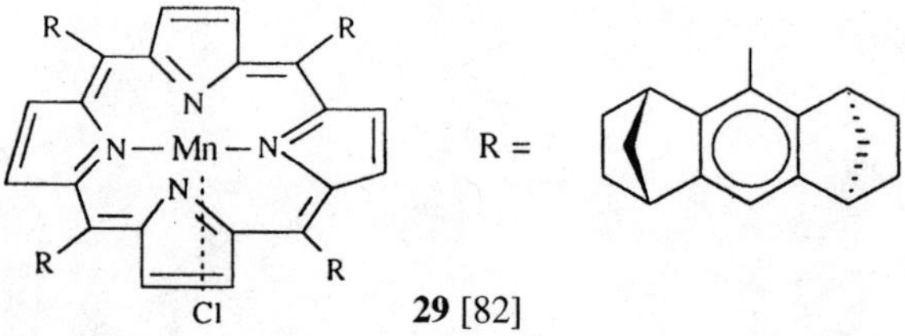

29 [82]

CHAPTER 6D

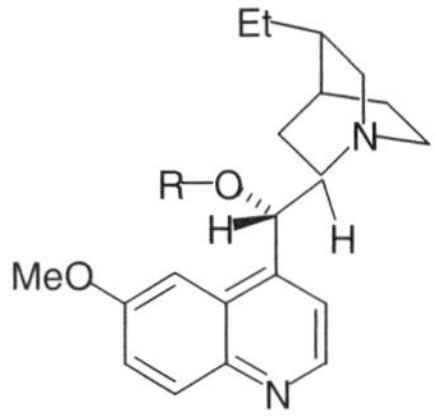

DHQD: R = H [39]

DHQD-OAc: R = —CCH₃ [17]

DHQD-CLB: R = —C—⟨ ⟩—Cl [17]

[(DHQD)₂-PHL: R = —⟨N-N⟩—ODHQD [29]

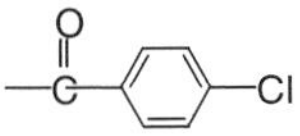

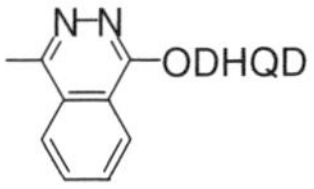

R = —⟨ ⟩ [41]

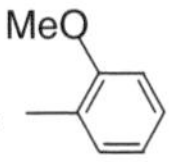

DHQD-PHN: R = — [41]

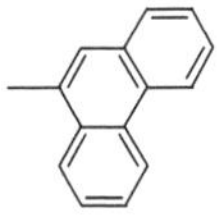

DHQD-MEQ: R = — [41]

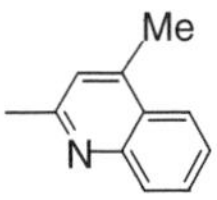

DHQD-IND: R = —C—N⟨ ⟩ [57b]

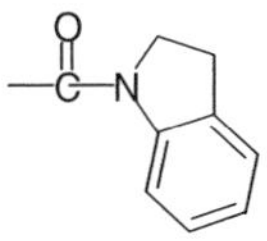

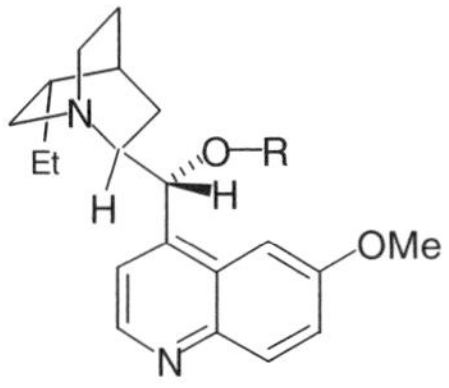

DHQ: R = H [39]

DHQ-OAc: R = —CCH₃ [17]

DHQD-CLB: R = —C—⟨ ⟩—Cl [17]

[(DHQ)₂-PHL: R = —⟨N-N⟩—ODHQ [29]

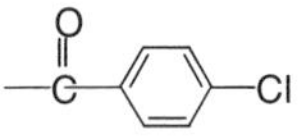

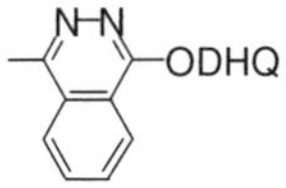

R = —⟨ ⟩ [41]

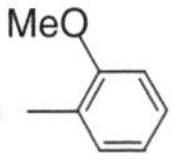

DHQ-PHN: R = — [41]

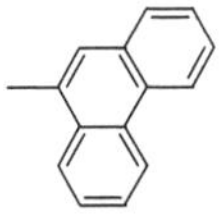

DHQ-MEQ: R = — [41]

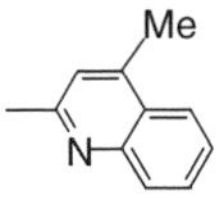

DHQ-IND: R = —C—N⟨ ⟩ [57b]

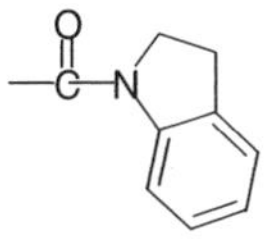

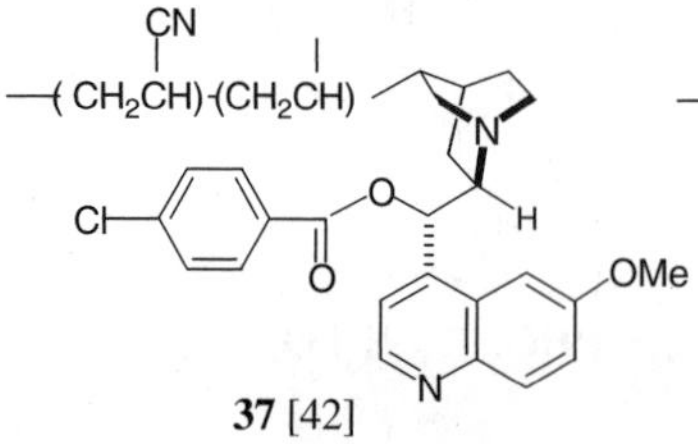

37 [42]

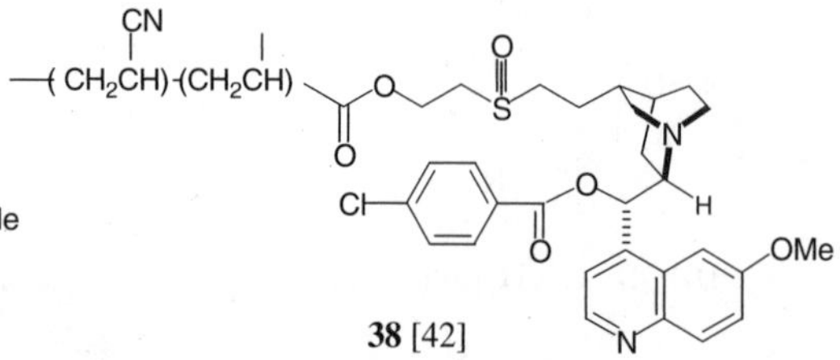

38 [42]

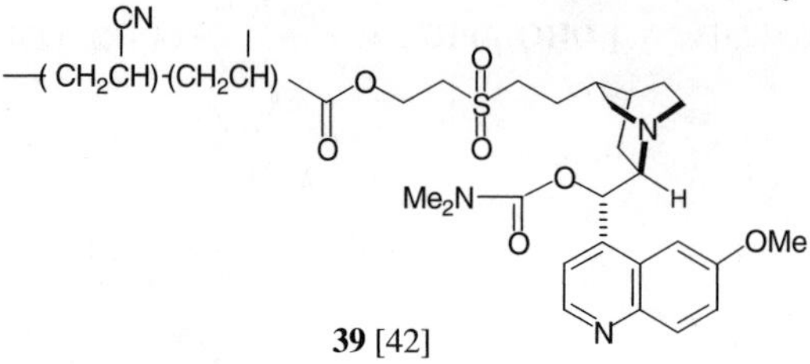

39 [42]

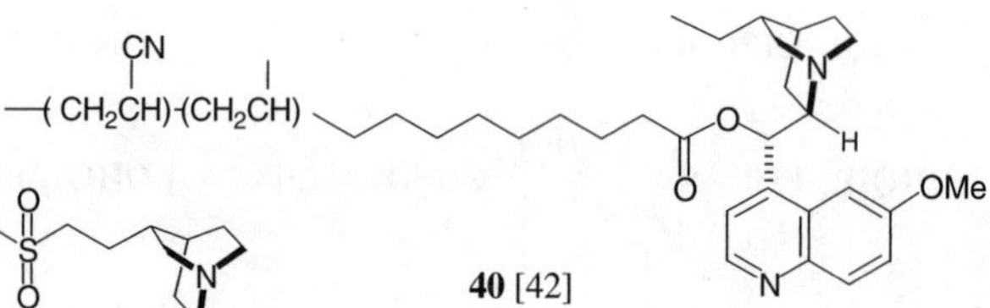

40 [42]

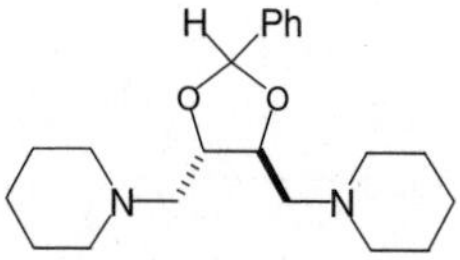

17 [20]

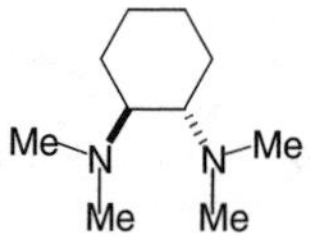

18 [21]

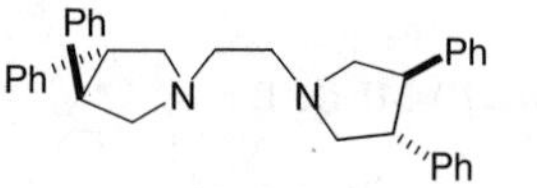

19 [22]

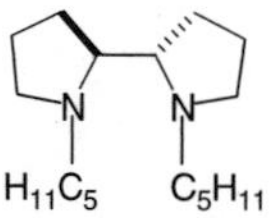

20 [23]

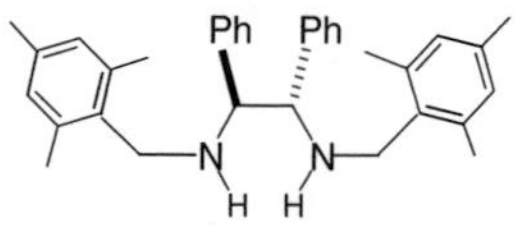

21 [24]

CHAPTER 6E

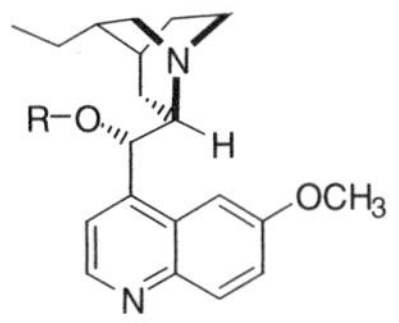

DHQD
Dihydroquinidine (R = H)

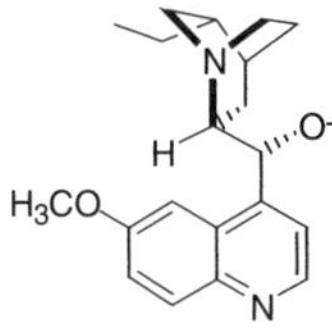

DHQ
Dihydroquinine (R = H)

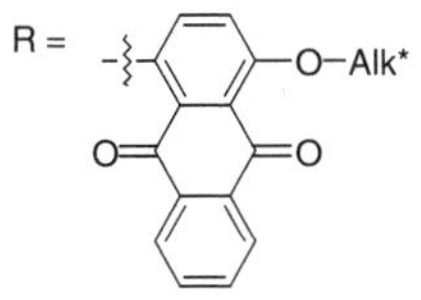

AQN Ligands [2]
Alk* = DHQ or DHQD

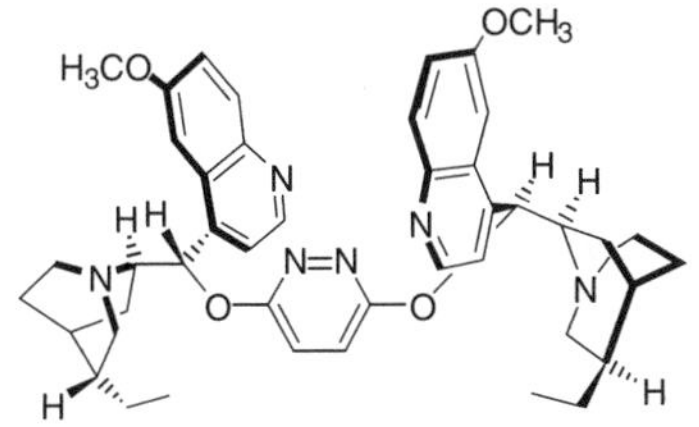

7 [31]

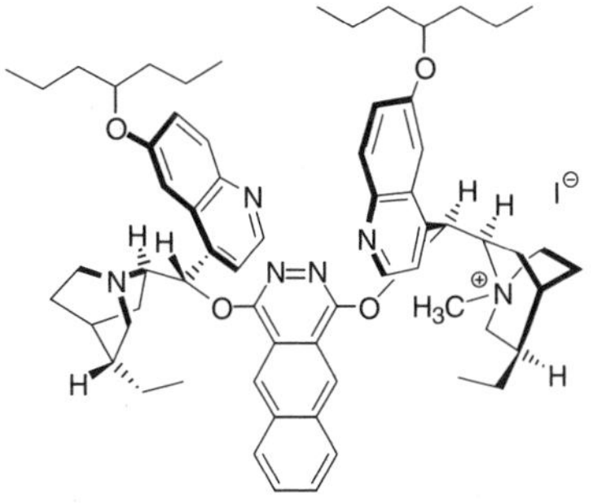

8 [5, 6]

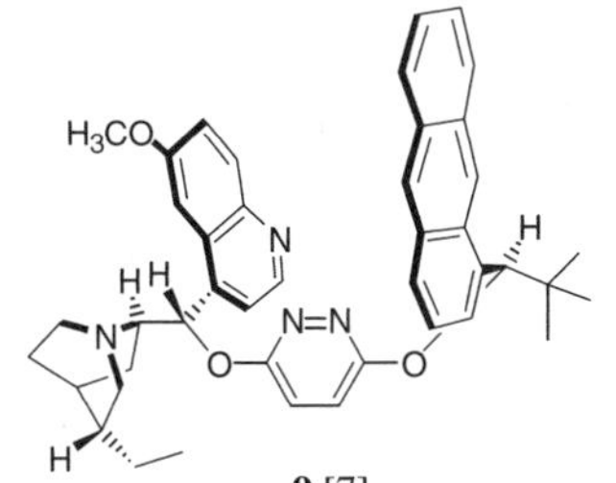

9 [7]

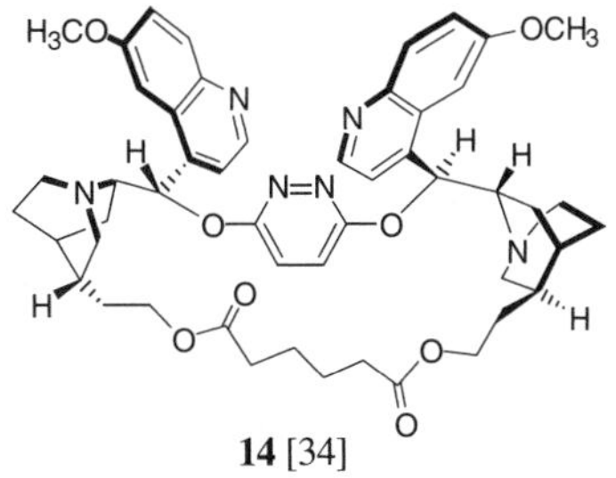

14 [34]

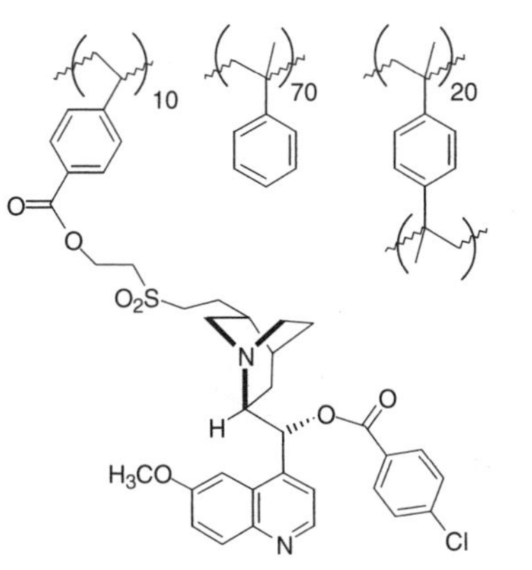

17 [54]

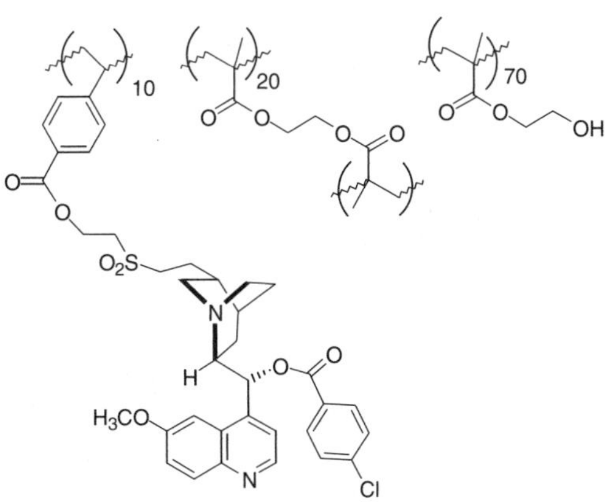

18 [55]

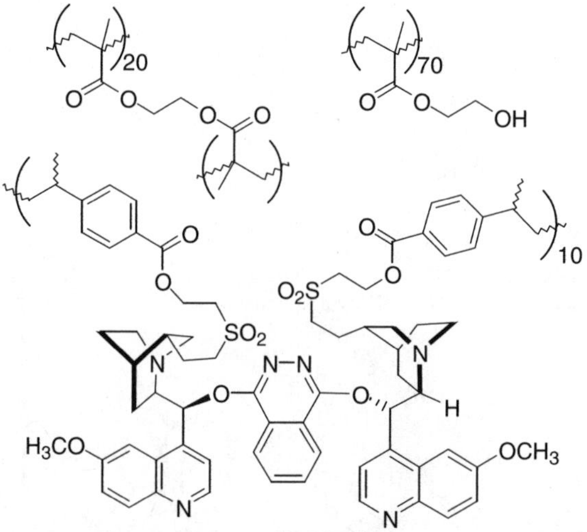

19 [57]

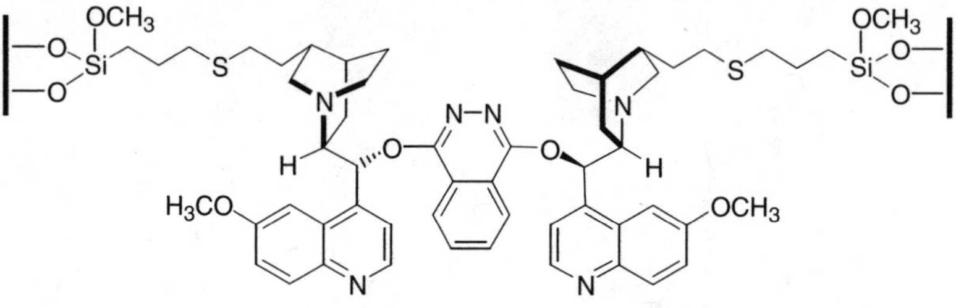

20 [62]

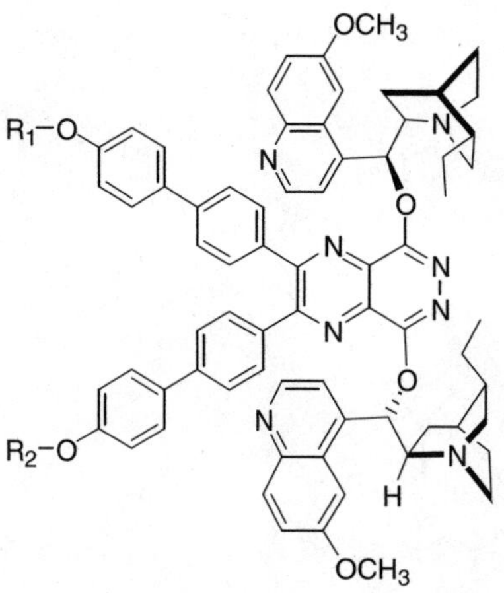

21a: R^1 =

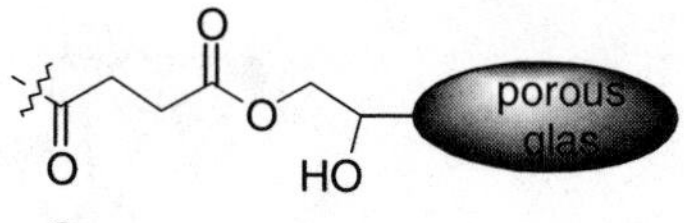

R^2 =

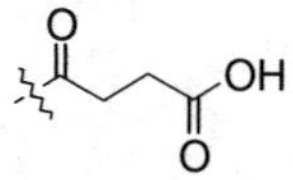

21b: R^1 =

R^2 = H

[63]

[63]

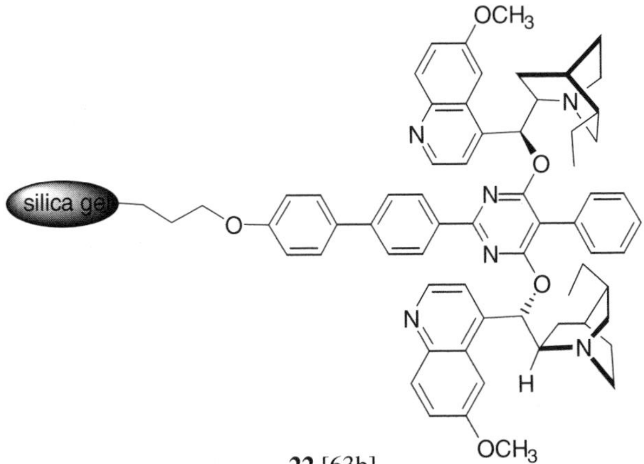

22 [63b]

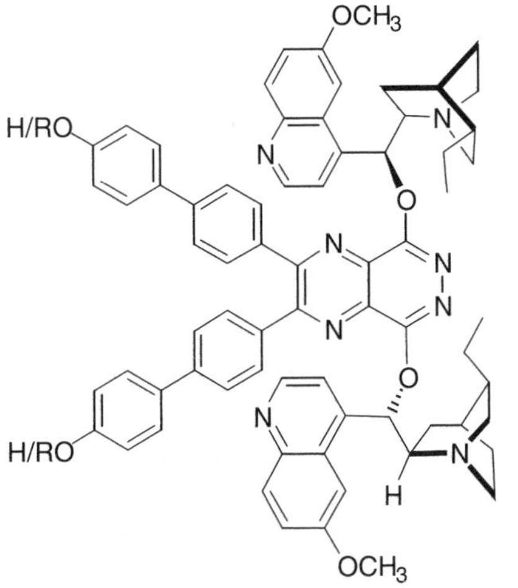

23a,b [68, 69]

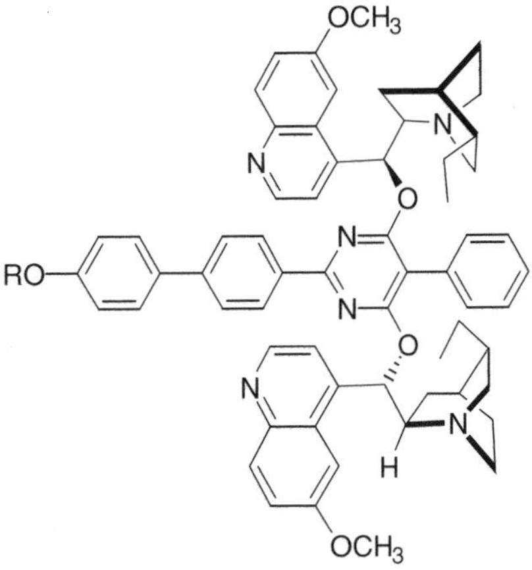

24a,b [68, 69]

a: R =

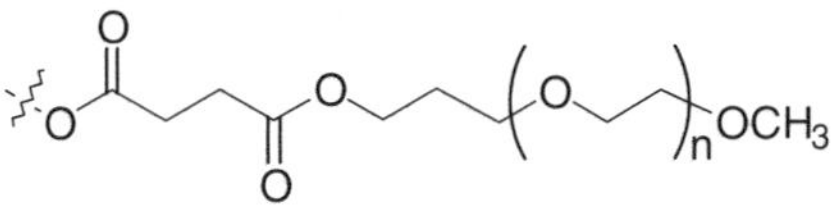

b: R =

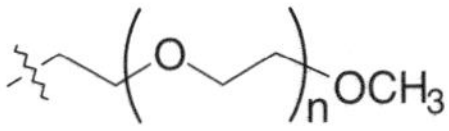

CHAPTER 7

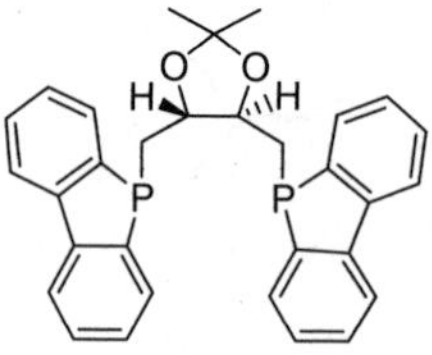

DBP-DIOP (**1**)
[8]

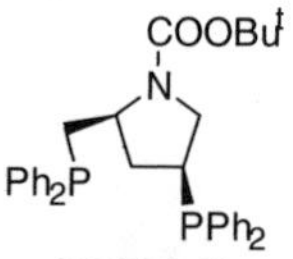

BPPM (**2**)
[10]

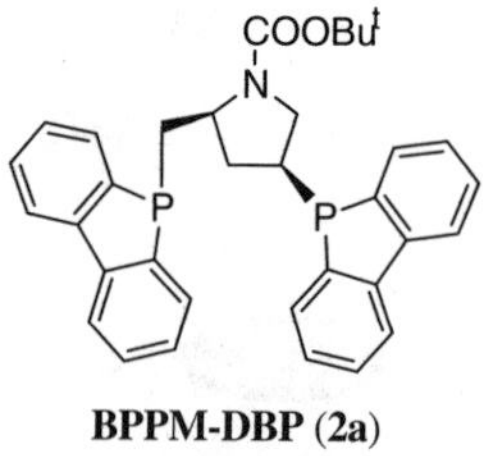

BPPM-DBP (**2a**)
[13]

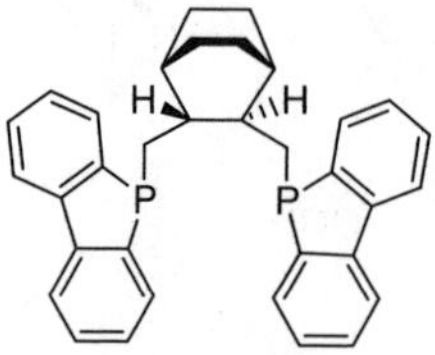

BCO-DBP (**3**) [11]

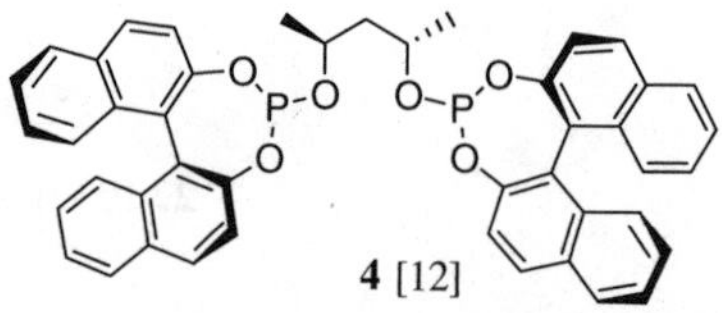

4 [12]

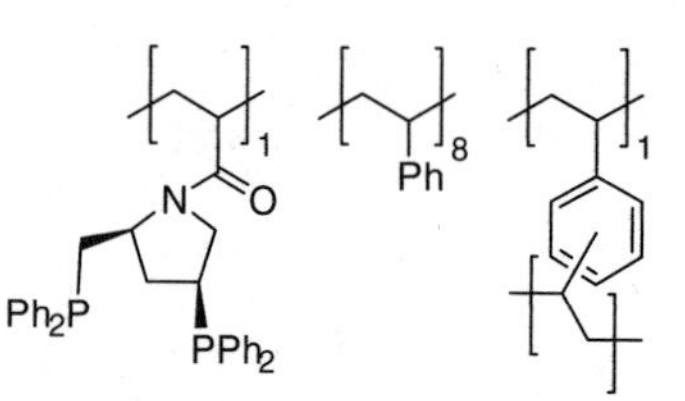

Polymer-**BPPM** (**5**)   [14]

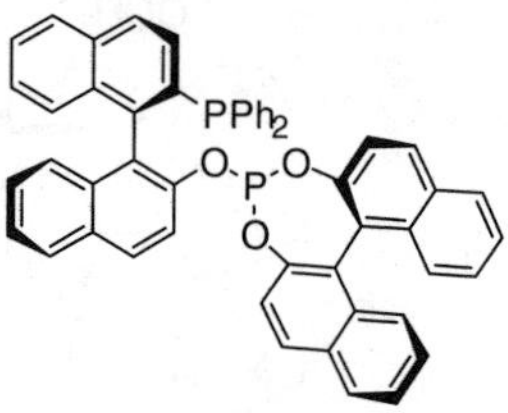

(*R,S*)-**BINAPHOS** (**6**)   [19]

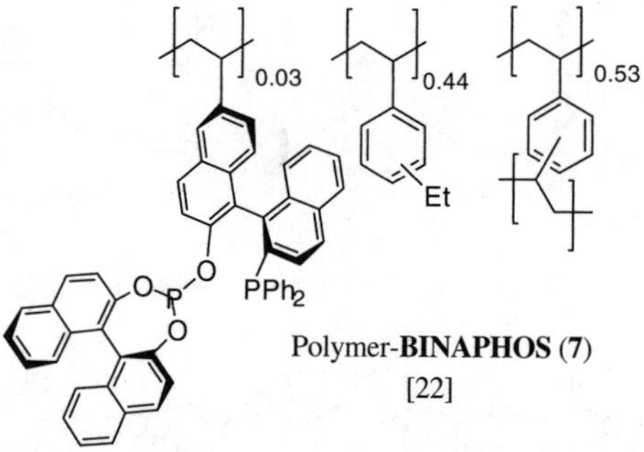

Polymer-**BINAPHOS** (**7**)
[22]

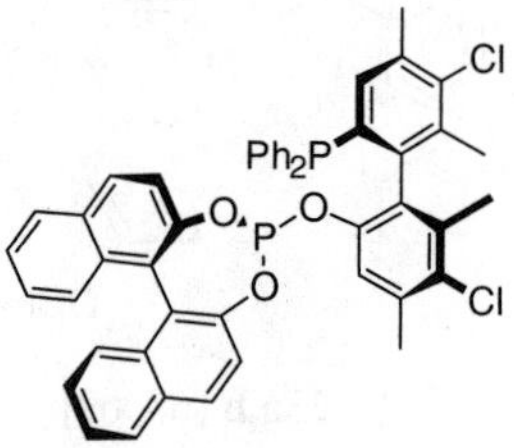

(*S,R*)-**BIPHEMPHOS** (**8**)
[23]

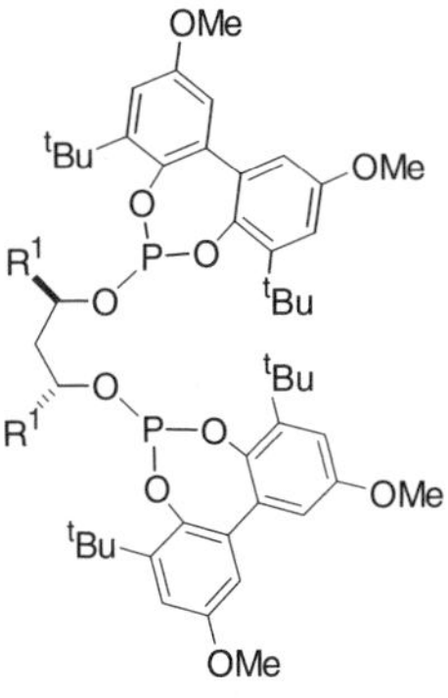

9 [25]

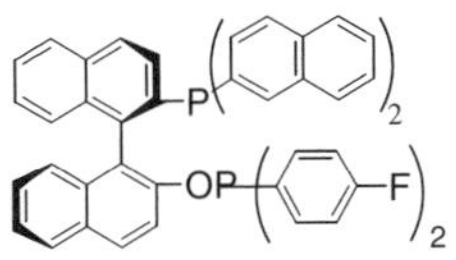

(*R*)-2-**Nap-BIPNITE**-*p*-**F** (**10**)
[33]

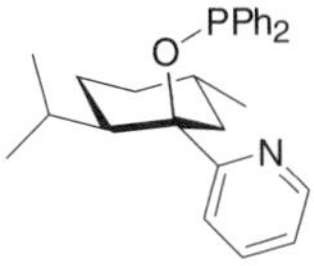

(-)-**menthoxy-PPy**
[87]

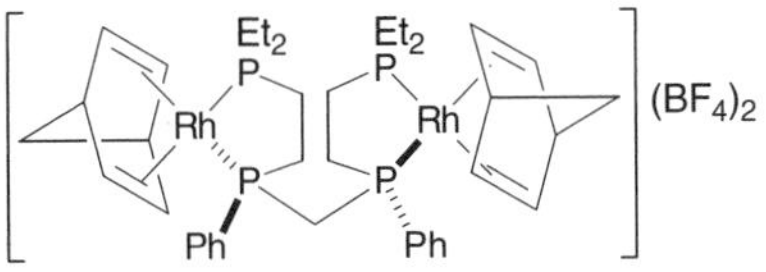

(*S,S*)-[Rh$_2$(nbd)$_2$(et,ph-P$_4$)](BF$_4$)$_2$ (**11**)
[34]

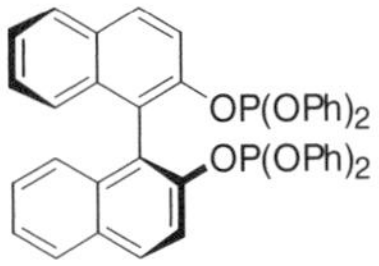

12 [31]

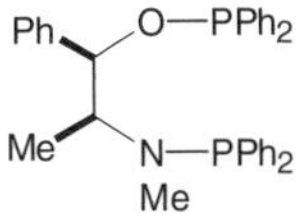

L-EPHOS (**13**)
[63]

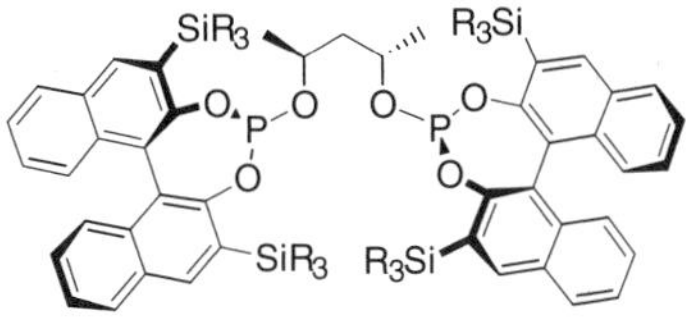

14 (R$_3$ = Me, Et, ^{t}BuMe$_2$) [30]

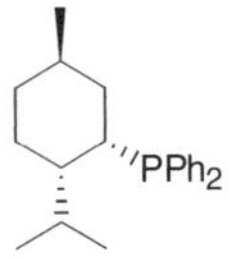

NMDPP (**48**)
[104]

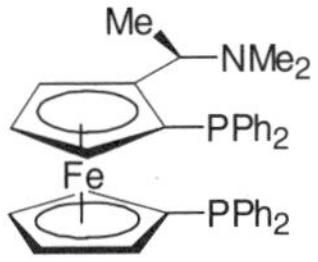

BPPFA (**49**)
[105]

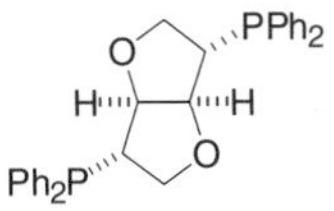

50 [106]

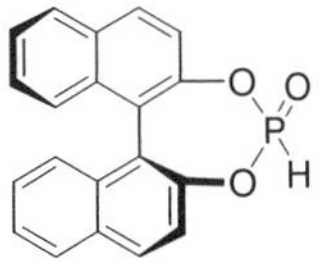

(*R*)-(−)-**BNPPA** (**51**)
[107]

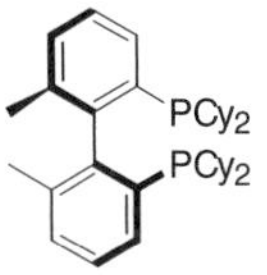

(*S*)-**BICHEP** (**59**)
Cy = cyclohexyl
[124]

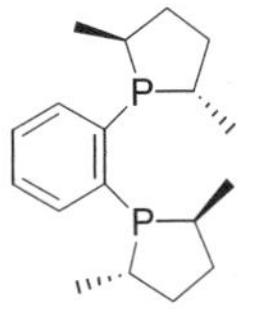

(*S,S*)-**Me-DUPHOS** (**60**)
[128]

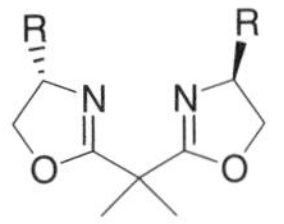

61a: R = i-Pr
61b: R = Me
[137]

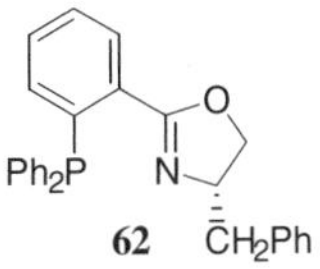

62
[138]

(*R,R*)-**DIOP**: See the Ligand List for Chapter 1

CHAPTER 8A

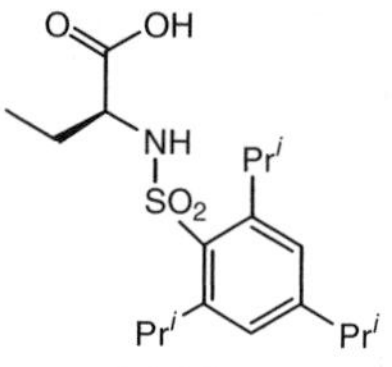

1, [3]

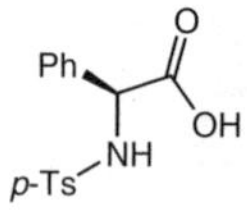

2, [4]

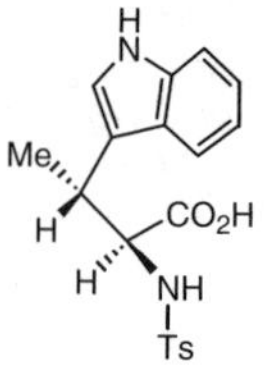

3, [7,8]

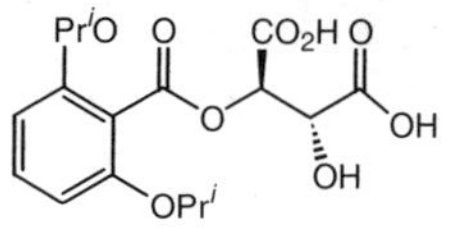

4, [9]

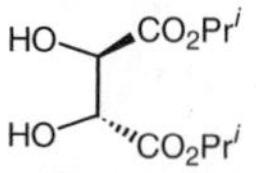

5, [10,70]

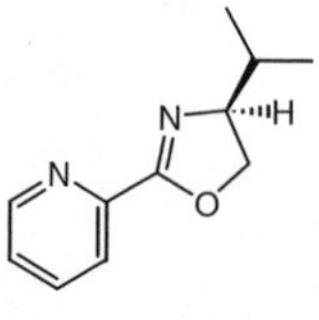

6, [11~13]

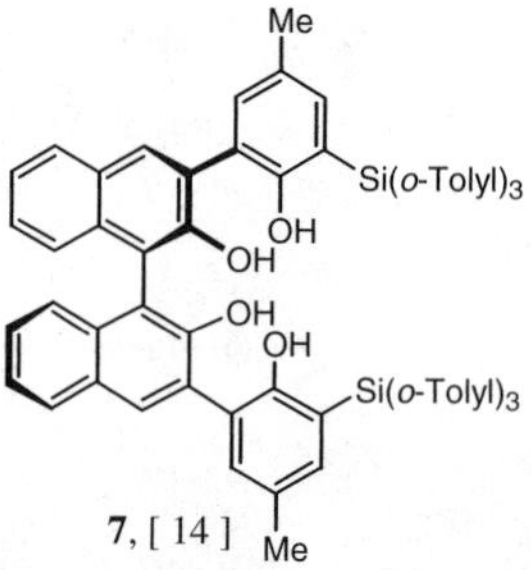

7, [14]

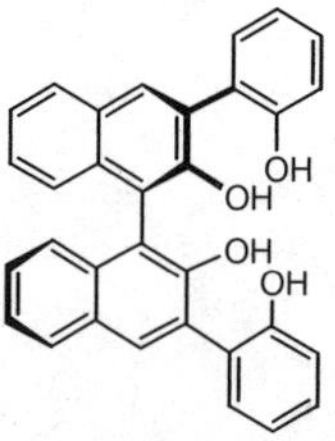

8, [15,17]

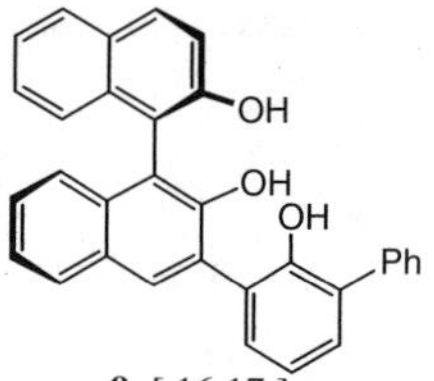

9, [16,17]

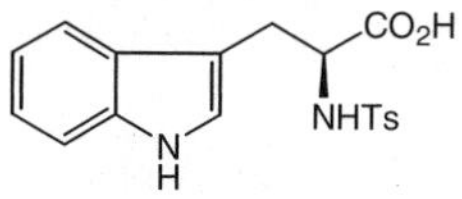

10, [18]

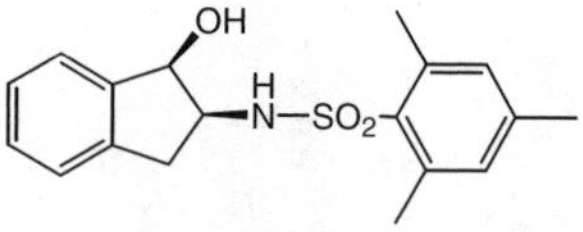

11, [19]

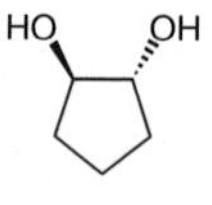

12, [20]

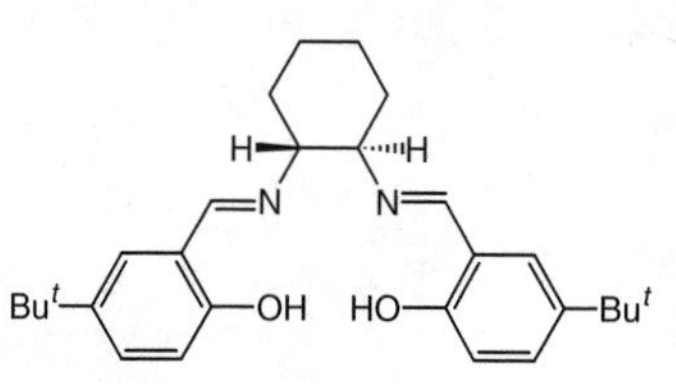

13, [21]

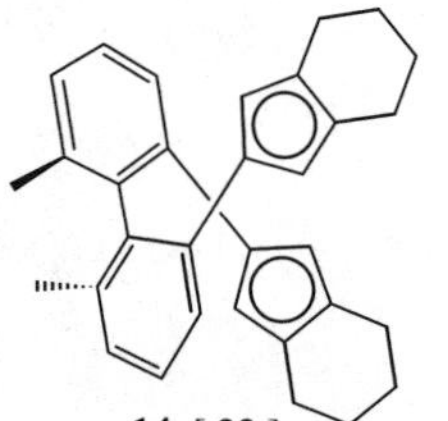

14, [22]

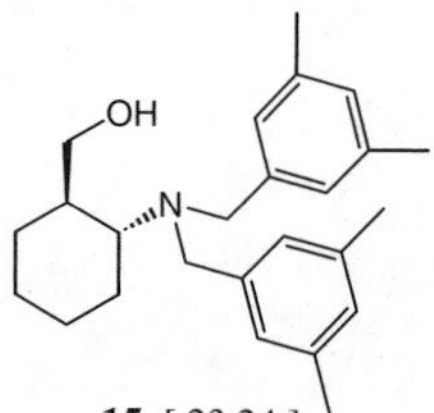

15, [23,24]

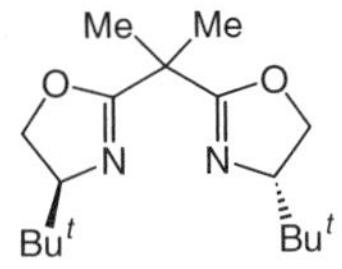

16, [25~29,33,61,63]

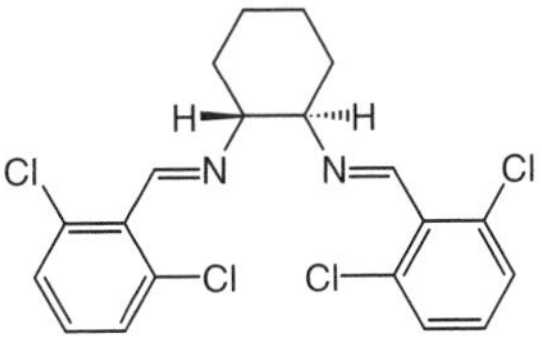

17, [30]

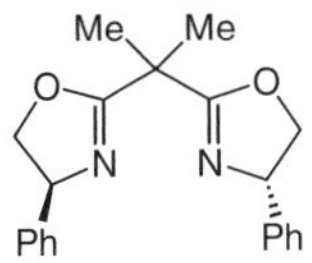

18, [31,32]

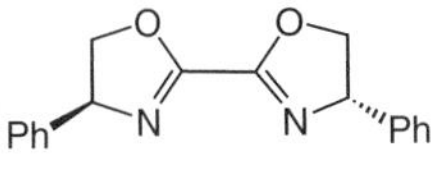

19, [33]

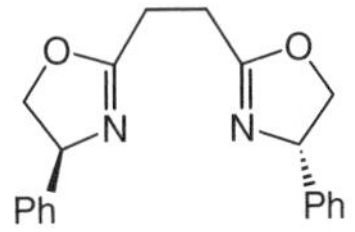

20, [33]

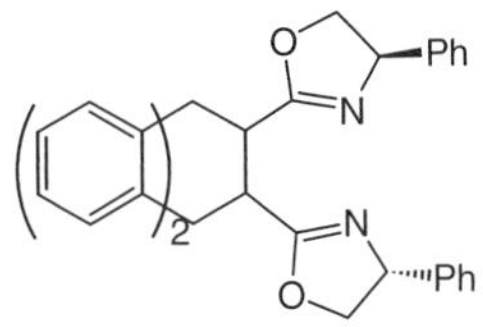

21, [34]

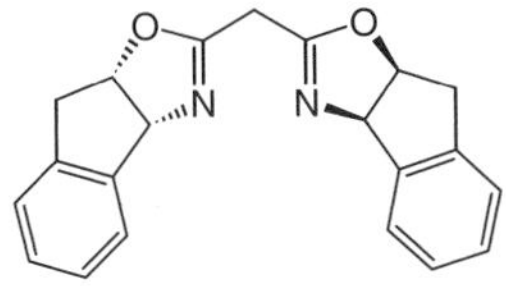

22, [35,62]

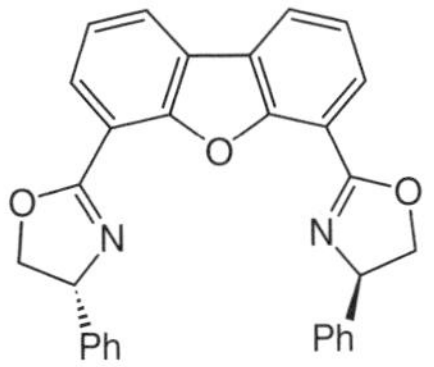

23, [36]

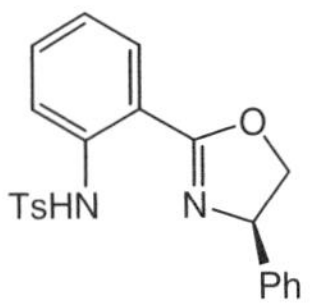

24, [37]

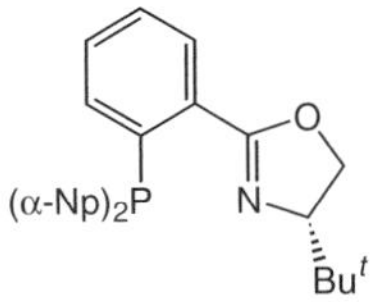

25, [38]

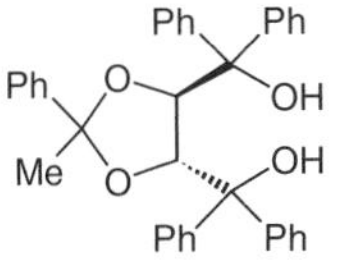

26, [39,67]

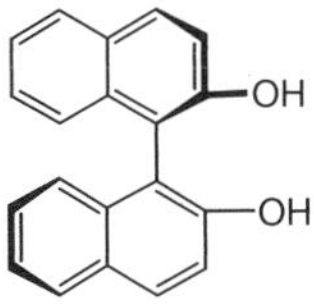

27, [43,44,66]

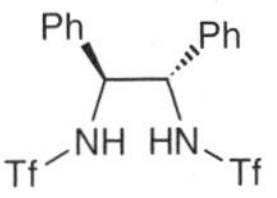

28, [45]

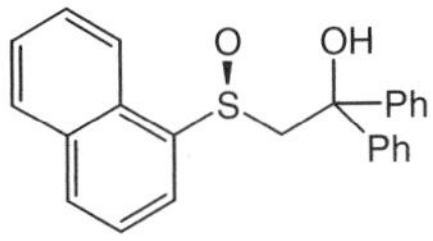

29, [46]

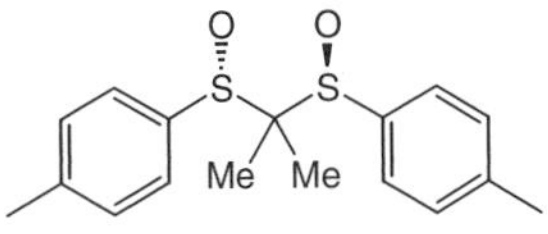

30, [47]

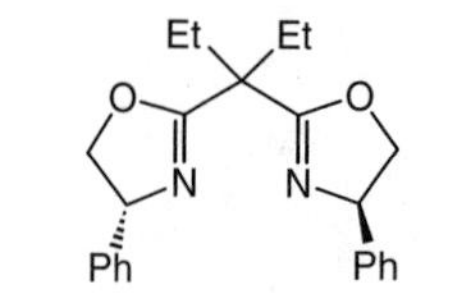

31, [48]

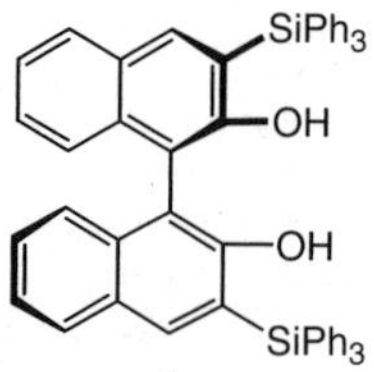

32, [49,50,68]

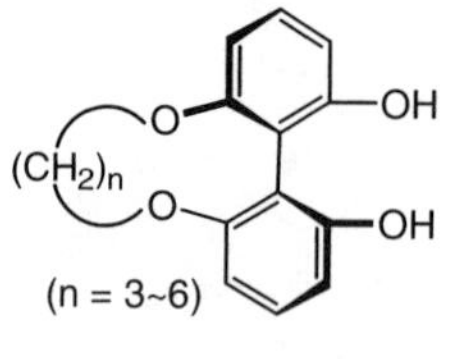

33, [51]
(n = 3~6)

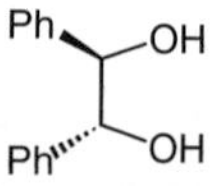

34, [52]

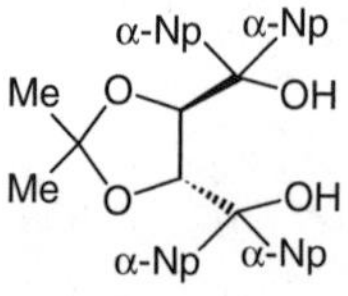

35, [53,54]

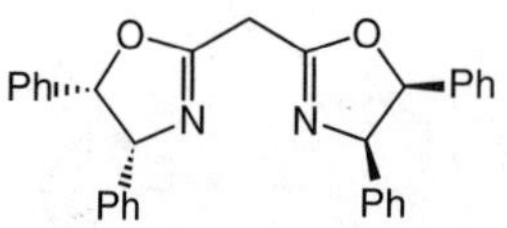

36, [55]

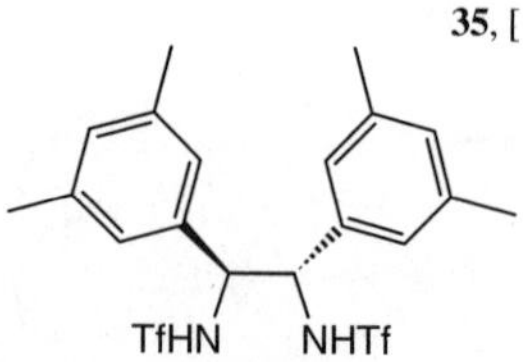

37, [56,57]

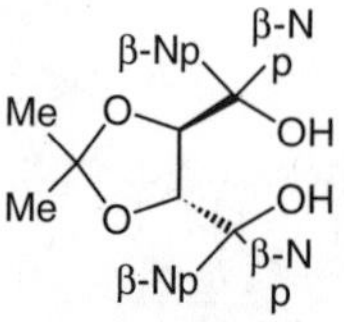

38, [58]

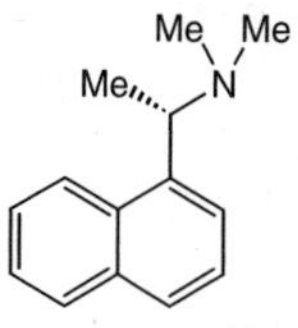

40, [60]

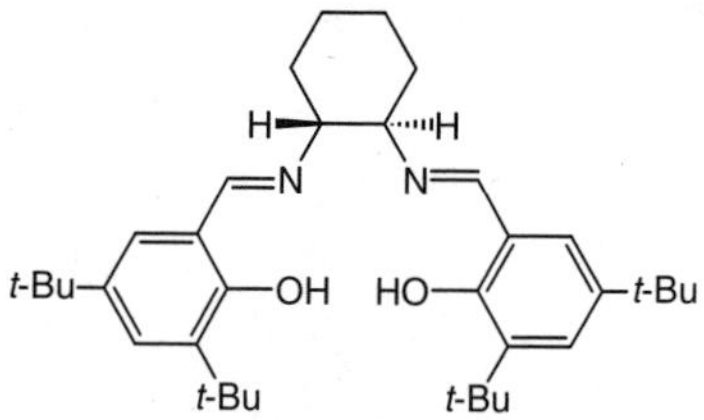

41, [64]

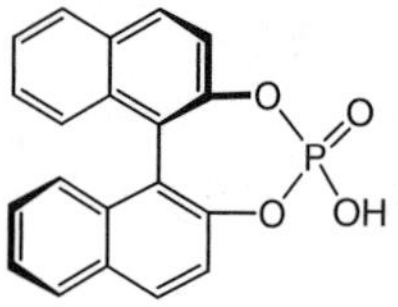

42, [65]

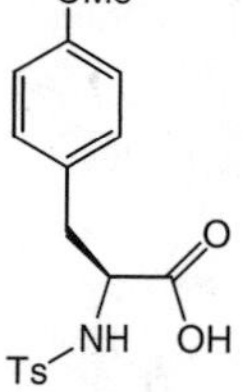

43, [69]

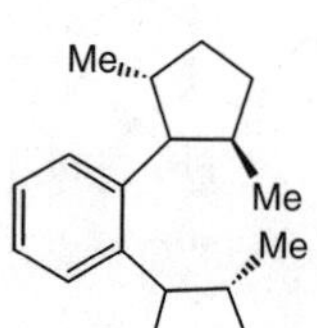

(R,R)-Me-DuPHOS
44, [71]

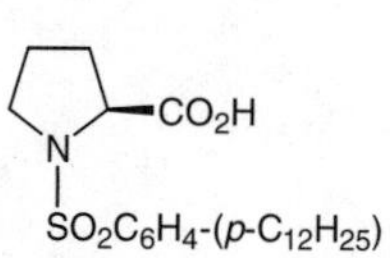

45, [72]

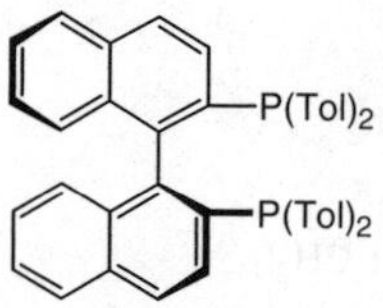

46, [73]

CHAPTER 8B.1

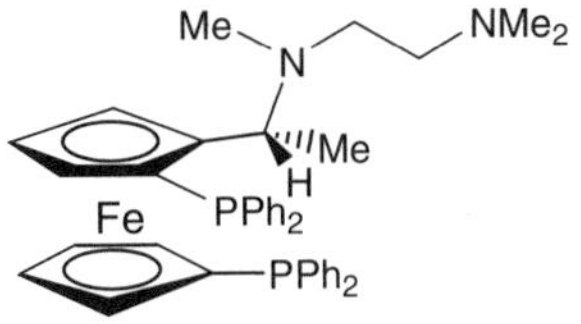

(*R*)-(*S*)-**BPPF**NMeCH₂CH₂NMe₂ [4]

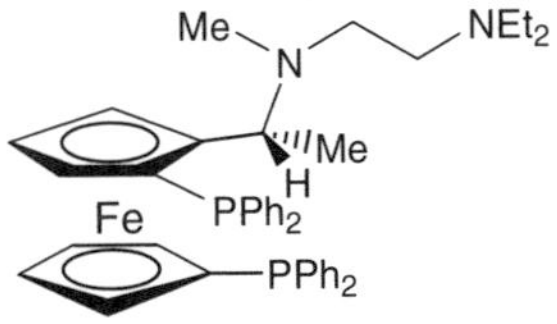

(*R*)-(*S*)-**BPPF**NMeCH₂CH₂NEt₂ [4]

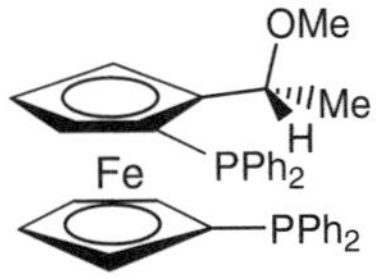

(*R*)-(*S*)-**BPPF**OMe [4]

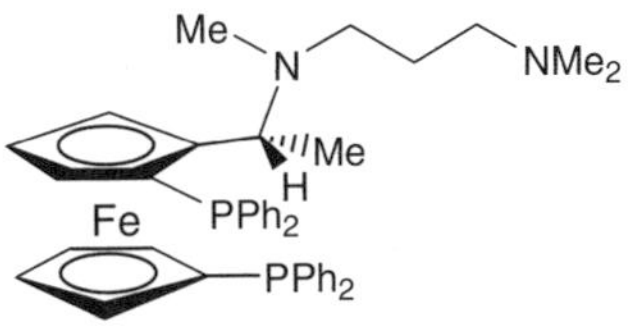

(*R*)-(*S*)-**BPPF**NMe(CH₂)₃NMe₂ [4]

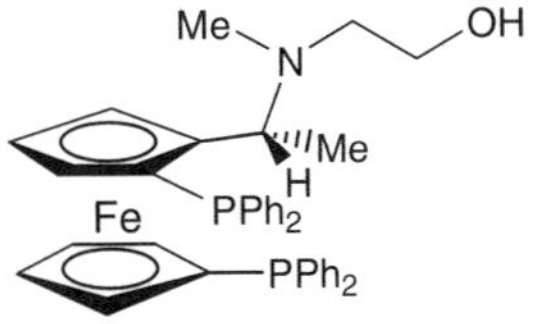

(*R*)-(*S*)-**BPPF**NMeCH₂CH₂OH [4]

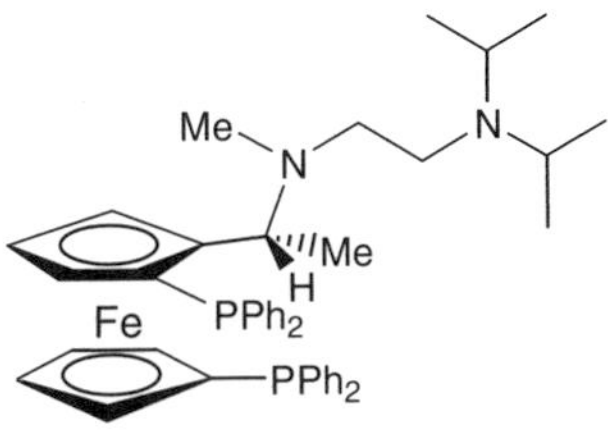

(*R*)-(*S*)-**BPPF**NMeCH₂CH₂N(*i*-Pr)₂ [5]

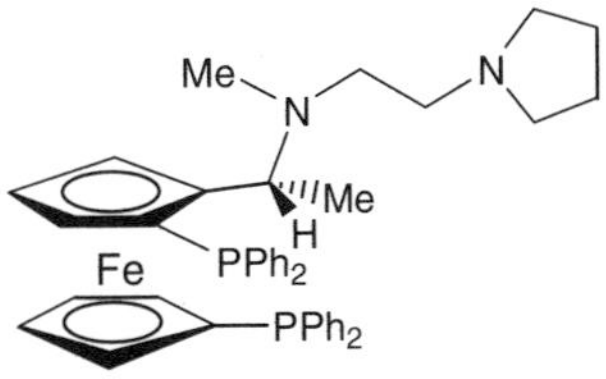

(*R*)-(*S*)-**BPPF**NMeCH₂CH₂(pyrrolidine) [5]

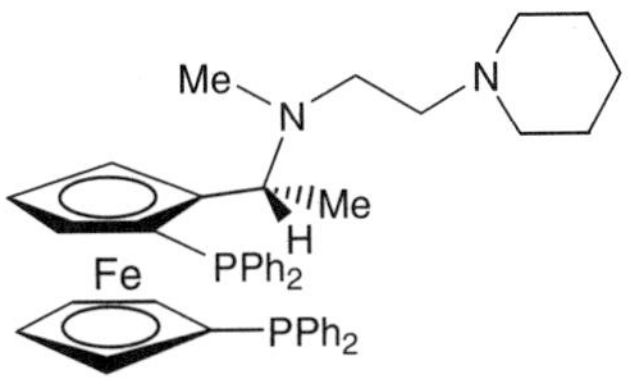

(*R*)-(*S*)-**BPPF**NMeCH₂CH₂(piperidine) [5]

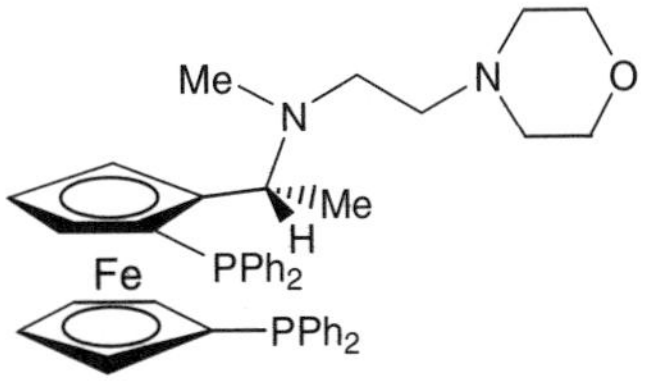

(*R*)-(*S*)-**BPPF**NMeCH₂CH₂(morpholine) [5]

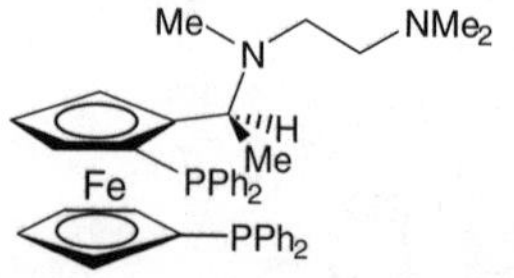

(*S*)-(*S*)-**BPPFNMeCH₂CH₂NMe₂** [23]

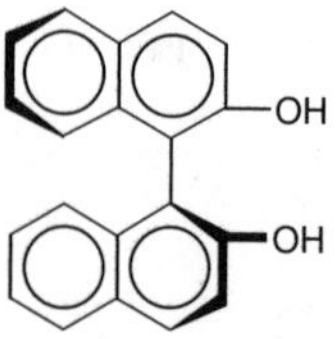

(*S*)-**Binaphthol**

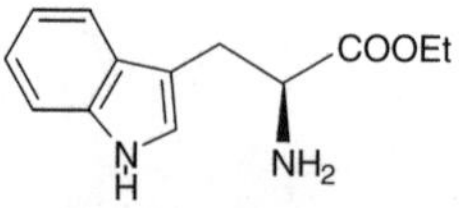

Tryptophan Ethyl Ester

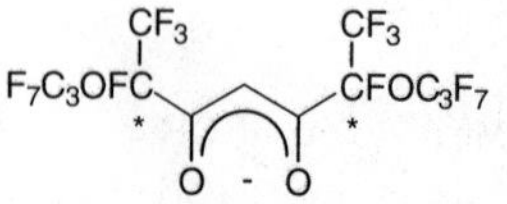

DPPM [30]

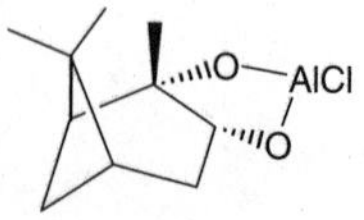

[33]

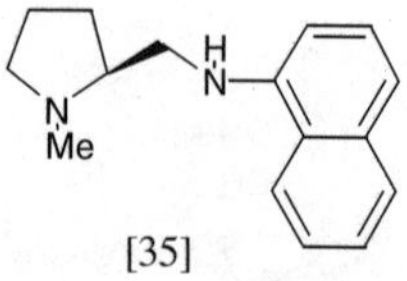

[35]

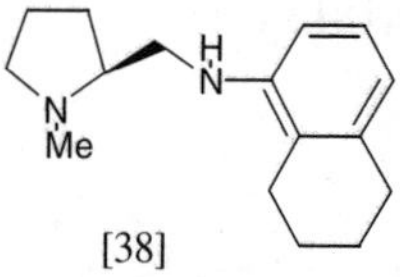

[38]

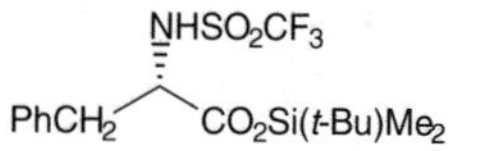

[42]

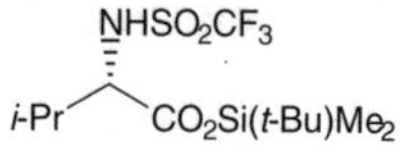

[42]

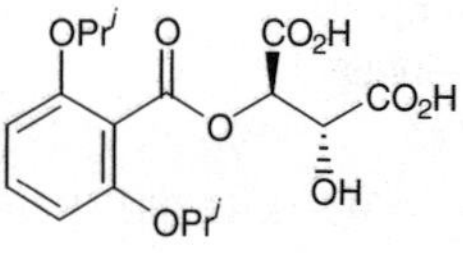

[43]

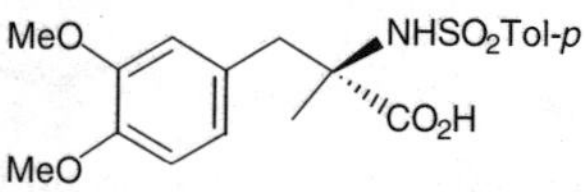

[45]

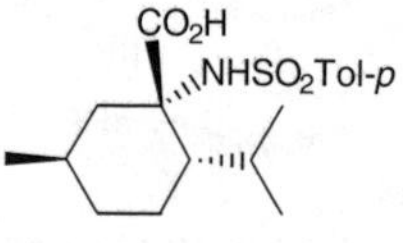

[45]

For **BPPFA**, **CHIRAPHOS**, **DIOP**, **TolBINAP**, and **NORPHOS**, see the Ligand list for Chapter 1.

CHAPTER 8B.2

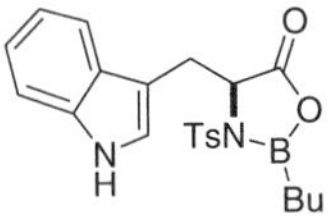

[5]

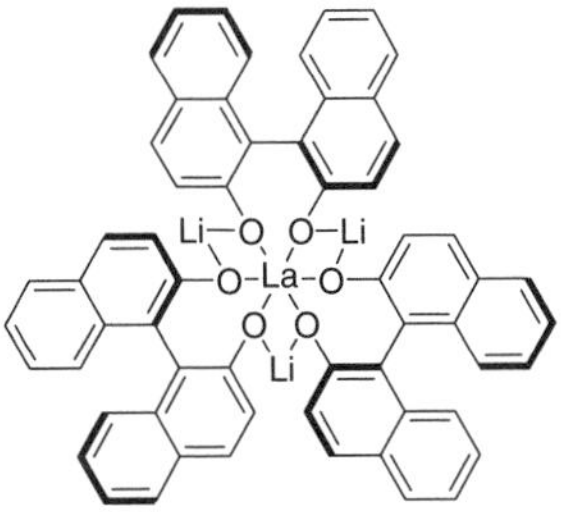

[9]

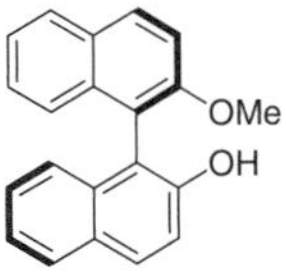

[11]

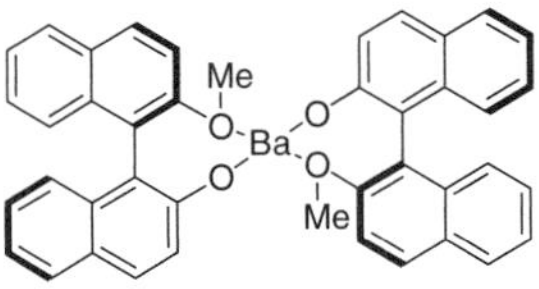

[11]

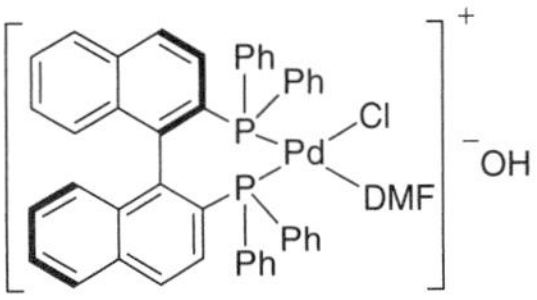

[13]

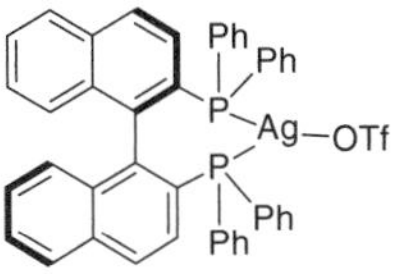

[14]

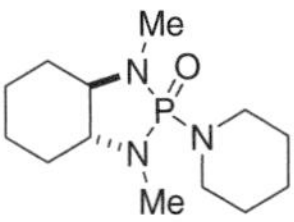

[15]

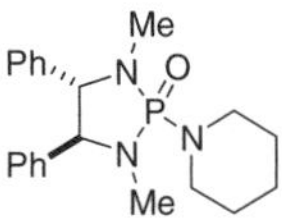

[15]

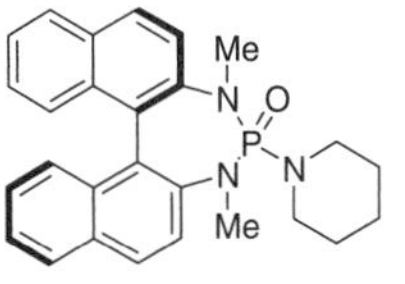

[15]

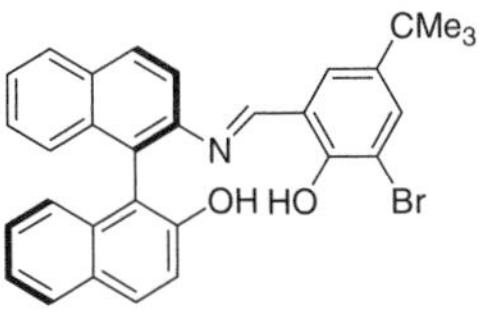

[16]

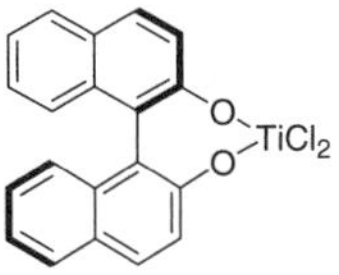

[17]

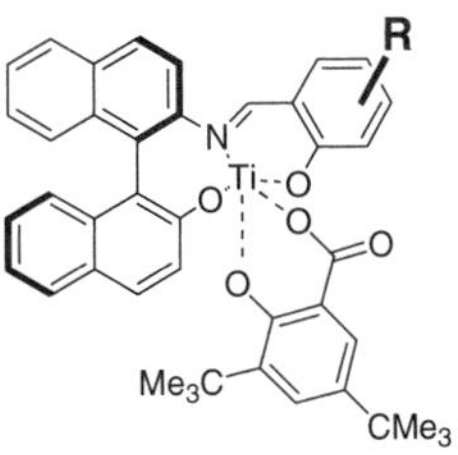

[16]

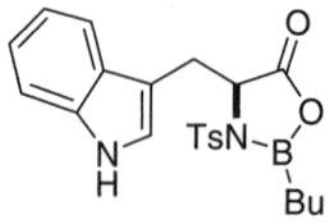

[5]

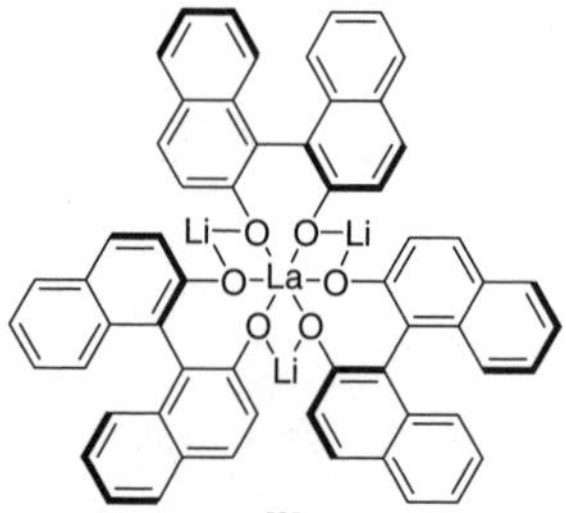

[9]

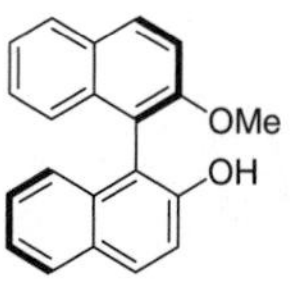

[11]

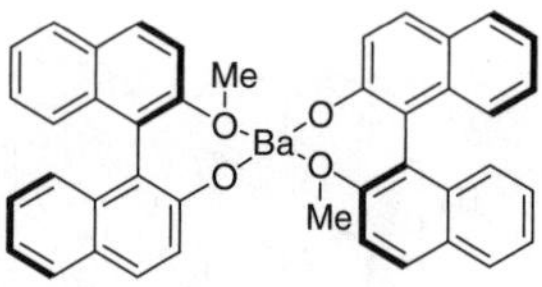

[11]

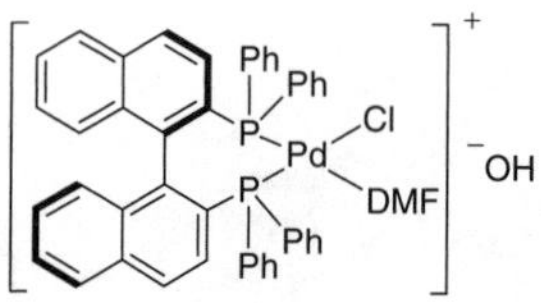

[13]

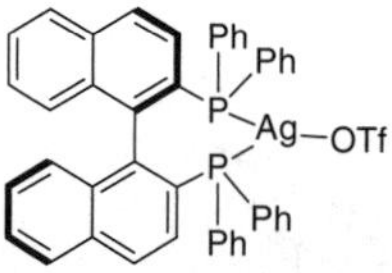

[14]

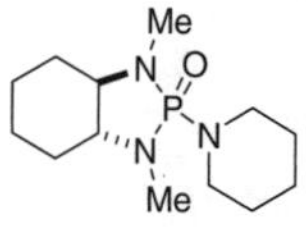

[15]

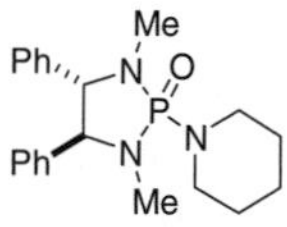

[15]

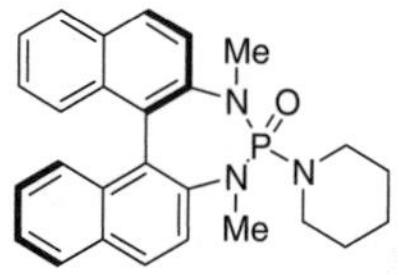

[15]

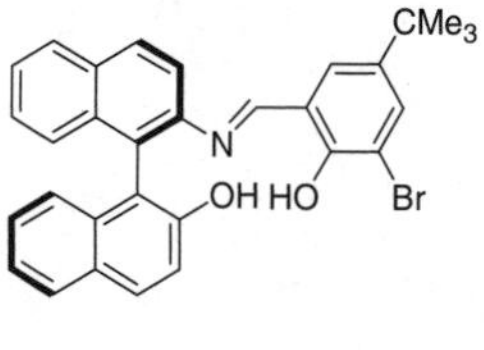

[16]

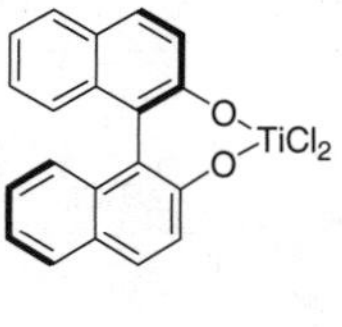

[17]

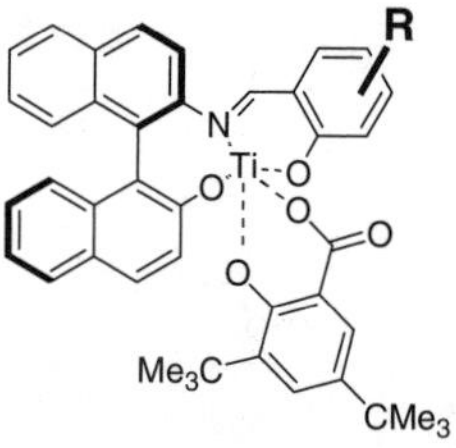

[16]

CHAPTER 8C

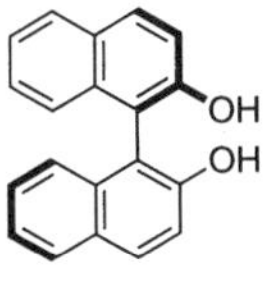

(*R*)-**BINOL** [9]

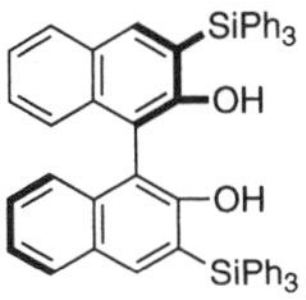

[6]

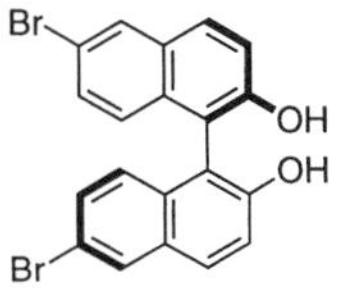

(*R*)-**6-Br-BINOL** [13]

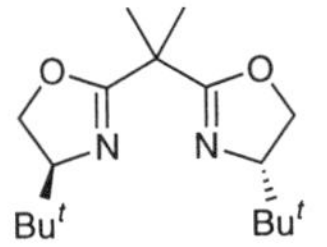

[26]

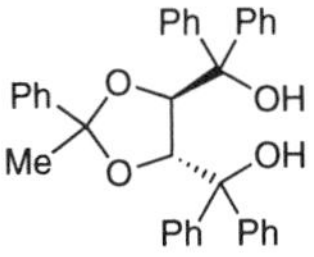

TADDOL [48]

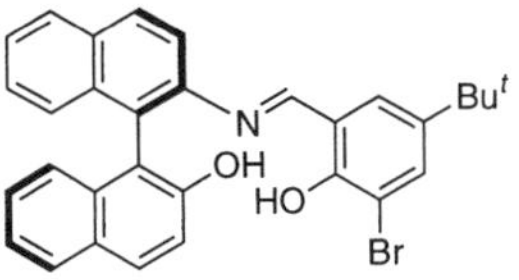

[64]

CHAPTER 8D

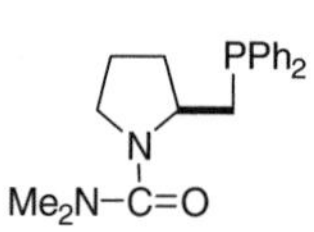

1 [4] 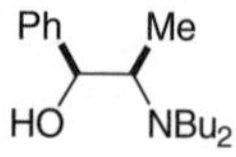   **2** [8] 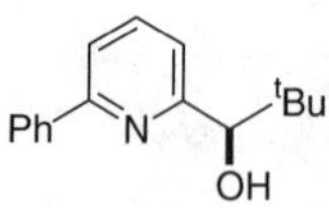   **3** [9] 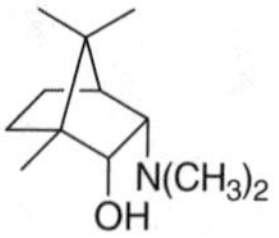   **4** [10] 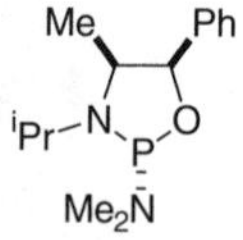   **5** [11]

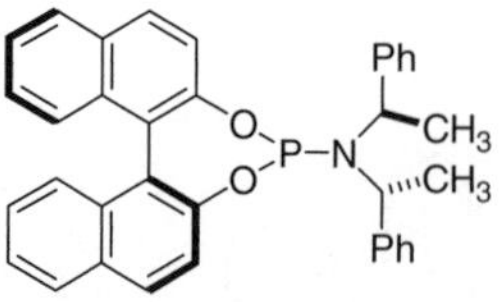

6 [12] 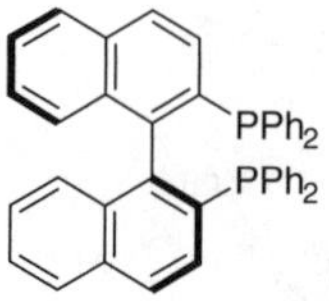   **BINAP** [13] 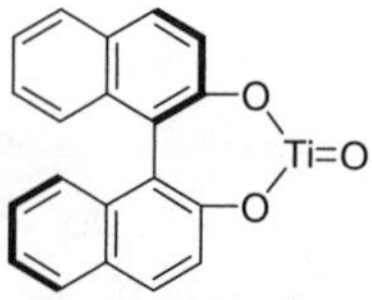   **7** [14]

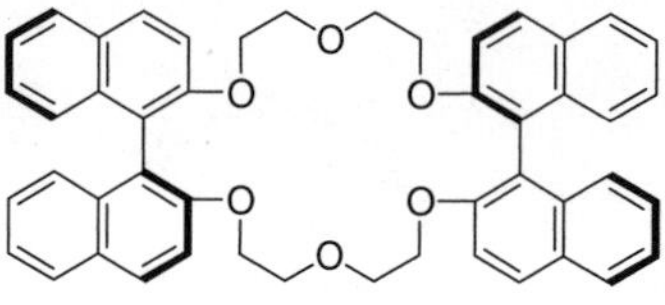

8 [16]

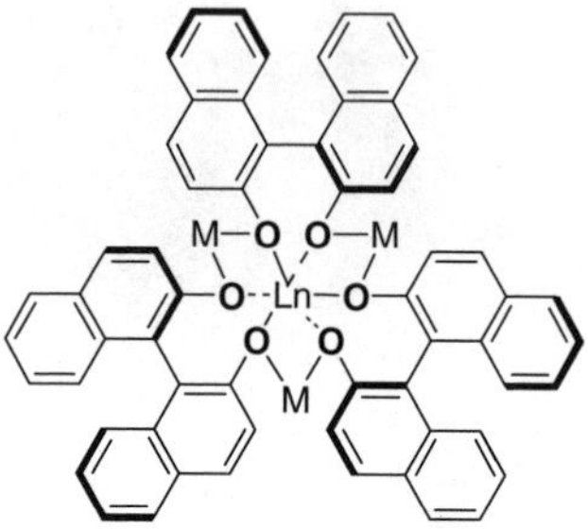

LnMB [17, 18]

9 (LSB): Ln = La, M = Na
(LLB): Ln = La, M = Li
(LPB): Ln = La, M = K
(PrSB): Ln = Pr, M = Na
(SmSB): Ln = Sm, M = Na
(GdSB): Ln = Gd, M = Na

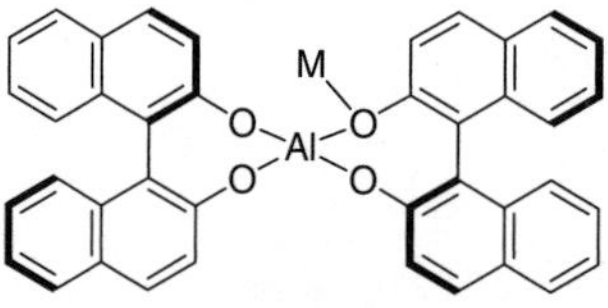

AMB [17, 23]

10 (ALB): M = Li
(ASB): M = Na
(APB): M = K

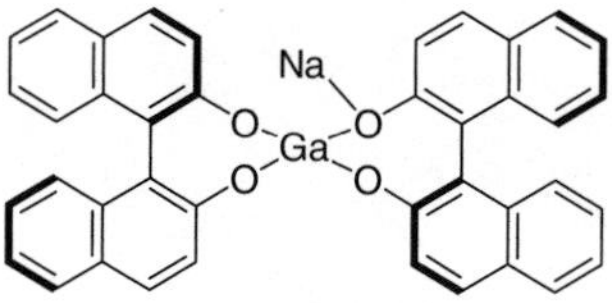

GaSB [17]

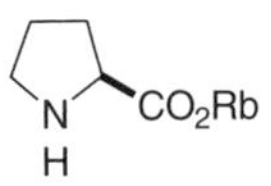

27 [32]

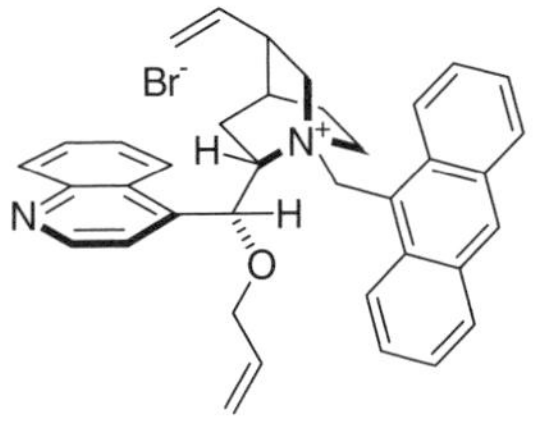

28 [33]

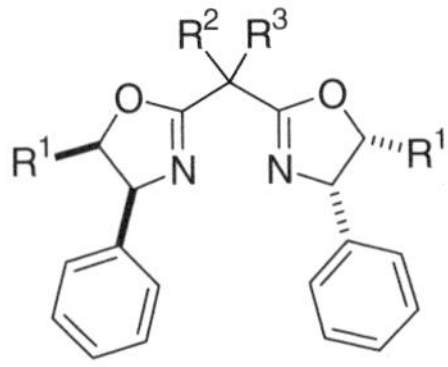

30a R₁ = H, R₂ = R₃ = CH₃
30b R₁ = H, R₂, R₃ = (CH₂CH₂)
30c R₁ = CH₃, R₂ = R₃ = CH₃
30d R₁ = CH₃, R₂, R₃ = (CH₂CH₂)
       [36, 38]

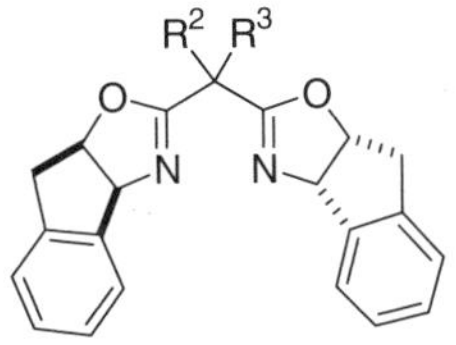

30e R₁ = R₂ = CH₃
30f R₁, R₂ = (CH₂CH₂)
30g R₁ , R₂ = (CH₂CH₂CH₂)
30h R₁ , R₂ = (CH₂CH₂CH₂CH₂)
       [36, 38]

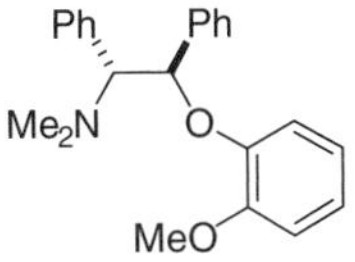

31 [39]

CHAPTER 8E

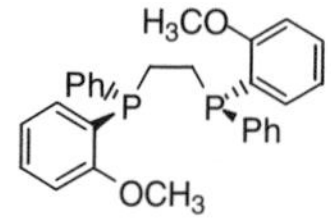

(R,R)-DiPAMP
(1) [2]

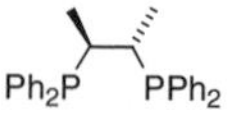

(S,S)-chiraphos
(2) [142, 206]

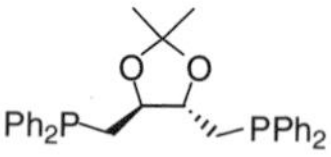

DIOP
(3) [12, 149]

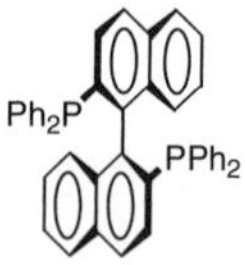

(R)-BINAP
(4) [43]

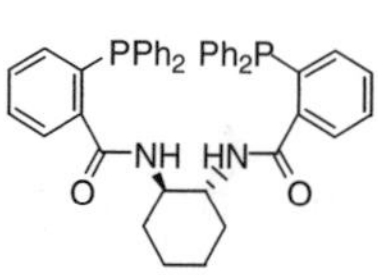

5 [56]

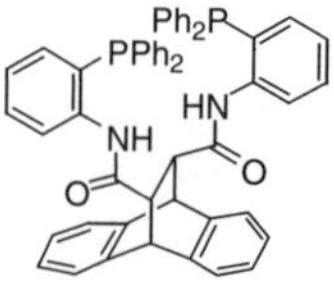

6 [57]

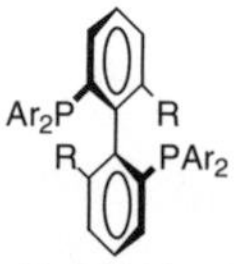

7a R = CH₃, Ar = Ph,
(R)-DiPHEMP [203]
7b R = OCH₃,
Ar = 2,4-t-Bu-C₆H₄

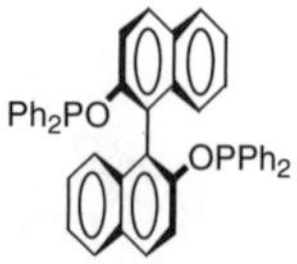

(R)-BINAPO
(8) [73, 65]

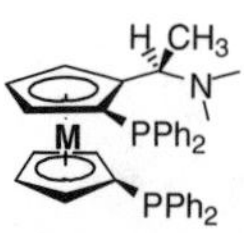

9a M = Fe, **(R,S)-BPPFA** [46]
9b M = Ru, **(R,S)-BPPRA** [194]

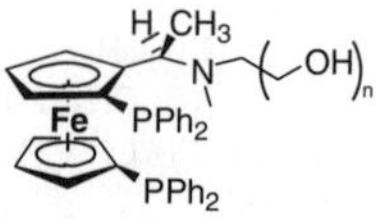

10a n = 1 [181]
10b n = 2 [79]
10c n = 3 [46]

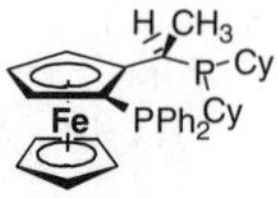

Josiphos
(11) [36]

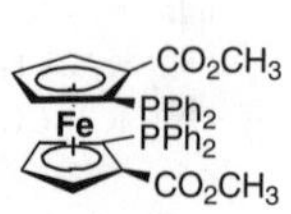

12 [97]

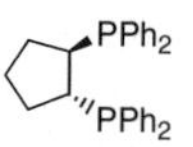

13 [203]

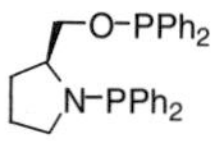

proliphos
(14) [193]

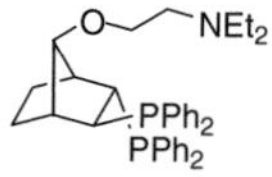

15 [101]

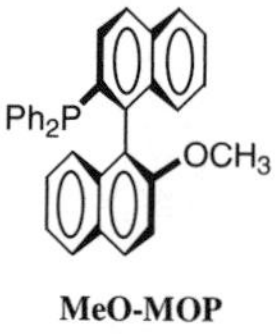

MeO-MOP
(16) [154]

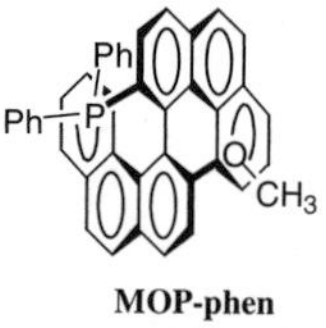

MOP-phen
(17) [196]

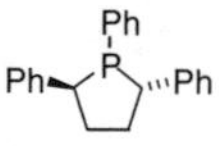

18 [192]

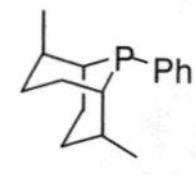

19 [110]

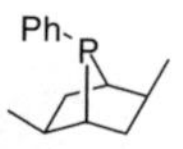

20 [111]

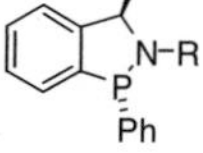

21 [89]

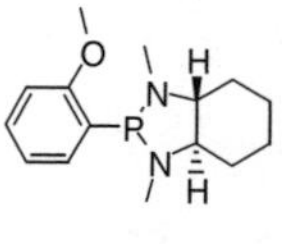

22 [109]

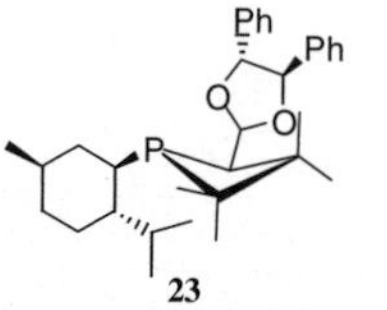

23
[*J. Organomet. Chem.* **1997**, *529*, 465]

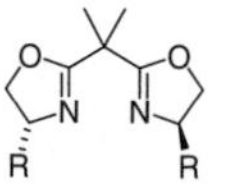

24a R = t-Bu
24b R = Bn [48]

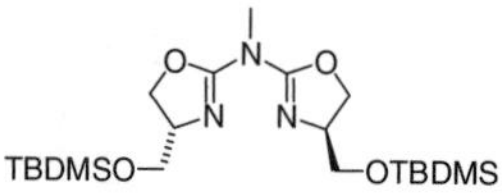

25 [85]

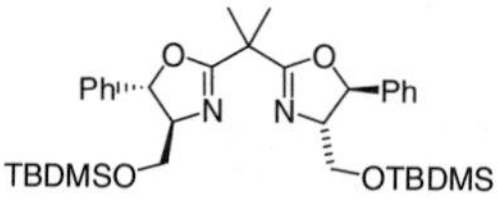

26 [48]

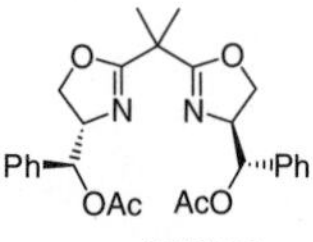

27 [102]

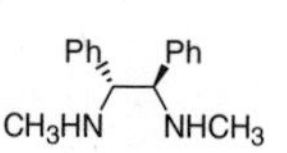

28 [86]

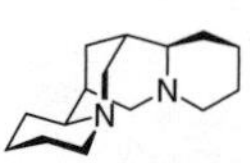

(−)-sparteine
(29) [77]

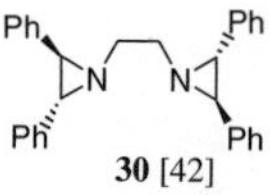

30 [42]

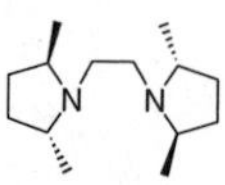

31 [90]

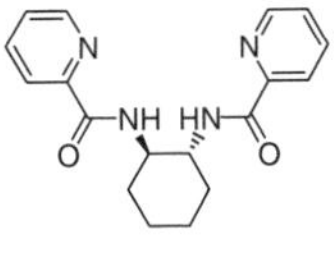

32 [210]

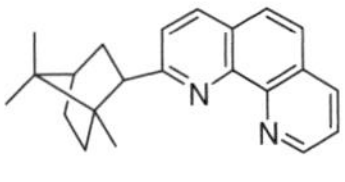

33 [117]

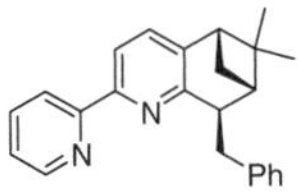

34
[*Tetrahedron: Asymmetry* **1998**, *9*, 531]

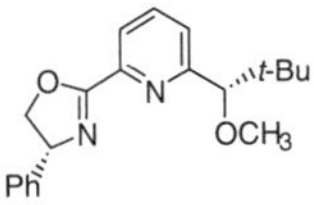

35 [107]

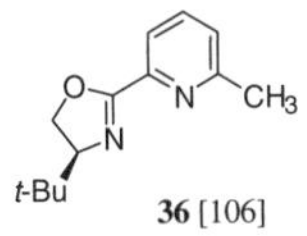

36 [106]

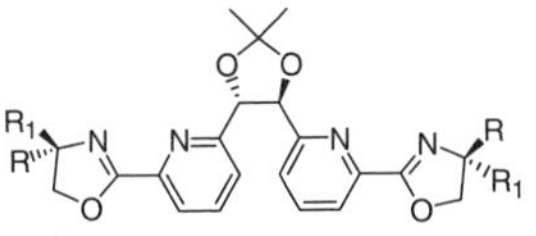

37a R = *i*-Pr, R₁ = H [105]
37b R = CH₃, R₁ = CH₃

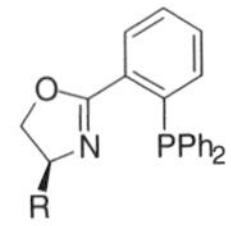

38a R = *i*-Pr [80-82]
38b R = Ph
38c R = *t*-Bu

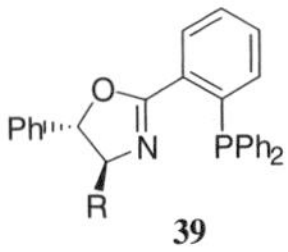

39
[*Synlett* **1994**, 551]

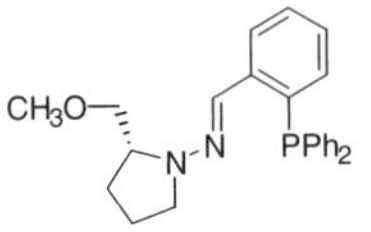

40 [96]

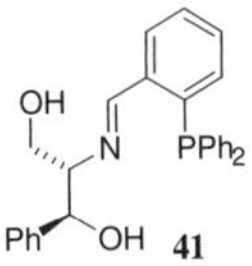

41
[*Inorg. Chim. Acta* **1993**, *220*, 63]

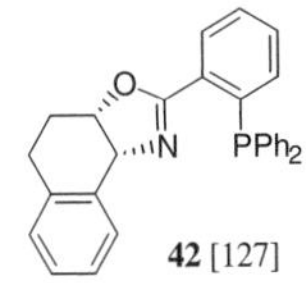

42 [127]

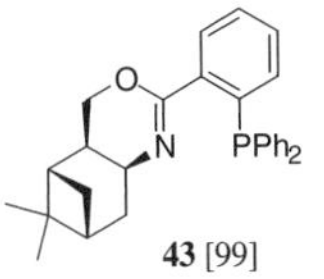

43 [99]

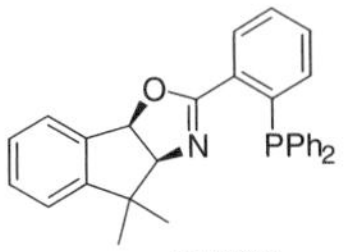

44 [122]

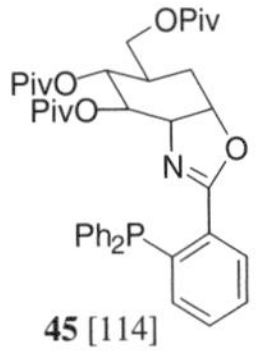

45 [114]

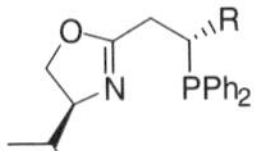

46 [100]
R = H, Ph

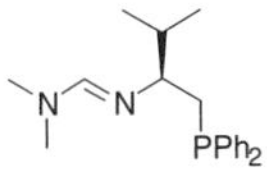

47 [98]

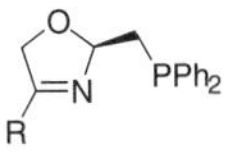

48 [91]
R = Adamantyl

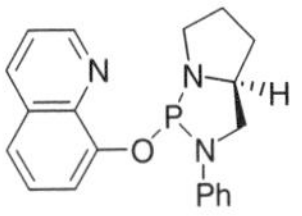

49 [70, 123]

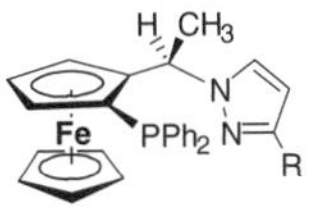

50 [118, 124]
R = Adamantyl

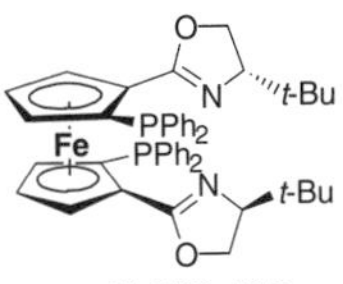

51 [103, 104]

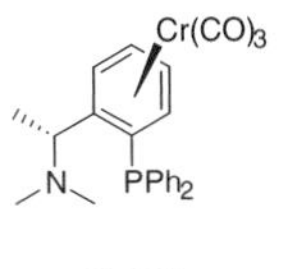

52 [93]

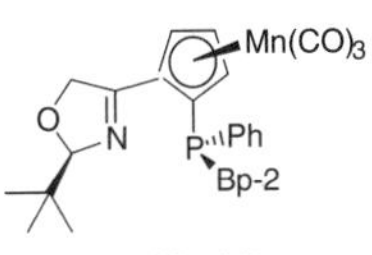

53 [134]

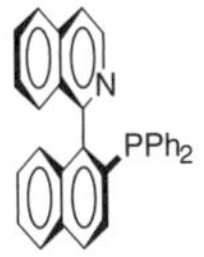

54 [43]

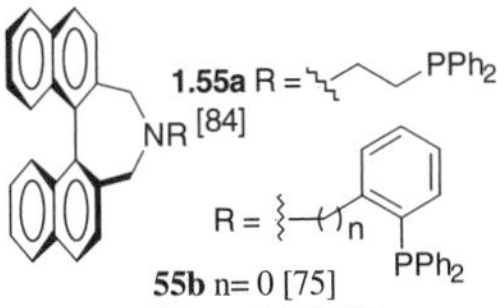

1.55a R = ⎔PPh₂ [84]

55b n= 0 [75]
55c n= 1 [76, 92]

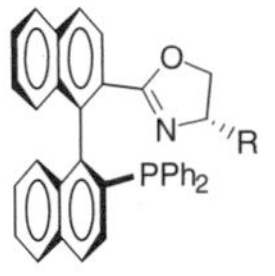

56a R = *i*-Pr [112]
56b R = *t*-Bu [113]

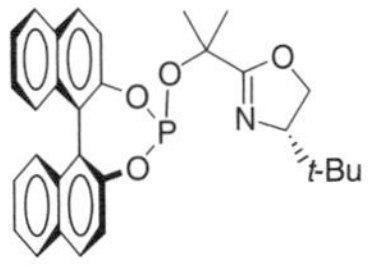

57 [152]

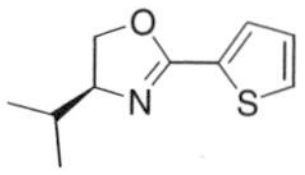

58 [131]

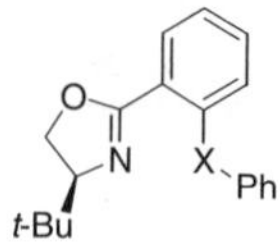

59a: S [83]
59b: SO [88]
59c: Se [35]

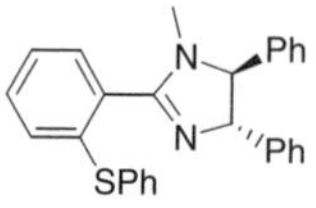

60 [108]

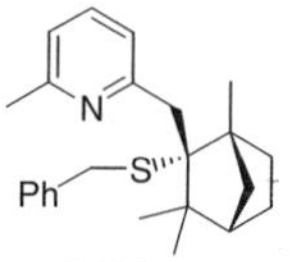

61 [94]

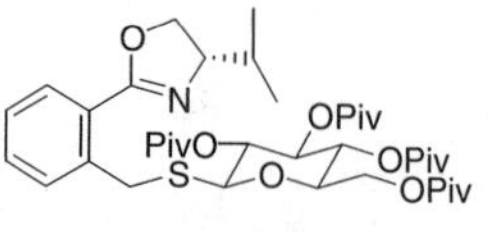

62 [95]

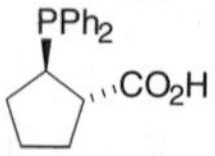

63
[*Tetrahedron: Asymmetry* **1991**, *2*, 667]

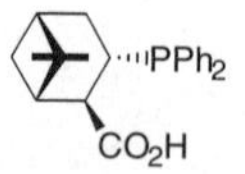

64 [130]

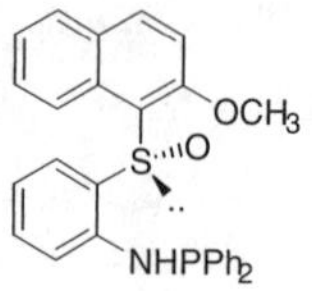

65 [115]

DPPBA and DPPA Derived Ligands [55-58]

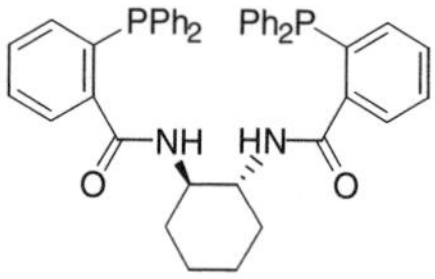

5

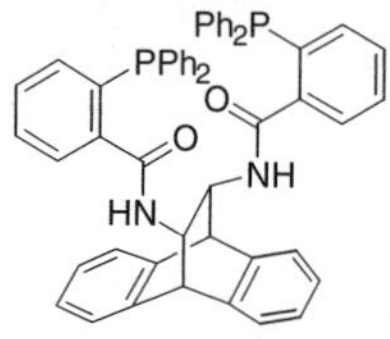

70

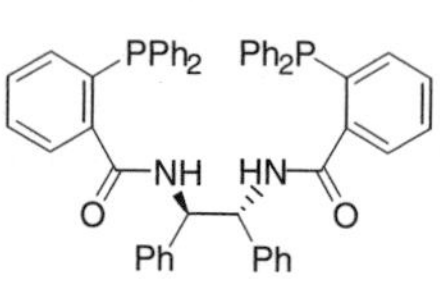

71

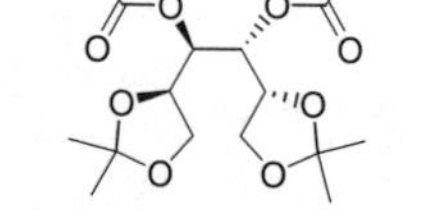

72

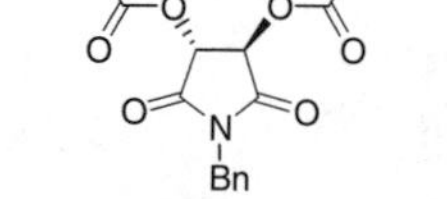

73

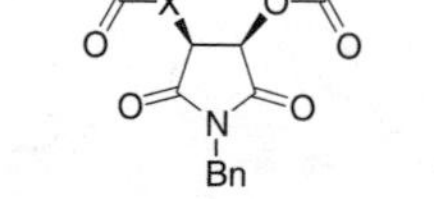

74a X = O
74b X = NH

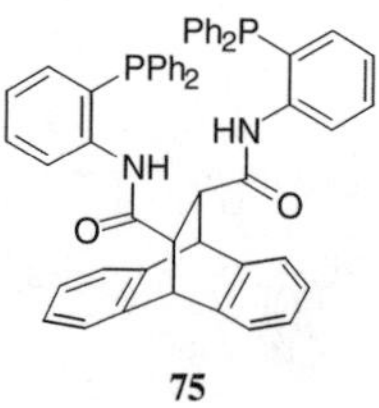

75

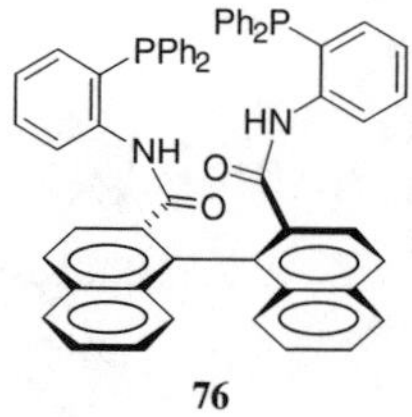

76

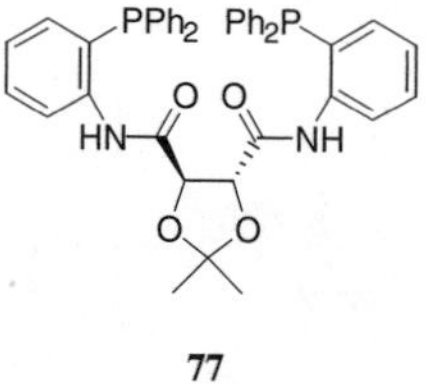

77

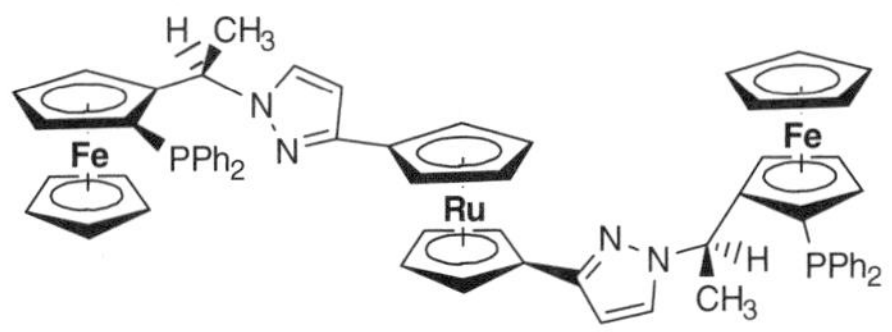

111 [119]

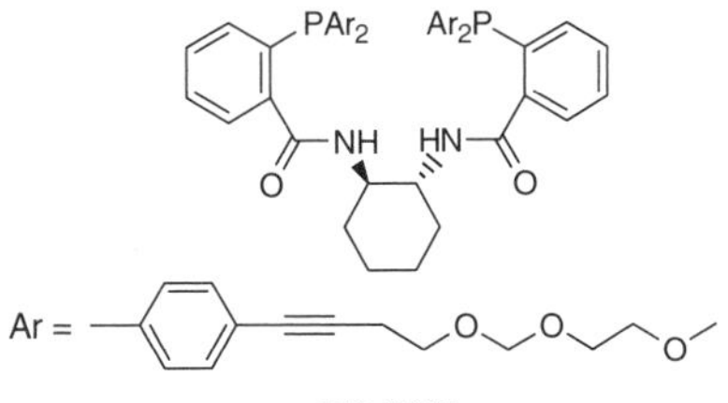

Ar =

112 [133]

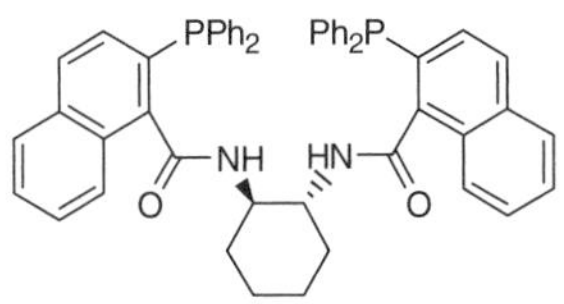

128 [157, 158]

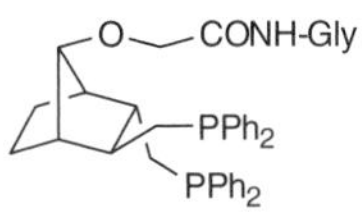

129 [161]

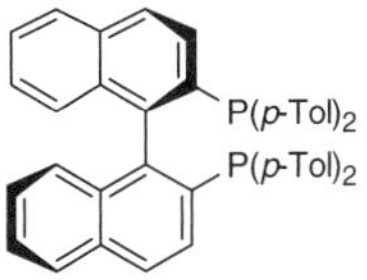

(*S*)-**TolBINAP** [166]

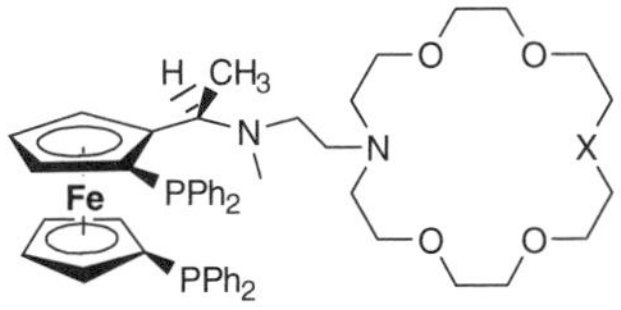

136a, X = NCH₃
136b, X = O [182]

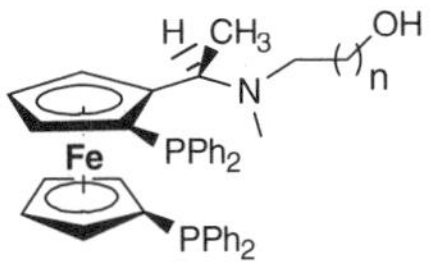

137a, n = 2 [183]
137b, n = 3 [184]

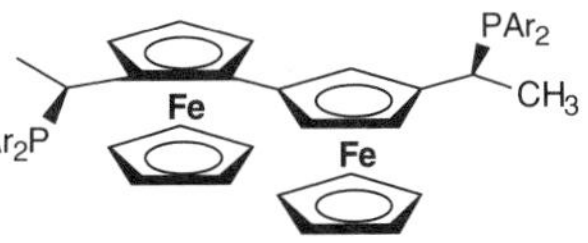

138, Ar = *p*-OMe-C₆H₄ [186]

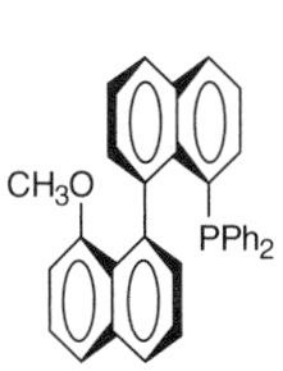

141 [197]

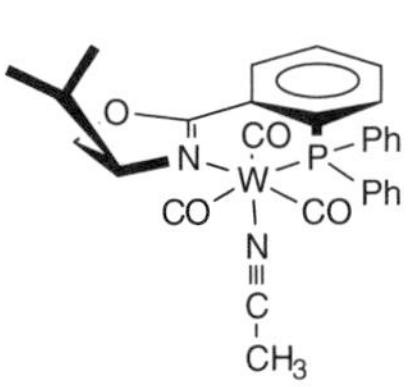

143 [209]

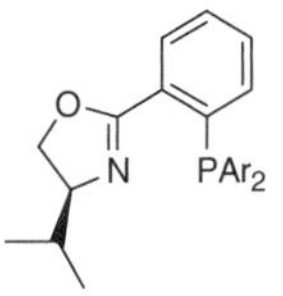

149 Ar = 4-CF₃C₆H₄
[217]

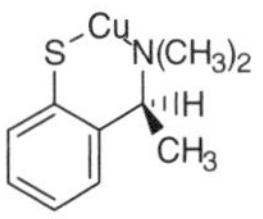

150 [219]

CHAPTER 8F

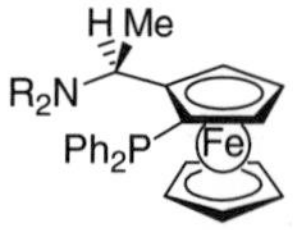

(S)-(R)-**PPFA** (**10a**): NR$_2$ = NMe$_2$ [16,17]

10b: NR$_2$ = N⟨⟩ [17]

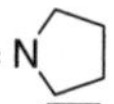

10c: NR$_2$ = N⟨⟩ [17]

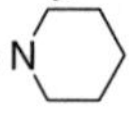

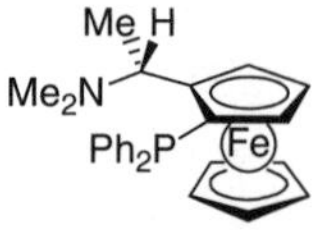

(R)-(R)-**PPFA** (**11**) [17]

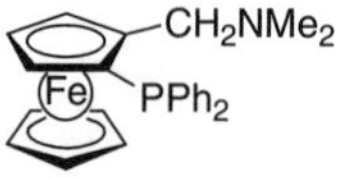

12 [16,17]

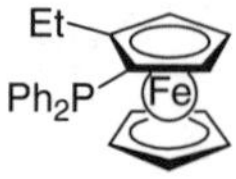

13 [16,17]

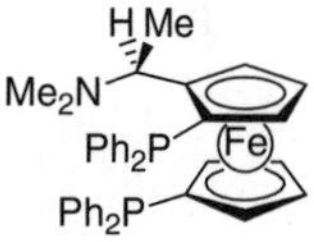

(S)-(R)-**BPPFA** (**14**) [17,19]

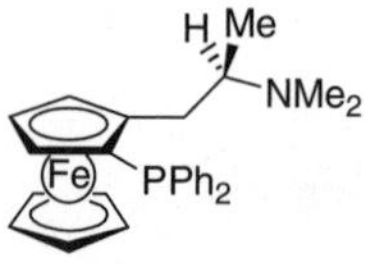

15 [18]

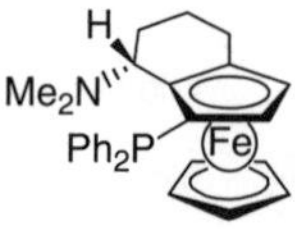

16 [20]

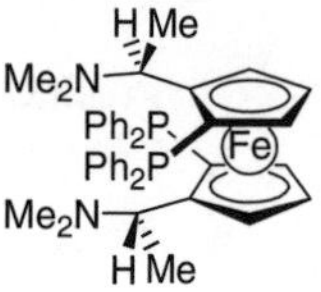

21 [23,24]

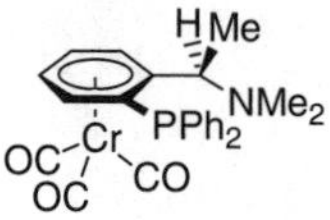

22 [25]

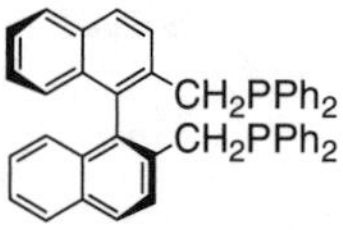

33 [31]

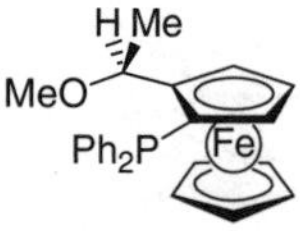

(S)-(R)-**37** [6]

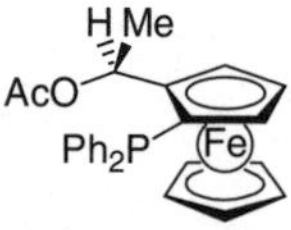

(S)-(R)-**48** [12]

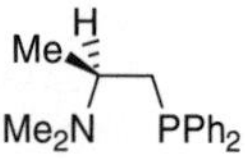

(S)-**alaphos** (**49a**) [33,34]

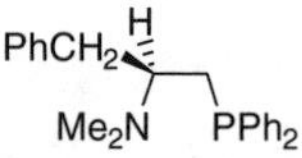

(S)-**phephos** (**49b**) [33,34]

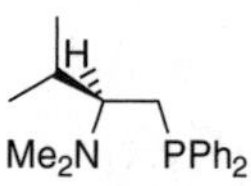

(S)-**valphos** (**49c**) [33,34]

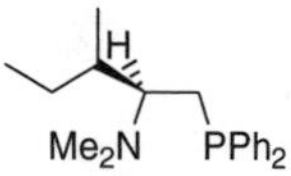

(S)-**ilephos** (**49d**) [33,34]

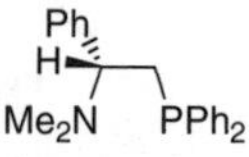

(R)-**phglyphos** (**49e**) [33,34]

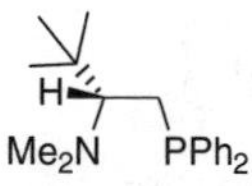

(R)-**t-leuphos** (**49f**) [33,34]

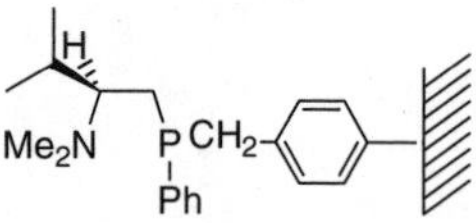

50 [35]

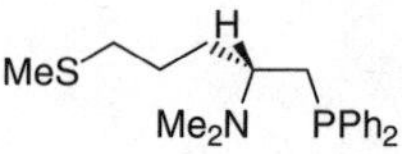

(R)-**51a** [36,37]

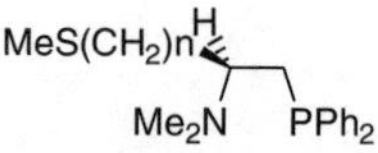

(S)-**51b**: n = 1 [36,37]
(S)-**51c**: n = 2 [36,37]

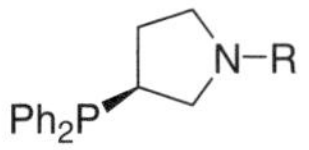

(*S*)-**67a**: R = CH₂Ph [41,42]
(*S*)-**67b**: R = CH₂C₆H₄OMe-*o* [41,42]
(*S*)-**67c**: R = Et [41,42]

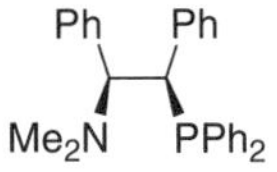

(1*R*,2*S*)-**68** [43]

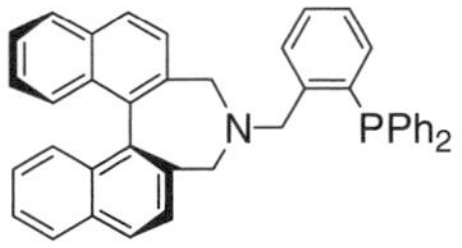

(*S*)-**69** [44,45]

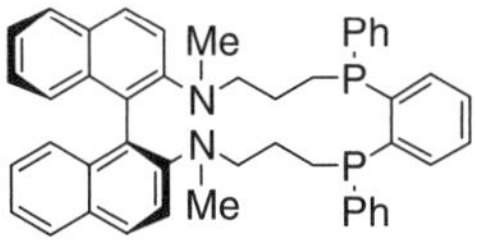

(*S*)-**70** [46]

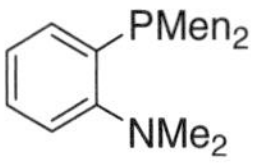

71 [47]

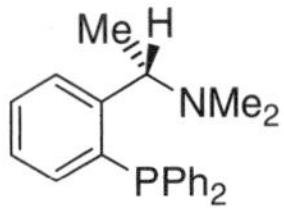

72 [48]

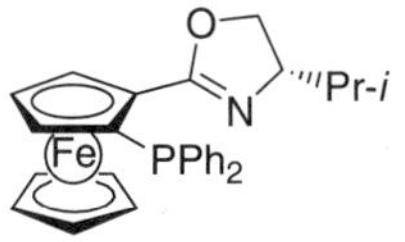

(*S*)-(*S*)-**73** [49]

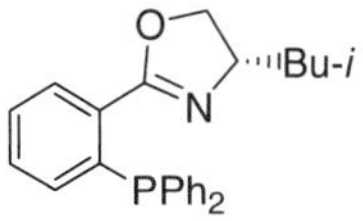

(*S*)-**74** [50]

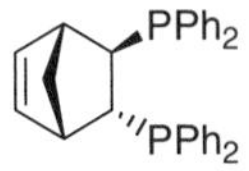

(*R*,*R*)-**norphos** (**75**) [51]

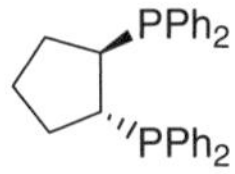

76 [19,52]

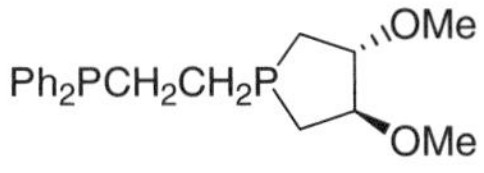

77 [53]

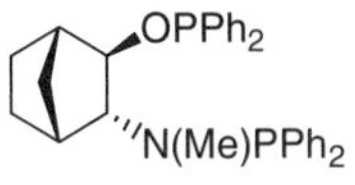

78 [54]

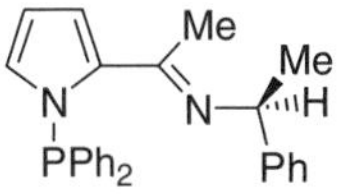

79 [55]

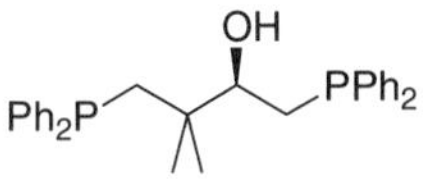

80 [56]

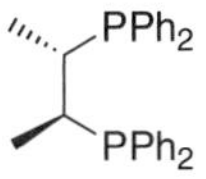

(*S*,*S*)-**chiraphos** (**81**) [57,58]

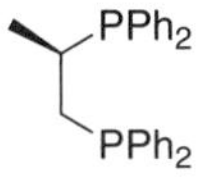

(*R*)-**prophos** (**82**) [57,58]

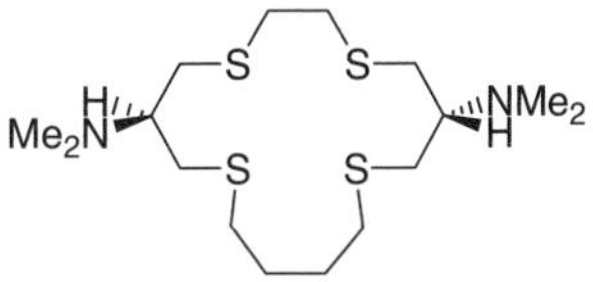

86 [61,62]

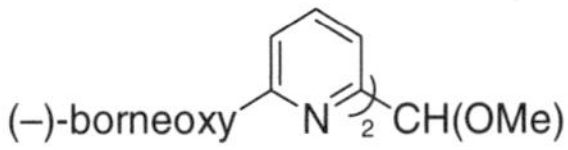

(−)-borneoxy   **87** [63]

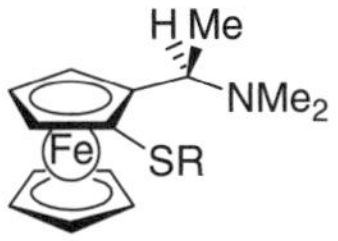

88 [64,65]

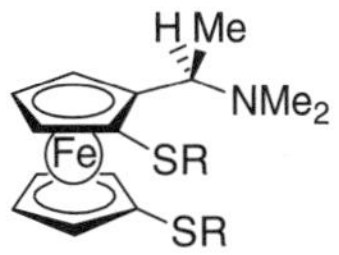

89 [64,65]

CHAPTER 8G

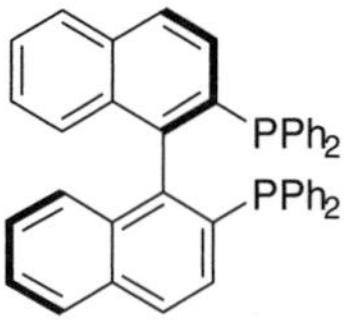

(*R*)-BINAP [11]

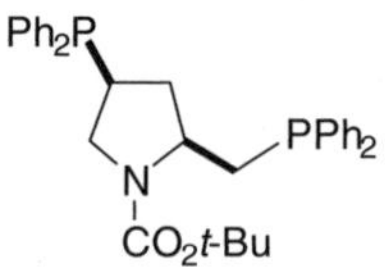

4.3

(*S,S*)-BPPM [5]

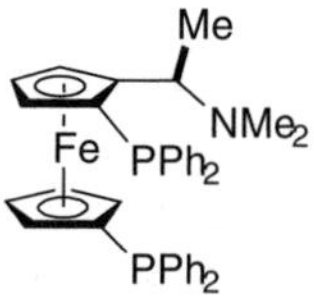

4.4

(*R,S*)-BPPFA [5]

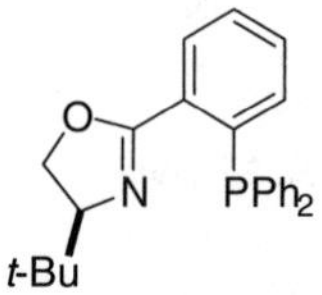

[28]

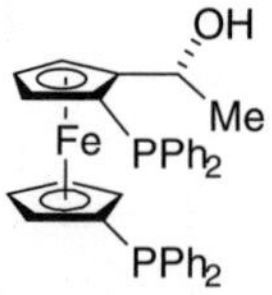

14.4

(*R,S*)-BPPFOH [34]

CHAPTER 9

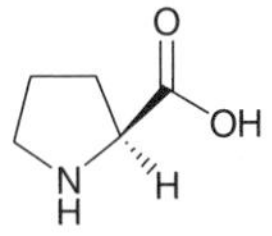

(*S*)-proline [5]

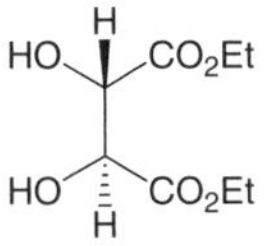

(*R,R*)-**DET** [5]

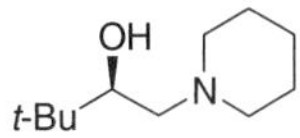

(-)-**5** [10]

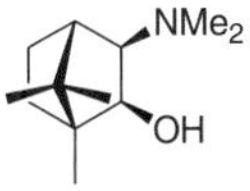

(-)-**6** [11]

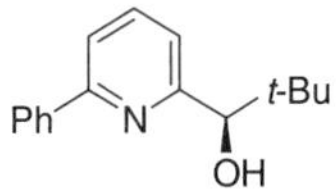

(*R*)-**7** [14]

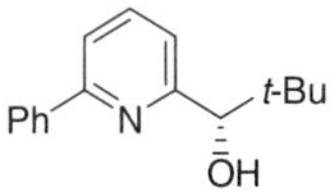

(*S*)-**7** [20]

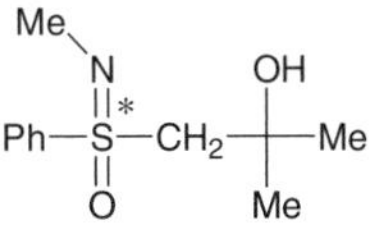

(*S*)-**8** [15]

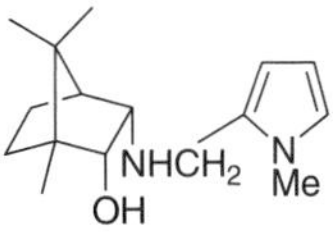

endo-(+)-**9** [16]

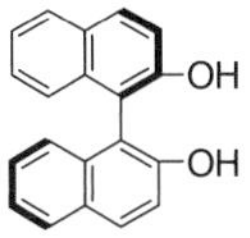

(*R*)-**BINOL**

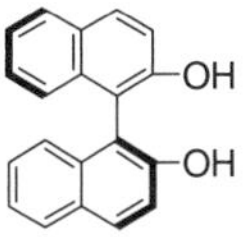

(*S*)-**BINOL**

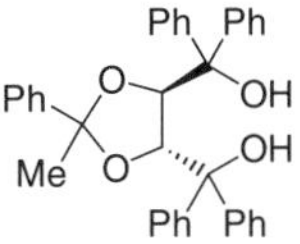

(*R,R*)-**18** [27,28]

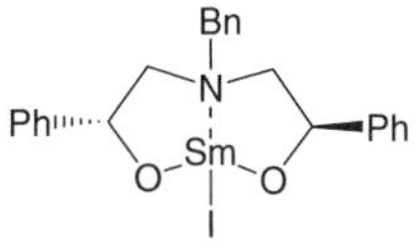

26 [34]

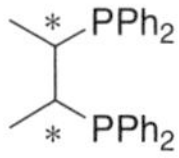

(*S,S*)- and (*R,R*)-
CHIRAPHOS (**28**) [35]

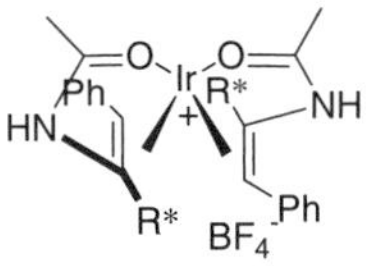

29 [35]
R*= CO$_2$((*R*)-menthyl)

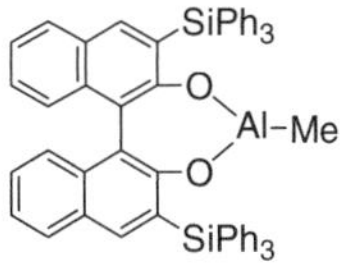

34 [36]

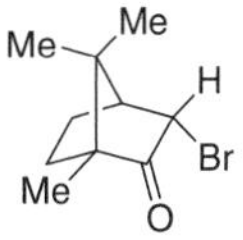

d-3-bromocamphor [36]

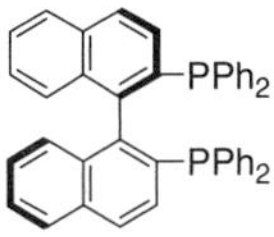

(*R*)-**BINAP** [38]

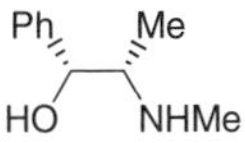

(1*R*,2*S*)-**ephedrine** [38]

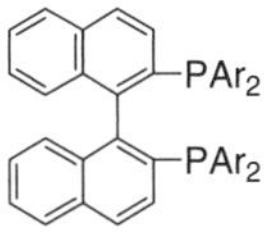

Ar=3,5-dimethylphenyl
39 [40]

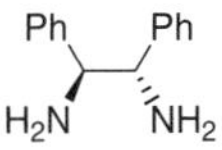

(*S,S*)-**40** [40]

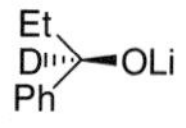

(*R*)-**41** [45]

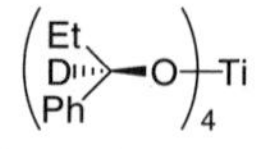

(*R*)-**43** [45]

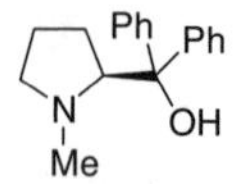

(*S*)-DPMPM (**45**) [48]

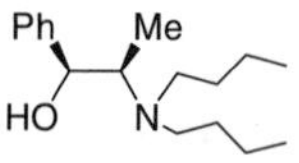

(1*S*,2*R*)-**DBNE** (**46**) [49]

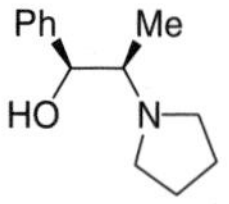

(1*S*,2*R*)-**47** [50]

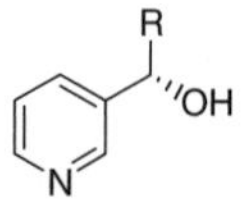

(*S*)-**49a**; R= *i*-Pr [43]
(*S*)-**49b**; R= Et [43]

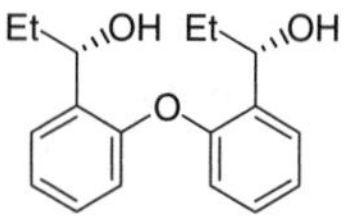

(*S*,*S*)-**52** [52]

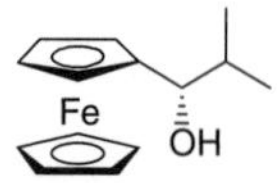

(*S*)-**54** [53]

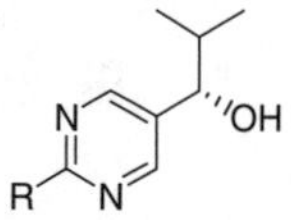

(*S*)-**56a**; R= H [54,56]
(*S*)-**56b**; R= Me [54,57]

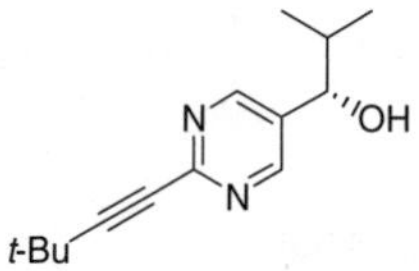

(*S*)-**56c** [55]

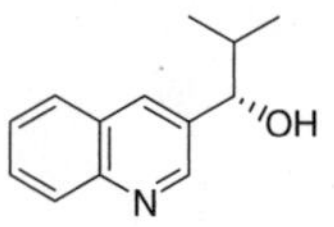

(*S*)-**59** [58,59]

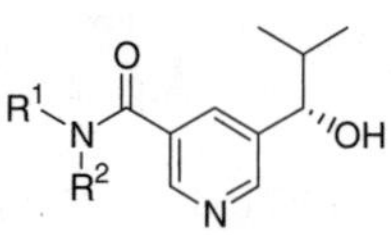

(*S*)-**61** (R^1=R^2= *i*-Pr) [60,61]

CHAPTER 10

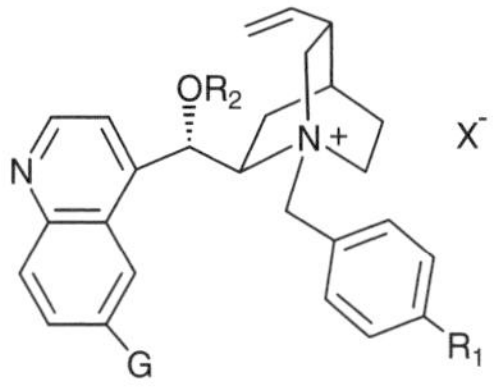

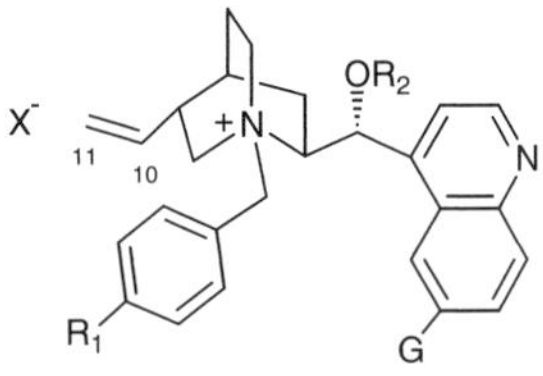

3a R$_1$ = 4-CF$_3$, R$_2$ = G = H, X = Br [9a] [36] [43] [52] [73] [79]
3b R$_1$ = 3,4-Cl$_2$, R$_2$ = G = H, X = Cl [10]
3c R$_1$ = H, R$_2$ = Allyl, G = H, X = Cl [5p]
3d R$_1$ = R$_2$ = G = H, X = F [32a]
3e R$_1$ = 4-CF$_3$, R$_2$ = G = H, X = F [31]
3f R$_1$ = R$_2$ = G = H, X = Cl [51] [77c]

6a R$_1$ = H, R$_2$ = Benzyl, G = H, X = Br [5p] [12c]
6b 6g R$_1$ = R$_2$ = G = H, X = Cl [12a] [68]
6c R$_1$ = 4-CF$_3$, R$_2$ = G = H, X = Br [37]
6d R$_1$ = 4-CF$_3$, R$_2$ = G = H, X = Br (10,11-dihydro) [11d]
6e R$_1$ = 4-CF$_3$, R$_2$ = H, G = OMe, X = Cl [53]
6f R$_1$ = R$_2$ = H, G = OMe, X = F [55]
6h 6i 6j R$_1$ = R$_2$ = H, G = OMe, X = Cl [47b] [69] [72]

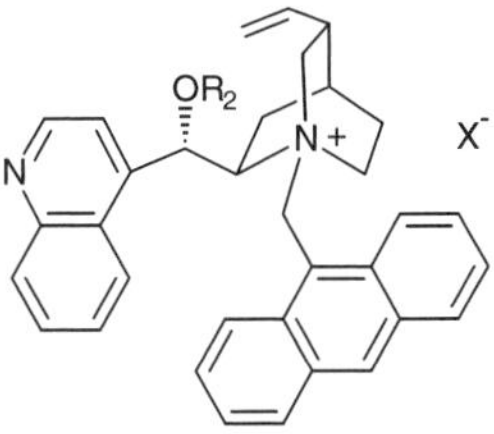

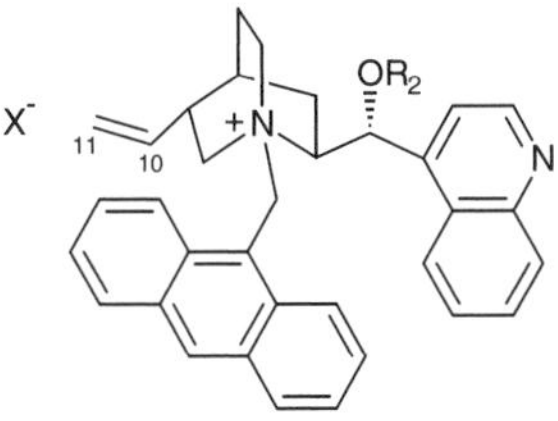

7a R$_2$ = Allyl, X = Br [15]
7b R$_2$ = H, X = Cl [13]
7c R$_2$ = Bn, X = OH [65]

8a R$_2$ = Allyl, X = Br [14] [21]
8b R$_2$ = H, X = Cl (10,11-dihydro) [13]
8c R$_2$ = Bn, X = OH (10,11-dihydro) [65]

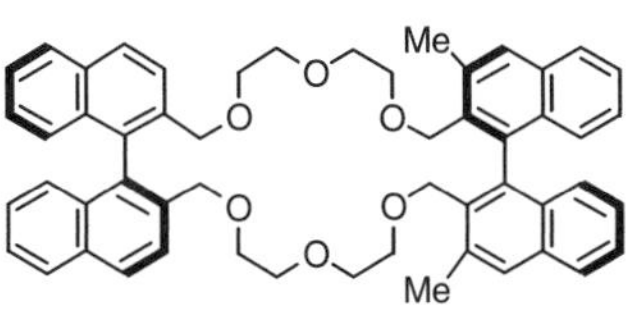

9 [35]

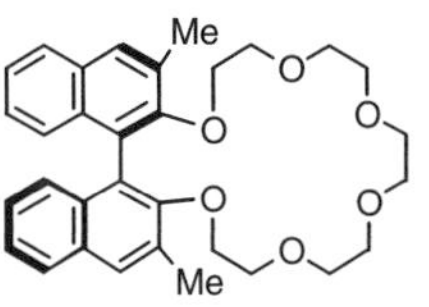

10 [35]

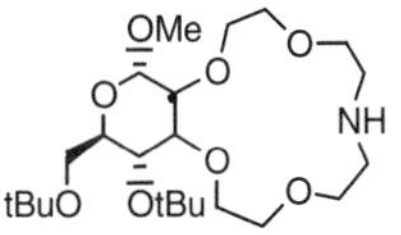

11 [38a]

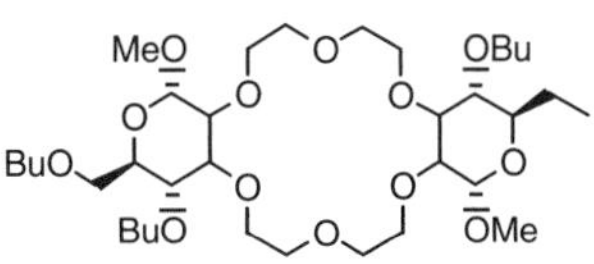

12 [39]

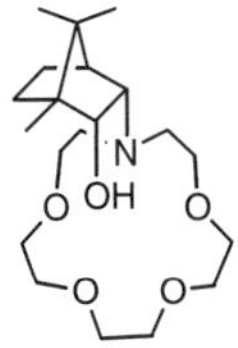

13 [40]

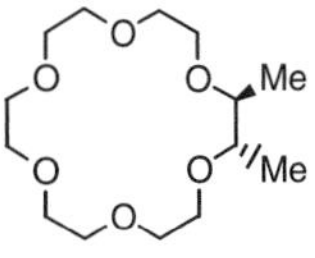

14 [41a]

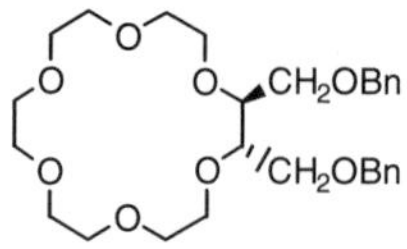

15 [24e]

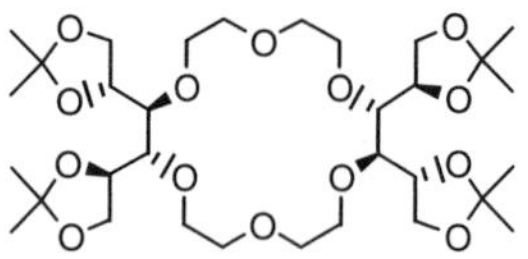

16 [41b]

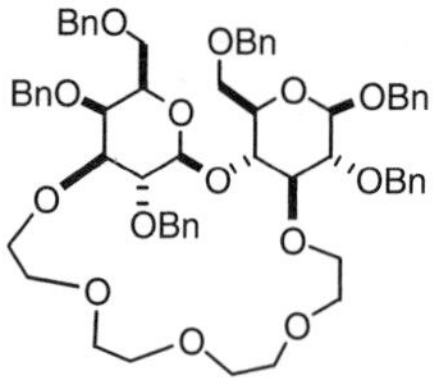

17 [42a]

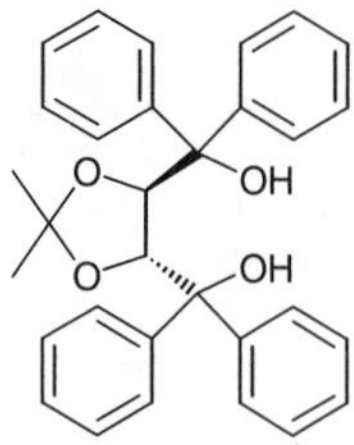

18 [23] [45a]

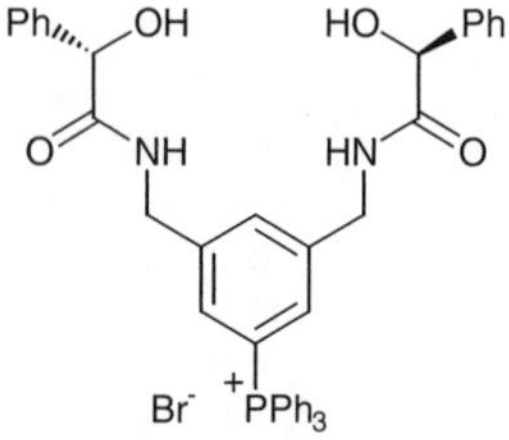

19 [11c]

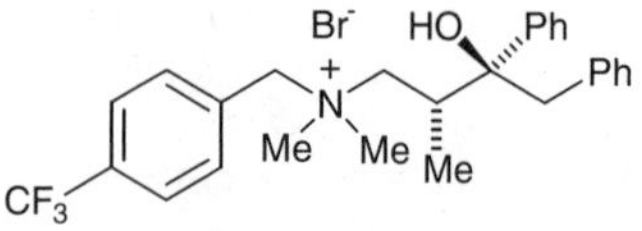

20 [24h]

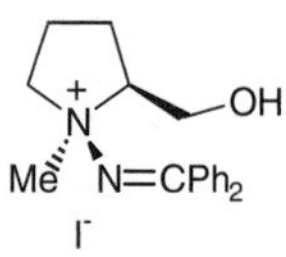

21 [29]

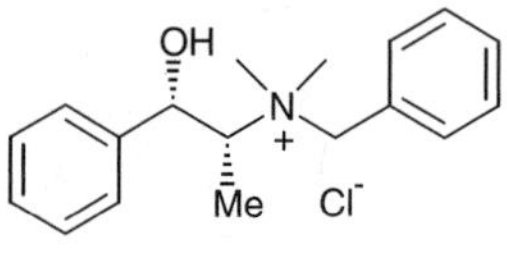

22 [30] [44b]

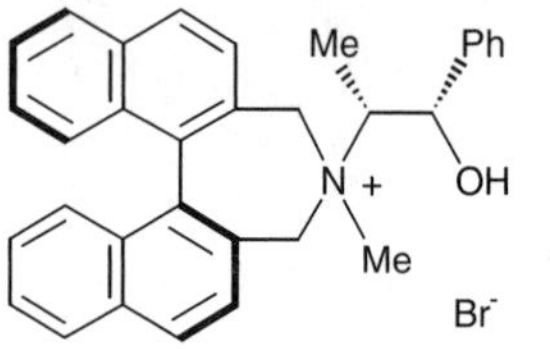

23 [44a]

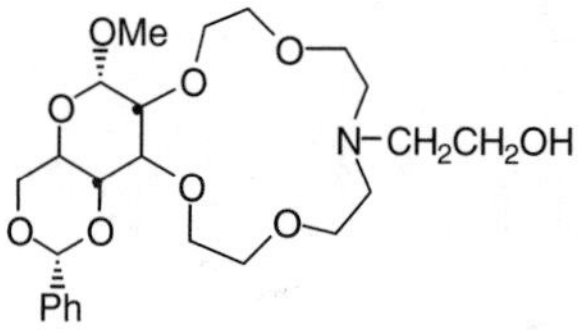

24 [38b]

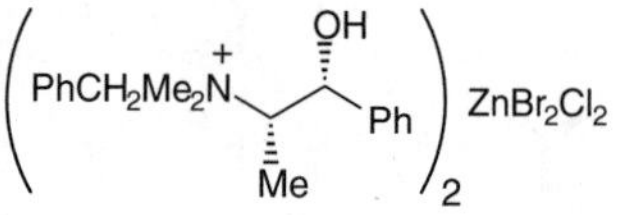

25 [56]

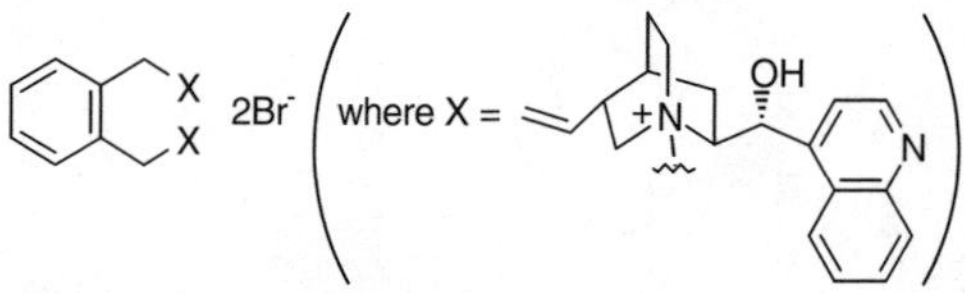

26 [71a]

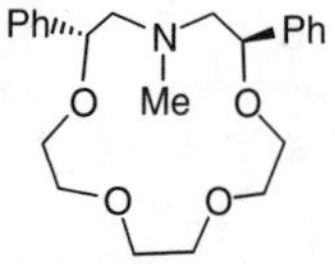

27 [74]

CHAPTER 11

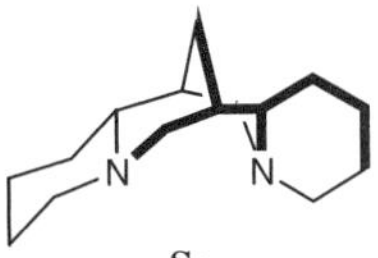

Sp
[20, 53,54,99, 100, 102, 103, 106, 175]

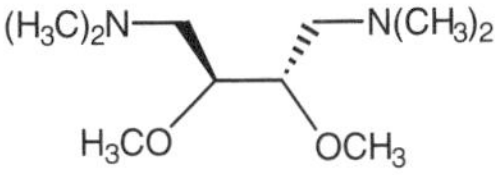

(+)-DDB
[20, 101, 103]

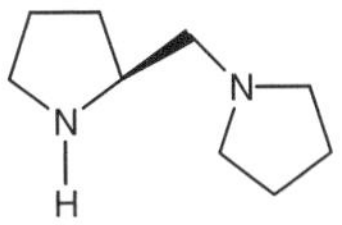

(+)-PMP
[20, 103,115, 116]

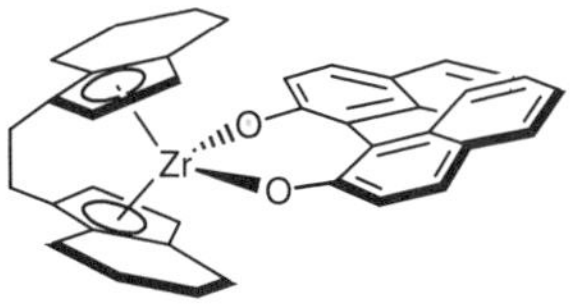

1
[21,22,58,59]

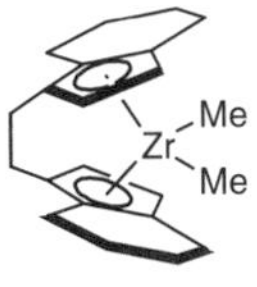

2
[21,22]

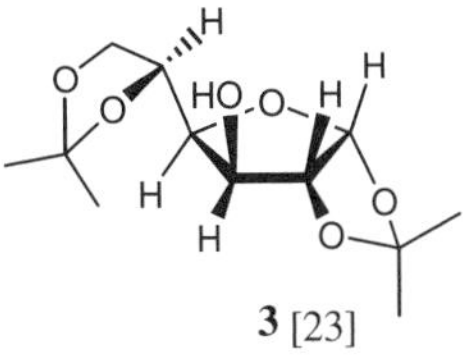

3 [23]

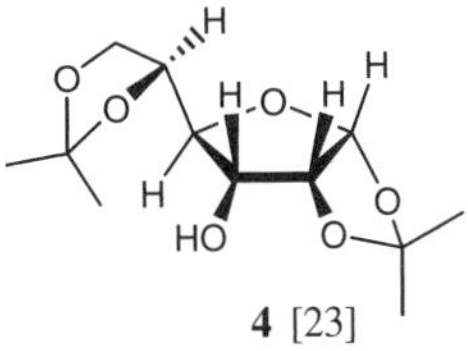

4 [23]

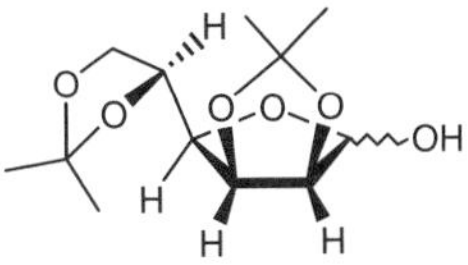

5 [23]

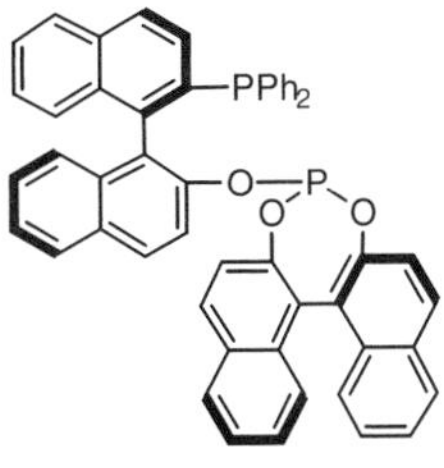

(R,S)-BINAPHOS
[24-26]

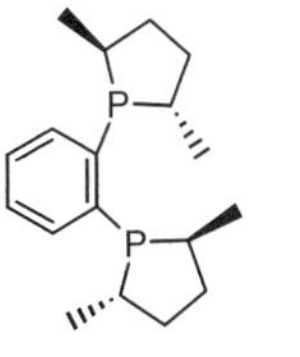

(S,S)-Me-DUPHOS
[27,28]

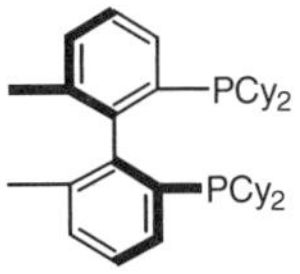

(S)-BICHEP
Cy = cyclohexyl
[29.30]

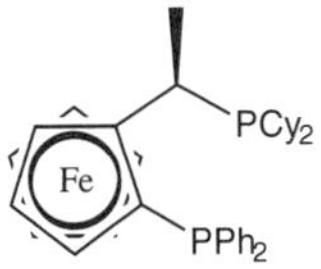

(R)(S$_p$)-JOSIPHOS
Cy = cyclohexyl
[31]

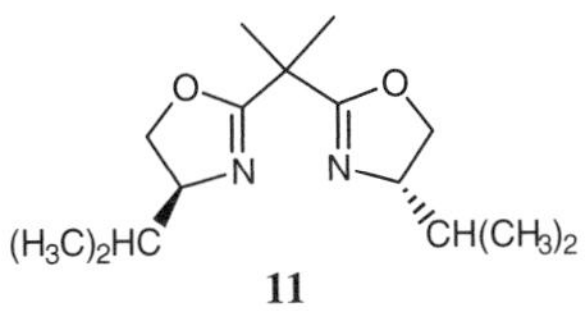

11
[33,34]

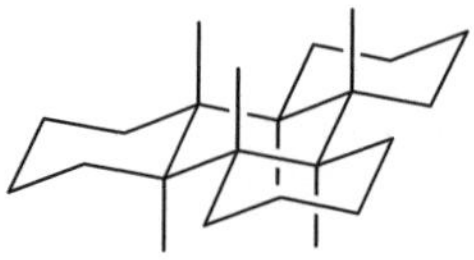

13
[42.43]

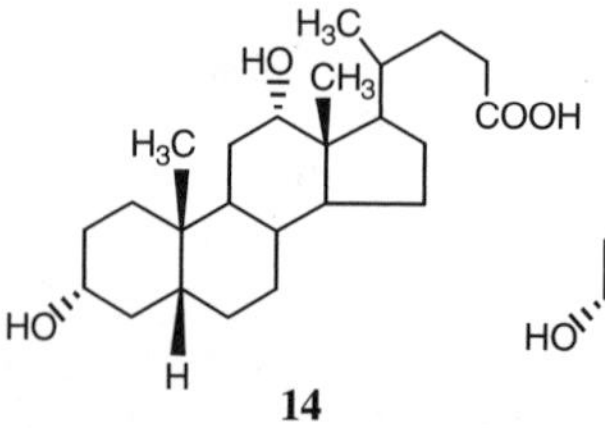

14
[44,45]

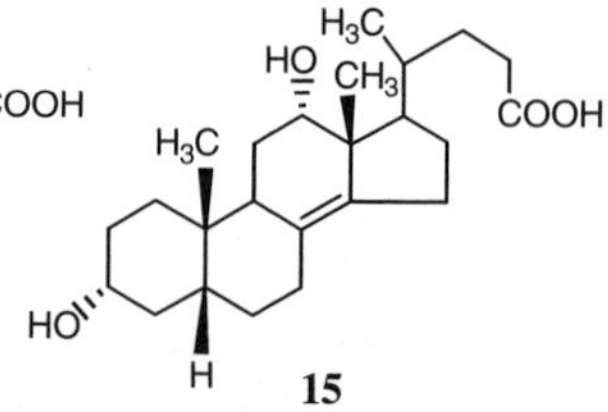

15
[46.47]

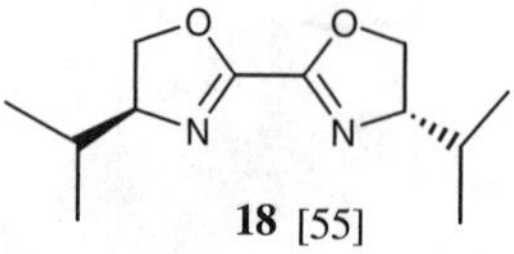

18 [55]

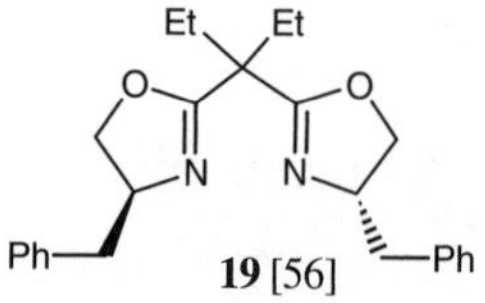

19 [56]

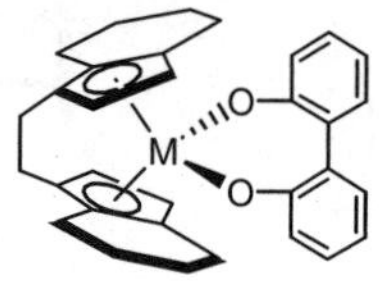

21: M = Zr
22: M = Hf [60-62]

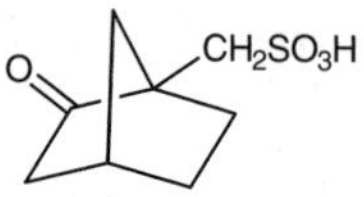

25 [64]

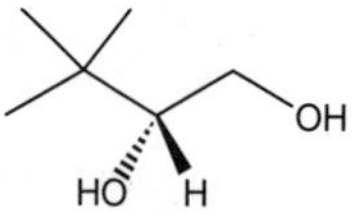

26
[65.66]

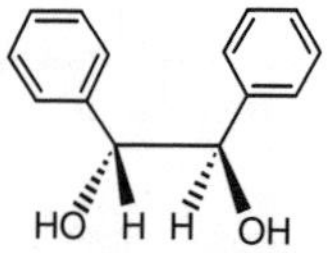

27
[65,66]

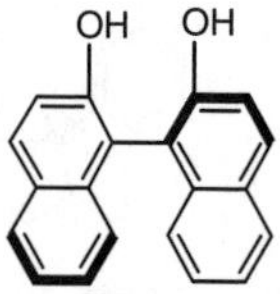

28 [65]

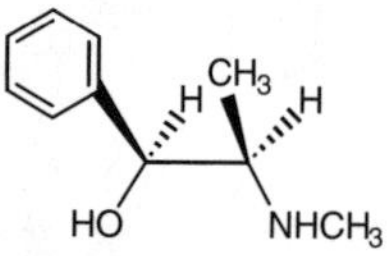

29 [67]

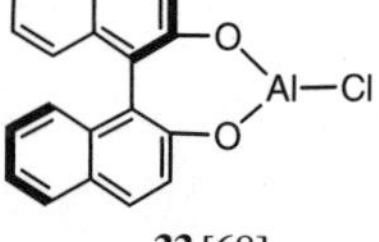

33 [68]

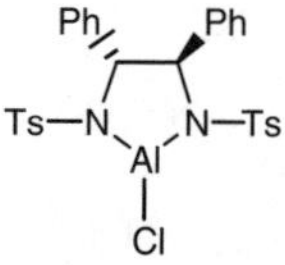

34 [68]

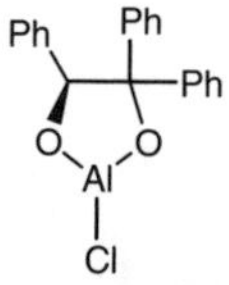

35 [68]

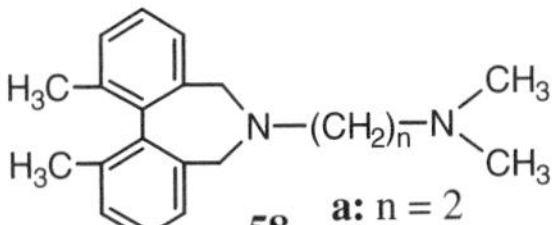

58   **a:** n = 2   **b:** n = 3
[104,105]

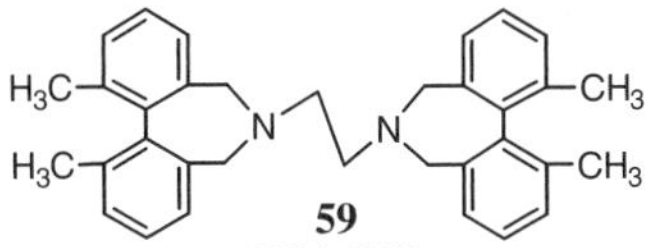

59
[104,105]

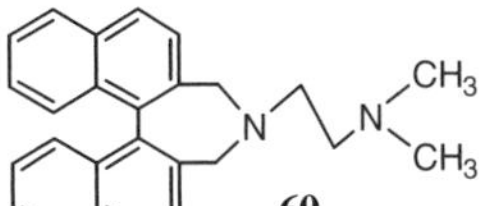

60
[104,105]

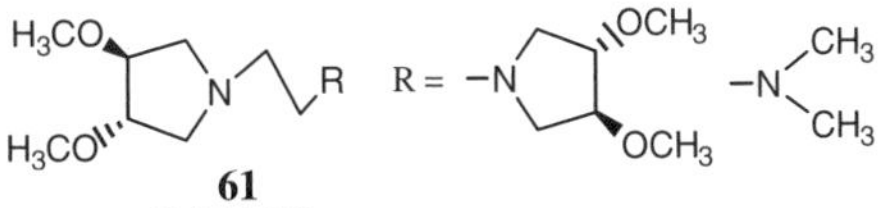

61
[104,105]

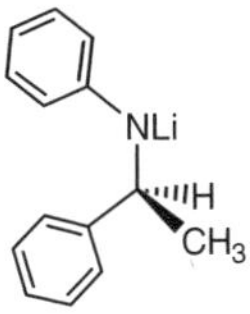

R =

62 [101]

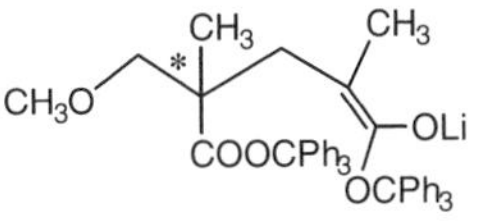

63 [107]

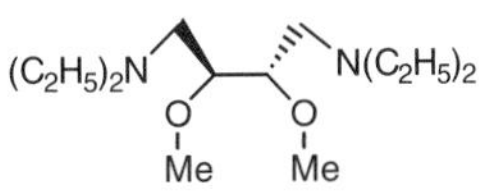

64 [116]

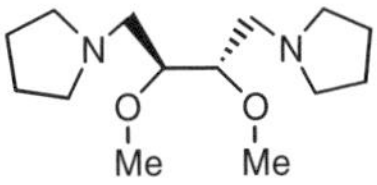

65 [116]

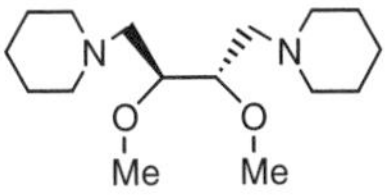

66 [116]

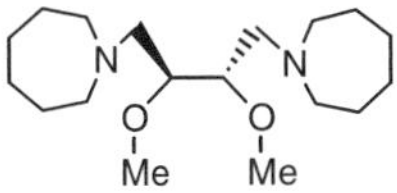

67 [116]

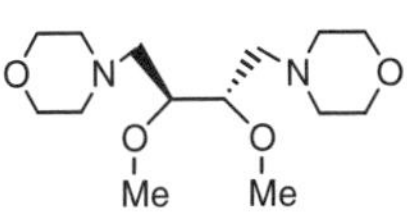

68 [116]

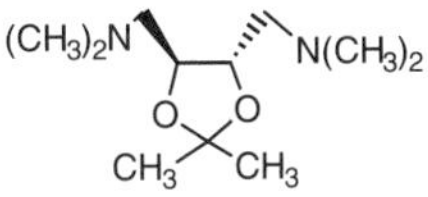

69 [116]

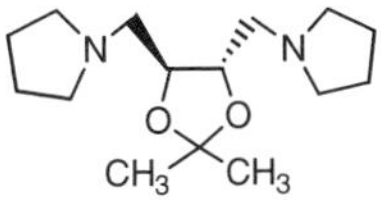

70 [116]

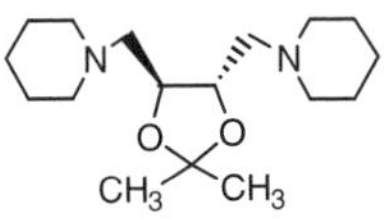

71 [116]

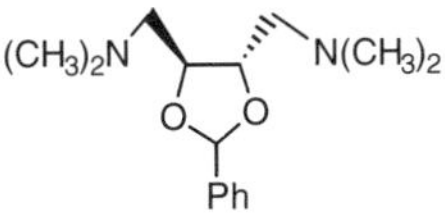

72 [116]

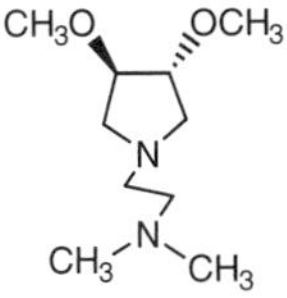

73 [116]

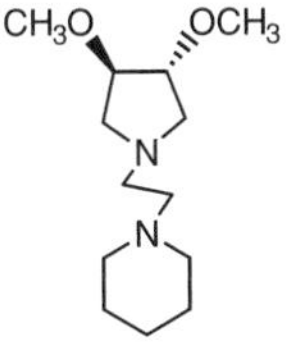

74 [116]

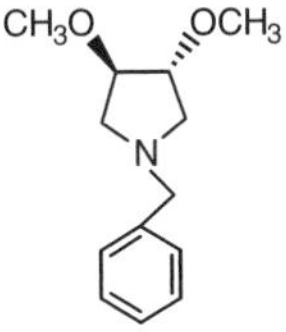

75 [116]

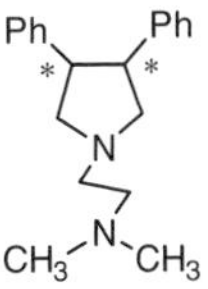

76 [116]

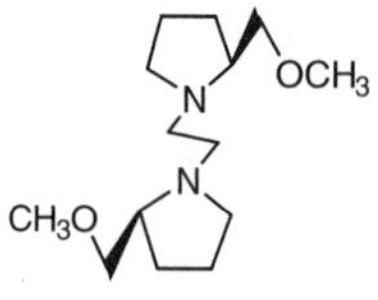

77 [116] 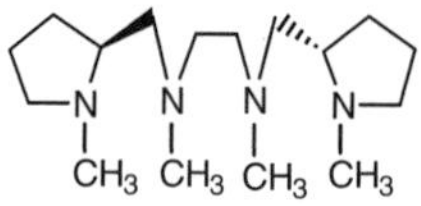   **78** [116] 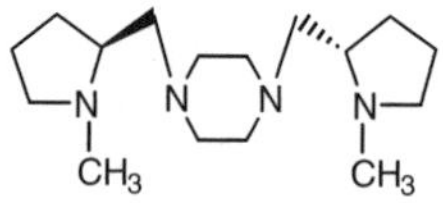   **79** [116]

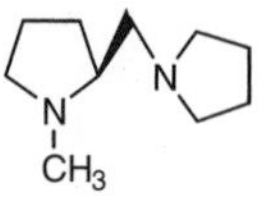

80 [116] 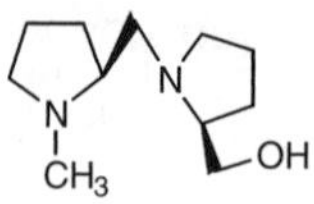   **81** [116] 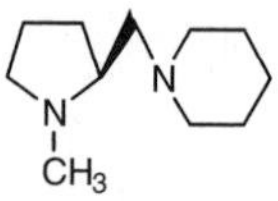   **82** [116]

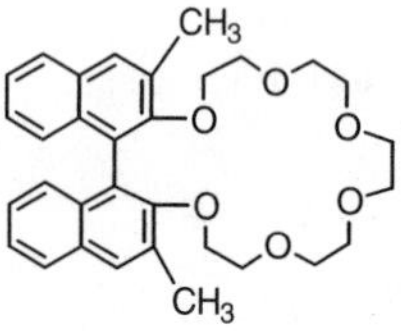

91 [138] 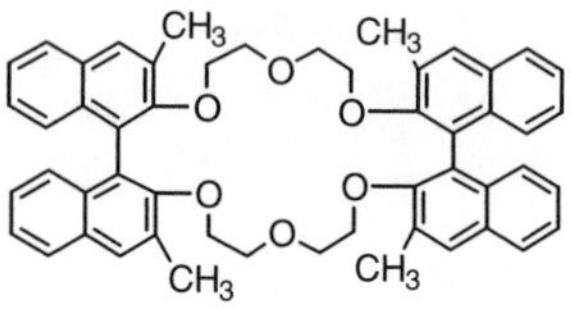   **92** [138] 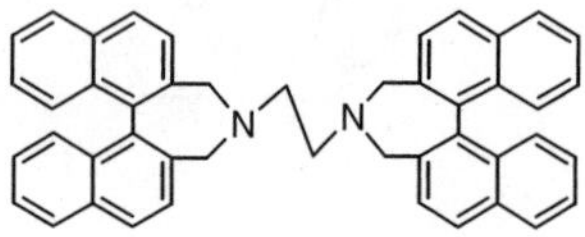   **93** [138]

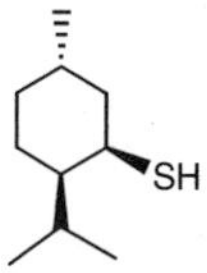

(+)-**110** [126] 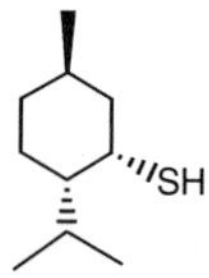   (-)-**110** [126] 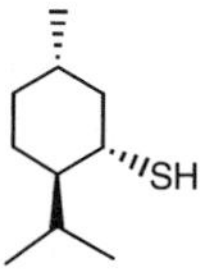   (-)-**111** [126]

(chain-transfer agents for free-radical polymeization)

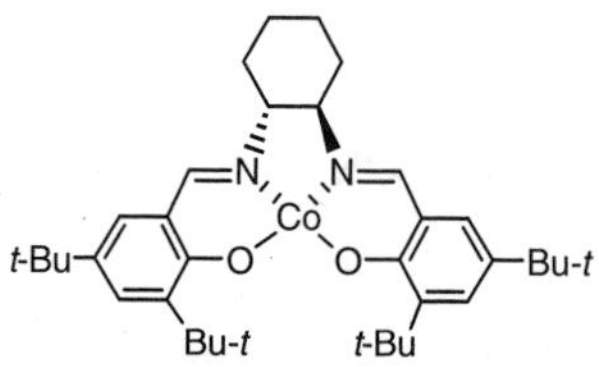

114 [153]

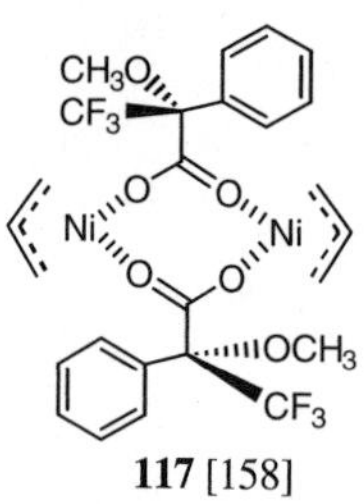

117 [158] 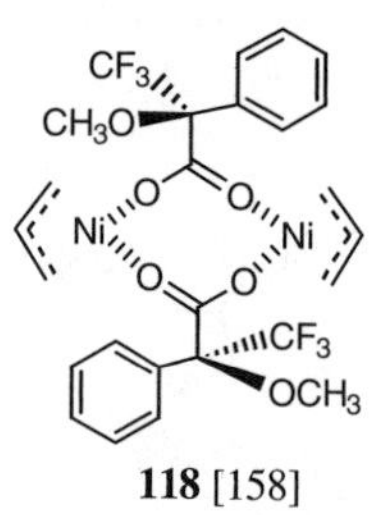   **118** [158] 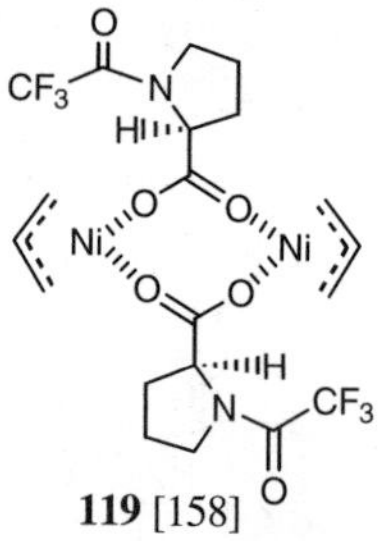   **119** [158]

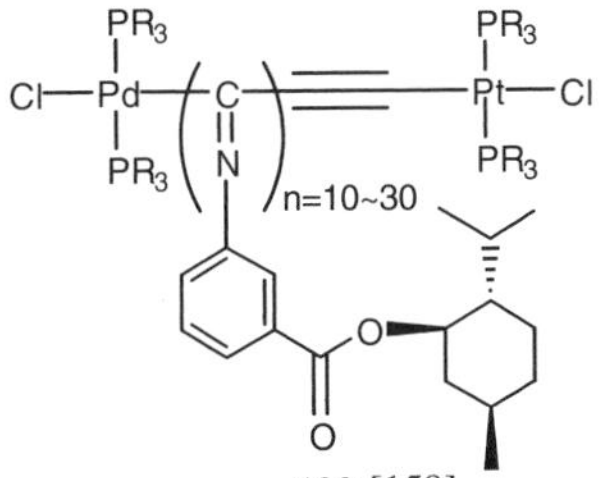

120 [159]

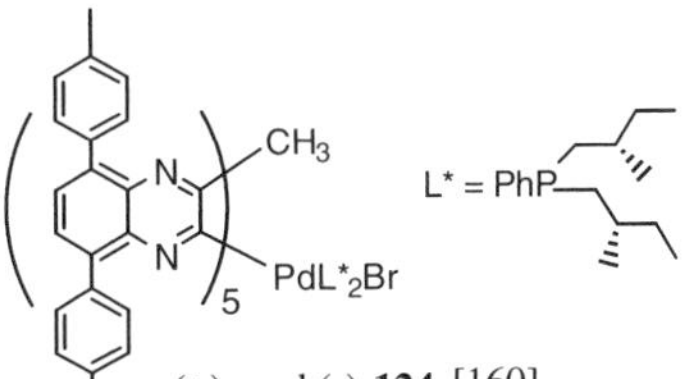

(+)- and (-)-**124** [160]

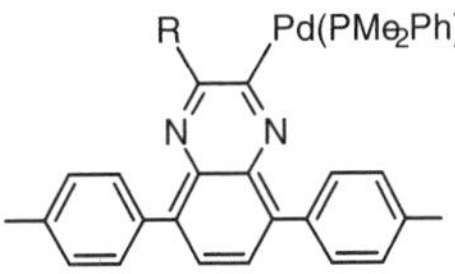

L* = PhP

126: R= 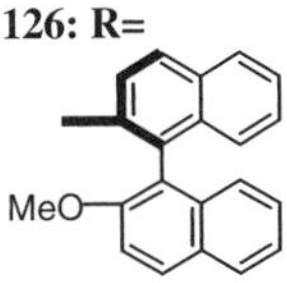 [163]

127: R= [163]

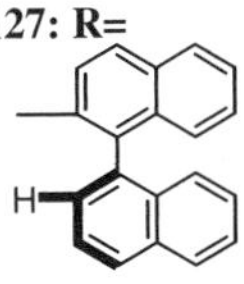

128: R= [164]

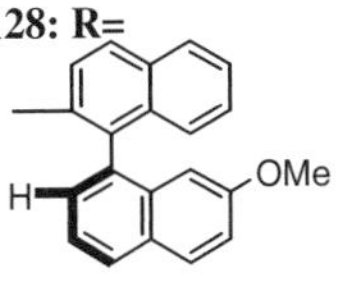

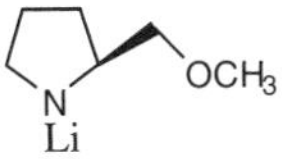

130 [167-169]

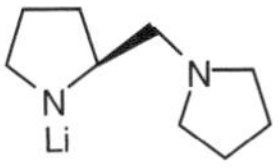

131 [167-169]

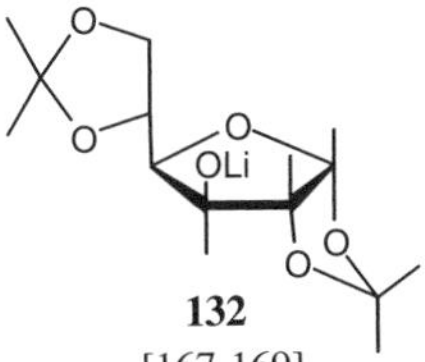

132 [167-169]

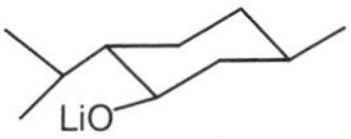

133 [167-169]

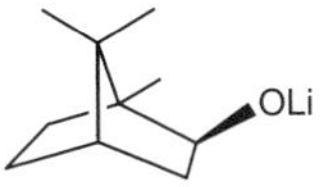

134 [167-169]

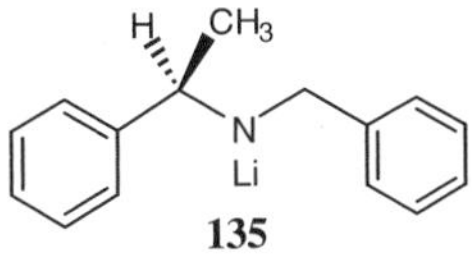

135 [167-169]

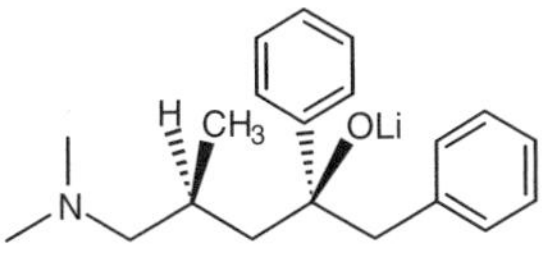

136 [167-169]

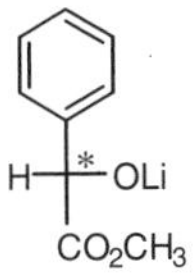

138 [171]

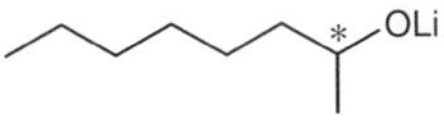

139 [171]

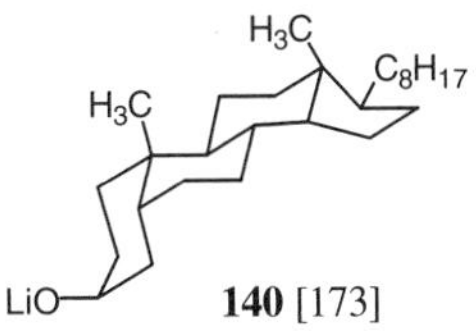

140 [173]

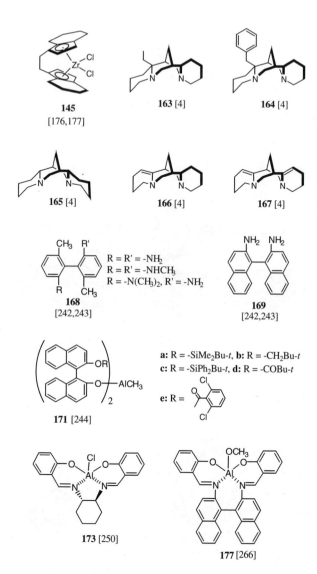

145
[176,177]

163 [4]

164 [4]

165 [4]

166 [4]

167 [4]

R = R' = -NH$_2$
R = R' = -NHCH$_3$
R = -N(CH$_3$)$_2$, R' = -NH$_2$

168
[242,243]

169
[242,243]

171 [244]

a: R = -SiMe$_2$Bu-t, b: R = -CH$_2$Bu-t
c: R = -SiPh$_2$Bu-t, d: R = -COBu-t

e: R =

173 [250]

177 [266]

INDEX